Solutions Manual and Study Guide to Accompany

INTRODUCTION TO ORGANIC CHEMISTRY

Revised Printing

Solutions Manual and Study Guide to Accompany

INTRODUCTION TO ORGANIC CHEMISTRY

FOURTH EDITION
Revised Printing

ANDREW STREITWIESER
University of California, Berkeley

CLAYTON H. HEATHCOCK
University of California, Berkeley

EDWARD M. KOSOWER
Tel Aviv University

Prepared by

Paul A. Bartlett
University of California, Berkeley

Judith G. Koch
Ithaca College, New York

PRENTICE HALL
UPPER SADDLE RIVER, NJ

A Simon & Schuster Company
Upper Saddle River, New Jersey 07458

Printed in the United States of America
10 9 8 7 6 5 4

ISBN 0-13-012990-9

Prentice-Hall International (UK) Limited, *London*
Prentice-Hall of Australia Pty. Limited, *Sydney*
Prentice-Hall Canada Inc., *Toronto*
Prentice-Hall Hispanoamericana, S.A., *Mexico*

Prentice-Hall of India Private Limited, *New Delhi*
Prentice-Hall of Japan, Inc., *Tokyo*
Simon & Schuster Asia Pte. Ltd., *Singapore*
Editora Prentice-Hall do Brasil, Ltda., *Rio de Janeiro*

TABLE OF CONTENTS

NOTES ON STUDYING

If you ask a group of college students which course is the most difficult, they will probably say "organic chemistry". This is intimidating to say the least, especially if you are about to start the course. There is no denying that a course in organic chemistry contains a great deal of material, including many names and facts and an extensive framework of basic principles. However, many students discover the logical foundation to the subject and thoroughly enjoy learning it. It requires work, but so do many other skills worth knowing and doing.

Learning organic chemistry requires logical reasoning as well as memorization. The importance of memorization is great, because the framework of organic chemistry rests on a foundation of facts, but you will find that memorization alone is not going to be enough. (If it were, organic chemistry would not have the reputation it has. Sheer memorization is easy – dull perhaps, but basically easy). To be able to discern a general principal from a collection of facts (reason deductively) and then to be able to predict the behavior of an unknown system (reason inductively) is really what is required to learn organic chemistry beyond the level of simple memorization. The subject can be compared to the game of chess: the chemical behavior of each class of compounds is the counterpart of the moves that different chess pieces are allowed to make. Each chess game becomes intricate and unique, but it always develops within the logical framework of the rules. Similarly, although the range and variety of chemical reactions is often bewildering, they too arise from a finite set of logical principles. We derive a great deal of satisfaction from organic chemistry, as do many of our friends from the game of chess. Hopefully we can help you to learn and enjoy the game of organic chemistry as well.

Despite what you may have heard, learning organic chemistry is not impossible, but it's not easy: it requires you to commit both time and intellectual effort. Learning anything requires reinforcement through repetition, especially repetition in different contexts. In a typical chemistry course repetition is provided in several different ways in order to involve the whole brain. Among these are:

<u>Lectures</u>. The lecture is often an underrated or misunderstood part of the learning process, but it shows you what aspects of the subject the instructor feels are most important, which reactions or mechanisms should receive emphasis, and what should be considered background information. Moreover, the lecturer will present the material in a different way than the text, provide different examples and thereby complement the text. Finally, you listen to a lecturer, and involve a different part of your brain than you do when reading.

<u>Lecture Notes</u>. Because the material that the lecturer presents both orally and on the blackboard is clearly important, it is necessary to recall it while studying. Many students in this electronic age use tape recorders for this purpose, but we feel that these are only a poor supplement to taking notes. Unless you videotape the lecture, you lose the blackboard material. Moreover, you can review a 50-minute lecture more effectively and quickly from notes than you can by spending 50 minutes listening to it again. All of which is to say that lecture notes are important. They help you focus your attention on the lecturer and the material (an important aid on Monday mornings if you or the lecturer is not at peak performance...), they force you to process the information (providing another form of repetition), and they make you use a different brain function as well.

Taking good notes does not end with the lecture. You should transcribe the rough notes into a more organized, legible, and complete form as soon as possible after the lecture. This further processing of information (more repetition...) should be done while the lecture is still fresh in your mind and before the short term memory has started to fade.

Textbook Study. As pointed out above, learning organic chemistry requires an intellectual commitment on top of the time commitment of studying. Mere reading is not study: study is work and requires self-discipline. It requires reading with concentration and thought. For instance, you should constantly ask yourself questions and formulate examples on your own as you proceed through the text. In studying a new reaction, run through specific examples in your mind, using those provided in the text as a guide. Interspersed throughout the text are exercises that ask you to do just this. You will find additional specific suggestions on what to look for and how to learn new reactions in this Study Guide.

Most reactions fall into a very few categories such as proton transfer, displacement, addition, elimination,

etc. In studying each reaction, see how these categories apply, and see how the mechanism of that reaction compares with similar ones of that class. The most important thing you should remember about a reaction is that it exists: that is, that a given functional group transformation or interconversion can be accomplished. It is secondarily useful, although still important, to know the overall reaction conditions and reagents required and any structural limitations involved. It is generally sufficient to know these aspects in general terms, such as the need for heating or cooling the reaction mixture, having acid or base present, etc., rather than memorizing a host of specific details.

Many students find it useful to construct "flash cards" for learning reactions. If you use this technique, be sure to make your own. Remember that the process of writing out the flash cards helps you to learn what's on them.

Problem Sets. Except for the first and last chapters, each chapter in the text is followed by a number of problems. **Working these problems is the most important way to learn organic chemistry**. Working a problem forces you to think, points out the things you don't know, and is good practice for exams. When you review your notes, or go back over the text, it is only human nature to say to yourself, "I know that reaction...I've seen that mechanism before...I'll remember that tomorrow morning..." But your professor is not going to be very sympathetic if you write, "I understood that yesterday" as an answer to an exam question. Working problems is really the only way to prove your mind to find out if you know something when you don't have it right in front of you. Multistep syntheses for example provide an excellent test of knowledge. You have to know each reaction, but that alone is not enough; you have to be able to put them together in the proper sequence.

It should be obvious that you have to work the problems on your own; that is, without thumbing back through the text to find the answer, or worse, simply looking it up in this Study Guide. You should make a serious attempt to work a problem before you look for the answer; if you honestly can't, then use the text and your notes to help you. Only when you are completely stymied by a problem, or you have solved it, should you look up the answer in this Study Guide. It will be a valuable check to see if you got the right answer, and it shows you how to solve the problem if you didn't. If you look the answer up right away, you'll just say to yourself, "Oh, I knew that", or "I'll remember that tomorrow" (when you probably didn't and won't).

If the answer given in the Study Guide doesn't make sense to you, check back through the text and your notes again. If you still are confused, by all means see your professor and/or teaching assistant. That is what they are paid for, and most organic chemistry classes are too large for the instructor to come to you. Most instructors like to teach, and they will be pleased if you use their office hours judiciously.

Laboratory. Most introductory organic chemistry courses have a laboratory associated with them. The primary purpose of the laboratory part of the course is for you to learn the experimental operations of organic chemistry and to gain facility in the associated physical manipulations. It takes actual practice to set up an experimental apparatus and carry out the preparation of a compound, just as it takes dexterity and practice to play the piano or make a sauce béarnaise. Moreover, your laboratory experience provides still another reinforcement route to learning organic chemistry. Organic compounds are stuff that melt, boil, smell, have colors, and crystalline form; modern organic chemistry has many abstract theories and principles, but they apply to real physical substances. This character of organic compounds is emphasized in the text, but will be reinforced by your laboratory work.

Further Suggestions. Finally, but most importantly, you should study frequently and in continuous pieces. If you limit your study to "cramming" for examinations, you will not learn organic chemistry. There is a great deal of structural hierarchy in the science, and you should understand each level thoroughly before you go on to the next one.

WHAT THIS GUIDE CONTAINS

The Study Guide is organized in chapters that parallel those of the text. Each chapter contains the following sections:

A. Outline of the chapter in the text, with keywords (bold-faced terms), important ideas introduced.
B. A brief equation illustrating each reaction discussed in the chapter, and where appropriate the major points to keep in mind for these reactions.
C. A discussion section which: 1) provides an overview of the chapter and a look at the important reactions from a different point of view, and 2) gives specific hints on how to approach certain topics and how to solve certain types of problems.
D. Answers to the exercises which are interspersed throughout the chapter in the text.
E. Answers and explanations to the problems at the end of the chapter in the text.
F. Supplementary problems to provide you with additional practice. These problems tend to be a bit harder than those in the text, and they often incorporate ideas from previous chapters as well. This can be valuable from the point of view of review, of course.
G. Answers to the supplementary problems. Although these are very near the questions themselves, don't look until you've worked the problems!

We hope that this Study Guide is useful to you in learning organic chemistry. Even more importantly, we hope that it helps you to appreciate the beauty and excitement of this area of science, and that you will be one of the many students who enjoy the subject. We would appreciate any comments or suggestions that you might have for this Study Guide; we apologize for any mistakes and will be grateful if you bring them to our attention.

2. ELECTRONIC STRUCTURE AND BONDING

2.A. Chapter Outline and Important Terms Introduced

2.1 **Periodic Table** (important elements in organic chemistry)

2.2 **Lewis Structures** (where the electrons are)

octet configuration
valence electrons
electropositive and electronegative
ionization potential and electron affinity
Lewis structures
Kekulé structures
covalent bonds
formal charges
lone pair electrons

2.3 **Geometric Structures** (bond angles and bond lengths)

single bond
double bond

2.4 **Resonance Structures** (when a single Lewis structure is inadequate)

resonance hybrids
bond orders
move electrons, not nuclei
importance of octet configuration
minimize charge separation
oxonium ions
carbocations

2.5 **Atomic Orbitals** (electrons as waves)

Heisenberg Uncertainty Principle
wave function
atomic function
probability function
electron density distribution
quantum number
node
shape of orbitals

2.6 **Electronic Structure of Atoms** (1s, 2s, 2p, etc.)

2.7 **Bonds and Overlap** (how orbitals are combined)

reinforcement and interference
covalent bond
molecular orbital
bonding and antibonding orbitals
overlap of orbitals
bent bonds

2.8 **Hybrid Orbitals and Bonds**

hybrid orbital
hybridization (sp, sp^2, sp^3)
bond angle

2.B. Important Reactions Introduced:

none yet

2.C. Important Concepts and Hints

The topics of electronic structure and bonding are presented from two different points of view in Chapter 2. The first is concerned with the accounting of electrons, predicting how many covalent bonds an atom will make, and assigning formal charges. All of this falls under the heading of **Lewis**, or "electron dot", structures. Being familiar with this accounting procedure is important, because as you progress in organic chemistry, you will be expected to keep track mentally of the information provided by Lewis structures, although you will usually write the simpler Kekulé structures. Another important concept in the accounting of electrons and bonds is that of **resonance**, which enables you to understand the structure and reactivity of a molecule that does not conform to any single Lewis structure. Students frequently describe resonance as the molecule or ion "flipping" between the different Lewis structures used. This is no more true than saying that a mule "flips" between being a horse and a donkey. A mule is not a horse one instant and a donkey the next. In the same way, the molecule exhibiting resonance is not one Lewis structure one instant and another the next. Lewis structures are valuable tools, but necessarily localize electrons in specific bonds between two atoms. For some species, this localization doesn't exist and the electron pair can probably be found over a greater part of the molecule or ion. It is this delocalization of the electrons that we are trying to represent in resonance structures.

In parallel with an understanding of Lewis structures and resonance, you will also need to be familiar with the concept of an **orbital**, and many of the topics pertaining to it. Because modern organic chemists describe almost all reaction mechanisms in terms of the orbitals involved, you should understand

1) what an orbital is
2) why it can be filled or empty
3) how a covalent bond is formed from the overlap of two atomic orbitals
4) what **molecular** and **bonding** orbitals are
5) what **hybridized** orbitals are, how and why they are formed and what their shape and orientation are
6) how the geometry of a molecule depends on this hybridization.

We don't exaggerate when we say that these concepts are the foundation of the modern approach to understanding organic chemistry. You would be wise, therefore, to understand them from the start.

HINTS: For determining the hybridization of an atom, a convenient approach is to count up the number of things attached to that atom. By "things" we mean other atoms **and** unshared electron pairs. For example, methane, ammonia, and water all have four things attached to the central atom; they are all, therefore, sp^3-hybridized.

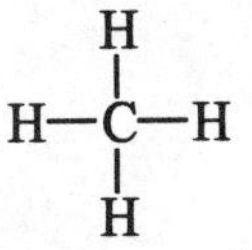

methane 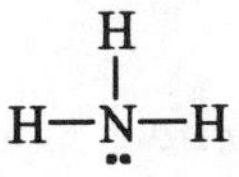 *ammonia* 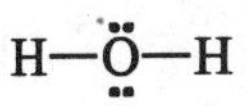 water

The carbon in methyl cation, the carbon and oxygen in formaldehyde, the nitrogens in diimide, and the boron in borane are all connected to only three things and are sp^2-hybridized. The carbons and boron in these examples are bonded to three atoms only, although in formaldehyde the carbon still has four bonds. Each nitrogen in diimide is attached to a hydrogen, another nitrogen, and a lone pair. The oxygen in formaldehyde is attached to a carbon and two lone pairs.

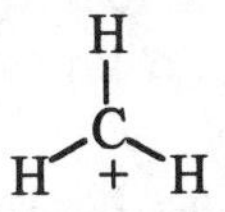

methyl cation 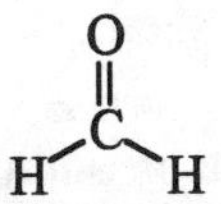 *formaldehyde* 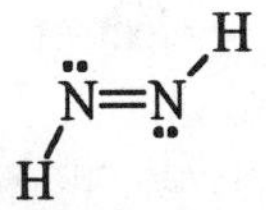 *diimide* 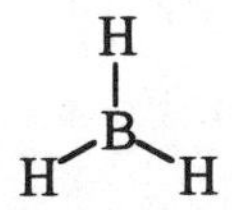 *borane*

Using these criteria, it is clear why all the non-hydrogen atoms in the structures below are sp-hybridized:

H–Be–H H–C≡N: $^{-}$:C≡C–H $^{+}$:O≡N:

The only difficulties arise in attempting to predict the hybridization of the halogens, and of elements below the second row of the period table. We would predict by the method outlined above that the halogens are sp^3-hybridized. Because the significance of evaluating hybridization lies in the prediction of bond angles, the hybridization of the monovalent halogens is unimportant. Nevertheless, in some instances (e.g., the fluorine in HF), they are considered to be unhybridized (see Section 8.1 later on in the text). For elements in the third row of the periodic table and below, the question of hybridization can become more complicated (see the discussion at the end of Section 2.4 in the text, as well as Exercise 2.8).

2.D. Answers to Exercises

2.1

a. $:\ddot{\underset{\cdot\cdot}{Cl}}:$

Cl: 7 — valence electrons for Cl
- charge: <u>1</u> — additional electron for negative charge
Sum: 8

b. $H{-}\ddot{\underset{\cdot\cdot}{O}}{-}H$

2H: 2
O: <u>6</u>
Sum: 8

c. $H:\ddot{\underset{\cdot\cdot}{O}}:^-$

H: 1
O: 6
- charge: <u>1</u>
Sum: 8

You can arrive at the structure for hydroxide by combining the hydrogen atom, the oxygen atom and the negative charge, as shown, or by taking the structure for the water molecule, shown above, and subtracting H^+.

d. $:\ddot{\underset{\cdot\cdot}{Cl}}:\ddot{\underset{\cdot\cdot}{O}}:^-$

Cl: 7
O: 6
- charge: <u>1</u>
Sum: 14

To make an octet for each atom using only 14 electrons, one pair has to be shared between the atoms. The oxygen has the negative charge because it controls seven electrons (6 + 1/2 of the 2 shared = 7), one more than in a free oxygen atom.

e. $H:\ddot{\underset{\cdot\cdot}{N}}:H$
$\quad\;\; H$

3 H: 3
N: <u>5</u>
Sum: 8

f. $:\ddot{\underset{\cdot\cdot}{F}}:C\vdots\vdots N:$

C: 4
N: 5
F: <u>7</u>
Sum: 16

In this case, with 16 electrons to make octets for three atoms, (3 x 8) - 16 = 4 pairs of electrons have to be shared. This is accomplishedby putting one pair between F and C (single bond) and three pairs between C and N (triple bond). All of the atoms are neutral since they control the same number of electrons as they do in the atomic state (F:6 + 1/2 of 2 = 7; C: 1/2 of 8 = 4; N: 2 + 1/2 of 6 = 5).

g. $:\dot{N}::\ddot{O}:$

N: 5
O: <u>6</u>
Sum: 11

With an odd number of valence electrons, it is impossible to have an octet for each atom. The best that can be achieved is to share two pairs to give oxygen an octet and nitrogen 7 electrons in the valence shell. There are no formal charges.

h. $H:\ddot{\underset{\cdot\cdot}{O}}{}^{+}:H$
$\quad\;\; H$

3 H: 3
O: 6
+ charge: <u>-1</u>
Sum: 8

Note that you can get this structure by adding a proton, H^+, to the water molecule. Oxygen has the positive charge because it controls only 5 electrons (2 + 1/2 of the 6 are shared.

i. Na^+

Na: 1
+ charge: <u>-1</u>
Sum: 0 — electrons in valence shell

The next shell lower in energy than the valence shell has a complete octet; the sodium cation in fact has the same number of electrons as neon.

j. He: k. H:Ö:Ö:H l. Ö::C::Ö

C: 4
2 O: 12
Sum: 16

Three octets from 16 electrons requires four pairs to be shared.

2.2 :Ṅ::Ö:

The Lewis dot structure indicates that two pairs of electrons are shared and that nitric oxide has a double bond. As indicated in the text, the N=O double bond distance is 1.15Å (0.115 nm); the experimentally determined value for nitric oxide is in fact 1.15Å.

2.3 In each case, the double bond will be shorter than the single bond.

a. double bond, shorter — H–C(=O:)–Ö–H — single bond, longer

b. double bond ⟶ :O: ‖ ·N–Ö–H — single bond

c. H–C(H)(H)–Ö–H — single bond

H₂C=Ö — double bond

2.4 Ozone: $^{-}$:Ö:Ö$^{+}$::Ö ⟷ Ö::Ö$^{+}$:Ö:$^{-}$ Hydrogen peroxide: H:Ö:Ö:H

Each O–O bond has a bond order of 1½ in ozone (50% single bond and 50% double bond). These bonds are therefore shorter than the O–O bond in H–O–O–H, bond order 1(100% single bond).

2.5

:Ö:$^{-}$:C̈l:N 2+ :Ö:$^{-}$	:Ö:$^{-}$ H:C + :Ö:$^{-}$	:Ö:$^{-}$ $^{-}$:Ö:C + :Ö:$^{-}$
Nitryl chloride	Formate	Carbonate

2.6 In order to have no more than an octet for P in PCl_6^-, a resonance form with no bonds between P and 2 Cl's must be invoked.

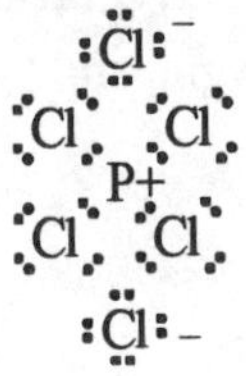

Because of the octahedral symmetry of PCl_6^-, unlike the trigonal bipyramid of PCl_5, all the Cl's are equivalent. If phosphorus is to have only an octet, 2 of the Cl's will be "no-bond", and the bond order will be 4 bonds/6 Cl's or 2/3 or longer than the normal P-Cl bond length.

2.7 **Resonance Structure** **Hybrid**

nitric acid: [H–O–N⁺(=O)(–O⁻) ⟷ H–O–N⁺(=O)(–O⁻)] ≡ hybrid with "single bond" to the O–H oxygen and "$1^1/_2$ bond" to each other oxygen, charge $-1/2$ on each of those oxygens

nitrate ion: [three equivalent resonance structures, each with one N=O bond and two N–O⁻ bonds, N⁺] ≡ hybrid with charge $-2/3$ on each oxygen

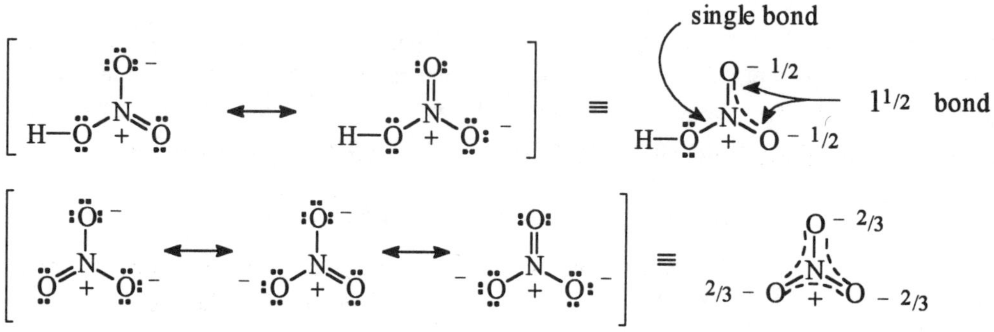

each N–O bond is a "single-and-a-third"

The single bond in nitric acid is the longest, and the "single-and-a-half" bonds in nitric acid are the shortest.

2.8 The fact that the H–S–H angle is closer to 90° than to 109° suggests that the sulfur atom uses unhybridized p orbitals to bond to the hydrogens, rather than sp^3 orbitals.

2.E. Answers and Explanations for Problems

1 a. [Resonance structures of the hydrogen sulfate ion HSO_4^-: H:O:S^{2+} with three O⁻ ⟷ H:O:S⁺ with one S=O ⟷ … ⟷ H:O:S with two S=O bonds and one O⁻ (seven structures in all)]

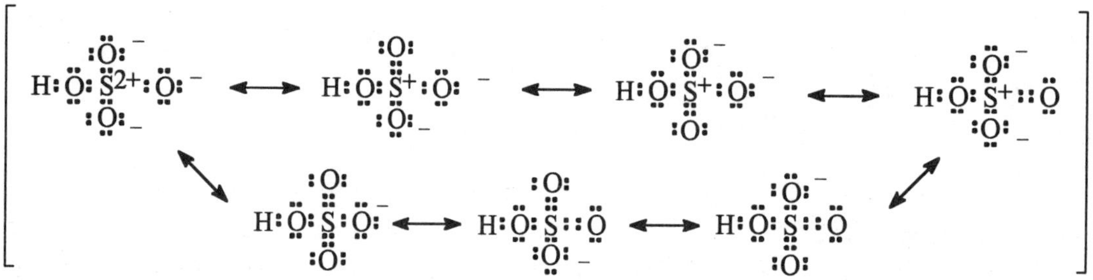

See the Lewis structure of the sulfate ion given in Section 2.2 of the text, and the discussion of the resonance hybrids with expanded valence shells in Section 2.4. Note the equivalence of the three resonance structures at the top right and at the bottom of the diagram above.

b. 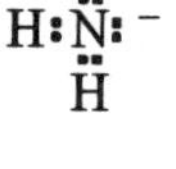 H:N̈:⁻ (with H below N) Nitrogen contributes five valence electrons, each hydrogen contributes one, and the negative charge represents one more for a total of eight.

c.  [O::N:O:⁻ ⟷ ⁻:O:N::O] Nitrogen contributes five valence electrons, the two oxygen contribute twelve, and the negative charge adds one for a total of eighteen.

Incorrect: ⁻:O:N⁺:O:⁻ 6 electrons around N   :O::N::O 10 electrons around N

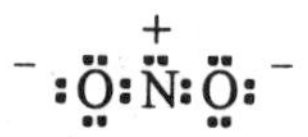

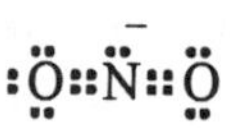

The first structure is less important since thare are fewer than 8 electrons around N; the second structure is wrong because nitrogen can not have more than an octet.

d. 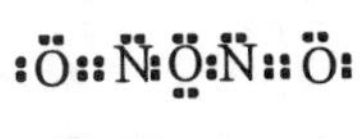

e. 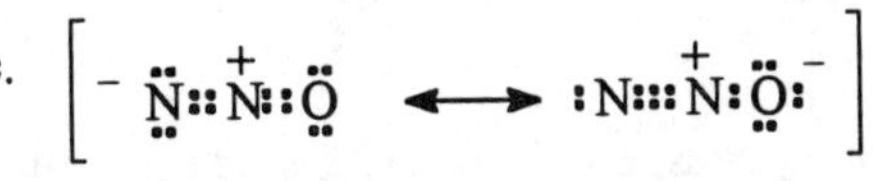

f. 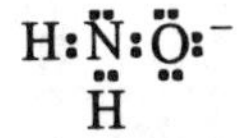

g. 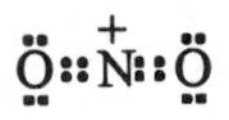 Compare with nitrite ion (ONO^-, 1 c.), which has two more electrons.

h. 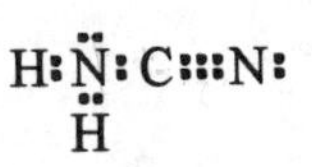

i. 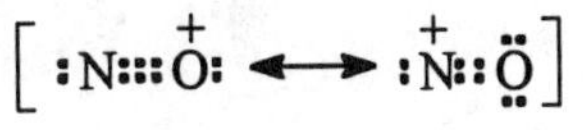

Note the NO triple bond. Compare with N_2, in Section 2.2 of the text.

j. 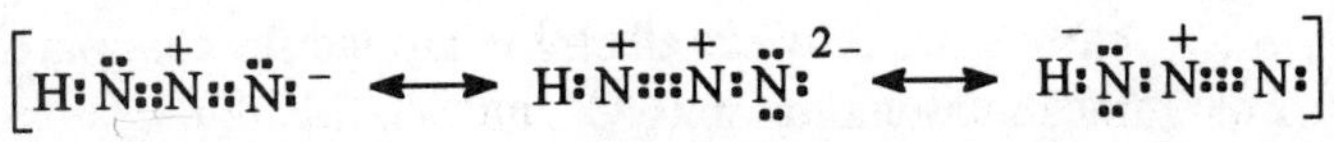

The middle structure with two + and one 2− can be ignored.

k. 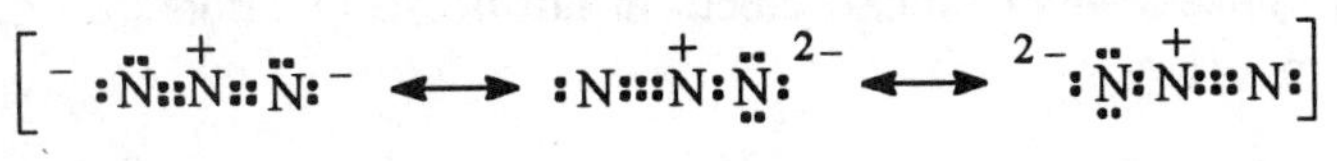

Two of the possible resonance structures put two formal charges on the nitrogen, a second row element. These can be ignored.

l. 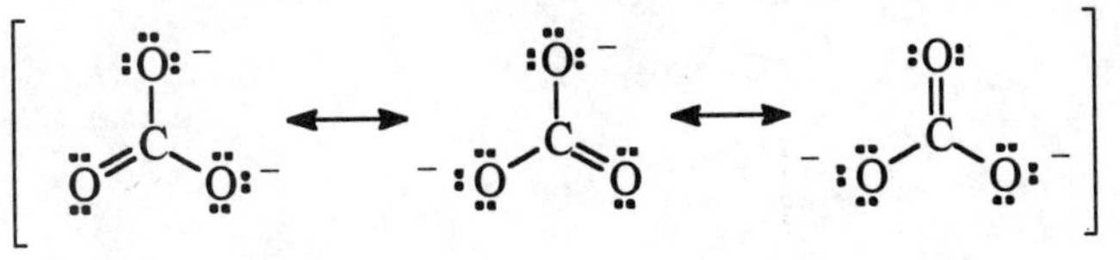

m. 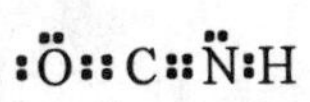

2. a. 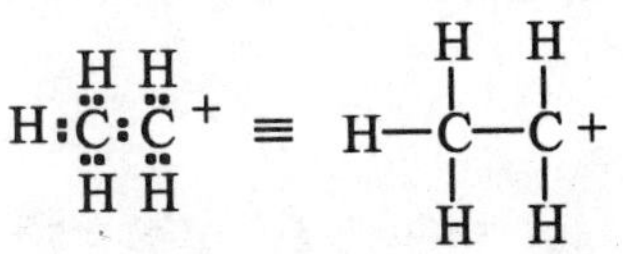

b. 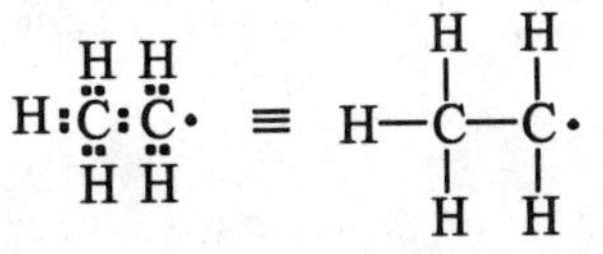

c. 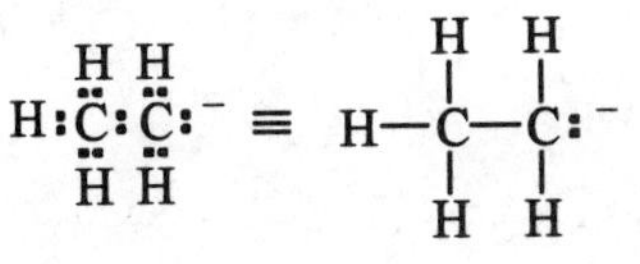

Note the change in charge on carbon as it goes from six (2a.) to seven (2b.) to eight (2c.) valence electrons.

d. 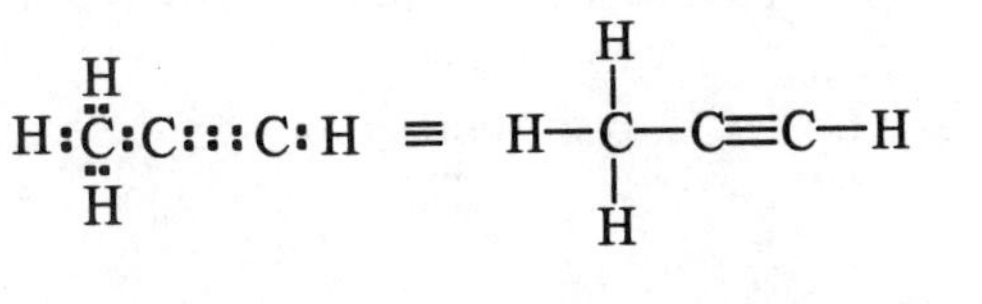

e. 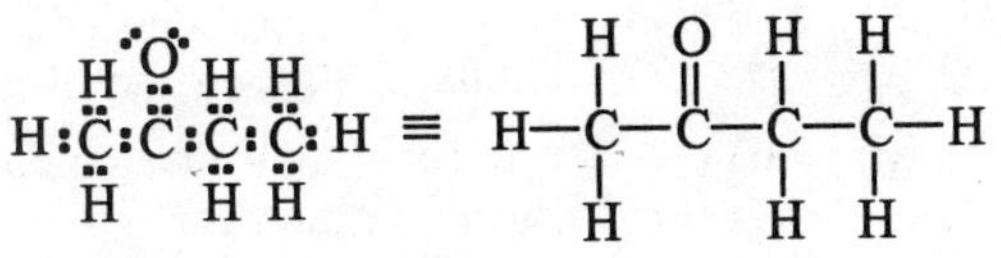

f. 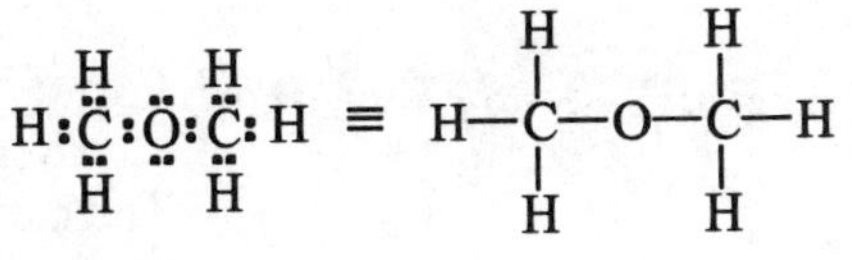

g. 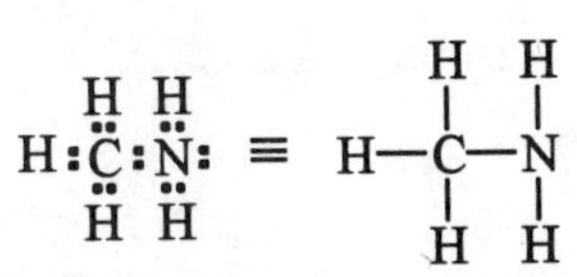

h. 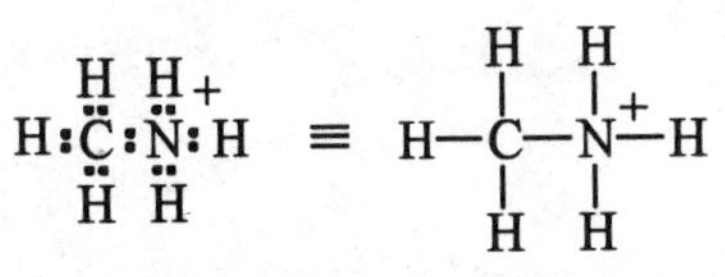

i. 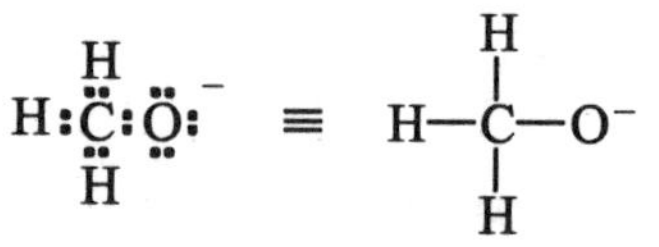 $H:\overset{H}{\underset{H}{C}}:\ddot{O}:^{-} \equiv H\!-\!C\!-\!O^{-}$ (with H above and below C)

j. 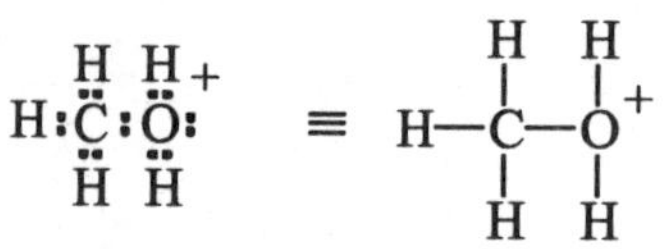 $H:\overset{H}{\underset{H}{C}}:\overset{H}{\underset{H}{O}}:^{+} \equiv H\!-\!C\!-\!O^{+}$ (with H above and below C and O)

Compare the electronic arrangement in the series: ethyl anion 2 c., methylamide 2 g., and methyl oxonium ion 2 j.

k. 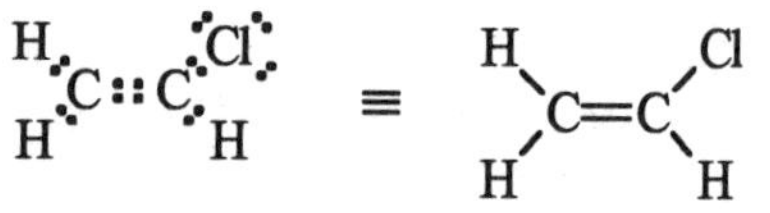 H, H, :Cl:, H around C::C ≡ H(H)C=C(Cl)(H)

l. 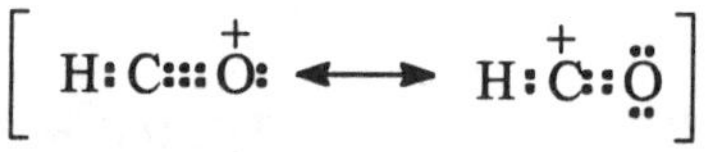 [H:C:::Ö:⁺ ⟷ H:Ċ⁺::Ö]

Note the non-equivalent resonance structures. The one on the right is less important than the other because the carbon has only six electrons around it. Compare with the nitrosonium ion (NO^+, i.)

NOTE: for many applications it is convenient to use lines to indicate electron-pair bonds and dots to show lone pairs; for instance:

2c. 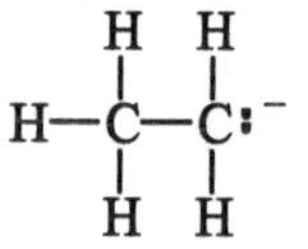 H—C(H)(H)—C(H)(H):⁻ 2e. 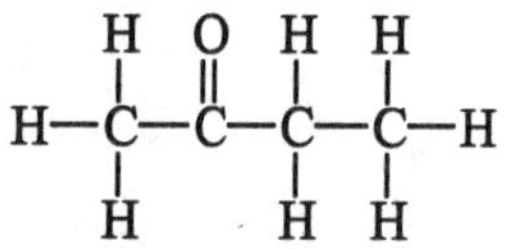 H—C(H)(H)—C(=O)—C(H)(H)—C(H)(H)—H 2j. 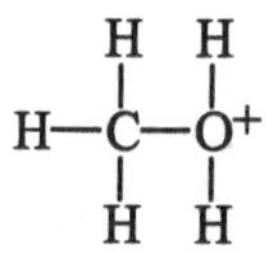 H—C(H)(H)—O⁺(H)(H)

3.

a. 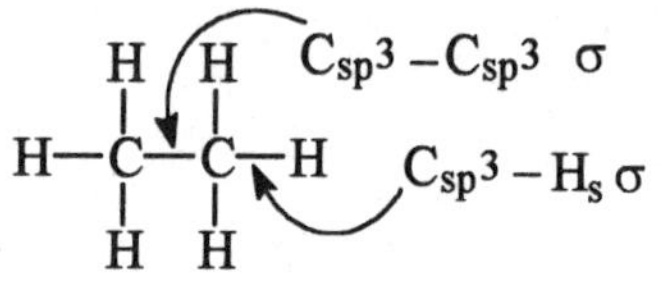 H—C(H)(H)—C(H)(H)—H ; $C_{sp^3}-C_{sp^3}$ σ ; $C_{sp^3}-H_s$ σ

Each carbon has four things bonded to it, and therefore is sp^3 hybridized.

b. 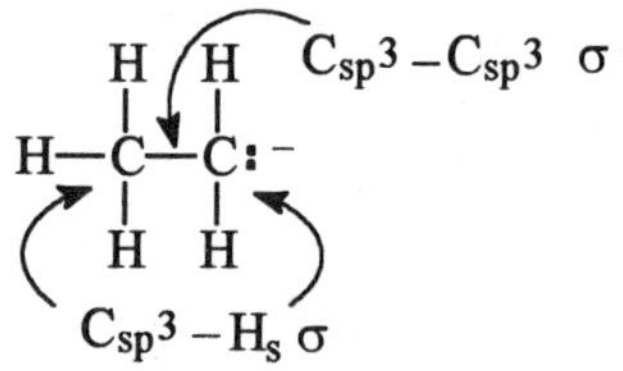 H—C(H)(H)—C(H)(H):⁻ ; $C_{sp^3}-C_{sp^3}$ σ ; $C_{sp^3}-H_s$ σ

Each carbon has four things bonded to it; therefore it is sp^3 hybridized.

c. 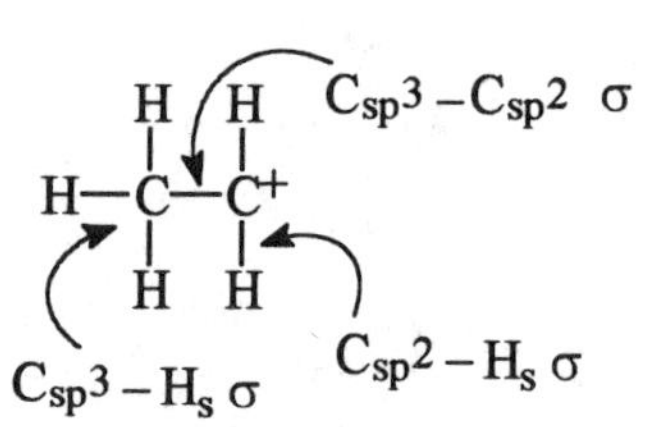 H—C(H)(H)—C⁺(H)(H) ; $C_{sp^3}-C_{sp^2}$ σ ; $C_{sp^3}-H_s$ σ ; $C_{sp^2}-H_s$ σ

The carbon with only three things bonded to it is planar.

d.

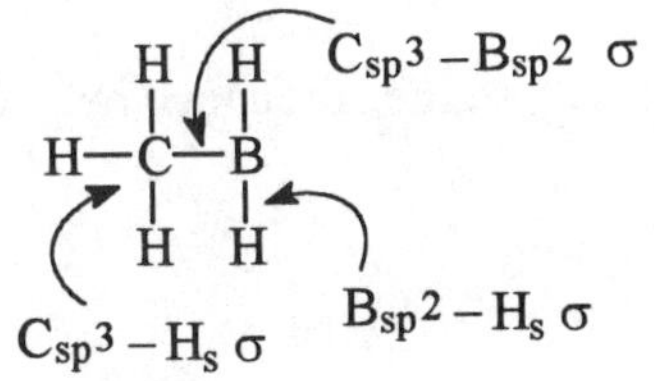

$$\begin{array}{c} \text{H} \quad \text{H} \\ | \quad | \\ \text{H—C—B} \\ | \quad | \\ \text{H} \quad \text{H} \end{array}$$

$C_{sp^3}-B_{sp^2}\ \sigma$

$C_{sp^3}-H_s\ \sigma$ $B_{sp^2}-H_s\ \sigma$

Methylborane has the same arrangement of electrons as ethyl cation (3c).

e.

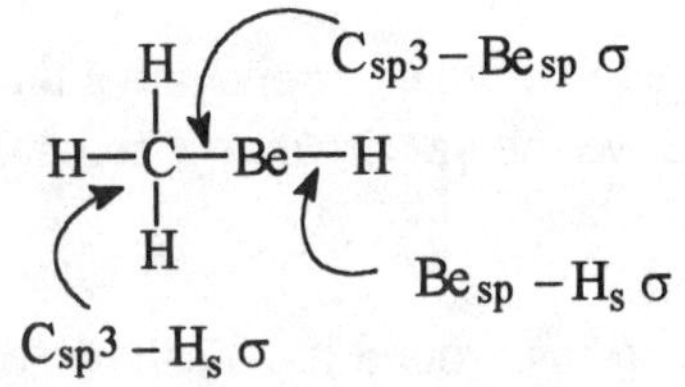

$$\begin{array}{c} \text{H} \\ | \\ \text{H—C—Be—H} \\ | \\ \text{H} \end{array}$$

$C_{sp^3}-Be_{sp}\ \sigma$

$Be_{sp}-H_s\ \sigma$

$C_{sp^3}-H_s\ \sigma$

Beryllium, bonded to only two other atoms, is linear.

f.

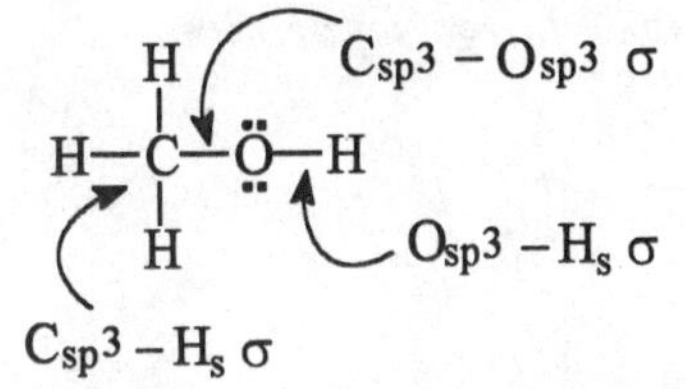

$$\begin{array}{c} \text{H} \\ | \\ \text{H—C—}\ddot{\underset{..}{\text{O}}}\text{—H} \\ | \\ \text{H} \end{array}$$

$C_{sp^3}-O_{sp^3}\ \sigma$

$O_{sp^3}-H_s\ \sigma$

$C_{sp^3}-H_s\ \sigma$

Oxygen, with four things bonded to it (a carbon, a hydrogen, and two lone electron pairs), is approximately sp^3-hybridized.

4. Resonance structures involve a shift in the position of electrons only. They do not involve changes in the position of the atomic nuclei. Therefore, b., c., f., g., j., k. and o. are not pairs of resonance structures because one or more atoms change positions.

5.

a. 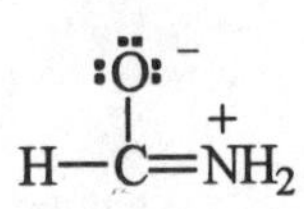 $:\ddot{O}:^-$ / $H-\overset{|}{C}=\overset{+}{N}H_2$ is less important because charge separation leads to higher energy.

b. 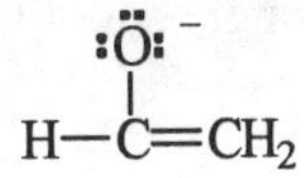 $:\ddot{O}:^-$ / $H-\overset{|}{C}=CH_2$ is more important (lower energy) because the negative charge is on a more electronegative atom (oxygen rather than carbon).

c. 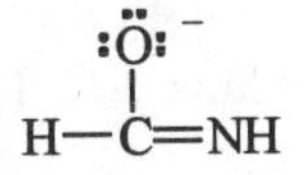 $:\ddot{O}:^-$ / $H-\overset{|}{C}=NH$ is more important for the same reason in 5b.: oxygen is more electronegative than nitrogen.

d. 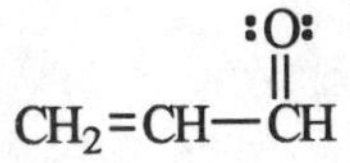 $:O:$ / $CH_2=CH-\overset{||}{C}H$ is the most important because of the absence of charge separation.

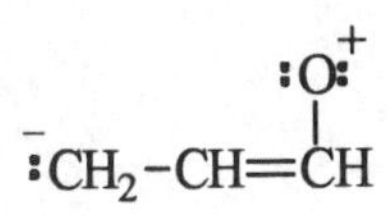

$:O:^+$ / $\bar{:}CH_2-CH=\overset{|}{C}H$ is the **least** important (highest energy) because an electronegative atom (oxygen) is forced to accept only six valence electrons, while a much less electronegative atom (carbon) must carry a negative charge.

e. These two structures are exactly equivalent, and therefore contribute equally to the resonance hybrid.

f. These two structures are also equivalent and contribute equally.

g. $CH_3–N^+≡C–O^-$ is more important than $CH_3–N^-–C≡O^+$ because oxygen is more electronegative than nitrogen. Neither resonance structure is as important as $CH_3–N=C=O$, which has no charge separation.

h. O=O=O is invalid because the central oxygen atom has ten valence electrons around it, as can be seen by drawing the full Lewis dot structure: :Ö::Ö::Ö: (three oxygen x 6= 18 valence electrons)

***NOTE** that **no** octet structure can be written for ozone that does not involve charge separation.*

i. ·N=O is more important because it has no charge separation and $^-$N=O·$^+$ does.

j. H—Ö—N̈=C: is less important because there are only six electrons around the carbon. There is charge separation in $HON^+≡C^-$, but all of the atoms have a filled octet.

k. In both structures, the atoms all have filled octets. However, $CH_2=C=O$ is more important because there is no charge separation.

6. $^-$:C≡O:$^+$ ↔ :C=Ö: will have a bond length between double and triple carbon-oxygen and will be shorter than the C-O bond in $(CH_3)_2C=O$, a simple carbon-oxygen double bond. $CH_3CO_2^-$ will have a carbon-oxygen bond length between that of a double and single bond; CH_3OH will have a simple carbon-oxygen single bond.

7. $CH_3C≡N$ will have the shortest C–N bond length; $HC≡N^+H$ ↔ $HC^+=NH$ will have a bond length between that of a double and triple bond; $(CH_3)_2C=NOH$ will have a C–N double bond and CH_3NH_2 will have a C–N single bond.

8. Na· ⟶ $Na^+ + e^-$ 118.0 kcal/mole *required*
Cl· + e^- ⟶ Cl^- -83.3 kcal/mole *liberated*

Na· + Cl· ⟶ $Na^+ + Cl^-$ 34.7 kcal/mole energy *required.*

From Coulomb's law, electrostatic attraction is proportional to q_1q_2/r^2; therefore if a positive and negative charge separated by 1Å have an energy of 330 kcal/mole, Na^+ and Cl^- at a distance of 2.36Å will have an energy of roughly $1/(2.36)^2 \times 330 = 59$ kcal/mole. This is enough to overcome the unfavorable ionization energy calculated above.

2.F Supplementary Problems

S1. a. Write Lewis structures for BH_4^-, CH_4, and NH_4^+. Do you expect the structures of these molecules to be similar or different?

b. Do the same for AlH_4^- and GaH_4^-. How should these structures compare to BH_4^-? to PH_4^+?

c. Write the Lewis structures of BH_3 and NH_3. Do these molecules have similar structures?

S2. Write Lewis structures for the following molecules.

a. CH_3F b. O_2^- c. ClNO

d. CH_3CHNH e. H_2CCCH_2 f. H_3BNH_3 (assign formal charges)

S3. For each of the compounds in problem #S2 above, show the hybridization of all atoms except hydrogen or halogens.

S4. Write out the Lewis structures and corresponding Kekulé structures for at least two resonance structures of each of the following compounds. Circle the one that represents the most important contributor to the structure of each compound, and justify your choice.

a. $[CH_2CHNH_2]$ b. $[CH_2NO]^-$ c. $[CH_2NCH_2]^+$ d. $[CH_2COCH_3]^-$ (with O above the first C of COCH₃)

2.G Answers to Supplementary Problems

S1. a.

```
 H -        H         H +
H:B:H     H:C:H     H:N:H
 H          H         H
```

Since the electronic structure is the same for all three molecules (they are isoelectronic), the structures will be similar: they are tetrahedral with sp^3-hybridized central atoms.

b.

```
 H -         H -
H:Al:H     H:Ga:H
 H           H
```

Aluminum and gallium are in the same column of the periodic table as boron, hence their valence shells are similar. BH_4^-, AlH_4^-, and GaH_4^- are all tetrahedral in shape (sp_3-hybridized), although the molecules get bigger on going from B→Al→Ga. PH_4^+ is isoelectronic with AlH_4^- and is tetrahedral as well.

c.

```
H:B:H      ..
 H        H:N:H
            H
```

BH_3 has six valence electrons, the boron is sp_2-hybridized and the molecule is therefore planar; NH_3 has eight valence electrons, the nitrogen is sp^3-hybridized, and the molecule is therefore pyramidal.

S2,3. a. 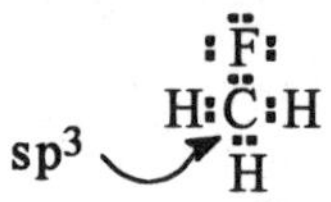 H:C(:F:):H with H below; carbon is sp^3

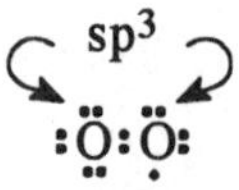

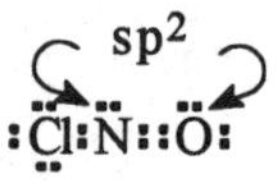

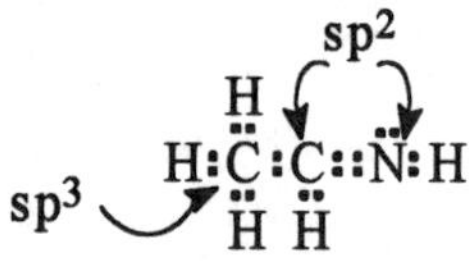

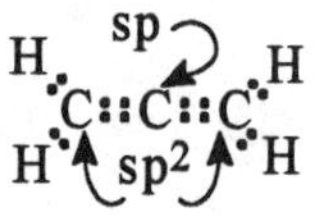

b. :Ö:Ö: — each oxygen sp^3

c. :Cl:N::O: — nitrogen and oxygen sp^2

d. H:C:C::N:H (with H's) — first carbon sp^3; second carbon and nitrogen sp^2

e. H₂C::C::CH₂ — terminal carbons sp^2, central carbon sp

f. $^{-}$H:B:N:H^{+} (with H's) — boron sp^3, nitrogen sp^3

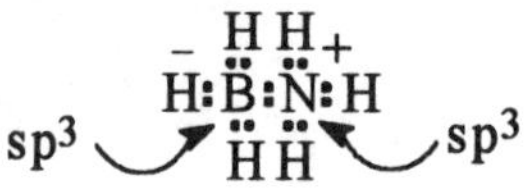

S4.

a.

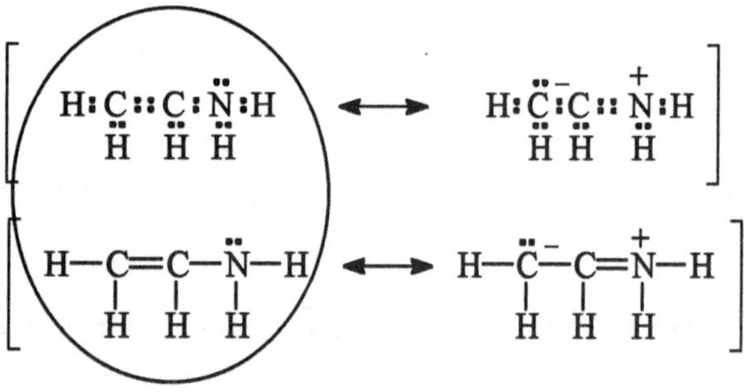

[H:C::C:N̈:H (H H H) ⟷ H:C̈⁻:C::N⁺:H (H H H)]

[H—C=C—N̈—H (H H H) ⟷ H—C̈⁻—C=N⁺—H (H H H)]

(left structures circled)

The structure on the left has no charge separation.

b.

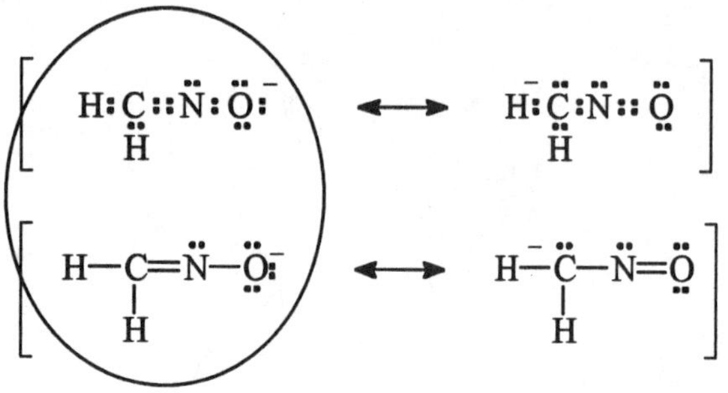

[H:C::N̈:Ö:⁻ (H) ⟷ H:C̈⁻:N̈::Ö (H)]

[H—C=N̈—Ö:⁻ (H) ⟷ H—C̈⁻—N̈=Ö (H)]

(left structures circled)

In the structure on the left, the more electronegative atom carries the negative charge.

c.

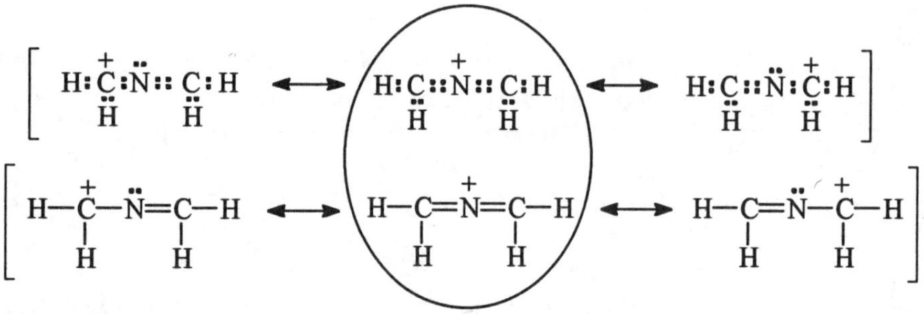

[H:C⁺:N̈::C:H (H H) ⟷ H:C::N⁺::C:H (H H) ⟷ H:C::N̈:C⁺:H (H H)]

[H—C⁺—N̈=C—H (H H) ⟷ H—C=N⁺=C—H (H H) ⟷ H—C=N̈—C⁺—H (H H)]

(center structures circled)

The structure in the center has filled octets.

d.

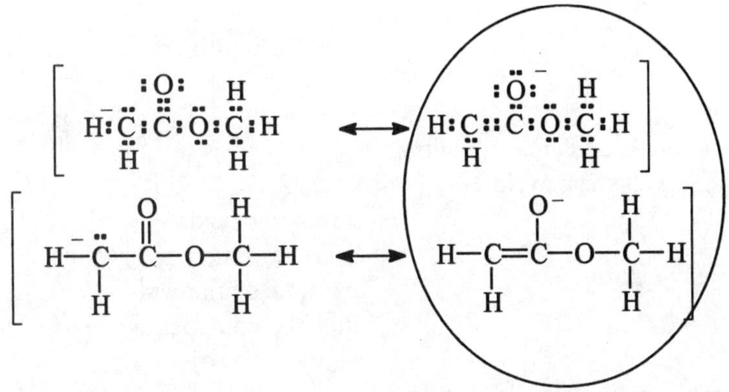

$$\left[\mathrm{H{:}\overset{-}{\ddot{C}}{:}\overset{:\ddot{O}:}{C}{:}\ddot{O}{:}\overset{H}{\underset{H}{C}}{:}H}\right] \longleftrightarrow \left[\mathrm{H{:}\underset{H}{C}{::}\overset{:\ddot{O}:^{-}}{C}{:}\ddot{O}{:}\overset{H}{\underset{H}{C}}{:}H}\right]$$

$$\left[\mathrm{H{-}\underset{H}{\overset{-\,\cdot\cdot}{C}}{-}\overset{O}{\overset{\|}{C}}{-}O{-}\underset{H}{\overset{H}{C}}{-}H}\right] \longleftrightarrow \left[\mathrm{H{-}\underset{H}{C}{=}\overset{O^{-}}{C}{-}O{-}\underset{H}{\overset{H}{C}}{-}H}\right]$$

In the structure on the right, the more electronegative element carries the negative charge.

3. ORGANIC STRUCTURES

3.A. Chapter Outline and Important Terms Introduced

3.1 Introduction (historical)

- isomerism
- structural formulas
- condensed formulas

3.2 Functional Groups (commonly-encountered subunits)

- hydrocarbons (alkanes, alkenes, alkynes, cyclo-)
- alcohols
- ethers
- aldehydes
- ketones
- carboxylic acids
- (all other functional groups)
- aromatic compounds
- aliphatic compounds

3.3 The Shape of Molecules (their three-dimensionality)

- molecular models

3.4 The Determination of Organic Structure (how to figure out the formula)

- characterization
- combustion analysis
- empirical formula
- molecular formula
- spectroscopy

3.5 n-Alkanes, the Simplest Organic Compounds (the backbone of organic structure)

- homologous series
- saturated and unsaturated hydrocarbons

3.6 Systematic Nomenclature (naming compounds)

- IUPAC system
- isomers
- common names (iso-, neo-)
- chemical literature

3.B. Important Reactions Introduced: still none yet

3.C. Important Concepts and Hints

Chapter 3 continues to discuss structure, but emphasizes the molecular level as opposed to the electronic aspects which were presented in Chapter 2. A good way to organize your thoughts on this topic is to consider the chart on the following page, in which each level down the scheme corresponds to a more specific or detailed description of the structure of a molecule. You should know the difference between empirical and molecular formulas, and understand the concept of **isomerism.** For instance, you should be able to draw all the structural isomers of a compound, given its molecular formula. In this regard, you should also be able to tell that compounds A,B, and C below are structural isomers of each other, while structures D,E, and F represent the same compound.

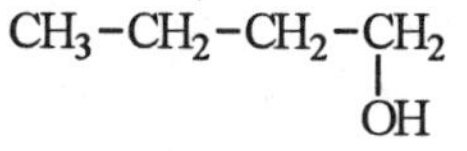

$CH_3-CH_2-CH_2-CH_2-OH$ (OH on the terminal CH_2)

<u>A</u>

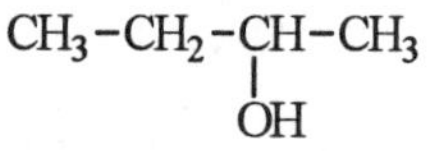

$CH_3-CH_2-CH(OH)-CH_3$ (OH on the CH)

<u>B</u>

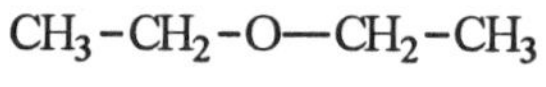

$CH_3-CH_2-O-CH_2-CH_3$

<u>C</u>

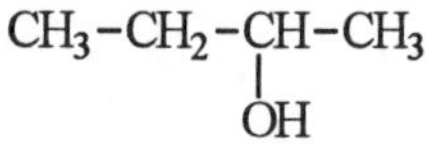

$CH_3-CH_2-CH(OH)-CH_3$ (OH on the CH)

<u>D</u>

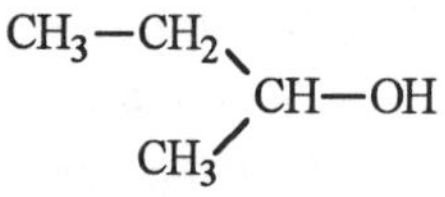

$(CH_3-CH_2)(CH_3)CH-OH$

<u>E</u>

$CH_3-CH(OH)-CH_2-CH_3$ (OH on the CH)

<u>F</u>

Another important concept is the idea of functional groups, subunits of molecular structure thet are larger than single atoms. These are important not only in considering structural isomerism, but (as you will discover as the course progresses) also in understanding and predicting the reactions compounds will undergo.

After discussing the various levels of molecular structure, how this information can be obtained for a compound is presented. In this chapter you should learn how to determine empirical and molecular formulas, and you should understand the concept of functional group tests. The more detailed methods of studying a molecule's structure are presented in the chapters that discuss spectroscopy (Chapters 13 and 17).

In addition to molecular and structural formulas, we also have to have names for compounds, so that we can discuss them verbally, index them, etc. Although nomenclature seems like a sideline to reactions when studying chemistry, it is important to learn it at the outset of the course so you will know what is being discussed the rest of the time. The systematic nomenclature (IUPAC) is the most important; it does become complex (as the molecules to be named become complex), but it is systematic and the step-by-step rules to follow are easily remembered. The simpler aspects of common nomenclature are useful too, just as colloquialisms and contractions are necessary in our everyday language.

Level	**Example**
Empirical Formula	$(C_3H_6O)_n$ (this excludes CH_4O, C_3H_8O, etc.)
Molecular Formula	$C_6H_{12}O_2$ (as opposed to C_3H_6O or $C_9H_{18}O_3$, etc.)
Structural Formula	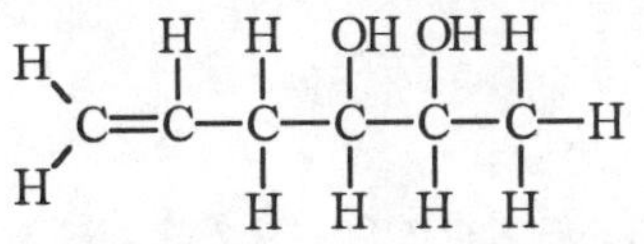 $H_2C{=}CH{-}CH_2{-}CH(OH){-}CH(OH){-}CH_3$ ≡ $CH_2{=}CHCH_2CHOHCHOHCH_3$

at this level, one can rule out molecules having the same molecular formula (isomers) but with different topology; i.e., with the atoms connected in a different order, such as

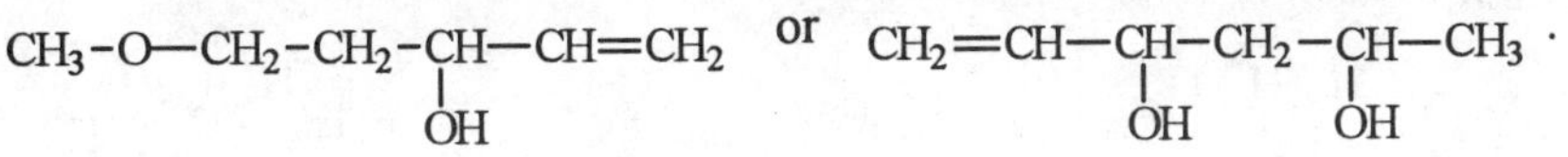

$CH_3{-}O{-}CH_2{-}CH_2{-}CH(OH){-}CH{=}CH_2$ or $CH_2{=}CH{-}CH(OH){-}CH_2{-}CH(OH){-}CH_3$.

[Stereochemical Formula (Newman projection: front H, HO, HO; back $CH_2CH{=}CH_2$, CH_3, H) = (zigzag drawing with H, OH; H, OH)

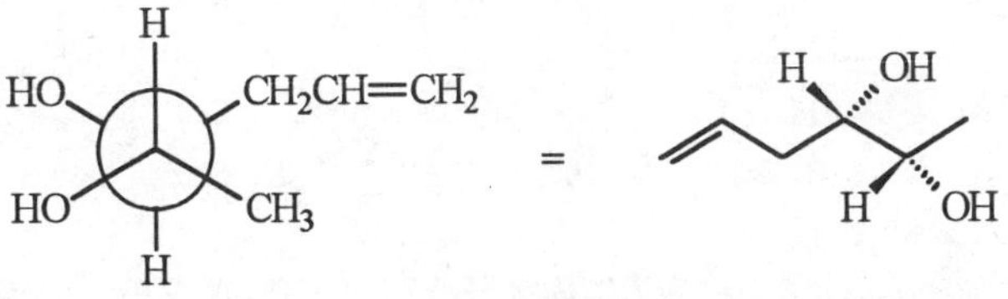

NOTE: the concept of stereochemistry is not discussed in the text until Chapter 7, Stereoisomerism]

HINTS: *BUY A SET OF MODELS*! As pointed out in Section 3.3, molecules are three-dimensional and much of organic structure and chemical reactions cannot be understood without thinking in three dimensions. You may not like to admit it, but most people spend their lives on the two-dimensional surface of the earth and have little experience in thinking in three dimensions. Using models is the best way for you to really get an idea of what the two-dimensional illustrations in the text or on the blackboard represent in three dimensions. Other aspects of structure, such as rotation around single bonds (and therefore the floppiness of big molecules), are easily understood using models, too.

Unless the three-dimensional aspects of structure are specifically indicated (for instance, as will be introduced in Chapter 7), structural formulas only show "connectedness", i.e. what atom is bonded to what atom. A common trap that students fall into arises from a misunderstanding of this point. Although the following drawings look different, they *all* represent the same molecule:

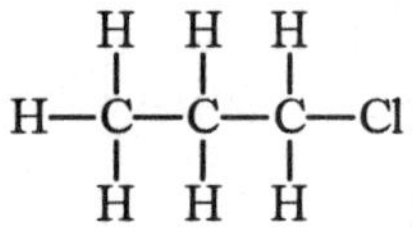

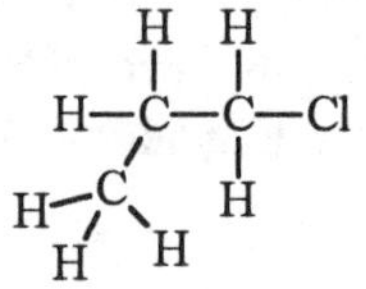

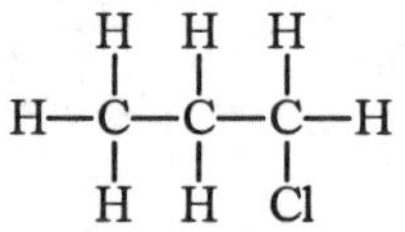

```
   H  H  H                 H  H                    H  H  H
   |  |  |                 |  |                    |  |  |
H—C——C——C—Cl            H—C——C—Cl              H—C——C——C—H
   |  |  |                /   |                    |  |  |
   H  H  H             H—C    H                    H  H  Cl
                        H  H
```

Similarly, the three drawings below represent the same molecule, because of the ability of groups joined by single bonds to rotate relative to each other. The best way to convince yourself of this may be to make models... However, when working on a plane surface, like a piece of paper during a quiz or exam, it is often helpful to find the longest continuous chain, just as if you were going to name the compound, then locate the substituents. In the examples below, the longest continuous chain has 6 carbons with a methyl and an OH both on C–3; therefore the three structures represent the same compound.

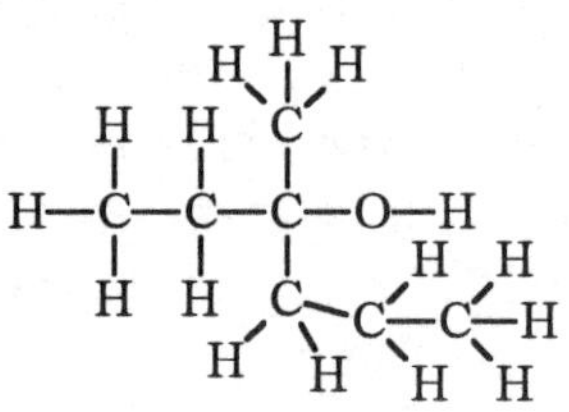

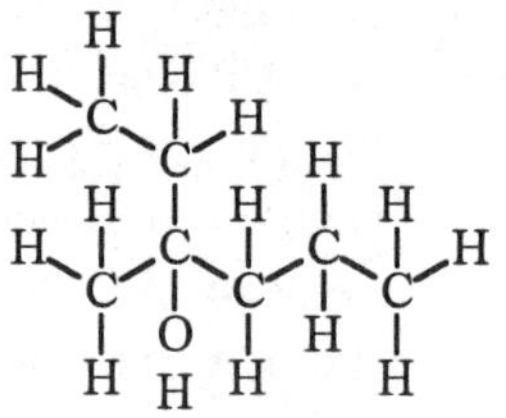

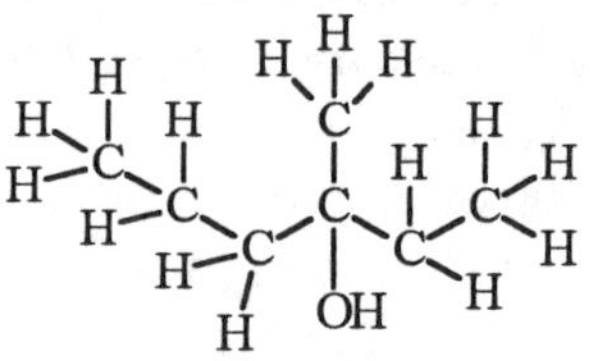

```
      H H H
       \|/
  H  H  C
  |  |  |
H—C——C——C—O—H
  |  |  |   H  H
  H  H  C—C——C—H
       H  H  H  H
```

3.D. Answers to Exercises

3.1 C_4H_{10} :

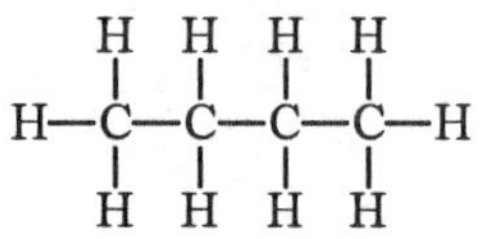

```
   H  H  H  H
   |  |  |  |
H—C——C——C——C—H   ≡ CH3CH2CH2CH3
   |  |  |  |
   H  H  H  H
```

$\equiv CH_3CH_2CH_2CH_3$

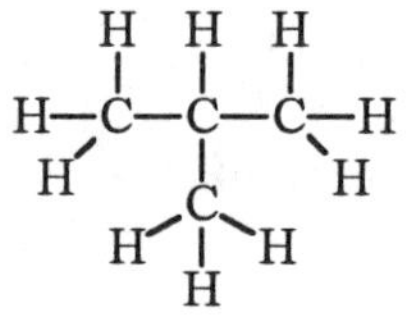

```
   H  H  H
   |  |  |
H—C——C——C—H
 H    |    H
      C
    H | H
      H
```

$\equiv (CH_3)_3CH$

C_5H_{12} :

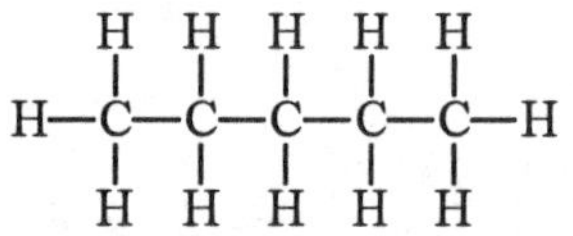

```
   H  H  H  H  H
   |  |  |  |  |
H—C——C——C——C——C—H
   |  |  |  |  |
   H  H  H  H  H
```

$\equiv CH_3CH_2CH_2CH_2CH_3$;

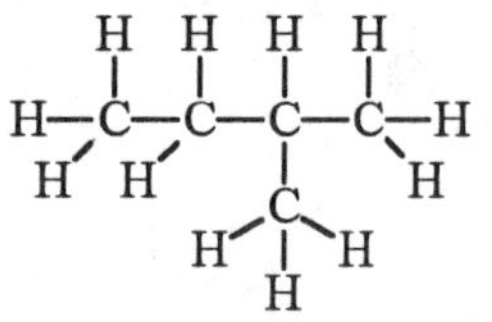

```
   H  H  H  H
   |  |  |  |
H—C——C——C——C—H
 H  H     |   H
          C
        H | H
          H
```

$\equiv CH_3CH_2CH(CH_3)_2$;

```
         H
       H | H
         C
   H     |     H
H—C——C——C—H
   H     |     H
         C
       H | H
         H
```

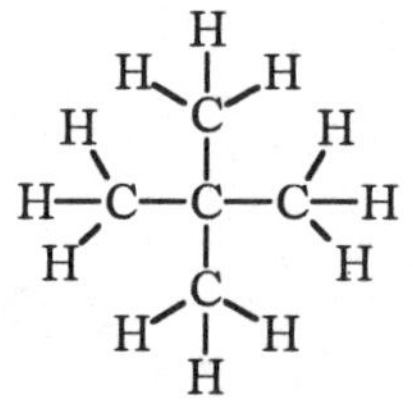

$\equiv (CH_3)_4C$

3.2

Class	*Structural*	*Condensed*
alkene	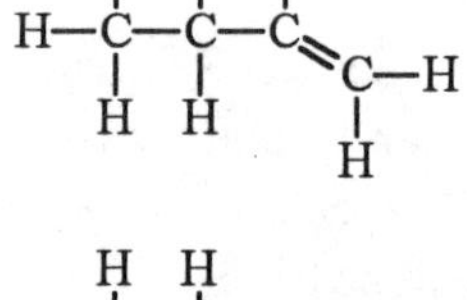 H—C(H)(H)—C(H)(H)—C(H)=C(H)—H	$CH_3CH_2CH{=}CH_2$
alkyne	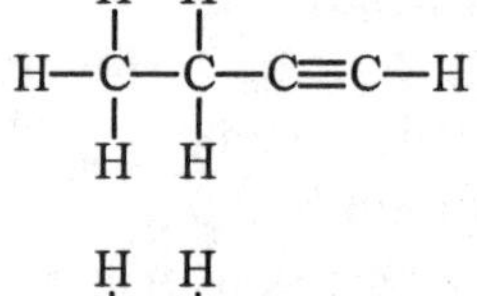 H—C(H)(H)—C(H)(H)—C≡C—H	$CH_3CH_2C{\equiv}CH$
alcohol	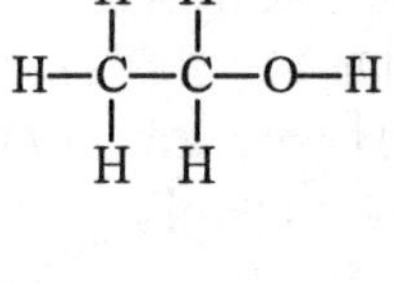 H—C(H)(H)—C(H)(H)—O—H	CH_3CH_2OH

ether

```
   H  H     H  H
   |  |     |  |
H—C—C—O—C—C—H
   |  |     |  |
   H  H     H  H
```

$(CH_3CH_2)_2O$

aldehyde

```
   H  H   O
   |  |  //
H—C—C—C
   |  |   \
   H  H    H
```

CH_3CH_2CHO

ketone

```
   H  H  O  H  H
   |  |  ||  |  |
H—C—C—C—C—C—H
   |  |     |  |
   H  H     H  H
```

$(CH_3CH_2)_2CO$

carboxylic acid

```
   H  H   O
   |  |  //
H—C—C—C
   |  |   \
   H  H    OH
```

$CH_3CH_2CO_2H$

amine

```
   H  H
   |  |
H—C—C—N—H
   |  |  |
   H  H  H
```

$CH_3CH_2NH_2$

nitrile

```
   H  H
   |  |
H—C—C—C≡N
   |  |
   H  H
```

$CH_3CH_2C{\equiv}N$

3.3 Model building exercise - no answer.

3.4 Molecular weight of $C_6H_{12}O$:

$$\begin{aligned} 6 \times 12 &= 72 \\ 12 \times 1 &= 12 \\ 1 \times 16 &= \underline{16} \\ & 100 \end{aligned}$$

Therefore 3.74 mg = $(3.74 \times 10^{-3}$ g)/100 g/mole = 3.74×10^{-5} mole

Overall reaction: $C_6H_{12}O + 8\ 1/2\ O_2 \longrightarrow 6\ CO_2 + 6\ H_2O$

3.74×10^{-5} mole of $C_6H_{12}O$ will produce $6 \times 3.74 \times 10^{-5} = 2.24 \times 10^{-4}$ mole of CO_2 and H_2O.
2.24×10^{-4} x [molecular weight of CO_2 = 44.0] = 9.87 mg of CO_2
2.24×10^{-4} x [molecular weight of H_2O = 18.0] = 4.04 mg of H_2O

3.5 0.677 g of CO_2 contains 0.677/44.0 = 0.0154 mole of C
0.311 g of H_2O contains 0.311/18.0 = 0.0173 mole of H_2 = 0.0346 mole of H

Together, this amount of carbon and hydrogen account for:
0.0154 mole x 12.0 g/mole = 0.185 g of carbon
0.0346 mole x 1.01 g/mole = 0.035 g of hydrogen
total = 0.220 g of the 0.250 g of di-*n*-butyl ether

The rest must be 0.250 – 0.220 = 0.030 g = 0.0019 mole of oxygen

Divide each of the molar amounts by the smallest number of moles, i.e., 0.0019 mole of O. This gives: 8.1 C, 18 H and 1.0 O. Therefore, di-*n*-butyl ether is $C_8H_{18}O$.

3.6

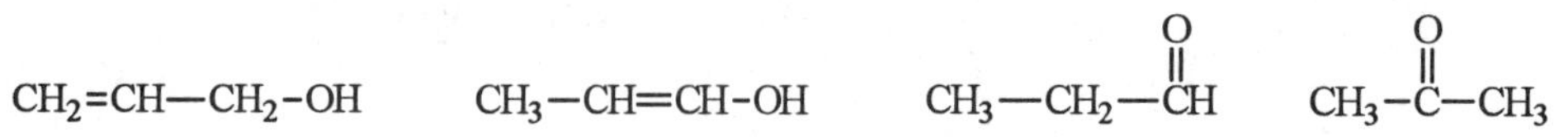

$CH_2{=}CH{-}CH_2{-}OH$ [1]

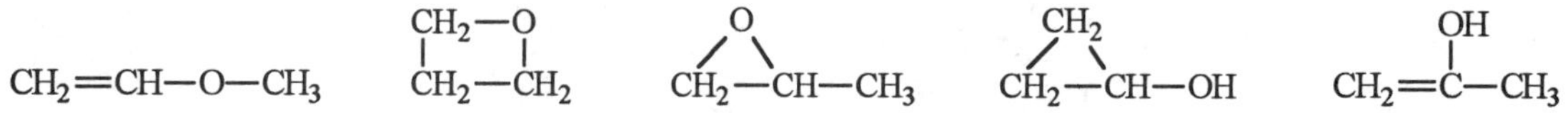

$CH_3{-}CH{=}CH{-}OH$ [2]

$CH_3{-}CH_2{-}CH{=}O$ [3]

$CH_3{-}C(=O){-}CH_3$ [4]

$CH_2{=}CH{-}O{-}CH_3$ [5]

$CH_2{-}O$ / $CH_2{-}CH_2$ (four-membered ring) [6]

$CH_2{-}CH{-}CH_3$ with O bridging (epoxide) [7]

$CH_2{-}CH{-}OH$ with CH_2 bridging (cyclopropanol) [8]

$CH_2{=}C(OH){-}CH_3$ [9]

The only structures with the formula C_3H_6O that also have C=O are [3] and [4].
(NOTE: You will learn in Chapter 15 that [2] and [9] are unstable structures).

3.7 Pentyl group: $CH_3CH_2CH_2CH_2CH_2{-}$

Pentyl iodide: $CH_3CH_2CH_2CH_2CH_2I$

3.8 3-ethylhexane:

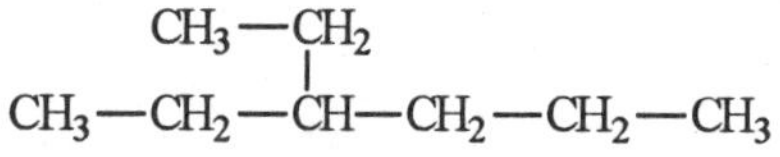

$$CH_3{-}CH_2{-}CH(CH_2{-}CH_3){-}CH_2{-}CH_2{-}CH_3$$

"2-ethylhexane":

$$CH_3{-}CH(CH_2{-}CH_3){-}CH_2{-}CH_2{-}CH_2{-}CH_3$$

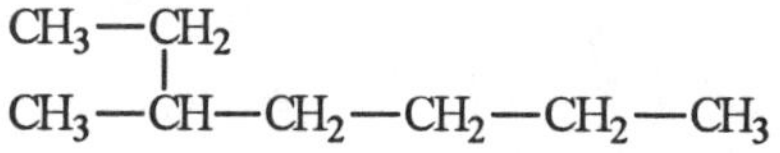

"2-ethylhexane" is actually 3-methylheptane.

3.9

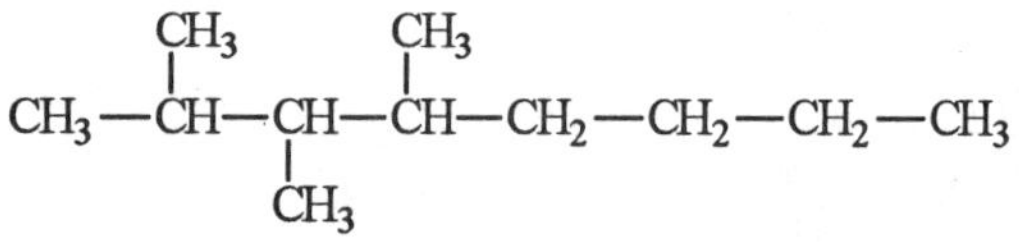

$$CH_3{-}CH(CH_3){-}CH(CH_3){-}CH(CH_3){-}CH_2{-}CH_2{-}CH_2{-}CH_3$$

2,3,4-trimethyloctane

$$CH_2(CH_3){-}CH_2{-}CH_2{-}CH_2(CH_3)$$

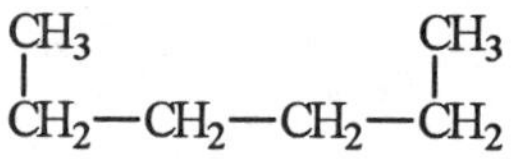

"1,4-dimethylbutane" is simply hexane

3.10

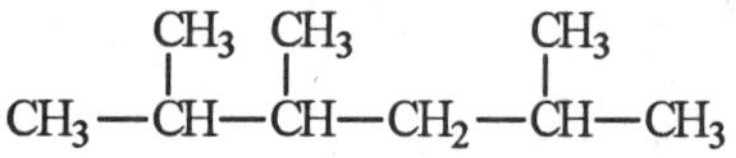

$$CH_3{-}CH(CH_3){-}CH(CH_3){-}CH_2{-}CH(CH_3){-}CH_3$$

1	2	3	4	5	6
6	5	4	3	2	1

2,3,5-trimethylhexane (correct name)

2,4,5-trimethylhexane (numbering from wrong end)
The first point of difference is at C–3 (C–4 numbering the wrong way)

3.11

a.

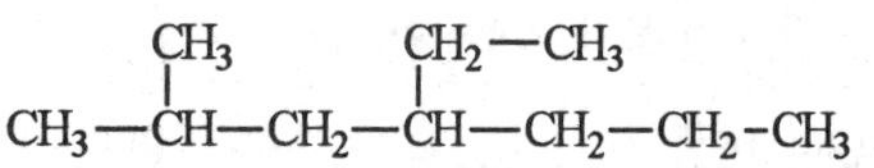

$$CH_3{-}CH(CH_3){-}CH_2{-}CH(CH_2{-}CH_3){-}CH_2{-}CH_2{-}CH_3$$

Incorrect name: "2-methyl-4-ethylheptane" (ethyl comes before methyl in alphabet)
Correct name: 4-ethyl-2-methylheptane (note that alphabetization doesn't change numbering scheme.)

b.
```
       CH3 CH2CH3
        |   |
CH3—C—CH—CH2—CH2—CH3
        |
       CH3
```

Correct name: 3-ethyl-2,2-dimethylhexane

c.
```
H3C    CH3
   \  /
    CH
    |
CH3—CH-CH—CH2—CH2—CH3
           |
          CH3
```

Incorrect name: 2-isopropyl-3-methylhexane (longest chain has seven carbons)
Correct: 2,3,4-trimethylheptane

d.
```
              H3C    CH3
                 \  /
                  CH
                  |
CH3—CH2—CH-CH—CH2—CH2—CH3
             |
      CH3—CH2
```

Incorrect name: 4-isopropyl-3-ethylheptane (ethyl comes before isopropyl)
Correct: 3-ethyl-4-isopropylheptane

3.12
```
       CH3        Cl
        |         |
CH3—CH—CH2-CH—CH3
```

Incorrect name: 4-chloro-2-methylpentane (chloro comes before methyl and, other things being equal, should get the lower number)
Correct: 2-chloro-4-methylpentane

```
                   CH3
                    |
         X          CH2         X
         |          |           |
CH3—CH—CH2-CH—CH2—CH—CH3
```

X = Cl: 2,6-dichloro-4-ethylheptane
X = I: 4-ethyl-2,6-diiodoheptane

3.E. Answers and Explanations for Problems

1. $CH_3CH_2CH_2CH_2CH_2Cl$ 1-chloropentane

   $CH_3CH_2CH_2CHClCH_3$ 2-chloropentane (this is the same as $CH_3CHClCH_2CH_2CH_3$; also, 4-chloropentane is just an incorrect name for 2-chloropentane.)

   $CH_3CH_2CHClCH_2CH_3$ 3-chloropentane

$(CH_3)_2CHCH_2CH_2Cl$	1-chloro-3-methylbutane
$(CH_3)_2CHCHClCH_3$	2-chloro-3-methylbutane
$(CH_3)_2CClCH_2CH_3$	2-chloro-2-methylbutane (note that the direction of numbering has changed; 3-chloro-3-methylbutane is an incorrect name)
$ClCH_2CH(CH_3)CH_2CH_3$	1-chloro-2-methylbutane
$CH_3C(CH_3)_2CH_2Cl$	1-chloro-2,2-dimethylpropane

2. $CH_3CH_2CH_2CH_2OH$ $CH_3OCH_2CH_2CH_3$
$CH_3CH_2CHOHCH_3$ $CH_3OCH(CH_3)_2$
$(CH_3)_2CHCH_2OH$ $CH_3CH_2OCH_2CH_3$
$(CH_3)_3COH$

3.

$CH_3CH_2CH_2CH_2CH_2CH_2CH_3$	heptane
$(CH_3)_2CHCH_2CH_2CH_2CH_3$	2-methylhexane
$CH_3CH_2CH(CH_3)CH_2CH_2CH_3$	3-methylhexane
$(CH_3)_3CCH_2CH_2CH_3$	2,2-dimethylpentane
$CH_3CH_2C(CH_3)_2CH_2CH_3$	3,3-dimethylpentane
$(CH_3)_2CHCH(CH_3)CH_2CH_3$	2,3-dimethylpentane
$(CH_3)_2CHCH_2CH(CH_3)_2$	2,4-dimethylpentane
$(CH_3CH_2)_3CH$	3-ethylpentane (this is the one most people miss)
$(CH_3)_3CCH(CH_3)_2$	2,2,3-trimethylbutane

4. There are **four** monobromo derivatives of heptane (1-bromo, 2-bromo, 3-bromo and 4-bromoheptane) 4

Six from 2-methylhexane (1-bromo-, 2-bromo-, 3-bromo- and 4-bromo-2-methylhexane; 1-bromo- and 2-bromo-5-methylhexane are all different) 6

Seven from 3-methylhexane (1-bromo-, 2-bromo- and 3-bromo-3-methylhexane; 1-bromo-, 2-bromo- and 3-bromo-4-methylhexane; 3-bromomethylhexane) 7

Four from 2,2-dimethylpentane (1-bromo-, 3-bromo- and 4-bromo-2,2-dimethylpentane; 1-bromo-4,4-dimethylpentane) 4

Three from 3,3-dimethylpentane (1-bromo- and 2-bromo-3,3-dimethylpentane and 3-bromomethyl-3-methylpentane) 3

Six from 2,3-dimethylpentane (1-bromo-, 2-bromo-, 3-bromo-2,3-dimethylpentane; 1-bromo- and 2-bromo-3,4-dimethylpentane as well as 3-bromomethyl-2-methylpentane) [Note: you might think that "2-bromomethyl-3-methylpentane" is a 7th isomer, but this is an incorrect name for one of the others. Which one?] 6

Three only from 2,4-dimethylpentane (1-bromo-, 2-bromo- and 3-bromo-2,4-dimethyl-pentane) 3

Three from 3-ethylpentane (1-bromo-, 2-bromo- and 3-bromo-3-ethylpentane) 3

Three from 2,2,3-trimethylbutane (1-bromo-2,2,3-trimethylbutane and 1-bromo- and 2-bromo-2,3,3-trimethylbutane; notice how the numbering scheme for the last two isomers changes from that of the hydrocarbon.) 3

Total is thirty-nine structural isomers* 39

*NOTE: You will learn in Chapter 8 that stereoisomers exist for some of these compounds as well.

5. a. alkene b. ether c. alcohol d. alkyl halide e. carboxylic acid f. ketone g. aldehyde h. disulfide i. thiol j. sulfide k. aromatic ring l. primary amine m. alkyne n. organometallic

6.

```
        CH3                          CH3                   CH3
a.      |               b.          /             c.       |
   CH3-C-CH3              CH3-CH                     CH3-C-Br  ≡ (CH3)3CBr
        |                            \                     |
        CH3                          CH3                   CH3
```

d. $(CH_3)_2CHCH_2I$ e. $(CH_3)_2CHCl$ f. $CH_3CH_2CHBrCH_3$

```
                                                                              H     H
                                                                               \   /
      H    H  H  H  H  H              H  H   CH2—CH2                            C—C       O
       \   |  |  |  |  |              |  |  /       \                          //   \\    //
7. a.   C=C—C—C—C—C—H        b. H—C—C—CH          C=O        c. H—C       C—C
       /      |  |  |  |              |  |  \       /                          \    /     \
      H       H  H  H  H              H  H   CH2—CH2                            C=C       OH
                                                                               /   \
                                                                              H     H

        H  H  H  O                         H  H  O     H
        |  |  |  ||                        |  |  ||    |
d. i. H—C—C—C—C—N—H              ii.  H—C—C—C—N—C—H
        |  |  |     |                      |  |     |  |
        H  H  H     H                      H  H     H  H
```

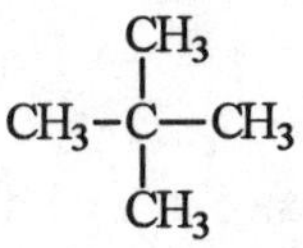

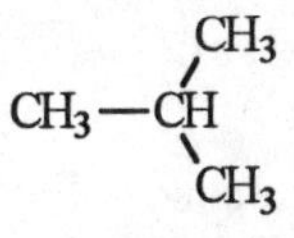

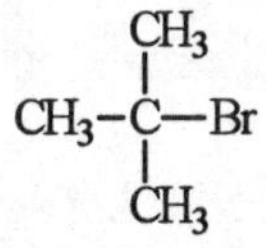

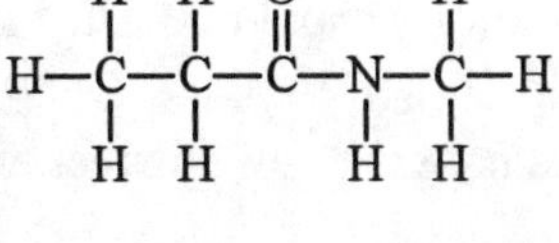

There are many other structures that correctly answer this question.

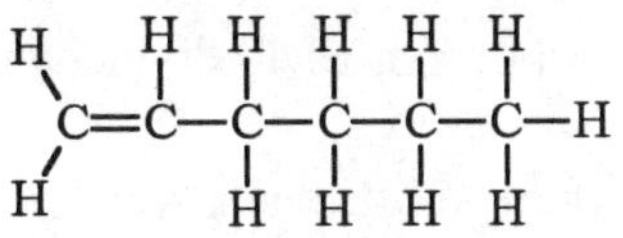

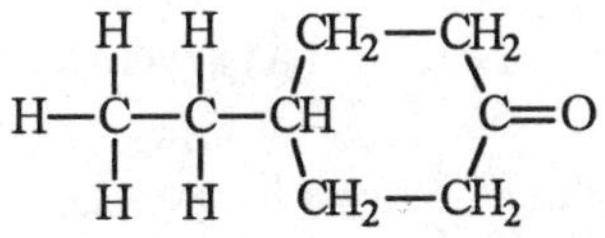

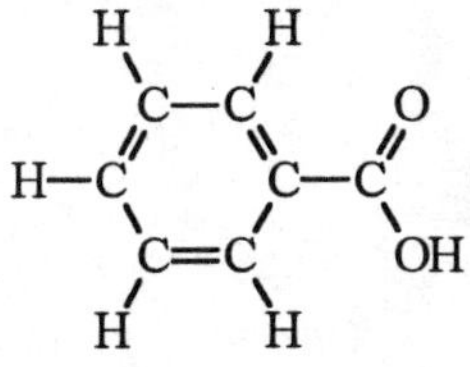

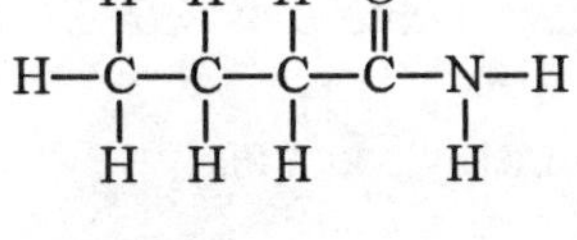

8.

```
                 CH2—CH2—CH3                                   CH2—CH3
                 |                                             |
a. CH3—CH2—CH-C—CH—CH2—CH2—CH3         b. CH3—CH2—C—CH2—CH2—CH3
            /    |   \                                         |
         CH3     CH3  CH3                                      F
```

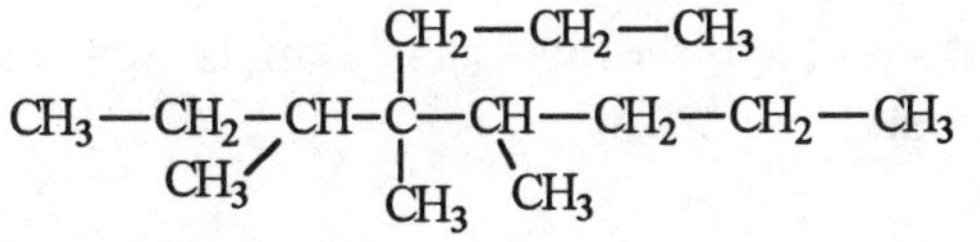

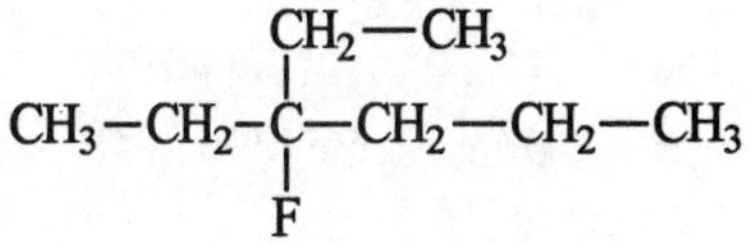

c.
$$\begin{array}{c} CH_3 \\ | \\ CH_2CH_2CHCH_3 \\ | \\ CH_3CH_2CH_2CH_2CH_2CHCH_2CH_2CH_2CH_2CH_3 \end{array}$$

d.
$$\begin{array}{c} CH_3 \\ | \\ CH_3-C-CH_3 \\ | \\ CH_3CH_2CH_2CHCH_2CH_2CH_3 \end{array}$$

e. $CH_3CHCH_2CH_2CH_2CH_2CH_2CH_2CH_2CH_2CH_2CH_2CH_2CH_2CH_2CH_2CH_3$ (with CH_3 on C-2) $\equiv (CH_3)_2CH(CH_2)_{14}CH_3$

f.
$$\begin{array}{c} H_3C\ \ CHClCH_3 \\ |\quad\ | \\ CH_3CH_2C-CHCH_2CH_2CH_3 \\ | \\ CH_3 \end{array}$$

g.
$$\begin{array}{c} H_3C\diagdown CHC(CH_3)_3 \\ | \\ CH_3CH_2CH_2CH_2CHCCH_2CH_2CH_2CH_2CH_2CH_3 \\ H_3C\diagup\ \diagdown CH_3 \end{array}$$

h. $(CH_3)_2CHCH_2CH_2C(CH_2CH_3)_3$

9. a. 2,5-dimethylhexane
   b. 3-ethyl-5,5,7-trimethylnonane
   c. 4-ethyl-3-methylheptane
   d. 1-bromo-4-chloro-2-methylpentane
   e. 5-ethyl-4-iodo-2,2-dimethyloctane
   f. 3,6-diethyl-2,6-dimethyloctane
   g. 7-(4,4-dimethylhexyl)-3,3,11,11-tetramethyltridecane
   h. 3,3-diethylpentane
   i. 3-ethyl-4-methylhexane
   j. 4-ethyl-3,3-dimethylhexane

10. a. Where on the heptane backbone is the methyl attached? This has to be specified in a complete name.

    b. "4-Methylhexane" should be numbered from the other end to give 3-methylhexane.

    c. The longest chain in "3-propylhexane" has seven carbons; it should be named 4-ethylheptane.

    d. When choosing between chains of equal length, the one that has more substituents should be chosen: 3-ethyl-2,5,5-trimethyloctane.

    e. Alkyl and halo substituents should be listed in alphabetical order: 3-chloro-4-methylhexane.

    f. The prefix "di-" does not count in alphabetizing: 3-ethyl-2,2-dimethylpentane.

    g. Numbered from the wrong end: 3,4,5,7-tetramethylnonane.

    h. A position must be specified for every substituent, even if the position is the same: 2,2-dimethylpropane.

11. The formula C_5H_{12} represents a fully saturated hydrocarbon; if the molecular formula were a multiple of this, there would be too many hydrogens to go around.

12. Compounds containing only C, H and O will always have an even number of H's since the valence of carbon is 4 and that of oxygen is 2. Since the atomic weights of C and O are even, 12 and 16

respectively, the molecular weight must be even. Since the valence of nitrogen is 3, there will always be an odd number of H's in molecules with odd numbers of N's. Since the atomic weight of N is even, the molecular weight will always be odd.

13. a. 3,3-dimethylheptane (C_9H_{20})

molecular weight = (9 x 12) + (20 x 1) = 128
% carbon = (9 x 12)/128 = 84.4%
% hydrogen = (20 x 1)/128 = 15.6%

b. ethyl acetate ($C_4H_8O_2$)

molecular weight = (4 x 12) + (8 x 1) + (2 x 16) = 88
% carbon = (4 x 12)/88 = 54.5%
% hydrogen = (8 x 1)/88 = 9.1%
% oxygen = (2 x 16)/88 = 36.4%

c. nitromethane (CH_3NO_2): 19.7% C, 4.9% H, 23.0% N, 52.5% O

d. trinitrotoluene ($C_7H_5N_3O_6$): 37.0% C, 2.2% H, 18.5% N, 42.3% O

14. a. 15.73 mg of CO_2 contain 15.73/44.01 = 3.574×10^{-4} mole of C
6.38 mg of H_2O contain 2 x 6.38/18.0 = 7.09×10^{-4} mole of H

$3.574 \times 10^{-4} \times 12.01 = 4.293$ mg of C
$7.09 \times 10^{-4} \times 1.01 = .716$ mg of H

The rest of the sample is assumed to be oxygen:

5.72 – 4.293 – .716 = 0.71 mg = 4.4×10^{-5} mole of O
Divide each of the moles by the smallest number, i.e., 4.4×10^{-5} mole O. This gives 8.0 moles C to 16 moles H to 1.0 mole O. Empirical Formula: $C_8H_{16}O$.

b. 12.3 mg of CO_2 = 3.36 mg C = 0.280 mmole of C (a mmole is 10^{-3} mole)
3.90 mg of H_2O = 0.433 mg H = 0.433 mmole of H

3.81 mg of sample – 3.355 mg of C – 0.433 mg of H = 0.02 mg (~ 0) unaccounted for

0.280 mmole C and 0.433 mmole H are compatible with the empirical formulas: C_2H_3, C_4H_6, etc.

C_2H_3 is excluded as a stable molecule by the rules of valence, hence the sample has the formula C_4H_6.

c. 5.87 mg of CO_2 = 1.60 mg = 0.133 mmole of C

2.40 mg of H_2O = 0.267 mg = 0.267 mmole of H

2.58 – 1.60 – 0.27 = 0.71 mg = 0.0444 mmole of O

$C_{0.133}H_{0.267}O_{0.0444} = C_3H_6O$

d. 4.99 mg of CO_2 = 1.36 mg = 0.113 mmole of C

2.05 mg of H_2O = 0.228 mg = 0.228 mmole of H

3.41 – 1.36 – .228 = 1.82 mg = 0.114 mmole of O

Empirical Formula: $C_{.113}H_{.228}O_{.114}$ or CH_2O

15. a. "70.4% C, 13.9% H" means: of 100 g of the compound, 70.4 g is carbon, 13.9 g is hydrogen, and the rest (15.7 g) is assumed to be oxygen. A 100-g sample would then contain 70.4/12 = 5.87 moles of carbon, 13.9/1 = 13.9 mole of hydrogen, and 15.7/16 = 0.98 moles of oxygen. This empirical formula of $C_{5.87}H_{13.9}O_{0.98}$ is clearly $C_6H_{14}O$ in integral values. Because this formula represents a fully saturated compound, no multiple of it is possible (see problem #11).

	Weight Ratio	Mole Ratio	Empirical Formula
b.	92.1% C	92.1/12 = 7.68	
	7.9% H	7.9/1 = 7.9	$C_{7.68}H_{7.9}$ = $(CH)_n$ (Benzene is C_6H_6)
c.	71.6% C	71.6/12 = 5.97	
	7.5% H	7.5/1 = 7.5	$C_{5.97}H_{7.5}N_{1.49}$, divide by the smallest number gives
	20.9% N	20.9/4 = 1.49	$(C_4H_5N)_n$ (Pyrrole is C_4H_5N)
d.	71.6% C	71.6/12 = 5.97	$C_{5.97}H_{6.7}N_{0.35}O_{1.05}$, divide by smallest number
	6.7% H	6.7/1 = 6.7	$(C_{17}H_{19}NO_3)_n$ (Morphine is $C_{17}H_{19}NO_3$)
	4.9% N	4.9/14 = 0.35	
	(16.8% O)	16.8/16 = 1.05	the weight not accounted for is assumed to be oxygen
e.	74.1% C	74.1/12 = 6.18	$C_{6.18}H_{7.5}N_{0.614}O_{0.613}$, divide by the smallest number
	7.5% H	7.5/1 = 7.5	$(C_{10}H_{12}NO)_n$ (Quinine is $C_{20}H_{24}N_2O_2$)
	8.6% N	8.6/14 = 0.614	
	(9.8% O)	9.8/16 = 0.613	the weight not accounted for is assumed to be oxygen
f.	47.4% C	3.95	divide by smallest: $C_{2.8}H_{1.85}Cl$
	2.6% H	2.6	take multiples until all subscripts are integers in this
	50.0% Cl	1.41	case, x 5: $(C_{14}H_9Cl_5)_n$ (DDT is $C_{14}H_9Cl_5$)
g.	38.4% C	3.2	
	4.9% H	4.9	C_2H_3Cl (Vinyl chloride is C_2H_3Cl)
	56.7% Cl	1.6	
h.	23.4% C	1.9	
	1.4% H	1.4	
	65.3% I	65.3/127 = 0.514	$(C_{15}H_{11}I_4NO_4)_n$ (Thyroxine is $C_{15}H_{11}I_4NO_4$)
	1.8% N	0.13	
	(8.1% O)	0.51	

16. a. 0.0132 g of camphor gives 0.0382 g of CO_2, which is equivalent to 8.68 x 10^{-4} mole (0.0382 divided by 44). Therefore, 0.0132 g of camphor contains 8.68 x 10^{-4} mole = 0.0104 g of carbon.

0.0132 g of camphor gives 0.0126 g of H_2O, which is equivalent to 7.0 x 10^{-4} mole (0.0126 divided by 18). Therefore 0.0132 g of camphor contains 14 x 10^{-4} mole = 0.0014 g of hydrogen.

0.0104 g of carbon plus 0.0014 g of hydrogen leaves 0.0014 g unaccounted for from the 0.0132 g sample of camphor; this is assumed to be oxygen (0.0014 divided by 16 = 0.875 x 10^{-4} mole).

mole ratio in camphor: $C_{8.68}H_{14}O_{0.875}$, divide by the smallest number: $(C_{10}H_{16}O)_n$
(Camphor is $C_{10}H_{16}O$)

b. <u>1.56 mg of sex attractant</u>:

3.73 mg = 0.0848 mmole of CO2
0.0848 mmole of carbon = 1.018 mg C
1.22 mg = 0.0678 mmole of H2
0.1356 mmole of hydrogen = <u>0.136</u> mg H
1.154 mg accounted for

The remaining 0.41 mg is assumed to be oxygen = 0.0256 mmole
$C_{0.0848}H_{0.136}O_{0.0256}$, divide by 0.0256 gives $C_{3.31}H_{5.31}O$, multiply by 3 gives $(C_{10}H_{16}O_3)_n$
(the molecular formula for this compound is actually $C_{10}H_{16}O_3$)

c. <u>2.16 mg of benzo[a]pyrene</u>:
7.5 mg = 0.17 mmole of CO_2
= 0.17 mmole of carbon = 2.05 mg C

0.92 mg = 0.051 mmole of H_2O
= 0.102 mmole of hydrogen = <u>0.10 mg H</u>
2.15 mg (this accounts for all of the benzo[a]pyrene)

$C_{0.17}H_{0.102}$: divide by 0.102 gives $C_{1.66}H$; multiply x 3 $(C_5H_3)_n$

(benzo[a]pyrene is $C_{20}H_{12}$)

17. <u>2.03 mg of sample:</u>

4.44 mg = 0.101 mmole of CO_2
= 0.101 mmole of carbon = 1.21 mg = 59.7% C

0.91 mg = 0.051 mmole of H_2O
0.102 mmole of hydrogen = 0.10 mg = 5.0% H

5.31 mg of sample gives X mmole of Cl^- in solution. Because one mmole of $AgNO_3$ is required for each mmole of Cl^-, X is equal to 4.80 x 0.110 = 0.0528 mmole of Cl. This is equivalent to 1.87 mg of Cl (0.0528 x 35.5) in 5.31 mg of sample = 35.2% Cl.

59.7% C	59.7/12 = 4.98	
5.0% H	5.0/1 = 5.0	C_5H_5Cl
35.2% Cl	35.2/35.5 = 0.99	

3.F. Supplementary Problems

S1. From the analytical values for each compound, derive its empirical formula.

a. Cecropia moth juvenile hormone: 73.6% C, 10.1% H

b. Valium: 67.5% C, 4.6% H, 12.5% Cl, 9.8% N

c. Nicotine: 74.0% C, 8.7% H, 17.3% N

d. Sarin (a nerve gas): 34.3% C, 7.2% H, 13.6% F, 22.1% P

S2. The following compounds were shown to contain only carbon, hydrogen, oxygen, and (if indicated) nitrogen. Calculate the empirical formula for each case.

a. Combustion of 5.63 mg of aspirin gave 12.39 mg of CO_2 and 2.27 mg of H_2O.

b. Combustion of 1.87 mg of vitamin E gave 5.55 mg of CO_2 and 1.97 mg of H_2O.

c. Combustion of 2.79 mg of caffeine gave 5.06 mg of CO_2 1.30 mg of H_2O, and 0.80 mg of N_2.

d. Combustion of 1.07 mg of epinephrine (adrenaline) gave 2.31 mg of CO_2 0.69 mg of H_2O, and 0.08 mg of N_2.

S3. On oxidation of sulfur-containing compounds, the sulfur is oxidized to sulfate, which can be determined by conversion to the very insoluble barium salt $BaSO_4$. Combustion of 3.27 mg of saccharin gave 5.50 mg of CO_2, 0.81 mg of H_2O, and 0.25 mg of N_2. Oxidation of a 6.73 mg sample of saccharin and conversion of the sulfate to the barium salt gave 8.59 mg of $BaSO_4$. What is the empirical formula for saccharin?

S4. a. What is the percent elemental composition of vitamin C ($C_6H_8O_6$)?

b. How much CO_2 and H_2O do you expect to obtain on combustion of a 3.97 mg sample of vitamin C?

S5. Write out condensed formulas and IUPAC names for all of the isomers of each of the following formulas.

a. $C_3H_5Cl_3$

b. C_6H_{14}

c. C_4H_8ClI

S6. Write condensed structural formulas for each of the following compounds.

a. 2,5,5-trimethylheptane

b. 1-bromo-3-ethylpentane

c. neopentyl bromide

d. 4,4-di(2,2-dimethylpropyl)-2,2-6,6-tetramethylheptane

e. 1,2-dichloro-1,1,2,2-tetrafluoroethane ("Freon 114")

f. 4-(1,1-dimethylpropyl)-2,2,3-trimethyloctane

S7. Give the IUPAC name for each of the following compounds

a.
```
                 CH3
                 |
CH3—CH—C—CH3
        |      |
       CH3  CH3
```

d. $((CH_3)_3C)_2CHCH_3$

b. $(CH_3CH_2)_2CHCH_2CH_2Cl$

e. $(CH_3)_2CHCH_2CH(CH_3)CClF_2$

c. 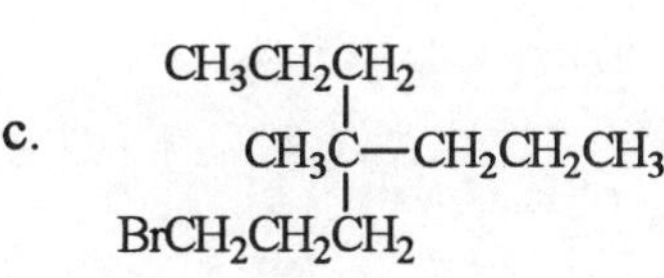

f. 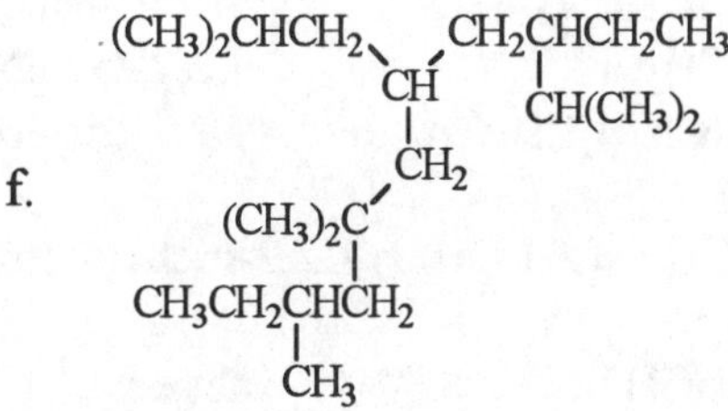

3.G. Answers to Supplementary Problems

S1. a. $C_{6.13}H_{10.1}O_{1.02} = (C_6H_{10}O)_n$ (Cecropia juvenile hormone is $C_{18}H_{30}O_3$)

b. $C_{5.63}H_{4.6}Cl_{0.35}N_{0.70}O_{0.35} = (C_{16}H_{13}ClN_2O)_n$ (Valium is $C_{16}H_{13}ClN_2O$)

c. $C_{6.17}H_{8.7}N_{1.24} = (C_5H_7N)_n$ (Nicotine is $C_{10}H_{14}N_2$)

d. $C_{2.86}H_{7.2}F_{0.72}O_{1.43}P_{0.71} = (C_4H_{10}FOP)_n$ (Sarin is $C_4H_{10}FO_2P$)

S2. a. (12.39/44.0) x 12/5.63 x 100 = 60.0% C(2.27/18.0) x 2/5.63 x 100 = 4.5% H

100% – 60.0% – 4.5% = 35.5 % O

$C_{60/12}H_{4.5/1}O_{35.5/16} = C_5H_{4.5}O_{2.2} = C_{2.27}H_2O = (C_9H_8O_4)n$ (Aspirin is $C_9H_8O_4$)

b. 80.9% C, 11.7% H 100% – 80.9% – 11.7% = 7.4% O

$C_{6.74}H_{11.6}O_{0.46} = C_{14.6}H_{25.1}O = (C_{29}H_{50}O_2)_n$ (Vitamin E is $C_{29}H_{50}O_2$)

c. 49.5% C, 5.2% H, 28.7% N 100% – 49.5% – 5.2% – 28.7% = 16.6% O

$C_{4.13}H_{5.2}N_{2.05}O_{1.04} = (C_4H_5N_2O)_n$ (Caffeine is $C_8H_{10}N_4O_2$)

d. 58.9% C, 7.2% H, 7.5% N 100% – 58.9% – 7.2% – 7.5% = 26.4% O

$C_{4.9}H_{7.2}N_{0.54}O_{1.65} = C_{9.07}H_{13.33}NO_{3.05} = (C_9H_{13}NO_3)_n$ (Epinephrine is $C_9H_{13}NO_3$)

S3. 45.9% C, 2.75% H, 7.6% N (8.59/233.3) x 32/6.73 x 100 = 17.5% S

100% – 45.9% – 2.75% – 7.6% – 17.5% = 26.25% O

$C_{3.38}H_{2.75}N_{0.54}O_{1.64}S_{0.55}$ = $(C_7H_5NO_3S)_n$ (Saccharin is $C_7H_5NO_3S$)

S4. a. MW = (6 x 12) + (8 x 1) + (6 x 16) = 176

(6 x 12)/176 x 100 = 40.9% C (8 x 1)/176 x 100 = 4.5% H (6 x 16)/176 x 100 = 54.5%

b. (0.409 x 3.97)/12 x 44 = 5.95 mg of CO_2 (0.045 x 3.97)/2 x 18 = 1.61 mg of H_2O

S5. a.

$CH_3CH_2CCl_3$	1,1,1-trichloropropane
$CH_3CHClCHCl_2$	1,1,2-trichloropropane
$CH_2ClCH_2CHCl_2$	1,1,3-trichloropropane
$CH_3CCl_2CH_2Cl$	1,2,2-trichloropropane
$CH_2ClCHClCH_2Cl$	1,2,3-trichloropropane

(NOTE: $CH_2ClCCl_2CH_3$ is 1,2,2-trichloropropane)

b.

$CH_3(CH_2)_4CH_3$	hexane
$(CH_3)_2CH(CH_2)_2CH_3$	2-methylpentane
$CH_3CH_2CH(CH_3)CH_2CH_3$	3-methylpentane
$(CH_3)_3CCH_2CH_3$	2,2-dimethylbutane
$(CH_3)_2CHCH(CH_3)_2$	2,3-dimethylbutane

c.

$CH_3CH_2CH_2CHClI$	1-chloro-1-iodobutane
$CH_3CH_2CHClCH_2I$	2-chloro-1-iodobutane
$CH_3CHClCH_2CH_2I$	3-chloro-1-iodobutane
$ClCH_2CH_2CH_2CH_2I$	1-chloro-4-iodobutane
$CH_3CH_2CHICH_2Cl$	1-chloro-2-iodobutane
$CH_3CHICH_2CH_2Cl$	1-chloro-3-iodobutane
$CH_3CH_2CClICH_3$	2-chloro-2-iodobutane
$CH_3CHClCHICH_3$	2-chloro-3-iodobutane
$(CH_3)_2CHCHClI$	1-chloro-1-iodo-2-methylpropane
$(CH_3)_2CClCH_2I$	2-chloro-1-iodo-2-methylpropane
$(CH_3)_2CICH_2Cl$	1-chloro-2-iodo-2-methylpropane
$ClCH_2CH(CH_3)CH_2I$	1-chloro-3-iodo-2-methylpropane

S6. a. $(CH_3)_2CHCH_2CH_2C(CH_3)_2CH_2CH_3$ b. $BrCH_2CH_2CH(CH_2CH_3)_2$ c. $(CH_3)_3CCH_2Br$

d.
```
            CH2C(CH3)3
            |
(CH3)3CCH2CCH2C(CH3)3
            |
            CH2C(CH3)3
```

e. $F_2ClCCClF_2$

f.
```
                 CH3
                 |
          CH3—CCH2CH3
                 |
(CH3)3CCHCHCH2CH2CH2CH3
          |
          CH3
```

S7. a. 2,2,3-trimethylbutane

b. 1-chloro-4-ethylpentane

c. 1-bromo-4-methyl-4-propylheptane

d. 2,2,3,4,4-pentamethylpentane

e. 1-chloro-1,1-difluoro-2,4-dimethylpentane

f. 3-ethyl-2,7,7,9-tetramethyl-5-(2-methylpropyl)undecane

4. ORGANIC REACTIONS

4.A. Chapter Outline and Important Terms Introduced

4.1 **Introduction** (distinction between equilibrium and rate)

concept of a reaction mechanism

4.2. **An Example of an Organic Reaction: Equilibria** (how far it goes)

"go to completion"
$\Delta G° = -RT \ln K$
Gibbs Standard Free Energy change, $\Delta G°$
driving force

enthalpy, $\Delta H°$
entropy, $\Delta S°$
exothermic and endothermic

4.3 **Reaction Kinetics** (how fast it goes)

energy barrier
activation energy (enthalpy), $\Delta H^{\ddagger}$
rate constant

first- and second-order reactions
pseudo first-order reactions

4.4 **Reaction Profiles and Mechanism** (graphs of energy changes in a reaction mechanism)

transition state
maximum energy
activation energy
reaction coordinate
theory of absolute rates ($\Delta G^{\ddagger} = -RT \ln k + \text{constant}$)

reaction mechanism
rate-determining
reaction intermediate
multi-step reaction

4.5 **Acidity and Basicity** (an important review - back to the basics)

solvation
dissociation constant

pK_a
dependence of pK_a on structure

4.B. Important Reactions Introduced: still none yet

4.C. Important Concepts and Hints

Whereas Chapters 2 and 3 present the underlying principles of structure, Chapter 4 discusses the principles of reactivity. Two quite separate concepts are those of **equilibrium** (a measure of how completely a reaction proceeds) and **rate** (a measure of how fast a reaction proceeds). Although they are independent, both depend on potential energy differences that organic chemists like to display pictorially with "reaction profile" or "reaction coordinate" diagrams, such as the simple one sketched below. These diagrams are seldom used in a quantitative sense, but they frequently provide a picture of the energy changes associated with a reaction.

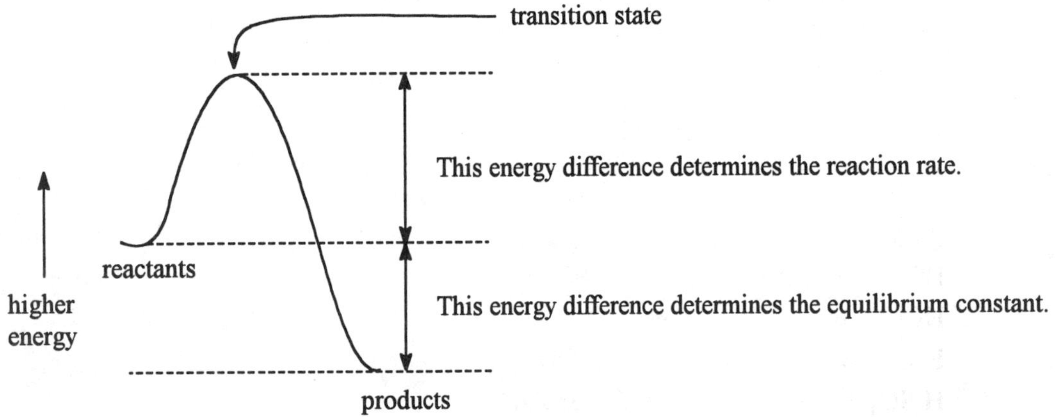

A helpful analogy can be drawn between molecules passing over the energy barrier (through the transition state) leading to product, and water passing over a physical barrier and running downhill. The higher the barrier, the more vigorously the water must be agitated before very much splashes over the edge. This is analogous to raising the temperature of a reaction mixture until enough molecules have the kinetic energy necessary to overcome the activation energy $\Delta G^‡$. The energy released as the water runs downhill, on the other hand, depends only on the difference in height between the upper and lower reservoirs, just as the energy released in a reaction depends only on the overall free energy of reaction, $\Delta G°$.

Reaction coordinate diagrams can be useful for an intuitive understanding of the energetics of reaction kinetics and equilibria. However, you should recognize their qualitative nature as well as some pitfalls that you may encounter when using them. Because they really refer only to standard states (approximately 1 M for liquid-phase reactions), they cannot be rigorously applied to reactions in which relative rates and equilibria depend on concentration. For example, see problem #S9 in this Chapter of the Study Guide.

Equilibrium: The equilibrium constant, K = [product(s)]/[reactants(s)], is a measure of how far a reaction will proceed before there is no further change in the concentration of reactants and products. It depends on the potential energy difference between the reactants and products, as expressed by the equation $\Delta G° = -RT \ln K$. This is a useful equation to remember, as it can be applied in many situations. Also useful to remember is a particular consequence of this equation: at normal temperatures ($T \approx 300K$), every 1.37 kcal change in $\Delta G°$ corresponds to a ten-fold change in the equilibrium constant K (and vice versa).

Rate: The rate of a reaction depends on a constant that is characteristic to each reaction (the rate constant, k) and on the concentration of the reactants. In this regard, be sure you understand the difference between **rate** and **rate constant**, and **first-** and **second-order reactions**. The rate constant, k, depends on the activation energy for the reaction, $\Delta G^‡$, as shown by the equation:

$$k = \text{constant} \times \exp(-\Delta G^‡/RT)$$

which rearranges to: $\Delta G^‡ = -RT \ln k + \text{another constant}$.

This equation is similar to the one that relates the standard free energy of reaction, $\Delta G°$, and the equilibrium constant, K, so the same "1.37 kcal $mole^{-1}$ = a factor of 10" rule-of-thumb applies. Although these equations are similar, don't get $\Delta G^‡$ and $\Delta G°$ mixed up.

Chapter 4 also discusses acid/base equilibria fairly thoroughly, although you may feel that the topic belongs more to General Chemistry courses. In fact, the fundamental aspects of acid/base equilibria will be referred to repeatedly throughout your course in organic chemistry. These concepts will be useful for explaining why reactions take place, what mechanisms they proceed by, and so on. You should be familiar with the idea that any molecule that donates a proton in the course of a reaction is an "acid", and any molecule that accepts a proton is a "base". All of the equilibria below are acid/base reactions. Note that some molecules can act either as acids or bases depending on the situation.

Acids		Bases		Bases		Acids
HCl	+	NH_3	⇌	Cl^-	+	NH_4^+
HCl	+	H_2O	⇌	Cl^-	+	H_3O^+
H_2O	+	CH_3O^-	⇌	OH^-	+	CH_3OH
H_2SO_4	+	$O{=}CH_2$	⇌	HSO_4^-	+	$HO^+{=}CH_2$
NH_4^+	+	OH^-	⇌	NH_3	+	H_2O
NH_3	+	CH_3^-	⇌	NH_2^-	+	CH_4

The most useful measure of how easily a compound gives up a proton is its pK_a (frequently called simply "pK"). The easiest way to remember what the numbers mean is the following: in aqueous solution, when the pH equals the pK_a of a compound, it is half-ionized (i.e., half of it is protonated and half of it is unprotonated). Weaker acids have higher pK_a's and therefore require higher pH's before they are half-ionized, and vice versa for stronger acids. Obviously, this relationship is only realistic for pK_a's in the range attainable in water (0-14), but the device is still useful for remembering what pK_a's outside this range mean; for instance, a negative pK_a means that the compound is a very strong acid. On the other hand, pK_a of 34 means that the compound is a very weak acid, but that the deprotonated form (called the "conjugate base") is a very strong base. Each pK_a unit represents a factor of 10 in equilibrium because it is a logarithmic scale. Remember: a factor of 10 = 1.37 kcal mole^{-1} in energy difference.

> **NOTE:** Later in your course in organic chemistry you may hear: "the pK_a of ammonia is 9". This is a careless statement, although a common one; what is intended is: "the pK_a of the ammonium ion (NH_4^+) is 9". Although we often think of both the acidity of a protonated compound and the basicity of the unprotonated form in terms of pK_a, we should always be aware that pK_a refers to the ability of the protonated species to give up its proton; i.e., for the compound to function as an acid. The pK_a of ammonia ($NH_3 \rightleftharpoons NH_2^- + H^+$) is actually 34.

If the whole concept of pH, and of acids and bases in general, has receded from your memory since you took introductory chemistry, you should definitely go back and review this subject....

4.D. Answers to Exercises

4.1 $\Delta G° = \Delta H° - T\Delta S°$

a. At 27°C (300K): $\Delta H° = -10$ kcal mole^{-1}
$T = 300°K$
$\Delta S° = -22$ e.u. $= -22$ cal deg^{-1}mole^{-1}

Don't overlook the difference between cal and kcal!

$\therefore \Delta G° = -10{,}000 - 300 \cdot (-22) = -3400$ cal mole^{-1}

$\Delta G° = -RT \ln K$

$\Delta G° = -3.4$ kcal mole^{-1}, R = 1.987 cal deg^{-1} mole^{-1} *(cal, not kcal)*, T = 300K.

$$\therefore \ln K = \frac{-3400}{-1.987 \times 300} = 5.70 \quad \text{and } K = e^{5.7} = 300 \text{ at } 300K$$

b. At 227°C (500K): $\Delta G° = -10{,}000 - 500 \cdot (-22) = +1000$ cal mole^{-1} = +1.0 kcal mole^{-1}

$$\ln K = \frac{1000}{-1.987 \times 500} = -1.01 \quad \text{and} \quad K = e^{-1.01} = 0.36$$

4.2 Rate = $k[OH^-][CH_3Cl]$; $k = 6 \times 10^{-6}\, M^{-1}\, sec^{-1}$

Initial reaction rate:

a. $k \times 1.0\,M \times 0.1\,M = 6 \times 10^{-7}\,M\,sec^{-1}$
b. $k \times 0.1\,M \times 0.1\,M = 6 \times 10^{-8}\,M\,sec^{-1}$
c. $k \times 0.01\,M \times 0.01\,M = 6 \times 10^{-10}\,M\,sec^{-1}$

After 90% of the methyl chloride has reacted:

a. $k \times 0.91\,M \times 0.01\,M = 6 \times 10^{-8}\,M\,sec^{-1}$
b. $k \times 0.01\,M \times 0.01\,M = 6 \times 10^{-10}\,M\,sec^{-1}$
c. $k \times 0.001\,M \times 0.001\,M = 6 \times 10^{-12}\,M\,sec^{-1}$

4.3

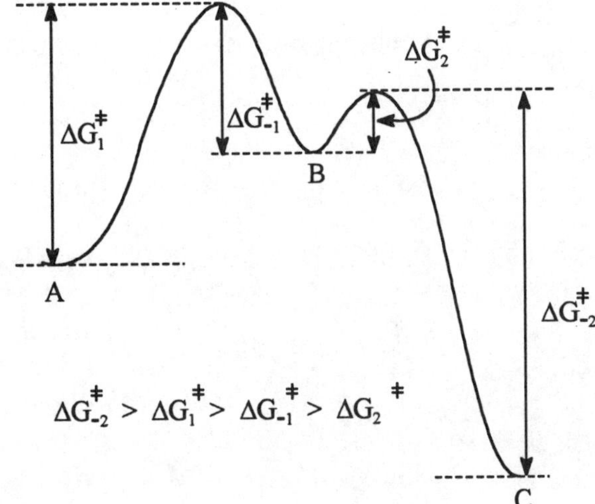

$\Delta G^{\ddagger}_{-2} > \Delta G^{\ddagger}_{1} > \Delta G^{\ddagger}_{-1} > \Delta G^{\ddagger}_{2}$

Rate-determining step is A → B. The rate of B → C is the faster one. For a two-step reaction A ⇌ B → C, in which B is an unstable intermediate (i.e., A → B is endothermic), the rate of A → B can become faster than B → C only if the first transition state is lower than the second (i.e., if $k_{-1} > k_2$).

4.4 $HA \rightleftharpoons H^+ + A^-$, $\quad K = \frac{[H^+][A^-]}{[HA]} \quad pK_a = -\log K$ and $K = 10^{-pKa}$

$[H^+] \cong [A^-] = 1 - [HA] \qquad K = \frac{(1 - [HA])^2}{[HA]}$

For HI and HCl: when K is very large, $[HA] \cong 0$ and $[H^+] \cong 1\,M$ pH=0

For HF: $K = 10^{-3.2} = 6.3 \times 10^{-4}$. In this case, K is small, $[HF] \cong 1\,M$ and $\frac{[H^+][F^-]}{1} \cong 6.3 \times 10^{-4}$;

$[H^+] = [F^-] = \sqrt{6.3 \times 10^{-4}} = 2.5 \times 10^{-2}\,M$ and pH = 1.6

For acetic acid: $K = 10^{-4.76} = 1.74 \times 10^{-5}$. As for HF, very little of HA dissociates: $[HA] \cong 1\,M$

$\frac{[H^+][A^-]}{1} \cong 1.74 \times 10^{-5}$; $[H^+] = \sqrt{1.74 \times 10^{-5}} = 4.17 \times 10^{-3}\,M$ and pH = 2.4.

For H_2S: $K = 10^{-6.97} = 1.1 \times 10^{-7}$. Again, very little of H_2S ionizes: $[H_2S] \cong 1\,M$ and $[H^+] = [HS^-] = \sqrt{1.1 \times 10^{-7}} = 3.3 \times 10^{-4}\,M$ and pH = 3.5

4.5 $\frac{[H^+][A^-]}{[HA]} = 10^{-pK_a} = 100 \qquad [H^+] = [A^-] = 1 - [HA]$

$[H^+] = [A^-] = 0.99\,M$; $[HA] = 0.01\,M$

4.6

Acid	Conjugate Base	Strength of Conjugate Base
CH_3CO_2H	$CH_3CO_2^-$	Weak
NH_4^+	$:NH_3$	Moderate
HI, HBr, HCl	I^-, Br^-, Cl^-	Very Weak
HCN	CN^-	Moderate
HF	F^-	Weak
H_2Se	HSe^-	Weak
H_2S	HS^-	Weak-Moderate
CH_3OH	CH_3O^-	Strong
HNO_3	NO_3^-	Very Weak
HNO_2	NO_2^-	Weak
H_3PO_4	$H_2PO_4^-$	Weak
(Second ionization:	HPO_4^{2-}	Weak-Moderate)
(Third ionization:	PO_4^{3-}	Strong)
C_6H_5OH	$C_6H_5O^-$	Moderate
H_2SO_4	HSO_4^-	Very Weak
(Second ionization:	SO_4^{2-}	Weak)
H_2O	OH^-	Strong

4.E. Answers and Explanations for Problems

1. a.

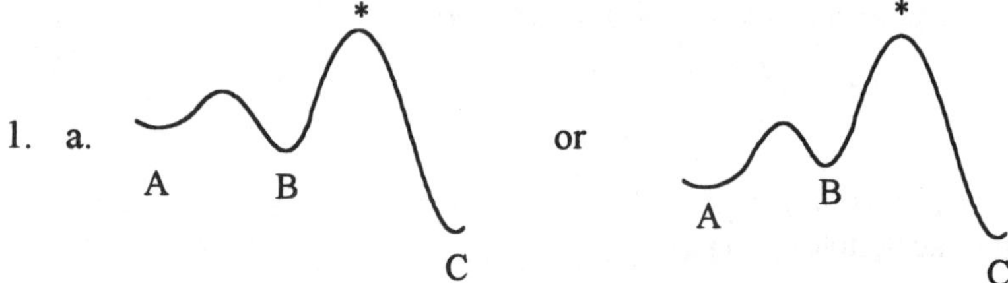

b. 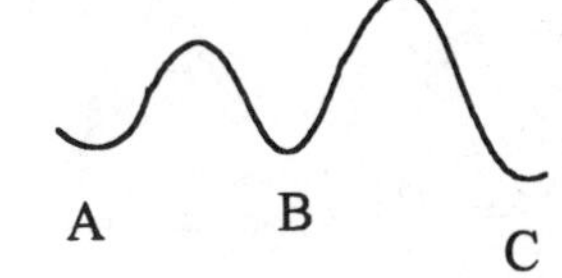

or 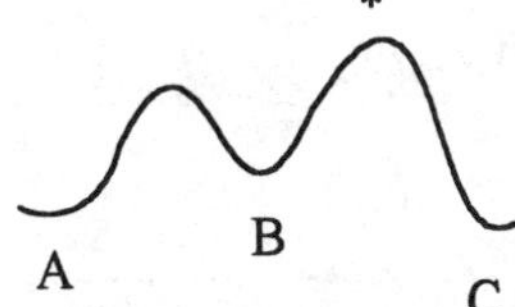

c. 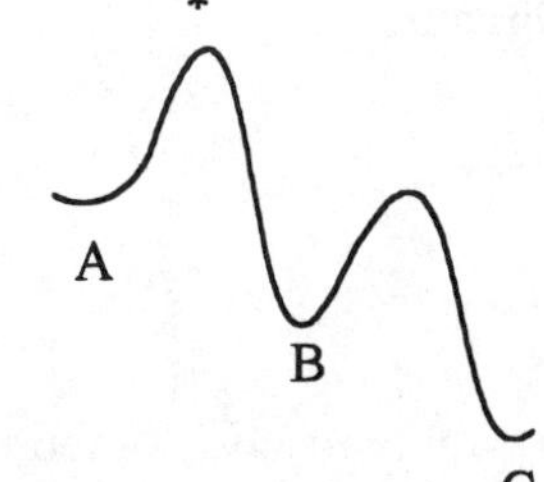

d. 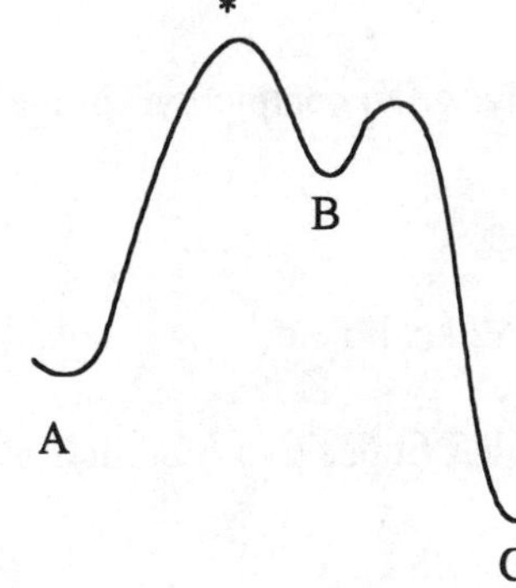

In each case, an asterisk (*) indicates the rate-determining transition state.

2. a. Endothermic. C is higher in energy than A. An "uphill" reaction requires you to put energy <u>in</u>.

b. 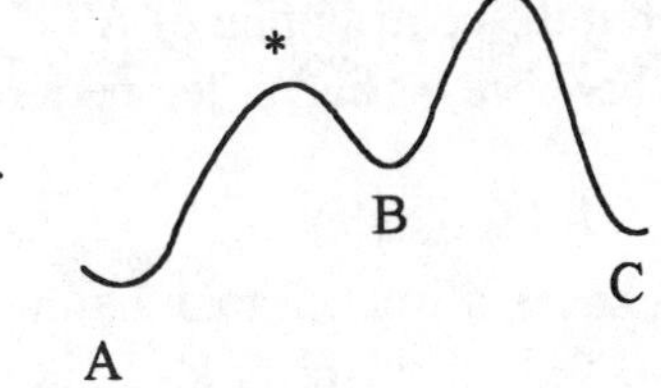

,' = transition states
*' = rate-determining transition state

c. $k_2 > k_3 > k_1 > k_4$ d. A is most stable e. B is least stable

3. a. 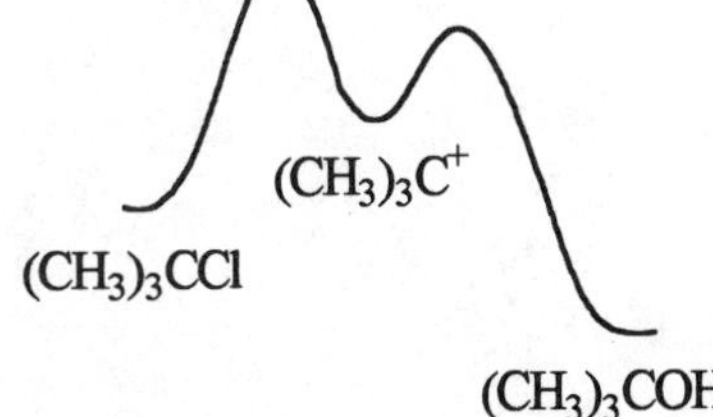

b. endothermic; uphill to the carbocation
c. exothermic; downhill overall
d. the first step, because that is the highest barrier to cross.

4. a. $\Delta H° = 7.3 \text{ kcal mole}^{-1}$ *Note difference between <u>kcal</u> and <u>cal</u>*

$T = 298K$

$\Delta S° = 0.3 \text{ cal deg}^{-1} \text{ mole}^{-1}$

$\Delta G° = \Delta H° - T\Delta S° = 7300 - (298 \times 0.3)$

$= 7211 \text{ cal mole}^{-1} = 7.21 \text{ kcal mole}^{-1}$

b. $\Delta G° = 7210$ cal mole^{-1}
$R = 1.987$ cal deg^{-1} mole^{-1} *(cal, not kcal)*
$T = 298$ K

$\Delta G° = -RT \ln K$, so that $\ln K = \dfrac{7210}{-1.987 \times 298} = -12.2$ and $K = e^{-12.2} = 5 \times 10^{-6}$

c. No; the reaction would actually "go to completion" in the opposite direction.

5. a. $\Delta G° = \Delta H° - T\Delta S°$
$= 22{,}200 - (298 \times 33.5)$
$= 12{,}200$ cal mole^{-1} = 12.2 kcal mole^{-1}

The equilibrium lies far to the left (a positive $\Delta G°$ indicates an unfavorable reaction in the direction written).

b. $\Delta G° = \Delta H° = T\Delta S°$
At 800 K, $\Delta G = 22{,}200 - (800 \times 33.5)$
$= -4{,}600$ cal mole^{-1} = -4.6 kcal mole^{-1}

At this temperature, the equilibrium lies to the right

c. The contribution of $\Delta H°$ to $\Delta G°$ is unaffected by temperature, but the contribution of $\Delta S°$ depends directly on temperature because of the $-T\Delta S°$ term. Therefore, at higher temperatures the entropy term $\Delta S°$ becomes more important.

6. Hydrogen is a very, very weak acid (i.e. hydride ion is a very strong base), a weaker acid than water. The products of the reaction of NaH with H_2O would be: $NaH + H_2O \rightarrow Na^+ + OH^- + H_2$. The reaction would be similar: $NaH + CH_3CO_2H \rightarrow CH_3CO_2Na + H_2$

7. Because NH_2^- is such a strong base, the equilibrium with water lies completely to the right: $H_2O + NH_2^- \rightarrow HO^- + NH_3$, and $[OH^-]$ will equal 0.1 *M*

$[H^+][OH^-] = 10^{-14} \rightarrow [H^+] = 10^{-13}$ and pH = 13.

This result is essentially the same as adding 0.1 *M* of NaOH itself.

PH_3 is expected to be a stronger acid because the P–H bonds are weaker than N–H bonds (P is below N in the periodic table).

8. At 50% reaction, $[A] = \frac{1}{2}[A]_{initial}$ and $[B] = \frac{1}{2}[B]_{initial}$

rate = $k \frac{1}{2}[A]_{initial} \times \frac{1}{2}[B]_{initial} = \frac{1}{4}$(initial rate)

9. a. and g. The presence of one negative charge makes it more difficult to generate another.

b., d., f., and h. The more electronegative oxygen substituents present, the better the negative charge can be stabilized.

c. The bonds become weaker farther down the periodic table.

d. Nitrite ion is also stabilized by resonance relative to the hydroxylamine ion.

e. Because it lies farther to the right in the periodic table, sulfur is more electronegative than phosphorus.

10. a. At room temperature (300 K), compare ΔG° for K and for 10K:

$\Delta G^\circ_A = -RT \ln 10K \qquad \Delta G^\circ_B = -RT \ln K$

$$\Delta G^\circ_A - \Delta G^\circ_B = -RT(\ln 10K - \ln K) = -RT \ln 10$$
$$= -1.987 \times 300 \times 2.303$$
$$= -1370 \text{ cal mole}^{-1} = -1.37 \text{ kcal mole}^{-1}$$

For a factor of 100, $\Delta G^\circ_A - \Delta G^\circ_B = -RT \ln 100 = -2.75 \text{ kcal mole}^{-1}$

b. A factor of 10 in K equals a change in ΔG° of 1.37 kcal mole^{-1} at room temperature. This could arise from a change in ΔH° of 1.37 kcal mole^{-1} if ΔS° remained constant, or a change in ΔS° of

$-\dfrac{1370}{298} = -4.6$ e.u., if ΔH° remained constant.

11. a. $k = 6 \times 10^{-6}\, M^{-1}\,\text{sec}^{-1}$ $\qquad [OH^-] = 0.10\,M$ $\qquad [CHCl_3] = 0.05\,M$

rate $= k[OH^-][CHCl_3]$

$= 6\times10^{-6}\,M^{-1}\,\text{sec}^{-1} \times 0.10\,M \times 0.05\,M$
$= 3.0 \times 10^{-8}\,M\,\text{sec}^{-1}$ at the start of reaction

b. and c. Each 10% of reaction means a change in $[CHCl_3]$ of 0.005 *M*.

% Reaction	k, $M^{-1}\text{sec}^{-1}$	$[OH^-]$, *M*		$[CHCl_3]$, *M*		Rate, $M\,\text{sec}^{-1}$	Time for 10% Reaction
0	6×10^{-6}	0.10	x	0.050	=	3.0×10^{-8}	
							1.67×10^5sec
10	6×10^{-6}	0.095	x	0.045	=	2.6×10^{-8}	
							1.92×10^5 sec
20	6×10^{-6}	0.090	x	0.040	=	2.2×10^{-8}	
							2.27×10^5 sec
30	6×10^{-6}	0.085	x	0.035	=	1.8×10^{-8}	
							2.78×10^5 sec
40	6×10^{-6}	0.080	x	0.030	=	1.4×10^{-8}	
							3.57×10^5 sec
50	6×10^{-6}	0.075	x	0.025	=	1.1×10^{-8}	
						Total time	$= 1.22 \times 10^6$ sec
							= 339 hours

12. a. The entropy is negative because more order is introduced into the system (less freedom of motion) when two molecules combine to give one.

b. $\Delta G° = \Delta H° - T\Delta S°$
$= -15.5 \text{ kcal mole}^{-1} - (298 \text{ deg x } -31.3 \text{ cal deg}^{-1} \text{ mole}^{-1})$
$= -6.17 \text{ kcal mole}^{-1}$

c. $\Delta G° = -RT \ln K$, so that $\ln K = \dfrac{-6170 \ cal \ mole^{-1}}{-1.987 \ cal \ deg^{-1} \ mole^{-1} \ x \ 298 \ K} = 10.4$

and $K = e^{10.4} = 3.29 \times 10^4$

$$K = \frac{[C_2H_5Cl]}{[HCl] \ [C_2H_4]} = \frac{P_{C_2H_5Cl}}{P_{HCl} \ x \ P_{C_2H_4}}$$

Because the reaction essentially goes to completion, $P_{C_2H_5Cl}$ at equilibrium = 1 atm (1 atm HCl

+ 1 atm $C_2H_4 \rightleftharpoons$ 1 atm C_2H_5Cl). Furthermore, because $P_{HCl} = P_{C_2H_4}$ at the start of the reaction,

and
one molecule of HCl is consumed for every molecule of C_2H_4. P_{HCl} will always equal $P_{C_2H_4}$.

$$K = \frac{1}{(P_{HCl})^2} = 3.29 \times 10^4 \Rightarrow P_{HCl} = P_{C_2H_4} = 5.5 \times 10^{-3} \text{ atm.}$$

d. For all three components to be present in equal amounts, i.e., at equal pressures of each component;

$$K = \frac{P}{P \cdot P} \quad \frac{1}{P} = 3.29 \times 10^4 \rightarrow P = 3 \times 10^{-5} \text{ atm.}$$

With each component at this pressure, the total pressure will be 9×10^{-5} atm.

13. a. $[H_2O]$ in water is $\dfrac{1000 \ g \ liter^{-1}}{18 \ g \ mole^{-1}} = 55 \ M$

b. $k_1 \ [CH_3Cl] = k_2 \ [CH_3Cl] \ [H_2O]$

$$k_1 = k_2[H_2O] \Rightarrow k_2 = \frac{3 \ x \ 10^{-10} \ \text{sec}^{-1}}{55 \ M} = 5.45 \times 10^{-12} M^{-1} \text{ sec}^{-1}.$$

The rate constant for the reaction with OH^- is $6 \times 10^{-6} M^{-1} \text{ sec}^{-1}$, i.e., much faster.

c. $\dfrac{0.69}{3 \ x \ 10^{-10} \ \text{sec}^{-1}} = 2.3 \times 10^9 \text{ sec} = 73 \text{ years.}$

4.F Supplementary Problems

S1. a. At 25 °C, the equilibrium constant K for the addition of water to ethylene in the gas phase is 23.1 M^{-1}. Calculate ΔG° for this reaction at 25 °C.

$$CH_2{=}CH_2 + H_2O \rightleftharpoons CH_3CH_2OH \quad K = 23.1\ M^{-1} \text{ at } 25°C$$

b. At 400 K, the equilibrium constant is 0.213 M^{-1}. Calculate ΔH° and ΔS° for the reaction.

c. . Why does the equilibrium constant decrease at higher temperature?

S2. The rate constant k for the reaction: $CH_3CH_2Br + CH_3CH_2O^- \longrightarrow CH_3CH_2OCH_2CH_3 + Br^-$ in ethanol (CH_3CH_2OH) solvent at 25 °C is 7.6 x 10^{-5} M^{-1} sec^{-1}, and the rate equation is: rate = $k[CH_3CH_2Br][CH_3CH_2O^-]$.

a. If 0.05 mole of CH_3CH_2Br and 0.05 mole of $CH_3CH_2O^-$ Na^+ are dissolved in 250 ml of ethanol, to a first approximation how long will it take before 10% of the CH_3CH_2Br has reacted?
b. If the same quantities are dissolved in one liter of ethanol?
c. If 0.05 mole of CH_3CH_2Br is dissolved in one liter of 0.2 M $CH_3CH_2O^-Na^+$ in ethanol?

S3. The rate constant k′ for the reaction

$$(CH_3)_3CBr + CH_3CH_2O^- \longrightarrow (CH_3)_3COCH_2CH_3 + Br^-$$

in ethanol solvent at 25 °C is: 5 x 10^{-4} sec^{-1}, and the rate equation is: rate = $k'[(CH_3)_3CBr]$.

a. If 0.05 mole of $(CH_3)_3CBr$ and 0.05 mole of $CH_3CH_2O^-Na^+$ are dissolved in 250 ml of ethanol, to a first approximation how long will it take before 10% of the $(CH_3)_3CBr$ has reacted?
b. If the same quantities are dissolved in one liter of ethanol?
c. If 0.05 mole of $(CH_3)_3CBr$ is dissolved in one liter of 0.2 M $CH_3CH_2O^-Na^+$ in ethanol?

S4. a. When 0.05 mole of CH_3CH_2I and 0.05 mole of $CH_3CH_2O^-Na^+$ are dissolved in 250 ml of ethanol solvent at 25 °C, it requires 65 minutes before 10% of the starting material has reacted according to the following equation:

$$CH_3CH_2I + CH_3CH_2O^-Na^+ \longrightarrow CH_3CH_2OCH_2CH_3 + Na^+I^-$$

What is the approximate rate of the reaction under these conditions?

b. If the same amounts of starting material are dissolved in 500 ml of the solvent at 25 °C, 130 minutes are required before 10% reaction has occurred. What is the rate of reaction this time?
c. What is the form of the rate equation? Is it a first- or second-order reaction?
d. What is the rate constant?

S5. Construct a reaction profile diagram for the following reaction sequence:

a. $$A \underset{k_{-1}}{\overset{k_1}{\rightleftharpoons}} B \underset{k_{-2}}{\overset{k_2}{\rightleftharpoons}} C \underset{k_{-3}}{\overset{k_3}{\rightleftharpoons}} D$$

$k_1 = 3 \times 10^{-3}\ sec^{-1}$ $k_2 = 10^{-1}\ sec^{-1}$ $k_3 = 6 \times 10^5\ sec^{-1}$
$k_{-1} = 4 \times 10^3\ sec^{-1}$ $k_{-2} = 10^{-3}\ sec^{-1}$ $k_{-3} = 10^{-5}\ sec^{-1}$

b. Which compound has the highest potential energy?
c. Which compound reacts the fastest?

d. What is the equilibrium constant $K = \frac{[D]}{[A]}$?

e. What is the rate-limiting step in the conversion of A to D?
f. Is the conversion of A to B endothermic or exothermic?
g. What is the most exothermic step?

S6. For each of the following pairs, choose the compound with the higher pK_a.

a. HCl, H_2S b. H_2O, HF c. H_3PO_4, $H_2PO_4^-$

d. H_2O, NH_3 e. H_3O^+, H_2O f. H_3O^+, NH_4^+

g. HOOH, H_2O

S7. a. Why can there be no stronger base in water than hydroxide ion?
b. What is the strongest acid possible in water?

S8. From the pK_a's given below, calculate the pH of a solution obtained when one mole of each substance is dissolved in one liter of water.

a. $NH_4^+Cl^-$ $pK_a = 9.2$ b. NH_3 (pK_a of $NH_4^+ = 9.2$)

S9. Consider the following reaction sequence:

$$A \underset{k_{-1}}{\overset{k_1}{\rightleftharpoons}} B, \text{ then } B + C \underset{k_{-2}}{\overset{k_2}{\rightleftharpoons}} P$$

$k_1 = 10^{-5}\ sec^{-1}$ $k_2 = 2 \times 10^{-2}\ M^{-1}\ sec^{-1}$
$k_{-1} = 10^{-3}\ sec^{-1}$ $k_{-2} = 10^{-8}\ sec^{-1}$

a. Draw a reaction coordinate diagram for the overall process $A + C \rightleftharpoons P$.

b. The rate equation for formation of P is: rate = $\frac{d[P]}{dt} = \frac{k_1 k_2 [A][C]}{k_{-1} + k_2[C]}$

Under standard state conditions

$[A] = [C] = 1\ M$; $k_{-1} + k_2[C]$ is approximately equal to $k_2[C]$, and the expression reduces to:

$$\frac{d[P]}{dt} \approx \frac{k_1 k_2 [A][C]}{k_2[C]} = k_1[A].$$

What is the rate-limiting step under these conditions?

c. What is the approximate form of the rate equation if $[A] = [C] = 0.001\ M$? What is now the rate-limiting step? What does this suggest about the limitations of reaction coordinate diagrams?

S10. Consider the following reaction sequence:

$$A \underset{k_{-1}}{\overset{k_1}{\rightleftharpoons}} B + C, \text{ then } B + D \underset{k_{-2}}{\overset{k_2}{\rightleftharpoons}} P$$

$k_1 = 2 \times 10^{-5}\ \text{sec}^{-1}$ $\qquad$ $k_2 = 3 \times 10^{-2} M^{-1}\ \text{sec}^{-1}$

$k_{-1} = 10^{-2} M^{-1}\ \text{sec}^{-1}$ $\qquad$ $k_{-2} = 10^{-8}\ \text{sec}^{-1}$

a. Draw a reaction coordinate diagram for the overall process $A + D \rightleftharpoons C + P$.

b. The complete rate equation for this reaction is : rate = $\dfrac{d[P]}{dt} = \dfrac{k_1 k_2 [A][D]}{k_{-1}[C] + k_2[D]}$.

If [A] = [D] = 0.1 M and [C] = 0 at the start of the reaction, what is the approximate form of the rate equation and what is the rate-determining step?

c. What is the approximate form of the rate equation if [C] = 2 M?
What is now the rate-determining step?

4.G. Answers to Supplementary Problems

S1. a. $\Delta G° = -RT \ln K$; $\Delta G°_{298} = -1.86$ kcal mole^{-1}

b. To calculate $\Delta H°$ and $\Delta S°$, you need $\Delta G°$ at two different temperatures:
$\Delta G° = \Delta H° - T\Delta S°$. If K = 0.213 at 400 K, $\Delta G°_{400} = +1.23$ kcal mole^{-1}.

$\Delta G°_{298} = \Delta H° - (298 \times \Delta S°) = -1.86$

$\Delta G°_{400} = \Delta H° - (400 \times \Delta S°) = +1.23$

$\Delta G°_{298} - \Delta G°_{400} = (-298 + 400)\ \Delta S° = -3.09$ kcal mole^{-1};
$\Delta S° = -30.3$ cal deg^{-1} mole^{-1} (e.u.)

$\Delta H° = \Delta G° + T\Delta S° = -10.9$ kcal mole^{-1}

c. The equilibrium constant decreases at higher temperatures because the unfavorable entropy term ($T\Delta S°$) becomes more important. The entropy for this reaction is unfavorable (negative) because two molecules are combining to form one.

S2. rate (moles liter^{-1} sec^{-1}) x time (sec) = amount (moles liter^{-1})

$k[CH_3CH_2Br][CH_3CH_2O^-]t = [CH_3CH_2OCH_2CH_3]$

(Note that this approximation is valid only for very short reaction times. As soon as $[CH_3CH_2Br]$ and $[CH_3CH_2O^-]$ change, the rate changes too.)

For 10% reaction, $[CH_3CH_2OCH_2CH_3] = 0.1 \times [CH_3CH_2Br]$

and the equation reduces to: $k[CH_3CH_2Br]\,[CH_3CH_2O^-]t \cong 0.1 \times [CH_3CH_2Br]$

For 10% reaction, $t \cong \dfrac{0.1}{k[CH_3CH_2O^-]}$; $k = 7.6 \times 10^{-5}\,M^{-1}\,sec^{-1}$

a. $[CH_3CH_2O^-] = \dfrac{0.05\ mole}{0.25\ liter} = 0.2\,M \rightarrow t = \dfrac{0.1}{7.6 \times 10^{-5} \times 0.2} = 6580\ sec = 1.83\ hr$

b. $[CH_3CH_2O^-] = \dfrac{0.05\ mole}{1.0\ liter} = 0.05\,M \rightarrow t = \dfrac{0.1}{7.6 \times 10^{-5} \times 0.05} = 26{,}300\ sec = 7.31\ hr$

c. $[CH_3CH_2O^-] = 0.2\,M \rightarrow t = 1.83$ hr (as in part a.)

S3. $k'[(CH_3)_3CBr]t = [(CH_3)_3COCH_2CH_3]$

For 10% reaction, $[(CH_3)_3COCH_2CH_3] = 0.1 \times [(CH_3)_3CCBr]$

and the equation above reduces to: $k'[(CH_3)_3CBr]t \cong 0.1 \times [(CH_3)_3CBr]$

For 10% reaction, $t \cong \dfrac{0.1}{k'} = \dfrac{0.1}{5 \times 10^{-4}\ sec^{-1}} = 200\ sec = 3.3\ min.$

Because the rate of the reaction is *independent* of $[CH_3CH_2O^-]$ (a first-order reaction), the length of time for 10% reaction is the same for all three cases a., b., and c. As far as each molecule of $(CH_3)_3CBr$ is concerned, the presence or absence of a nearby $CH_3CH_2O^-$ ion is unimportant in determining how fast it is going to react.

S4. a. 10% reaction = $\dfrac{0.005\ mole}{0.25\ liter} = 0.02\,M$ (change in $[CH_3CH_2I]$)

rate = $\dfrac{0.02\ M}{65\ min} = 3.08 \times 10^{-4}\,M\,min^{-1}$ ($5.13 \times 10^{-6}\,M\,sec^{-1}$)

b. 10% reaction = $\dfrac{0.005\ mole}{0.5\ liter} = 0.01\,M$ (change in $[CH_3CH_2I]$)

rate = $\dfrac{0.01\ M}{130\ min} = 7.69 \times 10^{-5}\,M\,min^{-1}$ ($1.28 \times 10^{-6}\,M\,sec^{-1}$)

c. The two likely possibilities are:

rate = $k[CH_3CH_2I][CH_3CH_2O^-]$ or rate = $k'[CH_3CH_2I]$

As shown in the preceding problem, the rate of a first order reaction is independent of concentration. Since the rate of this reaction changes with concentration, this is a second-order reaction, and the rate equation is:

$$\text{rate} = k[CH_3CH_2I][CH_3CH_2O^-]$$

d. From a., rate = $5.13 \times 10^{-6}\ M\ \text{sec}^{-1} = k\ [0.2\ M][0.2\ M]$

$k = 1.28 \times 10^{-4}\ M^{-1}\ \text{sec}^{-1}$

or

From b., rate = $1.28 \times 10^{-6}\ M\ \text{sec}^{-1} = k\ [0.1\ M][0.1\ M]$

$k = 1.28 \times 10^{-4}\ M^{-1}\ \text{sec}^{-1}$

S5. a.

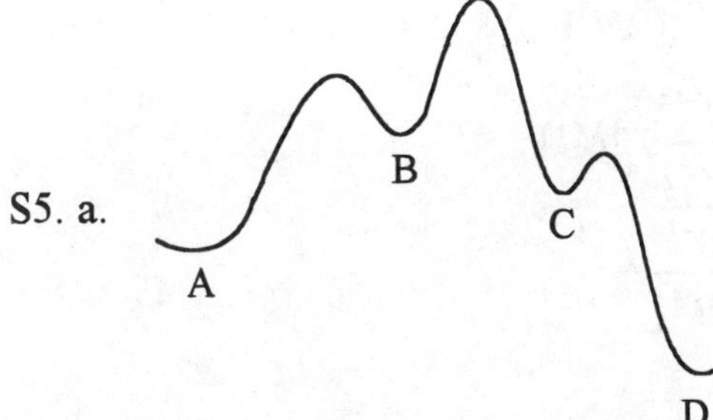

b. B

c. C (to go to D), because it has the lowest activation energy and thus the fastest rate constant of any of the reactions.

d. $K = \dfrac{[D]}{[A]} = \dfrac{k_1 \times k_2 \times k_3}{k_{-1} \times k_{-2} \times k_{-3}} = 4.5 \times 10^6$

e. B to C

f. endothermic

g. C to D

S6. a. H_2S is less basic = (higher pK_a) because S lies to the left of Cl in the periodic table (it is less electronegative).

b. H_2O (same reason as for a.)

c. $H_2PO_4^-$, because one negative charge destabilizes a second.

d. NH_3 (same reason as for a.)

e. H_2O, because loss of a proton from a cationic molecule is easier than from a neutral one, other factors being equal.

f. NH_4^+ (same reason as for a.)

g. H_2O, because in HOOH, one oxygen acts acts as an electronegative substituent on the other.

S7. a. The equilibrium: $Base^- + H_2O \rightleftharpoons Base\text{-}H + OH^-$ will always take place.
A base stronger than OH^- (pK_a of Base-H > 14) will simply react with water to give OH^-.

b. For a similar reason, H_3O^+ is the strongest acid possible in water.

S8. a. $$\frac{[NH_3]\,[H^+]}{[NH_4^+]} \approx \frac{[NH_3]\,[H^+]}{1} = 10^{-9.2}$$

so that $[H^+] = \sqrt{10^{-9.2}} = 2.51 \times 10^{-5}$ and pH = 4.6

(compare with Exercise at the end of Section 4.5)

b. $NH_3 + H_2O \rightleftharpoons NH_4^+ + OH^-$ $[NH_3] = 1 - [NH_4^+] = 1 - [OH^-]$

$$[NH_4^+] = \frac{10^{-14}}{[H^+]} \text{ so that } \frac{[NH_3]\,[H^+]}{[NH_4^+]} = \frac{(1 - \frac{10^{-14}}{[H^+]})\,[H^+]}{\frac{10^{-14}}{[H^+]}}$$

$10^{14} \times [H^+]^2 - [H^+] - K = 0$

$[H^+] = 2.5 \times 10^{-12}$ and pH = 11.6

S9. a.

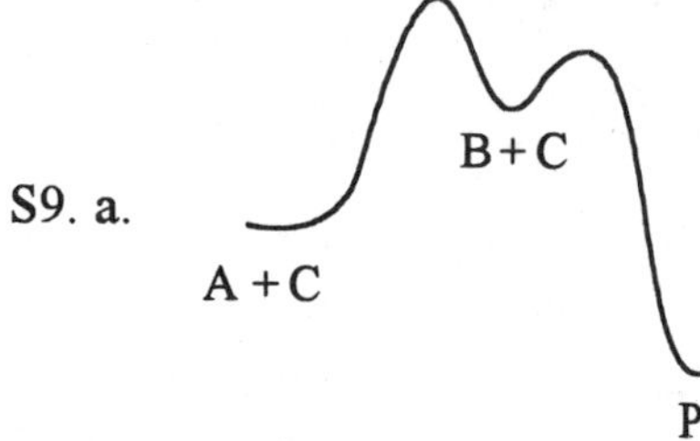

b. Under standard state conditions, the conversion of A → B is the slow step.

c. If [C] = 0.001, k_2 [C] (2×10^{-5}) is small relative to k_{-1} (10^{-3}) and the rate equation reduces to:

$$\frac{d[P]}{dt} = \frac{k_1 k_2}{k_{-1}}[A][C]$$

The reaction is now second-order and B + C → P is the rate-determining step. Notice how the rate-determining step can change with concentration. The reaction coordinate diagram does not provide a valid picture for concentrations other than standard state, unless all the transformations involved are first-order.

S10. a.

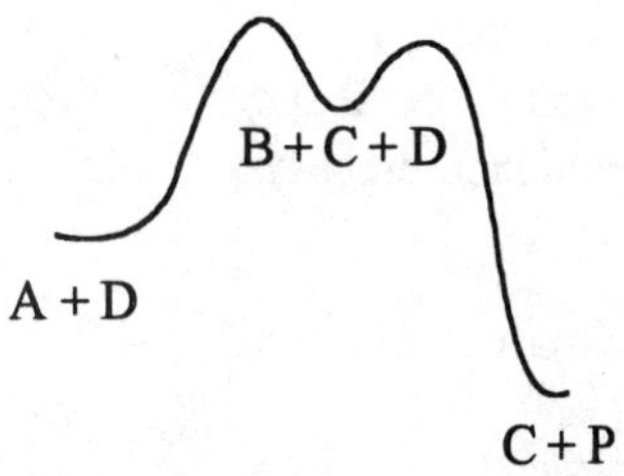

B + C + D

A + D

C + P

b. $k_{-1}[C] = 0$ during the early part of the reaction, and

$$\frac{d[P]}{dt} = \frac{k_1 k_2[A][D]}{k_2[D]} = k_1[A]$$

The first step is rate-limiting.

c. If [C] = 2, then $k_{-1}[C] + k[D] = (10^{-2} \times 2) + (3 \times 10^{-2}) \times 0.1$
$= 2.3 \times 10^{-2} \cong k_{-1}[C]$

so that $$\frac{d[P]}{dt} = \frac{k_1 k_2[A][D]}{k_{-1}[C]}$$

The second step is now rate-determining.

5. ALKANES

5.A. Chapter Outline and Important Terms Introduced

5.1 n-Alkanes; Physical Properties (boiling points, melting points, and density)

non-polar forces of attraction
Van der Waals forces
London force, dispersion force
dependence on surface area

5.2 n-Alkanes; Barriers to Rotation (rotation about single bonds)

conformations
eclipsed
staggered
Newman projections
anti
gauche

5.3 Branched Chain Alkanes (more on physical properties, conformations)

conformational "isomers" - conformers - versus structural isomers
line structures

5.4 Cycloalkanes (ring structures, physical properties, nomenclature)

5.5 Heats of Formation (how to calculate relative stabilities)

heats of formation, ΔH_f°
standard state
distinction from free energy, ΔG_f°

5.6 Cycloalkanes; Ring Strain (bending bonds in small rings; conformations of cyclic molecules)

cyclopropane and cyclobutane: bent bonds
cyclopentane: envelope
cyclohexane: chair ⇌ chair interconversion; axial vs. equatorial
larger-ring cycloalkanes

5.7 Occurrence of Alkanes (or lack thereof....)

crude oil
natural (living) sources

5.B. Important Reactions Introduced:

still none yet

5.C. Important Concepts and Hints

Chapter 5 presents the structural and physical properties of a specific class of organic compounds, the saturated hydrocarbons. Although there are only a few chemical reactions to be discussed for the alkanes (Chapter 6), carbon chains form the backbone of all other organic molecules, and it is therefore important to know what their shapes are, how their conformations differ in three dimensions, and how much more stable one isomer is compared to another. In this connection, the following terms are used over and over, and you should know what they mean: **anti** and **gauche**, **eclipsed** and **staggered**, **ring strain**, **chair conformation**, and **axial** and **equatorial**. Be sure you also understand the concept of **heat of formation** as an indication of relative thermodynamic stability.

By now you should be completely familiar with the relationship between free energy differences and relative amounts ($\Delta G^\circ = -RT \ln K$), and also quite convinced of the value of your molecular model kit.

HINTS: In problems that ask you to calculate energy differences (ΔH° or ΔG°), the best way to figure out what to subtract from what is:

(1) Write the equation for the reaction, with the products on the right;
(2) Write the thermodynamic value you are interested in comparing under the formula, or name of each component, keeping the signs straight;
(3) <u>ADD</u> the values for the products and <u>SUBTRACT</u> the values for the starting materials, again paying attention to the signs.

As an example, calculate ΔH° for the following reaction:

(1) $CH_2{=}CH_2$ + HBr ⟶ CH_3CH_2Br
(2) ΔH_f° = 12.5 -8.7 -15.2
(3) $\Delta H^\circ = +(-15.2) - (12.5 - 8.7) = -19.0$ kcal $mole^{-1}$

This will give you the thermodynamic value for the overall transformation, with the correct sign for the direction in which you have written the equation. If you write the reverse reaction, you change the sign of the value that you calculate.

If you like to remember things visually, the following scheme can also help you keep track of when to add and when to subtract.

(1) Across the page, draw a line representing zero energy.
(2) To determine the energy content of the reactants, go **UP** for positive values and **DOWN** for negative values, moving sequentially. For instance, "$CH_2{=}CH_2$ ($\Delta H_f^\circ = +12.5$) + HBr ($\Delta H_f^\circ = -8.7$)" would result in:

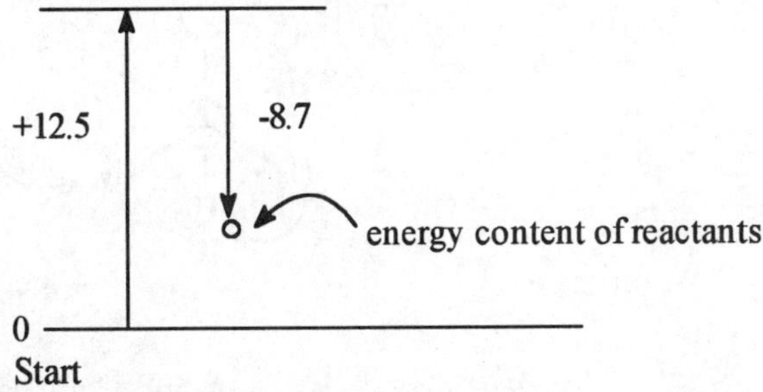

(3) Determine the energy content of the products the same way:

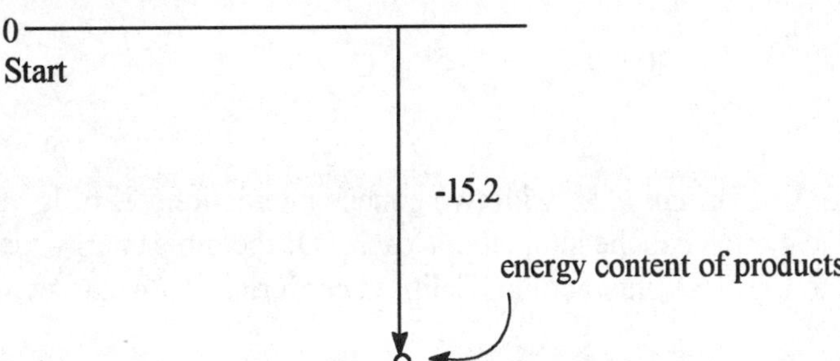

(4) Visually, then it is easy to see whether going from reactants to products is uphill (endothermic, ΔH° > 0) or downhill (exothermic, ΔH° < 0):

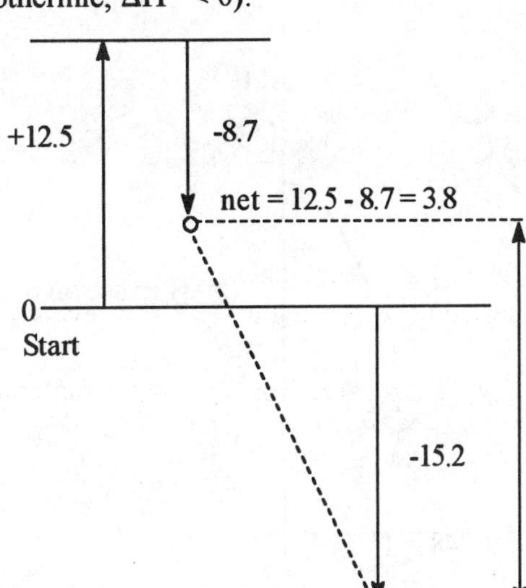

total energy difference = 19.0 and it is downhill; ΔH° = -19.0 kcal/mole.

Also, remember that ΔH and ΔG are usually presented as "**kcal** mole^{-1}", whereas ΔS and the gas constant R are given as "**cal** deg^{-1} mole^{-1}" (the same as an entropy unit, e.u.). A kcal is 1000 cal.

5.D. Answers to Exercises

5.1 The distance between eclipsed hydrogens in ethane is about 2.3Å, as indicated by molecular models (trigonometry with r(C-H) = 1.10Å, r(C-C) = 1.54Å, and tetrahedral angles gives 2.27Å). The distance between staggered hydrogens in ethane is about 2.5Å. By contrast, in gauche-butane one pair of 1,4-hydrogens is only 2.0Å apart. It is this close approach of these two hydrogens that probably accounts for most of the relative instability of the gauche conformation.

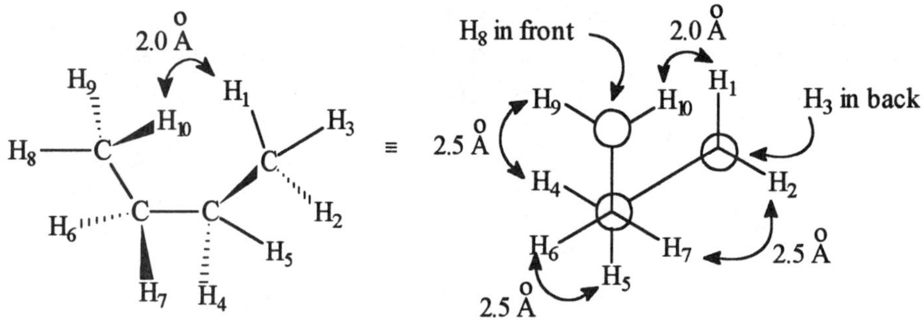

5.2 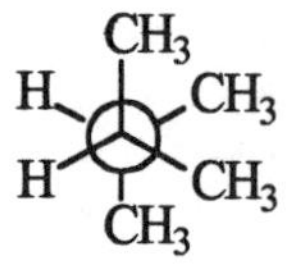 A and C are equivalent and have equal energy

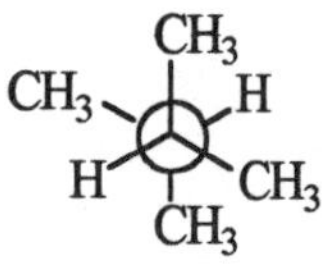

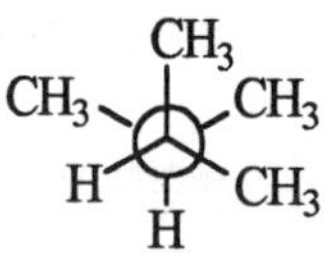

5.3 **2,3-Dimethylbutane**: Conformation D, with two gauche interactions, is of lower energy than the other two minima, B and F, which have three gauche interactions each. Of the three energy maxima, C and E each have two CH_3-H interactions and one CH_3-CH_3 interaction. Eclipsed conformation A has two CH_3-CH_3 interactions and one H-H eclipsed interaction.

[Me = CH_3]

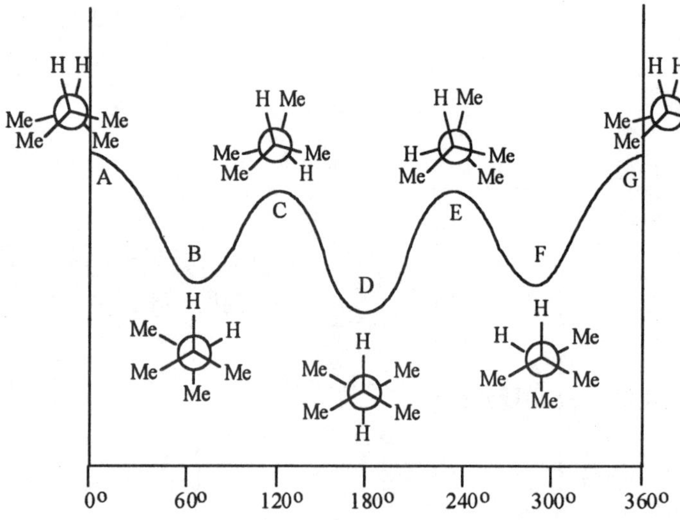

(NOTE: see answer to problem #5, too.)

5.4 a. CH_3 / $CH_3-CH-CH_2-CH_2-CH_2-CH_2-CH_2-CH_2-CH_3$

b. CH_3 / CH_3-CH, CH_2-CH_3, C, CH_3, CH_3

c. CH_3 / $CH_3-CH_2-CH-CH-CH_2-CH_2-CH_2-CH_3$ / CH_2 / CH_3

d. $CH_3-CH_2-C-CH_2-CH_2-CH_3$ / CH_2, CH_3 / Cl

5.5 a. H_3C, CH_3, C, H_2C, CH_2, H_2C, $CH-CH_3$, CH_2

b. $H_3C-CH_2-CH-CH_2-CH_3$, CH, H_2C, CH_2, H_2C-CH_2

c. Cl, CH, H_2C, CH_2, H_2C, CH_2, CH, H_2C, Cl

d. H_3C, CH_3, C, C, H_3C, CH_2, CH_3

5.6

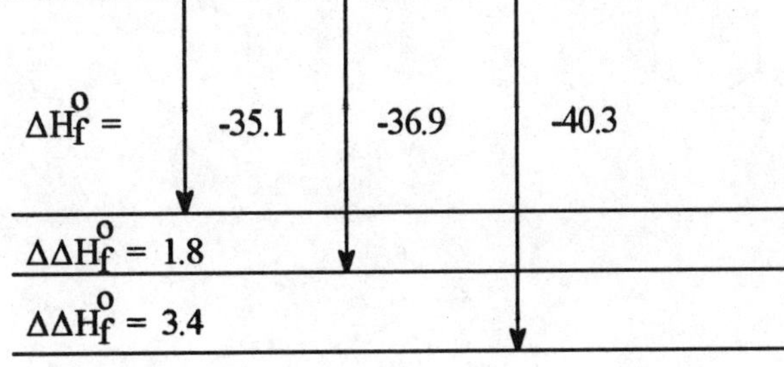

Standard States

pentane

2-methylbutane

2,2-dimethylpropane

5.7

$$(CH_3)_2CHCH_2CH_3 + H_2 \longrightarrow CH_3CH_3 + CH_3CH_2CH_3$$

$\Delta H_f^o =$ −36.9   0   −20.2   −24.8

$\Delta H° = -(-36.9 + 0) + (-20.2 - 24.8) = -8.1$ kcal mole^{-1}

5.8 Cyclohexane is strain-free, and its heat of formation per CH_2 group ($\Delta H^\circ_f/6 = -4.92$ kcal mole^{-1}) is taken as the standard. If cyclooctane were also strain-free, it would have $\Delta H^\circ_f = -4.92 \times 8 = -39.36$ kcal mole^{-1}. The actual ΔH°_f of -29.7 kcal mole^{-1} indicates that the strain energy of cyclooctane is $-29.7 - (-39.4) =$ 9.7 kcal mole^{-1}.

5.9 Model building. No solution required.

5.10 Although you may think that we're trying to teach you art in this question, consider this point: many exam answers are marked "wrong" because the structures drawn are incomprehensible to the grader....

5.11

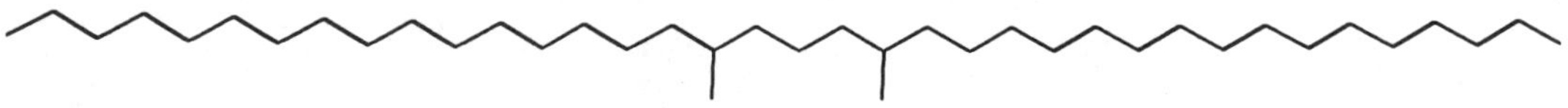

17,21-dimethylheptatriacontane

5.E Answers and Explanations for Problems

1. a. 1-ethyl-1-methylcyclohexane
   b. 1-isopropyl-3-methylcyclohexane or 1-methyl-3-(1-methylethyl)cyclohexane
   c. isopropylcyclodecane or (1-methylethyl)cyclodecane
   d. 1,1-dimethylcyclopropane
   e. isobutylcyclopentane or (2-methylpropyl)cyclopentane
   f. 1-cyclobutyl-3-methylpentane or (3-methylpentyl)cyclobutane
   g. 1-bromo-3-methylcyclohexane
   h. 1-ethyl-2-iodocyclopentane

2. a.

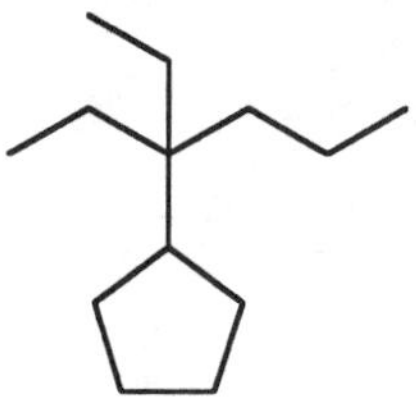

b. 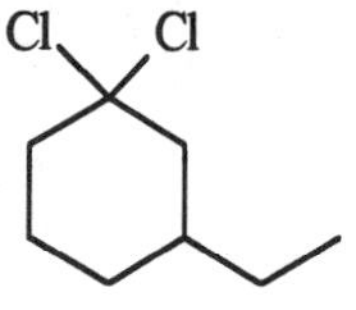 Cl Cl

c.

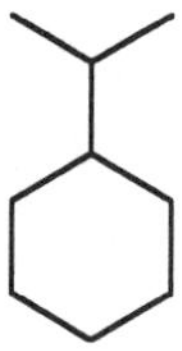

d.  Cl Cl

e.

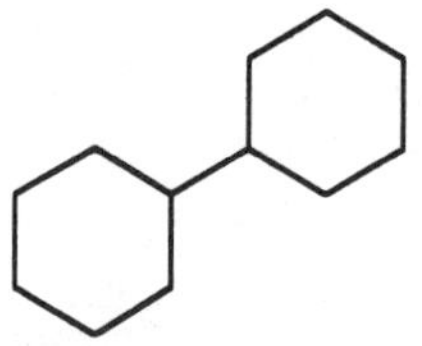

f.

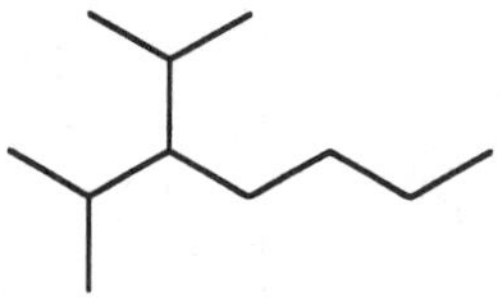

3.

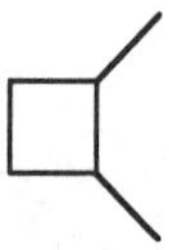

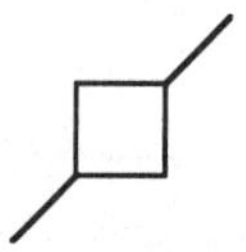

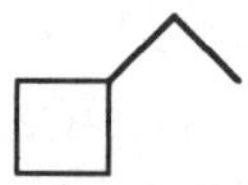

1,1-dimethylcyclobutane 1,2-dimethylcyclobutane 1,3-dimethylcyclobutane ethylcyclobutane

There are 2 more isomers that you will learn about in Chapter 7. These occur for 1,2 and 1,3 disubstituted cyclobutanes depending on whether the substituents are both on the same side with respect to the plane of the ring or on opposite sides.

4. The extra CH_2 group in heptane relative to hexane raises the b.p. by 98.4 − 68.7 = 30°. To estimate the b.p.'s of the heptane isomers, add 30° to the most similar branched hexane.

Name	Line Structure	Estimated b.p.
heptane		(98.4°C)
2-methylhexane	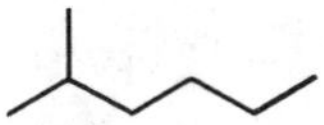	90
3-methylhexane	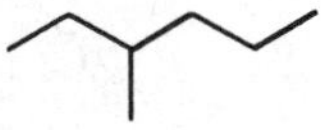	93
2,2-dimethylpentane	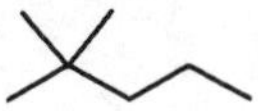	80
2,3-dimethylpentane	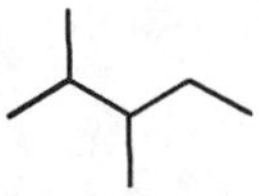	88
2,4-dimethylpentane	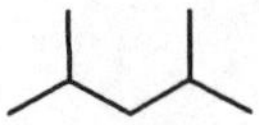	82

The structural difference between hexane and 2-methylpentane (−8.4°C in b.p.) is the same as that between 2-methylhexane and 2,4-dimethylpentane. This explains the estimate of 90 − 8 = 82°C for the b.p. of the latter compound.

3,3-dimethylpentane		80
3-ethylpentane	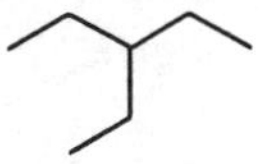	90
2,2,3-trimethylbutane		80

The b.p. for 2,2,3-trimethylbutane was estimated by adding a methyl to 2,2-dimethylbutane rather than 2,3-dimethylbutane.

5. **C_2-C_3 anti**

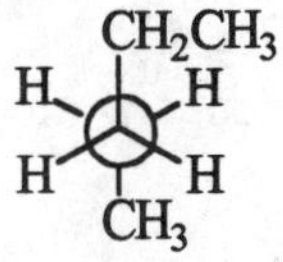

CH_2CH_3, H, H, H, H, CH_3

This corresponds to a. in Figure 5.11

C_2-C_3 gauche

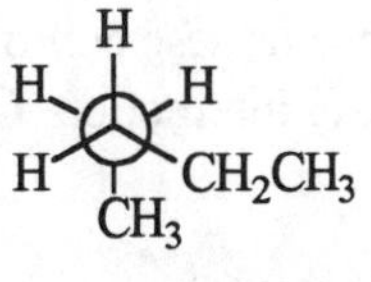

H, H, H, H, CH_2CH_3, CH_3 and H, H, H, CH_3CH_2, H, CH_3

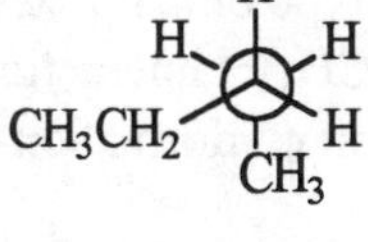

This corresponds to b. in Figure 5.11

6. Adamantane is $C_{10}H_{16}$ (mw = 136) and is very symmetrical – almost spherical. For comparison, 2,2,3,3-tetramethylbutane (C_8H_{18} ; mw = 114) has bp 106°C and mp 100°C. Adamantane would be expected to have a similar bp (it is even more symmetrical than tetramethylbutane, but a little larger). The symmetrical structure would suggest a high mp. In fact, adamantane melts at 270°C (in a sealed tube). At atmospheric pressure, adamantane sublimes instead of melting; that is, it goes directly from the solid to the vapor state.

7. most stable; all anti

one gauche interaction

two gauche interactions

least stable; methyls colliding

8. **2-Methylbutane:**

[Me = CH_3]

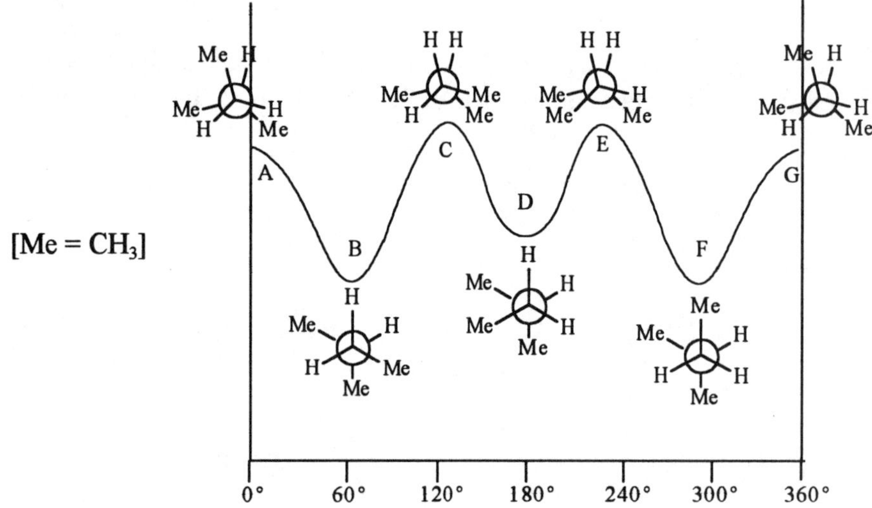

The three energy minima are B, D, and F. Conformations B and F are equivalent and equal in energy; they both have one gauche interaction. Conformation D, with two gauche interactions, is less stable. In rotating about the C_2-C_3 bond from B to F, the molecule passes through an eclipsed conformation (energy maximum) in which there are three CH_3-H interactions. In passing from B to D and also from D back to F, the molecule passes through another type of eclipsed conformation. In these conformations (C and E), there is one CH_3-CH_3 interaction, one CH_3-H interaction, and one H-H interaction. Conformations C and E are equivalent and are of higher energy than conformation A.

2,2-Dimethylbutane:

0°
60°
120°
180°
240°
300°
360°

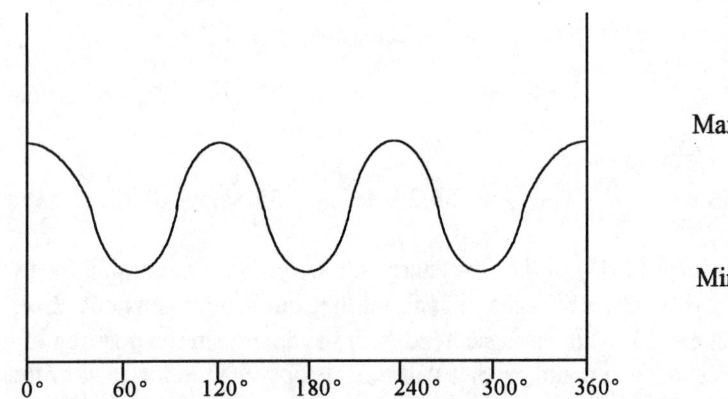

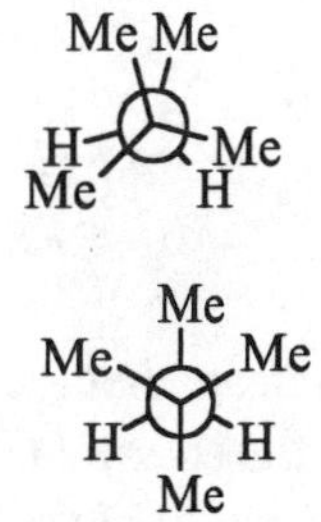

Maxima:
Me Me
H
Me
Me
H
Minima:
Me
Me
Me
H
H
Me

All minima are equal, and all maxima are equal.

2,2,3,3-Tetramethylbutane:

0°
60°
120°
180°
240°
300°
360°

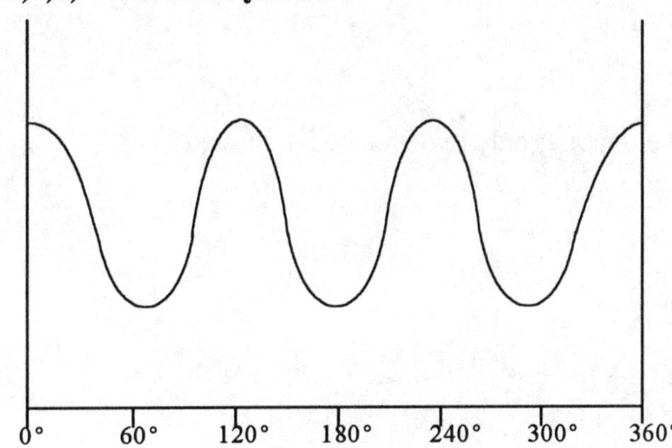

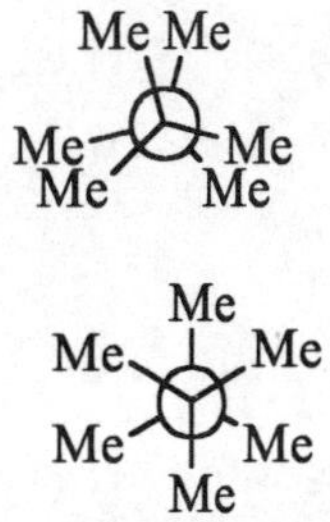

Maxima:
Me Me
Me
Me
Me
Me
Minima:
Me
Me
Me
Me
Me
Me

All minima are equal, and all maxima are equal.

9.

1
3
2
≡
H
anti
H_2C
CH_2
H
H_2C
CH_2
CH_3
H
1
3
2
≡
CH_3
gauche, therefore less stable
H_2C
CH_2
H
H_2C
CH_2
H
H

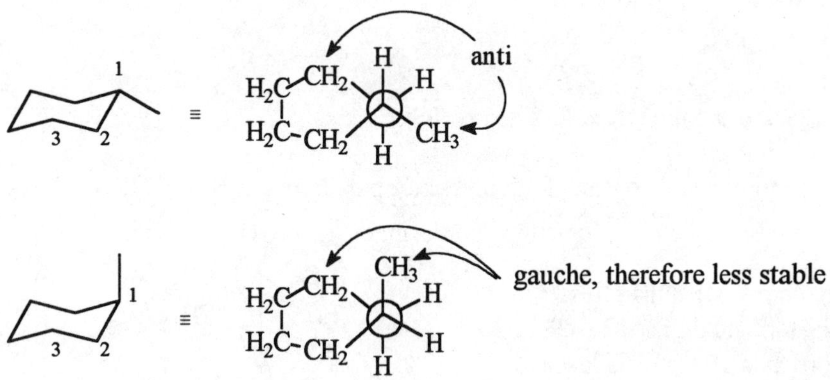

10. a. $CH_2{=}CH_2$ + H_2 ⟶ CH_3CH_3
ΔH°_f = +12.5 (0) −20.2 $\Delta H^\circ = -20.2 - 12.5 = -32.7$ kcal $mole^{-1}$

b. $CH_2{=}CH_2$ + HCl ⟶ CH_3CH_2Cl
ΔH°_f = +12.5 −22.1 −26.1 $\Delta H^\circ = -26.1 - (12.5 - 22.1) = -16.5$ kcal $mole^{-1}$

c. $CH_2{=}CH_2$ + H_2O ⟶ CH_3CH_2OH
ΔH°_f = +12.5 −57.8 −56.2 $\Delta H^\circ = -56.2 - (12.5 - 57.8) = -10.9$ kcal $mole^{-1}$

These calculations indicate that the enthalpy term (ΔH°) of the free energy change ΔG° is favorable (negative), but without knowing the entropy term (ΔS°) you can't say very much about the equilibrium constant. See, for example, Supplementary Problem #S1 in Chapter 4, which is concerned with the change in the position of equilibrium c. above with temperature. However, you could predict that the entropy change for each of these reactions would be about the same (two molecules ⟶ one molecule), and because of the enthalpy differences between a., b. and c., the equilibrium constant for reaction a. would be greater than that for b., and b. greater than c.

The enthalpy change of the overall reaction bears no particular relationship to the energy of activation, which is what governs the **rate** of reaction.

11. 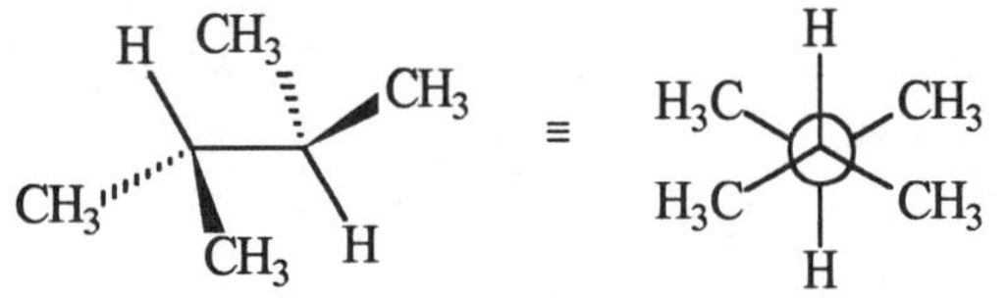 [Sawhorse structure A] ≡ [Newman projection: H top, H_3C, CH_3, H_3C, CH_3, H bottom] Lower energy; only two gauche interactions

A

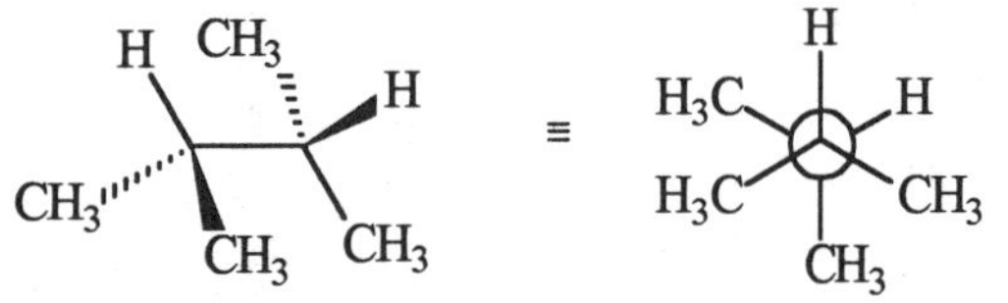

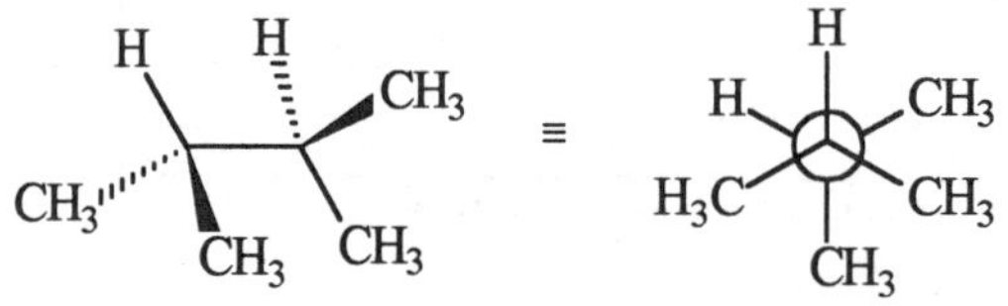

[Sawhorse structure B] ≡ [Newman projection: H, H_3C, H, H_3C, CH_3, CH_3] [Sawhorse structure B′] ≡ [Newman projection: H, H, CH_3, H_3C, CH_3, CH_3]

B B′

Both B and B′ have three gauche interactions, and so are expected to be 0.9 kcal $mole^{-1}$ less stable than A.

$$K = \frac{[A]}{[B]} = \frac{[A]}{[B']} \quad \text{and } \Delta G^\circ = -RT\ \ln K$$

$$-0.9 \text{ kcal mole}^{-1} = -1.987 \text{ cal deg}^{-1} \text{ mole}^{-1} \times 298 \text{ deg} \times \ln K$$

Based on this prediction, the mixture would consist of 4.5 parts A:1 part B:1 part B′ = 69% A, 31% B + B′

[Newman projection: H top, H_3C, CH_3, H_3C, CH_3, H bottom; labels: *worse interaction*, *repulsion*]

The situation is actually more complicated than this:
The geminal methyls undergo steric hindrance as well, and the angle between them is more than 109°. This leads to greater interaction between the vicinal methyls in the anti conformation than predicated above, and a much smaller difference in energy between the two staggered conformations of 2,3-dimethylbutane.

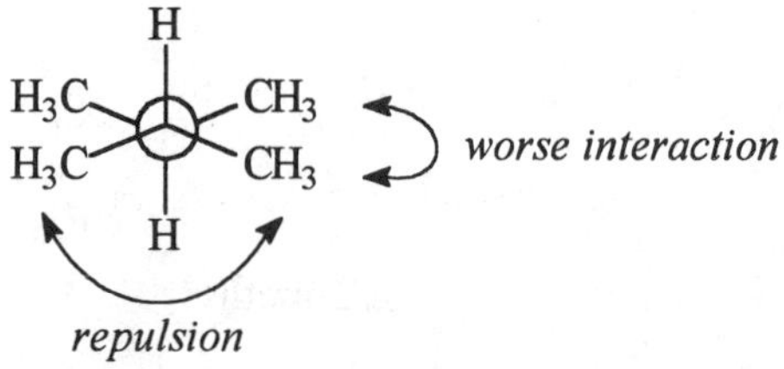

12. cyclooctane $\rightleftharpoons$ ethylcyclohexane (CH_2CH_3)

$$K = \frac{[ethylcyclohexane]}{[cyclooctane]}$$

$\Delta H^\circ_f = -29.7$ $\quad -41.0 \quad$ $\Delta H^\circ = -11.3$ kcal mole^{-1}

If $\Delta H^\circ = \Delta G^\circ = -RT \ln K$, then $K = 1.9 \times 10^8$

The fact that the equilibrium is even more in favor of ethylcyclohexane results from the favorable contribution from entropy to the free energy of this isomerization ($\Delta G^\circ = \Delta H^\circ - T\Delta S^\circ$). There is more freedom of motion in ethylcyclohexane than in cyclooctane (the ethyl group can spin relative to the cyclohexane ring, for instance); this results in a positive ΔS° and therefore a negative contribution to ΔG°.

13. Boiling points and melting points have nothing to do with relative thermodynamic stability. M.p. and b.p. are only indications of the stability of the solid vs. liquid vs. gaseous states of a molecule, whereas thermodynamic stability is in relation to possible reactions or decompositions. For example, hexahydro-1,3,5-trinitro-1,3,5-triazine (cyclonite or RDX) is a high-melting solid (m.p. 204°C) which on decomposition releases over 300 kcal mole^{-1}. (This substance is used as a high explosive).

14. $\Delta G^\circ = \Delta H^\circ - T\Delta S^\circ$

$-0.89 = -2.05 - (298 \times \Delta S^\circ)$, so that $\Delta S^\circ = -3.89$ cal deg^{-1} mole^{-1}

A negative entropy change indicates more "restriction of freedom". In this instance, it arises because there are fewer different conformations possible for isobutane than for butane.

$\Delta G^\circ = -RT \ln K = -0.89$ kcal mole^{-1}; therefore $K = 4.5$

15. pentane $\rightleftharpoons$ 2-methylbutane

$$\Delta G^\circ = -1.54 \text{ kcal mole}^{-1} \text{ and } K_1 = \frac{[2\text{-}methylbutane]}{[pentane]} = 13.5$$

2-methylbutane $\rightleftharpoons$ 2,2-dimethylpropane

$$\Delta G^\circ = -0.10 \text{ kcal mole}^{-1} \text{ and } K_2 = \frac{[2,2\text{-}dimethylpropane]}{[2\text{-}methylbutane]} = 1.2$$

Ratio of pentane/2-methylbutane/2,2-dimethylpropane = 1 : 13.5 : (13.5 x 1.2), which is equivalent to 3% /44% /53%.

16. Compare:

	ethane		propane		butane		pentane
ΔH_f° =	-20.2		-24.8		-30.4		-35.1
difference		-4.6		-5.6		-4.7	

average = -5.0 kcal $mole^{-1}$ per CH_2 group

ethane: two CH_3 ; ΔH_f° = -20.2, which is equivalent to -10.1 per CH_3

propane: two CH_3 + one CH_2 (-5.0) $\Delta H_f^\circ = -24.8 \Rightarrow$ - 9.9 per CH_3

butane: two CH_3 + two CH_2 (-10.0) $\Delta H_f^\circ = -30.4 \Rightarrow$ -10.2 per CH_3

pentane: two CH_3 + three CH_2 (-15.0) $\Delta H_f^\circ = -35.1 \Rightarrow$ -10.1 per CH_3

average = -10.1 kcal $mole^{-1}$ per CH_3 group

isobutane: three CH_3 + one CH (-30.3) $\Delta H_f^\circ = -32.4 \Rightarrow$ -2.1 per CH

isopentane: three CH_3 + one CH_2 + one CH (-30.3) (-5.0) $\Delta H_f^\circ = -36.9 \Rightarrow$ -1.6 per CH

average = -1.9 kcal $mole^{-1}$ per CH group

neopentane: four CH_3 + one C (-40.4) $\Delta H_f^\circ = -40.3 \Rightarrow$ +0.1 per C

PREDICTIONS:

		ΔH_f° (kcal $mole^{-1}$) Estimated	Found
hexane:	two CH_3 + four CH_2 (-20.2) (-20.0)	-40.2	-39.9
2-methylpentane: 3-methylpentane:	three CH_3 + two CH_2 + one CH (-30.3) (-10.0) (-1.9)	-42.2	-41.8 -41.1
2,2-dimethylbutane:	four CH_3 + one CH_2 + one C (-40.4) (-5.0) (+0.1)	-45.3	-44.3
2,3-dimethylbutane:	four CH_3 + two CH (-40.4) (-3.8)	-44.2	-42.6
nonane:	two CH_3 + seven CH_2 (-20.2) (-35.0)	-55.2	-54.7
2,2,4,4-tetramethylpentane:	six CH_3 + one CH_2 + two C (-60.6) (-5.0) (+0.2)	-65.4	-57.8

The steric strain of 2,2,4,4-tetramethylpentane makes it difficult to put the molecule together; i.e., the compound is less stable than estimated.

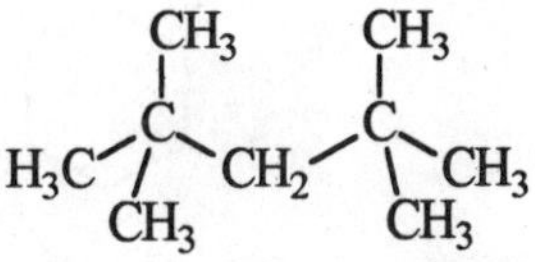

$$H_3C-C(CH_3)_2-CH_2-C(CH_3)_2-CH_3$$

NOTE: from a more extensive comparison of hydrocarbon heats of formation, the ΔH_f° incremental values below have been calculated:

CH_3	-10.12 kcal mole^{-1}
CH_2	-4.93
CH	-1.09
C	$+0.80$

5.F. Supplementary Problems

S1. Write the structures for all of the isomers of C_5H_{10} that have only C-C single bonds, and provide IUPAC names for each structure.

S2. Write line structure for each of the following compounds:

a. 1-ethyl-2-propylcyclopentane
b. 1,1,4-tribromocyclohexane
c. 3-chloro-1,1-dicyclopropylcycloheptane
d. 1-(3-chloropropyl)-4-*t*-butylcyclohexane
e. cyclotetradecane

S3. Using the heats of formation given in Appendix I, calculate or estimate ΔH° for the following reactions:

a. △ $+ H_2 \longrightarrow CH_3CH_2CH_3$

b. □ $+ H_2 \longrightarrow CH_3CH_2CH_2CH_3$

c. □ $\rightleftharpoons$ ▷$-CH_3$ (estimate)

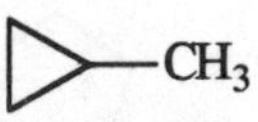

d. $CH_3CH_3 + HCl \longrightarrow CH_4 + CH_3Cl$

e.  △ $+ HCl \longrightarrow CH_3CH_2CH_2Cl$

f.  ⬡ $+ HCl \longrightarrow CH_3(CH_2)_4CH_2Cl$ (estimate)

S4. From the data provided in Appendix I, calculate ΔH° for the following isomerizations:

$$CH_3CH_2CH_2CH_2CH_3 \rightleftharpoons CH_3CH_2CH(CH_3)_2$$

CH_3

How do you account for the fact that increased branching is favorable in one case and not the other?

S5. Make a model of *t*-butylcyclobutane. Which conformation do you expect would be the most stable? Do the same for *t*-butylcyclopentane.

S6. Which of the following pairs represent structural isomers? Which are conformational isomers? Which are not isomers at all?

a.  H_3C CH_3 and CH_3 CH_3

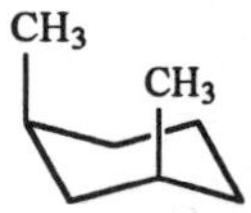

b. 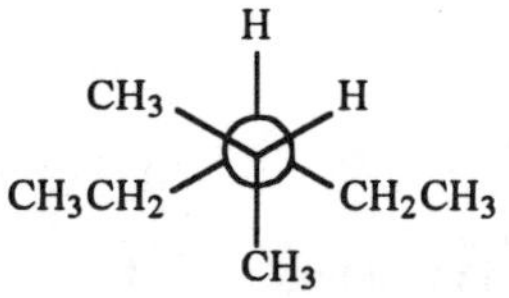 H CH_3 H CH_3CH_2 CH_2CH_3 CH_3 and CH_3 H H CH_3CH_2 CH_2CH_3 CH_3

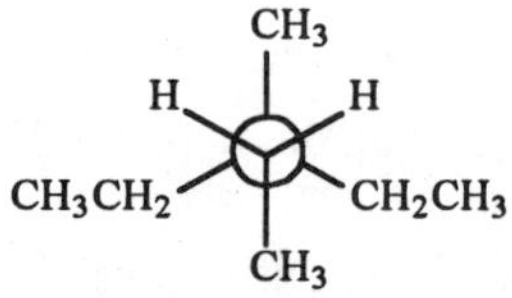

c. 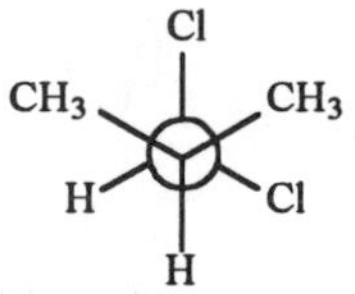 Cl CH_3 CH_3 H Cl H and H CH_3 CH_3 Cl Cl H

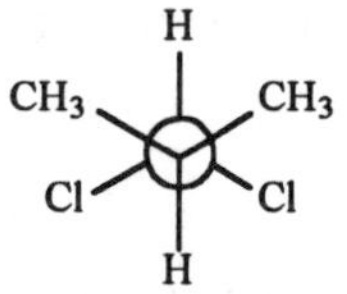

d.  Br Br and Br Cl

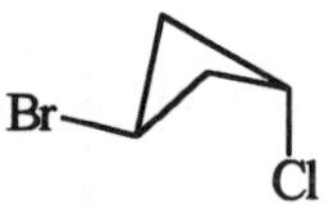

e.  H_3C and CH_3

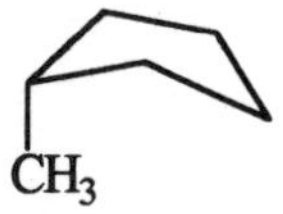

f. 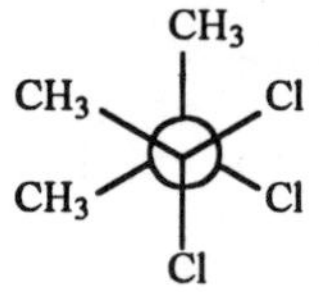 CH_3 CH_3 Cl CH_3 Cl Cl and CH_3 Cl Cl CH_3 CH_3 Cl

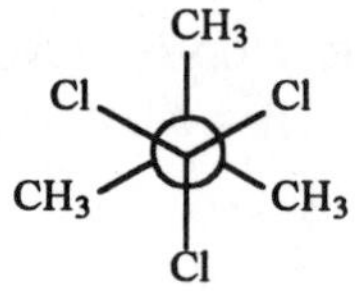

5.G Answers to Supplementary Problems

S1. 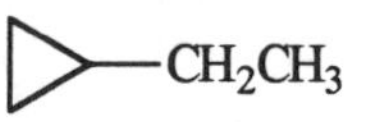 $-CH_2CH_3$ ethylcyclopropane; H_3C CH_3 1,2-dimethylcyclopropane

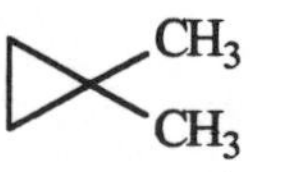

CH_3 CH_3 1,1-dimethylcyclopropane; $-CH_3$ methylcyclobutane; 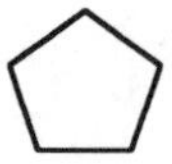 cyclopentane

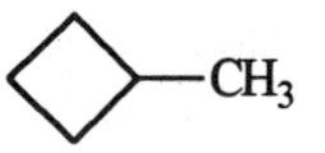

As mentioned in problem 3, 1,2-dimethylcyclopropane can exist as two geometric isomers. This will be discussed in chapter 7.

S2. a.

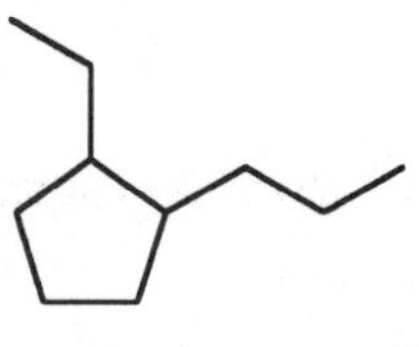

b. 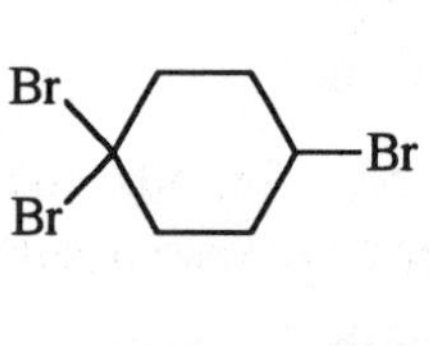 Br Br -Br

c. 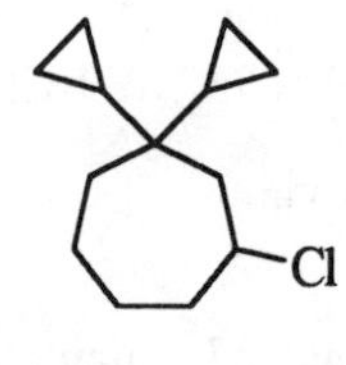 Cl

d. Cl

e.

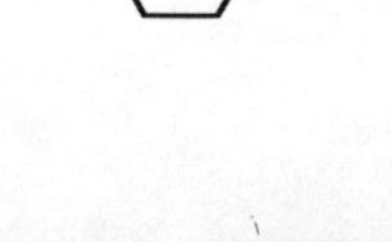

S3. ΔH_f° (products) − ΔH_f° (starting material) = ΔH° of the reaction

NOTE: ΔH_f° for $H_2 = 0$

a. $-24.8 - 12.7 = -37.5$ kcal mole^{-1}
b. $-30.4 - 6.8 = -37.2$ kcal mole^{-1}
c. To calculate the change in ΔH_f° caused by the addition of a methyl group:

ΔH_f° (isobutane) − ΔH_f°(propane) = -7.6 kcal mole^{-1}
ΔH_f° (methylcyclopentane) − ΔH_f° (cyclopentane) = -6.9 kcal mole^{-1}
ΔH_f° (methylcyclohexane) − ΔH_f° (cyclohexane) = -7.5 kcal mole^{-1}

Average is -7.3

Using this calculated average, the estimated ΔH_f° (methylcyclopropane) is $12.7 - 7.3 = 5.4$ kcal mole^{-1}, and ΔH° for the conversion cyclobutane ⟶ methylcyclopropane is $5.4 - 6.8 = -1.4$ kcal mole^{-1}.

d. $-17.9 - 20.6 - (-20.2) - (-22.1) = +3.8$ kcal mole^{-1} (an endothermic reaction)

e. $-31.0 - 12.7 - (-22.1) = -21.6$ kcal mole^{-1}

f. Estimate that ΔH_f° ($CH_3(CH_2)_4CH_2Cl$) is approx. ΔH_f° ($CH_3CH_2CH_2Cl$) + $3\Delta H_f^\circ$ (CH_2): = -46.0 kcal mole^{-1} (see problem #16 in this Chapter)

ΔH° for the reaction = $-46.0 - (-29.5 - 22.1) = +5.6$ kcal mole^{-1}

S4. pentane ⇌ 2-methylbutane
$\Delta H^\circ = -36.9 - (-35.1) = -1.8$ kcal mole^{-1}

cyclohexane ⇌ methylcyclopentane
$\Delta H^\circ = -25.3 - (-29.5) = +4.2$ kcal mole^{-1}

There is some strain in cyclopentane (6 kcal mole^{-1}, in fact), and none in cyclohexane, and this overcomes the small difference in ΔH_f° which is attributed to branching.

S5.

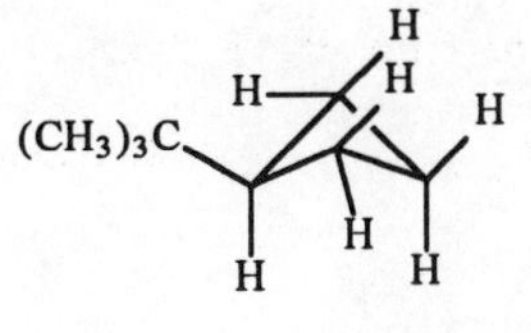

"equatorial" better than "axial"

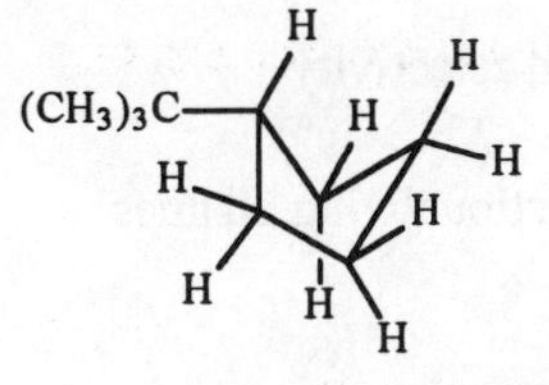

when *t*-butyl is attached to the "flap" of the envelope, it is in the least crowded position.

S6. Structural isomers: b., f.
Conformational isomers: a., c., e.
Not isomers: d.

6. REACTIONS OF ALKANES

6.A. Chapter Outline and Important Terms Introduced

6.1 **Bond Dissociation Energies** (what it takes to break a bond)

- vibrational energy levels
- zero point energy
- free radicals
- stability of radicals: tertiary > secondary > primary > methyl

6.2 **Pyrolysis of Alkanes: Cracking** (reactions of alkanes at high temperature)

disproportionation:

$$CH_3CH_2CH_2CH_2CH_3 \xrightarrow{\Delta} CH_3CH_3 + CH_2{=}CHCH_3$$

6.3 **Halogenation of Alkanes** (the first thorough analysis of a reaction mechanism)

Chlorination

$$RH + Cl_2 \longrightarrow RCl + HCl \text{ (can use } SO_2Cl_2 \text{, too)}$$

- homolysis vs. heterolysis
- chain reaction
- initiation, propagation and termination steps
- principle of microscopic reversibility
- relative reactivity
- statistical factor

Halogenation with Other Halogens

- selectivity of bromination

6.4 **Combustion of Alkanes** (how to **determine** relative stabilities)

- heat of combustion
- alkylperoxy radical

6.5 **Average Bond Energies** (to generalize between compounds)

- heat of atomization

6.B Important Reactions Introduced

Note: Starting with this chapter of the Study Guide, we will present an outline of the important reactions introduced in each chapter of the text, indicating some of the key aspects that you should remember.

Free radical halogenation (6.3)

Equation: $R_3C{-}H + X_2 \longrightarrow R_3C{-}X + HX$

Generality: R = alkyl or H

X =	
F	(seldom used)
Cl	(poor selectivity)
Br	(very selective for tertiary C–H)

Key Features: 3° > 2° > 1° C-H bond selectivity
(Cl: 5:4:1; Br: 3° >>> 2° >> 1°)
Useful reaction for functionalizing alkanes

6.C Important Concepts

In Chapter 6 you are introduced for the first time to a detailed description of a reaction: free radical halogenation. Although many industrial processes and other important reactions involve free-radical chemistry, such reactions usually receive less attention in introductory organic courses than do reactions that proceed by ionic mechanisms (those involving charged intermediates). Nevertheless, free-radical halogenation provides you with the opportunity to analyze closely the various aspects of chemical reactivity.

Many students approach each new reaction as another group of facts to be memorized. As we said in the introduction, memory is important, but it is not enough by itself. You should approach each new reaction with a series of questions, such as: "What sort of functional group transformation does it accomplish? What is the mechanism? What is the generality of the reaction? What are the stereochemical features to keep in mind? What limitations does it have?" These are general questions that will help you to understand the chemistry involved and fit the reaction into your chemical knowledge. Only after you have asked these questions should you ask: "What is unusual or unique about the reaction?" During the process of answering the first questions, you will probably have found the answer to the last one.

One of the essential facts to keep in mind is that all of organic chemistry makes logical sense. For free-radical halogenation, the details of each step are discussed thoroughly and it is pointed out how the mechanism makes sense in light of the basic principles discussed in previous chapters. Many of the subsequent reactions presented in the text will be discussed in as much detail as this one, many more in less detail, but they are all sensible. Don't pretend that organic reactions are magic; ask yourself questions about the reactions you see and, if you can't make sense of them, ask your teachers.

6D. Answers to Exercises

6.1

	$CH_4 \longrightarrow$	$CH_3\cdot$	+ $H\cdot$	$\Delta H°$ (kcal mole^{-1})
$\Delta H_f°$:	−17.9	+35	+52	35 + 52 − (−17.9) = 105

Note: Be aware of significant figures. A sum cannot be more accurate than the least acurate of the numbers that go into it. $\Delta H_f°$ for the radicals are known less precisely than for methane. The overall enthalpy change cannot be determined more precisely than $\Delta H_f°$ for the radicals: 35 + 52 − (−17.9) = 104.9 is therefore incorrect.

	CH_3CH_3	$\longrightarrow$	2 $\cdot CH_3$		
$\Delta H_f°$:	−20.2		2 x 35		90
	CH_3CH_3	$\longrightarrow$	$CH_3CH_2\cdot$	+ $H\cdot$	
	−20.2		29	52	101
	$CH_3CH_2CH_3$	$\longrightarrow$	$(CH_3)_2CH\cdot$	+ $H\cdot$	
	−24.8		21	52	98
	$(CH_3)_3CH$	$\longrightarrow$	$(CH_3)_3C\cdot$	+ $H\cdot$	
	−32.4		12	52	96
	$(CH_3)_4C$	$\longrightarrow$	$(CH_3)_3C\cdot$	+ $CH_3\cdot$	
	−40.3		12	35	87
	$(CH_3)_3CH$	$\longrightarrow$	$(CH_3)_2CHCH_2\cdot$	+ $H\cdot$	
	−32.4		17	52	101

6.2

$$(CH_3)_2CHCH_2CH_3 \xrightarrow{A} \dot{C}H_3 + CH_3\dot{C}HCH_2CH_3 \longrightarrow CH_4 + (CH_2{=}CHCH_2CH_3 + CH_3CH{=}CHCH_3)$$

$$(CH_3)_2CHCH_2CH_3 \xrightarrow{B} (CH_3)_2\dot{C}H + \dot{C}H_2CH_3 \longrightarrow (CH_2{=}CHCH_3 + C_2H_6) + (C_3H_8 + CH_2{=}CH_2)$$

$$(CH_3)_2CHCH_2CH_3 \xrightarrow{C} (CH_3)_2CH\dot{C}H_2 + \dot{C}H_3 \longrightarrow (CH_3)_2C{=}CH_2 + CH_4$$

Path A:

					ΔH° (kcal/mole)
	$(CH_3)_2CHCH_2CH_3$	⟶	$\dot{C}H_3$ +	$CH_3\dot{C}HCH_2CH_3$	
ΔH_f° =	-36.9		+35	+15	87

Path B:

	$(CH_3)_2CHCH_2CH_3$	⟶	$(CH_3)_2\dot{C}H$ +	$\dot{C}H_2CH_3$	
ΔH_f° =	-36.9		+21	+29	87

Path C:

	$(CH_3)_2CHCH_2CH_3$	⟶	$(CH_3)_2CH\dot{C}H_2$ +	$\dot{C}H_3$	
ΔH_f° =	-36.9		+17	+35	89

The easiest fragmentation is depicted in either Path A or B. Therefore the major products are expected to be methane, 1-butene and 2-butene from Path A and ethane, propane, ethene and propene from Path B.

6.3

Initiation:	Cl_2	⟶	2 Cl·	58
Propagation:	Cl· + CH_3CH_3	⟶	HCl + CH_3CH_2·	
ΔH_f° =	29 -20.2		-22.1 29	-2
	CH_3CH_2· + Cl_2	⟶	CH_3CH_2Cl + Cl·	
	29 0		-26.1 29	-26
Termination:				
	2 CH_3CH_2·	⟶	$CH_3CH_2CH_2CH_3$	
ΔH_f° =	2 x 29		-30.4	-88
	2 Cl·	⟶	Cl_2	-58
	CH_3CH_2· + Cl·	⟶	CH_3CH_2Cl	
	29 29		-26.1	-84

ΔH° for the overall reaction:

	CH_3CH_3 + Cl_2	⟶	CH_3CH_2Cl + HCl	
ΔH_f° =	-20.2 0		-26.1 -22.1	ΔH°= -28.0 kcal/mole

Note that the initiation and termination steps are not part of the overall reaction.

6.4

CH_3 (arrow) / CH_3 (arrow) $\diagdown$ $CHCH_2CH_3$ ← These 3 hydrogens are different from the other primary ones.

These 6 hydrogens are equivalent.

Product	Statistical Factor		Relative Reactivity		Relative Amount	Percent of Mixture
$ClCH_2CH(CH_3)CH_2CH_3$	6	x	1	=	6	6/22 ≅ 27%
$(CH_3)_2CClCH_2CH_3$	1	x	5	=	5	5/22 ≅ 23%
$(CH_3)_2CHCHClCH_3$	2	x	4	=	8	8/22 ≅ 36%
$(CH_3)_2CHCH_2CH_2Cl$	3	x	1	=	3	3/22 ≅ 14%
					22	100%

6.5

a. $(CH_3)_3CC(CH_3)_3$ b. cyclooctane c. "cubane"

6.6 $\Delta\Delta G^{\ddagger} = -RT\ln(k_2/k_1)$ $\Delta\Delta G^{\ddagger}$ = the difference in activation energy; k_2/k_1 = relative reactivity

$$-3000 \text{ cal/mole} = -1.987 \text{ cal/K} \times 600 \text{ K} \times \ln(k_2/k_1)$$
$$\ln(k_2/k_1) = 2.5$$
$$k_2/k_1 = 12.2$$

6.7

	ΔH°_f (t-$C_4H_9\cdot$)	+	ΔH°_f ($X\cdot$)	-	DH° (t-C_4H_9X)	=	ΔH°_f (t-C_4H_9X)
X = F	12	+	18.9	-	113	=	-82
X = Cl	12	+	28.9	-	84	=	-43
X = Br	12	+	26.7	-	70	=	-31
X = I	12	+	25.5	-	55	=	-18

	$(CH_3)_3CH$ +	X_2 ⟶	$(CH_3)_3CX$ +	HX	ΔH° (kcal/mole)
ΔH°_f X = F	-32.4	0	-82	-65.0	-115
ΔH°_f X = Cl	-32.4	0	-43	-22.1	-33
ΔH°_f X = Br	-32.4	7.4	-31	-8.7	-15
ΔH°_f X = I	-32.4	14.9	-18	+6.3	+6

Note: ΔH°_f for Br_2 and I_2 in the gas phase are not zero since the standard states of these elements are the liquid and solid phases, respectively. It takes energy to transfer them to the gas phase where the comparison between reactants and products is made.

6.8

butane: C_4H_{10} + $6\frac{1}{2}\ O_2$ ⟶ $4\ CO_2$ + $5\ H_2O$ $\Delta H°_{comb.}$ = -634.82 kcal/mole

isobutane: C_4H_{10} + $6\frac{1}{2}\ O_2$ ⟶ $4\ CO_2$ + $5\ H_2O$ $\Delta H°_{comb.}$ = -632.77 kcal/mole

graphite &
hydrogen: $4\ C + 5\ H_2 + 6\frac{1}{2}\ O_2$ ⟶ $4\ CO_2$ + $5\ H_2O$

$\Delta H_f°$: 0 0 0 4 x (-94.05) + 5 x (-57.80) = -665.20 kcal/mole

Therefore, for butane, $\Delta H_f°$ = -665.20 - (-634.82) = -30.38 kcal/mole
isobutane, $\Delta H_f°$ = -665.20 - (-632.77) = -32.43 kcal/mole

6.9

Pentane: $\Delta H° = \Delta H_{combustion}$ kcal/mole

	$CH_3CH_2CH_2CH_2CH_3$	+ $8\ O_2$	⟶ $5\ CO_2$	+	$6\ H_2O$	
$\Delta H_f°$ =	-35.1	0	5 x (-94.1)	+	6 x (-57.8)	-782.2

2-Methylbutane:

	$(CH_3)_2CHCH_2CH_3$	+ $8\ O_2$	⟶ $5\ CO_2$	+	$6\ H_2O$	
$\Delta H_f°$ =	-36.9	0	5 x (-94.1)	+	6 x (-57.8)	-780.4

2,2-dimethylpropane:

	$(CH_3)_4C$	+ $8\ O_2$	⟶ $5\ CO_2$	+	$6\ H_2O$	
$\Delta H_f°$ =	-40.3	0	5 x (-94.1)	+	6 x (-57.8)	-777.0

6.10

Using average bond energies: CH_3-H + Cl-Cl ⟶ CH_3-Cl + H-Cl
Bond energies 99 58 81 103
Bonds broken - bonds gained: (99 + 58) - (81 + 103) = –27 kcal/mole

Using heats of formation:
$\Delta H_f°$ = -17.9 0 -20.6 -22.1 = -24.8 kcal/mole

6.11

Butane: $CH_3CH_2CH_2CH_3$ ⟶ 4 C atoms + 10 H atoms

Using heats of formation: $\Delta H_{atomization}$
$\Delta H_f°$ = -30.4 4 x (170.9) 10 x (52.1) 1235 kcal/mole

Using average bond energies:
Bonds broken: 10 C-H bonds @ 99 + 3 C-C bonds @ 83 1239 kcal/mole

2-Methylpropane: $(CH_3)_3CH$ ⟶ 4 C atoms + 10 H atoms

Using heats of formation:
$\Delta H_f°$ = -32.4 4 x (170.9) 10 x (52.1) 1237 kcal/mole

Using average bond energies, you get the same answer as for butane:

Bonds broken: 10 C–H bonds @ 99 + 3 C–C bonds @ 83 — 1239 kcal/mole

6.E Answers and Explanations for Problems

1. a. $CH_3CH_2CH_2CH_2CH_3 \xrightarrow{\Delta} CH_4 + CH_3CH_3 + CH_2{=}CH_2 + CH_3CH_2CH_3 + CH_3CH{=}CH_2 + CH_3CH_2CH_2CH_3 + CH_3CH_2CH{=}CH_2$

   b. $CH_3CH_2CH_2CH_2CH_3 \xrightarrow{\Delta} CH_3\cdot + CH_3CH_2CH_2CH_2\cdot$
   $\longrightarrow \cdot CH_2CH_3 + CH_3CH_2CH_2\cdot$

$2\ CH_3\cdot \longrightarrow CH_3CH_3$

$CH_3\cdot + CH_3CH_2\cdot \longrightarrow CH_3CH_2CH_3$
$\longrightarrow CH_4 + CH_2{=}CH_2$

$2\ CH_3CH_2\cdot \longrightarrow CH_3CH_2CH_2CH_3$
$\longrightarrow CH_3CH_3 + CH_2{=}CH_2$

$CH_3\cdot + CH_3CH_2CH_2\cdot \longrightarrow CH_3CH_2CH_2CH_3$
$\longrightarrow CH_4 + CH_3CH{=}CH_2$

$CH_3CH_2\cdot + CH_3CH_2CH_2\cdot \longrightarrow CH_3(CH_2)_3CH_3$
$\longrightarrow CH_2{=}CH_2 + CH_3CH_2CH_3$
$\longrightarrow CH_3CH_3 + CH_3CH{=}CH_2$

$CH_3\cdot + CH_3CH_2CH_2CH_2\cdot \longrightarrow CH_3(CH_2)_3CH_3$
$\longrightarrow CH_4 + CH_3CH_2CH{=}CH_2$

Note that longer alkanes can also be produced by recombination of $CH_3CH_2CH_2\cdot$ and $CH_3CH_2CH_2CH_2\cdot$ radicals.

c.

Reactants		Products	ΔH°, kcal/mole
$CH_3CH_2CH_2CH_2CH_3$	→	$CH_3\cdot + CH_3CH_2CH_2CH_2\cdot$	+89
	→	$CH_3CH_2\cdot + CH_3CH_2CH_2\cdot$	+88
$2\ CH_3\cdot$	→	CH_3CH_3	-90
$CH_3\cdot + CH_3CH_2\cdot$	→	$CH_3CH_2CH_3$	-89
	→	$CH_4 + CH_2{=}CH_2$	-69
$2\ CH_3CH_2\cdot$	→	$CH_3CH_3 + CH_2{=}CH_2$	-66
$CH_3\cdot + CH_3CH_2CH_2\cdot$	→	$CH_3CH_2CH_2CH_3$	-89
	→	$CH_4 + CH_3CH{=}CH_2$	-72
$CH_3CH_2\cdot + CH_3CH_2CH_2\cdot$	→	$CH_2{=}CH_2 + CH_3CH_2CH_3$	-65
	→	$CH_3CH_3 + CH_3CH{=}CH_2$	-68
$CH_3\cdot + CH_3CH_2CH_2CH_2\cdot$	→	$CH_4 + CH_3CH_2CH{=}CH_2$	-72

OVERALL REACTIONS: $\underline{\Delta H°,\ kcal/mole}$

$$\nearrow \quad CH_4 + CH_3CH_2CH{=}CH_2 \qquad \begin{matrix} -17.9 & -0.2 \end{matrix} \qquad +17.0$$

$$CH_3CH_2CH_2CH_2CH_3 \;(\Delta H_f^\circ = -35.1) \rightarrow CH_3CH_3 + CH_3CH{=}CH_2 \qquad \begin{matrix} -20.2 & 4.9 \end{matrix} \qquad +19.8$$

$$\searrow \quad CH_3CH_2CH_3 + CH_2{=}CH_2 \qquad \begin{matrix} -24.8 & 12.5 \end{matrix} \qquad +22.8$$

2.

a.

1) $Br_2 \longrightarrow 2\ Br\cdot$ $\quad \Delta H° = DH° = 2\ \Delta H°_f(Br\cdot) - \Delta H_f^\circ(Br_2) = 46$ kcal/mole

2) $CH_3CH_3 + Br\cdot \longrightarrow CH_3CH_2\cdot + HBr$

To use DH° values, you must break this step into two reactions:

$$\begin{array}{lll} CH_3CH_3 \longrightarrow CH_3CH_2\cdot + H\cdot & & DH° = 101 \\ \underline{H\cdot + Br\cdot \longrightarrow HBr} & & \underline{-DH° = -87.5} \\ CH_3CH_3 + \cancel{H}\cdot + Br\cdot \longrightarrow CH_3CH_2\cdot + \cancel{H}\cdot + HBr & & \Delta H° = 14 \text{ kcal/mole} \end{array}$$

The use of heats of reaction gives a similar answer:

$$\begin{array}{lllll} & CH_3CH_3 & + Br\cdot \longrightarrow & CH_3CH_2\cdot & + HBr \\ \Delta H_f^\circ = & -20.2 & 26.7 & 29 & -8.7 \end{array} \qquad \Delta H° = 14 \text{ kcal/mole}$$

3) $CH_3CH_2\cdot + Br_2 \longrightarrow CH_3CH_2Br + Br\cdot$ $\quad \Delta H° = DH°(Br{-}Br) - DH°(CH_3CH_2{-}Br)$
$= 46 - 71 = -25$ kcal/mole

b. (2) + (3) $= \Delta H°$ for the overall reaction
$= 14 + (-25) = -11$ kcal/mole

c. To check:

$$\begin{array}{lllll} & Br_2 & + CH_3CH_3 \longrightarrow & CH_3CH_2Br & + HBr \\ \Delta H_f^\circ = & 7.4 & -20.2 & -15.2 & -8.8 \end{array} \qquad \Delta H° = -11.2 \text{ kcal/mole}$$

The values are the same except for the precision. ΔH_f°'s for radicals are not known as precisely as for normal compounds. *(see explanatory Note accompanying answer to Exercise 6.1)*

3. The reaction of Br· with ethane is an endothermic reaction; the $\Delta H^\ddagger$ is $\geq \Delta H° = +14$ kcal/mole. It is therefore likely to be slow. In contrast, the reaction of ethyl radical with Br_2 is very exothermic ($\Delta H° = -25$ kcal/mole). Because radical hydrogen abstraction has a relatively low $\Delta H^\ddagger$ for exothermic reactions, the reaction of $C_2H_5\cdot$ with Br_2 is expected to be fast. Therefore $[Br\cdot] > [C_2H_5\cdot]$.

4. a. All of the hydrogens in spiropentane are equivalent. Thus, there is only one possible monochlorospiropentane. Furthermore, the dichloro compounds have higher boiling points and can be separated by distillation.

b. Mechanism:

$Cl_2 \longrightarrow 2\ Cl\cdot$ (initiation)

spiropentane (H_2C, CH_2, H_2C, CH_2 around central C) $+ Cl\cdot \longrightarrow$ spiropentyl radical (H_2C, CH_2, H_2C, $CH\cdot$ around central C) $+ HCl$

(propagation)

spiropentyl radical (H_2C, CH_2, H_2C, $CH\cdot$) $+ Cl_2 \longrightarrow$ chlorospiropentane (H_2C, CH_2, H_2C, $CHCl$) $+ Cl\cdot$

$2\ R\cdot \longrightarrow R\text{–}R$ ($R\cdot$ = spiropentyl radical or $Cl\cdot$) (termination)

5. In section 6.3, we found the relative reactivities of primary (1°), secondary (2°), and tertiary (3°) hydrogens to be 1 : 4.0 : 5 in chlorination reactions.

a. In $CH_3CH_2CH_2CH_3$, the six hydrogens of the methyl groups are all equivalent and so are the four hydrogens of the CH_2 groups.

Product	Statistical Factor		Relative Reactivity	Relative Amount	Percent of Mixture
$CH_3CH_2CH_2CH_2Cl$	6	x	1	6	6/22 ≅ 27%
$CH_3CH_2CHClCH_3$	4	x	4	16	16/22 ≅ 73%
				22	100%

b. For $(CH_3)_2CHCH(CH_3)_2$, all twelve of the hydrogens in the four methyl groups are equivalent. The two H's in the two CH groups are equivalent.

Product	Statistical Factor		Relative Reactivity	Relative Amount	Percent of Mixture
$(CH_3)_2CHCH(CH_3)CH_2Cl$	12	x	1	12	12/22 ≅ 55%
$(CH_3)CHCCl(CH_3)_2$	2	x	5	10	10/22 ≅ 45%
				22	100%

c. For $(CH_3)_3CCH_2CH(CH_3)_2$, the nine H's in the three methyls of the *t*-butyl group are all equivalent, the six H's in the two methyls of the isopropyl ($-CH(CH_3)_2$) are equivalent.

Product	Statistical Factor		Relative Reactivity	Relative Amount	Percent of Mixture
$CH_2ClC(CH_3)_2CH_2CH(CH_3)_2$	9	x	1	9	9/28 ≅ 32%
$(CH_3)_3CCHClCH(CH_3)_2$	2	x	4	8	8/28 ≅ 29%
$(CH_3)_3CCH_2CCl(CH_3)_2$	1	x	5	5	5/28 ≅ 18%
$(CH_3)_3CCH_2CH(CH_3)CH_2Cl$	6	x	1	6	6/28 ≅ 21%
				28	100%

d. $(CH_3)_3CCH(CH_3)_2$

Product	Statistical Factor		Relative Reactivity	Relative Amount	Percent of Mixture
$CH_2ClC(CH_3)_2CH(CH_3)_2$	9	x	1	9	9/20 ≅ 45%
$(CH_3)_3CCCl(CH_3)_2$	1	x	5	5	5/20 ≅ 25%
$(CH_3)_3CCH(CH_3)CH_2Cl$	6	x	1	6	6/20 ≅ 30%
				20	100

e. $CH_3CH_2CH_2CH_2CH_3$ There are two different kinds of CH_2 groups here; the CH_2 in the middle of the molecule is next to two other CH_2 groups; the others are next to a CH_3 and a CH_2. Remember the rules for nomenclature; since the molecule is symmetrical, the C at position 3 is unique, but the C's at 2 and 4 are the same since you can number the chain starting from either end.

Product	Statistical Factor		Relative Reactivity	Relative Amount	Percent of Mixture
$CH_3CH_2CH_2CH_2CH_2Cl$	6	x	1	6	6/30≃ 20%
$CH_3CHClCH_2CH_2CH_3$	4	x	4	16	16/30≃ 53%
$CH_3CH_2CHClCH_2CH_3$	2	x	4	8	8/30≃ 27%
				30	100%

Note how chlorination is generally impractical as a synthetic method when the molecule contains non-equivalent hydrogens.

6. a.

$BrCH_2CH_2CH_2CH_3$	6 x 1	=	6	6/886 ≃ 0.7%
$CH_3CHBrCH_2CH_3$	4 x 220	=	880	880/886 ≃ 99.3%
			886	100 %

b.

$(CH_3)_2CHCH(CH_3)CH_2Br$	12 x 1	=	12	12/38012 ≃ 0.03%
$(CH_3)_2CBrCH(CH_3)_2$	2 x 19000	=	38000	38000/38012 ≃ 99.97%
			38012	100.0%

c.

$BrCH_2C(CH_3)_2CH_2CH(CH_3)_2$	9 x 1	=	9	9/19455 ≃ 0.05%
$(CH_3)_3CCH_2CBr(CH_3)_2$	1 x 19000	=	19000	19000/19455 ≃ 97.7%
$(CH_3)_3CCHBrCH(CH_3)_2$	2 x 220	=	440	440/19455 ≃ 2.3%
$(CH_3)_3CCH_2CH(CH_3)CH_2Br$	6 x 1	=	6	3/19455 ≃ 0.03%
			19455	100.0%

d.

$BrCH_2C(CH_3)_2CH(CH_3)_2$	9 x 1	=	9	9/19015 ≃ 0.05%
$(CH_3)_3CCBr(CH_3)_2$	1 x 19000	=	19000	19000/19015 ≃ 99.9%
$(CH_3)_3CCH(CH_3)CH_2Br$	6 x 1	=	6	6/19015 ≃ 0.03%
			19015	100.0%

e.

$BrCH_2CH_2CH_2CH_2CH_3$	6 x 1	=	6	6/1326 ≃ 0.5%
$CH_3CHBrCH_2CH_2CH_3$	4 x 220	=	880	880/1326 ≃ 66.4%
$CH_3CH_2CHBrCH_2CH_3$	2 x 220	=	440	440/1326 ≃ 33.2%
			886	100 %

7. a.

ΔH°, kcal/mole

$HCl + CH_3CH_2Cl$

↗ −22.1 −26.1   −28.0

$C_2H_6 + Cl_2$

$\Delta H^\circ_f = -20.2 \quad 0$

↘

$2\ CH_3Cl$

$2 \times (-20.6)$ −21.0

b.

				$\Delta H°$
	C_2H_6 + $Cl\cdot$	⟶	CH_3Cl + $CH_3\cdot$	
ΔH_f° =	−20.2 26.7		−20.6 35	+8
	$CH_3\cdot$ + Cl_2	⟶	CH_3Cl + $Cl\cdot$	
ΔH_f° =	35 0		−20.6 26.7	−29

c. Although there is no thermodynamic difficulty with either step, the reaction of ethane with the chlorine atom to give ethyl radical and HCl is so much more favorable ($\Delta H°$ = −2 kcal/mole; see exercise 6.3) than the reaction to give $CH_3\cdot$ and CH_3Cl that the latter is not observed.

8. Cyclopropane: CH_2 / $H_2C{-}CH_2$ (ring) ⟶ 3 C atoms and 6 H atoms

Using average bond energies:
3 C−C bonds (3 x 83) and 6 C−H bonds (6 x 99) = 843 kcal/mole

Using heats of formation:
$-\Delta H_f^\circ$ (cyclopropane) + 3 x ΔH_f° (C atom) + 6 x ΔH_f° (H atom)
−12.7 3 x 170.9 6 x 52.1 = 812.6 kcal/mole

The calculations using the average bond energies overestimate the heat of atomization by 30 kcal/mole because it does not take the ring strain into account. This difference is in fact a good approximation of the ring strain (see Table 5.5 in the text.)

Cubane: (cube structure with 8 H) ⟶ 8 C atoms and 8 H atoms

Using average bond energies:
12 C−C bonds (12 x 83) and 8 C−H bonds (8 x 99) = 1788 kcal/mole

Using heats of formation:
$-\Delta H_f^\circ$ (cubane) + 8 x ΔH_f° (C atom) + 8 x ΔH_f° (H atom)
−148.7 8 x 170.9 8 x 52.1 = 1635.3 kcal/mole

The difference between the two answers, 1788 − 1635.3, is an approximate measure of the strain, ≈ 153 kcal/mole.

9. △ + 4½ O_2 ⟶ 3 CO_2 + 3 H_2O

ΔH_f° = 12.7 0 3 x (−94.05) 3 x (−57.80) $\Delta H°_{combustion}$ = −468.3 kcal/mole

⬡ + 9 O_2 ⟶ 6 CO_2 + 6 H_2O

ΔH_f° = −29.5 0 6 x (−94.05) 6 x (−57.80) $\Delta H°_{combustion}$ = −881.6 kcal/mole

Since the molecular weight of cyclohexane is twice that of cyclopropane, on a weight basis cyclopropane releases
(2 x 468)/882 = 1.06 times as much energy on combustion as cyclohexane does. If the two fuels cost the same, cyclopropane would be the more economical.

10. a. In the free radical fluorination of octane, the chain propagation steps are:

1) $C_8H_{18} + F\cdot \longrightarrow HF + C_8H_{17}\cdot$
2) $C_8H_{17}\cdot + F_2 \longrightarrow C_8H_{17}F + F\cdot$

Using average bond energies, ΔH°_f for the two steps are:

1) ΔH°_f = 1 C−H bond broken and 1 H−F bond formed = +99 − 135 = −36 kcal/mole
2) ΔH°_f = 1 F−F bond broken and 1 C−F bond formed = +37 − 110 = −73 kcal/mole

Not only is the overall reaction highly exothermic, the individual steps are both very exothermic. Neither step will be particularly slow and the heat generated will further increase the rate.

b. By diluting the F_2, the number of effective collisions will be reduced, slowing the reaction to a manageable rate. See section 4.3 in the text.

c. The relative stability of alkyl radicals is tertiary > secondary > primary. This reflects the weaker 3° C-H bond energy. See Table 6.1 in the text.

11. a.

			HF + CH_3Cl	
		↗	−65.0 −20.6	ΔH° = −55.5 kcal/mole
	CH_4 + ClF			
ΔH°_f =	−17.9 −12.2	↘		
			CH_3F + HCl	
			−56.8 −22.1	ΔH° = −48.8 kcal/mole

b. CH_3Cl and HF, because their formation releases the greatest amount of energy.

c.

			$F\cdot$ + CH_3Cl	
		↗	18.9 −20.6	ΔH° = −25 kcal/mole
	$CH_3\cdot$ + ClF			
ΔH°_f =	35 −12.2	↘		
			CH_3F + $Cl\cdot$	
			−56.8 28.9	ΔH° = −51 kcal/mole

One would predict from this comparison that CH_3F would be formed faster.

This case provides a good illustration of a situation that is frequently encountered: the product that is formed more quickly (called the product of "kinetic control") is not always the most stable one (which is called the product of "thermodynamic control"). This sort of situation results from the independence of the activation energy ($\Delta G^{\ddagger}$) and the overall energy change ($\Delta G°$) in a reaction.

12. First calculate a value for ΔH°_f for ClO based on bond energies; $\Delta H^\circ_{f\,ClO\cdot} = \Delta H°$ for the reaction:

	$\frac{1}{2}O_2$	+	$\frac{1}{2}Cl_2$	⟶	$ClO\cdot$	
	½(119)		½(58)		-52	= 37 kcalmole
Step 1:	$Cl\cdot$	+	O_3	⟶	$ClO\cdot$ + O_2	
ΔH°_f:	29		34		37 0	$\Delta H°$ = -26 kcal/mole
Step 2:	O	+	$ClO\cdot$	⟶	O_2 + $Cl\cdot$	
ΔH°_f:	60		37		0 29	$\Delta H°$ = -68 kcal/mole
Overall:	O	+	O_3	⟶	2 O_2	$\Delta H°$ = -94 kcal/mole

Not only are both reactions energetically favorable, the Cl atom functions as a catalyst so that it is not consumed in the overall reaction, which converts ozone and O atoms to O_2.

The UV radiation from the sun is absorbed by O_3 and used as a source of energy for vibrational and electronic transitions within the molecule.

13.

							$\Delta H°$ (kcal/mole)	
	CH_4	+ HNO_3	⟶	CH_3NO_2	+	H_2O		
ΔH°_f =	-17.9	-32.1		-17.9		-57.8	-25.7	
	CH_4	+ $\cdot NO_2$	⟶	$CH_3\cdot$	+	HNO_2		this step is highly endothermic
and	-17.9	7.9		35		-18.4	+27	is therefore slow; it requires high temperature.
	HNO_2	+ HNO_3	⟶	2 $NO_2\cdot$	+	H_2O		
	-18.4	-32.1		2 x 7.9		-57.8	+8.5	
	$CH_3\cdot$	+ $\cdot NO_2$	⟶	CH_3NO_2				
	35	7.9		-17.9			-61	

Possible alternatives for $CH_3\cdot$:

			$\cdot NO_2$ +	CH_3OH	
		↗	7.9	-48.1	-43
	$CH_3\cdot$ + HNO_3				
ΔH°_f =	35 -32.1	↘			
			CH_3NO_2 +	$HO\cdot$	
			-17.9	9.4	-11

Resonance structures for $NO_2\cdot$: $[\ \ddot{O}{=}\overset{+}{N}{-}\ddot{O}{:}^{-} \longleftrightarrow {}^{-}{:}\ddot{O}{-}\overset{+}{N}{=}\ddot{O} \longleftrightarrow \cdot\ddot{O}{-}\ddot{N}{=}\ddot{O} \longleftrightarrow \ddot{O}{=}\ddot{N}{-}\ddot{O}\cdot\ ]$

$$\begin{array}{llll} & & CH_3ONO & \\ & \nearrow & -15.8 & \Delta H° = -59 \text{ kcal/mole} \\ & CH_3\cdot + NO_2\cdot & & \\ \Delta H_f^\circ = & 35 \quad\ 7.9 & & \\ & \searrow & CH_3NO_2 & \\ & & -17.9 & \Delta H° = -61 \text{ kcal/mole} \end{array}$$

14. At high concentrations of CH_4 and low concentrations of Cl_2, the rate at which chlorine atoms find and react with CH_4 molecules will become faster than the rate at which methyl radicals find and react with Cl_2. At equal concentrations of CH_4 and Cl_2, the reaction of chlorine atoms with CH_4 will be rate-determining. The chlorine atoms will be present at higher concentration (see problem #3). The relative rates of the propagation steps determine which radical is in excess, which in turn determines which termination process is faster. Termination processes involving $Cl\cdot$ will predominate if $Cl\cdot + CH_4$ is the slow step.

15.

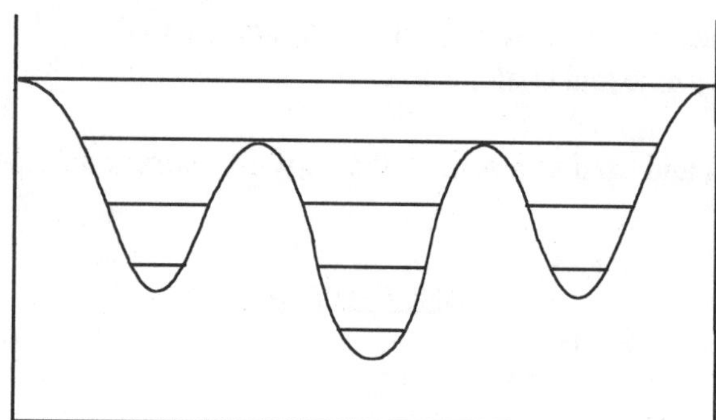

Free rotation at the 5th or 6th quantum level.

6.F Supplementary Problems

S1. Cracking propane to give a mixture of methane and ethylene is an endothermic process (compare with problem #1 in this chapter). However, from the point of view of entropy, splitting one molecule into two is a favorable process, as reflected by the standard entropy change, $\Delta S°$, for the reaction in question:

$$CH_3CH_2CH_3 \longrightarrow CH_4 + CH_2{=}CH_2 \qquad \Delta S° = +33 \text{ e.u.}$$

a. Using this number and Appendix I, calculate the temperature at which equilibrium favors the smaller molecules; i.e., $K > 1$.

b. In a more efficient cracking process, called hydrocracking, hydrogen is mixed with the hydrocarbons to be cracked. What products would you expect from the cracking of propane in the presence of hydrogen?

c. Calculate the $\Delta H°$ for the reaction you have written for part b.

S2. From the data in Table 6.2 and Appendix I, calculate or predict DH° for the weakest bond in each of the compounds listed below.

a. $(CH_3)_2CHCH(CH_3)_2$

b. $(CH_3)_3CC(CH_3)_3$ ($\Delta H_f^\circ = -54.0$ kcal/mole)

c. □

d. △

S3. Free radical halogenation can also be accomplished using sulfuryl chloride, SO_2Cl_2, via the following propagation steps:

$$\left.\begin{array}{l} R\cdot + ClSO_2Cl \longrightarrow RCl + \cdot SO_2Cl \\ \cdot SO_2Cl + RH \longrightarrow R\cdot + HSO_2Cl \end{array}\right\} \text{propagation}$$

$$(HSO_2Cl \xrightarrow{fast} SO_2 + HCl)$$

a. From the product compositions below, calculate the selectivity of the $\cdot SO_2Cl$ radical for primary, secondary and tertiary hydrogens.

$$(CH_3)_2CHCH_3 + ClSO_2Cl \xrightarrow{h\nu} \underset{31\%}{(CH_3)_3CCl} + \underset{69\%}{(CH_3)_2CHCH_2Cl}$$

$$CH_3CH_2CH_2CH_2CH_3 + ClSO_2Cl \rightarrow \underset{48\%}{CH_3CH_2CH_2CHClCH_3} + \underset{28\%}{CH_3CH_2CH_2CH_2CH_2Cl} + \underset{24\%}{CH_3CH_2CHClCH_2CH_3}$$

b. Predict the relative amounts of monochlorinated products obtained from this reaction with 2,4-dimethylpentane.

S4. On heating with HI, alkyl iodides react to give alkanes and I_2.

a. Using $CH_3I + HI \longrightarrow CH_4 + I_2$ as an example, calculate ΔH° for the reaction.

b. Propose a mechanism for this reaction and justify the steps involved by calculating ΔH° for each one. (*HINT*: take advantage of the principle of microscopic reversiblity.)

S5. One of the mechanisms by which HBr can add to alkenes is illustrated below for ethylene:

$$Br\cdot + CH_2{=}CH_2 \longrightarrow Br{-}CH_2{-}CH_2\cdot$$

$$Br{-}CH_2{-}CH_2\cdot + H{-}Br \longrightarrow BrCH_2CH_3 + Br\cdot$$

a. Using the data in Table 6.2 and in Appendices I and II, estimate ΔH° for each of the steps above and for the overall reaction.

b. Do the same calculations for the reaction of HCl by the same mechanism. What do you conclude about the relative rates of the two reactions?

c. Two isomeric products are possible for the addition of HBr to 2-methylpropene. Write out the steps involved in the formation of each one and estimate $\Delta H°$ for each of them. Which isomer is more stable? Which one will be the major product (formed faster)?

S6. a. Calculate the average bond dissociation energy for CCl_4 using the data from Appendix I.

b. Calculate the C–H bond dissociation energy for $CHCl_3$ (chloroform).

c. Is the reaction $CH_4 + CCl_4 \longrightarrow CH_3Cl + CHCl_3$ likely to take place on thermodynamic grounds?

d. Is the reaction likely to take place on kinetic grounds? In other words, are any of the steps in the mechanism you wrote very endothermic?

6.G Answers to Supplementary Problems

S1. a. $C_3H_8 \longrightarrow CH_4 + C_2H_4$ $\Delta H° = +19.4$ kcal/mole

$\Delta S° = +33$ e.u.

$K > 1$ when $\Delta G° < 0$; since $\Delta G° = \Delta H° - T\Delta S°$; $19{,}400 - T \times 33 < 0$, so that $T > 588$ K.

b. $C_3H_8 + H_2 \longrightarrow CH_4 + C_2H_6$

c. $\Delta H_f°$: −24.8 0 −17.9 −20.2 $\Delta H° = -13.3$ kcal/mole

S2. a. $(CH_3)_2CH{-}CH(CH_3)_2 \longrightarrow 2\ (CH_3)_2CH\cdot$

$\Delta H_f°$: −42.6 2 x 21 $\Delta H° = DH° = 85$ kcal/mole

b. $(CH_3)_3C{-}C(CH_3)_3 \longrightarrow 2\ (CH_3)_3C\cdot$

$\Delta H_f°$: −54 2 x 12 $\Delta H° = DH° = 78$ kcal/mole

c. To calculate DH° for □ $\longrightarrow \cdot CH_2CH_2CH_2CH_2\cdot$, follow the sequence below:

(*1*) H_2 + □ $\longrightarrow CH_3CH_2CH_2CH_3$ $\Delta H° = -37.2$ kcal/mole

(*2*) $CH_3CH_2CH_2CH_3 \longrightarrow 2\ H\cdot + \cdot CH_2CH_2CH_2CH_2\cdot$ $\Delta H° = 2\ DH°$(primary C–H) $= +202$ kcal/mole

(*3*) $2\ H\cdot \longrightarrow H_2$ $\Delta H° = -DH°(H_2) = -104$ kcal/mole

DH° for the ring opening of cyclobutane is the sum of (*1*), (*2*) and (*3*):
$DH° = -37.2 + 202 + (-104) = 61$ kcal/mole.

d. The same sequence may be applied to △ $\longrightarrow \cdot CH_2CH_2CH_2\cdot$; the value for $\Delta H°$ is different only for step (*1*):

(*1'*) H_2 + △ $\longrightarrow CH_3CH_2CH_3$ $\Delta H° = -37.5$ kcal/mole

DH° for the ring opening of cyclopropane is calculated to be +61 kcal/mole also. For your interest, compare this figure with the values you calculate for the cleavage of the cyclopentane and cyclohexane rings.

S3. a. Reaction of $(CH_3)_3CH$ results in 31% tertiary and 69% primary chlorination.

Relative Amount		Statistical Factor		Relative Reactivity
31%	=	1	x	3° reactivity
69%	=	9	x	1° reactivity

Relative reactivity: 3°/1° $= \frac{31}{69/9} = 4$

Reaction of $CH_3CH_2CH_2CH_2CH_3$ results in 48% 2-chloro isomer, 24% 3-chloro isomer and 28% 1-chloro isomer.

48%	=	4	x	2° reactivity
24%	=	2	x	2° reactivity
28%	=	6	x	1° reactivity

Relative reactivity: 2°/1° $= \frac{48/4}{28/6}$ or $\frac{24/2}{28/6} = 2.6$

Therefore, the ratio of 3° reactivity: 2° reactivity: 1° reactivity = 4:2.6:1

b. $(CH_3)_2CHCH_2CH(CH_3)_2 \longrightarrow$
$ClCH_2CH(CH_3)CH_2CH(CH_3)_2 + (CH_3)_2CClCH_2CH(CH_3)_2 + (CH_3)_2CHCHClCH(CH_3)_2$

Statistical factor:	12	2	2
Relative reactivity:	1	4	2.6
Relative amount	12 (= 12 x 1)	8 (= 2 x 4)	5.2 (= 2 x 2.6)
<u>Percent of Mixture</u>	48%	32%	20%

S4. a. $CH_3I + HI \longrightarrow CH_4 + I_2$
ΔH_f°: 3.4 6.3 −17.9 0 $\Delta H^\circ = -27.6$ kcal/mole

b. $CH_3\text{-}I \longrightarrow CH_3\cdot + I\cdot$ DH° = 56 kcal/mole

But many other initiation steps are possible; for example, from traces of organic peroxide, light, etc.

propagation

$I\cdot + CH_3I \longrightarrow CH_3\cdot + I_2$
ΔH_f°: 25.5 3.4 35 0 $\Delta H^\circ = +6$ kcal/mole

$CH_3\cdot + HI \longrightarrow CH_4 + I\cdot$
ΔH_f°: 35 6.3 −17.9 25.5 $\Delta H^\circ = -34$ kcal/mole

S5. a. $Br\cdot + CH_2{=}CH_2 \longrightarrow BrCH_2CH_2\cdot$

Calculate from the sequence: ΔH° (kcal/mole)

(*1*) $H\cdot + Br\cdot \longrightarrow HBr$	$\Delta H° = -DH°(HBr)$	-87.5
(*2*) $HBr + CH_2{=}CH_2 \longrightarrow CH_3CH_2Br$		-19.0
(*3*) $CH_3CH_2Br \longrightarrow H\cdot + \cdot CH_2CH_2Br$	$\Delta H° = DH°(1° \text{ C-H})$:	101
The sum of (*1*), (*2*) and (*3*) = ΔH° for the first step:		-5.5

$\cdot CH_2CH_2Br + HBr \longrightarrow CH_3CH_2Br + Br\cdot$

Calculate from the sequence:

(*4*) $HBr \longrightarrow H\cdot + Br\cdot$	$\Delta H° = DH°(HBr)$	87.5
(*5*) $\cdot CH_2CH_2Br + H\cdot \longrightarrow CH_3CH_2Br$	$\Delta H° = -DH°(1° \text{ C-H})$:	-101
The sum of (*4*) and (*5*) = ΔH° for the second step:		-13.5

ΔH° for the overall reaction was calculated as eq. (*2*) above. -19.0

b. For HCl addition, the following changes in the sequence of equations are made:

(*1'*) $H\cdot + Cl\cdot \longrightarrow HCl$	$\Delta H° = -DH°(HCl)$	-103.2
(*2'*) $HCl + CH_2{=}CH_2 \longrightarrow CH_3CH_2Cl$		-16.5
(*3'*) $CH_3CH_2Cl \longrightarrow H\cdot + \cdot CH_2CH_2Cl$	$\Delta H° = DH°(1° \text{ C-H})$:	101
The sum of (*1'*), (*2'*) and (*3'*) = ΔH° for the first step:		-18.7
(*4'*) $HCl \longrightarrow H\cdot + Cl\cdot$	$\Delta H° = DH°(HCl)$	103.2
(*5'*) $\cdot CH_2CH_2Cl + H\cdot \longrightarrow CH_3CH_2Cl$	$\Delta H° = -DH°(1° \text{ C-H})$:	-101
The sum of (*4'*) and (*5'*) = ΔH° for the second step:		+2.2

Again, the ΔH° for the overall reaction was calculated as eq. (*2'*) above. -16.5 kcal/mole

Because the second step in the chain reaction involving HCl is endothermic (ΔH° = +2.2 kcal/mole), this reaction proceeds much more slowly than the HBr addition.

c. $(CH_3)_2CBrCH_2\cdot \xleftarrow[\text{Path A}]{Br\cdot} (CH_3)_2C{=}CH_2 \xrightarrow[\text{Path B}]{Br\cdot} (CH_3)_2\dot{C}CH_2Br$

$\downarrow$ HBr (Path A) $\qquad \downarrow$ HBr (Path B)

$(CH_3)_2CBrCH_3 + Br\cdot \qquad (CH_3)_2CHCH_2Br + Br\cdot$

Path A:

$\Delta H^\circ_f((CH_3)_3CBr) = \Delta H^\circ_f(Br\cdot) + \Delta H^\circ_f((CH_3)_3C\cdot) - DH^\circ(CH_3)_3C{-}Br) = -31$ kcal/mole

		ΔH° (kcal/mole)
(*6*) $H\cdot + Br\cdot \longrightarrow HBr$	$\Delta H^\circ = DH^\circ(HBr)$	-87.5
(*7*) $HBr + (CH_3)_2C{=}CH_2 \longrightarrow (CH_3)_3CBr$ ΔH°_f: -8.7 -4.3 -31 (*from above*)		-18
(*8*) $(CH_3)_3CBr \longrightarrow H\cdot + (CH_3)_2CBrCH_2\cdot$	$\Delta H^\circ = DH^\circ(1^\circ\ C{-}H)$:	101
The sum of (*6*), (*7*) and (*8*) = ΔH° for the first step: $(Br\cdot + (CH_3)_2C{=}CH \longrightarrow (CH_3)_2CBrCH_2\cdot)$:		-4.5

For the second step, $(CH_3)_2CBrCH_2\cdot + HBr \longrightarrow (CH_3)_3CBr + Br\cdot$

$\Delta H^\circ = -DH^\circ(1^\circ\ C{-}H) + DH^\circ(HBr) = -101 + 87.5 =$ -13.5

Path B:

$\Delta H^\circ_f((CH_3)_2CHCH_2Br) = \Delta H^\circ_f(Br\cdot) + \Delta H^\circ_f((CH_3)_2CHCH_2\cdot) - DH^\circ(CH_3)_2CHCH_2{-}Br)$
$= -27$ kcal/mole

		ΔH° (kcal/mole)
(*6*) same as for Path A		-87.5
(*9*) $HBr + (CH_3)_2C{=}CH_2 \longrightarrow (CH_3)_2CHCH_2Br$ ΔH°_f: -8.7 -4.3 -27 (*from above*)		-14
(*10*) $(CH_3)_2CHCH_2Br \longrightarrow H\cdot + (CH_3)_2\dot{C}CH_2Br$	$\Delta H^\circ = DH^\circ$(tertiary C-H):	96
The sum of (*6*), (*9*) and (*10*) = ΔH° for the first step: $(Br\cdot + (CH_3)_2C{=}CH_2 \longrightarrow (CH_3)_2\dot{C}CH_2Br)$:		-5.5

For the second step $(CH_3)_2\dot{C}CH_2Br + HBr \longrightarrow (CH_3)_2CHCH_2Br + Br\cdot$

$\Delta H^\circ = -DH^\circ(\text{tertiary C-H}) + DH^\circ(HBr)$: -8.5

The overall ΔH° for Path A (-18 kcal/mole), equation *(7)*, is more favorable than for Path B (-14 kcal/mole), equation *(9)*, but the energy for the first step in Path B is more favorable (-5.5 kcal/mole), than for Path A (-4.5 kcal/mole). This is another instance where the less thermodynamically favored product is formed faster.

S6. a. $CCl_4 \longrightarrow \cdot\dot{\underset{\cdot}{C}}\cdot + 4\ Cl\cdot$

ΔH_f° : -25.2 170.9 4 x 28.9 $\Delta H^\circ = 311.7$ kcal/mole

average per C−Cl bond = 77.9 kcal/mole

b. $CHCl_3 \longrightarrow H\cdot + \cdot CCl_3$

Calculate from the sequence: ΔH°(kcal/mole)

(*1*) $CHCl_3 + Cl\cdot \longrightarrow CCl_4 + H\cdot$

ΔH_f° : -24.6 28.9 -25.2 52.1 22.6

(*2*) $CCl_4 \longrightarrow \cdot CCl_3 + Cl\cdot$ ΔH° = DH°(C-Cl) 78 (*from (a) above*)

The sum of (*1*) and (*2*) = ΔH° for the overall reaction: 101

c. $CH_4 + CCl_4 \longrightarrow CH_3Cl + CHCl_3$

ΔH_f° : -17.9 -25.2 -20.6 -24.6 $\Delta H^\circ = -2.1$ kcal/mole

On the basis of a negative overall enthalpy, one would assume that the reaction would take place.

d. Assuming that an initiation step can occur, the most reasonable propagation steps are:

(*I*) $\cdot CCl_3 + CH_4 \longrightarrow CHCl_3 + \cdot CH_3$

ΔH_f° : 24.3 -17.9 -24.6 35 $\Delta H^\circ = +4$ kcal/mole

($\Delta H_f^\circ(\cdot CCl_3) = \Delta H_f^\circ(CHCl_3) + DH^\circ(Cl_3C\text{-}H) - \Delta H_f^\circ(H\cdot)$ = 24.3 kcal/mole)

DH°(Cl_3C-H) calculated in (b) above

(*II*) $\cdot CH_3 + CCl_4 \longrightarrow CH_3Cl + \cdot CCl_3$

ΔH_f° : 35 -25.2 -20.6 24.3 $\Delta H^\circ = -6$ kcal/mole

Neither of these steps is unacceptably endothermic.

7. STEREOISOMERISM

7.A. Chapter Outline and Important Terms Introduced

7.1 Chirality and Enantiomers (mirror images that are different)

chirality
chiral and achiral
non-superimposable
enantiomers
stereoisomers
stereocenter

7.2 Physical Properties of Enantiomers: Optical Activity (the only physical difference between enantiomers)

plane polarized light
dextrorotatory and levorotatory
specific rotation, [α]
polarimeter

7.3 Nomenclature of Enantiomers: the R-S Convention (describing three dimensions using one dimension)

absolute configuration
sequence rule
higher atomic number
first point of difference

7.4 Racemates (chiral molecules, optically **inactive** mixtures)

racemate
racemic mixture
racemic compound
racemization

7.5 Compounds Containing More than One Stereocenter, Diastereomers

(stereoisomers with <u>**different**</u> physical properties)

n stereocenters → 2^n stereoisomers (but look out for meso compounds)

7.6 Stereoisomeric Relationships in Cyclic Compounds (looking for planes and points of symmetry)

symmetry plane

7.7 Conformations of Substituted Cyclohexanes (chair conformations revisited)

equilibria between two possible chair conformations
boat
skew-boat (also called twist-boat)

7.8 Chemical Reactions and Stereoisomerism (what happens when a stereocenter is involved)

inversion of absolute configuration
enantiomeric transition states

7.B. Important Reactions Introduced: none

7.C. Important Concepts and Hints

Stereoisomerism can be the most exciting aspect of organic chemistry. If you can take a molecule off a two-dimensional page and see it in three dimensions with your mind's eye, you will have a lot of fun studying Chapter 7 and discovering the details of stereochemistry throughout the course. Stereoisomerism can also be a very difficult and challenging subject, simply because you have to be able to imagine three-dimensional objects when confronted with two-dimensional pictures. This takes practice, which very few individuals have had (remember, most people have been stuck on the two-dimensional surface of the earth all their lives....). However, it is easy for you to have this practice: **USE YOUR MODELS!** As you go through the chapter, make models of the structures and compounds described, so you can see a three-dimensional representation alongside the two-dimensional representation in the text. For instance, make a model of the four 2-chloro-3-iodobutane stereoisomers and match them up with the structures depicted in Figure 7.9. Work the exercises and problems with models until you are confident of your understanding. Then, go through them again without models so that you can practice and develop your ability to make the mental transition between two and three dimensions.

Unfortunately, for the purposes of quizzes and exams, you often are not permitted visual aids like molecular models. There are a few other aids that you can use to help you manipulate 3-dimensional structures on a 2-dimensional surface, like an exam paper. Keep in mind that if you interchange 2 of the four groups at a chiral carbon, you will have the enantiomer of the original structure. If you interchange 2 other groups at that carbon, you will be back to the original configuration. Obviously, if you interchange the groups you interchanged to begin with, you will be back where you started; you may feel intuitively that if you interchange the other 2 groups, you will be back to the original structure, but any 2 groups will do it as long as they are directly attached to the chiral carbon of interest. (Be careful with molecules containing more than one chiral center.) This is a useful device for rotating a 3-dimensional structure on a 2-dimensional surface and being sure in your own mind that you have not inadvertently inverted the configuration. Suppose you want to rotate a given representation of a molecule so that the group of lowest priority is at the back of the molecule. Interchange that group with the group already at the back. Then interchange any other 2 groups and voilà, the molecule is in a representation where R and S can be readily determined. Use the following representation of the carbohydrate glyceraldehyde: $CH_2OHCHOHCHO$. The central carbon has 4 different groups attached to it: $-CH_2OH$, -H, -OH, -CHO, and is a chiral carbon. First arrange the groups in order of priority: -OH > -CHO > $-CH_2OH$ > H. Then switch two of the groups at the chiral carbon so that the group of lowest priority is at the back. **You now have the enantiomer of the molecule you started with.** So interchange any other two groups (not the lowest priority one). This will get the original enantiomer back again in an orientaion where you can readily determine whether it is R or S.

i.e. H, ·····CHO, OH, $HOCH_2$ one interchange → CHO, ·····H, OH, $HOCH_2$ this is the enantiomer of the original, so do

one more switch → HOH_2C, ·····H, OH, CHO The configuration is *R*.

This is a useful device for manipulating 3-dimensional structures on plane surfaces. Use it with care. If there is more than one stereocenter in the molecule, remember that **both switches must be done at the same stereocenter.** Otherwise, you have inverted configuration at each center where you have made only one interchange.

The most important concepts in Chapter 7 are those of **chirality** and **optical activity**, and how they differ (see the discussion under **Chiral** in the Glossary); how to describe **absolute configurations** using the **R-S convention**; how to draw molecules so as to depict their absolute configuration (**Wedged** and **Dashed** bonds; **Newman projections**) and how to manipulate these pictures in your mind and on paper; and the relationships between **enantiomers, diastereomers,** and **meso** compounds.

7.D. Answers to Exercises

7.1 Items c, d, f, g, h, j, k, and l are clearly chiral; a, b, e, i, and m are not chiral, unless one takes into account the printing or design on them. A portrait (n) is probably chiral, unless it is perfectly symmetrical; i.e. a face-on view, hair parted in the middle, arms not folded, etc.

7.2 b and c

7.2 Model building exercise.

7.4 $\alpha = [\alpha] \times \ell \times c$ $\quad V = \pi r^2 \ell$ $\quad \ell = \dfrac{V}{\pi r^2} = \dfrac{300 cm^2}{\pi \; x \; 6.25 \; cm^2} = 15.3$ cm

$[\alpha] = 66°$; $\ell = 15.3$ cm = 1.53 dm; c = 60 g/300 mL = 0.2 g mL^{-1}; therefore $\alpha = +20.2°$

7.5 H····C(CH₂CH₃)(F)(Cl) or CH₃CH₂–C(····Cl)(H)(F)

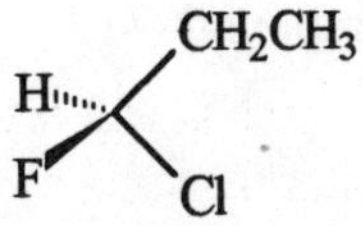

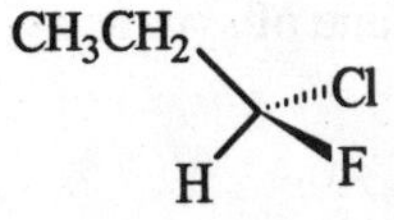

Be sure to note that there are other ways you can draw this compound.

7.6 (S)-3-chloro-2,6-dimethylheptane

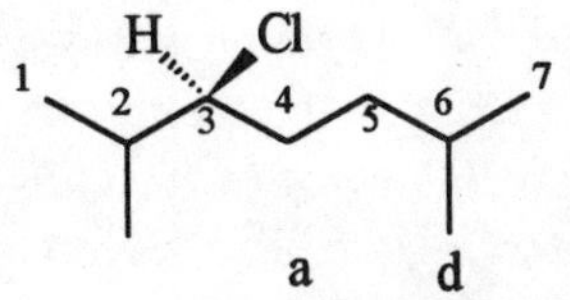

7.7 a. 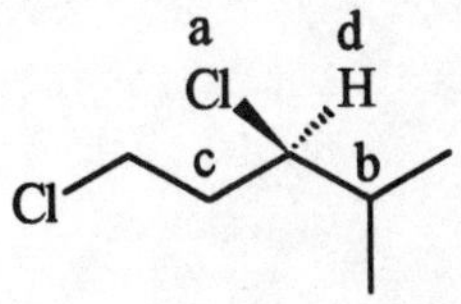 (R)

$(CH_3)_2CH$- has higher rank than Cl-CH_2-CH_2 because the *first difference* in going out the chain is two carbons (two methyls) vs. one carbon (Cl-CH_2); the fact that there is a chlorine farther out doesn't matter.

b. 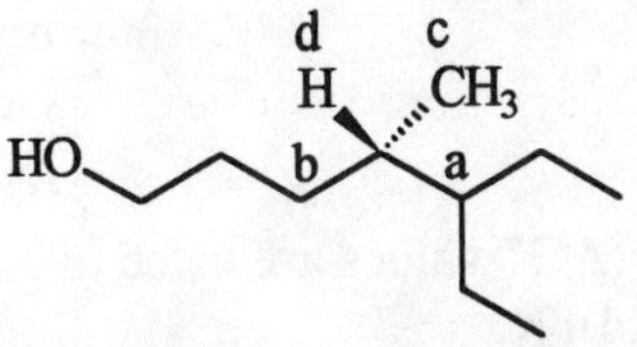 ≡ 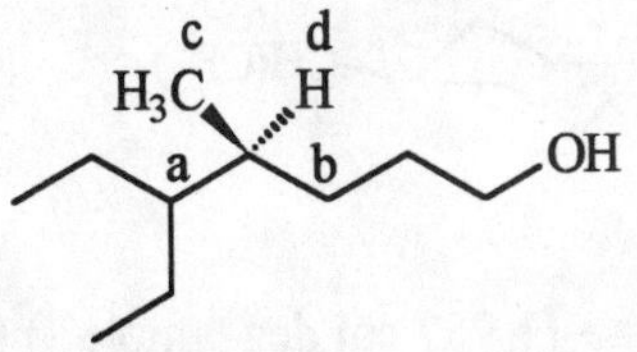 (S)

In the first drawing, the lowest priority group, (d), is in front. To rotate the structure, first exchange the group of lowest priority with the group in the back. This has inverted configuration, therefore to get back to the original structure, reverse any other two groups a & b, a & c or c & b. We have arbitrarily chosen a & b.

7.8 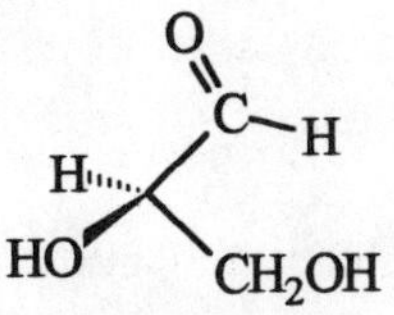 ≡ 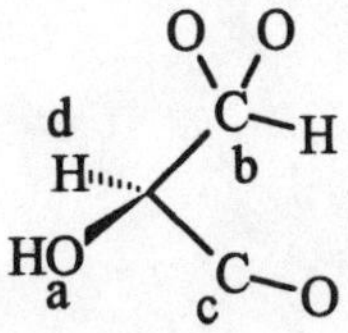 (R)

7.9 a. 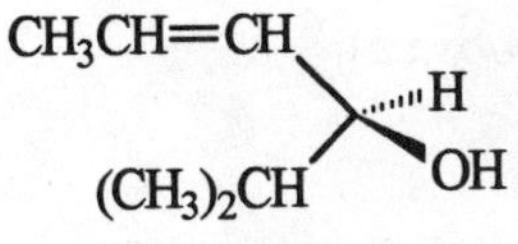 ≡ 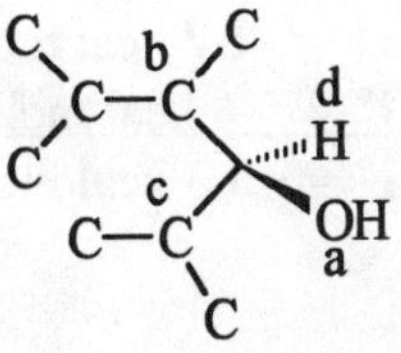 (S) b. 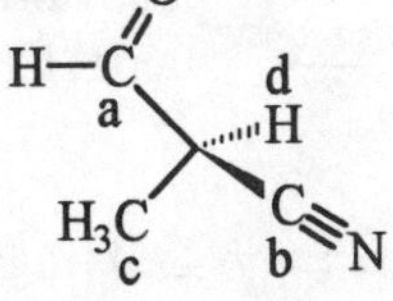 (R)

7.10

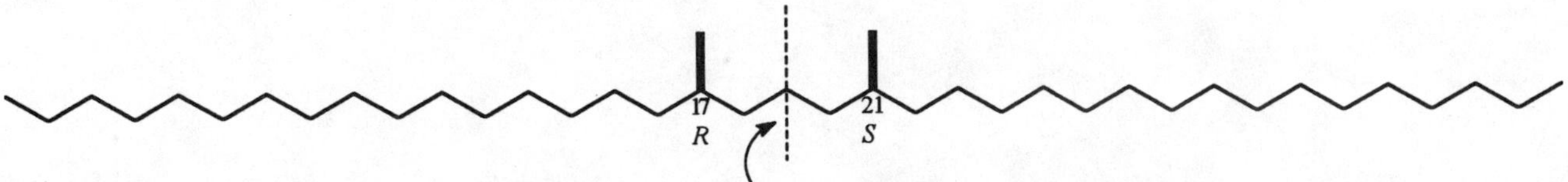

Plane of symmetry; compound is *meso*.

7.11

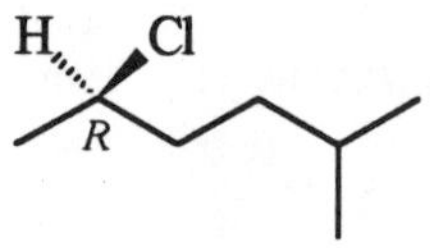

Plane of symmetry; *meso* compound.

(*R*)-2-chloro-5-methylhexane

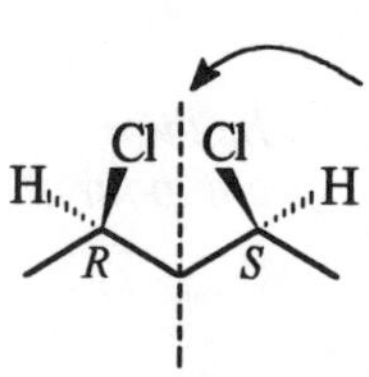

2(*R*),4(*S*)-2,4-dichloropentane

7.12, 13, 14 Model building exercises

7.15

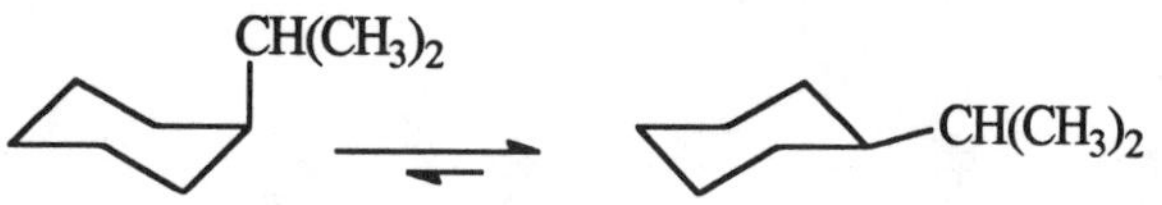

$\Delta G° = -RT \ln K$ ($\Delta G°$ values are listed in Table 7.1)

$-2.1 \text{ kcal mole}^{-1} = -2100 \text{ cal mole}^{-1} = -(1.987 \text{ cal deg}^{-1} \text{ mole}^{-1})(298 \text{ deg}) \ln K$

$K = 35$

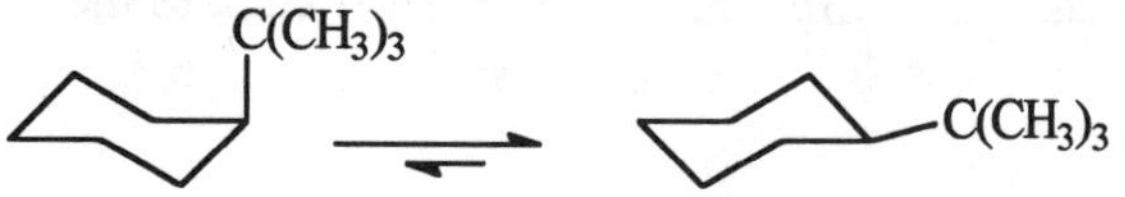

$(-5000 \text{ to } -6000) = 1.987 \times 298 \times \ln K$

$K \approx 4600$ to 25,000

7.16

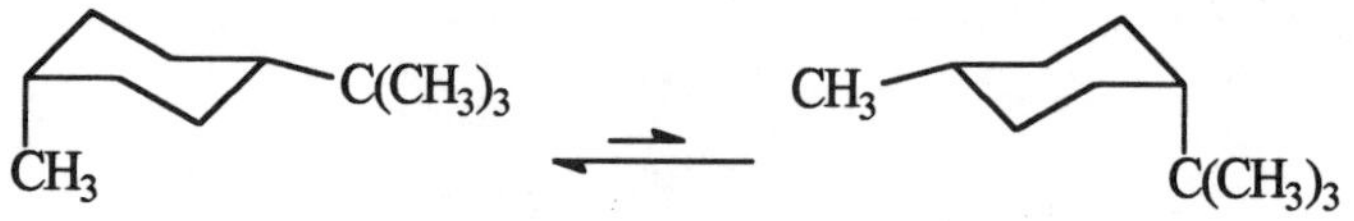

axial CH_3	⟶	equatorial CH_3:	$\Delta G° = -1.7 \text{ kcal mole}^{-1}$
equatorial $C(CH_3)_3$	⟶	axial $C(CH_3)_3$:	$\Delta G° \approx 5\text{-}6 \text{ kcal mole}^{-1}$ (*choose 5*)
		Sum:	$\Delta G° \approx +3.3 \text{ kcal mole}^{-1}$

$\Delta G° = -RT \ln K = +3{,}300 = -(1.987)(298) \ln K$

$K = 0.0038$ (i.e., conformation on the left is strongly favored)

7.E Answers and Explanations for Problems

1. a. 1 M $C_5H_{11}Cl$ = 106.5 g liter^{-1} = 0.1065 g ml^{-1} = c
ℓ = 10 cm = 1 dm

$$\alpha = +3.64°;\ [\alpha]_D = \frac{\alpha}{\ell \times c} = \frac{3.64}{1 \times 0.1065} = +34.2°$$

b. c = 0.096 g ml^{-1}, ℓ = 0.5 dm, α = −1.80°; therefore $[\alpha]_D$ = −37.5

2. a. three: RR, SS, and meso (RS)
b. and c. four: RR, SS, RS, and SR
d. and e., eight: RRR, RRS, RSR, SRR, RSS, SRS, SSR, and SSS

When no possibility of meso compounds exists, as in these cases, the number of possible stereoisomers is equal to 2^n, where n = the number of stereocenters in the molecule.

f. is achiral.

3. a. H OH HO H *R R* H H HO OH *S S* H H HO OH *R S* ≡ H H HO OH *S R*

enantiomers

If you imagine turning one of the pictures of the meso compounds over 180°, you can see that they represent the same structure.

H H HO OH *R S*

b. H OH *R R* H OH (1) H OH *R S* H OH (2) H OH *S S* H OH (3) H OH *S R* H OH (4)

(1) and (3) are enantiomers, as are (2) and (4). All other pairings: (1) and (2), (1) and (4), (2) and (3), (3) and (4), are diasteriomeric. There is no meso isomer in this case because the two ends of the molecules are different.

c. H Cl *R R* H OH (1) H Cl *R S* H OH (2) H Cl *S S* H OH (3) H Cl *S R* H OH (4)

(1) and (3) are enantiomers, as are (2) and (4). All other pairings: (1) and (2), (1) and (4), (2) and (3), (3) and (4), are diasteriomeric. There is no meso isomer because the substituents at C-2 and C-3 are different.

d. 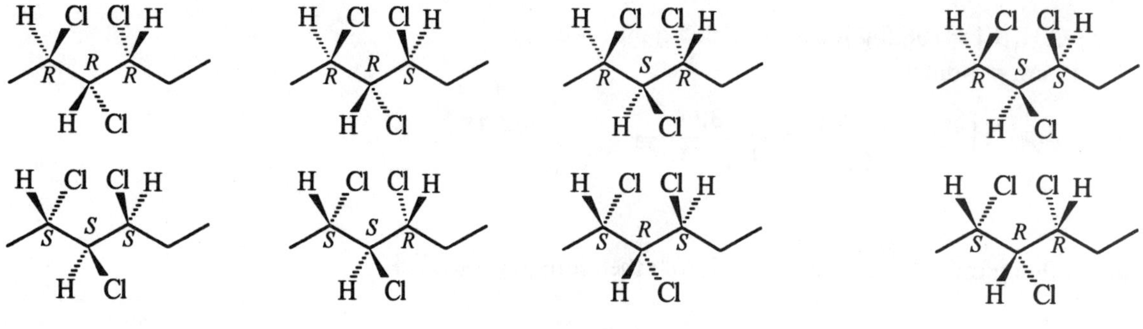 [Structures: eight stereoisomers labeled R,R,R; R,R,S; R,S,R; R,S,S (first row) and S,S,S; S,S,R; S,R,S; S,R,R (second row)]

The enantiomers of the compounds in the first row are given directly below. All other pairs are diasteriomers.

e. 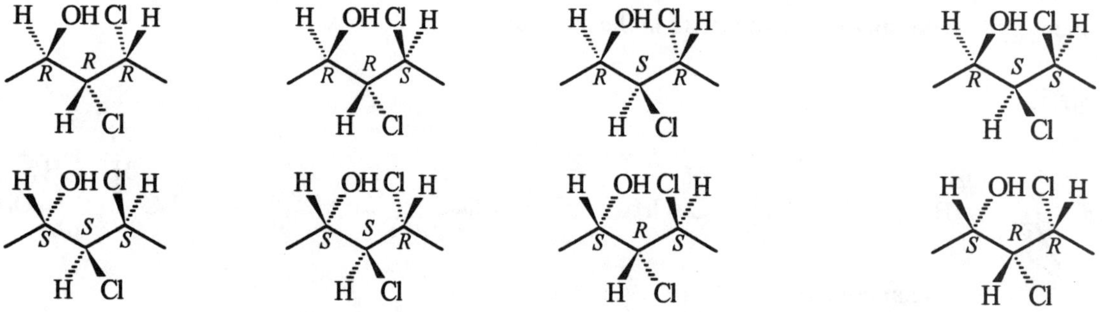 [Structures: eight stereoisomers labeled R,R,R; R,R,S; R,S,R; R,S,S (first row) and S,S,S; S,S,R; S,R,S; S,R,R (second row)]

The enantiomers of the compounds in the first row are given directly below. All other pairs are diasteriomers.

4. a. 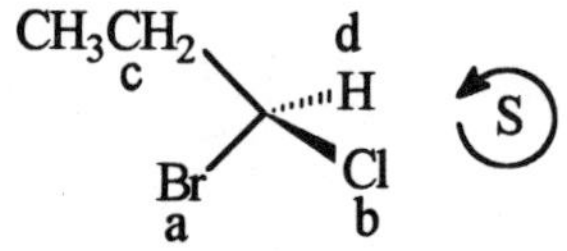 CH_3CH_2 (c), H (d), Cl (b), Br (a) — S — lowest priority in back

b. 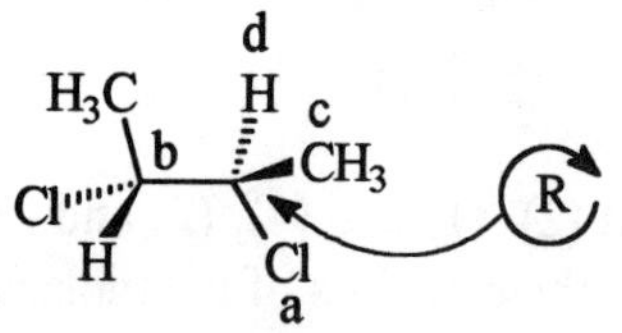 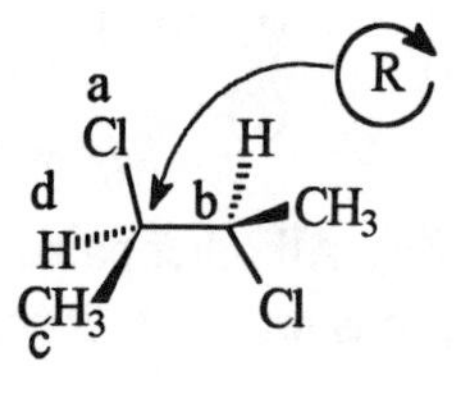 H_3C, H (d), CH_3 (c), Cl (a) — R — put the group of lowest priority in back.

This is the R,R-diastereomer

c. 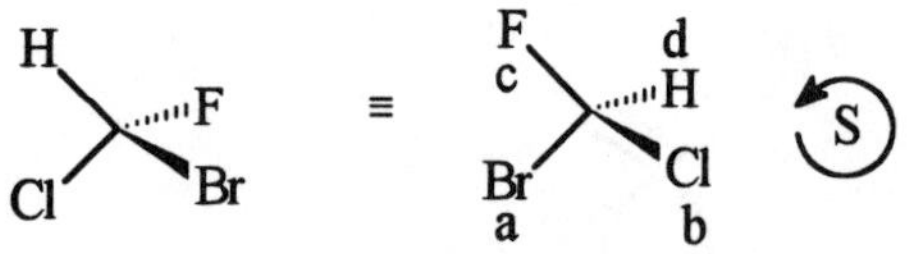 H, F, Cl, Br ≡ F (c), H (d), Cl (b), Br (a) — S

d. H_3C (b), H (d), OH (a), D (c) — S

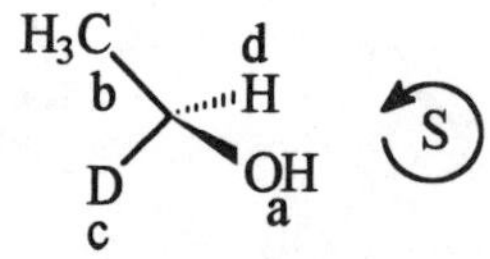

e. 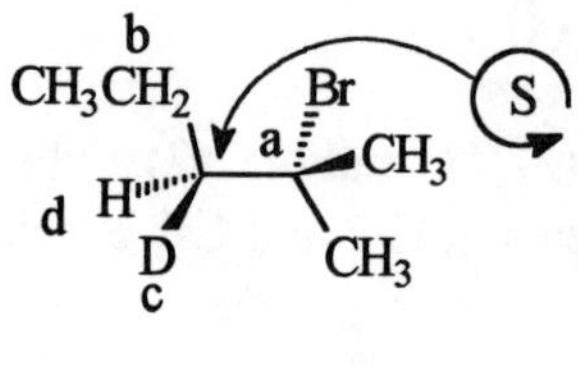 CH_3CH_2 (b), H (d), D (c), Br (a), CH_3, CH_3 — S

The carbon bearing the bromine is not chiral since two of the groups attached to it are the same (CH_3).

f. 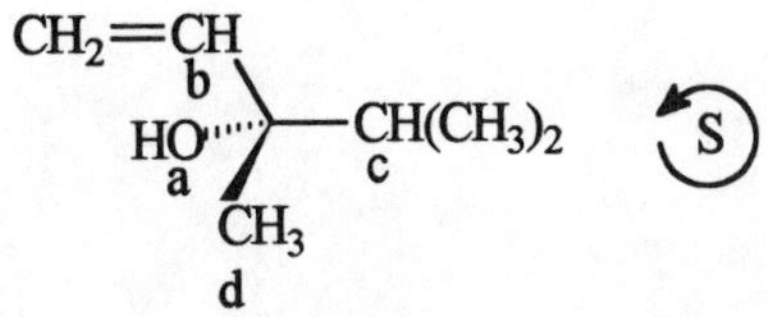 $CH_2{=}CH$ (b), HO (a), $CH(CH_3)_2$ (c), CH_3 (d) — (S) -CH=CH$_2$ counts as $-\underset{C}{C}-\underset{C}{C}$ and therefore has priority over $-\underset{C}{C}-C$

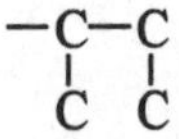

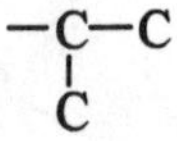

g. 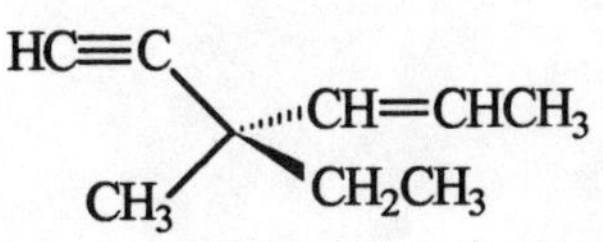 $HC{\equiv}C$, $CH{=}CHCH_3$, CH_3, CH_2CH_3 ≡ $HC{\equiv}C$ (a), CH_3 (d), CH_3CH_2 (c), $CH{=}CHCH_3$ (b) — *R*

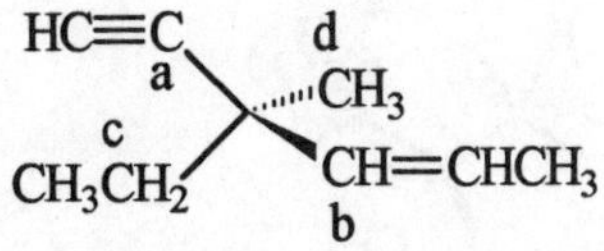

-C≡CH counts as $-\overset{C}{\underset{C}{C}}-\overset{C}{\underset{C}{C}}$ and has priority over -CH=CHCH$_3$, equivalent to $-\underset{C}{C}-\overset{C}{\underset{C}{C}}$

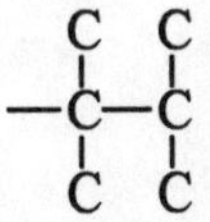

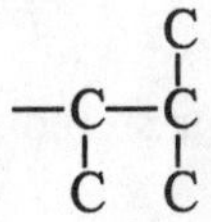

h. 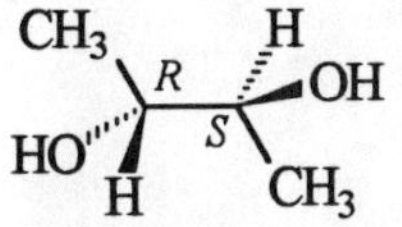 CH_3, H, OH, HO, H, CH_3 — *R*, *S* (meso)

5\. a. 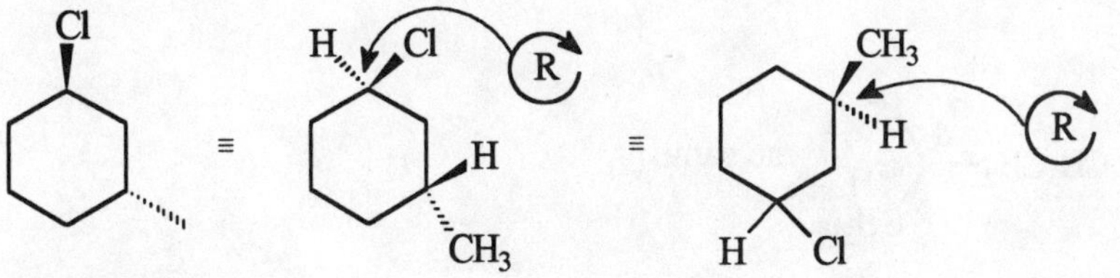 Cl … ≡ H, Cl, H, CH_3 (R) ≡ CH_3, H, H, Cl (R)

b. 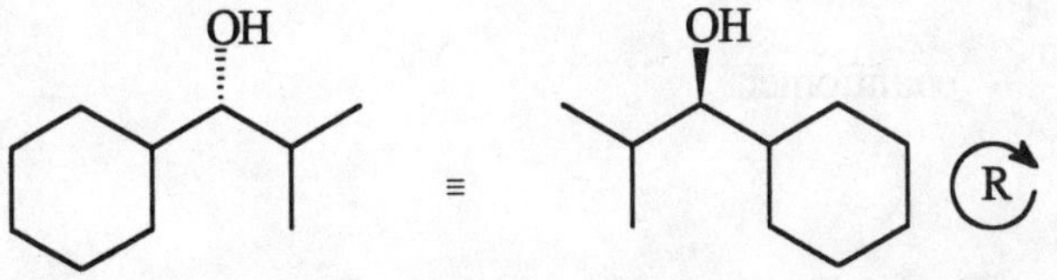 OH ≡ OH (R) c. 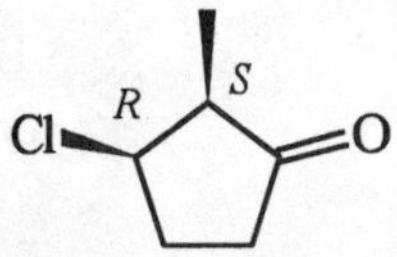 Cl, *R*, *S*, O

d. Determining the stereochemistry at C-6, *S*, is straightforward. It is harder for C-1. Write the two paths along the ring in terms of the equivalent carbon chains.

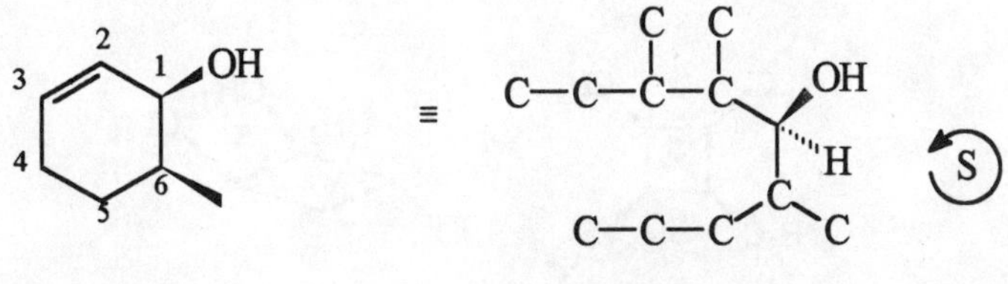

1, 2, 3, 4, 5, 6, OH ≡ C–C–C–C (C, C) OH, H, C–C–C–C, C (S)

6\. a. 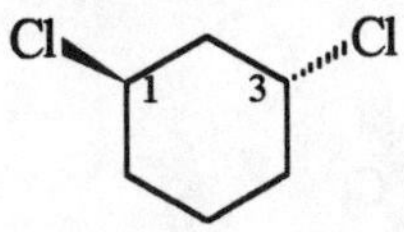 Cl, 1, 3, Cl b. 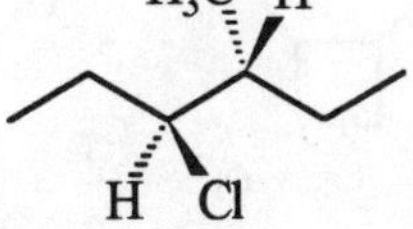 H_3C, H, H, Cl

c. 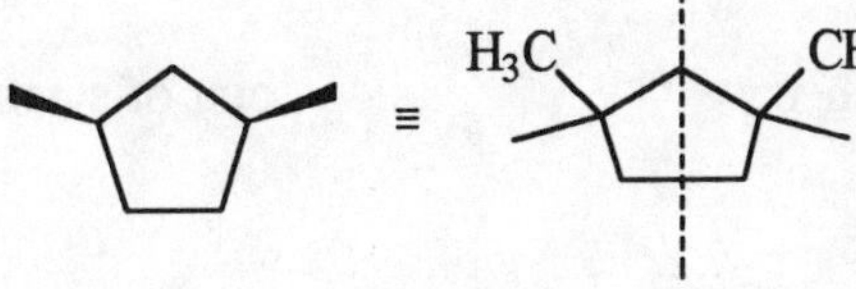 ≡ H_3C, CH_3 d. 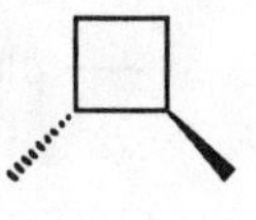 ≡

7. 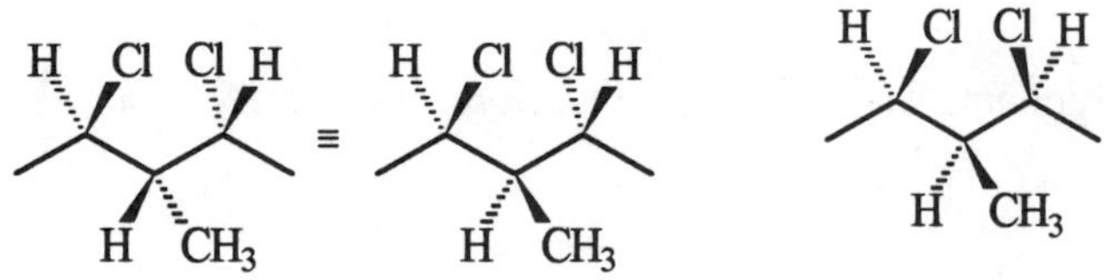 H Cl Cl H ≡ H Cl Cl H   H Cl Cl H
H CH_3 H CH_3 H CH_3
meso

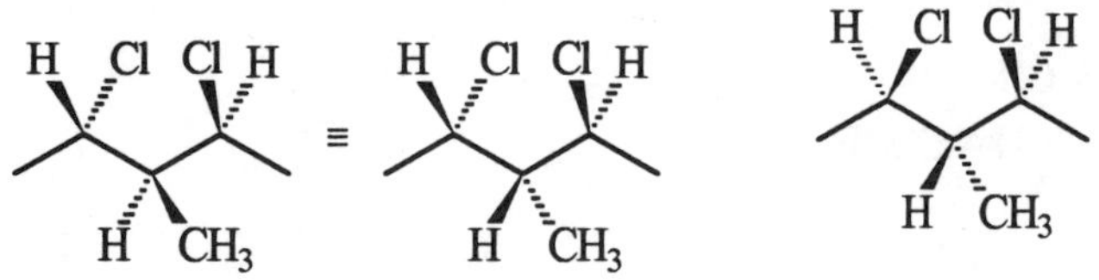

H Cl Cl H ≡ H Cl Cl H   H Cl Cl H
H CH_3 H CH_3 H CH_3
meso

There are only four stereoisomers of this compound.

8.  CH_3, H_3C, H, H, H, Br ≡ CH_3 CH_3 H S Br H H   compare to:

a.  Br, H, CH_3, H_3C, H, H ≡ CH_3CH_2 S Br H CH_3 the same

b.  CH_2CH_3, H, Br, CH_3 ≡ H R CH_2CH_3 Br CH_3 enantiomer

c. 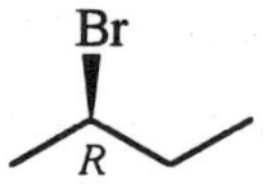 Br R   enantiomer   d. 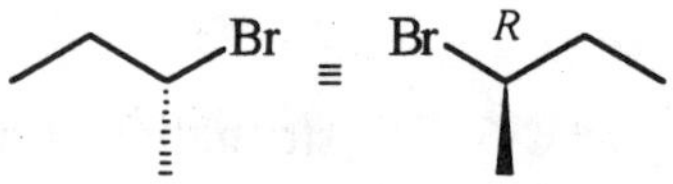 Br ≡ Br R   enantiomer

9. 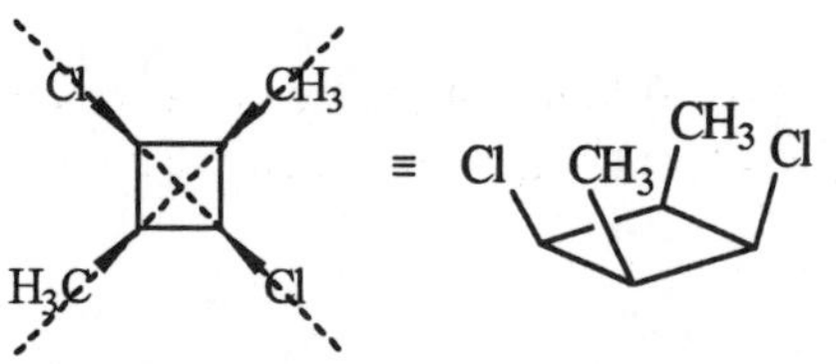 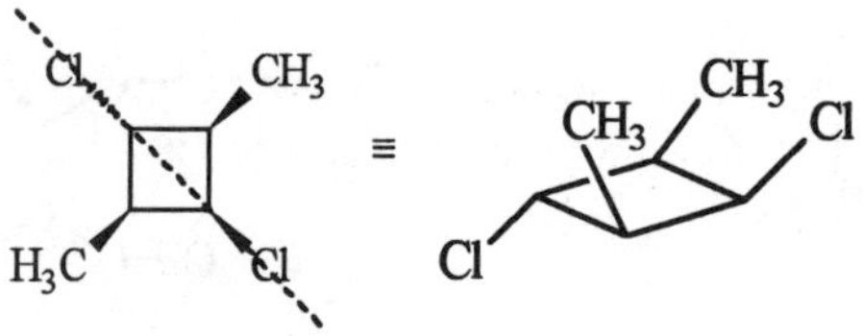 Cl CH_3 H_3C Cl ≡ Cl CH_3 CH_3 Cl   Cl CH_3 H_3C Cl ≡ CH_3 CH_3 Cl Cl

Cl CH_3 H_3C Cl ≡ Cl CH_3 Cl CH_3 Cl CH_3 H_3C Cl ≡ Cl CH_3 Cl CH_3

Cl CH_3 H_3C Cl ≡ CH_3 CH_3 Cl Cl

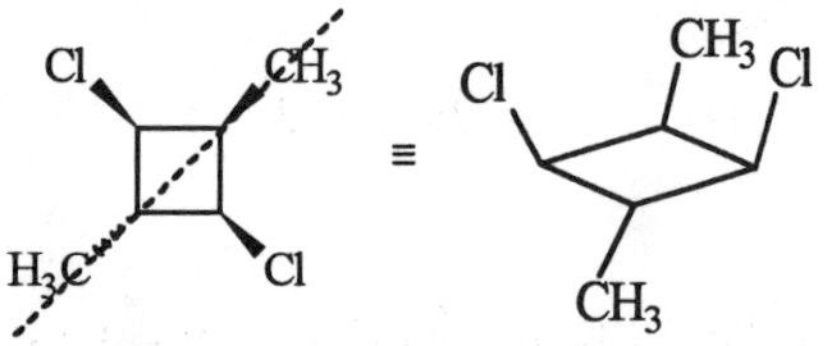

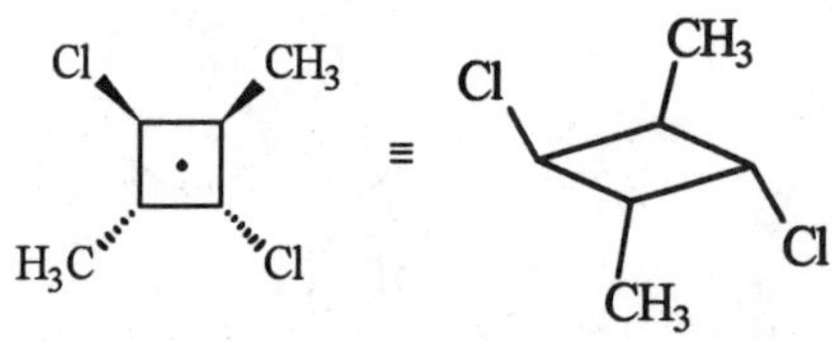

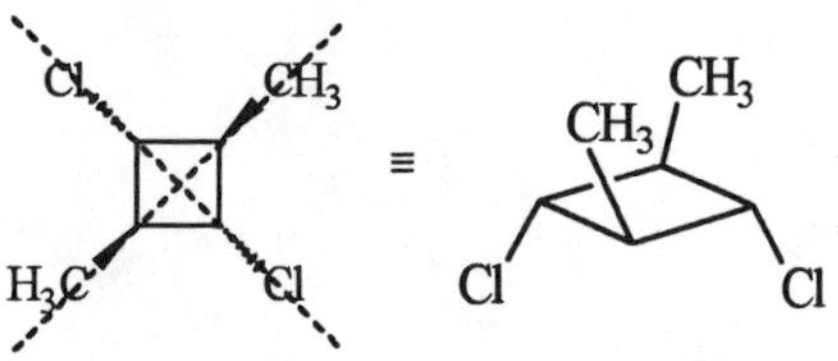

----- plane of symmetry   · point of symmetry

10. a. 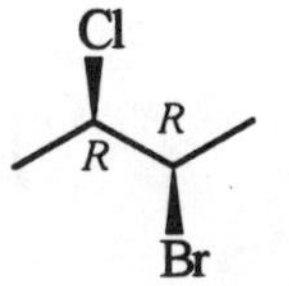 b. 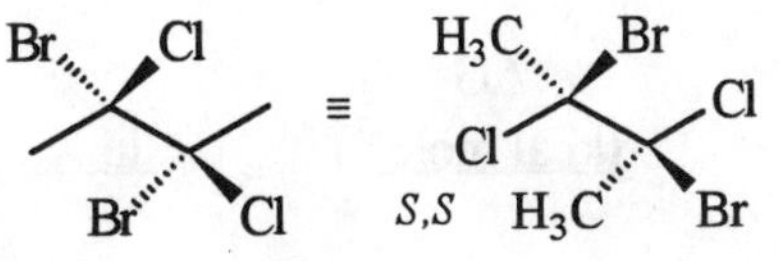 c.

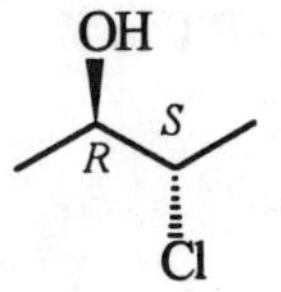

d. 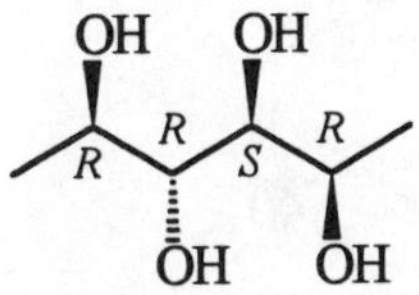 e. 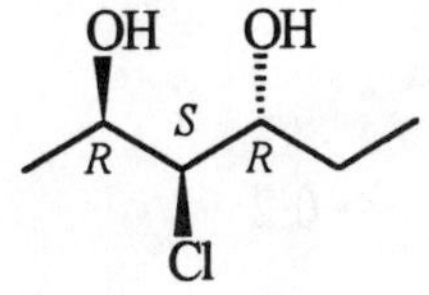 f.

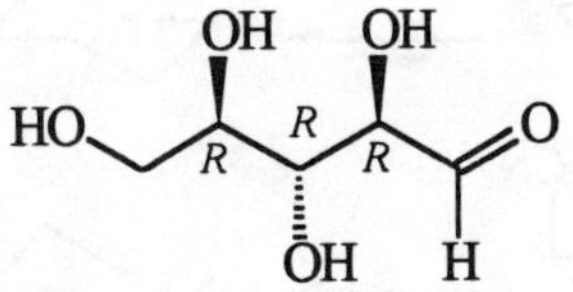

11. a. 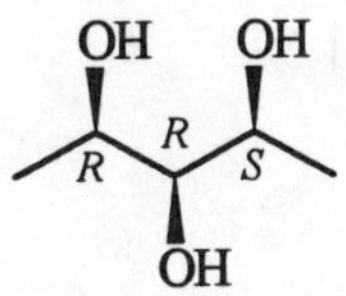 b. 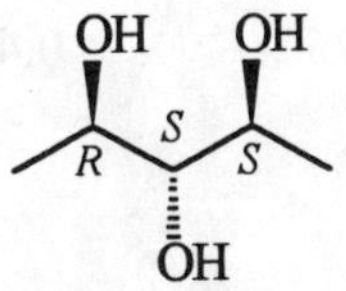 c.

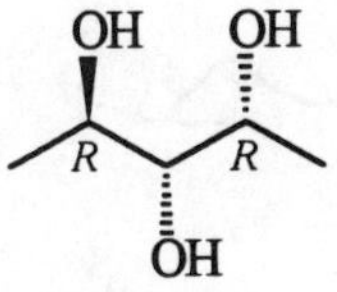

The central carbon in c. is **not** a stereocenter since two of the attached groups are identical: (*R*)-$CHOHCH_3$. See problem #7.

12.

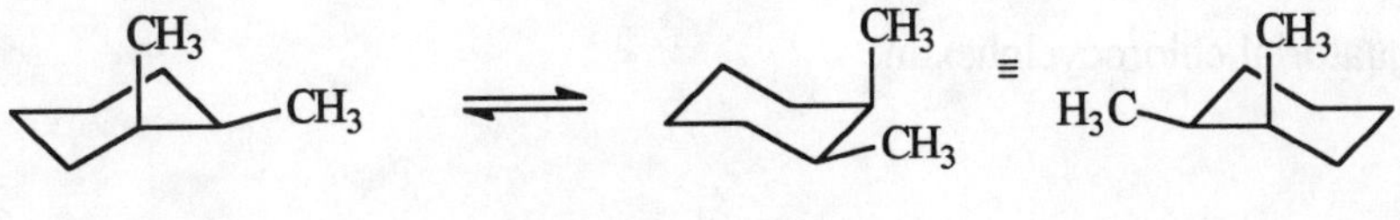

These are mirror images of each other. Be careful not to confuse **conformation** and **configuration. Conformation** changes by rotating, bending, or flexing the molecule. **Configuration** changes by breaking a bond and putting a substituent at a different place on the molecule. That is, the two conformations are enantiomeric. The two enantiomers of *cis*-1,2-dimethylcyclohexane are simply different conformations of the same molecule. They are not different configurations. Because the chair⇌chair interconversion is fast at ordinary temperatures, *cis*-1,2-dimethylcyclohexane behaves, on the time-average, as an achiral, meso compound.

13. a. Look at axial conformations for the two compounds; with the isopropyl substituent, there is one rotamer with only an H to interact with the other axial H's. With the *t*-butyl substituent, there is always a CH_3 group interacting with axial H's:

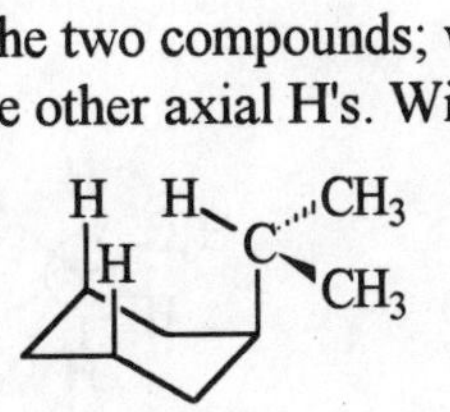

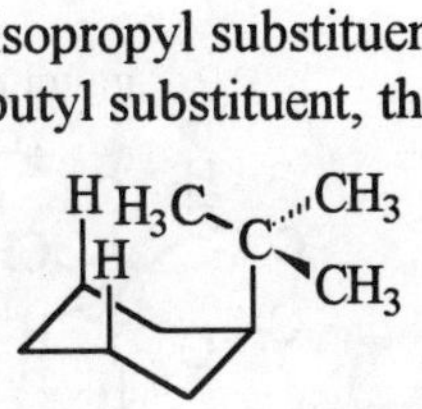

Consequently, the change in ΔG in going from CH_3CH_2- to $(CH_3)_2CH-$ is only about 0.3 kcal/mole, while the change from $(CH_3)_2CH-$ to $(CH_3)_3C-$ is about 3 kcal/mole.

b. Look at the Lewis structure for $-C\equiv CH$; the linear structure gives very little interference with axial H's.

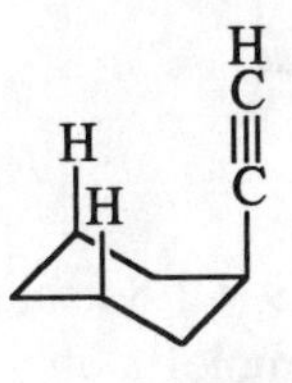

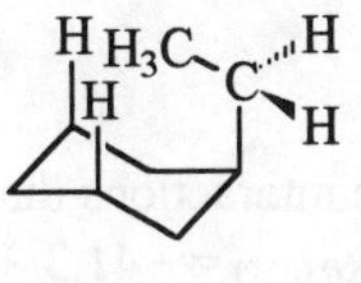

14. $\Delta G° = -RT \ln K$

	$\Delta G°$ (kcal mole^{-1})	ln K	K	Percent Equatorial
a. (OH, axial ⇌ equatorial)	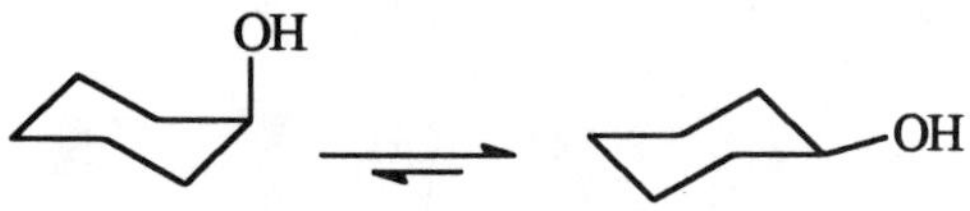 -1.0	1.69	5.42	84.4
b. (CN, axial ⇌ equatorial)	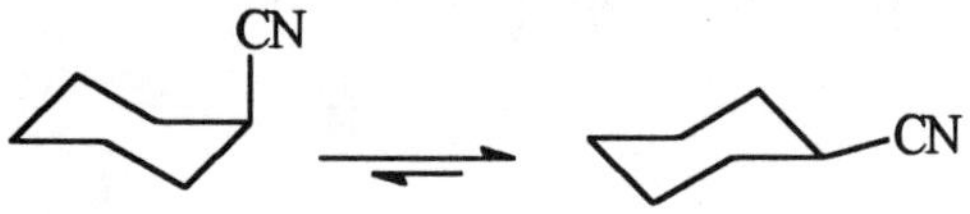 -0.2	0.338	1.40	58.3
c. (F, axial ⇌ equatorial)	 -0.25	0.422	1.53	60.5

15.

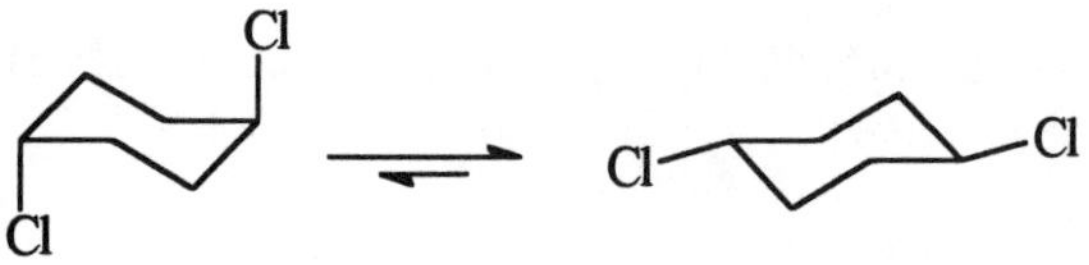

$\Delta G°$ ≅ 2 x $\Delta G°$ for axial ⟶ equatorial chlorocyclohexane
= 2 x 0.5 = 1.0 kcal mole^{-1}

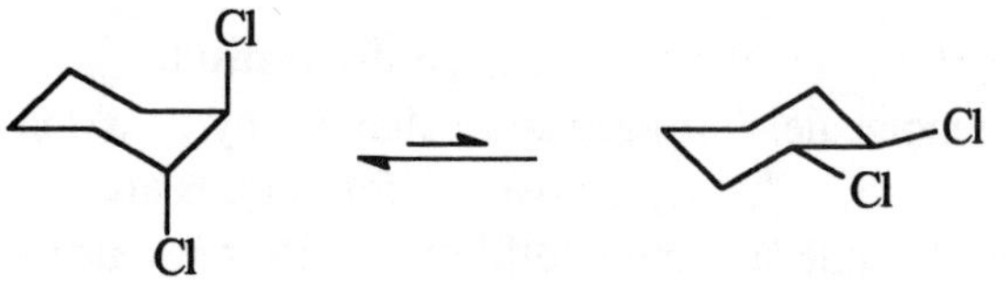

the gauche interaction between the adjacent, chlorines is more unfavorable than the axial equatorial interactions.

16. C-1 C-2 C-1 C-6 C-2 C-3

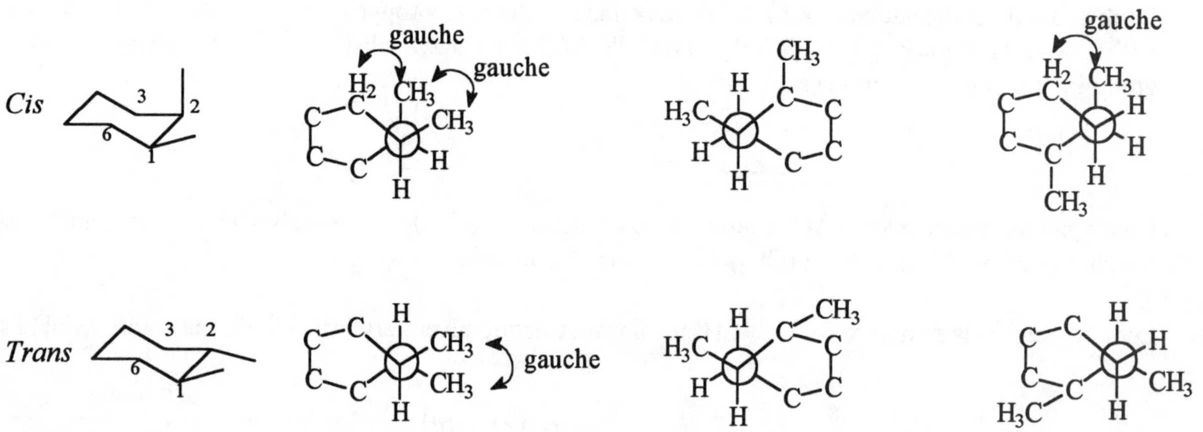

Cis: gauche, gauche (C-1 C-2); gauche (C-2 C-3)

Trans: gauche (C-1 C-2)

The *cis* isomer has two more gauche interactions than the *trans*: 2 x 0.9 = 1.8 kcal mole^{-1}
From Appendix I: $\Delta H^{\circ}_{f}(cis) - \Delta H^{\circ}_{f}(trans) = -41.3 - (-43.0) = 1.7$ kcal mole^{-1}.

17. 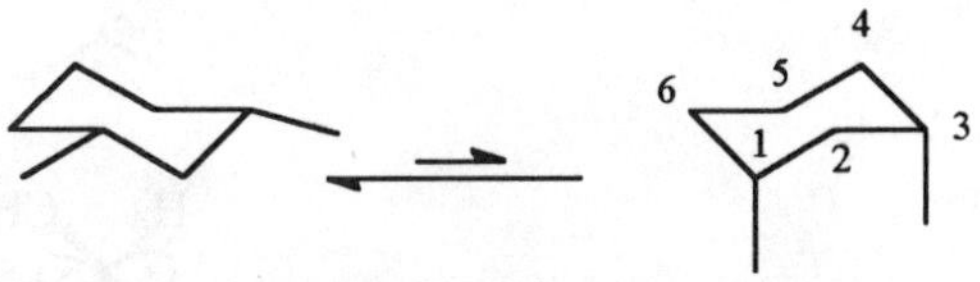 [structures: e,e and a,a conformations of cis-1,3-dimethylcyclohexane, carbons numbered 1–6]

There are no gauche interactions for the *e,e* conformation. However, for the *a,a* conformation, there are four gauche interactions, 2 between the CH_3-C_1 bond and the C_2-C_3 and C_5-C_6 bonds and 2 between the CH_3-C_3 bond and the C_1-C_2 and C_4-C_5 bonds: $4 \times 0.9 = 3.6$ kcal mole^{-1} less stable.

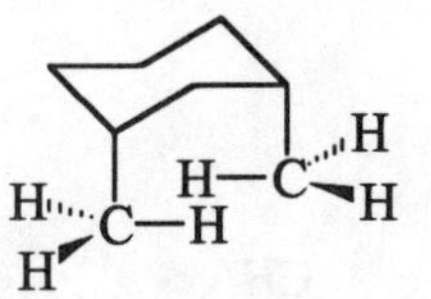

The diaxial conformation of *cis*-1,3-dimethylcyclohexane is even more unstable because the two methyl groups are so close that they are actually trying to occupy the same space:

18. a. 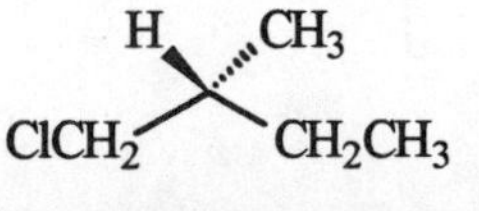 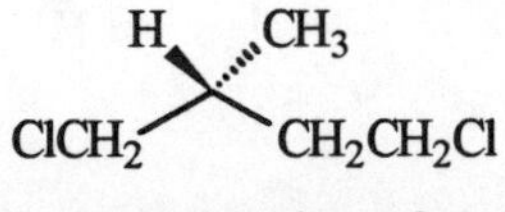 [structures: $ClCH_2C(H)(CH_3)CH_2CH_3$ and $ClCH_2C(H)(CH_3)CH_2CH_2Cl$]

There is no *a priori* relationship between sign of rotation, (+) or (−), and configuration (R or S).

b. $ClCH_2CCl(CH_3)CH_2CH_3$ is produced *via* the free radical $ClCH_2C(CH_3)CH_2CH_3$ intermediate. Since the product from this intermediate is racemic, the intermediate itself is probably achiral, either because it is achiral by symmetry (planar radical), or (if the radical is pyramidal) because it inverts much faster than it reacts. Other evidence is required to distinguish between these two alternatives.

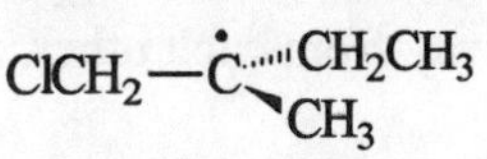

planar free radical (achiral)

rapidly equilibrating pyramidal free radicals (effectively achiral)

19. 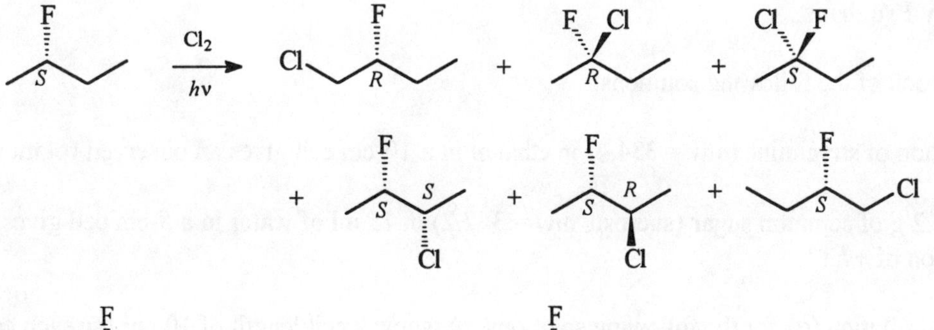 [reaction scheme: (S)-2-fluorobutane with Cl_2, $h\nu$ giving chlorinated products with configurations R, R, S, S/S, S/R, S]

a. 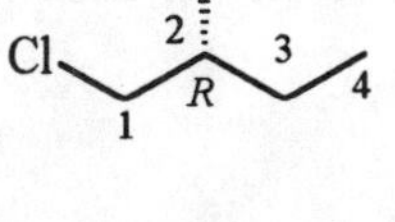 [structure: (R)-1-chloro-2-fluorobutane, carbons numbered 1–4]

b. 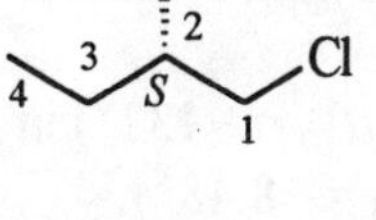 [structure: (S)-1-chloro-2-fluorobutane, carbons numbered 1–4]

The stereocenter has been unaffected in formation of these two products. The 1-chloro-2-fluoro-isomer is "R" only because of a change in the priority of the substituents.

c. 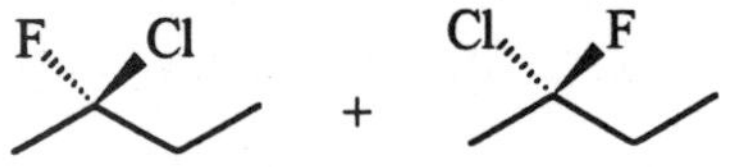 + Equal amounts formed via the achiral intermediate radical: 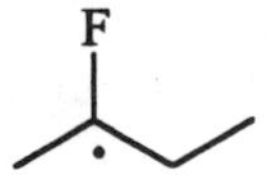 (see problem #18b.)

d. 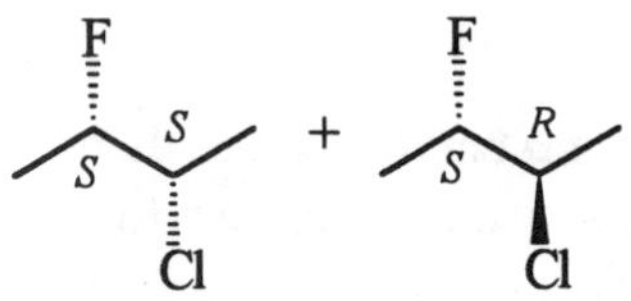 + The radical intermediate leading to these isomers is still chiral:

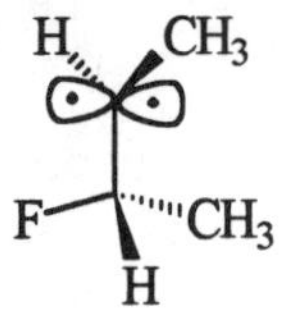

16% 24%

The transition states leading to the two products are therefore *diastereomeric*:

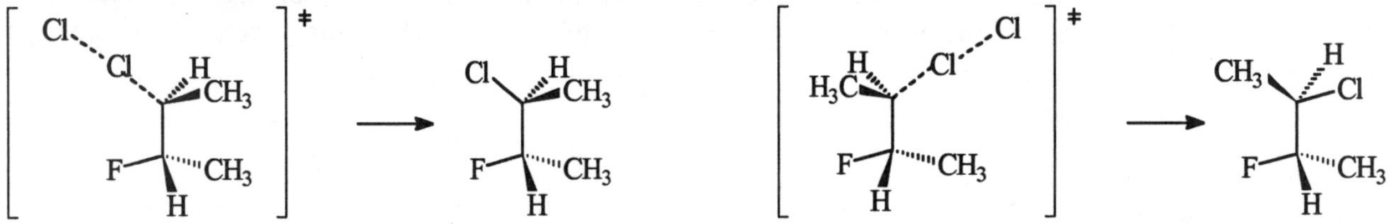

(2S,3S) (2S,3R)

Since the transition states are diastereomeric, they will have different physical properties, including different free energies of formation. Since the two competing reactions start at the same place and pass through transition states of different energies, the two activation energies are different. Therefore one diastereomer will be formed in greater amount than the other.

20. Bicyclo[2.2.1]heptane is a stable compound. The eclipsing energy present in boat cyclohexane is essentially unchanged in the bicyclo compound. The twist-boat (skew-boat) conformer is of much higher energy. Since the ends of the cyclohexane ring are tied together, they are prevented from moving away from each other.

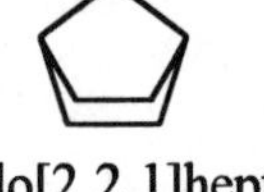

bicyclo[2.2.1]heptane

7.F. Supplementary Problems

S1. Calculate $[\alpha]_D$ for each of the following solutions:

a. A 0.13 *M* solution of strychnine (mw = 334.4) in ethanol in a 10-cm cell gives an observed rotation of $-2.26°$.
b. A solution of 3.2 g of common sugar (sucrose, mw = 342.2) in 15 ml of water in a 5-cm cell gives an observed rotation of $+7.1°$.

S2. Predict the observed rotation (α) for the following solutions. Assume a cell length of 10 cm for each case.

a. 3 g of morphine hydrate (mw = 303.3, $[\alpha]_D = -132°$) in 50 ml of methanol.
b. Pure (−)-2-chlorobutane (d = 0.87, $[\alpha]_D = -8.48°$).

S3. a. Pure (−)-α-pinene has a specific rotation, $[\alpha]_D = -51.3°$. What is the optical purity of a sample of α-pinene which shows a specific rotation of $-35.9°$ (i.e., how much of each enantiomer is present)?
b. Predict the specific rotation of a mixture of 30% of (−)-2-bromobutane (d = 1.254, $[\alpha]_D = -23.13°$) and 70% of the (+)-enantiomer.

S4. For each of the following compounds, make a model and label each stereocenter R or S as appropriate.

a. 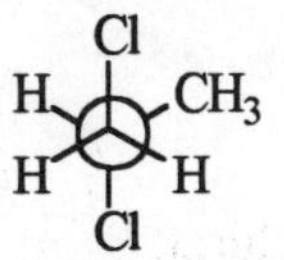 Cl, H, CH_3, H, H, Cl   b. 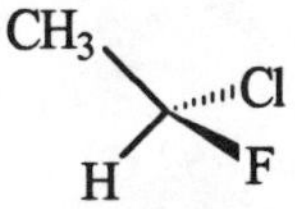 CH_3, Cl, H, F   c. 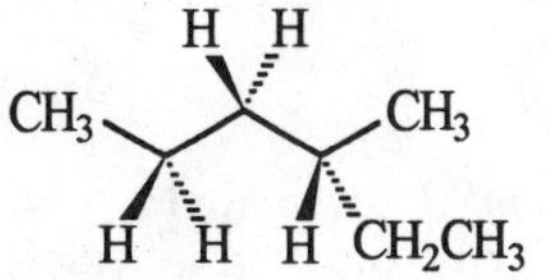 CH_3, H H, CH_3, H H, H CH_2CH_3 d. 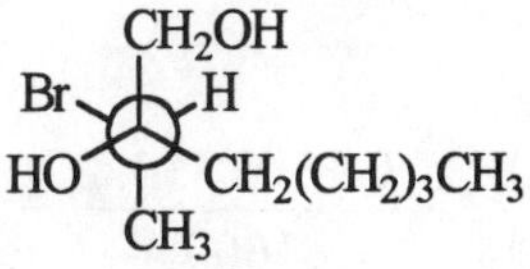 CH_2OH, Br, H, HO, $CH_2(CH_2)_3CH_3$, CH_3

S5. What is the relationship between the molecules of each of the following pairs (i.e., are they the same, enantiomers, diastereomers, structural isomers, etc.)? For each stereocenter in these molecules, assign R or S as appropriate.

a. 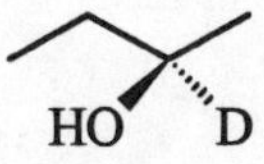 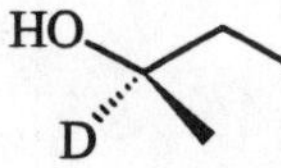 HO D; HO, D   b. 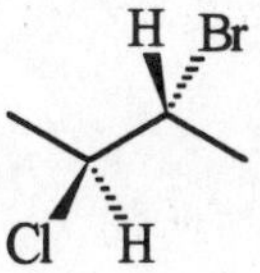 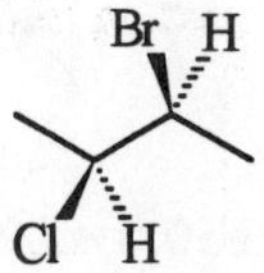 H Br, Cl H; Br H, Cl H

c. D, H, CH_3, F; CH_3, F, H, D   d. CH_3, H, Cl, Cl, H, CH_3; H Cl, Cl H

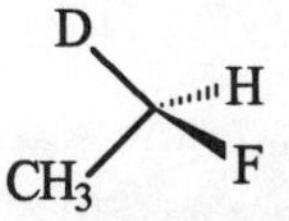

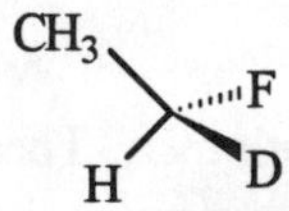

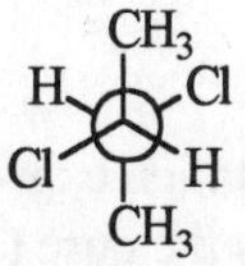

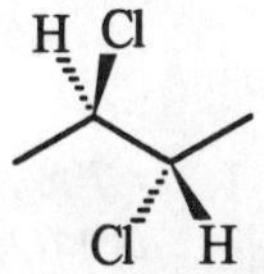

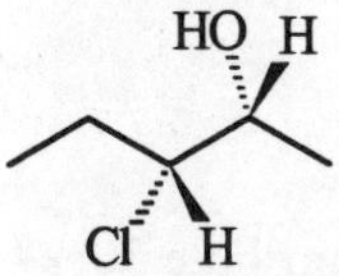

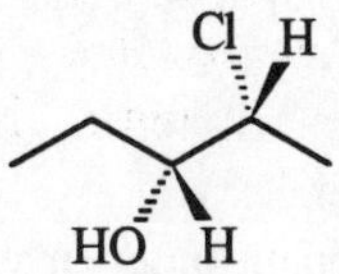

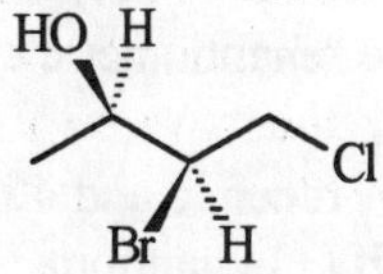

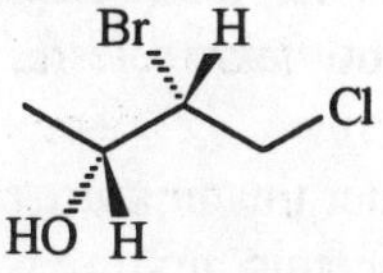

e. HO H, Cl H; Cl H, HO H   f. HO H, Cl, Br H; Br H, Cl, HO H

g. CH_3, H, CH_3, F, CH_2CH_3, F; H, F, F   h. $H_2C{=}CH$ H, H, Cl; Cl, H, H, $H_2C{=}CH$

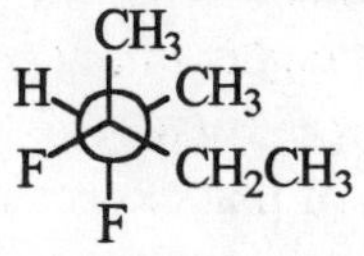

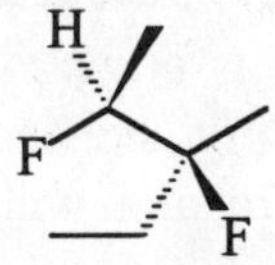

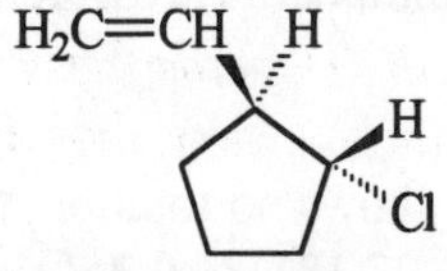

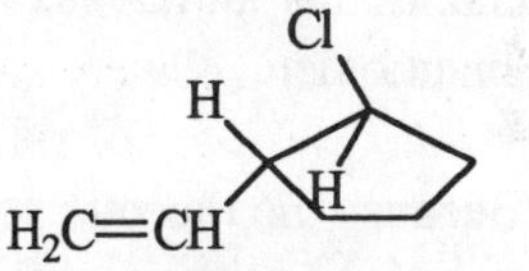

S6. For each of the compounds illustrated below, draw: (1) the enantiomer, (2) a diastereomer, (3) a Newman projection for a. and a line drawing for b.

a. 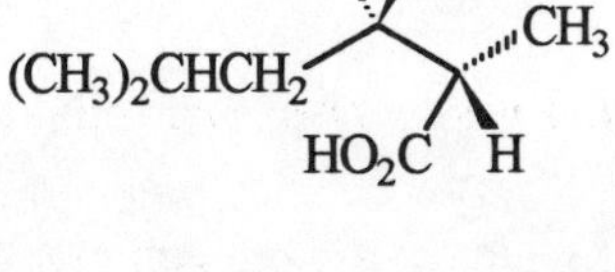 H Br, $(CH_3)_2CHCH_2$, CH_3, HO_2C H   b. 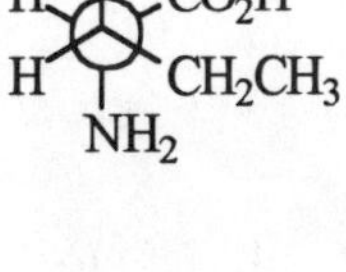 CH_3, H, CO_2H, H, CH_2CH_3, NH_2

S7. Write structures for all of the isomers of trimethylcyclopentane. Which ones are chiral and which ones are achiral?

S8. Consider the free-radical chlorination of 1,1,4,4-tetramethylcyclohexane with Cl_2.

a. Write the structures of all of the possible monochloro isomers.
b. Assign the configuration of each stereocenter as you have drawn it.
c. Predict the relative amount of each isomer.

7.G Answers to Supplementary Problems

S1. a. $\frac{0.13 \times 334.4}{1000} = 0.0435\ g\ ml^{-1}$; $[\alpha]_D = \frac{-2.26}{1 \times 0.0435} = -52.0°$

b. $[a]_D = \frac{+7.1}{0.5 \times 3.2/15} = +66.6°$

S2. a. $\alpha = [\alpha]_D \times \ell \times c$
$= -132 \times 1 \times 3/50 = -7.92°$
$\alpha = -8.48 \times 1 \times 0.87 = -7.38°$

S3. a. −35.9° is 35.9/51.3 = 70% of the rotation expected for the pure enantiomer. Therefore "30%" of the mixture is racemic and "70%" of it is the pure (−)-enantiomer. For the whole mixture, then, 70% + 15% = 85% of it is the (−)-enantiomer and 15% of it is the (+)-enantiomer. The optical purity is 70%; this is frequently also referred to as the "enantiomeric excess".

b. Consider this mixture to be 60% racemic and 40% (+)-enantiomer:
The racemic mixture will be half (+)-enantiomer and half (−)-enantiomer or will contribute 30% (+)-enantiomer and 30% (−)-enantiomer. The overall composition is 30% (−)-enantiomer and 70% (+)-enantiomer.

Because the racemic portion shows no rotation, the observed rotation will be 40% of that expected for the pure (+)-enantiomer: 0.4 x (−23.13) = −9.25°.

S4. a. ≡ [structure: Cl, H, Cl] R   b. R   c. R   d. 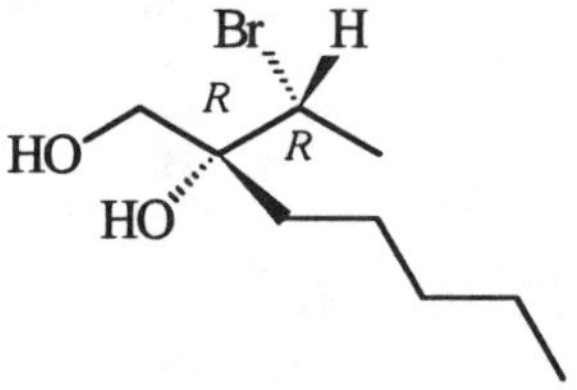 [structure: HO, HO, Br, H; *R*, *R*]

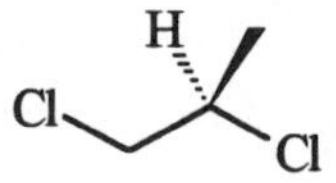

S5. a. identical: R   b. diastereomers: [structures: H, Br, Cl, H; *S*, *R*] [Br, H, Cl, H; *S*, *S*]

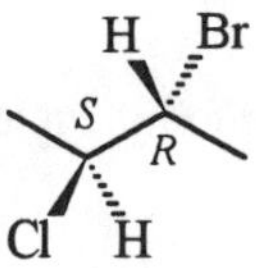

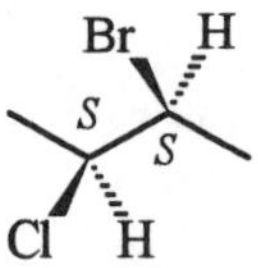

c. enantiomers: [structure: D, H, F, CH_3; *R*]   d. identical: [structure: H, Cl, Cl, H; *R*, *S*] *meso isomer*

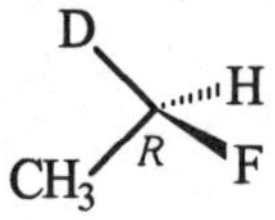

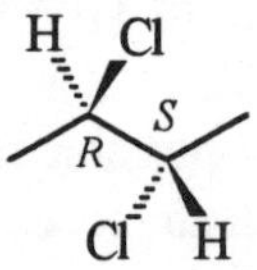

e. structural isomers:

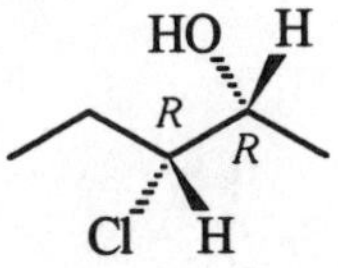

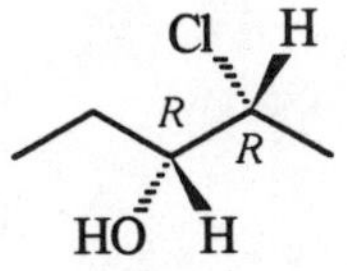

f. identical:

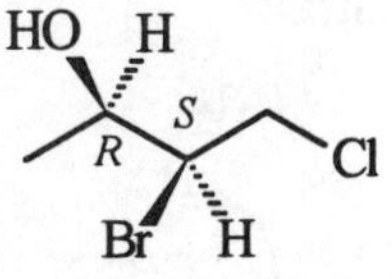

g. diasteriomers:

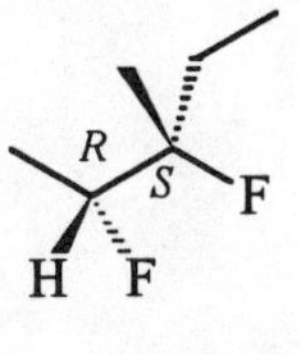

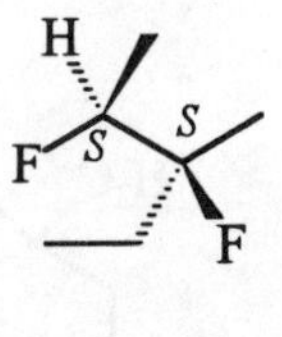

h. enantiomers:

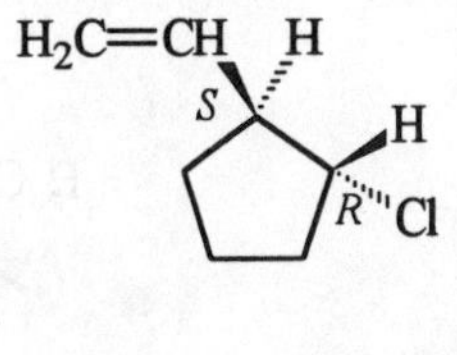

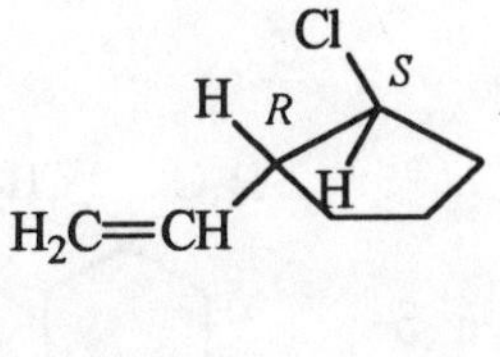

S6.

a. (1) 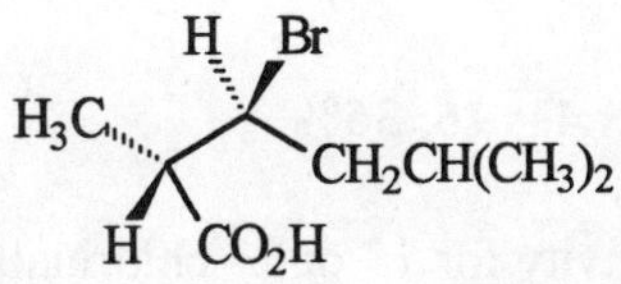 (2) 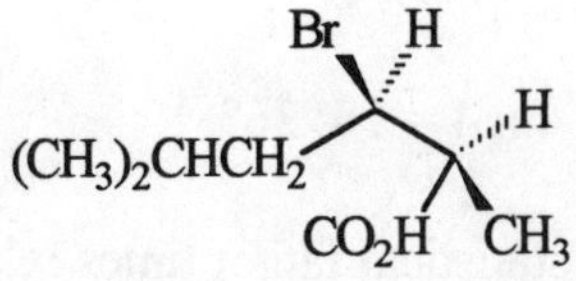 or

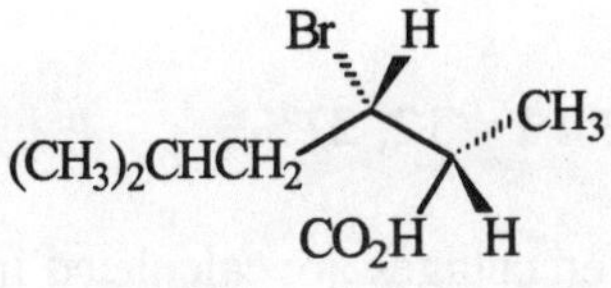

(3)

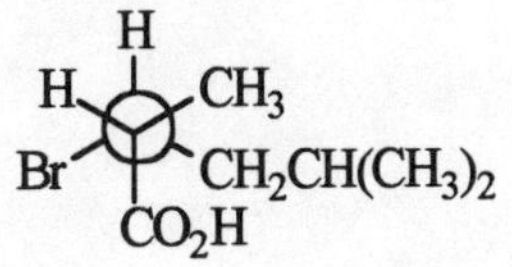

b.

(1) 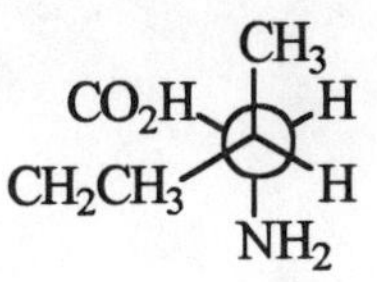 (2) 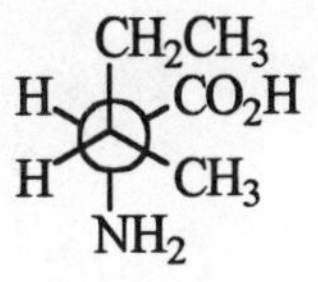 or 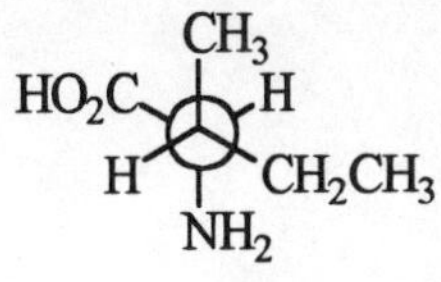 (3)

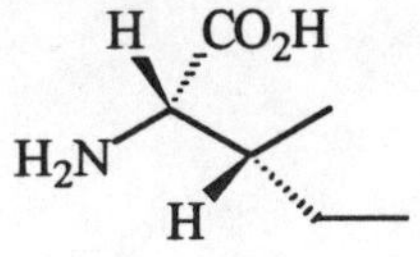

S7. Chiral

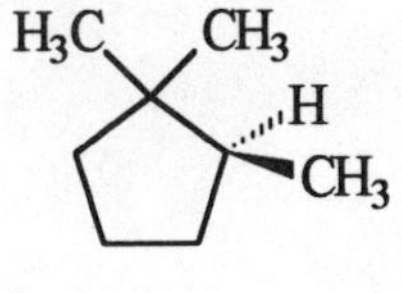

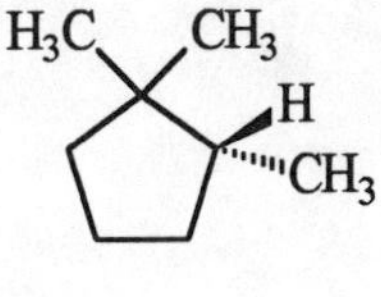

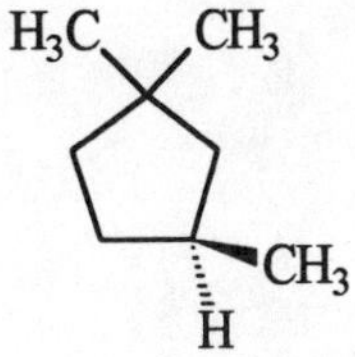

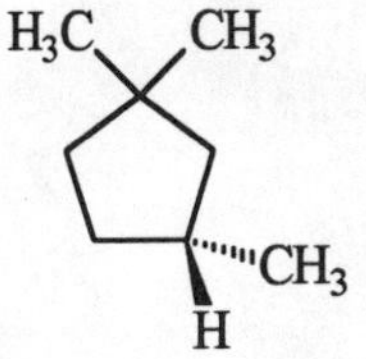

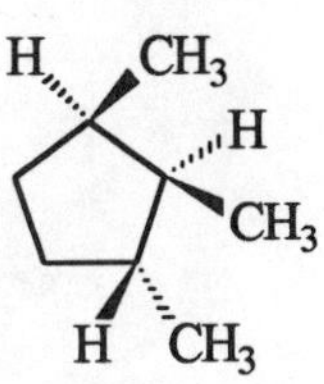

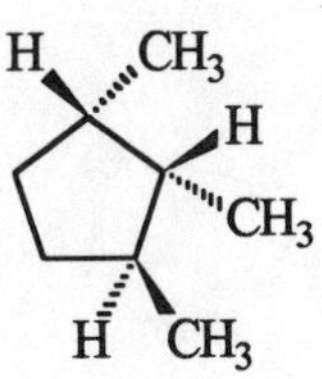

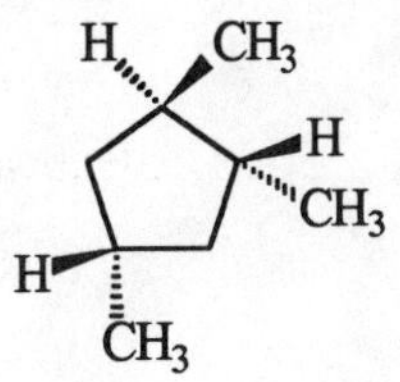

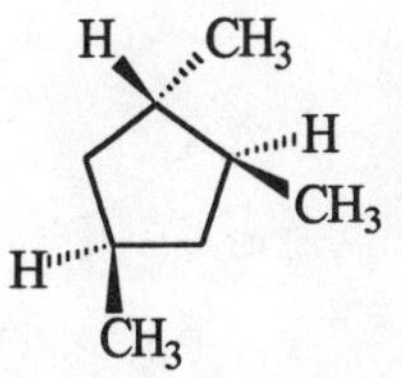

Achiral:

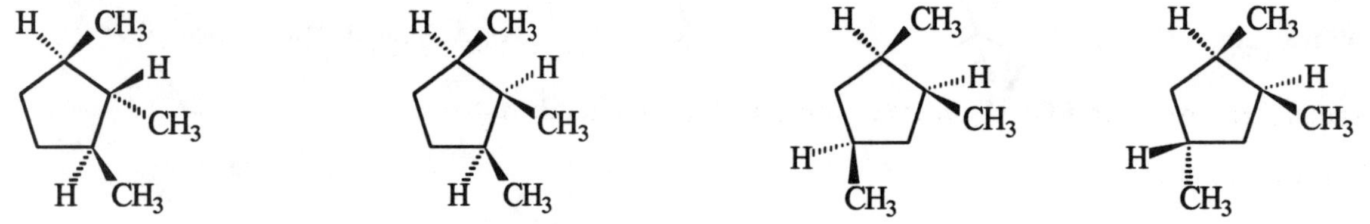

S8. a.,b.

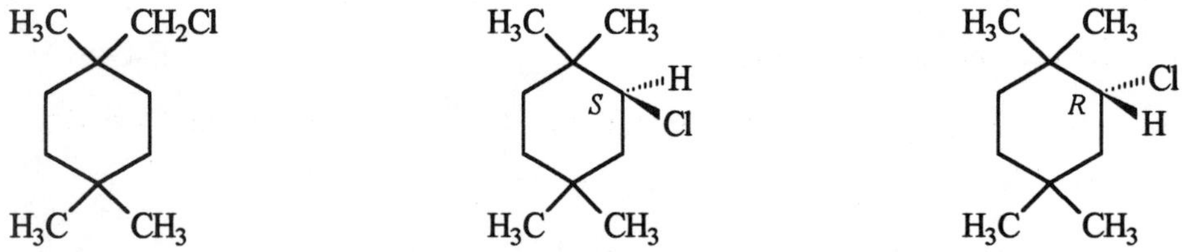

c. 12 x 1 = 12; 27% 4 x 4 = 16; 36% 4 x 4 = 16; 36%

(Percentages are calculated from the statistical factor times relative reactivity for 1° or 2° chlorination; see Chapter 6.)

8. ALKYL HALIDES AND ORGANOMETALLIC COMPOUNDS

8.A. Chapter Outline and Important Terms Introduced

8.1 Structure of Alkyl Halides (size of halogens, bond lengths)

dipole moment
van der Waals radius

8.2 Physical Properties of Alkyl Halides (boiling points, melting points)

polarizability

8.3 Conformations of Alkyl Halides

barriers to rotation

8.4 Some Uses of Halogenated Hydrocarbons (from dry cleaning to DDT)

8.5 Nomenclature of Organometallic Compounds

e.g. butyllithium, ethylmagnesium chloride (without a space between the organic prefix and the name of the metal)

8.6. Structures of Organometallic Compounds

electropositive metals
carbanion (carbon anion)
three-center, two-electron bond
Grignard reagent

8.7. Physical Properties of Organometallic Compounds (sensitive to water and air)

8.8 Preparation of Organometallic Compounds

A. Reaction of an Alkyl Halide with a Metal
   Grignard reaction
   formation of alkyllithium
B. Reaction of Organometallic Compounds with Salts
   electropositivity

8.9 Reactions of Organometallic Compounds

A. Hydrolysis: specific deuteration
B. Reaction with Halogens
C. Reaction with Oxygen

8.B. Important Reactions Introduced

Formation of organometallic compounds from alkyl halides (8.8.A)

Equation:
$R\text{-}X + 2\,M \longrightarrow R\text{-}M + MX$ (M = monovalent metal, such as Li or Na)
$R\text{-}X + Mg \longrightarrow R\text{-}Mg\text{-}X$ (formation of Grignard reagent)

Generality:
R = alkyl or aryl
M = highly electropositive metal (such as Li, Na, K, Mg)
X = Cl, Br, I (F not usually reactive enough)
an ether is the usual solvent

Key features: important way to prepare <u>anionic</u> carbon derivatives

Reaction of organometallic compounds with salts (8.8.B)

Equation: n RM + $M'X_n$ ⟶ R_nM' + n MX where M and M′ are both metals

Generality: equilibrium process
M′ must be less electropositive (*i.e.*, more electronegative) than M

Key features: important way to prepare organometallic compounds with metals that are not electropositive enough to react directly with R-X

Reactions of organometallic compounds with water (8.9.A)

Equation: R-M + H_2O ⟶ R-H + M-OH
Generality: M = Li, Na, K, Mg, other metals with electronegativities less than 1.7

Key features: Useful for preparing specifically deuterated or tritiated compounds

8.C. Important Concepts and Hints

In this chapter two important classes of organic compounds are introduced: alkyl halides and organometallic compounds. They are discussed primarily from the point of view of structure and physical properties, as well as the fact that alkyl halides are the most important precursors to organometallic compounds.

One of the important concepts to realize from this chapter is the difference in the electronic character between these two classes of compounds. In the alkyl halides, the halogen is **more electronegative** than carbon, and the bond between the two elements is polarized toward the halogen. That is, the electrons that occupy the bonding orbital between the two atoms are pulled toward the halogen atom, leaving the carbon atom with a partial positive charge. The situation is exactly reversed in the organometallic compounds: metals are **more electropositive** (= less electronegative) than carbon, with the result that there is partial negative charge on carbon.

As you might expect, and as you will see more dramatically in Chapter 9, where the major reactions of alkyl halides are presented, the fact that alkyl halides and organometallic compounds have opposite electronic characteristics means that they have the opposite behavior when it comes to the types of reactions that they undergo. Because the carbon atom in organometallic compounds is partially negative, it comes as no surprise that these compounds react with molecules that are positively charged. The simplest case is the reaction with a proton, as described in the section on hydrolysis of organometallic species. Other examples are the equilibrium reactions between organometallic compounds and metal salts. In contrast, alkyl halides react with negatively charged molecules, or those that can at least furnish a pair of electrons for the new bond, as described in Chapter 9.

8.D. Answers to Exercises

8.1 $H_3C^{\delta+}–X^{\delta-}$ d = C–X distance, q = δ± $\mu = q \cdot d$

For X = F : $$q = \frac{\mu}{d} = \frac{1.82 \times 10^{-18}\ esu{\cdot}cm}{1.39 \times 10^{-8}\ cm} = 1.31 \times 10^{-10}\ esu$$

One electronic charge is 4.8×10^{-10} esu, so 1.3×10^{-10} esu corresponds to 1.3/4.8 = 0.27 of an electronic charge.

X	μ (D= 10^{-18} esu·cm)	d (Å = 10^{-8} cm)	q (esu)	Fractional charge ($q/4.8 \times 10^{-10}$ esu)
F	1.82	1.39	1.31×10^{-10}	0.27
Cl	1.94	1.78	1.09×10^{-10}	0.23
Br	1.79	1.93	0.927×10^{-10}	0.19
I	1.64	2.14	0.766×10^{-10}	0.16

8.2 a. *t*-Butyl bromide has a higher melting point that *n*-butyl bromide. Little internal freedom of motion needs to be lost on incorporation of the *t*-butyl bromide molecule into the crystal, whereas the *n*-butyl bromide molecule loses its freedom to rotate about the C_1-C_2 and C_2-C_3 bonds. Therefore entropy makes it more difficult for *n*-butyl bromide to crystallize (see Section 5.3).

b. To estimate bp of Cl

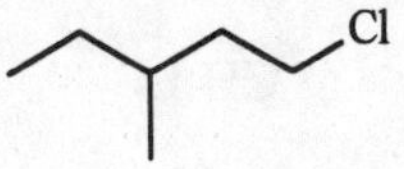

Compare  , bp 36° with 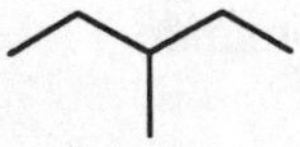 , bp 63°; branching increased bp 27°

From Cl bp 108°, estimate for Cl bp = 108° + 27° = 135°C.

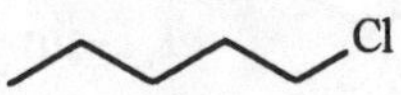

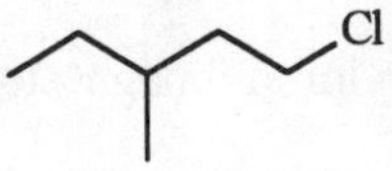

To estimate bp of Cl

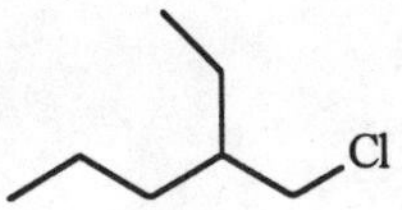

Compare  , bp 36° with Cl , bp 108°; Cl for H increased bp 72°

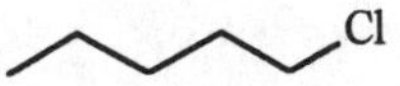

Compare  , bp 36° with 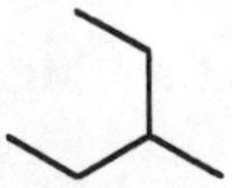 , bp 63°; branching increased bp 27°

From 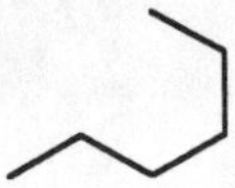 bp 69°, estimate for 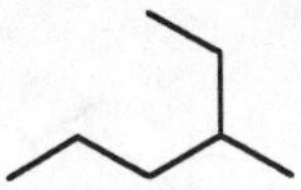 bp = 69° + 27° = 96°C.

Estimate for Cl bp = 96° + 72° = 168°C.

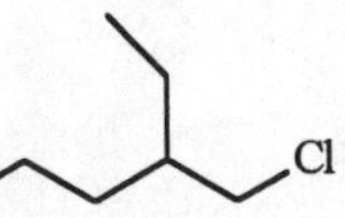

8.3 a. isopropylmagnesium chloride
b. tetramethylsilane
c. butyllithium
d. tetramethyllead
e. diethyldimethylstannane

8.4 Lewis structure | Dative bonds | Ionic Form

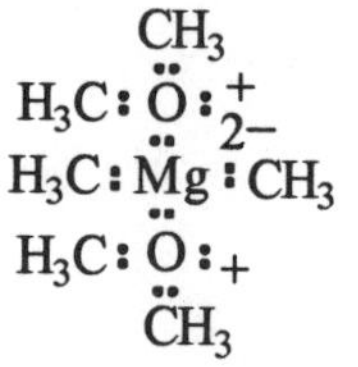

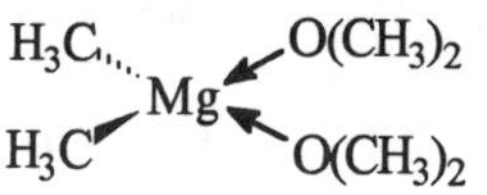

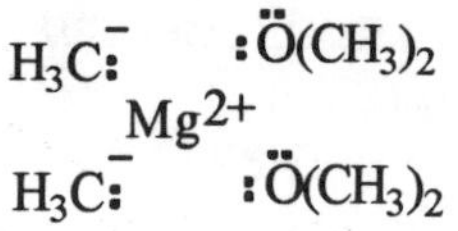

Lewis structure: $(CH_3)_2\ddot{O}^{+}$: $H_3C:\ddot{M}g^{2-}:CH_3$: $^{+}\ddot{O}(CH_3)_2$

Dative bonds: $(H_3C)_2Mg \leftarrow O(CH_3)_2$, $\leftarrow O(CH_3)_2$

Ionic Form: $H_3C:^-$ $H_3C:^-$ Mg^{2+} $:\ddot{O}(CH_3)_2$ $:\ddot{O}(CH_3)_2$

According to the Lewis structure, each oxygen has one formal positive charge (2 + 1/2 of 6 shared = 5, one less than in the element) and the magnesium is −2 (1/2 of 8 shared = 4, two more than in the element). The ionic form shows an electron distribution that is more consistent with the chemical behavior of the compound.

8.5 Lewis structure | Ionic form | Dative structure

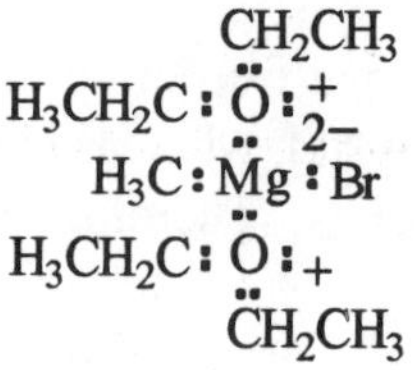

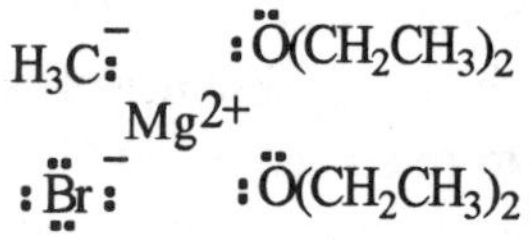

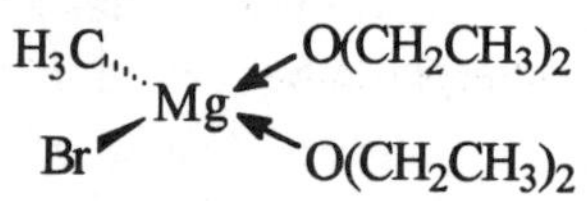

Lewis structure: $(CH_3CH_2)_2\ddot{O}^{+}$: $H_3C:\ddot{M}g^{2-}:Br$: $^{+}\ddot{O}(CH_2CH_3)_2$

Ionic form: $H_3C:^-$ $:\ddot{B}r:^-$ Mg^{2+} $:\ddot{O}(CH_2CH_3)_2$ $:\ddot{O}(CH_2CH_3)_2$

Dative structure: $H_3C(Br)Mg \leftarrow O(CH_2CH_3)_2$, $\leftarrow O(CH_2CH_3)_2$

8.6

CH_3MgI	methylmagnesium iodide	CH_3Li	methyllithium
cyclohexyl–MgBr	cyclohexylmagnesium bromide	cyclohexyl–Li	cyclohexyllithium
$CH_3CH_2C(CH_3)_2$–MgCl	1,1-dimethylpropylmagnesium chloride	$CH_3CH_2C(CH_3)_2$–Li	1,1-dimethylpropyllithium

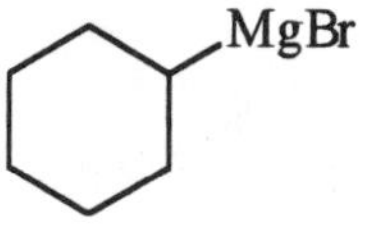

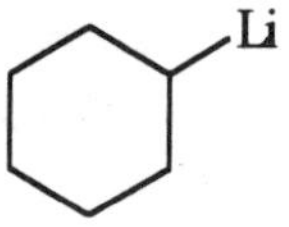

			Electronegativity
8.7 a. $2\ (CH_3)_3Al + 3\ ZnCl_2$	$\rightleftharpoons$	$3\ (CH_3)_2Zn + 2\ AlCl_3$	Al < Zn (1.5) (1.6)
b. $2\ (CH_3)_2Hg + SiCl_4$	$\leftrightharpoons$	$(CH_3)_4Si + 2\ HgCl_2$	Hg > Si (1.9) (1.7)
c. $(CH_3)_2Mg + CaBr_2$	$\leftrightharpoons$	$(CH_3)_2Ca + MgBr_2$	Mg > Ca (1.2) (1.0)

8.8 a. 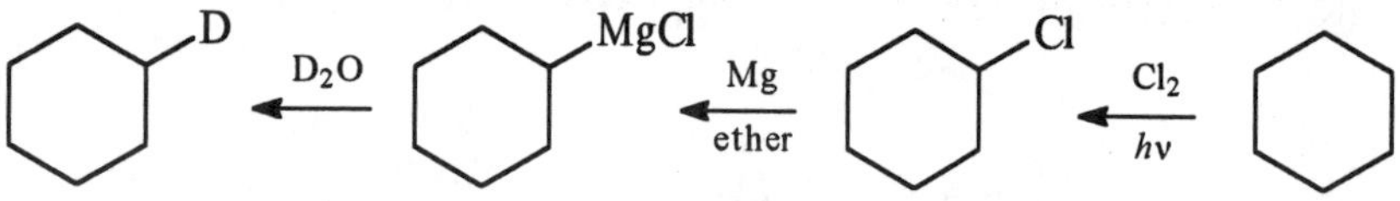 cyclohexyl–D $\xleftarrow{D_2O}$ cyclohexyl–MgCl $\xleftarrow[\text{ether}]{Mg}$ cyclohexyl–Cl $\xleftarrow[h\nu]{Cl_2}$ cyclohexane

b. $(CH_3)_3CD \xleftarrow{D_2O} (CH_3)_3CMgBr \xleftarrow[\text{ether}]{Mg} (CH_3)_3CBr \xleftarrow[h\nu]{Br_2} (CH_3)_3CH$

c. $(CH_3)_3CCH_2D \xleftarrow{D_2O} (CH_3)_3CCH_2Li \xleftarrow[\text{ether}]{Li} (CH_3)_3CCH_2Cl \xleftarrow[h\nu]{Cl_2} (CH_3)_4C$

8.E Answers and Explanations for Problems

1. a. 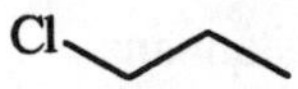
Cl
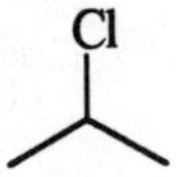
Cl

1-chloropropane 2-chloropropane

b. 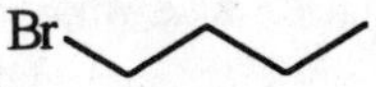
Br
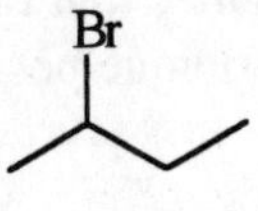
Br
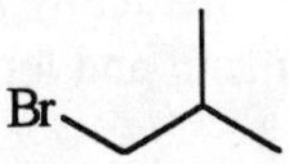
Br
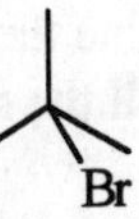
Br

1-bromobutane 2-bromobutane 1-bromo-2-methylpropane 2-bromo-2-methylpropane

c. 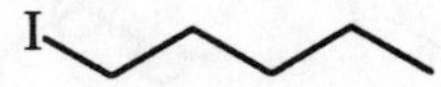
I
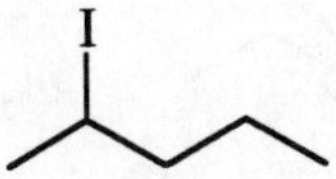
I
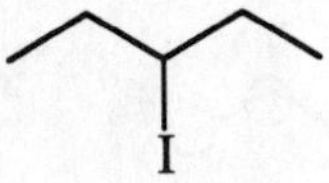
I
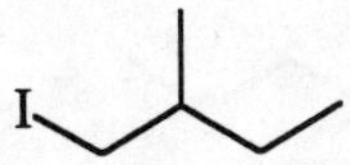
I

1-iodopentane 2-iodopentane 3-iodopentane 1-iodo-2-methylbutane

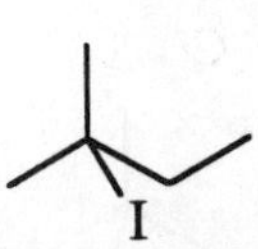
I
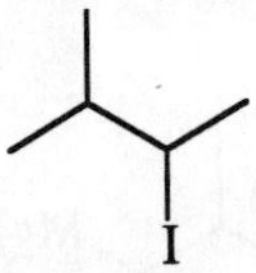
I
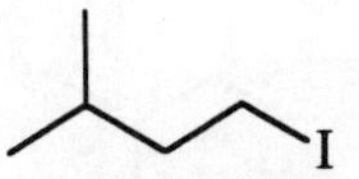
I
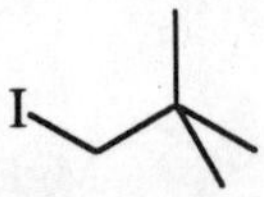
I

2-iodo-2-methylbutane 2-iodo-3-methylbutane 1-iodo-3-methylbutane 1-iodo-2,2-dimethylpropane

d. 1: 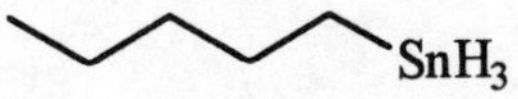
SnH3
2: 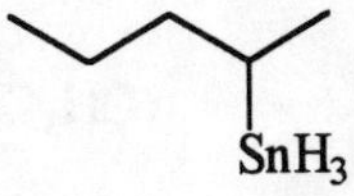
SnH3
3: 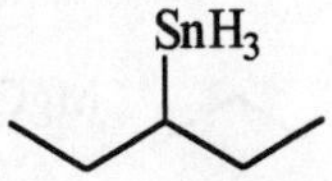
SnH3
4: 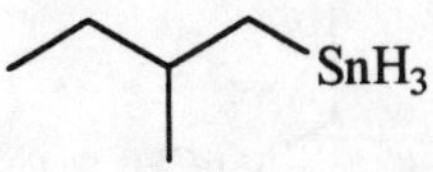
SnH3

pentylstannane 1-methylbutylstannane 1-ethylpropylstannane 2-methylbutylstannane

5: 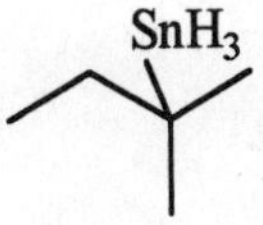
SnH3
6: 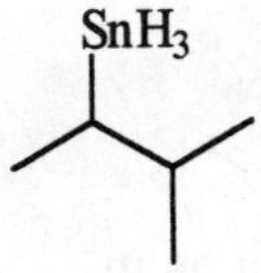
SnH3
7: 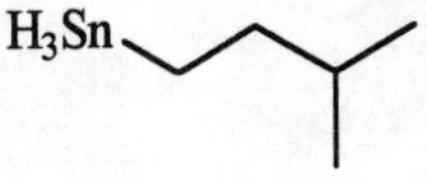
H3Sn
8: 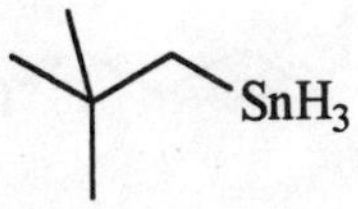
SnH3

1,1-dimethylpropyl-stannane 1,2-dimethylpropylstannane 3-methylbutylstannane 2,2-dimethylpropyl stannane

9: 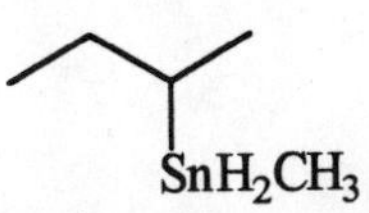
SnH2CH3
10: 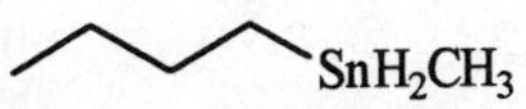
SnH2CH3
11: 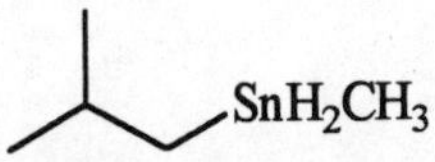
SnH2CH3
12: 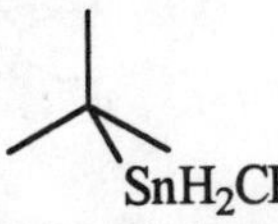
SnH2CH3

sec-butylmethylstannane butylmethylstannane isobutylmethylstannane *t*-butylmethylstannane

13: 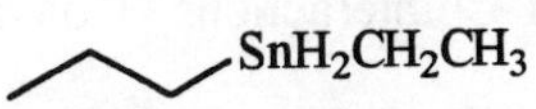
SnH2CH2CH3
14: 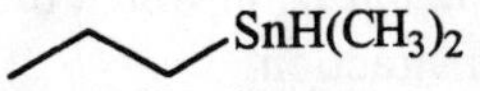
SnH(CH3)2
15: 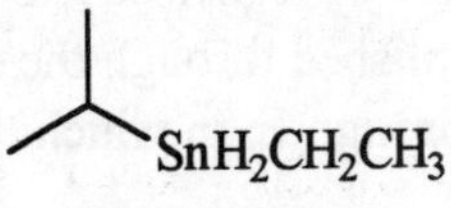
SnH2CH2CH3

ethylpropylstannane dimethylpropylstannane ethylisopropylstannane

16: $SnH(CH_3)_2$ dimethylisopropylstannane

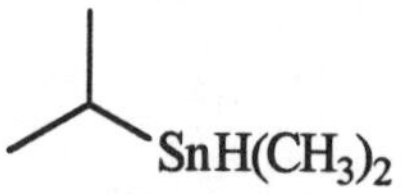

17: $(CH_3CH_2)_2SnHCH_3$ diethylmethylstannane

18: $CH_3CH_2Sn(CH_3)_3$ ethyltrimethylstannane

Another way to determine these isomers systematically is to recognize that tin has the same valence as carbon: write the skeletons of all the six-carbon isomers, and look for the unique positions where a C can be replaced with an Sn:

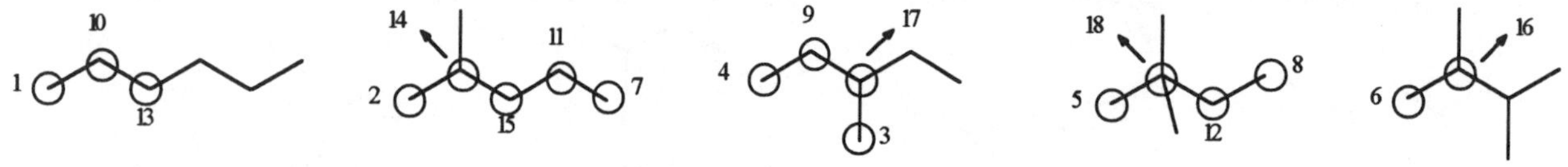

e.  MgH butylmagnesium hydride; MgH *sec*-butylmagnesium hydride; MgH isobutylmagnesium hydride; MgH *t*-butylmagnesium hydride

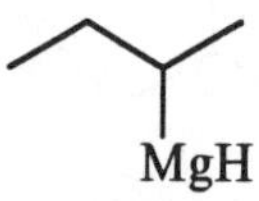

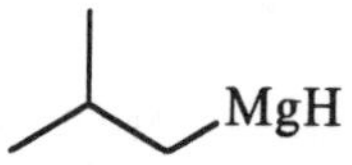

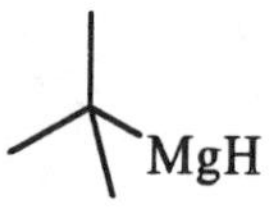

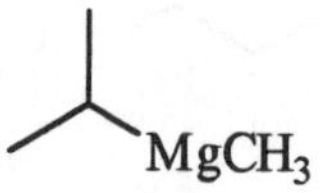

$MgCH_3$ isopropylmethylmagnesium; $MgCH_3$ methylpropylmagnesium; $(CH_3CH_2)_2Mg$ diethylmagnesium

f.  Li butyllithium; Li *sec*-butyllithium; Li isobutyllithium; Li *t*-butylithium

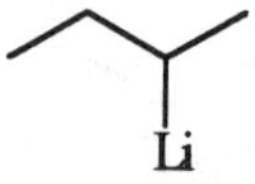

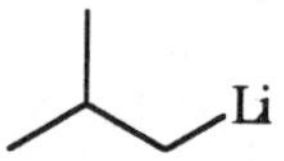

2. A: 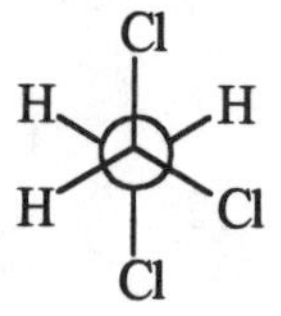 B: 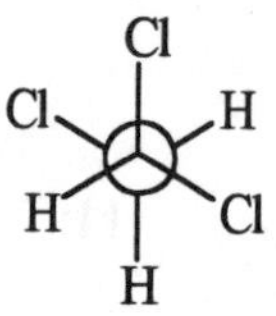 C:

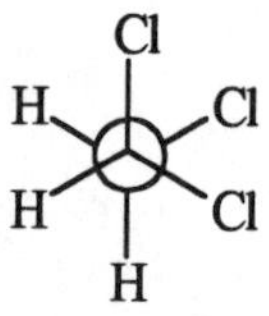

A and B are equivalent and equal in energy, and are more stable than C. A and B have one gauche and one anti interaction; C has two gauche interactions. Anti is more stable in the gas phase where dipole-dipole interactions are more important. Such interactions are less important in the liquid phase because they are masked by the dielectric effect of the liquid, and other interactions can dominate. Interconversion of A and B is accomplished through the eclipsed conformation D, in which there are three H-Cl interactions. Conversion to C goes through E, in which there is a Cl-Cl interaction.

D: 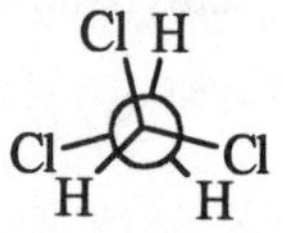 E:

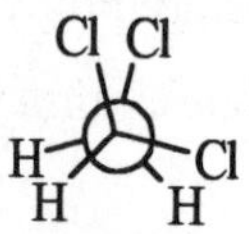

3. a. $(CH_3)_3CCD(CH_3)_2 \xleftarrow{D_2O} (CH_3)_3CCMgBr(CH_3)_2 \xleftarrow[ether]{Mg} (CH_3)_3CCBr(CH_3)_2 \xleftarrow[h\nu]{Br_2} (CH_3)_3CCH(CH_3)_2$

Bromination and not chlorination is used here since bromination is more selective and there are 15 1 ° H's as opposed to the 1 3 ° H where the reaction has to occur.

b. (skeletal structures) tetraalkyltin (R_4Sn) $\xleftarrow{SnCl_4}$ 4 alkyllithium (RLi) $\xleftarrow{8\ Li}$ 4 alkyl chloride (RCl)

c. $(CH_3)_3CCH_2Br \xleftarrow{Br_2} (CH_3)_3CCH_2MgCl \xleftarrow[ether]{Mg} (CH_3)_3CCH_2Cl$

d. $(CH_3CH_2)_2Cd \xleftarrow{CdCl_2} 2\ CH_3CH_2MgCl \xleftarrow[ether]{2\ Mg} 2\ CH_3CH_2Cl$

e. $[(CH_3)_3C]_2Hg \xleftarrow{HgBr_2} 2\ (CH_3)_3CMgCl \xleftarrow[ether]{2\ Mg} 2\ (CH_3)_3CCl$

4. This problem, like Exercise 8.6, can be answered by comparing the ionization potentials of the metals (Table 8.6). The more electropositive (= less electronegative) element prefers to exist as the cation in an inorganic salt rather than bound to carbon with some covalent bonding.

a. Al (1.5) < Cd (1.7) K > 1
b. Hg (1.9) > Zn (1.6) K < 1
c. Mg (1.2) < Si (1.7) K > 1
d. Li (1.0) < H (2.2) K » 1
e. Zn (1.6) > Li (1.0) K < 1

5. [Polymeric structure: –Be–Be–Be– chain bridged by $C(CH_3)_3$ groups; labels: CH_3, H_3C, C, CH_3, H_3C, CH_3, CH_3, Be, Be, Be, C, H_3C, CH_3, CH_3, H_3C, C, CH_3, CH_3]

$(CH_3)_3C{-}Be{-}C(CH_3)_3$ (180°)

a. For di-*t*-butylberyllium, the *t*-butyl group is too bulky to permit the polymeric structure; instead, it exists as a monomer with sp hybridization (see drawing).

b. The molecule is linear about Hg: $CH_3{-}Hg{-}CH_3$. The bonding is C_{sp^3}-Hg_{sp}. Mercury has two valence electrons; hence, bonds use sp hydrids, much as in beryllium.

6. $(CH_3)_3B + CH_3Li \rightarrow (CH_3)_4B^-Li^+$

The compound $(CH_3)_4B^-Li^+$ is tetrahedral about boron; that is, C_{sp^3}-B_{sp^3}; $(CH_3)_3B$ is trigonal planar, i.e., C_{sp^3}-B_{sp^2}.

7. D D D D D D D D

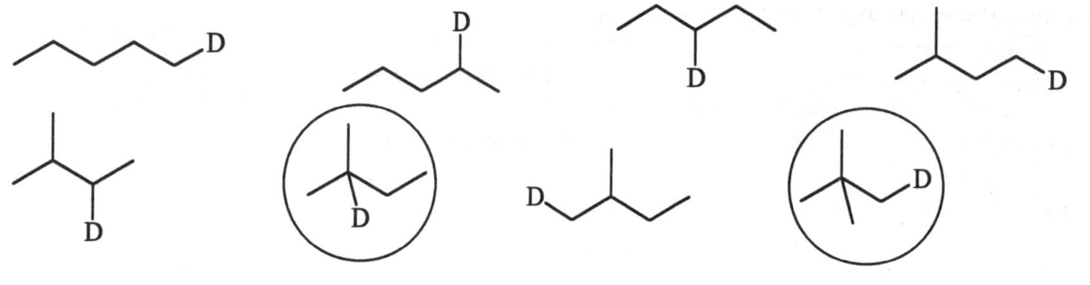

Only the circled isomers can reasonably be prepared by the route:

$$\underset{\text{(alkane)}}{RH} \xrightarrow[h\nu]{Br_2} RBr \xrightarrow[\text{ether}]{Mg} RMgBr \xrightarrow{D_2O} RD$$

The halides that would lead to the other isomers cannot be prepared *selectively* by free radical halogenation of the alkanes.

8. Dimerization and disproportionation are typical reactions which occur between two alkyl radicals (see Section 6.2 of the text). Free radical intermediates are involved in the formation of Grignard reagents (see Section 8.8.A.)

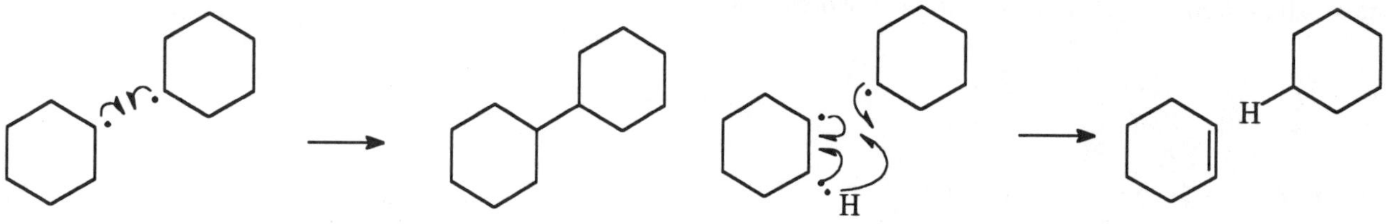

H H

8.F. Supplementary Problems

S1. Write Lewis structures for the trimethylaluminum dimer (Figure 8.3 in the Text) and trimethylantimony.

S2. The reaction between trimethylaluminum and trimethylantimony leads to the formation of a compound with the formula $C_6H_{18}AlSb$. Propose a structure for this material and write a Lewis structure to account for the bonding scheme.

S3. 1-Bromo-6-methoxyhexane can be converted to the Grignard reagent, whereas the corresponding alcohol cannot. Explain.

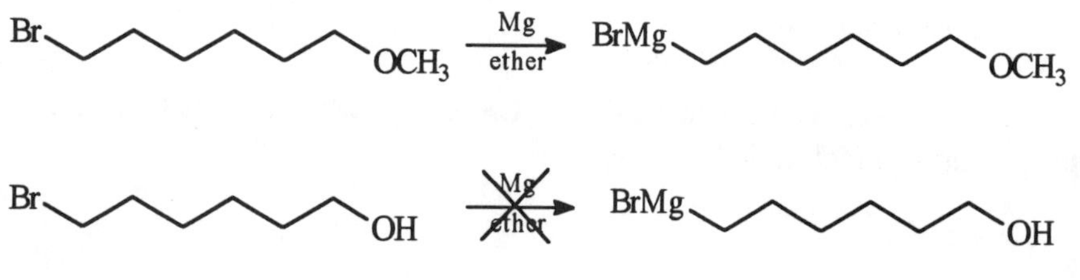

Br–(CH₂)₆–OCH₃ $\xrightarrow[\text{ether}]{Mg}$ BrMg–(CH₂)₆–OCH₃

Br–(CH₂)₆–OH ~~$\xrightarrow[\text{ether}]{Mg}$~~ BrMg–(CH₂)₆–OH

S4. Construct a diagram showing qualitatively the potential energy of 1,2-dichloroethane as a function of rotation about the C-C bond.

8.G. Answers to Supplementary Problems

S1. Trimethylaluminum dimer:

$$\begin{array}{ccc} H_3C\!: & \overset{H_3}{C} & :CH_3 \\ & Al :: Al & \\ H_3C\!: & \underset{H_3}{C} & :CH_3 \end{array}$$

Each electron pair in the center is shared by two aluminums and one carbon.

Trimethylantimony:

$$H_3C : \ddot{Sb} : CH_3$$
$$\underset{H_3}{\ddot{C}}$$

S2.

$$Me : \overset{Me}{\underset{Me}{Al}} : \overset{Me}{\underset{Me}{Sb}} : Me \quad (- \text{ on Al}, + \text{ on Sb}) \qquad Me = CH_3$$

S3. The alcohol function reacts with the Grignard reagent as soon as it is formed:

BrMg~~~~~~OH ⟶ ~~~~~~O^- $^{2+}$Mg Br^-

S4.

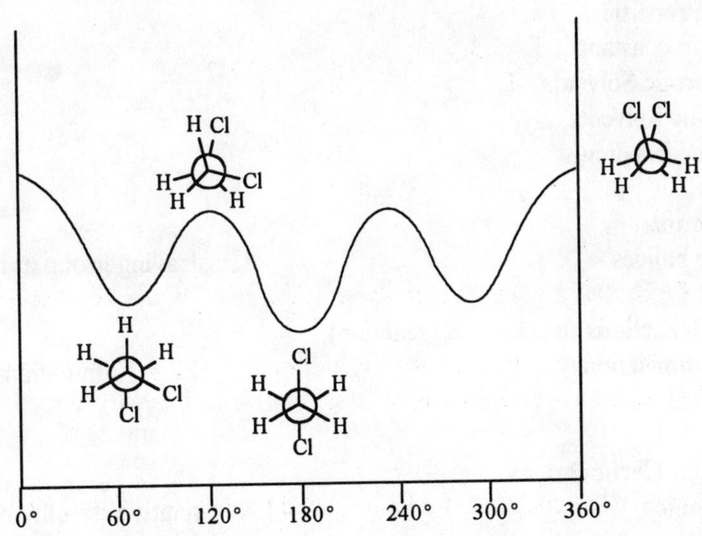

9. NUCLEOPHILIC SUBSTITUTION

9.A. Chapter Outline and Important Terms Introduced

9.1. **The Displacement Reaction** (the first "ionic" reaction discussed in detail)
nucleophile

9.2 **Mechanism of the Displacement Reaction** (how bonds are made and broken, and stereochemistry of reactions)
reaction kinetics
stereochemistry
inversion of configuration
S_N2 = <u>S</u>ubstitution <u>N</u>ucleophilic bimolecular(<u>**2**</u>)
second-order kinetics
bimolecular mechanism

9.3 **Effect of Alkyl Structure on Displacement Reactions**
steric hindrance
α-branching
β-branching
"neopentyl-type" systems

9.4 **Effect of Nucleophile Structure on Displacement Reactions**
nucleophilicity vs basicity
ambident nucleophiles
2nd and 3rd row nucleophiles better
solvent effects (*sec 9.5*)

9.5 **Nucleophilicity and Solvent Effects** (factors affecting the nucleophile)
A. Solvent Properties
dielectric constant
B. Polar Aprotic Solvents
C. Hydroxylic Solvents
hydrogen-bonding

9.6. **Leaving Groups**
not limited to halides
leaving group ability vs. basicity

9.7. **Elimination Reactions** (a competing reaction)
bimolecular elimination
steric hindrance
E2 = <u>**E**</u>limination-bimolecular (<u>**2**</u>)

9.8 **S_N1 Reactions: Carbocations**
solvolysis reaction
unimolecular,
first-order kinetics
carbocation
S_N1 = <u>S</u>ubstitution <u>N</u>ucleophilic unimolecular (<u>**1**</u>)
relative stability of carbocations:
tertiary > secondary > primary > methyl
hyperconjugation

9.9 **Ring Systems** (special aspects in cyclic molecules)
bond angle strain
intramolecular vs. intermolecular

9.B. Important Reactions Introduced

Displacement reactions (9.1, 9.2)

Equation: $Nu:^- + R\text{-}X \longrightarrow Nu\text{-}R + X^-$

Generality: X^- = good leaving group (= HX is strong acid)

for S_N2: R = unhindered alkyl (methyl > 1° > 2° >> 3°)
$Nu:^-$ = good nucleophile

for S_N1: R = substituted alkyl (3° > 2° >> 1° or methyl)
X = good leaving group

Key features: S_N2 proceeds with inversion of configuration and involves bimolecular process.
S_N1 proceeds through planar carbocation and involves unimolecular process.
E2 (elimination) reactions interfere if $Nu:^-$ is strong base and R is sterically hindered.
Influence of solvent (dielectric constant and hydrogen bonding) on reaction rate.

9.C. Important Concepts and Hints

This chapter is your introduction to a detailed analysis of an organic reaction that proceeds via an ionic (heterolytic) mechanism instead of by a free-radical (homolytic) mechanism. The displacement reaction is the primary reaction of interest for alkyl halides and a reaction of major importance in organic chemistry. It is analyzed in detail in Chapter 9 because it provides an opportunity to examine all of the factors that are generally important in understanding any reaction. In Section 6.C of this Study Guide, we suggested that you approach each new reaction with a series of questions to help you organize your thinking. As an example, the sort of questions you should ask and the sort of answers you should give yourself are illustrated below for displacement reactions.

(1) **Q:** What functional group transformation does the reaction accomplish?

ANS: In general terms, $Nu^- + R\text{-}X \longrightarrow R\text{-}Nu + X^-$
where Nu = nucleophile, R = alkyl group, and X = leaving group.

(2) **Q:** What is the mechanism of the reaction?

ANS: For some combination of reactants, the bimolecular S_N2 mechanism is involved, with simultaneous attack of Nu and departure of X from opposite sides of the carbon atom; for other combinations of reactants, the stepwise S_N1 mechanism is involved, with initial loss of X to give a carbocation intermediate (slow) and subsequent attack by Nu to give product (fast).

(3) **Q:** What is the generality of the reaction? That is, what characteristics must the reactants have?

ANS: **Nu**, the nucleophile, must have a lone pair of electrons (i.e., must be a Lewis base) in order to form a bond to the carbon. **Nucleophilicity** increases with basicity (to the left in the Periodic Table, other factors being equal) and with **polarizability** (down in the Periodic Table, other factors being equal), and helps to determine whether the reaction will proceed via the S_N2 or S_N1 mechanism. Stronger nucleophiles favor S_N2 reactions (other factors being equal).

R, the alkyl group, plays the major role in determining the mechanism (S_N1 or S_N2) and rate of the reaction. Displacement reactions at tertiary carbon occur by the S_N1 mechanism and at primary carbon by the S_N2 mechanism. Both situations reflect the combined influences of steric hindrance and carbocation stability. Substitution at secondary carbon is the gray area, and whether these reactions proceed by the S_N1 or S_N2 mechanism, or both simultaneously, depends on other factors. Steric effects are not limited to substitution on the carbon undergoing reaction (**α-branching**), but are seen at more remote positions as well (**β-branching**, effects of cyclic systems).

X, the leaving group, must be stable when it departs with two electrons (usually as an anion). This is most easily evaluated by considering the pK_a of HX: a low pK_a indicates that HX is a strong acid, which means that X^- is a weak base, which in turn says that X^- is stable and a good leaving group. Under most circumstances, the pK_a of HX should be less than 2 or 3 for the reaction to occur at a reasonable rate. ($R'O^-$, where R′=alkyl or H, is essentially *never* a leaving group in an S_N1 or S_N2 reaction.) The "better" the leaving group is (i.e., the more stable X^- is), the faster the displacement reaction occurs, but the influence is greater on S_N1 reactivity than on S_N2 reactivity.

Solvent plays a role in the way it stabilizes the reactants in comparison to intermediates and transition states. For instance, if a polar, hydroxylic solvent (such as methanol) can form hydrogen bonds to the nucleophile (such as chloride ion) more strongly than it can to the transition state, the S_N2 reaction will be slowed; if a neutral alkyl halide (such as *t*-butyl bromide) must ionize in order to react (by the S_N1 mechanism), then a polar solvent will speed up the reaction.

(4) **Q:** What are the stereochemical features to keep in mind?

ANS: **S_N2** Backside attack and *inversion* of configuration at the carbon undergoing substitution.
S_N1 Planar, carbocation intermediate which the nucleophile can approach from either side. Leads to *racemic products* if that carbon is the only stereocenter in the molecule.

(5) **Q:** What are the limitations, that is, what possible side reactions should be kept in mind?

ANS: Elimination reactions (E2) compete with substitution when the nucleophile is fairly basic (RO^- [R = alkyl or H], CN^-, RS^-, NH_2^-, etc.) and S_N2 displacement is slowed because of steric hindrance. Elimination competes when the nucleophile is impatient for reaction (basic; i.e., unstable as its anion Nu^-) and doesn't want to wait around for S_N1-type ionization to occur. This is always an important point to keep in mind: very basic nucleophiles give mostly elimination products, except with unhindered primary alkyl halides.

This Question-and-Answer outline for displacement reactions is quite lengthy and detailed for two reasons: first, because it covers a complex topic, and second, because we want to provide you with a comprehensive example. A good exercise for you would be to make a similar outline for free-radical halogenation (Chapter 6). A more skeletal outline of the questions and answers above, using key words to trigger your memory, would be:

(1) Functional Group Transformation

$$Nu^- + R\text{-}X \rightarrow R\text{-}Nu + X^-$$

(2) Mechanism

S_N2: $Nu^- + \;\; C{-}X \longrightarrow \left[\overset{-\delta}{Nu}\text{-----}C\text{-----}\overset{-\delta}{X} \right]^- \longrightarrow Nu{-}C \;\; + X^-$

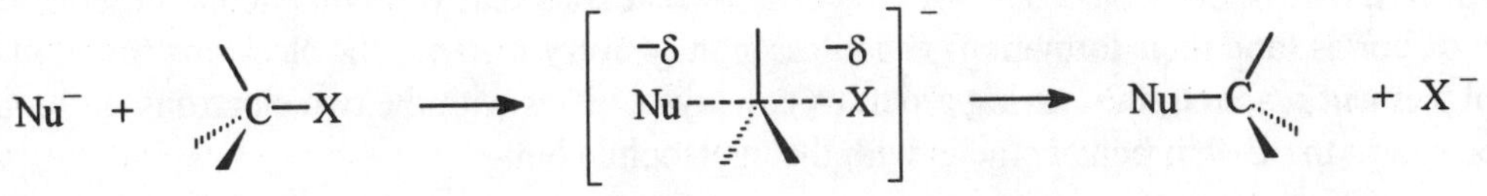

S_N1: $R{-}X \longrightarrow R^+ + X^- \xrightarrow{Nu^-} R{-}Nu$

(3) Generalizations

Nu: Lewis base; more basic, more polarizable → better nucleophile
R: S_N2 methyl > 1° > 2° > 3°; β-branching slows reaction, too
S_N1 3° > 2° > 1° >> methyl; decision hard only for 2°
X: X^- less basic ⇒ better leaving group
Solvent: dipolar solvents speed reactions
H-bonding slows S_N2 by tying up Nu^-

(4) Stereochemistry

S_N2: inversion
S_N1: loss of configuration via planar carbocation

(5) Side Reactions

Elimination (E2) if Nu is basic and/or R−X is sterically hindered.

HINTS: Organic chemists love to show what is going where in a reaction mechanism by drawing arrows, and writing these mechanisms often involves a lot of "arrow-pushing". A brief catalog of arrows that you will encounter in organic chemistry follows, as well as a description of those used when writing mechanisms.

A single-headed arrow written in an equation (for example, A + B → C) is used by organic chemists instead of the = sign common to general chemistry equations (for example, $P_4 + 3\ OH^- + 3\ H_2O = PH_3 + 3\ H_2PO_2^-$). Reversible reactions are written with two single-headed arrows pointing in opposite direction. Often, an idea of the position of equilibrium is given by the relative length of the arrows:

$$HCN + OH^- \rightleftharpoons CN^- + H_2O$$

These double arrows often connect different conformations of the same molecule, for example:

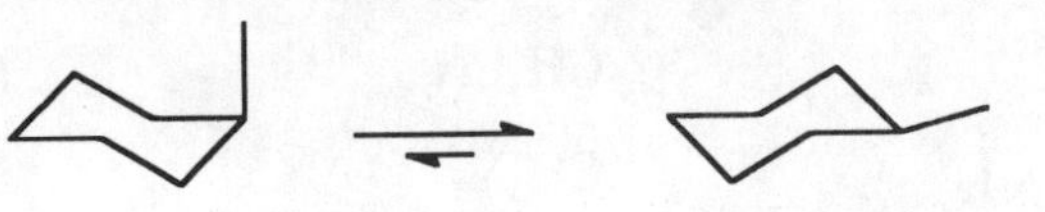

[cyclohexane chair conformations ⇌]

A double-headed arrow, ↔, is distinct from the equilibration idea of ⇌, and is specifically used between resonance structures, for example:

$[\,^-{:}\ddot{\underset{..}{O}}{-}C{\equiv}N{:} \longleftrightarrow {:}\ddot{O}{=}C{=}\ddot{N}{:}^-\,]$ or [carbonate resonance structures: $O{=}C(O^-)_2 \longleftrightarrow \longleftrightarrow$]

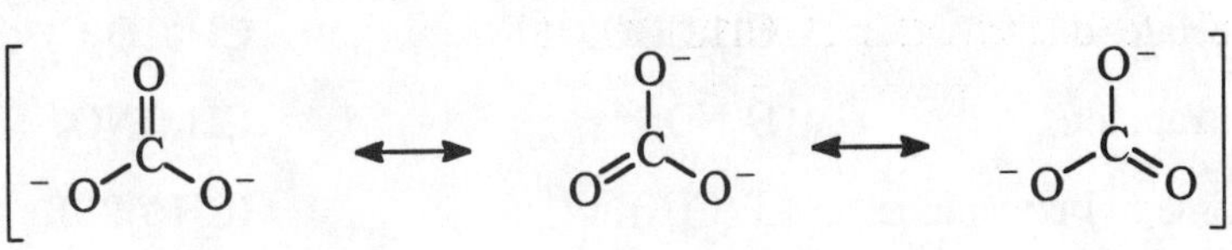

The curved arrows that you see leading molecules around in describing reaction mechanisms represent the movement of a pair of electrons. The vast majority of reactions that you will encounter involve hetereolytic cleavage of bonds (and their formation) and in essentially every instance the electrons travel in pairs. For instance, in a displacement reaction, the leaving group (X) always departs with the two electrons in the C-X bond, and the two electrons in the C-Nu bond come in with the nucleophile Nu:

$$HO:^- + H_3C{-}Cl: \longrightarrow HO{-}CH_3 \quad :Cl:^-$$

$$HO^- + CH_3{-}Cl \longrightarrow HO{-}CH_3 + Cl^-$$

The situation is more complicated for an E2 elimination reaction, because more bonds are being formed and broken, but the arrows help to keep everything organized.

$$CH_3O^- + H_3C{-}C(CH_3)_2Br \longrightarrow CH_3OH \quad H_2C{=}C(CH_3)_2 \quad Br^-$$

Pushing arrows is valuable because it helps you to keep track of electrons and charges and often prevents you from writing absurd mechanisms. Always bear in mind that an arrow represents the movement of a pair of electrons.

Sometimes, when describing free-radical reactions in which odd electron species are reacting, it is useful to show the movement of a single electron. Organic chemists often use "fish hook": or single-headed arrows to represent this:

$$Cl\cdot \;\; H{-}CH_3 \longrightarrow Cl{-}H + \cdot CH_3$$

9.D. Answers to Exercises

9.1 Displacment reactions with methyl iodide: CH_3I

Attacking Reagent		*Product*	
Name	Structure	Structure	Name
hydroxide ion	$HO:^-$	CH_3OH	methanol
ethoxide ion	$CH_3CH_2O:^-$	$CH_3CH_2OCH_3$	ethyl methyl ether
hydrosulfide ion	$HS:^-$	CH_3SH	methanethiol
thiocyanate ion	$^-:N{=}C{=}S \leftrightarrow N{\equiv}C{-}S:^-$	CH_3SCN	methyl thiocyanate
cyanide ion	$N{\equiv}C:^-$	CH_3CN	(methyl cyanide) acetonitrile
azide ion	$[:N{=}N{=}N:]^-$	CH_3N_3	methyl azide
ammonia	$:NH_3$	$CH_3N^+H_3\ I^-$	methylammonium iodide
water	$H_2O:$	$CH_3O^+H_2\ I^-$	methyloxonium iodide
acetate ion	$CH_3COO:^-$	$CH_3CO_2CH_3$	methyl acetate
nitrate ion	$O_2NO:^-$	CH_3ONO_2	methyl nitrate
trimethyl phosphine	$:P(CH_3)_3$	$(CH_3)_4P^+\ I^-$	tetramethylphosphonium iodide
triethylamine	Et_3N: (Et = CH_2CH_3)	$CH_3N^+Et_3\ I^-$	triethylmethylammonium iodide
diethyl sulfide	$Et_2S:$	$CH_3SEt_2\ I^-$	diethylmethylsulfonium iodide

9.2, 9.3

a. 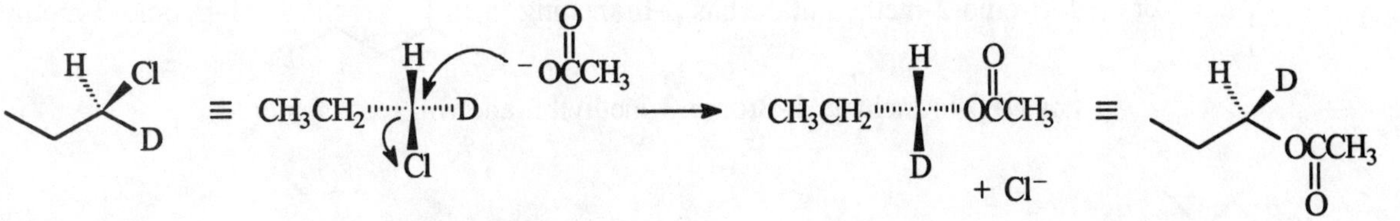 $\equiv CH_3CH_2$–C(H)(D)(Cl) + $^-OCCH_3$ (O) ⟶ CH_3CH_2–C(H)(D)–$OCCH_3$ (O) $\equiv$ + Cl^-

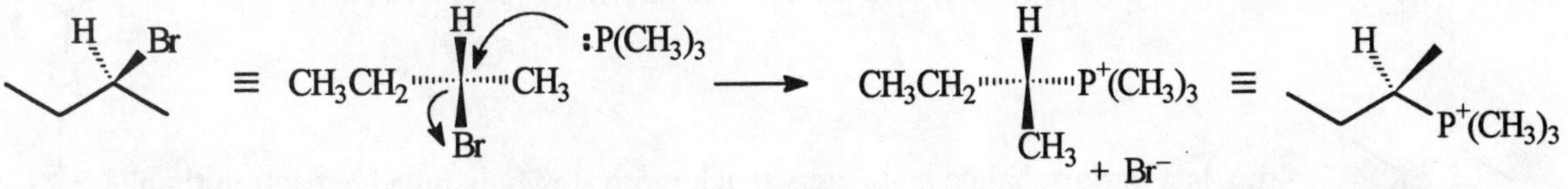

b. $\equiv CH_3CH_2$–C(H)(CH_3)(Br) + $:P(CH_3)_3$ ⟶ CH_3CH_2–C(H)(CH_3)–$P^+(CH_3)_3$ $\equiv$ + Br^-

c. H_3C, H, Br, H + $^-:C{\equiv}N$ ⟶ H_3C, H, H, CN Br^-

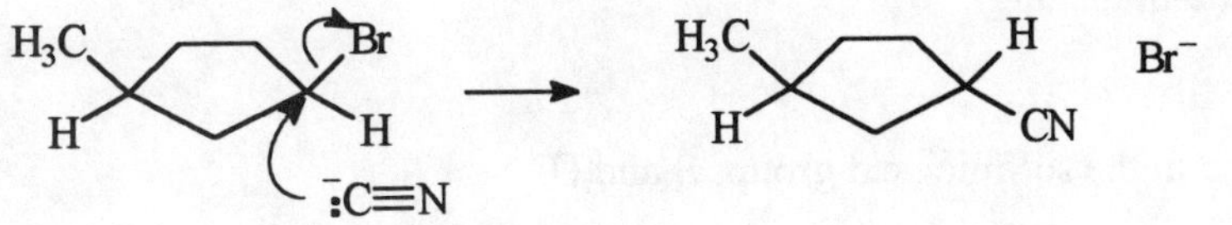

9.4

B ⇌ A; C ⇌ D; A $\overset{I^-}{\rightleftharpoons}$ D

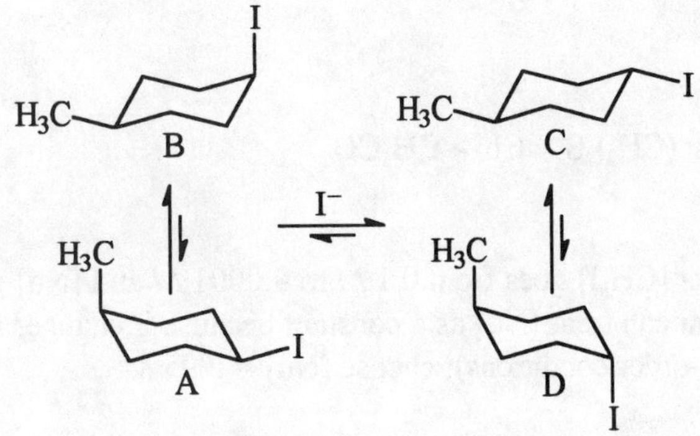

ΔG°_f difference for isomers: A: axial methyl: +1.7 kcal mole[-1]
B: axial iodine: +0.45 kcal mole[-1]
C: no axial substituents
D: axial methyl and axial iodine: +2.15 kcal mole[-1]

A : B : C : D = 0.06 : 0.47 : 1 : 0.03

cis	*trans*
34%	66%

9.5

a. 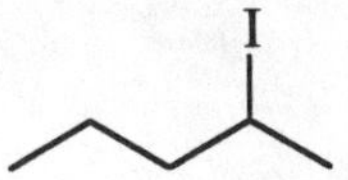 2-Iodopentane is a 2° alkyl halide with **one** β substituent (the C-4 and C-5 part of the chain).

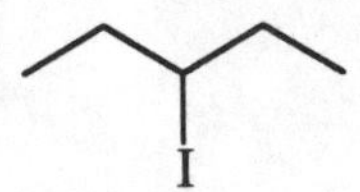

3-Iodopentane is a 2° alkyl halide with **two** β substituents (C-1 and C-5). 2-Iodopentane will react faster.

b. 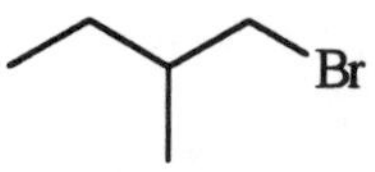 1-Bromo-2-methylbutane has β-branching. 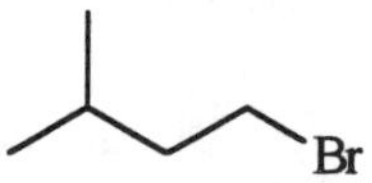 1-Bromo-3-methylbutane has no β-branching. 1-Bromo-3-methylbutane will react faster.

c. 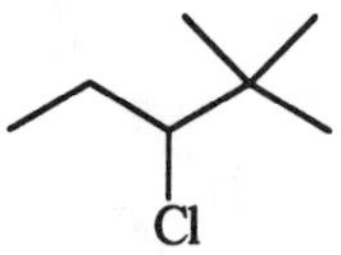 this is a "neopentyl-like" compound and will react more slowly than

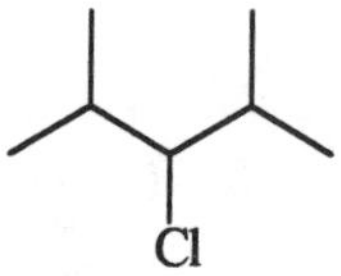

d. 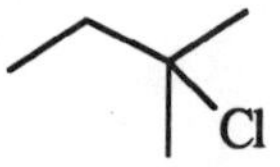 this is a tertiary halide and reacts much more slowly in an S_N2 reaction than a secondary halide (α-branching).

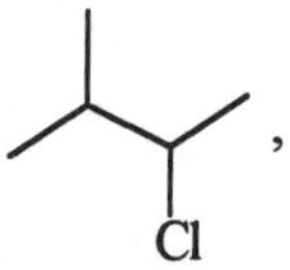

9.6 There are two nucleophilic atoms in the sulfinic acid group, S and O:

$$CH_3-S(=O)_2-CH_3 \qquad CH_3-S(=O)-OCH_3$$

9.7 $I^- > {}^-CN > {}^-SCN > NO_2^- > N_3^- = Br^- > (CH_3)_2S > Cl^- > CH_3CO_2^-$

9.8 Rate = $k_2[CH_3I]$ [Nu]; 99.9% reaction means $[CH_3I]$ goes from 0.1 *M* to 0.0001 *M* and [Nu] goes from 1.0 *M* to 0.9 *M*. To simplify the calculation, you can treat [Nu] as a constant because it changes by only 10% over the course of the reaction (pseudo-first-order conditions); choose [Nu] = 0.95 *M*.

$$-\frac{d[CH_3I]}{dt} = k_2[CH_3I][Nu] \qquad -\int_{0.10}^{0.0001} \frac{d[CH_3I]}{[CH_3I]} = k_2[Nu]\int_0^t dt \qquad -\ln\ [CH_3I]\Big]_{0.1}^{0.0001} = k_2\ [Nu]t\ \Big]_0^t$$

$$-(-2.303 - (-9.21) = 10^{-2} \times 0.95(0 - t)$$

$$t = 7.3 \times 10^2 \text{ sec} = 12 \text{ min}$$

9.9 $CH_3-S(=O)_2-O-CH_3 \quad :N(CH_3)_3 \longrightarrow CH_3-S(=O)_2-O^- + (CH_3)_4N^+$

$(H_3C)_2S^+-CH_3 \quad :N(CH_3)_3 \longrightarrow (CH_3)_2S + (CH_3)_4N^+$

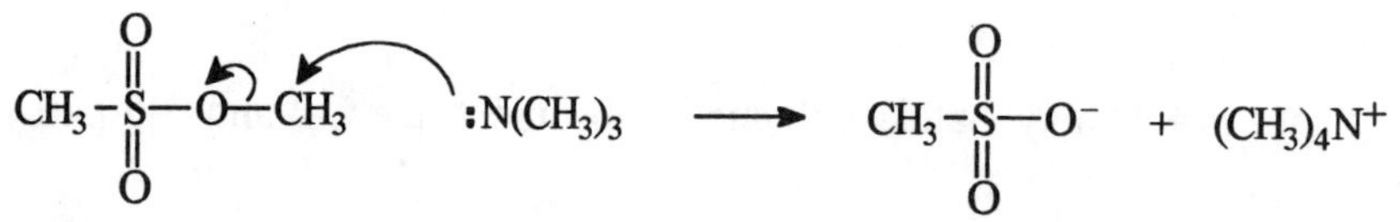

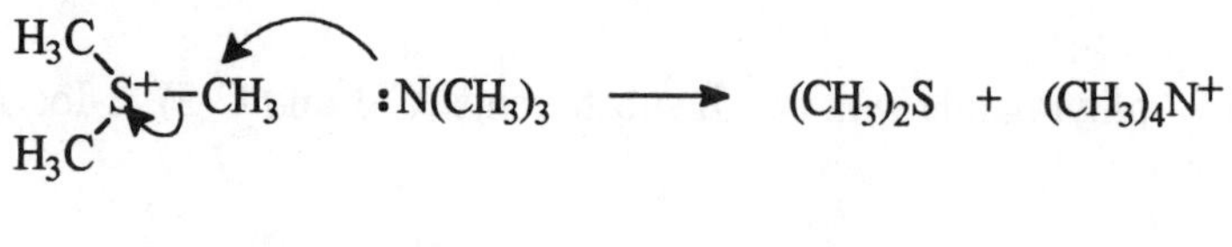

9.10

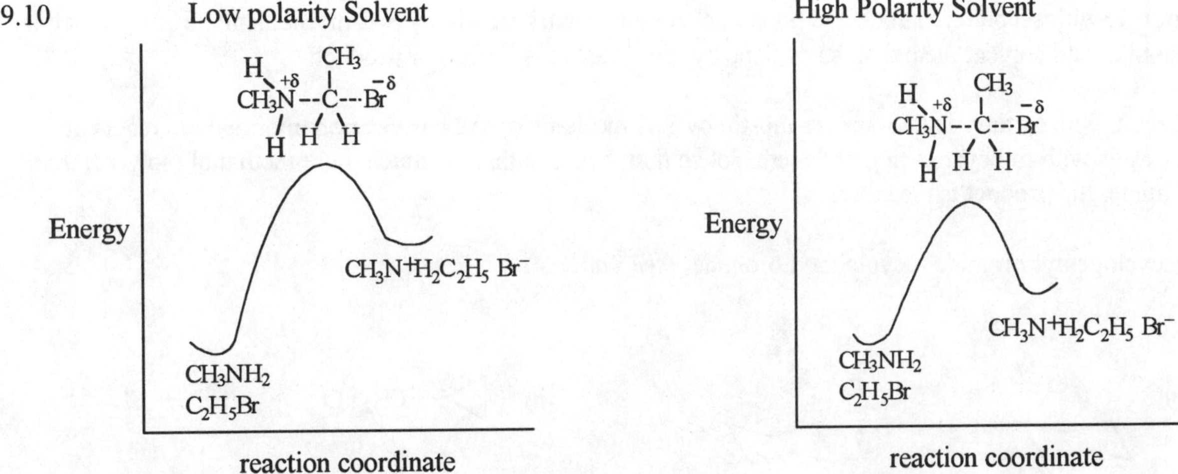

Since the transition state involves substantial charge formation compared with the neutral reactants, the solvent of higher polarity will stabilize the transition state and products to a greater degree than will a low polarity solvent. This lowering of the activation energy will enhance the rate for the reaction in a high polarity solvent.

9.11 a. 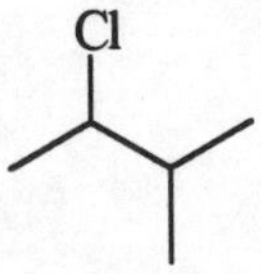 Cl — β-branching slows S_N2 reaction, therefore E2 reaction occurs more frequently.

b. $CH_3CH_2O^-$ — It's a stronger base than $CH_3CO_2^-$.

c. 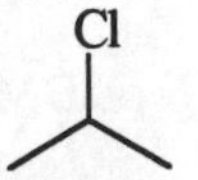 Cl — Primary halides react more readily via the S_N2 mechanism; secondary halides are slower S_N2 and faster E2 than primary.

d. 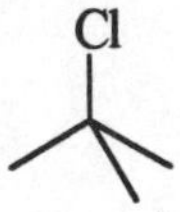 Cl — Tertiary halides are much slower in S_N2 reactions than secondary and hence faster E2.

9.12 1. Cl, H, $^-$:SCN → H, SCN, Cl^-

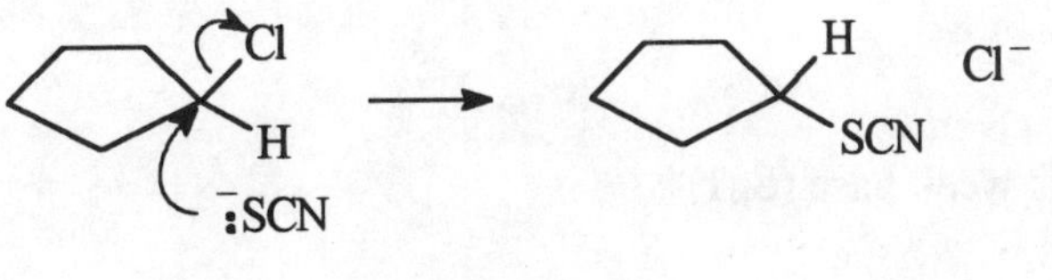

2. Cl, CH_3 → $^+$ $-CH_3$, Cl^-, H_3C-O-H → CH_3, $\overset{+}{O}CH_3$, H → CH_3, OCH_3, H^+

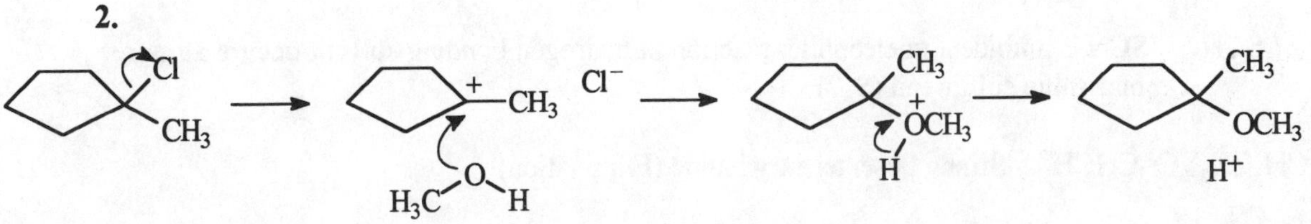

Reaction **1**, with secondary halide and good nucleophile, occurs mostly by S_N2 mechanism: ^-SCN is a better nucleophile than solvent methanol, so cyclopentyl thiocyanate is the major product.

Reaction **2**, with tertiary halide, occurs mostly by S_N1 mechanism: the carbocation intermediate reacts at similar rates with thiocyanate ion and methanol so that, because there is much more methanol (solvent) than thiocyanate, the product is the ether.

9.13 a. cyclopentyl bromide > cyclobutyl bromide *(see Table 9.9)*

b.

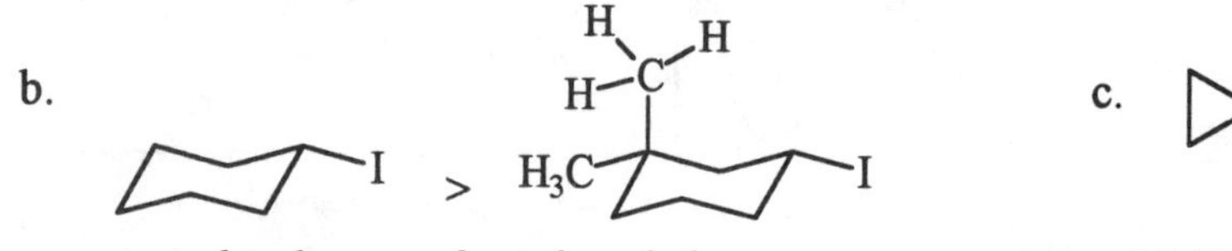

steric hindrance of axial methyl *1° vs 2° RX (also see Figure 9.12 and Table 9.9)*

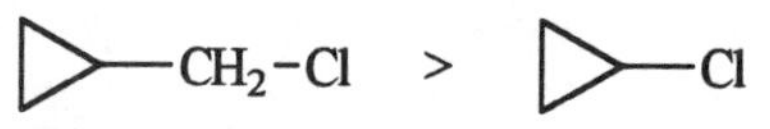

9.E. Answers and Explanations for Problems

1. a. $CH_3CH_2CH_2Br + Na^+SH^- \longrightarrow CH_3CH_2CH_2SH + NaBr$

   b. $(CH_3)_2CHCH_2CH_2Br + Na^+CN^- \longrightarrow (CH_3)_2CHCH_2CH_2CN + NaBr$

   c. $CH_3CH_2CH_2Br + Na^+OCH_3^- \longrightarrow CH_3CH_2CH_2OCH_3 + NaBr$

   d. $CH_3CH_2CH_2I + Na^+OH^- \longrightarrow CH_3CH_2CH_2OH + NaI$

   e. $CH_3I + Na^+NO_3^- \longrightarrow CH_3ONO_2 + NaI$

   f. $CH_3CH_2CH_2CH_2Br + Na^+N_3^- \longrightarrow CH_3CH_2CH_2CH_2N_3 + NaBr$

   All of these reactions are S_N2 and should be run in an aprotic, polar solvent such as DMF or DMSO where practical. Reactions with $Na^+CH_3O^-$ are usually run in CH_3OH. For cost or solubility reasons, a suitable alcohol is frequently used as solvent.

2. a.  (iodocyclopentane) secondary halide, good nucleophile, weak base, aprotic polar solvent, S_N2.

   b. 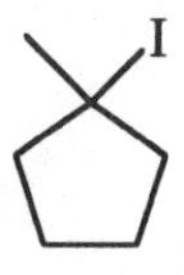 (1-iodo-1-methylcyclopentane) tertiary halide, nucleophile is weak base (S_N1).

   c. CH_3SCN ^-SCN is ambident nucleophile: reaction in hydrogen bonding solvent occurs at more polarizable sulfur end (S_N2)

   d. $(CH_3CH_2)_2C{=}CHCH_3$ Strong base, tertiary halide (Elimination)

   e. 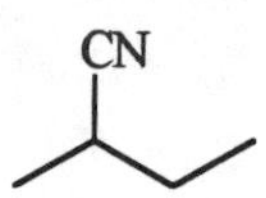 CN secondary halide, good nucleophile, aprotic polar solvent, (S_N2)

f. (N,N-dimethylpyrrolidinium) Br^- Intramolecular S_N2 reaction, primary alkyl halide, aprotic polar solvent.

g. $(CH_3)_3CCl + 2\ Na \xrightarrow{\text{ether}} (CH_3)_3CNa \xrightarrow{(CH_3)_3CCl} (CH_3)_3CH + CH_2{=}C(CH_3)_2 + NaCl$

The *t*-butylsodium formed in the organometallic reaction is such a strong base that it reacts with the halide starting material as soon as it is formed, by an elimination reaction.

h. $(CH_3)_2CHCH_2CH_2O_2CCH_3$ primary halide (S_N2)

i. $CH_3CH_2CH_2O{-}N{=}O$ NO_2^- is ambident nucleophile: reaction in dipolar, aprotic solvent occurs at atom bearing most of negative charge (O instead of N)

j. $(CH_3)_2CHCH_2N_3$ primary halide, excellent nucleophile (S_N2)

3. Because the C–O bond is not affected in the first step, the intermediate mesylate (*s*-butyl-$^{18}OSO_2CH_3$) still has the R configuration. The departure of the ^{18}O with the mesylate group shows that the substitution proceeds with cleavage of the C–O bond and by either an S_N2 or S_N1 mechanism (as opposed to cleavage of the S–O bond). Because OH^- is a strong nucleophile and the alkyl group is not tertiary, the reaction is expected to go by an S_N2 mechanism and inversion of configuration. Therefore the product is (S)-2-butanol.

<u>Note</u>: there is probably a lot of elimination taking place as a side reaction in this example, but the question only concerns the stereochemistry of the 2-butanol that <u>is</u> formed.

4. $(CH_3)_2CXCH_2CH_3 \longrightarrow X^- + (CH_3)_2C^+CH_2CH_3$

$(CH_3)_2C^+CH_2CH_3 \xrightarrow{-H^+} CH_2{=}C(CH_3)CH_2CH_3$

$(CH_3)_2C^+CH_2CH_3 \xrightarrow{-H^+} (CH_3)_2C{=}CHCH_3$

$(CH_3)_2C^+CH_2CH_3 \xrightarrow{CH_3OH} (CH_3)_2C(HO^+CH_3)CH_2CH_3 \xrightarrow{-H^+} (CH_3)_2C(OCH_3)CH_2CH_3$

The k_{-x} step is rate-determining and is different in rate for X = Cl, Br, I. The same carbocation is produced from each halide, and gives the same mixture of products.

5. a. A tertiary halide is sterically hindered and leads to a relatively stable carbocation; water is a weak nucleophile. Both imply an S_N1 substitution mechanism, by way of a planar, <u>achiral</u> carbocation:

(Structure: CH_3, CH_2CH_3, $CH_3CH_2CH_2$ and Br on a stereocenter $\longrightarrow$ planar carbocation CH_3, CH_2CH_3, $CH_3CH_2CH_2$ on C^+, attacked by H_2O from both lobes)

reaction is equally likely on both sides of the molecule, resulting in racemic product.

b. In this case, the chiral center is not involved in the reaction and is therefore unchanged:

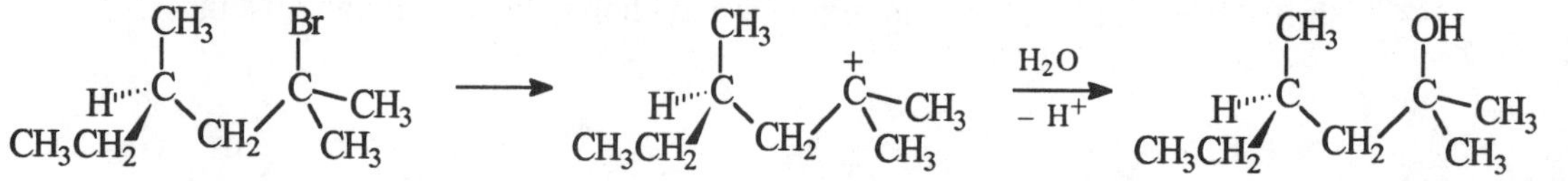

$$\underset{\text{(H, } CH_3\text{, } CH_3CH_2\text{)}}{C}\text{–}CH_2\text{–}C(Br)(CH_3)CH_3 \longrightarrow \underset{\text{(H, } CH_3\text{, } CH_3CH_2\text{)}}{C}\text{–}CH_2\text{–}\overset{+}{C}(CH_3)CH_3 \xrightarrow[-\,H^+]{H_2O} \underset{\text{(H, } CH_3\text{, } CH_3CH_2\text{)}}{C}\text{–}CH_2\text{–}C(OH)(CH_3)CH_3$$

6. a. 2nd In general, the more stable the leaving group is as the free Lewis base, the faster the S_N2 reaction. This is conveniently estimated by considering how basic the leaving group is (or how acidic the protonated compound - the conjugate acid - would be). The leaving group acquires additional electron density in the S_N2 transition state, and its basicity is a good indication of how easily it can accommodate (stabilize) this increased negative charge. In this case, I^- is a weaker base than Cl^- (HI is a stronger acid than HCl; see Appendix IV), and alkyl iodides are generally more reactive in displacement reactions than alkyl chlorides.

b. 1st Water is a weak nucleophile, therefore the S_N1 mechanism is the most probable; *t*-butyl bromide reacts faster by an S_N1 mechanism than isopropyl bromide does because the tertiary carbocation is more easily formed than a secondary one.

c. 2nd The methyl branch slows down the S_N2 reaction by steric hindrance; the straight-chain halide reacts faster.

d. 2nd ^-CN is stabilized by hydrogen-bonding solvation to methanol and its reactivity is reduced; ^-CN reacts faster in dimethylformamide because this polar aprotic solvent does not form hydrogen bonds to anions.

e. 2nd Reaction with $:NH_3$ is an S_N2 reaction which is faster with the unbranched, less sterically hindered, primary halide.

f. 1st Hydroxide is more basic than acetate ion, hence it is a better nucleophile.

g. 2nd Phosphorus is more nucleophilic than nitrogen. In general, 3rd-period atoms are more nucleophilic than their 2nd-period counterparts; hence, trimethylphosphine is more reactive toward methyl bromide than trimethylamine.

h. 1st The reaction giving CH_3CH_2SCN is faster. SCN^- is an ambident anion and may react on sulfur or nitrogen. Sulfur is the more nucleophilic end, and the faster reaction occurs there (see answer to Problem #1c.).

i. 2nd Even though the anions are about equal in basicity, sulfur is more nucleophilic because of its greater polarizability.

j. 2nd In a displacement reaction, the carbon atom undergoing substitution changes hybridization from sp^3 (bond angles 109°) to sp^2 (bond angles 120°) when proceeding to the transition state. Ring strain will oppose this "spreading apart" of the bonds to the carbon and will make the reaction more difficult. Therefore cyclobutyl chloride reacts much more slowly than cyclopentyl chloride. See Table 9.9.

k. 1st The three-membered and four-membered rings have about the same amount of strain (see Table 5.5). The likelihood that the ends of the chain will find each other to make a three-membered ring is higher than for the ends of a chain which forms a four-membered ring. Therefore O (oxirane, ethylene oxide) is formed faster than O (oxetane).

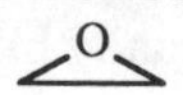

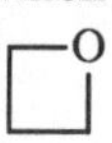

7. a. No; ^{-}CN is a poor leaving group. HCN is a relatively weak acid; recall that there is good correlation between acidity of H-Y and the leaving ability of Y^{-} in displacement reactions.

   b. Slow; F^{-} is a relatively strong base (HF is a weak acid) and a poor leaving group.

   c. No; ^{-}OH is a strong base and an exceptionally poor leaving group. The only reaction observed is an acid-base reaction:

$$(CH_3)_3COH + NH_2^{-} \rightarrow (CH_3)_3CO^{-} + NH_3$$

   d. Yes; $CH_3OSO_3^{-}$ is a weak base and a good leaving group. The corresponding acid, CH_3OSO_2OH, is a strong acid, comparable to H_2SO_4.

   e. No; $^{-}NH_2$ is a very strong base and a perfectly miserable leaving group. NH_3 is a very weak acid.

   f. Yes; I^{-} is a perfectly good leaving group.

   g. No; ^{-}OH is a strong base and an exceptionally poor leaving group.

   h. No; S_N2 reactions in cyclopropyl and cyclobutyl rings will not occur because of the large increase in ring strain that would build up in the transition state (see Figure 9.12).

8. a. False b. True *if* the S_N2 displacement takes place at the chiral center. c. False, second order.
   d. True e. True f. False g. True h. True

9. a. True b. False S_N1 reactions involve carbocations and give racemic mixtures. c. True
   d. False e. False, the reaction is at least 2 step. f. True
   g. False, the rate is independent of the nucleophile (see d.) h. True

10. a. $N_3^{-} + CH_3Cl \xrightarrow{\text{(slower)}} CH_3N_3 + Cl^{-}$ Cl^{-} more basic

   $N_3^{-} + CH_3I \xrightarrow{\text{(faster)}} CH_3N_3 + I^{-}$ I^{-} less basic

   Many other examples could have been cited, but this is the most common situation. If HX is more acidic than HY, X^{-} is less basic than Y^{-}, and RX is more reactive than RY.

   b. $CH_3Br + SCN^{-} \longrightarrow CH_3SCN + CH_3NCS$
   (major) *(minor)*

   c. $(CH_3)_3CCl \xrightarrow[\text{acetone}]{H_2O} (CH_3)_3COH$ *(faster)*

   $(CH_3)_2CHCl \xrightarrow[\text{acetone}]{H_2O} (CH_3)_2CHOH$ *(slower)*

d. $CH_3CH_2CH_2I + CH_3S^- \longrightarrow CH_3CH_2CH_2SCH_3$ *(faster)*

$(CH_3)_2CHI + CH_3S^- \longrightarrow (CH_3)_2CHSCH_3$ *(slower)*

Other examples could be chosen among primary halides with branching in the β-position. Recall that relative rates are: $CH_3CH_2CH_2- > (CH_3)_2CHCH_2- > (CH_3)_3CCH_2-$, entirely because of steric hindrance effects.

e. $CH_3CH_2Cl + SH^- \xrightarrow{C_2H_5OH} CH_3CH_2SH$ *(slower)*

$CH_3CH_2Cl + SH^- \xrightarrow{DMF} CH_3CH_2SH$ *(faster)*

In general, anions are less reactive in S_N2 reactions in hydroxylic solvents (such as alcohol) than in polar aprotic solvents (such as DMF, HMPT, and DMSO). In hydroxylic solvents, hydrogen bonds to the anion need to be broken in order to form the S_N2 transition state.

11. a. SCN^- b. I^- c. $P(CH_3)_3$ d. CH_3S^-

In each case, the nucleophile giving the larger substitution/elimination ratio contains an atom further down the Periodic Table, and is therefore more polarizable. Greater polarizability enhances nucleophilicity more than basicity; that is, polarizability is relatively more important in S_N2 reactions at carbon and is relatively less important in E2 reactions at hydrogen.

12.

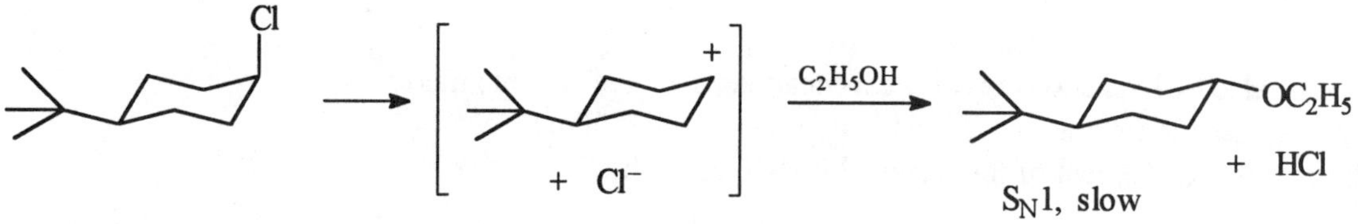

Cl

+

+ Cl^-

C_2H_5OH

OC_2H_5

+ HCl

S_N1, slow

This reaction proceeds through a carbocation intermediate, reacting with the solvent, a weak base.

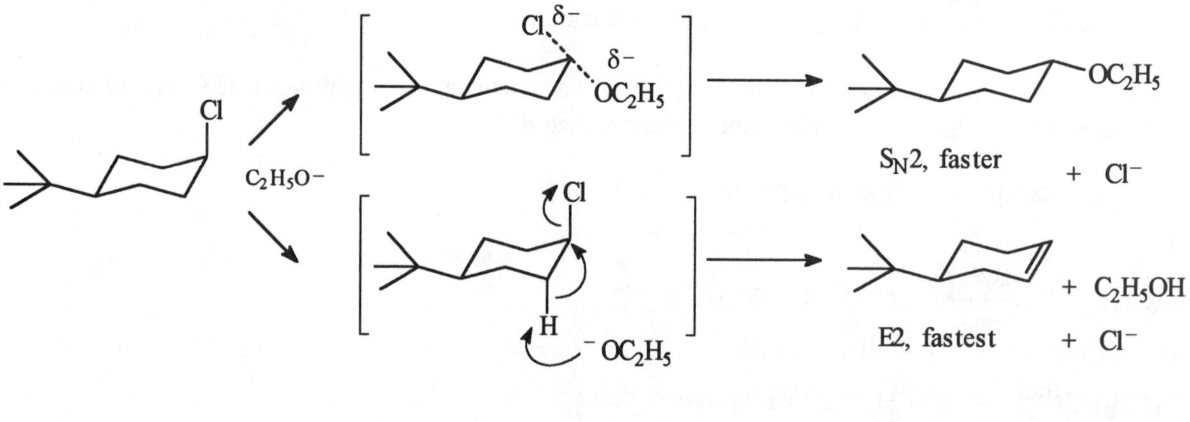

Cl

C_2H_5O-

Cl $\delta-$

$\delta-$

OC_2H_5

OC_2H_5

S_N2, faster

+ Cl^-

Cl

H

$^-OC_2H_5$

+ C_2H_5OH

E2, fastest

+ Cl^-

In the presence of a strong base, there is a competition between the S_N1 (slow) reaction, the S_N2 reaction at a 2° carbon and E_2, always a side reaction with 2° and 3° alkyl halides and a strong base.

13. For the cyclization reaction, both reactants are part of the same molecule, and the reaction can only have first-order kinetics. The reaction of the amino group of one molecule with the alkyl bromide of another (the <u>inter</u>molecular displacement rather than the <u>intra</u>molecular or cyclization reaction) has second-order kinetics; hence, the rate of this reaction is reduced much more than the other by reducing the concentration:

$$\frac{inter}{intra} = \frac{k_{inter}[H_2N(CH_2)_4Br]^2}{k_{intra}[H_2N(CH_2)_4Br]} = \frac{k_{inter}}{k_{intra}}[H_2N(CH_2)_4Br]$$

The ratio of the two is concentration-dependent, and the cyclization reaction is favored by "high dilution" methods.

14. In the equilibrium between carboxylate anion (RCO_2^-) and carboxylic acid (RCO_2H), in which a full negative charge is lost, the chloroacetate ion is $10^{1.9} = 79$ times less reactive than acetate ion:

$$RCO_2^- + H^+ \overset{1/K_a}{\rightleftharpoons} RCO_2H \qquad 1/K_a = 10^{(pK_a)}$$

In going to the transition state for the S_N2 reaction in methanol, the difference in reactivity is only 10, a reduction by a factor of 8. This suggests that only one-eighth of a charge is lost in going from the hydrogen-bonded carboxylate ion to the transition state:

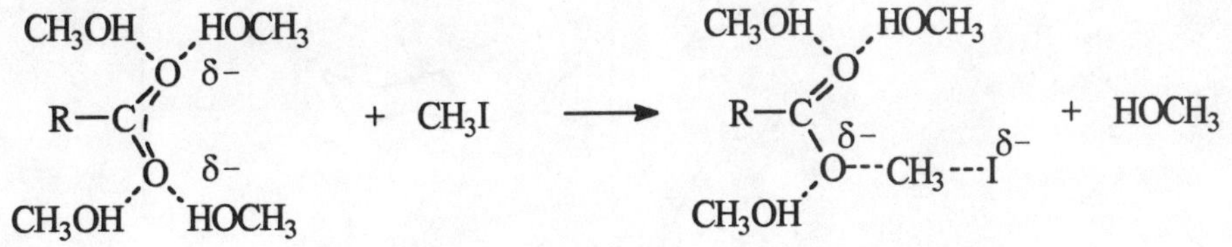

CH_3OH $HOCH_3$ / O δ− / R—C / O δ− / CH_3OH $HOCH_3$ + CH_3I ⟶ CH_3OH $HOCH_3$ / O / R—C δ− / O---CH_3---I δ− / CH_3OH + $HOCH_3$

15. 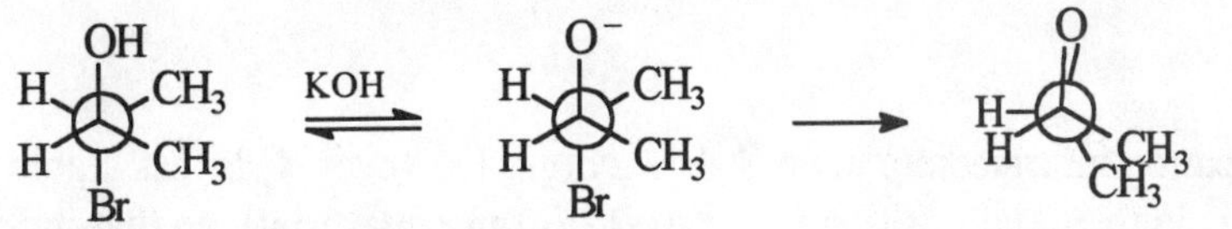 (Newman projections) OH, H, CH_3, H, CH_3, Br $\xrightleftharpoons{KOH}$ O⁻, H, CH_3, H, CH_3, Br ⟶ O, H, H, CH_3, CH_3

or the enantiomer

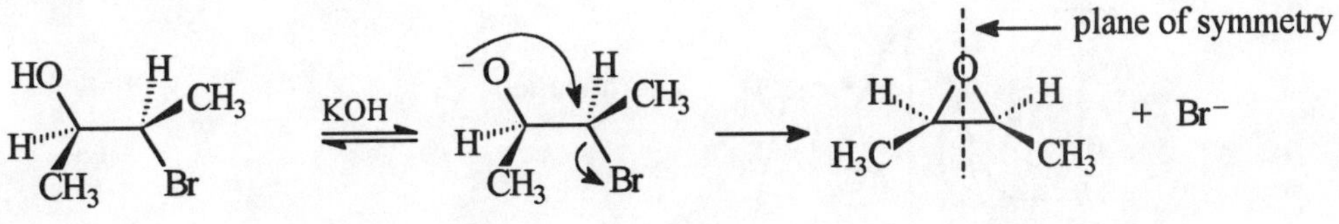

HO, H, CH_3, H, CH_3, Br $\xrightleftharpoons{KOH}$ ⁻O, H, CH_3, H, CH_3, Br ⟶ O, H, H, H_3C, CH_3 + Br⁻ (← plane of symmetry)

Although the cyclization reaction is unimolecular, it still proceeds by the "S_N2" mechanism involving backside attack and inversion of configuration.

Work out the stereochemistry of this reaction with the other <u>*enantiomer*</u> of the starting material and demonstrate that it gives the same product. Then do it for a <u>*diastereomer*</u>, and show that it gives a chiral product.

16. a. Look at models or pictures of chair structures:

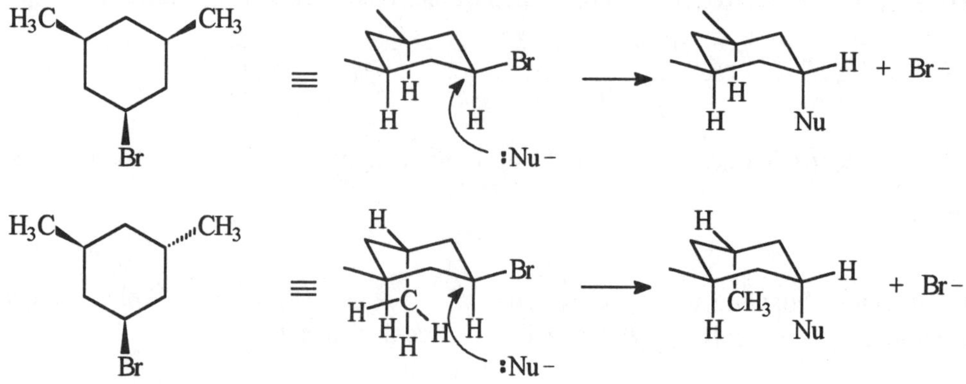

Axial attack is hindered even more by an axial methyl.

b.

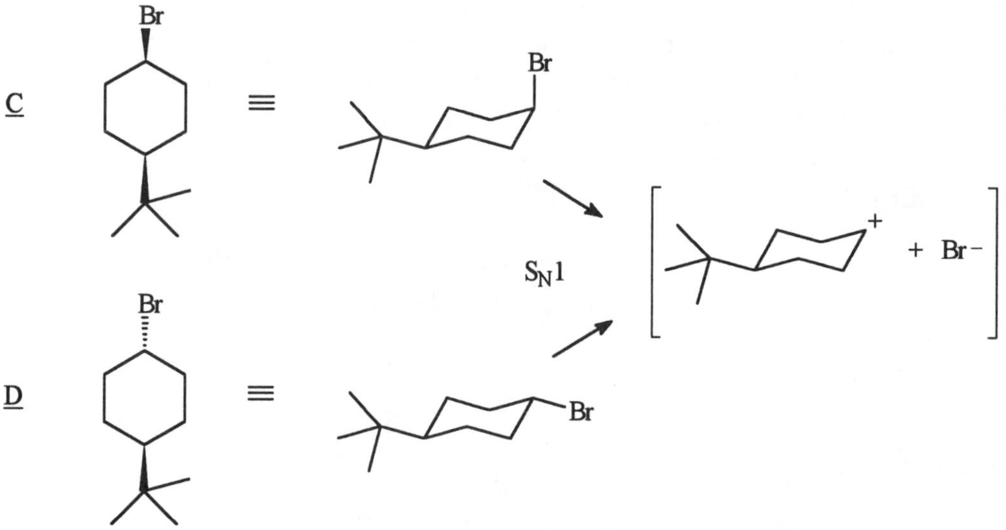

Both C and D lead to the same carbocation intermediate in an S_N1 reaction. However, C is less stable than D to begin with because the bromine is in the axial position (*t*-butyl) is *always* equatorial), so that it doesn't have to go as far "uphill" in order to react. This can be depicted in a reaction profile diagram:

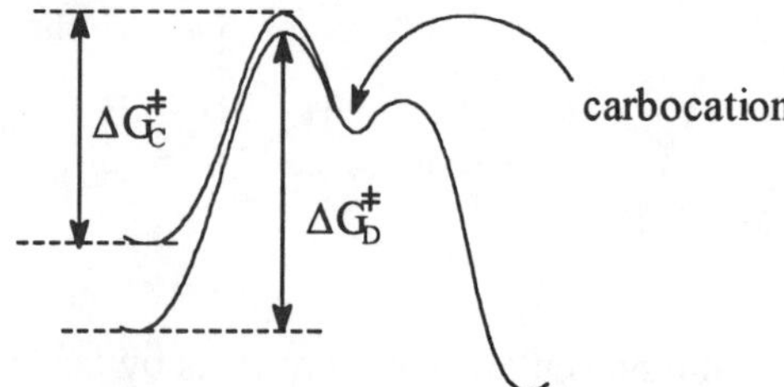

c. *t*-Butyl chloride reacts via the S_N1 mechanism involving the intermediate *t*-butyl cation. The higher the dielectric constant of a solvent, the more it can mask the attraction of a negative charge (the departing Cl^- ion) for a positive charge (the *t*-butyl cation), and the easier it is for the *t*-butyl chloride molecule to ionize. The dielectric constant of methanol is higher than that of ethanol, therefore S_N1 reactions proceed faster in methanol.

17. $Nu^- + CH_3I \longrightarrow Nu\text{-}CH_3 + I^-$

Nu^-	$\Delta H_f^\circ(Nu^-)$	- $\Delta H_f^\circ(CH_3I)$	+ $\Delta H_f^\circ(NuCH_3)$	+ $\Delta H_f^\circ(I^-)$	= ΔH° (kcal $mole^{-1}$)
CN^-	-16	- 3.4	+ 17.6	- 45.1	= -47
$CH_3CO_2^-$	-(-120.5)	- 3.4	+ -97.6	- 45.1	= -25.9
NO_2^-	-(-27.1)	- 3.4	+ -17.9	- 45.1	= -39.3
Cl^-	-(-54.5)	- 3.4	+ -20.6	- 45.1	= -14.6
Br^-	-(-50.9)	- 3.4	+ -9.1	- 45.1	= -6.7
I^-	-(-45.1)	- 3.4	+ 3.4	- 45.1	= 0

Little correlation between the values since the solvent affects the rate of the reaction in solution, which is determined by $\Delta G^\ddagger$, while the ΔH° is related to the thermodynamic stability of the products compared to the reactants, ΔG°, in the gas phase, not in solution.

9.F. Supplementary Problems

Sl. Predict the major product and mechanism of each of the following reactions:

a. $(CH_3)_2CHCH_2Cl + NaN_3 \xrightarrow{CH_3OH}$

b. $CH_3CH_2CH_2I + \left[H_3C\text{—}C(=O)S^- \longleftrightarrow H_3C\text{—}C(O^-)=S \right]$

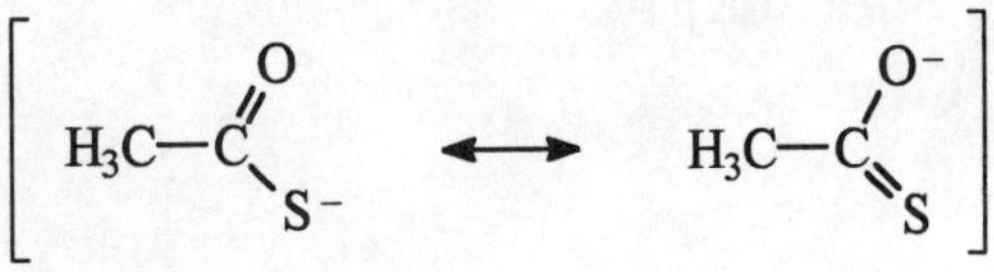

c. NaCN + 1-bromo-1-methylcyclohexane (ring carbon bearing Br and CH_3) $\longrightarrow$

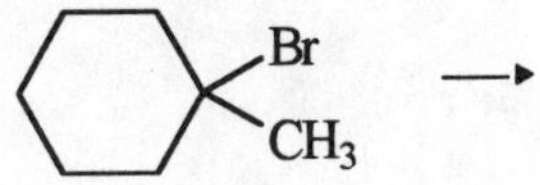

d. $(CH_3)_3CBr + NaCl \xrightarrow{CH_3OH}$

e. $(CH_3)_2COHCH_2CH_2CH_2Br + NaNH_2 \longrightarrow$

f. $CH_3CH_2CH_2Cl + :N(CH_3)_3 \longrightarrow$

g. $CH_3CH_2\text{—}C(CH_3)_2\text{—}CHClCH_3 + NaOH \longrightarrow$

h. 1-chloro-1-methylcyclopentane (ring carbon bearing Cl and CH_3) $+ CH_3SCH_3 \longrightarrow$

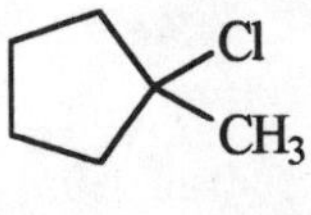

i. $BrCH_2CH_2CH_2\text{—}\ddot{N}(CH_3)\text{—}CH_2CH_2CH_2CH_2Br \xrightarrow{acetone}$

S2. What are the absolute configurations of the products of the following reaction sequences?

a. 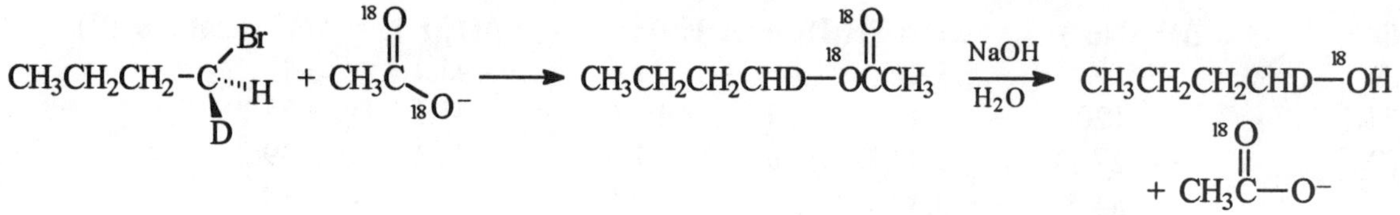 $CH_3CH_2CH_2C(Br)(H)(D) + CH_3C(=^{18}O)^{18}O^- \longrightarrow CH_3CH_2CH_2CHD{-}^{18}OC(=^{18}O)CH_3 \xrightarrow[H_2O]{NaOH} CH_3CH_2CH_2CHD{-}^{18}OH + CH_3C(=^{18}O){-}O^-$

b. 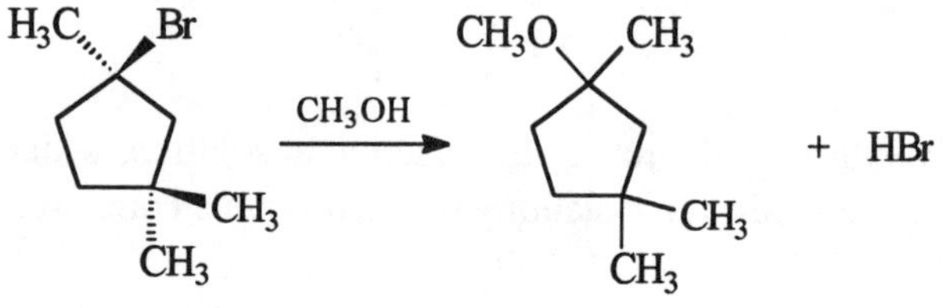 1-bromo-1,3,3-trimethylcyclopentane (H₃C, Br; CH₃, CH₃) $\xrightarrow{CH_3OH}$ 1-methoxy-1,3,3-trimethylcyclopentane (CH_3O, CH_3; CH_3, CH_3) + HBr

c. 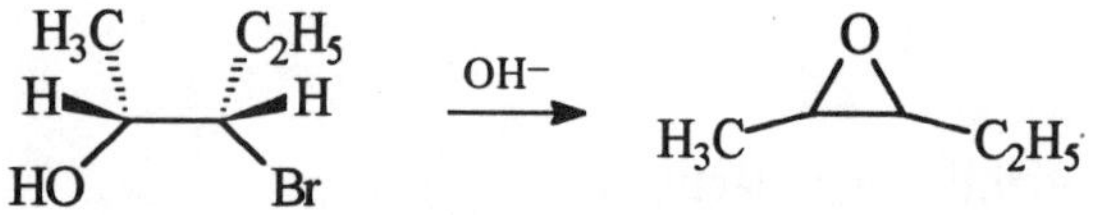 H₃C, H, HO / C₂H₅, H, Br $\xrightarrow{OH^-}$ epoxide (H_3C, O, C_2H_5)

d. Newman projection (Br, CH_2CH_3, CH_3, H, H_3C, H_3C) $\xrightarrow{CH_3OH}$ $CH_3CH_2CH(CH_3){-}C(CH_3)_2{-}OCH_3$ + HBr

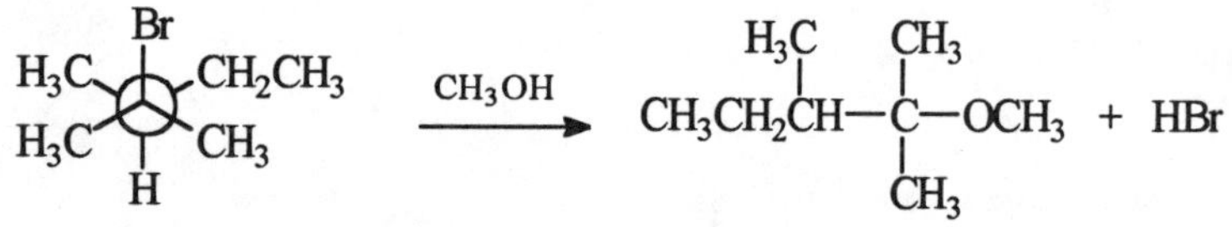

e. 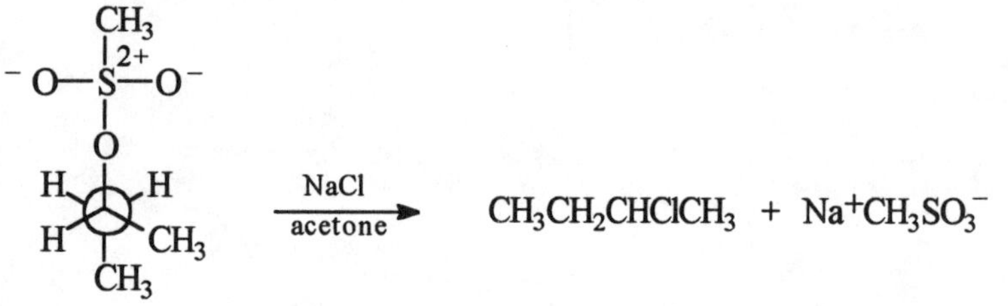 $CH_3{-}S^{2+}(O^-)_2{-}O{-}$ Newman projection (H, H, H, CH_3, CH_3) $\xrightarrow[acetone]{NaCl}$ $CH_3CH_2CHClCH_3 + Na^+CH_3SO_3^-$

S3. Rank the following nucleophiles in order of their rate of reaction with methyl sulfate in methanol:

Cl^-, OH^-, F^-, SH^-, H_2O

S4. Rank the following compounds in order of their rate of reaction with sodium azide in methanol.

CH_3Br, $(CH_3)_3CCH_2Cl$, $CH_3CH_2CHClCH_3$, $CH_3CH_2CHBrCH_3$, $(CH_3)_2CHCH_2CH_2Br$

S5. Rank the following compounds in order of their rate of reaction in refluxing (boiling) methanol:

$CH_3CH_2CHBrCH_3$, $CH_3CH_2CHClCH_3$, $(CH_3)_3CCH_2Br$, 1-bromo-1-methylcyclohexane (Br, CH_3)

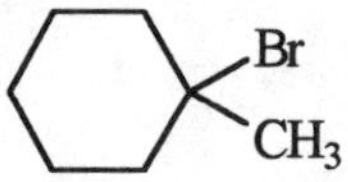

S6. For each of the sets of reactions and conditions below, choose that corresponding to the <u>faster</u> process. Indicate which explanations in the list of statements following are the most pertinent ones for each case. Note that more than one explanation may apply to each set of reactions.

a. $CH_3CH_2CH_2Br + Y^- \xrightarrow{CH_3OH} CH_3CH_2CH_2Y + Br^-$

$Y^- = CH_3O^-$ or $CH_3CO_2^-$

b. $(CH_3)_3CCl + CH_3O^- \xrightarrow{CH_3OH} A + Cl^-$

$A = (CH_3)_3COCH_3$ or $(CH_3)_2C{=}CH_2 + CH_3OH$

c. $R\text{-}Br + HS^- \longrightarrow R\text{-}SH + Br^-$

$R = (CH_3)_2CHCH_2CH_2Br$ or $CH_3CH_2CH(CH_3)CH_2Br$

d. $Cl^- + CH_3CH_2OSO_2CH_3 \xrightarrow{\text{solvent}} CH_3CH_2Cl + CH_3SO_3^-$

solvent = acetone or CH_3CH_2OH

e. $CH_3CH_2CHICH_3 + (CH_3)_3Z \longrightarrow CH_3CH_2CH(CH_3)Z^+(CH_3)_3 + I^-$

Z = N or P

f. $CH_3CH_2C(CH_3)_2CHBrCH_3 + CH_3O^- \xrightarrow{CH_3OH} D + Br^-$

$D = CH_3CH_2C(CH_3)_2CH(OCH_3)CH_3$ or $CH_3CH_2C(CH_3)_2C{=}CH_2 + CH_3OH_3$

g. 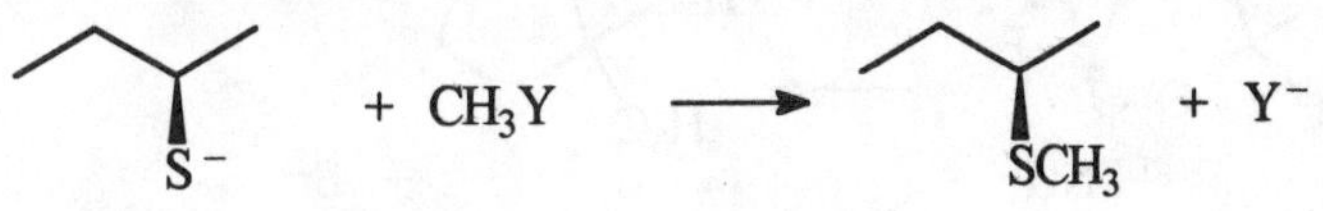 [sec-butyl thiolate, S⁻ on wedge] $+ CH_3Y \longrightarrow$ [sec-butyl methyl sulfide, SCH_3 on wedge] $+ Y^-$

$Y^- = F^-$ or Br^-

h. $R\text{-}Br + CH_3OH \longrightarrow R\text{-}OCH_3 + HBr$

R-Br = [1-bromo-1-methylcyclopentane (CH_3, Br)] or [cyclopentyl–CH_2Br]

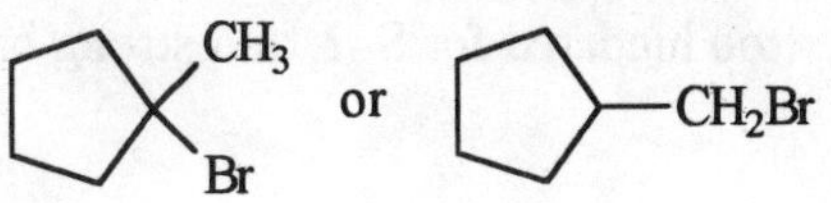

i. $CH_3CH_2CHBrCH_3 + Y^- \longrightarrow CH_3CH_2CHYCH_3 + Br^-$

$Y^- = N_3^-$ or NH_2^-

EXPLANATIONS

(A) Steric hindrance from α-branching results in slower S_N2 reactions.
(B) Steric hindrance from α-branching has little effect on S_N2 reactions.
(C) Steric hindrance from β-branching results in slower S_N2 reactions.
(D) Steric hindrance from β-branching has little effect on S_N2 reactions.
(E) The order of stability of carbocations is 3°>2°>1°.
(F) The order of stability of carbocations is 3°<2°<1°.
(G) E2 reactions are generally poor for tertiary systems.
(H) Stronger bases make poorer leaving groups.
(I) The reactivity of anions in protic solvents is diminished by hydrogen bonding.
(J) The reaction of cations in polar solvents is diminished by interaction with the solvent lone pair electrons.
(K) Very strong bases favor elimination over substitutions.
(L) Other factors being equal, stronger bases are generally better nucleophiles.
(M) Other factors being equal, the more polarizable reagent is the better nucleophile.
(N) Reagents with no lone pair electrons are relatively poor nucleophiles.
(O) Chiral molecules are generally more effective nucleophiles.
(P) Neopentyl-type systems are exceptionally slow in S_N2 reactions because of steric hindrance.

9.G. Answers to Supplementary Problems

S1. a. $(CH_3)_2CHCH_2N_3$ S_N2

b. $CH_3CH_2CH_2S-COCH_3$ S_N2

c. CH_3 E$_2$

d. $(CH_3)_3COCH_3 + HBr$ S_N1
(don't forget that the solvent can react!)

e. H_3C, H_3C, Br, O, H, $^-NH_2$ ⇌ H_3C, H_3C, O^-, Br + NH_3 → H_3C, H_3C, O + Br^-

Intramolecular S_N2

f. $CH_3CH_2CH_2N^+(CH_3)_3Cl^-$ S_N2

g. $CH_3CH_2C(CH_3)_2CH=CH_2$ E2
(too hindered for S_N2, and strong base)

h. H_3C, $S^+(CH_3)_2$ Cl^- S_N1

i. Br, H_3C, N, Br → Br, H_3C, N^+ Br^- Intramolecular S_N2

S2. a. 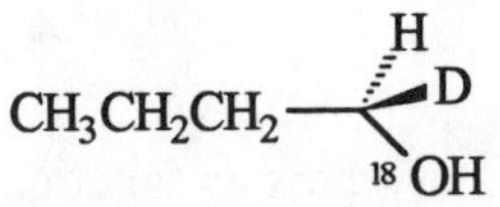 $CH_3CH_2CH_2-C(H)(D)(^{18}OH)$ (*R*); the first step is S_N2 with inversion; the second step doesn't affect the CHD-O bond.

b. 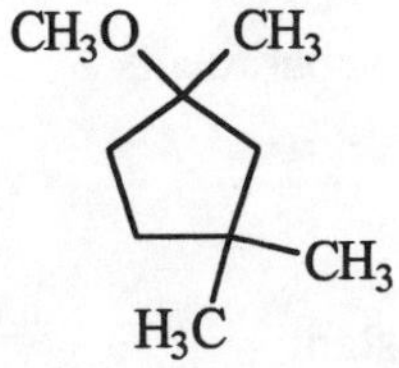 1-methoxy-1,3,3-trimethylcyclopentane (CH_3O, CH_3 on one carbon; CH_3, H_3C on the other) racemic; reaction proceeds via achiral carbocation (S_N1)

c. 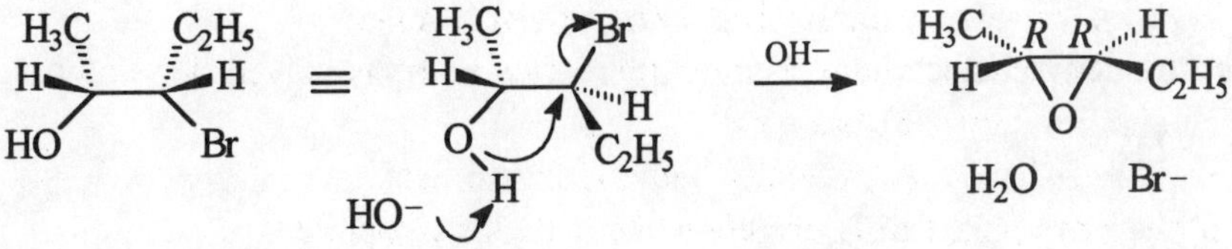 H₃C, H, HO / C₂H₅, H, Br ≡ H₃C, H, O–H ... Br, H, C₂H₅; HO⁻ $\xrightarrow{OH^-}$ H_3C, H (R R) H, C_2H_5 epoxide (O) + H_2O + Br^-

d. The reaction is S_N1, but does not involve the stereocenter: OCH_3, H_3C, CH_2CH_3, H_3C, CH_3, H (*S*)

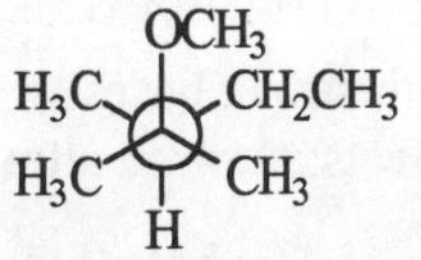

e. A mixture of S_N2 and S_N1 reactions, resulting in partial inversion, partial racemization.

Major enantiomer: H; H, H; Cl, CH_3; CH_3 = (*S*)-2-chlorobutane

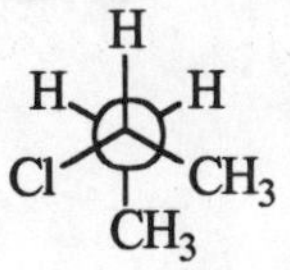

S3. $SH^- > OH^- > Cl^- > F^- > H_2O$

S4. S_N2: $CH_3Br > (CH_3)_2CHCH_2CH_2Br > CH_3CH_2CHBrCH_3 > CH_3CH_2CHClCH_3 > (CH_3)_3CCH_2Cl$

S5. 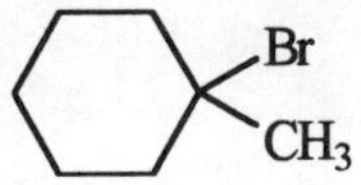 1-bromo-1-methylcyclohexane (Br, CH_3) $> CH_3CH_2CHBrCH_3 > CH_3CH_2CHClCH_3 > (CH_3)_3CCH_2Br$

S6. a. $Y^- = CH_3O^-$; L

b. $A = (CH_3)_2C{=}CH_2 + CH_3OH$; A (K)

c. $R = (CH_3)_2CHCH_2CH_2Br$; C

d. solvent = acetone; I

e. Z = P; M

f. $D = CH_3CH_2C(CH_3)_2CH{=}CH_2 + CH_3OH$; P

g. $Y^- = Br^-$; H

h. RBr = 1-bromo-1-methylcyclopentane (CH_3, Br) ; E

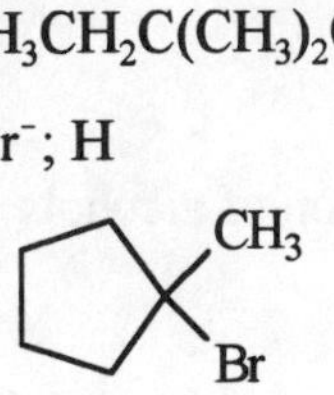

i. $Y^- = N_3^-$; K

NOTE: "Explanations" B,D,F, and O are *false* statements.

10. ALCOHOLS AND ETHERS

10.A. Chapter Outline and Important Terms Outlined

10.1 Introduction: Structure

sp^3-hybridization

alcohols and ethers as functional groups

10.2 Nomenclature of Alcohols

alkyl alcohol system
alkanols (IUPAC system)

hydroxyalkyl substituents

10.3 Physical Properties (solubility, boiling point, etc.)

dipole moment
hydrogen bonding

dielectric constant

10.4 Acidity of Alcohols: Inductive Effects

$$ROH \rightleftharpoons RO^- + H^+$$

$$ROH + B^- \text{ (strong base)} \rightleftharpoons RO^- + HB$$

reactions with strong bases
inductive effects: electron attracting or donating

reaction with alkali metals
ion pair

10.5 Preparation of Alcohols (functional group interconversions)

hydrolysis of alkyl halides

acetate displacement

10.6 Reactions of Alcohols

A. Reactions of Alkoxides with Alkyl Halides
B. Conversion of Alcohols into Alkyl Halides
   alkyloxonium salts
   inorganic esters: chlorosulfite, phosphite, sulfonate
C. Carbocation Rearrangements
   outline of best ways for ROH ⟶ RX
D. Dehydration of Alcohols: Formation of Ethers and Alkenes
   alkylsulfuric acid
E. Oxidation of Alcohols (loss of two protons <u>and</u> two electrons)
   chromic acid
   pyridinium chlorochromate (PCC)

chromate ester
(balancing oxidation/reduction reactions)

10.7 Nomenclature of Ethers

$alkyl_1$ $alkyl_2$ ether

alkoxyalkane

10.8 Physical Properties of Ethers

lower b.p. and solubility than comparably sized alcohols
THF

10.9 Preparation of Ethers

Williamson ether synthesis

reaction of alcohols with sulfuric acid

10.10 **Reactions of Ethers**

reactions with acids

autoxidation to peroxides

10.11 **Cyclic Ethers**

A. Epoxides: Oxiranes
heteroatom

glycols, glymes

B. Higher Cyclic Ethers
THF

crown ethers

10.12 **Multistep Synthesis** (how to design a sequence of reactions)

avoid isomers

work backwards

10.B. Important Reactions Discussed

A number of reactions are discussed that have been presented before in a more general context. The reactions of alkyl halides with water (hydrolysis) or another alcohol (ether formation) are simply nucleophilic substitution reactions, and proceed by either the S_N1 or S_N2 mechanisms as discussed in Chapter 9.

Conversion of alcohols to alkyl halides (10.5)

Acid-catalyzed:

Equation: $ROH + HX \xrightarrow{H^+} RX + H_2O$

Generality: easiest for 3° ROH; 1° and 2° okay if carbocation rearrangements can be avoided

Key features: mechanism involves oxonium ions (ROH_2^+); (X^- will **not** displace OH^-)
1° ROH react via S_N2; 2° and 3° ROH via S_N1
carbocation rearrangements common

Via inorganic esters:

Equation: $ROH + SOCl_2 \longrightarrow RCl + SO_2 + HCl$
$3\ ROH + PX_3 \longrightarrow 3\ RX + H_3PO_3$

Generality: good for 1°, 2°, and 3° ROH
$SOCl_2$ for RCl
PCl_3, PBr_3, or $P + I_2$ ($\rightarrow PI_3$) for RCl, RBr, and RI

Key features: milder conditions than HX; can be run with pyridine (non-acidic conditions) to minimize carbocation rearrangements

Note: The table at the end of Section 10.6.C in the Text is a very useful summary of the best conditions to use to convert ROH to RX.

Dehydration of alcohols to ethers (10.6.D)

Equation: $2\ ROH \xrightarrow{H_2SO_4} ROR + H_2O$

Generality: For formation of 1°, symmetrical ethers only

Key features: limited utility, strong acid conditions
dehydration occurs with 2° or 3° ROH

Oxidation of alcohols to ketones and aldehydes (10.6.E)

Equation: 3 RCH(OH)R' + 2 Cr(VI) ⟶ 3 RC(=O)R' + 2 Cr(III)

Generality: 1° and 2° alcohols, to give aldehydes and ketones, respectively
3° not oxidized under normal conditions

Key features: for 2° alcohols to ketone: $K_2Cr_2O_7$ in aqueous H_2SO_4 ("Jones reagent"); with 1° alcohol, aq. Cr(VI) reagents give overoxidation to carboxylic acid. For 1° alcohol to aldehyde: pyridinium chlorochromate (PCC) in CH_2Cl_2

Williamson ether synthesis (10.9)

Equation: $RX + R'O^- \longrightarrow ROR'$ to X^-(Formation of $R'O^-M+$, see 10.6A.)

Generality: best with CH_3X, 1° RX
3° RX give only elimination (E2)
$R'O^-$ can be 1°, 2°, or 3°

Key features: S_N2 mechanism
$R'O^-$ is strong base, so E2 is frequent side reaction
intramolecular reaction gives cyclic ethers

Reactions of epoxides (10.11.A)

Equation: for reactions with basic nucleophiles (base-catalyzed mechanism):

(epoxide C–C–O) + :Nu⁻ ⟶ O^-–C–C–Nu

for reactions under acid conditions (acid-catalyzed mechanism):

(epoxide) $\overset{H^+}{\rightleftharpoons}$ (protonated epoxide, H^+ on O) + :Nu⁻ ⟶ OH–C–C–Nu

Generality: useful nucleophiles are H_2O (OH^-) (to give glycols) or Grignard reagents (to form C-C bonds)
acid catalysis leads to reaction at more substituted end (S_N1-like mechanism)
basic reactions occur at less substituted end (S_N2-like)

Key features: reactivity is due to ring strain of three-membered ring

10.C. Important Concepts and Hints

Chapter 10 outlines the chemistry of alcohols and ethers in the way that other functional groups will be discussed throughout the text. First, the nomenclature and physical properties of molecules that contain the hydroxy group or the ether linkage are presented. Although these topics were introduced in preceding chapters, they must be expanded upon to show how they apply to each additional functional group. (Beginning with Chapter 14, spectroscopic properties will also be included here.)

The bulk of the chapter is devoted to the functional group interconversions in which alcohols and ethers take part. In other words, the ways you can make alcohols and ethers (preparation) and what you can make from them (reactions) are presented. As the course progresses, you will realize that a <u>lot</u> of organic reactions are being discussed (along with a lot of mechanisms). If you learn each reaction by itself, you will soon be overwhelmed by details, at the expense of understanding. You will always need to learn specific facts, but you should try to get an overview and an understanding of the relationship between these facts. Look for the underlying patterns in reactions. As an example, consider the following reactions:

$$(CH_3)_2CHCH_2CH_2OH \xrightarrow{PBr_3} (CH_3)_2CHCH_2CH_2Br$$

$$(CH_3)_2CHCH_2CH_2Br \xrightarrow{NaOH} (CH_3)_2CHCH_2CH_2OH$$

If you approach organic chemistry as a collection of unrelated reactions, you will learn the first reaction twice - first, as a way to prepare alkyl bromides and second, as a reaction of alcohols. Similarly, you will learn the second reaction twice - both as a way to prepare alcohols and as a reaction of alkyl bromides. On the other hand, if you try to get an overview, you will think of the reactions above as <u>one</u> interconversion of functional groups: alcohol $\rightleftharpoons$ alkyl halide. At this point you will have learned the most important aspect of these two reactions - that they exist. Your thinking of functional groups should grow to include ways that interconvert the indicated functional groups. With this as a framework, organizing the specific details in your mind becomes easier. At first you will know just a few reactions which interconvert alcohols and alkyl halides; as your knowledge grows you will learn more of them, their limitations and exceptions and side reactions, and so on.

Although an understanding of the framework outlined above is a starting point, you must still know the specific reactions involved. After all, a singer has to know some songs. Think of the functional group interconversions as music and the specific reactions as your repertoire. In this regard, Appendix II in this Study Guide is useful. It is a list of reactions, organized on the basis of functional group preparations and carbon-carbon bond forming reactions. You can try to memorize it if you want to, but it would be wiser to learn a little at a time, referring to it frequently to refresh your memory, using it in the manner suggested above.

10.D Answers to Exercises

10.1 a. neopentyl alcohol b. β-bromopropyl alcohol c. ω-chloropentyl alcohol

10.2 a. OH b. HO Br c. OH ≡ OH

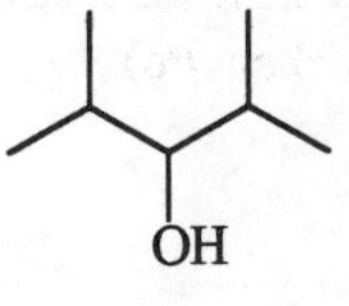

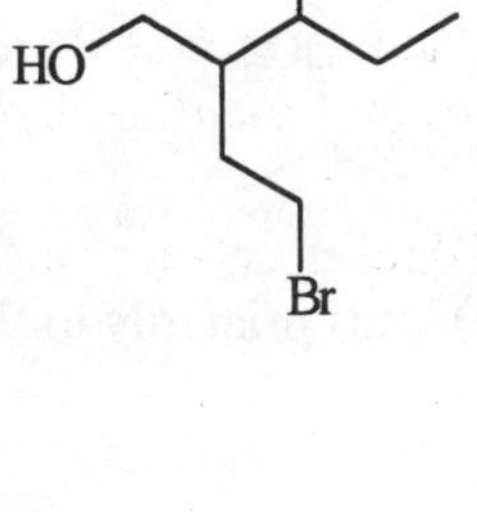

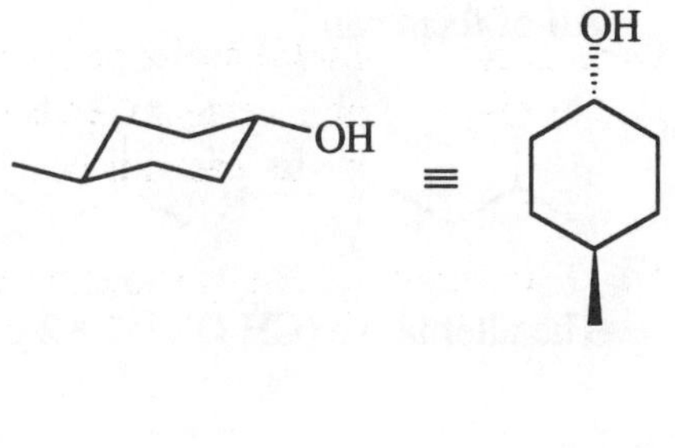

10.3 a. 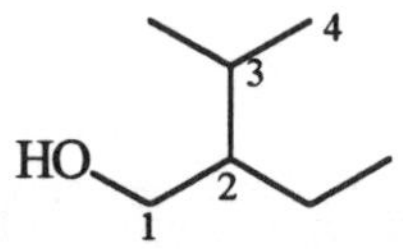 "2-Isopropyl-1-butanol" should be named 2-ethyl-3-methyl-1-butanol (choose the carbon chain that has more, smaller substituents).

b. 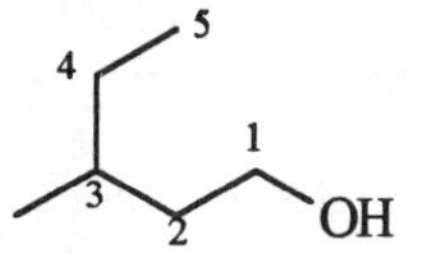 The name "2-ethyl-4-butanol" does not include the longest carbon chain in the root and doesn't give the lowest number to the OH function: 3-methyl-1-pentanol.

c. 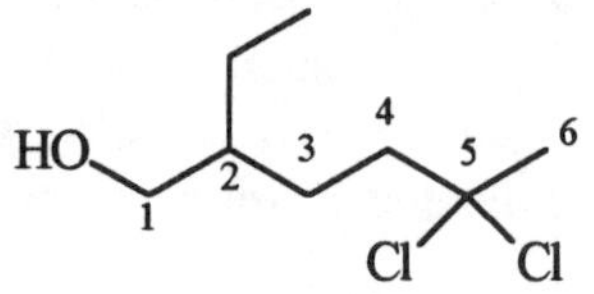 "2,2-Dichloro-5-hydroxymethylheptane" should be named as an alkanol: 5,5-dichloro-2-ethyl-1-hexanol. Note that the longest carbon chain doesn't contain the OH function.

10.4 a. 2,4,7-trimethyl-4-octanol b. 2,3,3-trimethyl-2-butanol c. 5-chloro-2-methyl-3-hexanol
d. *trans*-3-chlorocyclobutanol

10.5 a. $CH_3CHClCH_2OH$ is more acidic because the electronegative substituent is closer to the hydroxy group.

b. $CH_3OCH_2CH_2OH$ is more acidic because CH_3O is more electron attracting than ethyl.

c. The dichloro compound is more acidic because it has more electron attracting groups than the monochloro compound.

d. *cis*-3-Chlorocyclohexanol is more acidic than *trans*-4-chlorocyclohexanol for the same reason given in a. above.

10.6 a. 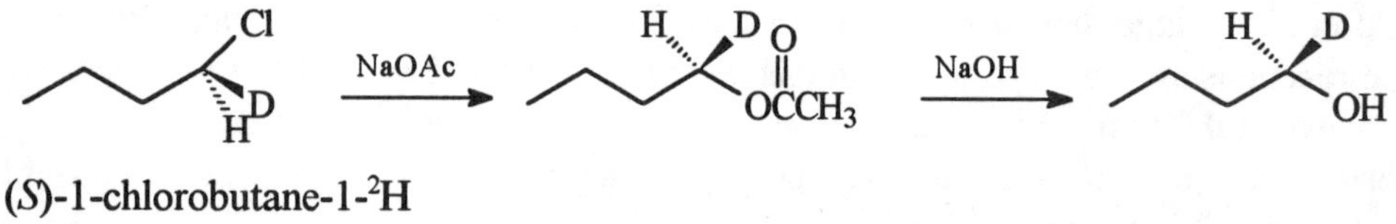 NaOAc, NaOH

(*S*)-1-chlorobutane-1-^{2}H

b. 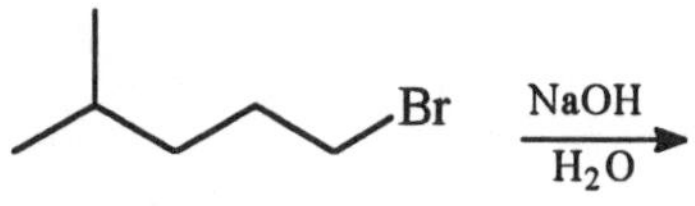 NaOH, H_2O

c. 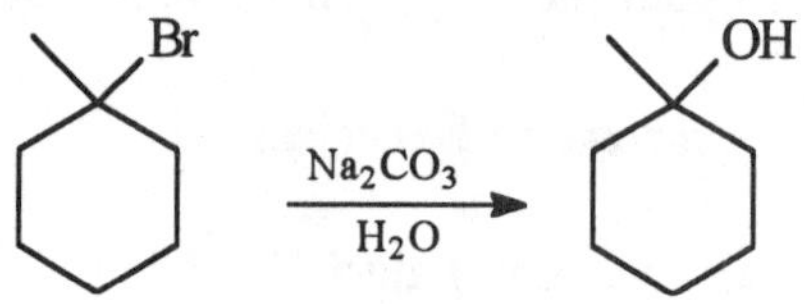 Na_2CO_3, H_2O

10.7 All of them can:

a. 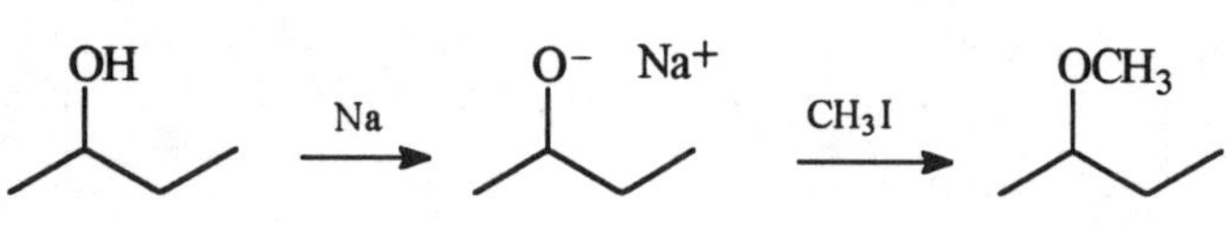 Na, CH_3I [S_N2]

The alternative ($CH_3O^-Na^+$ + 2-bromobutane) leads primarily to E2 elimination.

b. $(CH_3)_3CBr + HOCH(CH_3)_2 \xrightarrow{solvent} (CH_3)_3COCH(CH_3)_2$ [S_N1]

The reaction works because the halide is tertiary and can react via an S_N1 mechanism. A reaction between an alkoxide and an alkyl halide could not lead to *t*-butyl isopropyl ether; elimination would occur instead.

c. $CH_3CH_2Br + Na^+ \, ^-OCH_2C(CH_3)_3 \longrightarrow CH_3CH_2OCH_2C(CH_3)_3$ [S_N2]

The alternative [$CH_3CH_2O^-Na^+ + BrCH_2C(CH_3)_3$] will not work because of steric hindrance.

10.8 HO (*t*-butyl alcohol) + HBr ⇌ H_2O^+ (protonated *t*-butyl alcohol) Br^- ⟶ H_2O + (*t*-butyl cation, +) Br^- ⟶ Br (*t*-butyl bromide) + H_2O

10.9 The reaction between the alcohol and $SOCl_2$ leads first to the chlorosulfite ester and HCl. In the absence of water or a base, HCl is only slightly dissociated and there is only a low concentration of Cl^- ion to carry out the second step. As a mild base, pyridine converts the HCl to pyridinium ion and chloride ion, greatly increasing the concentration of the latter and thereby the rate of reaction.

10.10 a. $CH_2CH_2OCH_2CH_2OH + PBr_3 \longrightarrow CH_3CH_2OCH_2CH_2OPBr_2 + H^+ + Br^- \longrightarrow CH_3CH_2OCH_2CH_2Br + {}^-OPBr_2 + H^+ \longrightarrow HOPBr_2$

$HOPBr_2$ can react with two more alcohol molecules in a similar way, eventually yielding H_3PO_3.

b. OH (2-butanol) + $SOCl_2$ ⟶ (chlorosulfite ester, O–S(=O)–Cl) + H^+ + Cl^- ⟶ [O–S(=O)–Cl ester attacked by Cl^-] ⟶ Cl (2-chlorobutane) + SO_2 + HCl

c. $CH_3CH_2CH_2CH_2OH + PI_3 \longleftarrow \frac{1}{2}(2\,P + 3\,I_2)$

↓

$CH_3CH_2CH_2CH_2OPI_2 + H^+ + I^- \longrightarrow CH_3CH_2CH_2CH_2I + HOPI_2$

10.11 a. (*R*)-2-bromopentane b. *trans*-1-bromo-4-methylcyclohexane

10.12 Rearrangement of the secondary cation from ionization of 2-pentanol could at best only lead to another secondary cation. On the other hand, migration of the hydrogen from the 3-position on ionization of 3-methyl-2-butanol leads to a tertiary cation, which is more stable. Since there is some positive charge on both carbons in the transition state for rearrangement, the transition state that leads to the tertiary cation is of lower energy and that rearrangement occurs faster.

10.13 Model building exercise.

10.14 a. 3° alcohol with no fear of rearrangement: HCl at 0°C
b. 1° alcohol: $SOCl_2$ in pyridine
c. 1° alcohol, β-branching, small chance of rearrangement: PBr_3, < 0°C
d. 2° alcohol, complete inversion of configuration desired: *p*-toluenesulfonyl chloride to give tosylate, then NaBr
e. 2° alcohol, complete inversion of configuration desired: $SOCl_2$ and pyridine

10.15 The conditions necessary for the acid-catalyzed dehydration of a primary alcohol such as 1-butanol are so vigorous that the product will isomerize to a mixture of 1- and 2-butenes during the reaction.

10.16 Equation: 3 (cyclohexanol, HO, H) + $Na_2Cr_2O_7$ + 8 H^+ ⟶ 3 (cyclohexanone, O) + 2 Na^+ + 2 Cr^{3+} + 7 H_2O

70 g (0.70mole) of cyclohexanol requires 0.70/3 = 0.23 mole = 61 g of $Na_2Cr_2O_7$

10.17

	Common	IUPAC
a.	cyclohexyl ethyl ether	ethoxycyclohexane
b.	*sec*-butyl ethyl ether	2-ethoxybutane
c.	isobutyl isopropyl ether	2-methyl-1-(1-methylethoxy)propane or 1-isopropoxy-2-methylpropane

10.18 $(CH_3)_3CCH_2O^- Na^+ + CH_3O_3SC_6H_5 \longrightarrow (CH_3)_3CCH_2OCH_3 + Na^{+-}O_3SC_6H_5$ S_N2 displacement of CH_3X

$CH_3O^-Na^+ + (CH_3)_3CCH_2O_3SC_6H_5$ won't be useful since the β-branching on the neopentyl benzenesulfonate offers too much steric hindrance to S_N2.

10.19 a. (1-bromo-1-methylcyclohexane, Br) + CH_3OH *(solvent)* ⟶ (1-methoxy-1-methylcyclohexane, OCH_3) + HBr [S_N1]

b. $2\ CH_3CH_2CH_2CH_2OH \xrightarrow[\Delta]{H_2SO_4} (CH_3CH_2CH_2CH_2)_2O$

10.20 $CH_3OCH_2C(CH_3)_3 + HBr \longrightarrow CH_3Br + HOCH_2C(CH_3)_3$ S_N2 displacement by Br^- occurs much more rapidly at the methyl carbon of CH_3O than at the sterically hindered carbon of the neopentyl group of $(CH_3)_3CCH_2O$.

10.21 *cis:*

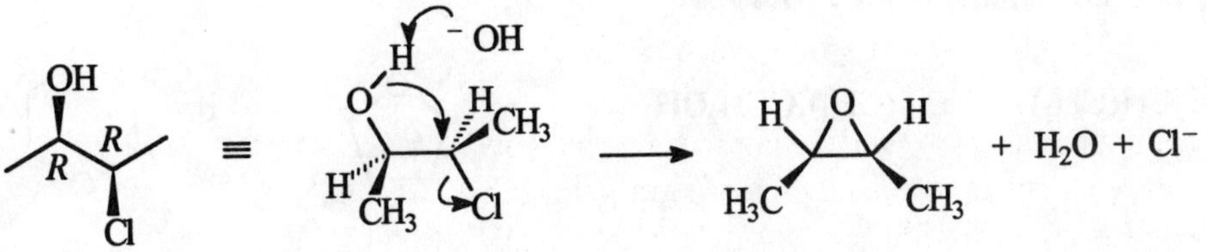

OH, R, R, Cl ≡ H, −OH, O, H, CH3, H, CH3, Cl → H, O, H, H3C, CH3 + H_2O + Cl^-

10.22

H+, H3C, O, H, H, CH3 → H, H3C, O+, H, H, CH3, HOCH3 → H, OH, H3C, H, CH3, +O, H3C, H → H, OH, H3C, H, CH3O, CH3, H+

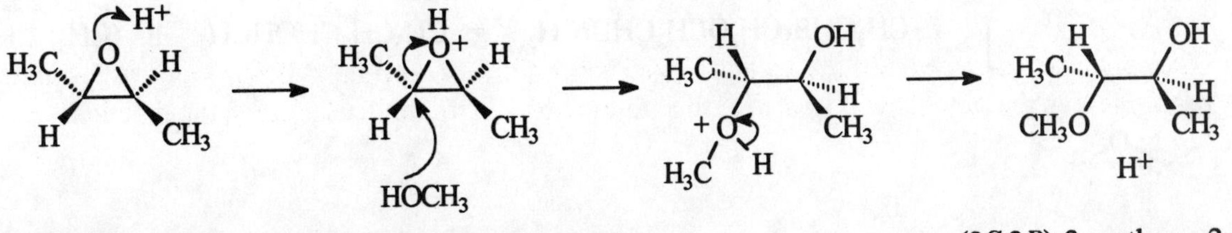

Attack at the other carbon leads to the same compound.
The enantiomeric epoxide leads to the (2*R*, 3*S*) product.

(2*S*,3*R*)-3-methoxy-2-butanol

10.23

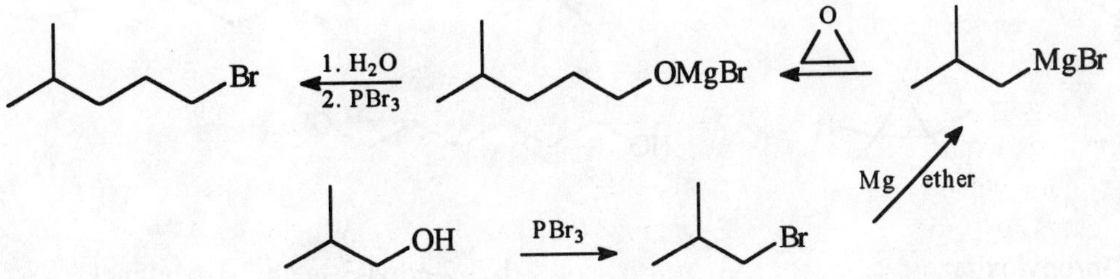

Br ← 1. H_2O 2. PBr_3 — OMgBr ← (epoxide) — MgBr

OH — PBr_3 → Br — Mg, ether → MgBr

10.24

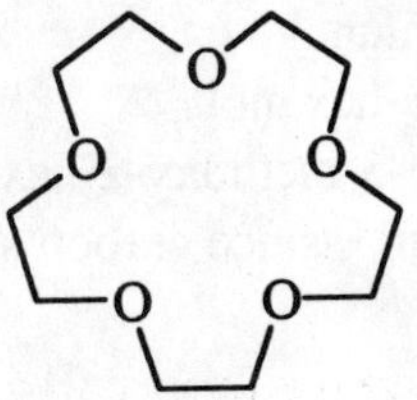

(crown ether: five O atoms in ring)

10.25 a. (cyclohexanone) ← $K_2Cr_2O_7$ / H_2SO_4 — OH (cyclohexanol) ← NaOH or Na_2CO_3 — Br (bromocyclohexane) ← Br_2 / $h\nu$ — (cyclohexane)

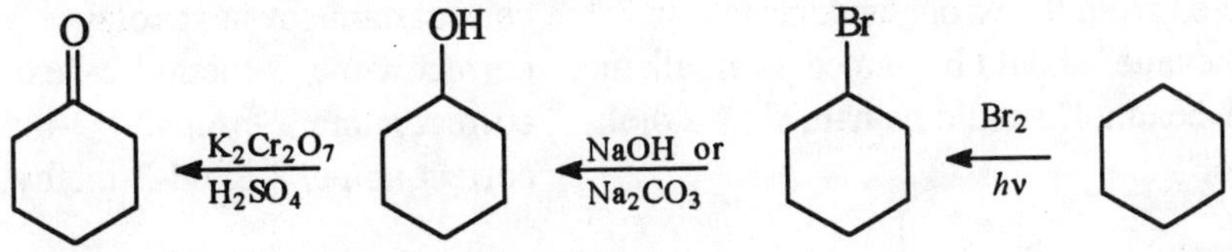

b. CH_3CH_2CN ← NaCN — CH_3CH_2Br ← Br_2 / $h\nu$ — CH_3CH_3

c. (cyclopentene) ← $(CH_3)_3CO^-\ K^+$ (E2) — Cl (chlorocyclopentane) ← Cl_2 / $h\nu$ — (cyclopentane)

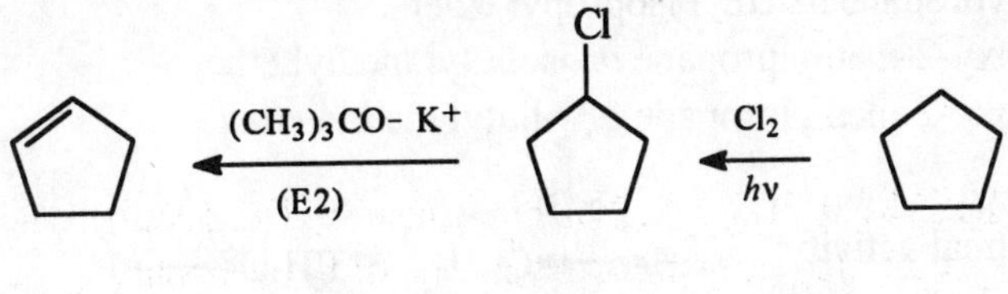

10E. Answers and Explanations for Problems

1. a. $CH_3OCH_2CH(CH_3)_2$ b. $(CH_3)_3CCH_2OH$ c.  d.

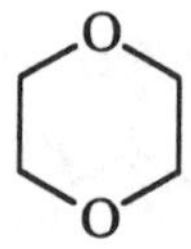

e. 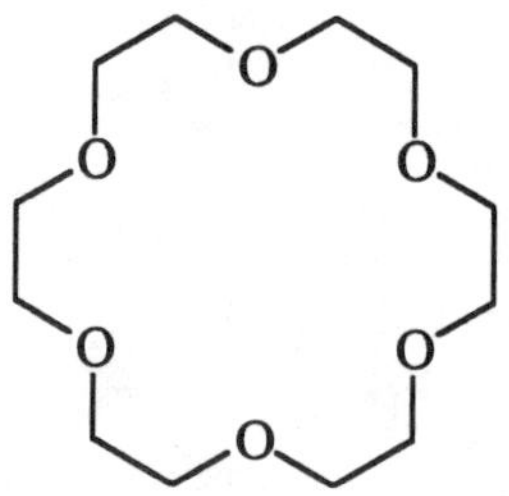 f. $CH_3CHBrCH_2OCH_2CHBrCH_3$ g. $CH_3CH_2CHOHCH_3$ h. $(CH_3)_2CHCH_2OH$

2. a. 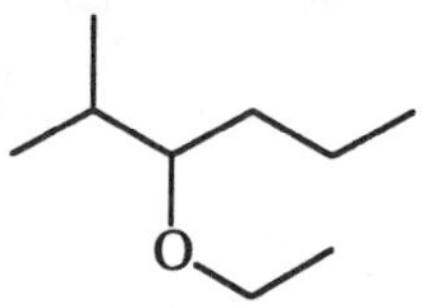 b. 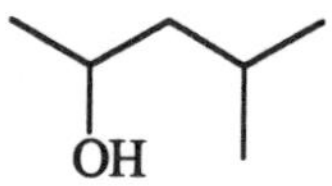 OH c. 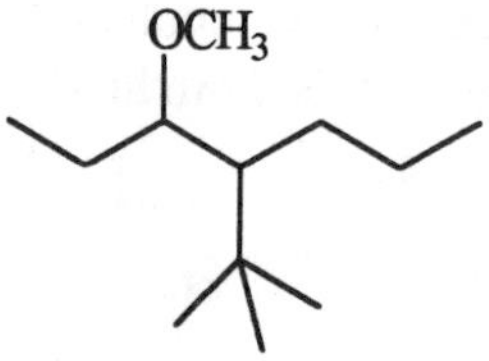 OCH_3 d. 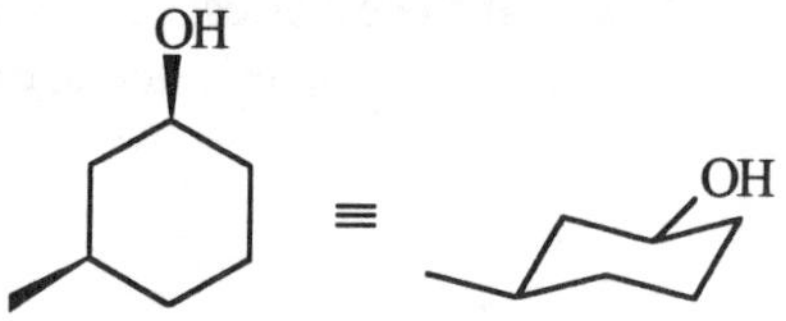 OH ≡ OH

e. 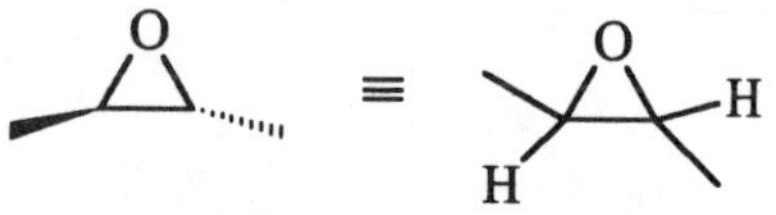 O ≡ O H H f. 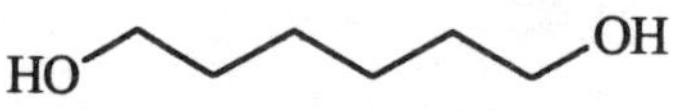 HO OH

3. a. 2-methyl-2-propyloxirane
   b. 2-ethyl-3-methyl-1-pentanol
   c. 6-chloro-2-ethyl-4-methyl-1-hexanol
   d. 3-methoxy-2-pentanol
   e. 3,3-dimethylcyclopentanol*
   f. 2,3,5-trimethyl-3-hexanol
   g. (2*R*,4*S*)-heptane-2,4-diol or (2*R*,4*S*)-2,4-heptanediol
   h. (2*S*,5*R*)-6-chloro-5-methoxy-2-hexanol

   *(NOTE: the position of the alcohol doesn't have to be specified in a cycloalkanol, since carbon #1 is automatically the one with the OH.)

4. a. "4-hexanol" is numbered from the wrong end; correct name: 3-hexanol
   b. "2-hydroxy-3-methylhexane" should be named as an alkanol; correct name: 3-methyl-2-hexanol
   c. "3-(hydroxymethyl)-1-hexanol" should be named as a diol; correct name: 2-propyl-1,4-butanediol
   d. "2-isopropyl-1-butanol": correct name: 2-ethyl-3-methyl-1-butanol.

5. 1) $CH_3OCH_2CH_2CH_2CH_3$ 1-methoxybutane *or* *n*-butyl methyl ether
   2) $CH_3CH(OCH_3)CH_2CH_3$* 2-methoxybutane* *or* *sec*-butyl methyl ether
   3) $CH_3CH_2OCH_2CH_2CH_3$ 1-ethoxypropane *or* ethyl propyl ether
   4) $CH_3CH_2OCH(CH_3)_2$ 2-ethoxypropane *or* ethyl isopropyl ether
   5) $CH_3OCH_2CH(CH_3)_2$ 1-methoxy-2-methylpropane *or* isobutyl methyl ether
   6) $CH_3OC(CH_3)_3$ 2-methoxy-2-methylpropane *or* *t*-butyl methyl ether

   * chiral; this is the only isomer capable of optical activity: CH_3 H OCH_3 C_2H_5 (*S*) CH_3 CH_3O H C_2H_5 (*R*)

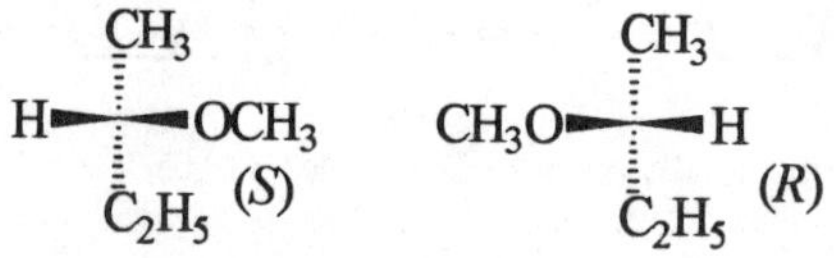

6.

1)	$CH_3CH_2CH_2CH_2CH_2CH_2OH$	1-hexanol	1°
2)	$CH_3CH_2CH_2CH_2CHOHCH_3$	2-hexanol	2°
3)	$CH_3CH_2CH_2CHOHCH_2CH_3$	3-hexanol	2°
4)	$(CH_3)_2CHCH_2CH_2CH_2OH$	4-methyl-1-pentanol	1°
5)	$(CH_3)_2CHCH_2CHOHCH_3$	4-methyl-2-pentanol	2°
6)	$(CH_3)_2CHCHOHCH_2CH_3$	2-methyl-3-pentanol	2°
7)	$(CH_3)_2COHCH_2CH_2CH_3$	2-methyl-2-pentanol	3°
8)	$HOCH_2CH(CH_3)CH_2CH_2CH_3$	2-methyl-1-pentanol	1°
9)	$CH_3CH_2CH(CH_3)CH_2CH_2OH$	3-methyl-1-pentanol	1°
10)	$CH_3CH_2CH(CH_3)CHOHCH_3$	3-methyl-2-pentanol	2°
11)	$CH_3CH_2C(CH_3)OHCH_2CH_3$	3-methyl-3-pentanol	3°
12)	$(CH_3)_2CHCH(CH_3)CH_2OH$	2,3-dimethyl-1-butanol	1°
13)	$(CH_3)_2CHC(CH_3)(OH)CH_3$	2,3-dimethyl-2-butanol	3°
14)	$CH_3CH_2CH(CH_2CH_3)CH_2OH$	2-ethyl-1-butanol	1°
15)	$(CH_3)_3CCH_2CH_2OH$	3,3-dimethyl-1-butanol	1°
16)	$(CH_3)_3CCHOHCH_3$	3,3-dimethyl-2-butanol	2°
17)	$HOCH_2C(CH_3)_2CH_2CH_3$	2,2-dimethyl-1-butanol	1°

7. a. The name is incorrect since it mixes common and systematic nomemclature.

Correct IUPAC: 2-methoxy-2-methylbutane
Common: methyl *t*-pentyl ether or *t*-amyl methyl ether

Amyl is an older common name for pentyl; e.g., *n*-amyl, isoamyl, *tert*-amyl. In modern usage, amyl is being replaced increasingly by pentyl, but the name is common in the older literature.

c. Since both molecules are comparable in size and shape, the deciding factor in determining physical properties is the presence of hydrogen bonding in the alcohol which will raise its boiling point with respect to the ether.

8. a. $CH_3CH_2CH_2OCH_2CH_2CH_3$ — *Sulfuric acid-catalyzed ether formation*

b. $CH_3CH_2CH_2OH + CH_2{=}C(CH_3)_2$ — *E2 elimination (remember: 3° halide and strong base)*

c. $(CH_3)_3COCH_2CH_2CH_3$ — *Williamson ether synthesis*

d. CH_3CH_2CHO — *Volatile aldehydes can be isolated under these conditions*

e. 4-*tert*-butylcyclohexanone (structure: cyclohexane ring with =O and a *tert*-butyl group) — *Standard chromic acid ("Jones") oxidation*

f. $CH_3CH_2CH_2I + Na^{+}\ {}^{-}O_3S{-}C_6H_4{-}CH_3$ — *S_N2 displacement, sodium tosylate precipitates*

g. $CH_3CH_2CH_2Br$ — *Acid-catalyzed formation of alkyl halide*

h. $(CH_3CH_2)_2CBrCH_2CH_2CH_3$ — *Reaction proceeds with rearrangement to give the tertiary carbocation as an intermediate*

i. (R)-$CH_3CHClCH_2CH_3$ — *Alkyl chloride formation with inversion of configuration*

j. $(CH_3CH_2)_2CHO^{-}K^{+} + H_2\uparrow$ — *Potassium hydride is a strong base*

k. $2\ CH_3CH_2CH_2I$

l. $CH_3CH_2CHBrCH_2CH_3 + CH_3Br$

(k and l) — *Acid-catalyzed formation of alkyl halide*

9. a. and b. Primary alcohols are best converted to primary alkyl chlorides with $SOCl_2$ in pyridine.

c. In this case, rearrangement is desired, so a reagent favoring carbocation intermediates should be chosen: $ZnCl_2 + HCl$.

d. The best way to make alkyl iodides from alcohols is to carry out a displacement on a sulfonate ester intermediate:

$$CH_3CH_2CH_2CHOHCH_3 + C_6H_5SO_2Cl \longrightarrow CH_3CH_2CH_2CH(OSO_2C_6H_5)CH_3 \xrightarrow[\text{acetone}]{\text{NaI}} CH_3CH_2CH_2CHICH_3$$

e. Rearrangement must be avoided: PBr_3 at 0°C.

f. Tertiary chloride from the alcohol, with no fear of rearrangement: HCl at 0°C.

10. a. Rearrangement is desired: $ZnCl_2 + HCl$

b. $(CH_3)_2CHCHOHCH_3 + PBr_3 \xrightarrow{\text{low temp.}} (CH_3)_2CHCHBrCH_3$

c. $(CH_3CH_2)_3COH\ +\ H_2SO_4 \xrightarrow{\Delta} CH_3CH{=}C(CH_2CH_3)_2$
A drop of sulfuric acid or a small amount of sulfonic acid is sufficient. Alternatively, the alcohol can be passed over hot Al_2O_3.

d. $CH_3CH_2CH_2CH_2OH + SOCl_2 \longrightarrow CH_3CH_2CH_2CH_2Cl \xrightarrow{CN^-} CH_3CH_2CH_2CH_2CN$

e. $(CH_3)_3CCl + H_2O \xrightarrow{\Delta} (CH_3)_3COH \xrightarrow{K} (CH_3)_3CO^-K^+ \xrightarrow[\Delta]{CH_3I} (CH_3)_3COCH_3$

Note that the reverse procedure of treating $(CH_3)_3CCl$ with $CH_3O^-K^+$ will not work, because the halide is tertiary and E2 elimination would dominate.

f. $(CH_3)_2CHCH_2OH \xrightarrow{PBr_3} (CH_3)_2CHCH_2Br \xrightarrow{CH_3S^-Na^+} (CH_3)_2CHCH_2SCH_3$

g. $(CH_3)_3COH + HBr \longrightarrow (CH_3)_3CBr \xrightarrow{Mg} (CH_3)_3CMgBr \xrightarrow[2.\ H_2O]{1.\ \text{ethylene oxide}} (CH_3)_3CCH_2CH_2OH$

11. a. $CH_4 + Br_2 \xrightarrow{h\nu} CH_3Br \xrightarrow{NaOH} CH_3OH$

b. cyclohexane $\xrightarrow[h\nu]{Br_2}$ bromocyclohexane (Br) $\xrightarrow{NaOH}$ cyclohexanol (OH) $\xrightarrow{Na}$ cyclohexoxide (O^-Na^+) $\xrightarrow[\text{(from a.)}]{CH_3Br}$ methoxycyclohexane (OCH_3)

c. $CH_3CH_2CH_3 + Br_2 \xrightarrow{h\nu} CH_3CHBrCH_3 \xrightarrow{(CH_3)_3CO^-K^+} CH_3CH{=}CH_2$ [E2]
[$(CH_3)_3CO^-K^+$ can be made from $(CH_3)_3CH$ by a series of steps similar to b. above.]

d. $CH_3CH_3 + Br_2 \xrightarrow{h\nu} CH_3CH_2Br \xrightarrow{NaSH} CH_3CH_2SH$

12. $CH_3CH_2C^+DCH_3 \rightleftharpoons CH_3CH^+CHDCH_3$ Equilibration of secondary carbocations has occurred.

$\updownarrow Br^-$ $\qquad$ $\updownarrow Br^-$

$CH_3CH_2CDBrCH_3 \rightleftharpoons CH_3CHBrCHDCH_3$

13. cyclopropyl–C(CH₃)₂–OH $+ H^+ \rightleftharpoons$ cyclopropyl–C(CH₃)₂–OH_2^+ $\rightleftharpoons$ cyclopropyl–$C^+(CH_3)_2$ (A) $\xrightarrow{Cl^-}$ cyclopropyl–C(CH₃)₂–Cl (observed product)

A $\updownarrow$ B (cyclobutyl cation, +) $\xrightarrow{Cl^-}$ chlorocyclobutane derivative (Cl) — not formed

The possible rearrangement A → B, is not observed. The ring strain of a cyclobutane ring is as high as that of a cyclopropane ring (see Table 5.5 in text), so that the isomerization A → B would involve conversion of a tertiary to a secondary carbocation with no relief of ring strain.

14. a. $3\ (CH_3)_2CHOH + 2\ KMnO_4 + 2\ H^+ \longrightarrow 3\ (CH_3)_2C{=}O + 2\ MnO_2 + 2\ K^+ + 4\ H_2O$

b. $ClCH_2CH_2CH_2OH + 4\ HNO_3 \longrightarrow ClCH_2CH_2CO_2H + 4\ NO_2 + 3\ H_2O$

c. (cyclohexane-1,2-diol, H and OH on each of two adjacent carbons) $+ K_2Cr_2O_7 + 8\ H^+ \longrightarrow HO_2C(CH_2)_4CO_2H + 2\ K^+ + 2\ Cr^{+3} + 5\ H_2O$

15. a. The rearrangement requires a carbocation intermediate. The formation of the high-energy primary carbocation cannot compete with S_N2 displacement on the oxonium ion.

$$(CH_3)_2CHCH_2OH + H^+ \rightleftharpoons (CH_3)_2CHCH_2O^+H_2 \xrightarrow{Br^-} (CH_3)_2CHCH_2Br + H_2O$$

$$(CH_3)_2CHCH_2O^+H_2 \not\longrightarrow (CH_3)_2CHCH_2^+ \longrightarrow (CH_3)_3C^+$$

The secondary carbocation forms far more readily:

$$\longrightarrow (CH_3)_2CHC^+HCH_3 \longrightarrow (CH_3)_2C^+CH_2CH_3 \xrightarrow{Br^-} (CH_3)_2CBrCH_2CH_3$$

b. Mixtures of ethers generally result:

$$R'CH_2OH + R''CH_2OH \xrightarrow{H_2SO_4} R'CH_2OCH_2R' + R'CH_2OCH_2R'' + R''CH_2OCH_2R''$$

Either alcohol forms $RCH_2O^+H_2$ and can be displaced by either alcohol. However *t*-butyl alcohol readily forms a carbocation,

$$(CH_3)_3COH + H^+ \rightleftharpoons (CH_3)_3CO^+H_2 \longrightarrow (CH_3)_3C^+ + H_2O$$

which reacts more readily with methanol than with the bulkier *t*-butyl alcohol:

$$(CH_3)_3C^+ + CH_3OH \longrightarrow (CH_3)_3C\underset{H}{O^+}CH_3 \longrightarrow (CH_3)_3COCH_3 + H^+$$

c. $C_2H_5OC_2H_5 + HI \rightleftharpoons CH_3CH_2\underset{H}{O^+}CH_2CH_3 + I^-$

$$I^- + CH_3CH_2\text{–}\overset{H}{\underset{+}{O}}CH_2CH_3 \longrightarrow CH_3CH_2I + CH_3CH_2OH$$

$$CH_3CH_2OH + HI \rightleftharpoons CH_3CH_2O^+H_2 + I^- \longrightarrow CH_3CH_2I + H_2O$$

16. $CH_3CH_2OCH_2CH_2CH_2CH_3 + HBr \rightleftharpoons CH_3CH_2\overset{H}{O^+}CH_2CH_2CH_2CH_3 + Br^-$

$CH_3CH_2\overset{H}{O^+}CH_2CH_2CH_2CH_3 + Br^- \rightleftharpoons CH_3CH_2Br + HOCH_2CH_2CH_2CH_3$

$CH_3CH_2\overset{H}{O^+}CH_2CH_2CH_2CH_3 + Br^- \rightleftharpoons CH_3CH_2OH + BrCH_2CH_2CH_2CH_3$

$(CH_3)_3COCH_2CH_3 + HBr \rightleftharpoons (CH_3)_3C\overset{H}{O^+}CH_2CH_3 + Br^-$

$(CH_3)_3C\overset{H}{O^+}CH_2CH_3 \rightleftharpoons (CH_3)_3C^+ + HOCH_2CH_3$

$(CH_3)_3C^+ + Br^- \rightleftharpoons (CH_3)_3CBr$

In the first case, product is formed by S_N2 attack on the protonated ether. Attack can occur at either primary carbon at comparable rates, so both ethyl and *n*-butyl bromide are produced. In the second case, the ether cleaves by an S_N1 process to give the *t*-butyl cation,and thus *t*-butyl bromide is the main product. The tertiary carbocation is sufficiently stable that it is produced readily by concentrated acid, even in the cold. Recall that tertiary alcohols rapidly give the halide with cold concentrated HCl or HBr.

17. $(CH_3CH_2)_3O^+ \; BF_4^-$ [structure: CH_3CH_2—O(+)(—CH_2CH_3)(—CH_2CH_3) with F—B(−)(F)(F)—F]

The O^+ is strongly electron accepting. The positive charge on the oxygen enhances its electronegativity making the CH_2 group of the ethyl electron deficient.

Mechanism of Ethylation:

$R\ddot{O}H + CH_3CH_2—\overset{+}{O}(CH_2CH_3)—CH_2CH_3 \longrightarrow R\overset{H}{\overset{+}{O}}CH_2CH_3 + O(CH_2CH_3)—CH_2CH_3$ (leaving group)

$R\overset{H}{\overset{+}{O}}CH_2CH_3 \rightleftharpoons ROCH_2CH_3 + H^+$

It is a very reactive compound because $(CH_3CH_2)_2O$ is a good leaving group, similar to H_2O. (The pK_a of $(CH_3CH_2)_2O^+H$ is about −3.6)

18.

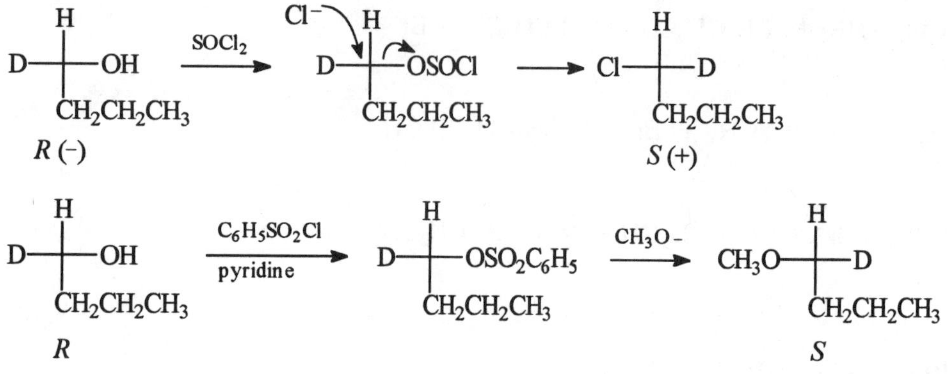

19.

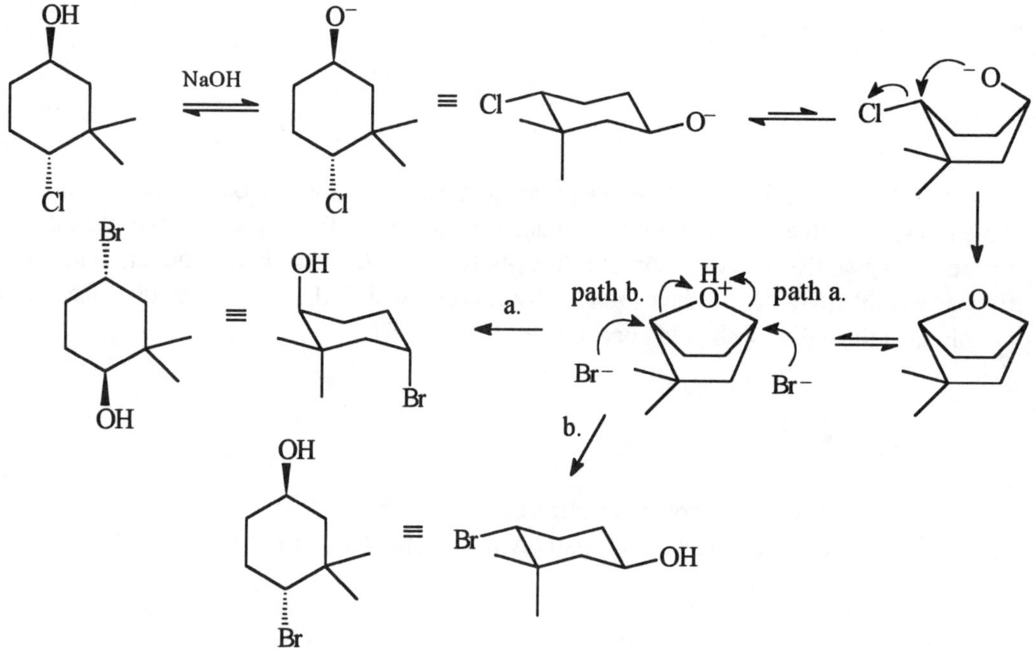

The clues here are the *trans* relationship of the OH and Br in both isomers of the product, and the fact that the bromine can end up at either end of the molecule.

20.

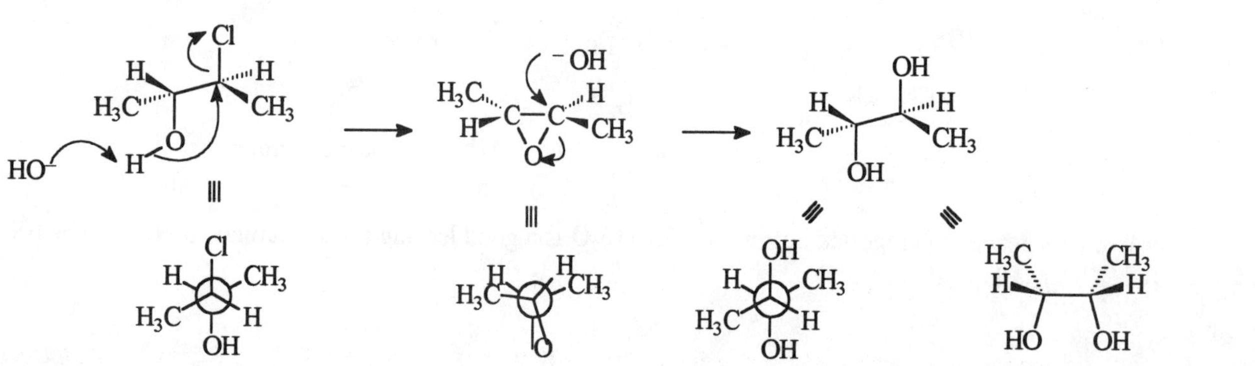

meso, optically inactive

(You will find that you get the same result regardless of which end of the epoxide is attacked in the hydrolysis step)

21. 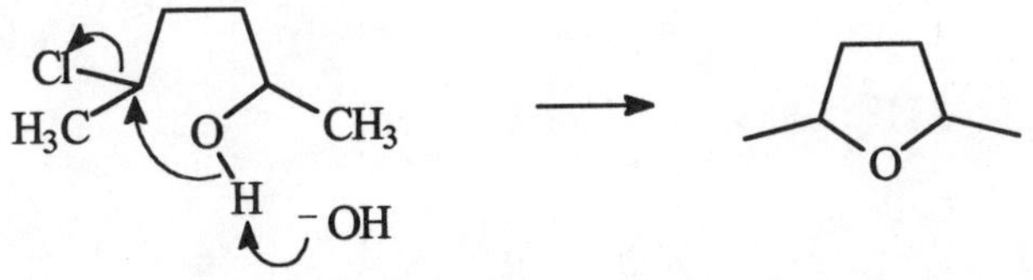 H_3C, Cl, O, CH_3, H, ^-OH → 2,5-dimethyltetrahydrofuran

The product could exist as an achiral, *cis* isomer: H_3C, CH_3, H, O, H — plane of symmetry

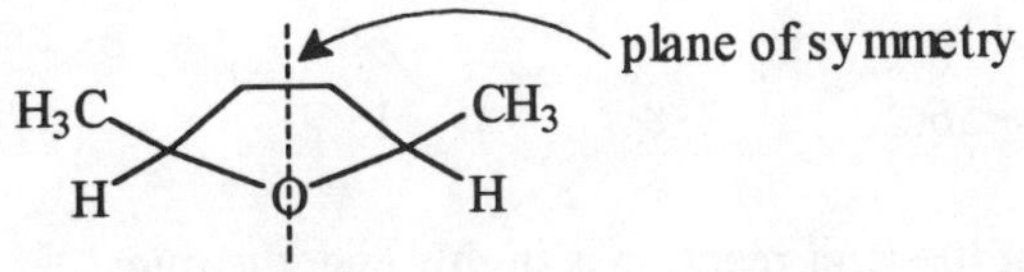

or as a chiral, *trans*, isomer: H_3C, H, H, O, CH_3 or H, CH_3, H_3C, O, H

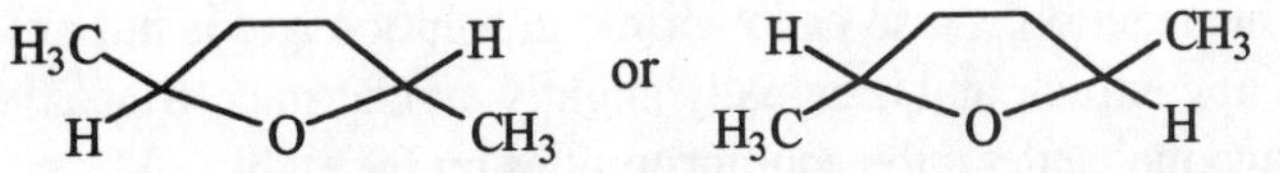

The starting material was **not** racemic (it was optically active), so if the *trans* isomer had been formed, it would not have been racemic. However, the product obtained is optically **inactive**, suggesting that the product is the achiral, **meso** isomer: the *cis* compound. This means that the starting material was either (2*R*,5*R*)-5-chloro-2-hexanol or the (2*S*,5*S*) isomer, and **not** the (2*R*,5*S*) or (2*S*,5*R*) isomers:

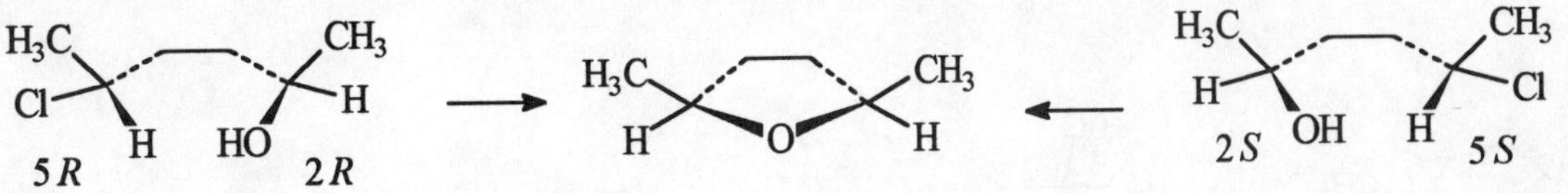

H_3C, CH_3, Cl, H, H, HO, 5*R*, 2*R* → H_3C, CH_3, H, O, H ← H_3C, CH_3, H, Cl, OH, H, 2*S*, 5*S*

22. The cavity inside 18-crown-6 is just the right size for K^+ to fit in (diameter of K^+ = 2.66Å) Na^+ is smaller (diameter = 1.96Å) and it is not coordinated so tightly. Therefore, 18-crown-6 solubilizes potassium salts much more effectively than sodium salts.

23. 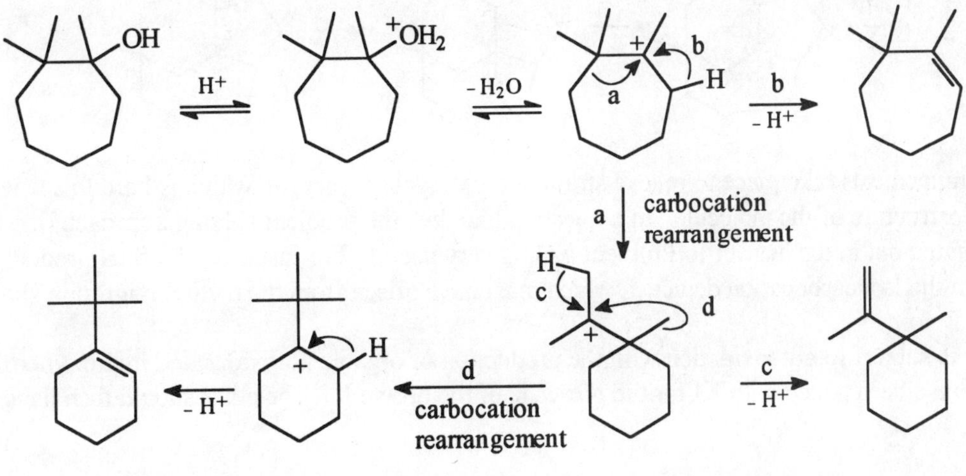 OH —H+→ $^+OH_2$ —$-H_2O$→ (carbocation, a, b, H) —b, $-H^+$→

a carbocation rearrangement ↓

(H, c, d, +) —c, $-H^+$→

(H, +) ←d, carbocation rearrangement— ; ←$-H^+$—

REMEMBER, cycloheptane is strained relative to cyclohexane (see Table 5.5 in text).

(One of the ways to solve mechanism problems is to work both backwards as well as forwards. For instance, 1-*t*-butylcyclohexene must have come from the tertiary carbocation shown next to it in the scheme; that in turn must have come from the same carbocation that led to 1-isopropenyl-1-methylcyclohexane, etc.)

24. CH_3CH_2OH + Br^- ⇌ CH_3CH_2Br + OH^-

ΔH_f°: -56.2 -50.9 -15.2 -32.7 $\Delta H^\circ = +59.2$ kcal mole^{-1}

CH_3CH_2OH + HBr ⇌ CH_3CH_2Br + H_2O

ΔH_f°: -56.2 -8.7 -15.2 -57.8 $\Delta H^\circ = -8.1$ kcal mole^{-1}

Note that the first reaction is highly endothermic; OH^- is a much stronger base than Br^- in the gas phase just as it is in solution. This reaction is also endothermic in solution and is not observed. The second reaction is almost thermoneutral in contrast, and is actually slightly exothermic. Since the entropy change is close to zero (two molecules give two molecules), the equilibrium lies on the right.

The difference between the two reactions can be seen from the following comparison: the O-H bond strength is much greater than the Br-H bond strength, and this **difference** is greater than for the C-O and C-Br bond strengths (see Appendix III).

25.

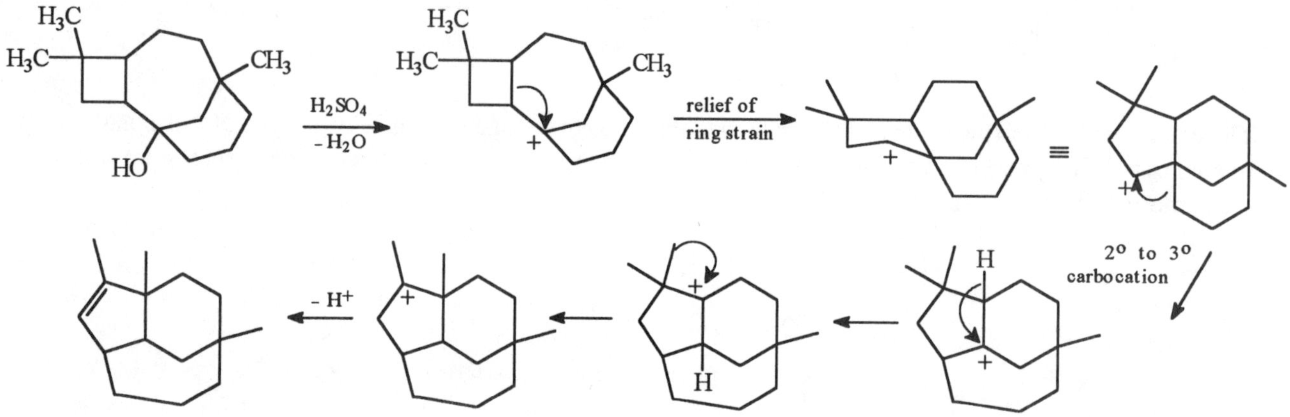

H₃C H₃C CH₃ HO $\xrightarrow[-H_2O]{H_2SO_4}$ H₃C H₃C CH₃ + $\xrightarrow{\text{relief of ring strain}}$ + ≡ +

2° to 3° carbocation

H + H + + $\xleftarrow{-H^+}$

The last two rearrangements take place to relieve strain in the tricyclic structure, which is hard to see without knowing the stereostructure of the molecule. In a question like this, the problem-solving approach of working backwards, as pointed out in the answer to Problem #23, is very useful. For instance, the final product must have resulted from the last carbocation depicted, which must have arisen from the methyl migration shown, etc.

26. Diisopropyl ether is subject to autooxidation with the production of organic peroxides and hydroxyperoxides. The ether should have been tested with KI for the presence of the brown I_3^-. The ether should then have been washed with Fe^{2+}.

10.F Supplementary Problems

S1. Provide the structure and IUPAC name for each of the following compounds:

a. α-chloroethyl ethyl ether
b. γ-chloropropyl alcohol
c. di(neopentyl) ether
d. ω-chloroheptyl alcohol

S2. All of the names below are incorrect. Provide a correct name, either common or IUPAC, for each:

a. isopropanol b. 2,2-dimethyl-5-hexanol c. β,β-dichloropropanol d. 2,3-dihydroxybutane

S3. Give the principal product(s) from each of the following reactions:

a. $CH_3CH_2CH_2OH \xrightarrow{H_3C-C_6H_4-SO_2Cl} \xrightarrow[\text{acetone}]{NaI}$

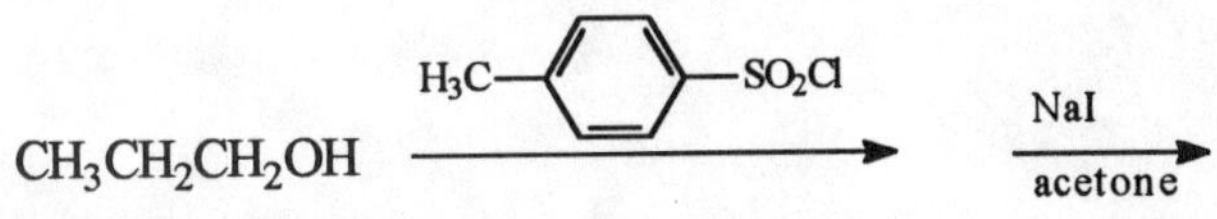

b. $ICH_2CH_2CH_2CH_2I \xrightarrow{NaOH}$

c. $(CH_3)_2CHOCH_3 \xrightarrow[\text{light}]{O_2}$

d. O (cyclopentene oxide) $\xrightarrow{H_3O^+}$

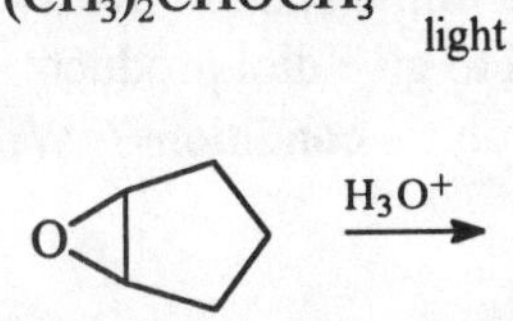

e. $(CH_3)_3COC(CH_3)_3 \xrightarrow[\Delta\ \text{(heat)}]{H_2SO_4}$

f. CH_3CH_2–C(OH)(H)(D) $\xrightarrow{PBr_3} \xrightarrow{NaOH}$

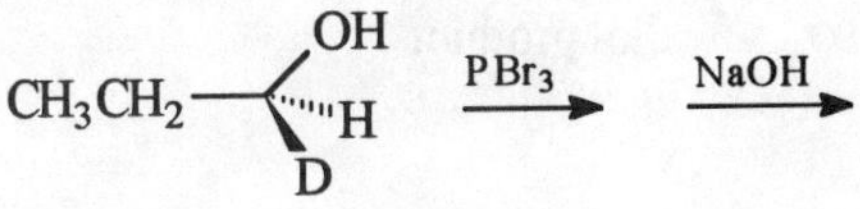

S4. Give the reagents and best conditions for carrying out the following transformations:

a. $CH_3CH_2C(CH_3)_2Br \longrightarrow CH_3CH_2C(CH_3)_2OH$

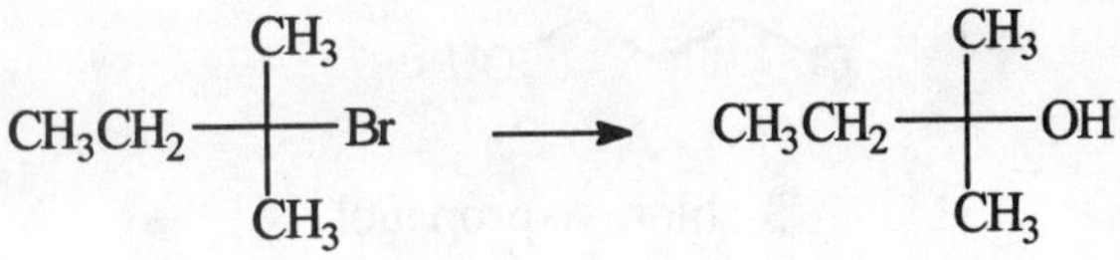

b. $CH_3CH_2CH{=}CH_2 \longrightarrow CH_3CH_2CH(OH)CH_2OH$

c. 2-methylcyclohexanol (CH_3, OH) $\longrightarrow$ 1-chloro-2-methylcyclohexane (CH_3, Cl)

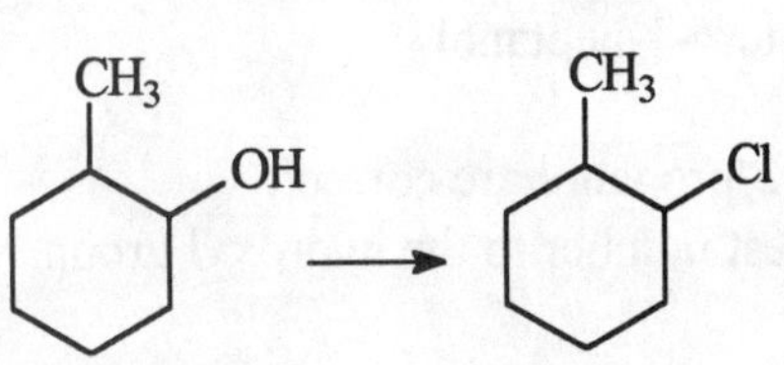

d. cyclopentanol (–OH) $\longrightarrow HO_2CCH_2CH_2CH_2CO_2H$

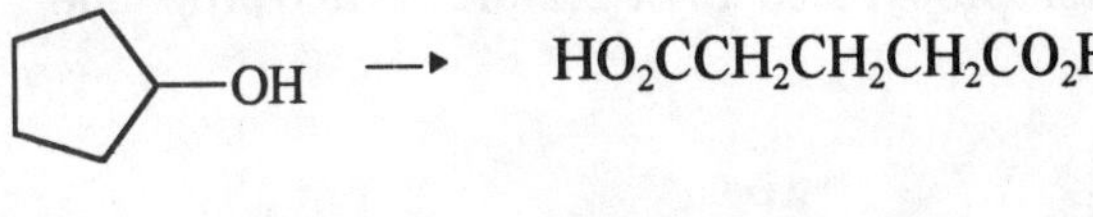

S5. Rank the following compounds in order of decreasing acidity.

a. CH_3CH_2OH b. $(CH_3)_3COH$ c. $ClCH_2CH_2OH$

d. $ClCH_2CH_2CO_2H$ e. CF_3CH_2OH f. CF_3OCH_3

S6. Which of the following reactions give(s) optically active products? Justify your answers.

a. 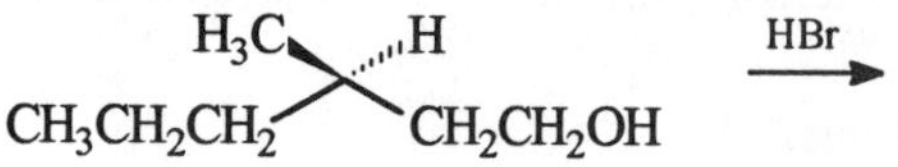 H_3C, H / $CH_3CH_2CH_2$, CH_2CH_2OH $\xrightarrow{HBr}$

b. 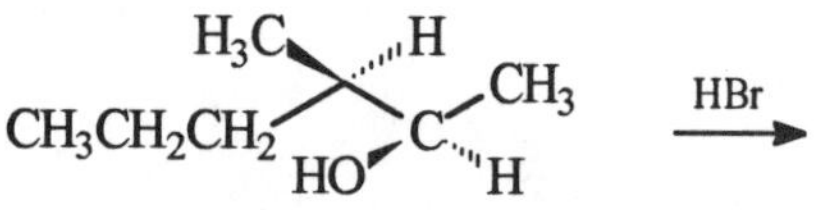 H_3C, H / $CH_3CH_2CH_2$, C(CH_3)(HO)(H) $\xrightarrow{HBr}$

c. H_3C, OH / $CH_3CH_2CH_2$, CH_2CH_3 $\xrightarrow{HBr}$

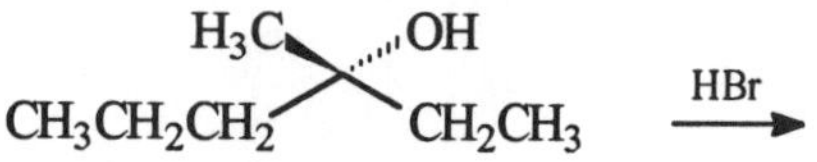

S7. (*S*)-1-Bromo-2-methylbutan-2-ol is converted to an optically active epoxide with dilute sodium hydroxide, as depicted below. The epoxide ring can be cleaved either in strong base or in acid to give diol products. What is the difference (if any) between the products formed by the acidic and basic hydrolysis conditions? Write a step-by-step mechanism to explain any differences you expect to see.

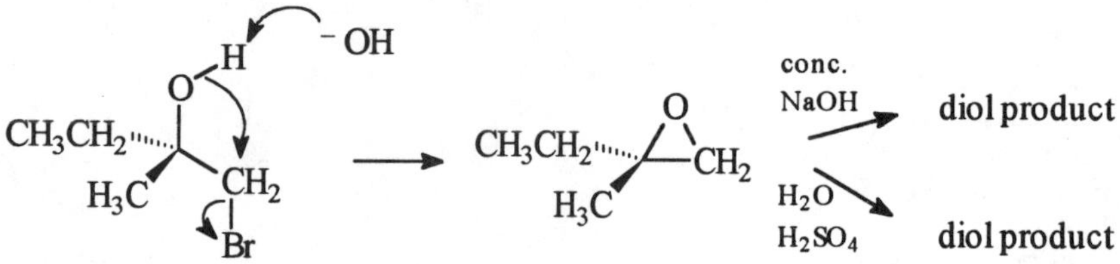

CH_3CH_2, H_3C, O–H, CH_2–Br, ^-OH → CH_3CH_2, H_3C, O, CH_2 (epoxide)

conc. NaOH → diol product

H_2O / H_2SO_4 → diol product

10.G Answers to Supplementary Problems

S1. a. 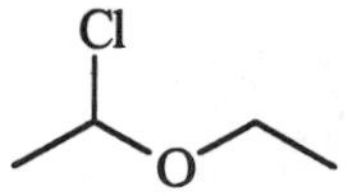 1-chloro-1-ethoxyethane

b.  3-chloro-1-propanol

c. 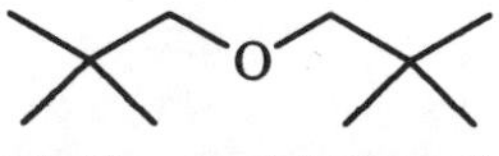 2,2-dimethyl-1-(2,2-dimethylpropoxy)propane

d.  7-chloro-1-heptanol

S2. a. Mixture of common and IUPAC usage: isopropyl alcohol and 2-propanol are correct.
b. As an "alkanol", the chain should be numbered to give the lowest number to the hydroxyl group: 5,5-dimethyl-2-hexanol.
c. Mixture of common and IUPAC usage: β,β-dichloropropyl alcohol or 2,2-dichloro-1-propanol.
d. Should be named as an alkanediol: 2,3-butanediol.

S3. a. $CH_3CH_2CH_2I$

b.  (tetrahydrofuran, O) After substitution of one of the iodides with hydroxyl, the second iodide is attacked faster <u>intra</u>molecularly by the alkoxide anion than it is displaced <u>inter</u>molecularly by hydroxide:

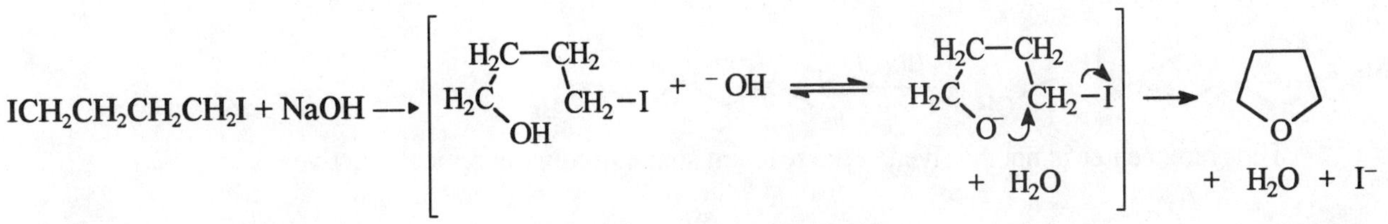

$$ICH_2CH_2CH_2CH_2I + NaOH \longrightarrow \left[\begin{matrix} H_2C{-}CH_2 \\ H_2C \quad CH_2{-}I \\ OH \end{matrix} + {}^{-}OH \rightleftharpoons \begin{matrix} H_2C{-}CH_2 \\ H_2C \quad CH_2{-}I \\ O^{-} \end{matrix} + H_2O \right] \longrightarrow \text{(tetrahydrofuran)} + H_2O + I^{-}$$

c. 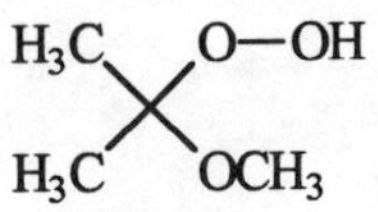 $(H_3C)_2C(O{-}OH)(OCH_3)$ (explosive!)

d. 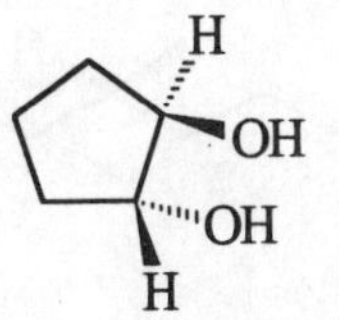 cyclopentane-1,2-diol (H, OH on each carbon, trans)

e. $2\ (CH_3)_2C{=}CH_2$

f. 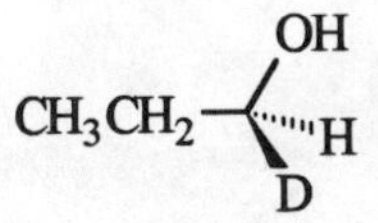 $CH_3CH_2{-}C(OH)(H)(D)$ (Two inversions give the same product back again)

S4.

a. 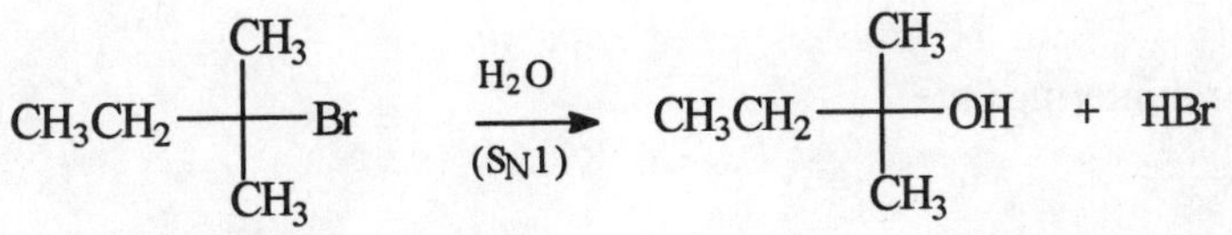 $CH_3CH_2C(CH_3)_2Br \xrightarrow[(S_N1)]{H_2O} CH_3CH_2C(CH_3)_2OH + HBr$

b. 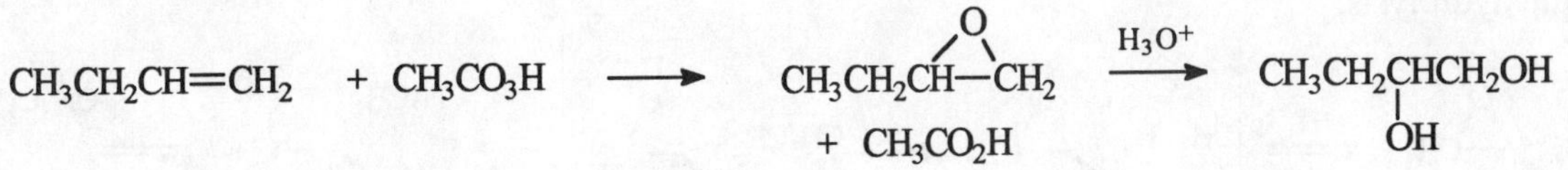 $CH_3CH_2CH{=}CH_2 + CH_3CO_3H \longrightarrow CH_3CH_2\overset{O}{CH{-}CH_2} + CH_3CO_2H \xrightarrow{H_3O^+} CH_3CH_2CH(OH)CH_2OH$

c. 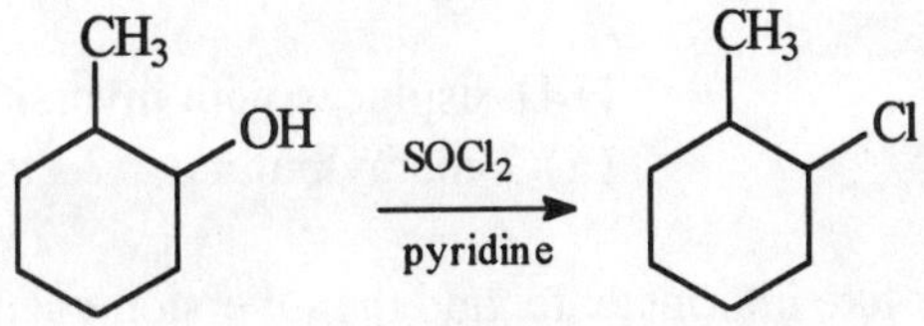 2-methylcyclohexanol $\xrightarrow[\text{pyridine}]{SOCl_2}$ 1-chloro-2-methylcyclohexane (Carbocation rearrangement must be avoided.)

d. 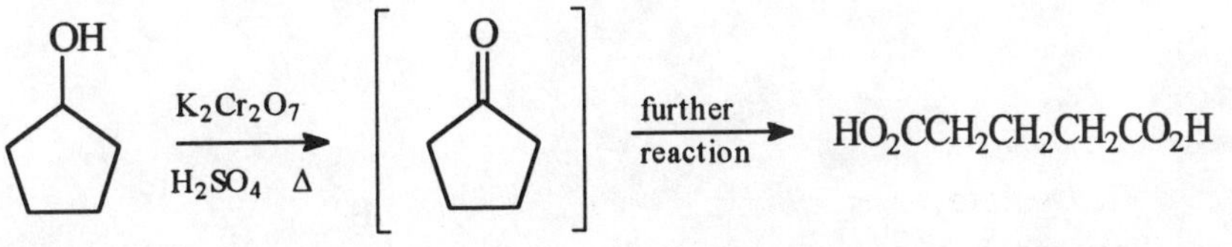 cyclopentanol $\xrightarrow[H_2SO_4\ \Delta]{K_2Cr_2O_7}$ [cyclopentanone] $\xrightarrow[\text{reaction}]{\text{further}}$ $HO_2CCH_2CH_2CH_2CO_2H$

S5. Most acidic: $ClCH_2CH_2CO_2H$ (it's a carboxylic acid)

CF_3CH_2OH (halogens are electron withdrawing, stabilize negative charge)

$ClCH_2CH_2OH$

CH_3CH_2OH

$(CH_3)_3COH$ (alkyl groups are elecron donating, don't stabilize negative charge)

Least acidic: CF_3OCH_3 (there's no hydrogen that can dissociate)

S6. a. [structure: alcohol with stereocenter, H, OH] $\xrightarrow{HBr}$ [structure: bromide with stereocenter, H, Br]

The stereocenter is not involved in the reaction so the product is optically active.

b. [structure: H, HO, H] $\xrightarrow[-\ H_2O]{H^+}$ [[carbocation, H, +] ⟶ [tertiary carbocation, +]] $\xrightarrow{Br-}$ [tertiary bromide, Br]

racemic

The reaction involves a rearrangement to give a planar, achiral carbocation intermediate and therefore a racemic product.

c. [structure: OH] $\xrightarrow[-\ H_2O]{H^+}$ [tertiary carbocation, +] $\xrightarrow{Br-}$ [tertiary bromide, Br]

racemic

This reaction involves the same achiral intermediate.

S7. Acidic hydrolysis:

[epoxide: CH_3CH_2, H_3C, O, CH_2] $\underset{}{\overset{H^+}{\rightleftharpoons}}$ [[protonated epoxide: H, +O, CH_3CH_2, H_3C, CH_2] ⟷ [H, O, CH_3CH_2, +, H_3C, CH_2]] ⟶ [CH_3CH_2, OH, CH_2, H_3C, $+OH_2$] ⇌ [CH_3CH_2, OH, CH_2, H_3C, OH]

H_2O displaces with inversion to give (*R*)-2-methylbutane-1,2-diol

The acid-catalyzed process favors cleavage via a carbocation-like intermediate, and the inversion occurs at the tertiary center.

Basic hydrolysis:

[epoxide: CH_3CH_2, O, CH_2, H_3C, $^-$OH] ⟶ [CH_3CH_2, O^-, CH_2OH, H_3C] $\xrightarrow{H_2O}$ [CH_3CH_2, OH, CH_2OH, H_3C] + $^-$OH

(*S*)-2-methylbutane-1,2-diol

The mechanism of the basic hydrolysis is like the S_N2 mechanism, and attack and inversion at the less sterically hindered, primary carbon occurs, leading to the enantiomer of the product obtained in acid.

11. ALKENES

11.A Chapter Outline and Important Terms Introduced

11.1 Electronic Structure

σ- and π-bonds
configurational vs. conformational isomers

11.2 Nomenclature of Alkenes

olefin
alkylenes
vinyl group
cis and *trans*
Z (zusammen) and E (entgegen)

11.3 Physical Properties of Alkenes

11.4 Relative Stabilities of Alkenes: Heats of Formation

steric hindrance
effects of substitution

11.5 Preparation of Alkenes

A. Dehydrohalogenation of Alkyl Halides (and sulfonates)
  rate depends on [RX] and [Base]
  isotope effects
  stereospecificity
  E2 mechanism
  rate depends on leaving group
  anti elimination stereochemistry

B. Dehydration of Alcohols
  acid-catalyzed
  alumina at 350-400°C
  alkene isomerization

C. Industrial Preparation of Alkenes
  cracking:

$$RCH_2CH_2R' \xrightarrow{700\text{-}900°C} RH + CH_2{=}CHR'$$

11.6 Reactions of Alkenes

A. Catalytic Hydrogenation
  syn addition stereochemistry

B. Addition of Halogens
  cyclic halonium ion
  vicinal dihalides
  anti addition stereochemistry

C. Addition of HX and Water
  Markovnikov's Rule
  addition of mercuric acetate
  hydration

D. Hydroboration
  organoboranes
  "anti-Markovnikov" hydration

E. Oxidation
  glycol formation
  syn addition of hydroxyls
  oxidative cleavage with ozone
  epoxidation with a peracid (*syn* addition)

F. Addition of Carbenes and Carbenoids: Preparation of Cyclopropanes
  syn addition

G. Free Radical Additions
  addition of HBr (not HCl or HI)
  addition of CX_4
  anti-Markovnikov orientation

11.B Important Reactions Introduced

Dehydrohalogenation of alkyl halides (and sulfonates) (11.5.A)

Equation:

Base:⁻ H–C–C–X → Base-H C=C X^-

Generality: Base = OH^- or RO^-, NH_2^-, or in some cases 3° amine (R_3N:)
X = good leaving group, such as Cl^-, Br^-, I^-, or sulfonate ester ($R'SO_3^-$) (**not** OH^- or RO^-)

Key features: concerted, bimolecular mechanism (E2)
faster if X is a better leaving group
anti relationship between C-H and C-X bonds that are broken
more stable isomer (more substituted > less substituted; *trans* > *cis*) usually favored, but bulky base (e.g. $(CH_3)_3CO^-$) forms less substituted isomer.

Dehydration of alcohols (11.5.B)

Equation: CH–C–OH $\xrightarrow{H^+}$ C=C + H_2O

Generality: ease of reaction: 3° > 2° > 1° alcohol
H^+ usually H_2SO_4; also Al_2O_3, heat

Key features: more stable olefin favored
mechanism involves carbocation intermediates (E1 mechanism)
possibility of carbocation rearrangements

Hydrogenation of alkenes (11.6.A)

Equation: C=C + H_2 $\xrightarrow{\text{catalyst}}$ H–C–C–H

Generality: catalyst = Pd/C, PtO_2, Ni, etc.
sometimes requires high pressure of H_2 gas (for hindered or very stable double bond)

Key features: *syn* addition stereochemistry
sometimes double bond rearrangement observed before H_2 addition

Electrophilic addition to alkenes (11.6.B)

Equation: E^+ + C=C ⇌ C–C bridged by E (+) ; :Nu → E–C–C–Nu

Generation:	**Reagent**	**E**	**Nu:**	
	X_2/CCl_4 solvent	X	X	[X = Cl, Br, I]
	X_2/ROH solvent	X	OR	[X = Cl, Br, I; R = H or alkyl]
	$Hg(OAc)_2$/ROH	Hg(OAc)	OR	[R = H, alkyl, or acetyl]

Key features: *anti* addition stereochemistry as result of cyclic halonium ion (or "mercurinium" ion)
Markovnikov orientation (Nu becomes attached to carbon best able to stabilize positive charge.)

Addition of HX and water to alkenes (11.6.C)

Equation:

$$\text{H-Nu} + \text{>C=C<} \underset{}{\overset{H^+}{\rightleftharpoons}} \text{>CH-C}^+\text{<} \;(\text{:Nu}^-) \longrightarrow \text{>CH-C(Nu)<}$$

Generality: Nu = H_2O, HOR, Cl^-, Br^-

Key features: acid-catalyzed, via carbocation intermediate (therefore rearrangements possible)
Markovnikov orientation (Nu becomes attached to carbon best able to stabilize positive charge.)
no specificity for *syn* or *anti* stereochemistry of addition
Note: HBr can add via a free radical mechanism (see below).

Hydroboration (11.6.D)

Equation:

$$\text{>C=C<} \xrightarrow{\text{H-BR}_2} \text{(H)C-C(BR}_2\text{)} \xrightarrow[\text{OH}^-]{\text{H}_2\text{O}_2} \text{(H)C-C(OH)}$$

Generality: normal alkenes
hydroboration reagent: R = H ($B_2H_6 \rightleftharpoons 2\ BH_3$) or alkyl

Key features: *syn* addition of H and B
retention of configuration in oxidation of C-B to C-O bond
boron becomes bonded to less substituted (= less hindered) carbon, therefore overall process is *syn*, anti-Markovnikov hydration of alkene

Oxidation of alkenes to vicinal diols (11.6.E)

Equation:

$$\text{>C=C<} \xrightarrow{\text{MO}_4} \text{cyclic } \text{C(O)-C(O)} \text{ (O}_2\text{M ester)} \xrightarrow{\text{H}_2\text{O}} \text{(HO)C-C(OH)}$$

Generality: MO_4 = OsO_4 or MnO_4^-

Key features: *syn* addition of hydroxyl groups
OsO_4 is toxic and expensive, but can be used catalytically with H_2O_2
aq. MnO_4^- ($KMnO_4$, dilute, cold) is cheap, but reaction often goes in low yield.

Oxidative cleavage of alkenes (11.6.E)

Equation: $>C=C< \xrightarrow{O_3} \xrightarrow{\text{reduction}} >C=O \quad O=C<$

Generality: very general, almost all types of alkenes undergo this reaction

Key features: mechanism involves cyclic peroxide intermediates (molozonides and ozonides: explosive) work-up involves reduction of ozonide: Zn/H_2O to give ketones and aldehydes, or $NaBH_4$ to give alcohol products.

Epoxidation of alkenes (11.6.E)

Equation: $>C=C< + RCO_3H \longrightarrow$ epoxide (C–C bridged by O) $+ RCO_2H$

Generality: RCO_3H usually CH_3CO_3H or Cl-C$_6$H$_4$-CO_3H (meta-chloroperoxybenzoic acid)

Key features: more substituted double bonds react faster
stereospecific *syn* addition

Carbene addition to alkenes (11.6.F)

Equation: $>C=C< + :CR_2 \longrightarrow$ cyclopropane (C–C bridged by CR_2)

Generality: $:CR_2 = :CCl_2$ or $:CBr_2$ from HCX_3 + strong base

$= :CH_2$ from $CH_2N_2 \xrightarrow{h\nu} :CH_2 + N_2$
or = ":CH_2" from $CH_2I_2 + Zn \longrightarrow IZnCH_2I$

Key features: stereospecific *syn* addition

Free radical addition to alkenes (11.6.G)

Equation: $Y\cdot + >C=C< \longrightarrow Y-\overset{|}{\underset{|}{C}}-\dot{C}< \quad X-Y \longrightarrow Y-\overset{|}{\underset{|}{C}}-\overset{|}{\underset{|}{C}}-X + Y\cdot$

Generality: X-Y = H-Br (not HCl, HI, or HOR), X-CX_3, and H-SR

Key features: "anti-Markovnikov" orientation (Y becomes bonded to **least** substituted carbon)
lack of stereoselectivity

11.C. Important Concepts and Hints

The two general methods for formation of carbon-carbon double bonds, E2 elimination of alkyl halides and acid-catalyzed dehydration of alcohols and ethers, have been presented briefly before (sections 9.6 and 10.6.D). In this chapter they are discussed in greater detail.

The major reactions of alkenes involve addition of a reagent to the two ends of the double bond. This results in rehybridization of the carbons involved, from sp^2 to sp^3, and conversion of the π-bond to two σ-bonds as shown schematically below:

π bond — sp^2-hybridized $\xrightarrow{X-Y}$ X–C–C–Y — σ bonds — sp^3-hybridized

A wide variety of reagents, X-Y, and products are possible, and a number of different mechanisms are observed. The discussion of addition reactions in the text is organized by type of reaction and product. To provide you with a different perspective, we have outlined them below according to the type of **mechanism**, with examples, an indication of the reagents for each and the section in the text where it is discussed.

One-Step Additions

1. Hydrogenation	11.6.A	4. Ozonide formation	11.6.E
2. Hydroboration	11.6.D	5. Epoxidation	11.6.E
3. Glycol formation with $KMnO_4$ or OsO_4/H_2O_2	11.6.E	6. Carbene addition	11.6.F

H_2 / Pt; $CHCl_3$ / NaOH; B_2H_6; H_2O_2 / OH^-; CH_3CO_3H; OsO_4 / H_2O_2 (or cold, dilute $KMnO_4$); O_3; Zn / H_2O

$H_3CDC(O–O)(O)CDCH_3 \xrightarrow[H_2O]{Zn}$ 2 CH_3CDO

Because they are all one-step additions, the two new σ-bonds are formed with the *syn* relationship, that is, from the same face of the double bond. (You may hear this referred to as "cis" addition, but the terms *cis* and *trans* should strictly speaking be applied only to cyclic systems or to those in which double bond stereochemistry remains in the product.)

Hydroboration is the only case above in which an unsymmetrical addition takes place (i.e., X ≠ Y). The boron group becomes attached to the sterically less congested end of the double bond, usually the less substituted end. Note also that the stereochemistry at that carbon is retained on replacing the C-B bond with C-OH.

Two-Step Additions

A. Radical Chain Reactions (11.6.G)

$$CH_3-CH=CH_2 + Y\cdot \longrightarrow CH_3-\dot{C}H-CH_2Y$$

$$CH_3-\dot{C}H-CH_2Y + X-Y \longrightarrow CH_3-CH(X)-CH_2Y + Y\cdot$$

X-Y = H-Br (**not** HCl, HI); X-CX_3 (X=Cl,Br); H-SR (R=H, alkyl, etc.); or (polymerization)

The first propagation step is addition of a free radical to one end of the π-bond to generate an alkyl radical. Note that the radical is **at the other end** of the original double bond. This reaction occurs in such a way as to generate the more stable free radical (tertiary > secondary > primary), so the Y· attacks the least substituted end of the double bond. There is no stereochemical preference in the second step, so mixtures of diastereomers are possible. For example:

$$(D)(H_3C)C=C(D)(CH_3) + HBr \xrightarrow{\text{(peroxides)}} H_3C-C(H)(D)-C(Br)(D)-CH_3 + H_3C-C(D)(H)-C(Br)(D)-CH_3$$

(each of these is produced in racemic form; that is, an equal amount of the enantiomer of each of the above is also produced)

B. Electrophilic Additions

1. Acid-catalyzed additions: HX and H_2O (11.6.C)

$$CH_3-CH=CH_2 \overset{H^+}{\rightleftharpoons} CH_3-\overset{+}{C}H-CH_3 \overset{H_2O}{\rightleftharpoons} CH_3-CH(\overset{+}{O}H_2)-CH_3 \overset{-H^+}{\rightleftharpoons} CH_3-CH(OH)-CH_3$$

$$CH_3-\overset{+}{C}H-CH_3 \overset{X^-}{\rightleftharpoons} CH_3CHXCH_3$$

$$CH_3-\overset{+}{C}H-CH_3 \xrightarrow{>C=C<} (CH_3)_2CH\left(-\overset{|}{\underset{|}{C}}-\overset{|}{\underset{|}{C}}-\right)_n$$

So-called electrophilic additions involve the attack of both an electrophile and a nucleophile, but the electrophile attacks first. An isolated double bond is itself weakly nucleophilic, so its preference for reaction with the electrophile is understandable. In acid-catalyzed additions, the first species to attack the double bond is a proton. It forms a C-H bond, using the two electrons that were in the π-bond and generating a carbocation at the other end. Because of the greater stability of 3° > 2° > 1° carbocations, the proton is attached to the less substituted end of the double bond. This generality is the original formulation of Markovnikov's rule. Two other generalities result from the intermediacy of the carbocation: 1) carbocation rearrangements are possible, obviously; and 2) stereochemical preference is usually not seen on attack by the nucleophile. For instance:

$$(D)(H_3C)C=C(D)(CH_3) \xrightarrow{H_3O^+} H_3C-C(HO)(D)-C(H)(D)-CH_3 + H_3C-C(D)(HO)-C(H)(D)-CH_3$$

Confusion can arise over the use of acid (typically H_2SO_4) to cause **both** the addition of water to a double bond to make an alcohol, and removal of water from an alcohol to make an alkene. Students often ask: how can the same reagent carry out opposing reactions? The situation is easily understood when you realize that there are not two reactions, only one, the equilibration of an alcohol with an alkene plus water, and that the acid is only a catalyst. It is not consumed or formed during the course of the reaction, and it will speed up the reaction in either direction.

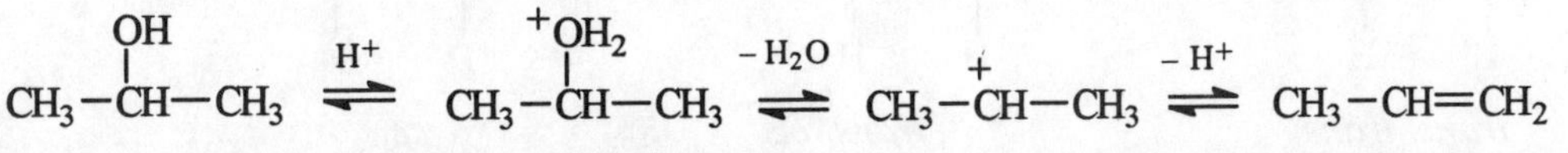

$$CH_3-CH(OH)-CH_3 \underset{}{\overset{H^+}{\rightleftharpoons}} CH_3-CH(\overset{+}{O}H_2)-CH_3 \overset{-H_2O}{\rightleftharpoons} CH_3-\overset{+}{C}H-CH_3 \overset{-H^+}{\rightleftharpoons} CH_3-CH{=}CH_2$$

The factors which control the direction that the reaction proceeds are the conditions: in dilute aqueous acid (a **lot** of water around), the equilibrium is driven to the left (as written above), and alcohol is formed from an alkene. Under these conditions an alcohol would not form appreciable amounts of alkene. On the other hand, in concentrated sulfuric acid (60% to 95%, depending on ease of dehydration), especially at higher temperatures, the equilibrium is driven to the right by distillation of the alkene from the reaction mixture and by protonation of the water formed (e.g., $H_2O + H_2SO_4 \rightleftharpoons H_3O^+ + H_2SO_4^-$).

2. Via bridged, cationic intermediates: X_2 (11.6.B) $Hg(OAc)_2$ (11.6.C)

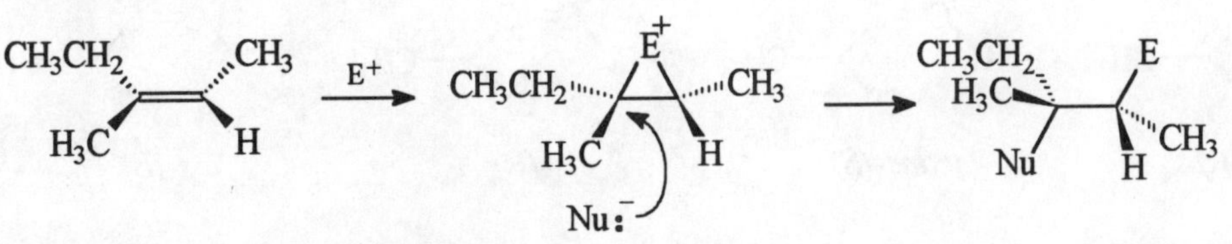

CH₃CH₂/H₃C alkene with CH₃/H $\xrightarrow{E^+}$ bridged E^+ intermediate (CH₃CH₂, H₃C, CH₃, H), attacked by Nu:⁻ $\longrightarrow$ product with E, Nu (CH₃CH₂, H₃C, CH₃, H)

$E^+ = Cl^+, Br^+, I^+$, or ^+HgOAc;
$Nu:^- = Cl^-, Br^-, I^-$, ROH, AcO^-, etc.)

Related reaction: hydrolysis of epoxides (10.11.A)

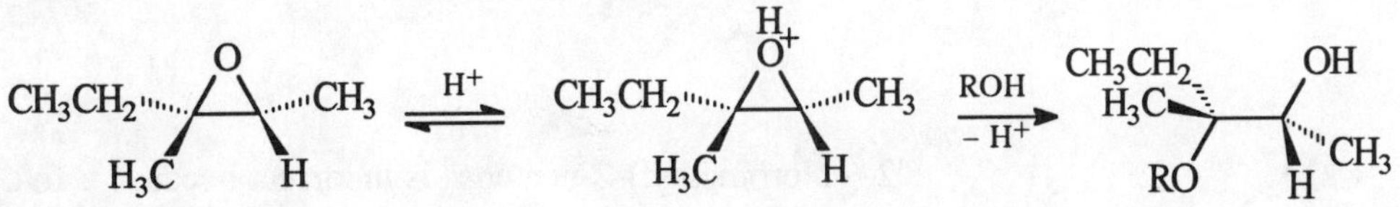

Epoxide (CH₃CH₂, H₃C, CH₃, H) $\underset{}{\overset{H^+}{\rightleftharpoons}}$ protonated epoxide ($\overset{+}{O}H$) $\xrightarrow[-H^+]{ROH}$ product with OH, RO (CH₃CH₂, H₃C, CH₃, H)

Because of the bridged nature of the cationic intermediate, there are two generalities for reactions of this type: 1) there is a stereochemical preference for *anti* attack by the nucleophile; and 2) Markovnikov's rule is followed in a broader definition: attack of an electrophile occurs in such a way as to form the more stable carbocationic intermediate. You can rationalize this by thinking of the resonance structures possible for the bridged intermediate: the one with a tertiary carbocation is more important than the secondary carbocation, and nucleophilic attack occurs faster at that position.

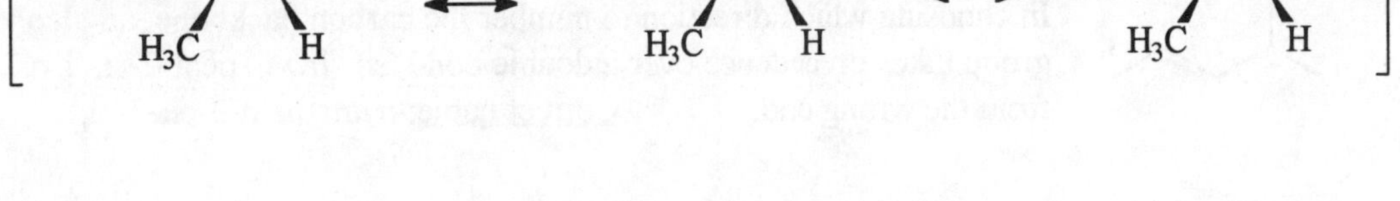

[tertiary cation (E, + on carbon bearing CH₃CH₂/H₃C) $\longleftrightarrow$ bridged E^+ $\longleftrightarrow$ secondary cation (E, + on carbon bearing CH₃/H)]

11.D. Answers to Exercises

11.1 a. Three:

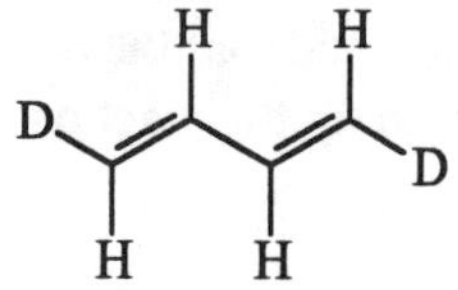

trans, trans 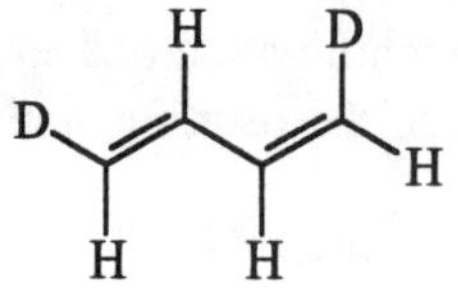 *trans, cis* 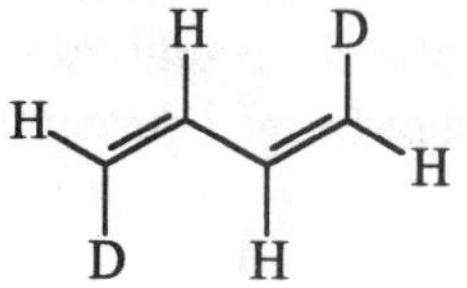 *cis, cis*

b. Two:

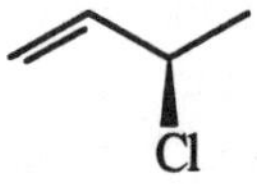

(*R*)-3-chloro-1-butene 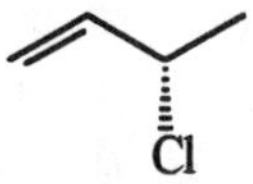 (*S*)-3-chloro-1-butene

c. 4 isomers: *cis* or *trans*, *R* or *S*

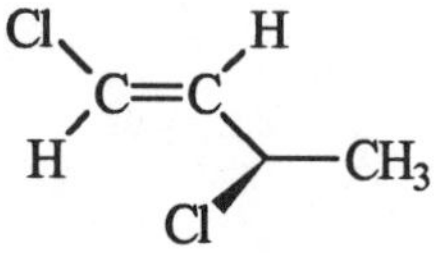

trans-(R) 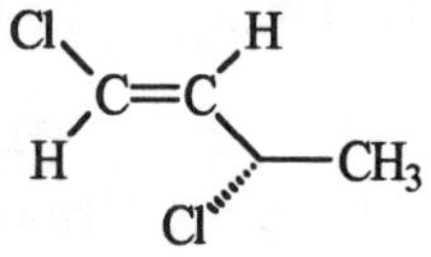 *trans-(S)* 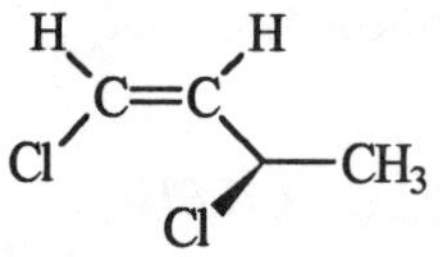 *cis-(R)* 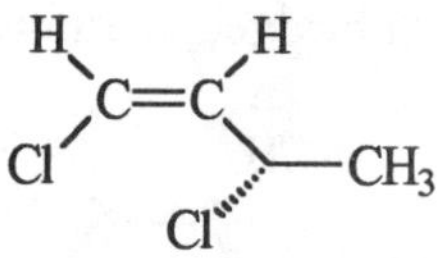 *cis-(S)*

11.2 a. *trans*-4-octene b. 1-hexene
c. *trans*-2,5-dimethyl-3-heptene d. 5-chloro-2-methyl-2-pentene

11.3 a. (*Z*)-2-chloro-2-hexene b. (*Z*)-3-(1-chloroethyl)-3-hexen-2-ol
c. (*E*)-3,4-dimethyl-3-hexene d. (*E*)-1-chloropropene

11.4 a. $ClCH_2(CH_3)C{=}CHCH_2CH_3$

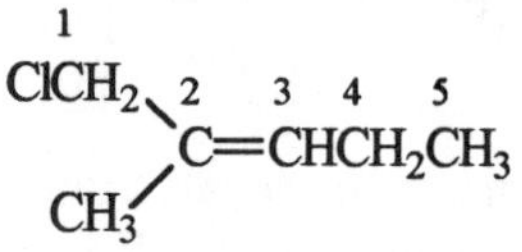

"2-(chloromethyl)-2-pentene" is incorrect because a five-carbon backbone can be chosen that includes the halogen.
Correct name: 1-chloro-2-methyl-2-pentene.

b.

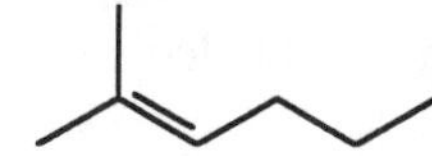

Both of the substituents on one end of the double bond are the same, so the *E* specification does not apply.
Correct name: 2-methyl-2-hexene.

c.

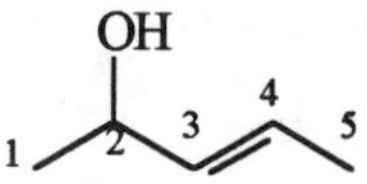

In choosing which direction to number the carbon backbone, an alcohol functional group takes precedence over a double bond, so "*trans*-pent-2-en-4-ol" is numbered from the wrong end. Correct name: *trans*-pent-3-en-2-ol.

11.5

	cis-2-pentene ⇌	*trans*-2-pentene	
ΔH°_f	-7.0	-7.9	$\Delta H^\circ = -0.9$ kcal mole^{-1}

Assuming that ΔS° is ~0, so that $\Delta H^\circ \approx \Delta G^\circ = -RT \ln K$:

at 25°C = 298K:	$-900/(-1.987 \times 298) = \ln K$;	K = 4.6	18% *cis* and 82% *trans*
at 300°C = 573K:	$-900/(-1.987 \times 573) = \ln K$;	K = 2.2	31% *cis* and 69% *trans*

11.6 *trans*-cyclooctene ⇌ *cis*-cyclooctene $\Delta G° = -9.1$ kcal mole^{-1}

$\Delta G° = -RT \ln K$: $-9100/(-1.987 \times 298) = \ln K$; $K = 4.72 \times 10^6$
% *trans* at equilibrium = 2.1×10^{-5}

11.7. If H_2O and C_2H_5OH have equal dissociation constants in C_2H_5OH solution, then the ratio of hydroxide (OH^-) to ethoxide ($C_2H_5O^-$) will be the same as the ratio of (water + hydroxide) to (ethanol + ethoxide). In one liter of ethanol solution, there are about 1000 mL x 0.789 g/mL (density of ethanol) = 789 g = 789 g/(46 g/mole) = 17.2 moles of ethanol. The amount of (water + hydroxide) is equal to the amount of hydroxide added (1 mole), therefore the ratio of hydroxide to ethoxide is about 1/17.

11.8 For a primary alkyl halide, the mechanism of a substitution reaction is S_N2, and the rate will be directly dependent on the concentration of the base. The mechanism for the elimination reaction is E2, hence this rate is also directly dependent on the concentration of base. Therefore to a first approximation, the **ratio** of substitution to elimination will be independent of base concentration.

11.9 The substitution reaction will be unaffected by the presence or absence of deuterium on the β-carbon and will therefore occur at the same rate for the two substrates. The elimination reaction will be slower for the deutero compound, however, since the carbon-deuterium bond is partially broken in the transition state. The ratio of substitution/elimination will therefore be higher for the deuterated substrate.

11.10 No answer. Model-building exercise.

11.11

H H Br Br R S H₃C CH₃ ≡ H Br H Br CH₃ CH₃ ≡ H ⁻OH H₃C Br H₃C H Br → H₃C Br H₃C H

11.12 Hot sulfuric acid would <u>not</u> be a good method in this case, since double bond isomerization and loss of the deuteriums would occur:

$$CD_3CH_2CH_2OH \xrightarrow[\Delta]{H_2SO_4} CD_3CH{=}CH_2 \overset{H^+}{\rightleftharpoons} CD_3\overset{+}{C}HCH_3 \overset{-D^+}{\rightleftharpoons} CD_2{=}CHCH_3 \overset{H^+}{\rightleftharpoons} CHD_2\overset{+}{C}HCH_3$$

11.13 1-Methylcyclohexene, via cationic rearrangement to the more stable tertiary cation and loss of a proton:

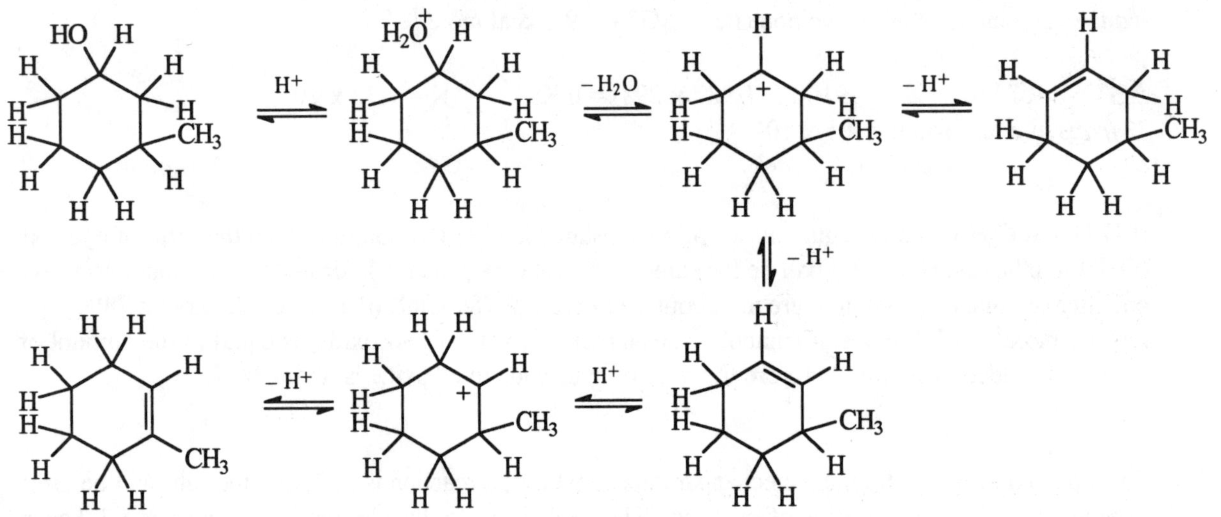

H^+ $-H_2O$ $-H^+$ $-H^+$ $-H^+$ H^+ CH_3

most stable product

NOTE: Many equilibria between various alkenes and carbocations are possible in this system. **<u>All</u>** of them can take place (for example, as depicted for the formation of 4-methylcyclohexene), but only the important ones, leading to the most stable product (1-methylcyclohexene), are shown.

11.14

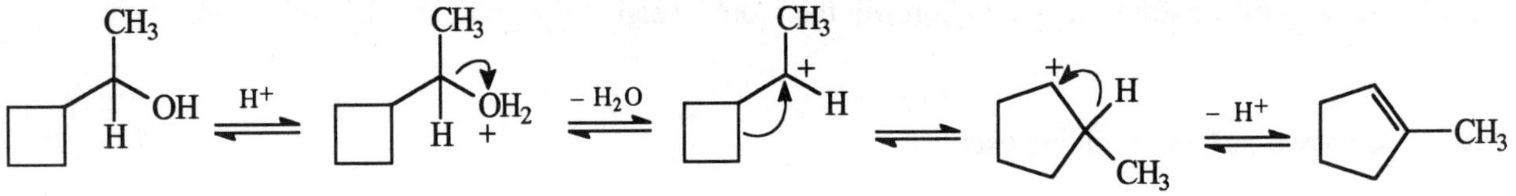

CH_3 OH H^+ OH_2 $-H_2O$ $-H^+$ CH_3

11.15 Using palladium as catalyst results in isomerization and loss of chirality before hydrogenation occurs:

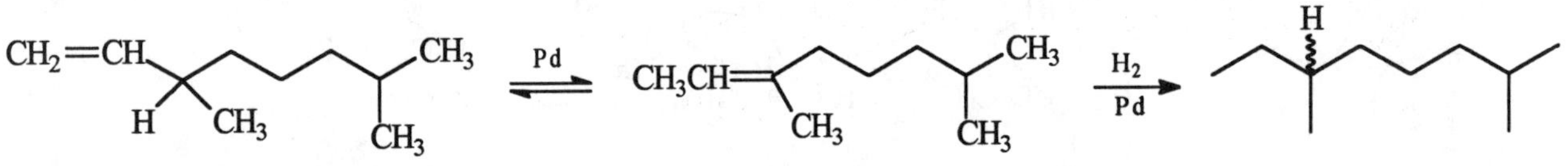

$CH_2{=}CH$ CH_3 Pd $CH_3CH{=}$ CH_3 H_2 / Pd

optically active   achiral   racemic

11.16

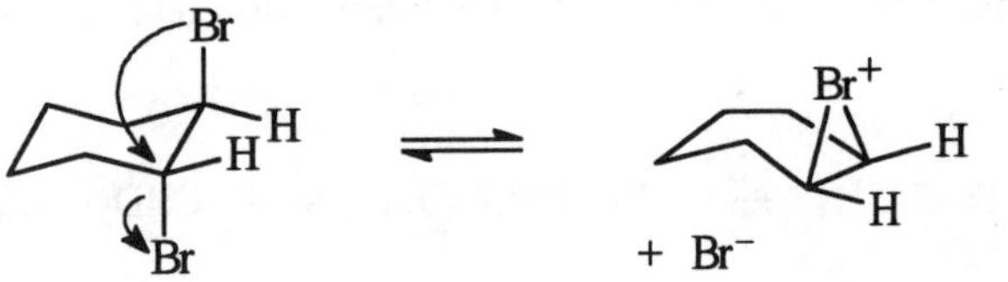

Br   H Br^+ + Br^-

This path can be followed in either direction.

This path can **not** be followed, in either direction.

11.17 *Immediate product:* *More stable product:*

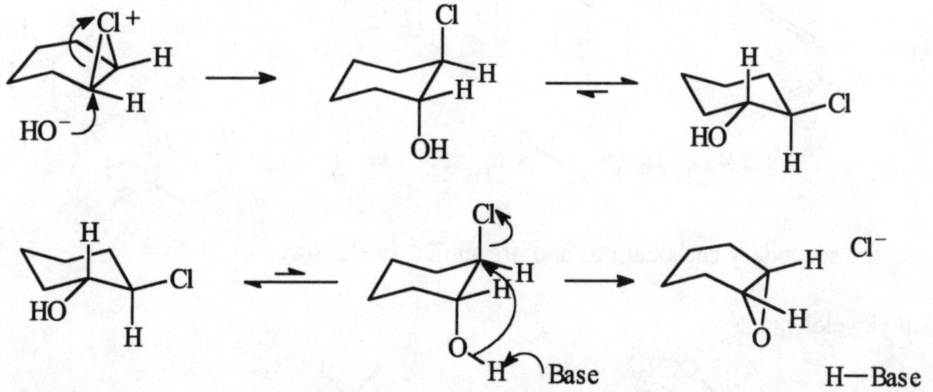

Cl+ H H HO⁻ → Cl H H OH ⇌ H Cl HO H

H Cl HO H ⇌ Cl H H O H Base → H Cl⁻ H O H—Base

11.18 a. *trans*-1,2-dibromo-1-methylcyclohexane and its enantiomer

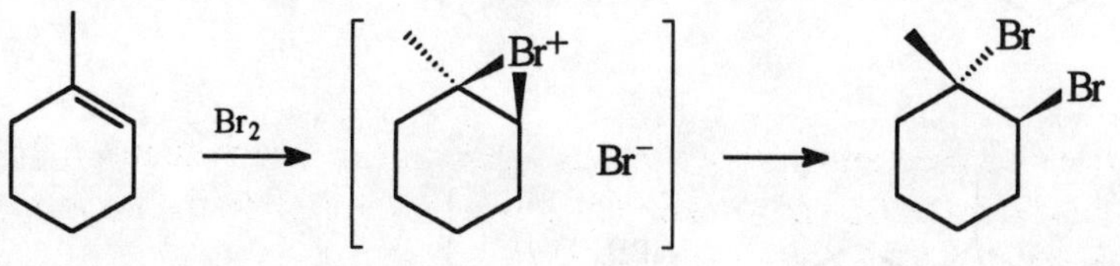

Br_2 → [Br+ Br⁻] → Br Br

b. (2*S*,3*S*)-3-chloro-2-butanol (and its enantiomer)

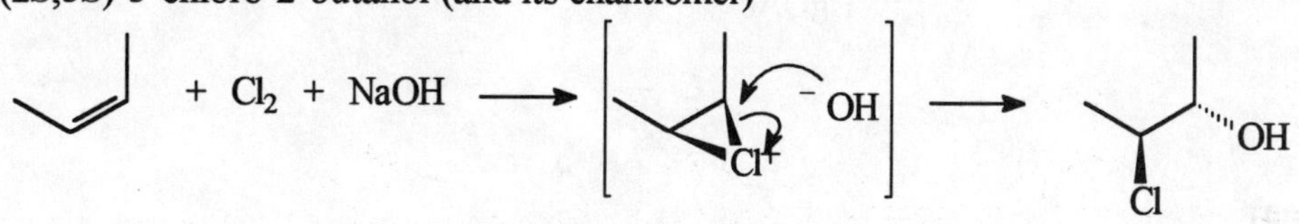

+ Cl_2 + NaOH → [⁻OH Cl+] → OH Cl

c. (2*R*,3*R*)-2,3-butanediol (and its enantiomer)

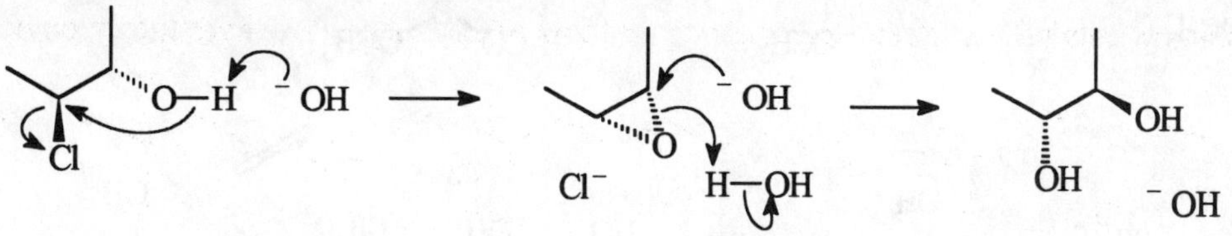

O—H ⁻OH Cl → ⁻OH O Cl⁻ H—OH → OH OH ⁻OH

11.19

H_3C H_3C C=CH_2 H—Cl ⇌ H_3C H_3C +C—CH_3 Cl⁻ ⇌ $(CH_3)_3CCl$

H—Br ⇌ + Br⁻ ⇌ Br

H—Cl ⇌ + Cl⁻ ⇌ Cl

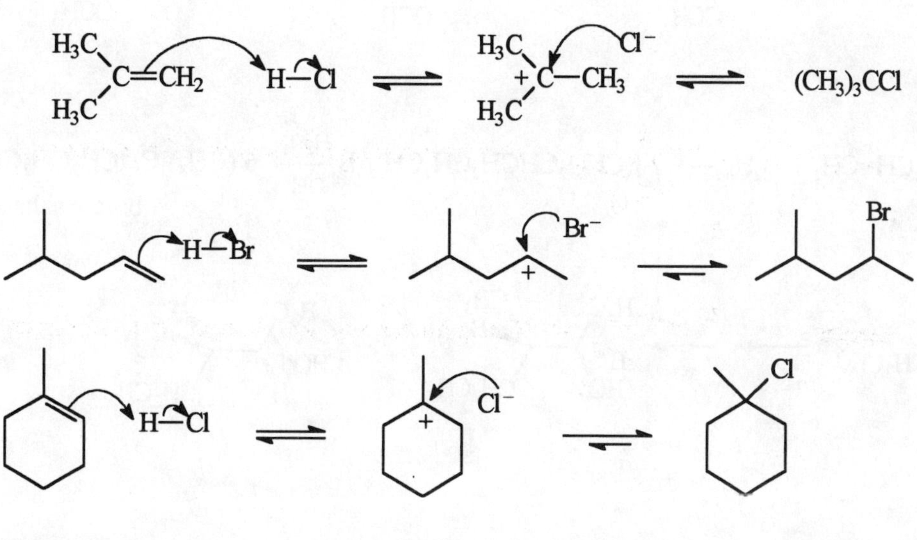

11.20

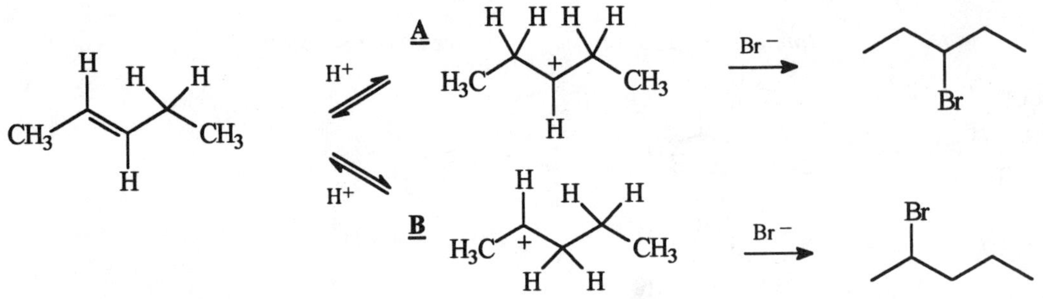

Both intermediates are secondary carbocations and are similar in stability.

11.21 a. 1-methoxy-1-methylcyclohexane:

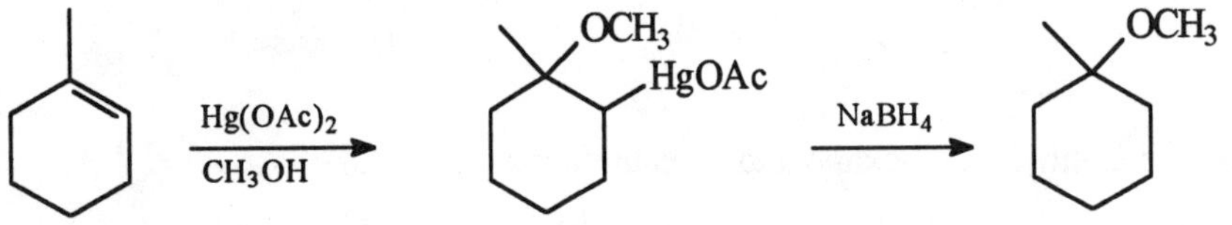

$Hg(OAc)_2$ / CH_3OH → $NaBH_4$ →

b. 2-methyl-2-pentanol:

$Hg(OAc)_2$ / H_2O / THF → $NaBH_4$ →

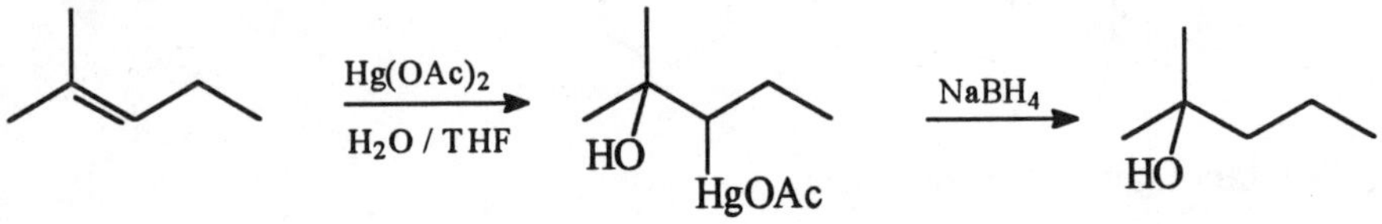

11.22

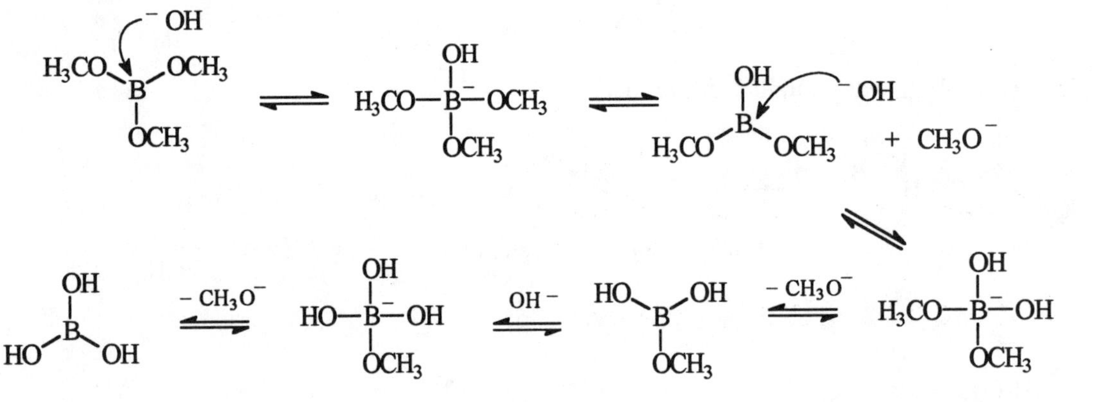

11.23

a. $6\ (CH_3)CHCH_2CH{=}CH_2 + B_2H_6 \longrightarrow 2\ [(CH_3)_2CHCH_2CH_2CH_2]_3B \xrightarrow[OH^-]{H_2O_2} 6\ (CH_3)_2CHCH_2CH_2CH_2OH +$ borate salts

b.

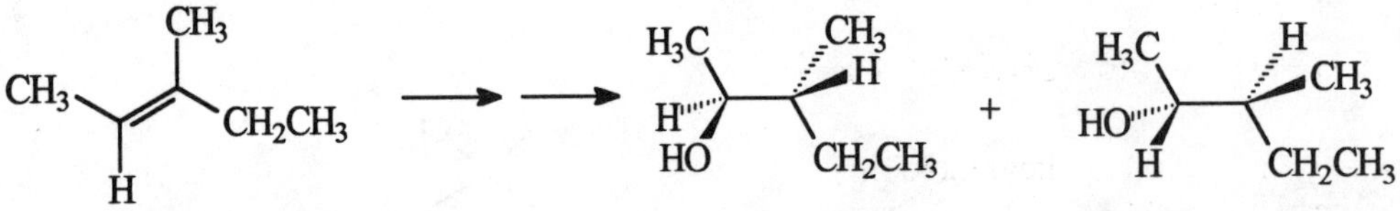

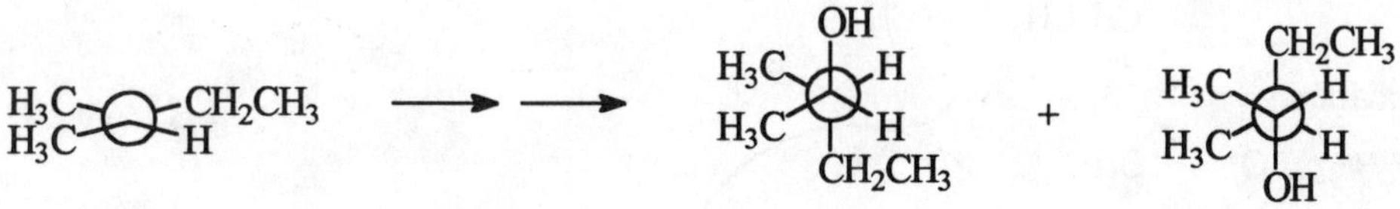

c. 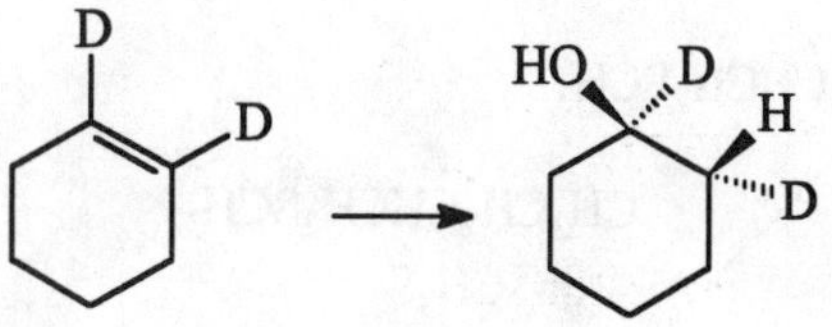 (*racemic*)

11.24 a. 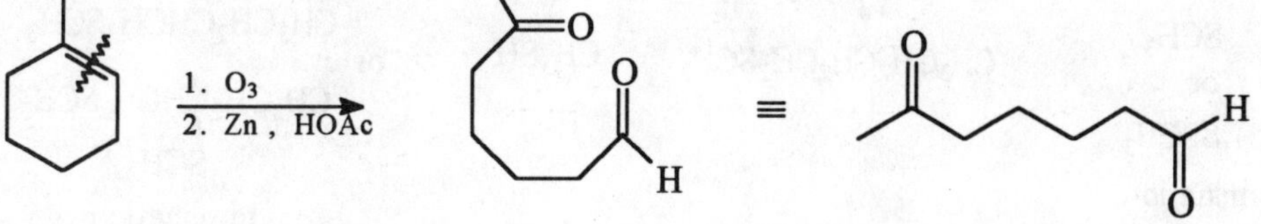 1. O_3 2. Zn , HOAc

b. 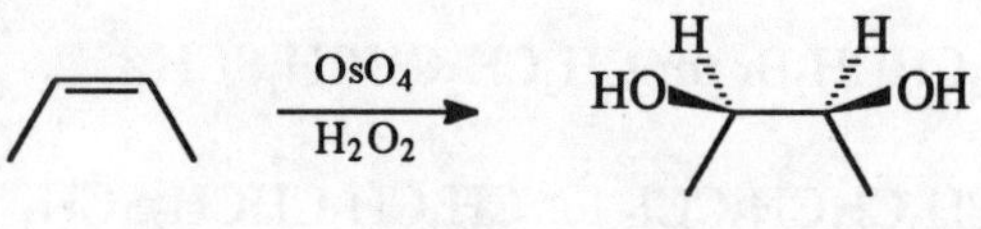 OsO_4 / H_2O_2 c. $KMnO_4$ / 5°C

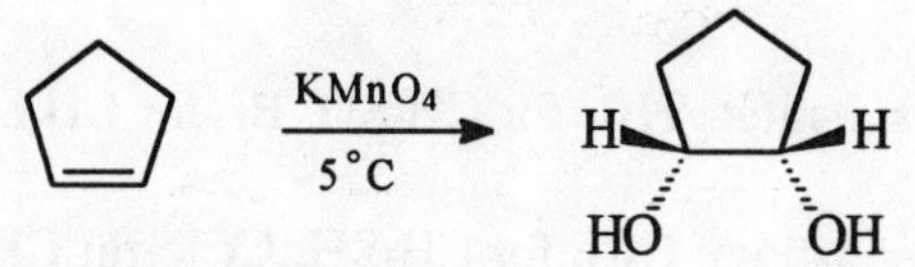

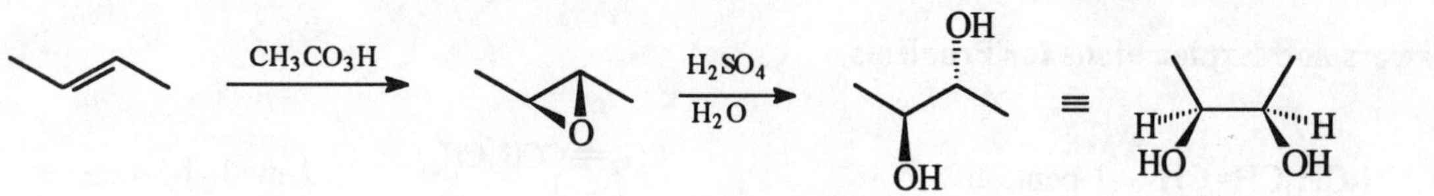

11.25 CH_3CO_3H → H_2SO_4 / H_2O → (*meso*)

11.26 a. 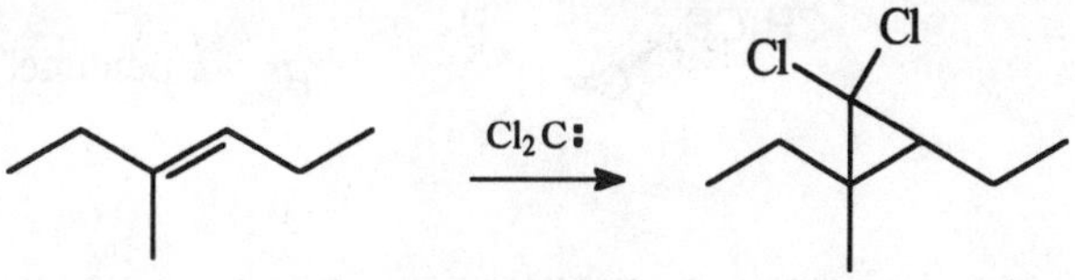 Cl_2C:

b. 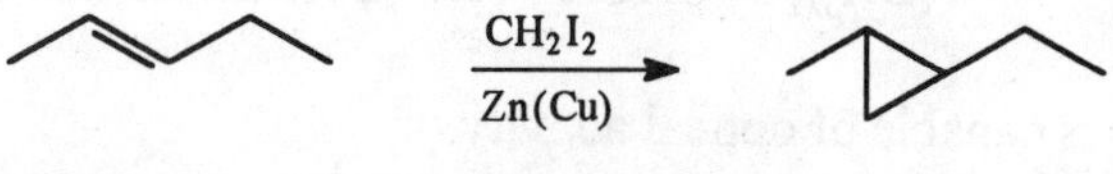 CH_2I_2 / Zn(Cu)

11.27 $CH_3S\cdot + CH_2{=}CH_2 \longrightarrow CH_3SCH_2CH_2\cdot$

Bonds formed:	C–S	−65	kcal mole^{-1}
	C–C	−83	
Bonds broken:	C=C	+150	
Overall change:		+2	kcal mole^{-1}

$CH_3SCH_2CH_2\cdot + CH_3SH \longrightarrow CH_3SCH_2CH_3 + CH_3S\cdot$

Bonds formed:	C–H	−101	kcal mole^{-1}
Bonds broken:	S–H	+83	
Overall change:		−18	kcal mole^{-1}

11.28

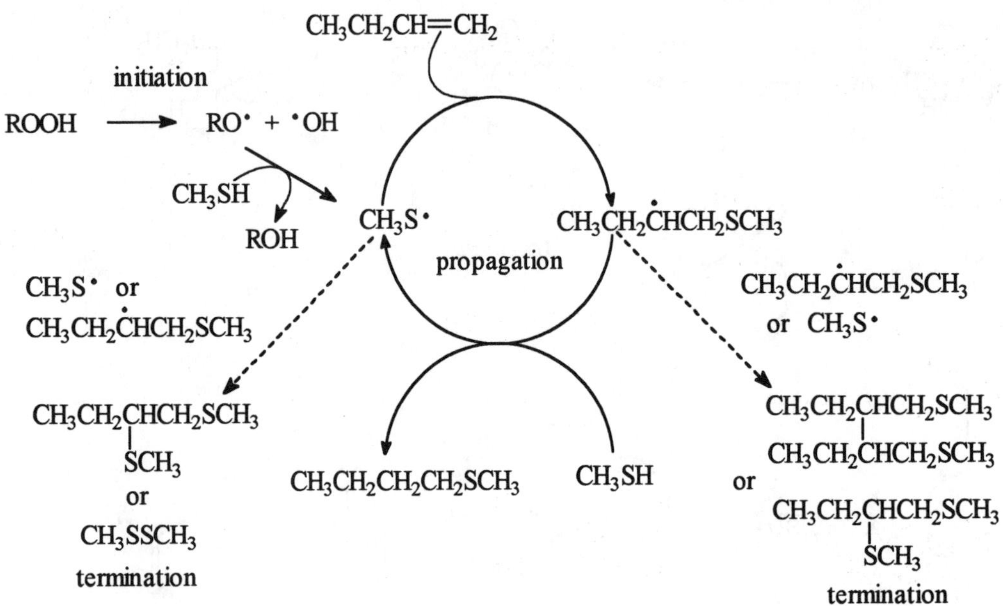

For HBr, substitute HBr for CH_3SH, Br· for $CH_3S\cdot$ and $CH_3CH_2\dot{C}HCH_2Br$ for $CH_3CH_2\dot{C}HCH_2SCH_3$;

for CCl_4, substitute CCl_4 for CH_3SH, $CCl_3\cdot$ for $CH_3S\cdot$ and $CH_3CH_2\dot{C}HCH_2CCl_3$ for $CH_3CH_2\dot{C}HCH_2SCH_3$.

11.E Answers and Explanations for Problems

1. a. $CH_3CH_2CH_2CH{=}CH_2$ 1-pentene

$CH_2{=}CCH_2CH_3$ (with CH_3 on C2) 2-methyl-1-butene

$CH_3CH_2(H)C{=}C(H)CH_3$ *cis*-2-pentene

$CH_3CH_2(H)C{=}C(CH_3)H$ *trans*-2-pentene

$(CH_3)_2CHCH{=}CH_2$ 3-methyl-1-butene

$(CH_3)_2C{=}CHCH_3$ 2-methyl-2-butene

There is only one pair of stereoisomers. None is capable of optical activity.

b. 1-methycyclopentene

4-methylcyclopentene

(R)-3-methylcyclopentene

(S)-3-methylcyclopentene

stereoisomers, capable of optical activity

2. a. H_3C, CH_3 / C=C / H, $CH(CH_3)_2$

   b. $HOCH_2CH_2CH{=}C(CH_3)_2$

   c. H_3C, $CH_2CH_2CH_3$ / C=C / H, CH_2CH_3

   d. $CH_2{=}CHF$

   e. Br

   f. CH_3

   g. H_3C, H / C=C / Br, CH_2CH_3

   h. H_3C, CH_3 / C=C / H, $CH_2CH_2CH_3$

   i. CH_3

   j. CH_3

   k. H, CH_3, H–, CH_3, H, H

   l. H_3C, CH_3 / C=C / H_3C, CH_3

3. a. "2-methylcyclopentene" is numbering the double bond from the wrong direction: correct name is 1-methylcyclopentene.

   b. The double bond takes precedence in numbering the carbon chain: correct name is 4-methyl-*cis*-2-pentene.

   c. There is no stereochemistry for unsubstituted 1-butene: correct name is simply 1-butene.

   d. Common and systematic nomenclature should not be mixed: correct name is 1-bromo-2-methylpropene.

   e. Same error as in **a**: correct name is 1-chlorocyclopentene.

   f. One end of the double bond has two identical substituents, therefore there is no stereochemistry to specify (see **c** above); also, "3-ethyl-3-pentene" is numbered from the wrong end: correct name is 3-ethyl-2-pentene.

   g. The hydroxyl group takes precedence over the double bond in numbering: correct name is *trans*-3-penten-2-ol.

   h. In case of two chains of identical length, the one with more, smaller substituents should be chosen: correct name is (*Z*)-3-ethyl-2-methyl-3-heptene.

4. a. (*R*)-3-methyl-1-hexene
   b. 1-methylcyclohexene
   c. (*E*)-3,4-dimethyl-3-heptene
   d. (*Z*)-4-isopropyl-3-methyl-3-heptene
   e. (*Z*)-1-bromo-1-chloro-2-fluoro-2-iodoethene
   f. 4-chloro-1-butene
   g. 4-chloro-2-ethyl-1-butene
   h. 2-(2-chloroethyl)-1-pentene
   i. (*Z*)-1-chloro-3-heptene
   j. 4-penten-1-ol
   k. (*S*,*E*)-hex-4-en-3-ol or (*S*,*E*)-4-hexen-3-ol
   l. (*R*)-2-ethyl-4-methyl-1-hexene

5. a. A 2-halo-4-methylpentane, $(CH_3)_2CHCH_2CHXCH_3$ will give $(CH_3)_2CHCH{=}CHCH_3$ (mix of *cis* and *trans* isomers) as the principal product. A 3-halo-2-methylpentane will give primarily 2-methyl-2-pentene.

   b. A 1-halo-4-methylpentane with a strong, sterically hindered base like $KOC(CH_3)_3$ is needed for a 1-alkene. A 2-halo-4-methylpentane will give a mix of 1- and 2-pentenes; $KOC(CH)_3$ will help increase the yield of the desired 1-alkene.

   c. The symmetrically substituted 1-halo-4-methylcyclohexane will yield only the desired product. The *cis* isomer will provide the correct anti geometry for the elimination with the lowest possible energy since the CH_3 will be equatorial.

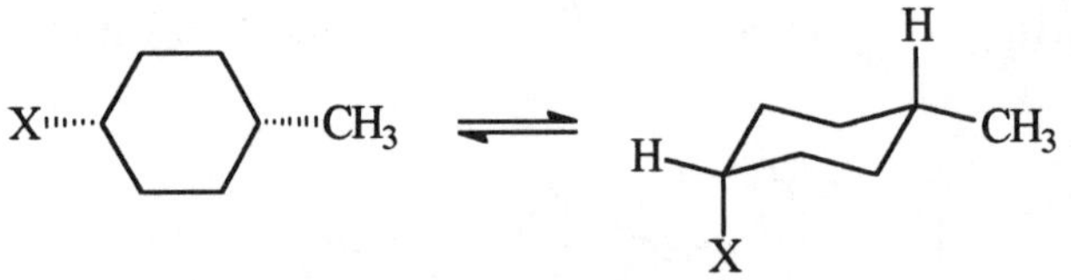

X CH3 ⇌ H H CH3 X

   d. The obvious choice here is 1-halo-1-methylcyclohexane, since only one product is possible. The other possibility is 1-halo-2-methylcyclohexane but the only possible isomer is the *cis*. With the *trans* isomer, the only *trans*-axial H available to eliminate with the axial X will be on C-3.

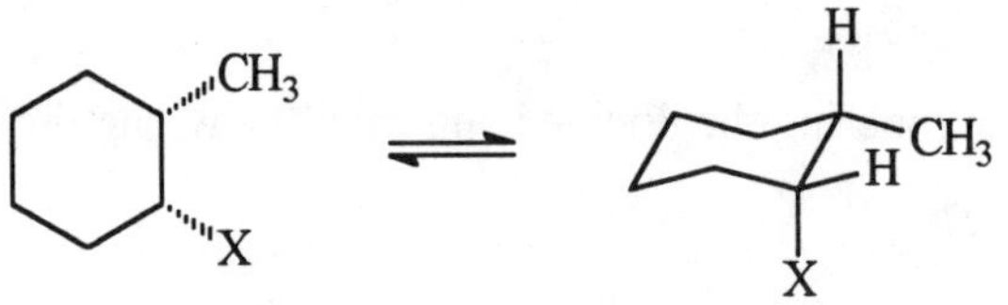

CH3 X ⇌ H CH3 H X

6. a. $(CH_3CH_2)_3CH$

3-ethylpentane

b. $CH_3CH_2C(CH_2CH_3)(OH)CHBrCH_3$

2-bromo-3-ethyl-3-pentanol

c. $CH_3CH_2C(CH_2CH_3)(Cl)CHClCH_3$

2,3-dichloro-3-ethylpentane

d. $CH_3CH_2C(CH_2CH_3)(OH)CHOHCH_3$

3-ethylpentane-2,3-diol

e. $CH_3CH_2CH(CH_2CH_3)CHOHCH_3$

3-ethyl-2-pentanol

f. $(CH_3CH_2)_3COH$

3-ethyl-3-pentanol

g. $CH_3CH_2COCH_2CH_3 + CH_3CHO$
diethyl ketone or 3-pentanone; acetaldehyde

h. $(CH_3CH_2)_3CBr$
3-bromo-3-ethylpentane

i. $(CH_3CH_2)_2CHCHBrCH_3$

2-bromo-3-ethylpentane

j. $CH_3CH_2C(CH_2CH_3)(OCH_3)CHBrCH_3$

2-bromo-3-ethyl-3-methoxypentane

k. $CH_3CH_2C(CH_2CH_3)$—$CHCH_3$ (epoxide, bridging O)

2,2-diethyl-3-methyloxirane

l. $CH_3CH_2C(CH_2CH_3)$—$CHCH_3$ (ring closed by CBr_2)

1,1-dibromo-2,2-diethyl-3-methylcyclopropane

m. $CH_3CH_2C(CH_2CH_3)$—$CHCH_3$ (ring closed by CH_2)

1,1-diethyl-2-methylcyclopropane

7. <u>Same</u> products from cis-$CH_3CH_2CH{=}CHCH_2CH_3$ and trans-$CH_3CH_2CH{=}CHCH_2CH_3$

a. hexane

e,f. $CH_3CH_2CHOHCH_2CH_2CH_3$ 3-hexanol

g. CH_3CH_2CHO propionaldehyde (propanal)

h,i. $CH_3CH_2CHBrCH_2CH_2CH_3$ 3-bromohexane

<u>Different</u> products:

b. *cis*-3-hexene gives bromonium ion (H, CH_3CH_2)C–C(H, CH_2CH_3) with bridging Br^+ $\xrightarrow{H_2O}$ Br(H)(CH_3CH_2)C–C(H)(CH_2CH_3)(OH)

(3R,4R)-4-bromo-3-hexanol plus an equal amount of the (3S,4S) enantiomer

To indicate that this diastereomer is present as a racemic mixture, the designation (3RS, 4RS) may be used.

trans-3-hexene gives Br(H)(CH_3CH_2)C–C(CH_2CH_3)(H)(OH) (3RS,4SR)-4-bromo-3-hexanol

c. *cis* gives (±) (or dl) 3,4-dichlorohexane; *trans* gives meso-3,4-dichlorohexane.

d. *cis* gives meso-3,4-hexanediol; *trans* gives (±)-3,4-hexanediol.

j. Same as b. , except CH_3O replaces OH: *cis* gives (3RS, 4RS)-3-bromo-4-methoxyhexane;
trans gives (3RS, 4SR)-3-bromo-4-methoxyhexane

k. *cis* gives CH_3CH_2, CH_2CH_3, H, O, H — *cis*-2,3-diethyloxirane

trans gives CH_3CH_2, H, H, O, CH_2CH_3 — *trans*-2,3-diethyloxirane

l. *cis* gives CH_3CH_2, CH_2CH_3, H, H, Br, Br — *cis*-1,1-dibromo-2,3-diethylcyclopropane

trans gives CH_3CH_2, H, H, CH_2CH_3, Br, Br — *trans*-1,1-dibromo-2,3-diethylcyclopropane

m. *cis* gives CH_3CH_2, CH_2CH_3, H, H — *cis*-1,2-diethylcyclopropane

trans gives CH_3CH_2, H, H, CH_2CH_3 — *trans*-1,2-diethylcyclopropane

8. a. cyclohexane

b. Br, OH — (±)-*trans*-2-bromo-cyclohexanol

c. Cl, Cl — (±)-*trans*-1,2-dichloro-cyclohexane

d. OH, OH — *cis*-1,2-cyclohexanediol

e,f. OH — cyclohexanol

g. 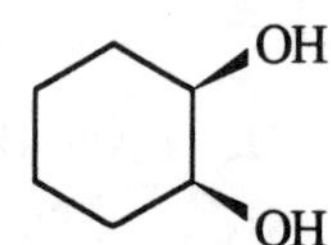

hexanedial

h,i. Br — bromocyclohexane

j. Br, OCH_3 — (±)-*trans*-1-bromo-2-methoxycyclohexane

k. O — epoxycyclohexane

l. Br, Br — 7,7-dibromo-bicyclo[4.1.0]heptane

m. bicyclo[4.1.0]heptane

9. a. $CH_3CH_2CH_2CH{=}CH_2$

b. 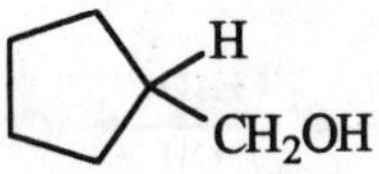 (cyclopentyl) H, CH_2OH

c. 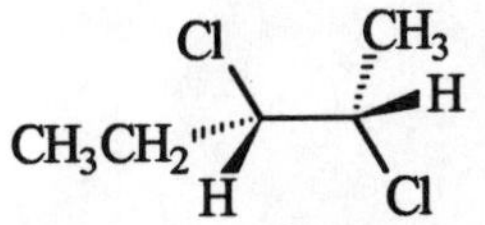 Cl, CH_3, H + enantiomer
CH_3CH_2, H, Cl

d. 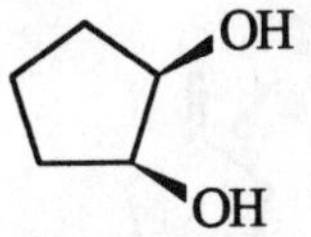 (cyclopentane) OH, OH

e. 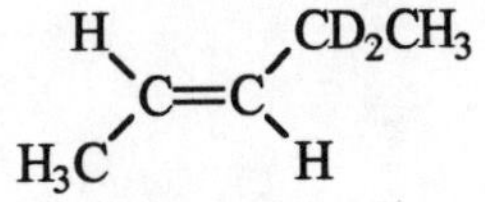 H, CD_2CH_3, C=C, H_3C, H

f. 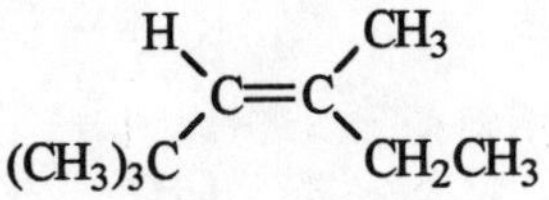 H, CH_3, C=C, $(CH_3)_3C$, CH_2CH_3

g. 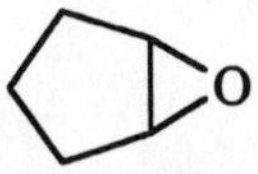 (cyclopentene oxide) O

h. $CH_3\overset{O}{\overset{\|}{C}}CH_2\underset{CH_3}{\overset{CH_3}{\overset{|}{C}}}CH_2\overset{O}{\overset{\|}{C}}H$

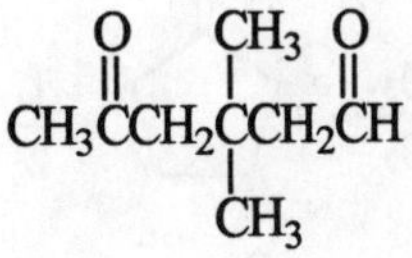

i. 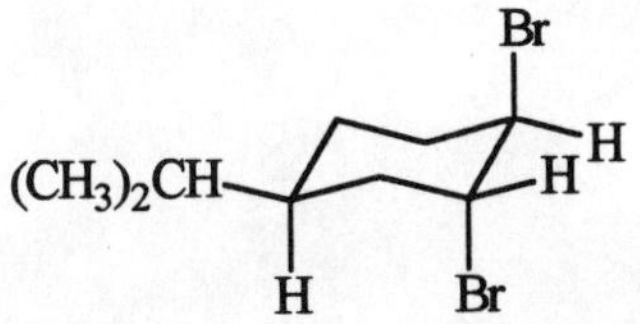 $(CH_3)_2CH$, Br, H, H, H, Br — product from *anti*, diaxial addition (*i*-propyl is almost always equatorial)

10. a. $CH_3CHBrCH_3 \xrightarrow[\text{or } t\text{-}C_4H_9OK/t\text{-}C_4H_9OH]{C_2H_5OK/C_2H_5OH,\ \Delta} CH_3CH{=}CH_2 \xrightarrow[\text{peroxides}]{HBr} CH_3CH_2CH_2Br$

b. $CH_3CHOHCH_3 \xrightarrow[\text{or } H_2SO_4,\ \Delta]{Al_2O_3,\ \Delta} CH_3CH{=}CH_2 \xrightarrow{B_2H_6} \xrightarrow{H_2O_2/OH^-} CH_3CH_2CH_2OH$

c. 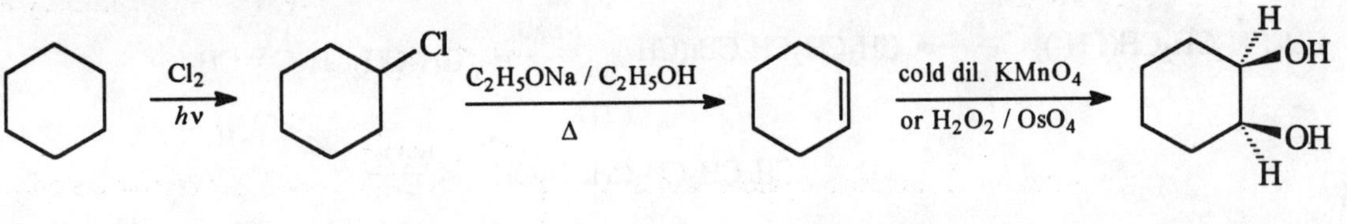 (cyclohexane) $\xrightarrow[h\nu]{Cl_2}$ (chlorocyclohexane) Cl $\xrightarrow[\Delta]{C_2H_5ONa\ /\ C_2H_5OH}$ (cyclohexene) $\xrightarrow[\text{or } H_2O_2\ /\ OsO_4]{\text{cold dil. } KMnO_4}$ (cyclohexane) H, OH, OH, H

d. (cyclohexene) $\xrightarrow{CH_3CO_3H}$ (cyclohexene oxide) O $\xrightarrow[H_2O]{H_2SO_4}$ (cyclohexane) OH, OH

e. $CH_3CH_2\overset{CH_3}{\overset{|}{C}}{=}CH_2 \xrightarrow[CH_3OH]{Hg(OAc)_2} CH_3CH_2\underset{OCH_3}{\underset{|}{\overset{CH_3}{\overset{|}{C}}}}{-}CH_2HgOAc \xrightarrow[NaOH]{NaBH_4} CH_3CH_2\underset{OCH_3}{\underset{|}{\overset{CH_3}{\overset{|}{C}}}}{-}CH_3$

or

$CH_3CH_2\overset{CH_3}{\overset{|}{C}}{=}CH_2 \xrightarrow{\text{HCl or HBr}} CH_3CH_2\underset{X}{\underset{|}{\overset{CH_3}{\overset{|}{C}}}}{-}CH_3 \xrightarrow[(S_N1\text{ reaction})]{CH_3OH} CH_3CH_2\underset{OCH_3}{\underset{|}{\overset{CH_3}{\overset{|}{C}}}}{-}CH_3$

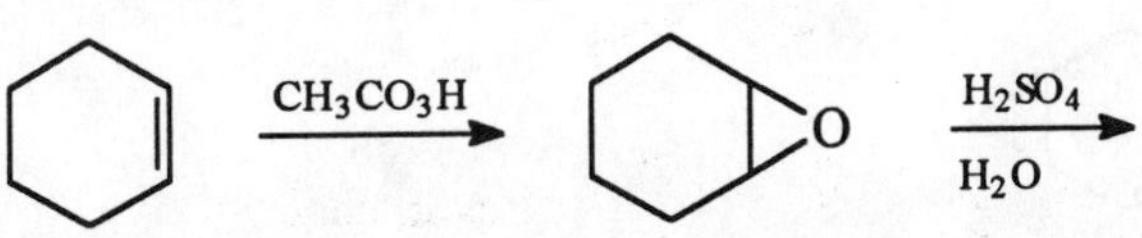

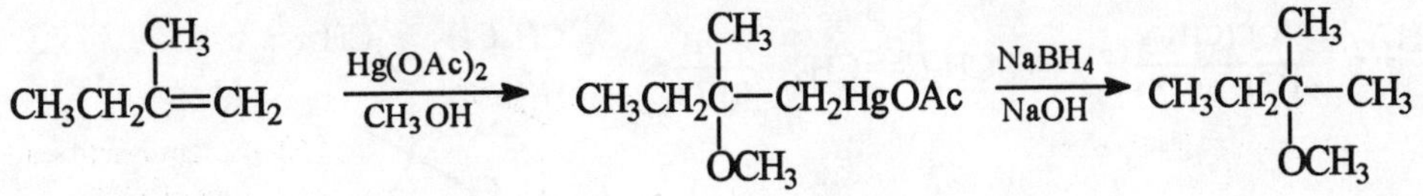

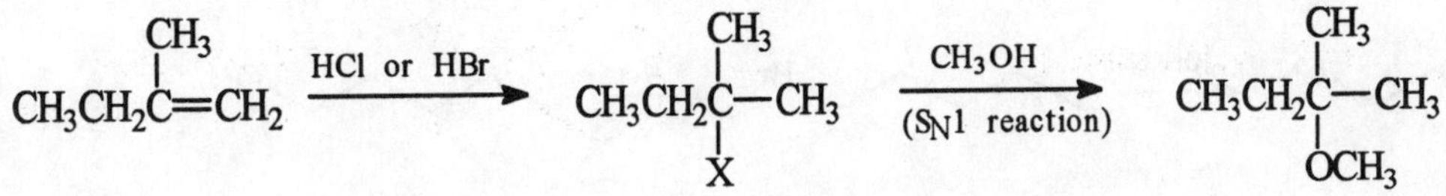

f. $CH_3CH_2C(CH_3){=}CH_2 \xrightarrow{B_2H_6} \xrightarrow[OH^-]{H_2O_2} CH_3CH_2CH(CH_3)CH_2OH \xrightarrow{K} \xrightarrow{CH_3I} CH_3CH_2CH(CH_3)CH_2OCH_3$

g. methylenecyclopentane ($=CH_2$) $\xrightarrow{O_3} \xrightarrow[HOAc]{Zn}$ cyclopentanone (C=O)

11. a. (bromomethyl)cyclopentane (CH_2Br) $\xrightarrow[HOC(CH_3)_3]{KOC(CH_3)_3}$ methylenecyclopentane ($=CH_2$) $\xrightarrow[2.\ NaBH_4,\ CH_3OH]{1.\ O_3,\ CH_2Cl_2}$ cyclopentanol (OH)

b. bromocyclopentane (Br) $\xrightarrow{alc.\ KOH}$ cyclopentene $\xrightarrow[H_2O]{Br_2}$ trans-2-bromocyclopentanol (Br wedged, OH hashed)

c. $CH_3CH_2CH_2CH_2CH_2OH \xrightarrow[H_2SO_4]{HBr} CH_3CH_2CH_2CH_2CH_2Br \xrightarrow{t\text{-}BuO^-} CH_3CH_2CH_2CH{=}CH_2$

or $CH_3CH_2CH_2CH_2CH_2OH \xrightarrow{Al_2O_3,\ \Delta} CH_3CH_2CH_2CH{=}CH_2 \xrightarrow[inhibitors\ (ionic)]{HBr} CH_3CH_2CH_2CHBrCH_3$

d. $CH_3CH_2C(CH_3)(D)CH_2OH \xrightarrow[\Delta]{Al_2O_3} CH_3CH_2C(CH_3){=}CH_2 \xrightarrow[cat.]{H_2} CH_3CH_2CH(CH_3)CH_3$

e. $CH_3CH_2CH_2CH(CH_3)_2 \xrightarrow{Br_2} CH_3CH_2CH_2CBr(CH_3)_2 \xrightarrow{t\text{-}BuO^-} CH_3CH_2CH_2C(CH_3){=}CH_2 \xrightarrow{B_2H_6} \xrightarrow[NaOH]{H_2O_2} CH_3CH_2CH_2CH(CH_3)CH_2OH$

f. cyclohexane $\xrightarrow[h\nu]{Cl_2} \xrightarrow[C_2H_5OH]{NaOC_2H_5} \xrightarrow[Zn(Cu)]{CH_2I_2}$ bicyclo[4.1.0]heptane

g. $CH_3CH_2C(CH_3)_2Br \xrightarrow[(CH_3)_3COH]{KOC(CH_3)_3} CH_3CH_2C(CH_3){=}CH_2 \xrightarrow[CHCl_3]{KOH}$ 1,1-dichloro-2-ethyl-2-methylcyclopropane (CH_3CH_2, CH_3; Cl, Cl)

12. a. hexanenitrile (CN) $\xleftarrow{NaCN}$ 1-bromopentane (Br) $\xleftarrow{PBr_3}$ 1-pentanol (OH)

b. Br, Br ← Br_2 / CCl_4 ← H_2SO_4 / Δ ← OH

c. HO OH ← 1. O_3 2. $NaBH_4$ ← H_2SO_4 ← OH

d. O ← CH_3CO_3H ← H_2SO_4 ← OH

13.

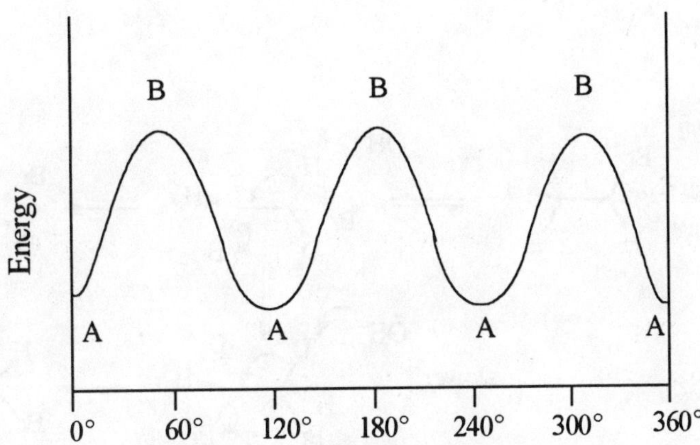

The angle plotted is the dihedral angle between the plane of the double bond and the $C-C_{methyl}-H$ plane.

14. The structure of 4,4-dimethyl-2-pentene is:

cis: $(H_3C)_3C-CH=CH-CH_3$ (H, H on same side) *steric hindrance*

trans: $(H_3C)_3C-CH=CH-CH_3$ (H, CH_3 on opposite sides) *less steric hindrance*

In the *cis* isomer, the terminal methyl groups are close in space, and the electron clouds interact and repel each other (steric hindrance).

15. $CH_3C(CH_3)(OH)CH_2CH_2CH_2CH_2CH_2OH$ ← primary OH; OH ← tertiary OH

The tertiary hydroxyl group dehydrates readily under acid conditions via the relatively stable *tert*-carbocation. Primary hydroxyl groups dehydrate much less readily.

$$CH_3C(CH_3)(OH)CH_2CH_2CH_2CH_2CH_2OH \xrightarrow{H^+} (CH_3)_2\overset{+}{C}(CH_2)_5OH \longrightarrow (CH_3)_2C{=}CH(CH_2)_4OH + CH_2{=}C(CH_3)(CH_2)_5OH$$

major product: $(CH_3)_2C{=}CH(CH_2)_4OH$; minor product: $CH_2{=}C(CH_3)(CH_2)_5OH$

16. Both isomers are produced via the carbocation, $CH_3\overset{+}{C}(CH_3)CH_2C(CH_3)_3$.

Loss of a primary hydrogen gives $CH_2{=}C(CH_3)CH_2C(CH_3)_3$, a disubstituted alkene with little steric hindrance.

Loss of a *sec*-H gives a trisubstituted alkene with much steric hindrance between the adjacent methyl and *t*-butyl.

$(CH_3)_3C(H)C{=}C(CH_3)_2$ Look at models!

17.

$\text{DHC}{=}\text{CDH} \xrightarrow{Br_2}$ bromonium ion $\rightarrow$ dibromide $\rightleftharpoons$ (OH^-) $\rightarrow$ $\text{BrDC}{=}\text{CDH}$

slower: (OH^-) $\rightarrow$ $\text{HBrC}{=}\text{CDH}$

$\text{HDC}{=}\text{CDH} \longrightarrow \longrightarrow \text{DHC}{=}\text{CDBr}$

18. $CH_3CH(OC_2H_5)CH_3 \xleftarrow{S_N2} CH_3CHBrCH_3 \xrightarrow[-\,HBr]{E2} CH_3CH{=}CH_2$

S_N2 reaction is almost unaffected by dueterium. Loss of deuterium in E_2 is slower than loss of hydrogen.

$CD_3CH(OC_2H_5)CD_3 \xleftarrow[\text{(same rate)}]{S_N2} CD_3CHBrCD_3 \xrightarrow[-\,DBr]{E2} CD_3CH{=}CD_2$ (slower rate)

The deuterium isotope effect, k_D/k_H (E2), $= \dfrac{1/2}{3/1} = \dfrac{1}{6}$

19.

Monosubstituted ethylenes	$-\Delta H^\circ_{hydrog}$	Disubstituted ethylenes	$-\Delta H^\circ_{hydrog}$
propene	29.7	*cis*-2-butene	28.5
1-butene	30.2	*cis*-2-pentene	28.1
1-pentene	29.8	*trans*-2-butene	27.4
		trans-2-pentene	27.2
		2-methylpropene	28.1
		2-methyl-1-butene	28.3

Alkyl substitution on a double bond has a stabilizing effect (see Section 11.4), therefore less energy is released on hydrogenation of the alkylethylenes than for ethylene itself. Similarly, *trans*-alkenes are more stable than *cis*-alkenes, and they have less negative ΔH°_{hydrog}.

20.

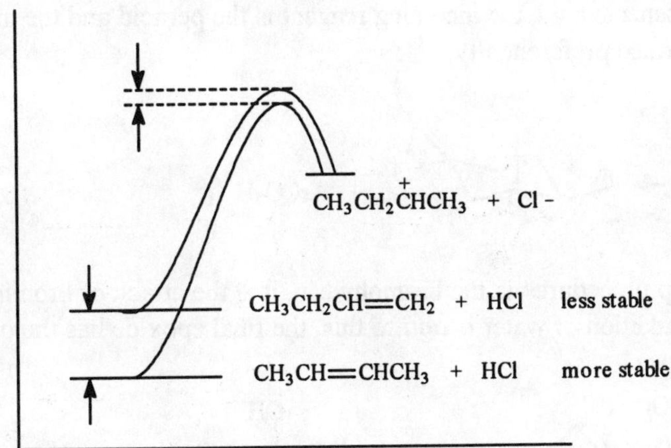

Reaction Coordinate

The difference in energies of the two transition states is probably less than the difference in energies of the two reactants. Thus, $\Delta H^\ddagger$ for 1-butene is smaller than $\Delta H^\ddagger$ for 2-butene, and 1-butene reacts more rapidly. On this basis, *cis*-2-butene should be more reactive than *trans*-2-butene. In general, when two isomers give the same intermediate or product via comparable transition states, the <u>less</u> stable isomer reacts <u>faster</u>, because it has a smaller energy barrier to overcome.

21.

Cl–cyclopentyl–CH_3 → cyclopentyl–CH_3 + cyclopentenyl–CH_3

optically active — *achiral, ∴optically inactive isomer* — *chiral, ∴ optically active isomer*

22. a. $HOCH_2CH_2CH_2CH{=}CH_2$ (I–I) → $H\ddot{O}CH_2CH_2CH_2CH{-}CH_2$ (bridged I^+) → (HCO_3^-) oxonium ion with CH_2I → tetrahydrofuran ring with CH_2I

b. 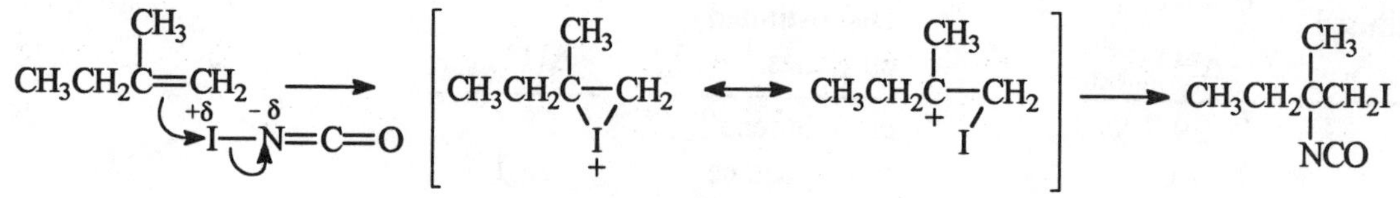 $CH_3CH_2C(CH_3)=CH_2 + I{-}N{=}C{=}O \longrightarrow [CH_3CH_2C(CH_3){-}CH_2\,(\text{bridged } I^+) \longleftrightarrow CH_3CH_2C^+(CH_3){-}CH_2I] \longrightarrow CH_3CH_2C(CH_3)(NCO)CH_2I$

c. $H_2C{-}CHCD_2Cl$ (epoxide) $+ CH_3O^- \longrightarrow CH_3OCH_2CH(O^-)CD_2{-}Cl \longrightarrow CH_3OCH_2CH{-}CD_2$ (epoxide)

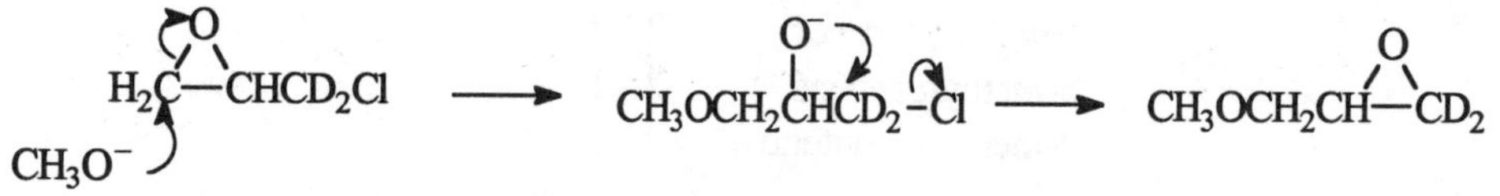

23. 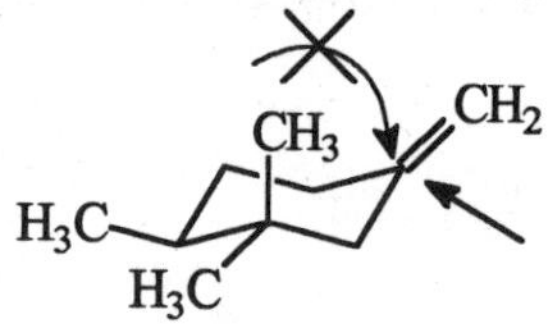 Attack from the top side of the molecule (as shown at the left) is sterically hindered, so the incoming reagent approaches from the bottom. With *m*-chloroperbenzoic acid, the incoming reagent is the peracid and the indicated epoxide is formed preferentially:

$+ ArCO_3H \longrightarrow$ (epoxide) $+ ArCO_2H$

When the epoxide is formed in the two-step procedure via the bromohydrin, it is the attack of bromine that determines the stereochemistry. Because addition of water is *anti* to this, the final epoxide has the opposite configuration:

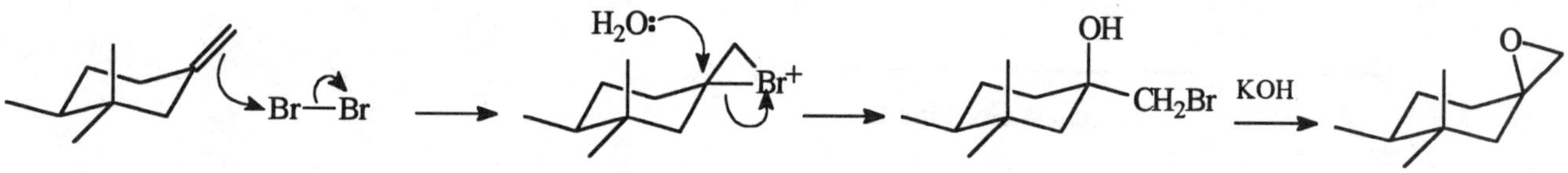

$Br{-}Br \longrightarrow$ ($H_2O{:}$, Br^+) $\longrightarrow$ (OH, CH_2Br) $\xrightarrow{KOH}$ (epoxide)

24. $\underline{H}$ ($C_{11}H_{24}O$) $\xrightarrow{PBr_3}$ $\underline{I}$ ($C_{11}H_{23}Br$) $\xrightarrow{KOCH_2CH_3}$ $\underline{J}$ *(major)* ($C_{11}H_{22}$) + $\underline{K}$ *(minor)* ($C_{11}H_{22}$)

J & K $\xrightarrow{O_3}\xrightarrow{NaBH_4}$ $(CH_3)_2CHCH_2CH_2OH + (CH_3)_2CHCH_2CH_2CH_2OH$

The sequence:
1. O_3, 2. $NaBH_4$ cleaves double bonds and reduces the products to alcohols, therefore J and K must be $(CH_3)_2CHCH_2CH{=}CHCH_2CH_2CH(CH_3)_2$.

J must be the *trans* isomer (formed more easily) and K the *cis* isomer. The 2,8-dimethyl-4-nonenes arose from an alkyl bromide by E2 elimination. Two bromides could have given J and K, but the 4-bromo isomer would have given 2,8-dimethyl-3-nonene as well. Therefore I must be 5-bromo-2,8-dimethylnonane and H must be the corresponding alcohol.

$$(CH_3)_2CHCH_2CHBrCH_2CH_2CH_2CH(CH_3)_2 \xrightarrow[HOC_2H_5]{KOC_2H_5} (CH_3)_2CHCH{=}CHCH_2CH_2CH_2CH(CH_3)_2 + \underline{J} \text{ and } \underline{K}$$

$$\underset{\underline{I}}{(CH_3)_2CHCH_2CH_2CHBrCH_2CH_2CH(CH_3)_2} \xrightarrow[HOC_2H_5]{KOC_2H_5} \underline{J} \text{ and } \underline{K} \text{ only}$$

$$\underset{\underline{H}}{(CH_3)_2CHCH_2CH_2CHOHCH_2CH_2CH(CH_3)_2}$$

25. $$CH_3CH_2CH_2\overset{Br}{\overset{|}{C}}HCH(CH_3)_2 \xrightarrow{base} \underset{\textbf{C}}{CH_3CH_2CH_2CH{=}C(CH_3)_2} + \underset{\textbf{D + E}}{\textit{cis/trans}\text{-}CH_3CH_2CH{=}CHCH(CH_3)_2}$$

$$\textbf{C} \xrightarrow[H_2O_2\text{ , NaOH}]{B_2H_6} \underset{\textbf{F}}{CH_3CH_2CH_2\overset{OH}{\overset{|}{C}}HCH(CH_3)_2} \qquad \textbf{D + E} \xrightarrow[H_2O_2\text{ , NaOH}]{B_2H_6} \textbf{F} + \underset{\textbf{G}}{CH_3CH_2\overset{OH}{\overset{|}{C}}HCH_2CH(CH_3)_2}$$

From the data given, we cannot determine whether D is *cis* or *trans*. When working this type of "roadmap" problem, it helps to summarize the data in the following manner:

$$C_7H_{15}Br \xrightarrow{base} \underset{(C_7H_{14})}{C + D + E} \xrightarrow{H_2/catalyst} (CH_3)_2CHCH_2CH_2CH_2CH_3$$

$$C \xrightarrow{B_2H_6} \xrightarrow[OH^-]{H_2O_2} F \text{ (alcohol)}$$

$$D + E \xrightarrow{B_2H_6} \xrightarrow[OH^-]{H_2O_2} F + G \quad \text{(isomeric alcohols, about equal amounts)}$$

We can rule out Br at C-1 or 6 (E2 would give a single alkene), C-2, (E2 would give two alkenes), C-4 (E2 would give four alkenes, two *cis-trans* pairs). For C-5, there are also 3 isomeric alkenes possible with E2:

$$(CH_3)_2CHCH_2CH_2CHBrCH_3 \longrightarrow \underset{\textit{cis and trans}}{(CH_3)_2CHCH_2CH{=}CHCH_3} + (CH_3)_2CHCH_2CH_2CH{=}CH_2$$

but with B_2H_6 and H_2O_2/OH^-, 5-methyl-1-hexene would give mainly 5-methyl-1-hexanol, and the 5-methyl-2-hexenes cannot give this alcohol. This leaves only $(CH_3)_2CHCHBrCH_2CH_2CH_3$, which reacts as shown above.

26. Both reactions must have $\Delta H°$ that is not too positive; otherwise, $\Delta E^‡$ is too high and reaction will be slow. For Y=Br, HS and $(CH_3)_3C$, both $\Delta H°$'s are negative (exothermic), and both steps should be facile. For the other compounds, one step or the other is no good; that is, one $\Delta H°$ is so positive that the reaction has a high activation energy and is slow.

27. ΔH°_f ($CH_3CH{=}CH_2$) = 4.9 kcal mole^{-1} *(Appendix I)*

ΔH°_f ($\cdot CN$) = 104 kcal mole^{-1} *(Appendix I)*

ΔH°_f (HCN) = ΔH°_f (H$\cdot$) + ΔH°_f ($\cdot$CN) - DH° (H-CN)

= 52 + 104 - 125 = 31 kcal mole^{-1} *(also given in Appendix I)*

ΔH°_f ($CH_3CH_2CH_2CN$) = ΔH°_f (CH_3CH_2CN) + ΔH°_f ($-CH_2-$)

= 12.1 - 5.0 = 7.1 kcal mole^{-1} *(see problem #16, Chapter 5)*

ΔH°_f ($CH_3\dot{C}HCH_2CN$) = ΔH°_f ($CH_3CH_2CH_2CN$) + DH° ($CH_3CH_2CH_2CN$) - ΔH°_f (H$\cdot$)

= 7 + 98 - 52 = 53 kcal mole^{-1}

a. $CH_3{-}CH{=}CH_2$ + HCN ⟶ $CH_3CH_2CH_2CN$
ΔH°_f = 4.9   31.2   7.1 ΔH°_f = -29.0 kcal mole^{-1}

b. *STEP 1:*
$CH_3CH{=}CH_2$ + $\cdot$CN ⟶ $CH_3\dot{C}HCH_2CN$
ΔH°_f = 4.9   104   53 ΔH°_f = -56 kcal mole^{-1}

STEP 2:
$CH_3\dot{C}HCH_2CN$ + HCN ⟶ $CH_3CH_2CH_2CN$ + $\cdot$CN
ΔH°_f = 53   31   7   104   ΔH° = +27 kcal mole^{-1}

c. The overall reaction is exothermic, but the second step is far too endothermic. The intermediate $CH_3\dot{C}HCH_2CN$ radicals would dimerize rather than abstract H from HCN.

11.F. Supplementary Problems

S1. Write structures for the following compounds:

a. *cis*-3-methyl-2-heptene
b. (*R*)-5-methylhex-4-en-2-ol
c. (*Z*)-4-methyl-2-pentene
d. (*E*)-5-chloro-3-isopropyl-5-methyl-2-hexene
e. (*Z*)-3-bromo-3-hexene
f. 1-chloro-6-methylcyclohexene
g. (*E*)-cyclododecene

S2. Show how to accomplish the following transformations in a practical manner (more than one step is necessary in each case).

a. $CH_3CH_2CH_2CH_2OH$ ⟶ $CH_3CH_2CHOHCH_3$

b. (cyclohexylidene)$=CH_2$ ⟶ (cyclohexyl)$-CH_2CN$

c. [cyclopentanol] → [cis-1,2-cyclopentanediol]

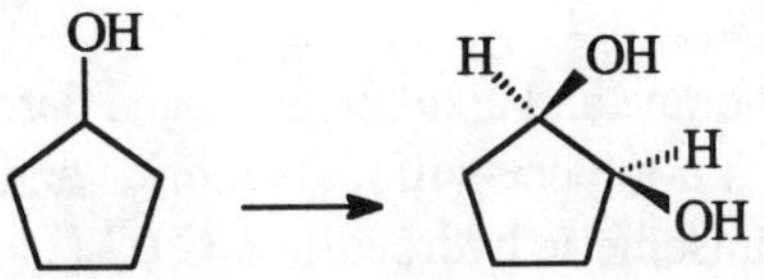

d. [1-methylcyclohexene] → [2-methylcyclohexanone]

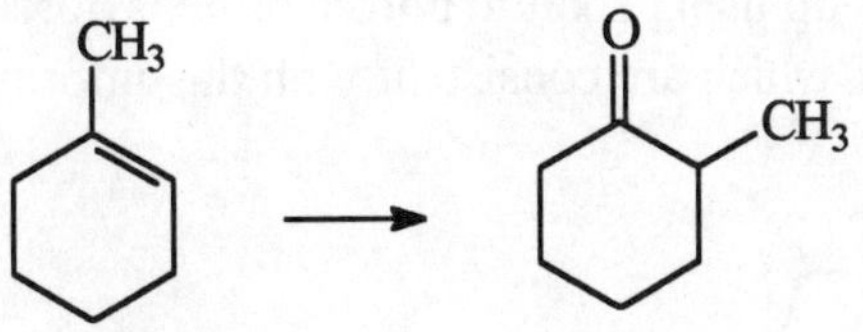

e. [methylcyclopentane] → [cyclopentanone]

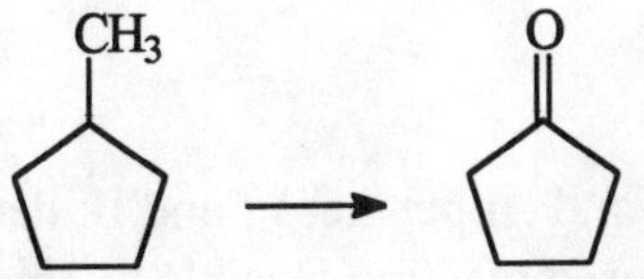

f. $(H_3C)_2C{=}CH_2$ → [3,3-dimethyl-1-bromocyclopropene]

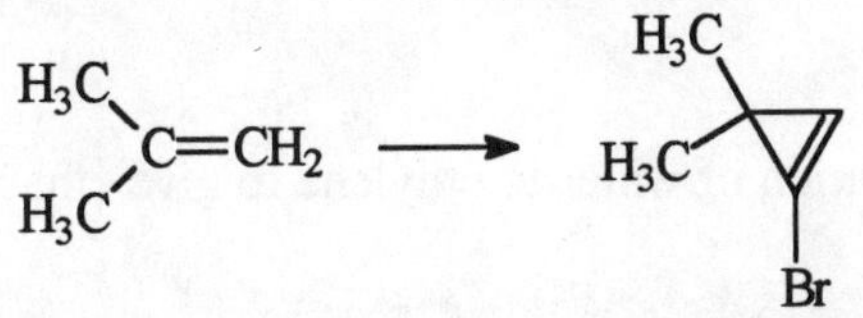

g. [cyclohexene] → [deuterated bromocyclohexane]

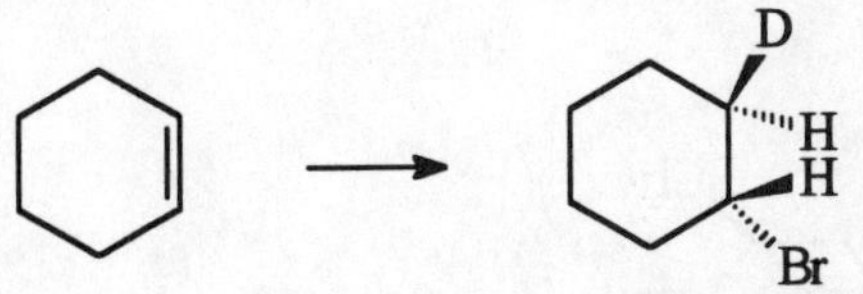

S3. Predict the product from the reaction of iodine chloride (ICl) with 1-methylcyclohexene, and justify your choice by writing a step-by-step mechanism for its formation.

S4. Predict the major products from the following reaction sequences:

a. $CH_2{=}C(CH_3)CH(OH)CH_3 \xrightarrow[CH_2I_2]{Zn(Cu)} \xrightarrow{HCl}$

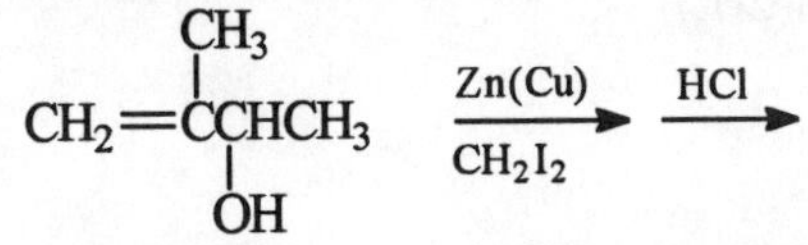

b. [1,2-dimethylcyclopentanol] $\xrightarrow[\Delta]{60\%\ H_2SO_4} \xrightarrow[25^0\ C]{KMnO_4}$

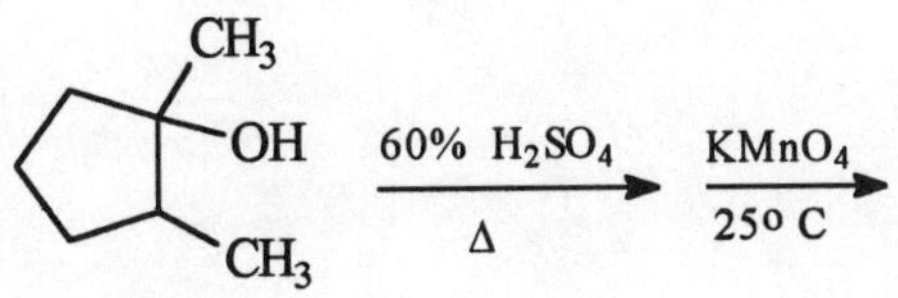

c. $CH_3CH_2CH{=}CHC(CH_3)_3 \xrightarrow[HOC_2H_5]{Hg(OAc)_2} \xrightarrow{NaBH_4}$

d. $(CH_3)_2C{=}CHCH_3 \xrightarrow[CH_3OH]{Br_2} \xrightarrow[HOC_2H_5]{KOC_2H_5} \xrightarrow[Pt]{D_2}$

S5. A hydrocarbon (A) of formula C_7H_{12} was treated successfully with diborane and alkaline hydrogen peroxide to provide compound **B** ($C_7H_{14}O$) as the only product. Reaction of **B** with *p*-toluene-sulfonyl chloride and pyridine, and then with potassium *t*-butoxide in *t*-butyl alcohol gave an isomeric hydrocarbon **C** (C_7H_{12}). Finally, treatment of **C** with ozone in methanol, followed by work-up using sodium borohydride, afforded 2-methyl-1,6-hexanediol. Write complete structures for **A**,**B**, and **C** which are consistent with this information.

S6. Write a reasonable mechanism for the following transformation:

$$CH_2{=}CHCH_2CH_2CH{=}CH_2 \xrightarrow[H_2O]{2\ Hg(OAc)_2} \xrightarrow{NaBH_4}$$ (2,5-dimethyltetrahydrofuran: H_3C–ring O–CH_3)

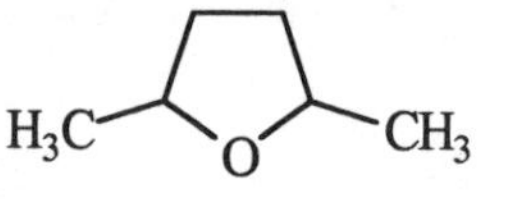

S7. The bond dissociation energy for H-SH is 90 kcal mole^{-1}. Using this value and Appendices I and II, determine H° for each step in the free radical addition of H_2S to ethylene. Is the proposed reaction feasible by this mechanism?

S8. Using Appendices I and II, determine whether the free radical addition of water to ethylene to give ethanol is feasible thermodynamically.

11.G. Answers to Supplementary Problems

S1. a. (H)(H_3C)C=C(CH_3)($CH_2CH_2CH_2CH_3$)

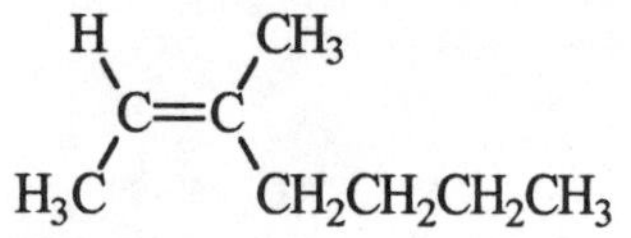

b. (H_3C)(H_3C)C=C(H)–CH_2–C(H)(CH_3)(OH)

c. (H_3C)(H)C=C($CH(CH_3)_2$)(H)

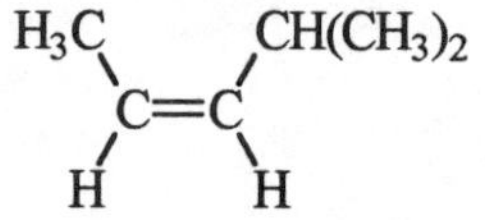

d. (H_3C)(H)C=C($CH_2C(Cl)(CH_3)_2$)($CH(CH_3)_2$)

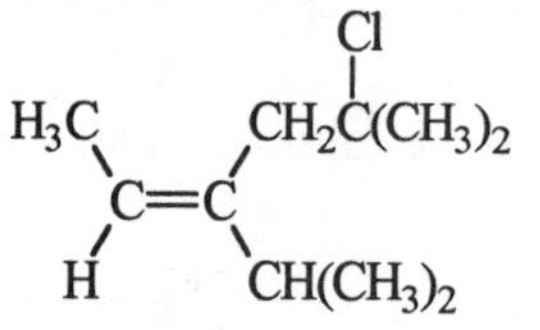

e. (H)(CH_3CH_2)C=C(CH_2CH_3)(Br)

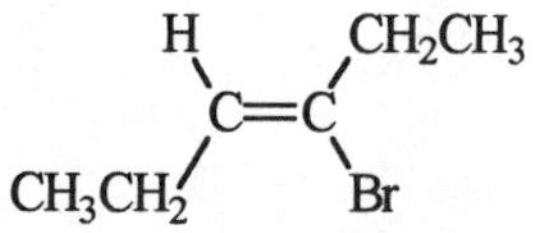

f. cyclohexene ring bearing CH_3 and Cl

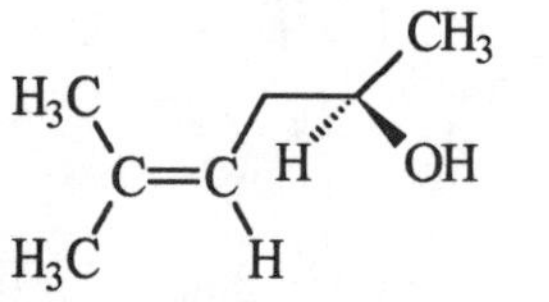

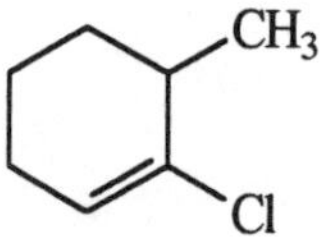

g. (cycloalkene ring)

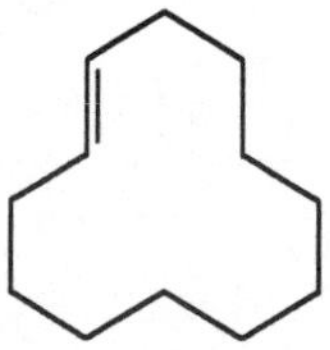

S2. a. $CH_3CH_2CH_2CH_2OH \xrightarrow[400°]{Al_2O_3} CH_3CH_2CH{=}CH_2 \xrightarrow[H_2O]{H_2SO_4} CH_3CH_2CHOHCH_3$

b. 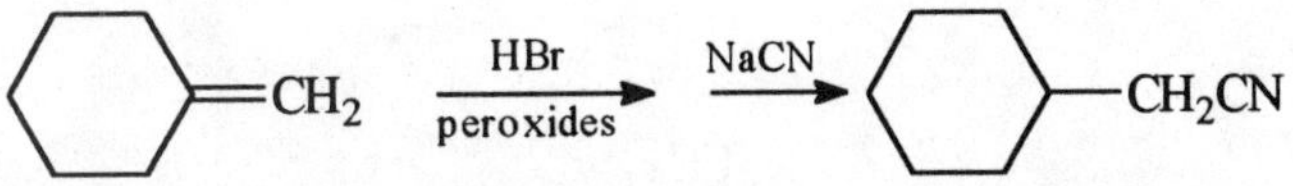
=CH2
HBr
peroxides
NaCN
-CH2CN

c. 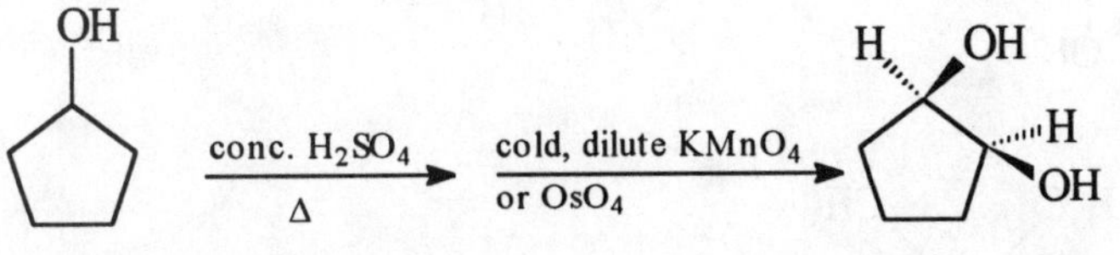
OH
conc. H2SO4
Δ
cold, dilute KMnO4
or OsO4
H
OH
H
OH

d. 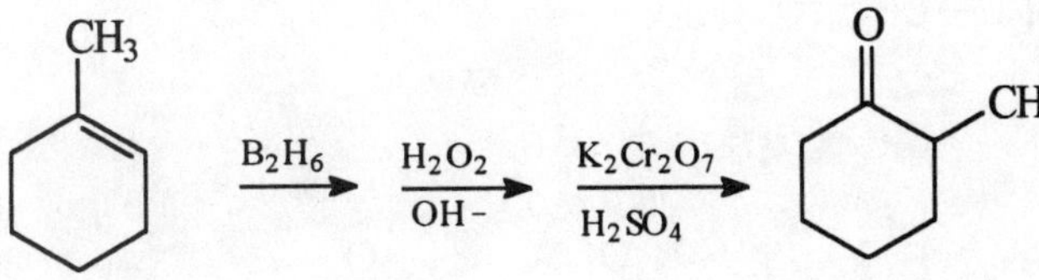
CH3
B2H6
H2O2
OH-
K2Cr2O7
H2SO4
O
CH3

e. 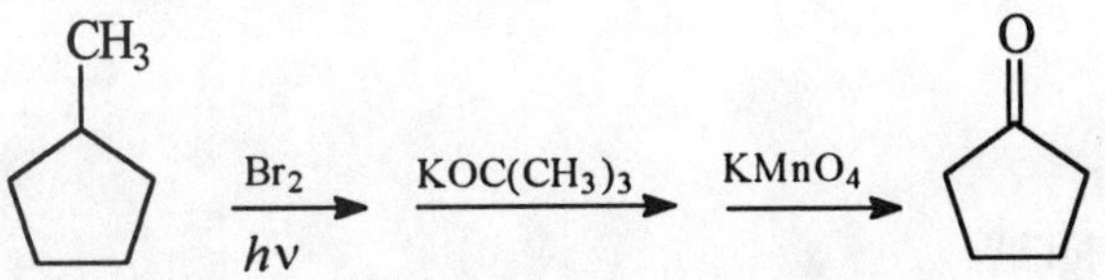
CH3
Br2
hν
KOC(CH3)3
KMnO4
O

f. 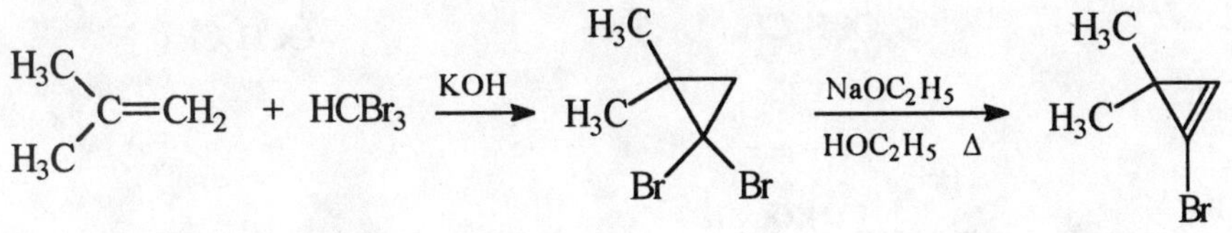
H3C
C=CH2
H3C
+ HCBr3
KOH
H3C
H3C
Br
Br
NaOC2H5
HOC2H5 Δ
H3C
H3C
Br

g. 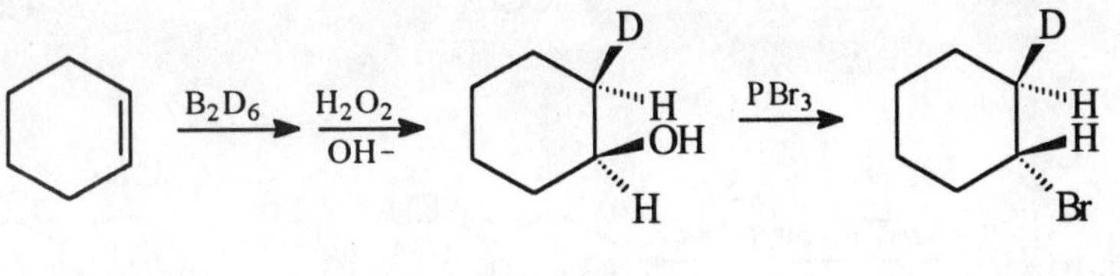
B2D6
H2O2
OH-
D
H
OH
H
PBr3
D
H
H
Br

S3. 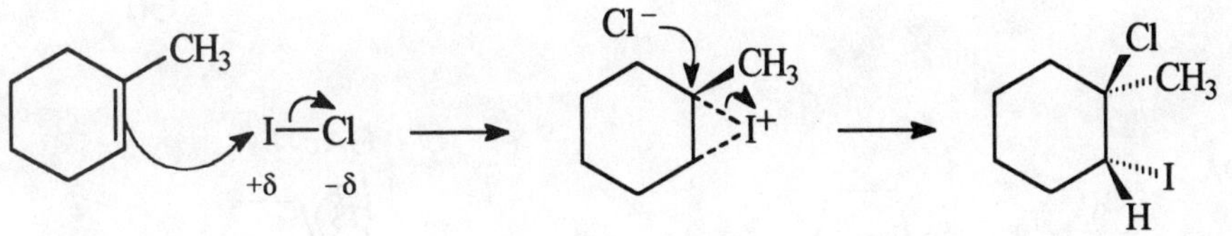
CH3
I—Cl
+δ -δ
Cl−
CH3
I+
Cl
CH3
I
H

S4. a.

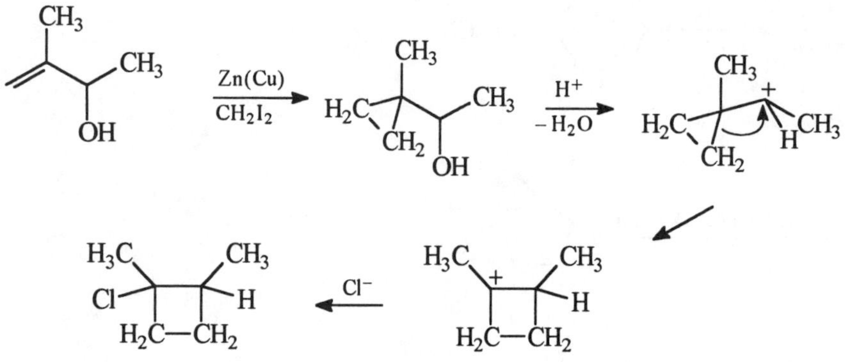

b.

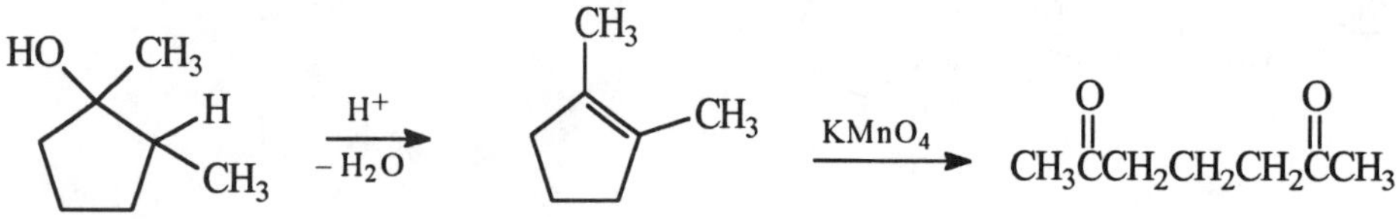

c.

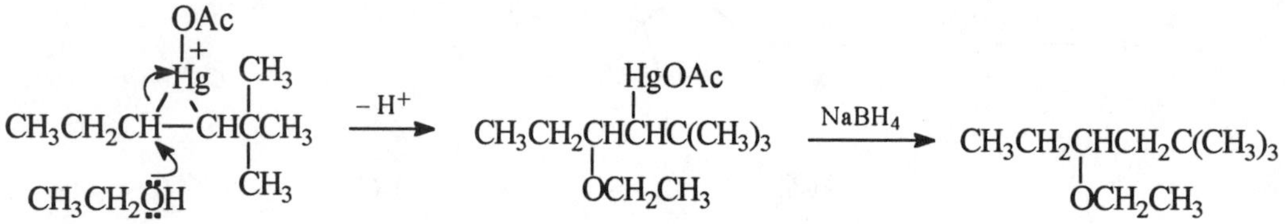

d.

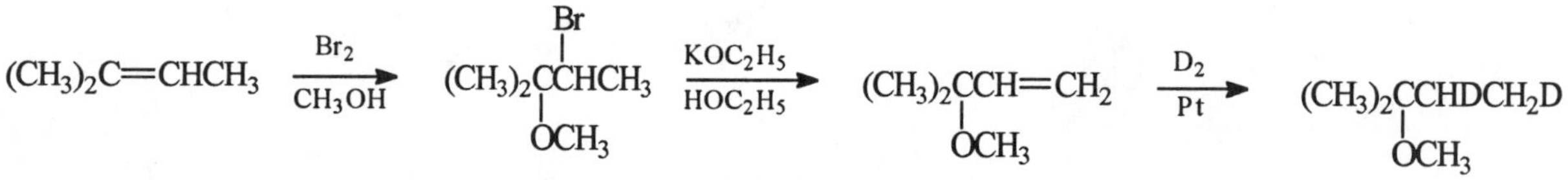

S5.

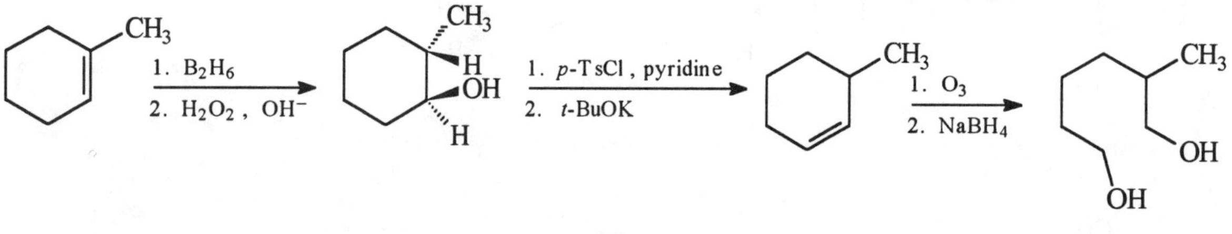

S6.

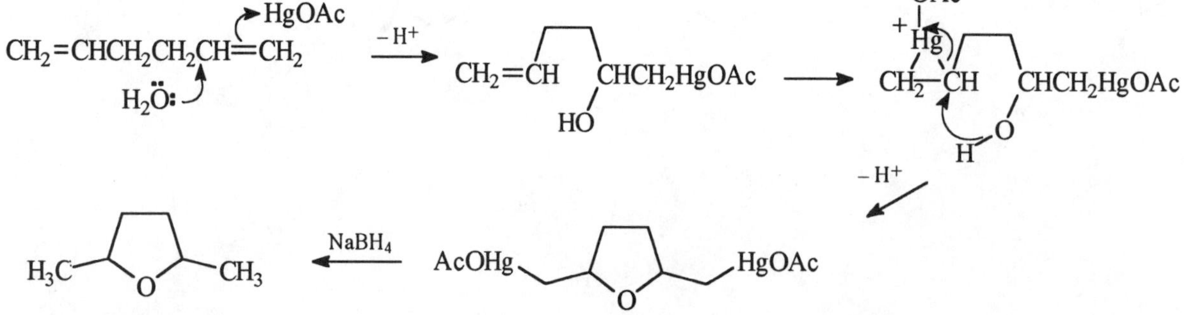

S7. ΔH°_f ($\cdot CH_2CH_2SH$) = ΔH°_f (CH_3CH_2SH) + DH° (primary C-H) − ΔH°_f (H·)
= −11 + 98 − 52 = 35 kcal mole^{-1}

$CH_2{=}CH_2$ + $\cdot SH$ ⟶ $\cdot CH_2CH_2SH$
ΔH°_f = −12.5 34 35 ΔH° = −12 kcal mole^{-1}

$\cdot CH_2CH_2SH$ + H_2S ⟶ CH_3CH_2SH + $HS\cdot$
ΔH°_f = 35 −4.8 −11 34 ΔH° = −7 kcal mole^{-1}

The proposed reaction is clearly feasible, because each step is exothermic.

S8. ΔH°_f ($\cdot CH_2CH_2OH$) ≅ ΔH°_f (CH_3CH_2OH) + DH° (primary C-H) − ΔH°_f (H·)

= −56 + 98 − 52 = −10 kcal mole^{-1}

$CH_2{=}CH_2$ + $\cdot OH$ ⟶ $\cdot CH_2CH_2OH$
ΔH°_f = 12.5 9.4 −10 ΔH° = −32 kcal mole^{-1}

$\cdot CH_2CH_2OH$ + H_2O ⟶ CH_3CH_2OH + $\cdot OH$
ΔH°_f = −10 −57.8 −56.2 9.4 ΔH° = +21 kcal mole^{-1}

The second step is too endothermic for this mechanism of hydration to be feasible.

12. ALKYNES AND NITRILES

12.A Chapter Outline and Important Terms Introduced

12.1 Electronic Structure

sp hybridization — triple bond, 1 σ, 2 π

isoelectronic, equivalent electron distribution, different nuclei

12.2 Nomenclature

alkyne — acetylenes - common

alkenyne — alkanenitrile, include the nitrile carbon

12.3 Physical Properties

higher μ and b.p. for nitriles than corresponding alkynes

12.4 Acidity of Alkynes

acetylene $pK_a = 25$

12.5 Preparation of Alkynes

A. Acetylene
- $Ca_2C + H_2O \longrightarrow HC{\equiv}CH$
- Flash pyrolysis of CH_4

B. Nucleophilic Substitution Reactions
- alkylation of acetylide anions

C. Elimination Reactions
- geminal dihalides — vicinal dihalides
- migration of the triple bond

12.6 Reaction of Alkynes and Nitriles

A. Reduction
- hydrogenation — dissolving metal reductions
- partial hydrogenation — radical anion, vinyl radical, vinyl anion intermediates

B. Electrophilic Additions (slower than alkenes because vinyl cations are high in energy)
- hydration — hydrohalogenation

C. Nucleophilic Additions
- vinyl alkoxides — amides (from nitriles)

D. Hydroboration
- vinyl boranes — "anti-Markovnikov" hydration

12.7 Vinyl Halides (haloalkenes)

12.B. Important Reactions Introduced

Alkylation of acetylide and cyanide (12.5.B)

Equation:

$$RC{\equiv}CH + M^+Base^- \longrightarrow Base\text{-}H + RC{\equiv}C{:}^-M^+ \xrightarrow{R'X} RC{\equiv}CR' + MX$$

$$N{\equiv}C{:}^-M^+ \xrightarrow{R''X} N{\equiv}CR'' + MX$$

Generality:

R = H, alkyl, aryl

M^+Base^- = $NaNH_2$, butyllithium or R"MgX

for acetylene alkylation: R′X = 1° alkyl halide or sulfonate (otherwise E2)
for cyanide alkylation: R″X = 1° or 2° alkyl halide or sulfonate
(E2 with R = 3° alkyl)

Key features: important carbon-carbon bond-forming process

Synthesis and isomerization of alkynes via elimination (12.5.C)

Equation:

$$\left.\begin{array}{c} RCHXCHXR' \\ \text{or} \\ RCX_2CH_2R' \end{array}\right\} \xrightarrow[\text{base}]{\text{strong}} RCX{=}CHR' \xrightarrow[\text{base}]{\text{strong}} RC{\equiv}CR'$$

Generality: R = H, alkyl, or aryl; X = Cl, Br, I
strong base = very hot KOH or KOR; $NaNH_2$ or KAPA (potassium 3-aminopropylamide)

Key features: reaction can be stopped after first elimination
triple bond isomerization often occurs:

$$RC{\equiv}CCH_3 \underset{KOH,\ \Delta}{\overset{1.\ NaNH_2 \text{ or KAPA},\ 2.\ H_2O}{\rightleftarrows}} RCH_2C{\equiv}CH$$

Partial hydrogenation of alkynes: cis (12.6.A)

Equation:

$$R{-}C{\equiv}C{-}R' \xrightarrow[\text{catalyst}]{H_2} \text{(R)(H)C}{=}\text{C(R')(H)} \text{ (cis: R and R' on same side)}$$

Generality: R = H, alkyl, or aryl
catalyst = $Pd/BaSO_4$ + quinoline = Lindlar's catalyst or Ni-B

Key features: *syn* addition of hydrogens give *cis* stereochemistry

Partial reduction of alkynes: trans (12.6.A)

Equation:

$$R{-}C{\equiv}C{-}R' \xrightarrow{[H]} \text{(R)(H)C}{=}\text{C(H)(R')} \text{ (trans: R and R' on opposite sides)}$$

Generality: R = H or alkyl
[H] = Na or Li in liquid NH_3, or $LiAlH_4$

Key features: *trans* stereospecificity

Reduction of nitriles (12.6.A)

Equation: $RC{\equiv}N \longrightarrow RCH_2NH_2$

Generality: R = alkyl or aryl
[H] = H_2/Pt, H_2/Ni or $LiAlH_4$

Hydration of alkynes: mercuric ion catalyzed addition of H_2O (12.6.B)

Equation:

$$R{-}C{\equiv}C{-}R' \xrightarrow[H_2SO_4]{Hg^{2+}} RC(=O)CH_2R'$$

Generality: R = alkyl or aryl; R′ = H, alkyl, or aryl

Key features: Markovnikov orientation
works best if R′ = H or same as R (otherwise gives mixture of isomers)
mechanism involves enol (vinyl alcohol) intermediate

Hydration of alkynes via hydroboration (12.6.D)

Equation:

$$R{-}C{\equiv}C{-}R' + HBR''_2 \longrightarrow (R)(H)C{=}C(R')(BR''_2) \xrightarrow[H_2O_2]{^-OH} RCH_2C(=O)R' + 2\ R''OH + B(OH)_3$$

Generality: R = alkyl or aryl, R′ = H, alkyl, or aryl
R" = H ($B_2H_6 \rightleftharpoons 2\ BH_3$) or alkyl or other alkenyl groups

Key features: "anti-Markovnikov" orientation
works best if R′ = H or same as R (otherwise gives mixture of isomers)
mechanism involves enol (vinyl alcohol) intermediate

Addition of HX to alkynes (12.6.B)

Equation: $RC{\equiv}CR' + HX \longrightarrow RCX{=}CHR' \xrightarrow{HX} RCX_2CH_2R'$

Generality: R, R′ = H, alkyl, or aryl
X = Cl, Br

Key features: Markovnikov orientation
can be stopped after one addition step

Hydration of nitriles (12.6.B and C)

Equation:

$$RC{\equiv}N \xrightarrow{H_2O} RC(=O)NH_2$$

Generality: requires either acid (e.g. aq. H_2SO_4) or base (eg. KOH in *t*-BuOH) catalysis

12.C. Important Concepts and Hints

Chemistry of Alkynes: Because the chemistry of alkynes is so similar to that of alkenes, it is useful to focus on the contrasts. Look particularly at reduction methods: you can hydrogenate alkynes with *syn* delivery of H_2, just as you can in alkene chemistry. However, with alkynes there is the added complication of stopping after the first addition or not, depending on the catalyst you use. Sodium in ammonia reduces alkynes (to *trans* alkenes), but does not reduce ordinary alkenes.

Electrophilic addition to alkynes is subject to the same orientation and stereochemical effects as alkenes (Markovnikov or anti-Markovnikov, *syn* or *anti* or neither). There are two complications: with alkynes the addition can occur **twice**, and the hydration reactions (Markovnikov-oriented with Hg^{2+}, dil. H_2SO_4, or anti-Markovnikov

can occur **twice**, and the hydration reactions (Markovnikov-oriented with Hg^{2+}, dil. H_2SO_4, or anti-Markovnikov with B_2H_6/H_2O_2, OH^-) do not lead to alcohols but to ketones or aldehydes instead.

$NaNH_2$ vs. Na, NH_3: Much of the chemistry of alkynes involves either sodium amide or sodium in ammonia as reagents, and students get confused over the difference between them. Sodium *amide* is a **base**; it is a salt, comprised of the sodium cation (Na^+) and amide anion (NH_2^-), and it is often used when a very strong base is required ($NH_3 \rightleftharpoons NH_2^- + H^+$; $pK_a = 34$). Benzene, or liquid ammonia itself, is frequently employed as the solvent for reactions involving $NaNH_2$.

Sodium *metal* (Na^0), dissolved in liquid ammonia, is a **reducing agent**; it ionizes to give a dark blue solution of sodium cations (Na^+) and solvated electrons (e^-). The solvated electron is a powerful reducing agent; hence it is for reduction that "sodium in ammonia" (Na, NH_3) is used.

Organic Synthesis: In the chapter on alkynes, you encounter the first reaction that is useful for formation of carbon-carbon bonds: alkylation of acetylide anions with alkyl halides. Reactions such as this are important because they allow you to build larger molecules, instead of just interchanging functional groups. There are limitations to be sure – the alkyl halide must be primary and unhindered, and there cannot be functional groups elsewhere in the molecule that are sensitive to strong base.

Now that you know a way to build large molecules from small ones, you will encounter a very important learning device – synthetic problems. These questions ask you to devise a way to make the target compound starting from simpler materials. From the beginning of alchemy to the present day, finding ways to turn simpler (or cheaper) compounds into more complicated (or expensive) ones has been a major pursuit of chemists. When you have learned a greater variety of carbon-carbon bond forming reactions and functional group interconversions, a whole chapter in the text will be devoted to Organic Synthesis (Chapter 16). At this time, it is still worthwhile to get you started on synthesis problems the right way by giving you an important hint: work the problems **backwards**.

When you want to go somewhere you've never been before, you find your destination on a map and work your way backward to where you are, before you actually set out. You don't just climb on the first bus that goes by your house, or drive down the first highway you see, and then determine whether it's going to your destination. The same is true for synthesis problems – don't take the starting materials and see what you can turn them into, hoping eventually to bump into the target. You should look at the target and think what its immediate precursor could be. Look for appropriate places for carbon-carbon bond-forming reactions, and work your way backwards to simpler compounds until you get to the starting materials (more will be said on this subject when you reach Chapter 16).

Road Map Problems: You will also encounter problems that give you an idea of how to figure out the structure of a compound by chemical methods. These are called road-map problems (sometimes they seem more like road-block problems...), and are actually just puzzles that need to be solved logically from the clues given. The best way to approach them is to be organized: write down schematically all the information presented in words in the problem. Then, look for the compound about which the most is known relative to the possibilities for it. If the structure of this key compound is actually provided, or if you can figure out what it is, work your way outward from that point as logically as you can. The best advice is to be systematic. Problem #17 and its answer are a good example of this approach.

Degrees of Unsaturation: In solving road-map problems, one particular point is important – using the formula to deduce "degrees of unsaturation". A degree of unsaturation can be a ring or a π-bond. Both C=C and C=O double bonds are unsaturations; a C≡C triple bond is **two** unsaturations. To find the number of degrees of unsaturation, find the maximum number of H's for the C's in the compound; i.e., 2n + 2. Subtract the actual number of H's. Divide the result by 2. For hydrocarbons and oxygen-containing compounds, the formula tells you directly the number of degrees of unsaturation: for every two hydrogens less than that (2 x number of carbons + 2) there is one unsaturation. For instance, $C_7H_{10}O_2$ has three degrees of unsaturation, $[(2 \times 7 + 2) - 10 \div 2]$, and it could be represented by the following possibilities (as well as many others):

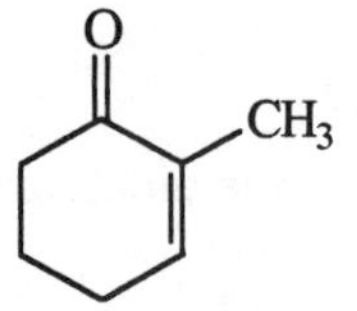

$CH_3C{\equiv}CCH_2CH_2CH_2CH$ (with C=O)

CH_2 ... CH_2OH (H H H / H H)

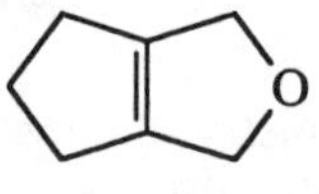

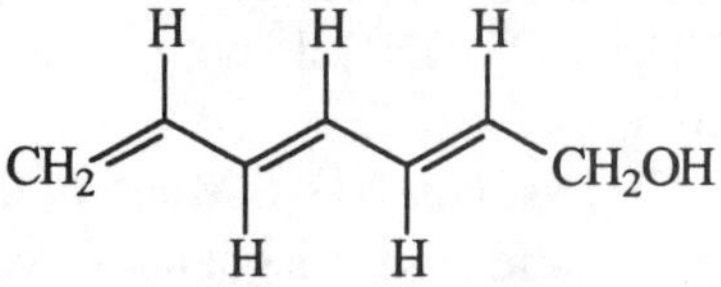

When halogens are present in the molecule, simply count them as hydrogens: e.g., $C_3H_5Br_2Cl$ is saturated ($CH_3CHClCHBr_2$ is one possibility). When nitrogens are present, add one hydrogen for each nitrogen to the theoretical maximum number of hydrogens; $C_5H_{11}N$: $(2x5 + 2 + 1) - 11 \div 2$ = one unsaturation.

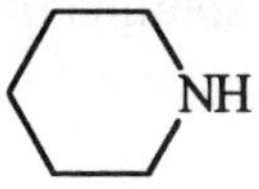

NH and $CH_3CH_2CH_2CH_2CH{=}NH$ are possibilities.

You can't distinguish types of unsaturation (π-bond or ring) from the formula, but there are often clues given from reactions that a compound undergoes. For example, hydrogenation usually removes only C=C or C≡C unsaturation, not rings or C=O. (We say usually because cyclopropane rings and ketones and aldehydes can be reduced under vigorous hydrogenation conditions.)

12.D. Answers to Exercises

12.1 $HC{\equiv}CCH_2CH_2CH_2OH$
4-pentyn-1-ol

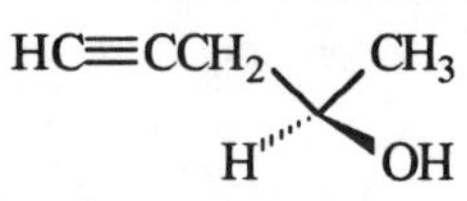

$HC{\equiv}CCH_2$, CH_3, H, OH — (*R*)-4-pentyn-2-ol

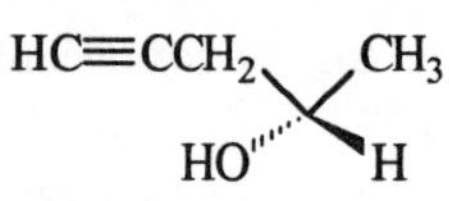

$HC{\equiv}CCH_2$, CH_3, HO, H — (*S*)-4-pentyn-2-ol

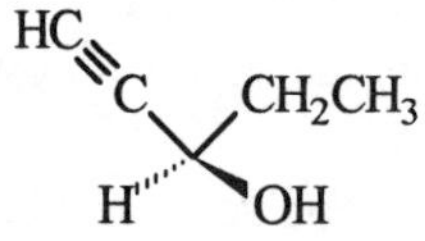

HC≡C, CH_2CH_3, H, OH — (*R*)-1-pentyn-3-ol

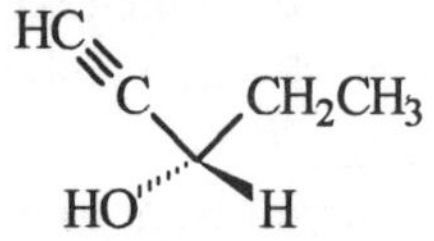

HC≡C, CH_2CH_3, HO, H — (*S*)-1-pentyn-3-ol

$HOC{\equiv}CCH_2CH_2CH_3$
1-pentyn-1-ol
(this molecule is unstable)

$CH_3C{\equiv}CCH_2CH_2OH$
3-pentyn-1-ol

$HOCH_2C{\equiv}CCH_2CH_3$
2-pentyn-1-ol

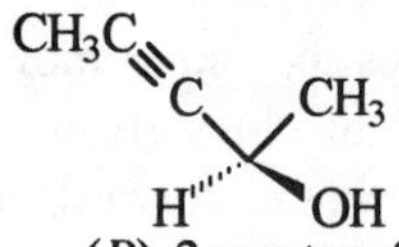

$CH_3C{\equiv}C$, CH_3, H, OH — (*R*)-3-pentyn-2-ol

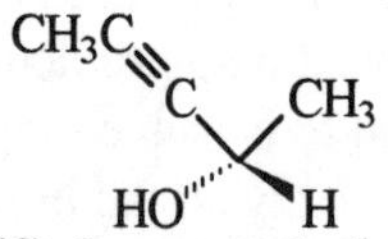

$CH_3C{\equiv}C$, CH_3, HO, H — (*S*)-3-pentyn-2-ol

12.2 a. $CH_3C{\equiv}CCH_3$ b. 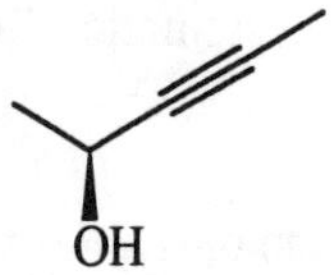 OH c. $CH_3CH_2CHClC{\equiv}N$

12.3. The equilibrium constant K for the equilibrium: $RH \rightleftharpoons R^- + H^+$ is 10^{-pKa} (see Section 4.5).

The equilibrium constant for the acid-base reaction:

$$RH + B^- \rightleftharpoons R^- + BH$$

can be calculated by dissecting the reaction into two components:

$$RH \rightleftharpoons R^- + H^+ \qquad K_1 = 10^{-pKa}$$

$$B^- + H^+ \rightleftharpoons BH \qquad K_2 = 1/10^{-pKa'} = 10^{pKa'}$$

Sum: $RH + B^- \rightleftharpoons R^- + BH \quad K = \dfrac{[R^-][BH]}{[RH][B^-]} = \dfrac{[R^-][H^+]}{[RH]} \; x \; \dfrac{[BH]}{[H^+][B^-]}$

$$K = K_1K_2 = 10^{(pKa' - pKa)}$$

For $HC{\equiv}CH + NH_2^- \rightleftharpoons HC{\equiv}C^- + NH_3$:
$pK_{a'} - pK_a = 34 - 25 = 9$; therefore $K_{eq} = 10^9$

For $H_2C{=}CH_2 + NH_2^- \rightleftharpoons H_2C{=}CH^- + NH_3$:
$pK_{a'} - pK_a = 34 - 44 = -10$; therefore $K_{eq} = 10^{-10}$

For $CH_4 + NH_2^- \rightleftharpoons CH_3^- + NH_3$:
$pK_{a'} - pK_a = 34 - 50 = -16$; therefore $K_{eq} = 10^{-16}$

12.4 pK_a (HCN) = 9.2; pK_a (H_2O) = 15.7

For $HCN + OH^- \rightleftharpoons CN^- + H_2O$ $\quad K_{eq} = K_a(HCN)/K_a(H_2O) = 10^{(15.7-9.2)} = 10^{6.5} = 3.2 \times 10^6$

12.5 1-pentyne will form a precipitate with $AgNO_3$, whereas 2-pentyne will not.

12.6. Cyanide ion (pK_a of HCN = 9.2) is much less basic than an acetylide ion (pK_a of $HC{\equiv}CH$ = 25), therefore it is less likely to cause elimination reactions.

12.7 a. $CH_3CH_2Br \xrightarrow[t\text{-BuOH}]{t\text{-BuOK}} CH_2{=}CH_2 \xrightarrow{Br_2} BrCH_2CH_2Br \xrightarrow[NH_3]{NaNH_2} HC{\equiv}C^-\,Na^+ \xrightarrow{CH_3(CH_2)_5Br} CH_3(CH_2)_5C{\equiv}CH$

b. $\underset{\textit{from a.}}{HC{\equiv}C^-\,Na^+} + CH_3(CH_2)_4Br \longrightarrow CH_3(CH_2)_4C{\equiv}CH \xrightarrow[NH_3]{NaNH_2} CH_3(CH_2)_4C{\equiv}C^-Na^+$

$\qquad CH_3I \downarrow$

$CH_3(CH_2)_4C{\equiv}CCH_3$

c. $(CH_3)_2CHCH_2CH_2Cl + NaCN \xrightarrow[DMF]{} (CH_3)_2CHCH_2CH_2CN$

12.8 $CH_3CH_2C\equiv CH \rightleftharpoons CH_3C\equiv CCH_3$

ΔH°_f 39.5 34.7 $\Delta H^\circ = 34.7 - 39.5 = -4.8$ kcal mole^{-1}

$\Delta H^\circ \sim \Delta G^\circ = -RT\ln K$ $\ln K = -4800/(-1.987 \times 373)$ $K = 6.5$

12.9

$$CH_3CH_2CH{=}CHCH_3 + Br_2 \xrightarrow{CCl_4} CH_3CH_2CHBrCHBrCH_3 \xrightarrow[C_2H_5OH,\ \Delta]{KOH} CH_3CH_2C{\equiv}CCH_3 + 2\ KBr \xrightarrow[150^\circ C]{NaNH_2} CH_3CH_2CH_2C{\equiv}CH$$

Isomerization of an internal alkyne to the terminal isomer involves the following steps:

$$\underset{\underline{1}}{CH_3CH_2C{\equiv}CCH_2{-}H}\ :B^- \rightleftharpoons CH_3CH_2C{\equiv}C{-}\overset{-}{C}H_2 \ (B{-}H) \rightleftharpoons CH_3CH_2CH{=}C{=}\overset{-}{C}H \ (:B^-)$$

$$CH_3CH_2CH{=}\overset{-}{C}{=}CH \ (B{-}H) \rightleftharpoons \underset{\underline{2}}{CH_3CH_2CH_2C{\equiv}C{-}H}\ :B^- \rightleftharpoons \underset{\underline{3}}{CH_3CH_2CH_2C{\equiv}C^-}\ \ B{-}H$$

When hydroxide is used as the base (KOH in ethanol, for example), the equilibrium $\underline{2} \rightleftharpoons \underline{3}$ favors $\underline{2}$; that is, hydroxide is a **weaker** base than the acetylide ion. Therefore, the equilibration that takes place is between the internal alkyne $\underline{1}$ and the terminal alkyne $\underline{2}$, and the former predominates because it is more stable.

When amide ion is the base ($NaNH_2$, for example), the equilibrium $\underline{2} \rightleftharpoons \underline{3}$ favors $\underline{3}$; that is amide ion is a stronger base than the acetylide ion. Therefore, the equilibration that takes place favors $\underline{3}$, and when the reaction is worked up the terminal alkyne is obtained.

The ability to obtain either the internal or the terminal alkyne by choosing the appropriate base is a useful aspect of alkyne chemistry.

12.10 a. $CH_3CH_2CH_2CH_2Br \xrightarrow[t\text{-BuOH}]{t\text{-BuOK}} CH_3CH_2CH{=}CH_2 \xrightarrow{Br_2} CH_3CH_2CHBrCH_2Br \xrightarrow[NH_3]{NaNH_2} CH_3CH_2C{\equiv}C^-\ Na^+ \xrightarrow{H_2O} CH_3CH_2C{\equiv}CH$

b. $CH_3CH_2C{\equiv}CH$ (from a.) $\underset{\Delta}{\rightleftharpoons} CH_3C{\equiv}CCH_3$

c. $CH_3CH_2C{\equiv}C^-\ Na^+$ (from a.) $\xrightarrow{CH_3(CH_2)_3Br} CH_3CH_2C{\equiv}CCH_2CH_2CH_2CH_3$

d. $CH_3CH_2C{\equiv}C(CH_2)_3CH_3 \xrightarrow[K^{+-}NH(CH_2)_3NH_2]{NaNH_2\text{ at }150^\circ C\text{ or}} Na^{+}\,{}^-C{\equiv}C(CH_2)_5CH_3 \xrightarrow{H_2O} HC{\equiv}C(CH_2)_5CH_3$

12.11 a.

$$CH_3(CH_2)_3Cl \xrightarrow{t\text{-BuOK}} CH_3CH_2CH{=}CH_2 \xrightarrow{Br_2} CH_3CH_2CHBrCH_2Br \xrightarrow{NaNH_2} CH_3CH_2C{\equiv}C^-\ Na^+$$

$$\xrightarrow{CH_3(CH_2)_3Cl} CH_3CH_2C{\equiv}C(CH_2)_3CH_3 \xrightarrow[Pd(BaSO_4)]{H_2} cis\text{-}CH_3CH_2CH{=}C(CH_2)_3CH_3$$

b. $$CH_3CH_2C{\equiv}C(CH_2)_3CH_3 \text{ (from a.)} \xrightarrow[NH_3]{Na} trans\text{-}CH_3CH_2CH{=}CH(CH_2)_3CH_3$$

c. $$CH_3CH_2C{\equiv}C(CH_2)_3CH_3 \xrightarrow[NH_2(CH_2)_3NH_2]{KNH(CH_2)_3NH_2} K^{+}\,{}^{-}C{\equiv}C(CH_2)_5CH_3 \xrightarrow{H_2O} HC{\equiv}C(CH_2)_5CH_3$$

12.12 $$CH_3CH_2CH_2CH_2Cl \xrightarrow{NaCN} CH_3CH_2CH_2CH_2CN \xrightarrow[ether]{LiAlH_4} CH_3(CH_2)_4NH_2$$

12.13

$$CH_3C{\equiv}CCH_2CH_3 \xrightarrow{H^+} CH_3CH{=}\overset{+}{C}CH_2CH_3 \xrightarrow{H_2\ddot{O}:} CH_3CH{=}\overset{H\overset{+}{O}H}{C}CH_2CH_3 \rightleftharpoons CH_3CH{=}\overset{:\ddot{O}H}{C}CH_2CH_3 \rightleftharpoons CH_3CH_2\overset{\overset{+}{O}H}{\overset{\|}{C}}CH_2CH_3 \xrightleftharpoons{-H^+} CH_3CH_2\overset{O}{\overset{\|}{C}}CH_2CH_3$$

$$CH_3C{\equiv}CCH_2CH_3 \xrightarrow{H^+} CH_3\overset{+}{C}{=}CHCH_2CH_3 \xrightarrow{H_2\ddot{O}:} CH_3\overset{H\overset{+}{O}H}{C}{=}CHCH_2CH_3 \rightleftharpoons CH_3\overset{H\ddot{O}}{C}{=}CHCH_2CH_3 \rightleftharpoons CH_3\overset{\overset{+}{O}H}{\overset{\|}{C}}CH_2CH_2CH_3 \xrightleftharpoons{-H^+} CH_3\overset{O}{\overset{\|}{C}}CH_2CH_2CH_3$$

The two cations differ only in substitution with ethyl versus methyl, which is a very minor difference. They will therefore be very similar in stability and both products will be formed in similar amounts.

12.14

$$RC{\equiv}CH \xrightarrow{H^+} R\overset{+}{C}{=}CH_2 \longrightarrow \longrightarrow \longrightarrow R\overset{O}{\overset{\|}{C}}CH_3$$

$$RC{\equiv}CH \xrightarrow{H^+} RCH{=}\overset{+}{C}H \longrightarrow \longrightarrow \longrightarrow RCH_2\overset{O}{\overset{\|}{C}}H$$

less substituted, therefore less stable *not formed*

In the case of a terminal alkyne, the two possible cations do differ substantially in terms of alkyl substitution. The more substituted cation will be formed more easily, and the methyl ketone will be the major product.

12.15 a. $$CH_3(CH_2)_3Cl \xrightarrow{NaCN} CH_3(CH_2)_3C{\equiv}N \xrightarrow[\Delta]{NaOH} CH_3(CH_2)_3\overset{O}{\overset{\|}{C}}NH_2$$

b. $$CH_3(CH_2)_3Cl \xrightarrow{NaC{\equiv}CH} CH_3(CH_2)_3C{\equiv}CH \xrightarrow[HgSO_4]{H_2SO_4} CH_3(CH_2)_3\overset{O}{\overset{\|}{C}}CH_3$$

12.16

$$3\ CH_3CH_2C{\equiv}CH + \tfrac{1}{2}\ B_2H_6 \longrightarrow \left(\begin{matrix} CH_3CH_2 & & H \\ & C{=}C & \\ H & & \end{matrix}\right)_3 B \xrightarrow[OH^-]{H_2O_2} 3\left[\begin{matrix} CH_3CH_2 & & H \\ & C{=}C & \\ H & & OH \end{matrix}\right] \longrightarrow 3\ CH_3(CH_2)_2\overset{O}{\overset{\|}{C}}H$$

$$3\ CH_3C{\equiv}CCH_3 + \tfrac{1}{2}\ B_2H_6 \longrightarrow \left(\begin{matrix} CH_3 & & CH_3 \\ & C{=}C & \\ H & & \end{matrix}\right)_3 B \xrightarrow[OH^-]{H_2O_2} 3\left[\begin{matrix} CH_3 & & CH_3 \\ & C{=}C & \\ H & & OH \end{matrix}\right] \longrightarrow 3\ CH_3CH_2\overset{O}{\overset{\|}{C}}CH_3$$

$$3\ CH_3CH_2C{\equiv}CCH_3 + \tfrac{1}{2}\ B_2H_6 \nearrow \left(\begin{matrix} CH_3CH_2 & & CH_3 \\ & C{=}C & \\ H & & \end{matrix}\right)_3 B \xrightarrow[OH^-]{H_2O_2} 3\ CH_3CH_2CH_2\overset{O}{\overset{\|}{C}}CH_3$$

+

$$3\ CH_3CH_2C{\equiv}CCH_3 + \tfrac{1}{2}\ B_2H_6 \searrow \left(\begin{matrix} CH_3 & & CH_2CH_3 \\ & C{=}C & \\ H & & \end{matrix}\right)_3 B \xrightarrow[OH^-]{H_2O_2} 3\ CH_3CH_2\overset{O}{\overset{\|}{C}}CH_2CH_3$$

12.17

$$\begin{matrix} CH_3CH_2 & & Cl \\ & C{=}C & \\ CH_3CH_2 & & H \end{matrix} \xleftarrow{-\,H_bCl_b} \text{(3R,4S)-3,4-dichlorohexane: } H_b,\ Cl_a,\ H_a,\ Cl_b,\ CH_3CH_2,\ CH_3CH_2 \rightleftharpoons Cl_a,\ H_b,\ CH_2CH_3,\ Cl_b,\ CH_2CH_3,\ H_a \xrightarrow{-\,H_aCl_a} \begin{matrix} H & & CH_2CH_3 \\ & C{=}C & \\ Cl & & CH_2CH_3 \end{matrix}$$

(3*R*,4*S*)-3,4-dichlorohexane

$$CH_2CH_2C{\equiv}CCH_2CH_3 \xrightarrow[Pd(BaSO_4)]{H_2} \begin{matrix} CH_3CH_2 & & CH_2CH_3 \\ & C{=}C & \\ H & & H \end{matrix} \xrightarrow{Cl_2} \text{Cl, H, CH}_2\text{CH}_3\text{, H, CH}_3\text{CH}_2\text{, Cl}$$

$$\begin{matrix} CH_3CH_2 & & H \\ & C{=}C & \\ Cl & & CH_2CH_3 \end{matrix} \xleftarrow{KOC_2H_5} \text{H, H, CH}_2\text{CH}_3\text{, CH}_3\text{CH}_2\text{, Cl, Cl}$$

12.E. Answers and Eplanations for Problems

1. a. $CH_3C{\equiv}CCH(CH_3)_2$ b. $CH_3C{\equiv}N$ c. $CH_2{=}CHC{\equiv}CH$ d. $(CH_3)_3CC{\equiv}CC(CH_3)_3$
   e. $CH_2{=}CHBr$ f. $(CH_3)_2CHCH_2C{\equiv}CH$ g. $CH_3OC{\equiv}CH$

2. a. 10 1 2 3 9 4 8 7 6 5 b. c. HO d. $CH_3OC{\equiv}CH$

   e. OCH_3 f. $HOCH_2C{\equiv}CCH_2CH_3$ g. CN Cl h. $CH_2{=}CHC{\equiv}CCH_3$

   i. $(CH_3)_3CC{\equiv}CC(CH_3)_3$ j. $(CH_3)_2CHC{\equiv}CCH_3$

3. a. 6-bromo-1-hexyne b. 2,2-dimethyl-3-hexyne
   c. propanenitrile d. cyclopropylacetylene or ethynylcyclopropane
   e. (*R*)-pent-1-en-4-yn-3-ol f. 2-butyn-1,4-diol
   g. hex-1-en-5-yne h. (*Z*)-2-chloro-2-butene

4. a. $CH_3CH_2C{\equiv}C^-Li^+$ b. $CH_3CH_2CH_2CH_3$ c. $CH_3CH_2CH{=}CH_2$ d. No reaction
   e. $CH_3CH_2C{\equiv}CAg$ f. $CH_3CH_2CH_2CHO$ g. $CH_3CH_2COCH_3$ h. $CH_3CH_2CH{=}CH_2$

5. a. No reaction b. $CH_3CH_2CH_2CH_3$ c. *trans*-$CH_3CH{=}CHCH_3$ d. No reaction
   e. No reaction f. $CH_3CH_2COCH_3$ g. $CH_3CH_2COCH_3$ h. *cis*-$CH_3CH{=}CHCH_3$

6. a. $CH_3CH_2CH_2CH_2NH_2$ (as HCl salt)

   b. and c. $CH_3CH_2CH_2C(=O)NH_2$ d. $CH_3CH_2CH_2CH_2NH_2$

7. a. $(CH_3)_2CHBr \xleftarrow[\text{(radical inhibitors)}]{HBr} CH_2{=}CHCH_3 \xleftarrow[\text{poisoned Pd catalyst}]{H_2} HC{\equiv}CCH_3$

   b. $CH_3COCH_3 \xleftarrow[HgSO_4]{H_2SO_4} HC{\equiv}CCH_3$

c. $CH_3CH_2CH{=}O \xleftarrow[2.\ H_2O_2,\ OH^-]{1.\ B_2H_6} CH_3C{\equiv}CH$

d. $CH_3(CH_2)_4CH_3 \xleftarrow[Pt]{H_2} CH_3CH_2CH_2C{\equiv}CCH_3 \longleftarrow$

$Na^+\ {}^-C{\equiv}CCH_3 \xleftarrow{NaNH_2} HC{\equiv}CCH_3$

$CH_3CH_2CH_2Br \xleftarrow[peroxides]{HBr} CH_2{=}CHCH_3$ (from a.)

e. *trans*-$CH_3CH_2CH_2CH{=}CHCH_3 \xleftarrow[NH_3]{Na} CH_3CH_2CH_2C{\equiv}CCH_3$ (from d.)

f. $CH_3CCl{=}CH_2 \xleftarrow{HCl\ (one\ mole)} CH_3C{\equiv}CH$

g. $CH_3CH_2CH_2CN \xleftarrow[DMF]{NaCN} CH_3CH_2CH_2Br$ (from d.)

h. $CH_3CH_2CH_2CH_2NH_2 \xleftarrow[ether]{LiAlH_4} CH_3CH_2CH_2CN$ (from g.)

8. a. $CH_3(CH_2)_2CH{=}CH_2 \xrightarrow{Cl_2} CH_3(CH_2)_2CHClCH_2Cl \xrightarrow[NH_3]{NaNH_2} CH_3(CH_2)_2C{\equiv}C^-\ Na^+ \xrightarrow{H_3O^+} CH_3(CH_2)_2C{\equiv}CH$

b. $CH_3(CH_2)_2Br \xrightarrow[t\text{-}BuOH]{t\text{-}BuO^-} CH_3CH{=}CH_2 \xrightarrow{Br_2} CH_3CHBrCH_2Br \xrightarrow[\Delta]{KOH} CH_3C{\equiv}CH \xrightarrow{NaNH_2} \xrightarrow{CH_3(CH_2)_2Br} CH_3(CH_2)_2C{\equiv}CCH_3$

c. $HC{\equiv}CH \xrightarrow[Lindlar\ catalyst]{H_2} CH_2{=}CH_2 \xrightarrow{HBr} CH_3CH_2Br \xrightarrow{NaC{\equiv}CH} CH_3CH_2C{\equiv}CH \xrightarrow[Lindlar]{H_2} CH_3CH_2CH{=}CH_2$

$\downarrow$ 1. B_2H_6 2. H_2O_2, OH^-

$CH_3CH_2CH_2CH_2OH$

d. $CH_3CH_2C{\equiv}CH$ (from c.) $\xrightarrow[2.\ C_2H_5Br]{1.\ NaNH_2}$ (from c.) $CH_3CH_2C{\equiv}CCH_2CH_3$ $\xrightarrow{1.\ H_2\ /\ Lindlar}$ $CH_3CH_2CH(OCH_3)CH_2CH_2CH_3$

2. $Hg(OAc)_2$, CH_3OH (see below)
3. $NaBH_4$

$HC{\equiv}CH \xrightarrow{[H]} CH_2{=}CH_2 \xrightarrow[2.\ NaBH_4]{1.\ O_3} 2\ CH_3OH$

e. $trans\text{-}(CH_3)_3CCH{=}CHC(CH_3)_3 \xrightarrow{Br_2} \xrightarrow{2\ NaNH_2} (CH_3)_3CC{\equiv}CC(CH_3)_3 \xrightarrow[\text{Lindlar}]{H_2} cis\text{-}(CH_3)_3CCH{=}CHC(CH_3)_3$

f. $CH_3CH_2CH_2OH \xrightarrow[\Delta]{H_2SO_4} CH_3CH{=}CH_2 \xrightarrow[2.\ 2\ NaNH_2]{1.\ Br_2} CH_3C{\equiv}CH \xrightarrow[Hg^{2+}]{H_2SO_4} CH_3COCH_3$

g. $CH_3CH_2C{\equiv}CH \xrightarrow{n\text{-BuLi}} CH_3CH_2C{\equiv}CLi \xrightarrow{D_2O} CH_3CH_2C{\equiv}CD \xrightarrow[2.\ CH_3COOH]{1.\ B_2H_6} (E)\text{-}CH_3CH_2CH{=}CDH$

Note that in the final step of this sequence, catalytic hydrogenation is not recommended because of the possibility of some H-D exchange.

h. $HC{\equiv}CH \xrightarrow[2.\ C_2H_5Br\ \text{from c.}]{1.\ NaNH_2} CH_3CH_2C{\equiv}CH \xrightarrow{H_2/Pt} CH_3CH_2CH_2CH_3$

9. a. $(CH_3)_2CHC(=O)NH_2 \xleftarrow{KOH} (CH_3)_2CHC{\equiv}N \xleftarrow[DMSO]{NaCN} (CH_3)_2CHBr$

b. $CH_3(CH_2)_4NH_2 \xleftarrow[\text{ether}]{LiAlH_4} CH_3(CH_2)_3C{\equiv}N \xleftarrow[DMSO]{NaCN} CH_3(CH_2)_3Cl$

c. $CH_3CH_2I \xrightarrow[t\text{-BuOH}]{t\text{-BuOK}} CH_2{=}CH_2 \xrightarrow{Br_2} BrCH_2CH_2Br \xrightarrow{3\ NaNH_2} HC{\equiv}C^-Na^+$

$(CH_3)_2CBr \xrightarrow[t\text{-BuOH}]{t\text{-BuOK}} CH_3CH{=}CH_2 \xrightarrow[\text{peroxides}]{HBr} CH_3CH_2CH_2Br \xrightarrow[\text{from above}]{HC{\equiv}C^-Na^+} CH_3CH_2CH_2C{\equiv}CH \xrightarrow{NaNH_2}$

$CH_3(CH_2)_2C{\equiv}C^-Na^+ \xrightarrow{C_2H_5I} CH_3(CH_2)_2C{\equiv}CCH_2CH_3 \xrightarrow[NH_3]{Na} trans\text{-}CH_3(CH_2)_2CH{=}CHCH_2CH_3$

d. $CH_2{=}CClCH_2CH_2CH_3 \xleftarrow{HCl} CH_3CH_2CH_2C{\equiv}CH$ (from c.)

10. a. $CH_3CH_2C(=O)CH_3 \xleftarrow[HgSO_4]{H_2SO_4} CH_3CH_2C{\equiv}CH \leftarrow$ {$Na^+\ {}^-C{\equiv}CH \xleftarrow{NaNH_2} HC{\equiv}CH$; $CH_3CH_2Br \xleftarrow{HBr} CH_2{=}CH_2$}

$HC{\equiv}CH \xrightarrow{H_2,\ Pd(BaSO_4)} CH_2{=}CH_2$

b. $CH_3O(CH_2)_4CN \xleftarrow{NaCN} CH_3O(CH_2)_4Br \xleftarrow[\text{peroxides}]{HBr} CH_3OCH_2CH_2CH{=}CH_2 \xleftarrow[\text{(poisoned Pd cat.)}]{H_2} CH_3OCH_2CH_2C{\equiv}CH$

c. $HC{\equiv}CH + NaNH_2 \longrightarrow HC{\equiv}CNa \xrightarrow{C_2H_5I} CH_3CH_2C{\equiv}CH \xrightarrow{NaNH_2} CH_3CH_2C{\equiv}CNa \xrightarrow{CH_3CH_2CH_2I} CH_3CH_2C{\equiv}CCH_2CH_2CH_3$

d. $(CH_3)_2COHCH_2CH_3 \xrightarrow[\Delta]{H_2SO_4} (CH_3)_2C{=}CHCH_3 \xrightarrow[\text{peroxides}]{HBr} (CH_3)_2CHCHBrCH_3 \xrightarrow{t\text{-BuO}^-} (CH_3)_2CHCH{=}CH_2 \xrightarrow{Br_2} (CH_3)_2CHCHBrCH_2Br \xrightarrow{NaNH_2} (CH_3)_2CHC{\equiv}CNa \xrightarrow{CH_3I} (CH_3)_2CHC{\equiv}CCH_3$

11. $(CH_3)_2CHCH_2CH_2OH \xrightarrow[\text{or } PBr_3]{HBr/H_2SO_4} (CH_3)_2CHCH_2CH_2Br$

$HC{\equiv}CH \xrightarrow[\text{liq. } NH_3]{NaNH_2} HC{\equiv}C^- Na^+ \xrightarrow{(CH_3)_2CHCH_2CH_2Br} (CH_3)_2CHCH_2CH_2C{\equiv}CH \xrightarrow{NaNH_2} (CH_3)_2CH(CH_2)_2C{\equiv}C^-Na^+ \xrightarrow{CH_3(CH_2)_9CH_2Br} (CH_3)_2CHCH_2CH_2C{\equiv}C(CH_2)_{10}CH_3 \xrightarrow{H_2/Pt} (CH_3)_2CH(CH_2)_{14}CH_3$

12. $CH_3(CH_2)_xOH \xrightarrow[\text{or } PBr_3]{HBr/H_2SO_4} CH_3(CH_2)_xBr$

$HC{\equiv}CH \xrightarrow[\text{liq. } NH_3]{NaNH_2} HC{\equiv}C^- \xrightarrow{CH_3(CH_2)_7Br} CH_3(CH_2)_7C{\equiv}CH \xrightarrow{NaNH_2} \xrightarrow{CH_3(CH_2)_{12}Br} CH_3(CH_2)_7C{\equiv}C(CH_2)_{12}CH_3$

$\xrightarrow[\text{quinoline or}]{Pd/BaSO_4/H_2}$ *cis*-$CH_3(CH_2)_7CH{=}CH(CH_2)_{12}CH_3$
quinoline or
1. B_2H_6
2. CH_3COOH

13.

$RC{\equiv}CR + Hg^{2+} \rightleftharpoons \left[\text{RC=CR (}\pi\text{-complex with } Hg^{2+}\text{)} \leftrightarrow \overset{+}{R}C{=}C(Hg^+)R \right] \overset{H_2O}{\rightleftharpoons}$ R(H–$\overset{+}{O}$H)C=C(Hg$^+$)R $\rightleftharpoons$ R(HÖ:)C=C(Hg$^+$)R + H^+ $\rightleftharpoons$ R(H$\overset{+}{O}$=)C–C(Hg$^+$)(H)–R

$\rightleftharpoons$ +H^+

R(HÖ:)C=C(H)R + H^+ $\rightleftharpoons$ Hg^{2+} + R(H$\overset{+}{O}$=)C–CH(H)–R $\rightleftharpoons$ H^+ + $RCCH_2R$ (C=O)

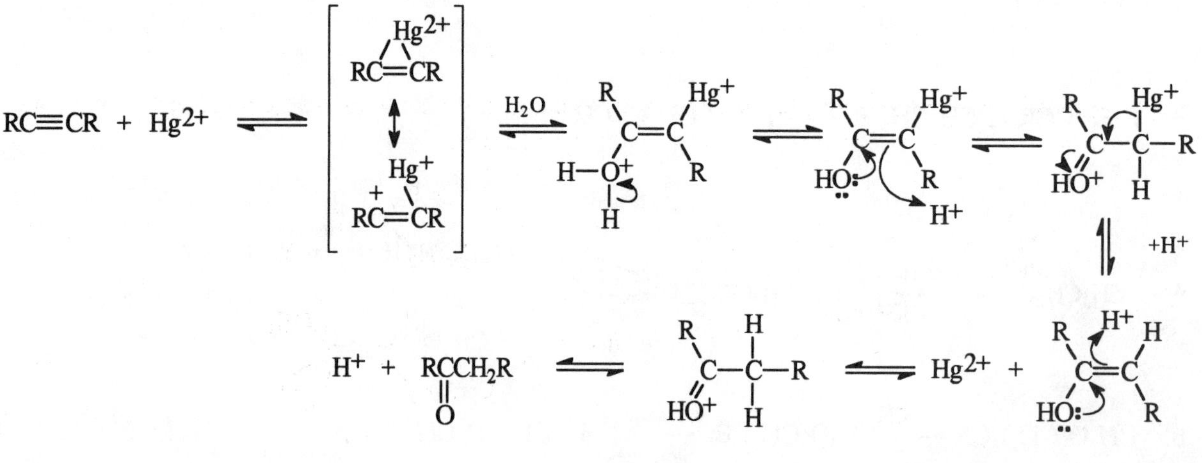

The interconversion of HO–C=C $\rightleftharpoons$ –C(=O)–CH will be discussed in greater detail in Chapter 15.

14. a. Solutions of C_2H_6 or C_2H_4 in $NH_3/NaNH_2$ have so little carbanion that there is no reaction with CH_3I (see exercise 12.3). $NaNH_2$ will react instead to give CH_3NH_2. C_2H_2 is converted completely into $HC{\equiv}C^-$, which can react with CH_3I.

b. We must consider the following equations:

$RH = R\cdot + H\cdot \quad \Delta H° = DH°$

$H\cdot = H^+ + e \quad \Delta H° = \text{ionization potential of } H\cdot\text{, constant for all RH}$

$\underline{R\cdot + e = R^-} \quad \Delta H° = -E.A. \text{ (electron affinity of } R\cdot) \quad$ (**Note**: positive E.A. corresponds to negative ΔH)

$RH = H^+ + R^- \quad \Delta H° = DH° + I.P.(H\cdot) - E.A.(R\cdot)$

The difference in enthalpy, $\Delta H_2° - \Delta H_1° \equiv \Delta\Delta H°$, for two different hydrocarbons, R_2H and R_1H, is therefore given by:

$$\Delta\Delta H° = [DH_2° - DH_1°] - [E.A.(R_2\cdot) - E.A.(R_1\cdot)]$$

For $R_2H = HC{\equiv}CH$ and $R_1H = CH_3CH_3$ the negative $\Delta\Delta H°$ despite the positive $(DH_2° - DH_1°)$ means that
$$[E.A.(HC{\equiv}C\cdot) - E.A.(CH_3CH_2\cdot)] > [DH°(HC{\equiv}C\text{-}H) - DH°(C_2H_5\text{-}H)].$$

The E.A. of $HC{\equiv}C\cdot$ corresponds to putting an electron in an sp hybrid orbital. The greater s-character of the sp-hybrid gives ethynyl radical a high electron affinity. The experimental values of these electron affinities are not known accurately, but the available data give $[E.A.(HC{\equiv}C\cdot) - E.A.(C_2H_5\cdot)] = 50$ kcal mole^{-1}, a value substantially higher than the difference in bond dissociation energies.

In short, although increasing s-character in an orbital increases the stability of a bond involving the orbital, it increases the stability of a lone pair still more. An electron pair in a bond involves two orbitals, whereas a lone pair involves a single orbital.

15. a. Overall retention of configuration shows that the vinyl anion is not linear and that protonation of the vinyl anion by NH_3 is faster than inversion of the carbanion carbon:

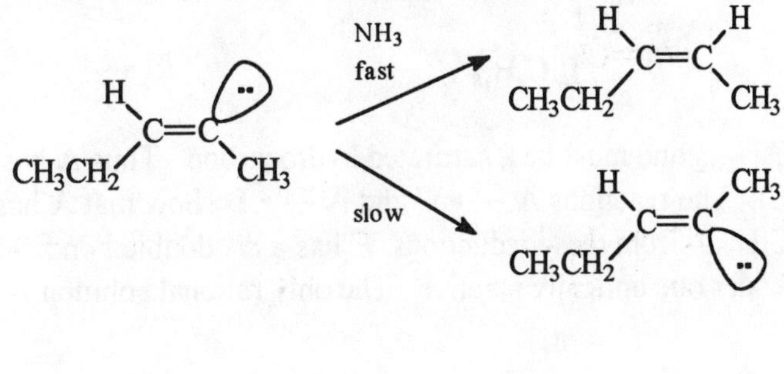

(H)(CH_3CH_2)C=C(:⁻)(CH_3) —NH_3, fast→ (H)(CH_3CH_2)C=C(H)(CH_3)

(H)(CH_3CH_2)C=C(:⁻)(CH_3) —slow→ (H)(CH_3CH_2)C=C(CH_3)(:⁻)

b. $C_2H_5(H)C{=}C(H)C_2H_5$ $\xrightarrow{Cl_2}$ CH_3CH_2–CHCl–CHCl–CH_2CH_3 $\rightleftharpoons$ (rotated conformer) $\xrightarrow{C_2H_5OK \text{ or } t\text{-BuOK}}$ $CH_3CH_2(H)C{=}C(Cl)CH_2CH_3$ $\xrightarrow{Na / liq.\ NH_3}$ $CH_3CH_2(H)C{=}C(H)CH_2CH_3$

16. $Br(H)C{=}C(H)CH_2CH_2CH_2Br$ $\xrightarrow{EtO^-}$ $Br(H)C{=}C(H)CH_2CH_2CH_2OEt$ + $CH_2{=}C(OEt)CH_2CH_2CH_2OEt$ + $HC{\equiv}CCH_2CH_2CH_2OEt$

but no $EtOCH{=}CHCH_2CH_2CH_2OEt$

The terminal -CH_2OEt comes from a normal S_N2 reaction on the primary bromide. The terminal $HC{\equiv}C$- group results from E2 elimination of the vinyl bromide. The vinyl ether results from a subsequent reaction of the triple bond:

$$HC{\equiv}CCH_2CH_2CH_2OEt + EtO^- \longrightarrow {}^-CH{=}C(OEt)CH_2CH_2CH_2OEt \xrightarrow{EtOH} CH_2{=}C(OEt)CH_2CH_2CH_2OEt$$

Ethoxide ion adds to the triple bond to give the primary carbanion, $=C^-{-}H$, rather than the less stable secondary carbanion $=C^-{-}C$. This latter mode of addition, if it occurred, would give rise to $EtOCH{=}CHCH_2CH_2CH_2OEt$, but cannot compete with the alternative mode of addition.

17. The information in this roadmap problem is summarized as follows:

C C_8H_{14} optically active $\xleftarrow{H_2/Lindlar}$ **A** C_8H_{12} optically active $\xrightarrow{Na/NH_3}$ **D** C_8H_{14} optically inactive

A $\xrightarrow{H_2,\ Pt}$ **B** C_8H_{18}

B, C_8H_{18}, corresponds to C_nH_{2n+2} and must be a saturated hydrocarbon. Thus **A** has three units of unsaturation (3 C=C or 1 C=C + 1 C≡C). The reactions **A** ⟶ **C** and **A** ⟶ **D** show that **A** has a triple bond. Our partial structure is (C=C) (C≡C)C_4H_{12}. From these reductions, **C** has a *cis* double bond and **D** has a *trans*, yet this difference is sufficient to render one optically inactive. The only rational solution is therefore:

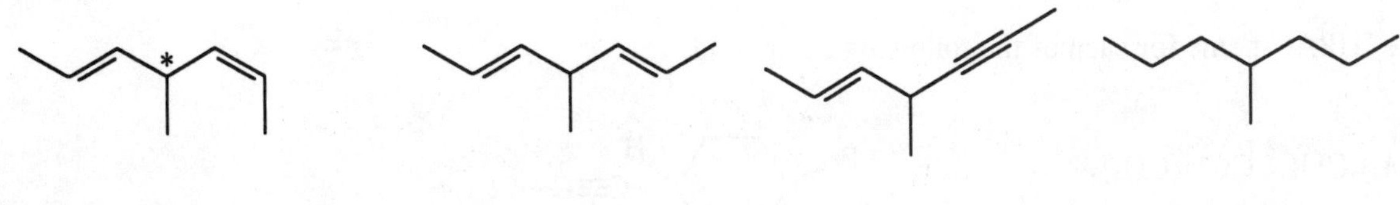

optically active	optically inactive		
C	**D**	**A**	**B**

18. **E** $\xrightarrow[-20°]{\text{HCl}}$ **F** $\xrightarrow{t\text{-BuOK}}$ **E** + **G** $\xrightarrow{O_3}$ (cyclohexanone)=O + CH_2=O

C_7H_{12} $C_7H_{13}Cl$ C_7H_{12}

From the last reaction, **G** must be (cyclohexylidene)=CH_2 . Then **F** must be (1-chloro-1-methylcyclohexane, Cl) and **E** is (1-methylcyclohexene) .

F cannot be (cyclohexylmethyl chloride, Cl) because this chloride cannot be formed by addition of HCl to an alkene.

19.

	DH°		
CH_3CH_2-F	107	CH_2=CH-F	?
CH_3CH_2-Cl	81	CH_2=CH-Cl	88
CH_3CH_2-Br	68	CH_2=CH-Br	76
CH_3CH_2-I	53	CH_2=CH-I	?

For the chloride and bromide, DH° for the vinyl compound is 7-8 kcal mole^{-1} higher than for the ethyl compound. Rough estimates for CH_2=CHF and CH_2CHI are 115 and 61 kcal mole^{-1}, respectively. The vinyl-halide bond is stronger in part because of increased s character and in part because of delocalization of a halide lone pair electron with the double bond.

C_2H_5—X ($C_{sp}3$—X) C_2H_3—X ($C_{sp}2$—X)

$$\left[\,CH_2{=}CH{-}\ddot{X} \longleftrightarrow \bar{:}CH_2{-}CH{=}X^{+}\right]$$

This structure contributes a small but significant amount.

12.F. Supplementary Problems

S1. Write out the structure corresponding to each of the following names:

a. neopentylacetylene
b. 1,3,3-tribromopropyne
c. sodium acetylide
d. 1,5,9-cyclododecatriyne
e. 2,2,7,7-tetramethyl-3,5-octadiyne
f. ethyne

S2. Give the IUPAC name for each of the following compounds:

a. $(CH_3)_2CHC{\equiv}CCH(CH_3)_2$

b.  [structure: cyclohexane with Br and H at one carbon, H and C≡C—Cl at the opposite carbon]

c. [structure: H_3C, H_3C C=C H; stereocenter with H, CH_3, C≡CH]

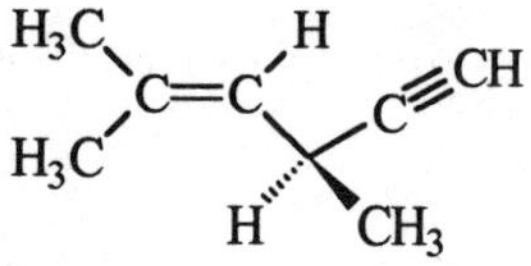

d.  [structure: OH, alkyne chain ending in Br]

S3. Give the principal product of reaction of compound (c), problem #S2, under each of the following conditions.

a. $\xrightarrow{H_2\text{/Lindlar catalyst}}$

b. $\xrightarrow{Na/NH_3}$

c. $\xrightarrow{NaNH_2}$ $\xrightarrow{CH_3I}$

d. $\xrightarrow{NaNH_2}$ $\xrightarrow{\text{cyclohexyl-Br}}$

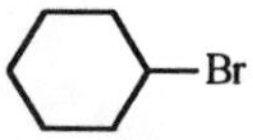

e. $\xrightarrow{HCl\ (1\ eq)}$

S4. Show how each of the following conversions can be accomplished in good yield. You may use other organic compounds if necessary.

$(CH_3)_2CHCH_2CH{=}CH_2 \longrightarrow$

a. $(CH_3)_2CH\overset{O}{\overset{\|}{C}}CH_3$

b. $(CH_3)_2CHCH_2\overset{O}{\overset{\|}{C}}CH_3$

c. $(CH_3)_2CHCH_2CH_2\overset{O}{\overset{\|}{C}}H$

d. $(CH_3)_2CHCH_2CH_2\overset{O}{\overset{\|}{C}}CH_3$

e. $(CH_3)_2CH\overset{O}{\overset{\|}{C}}CH_2CH_3$

Do you anticipate any problems with the route you would most likely select for conversion (e)?

S5. The sex attractant of the galechiid moth *Bryotopha similis* is a derivative of *trans*-9-tetra-decen-1-ol, and one of the components of the sex attractant of the butterfly *Lycorea ceresceres* is a derivative of *cis*-11-octadecen-1-ol. Show how to synthesize these two alcohols from 8-bromo-1-octanol and any other compound of four carbons or less. (HINT: protect the hydroxy group during your synthesis by forming the *t*-butyl ether; see end of Section 10.9 in the Text.)

S6. a. Using Appendix I, calculate the heat of hydrogenation, $\Delta H^\circ_{hydrog.}$ for each step in the hydrogenation of 2-butyne to butane.

b. Do you expect this to be a good model for calculating $\Delta H^\circ_{hydrog.}$ of cyclodecyne? If not, in which step(s) do you think the biggest difference will be seen?

S7. Show how to synthesize *cis* and *trans*-1,2-dibromocyclododecane stereospecifically from cyclododecane.

S8. Treatment of *cis*-3-hexene with bromine and then KOH in ethanol gives the vinyl halide, *Z*-bromo-3-hexene. However, when the same sequence of reactions is applied to cyclohexene, no vinyl halide (1-bromocyclohexene) is produced. Instead, 1,3-cyclohexadiene is obtained. Provide an explanation for the difference in behavior of these two alkenes.

12.G. Answers to Supplementary Problems.

S1. a. $(CH_3)_3CCH_2C\equiv CH$

b. $BrC\equiv CCHBr_2$

c. $HC\equiv CNa$

d. 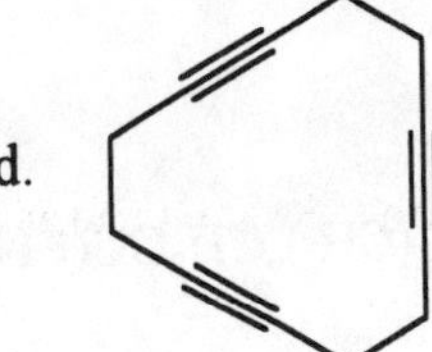

e. $(CH_3)_3CC\equiv C-C\equiv CC(CH_3)_3$

f. $HC\equiv CH$

S2. a. 2,5-dimethyl-3-hexyne

b. *trans*-1-bromo-4-(2-chloroethynyl)cyclohexane

c. (*S*)-3,5-dimethylhex-4-en-1-yne

d. (*S*)-6-bromo-4-hexyn-2-ol

S3. a. and b. (*R*)-$(CH_3)_2C=CHCH(CH_3)CH=CH_2$

c. (*S*)-$(CH_3)_2C=CHCH(CH_3)C\equiv CCH_3$

d. (*S*)-$(CH_3)_2C=CHCH(CH_3)C\equiv CH$ + [cyclohexene] + NaBr *(2° RX undergoes E2)*

e. (*S*)-$(CH_3)_2CClCH_2CH(CH_3)C\equiv CH$ *(electrophilic addition to alkenes is faster than to alkynes)*

S4. a. $(CH_3)_2CHCH_2CH{=}CH_2 \xrightarrow[\text{2. NaBH}_4]{\text{1. O}_3} (CH_3)_2CHCH_2CH_2OH \xrightarrow[\Delta]{Al_2O_3} (CH_3)_2CHCH{=}CH_2$ $\downarrow Br_2$

$(CH_3)_2CHCOCH_3 \xleftarrow[H_2SO_4]{\text{aq. }Hg^{2+}} (CH_3)_2CHC{\equiv}CH \xleftarrow{NaNH_2} (CH_3)_2CHCHBrCH_2Br$

b. $(CH_3)_2CHCH_2CH{=}CH_2 \xrightarrow[\text{2. NaNH}_2]{\text{1. Br}_2} (CH_3)_2CHCH_2C{\equiv}CH \xrightarrow[H_2SO_4]{\text{aq. }Hg2+} (CH_3)_2CHCH_2\overset{O}{\overset{\|}{C}}CH_3$

c. $(CH_3)_2CHCH_2C{\equiv}CH \xrightarrow{B_2H_6} ((CH_3)_2CHCH_2CH{=}CH)_3B \xrightarrow[NaOH]{H_2O_2} (CH_3)_2CHCH_2CH_2\overset{O}{\overset{\|}{C}}H$
(from b.)

d. $(CH_3)_2CHCH_2C{\equiv}CH \xrightarrow{NaNH_2} \xrightarrow{CH_3I} (CH_3)_2CHCH_2C{\equiv}CCH_3 \xrightarrow[150°]{NaNH_2} (CH_3)_2CHCH_2CH_2C{\equiv}C^- \; Na^+$
(from b.)

$\xrightarrow{H_2O} \xrightarrow[H_2SO_4]{Hg2+} (CH_3)_2CHCH_2CH_2\overset{O}{\overset{\|}{C}}CH_3$

e. $(CH_3)_2CHCH_2C{\equiv}CH \xrightarrow[C_2H_5OH]{KOH} (CH_3)_2CHC{\equiv}CCH_3 \xrightarrow[H_2SO_4]{Hg2+} (CH_3)_2CHCH_2\overset{O}{\overset{\|}{C}}CH_3 + (CH_3)_2CH\overset{O}{\overset{\|}{C}}CH_2CH_3$
(from b.)

The trouble with route (e) is that it will produce a lot of the isomeric ketone as well.

S5. $BrCH_2(CH_2)_6CH_2OH \xrightarrow[H^+]{(CH_3)_2C{=}CH} BrCH_2(CH_2)_6CH_2OC(CH_3)_3 \xrightarrow{HC{\equiv}CNa} HC{\equiv}C(CH_2)_8OC(CH_3)_3$ **A**

A $\xrightarrow{NaNH_2} \xrightarrow{BuBr} CH_3(CH_2)_3C{\equiv}C(CH_2)_8OC(CH_3)_3 \xrightarrow{Na/NH_3} \xrightarrow{H^+} CH_3(CH_2)_3CH{=}CH(CH_2)_8OH$
trans-9-tetradecen-1-ol

$BuBr + NaC{\equiv}CC_2H_5 \longrightarrow CH_3(CH_2)_3C{\equiv}CCH_2CH_3 \xrightarrow{NaNH_2,\ 150°} CH_3(CH_2)_5C{\equiv}CNa$ **B**

A $\xrightarrow{H_2,\ Lindlar} CH_2{=}CH(CH_2)_8OC(CH_3)_3 \xrightarrow{HBr,\ h\nu} Br(CH_2)_{10}OC(CH_3)_3$ **C**

B + **C** $\longrightarrow CH_3(CH_2)_5C{\equiv}C(CH_2)_{10}OC(CH3_3 \xrightarrow{H_2,\ Lindlar} \xrightarrow{H^+} CH_3(CH_2)_5CH{=}CH(CH_2)_{10}OH$
cis-11-octadecen-1-ol

S6. a. $CH_3C{\equiv}CCH_3 + H_2 \longrightarrow$ *cis*-$CH_3CH{=}CHCH_3$

ΔH°_f: 34.7 0 −1.9 $\Delta H^\circ_{hydrog(1)} = -36.6$ kcal mole^{-1}

cis-$CH_3CH{=}CHCH_3 + H_2 \longrightarrow CH_3CH_2CH_2CH_3$

ΔH°_f: −1.9 0 −30.4 $\Delta H^\circ_{hydrog(2)} = -28.5$ kcal mole^{-1}

b. In cyclodecyne there is a lot of strain because the C−C≡C−C group of atoms must bend to fit into the ring structure. This strain will be released in going from cyclodecyne to *cis*-cyclodecene, so $\Delta H^\circ_{hydrog(1)}$ will be more negative than predicted in a. The second step should be similar to the acyclic model.

S7.

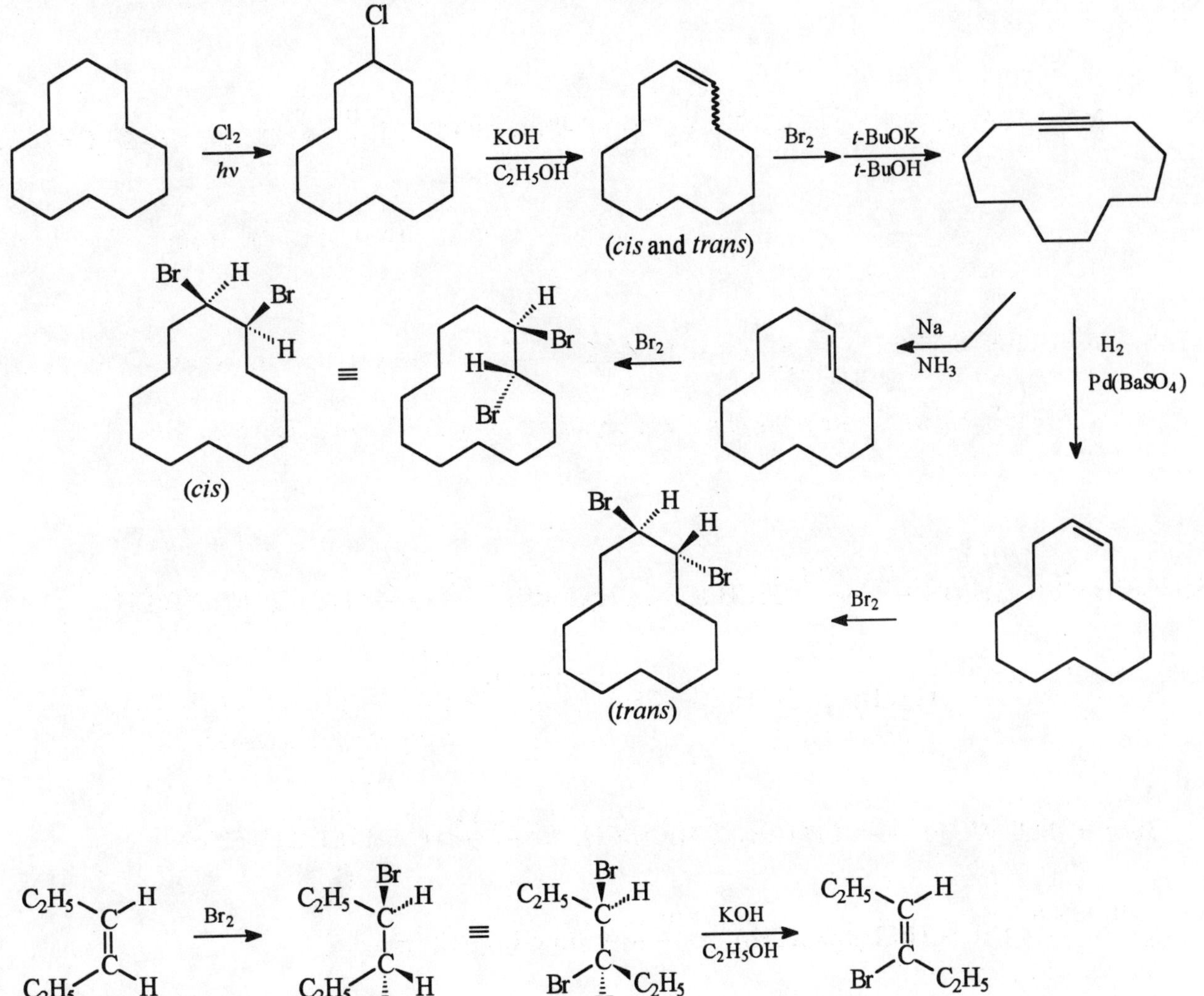

Cl

Cl_2, *hv* → KOH, C_2H_5OH → (*cis* and *trans*) → Br_2 → *t*-BuOK, *t*-BuOH →

Na, NH_3 → ; Br_2 → Br, H, Br, H (*cis*) ≡ H, Br, H, Br

H_2, Pd($BaSO_4$) → ; Br_2 → Br, H, H, Br (*trans*)

S8. C_2H_5, H, C_2H_5, H → Br_2 → Br, C_2H_5, H, C_2H_5, H, Br ≡ Br, C_2H_5, H, Br, C_2H_5, H → KOH, C_2H_5OH → C_2H_5, H, Br, C_2H_5

via anti elimination

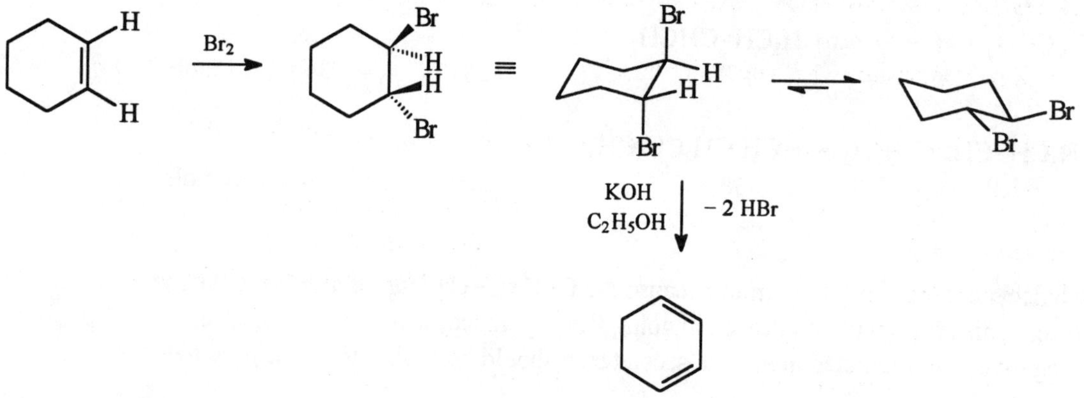

Br$_2$ KOH C_2H_5OH $-$ 2 HBr

A vinyl halide cannot be formed from this compound by anti elimination.

13. NUCLEAR MAGNETIC RESONANCE SPECTROSCOPY

13.A. Chapter Outline and Important Terms Introduced

13.1 Structure Determination (what does it involve?)

13.2 Introduction to Spectroscopy (the most important methods of structure determination)

quantization
quantum states
energy level differences
$\Delta E = h\nu$

microwave spectroscopy
infrared spectroscopy
ultraviolet-visible spectroscopy

13.3 Nuclear Magnetic Resonance (the physics behind the technique)

nuclear spin
magnetic moment
α- and β-spins, spin "flipping"

magnetic field strength, H
magnetogyric ratio, γ
$\nu = \gamma H/2\pi$

13.4 Chemical Shift (information about the environment of the magnetic nucleus)

shielding
diamagnetic shielding
resonance
upfield-downfield

tetramethylsilane (TMS)
parts per million (ppm)
δ scale

13.5 Carbon NMR Spectroscopy (CMR the number of different carbons in the molecule)

natural abundance
Fourier transform instrumentation
chemical shift prediction
symmetry

proton decoupling and off-resonance decoupling
α-, β-, and γ-effects
stereochemical effects

13.6 Relative Peak Areas (information on numbers of protons)

integration of spectra

saturation/relaxation

13.7 Spin-Spin Splitting (information on adjacent protons)

magnetic non-equivalence
applied vs. effective field
singlet, doublet, triplet, quartet, multiplet

coupling constant, J
polynomial pattern, Pascal's triangle

13.8 More Complex Splitting (the real world....)

"first-order" vs. non-first order spectra, $\Delta\nu \gg J$ vs $\Delta\nu \sim J$
$J_{ab} \neq J_{ac}$, overlap of patterns
"tree" diagrams

13.9 **Effect of Conformation on Coupling Constants**
dihedral angle dependence
Karplus curve

13.10 **Remote Shielding by Multiple Bonds: Magnetic Anisotropy** (why alkenes are downfield and alkynes upfield)
induced magnetic field
π-electron circulation
upfield shift of acetylenic hydrogens
long range coupling

13.11 **Dynamical Systems** (slow or fast on the "NMR time scale")
Heisenberg Uncertainty Principle
NMR time scale
magnetic equivalence
chair-chair interconversion
variable temperature NMR

13.12 **Chemical Exchange of Hydrogens Bonded to Oxygen** (why the coupling sometimes disappears)

13.13 **Solving Spectral Problems** (useful generalizations and hints)

13.B. Important Reactions Introduced: none in this chapter.

13.C. Important Concepts and Hints

NMR spectroscopy has become the most important method available to organic chemists for determining the structure of a compound. Chapter 13 presents the physical principles that underly the technique, and describes the spectra observed for alkanes, alkenes, alkynes, and alkyl halides. As other functional groups are introduced in subsequent chapters, their characteristic NMR resonances will be discussed. The facts that make up the topic of NMR form a logical framework and are as intuitively understandable as the rest of organic chemistry. Nevertheless, a common tendency for many students when presented by tables of numbers is to **memorize**. A certain amount of instant recall will help you in solving NMR problems, although as the Text points out, it is really important to know only the general regions in the NMR spectrum where various types of hydrogens or carbons appear, rather than memorizing exact resonance positions. The following outline is a general guide to the things that you should definitely "know" about NMR:

I. "To the right" in an NMR spectrum is:
  A. more "shielded"
  B. "upfield" (= higher field strength for a given resonance frequency)
  C. "lower frequency" (to reach resonance for a given magnetic field)
  [and vice versa: "to the left" = "deshielded" = "downfield" = "higher frequency"]

II. For proton spectra (**not CMR**), **area** is proportional to **number of protons**

III. Chemical shift depends on substituents:

  A. For proton spectra, the following generalizations are very useful:
    1. Alkyl hydrogens have δ ~ 1 ppm
    2. δ (tertiary C-H) > δ (secondary C-H) > δ (primary C-H)
    3. A halogen atom on the same carbon causes a downfield shift by 2-3 ppm
    4. A halogen on the next carbon still has an effect of about 0.5 ppm
    It is useful to learn the NMR characteristics of each functional group as they are introduced in subsequent chapters.

B. For carbon spectra:
   1. Learn some basic shifts (e.g. methane-butane)
   2. Learn the substituent effects (α-, β-, γ-effects)
      When it comes to solving spectral problems, it helps to know right away what chemical shift corresponds to what possibilities.

IV. For CMR spectra

The most important piece of information you can obtain from a CMR spectrum is how many different types of carbon atoms there are in the molecule. In many cases, this will tell you whether the structure is symmetrical or non-symmetrical in some way. If given, the multiplicity of the carbon resonance in the proton-**coupled** CMR spectrum will tell you directly if the carbon is a CH_3, CH_2, CH, or quaternary C.

Secondarily, chemical shift information can tell you:

A: Whether the signal corresponds to an alkene or alkane carbon, or if there is a heteroatom substituent

B: What nearby regions of the carbon skeleton look like (α-, β- and γ-steric effects, *cis/trans* stereochemistry of alkenes, etc.)

V. For proton spectra, as a basic rule, **n** adjacent protons lead to a splitting of **n** + 1 peaks. For first order spectra, sp^3 C and no rings, this is usually true. This can easily become more complicated, however. J for neighboring hydrogens on an alkane chain is usually 4-10 Hz.

VI. You should also be aware of complications in splitting patterns that arise from:

A. non-first order spectra

B. non-equivalent coupling constants

C. dihedral angle dependence

NMR spectral problems are usually one of two types: given the structure of the compound, predict the spectrum it would give; or, given the spectral data, deduce the structure of the compound. Solving the first type of question is fairly straightforward.

FIRST: Pick out the non-equivalent nuclei (nuclei can be equivalent either through symmetry alone, or through conformational interconversions and symmetry);

SECOND: Predict the chemical shift for each group;

THIRD: For proton spectra, (a) assign relative area (number of nuclei) in each group; (b) calculate splitting patterns.
For carbon spectra, sometimes you are asked to predict the multiplicity of the off-resonance decoupled spectrum.

FOURTH: Look for complications....

Solving the second type of problem is like solving a puzzle. It can be fun and there is a system. It is most complex for proton spectra, where there are the added complications of splitting patterns and relative areas to deal with, so we will point out a systematic way to approach this kind of question. Take the following example:

formula: $C_6H_{12}Br_2$
NMR spectrum, δ, ppm: 0.9 (t, 3H), 1.4 (s, 6H), 1.8 (m, 2H), 4.6 (t, 1H)

1. Write out, **underlined**, groups that correspond to the relative areas of each resonance (starting with the most probable combinations), including enough unprotonated carbons and other substituents to satisfy the formula:

"3H" is usually a methyl:	$\underline{CH_3}$–
"6H" is usually two equivalent methyls:	$\underline{CH_3}$–, $\underline{CH_3}$– (but it could be 3 x –$\underline{CH_2}$–)
"2H" is usually a CH_2:	–$\underline{CH_2}$–, (sometimes –$\underline{NH_2}$)
"1H" is, of course:	>$\underline{CH}$–, (or –$\underline{OH}$ or >$\underline{NH}$)
other atoms:	>$\underline{C}$<, 2 $\underline{Br}$– to satisfy $C_6H_{12}Br_2$

Make sure the number of protons you have written adds up to the number in the molecule; if it doesn't, multiply everything by an integer (if that doesn't work, then the entire spectrum was not given).

2. Look at the splitting pattern, and include in your part structures the adjacent carbons with an appropriate number of protons, **not underlined**. Simply circle the multiplets:

(t, 3H): $\underline{CH_3}$–CH_2–

(s, 6H): $\underline{CH_3}$–C(–)(–)–, $\underline{CH_3}$–C(–)(–)– (3 x –C(–)(–)–$\underline{CH_2}$–C(–)(–)– can be ruled out because there are not enough non-hydrogen-bearing substituents to fit)

(m, 2H): (–$\underline{CH_2}$–) (circled)

(t, 1H): –$\underline{CH}$(–)–CH_2– or –CH(–)–$\underline{CH}$(–)–CH(–)–

>$\underline{C}$<, 2 $\underline{Br}$–

3. Look for some correlation between the underlined and non-underlined groups.

A. The underlined and circled CH_2 group (third set, above) must be the non-underlined CH_2 group of the first and fourth sets:

CH_3–CH_2–CH(–)–

t, 3H m, 2H t, 1H

B. The underlined, quarternary carbon must be the non-underlined, quarternary carbon of the second set:

$(CH_3)_2C$<

s, 6H

C. There is only one way to hook these two pieces together with two bromine atoms:

$$CH_3CH_2CHBrC(CH_3)_2Br$$

4. Use the chemical shift information as a check, at least, or as further help if the coupling patterns can't solve the problem (as above):

CH_3 in CH_3CH_2 $\delta = 0.9$; CH_2 $\delta = 1.25 + 0.5$ for the adjacent Br = 1.75
CH $\delta = 4.1 + 0.5$ for the adjacent Br = 4.6; $(CH_3)_2$ $\delta = 0.9 + 0.5 = 1.4$

For many problems, you will be able to "see" the structure much more quickly than this stepwise process would suggest, but this methodical approach often avoids confusion in complex cases.

13.D. Answers to Exercises

13.1 With a spectrometer operating at 250 MHz, the difference between the TMS and CH_3 resonances is 557.5 Hz (2.23 ppm); the difference between the CH_3 and CH_2 resonances (4.00 - 2.23 = 1.77 ppm) is 442.5 Hz:

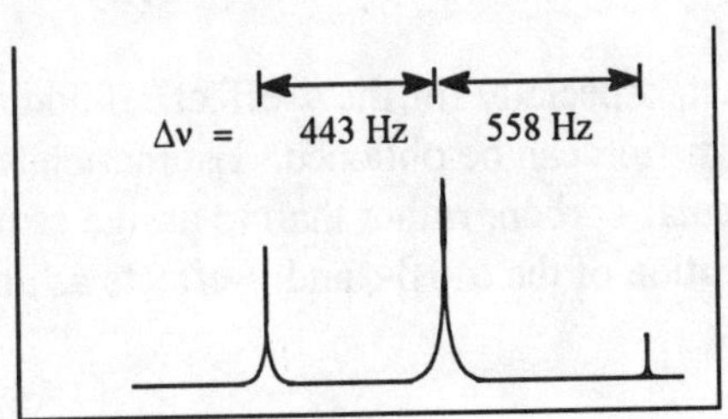

13.2 The fact that the CMR spectrum shows only four resonances indicates that the isomer is a meso compound, i.e., that there is a plane of symmetry that makes the C-1 equivalent to the C-5 carbon, C-2 equivalent to C-4, and the two methyls equivalent. The two isomers below are therefore eliminated by the CMR spectrum:

Cl ... Cl Cl ... Cl
OH OH

These meso compounds would give four-line CMR spectra:

Cl ... Cl Cl ... Cl
OH OH

13.3 $\underline{C}H_3OH$, δ = 49.3; each adjacent C adds about 9; therefore, for $CH_3\underline{C}H_2OH$ the calculated value for $\underline{C}$ is 49 + 9 = 58, actual 57.3; for $(CH_3)_2\underline{C}HOH$ the calculated value for $\underline{C}$ is 49 + 2 x 9 = 67, actual = 63.7; $(CH_3)_3\underline{C}OH$, the calculated value for $\underline{C}$ is 49 + 3 x 9 = 76. The actual value is 68.7.

13.4 $CH_3CH_2CH(CH_3)CH_2CH_2CH_3$

Carbon	δ for hexane	+ correction	= estimated δ	(actual δ)
C-1	13.9	γ, −2.5	11.4	10.0
C-2	22.9	β, +9.5	32.4	29.5
C-3	32.0	α, +9.0	41.0	34.3
C-4	32.0	β, +9.5	41.5	39.0
C-5	22.9	γ, −2.5	20.4	20.2
C-6	13.9	none	13.9	13.9

To calculate resonance for the C-3 methyl group:

13.7 [δ (CH_3 in <u>n</u>-pentane]
+9.5 [β-substituent]
−2.5 [γ-substituent]
= 20.7 ppm (actual = 18.8 ppm)

Note that the correspondance is not exact, especially for the α-effect. If additional effects such as branching are taken into account, a more accurate prediction can be obtained. Estimations of this sort are usually employed to assign a particular resonance to a particular carbon, rather than to assign structure of an unknown compound. From the practical point of view, application of the α-, β-, and γ-effects as above is enough to confirm or rule out various possibilities.

13.5 The CH_2 group of the (*Z*)-isomer will resonate at lower chemical shift values (will be shifted upfield) relative to that of the (*E*)-isomer, as a result of the γ-interaction with the methyl group at the other end of the double bond. For the same reason, the effect on the resonance which corresponds to the methyl substituent on C-3 will be opposite: it will occur at higher chemical shift (downfield) in the (*Z*)-isomer than in the (*E*)-isomer.

13.6 The CMR spectrum of cycloheptene will consist of 4 lines, three in the upfield region (δ 20-35 ppm) and one in the alkene region (δ 125-135 ppm).

13.7 a. $CHCl_2C(CH_3)_2CH_2Cl$

CH: area 1; $(CH_3)_2$: area 6; CH_2: area 2

b.

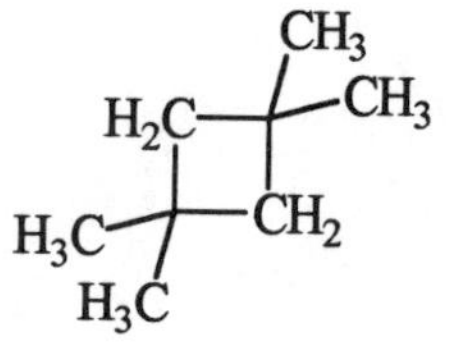

CH_3
H_2C — CH_3
H_3C — CH_2
H_3C

4 equivalent methyl groups = area 12; 2 equivalent CH_2's = area 4
Ratio of peaks = 12:4 = 3:1

c. $CH_3OCH_2CH_2OCH_3$ two equivalent CH_2 groups = area 4; two equivalent CH_3 groups = area 6
Ratio of peaks = 6:4 = 3:2

d. H; H_3C; Cl; H Ratio of peaks = 3:1:1

13.8

a.

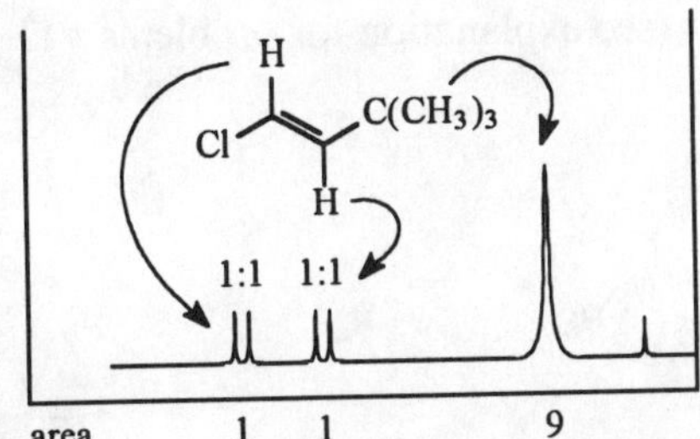

b.

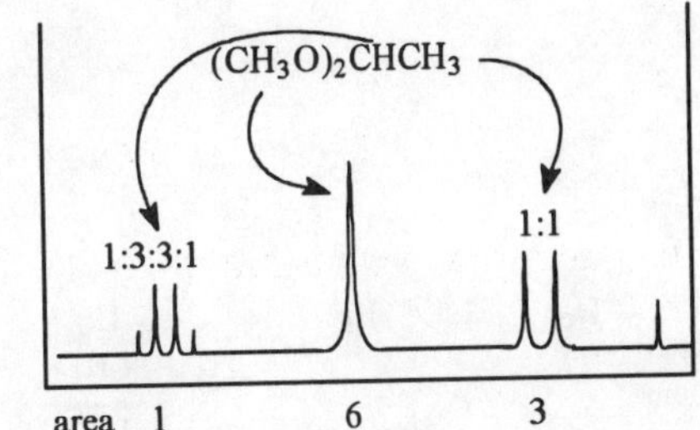

c.

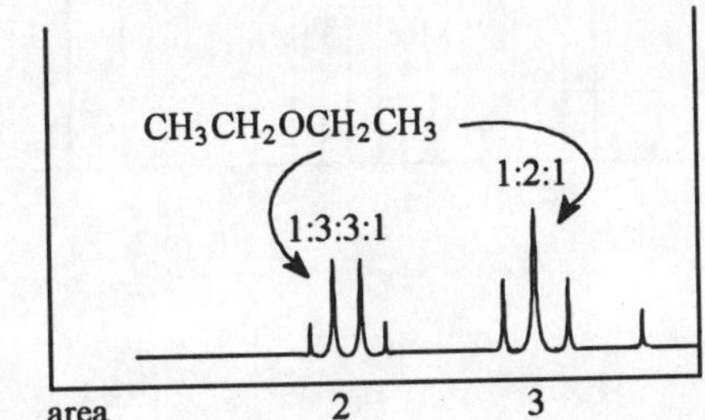

d.

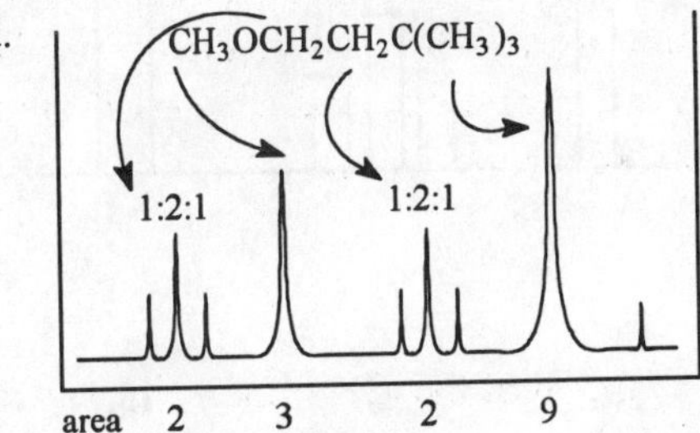

13.9

a.

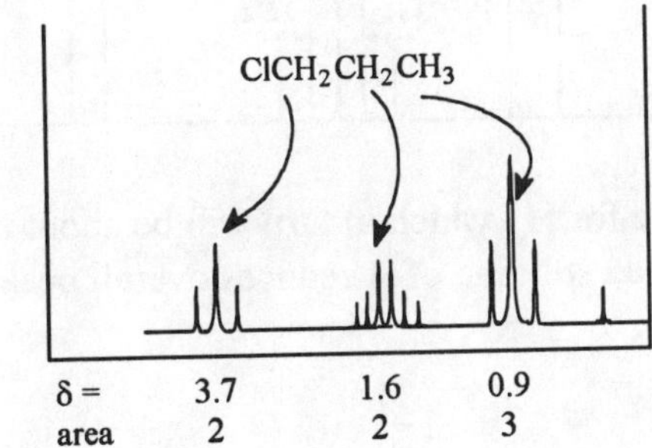

b.

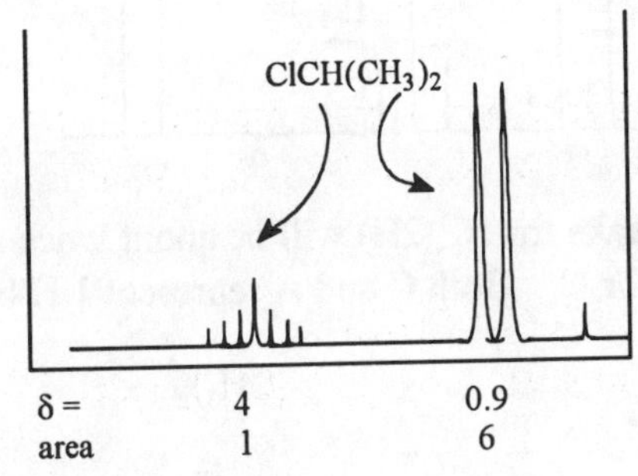

13.10

a. $ClCH_2CH_2CH_2Cl$

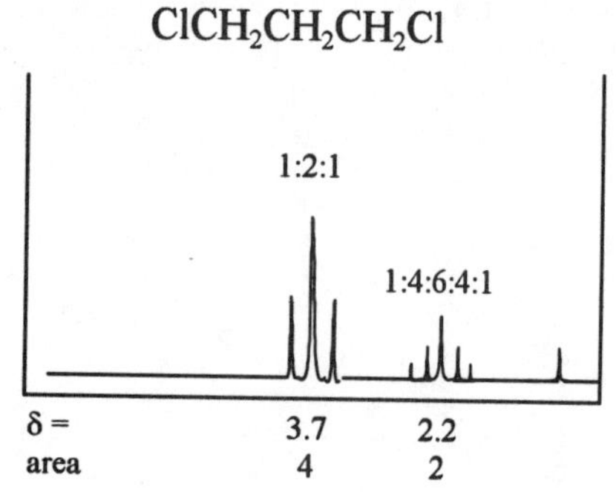

b. $CH_3OCH_2CH(OCH_3)CH_3$

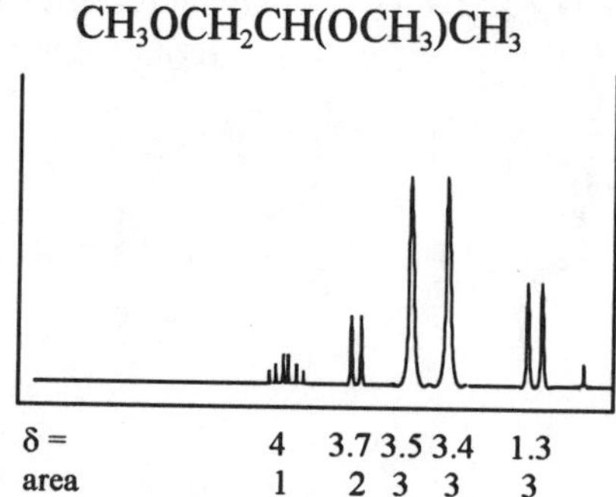

NOTE: the resonance for the CH_2 group will not actually be a simple doublet (see explanation for problems #11 b. and #12 f.

13.11

a.

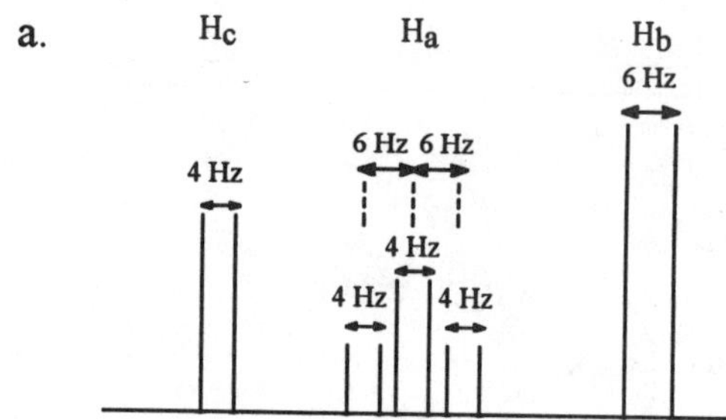

b.

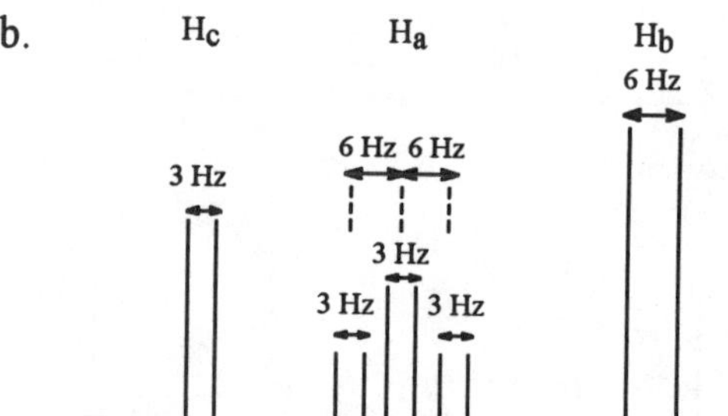

c.

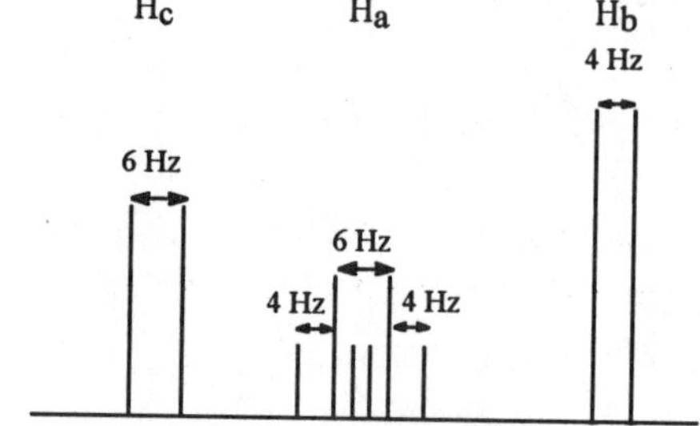

d.

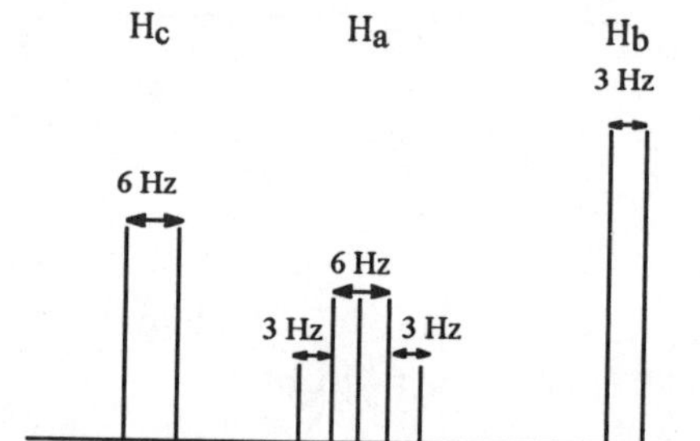

NOTE: the peaks for H_b (2H) will be about twice as tall as those for H_c, which in turn will be about twice as tall as those for H_a. Both C and A represent 1 H but the increased splitting of A reduces overall peak height but not area.

13.12

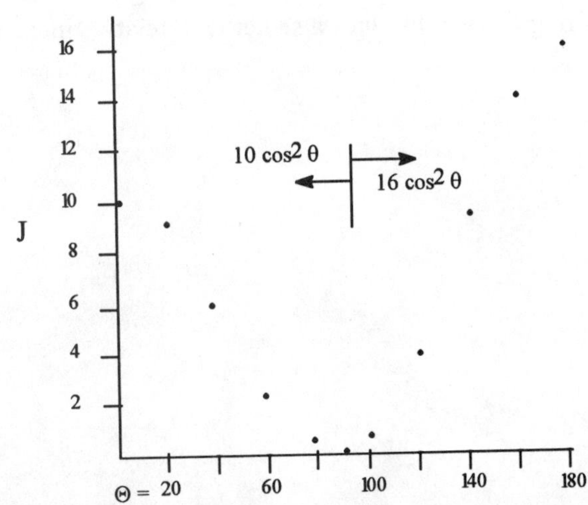

13.13 To a first approximation it will be a triple triplet, split equally by the two adjacent equatorial hydrogens ($J_{axial\text{-}equatorial} \approx 3$ Hz) and the two adjacent axial hydrogens ($J_{axial\text{-}axial} \approx 10$ Hz):

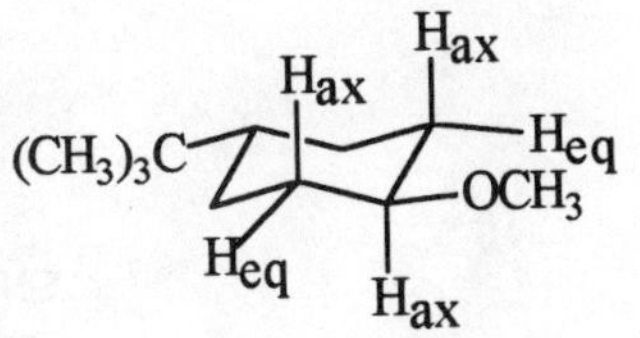

13.14 δ = 0.9 ppm, triplet, 3 hydrogens: CH_3 next to a CH_2 group
δ = 1.1 ppm, doublet, 3 hydrogens: CH_3 next to a CH group
δ = 1.5 ppm, multiplet, 2 hydrogens: CH_2 next to several hydrogens
δ = 1.8 ppm, doublet, J = 2.3 Hz, 1 hydrogen: CH coupled to another CH group; the coupling constant suggests that the coupling is long range, through an acetylene: $\underline{H}C{\equiv}C{-}CH$
δ = 2.3 ppm, multiplet, 1 hydrogen: CH next to several hydrogens

The pieces are therefore $C\underline{H_3}{-}CH_2$, $C\underline{H_3}{-}CH$, $CH_n{-}C\underline{H_2}{-}CH_n$, $\underline{H}C{\equiv}C{-}CH$, $-C\underline{H}(CH_n)$

From the formula, you can determine that there are 2 rings, 2 double bonds, 1 ring and 1 double bond, or 1 triple bond; the evidence above indicates that there is a triple bond. The structure which is consistent with all of the data is 3-methyl-1-propyne: $\mathbf{CH_3CH_2CH(CH_3)C{\equiv}CH}$

13.15 At room temperature, you expect a singlet because the resonances from axial and equatorial hydrogens are averaged rapidly by chair ⇌ chair interconversion.

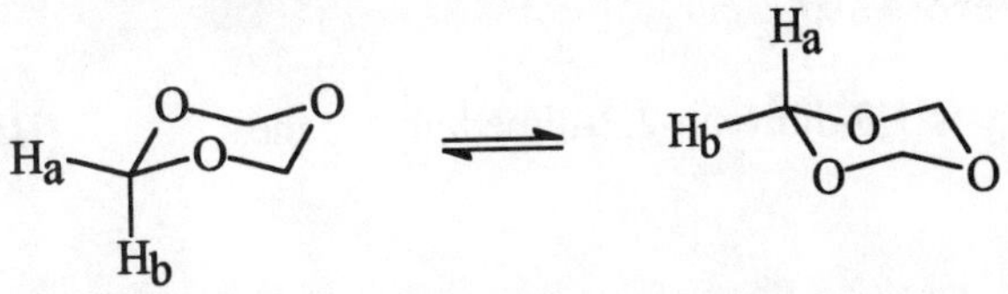

At −100°C, this interconversion is slow ("frozen out"), and separate resonances for axial and equatorial hydrogens will be seen. Because the axial and equatorial H's are not equivalent, they will split each other and a pair of doublets will be observed.

13.16 CH in diisopropyl ether: 6 equivalent neighboring H's will produce a septet with relative intensities of 1:6:15:20:15:6:1.

13.17

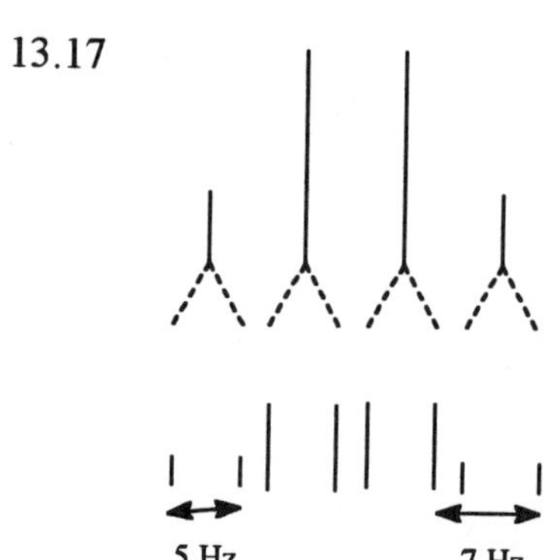

If the two J's were equal, the peak for the CH_2 resonance would be a simple quintet: 1:4:6:4:1.

13.18

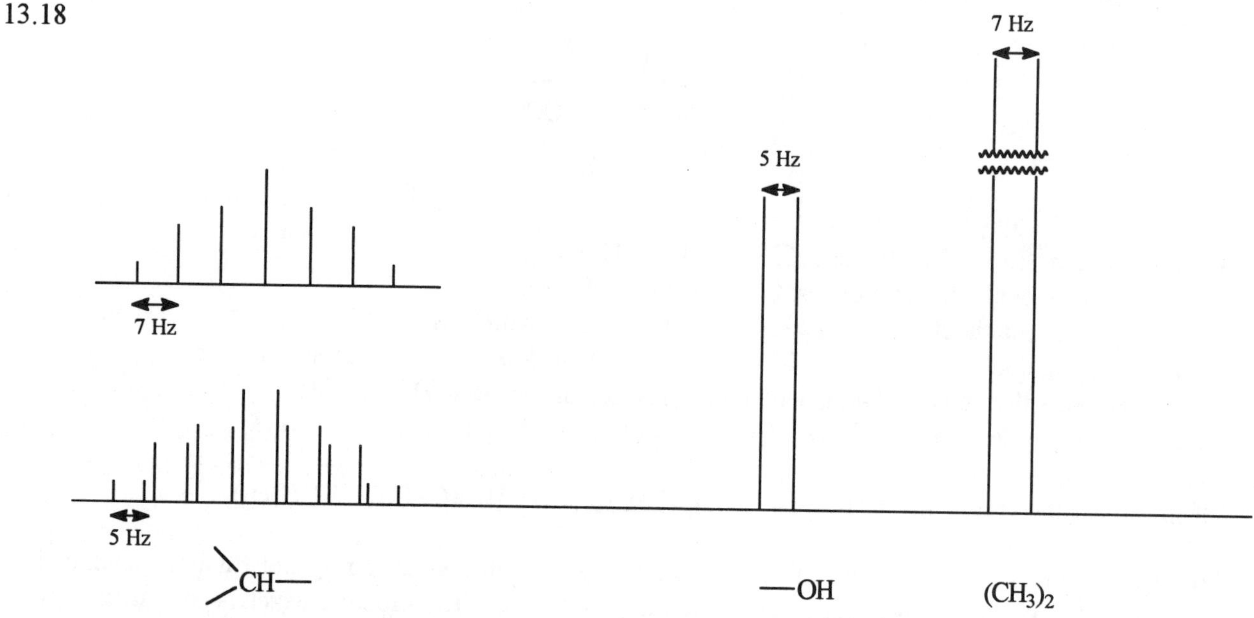

13.19 a. δ 1.0, s, 6H: 2 isolated CH_3's
δ 3.4, s, 4H: 2 isolated CH_2's

The only possibility is 1,3-dibromo-2,2-dimethylpropane: **$(BrCH_2)_2C(CH_3)_2$**

b. δ 1.0, t, 6H: 2 $C\underline{H_3}$-CH_2
δ 2.4, q, 4H: 2 CH_3-$C\underline{H_2}$-C

A symmetrical structure is again required; 3,3-dibromopentane: **$CH_3CH_2CBr_2CH_2CH_3$**

c. δ 0.9, d, 6H: $(C\underline{H}_3)_2CH$
δ 1.5, m, 1H: CH
δ 1.85, t, 2H: $CH\text{-}C\underline{H}_2\text{-}CH$ ($CH_2\text{-}C\underline{H}_2\text{-}C$ won't fit rest of the data)
δ 5.3, t, 1H: chemical shift indicates $CH_2\text{-}C\underline{H}Br_2$

All the pieces fit together as 1,1-dibromo-3-methylbutane: **$(CH_3)_2CHCH_2CHBr_2$**

d. δ 1.0, s, 9H: $(CH_3)_3C$
δ 5.3, s, 1H: $C\text{-}CHBr_2$

1,1-dibromo-2,2-dimethylpropane: **$(CH_3)_3CCHBr_2$**

e. δ 1.0, d, 6H: $(C\underline{H}_3)_2)CH$
δ 1.75, m, 1H: $C\underline{H}$
δ 3.95, d, 2H: $Br\text{-}C\underline{H}_2\text{-}CH$
δ 4.7, q, 1H: $CH\text{-}C\underline{H}Br\text{-}CH_2$

These fit together only one way: 1,2-dibromo-3-methylbutane, **$(CH_3)_2CHCHBrCH_2Br$**

f. δ 1.3, m, 2H: $C\underline{H}_2$
δ 1.85, m, 4H: 2 $CH_2\text{-}C\text{-}Br$
δ 3.35, t, 4H: 2 $CH_2\text{-}C\underline{H}_2\text{-}Br$

1,5-dibromopentane: **$BrCH_2CH_2CH_2CH_2CH_2Br$**

13.E. Answers and Explanations for Problems

1. In the NMR spectrum of ethyl bromide, the methyl hydrogens have δ = 1.7 ppm, the methylene hydrogens have δ = 3.3 ppm, and J = 7 Hz. The number of peaks given by the methyl hydrogens is **three,** with the approximate area ratio of **1:2:1**. These peaks are separated by **7** Hz. The number of peaks given by the methylene hydrogens is **four**, with an approximate area ratio of **1:3:3:1**. These peaks are separated by **7** Hz. The total area of the methyl peaks compared to the methylene peaks is in the ratio **3:2**. Of these two groups of peaks, the **methylene** peaks are farther downfield. The chemical shift difference between these peaks of 1.6 ppm corresponds in a 180-MHz instrument to **288** Hz and in a 250-MHz instrument to **400** Hz.

2. a. $CH_2BrCH_2CH_2CH_2Br$ b. $CH_2ClCH(CH_3)CH_2Cl$ c. $(CH_3)_2CBrCH_2Br$

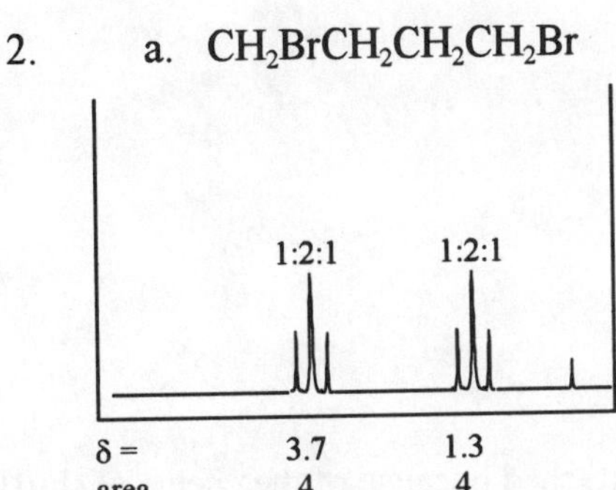

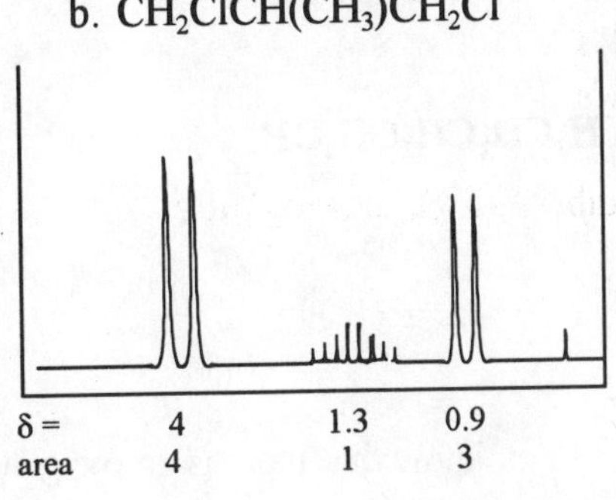

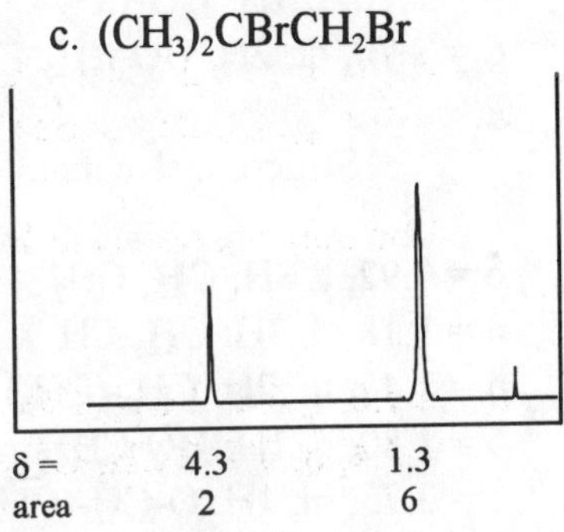

3. a. δ = 0.95, t, 3H: C**H**$_3$-CH$_2$
δ = 1.52, sextet, 2H: CH$_2$-C**H**$_2$-CH$_3$
δ = 3.30, s, 3H: C**H**$_3$-O (you can tell it's a methyl ether from the chemical shift)
δ = 3.40, t, 2H: CH$_2$-C**H**$_2$-O

Structure: methyl propyl ether, **$CH_3OCH_2CH_2CH_3$**

b. δ = 1.15, s, 1: this can only be an OH
δ = 1.28, s, 9: and this must be a *t*-butyl group

Structure: *t*-butyl alcohol, **$(CH_3)_3COH$**

c. δ = 1.20, t, 3H: C**H**$_3$-CH$_2$
δ = 3.45, q, 2H: CH$_3$-C**H**$_2$-O (you can tell the oxygen is attached to this carbon because of the chemical shift)

Although there appear to be only 5 hydrogens accounted for by the data given, remember that the areas are a relative ratio only: five hydrogens in a 3:2 ratio could just as well be ten hydrogens in a 6:4 ratio. The structure is diethyl ether: **$CH_3CH_2OCH_2CH_3$**

d. δ = 0.90, d, 6H: (C**H**$_3$)$_2$CH
δ = 1.78, m, 1H: C**H**-(CH$_n$)
δ = 2.45, t, 1H: could be **H**O-CH$_2$
δ = 3.45, t, 2H: HO-C**H**$_2$-CH

Structure: 2-methyl-1-propanol, **$(CH_3)_2CHCH_2OH$**

e. δ = 1.13, d, 6H: (C**H**$_3$)$_2$CH (because of the downfield shift of the methyl groups, there is probably an oxygen on the CH carbon)
δ = 3.30, s, 3H: C**H**$_3$-O
δ = 3.65, septet, 1H: (CH$_3$)$_2$C**H**-O

Structure: methyl isopropyl ether, **$(CH_3)_2CHOCH_3$**

f. δ = 0.95, t, 3H: C**H**$_3$-CH$_2$
δ = 1.50, m, 4H: four hydrogens must be 2 CH$_2$ groups
δ = 2.20, t, 1H: **H**O-CH$_2$
δ = 3.70, dt, 2H: HO-C**H**$_2$-CH$_2$

Structure: 1-butanol, **$CH_3CH_2CH_2CH_2OH$**

g. δ = 0.92, t, 3H: C**H**$_3$-CH$_2$
δ = 1.18, d, 3H: C**H**$_3$-CH
δ = 1.45, m, 2H: C**H**$_2$-(CH$_n$)
δ = 1.80, d, 1H: **H**O-CH
δ = 3.75, m, 1H: O-**CH**-(CH$_n$) (You know that there is an oxygen attached because of the chemical shift)

Structure: 2-butanol, **$CH_3CH_2CHOHCH_3$**

4. A: two methyls, $\mathbf{CH_3CCl_2CH_3}$

B: $\delta = 1.2$ (t, 3H) = $C\underline{H}_3\text{-}CH_2$
$\delta = 1.9$ (quint, 2H) = $CH_3\text{-}C\underline{H}_2\text{-}CH$ (<u>or</u> $\text{-}CH_2\text{-}C\underline{H}_2\text{-}CH_2\text{-}$ but not consistent with the other information)
$\delta = 5.8$ (t, 1H) = $\text{-}CH_2\text{-}C\underline{H}X_2$

This is all consistent with $\mathbf{CH_3CH_2CHCl_2}$.

C: $\delta = 1.4$ (d, 3H) = $C\underline{H}_3\text{-}CH$
$\delta = 3.8$ (d, 2H) = $\text{-}CH\text{-}C\underline{H}_2\text{-}$
$\delta = 4.3$ (sextet, 1H) = $CH_3\text{-}C\underline{H}\text{-}CH_2\text{-}$

This is all consistent with $\mathbf{CH_3CHClCH_2Cl}$.

D: $\delta = 2.2$ (quint, 2H) = $\text{-}CH_2\text{-}C\underline{H}_2\text{-}CH_2\text{-}$ (<u>or</u> $CH_3\text{-}C\underline{H}_2\text{-}CH\text{-}$ but this is inconsistent with the other information)
$\delta = 3.7$ (t, 4H) = $\text{-}C\underline{H}_2\text{-}CH_2\text{-}C\underline{H}_2\text{-}$

The structure must be $\mathbf{ClCH_2CH_2CH_2Cl}$.

5. a. The doublet at δ 1.7 ppm, corresponding to six hydrogens, suggests two equivalent $C\underline{H}_3\text{-}CH$ units. $(CH_3)_2CH\text{-}$ is inconsistent with other information. The quartet at δ 4.4 ppm corresponds therefore to the two equivalent C-H's. Hence, $\mathbf{CH_3CHBrCHBrCH_3}$.

b. Again, the doublet at δ 1.7 suggests $CH_3CH\text{-}$, but now there is only one. Hence, the molecule contains the unit: $CH_3CHBr\text{-}$. Since there is no other methyl, the second bromine must be at the other end of the chain. Thus the structure is $\mathbf{CH_3CHBrCH_2CH_2Br}$.

6. Since there are only three separate resonances, there must be some symmetry to the structure. The doublet in the proton-coupled CMR spectrum corresponds to a CH group, the triplet to a CH_2 and the quartet to CH_3. To add up to ten hydrogens, these must be present as two equivalent CH_3's, two equivalent CH's, and one CH_2 group:

2,4-dibromopentane, $\mathbf{CH_3CHBrCH_2CHBrCH_3}$.

7. E and F: (structure: H, Cl, Cl) and (structure: H, Cl, Cl) G: (structure: H, Cl, H_3C, CH_3, H, Cl)

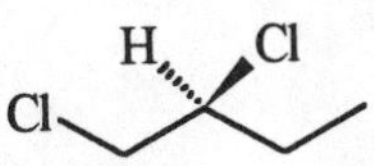

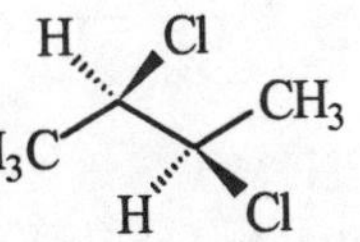

H: $CH_3CCl_2CH_2CH_3$ I: (structure: H, Cl, H_3C, CH_3, Cl, H) (meso)

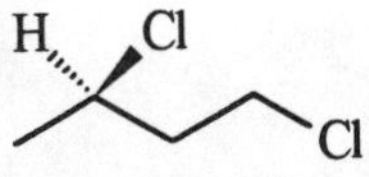

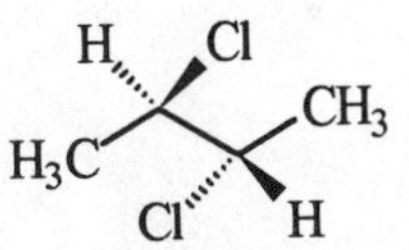

8. a. $CH_3OC(CH_3)_3$ b. CH_3CH_2I
c. CH_3CCl_3 d. $ClCH_2CH_2CH_2Cl$
e. $(CH_3)_2CHOH$ f. $ClCH_2C{\equiv}CH$ (Note the long range coupling across the triple bond.)

9. $CH_3CHClCHCl_2$ NOTE: the resonance at δ 0 ppm is due to tetramethylsilane as chemical shift reference.

10. The peaks 104.5 Hz above and below the main peak for $CHCl_3$ are the doublet due to $^{13}CHCl_3$. The separation of these two peaks corresponds to the carbon-proton coupling constant (209 Hz). In the proton-coupled CMR spectrum of $CHCl_3$, a doublet would appear (at 77 ppm) with the same coupling constant, i.e. with a separation of 209 Hz also.

11. a. With equal area under the two peaks in the NMR spectrum, and only two peaks in the CMR spectrum, the molecule must be quite symmetrical. The presence of an ether is suggested by the downfield position (3.6 ppm) of one of the peaks in the NMR: **tetrahydrofuran**

The spectrum of tetrahydrofuran does not show the pair of clean triplets that one would expect from a first order analysis. Although coupling is not observed between the two equivalent CH_2 groups at C-3 and C-4, their interaction has an effect on the patterns of the other CH_2 resonances. This "non-first order" effect, which is sometimes called "virtual coupling", results in a more complicated pattern for both sets of peaks.

b. Two peaks in the CMR spectrum: symmetrical structure again. The downfield position of the resonances in both spectra suggest that electronegative elements are present. In the NMR spectrum, the area ratio of 4:1, and the fact that the downfield resonance (δ 4.2) is a quintet, suggests that the compound is 1,2,3-trichloropropane: $\mathbf{ClCH_2CHClCH_2Cl}$

(NOTE: the resonances for the CH_2 groups are not simple doublets as you might expect. The two hydrogens in each CH_2 are **not** equivalent, and therefore they can have different chemical shifts as well as split each other. The easiest way to convince yourself that the two hydrogens are not equivalent is to make two different models of 1,2,3-trichloropropane, substituting first one and then the other hydrogen on one of the CH_2 groups with bromine (or anything other than hydrogen or chlorine). You will see that the two models you have made are **not** the same, and that they are not enantiomers either. They are diastereomers, hence the relationship between the two hydrogens that you substituted is referred to as **diastereotopic**.)

c. The CMR spectrum indicates four carbons, two that are probably double-bond carbons and two that are sp^3-hybridized; one of the latter contains an electron-withdrawing group because of its position around 63 ppm. The NMR spectrum indicates four types of hydrogens: two on a double bond (δ 5.6 ppm), one OH group (broad, 4.5 ppm), and CH_2 and CH_3 groups. Both of these are downfield from their position in alkanes because they are attached to the double bond, and because the OH is on the CH_2 group. The structure is 2-buten-1-ol: $\mathbf{CH_3CH{=}CHCH_2OH}$

12. a. δ = 1.0, d, 6H: $(C\underline{H_3})_2CH$
δ = 2.0, looks like a septet, but could be **nine** peaks, 1H: $(CH_3)_2C\underline{H}\text{-}CH_2$
δ = 3.4, d, 2H: $CH\text{-}C\underline{H_2}\text{-}Cl$

Structure: $\mathbf{(CH_3)_2CHCH_2Cl}$

b. $\mathbf{CH_3CH_2CH_2CH_2I}$

c. The downfield resonances (δ 5.1 and ~5.8 ppm, area 2:1) are characteristic for a vinyl group ($CH_2{=}CH\text{-}$), with a complex non-first order splitting pattern. The remaining two carbons, four hydrogens, and a bromine, must be arranged either as 3-bromo-1-butene or 4-bromo-1-butene:

$CH_3CHBrCH{=}CH_2$ or $BrCH_2CH_2CH{=}CH_2$

These are easily distinguished by the areas of 2:2 rather than 1:3 in the resonances around 3 ppm as well as the resonances in the upfield portion of the spectrum: the triplet at δ 3.4 and the quartet at δ 2.4 ppm are clearly the CH_2's next to the bromine and the double bond, respectively, in 4-bromo-1-butene. If you look closely at the quartet at δ 2.4 ppm, you can actually see long range coupling to the $=CH_2$ hydrogens: each peak in the quartet is barely perceptible as a triplet.

d. The formula C_5H_8O tells you that there are either two π bonds or two rings, or one of each. The NMR spectrum shows the following:

δ = 1.5, s, 6H: probably two equivalent, isolated methyl groups, with the chemical shift suggesting that the carbon they are attached to has the oxygen attached

δ = 2.2, s, 1H: could be an acetylenic hydrogen, ≡C-H

δ = 2.9, broad, 1H: because it is broad, it is most likely an OH

The structure is 2-methyl-3-butyn-2-ol: **$(CH_3)_2C(OH)C{\equiv}CH$**

e. The formula indicates that there is one double bond or a ring.

δ = 1.3, t, 3H: $C\underline{H}_3\text{-}CH_2$
δ = 3.7, q, 2H: $CH_3\text{-}C\underline{H}_2\text{-}O$
δ = 3.9, dd (see insert to discern pattern), 1H, and
δ = 4.0, dd, 1H: two hydrogens at the end of a double bond: $=CH_2$
δ = 6.4, dd, 1H: $O\text{-}C\underline{H}{=}CH_2$

The structure is ethyl vinyl ether: **$CH_3CH_2OCH{=}CH_2$**

Notice how the CH= hydrogen is shifted further downfield than a normal olefinic hydrogen because of the electron-withdrawing oxygen substituent. Conversely, the $=CH_2$ hydrogens are shifted upfield in comparison to normal because of the enol ether resonance structure (see below) which puts **more** electron density at that position.

$$CH_3CH_2\ddot{O}\text{—}CH{=}CH_2 \longleftrightarrow CH_3CH_2\overset{+}{O}{=}CH\text{—}CH_2^-$$

f. This is a spectacular spectrum, isn't it?!! Note that the peak areas are in the ratio of 6:1:4:1, which for 24 hydrogens is 12:2:8:2.

δ = 1.2, t, 12H: four equivalent methyl groups, attached to CH_2's
δ = 1.8, t, 2H: CH_2 next to another CH_2 or between two CH's
δ = 3.5, multiplet, 8H:

This pattern is actually quite interpretable; if you look closely, you can see that there are actually four quartets that are overlapping. The quartets arise from coupling to a methyl group. If we imagine that the quartet coupling is "removed", what would be left is a pattern of four lines, shown below, which is known as an "AB pattern". In other words, if you take an AB pattern and split every line into a quartet, you will get the observed multiplet:

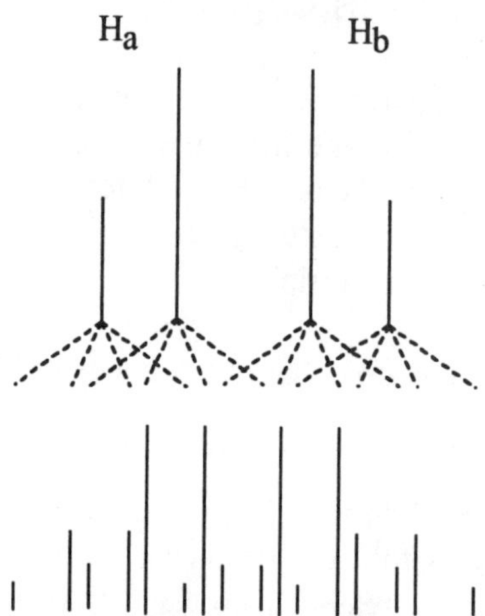

An AB pattern arises from two hydrogens that are strongly coupled to each other; that is, the chemical shift difference between them is not much greater than their coupling constant. Rather than appearing as a simple pair of doublets, the doublets of an AB pattern "lean" strongly toward each other. The two hydrogens of the CH_2 groups that give rise to this "quartet" of AB patterns are **diastereotopic** because, from their point of view, a carbon near to them is chiral. Again, you can convince yourself of this by making two models of the compound (shown below), substituting first one and then the other hydrogen of one of the CH_2 groups with something else, and comparing the two models that you have made. They will be diastereomers.

$\delta = 4.5$, 5, 2H: because of the far downfield position, it must be on a carbon with two oxygens: $(RO)_2C\underline{H}\text{-}CH_2$

The structure is 1,1,3,3-tetraethoxypropane: $\mathbf{(CH_3CH_2O)_2CHCH_2CH(OCH_2CH_3)_2}$

13. There are only four isomers of C_4H_9Br, shown below:

$CH_3CH_2CH_2CH_2Br$ $CH_3CH_2CHBrCH_3$ $(CH_3)_2CHCH_2Br$ $(CH_3)_3CBr$

1- and 2-Bromobutane will have four resonances in the CMR spectrum: *t*-butyl bromide will have only two. With only three carbon resonances, there is only one possibility: 1-bromo-2-methyl-propane. The proton NMR spectrum will be similar to that of the chloride; see the spectrum for Problem #12 a.

14. The four possible compounds are shown for the answer above, substituting Cl for Br. The fact that there are four resonances in the CMR spectrum of the compound indicates that it is either 1- or 2- chlorobutane. The proton-coupled CMR spectrum of 1-chlorobutane would show a quartet (for C-4) and three triplets (for carbons 1-3); that of the 2-chloro isomer would show two quartets (C-1 and C-4), a triplet (C-3) and a doublet (C-2).

15. On the NMR time scale, the two methyls are averaged by the chair ⇌ chair interconversion:

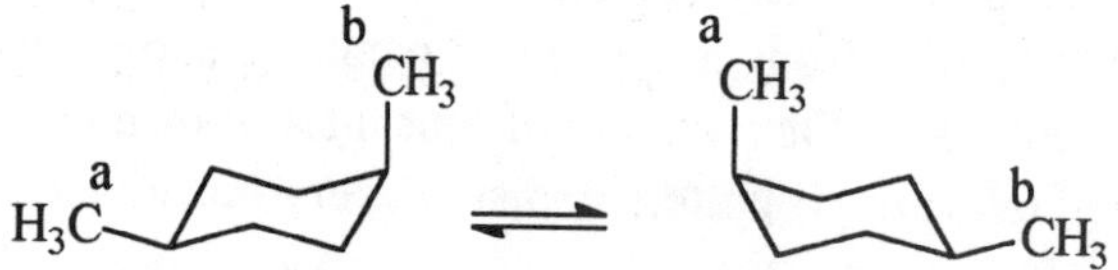

At lower temperatures where this interconversion occurs slowly, separate signals are seen for the two methyls.

16. a. All of the carbons are CH_2 groups since they are all triplets in the proton-coupled spectrum. The compound must therefore be 5-chloro-1-pentanol:

$ClCH_2CH_2CH_2CH_2CH_2OH$

b. The fact that there are only four carbon resonances for eight carbons indicates that the compound must be a dibutyl ether. The multiplicity of the four peaks (3 triplets and 1 quartet) shows that the compound is di-*n*-butyl ether:

$CH_3CH_2CH_2CH_2OCH_2CH_2CH_2CH_3$

c. $\delta = 75.1$ (d): $\underline{C}HOH$
$\delta = 35.3$ (s): $\underline{C}$, with no hydrogens
$\delta = 25.8$ (q): much more intense than the others: probably the $\underline{C}H_3$'s of a *t*-butyl group
$\delta = 18.2$ (q): $\underline{C}H_3$

The structure is 3,3-dimethyl-2-butanol: $CH_3CHOHC(CH_3)_3$

d. $\delta = 65.5$ (d): $\underline{C}H\text{-}O$
$\delta = 49.2$ (t): $\underline{C}H_2$
$\delta = 25.1$ (d): $\underline{C}H$
$\delta = 24.3$ and 22.7 (q): $\underline{C}H_3$'s.

Since there are 5 resonances and 6 carbons, it is likely that one of the methyl resonances represents two equivalent methyl groups.

The structure is 4-methyl-2-pentanol: $CH_3CHOHCH_2CH(CH_3)_2$

e. With two types of carbon, the possibilities are obviously limited. The formula indicates one ring or double bond. From the chemical shifts, there are no sp^2 carbons.

$\delta = 71.1$ (d): $\underline{C}H\text{-}O$
$\delta = 33.9$ (t): $\underline{C}H_2$

With only CH and CH_2 groups, the only reasonable structure is cyclohexane-1,4-diol (from the data given it is difficult to predict whether it is the *cis* or *trans* isomer):

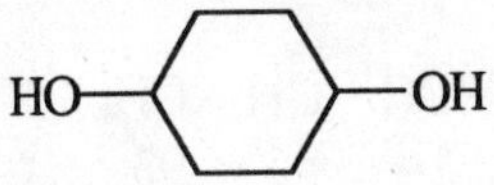

HO–(cyclohexane)–OH

17. The two isomers that can be formed are 3-methylcyclopentene 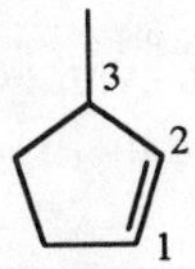 and 4-methylcyclopentene

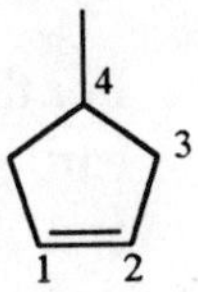

4-Methylcyclopentene is symmetrical and will show only four resonances in the CMR spectrum; every carbon in 3-methylcyclopentene is different, however, and six peaks appear in the CMR spectrum.

18. The remaining possibilities for simple elimination products from 3-bromo-2,3-dimethylpentane are *E* and *Z*-3,4-dimethyl-2-pentene. These can be readily distinguished from their CMR spectrum: the *Z*-isomer will have the resonance for the C-3 methyl group downfield from that for the *E*-isomer; conversely, the C-4 carbon in the *Z*-isomer will be upfield of that in the *E*-isomer:

Isomer J: 12.4, 20.6, 117.9, 28.5, 141.0, 17.8

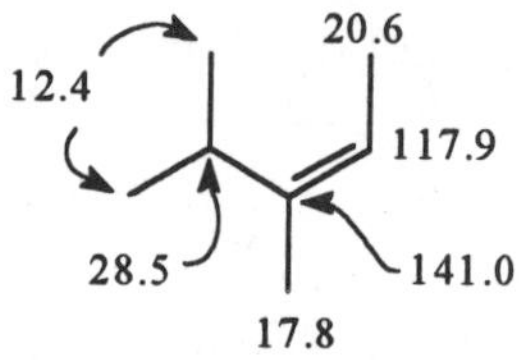

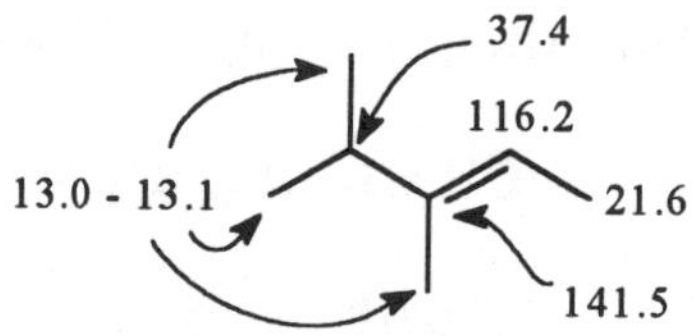

(*Z*-3,4-dimethyl-2-pentene)

Isomer K: 37.4, 116.2, 13.0 - 13.1, 21.6, 141.5

(*E*-3,4-dimethyl-2-pentene)

19. Isomers L and M: The CMR spectrum indicates one type of methyl and two types of alkene carbon. The structures are therefore the symmetrical stereoisomers of 2,4-hexadiene. Which one is *Z,Z*- and which is *E,E*- can be decided from the chemical shifts of the methyl groups: the methyl resonance in the *Z,Z*-isomer will come upfield of that of the *E,E*-isomer. Isomer L is therefore *Z,Z*-2,4-hexadiene, and isomer M is *E,E*-2,4-hexadiene. Not surprisingly, Isomer N is the unsymmetrical isomer.

Isomer L: *Z,Z*-2,4-hexadiene

Isomer M: *E,E*-2,4-hexadiene

Isomer N: *E,Z*-2,4-hexadiene

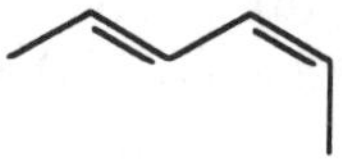

20.

Butane	δ	1-Butanol	δ	$\Delta\delta$
C-4	13.2	C-4	13.9	+0.7 (= δ-effect)
C-3	25.0	C-3	19.4	−5.6 (= γ-effect)
C-2	25.0	C-2	35.3	+10.3 (= β-effect)
C-1	13.2	C-1	61.7	+48.5 (= α-effect)

Using the known effect of Cl substitution, the following chemical shifts would be predicted:

$ClCH_2-CH_2-CH_2-CH_2OH$
45 30 30 62

$CH_3-CHCl-CH_2-CH_2OH$
25 50 46 57

$CH_3-CH_2-CHCl-CH_2OH$
9 30 66 61

($CH_3-CH_2-CH_2-CHClOH$ This molecule is unstable.)
14 14 46 93

4-Chloro-1-butanol certainly corresponds most closely to the observed values.

21. For the same reason that acetylenic hydrogens come in between alkane and alkene hydrogens, the sp-hybridized carbons in an acetylene are shielded by the diamagnetic anisotropy of the π-electron cloud (see Section 13.4 and 13.10 in the Text).

22. Isomer O, with one vinyl hydrogen in the proton NMR spectrum, must be 1-methylcyclohexene; isomers P and Q, with 2 vinyl hydrogens in the proton NMR, must be 3- and 4-methylcyclohexene. Isomer P must be 3-methylcyclohexene because one of the alkene carbons shows a downfield shift from the adjacent methyl group (β-effect). In the isomeric 4-methylcyclohexene (isomer Q), the methyl substituent is further removed from the double bond and has no effect on the resonances of the sp^2-hybridized carbons. It is only by coincidence that these carbons resonate exactly at the same frequency, since they are not equivalent.

23. The hydrogens are split by the fluorine spins, with J_{HF} = 50 Hz. The spectrum is a triplet since the two fluorines are equivalent. Note that in a different region in the NMR spectrum it is possible to determine the fluorine NMR spectrum, which would also show a triplet with the same coupling constant.

24. CH_3CH_2F: CH_3- $\delta = 14.6$; -CH_2- $\delta = 79.3$

Each resonance is split by ^{19}F (remember, the spectrum is only **proton** decoupled)

25. a. The overall appearance of a CMR spectrum of 2-pentanol doubly labeled at C-2 and C-3 will be of two doublets (^{13}C-2 and ^{13}C-3 will split each other). There will also be much weaker peaks (1% of the intensity of the labeled carbons) for C-1, C-4, and C-5; these minor peaks will be split into double doublets, since each of these carbons will be coupled to the two labeled carbons (and the two coupling constants will be different).

b. Since the 2-pentyl cation intermediate undergoes carbocation rearrangement during the course of the dehydration, and the alkene itself isomerizes in the reaction mixture, the *trans*-2-pentene that is isolated is a mixture of C-2, C-3-, and C-3, C-4-^{13}C-labeled material. That which has the ^{13}C-label in the C-2 and C-3 positions is responsible for the doublets at δ 123.6 (J = 70 Hz) and 133.2 (J = 70 Hz) ppm, and the isomer with the label at C-3 and C-4 for the doublets at 25.8 (J = 44 Hz) and 133.2 (J = 44 Hz).

26. The excess proportion of α-spins is 3×10^{-5} (Section 13.6) x 0.01 moles = 3×10^{-7} moles (of α-spin converted to β) x 0.017 cal mole^{-1} (Section 13.3) = 5.1×10^{-9} cal/mL.

$$\frac{5.1 \times 10^{-9}\ cal\ mL^{-1}}{1\ cal\ deg^{-1}\ mL^{-1}} = 5.1 \times 10^{-9} \text{ deg (not much!)}$$

13.F. Supplementary Problems

S1. From the molecular formulas and NMR spectra data given below, deduce the structure of each compound.

a. $C_4H_7Cl_3$; δ 1.4 (s, 3H); 4.0 (s, 4H)
b. $C_4H_7Cl_3$; δ 1.3 (d, 3H); 2.1 (s, 3H); 4.6 (q, 1H)
c. $C_4H_8Br_2$; δ 1.0 (d, 3H); 2.5 (m, 1H); 3.3 (d, 4H)
d. $C_4H_7Br_3$; δ 1.4 (d, 3H); 2.6 (t, 2H); 3.6 (m, 1H), 5.4 (t, 1H)

S2. Sketch the NMR spectra of the following compounds. Be sure to represent the expected δ for each group of peaks, the relative areas, and the splitting patterns.

a. $ICH_2CH_2CHCl_2$

b. $(CH_3)_3CCHClCH_2Cl$

c. $(CH_3)_2CHCHClCH_2Cl$

d. $(CH_3)_2CHCHCl_2$

e. $(CH_3CH_2)_2CHCH_2Br$

f. 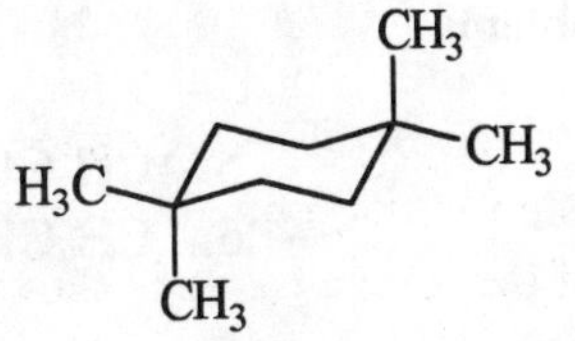 CH_3 CH_3 H_3C CH_3

S3. How many resonances do you expect to see in the CMR spectrum of each of the following compounds at 25°C? At −100°C?

a. Cl, Cl

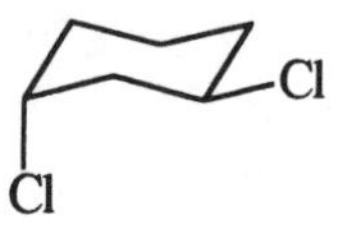

b. Cl, Cl, Cl, Cl

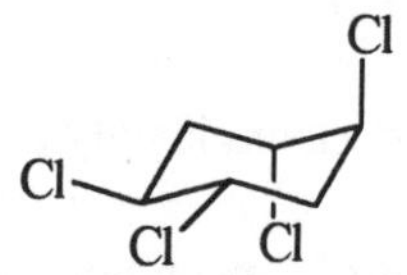

c. H_3C, CH_3, CH_3, CH_3

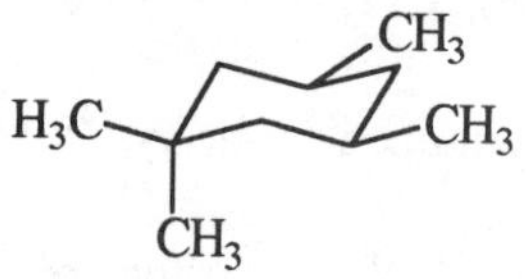

d.

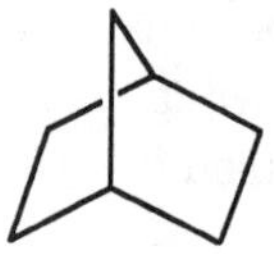

e. H_3C, Cl

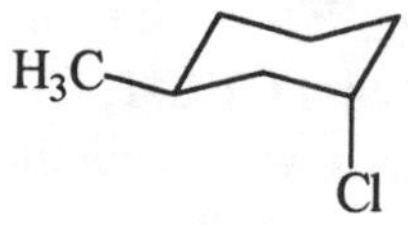

S4. The NMR spectrum of a rapidly (on the NMR time scale) interconverting mixture of conformational isomers is the weighted average of the spectra of the individual conformations. At low temperature, the chemical shifts of H-1 in axial- and equatorial-chlorocyclohexane are as shown below. Predict the observed chemical shift of this proton at room temperature.

δ C<u>H</u>Cl:	4.40 ppm	3.68 ppm

S5. For 1,1-diphenylpropane, J_{ab} = 7 Hz is in the expected range for acyclic alkanes.
In 2-methyl-1,1-diphenylpropane, however $J_{a',b'}$ is > 10 Hz.
Using the Karplus curve (Fig. 13.33) and Newman projections, offer an explanation for this increase in the coupling constant J.

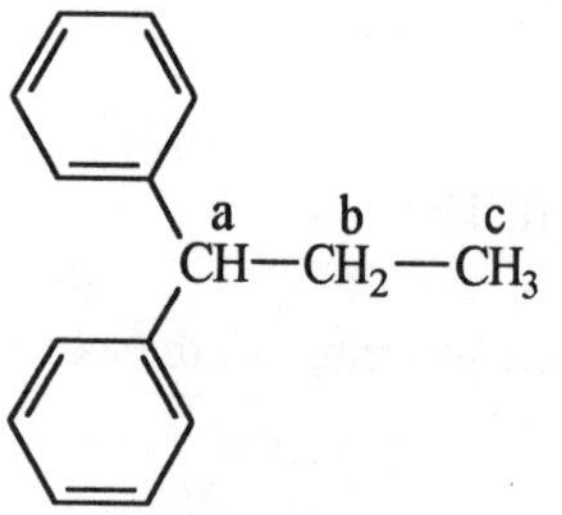

J_{ab} = 7 Hz 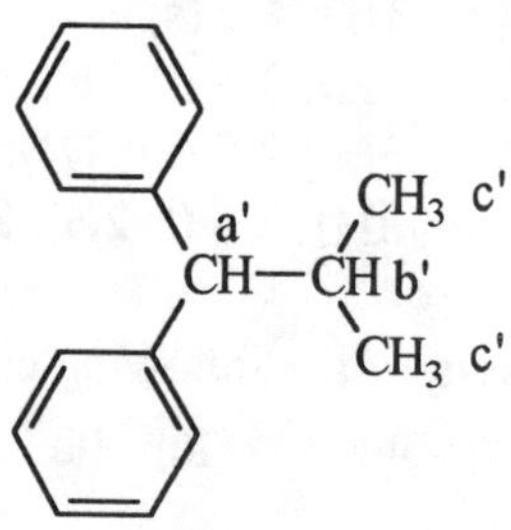 $\qquad$ $J_{a'b'}$ => 10 Hz

13.G Answers to Supplementary Problems

S1. a. $(ClCH_2)_2CClCH_3$ $\qquad$ b. $CH_3CCl_2CHClCH_3$
c. $CH_3CH(CH_2Br)_2$ $\qquad$ d. $CH_3CHBrCH_2CHBr_2$

S2. a.

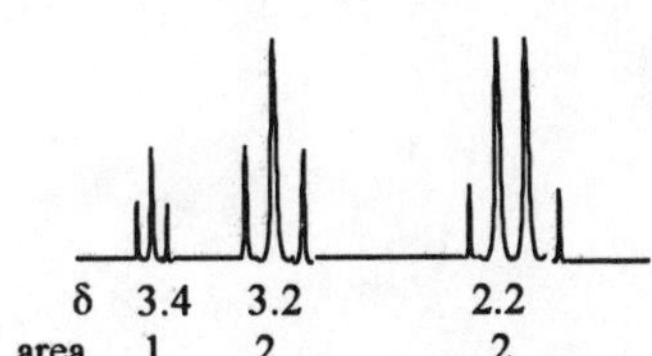

b.

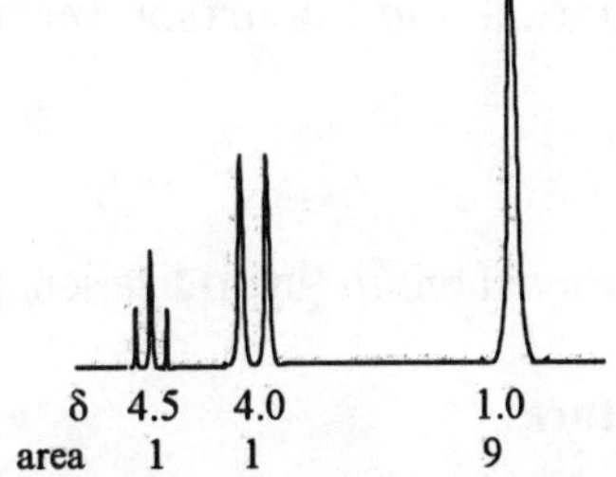

c.

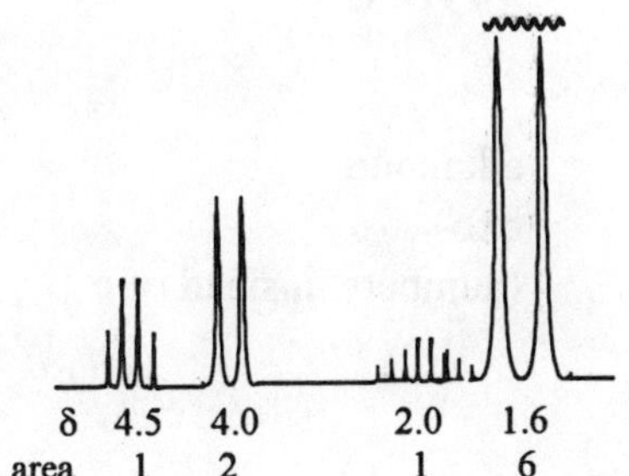

d.

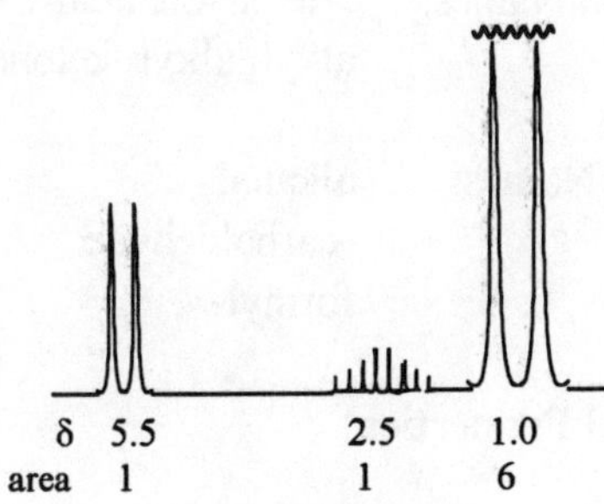

e.

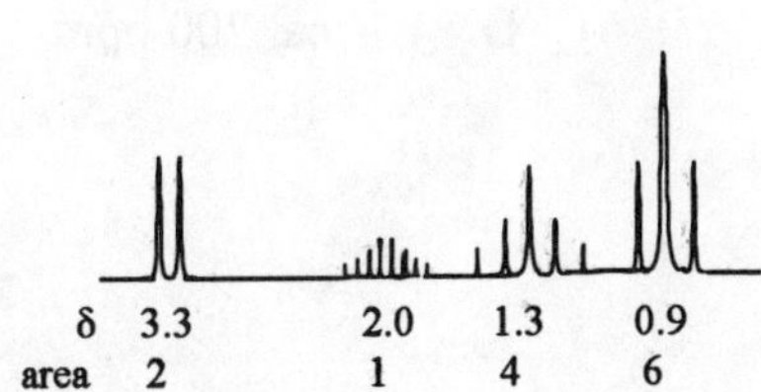

f.

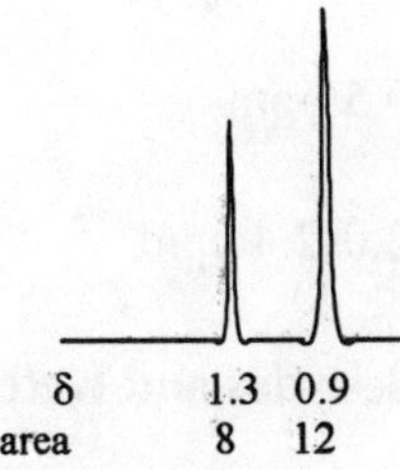

S3. a. 25°C: 4; −100°C: 6 b. 25°C: 2; −100°C: 3
c. 25°C: 7; −100°C: 7 (Too little of the other chair conformation is present at equilibrium to be seen.)
d. 25°C and −100°C: 3 e. 25°C: 7; −100°C: 14

S4. (axial-Cl chlorocyclohexane ⇌ equatorial-Cl chlorocyclohexane)

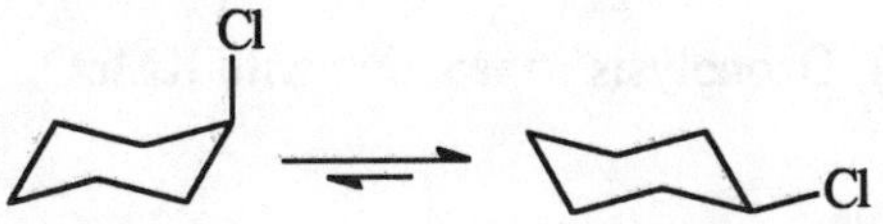

$$\Delta G = -0.5 \text{ kcal mole}^{-1} = -RT \ln K$$

$$K = \frac{[equatorial\text{-}Cl]}{[axial\text{-}Cl]} = 2.3$$

relative percent: $\frac{1}{3.3} = 30\%$; $\frac{2.3}{3.3} = 70\%$; $\delta_{average} = 0.3 \times 4.4 + 0.7 \times 3.7 = 3.9$ ppm

S5. In 1,1-diphenylpropane, the C_2-C_3 bond has two favored staggered conformations that are in rapid equilibrium, thus averaging J_{ab} to the usual ~7 Hz value for acyclic hydrocarbons. In the 2-methyl substituted derivative, only one staggered conformation is favored, with a 180 dihedral angle between $H_{a'}$ and $H_{b'}$.

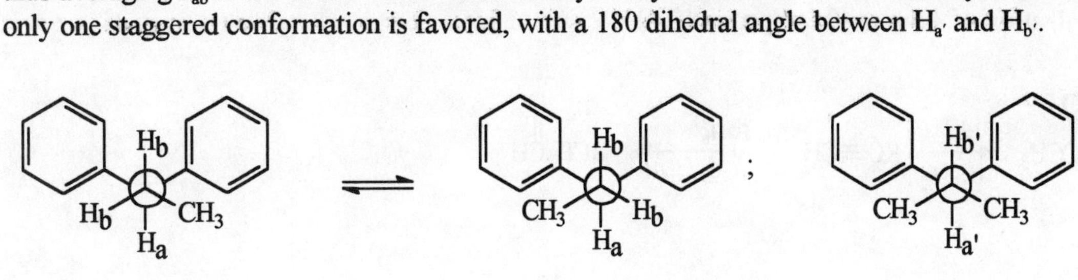

14. ALDEHYDES AND KETONES

14.A Chapter Outline and Important Terms Introduced

14.1 Structure

C=O carbonyl group (hybridization, polarization)

14.2 Nomenclature

Common Names: "-ic or -oic acid" → "-aldehyde" $\alpha,\beta,\gamma,...$
alkyl$_1$ alkyl$_2$ ketone

IUPAC Names: alkanal / alkanone
-carbaldehyde / oxo-
formyl- / (numbers, instead of $\alpha,\beta,\gamma,...$)

14.3 Physical Properties

14.4 Nuclear magnetic Resonance Spectra

C<u>H</u>=O ca. 9.5 ppm -<u>C</u>=O ca. 200 ppm

C<u>H</u>C=O ca. 2.0-2.4 ppm

14.5 Synthesis of Aldehydes and Ketones

Oxidation of Alcohols (see Section 10.6.E); Cr(VI) reagents:

$RCH_2COH \longrightarrow RCH{=}O$ (use CrO_3/HCl/pyridine (PCC) to avoid overoxidation)

$RCH(OH)R' \longrightarrow RC(=O)R'$ ($Na_2Cr_2O_7$ (or $K_2Cr_2O_7$)/H_2SO_4 (Jones reagent) is most common)

Oxidative Cleavage of Alkenes (see Section 11.6.E); Ozonolysis or reaction with $KMnO_4$:

$RCH{=}CR'R'' \longrightarrow RCH{=}O + O{=}CR'R''$

$RCH{=}CR'R'' \longrightarrow RCO_2H + O{=}CR'R''$

Hydration of Alkynes (see Section 12.6)

$$RC(=O)CH_3 \xleftarrow[H_2SO_4]{Hg^{2+}} RC\equiv CH \xrightarrow[2.\ H_2O_2,\ ^-OH]{1.\ B_2H_6} RCH_2CH{=}O$$

14.6 Addition of Oxygen and Nitrogen Nucleophiles

A. Carbonyl Hydrates: gem-Diols

^{18}O-exchange

$$R-\overset{^{16}O}{\overset{\|}{C}}-R' + H_2{}^{18}O \rightleftharpoons R'-\underset{^{18}OH}{\overset{^{16}OH}{C}}-R' \rightleftharpoons R-\overset{^{18}O}{\overset{\|}{C}}-R' + H_2{}^{16}O$$

acid-catalyzed vs. base-catalyzed mechanisms

inductive effects

B. Acetals and Ketals

hemiacetal and hemiketal formation (acid- or base-catalyzed)
acetal and ketal formation (acid-catalyzed only)
protecting group
enol ether formation and hydrolysis:

$$R-\underset{OR'}{\overset{OR'}{C}}-CH\langle \xrightarrow[\Delta]{} R-\overset{OR'}{C}=C\langle + R'OH \xrightarrow{H_3O^+} R-\overset{O}{\overset{\|}{C}}-CH\langle + R'OH$$

C. Imines and Related Compounds

imines = Schiff base
hemiaminal = carbinolamine

oximes, hydrazones, phenylhydrazones

14.7 Addition of Carbon Nucleophiles (carbon-carbon bond-forming reactions)

A. Addition of Organometallic Reagents: Synthesis of Alcohols

Grignard reagent
alkyllithium reagent

acetylide anion

B. Addition of HCN; cyanohydrin formation

C. Diastereomeric Transition States

formation of stereoisomers

attack from less sterically hindered side

D. The Wittig Reaction

phosphonium salts
ylide or phosphorane

oxaphosphetane
betaine

14.8 Oxidation and Reduction

A. Oxidation of Aldehydes and Ketones

autoxidation of aldehydes:

$$RCH{=}O \xrightarrow{\text{air}} RCO_2H$$

peroxycarboxylic acids
Baeyer-Villiger reaction
migratory aptitudes (H > 3° alkyl > 2° alkyl > phenyl > 1° alkyl > methyl)
lactones

B. Metal Hydride Reduction (addition of protons <u>and</u> electrons)

lithium aluminum hydride ($LiAlH_4$)
sodium borohydride ($NaBH_4$)
diborane (B_2H_6)

C. Catalytic Hydrogenation

D. Deoxygenation Reactions

Wolff-Kishner reduction
Clemmensen reduction

E. Enzymatic Reduction

enantiomeric excess (selectivity)

14.B Important Reactions Introduced

Hemiacetal (hemiketal) and acetal (ketal) formation (14.6.B)

Equation:

$$R{-}\overset{\overset{\displaystyle O}{\|}}{C}{-}R' \underset{}{\overset{R''OH}{\rightleftharpoons}} R{-}\overset{\overset{\displaystyle OH}{|}}{\underset{\underset{\displaystyle OR''}{|}}{C}}{-}R' \overset{R''OH}{\rightleftharpoons} R{-}\overset{\overset{\displaystyle OR''}{|}}{\underset{\underset{\displaystyle OR''}{|}}{C}}{-}R' + H_2O$$

Generality: R, R′ = H, alkyl, or aryl; R" = alkyl
Key features: first step acid- or base-catalyzed, reversible
second step occurs **only** with acid catalysis
equilibrium to acetal driven by removal of water
acetals useful as protecting groups

Formation of imines and derivatives (14.6.C)

Equation:

$$R{-}\overset{\overset{\displaystyle O}{\|}}{C}{-}R' + H_2NY \rightleftharpoons R{-}\overset{\overset{\displaystyle NY}{\|}}{C}{-}R' + H_2O$$

Generality: Y = H, alkyl, aryl, OH (product called an oxime), NH_2 (product called a hydrazone), NHPh (product called a phenylhydrazone)

Key features: reaction occurs via hemiaminal, usually with mild acid catalysis
equilibrium favors carbonyl compound if Y = H, alkyl
equilibrium favors imine if Y = aryl, OH, NH_2 or NHPh.

Addition of organometallic reagents to ketones and aldehydes (14.7.A)

Equation:

$$R-\overset{O}{\overset{\|}{C}}-R' + R''M \longrightarrow R-\underset{R'}{\underset{|}{\overset{O^- M^+}{\overset{|}{C}}}}-R'' \xrightarrow[\text{workup}]{H_2O} R-\underset{R'}{\underset{|}{\overset{OH}{\overset{|}{C}}}}-R''$$

Generality: M = MgX, Li

Key features: very important carbon-carbon bond-forming reaction
equation often written without specifically indicating aqueous workup

Cyanohydrin formation (14.7.B)

Equation:

$$R-\overset{O}{\overset{\|}{C}}-R' + HCN \longrightarrow R-\underset{CN}{\underset{|}{\overset{OH}{\overset{|}{C}}}}-R'$$

Generality: R, R′ = alkyl, aryl, H

Key features: addition catalyzed by cyanide salt
reaction is reversed in base

Wittig reaction (14.7.D)

Equation:

$$Ph_3P^+CHR_2\ X^- \xrightarrow[\text{(or other strong base)}]{\text{BuLi}} Ph_3P{=}CR_2 \xrightarrow{R'-\overset{O}{\overset{\|}{C}}-R'} R_2C{=}CR'_2 + Ph_3P{=}O$$

Generality: R, R′= various combinations of alkyl, aryl, or hydrogen

Key features: important reaction for forming carbon-carbon **double** bonds
reaction proceeds via oxaphosphetane
$R'_2C{=}O$ must be unhindered; i.e., mono-, di- and tri-substituted alkenes can be synthesized; tetra not.

Oxidation of aldehydes (14.8.A)

Equation:

$$R-\overset{O}{\overset{\|}{C}}-H \xrightarrow{[Ox]} R-\overset{O}{\overset{\|}{C}}-OH$$

Generality: [Ox] = air (O_2), Ag_2O, H_2O_2, $KMnO_4$, aq. CrO_3, RCO_3H, etc.

Key features: usually a reaction to be avoided, i.e., it often occurs as a side reaction in preparation of aldehydes or on their storage.

Baeyer-Villiger oxidation (14.8.A)

Equation:

$$R-\overset{O}{\overset{\|}{C}}-R' + R''CO_3H \longrightarrow R-\overset{O}{\overset{\|}{C}}-OR' + R''CO_2H$$

Generality: R groups = various combinations of alkyl, aryl, or H

Key features: migratory aptitude (i.e. R′ vs. R): H > 3° alkyl > 2° alkyl > phenyl > 1° alkyl > CH_3

Reduction of aldehydes and ketones (14.9.B and C)

Equation:

$$R-\overset{O}{\overset{\|}{C}}-R' \xrightarrow{[H]} R-\underset{H}{\overset{OH}{C}}-R'$$

Generality: [H] = $LiAlH_4$, $NaBH_4$; B_2H_6; H_2/catalyst (difficult; catalyst = Pd, Pt, etc.)

Key features: important preparation of alcohols

Deoxygenation of aldehydes and ketones (14.8.D)

Equation:

$$R\overset{O}{\overset{\|}{C}}R' \xrightarrow{[\text{deoxygenation}]} RCH_2R'$$

Generality: [deoxygenation] = KOH/H_2NNH_2/diethyleneglycol/240°C (Wolff-Kishner reduction) or Zn(Hg)/HCl (Clemmensen reduction)

Key features: for Wolff-Kishner reduction, aldehyde or ketone must be stable to strong base; for Clemmensen reduction, it must be stable to strong acid

14.C. Important Concepts and Hints

Reactions of Double Bonds: C=O vs. C=C

If you are developing an intuitive understanding of organic chemistry, the chapter on aldehydes and ketones should give you a view of its logical foundations as well as it complexity. Intuition is difficult to teach, but as you study this chapter, try to apply the Question-and-Answer outline suggested in Section 9.C of this Study Guide. It is hoped that you will not only see similarities among many of the reactions discussed in this chapter, but also recognize analogies between these reactions and those discussed earlier, in the chapter on alkenes for instance. As an illustration of the sort of analogy we want you to see, study the examples below, in which we've juxtaposed reaction mechanisms from carbonyl and alkene chemistry. There are great differences in the conditions under which these reactions occur, and in their rates and the position of equilibrium, but there is a lot of similarity among the mechanisms.

$$(H_3C)_2C{=}CH_2 \;\xrightleftharpoons{H^+}\; (H_3C)_2\overset{+}{C}{-}CH_3 \;\xrightleftharpoons{:OH_2}\; H_3C{-}C(\overset{+}{O}H_2)(CH_3){-}CH_3 \;\xrightleftharpoons{-H^+}\; H_3C{-}C(OH)(CH_3){-}CH_3$$

$$(H_3C)_2C{=}\ddot{O}: \;\xrightleftharpoons{H^+}\; \left[(H_3C)_2C{=}\overset{+}{O}H \longleftrightarrow (H_3C)_2\overset{+}{C}{-}OH\right] \;\xrightleftharpoons{H_2O}\; H_3C{-}C(\overset{+}{O}H_2)(OH){-}CH_3 \;\xrightleftharpoons{-H^+}\; H_3C{-}C(OH)(OH){-}CH_3$$

$$CH_3CH{=}O + HCN \rightleftharpoons \left[CH_3CH{=}\overset{+}{O}H \longleftrightarrow CH_3\overset{+}{C}HOH\right] \xrightleftharpoons{^-CN} CH_3CH(CN)OH$$

$$CH_3CH(OH)CH_3 \xrightleftharpoons{H^+} CH_3CH(\overset{+}{O}H_2)CH_3 \xrightleftharpoons{-H_2O} CH_3\overset{+}{C}HCH_3 \xrightleftharpoons{-H^+} CH_3CH{=}CH_2$$

$$CH_3CH(OH)NHNH_2 \xrightleftharpoons{H^+} CH_3CH(\overset{+}{O}H_2)NHNH_2 \xrightleftharpoons{-H_2O} \left[CH_3\overset{+}{C}HNHNH_2 \longleftrightarrow CH_3CH{=}\overset{+}{N}HNH_2\right] \xrightleftharpoons{-H^+} CH_3CH{=}NNH_2$$

$$CH_3CH_2CH{=}CH_2 \xrightleftharpoons{H^+} CH_3CH_2\overset{+}{C}HCH_3 \xrightleftharpoons{-H^+} CH_3CH{=}CHCH_3$$

$$CH_3CH_2CH{=}O \xrightleftharpoons{H^+} \left[CH_3CH_2CH{=}\overset{+}{O}H \longleftrightarrow CH_3CH_2\overset{+}{C}HOH\right] \xrightleftharpoons{-H^+} CH_3CH{=}CHOH$$

Deprotonation vs. Oxidation; Protonation vs. Reduction

When you convert cyclohexanol to cyclohexanone, all you do is remove two protons, right? Wrong! You also have to remove two electrons, which is why an oxidizing agent (electron acceptor) is required, instead of just a base. Similarly, to convert cyclohexanone back to cyclohexanol, two electrons need to be supplied by a reducing agent.

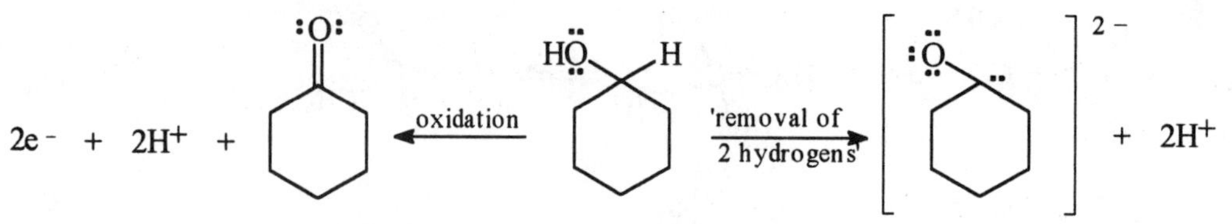

$2e^- + 2H^+ +$ [cyclohexanone] ← oxidation — [cyclohexanol] — 'removal of 2 hydrogens' → $[\ldots]^{2-} + 2H^+$

The simple addition of two protons by a strong acid (even if possible) would not give cyclohexanol:

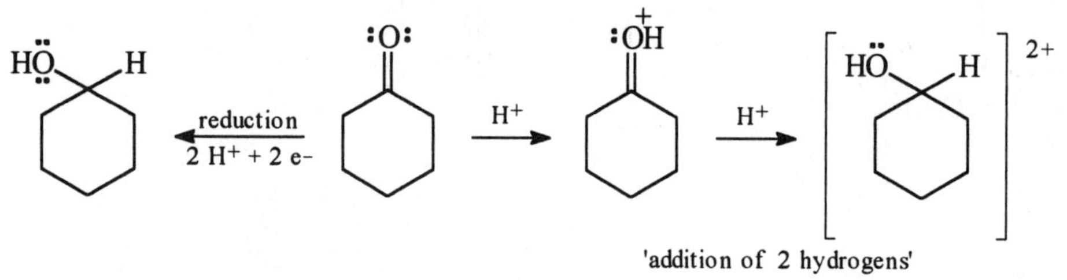

[cyclohexanol] ← reduction, $2H^+ + 2e^-$ — [cyclohexanone] —H^+→ [protonated cyclohexanone, :OH⁺] —H^+→ $[\text{HO, H}]^{2+}$

'addition of 2 hydrogens'

Neither oxidation nor reduction is accomplished with base or acid alone.

14.D. Answers to Exercises

14.1 a. 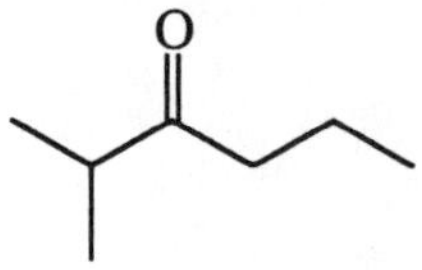 b.

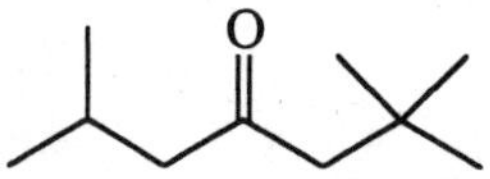

c. 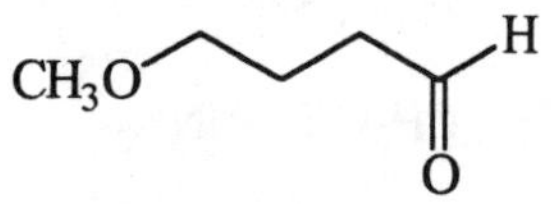 CH_3O ... H, O d.

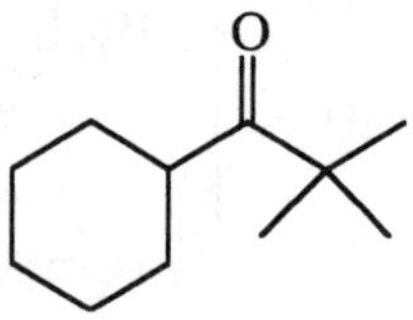

14.2 a. 2-methyl-3-hexanone b. 2,2,6-trimethyl-4-heptanone
c. 4-methoxybutanal d. 1-cyclohexyl-2,2-dimethyl-1-propanone

14.3 $CDCl_3$ used as the solvent. The carbon is split into a 1:1:1 triplet by the deuterium, I = 1.

14.4 3-methylbutanal: $(CH_3)_2CHCH_2CHO$
The small value of J is characteristic for aldehyde protons in aliphatic compounds. Four signals for 5 carbons indicate 2 equivalent carbons.

14.5 a. OH → PCC → O, H ← 1. B_2H_6 2. H_2O_2 , OH^-

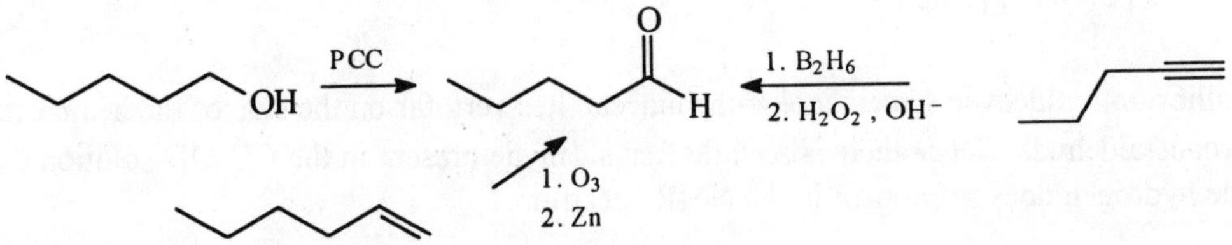

1. O_3 2. Zn

b. OH → $K_2Cr_2O_7$ H^+ → O ← 1. O_3 2. Zn

H_2SO_4 $HgSO_4$

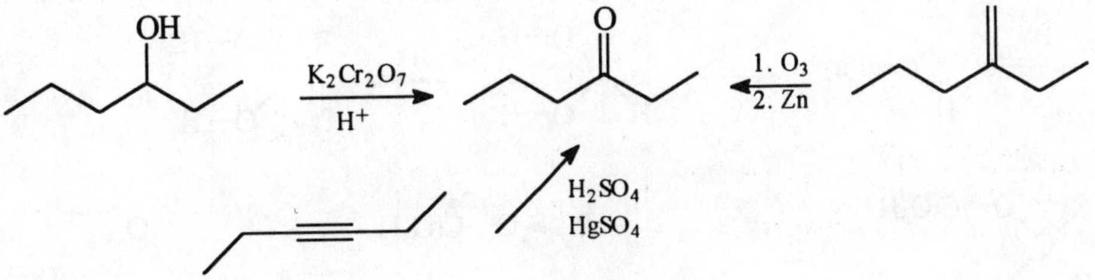

NOTE: 2-Hexyne would give a mixture of isomers.

c. OH → $K_2Cr_2O_7$ H^+ → O ← 1. O_3 2. Zn

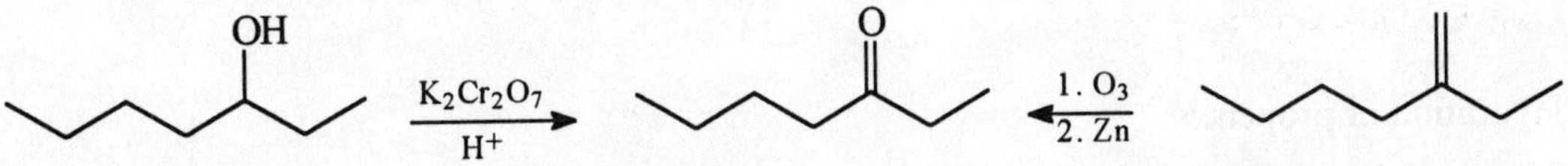

NOTE: Hydration of any appropriate alkyne would give a mixture of isomers.

d. 1. O_3 2. Zn → O, H ← PCC OH

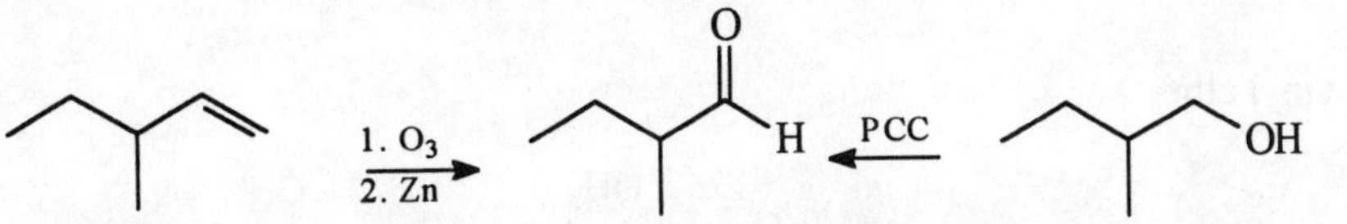

14.6 $$K = \frac{[C(OH)_2]}{[C{=}O]\ [H_2O]}$$ where $[C(OH)_2]$ = concentration of hydrate; $[C{=}O]$ = concentration of carbonyl

$[H_2O] = 55.5$; $[C{=}O] + [C(OH)_2] = 1$ M $[C{=}O] = 1 - [C(OH)_2]$

$$K = \frac{[C(OH)_2]}{(1-[C(OH)_2])\ x\ 55.5}$$

$K \times 55.5 = [(K \times 55.5) + 1] \times [C(OH)_2]$

For formaldehyde: K = 18; $[C(OH)_2]$ = 0.999 M; $[C{=}O]$ = 0.001 M
ratio $[C(OH)_2]/[C{=}O]$ = 999

For acetaldehyde: K = 0.01; $[C(OH)_2]$ = 0.36 M; $[C{=}O]$ = 0.64 M
ratio $[C(OH)_2]/[C{=}O]$ = 0.56

For acetone: K = 4 x 10^{-5}; $[C(OH)_2]$ = 2.2 x 10^{-3}; [C=O] = 1.0 M
ratio $[C(OH)_2]/[C{=}O]$ = 2.2 x 10^{-3}

14.7 The equilibrium: aldehyde + methanol ⇌ hemiacetal lies very far on the side of the hemiacetal in the case of chloroacetaldehyde. Hence there is so little free aldehyde present in the CD_3OD solution that the aldehyde hydrogen does not appear in the NMR spectrum.

14.8

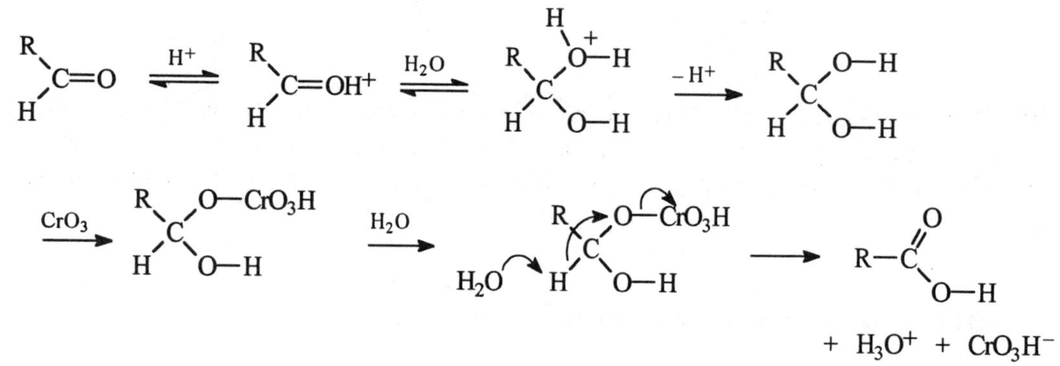

$$RHC{=}O \overset{H^+}{\rightleftharpoons} RHC{=}OH^+ \overset{H_2O}{\rightleftharpoons} RHC(OH)(O^+H_2) \xrightarrow{-H^+} RHC(OH)_2$$

$$\xrightarrow{CrO_3} RHC(OH)(O{-}CrO_3H) \xrightarrow{H_2O} [H_2O \text{ removes H}; O{-}CrO_3H \text{ leaves}] \longrightarrow R{-}C(=O){-}O{-}H + H_3O^+ + CrO_3H^-$$

a. R = H b. R = CH_3 c. R = $CHCl_2$ d. R = CF_3

14.9 Hydration of propene:

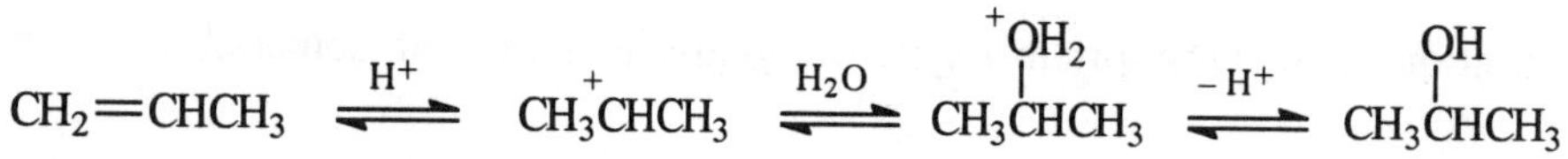

$$CH_2{=}CHCH_3 \overset{H^+}{\rightleftharpoons} CH_3\overset{+}{C}HCH_3 \overset{H_2O}{\rightleftharpoons} CH_3CH(\overset{+}{O}H_2)CH_3 \overset{-H^+}{\rightleftharpoons} CH_3CH(OH)CH_3$$

Hydrolysis of ethyl vinyl ether:

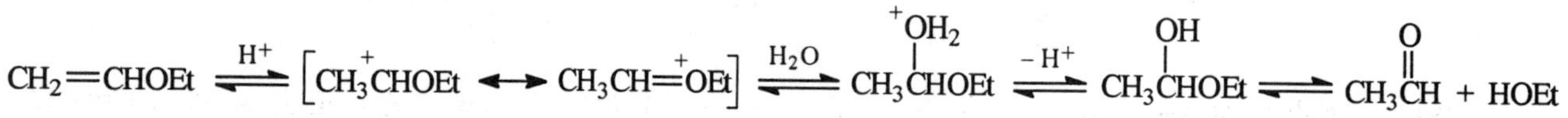

$$CH_2{=}CHOEt \overset{H^+}{\rightleftharpoons} [CH_3\overset{+}{C}HOEt \longleftrightarrow CH_3CH{=}\overset{+}{O}Et] \overset{H_2O}{\rightleftharpoons} CH_3CH(\overset{+}{O}H_2)OEt \overset{-H^+}{\rightleftharpoons} CH_3CH(OH)OEt \rightleftharpoons CH_3CH{=}O + HOEt$$

In both the hydration of an alkene and the acid-catalyzed hydrolysis of an enol ether, the hard step is formation of the initial carbocation intermediate. In the case of an enol ether, the carbocation is stabilized by resonance with the lone pair electrons of the oxygen substituent, therefore it is formed more easily.

14.10 The cyclic acetal will be produced. The ethyl group from the aldehyde and the substituent methyl from the diol can be either *cis* or *trans*:

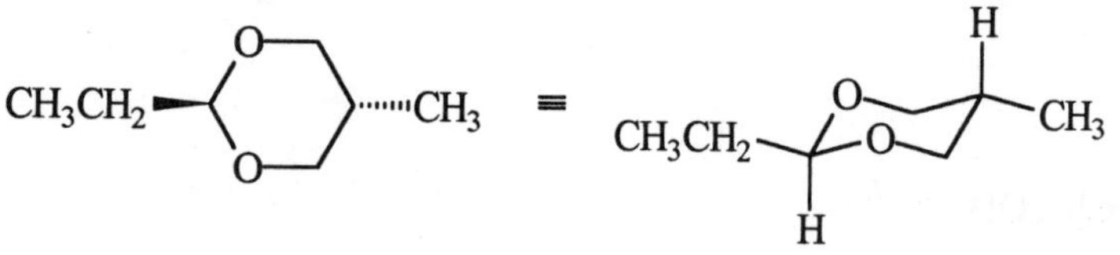

CH_3CH_2–(1,3-dioxane)····CH_3 ≡ CH_3CH_2, H / H, CH_3 (chair) *trans* isomer

CH_3CH_2–(1,3-dioxane)–CH_3 ≡ CH_3CH_2, H / CH_3, H (chair) *cis* isomer.

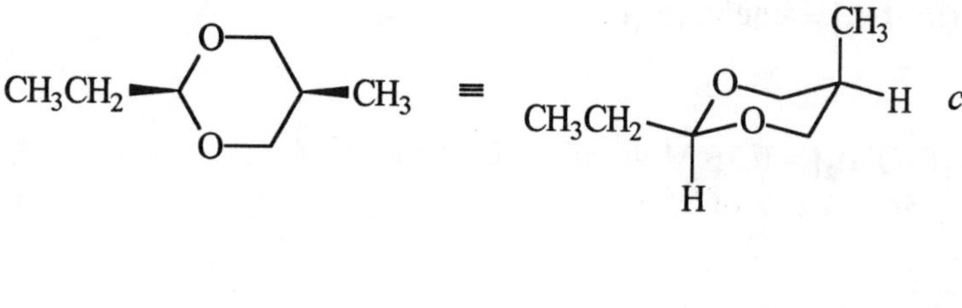

Of the two, the *trans* isomer is expected to be the major product since both substituents are able to assume an equatorial orientation.

14.11 In the all equatorial isomer, all of the methyls are the same, and all of the CH's are the same: two signals in the CMR spectrum.

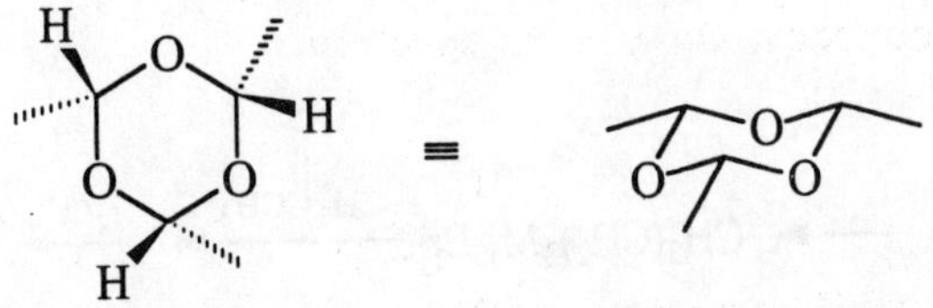

In the other isomer, two of the methyls will be *cis* to each other and different from the third which is *trans* to them; likewise the CH group that is between the two *cis* methyls is different from those that are between *trans* methyls; different types of methyl and two different types of CH groups: four signals in the CMR spectrum.

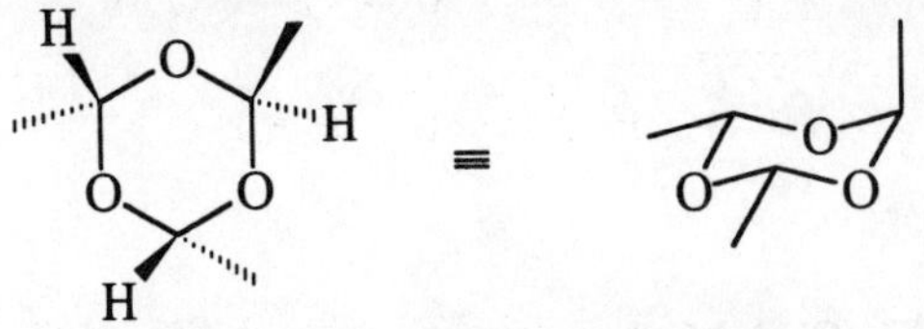

14.12 Because of the sp^2-hybridization of the nitrogen and the aldehyde carbon, there exist *cis* and *trans* isomers of the oxime product:

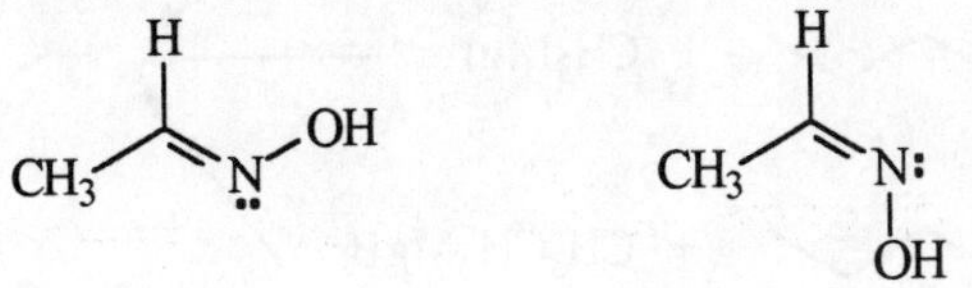

14.13 *rate-limiting step*

$$R_2C{=}O + H_2NR' \underset{}{\overset{K_1}{\rightleftharpoons}} R_2C(O^-)(^+NH_2R') \overset{K_2}{\rightleftharpoons} R_2C(OH)(NHR') \overset{K_3}{\rightleftharpoons} R_2C(^+OH_2)(NHR') \overset{k_1}{\rightleftharpoons} R_2C{=}\overset{+}{N}HR' \rightleftharpoons R_2C{=}NR'$$

ketone + amine | $NH_2^+O^-$ | NH-OH | $NH\text{-}OH_2^+$ | immonium ion | product

Rate = k_1 x $[NH\text{-}OH_2^+]$

$[NH\text{-}OH_2^+] = K_3$ x $[H^+]$ x $[NH\text{-}OH]$

$[NH\text{-}OH] = K_2$ x $[NH_2^+O^-]$

$[NH_2^+O^-] = K_1$ x [ketone] x [amine]

Therefore, Rate = k_1 x K_3 x K_2 x K_1 x $[H^+]$ x [ketone] x [amine]

If the first step were rate-limiting:

Rate = k' x [ketone] x [amine], independent of $[H^+]$

14.14 a. $CH_3(CH_2)_3Br$ + Mg ⟶ $CH_3(CH_2)_3MgBr$ $\xrightarrow{CH_2=O}$ $\xrightarrow{H_2O}$ HO

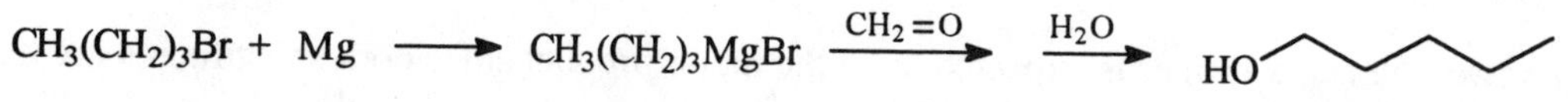

b. $CH_3(CH_2)_3Br$ + Mg ⟶ $CH_3(CH_2)_3MgBr$ $\xrightarrow{CH_3CH_2CH(=O)}$ $\xrightarrow{H_2O}$ OH

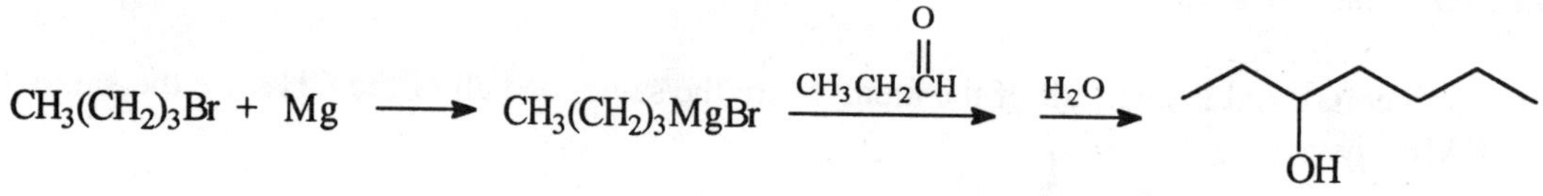

c. $CH_3(CH_2)_3Br$ + Mg ⟶ $CH_3(CH_2)_3MgBr$ $\xrightarrow{CH_3CCH_3 (=O)}$ $\xrightarrow{H_2O}$ OH

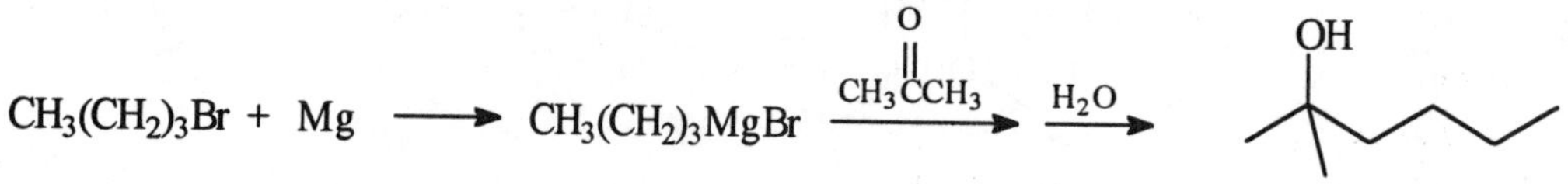

d. $CH_3(CH_2)_3MgBr$ from a. $\xrightarrow{H_2C\text{—}CH_2 \text{ (O)}}$ $\xrightarrow{H_2O}$ OH

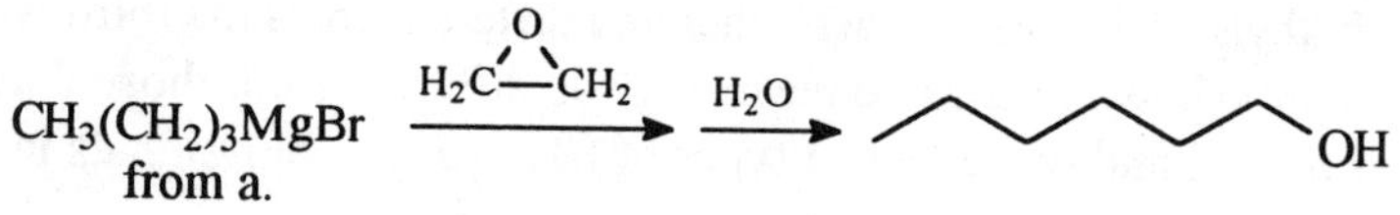

14.15

a. + $CH_3(CH_2)_5MgBr$

+ CH_3MgI ⟶ OH

+ CH_3CH_2MgBr

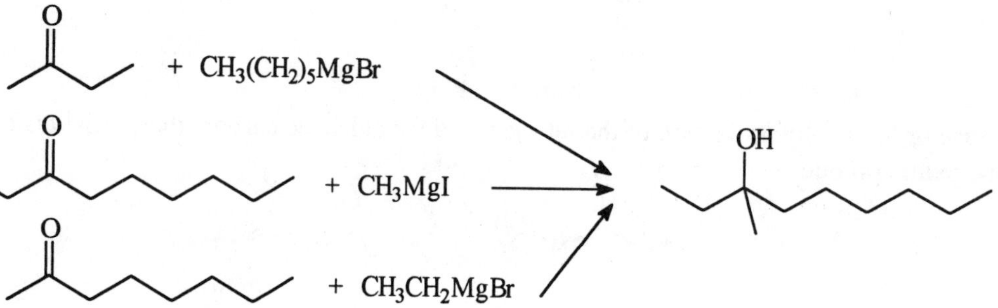

b. + MgBr

+ CH_3MgI ⟶ OH

+ CH_3CH_2MgBr

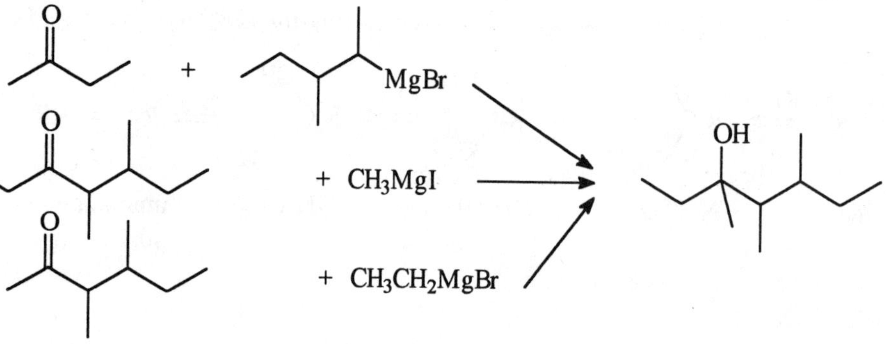

c. + $CH_3(CH_2)_3MgBr$

+ CH_3CH_2MgBr ⟶ OH

+ MgBr

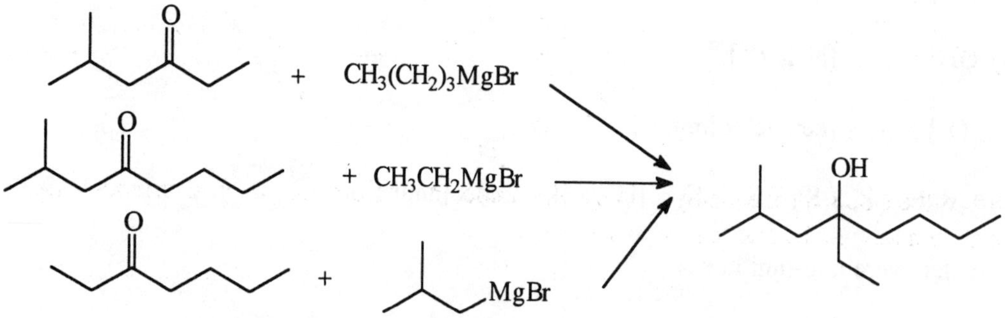

14.16 a. OH (1-ethynylcyclohexanol) $\xrightarrow[HgSO_4]{H_2SO_4}$ HO, O (1-acetylcyclohexanol)

b. OH $\xrightarrow[\text{Lindlar cat.}]{H_2}$ HO (1-vinylcyclohexanol)

c. OH $\xrightarrow[2.\ H_2O_2\ ,\ OH^-]{1.\ B_2H_6}$ HO, O, H

14.17 a. O $\xrightarrow{HCN}$ HO, CN $\xrightarrow[2.\ HCl]{1.\ H_2\ /\ PtO_2}$ HO, NH_2

b. O + $CH\equiv C^-$ ⟶ HO— $\xrightarrow[\text{Lindlar}]{H_2}$ HO

c. O + $CH_3C\equiv C^-$ ⟶ —OH $\xrightarrow[\text{Lindlar cat.}]{H_2}$ HO

14.18 CH_3, H, O ⇌ CH_3, H, HÖ, H, Br—Br ⟶ CH_3, H, H, O, Br + HBr

Attack from the side with the methyl is sterically hindered. The product, *trans*-2-bromo-3-methylcyclobutanone is favored both kinetically and thermodynamically.

14.19

a.

$$Ph_3P + (CH_3)_2CHI \longrightarrow Ph_3P^+CH(CH_3)_2I^- \xrightarrow{BuLi} Ph_3P{=}C(CH_3)_2 + CH_3CH_2\overset{O}{\overset{\|}{C}}H \longrightarrow (CH_3)_2C{=}CHCH_2CH_3$$

b.

$$Ph_3P + CH_3(CH_2)_3Br \longrightarrow \xrightarrow{BuLi} Ph_3P{=}CH(CH_2)_2CH_3 + (CH_3)_3C\overset{O}{\overset{\|}{C}}H \longrightarrow (CH_3)_3CCH{=}CH(CH_2)_2CH_3$$

c.

$$Ph_3P + CH_3CH_2I \longrightarrow \xrightarrow{BuLi} Ph_3P{=}CHCH_3 + CH_3\overset{O}{\overset{\|}{C}}CH_2CH_2CH_3 \longrightarrow CH_3CH{=}C(CH_3)CH_2CH_2CH_3$$

14.20

a. 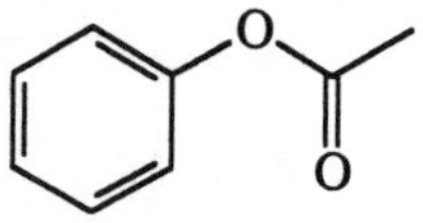 b. 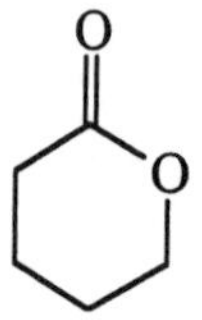 c. 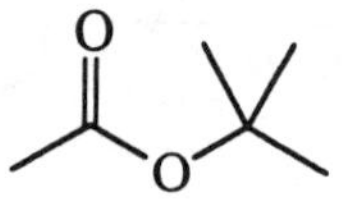 d.

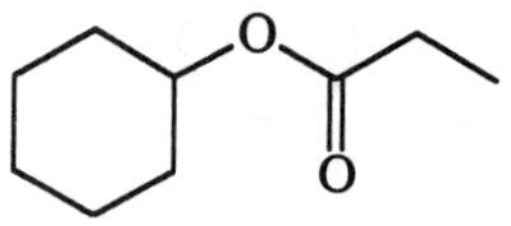

14.21

a. $CH{\equiv}C(CH_2)_5CH_3 \xrightarrow[\text{Lindlar cat.}]{H_2} CH_2{=}CH(CH_2)_5CH_3 \xrightarrow{H_2SO_{4(aq)}} CH_3CHOH(CH_2)_5CH_3$

or

$CH{\equiv}C(CH_2)_5CH_3 \xrightarrow[HgSO_4]{H_2SO_4} CH_3COCH(CH_2)_5CH_3 \xrightarrow{NaBH_4} CH_3CHOH(CH_2)_5CH_3$

b.

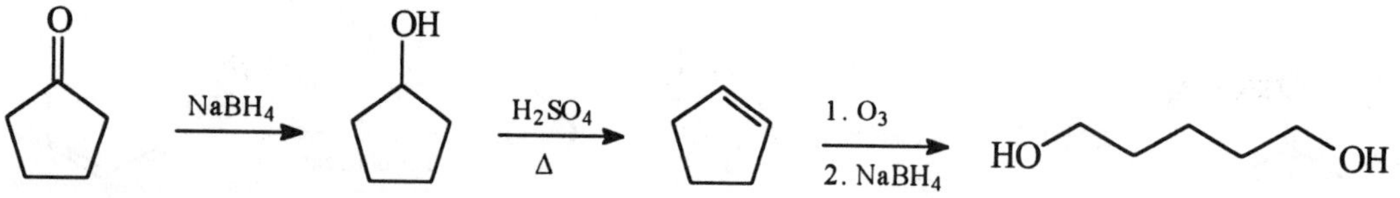

O NaBH$_4$ OH H_2SO_4 Δ 1. O_3 2. $NaBH_4$ HO OH

14.22 a. Clemmenson reduction (1° bromide would not survive the conditions of the Wolff-Kishner reduction).

b. Wolff-Kishner reduction (3° alcohol would not survive the Clemmenson reduction conditions).

c. Neither the Wolff-Kishner reduction nor the Clemmenson reduction would be appropriate for deoxygenation of the carbonyl group of c; the epoxide group would be destroyed by the strongly basic or acidic conditions.

d. Both methods would work.

E. Answers and Explanations for Problems

1. a. methyl ethyl ketone b. isopropyl *n*-propyl ketone c. methyl *t*-butyl ketone
   d. ethyl neopentyl ketone e. methyl cyclopentyl ketone f. isobutyraldehyde
   g. α-bromopropionaldehyde h. β-methoxybutyraldehyde i. ethyl vinyl ketone
   j. cyclopropylacetone

2. a. 2-butanone b. 2-methyl-3-hexanone c. 3,3-dimethyl-2-butanone
   d. 5,5-dimethyl-3-hexanone e. 1-cyclopentylethanone f. 2-methylpropanal
   g. 2-bromopropanal h. 3-methoxybutanal i. 1-penten-3-one
   j. 1-cyclopropyl-2-propanone

3. a. $CH_3\overset{O}{\overset{\|}{C}}CH_2CH(CH_3)_2$ b. $CH_3CH_2CH(OC_2H_5)_2$ c. $CH_3CHClCH_2CHO$ d. 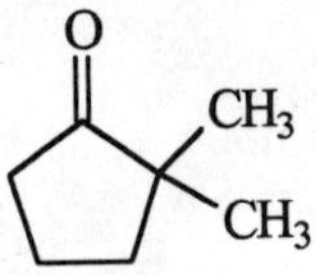 O, CH₃, CH₃

   e. 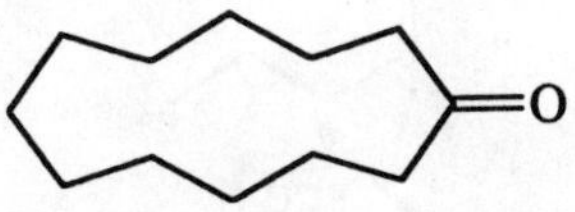 =O f. 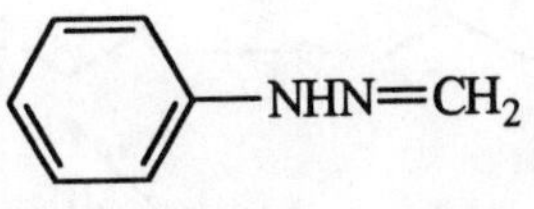 —NHN=CH₂ g. 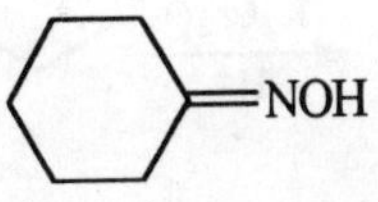 =NOH h. $(CH_3)_2C{=}NNH_2$

4. a. 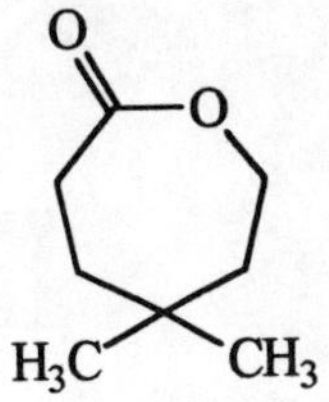 O, O, H₃C, CH₃ b. 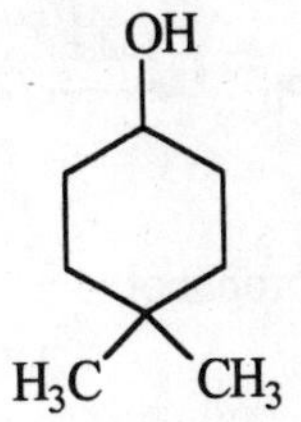 OH, H₃C, CH₃ c. 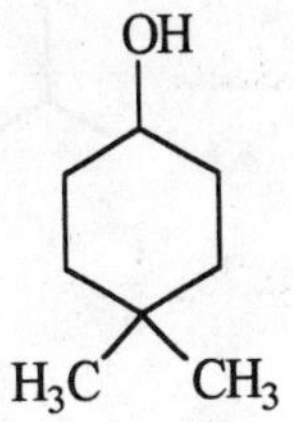 OH, H₃C, CH₃

   d. 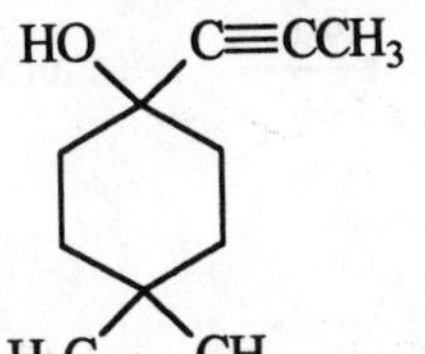 HO, C≡CCH₃, H₃C, CH₃ e. 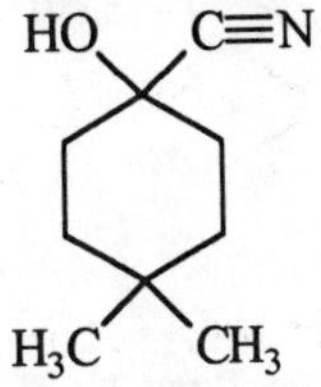 HO, C≡N, H₃C, CH₃ f. 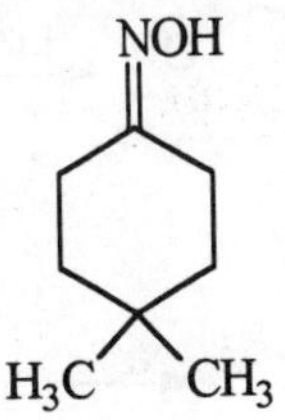 NOH, H₃C, CH₃

   g. 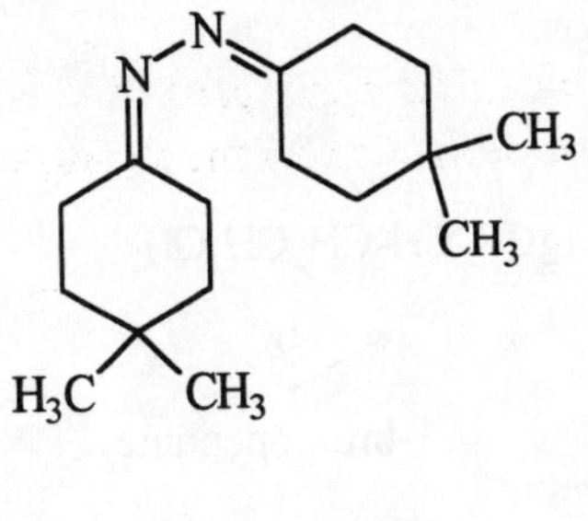 N, N, CH₃, CH₃, H₃C, CH₃ h,i. H₃C, CH₃ j. 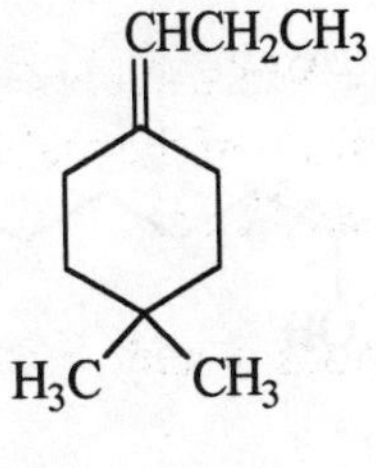 CHCH₂CH₃, H₃C, CH₃

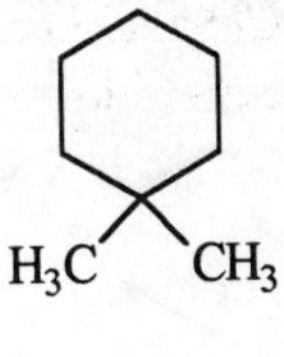

5. a. 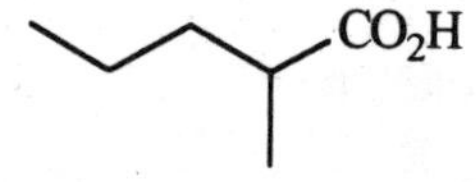 CO_2H b. 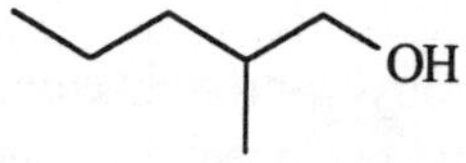 OH c. 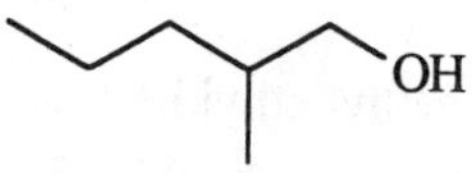 OH

d. 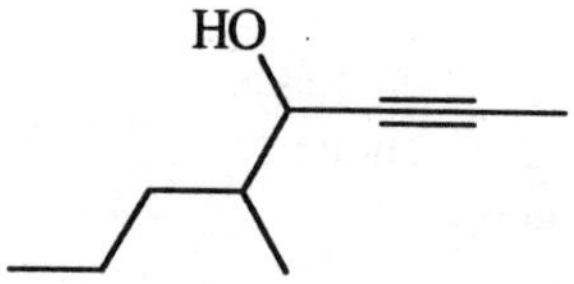 HO e. 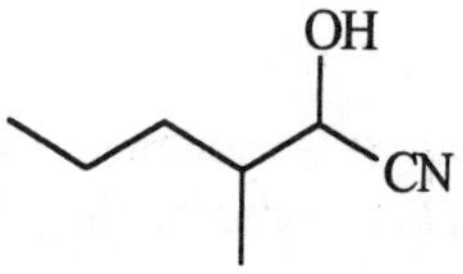 OH CN f. 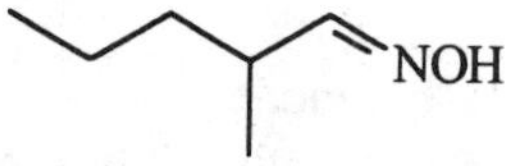 NOH

g. 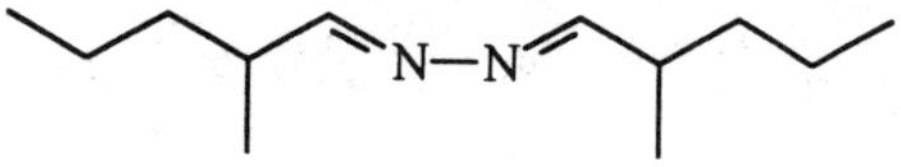 N—N h,i. 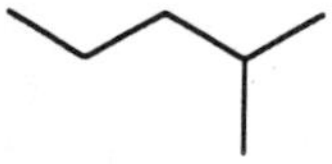 j.

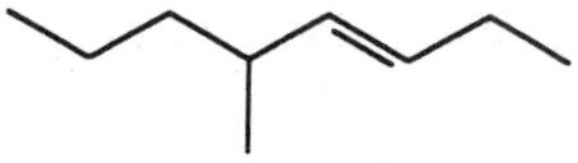

6. a. 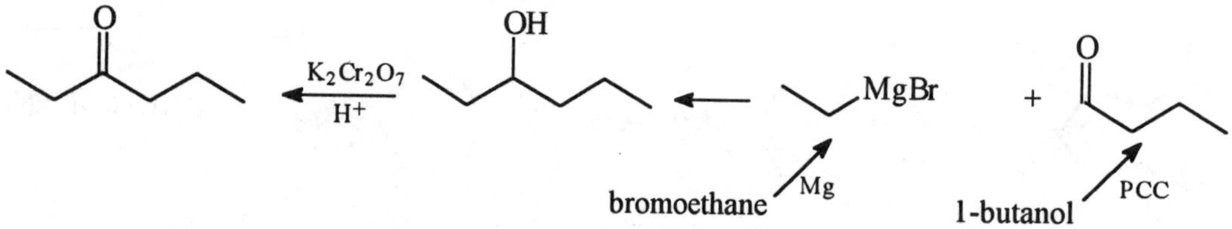 O $\xleftarrow[H^+]{K_2Cr_2O_7}$ OH ← MgBr + O

bromoethane $\xrightarrow{Mg}$ 1-butanol $\xrightarrow{PCC}$

b. 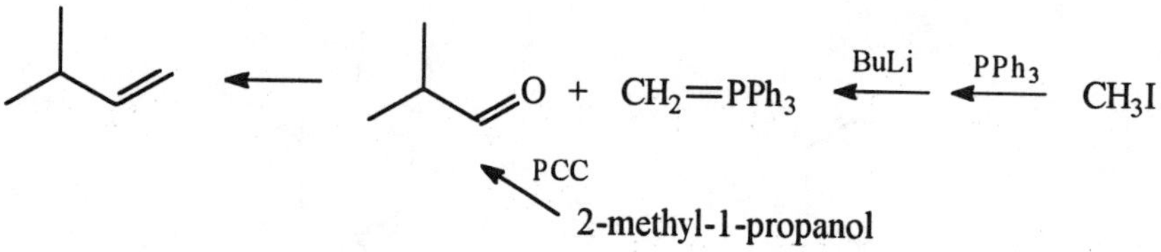 ← O + $CH_2{=}PPh_3$ $\xleftarrow{BuLi}$ $\xleftarrow{PPh_3}$ CH_3I

PCC 2-methyl-1-propanol

c. 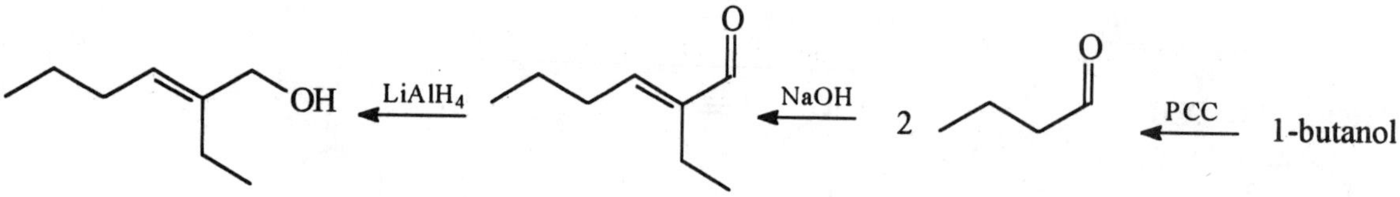 OH $\xleftarrow{LiAlH_4}$ O $\xleftarrow{NaOH}$ 2 O $\xleftarrow{PCC}$ 1-butanol

d. 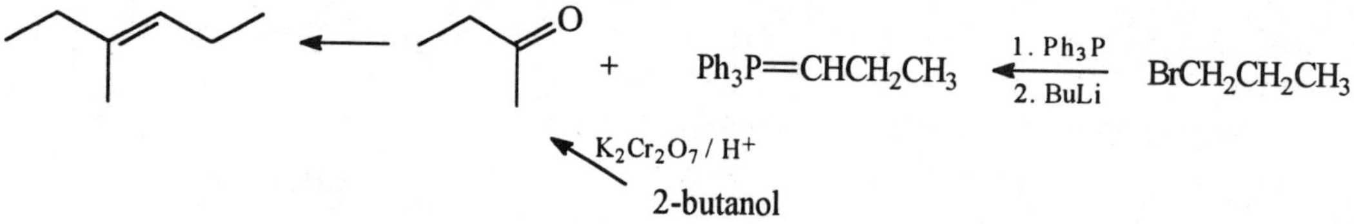 ← O + $Ph_3P{=}CHCH_2CH_3$ $\xleftarrow[2.\ BuLi]{1.\ Ph_3P}$ $BrCH_2CH_2CH_3$

$K_2Cr_2O_7$ / H^+ 2-butanol

e. 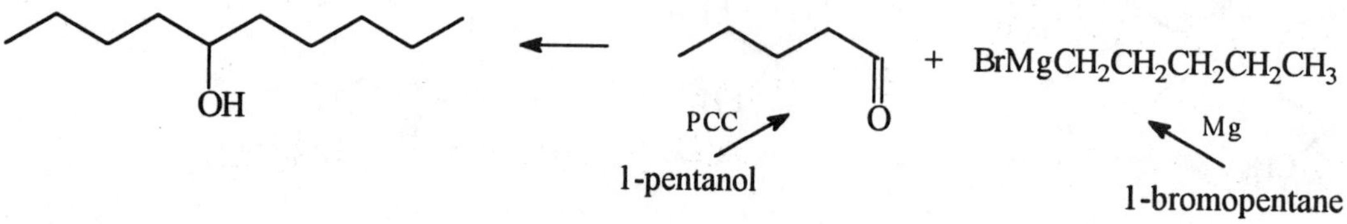 OH ← O + $BrMgCH_2CH_2CH_2CH_2CH_3$

PCC 1-pentanol Mg 1-bromopentane

$$CH_3(CH_2)_2Br + Mg \longrightarrow CH_3(CH_2)_2MgBr \xrightarrow{\text{(ethylene oxide)}} CH_3(CH_2)_4OH \xrightarrow{PBr_3} CH_3(CH_2)_4Br$$

f.

OH ← (ethylene oxide) MgBr ←Mg— Br ←HBr— OH

$CH_3\overset{OH}{C}HCH_2CH_2CH_3 \xrightarrow[H^+]{K_2Cr_2O_7}$ (2-pentanone) + $CH_3MgI \xleftarrow{Mg} CH_3I$

$CH_3(CH_2)_2Br + Mg \longrightarrow CH_3(CH_2)_2MgBr \xrightarrow{CH_3CHO} CH_3(CH_2)_2CHOHCH_3$

g. OH ← (acetone) + $ClMgCH_2CH_2CH_2CH_3 \xleftarrow{Mg} ClCH_2CH_2CH_2CH_3$

2-propanol $\xrightarrow{K_2Cr_2O_7 / H^+}$

h. OH $\xleftarrow{CH_3MgI}$ O $\xleftarrow[H^+]{K_2Cr_2O_7}$ OH

Br $\xrightarrow[2.\ CH_2=O]{1.\ Mg}$ OH $\xrightarrow{PCC}$ O + $CH_3(CH_2)_3MgCl \xleftarrow{Mg} CH_3(CH_2)_3Cl$

7. a. $CH_3CH_2CH_2CH_2CHO \xrightarrow[\Delta \text{ or } H_2NNH_2\text{-}NaOH\ (HOCH_2CH_2)_2O]{Zn(Hg)\text{-}HCl} CH_3CH_2CH_2CH_2CH_3$

b. $CH_3(CH_2)_3CHO + HC{\equiv}C^-Na^+ \longrightarrow \xrightarrow{H^+} CH_3(CH_2)_3CHOHC{\equiv}CH$

c. $CH_3(CH_2)_3CHO + KCN \xrightarrow{H_2SO_4} CH_3(CH_2)_3CHOHCN$

d. $CH_3(CH_2)_3CHO \xrightarrow{Ag_2O} \xrightarrow{H_3O^+} CH_3CH_2CH_2CH_2CO_2H$

e. $CH_3(CH_2)_3CHO + CH_3\overset{MgBr}{C}H(CH_2)_2CH_3 \longrightarrow \xrightarrow{H_2O} CH_3(CH_2)_3\underset{OH}{C}H\overset{CH_3}{C}H(CH_2)_2CH_3 \xrightarrow{Al_2O_3,\ \Delta}$

$CH_3(CH_2)_4\overset{CH_3}{C}HCH_2CH_2CH_3 \xleftarrow[Pd]{H_2} CH_3(CH_2)_3CH{=}\overset{CH_3}{C}(CH_2)_2CH_3$

f. $CH_3CH_2CH_2CH_2CHO \xrightarrow{NaBH_4} \xrightarrow{PBr_3} CH_3(CH_2)_4Br$

g. $CH_3(CH_2)_3CHO \xrightarrow{NaBH_4} CH_3(CH_2)_3CH_2OH \xrightarrow[100°C]{H_2SO_4} (CH_3CH_2CH_2CH_2CH_2)_2O$

h. $CH_3CH_2CH_2CH_2CHO + (C_6H_5)_3P{=}CH_2 \longrightarrow CH_3CH_2CH_2CH_2CH{=}CH_2$

8. a. CHO $\xrightarrow{(CH_3)_3CLi}$ OH

b. O $\xrightarrow{Ph_3P{=}C(CH_3)_2}$

c. CHO $\xrightarrow{NaBH_4} \xrightarrow{PBr_3} \xrightarrow{Mg} \xrightarrow{CH_2{=}O} \xrightarrow{H_2O}$ OH

d. CHO $\xrightarrow{CH_3C{\equiv}CNa} \xrightarrow[\text{Lindlar cat.}]{H_2}$ OH

e. O $\xrightarrow[H_2SO_4]{KCN}$ OH CN $\xrightarrow[PtO_2,\ HCl]{H_2}$ OH NH_2

9.

a. $CH_3(CH_2)_3CH_2Br \xleftarrow{PBr_3} CH_3(CH_2)_3CH_2OH \xleftarrow{H_2C{=}O} CH_3(CH_2)_3MgCl \xleftarrow{Mg}$ 1-chlorobutane

b. O $\xleftarrow[H^+]{K_2Cr_2O_7}$ OH $\longleftarrow$ $CH_3CH_2\overset{O}{\overset{\|}{C}}H$ + $BrMgCH_2CH_2CH_3$

or

CH_3CH_2MgI + $H\overset{O}{\overset{\|}{C}}CH_2CH_2CH_3$

c.

(skeletal structures) $\xleftarrow{H_2SO_4,\ \Delta}$ OH $\xleftarrow{\text{n-PrMgCl}}$ O

$\uparrow K_2Cr_2O_7,\ H^+$

$CH_3CH_2CH_2Cl \xrightarrow{Mg} CH_3CH_2CH_2MgCl + CH_3CH_2CH_2\overset{O}{\overset{\|}{C}}H \longrightarrow$ OH

d.

Br $\xrightarrow{Mg}$ MgBr + $CH_3CH_2\overset{O}{\overset{\|}{C}}H \longrightarrow$ OH

$\swarrow K_2Cr_2O_7,\ H^+$

OH $\longleftarrow$ Li + O

$(CH_3)_2CHBr \nearrow$ 2 Li

10. a. $CH_3(CH_2)_2CD_2OH \xleftarrow{HCl} \xleftarrow{NaBD_4} CH_3(CH_2)_2CD{=}O \xleftarrow[H_2O_2,\ OH^-]{B_2H_6} CH_3CH_2C{\equiv}CD$

$D_2O + CH_3CH_2C{\equiv}CNa$ ↑

b. OH, D, D, D, D $\xleftarrow[Pt]{D_2}$ HO

c. OH, NH_2, D, D $\xleftarrow[Pt,\ HCl]{D_2}$ HO, $-C{\equiv}N$

d. $CH_3CH_2CDOHCH_3 \longleftarrow$ 2-butanone + $NaBD_4$

11. a. It is not possible to make the phosphonium salt precursor to the ylide reagent, because neopentyl halides are too hindered for S_N2 attack.

b. The acidic conditions of the Clemmensen reduction would lead to loss of the alcohol group, with carbocation rearrangement and substitution by chloride.

12.

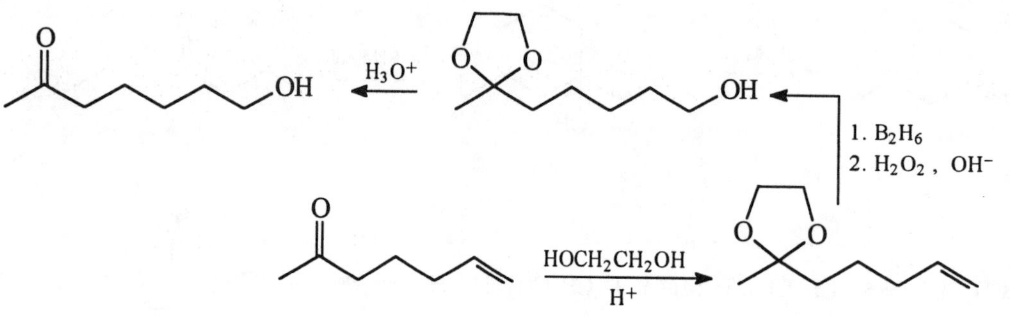

H_3O^+

OH

OH

1. B_2H_6
2. H_2O_2 , OH^-

$HOCH_2CH_2OH$
H^+

13.

14.

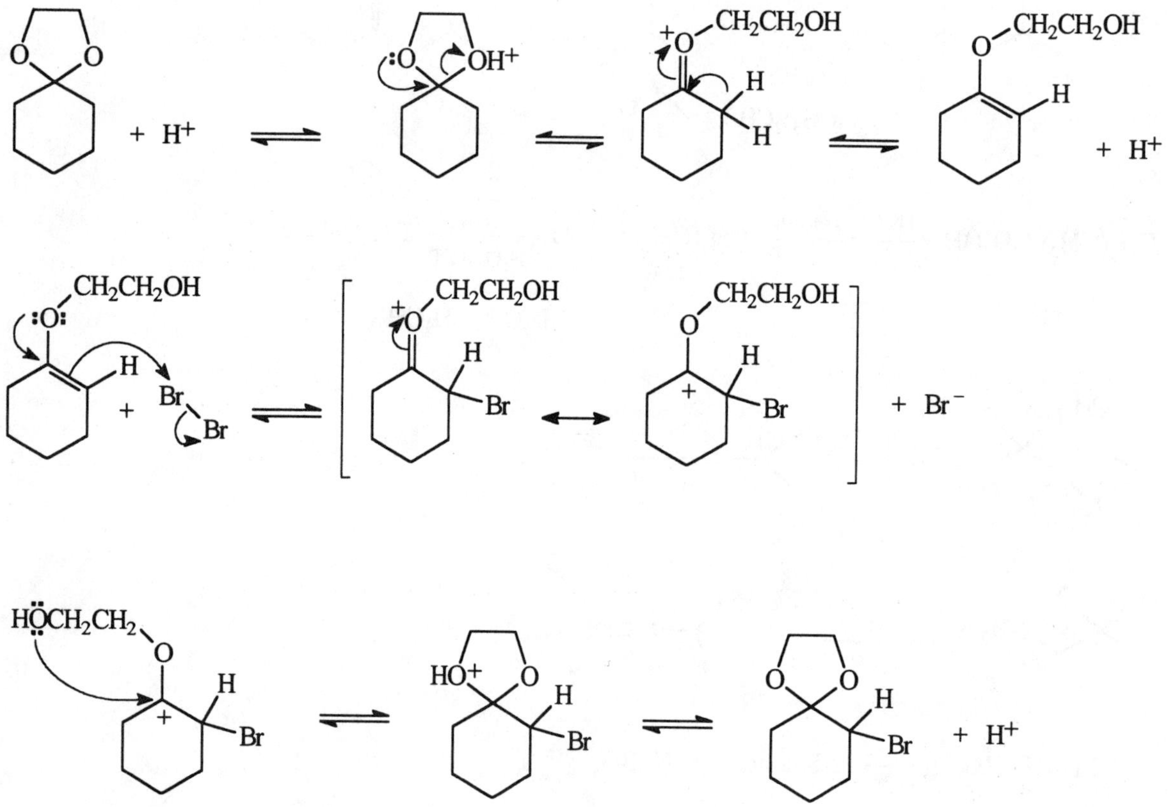

+ H^+

CH_2CH_2OH

+ H^+

+ Br^-

$H\ddot{O}CH_2CH_2$

+ H^+

15.

$$CH_3CH(OEt)OEt + H^+ \rightleftharpoons CH_3CH(\overset{+}{H}OEt)OEt \rightleftharpoons \left[CH_3\overset{+}{C}H-\ddot{O}Et \longleftrightarrow CH_3CH{=}\overset{+}{O}Et \right] + EtOH$$

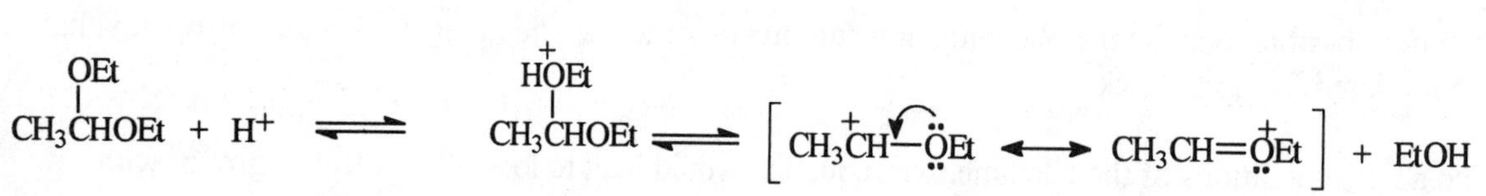

$$CH_3CH{=}\overset{+}{O}Et + H_2O \rightleftharpoons CH_3\overset{\overset{+}{O}H_2}{CH}OEt \rightleftharpoons CH_3\overset{OH}{CH}OEt + H^+ \rightleftharpoons CH_3\overset{OH}{CH}\overset{+}{O}(H)Et$$

$$CH_3\overset{OH}{CH}{-}\overset{+}{O}(H)Et \rightleftharpoons \left[CH_3\overset{+}{C}H{-}OH \longleftrightarrow CH_3CH{=}\overset{+}{O}H\right] + EtOH \rightleftharpoons CH_3CHO + EtOH + H^+$$

16. Greater than unity. The fluorinated carbonyl group is destabilized because the dipolar resonance structure is less stable.

$$CF_3\overset{O}{\overset{\|}{C}}H \longleftrightarrow F_3\overset{}{C}{-}\overset{O^-}{\underset{+}{C}}H$$

destabilized by electrostatic repulsions between C-F dipoles and C$^+$.

17. a. a cyclic acetal

b. $HOCH_2CH_2CH_2\overset{O}{\overset{\|}{C}}H + H^+ \rightleftharpoons H\ddot{O}CH_2CH_2CH_2\overset{+OH}{\overset{\|}{C}}H \rightleftharpoons$ (cyclic, $\overset{+}{O}$H, H, OH) $\rightleftharpoons$ (cyclic, O, H, OH) + H^+

(cyclic, O, H, OH) + $H^+ \rightleftharpoons$ (cyclic, $\ddot{O}$, H, $\overset{+}{O}H_2$) $\rightleftharpoons$ [(cyclic, $\overset{+}{O}$=C, H) $\longleftrightarrow$ (cyclic, O, $\overset{+}{C}$, H)] + H_2O

(cyclic, O, $\overset{+}{C}$, H) + $CH_3\ddot{O}H \rightleftharpoons$ (cyclic, O, H, $\overset{+}{O}(H)CH_3$) $\rightleftharpoons$ (cyclic, O, H, OCH_3) + H^+

c. a cyclic hemiacetal

d. See the first steps in part b.

18. $CH_3\overset{O}{\overset{\|}{C}}H + H^+ \rightleftharpoons CH_3\overset{+OH}{\overset{\|}{C}}H \qquad CH_3\overset{+OH}{\overset{\|}{C}}H + CH_3CHO \rightleftharpoons CH_3\overset{OH}{CH}{-}\overset{+}{O}{=}CHCH_3$

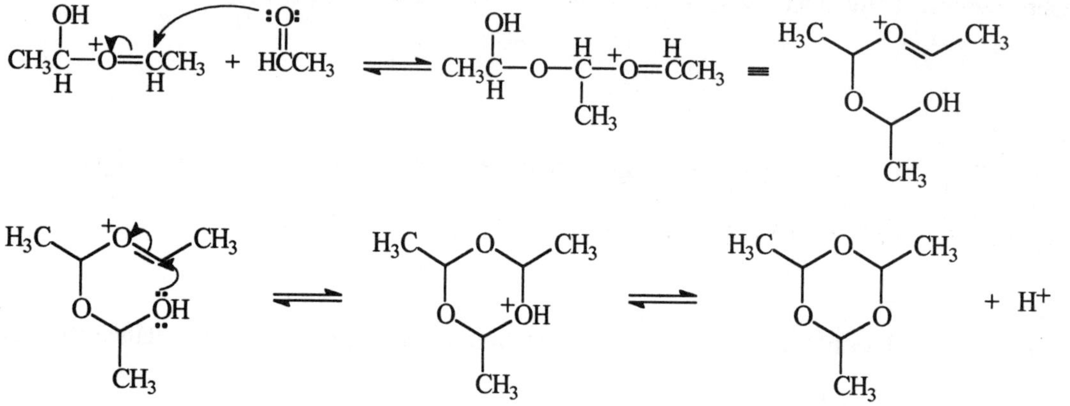

$$CH_3\overset{OH}{\underset{H}{C}}-\overset{+}{O}=\underset{H}{C}CH_3 + HCCH_3 \rightleftharpoons CH_3\overset{OH}{\underset{H}{C}}-O-\overset{H}{\underset{CH_3}{C}}-\overset{+}{O}=\overset{H}{C}CH_3$$

paraldehyde

The depolymerization of paraldehyde involves this same mechanism, starting from paraldehyde. The acetaldehyde is removed by distillation to displace the equilibrium

19. a. $CH_3(CH_2)_8CH_2OH \xrightarrow{PBr_3} \xrightarrow{Mg} CH_3(CH_2)_8CH_2MgBr \xrightarrow{H_2C=O} \xrightarrow{H_2O} CH_3(CH_2)_9CH_2OH \xrightarrow{PCC} CH_3(CH_2)_9CHO$

b. $CH_3(CH_2)_9CH_2CH_2OH \xrightarrow[\Delta]{Al_2O_3} CH_3(CH_2)_9CH=CH_2 \xrightarrow{O_3} \xrightarrow[AcOH]{Zn} CH_3(CH_2)_9CHO$

20.

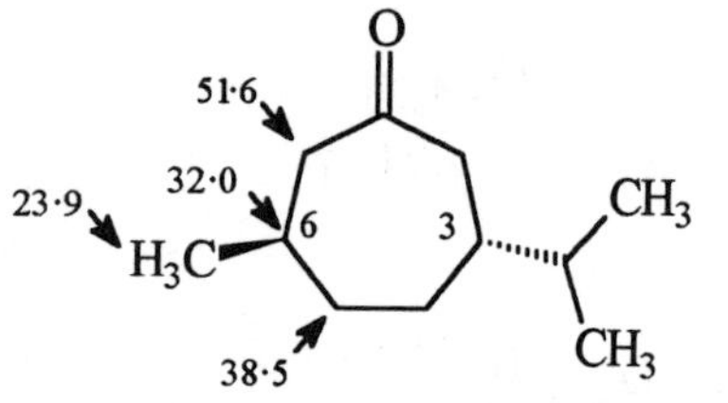

18.5 (q), 18.9 (q), and 23.9 (q) all belong to methyl groups. It is reasonable to assume that the isopropyl methyls are the ones with similar resonances. (Note: the resonances are not identical because the methyls are diastereotopic; see answer to problem #6, Chapter 17.)

32.0 (d), 34.9 (d), and 41.9 (d) are the CH's. C-3 clearly has the most β-branches and is therefore expected to come the furthest downfield.

32.5 (t), 38.5 (t), 47.0 (t), and 51.6 (t) are the CH_2's. C-2 and C-7 should be downfield from C-4 and C-5 (presence of electron withdrawing C=O group), therefore they are 47.0 and 51.6. C-2 and C-4 have 2 γ-substituents more than C-7 and C-5, respectively, because of the isopropyl vs. methyl groups. Therefore they should come ~ 5 ppm further upfield.

The signals at 212.1 and 213.4 correspond to the carbonyl carbons.

Since the signals of Isomer K correspond most closely to those of the *trans*-isopropyl compound, K is probably the *trans* isomer, J the *cis* isomer.

21. $CH_3\overset{O}{\overset{\|}{C}}CH_2CH_2C{\equiv}CCH_3 \xrightleftharpoons{Hg^{2+}} CH_3\overset{:O:}{\overset{\|}{C}}CH_2CH_2C{=}CCH_3 \cdot Hg^{2+} \longrightarrow$ (cyclic oxonium ion, Hg^+) + $H_2\ddot{O}:$; $\xrightarrow{-H^+}$ HO-substituted cyclic ether (Hg^+)

$\Big\updownarrow H_2O$ $\Big\updownarrow H_2O$

$CH_3C(OH)_2CH_2CH_2C{\equiv}CCH_3 \xrightleftharpoons{Hg^{2+}} CH_3C(OH)(\text{HO:})CH_2CH_2C{=}CCH_3 \cdot Hg^{2+} \xrightarrow{-H^+}$ HO-substituted cyclic ether (Hg^+)

$\Big\updownarrow H^+$

$CH_3\overset{O}{\overset{\|}{C}}CH_2CH_2\overset{+}{C}(OH){-}CH(Hg^+)CH_3 \xleftarrow{H^+} CH_3\overset{O}{\overset{\|}{C}}CH_2CH_2C(OH){=}C(Hg^+)CH_3 \xleftarrow{-H^+}$ protonated cyclic ether (Hg^+)

$\Big\downarrow -Hg^{2+}$

$CH_3\overset{O}{\overset{\|}{C}}CH_2CH_2C(OH){=}CHCH_3 \rightleftharpoons CH_3\overset{O}{\overset{\|}{C}}CH_2CH_2\overset{O}{\overset{\|}{C}}CH_2CH_3$

22. $CH_2{=}CHOEt \xrightleftharpoons{H^+} \left[CH_3\overset{+}{C}HOEt \leftrightarrow CH_3CH{=}\overset{+}{O}Et \right] \xrightleftharpoons{BuOH} CH_3{-}CH(\overset{+}{O}(H)Bu)(OEt) \xrightleftharpoons{-H^+} CH_3{-}CH(OBu)(OEt)$

$CH_3{-}CH(OR)(OR') \xrightleftharpoons{H^+} CH_3{-}CH(\overset{+}{O}(H)R)(OR') \xrightleftharpoons{-ROH} \left[CH_3\overset{+}{C}HOR' \leftrightarrow CH_3CH{=}\overset{+}{O}R' \right] \xrightleftharpoons{H_2O} CH_3{-}CH(\overset{+}{O}H_2)(OR')$

R, R' = Et, Bu

$\Big\updownarrow -H^+$

$CH_3CH{=}O \xleftharpoons{-H^+} \left[CH_3\overset{+}{C}HOH \leftrightarrow CH_3CH{=}\overset{+}{O}H \right] \xleftharpoons{-R'OH} CH_3{-}CH(OH)(\overset{+}{O}(H)R') \xleftharpoons{H^+} CH_3{-}CH(OH)(OR')$

The normal conditions for $LiAlH_4$ or $NaBH_4$ reduction of ketones to alcohols are alkaline, and the bromohydrin would cyclize to the epoxide: $H_2C{-}CH{\sim}$ (with O bridging, epoxide). Protect the alcohol by formation of the mixed acetal, by using ethyl vinyl ether:

$$HOCH_2CH(Br)CH_2CH_2CH_2C(=O)CH_3 \xrightarrow[H^+]{CH_2=CHOEt} EtOCH(CH_3)OCH_2CH(Br)CH_2CH_2CH_2C(=O)CH_3$$

$$\xrightarrow{NaBH_4} EtOCH(CH_3)OCH_2CH(Br)CH_2CH_2CH_2CH(OH)CH_3 \xrightarrow{H_3O^+} HOCH_2CH(Br)CH_2CH_2CH_2CH(OH)CH_3 + CH_3CHO + EtOH$$

23.

$$(CH_3)_2CHCH_2CH{=}O + {}^+NH_4 \rightleftharpoons (CH_3)_2CHCH_2CH{=}\overset{+}{O}H \;(+\,:NH_3) \rightleftharpoons (CH_3)_2CHCH_2CH(OH)(\overset{+}{N}H_3) \underset{H^+}{\overset{-H^+}{\rightleftharpoons}} (CH_3)_2CHCH_2CH(OH)(:NH_2)$$

$$\overset{H^+}{\rightleftharpoons} (CH_3)_2CHCH_2CH(\overset{+}{O}H_2)(:NH_2) \overset{-H_2O}{\rightleftharpoons} (CH_3)_2CHCH_2CH{=}\overset{+}{N}H_2 \;(+\,{}^-:CN) \rightleftharpoons (CH_3)_2CHCH_2CH(CN)(NH_2)$$

Heptane-2,6-dione $\xrightarrow[HCN]{NH_4Cl}$ [2-amino-2-cyano-6-oxo intermediate (CN, NH_2, O)] $\rightarrow$ 2,6-dimethylpiperidine-2,6-dicarbonitrile (CN, N–H, CN)

14.F. Supplementary Problems

S1. Provide IUPAC names for the following compounds:

a. $CH_3C(=O)CH{=}C(Cl)CH{=}O$

b. $HOCH_2$–CH(OH)–CH=O (wedge H, hashed OH)

c. cyclopropyl–CH=O

d. 3,5,5-trimethylcyclohex-2-enone (skeletal structure)

e. C_6H_5–C(H_3C)(Cl)–C(=O)CH_3

S2. Write the structure of the principal product formed in each of the following reaction sequences.

a. 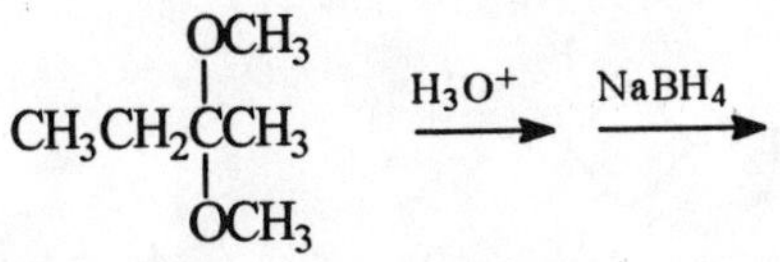 $CH_3CH_2C(OCH_3)_2CH_3 \xrightarrow{H_3O^+} \xrightarrow{NaBH_4}$

b. 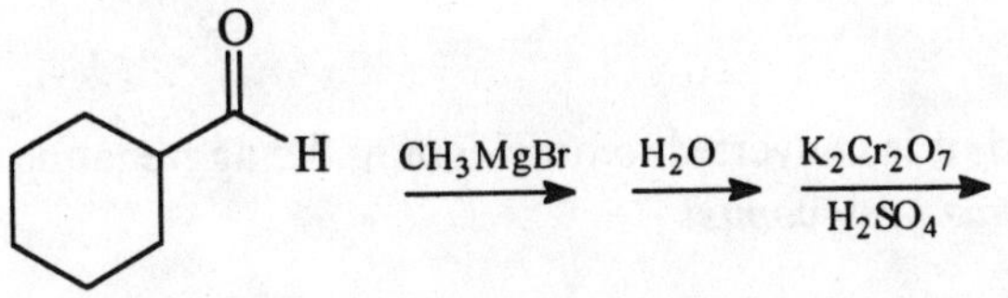 Cyclohexyl–CHO $\xrightarrow{CH_3MgBr} \xrightarrow{H_2O} \xrightarrow[H_2SO_4]{K_2Cr_2O_7}$

c. $CH_3CH_2CH_2C{\equiv}CH \xrightarrow[2.\ H_2O_2,\ OH^-]{1.\ B_2H_6} \xrightarrow{HCN}$

d. 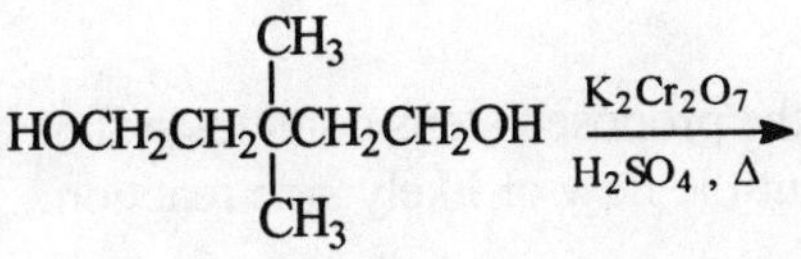 $HOCH_2CH_2C(CH_3)_2CH_2CH_2OH \xrightarrow[H_2SO_4\ ,\ \Delta]{K_2Cr_2O_7}$

e. 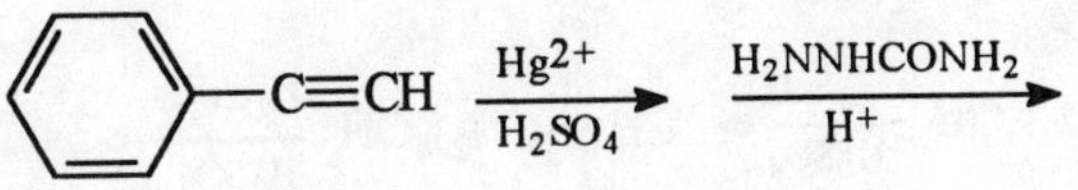 $C_6H_5C{\equiv}CH \xrightarrow[H_2SO_4]{Hg^{2+}} \xrightarrow[H^+]{H_2NNHCONH_2}$

f. 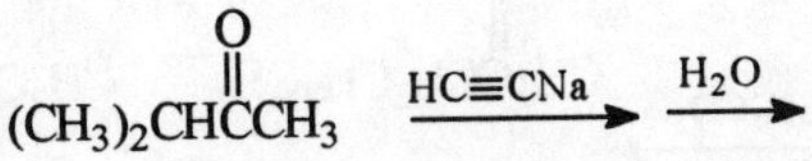 $(CH_3)_2CHCOCH_3 \xrightarrow{HC{\equiv}CNa} \xrightarrow{H_2O}$

S3. Show how to carry out each of the following transformations. More than one step is involved in each case.

a. 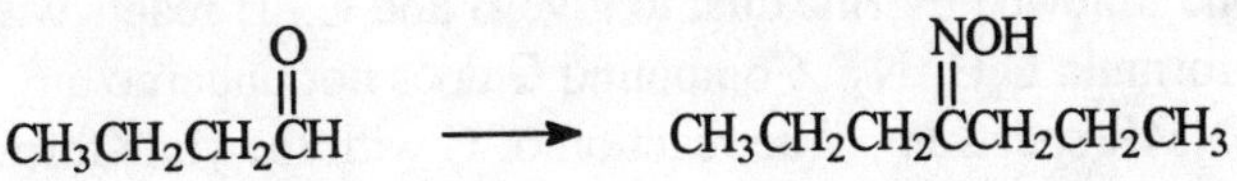 $CH_3CH_2CH_2CHO \longrightarrow CH_3CH_2CH_2C(=NOH)CH_2CH_2CH_3$

b. 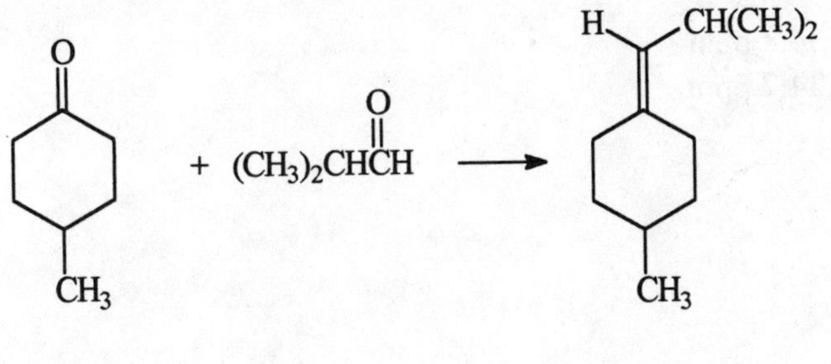 4-Methylcyclohexanone + $(CH_3)_2CHCHO \longrightarrow$ 4-methyl-1-($(CH_3)_2CH$–CH=)cyclohexane (ring C=CH–CH$(CH_3)_2$)

S4. Show how each of the following compounds can be prepared from materials containing four carbons or less.

a. 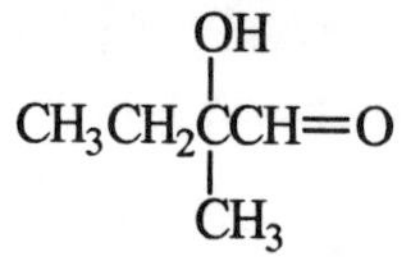 $CH_3CH_2C(OH)(CH_3)CH{=}O$

b. 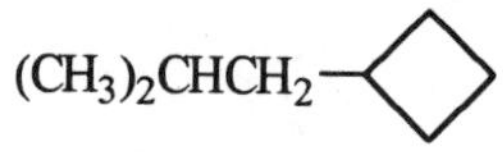 $(CH_3)_2CHCH_2$–cyclobutane

S5. On heating *cis*-6,7-epoxy-2-nonanone with a trace of acid, it is converted to brevicomin, the aggregating pheromone of the female western pine beetle *Dendroctomus brevicomis*:

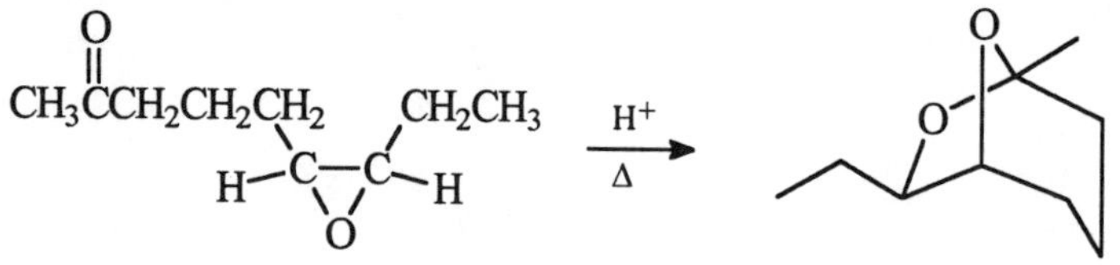

$$CH_3\overset{O}{\overset{\|}{C}}CH_2CH_2CH_2(H)C\overset{O}{-}C(H)CH_2CH_3 \xrightarrow[\Delta]{H^+} \text{brevicomin}$$

A proposed synthesis of this epoxyketone is shown below. Each of the proposed steps contains a flaw and will lead to a significant amount of side products. For each step, point out the flaw or likely side reaction. Then, suggest a synthesis which avoids these problems.

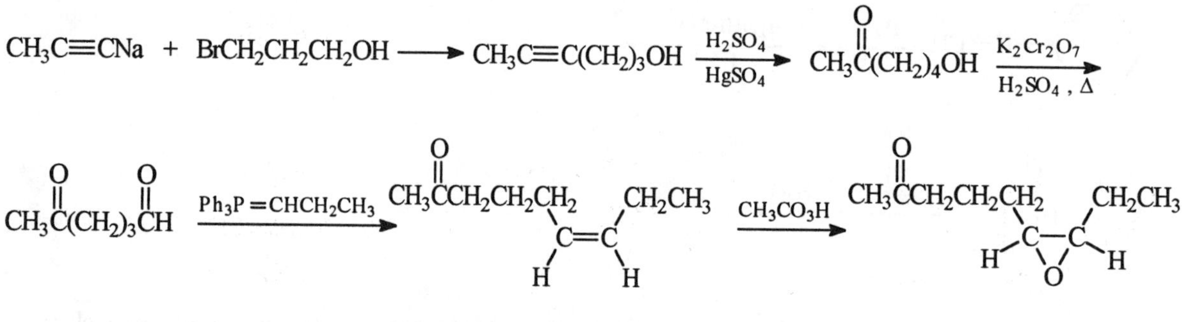

$$CH_3C{\equiv}CNa + BrCH_2CH_2CH_2OH \longrightarrow CH_3C{\equiv}C(CH_2)_3OH \xrightarrow[HgSO_4]{H_2SO_4} CH_3\overset{O}{\overset{\|}{C}}(CH_2)_4OH \xrightarrow[H_2SO_4,\ \Delta]{K_2Cr_2O_7}$$

$$CH_3\overset{O}{\overset{\|}{C}}(CH_2)_3\overset{O}{\overset{\|}{C}}H \xrightarrow{Ph_3P{=}CHCH_2CH_3} CH_3\overset{O}{\overset{\|}{C}}CH_2CH_2CH_2(H)C{=}C(H)CH_2CH_3 \xrightarrow{CH_3CO_3H} CH_3\overset{O}{\overset{\|}{C}}CH_2CH_2CH_2(H)C\overset{O}{-}C(H)CH_2CH_3$$

S6. Compound A, of formula C_7H_{14}, reacts with ozone followed by zinc dust to give B and C. B reacts with phenylhydrazine to give a crystalline product of formula $C_8H_{10}N_2$. Compound C does not undergo any reaction with Ag_2O, but it shows a singlet at δ 2.0 in the NMR spectrum. The reaction of C with the ylide formed from ethyltriphenylphosphonium bromide gives two compounds, A and an isomer D. The CMR spectra of A and D are given below:

A: δ 12.0, 12.8, 20.2, 22.1, 32.7, 118.3, 134.8 ppm
D: δ 12.0, 12.6, 14.2, 20.4, 41.2, 117.6, 134.7 ppm

What are A,B,C, and D?

S7. Suggest a mechanism for the following conversion:

O O (2,4-pentanedione) + H_2NNH–(phenyl) $\xrightarrow{H^+}$ N—N(phenyl) (3-methyl-1-phenylpyrazole)

S8. Write a mechanism which accounts for the formation of butyrolactone, (O=C lactone ring, O), when 1,4-butanediol is with CrO_3/pyridine.

treated

14.G Answers to Supplementary Problems

S1. a. (*Z*)-2-chloro-4-oxo-2-pentenal
b. (*R*)-2,3-dihydroxypropanal
c. cyclopropanecarbaldehyde
d. 2,4,4-trimethyl-2-cyclohexenone
e. (*S*)-3-chloro-3-phenyl-2-butanone

S2. a. $CH_3CH_2CH(OH)CH_3$ — OH over $CH_3CH_2\overset{|}{C}HCH_3$

b. cyclohexyl–$\overset{O}{\overset{\|}{C}}CH_3$

c. $CH_3CH_2CH_2CH_2CH(OH)C{\equiv}N$

d. $HO_2CCH_2C(CH_3)_2CH_2CO_2H$ (CH_3 above and below the central C)

e. phenyl–$\overset{NNHCONH_2}{\overset{\|}{C}}CH_3$

f. $(CH_3)_2CHC(OH)(CH_3)C{\equiv}CH$

S3. a.

$$CH_3(CH_2)_2\overset{O}{\overset{\|}{C}}H + CH_3(CH_2)_2MgBr \longrightarrow \xrightarrow{H_2O} CH_3(CH_2)_2\overset{OH}{\overset{|}{C}}H(CH_2)_2CH_3 \xrightarrow[H_2SO_4]{K_2Cr_2O_7}$$

$$CH_3(CH_2)_2\overset{O}{\overset{\|}{C}}(CH_2)_2CH_3 \xrightarrow{NH_2OH\text{ , }H^+} CH_3(CH_2)_2\overset{NOH}{\overset{\|}{C}}(CH_2)_2CH_3$$

b.

$$(CH_3)_2CH\overset{O}{\overset{\|}{C}}H \xrightarrow{NaBH_4} (CH_3)_2CHCH_2OH \xrightarrow{PBr_3} (CH_3)_2CHCH_2Br \xrightarrow{PhP_3} (CH_3)_2CHCH_2PPh_3{}^+Br^-$$

$$\xrightarrow{BuLi} (CH_3)_2CHCH{=}PPh_3 \xrightarrow{H_3C\text{–}C_6H_{9}{=}O\ (4\text{-methylcyclohexanone})} H_3C\text{–}C_6H_9{=}C(H)CH(CH_3)_2$$

S4.

a.

$$CH_3CH_2\overset{O}{\overset{\|}{C}}CH_3 + NaC{\equiv}CH \longrightarrow \xrightarrow{H_2O} CH_3CH_2\overset{OH}{\underset{CH_3}{C}}C{\equiv}CH \xrightarrow[\text{Lindlar cat.}]{H_2} CH_3CH_2\overset{OH}{\underset{CH_3}{C}}CH{=}CH_2$$

$$\xrightarrow[\text{2. Zn, HOAc}]{\text{1. } O_3} CH_3CH_2\overset{OH}{\underset{CH_3}{C}}CH{=}O$$

b.

$$(CH_3)_2CHCH_2Br + Ph_3P \longrightarrow \xrightarrow{BuLi} (CH_3)_2CHCH{=}PPh_3 \xrightarrow{\text{cyclobutanone}} \text{cyclobutylidene}{=}CHCH(CH_3)_2$$

$$\xrightarrow[Pt]{H_2} \text{cyclobutyl}{-}CH_2CH(CH_3)_2$$

S5. a. $CH_3C{\equiv}CNa$ will react with the alcohol: $CH_3C{\equiv}CNa + ROH \longrightarrow CH_3C{\equiv}CH + RO^-Na^+$

b. Two isomeric ketones can be formed: $CH_3\overset{O}{\overset{\|}{C}}CH_2CH_2CH_2CH_2OH$ and $CH_3CH_2\overset{O}{\overset{\|}{C}}CH_2CH_2CH_2OH$

c. Aldehyde will be overoxidized, to give acid: $CH_3\overset{O}{\overset{\|}{C}}CH_2CH_2CH_2CO_2H$

d. Wittig reagent can react with ketone, too: $CH_3\overset{CHCH_2CH_3}{\overset{\|}{C}}CH_2CH_2CH_2\text{---}$ and it can also give a mixture of *cis* and *trans* isomers:

$$\text{---}CH_2CH_2(H)C{=}C(H)CH_2CH_3$$

e. The ketone can undergo Baeyer-Villiger reaction in the presence of a peracid: $CH_3\overset{O}{\overset{\|}{C}}OCH_2CH_2CH_2\text{---}$

A synthesis that avoids these problems would make use of protecting groups:

$$BrCH_2CH_2CH_2OH + C_2H_5OCH{=}CH_2 \longrightarrow BrCH_2CH_2CH_2OCH(CH_3)OC_2H_5 \xrightarrow{NaC\equiv CH}$$

$$HC\equiv CCH_2CH_2CH_2OCH(CH_3)OC_2H_5 \xrightarrow[H_2SO_4]{HgSO_4} CH_3\overset{O}{\overset{\|}{C}}CH_2CH_2CH_2OH \xrightarrow[H^+]{HOCH_2CH_2OH} CH_3C(OCH_2CH_2O)CH_2CH_2CH_2OH$$

$$\xrightarrow[pyridine]{Tosyl\ Cl} CH_3C(OCH_2CH_2O)CH_2CH_2CH_2OTosyl \xrightarrow{NaC\equiv CCH_2CH_3} CH_3C(OCH_2CH_2O)CH_2CH_2CH_2C\equiv CCH_2CH_3 \xrightarrow[Lindlar]{H_2}$$

$$cis\text{-}CH_3C(OCH_2CH_2O)CH_2CH_2CH_2CH{=}CHCH_2CH_3 \xrightarrow{CH_3CO_3H} CH_3\overset{O}{\overset{\|}{C}}CH_2CH_2CH_2CH\overset{O}{-}CHCH_2CH_3$$

S6. 1) A, C_7H_{14}, CMR δ 118.3, 134.8: A is an alkene.

2) $A \xrightarrow[2.\ Zn]{1.\ O_3} B + C$; B + $C_6H_5NHNH_2 \longrightarrow C_8H_{10}N_2$, $C_6H_5NHN{=}CHCH_3$; B is acetaldehyde.

3) $C \xrightarrow{Ag_2O}$ N.R.; singlet at δ 2.0: C is a methyl ketone, either $CH_3\overset{O}{\overset{\|}{C}}CH_2CH_2CH_3$ or $CH_3\overset{O}{\overset{\|}{C}}CH(CH_3)_2$.

4) CMR of A shows 7 carbons: no isopropyl group present, so C is $CH_3\overset{O}{\overset{\|}{C}}CH_2CH_2CH_3$.

5) A and D are therefore the two stereoisomers of 3-methyl-2-hexene:

A: $(H)(H_3C)C{=}C(CH_3)(CH_2CH_2CH_3)$ — CH_3 δ 22.1; $CH_2CH_2CH_3$ δ 32.7

D: $(H_3C)(H)C{=}C(CH_3)(CH_2CH_2CH_3)$ — CH_3 δ 14.2; $CH_2CH_2CH_3$ δ 41.2

(A *cis* substituent causes an upfield shift in the CMR)

S7.

$CH_3CCH_2CCH_3$ (O, O) $\overset{H^+}{\rightleftharpoons}$ $H_2NNHC_6H_5$ + $CH_3CCH_2CCH_3$ (O, ^+OH) $\rightleftharpoons$ $CH_3CCH_2CCH_3$ ($H_2\overset{+}{N}$–NHC_6H_5, O, OH) $\overset{-H^+}{\rightleftharpoons}$ $CH_3CCH_2CCH_3$ (HN–NHC_6H_5, O, OH) $\overset{H^+}{\rightleftharpoons}$

$CH_3CCH_2CCH_3$ (:NH–NHC_6H_5, O, $^+OH_2$) $\overset{-H_2O}{\rightleftharpoons}$ $CH_3CCH_2CCH_3$ ($\overset{+}{N}H$–NHC_6H_5, O) $\overset{-H^+}{\rightleftharpoons}$ $CH_3CCH_2CCH_3$ (N–NHC_6H_5, O) $\overset{H^+}{\rightleftharpoons}$ $CH_3CCH_2CCH_3$ (N–NHC_6H_5, ^+OH) $\rightleftharpoons$

N—$\overset{+}{N}HC_6H_5$, OH, CH_3, H_2 $\overset{-H^+}{\rightleftharpoons}$ N—NC_6H_5, OH, CH_3, H_2 $\overset{H^+}{\rightleftharpoons}$ N—NC_6H_5, $^+OH_2$, CH_3, H_2 $\overset{-H_2O}{\rightleftharpoons}$ N—$\overset{+}{N}C_6H_5$, H, H

$\overset{-H^+}{\rightleftharpoons}$ N—NC_6H_5

S8.

$HOCH_2CH_2CH_2CH_2OH$ $\xrightarrow[\text{pyr.}]{CrO_3}$ [$HOCH_2CH_2CH_2CH$ (O) $\rightleftharpoons$ H, OH, O] $\xrightarrow[\text{pyr.}]{CrO_3}$ (O, O)

15. ALDEHYDES AND KETONES: ENOLS

15 A. Chapter Outline and Important Terms Introduced

15.1 Enolization (carbonyl compounds as nucleophiles)

acidity of the α-hydrogen

A. Keto-Enol Equilibria

enol, enolate ion (resonance structures of enolate anion)
tautomerism: (equilibrium that moves H from one atom to another

$$\rangle C-\overset{O}{\overset{\|}{C}}-C\langle-H \rightleftharpoons \rangle C-\overset{OH}{\overset{|}{C}}=C\langle$$

equilibrium (tautomerism), acid or base catalyzed

deuterium exchange:

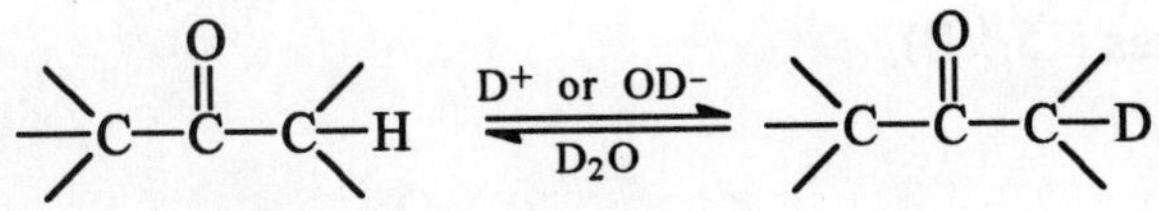

$$\rangle C-\overset{O}{\overset{\|}{C}}-C\langle-H \underset{D_2O}{\overset{D^+ \text{ or } OD^-}{\rightleftharpoons}} \rangle C-\overset{O}{\overset{\|}{C}}-C\langle-D$$

B. Enolate Ions

pK_a of acetone = 19
lithium diisopropylamide (LDA)
formation, alkylation, and silylation of enolates

ambident anions
degree of association

C. Racemization (via enol or enolate)

effect on chiral center α to carbonyl

D. Halogenation

autocatalytic

induction period

15.2 The Aldol Addition Reaction (an important C-C bond-forming reaction)

α,β-unsaturated aldehydes
intramolecular aldol condensations
mixed aldol condensations using preformed enolate

cyclic compounds

15.B. Important Reactions Introduced

Proton exchange of ketones and aldehydes (15.1.A)

Equation:

$$R-\overset{O}{\overset{\|}{C}}-CHR'_2 \xrightarrow{*H_2O} R-\overset{O}{\overset{\|}{C}}-C^{*}HR'_2$$

Generality: acid- or base-catalyzed
$*H = {}^1H$ (proton exchange), ^2H (deuterium), or ^3H (tritium)
all α-hydrogens can be exchanged

Key features: useful for determining number of α-hydrogens by exchanging in ^2H
useful for making labeled compounds
stereocenters α to carbonyl can be isomerized

Alkylation and silylation of ketone enolates (15.1.B)

Equation:

$$R-\overset{O}{\overset{\|}{C}}-CH_2R' \xrightarrow{LDA} R-\overset{O^- Li^+}{\overset{|}{C}}=CHR' \begin{cases} \xrightarrow{R''X} R-\overset{O}{\overset{\|}{C}}-CH(R')(R'') \\ \xrightarrow{Me_3SiCl} R-\overset{OSiMe_3}{\overset{|}{C}}=CHR' \end{cases}$$

Generality: LDA = lithium diispropylamide (or other hindered very strong base)
R" = methyl or 1° alkyl

Key features: important carbon-carbon bond-forming reaction
enolate is strong base, so elimination occurs with 2° or 3° R"X
silylation goes on oxygen to give silyl enol ether

Halogenation of ketones and aldehydes (15.1.D)

Equation:

$$R-\overset{O}{\overset{\|}{C}}-CHR'_2 \xrightarrow{X_2} R-\overset{O}{\overset{\|}{C}}-CXR'_2$$

Generality: X = Cl, Br, I
requires either acid or base catalysis

Key features: with acid catalysis, second and third halogens are introduced more slowly than first
with base catalysis, usually replace all α-hydrogens
uncatalyzed process is autocatalytic

Aldol addition reaction (15.2)

Equation:

$$R-\overset{O}{\overset{\|}{C}}-R + \underset{R'}{\underset{|}{CH_2}}-\overset{O}{\overset{\|}{C}}-R \longrightarrow R-\underset{R}{\underset{|}{\overset{OH}{\overset{|}{C}}}}-\underset{R'}{\underset{|}{CH}}-\overset{O}{\overset{\|}{C}}-R \longrightarrow (R)(R)C=C(R')-\overset{O}{\overset{\|}{C}}-R$$

Generality: all R's above = alkyl, aryl, or H
acid or base catalysis required

Key features: important carbon-carbon bond-forming reaction
important method for making cyclic compounds (intramolecular reaction)
often can isolate hydroxy aldehyde or hydroxy ketone intermediate
mechanism of addition reaction involves enolate attack on carbonyl component (base-catalyzed), or enol attack on protonated carbonyl component (acid-catalyzed)
mechanism of dehydration reaction involves loss of hydroxide ion from enolate (base-catalyzed), or loss of H_2O from enol (acid-catalyzed)
usually best for self-condensation of a single aldehyde or ketone (mixed aldol reactions can give difficult mixtures)
mixed aldol reactions can be successful in special circumstances, such as:
one carbonyl component = a non-enolizable aldehyde
mixed aldol reactions can also be carried out by preforming enolate with LDA, then adding carbonyl component

15.C. Important Concepts and Hints

Acid- vs. Base-Catalyzed Reactions

In Chapter 14, you were introduced to a reaction that could be either acid - or base-catalyzed:

$$\text{ketone + alcohol} \rightleftharpoons \text{hemiketal}$$

Chapter 15 develops this concept with the reaction at the α-carbon: bromination, deuteration and the related *aldol* addition reaction.

In an acid-catalyzed reaction involving a carbonyl, the reaction usually begins by attack of H^+ on C=O:

$$RCH_2\overset{:O:}{\overset{\|}{C}}R' \;(\curvearrowright H^+) \rightleftharpoons \left[RCH_2\overset{+OH}{\overset{\|}{C}}R' \longleftrightarrow RCH\overset{OH}{\overset{|}{C}}{}^{+}R' \right] \rightleftharpoons RCH=\overset{OH}{\overset{|}{C}}R' + H^+$$

(the α-H is shown below the α-carbon in the bracketed resonance form)

In the base-catalyzed reaction, it is loss of the weakly acidic α-hydrogen that initiates the reaction.

$$R\underset{H \curvearrowleft :OH^-}{\underset{|}{C}}H\overset{O}{\overset{\|}{C}}R' \rightleftharpoons \left[R\overset{-}{\underset{\cdot\cdot}{C}}H\overset{O}{\overset{\|}{C}}R' \longleftrightarrow RCH=\overset{O^-}{\overset{|}{C}}R' \right] + H_2O$$

In either case, it is the enol (acid-catalyzed) or enolate ion (base-catalyzed) that is important.

Acid-Catalyzed:

$$\text{RCH}{=}\overset{\text{O–H}}{\text{CR}'} + \text{Br–Br} \longrightarrow \text{RCHBr}\overset{\text{O}}{\overset{\|}{\text{C}}}\text{R}' + \text{H}^+ + \text{Br}^-$$

Base-Catalyzed:

$$\text{RCH}{=}\overset{:\ddot{\text{O}}:^-}{\text{CR}'} + \text{Br–Br} \longrightarrow \text{RCHBr}\overset{\text{O}}{\overset{\|}{\text{C}}}\text{R}' + \text{Br}^-$$

A few generalizations are useful to keep in mind when you are studying reactions like these, or trying to recall their mechanisms during an exam:

A. In acid-catalyzed reactions, the protons usually go on the molecule before other steps occur (nucleophilic attack, loss of a proton elsewhere (= tautomerization), etc.). *Anionic* intermediates are **almost never** observed (except counterions of strong acids such as Cl^-, tosylate ion, etc); enolates and alkoxide ions are not involved as intermediates.

B. In base-catalyzed reactions, proton **removal** usually precedes other steps. *Cationic* intermediates are rare, unless they are the protonated forms of good bases (ammonium ions, etc.); oxonium ions and carbocations are **not** involved as intermediates.

The equations below are examples of **incorrect** mechanisms:

$$\text{CH}_3\overset{\text{O}}{\overset{\|}{\text{C}}}\text{CH}_2\overset{\text{OH}}{\overset{|}{\text{C}}}(\text{CH}_3)_2 \underset{}{\overset{\text{H}^+}{\rightleftharpoons}} \text{CH}_3\overset{\text{O}}{\overset{\|}{\text{C}}}\text{CH}_2\overset{^+\text{OH}_2}{\overset{|}{\text{C}}}(\text{CH}_3)_2 \overset{-\text{H}^+}{\rightleftharpoons} \text{CH}_3\overset{\text{O}^-}{\overset{|}{\text{C}}}{=}\text{CH}\overset{^+\text{OH}_2}{\overset{|}{\text{C}}}\text{C}(\text{CH}_3)_2 \xrightarrow{-\text{H}_2\text{O}} \text{CH}_3\overset{\text{O}}{\overset{\|}{\text{C}}}\text{CH}{=}\text{C}(\text{CH}_3)_2$$

you can't have acid-catalysis (oxonium ion formation) and base-catalysis (enolate formation) at the same time

$$\text{CH}_3\overset{\text{O}}{\overset{\|}{\text{C}}}\text{CH}_3 \overset{\text{H}^+}{\rightleftharpoons} \text{CH}_3\overset{^+\text{OH}}{\overset{\|}{\text{C}}}\text{CH}_3 \overset{\text{CH}_3\text{O}^-}{\rightleftharpoons} \text{CH}_3\underset{\text{OCH}_3}{\overset{\text{OH}}{\text{C}}}\text{CH}_3$$

$$\text{CH}_3\overset{^+\text{OH}}{\overset{\|}{\text{C}}}\text{CH}_3 \overset{\text{HC}\equiv\text{C}^-}{\rightleftharpoons} \text{CH}_3\underset{\text{CH}_3}{\overset{\text{OH}}{\text{C}}}\text{C}\equiv\text{CH}$$

Both methoxide and acetylide ions (strong bases) will react with the acid used to protonate the ketone.

Remember from your course in general chemistry that a strong acid and a strong base cannot both exist in the same solution; in a strongly basic solution, the $[H^+]$ wil be very low, « 10^{-7} and vice versa.

Aldol vs Grignard in Carbon-Carbon Bond Formation

In a synthesis problem involving the formation of a carbon-carbon bond, in order to decide whether to use a Grignard reaction or an aldol addition, look carefully at the functionality of the desired product. The aldol addition will produce a compound with 2 functional groups at the reaction site, either a β-hydroxy carbonyl or an α,β-unsaturated carbonyl. The Grignard reaction produces only one functional group at the reaction site, the −OH function.

Also the Grignard allows more flexibility in the choice of side chains. A mixed aldol reaction is more difficult, though not impossible.

15.D. Answers to Exercises

15.1 Rate = k x [OH⁻] [ketone] (second-order reaction)

15.2 In one liter of D_2O, there are 2 x 55.5 = 111 moles of deuterium atoms. In one liter of 1 *M* solution of 3-pentanone, there are 4 x 1 = 4 moles of exchangeable hydrogen atoms. The ratio of deuterium to exchangeable hydrogen is therefore 111/4 = 27.75. The chance that any one exchangeable position will still have a proton after equilibrium is reached is 1/27.75; the chance that one out of the four exchangeable positions in 3-pentanone will still have a proton is 4 x (1/27.75) = 14.4%. Therefore 86% of the 3-pentanone molecules will contain four deuteriums and the average per molecule will be 4 x 0.86 = 3.44.

15.3 a. One: O D

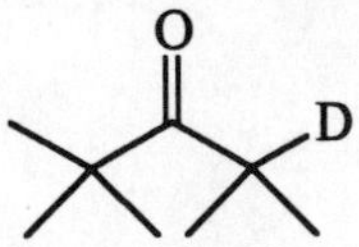

b. One: D O H

c. Four: O D D D D

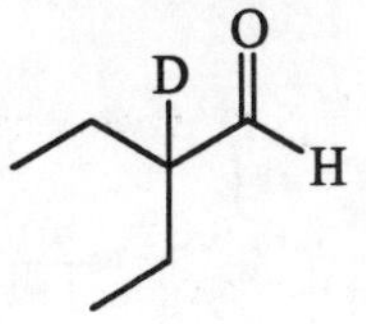

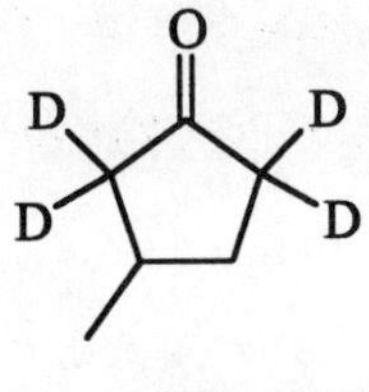

d. None: an alkene is not deprotonated by hydroxide

15.4 1) $CH_3\overset{O}{\overset{\|}{C}}CH_3 \rightleftharpoons CH_3\overset{O^-}{\overset{|}{C}}{=}CH_2 + H^+$ $K_1 = \dfrac{[\text{enolate}][H^+]}{[\text{ketone}]} = 10^{-19}$

2) $CH_3\overset{O}{\overset{\|}{C}}CH_3 \rightleftharpoons CH_3\overset{OH}{\overset{|}{C}}{=}CH_2$ $K_2 = \dfrac{[\text{enol}]}{[\text{ketone}]} = 1.5 \times 10^{-7}$

3) $CH_3\overset{OH}{\overset{|}{C}}{=}CH_2 \rightleftharpoons CH_3\overset{O^-}{\overset{|}{C}}{=}CH_2 + H^+$ $K_3 = \dfrac{[\text{enolate}][H^+]}{[\text{enol}]}$

$$K_3 = \frac{[\text{enolate}][H^+]}{[\text{ketone}]} \times \frac{[\text{ketone}]}{[\text{enol}]} \qquad K_3 = \frac{K_1}{K_2} = \frac{10^{-19}}{1.5 \times 10^{-7}}$$

so that pK_a (enol) = 12.2 (more acidic than an alcohol)

The difference in ΔH° for these two reactions is precisely equal to the difference ΔH_f° for the two isomeric forms of acetone. *Since the enol form is much less stable, it is more acidic.*

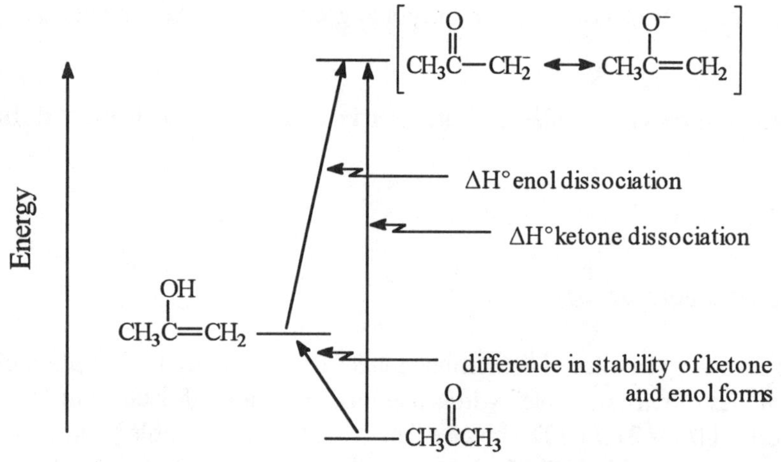

$$\left[CH_3\overset{O}{\overset{\|}{C}}-\bar{C}H_2 \longleftrightarrow CH_3\overset{O^-}{\overset{|}{C}}=CH_2 \right]$$

Energy

ΔH° enol dissociation

ΔH° ketone dissociation

$CH_3\overset{OH}{\overset{|}{C}}=CH_2$

difference in stability of ketone and enol forms

$CH_3\overset{O}{\overset{\|}{C}}CH_3$

15.5

a. (1-cyclohexenyloxy)triethylsilane (OSiEt$_3$) ← Et_3SiCl — cyclohexanone enolate (O^-) ← LDA, THF, -78° — cyclohexanone

b. 2-methylcyclohexanone ← CH_3I — cyclohexanone enolate

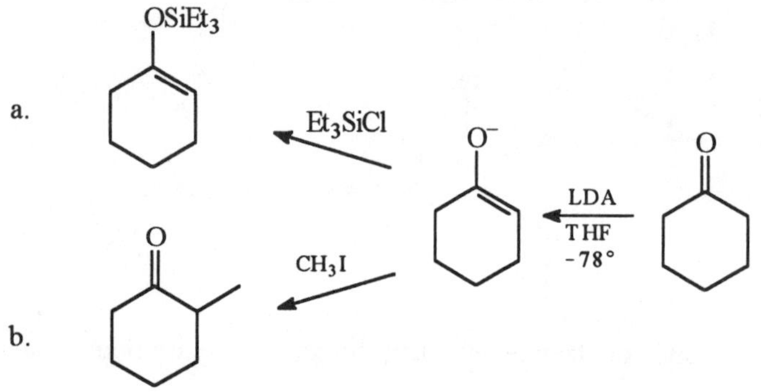

15.6 The only ones that will racemize are those that have an enolizable stereocenter; i.e., a stereocenter that (1) is next to a carbonyl group and (2) has a hydrogen substituent:

(a) and (c) will racemize: CHO (H) ⇌ [enolate O^-] ⇌ CHO (H)

ketone (O, H) ⇌ [enolate O^-] ⇌ ketone (O, H)

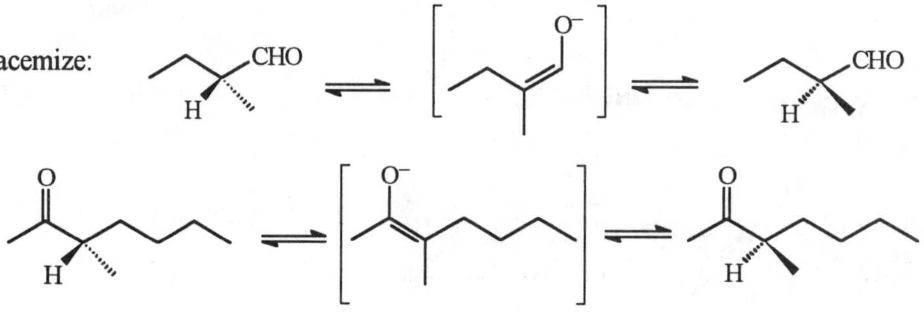

(b) and (d) will not:

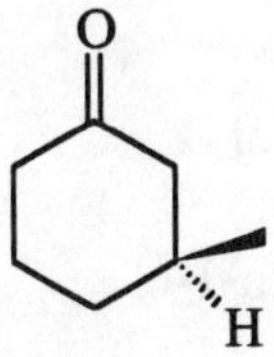

stereocenter not next to ketone

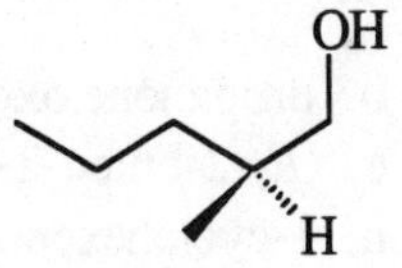

alcohol unaffected by base

15.7 a. $\xrightarrow[\text{HOAc}]{Br_2}$

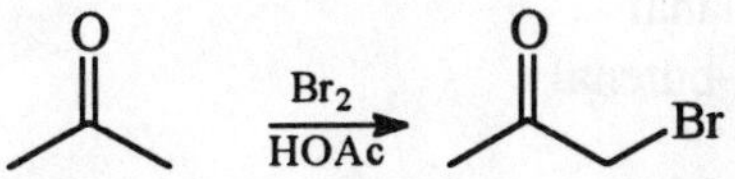

b. $\xrightarrow{NaBH_4}$ $\xrightarrow{NaOH}$

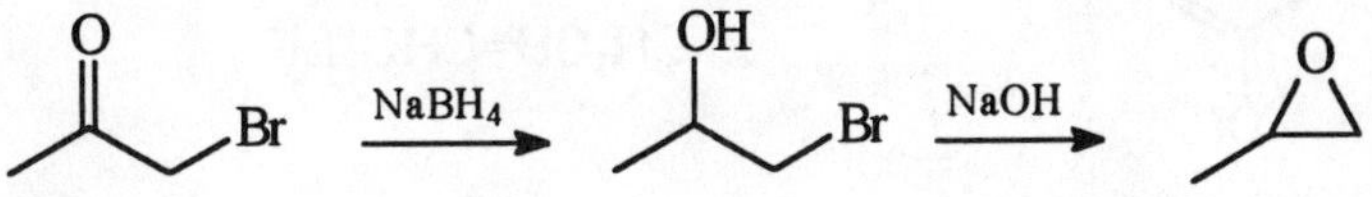

[(from (a)]

15.8 a.

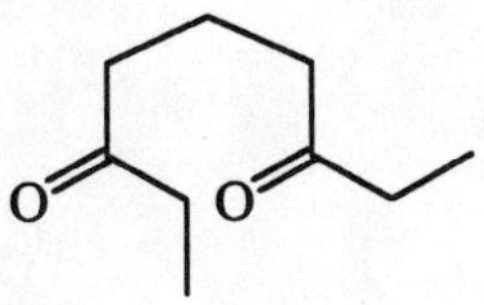

b.

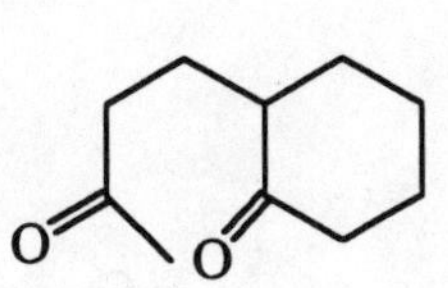

15.9 a. + CHO $\xrightarrow{NaOH}$ CHO

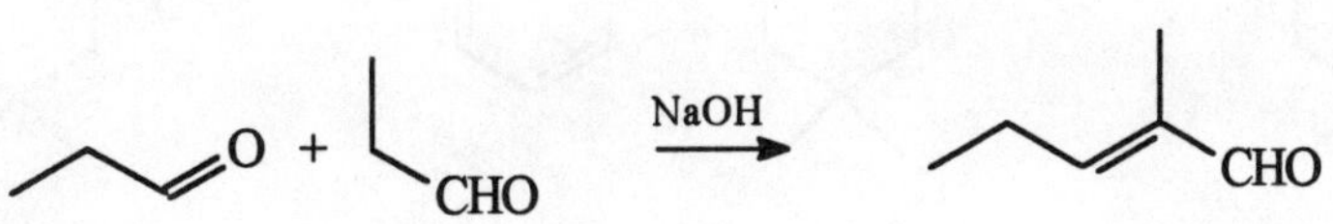

b. $\xrightarrow[\text{THF}]{\text{LDA}}$ $\xrightarrow{}$ $\xrightarrow{H_2O}$ $\xrightarrow[\Delta]{H^+}$

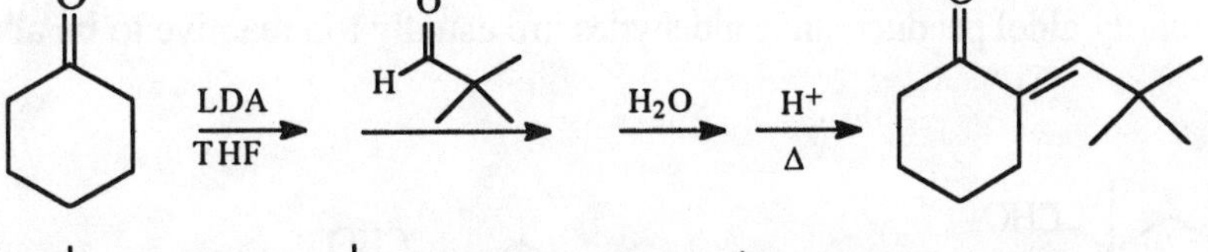

c. + CHO $\xrightarrow{NaOH}$ CHO OH

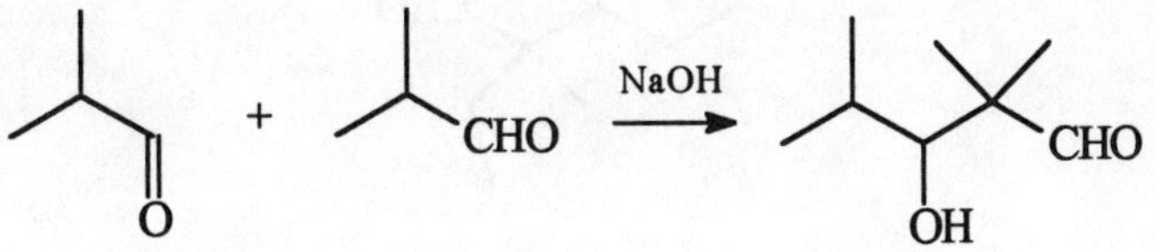

d. $\xrightarrow[\text{THF, -78°}]{\text{LDA}}$ Li^+O^- $\xrightarrow{}$ O OH

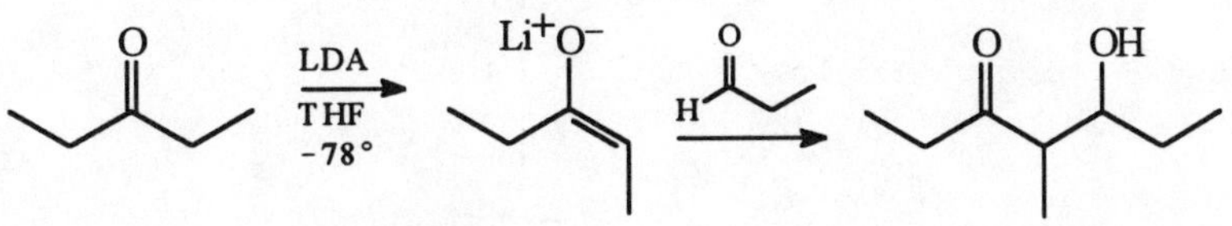

15.E. Answers and Explanations for Problems

1. a. 1-propen-2-ol b. propanone oxonium ion c. (*E*)-2-buten-2-ol
   d. (*Z*)-2-buten-2-ol e. (*E*)-2-buten-2-olate ion f. ethenol
   g. 1-cyclopentylidenethanol h. 1-cyclohexenol i. (*Z*)-1,3-butadien-1-ol
   j. (*E*)-1,3-butadien-1-ol

2. a. propanone c,d. butanone f. ethanal
   g. 1-cyclopentylethanone h. cyclohexanone i,j. 3-butenal

3. a. $CH_3CH_2CH{=}CHOH$ b. 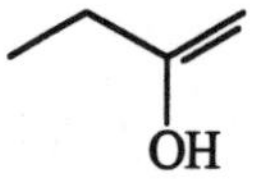 OH c. $CH_3CH{=}CHO^-Li^+$

   d. 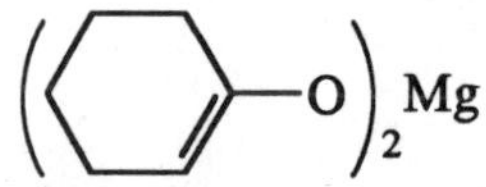 O Mg 2 e. $Li^+ \, ^-N[CH(CH_3)_2]_2$ f. 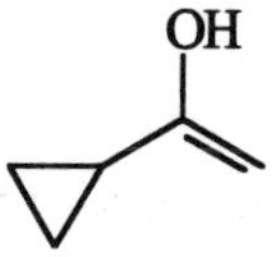 OH

4. a. 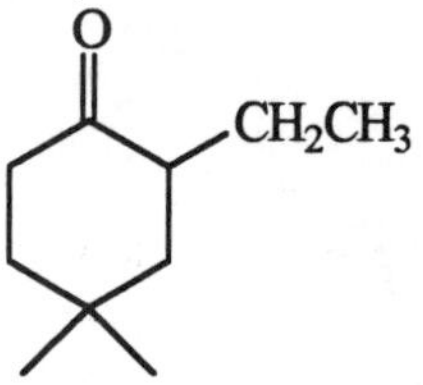 O CH_2CH_3 b. 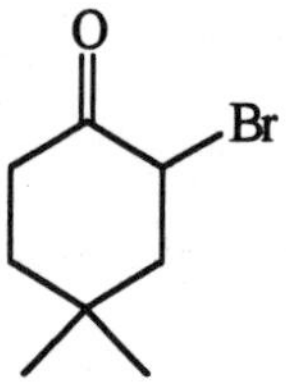 O Br c. 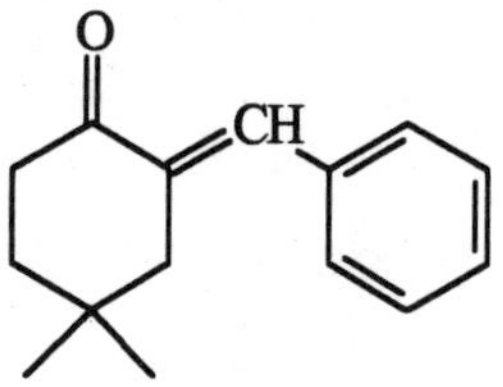 O CH d. O D D D D

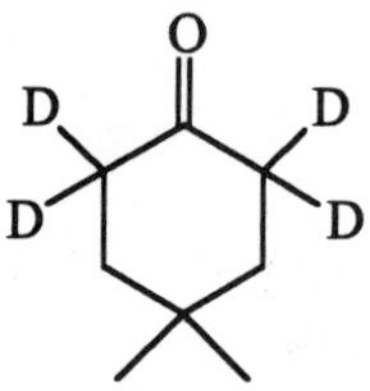

5. a. This reaction will yield primarily aldol product since aldehydes are usually too reactive to be alkylated by this method.

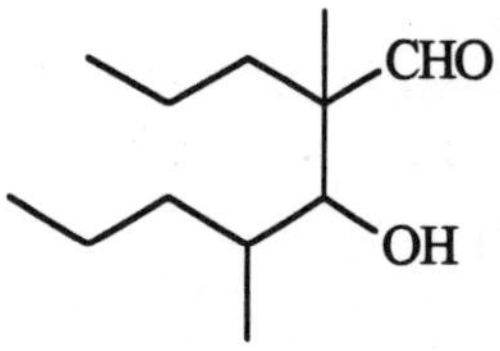

   CHO OH CHO

   aldol product 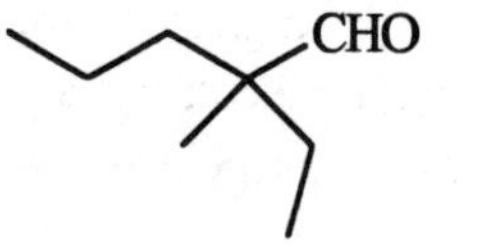 *alkylation product*

   b. 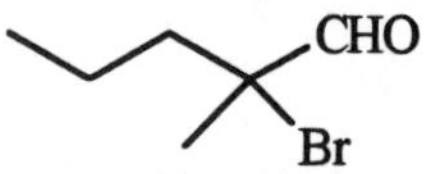 CHO Br

c. Depending on the reaction conditions, this will give a mixture of aldol products including the one from a. above and the mixed aldol shown at right or it can give primarily the mixed aldol.

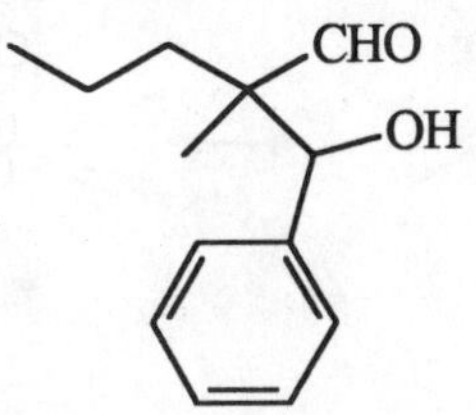

CHO
OH

mixed aldol product

d. CHO
D

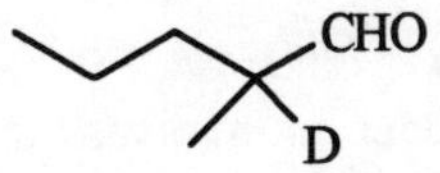

6. a. The four-membered ring is too strained to be formed by this reversible reaction:

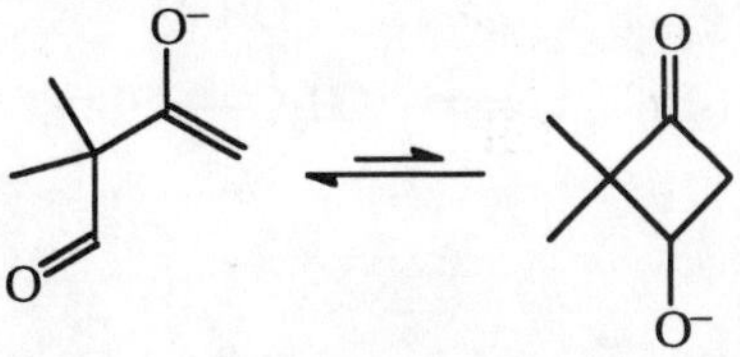

b. There would be three other products formed in comparable amounts since both aldehydes that would have to be used have enolizable hydrogens.

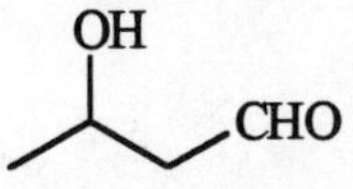

3-hydroxybutanal 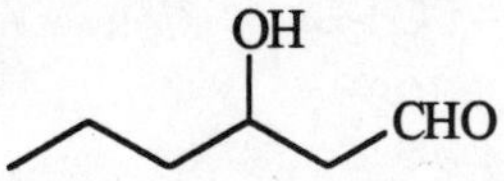 *3-hydroxyhexanal* 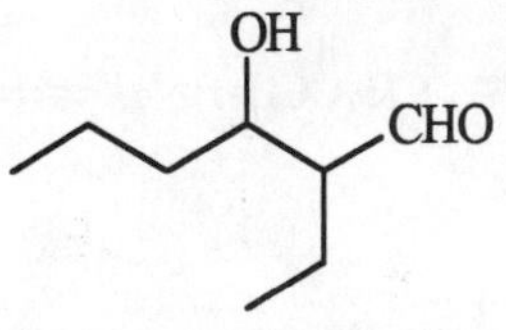 *2-ethyl-3-hydroxyhexanal* 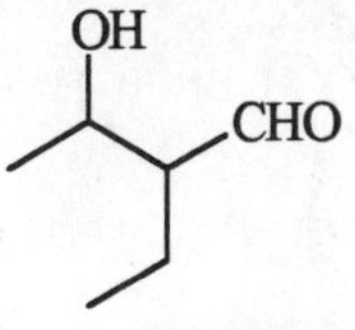 *2-ethyl-3-hydroxybutanal*

7. Consider the reaction of a base at the two sites:

Reaction on oxygen:

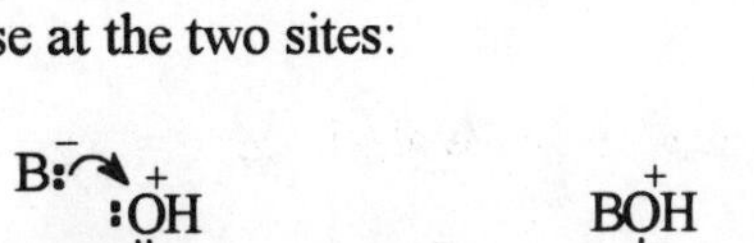

$$\mathrm{B:^- + CH_3C(=\overset{+}{O}H)CH_3 \longrightarrow CH_3\overset{-}{C}(B\overset{+}{O}H)CH_3}$$

Since the oxygen already has eight electrons, one pair in the double bond must be displaced onto carbon as the new bond is formed using the electron pair as the base. The resulting intermediate would be a high-energy charge-separated species.

Reaction on carbon:

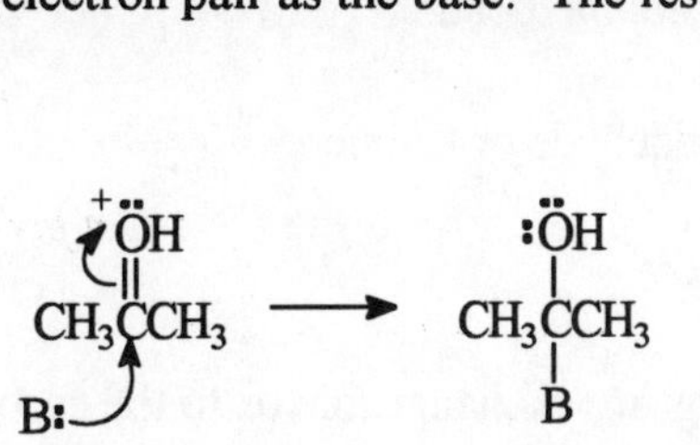

$$\mathrm{CH_3C(=\overset{+}{O}H)CH_3 + B: \longrightarrow CH_3C(OH)(B)CH_3}$$

Again, the carbon already has eight electrons, so one pair in the double bond must be displaced onto oxygen.

However, in this case, the resulting product is neutral, and of much lower energy.

$$\overset{+}{O}H=C(CH_3)_2 \longleftrightarrow HO-\overset{+}{C}(CH_3)_2$$

This example points up the value of using resonance structures. Of the two contributions to the resonance hybrid, the former is more important, since both C and O have octet configurations. However, the latter structure shows that C also has positive character and, to the extent that it has positive character, is electron deficient and can react with a base.

8. This example is another case of kinetic vs. equilibrium control. The acid-catalyzed reaction involves the more highly substituted enol which is formed faster than its isomer, but 1,3-dibromoacetone is apparently more stable than 1,1-dibromoacetone, perhaps because of steric effects. The important implication in these results is that the acid-catalyzed bromination reaction is reversible.

$$CH_3\overset{O}{\overset{\|}{C}}CH_2Br + H^+ \rightleftharpoons CH_3\overset{\overset{+}{O}H}{\overset{\|}{C}}CH_2Br \rightleftharpoons CH_3\overset{OH}{\overset{|}{C}}=CHBr \quad \textit{faster}$$

$$CH_3\overset{OH}{\overset{|}{C}}=CHBr \overset{Br_2}{\rightleftharpoons} CH_3\overset{\overset{+}{O}H}{\overset{\|}{C}}CHBr_2 + Br^- \rightleftharpoons CH_3\overset{O}{\overset{\|}{C}}CHBr_2 + HBr$$

less stable

$$CH_3\overset{O}{\overset{\|}{C}}CH_2Br + H^+ \rightleftharpoons CH_3\overset{\overset{+}{O}H}{\overset{\|}{C}}CH_2Br \rightleftharpoons CH_2=\overset{OH}{\overset{|}{C}}CH_2Br \quad \textit{slower}$$

$$CH_2=\overset{OH}{\overset{|}{C}}CH_2Br \overset{Br_2}{\rightleftharpoons} BrCH_2\overset{\overset{+}{O}H}{\overset{\|}{C}}CH_2Br + Br^- \rightleftharpoons BrCH_2\overset{O}{\overset{\|}{C}}CH_2Br + HBr$$

more stable

9. $$\underset{C_7H_{16}O}{\mathbf{A}} \xrightarrow[H_2SO_4]{Na_2Cr_2O_7} \underset{C_7H_{14}O}{\mathbf{B}} \xrightarrow{NaOD} C_7H_{12}D_2O; \quad \mathbf{B} \xrightarrow{Ag_2O} \text{no reaction}$$

i) **A**, $C_7H_{16}O$: formula indicates no rings or double bonds

ii) **A** $\xrightarrow{Cr(VI)}$ **B:** A is an alcohol, B is an aldehyde or ketone

iii) **B** $\xrightarrow{Ag_2O}$ no reaction: B is not an aldehyde

iv) **B** $\xrightarrow{NaOD}$ $C_7H_{12}D_2O$: only two hydrogens are exchangeable (α to the carbonyl group of B)

The only possibilities for this latter case are: $(CH_3)_2CHC(=O)CH(CH_3)_2$ or $(CH_3)_3CC(=O)CH_2CH_3$

A is therefore $(CH_3)_2CHCH(OH)CH(CH_3)_2$ or $(CH_3)_3CCH(OH)CH_2CH_3$.

To distinguish between these two possibilities, further information is needed.

10. **D_1 and D_2** $\xleftarrow{H_2/Pt}$ **C** $\xrightarrow{O_3}$ **E** $\xrightarrow{H_2NOH}$ **F**

$C_{12}H_{22}$ — $C_{12}H_{20}$ (optically active) — $C_6H_{10}O$ (optically active) — $C_6H_{11}NO$

E $\xrightarrow{DCl}$ mw 101

NMR: only one methyl, as a doublet

i) **C**, $C_{12}H_{20}$: formula indicates three "degrees of unsaturation"; that is: three rings, two rings and one double bond; one ring and two double bonds, or one ring and one triple bond, etc.

ii) **C** $\xrightarrow{H_2/Pt}$ **D_1** and **D_2**, $C_{12}H_{22}$: hydrogenation adds only two H's, so unsaturation is two rings and one double bond

iii) **C** $\xrightarrow{O_3}$ **E**, $C_6H_{10}O$: C has been cleaved into two molecules of the same formula, E, which is either a ketone or an aldehyde.

iv) **E**, $C_6H_{10}O$: formula indicates mw 98 and two unsaturations: one C=O and one ring.

v) **E** $\xrightarrow{DCl}$ mw 101: there are three exchangeable protons, so E cannot be an aldehyde.

vi) **E**, only one methyl group, as a doublet in NMR: >$CHCH_3$ is part of the structure.

vii) There are many ketones of formula $C_6H_{10}O$ with one ring, but only one which satisfies iv), v) and vi) above: [structure: cyclopentanone with H, H on one α-carbon and H, CH_3 on the other α-carbon]

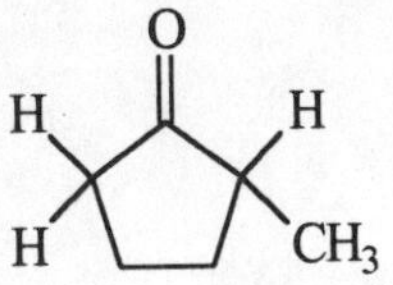

viii) **E** is optically active: it is not a racemic mixture, and the two halves of C must have the same configuration.

ix) **C** could be either:

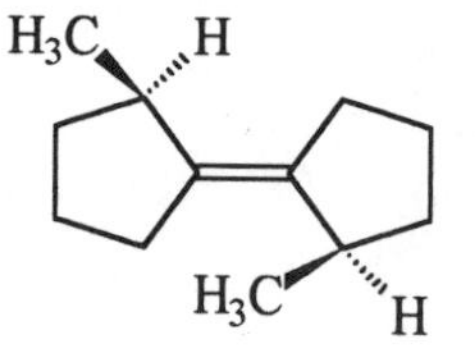

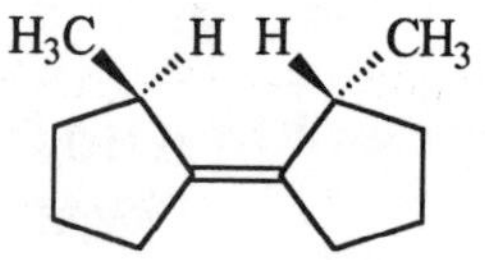

(or the (S,S)-enantiomers)

x) C $\xrightarrow[Pt]{H_2}$ **D_1 and D_2**: C is not the *cis* isomer in ix); only the *trans* isomer can give two isomers on hydrogenation.

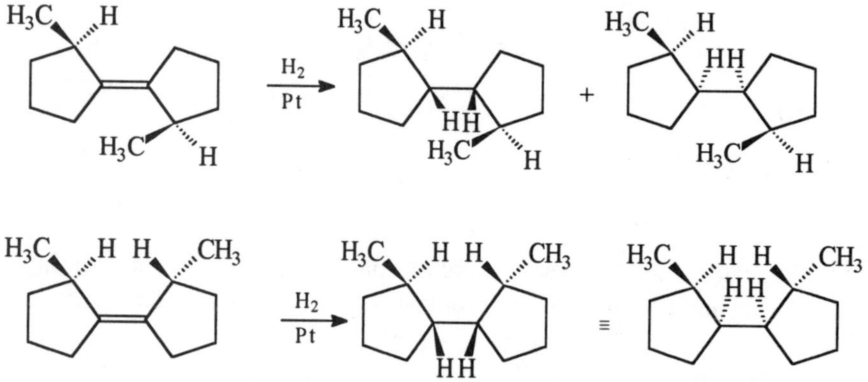

(You can see this easily with models)

xi) Finally, **F** is the oxime:

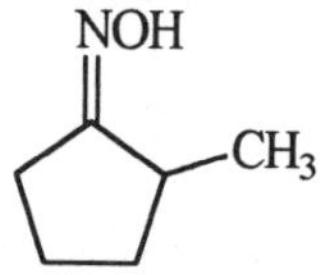

11. **G** $\xrightarrow{H_2/Pd}$ **H** $\xrightarrow{CH_3CO_3H}$ **I**

C_6H_8O $C_6H_{10}O$ $C_6H_{10}O_2$ NMR: only one methyl, as a doublet, δ 1.2

H $\xrightarrow{NaOD}$ $C_6H_7D_3O$

i) **G**, C_6H_8O: formula indicates three degrees of unsaturation

ii) **G** $\xrightarrow{H_2/Pd}$ **H**, $C_6H_{10}O$: only one saturation is a C=C; that leaves one ring and one C=O.

iii) **H** $\xrightarrow{NaOD}$ $C_6H_7D_3O$: three hydrogens α to C=O

iv) **H** $\xrightarrow{CH_3CO_3H}$ **I**, $C_6H_{10}O_2$:

H undergoes the Baeyer-Villiger Reaction: $-\overset{O}{\overset{\|}{C}}-C \longrightarrow -\overset{O}{\overset{\|}{C}}-O-C$

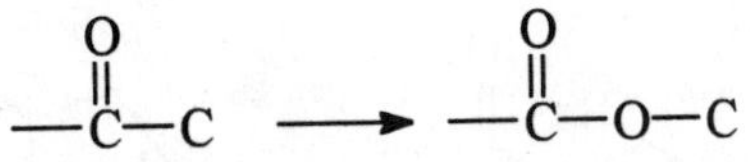

v) **I**, NMR: one methyl, doublet at δ 1.2: $-O-\overset{H}{\overset{|}{\underset{|}{C}}}-CH_3$

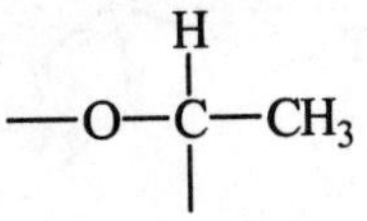

vi) From 3) and 4) above, **I** must be: 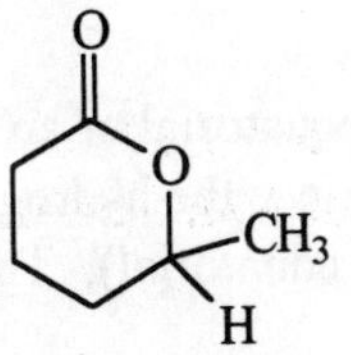 *any smaller ring would require a 2nd CH_3.*

vii) ∴ **H** must be:

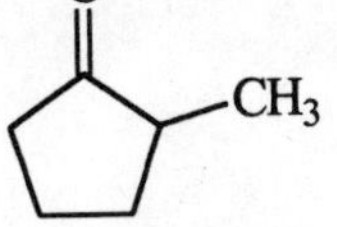

viii) **G**, nmr: methyl is a singlet: **G** is

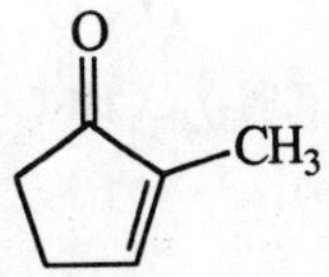

12.

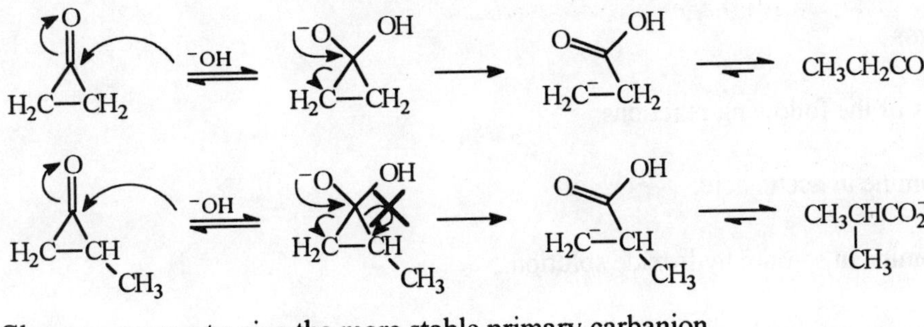

H_2C-CH_2 (cyclopropanone) $\overset{^-OH}{\rightleftharpoons}$ H_2C-CH_2 (adduct) $\longrightarrow$ $H_2\overset{-}{C}-CH_2$ (carboxylic acid) $\rightleftharpoons$ $CH_3CH_2CO_2^-$

$H_2C-CH(CH_3)$ $\overset{^-OH}{\rightleftharpoons}$ $H_2C-CH(CH_3)$ $\longrightarrow$ $H_2\overset{-}{C}-CH(CH_3)$ $\rightleftharpoons$ $CH_3\underset{CH_3}{CH}CO_2^-$

Cleavage occurs to give the more stable primary carbanion.

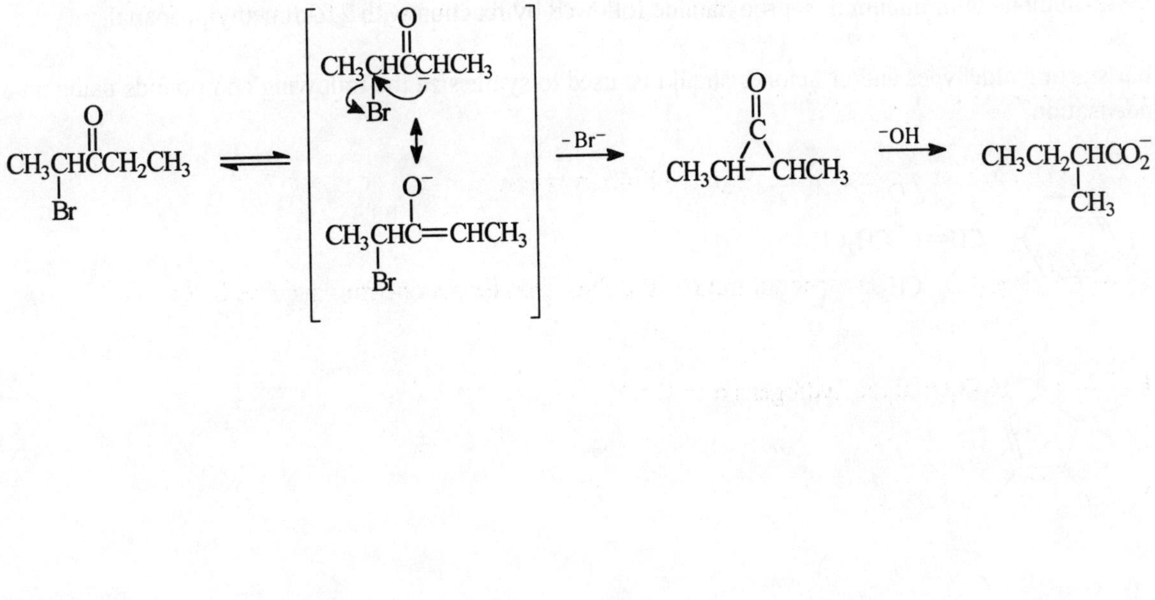

$CH_3\underset{Br}{CH}\overset{O}{\overset{\|}{C}}CH_2CH_3 \rightleftharpoons \left[CH_3\underset{Br}{CH}\overset{O}{\overset{\|}{C}}\overset{-}{C}HCH_3 \longleftrightarrow CH_3\underset{Br}{CH}\overset{O^-}{\overset{|}{C}}=CHCH_3 \right] \xrightarrow{-Br^-} CH_3CH-CHCH_3$ (cyclopropanone) $\xrightarrow{^-OH} CH_3CH_2\underset{CH_3}{CH}CO_2^-$

13. a. 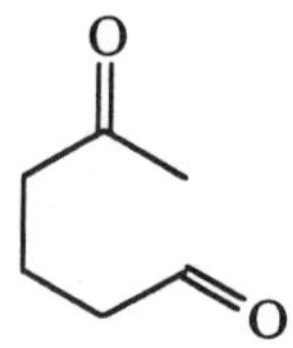 b. 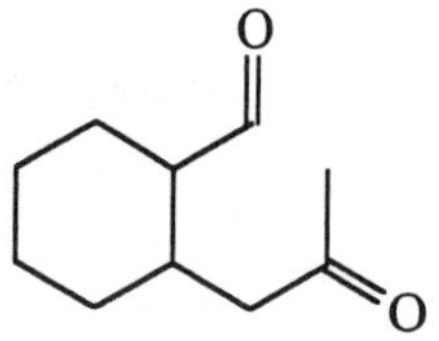 c. 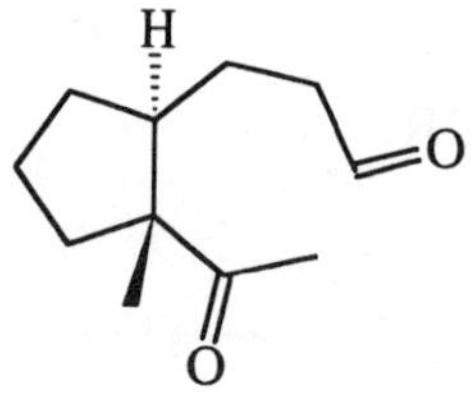 d.

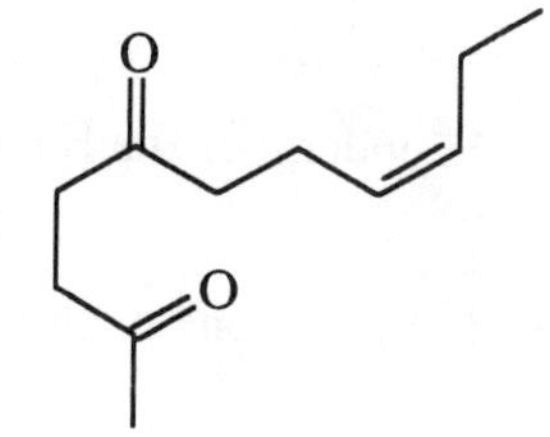

14. The 2.8 Hz coupling corresponds to an equatorial-equatorial or an axial-equatorial H-C-C-H coupling; the 11.8 Hz coupling corresponds to axial-axial coupling. Since the hydrogen at C-5 is axial, the initial bromo compound must have bromine axial (C-6 hydrogen equatorial). This isomerizes to the more stable product with bromine equatorial.

15. The first step is just like an aldol condensation:

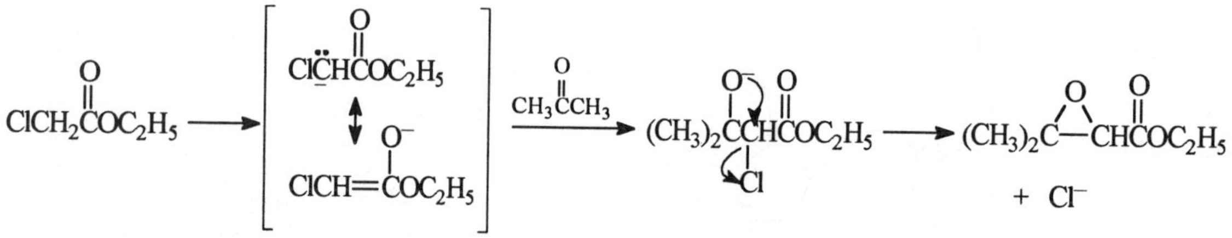

$$ClCH_2\overset{O}{\overset{\|}{C}}OC_2H_5 \longrightarrow \left[Cl\ddot{C}H\overset{O}{\overset{\|}{C}}OC_2H_5 \updownarrow ClCH{=}\overset{O^-}{\overset{|}{C}}OC_2H_5 \right] \xrightarrow{CH_3\overset{O}{\overset{\|}{C}}CH_3} (CH_3)_2\overset{O^-}{\overset{|}{C}}\underset{Cl}{\underset{|}{C}}H\overset{O}{\overset{\|}{C}}OC_2H_5 \longrightarrow (CH_3)_2\overset{O}{C{-}}CH\overset{O}{\overset{\|}{C}}OC_2H_5 + Cl^-$$

15.F. Supplementary Problems

S1. Give the anticipated products of the following reactions:

a. 3-methylbutanal and bromine in acetic acid.

b. 3-methylbutanal and bromine in sodium hydroxide solution.

c. 3-methylbutanal heated in aqueous sodium hydroxide.

d. 3-pentanone with lithium diisopropylamide followed by reaction with 2,2-dimethylpropanal.

S2. What starting aldehydes and/or ketones should be used to synthesize the following compounds using an aldol condensation?

a. 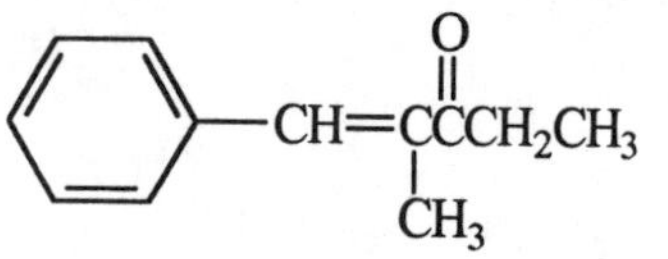 $C_6H_5{-}CH{=}\underset{CH_3}{\underset{|}{C}}\overset{O}{\overset{\|}{C}}CH_2CH_3$

b.

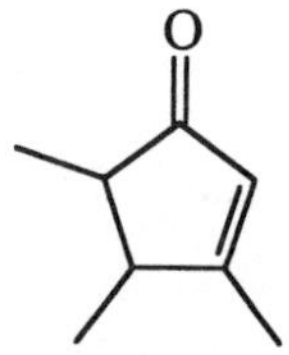

c. 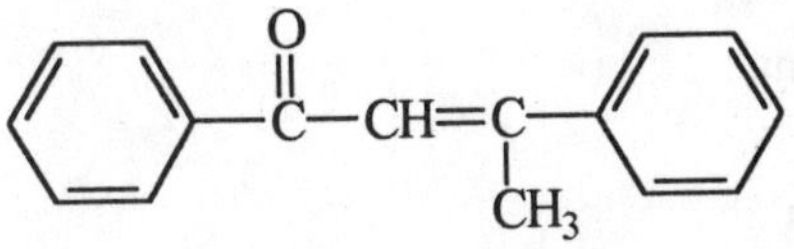 $C_6H_5-\overset{O}{\overset{\|}{C}}-CH=\underset{CH_3}{\underset{|}{C}}-C_6H_5$

S3. Write the structure of the principal product formed in each of the following reaction sequences.

a. 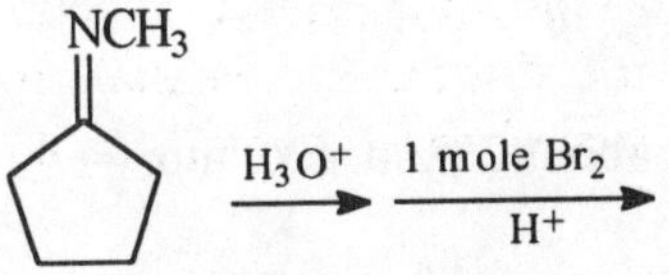 (cyclopentanone N-methylimine, $C=NCH_3$) $\xrightarrow{H_3O^+}$ $\xrightarrow[H^+]{\text{1 mole } Br_2}$

b. 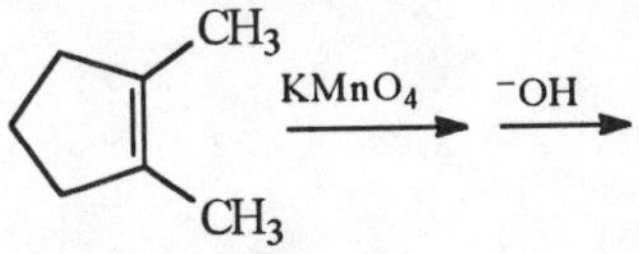 (1,2-dimethylcyclopentene) $\xrightarrow{KMnO_4}$ $\xrightarrow{^-OH}$

c. (cyclohexanone) $\xrightarrow{LDA}$ $\xrightarrow{CH_3CH_2I}$ $\xrightarrow[KOH,\ \Delta]{H_2NNH_2}$

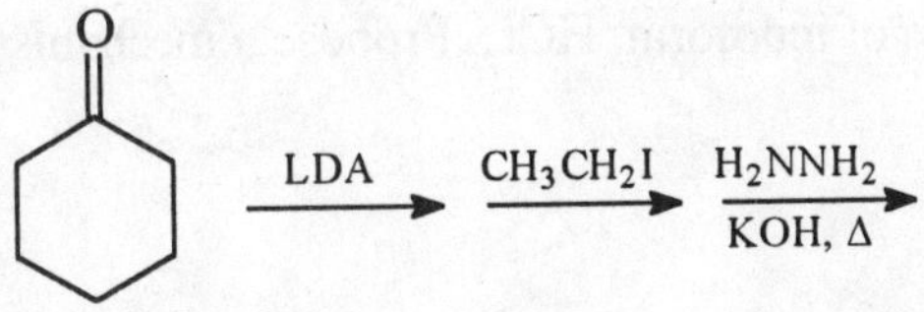

S4. Show how each of the following compounds can be prepared from materials using four or fewer carbons.

a. 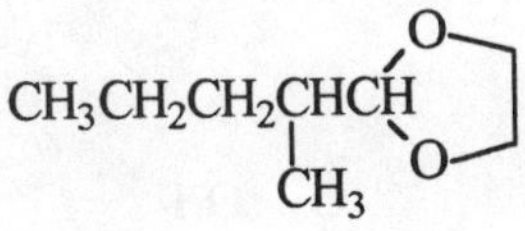 $CH_3CH_2CH_2\underset{CH_3}{\underset{|}{C}}HCH$ (1,3-dioxolane ring: $-CH<\genfrac{}{}{0pt}{}{O-}{O-}$)

b. $CH_3\overset{OH}{\overset{|}{C}}HCH=CHCH(CH_3)_2$

S5. Show how to carry out each of the following transformations. More than one step is involved in each case.

a. 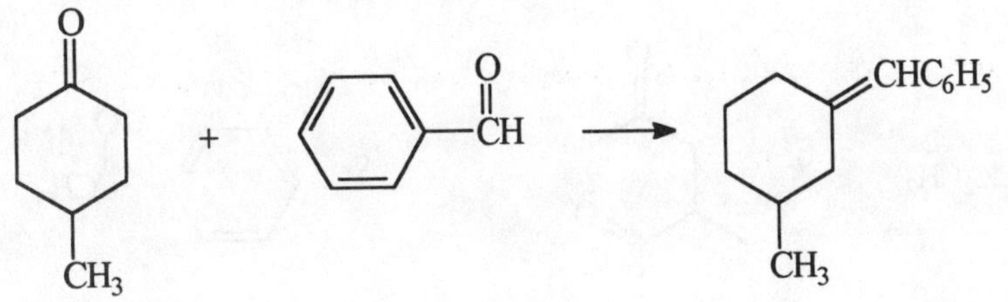 (4-methylcyclohexanone) + $C_6H_5\overset{O}{\overset{\|}{C}}H$ $\longrightarrow$ (3-methylcyclohexylidene)$=CHC_6H_5$

b. $CH_3CH_2\overset{O}{\overset{\|}{C}}H \longrightarrow HOCH_2\underset{CH_3}{\underset{|}{C}}H\overset{OH}{\overset{|}{C}}HCH_2CH_3$

S6. Show all the intermediates involved in the following reaction sequence

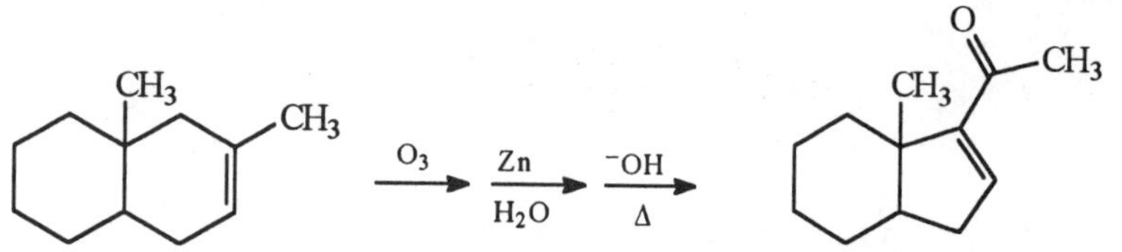

[bicyclic alkene with CH_3 groups] $\xrightarrow{O_3}$ $\xrightarrow[H_2O]{Zn}$ $\xrightarrow[\Delta]{^-OH}$ [bicyclic enone product with CH_3 and $COCH_3$]

S7. The NMR of 2,4-pentanedione shows two signals of different intensity at about 2 ppm. Explain.

S8. A reaction that can be used to identify methyl ketones is the iodoform test:

$$RCOCH_3 + 3\,I_2 + 4\,OH^- \longrightarrow RCO_2^- + HCI_3 + 3\,I^- + 3\,H_2O$$

A positive test is the appearance of the pale yellow precipitate of iodoform, HCI_3. Propose a mechanism for this reaction.

15.G. Solutions to the Supplementary Problems

S1. a. $(CH_3)_2CHCHBrCHO$ b. $(CH_3)_2CHCBr_2CHO$

c. $(CH_3)_2CHCH_2CH(OH)CH(CH(CH_3)_2)CHO$ or $(CH_3)_2CHCH_2CH{=}C(CH(CH_3)_2)CHO$

d. $CH_3CH_2COCH(CH_3)CH(OH)C(CH_3)_3$

S2. a. 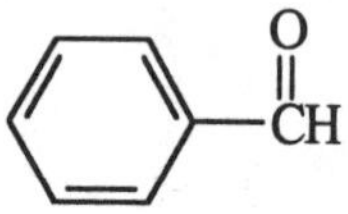 C_6H_5CHO and $CH_3CH_2COCH_2CH_3$ b. 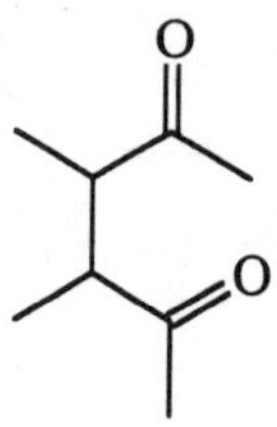 [diketone structure] c. 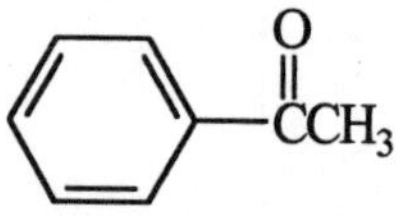 $C_6H_5COCH_3$

S3. a. 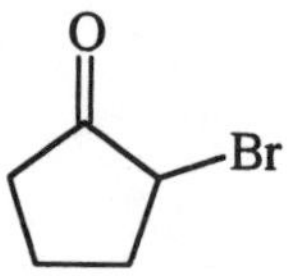 [2-bromocyclopentanone] b. 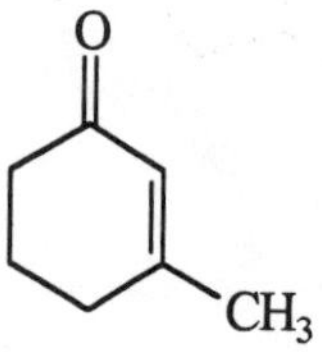 [3-methylcyclohex-2-enone] c. 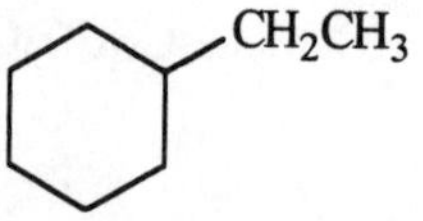 [ethylcyclohexane, CH_2CH_3]

S4.

a. 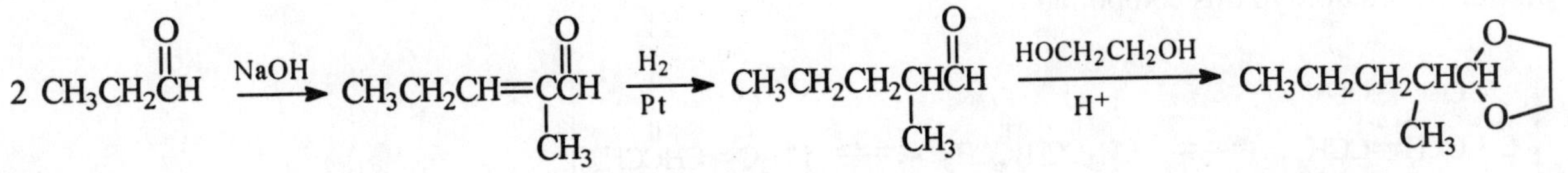 $2\ CH_3CH_2\overset{O}{\overset{\|}{C}}H \xrightarrow{NaOH} CH_3CH_2CH{=}\underset{CH_3}{\underset{|}{C}}\overset{O}{\overset{\|}{C}}H \xrightarrow[Pt]{H_2} CH_3CH_2CH_2\underset{CH_3}{\underset{|}{C}H}\overset{O}{\overset{\|}{C}}H \xrightarrow[H^+]{HOCH_2CH_2OH} CH_3CH_2CH_2\underset{CH_3}{\underset{|}{C}H}CH(OCH_2CH_2O)$

b. $CH_3\overset{O}{\overset{\|}{C}}CH_3 \xrightarrow{BuLi} CH_3\overset{OLi}{\overset{|}{C}}{=}CH_2 \xrightarrow{(CH_3)_2CH\overset{O}{\overset{\|}{C}}H} CH_3\overset{O}{\overset{\|}{C}}CH_2\overset{OH}{\overset{|}{C}H}CH(CH_3)_2 \xrightarrow{H^+} CH_3\overset{O}{\overset{\|}{C}}CH{=}CHCH(CH_3)_2$

$\xrightarrow{NaBH_4} CH_3\overset{OH}{\overset{|}{C}H}CH{=}CHCH(CH_3)_2$

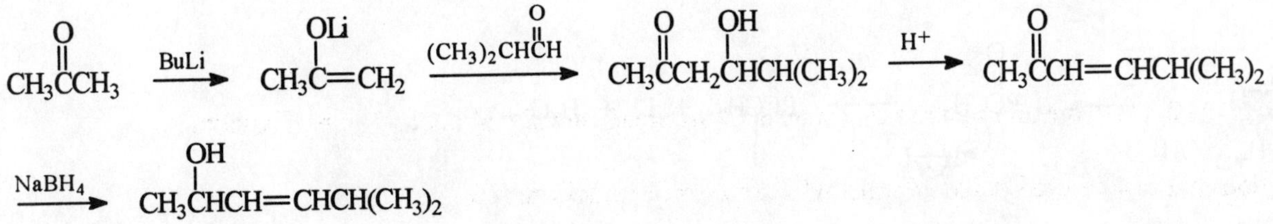

S5.

a. 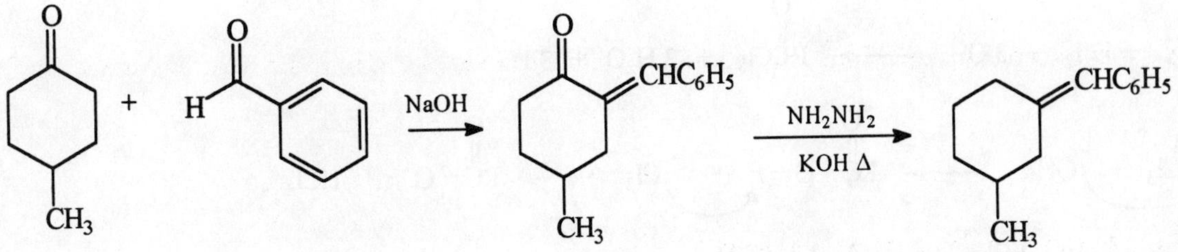 4-methylcyclohexanone + C_6H_5CHO (H–C(=O)–C$_6$H$_5$) $\xrightarrow{NaOH}$ 2-($=CHC_6H_5$)-4-methylcyclohexanone $\xrightarrow[KOH\ \Delta]{NH_2NH_2}$ 1-($=CHC_6H_5$)-3-methylcyclohexane (CH_3)

b. 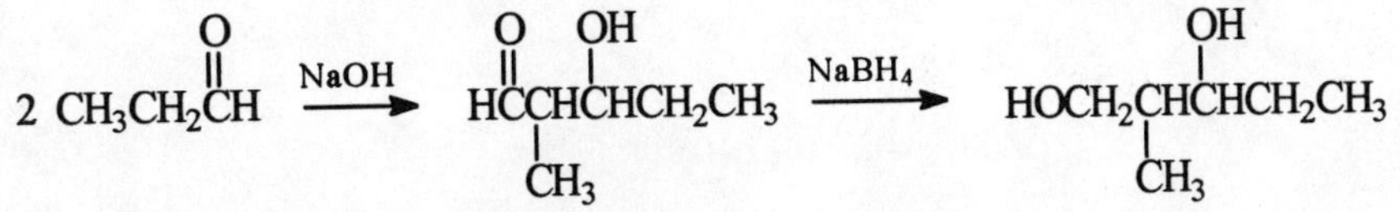 $2\ CH_3CH_2\overset{O}{\overset{\|}{C}}H \xrightarrow{NaOH} H\overset{O}{\overset{\|}{C}}\underset{CH_3}{\underset{|}{C}H}\overset{OH}{\overset{|}{C}H}CH_2CH_3 \xrightarrow{NaBH_4} HOCH_2\underset{CH_3}{\underset{|}{C}H}\overset{OH}{\overset{|}{C}H}CH_2CH_3$

S6.

Bicyclic alkene $\xrightarrow{O_3}$ ozonide $\longrightarrow$ keto aldehyde (H, O) $\underset{}{\overset{^-OH}{\rightleftharpoons}}$ enolate (^-O, H, O) $\rightleftharpoons$ alkoxide (O, O^-) $\underset{}{\overset{H_2O}{\rightleftharpoons}}$ β-hydroxy ketone (O, OH) $\overset{^-OH}{\rightleftharpoons}$ enolate (^-O, OH) $\xrightarrow{-OH^-}$ α,β-unsaturated ketone (O)

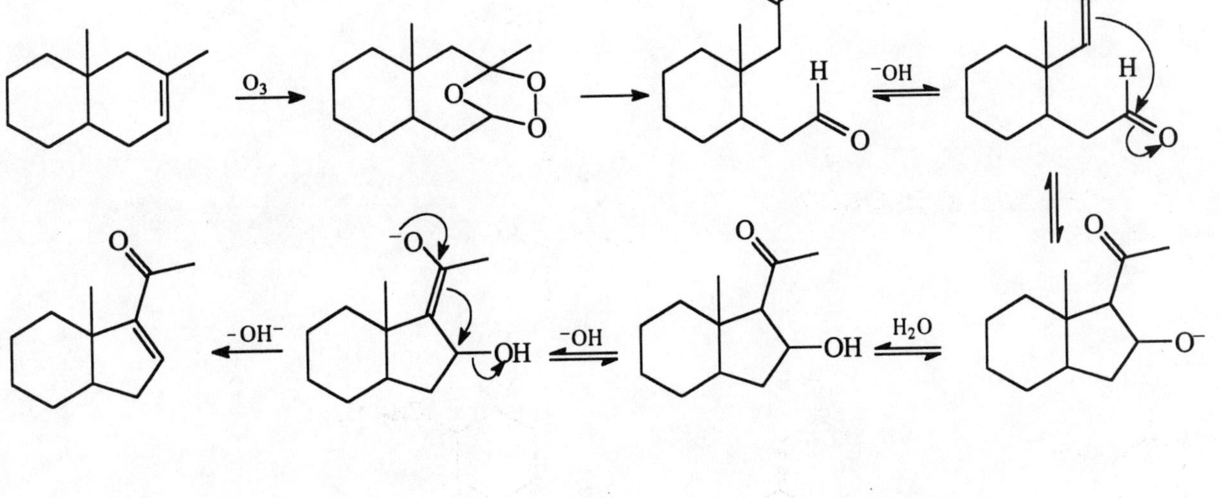

S7. The two signals arise from the non-equivalent methyls from the keto and enol forms of the dione. The enol is particularly stable in this compound.

$$\left[CH_3\overset{O}{\overset{\|}{C}}CH{=}\overset{OH}{\overset{|}{C}}CH_3 \rightleftharpoons CH_3\overset{O}{\overset{\|}{C}}CH_2\overset{O}{\overset{\|}{C}}CH_3 \rightleftharpoons CH_3\overset{OH}{\overset{|}{C}}{=}CH\overset{O}{\overset{\|}{C}}CH_3 \right]$$

S8. $$R\overset{O}{\overset{\|}{C}}\underset{H\ \ {}^{-}OH}{CH_2} \longrightarrow R\overset{O}{\overset{\|}{C}}\overset{-}{\ddot{C}}H_2 \ \ I{-}I \longrightarrow R\overset{O}{\overset{\|}{C}}\underset{I}{\underset{|}{C}}H_2 + I^- + H_2O$$

This step is repeated twice more. In basic solution, the more stable anion is the methyl. Once one H has been substituted by I, the other H's on that carbon are even more rapidly replaced.

$$R\overset{O}{\overset{\|}{C}}CH_3 + 3\,I_2 + 3\,OH^- \longrightarrow R\overset{O}{\overset{\|}{C}}CI_3 + 3\,H_2O + 3\,I^-$$

$$R\overset{O}{\overset{\|}{C}}{-}CI_3 + {}^{-}\!:OH \longrightarrow R\overset{O}{\overset{\|}{C}}{-}O{-}H + {}^{-}\!:CI_3 \longrightarrow R\overset{O}{\overset{\|}{C}}{-}O^- + HCI_3$$

16. ORGANIC SYNTHESIS

16.A. Chapter Outline and Important Terms Introduced

16.1 Introduction

16.2 Considerations in Synthesis Design

1. Construction of the proper carbon skeleton
2. Placement of desired functional groups in their proper place
3. Control of stereochemistry where relevant

16.3 Planning a Synthesis

retrosynthesis
synthetic tree
synthetic equivalents: synthons

16.4 Protecting Groups

16.5 Industrial Syntheses

process development

16.B Important Reactions Introduced: (none in this chapter)

16.C Important Concepts and Hints

To many organic chemists, the best examples of the elegance and creativity of chemistry are found in synthesis. It is in designing a synthesis of a complex molecule, and carrying it out in the laboratory, that many chemists find their greatest fulfillment. Although you are introduced to this topic with relatively simple molecules as targets, the field has advanced to the extent that we can confidently design syntheses of molecules that are bewilderingly complex. The successful synthesis of Vitamin B_{12} by the late R.B. Woodward, A. Eschenmoser, and their research groups is a spectacular example of what the chemist can now accomplish.

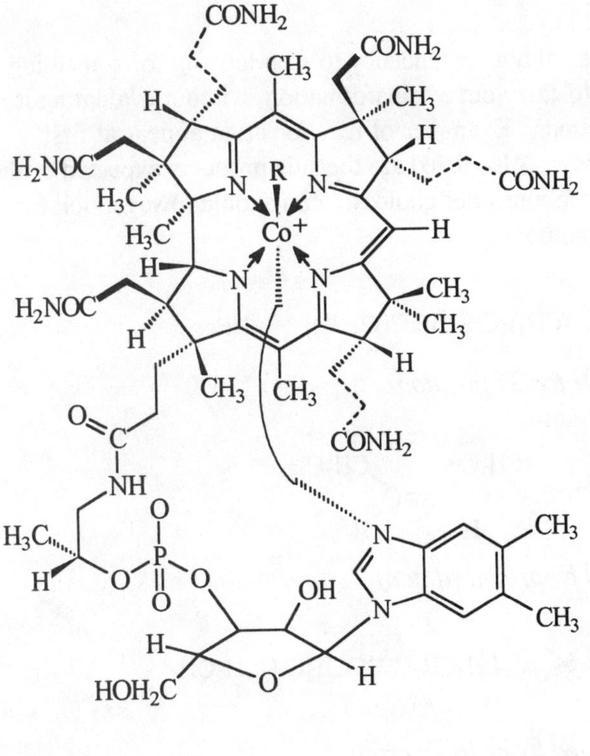

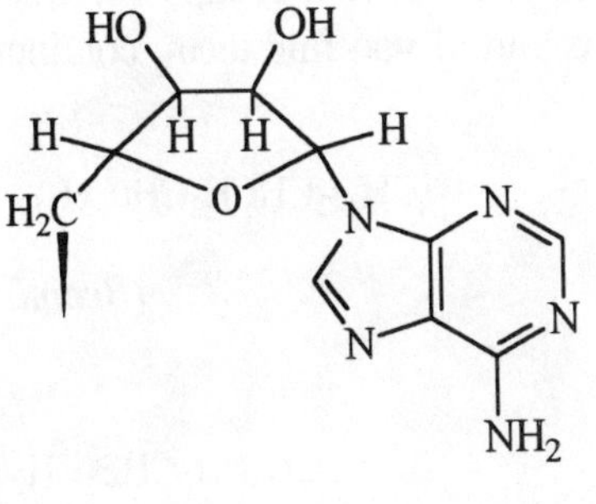

For the most part, organic chemists choose as their synthetic targets compounds isolated from natural sources, or compounds similar to them. Such syntheses are pursued in order to produce compounds of potentially valuable biological activity, to assist in structure determination, to demonstrate the usefulness of a new reaction sequence, or, one suspects in some cases, for the fun of it. The development of efficient syntheses of less complex, commercially important compounds is an equally challenging aspect of organic synthesis. In this regard, elegance and creativity are measured by the simplicity and economy of the processes, their energy efficiency, the absence of byproducts, etc.

Regardless of the type of synthesis you are trying to develop, you must be able to think logically and you must know a broad range of organic reactions. With these two abilities, as well as practice, you will develop an intuition for organic synthesis and be able to appreciate the creative possibilities. Chapter 16 outlines the logical processes involved in synthesis design, the rest of the text provides you with the background knowledge of reactions (your repertoire). It's up to you to practice synthesis problems and develop the necessary intuition.

The suggestion that you should work synthesis problems backwards was made in Section 12.B of this Study Guide, and is discussed in more detail in Chapter 16 of the Text. An example of the sort of thing that happens when you work a synthesis problem **forward** is the following scheme, taken from an actual answer to a midterm exam question:

cyclohexane $\xrightarrow[hv]{Br_2}$ bromocyclohexane (Br) $\xrightarrow[C_2H_5OH]{KOH}$ cyclohexene $\xrightarrow[H_2SO_4]{dilute}$ cyclohexanol (OH) $\xrightarrow{PBr_3}$ bromocyclohexane (Br) $\xrightarrow{NaCN}$ cyclohexanecarbonitrile (CN)

Because the student was thinking "What can I make from this compound?" at each stage, he wasted three steps converting bromocyclohexane into bromocyclohexane!

A more subtle mistake that we all make arises from the usual human tendency to stop looking for something once we think we've found it. Often you will think of a way to carry out a transformation, without evaluating it closely to determine whether there are any potential side reactions. Examples of reactions that appear at first glance to accomplish their intended conversion are shown below. Also listed are the side products expected or the incompatibility of functional groups that make these transformations poor choices. You should always look for such problems and, if you find them, continue to search for another route.

$$(CH_3)_2CHCH_2CH(CH_3)_2 + Br_2 \xrightarrow[hv]{} \times\ (CH_3)_2CHCHBrCH(CH_3)_2 + HBr$$

(Bromination is selective for 3° position)

$$CH_3CH_2CHBrCH_2CH_2CH_3 \xrightarrow[EtOH]{KOH} \times\ \text{(E)-}CH_3CH_2CH{=}CHCH_2CH_3$$

(product drawn as CH_3CH_2 and H on one carbon, CH_2CH_3 and H on the other, $C{=}C$)

(Three other isomers will be produced, too)

$$CH_3CH_2C{\equiv}CNa + BrCH_2CH_2CH_2OH \longrightarrow \times\ CH_3CH_2C{\equiv}CCH_2CH_2CH_2OH$$

(Acetylide ion will be protonated by hydroxy proton)

$$2\ \overset{O}{\overset{\|}{H C}}CH_2CH_2Br \xrightarrow{NaOH} \!\!\!\!\!\times\ \overset{O}{\overset{\|}{HC}}\overset{OH}{\overset{|}{CH}}\underset{CH_2Br}{\underset{|}{CH}}CH_2CH_2Br$$

(Bromide will undergo substitution or elimination under the basic conditions)

$$CH_3\overset{O}{\overset{\|}{C}}CH_2CH_2CH{=}CH_2 \xrightarrow[HBr]{Br_2} \!\!\!\!\!\times\ CH_3\overset{O}{\overset{\|}{C}}\underset{Br}{\underset{|}{CH}}CH_2CH{=}CH_2$$

(Double bond will undergo bromine addition)

$$CH_3\overset{Br}{\overset{|}{CH}}CH_2CH_2CH_2Br \xrightarrow{Mg} \!\!\!\!\!\times\ CH_3\overset{Br}{\overset{|}{CH}}CH_2CH_2CH_2MgBr$$

(The other bromide will react, too)

These are only examples to show you the sort of difficulties to look out for.

16.D. Answers to Exercises

16.1

$$R\text{-}Y \xrightarrow{?} R\text{-}Z$$

NOTE: this matrix was constructed based on reactions presented in Chapters 1-15; parentheses indicate entries that would be changed based on reactions introduced in subsequent chapters.

RY↓RZ→	RH	RBr	ROH	RCH_2OH	RCHO	$RCOCH_3$	RCN
RH		–*	–	–	–	–	–
RBr	1[*]		1	1[*]	1[*]	+	1
ROH	+	1		+	+	+	+
RCH_2OH	+	(-)	(-)		1	+	(-)
RCHO	1	(-)	(-)	1		+	(-)
$RCOCH_3$	+	(-)	+	(-)	(-)		–
RCN	+	(-)	(-)	(-)	1	(-)	

*Although some RH can be converted to RBr by free radical bromination, the reaction is not general because some RH do not react and others give impractical mixtures.

[*]One step from Grignard

16.2 4-Methylpentanenitrile: the possible carbon-carbon connections to build up the six-carbon skeleton are indicated below, with the asterisk showing where the functional group is located in the product:

1. (C)(C)C–C–C–C* 2. (C)(C)C C–C–C* 3. (C)(C)C–C C–C* 4. (C)(C)C–C–C C*

The most logical connection is 4: making the new bond next to the nitrile functional group. You can use the displacement reaction of an alkyl halide with cyanide ion to accomplish this:

$$(CH_3)_2CHCH_2CH_2Br + NaCN \longrightarrow (CH_3)_2CHCH_2CH_2CN$$

16.3 a. Acetylene is a synthon for the -$COCH_3$ group, via the following sequence:

$$HC{\equiv}CH \xrightarrow{NaNH_2} HC{\equiv}CNa \xrightarrow{RX} HC{\equiv}CR \xrightarrow[HgSO_4]{H_2SO_4} R\overset{O}{\overset{\|}{C}}CH_3$$

b. Cyanide is the best synthon for -CH_2NH_2, via the following reactions:

$$RX + Na^{+}\,{}^{-}C{\equiv}N \rightarrow RC{\equiv}N \xrightarrow[PtO_2,\ H^+]{H_2} RCH_2NH_2$$

c. Formaldehyde: $RMgX + CH_2{=}O \rightarrow \xrightarrow{H_2O} RCH_2OH$

d. Propyne: $HC{\equiv}CCH_3 \xrightarrow{NaNH_2} Na^{+}\,{}^{-}C{\equiv}CCH_3 \xrightarrow{RX} RC{\equiv}CCH_3$

16.4 **1-Heptene:** If we are only going to consider combinations that will give the seven-carbon skeleton by formation of **one** carbon-carbon bond (with starting materials of five carbons or less), the possibilities are indicated below, with the asterisk showing where the functional group is located in the product:

1. C–C C–C–C–C*–C 2. C–C–C C–C–C*–C

3. C–C–C–C C–C*–C 4. C–C–C–C–C C*–C

Review the methods available for introduction of a carbon-carbon double bond:

a. Dehydration of an alcohol b. Elimination of an alkyl halide

Both of these can be ruled out for the same reasons as in Example 16.3 of the Text.

c. Partial hydrogenation of an alkyne: $RC{\equiv}CH \xrightarrow[\text{poisoned Pd catalyst}]{H_2} RCH{=}CH_2$

This is particularly attractive, since the alkyne can be used to form the carbon-carbon bond via combination 4.

d. Wittig reaction: this will form the carbon-carbon double bond, but it does not correspond to any of the disconnections given above, and would necessitate **two** C-C bond forming reactions.

As indicated in (c), the route via 1-heptyne is desirable, since alkylation of a primary halide with sodium acetylide will afford this in one step:

$$CH_3(CH_2)_4Br + Na^{+\,-}C{\equiv}CH \longrightarrow CH_3(CH_2)_4C{\equiv}CH \longrightarrow CH_3(CH_2)_4CH{=}CH_2$$

16.5 Non-3-yn-1-ol: as pointed out in the text, oxirane is a synthon for the CH_2CH_2OH group, via the following reaction:

$$R{-}M + \text{(oxirane)} \longrightarrow \xrightarrow{H_2O} RCH_2CH_2OH$$

The appropriate precursor to "R-M" in this case would be 1-heptyne, which was an intermediate in the 1-heptene synthesis that was outlined in Exercise 16.4:

$$CH_3(CH_2)_4C{\equiv}CH \longrightarrow CH_3(CH_2)_4C{\equiv}C^-Na^+ \longrightarrow \longrightarrow CH_3(CH_2)_4C{\equiv}CCH_2CH_2OH$$

16.6 2,9-dimethyldecan-5-ol: with twelve carbons in the target, at least two carbon-carbon bond-forming reactions will be required. If we consider only combination strategies that use two five-carbon units and a two-carbon unit, the most reasonable connection we look for is:

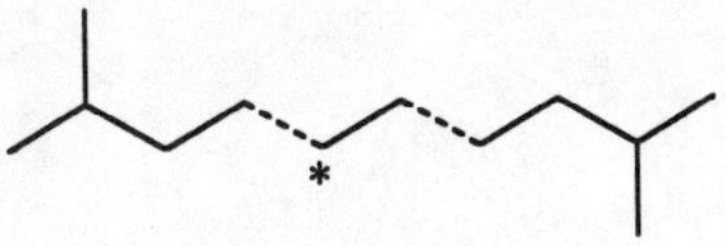

*

Review methods for introduction of an alcohol group:

a. Hydration of an alkene:

$$RCH{=}CHR' \xrightarrow[\text{or } 1.\ B_2H_6 \quad 2.\ H_2O_2,\ OH^-]{\text{aq. } H_2SO_4} RCHOHCH_2R'$$

b. Reduction of a ketone:

$$RC(=O)R' \longrightarrow RCH(OH)R'$$

c. Hydrolysis of an alkyl halide:

$$RCH(X)R' \xrightarrow{OH^-} RCH(OH)R'$$

d. Grignard (or other organometallic reaction):

$$RMgX + HC(=O)R' \longrightarrow \xrightarrow{H_2O} RCH(OH)R'$$

$$\mathrm{RMgX} + \mathrm{H\overset{O}{\overset{\|}{C}}R'} \longrightarrow \xrightarrow{H_2O} \mathrm{R\overset{OH}{\overset{|}{C}}HR'}$$

This problem is a particularly good one for illustrating the fact that there are often several good routes to a given target. Compare the following routes, which rely on either (a), (b), or (d) above. We can rule out (c), because from the point of view of synthesis, halides are usually made from alcohols, and not vice versa.

A. [via (a)]

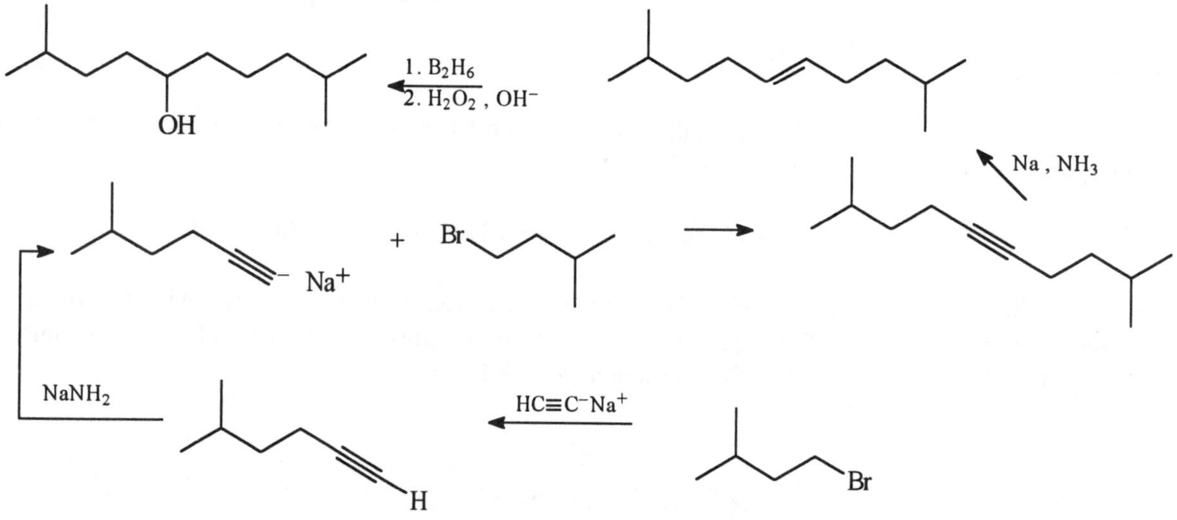

1. B_2H_6
2. H_2O_2, OH^-

OH

Na , NH_3

+ Br

Na^+

$NaNH_2$

$HC{\equiv}C^-Na^+$

H

Br

In this route, we recognize that the alkene precursor to the alcohol can be **symmetrical**, therefore it doesn't matter which end the OH group ends up on. The alkene in turn is available from the alkyne, which allows us to make easily the carbon-carbon bonds required by our connection strategy. Note that several equally good alternatives are available for partial reduction of the triple bond and hydration of the alkene.

B. [via (b)]

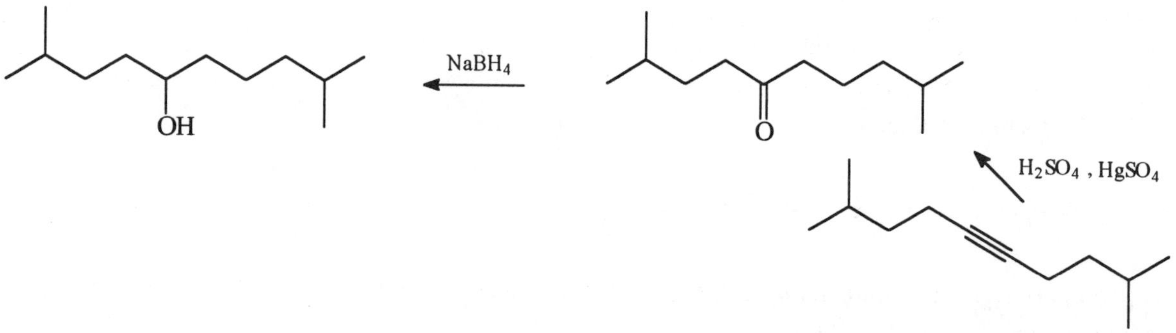

$NaBH_4$

OH

O

H_2SO_4 , $HgSO_4$

In terms of number of steps, this method is the same as A., and has a similar advantage in that it uses the same starting materials.

C. [via (d)]

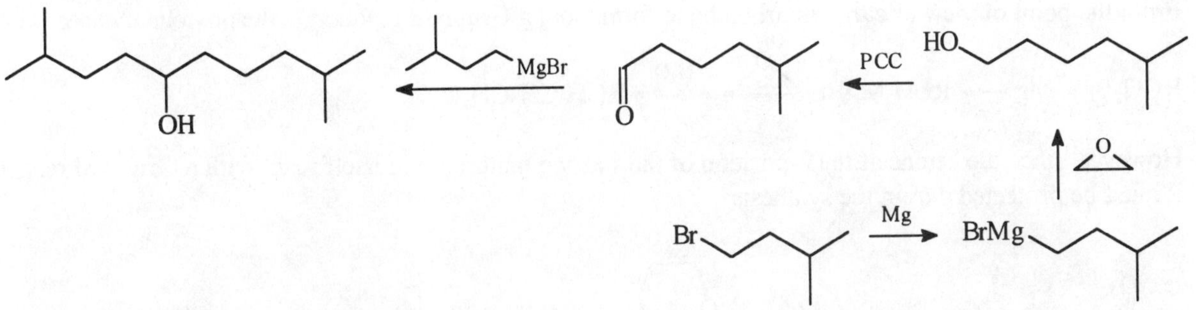

This sequence relies on Grignard reactions for the carbon-carbon bond forming reactions, and takes advantage of oxirane as a synthon for the O-C-C- unit.

D. The following synthesis is given as an example of a route that is workable, but since it involves quite a few more steps than the ones above, it is a poor answer.

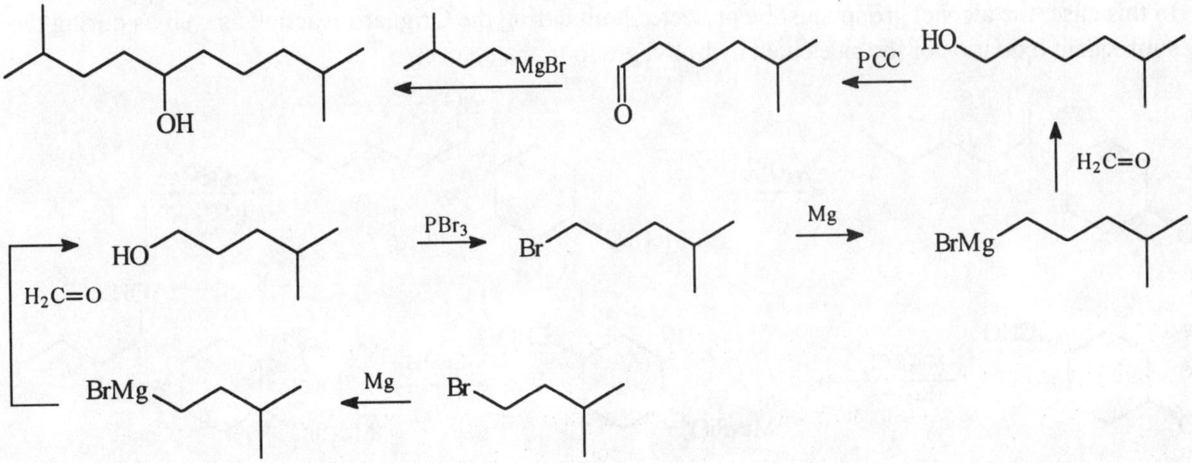

16.7 4-(Deuteriomethyl)nonane: In this case, we need to introduce the deuterium atom specifically on the sidechain methyl group, so we have to consider connection strategies that will leave functionality on that carbon. Working the synthesis backward, we realize that 2-propyl-1-heptanol is a key intermediate.

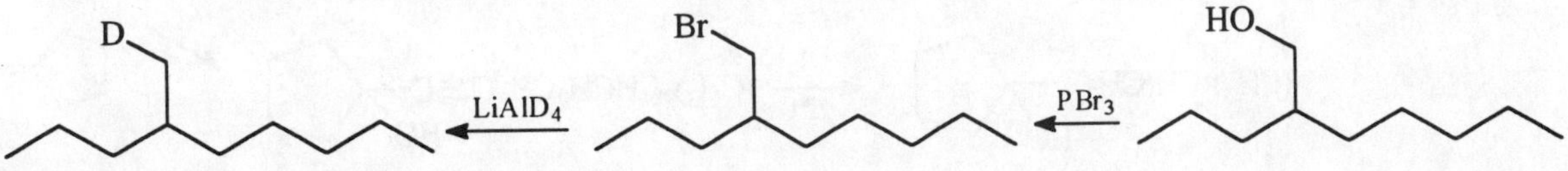

A reaction that forms a carbon-carbon bond **adjacent** to an oxygenated carbon is required, and the most appropriate is the aldol condensation:

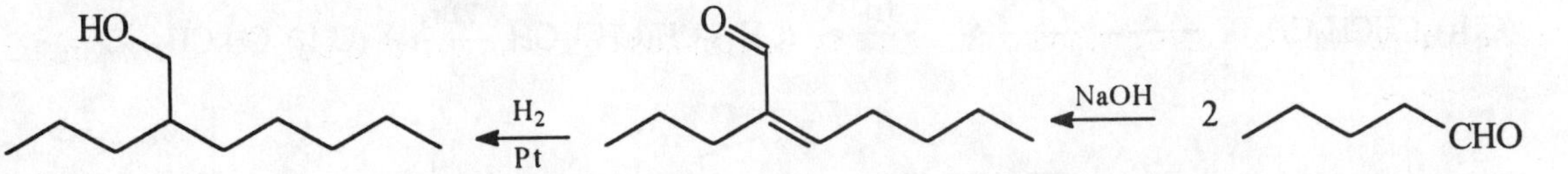

16.8 From the point of view of carbon-carbon bond formation, a Grignard reaction is the obvious choice:

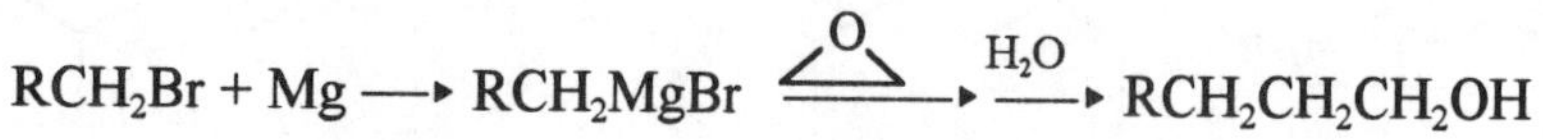

$$RCH_2Br + Mg \longrightarrow RCH_2MgBr \xrightarrow{\text{ethylene oxide}} \xrightarrow{H_2O} RCH_2CH_2CH_2OH$$

However, since the ketone at the 2-position of the starting material will itself react with a Grignard reagent, it must be protected during the synthesis:

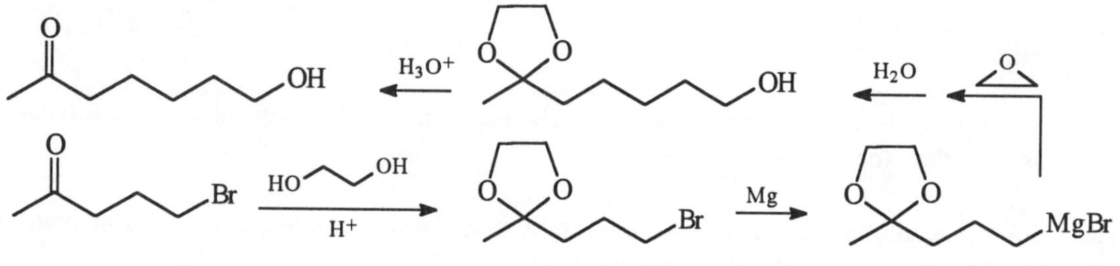

O, OH ← H_3O^+ — O, O, OH ← H_2O ← ethylene oxide

O, Br — HO⌒⌒OH / H^+ → O, O, Br — Mg → O, O, MgBr

16.9 In this case, the alcohol group must be protected both during the Grignard reaction as well as during the subsequent oxidation of the side chain hydroxyl group to the ketone.

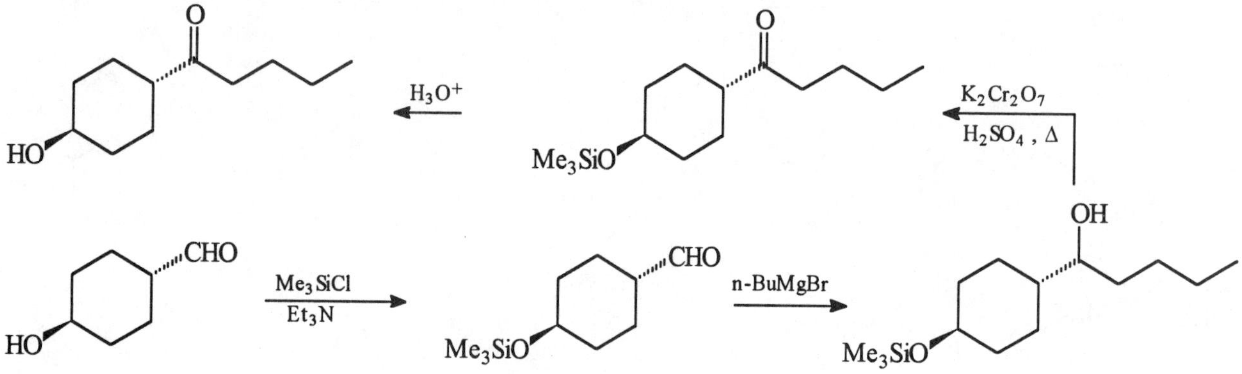

O, HO ← H_3O^+ — O, Me_3SiO ← $K_2Cr_2O_7$, H_2SO_4, Δ — OH

CHO, HO — Me_3SiCl / Et_3N → CHO, Me_3SiO — n-BuMgBr → OH, Me_3SiO

16.10

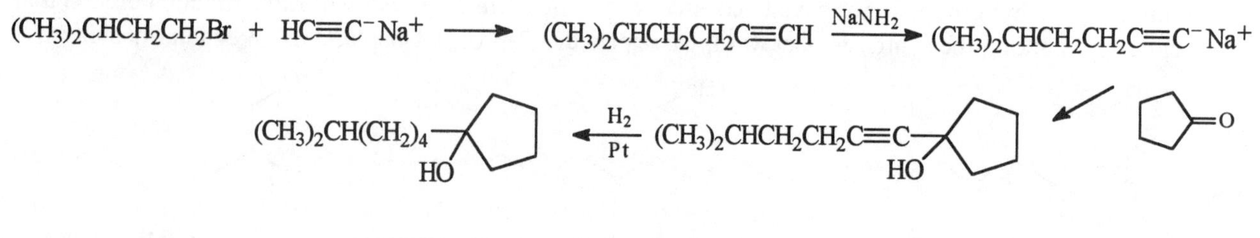

$$(CH_3)_2CHCH_2CH_2Br + HC{\equiv}C^-Na^+ \longrightarrow (CH_3)_2CHCH_2CH_2C{\equiv}CH \xrightarrow{NaNH_2} (CH_3)_2CHCH_2CH_2C{\equiv}C^-Na^+$$

$(CH_3)_2CH(CH_2)_4$–(cyclopentyl)–HO ← H_2 / Pt — $(CH_3)_2CHCH_2CH_2C{\equiv}C$–(cyclopentyl)–HO ← cyclopentanone (=O)

or

$$(CH_3)_2CHCH_2CH_2Br \xrightarrow{Mg} \xrightarrow{\text{ethylene oxide}} \xrightarrow{H_2O} (CH_3)_2CH(CH_2)_4OH \xrightarrow{PBr_3} (CH_3)_2CH(CH_2)_4Br$$

↓ Mg, ether

$(CH_3)_2CH(CH_2)_4$–(cyclopentyl)–HO ← H_2O — ← cyclopentanone (=O)

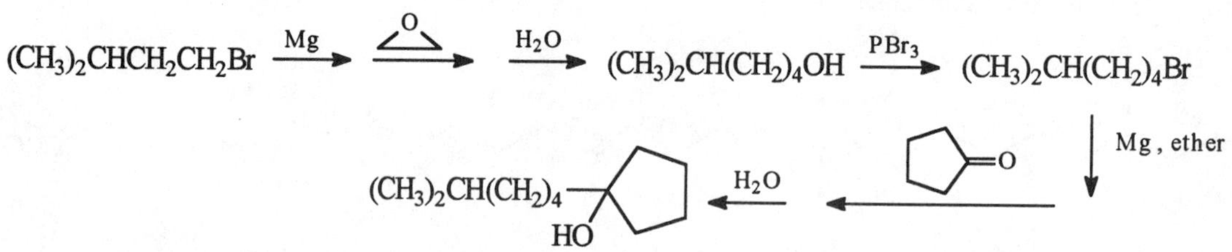

16.11 Since (*S*)-active amyl alcohol is commercially available, it can be brominated using PBr_3 to produce (*S*)-1-bromo-2-methylbutane. The reaction sequences then will be the same as those shown in 16.10. Since the methyl group is now on C-2 rather than C-3, there will be β-branching to slow down the reaction with acetylide.

16.12 HC≡CH and oxirane are the synthons for a two carbon fragment with a functional group. Cyclopentanone is the synthon for cyclopentanol; the (*S*)-active amyl alcohol and 1-bromo-3-methylbutane are the synthons for the five carbon branched chain.

16.E. Answers and Explanations for Problems

NOTE: for most synthesis problems, especially the more complex ones, there is more than one "correct" answer. Two or three routes can be devised which will lead to the target compound. Often one route will stand out above the others because it is shorter, or avoids low-yield reactions or isomeric products. Always try to find such a route, because that one is "more correct" as an answer. In this Study Guide, we cannot list every conceivable way to make the target compound, and your answer may differ from the one depicted here. Look at our explanations carefully, and learn to evaluate possible syntheses by comparison of yours with ours.

1\. a.

$$CH_3(CH_2)_5OH \xleftarrow{H_2O} CH_3(CH_2)_5OMgBr \xleftarrow{H_2C{=}O} CH_3(CH_2)_4MgBr \xleftarrow[\text{ether}]{Mg} CH_3(CH_2)_4Br$$

$CH_3(CH_2)_5OMgBr$ ← oxirane + $CH_3(CH_2)_3MgBr$

b.

$$CH_3(CH_2)_3CH(OH)CH_3 \xleftarrow{H_2O} CH_3(CH_2)_3CH(OMgBr)CH_3$$

← $CH_3(CH_2)_3CHO + CH_3MgI$

or

← $CH_3(CH_2)_3MgBr + HC(=O)CH_3$

c.

$$CH_3(CH_2)_2CH(OH)CH_2CH_3 \xleftarrow{H_2O} CH_3(CH_2)_2CH(OMgBr)CH_2CH_3$$

← $CH_3(CH_2)_2CHO + C_2H_5MgI$

or

← $CH_3(CH_2)_2MgBr + HC(=O)C_2H_5$

d.

$$CH_3(CH_2)_4C{\equiv}CH \longleftarrow CH_3(CH_2)_4Br + NaC{\equiv}CH$$

e.

$$\text{trans-}CH_3CH_2CH{=}CHCH_2CH_2CH_3 \xleftarrow[NH_3]{Na} CH_3CH_2C{\equiv}CCH_2CH_2CH_3$$

$CH_3CH_2C{\equiv}CCH_2CH_2CH_3 \xleftarrow{CH_3CH_2Br} \xleftarrow{NaNH_2} HC{\equiv}CCH_2CH_2CH_3$

or

$CH_3CH_2C{\equiv}CCH_2CH_2CH_3 \xleftarrow{BrCH_2CH_2CH_3} \xleftarrow{NaNH_2} CH_3CH_2C{\equiv}CH$

(Any route involving elimination of an alcohol or a halide would give some *cis* isomer, too.)

f.

$$CH_3CH_2CH_2CH_2\overset{O}{\overset{\|}{C}}H \xrightarrow[2.\ H_2O]{1.\ CH_3(CH_2)_3MgBr} \xrightarrow[H_2SO_4\ ,\ \Delta]{K_2Cr_2O_7} \xrightarrow{CH_3MgI} (CH_3CH_2CH_2CH_2)_2\overset{OMgX}{\overset{|}{C}}CH_3$$

$$CH_3CH_2CH_2CH_2\overset{O}{\overset{\|}{C}}H \xrightarrow[2.\ H_2O]{1.\ CH_3MgBr} CH_3CH_2CH_2CH_2\overset{OH}{\overset{|}{C}}HCH_3$$

$$CH_3CH_2CH_2CH_2Br \xrightarrow{1.\ Mg} \xrightarrow[3.\ H_2O]{2.\ CH_3CHO} CH_3CH_2CH_2CH_2\overset{OH}{\overset{|}{C}}HCH_3$$

$$CH_3CH_2CH_2CH_2\overset{OH}{\overset{|}{C}}HCH_3 \xrightarrow[H_2SO_4\ ,\ \Delta]{K_2Cr_2O_7} \xrightarrow{CH_3(CH_2)_3MgBr} (CH_3CH_2CH_2CH_2)_2\overset{OMgBr}{\overset{|}{C}}CH_3$$

$$(CH_3CH_2CH_2CH_2)_2\overset{OMgX}{\overset{|}{C}}CH_3 \xrightarrow{H_2O} (CH_3CH_2CH_2CH_2)_2\overset{OH}{\overset{|}{C}}CH_3$$

Note that in many instances a number of similar, equally valid routes are available. In Section 19.7.D of the Text, you will find the reaction depicted here, which would be the best way to prepare 5-methyl-5-nonanol.

$$2\ RMgX + R'\overset{O}{\overset{\|}{C}}OR'' \longrightarrow \xrightarrow{H_2O} R'\overset{OH}{\overset{|}{\underset{R}{\underset{|}{C}}}}R$$

g. **1st decision:** which C-C bond to form

$$\underbrace{(CH_3)_2\overset{OH}{\overset{|}{C}}CH_2}_{4\ carbons}\text{----}\underbrace{CH_2CH_2CH(CH_3)_2}_{5\ carbons} \quad \text{or} \quad \underbrace{(CH_3)_2\overset{OH}{\overset{|}{C}}CH_2CH_2}_{5\ carbons}\text{----}\underbrace{CH_2CH(CH_3)_2}_{4\ carbons}$$

The first choice is the best, because the functional group in the product is closest to the bond to be formed.

2nd decision: what C-C bond-forming reaction to choose

$$(CH_3)_2CH\overset{O}{\overset{\|}{C}}H + MCH_2CH_2CH(CH_3)_2 \longrightarrow \xrightarrow{H_2O} (CH_3)_2CH\overset{OH}{\overset{|}{C}}HCH_2CH_2CH(CH_3)_2$$

This reaction would put the hydroxy group in the wrong place. You could correct this, but the sequence would involve several steps and would therefore be inefficient.

$$(CH_3)_2CHCHOHCH_2CH_2CH(CH_3)_2 \xrightarrow{H_2SO_4} \underset{(major)}{(CH_3)_2C{=}CHCH_2CH_2CH(CH_3)_2} \xrightarrow{dil.\ H_2SO_4} (CH_3)_2COH(CH_2)_3(CH_3)_2$$

$$+$$

$$(CH_3)_2CHCH{=}CHCH_2CH(CH_3)_2\ \ (minor)$$

Alternatively,

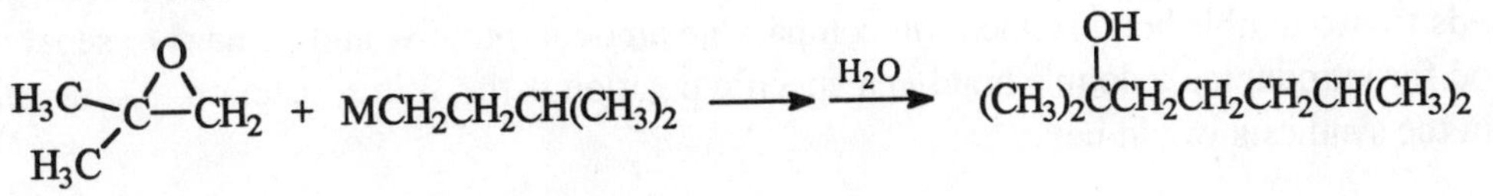

$$H_3C\text{-}\overset{O}{C}(CH_3)\text{-}CH_2 + MCH_2CH_2CH(CH_3)_2 \longrightarrow \xrightarrow{H_2O} (CH_3)_2C(OH)CH_2CH_2CH_2CH(CH_3)_2$$

This reaction gives the desired products directly.

FINAL DECISION: what M should be: MgX or Li

With an epoxide such as this one, Grignard reactions often induce rearrangement prior to addition. The organolithium reaction is therefore the better choice.

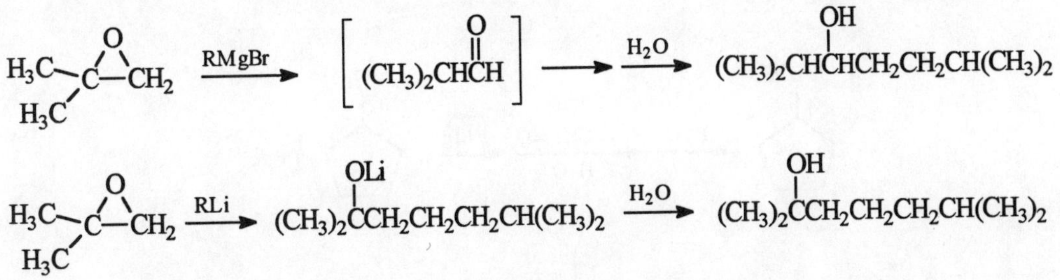

$$H_3C\text{-}\overset{O}{C}(CH_3)\text{-}CH_2 \xrightarrow{RMgBr} \left[(CH_3)_2CHCH{=}O \right] \longrightarrow \xrightarrow{H_2O} (CH_3)_2CHCH(OH)CH_2CH_2CH(CH_3)_2$$

$$H_3C\text{-}\overset{O}{C}(CH_3)\text{-}CH_2 \xrightarrow{RLi} (CH_3)_2C(OLi)CH_2CH_2CH_2CH(CH_3)_2 \xrightarrow{H_2O} (CH_3)_2C(OH)CH_2CH_2CH_2CH(CH_3)_2$$

Many other routes can be imagined, such as the multistep sequence below, but the extra steps required make them very inefficient, and therefore poor solutions to the problem.

$$BrMg(CH_2)_2CH(CH_3)_2 \xrightarrow[2.\ H_2O]{1.\ CH_2{=}O} HO(CH_2)_3CH(CH_3)_2 \xrightarrow[2.\ Mg]{1.\ PBr_3} BrMg(CH_2)_3CH(CH_3)_2$$

$$\xrightarrow[2.\ H_2O]{1.\ (CH_3)_2C{=}O} (CH_3)_2COH(CH_2)_3CH(CH_3)_2$$

h. 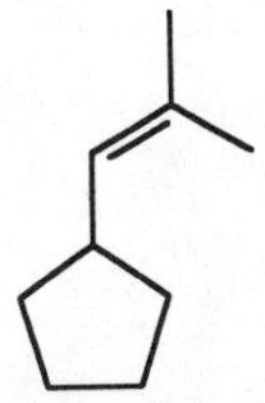 First decision: which C-C bond to form

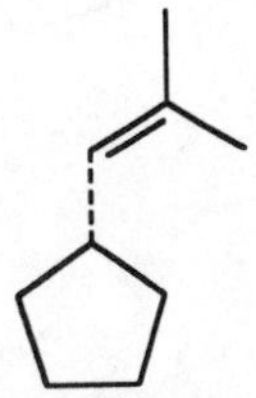

4 carbons

5 carbons

This would be efficient from the point of view of requiring only one C-C bond-forming step. The most obvious way to do this would involve a Grignard reaction and dehydration sequence:

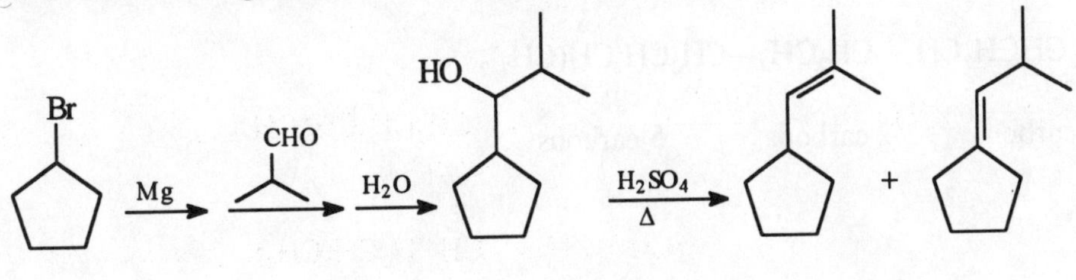

Br — Mg — CHO — H_2O — HO — H_2SO_4, Δ — +

However, this route leads to two double bond isomers (in comparable amounts) that would be hard to separate. The best general method for introducing a double bond in a specific position is the Wittig reaction. Using this reaction, the last step in the synthesis would be:

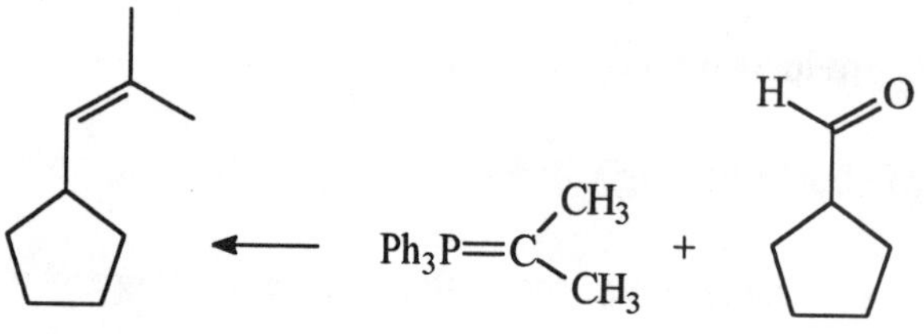

$Ph_3P{=}C(CH_3)_2$ + H, O (cyclopentanecarbaldehyde)

The aldehyde would be synthesized as follows:

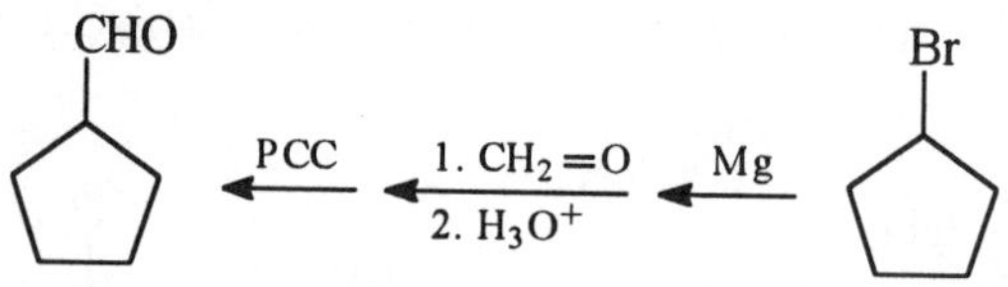

CHO ← PCC ← 1. $CH_2{=}O$ 2. H_3O^+ ← Mg ← Br

2. a. ← H_2 / Pt ← (mixture of alkenes) ← H_2SO_4, Δ ← OH

CHO + BrMg →

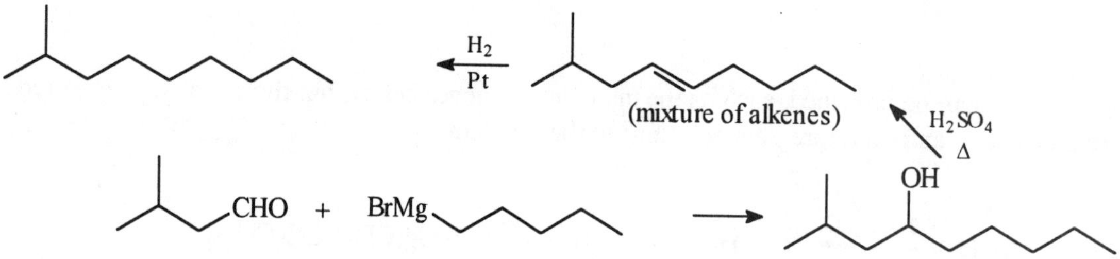

b. ← H_2 / Pt ← H_2SO_4 / Δ ← OH ← H_2O ←

=O + BrMg—

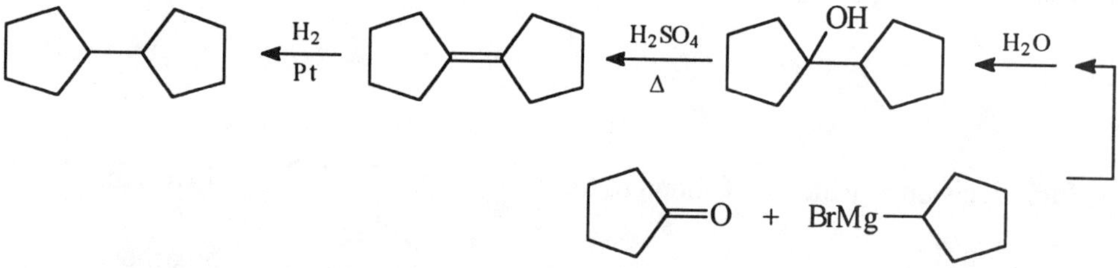

c. $(CH_3)_2CH(CH_2)_6CH(CH_3)_2$

1st decision: which C-C bonds to form:

$(CH_3)_2CHCH_2CH_2$---CH_2CH_2---$CH_2CH_2CH(CH_3)_2$

5 carbons 2 carbons 5 carbons

2nd decision: what reaction to use (take the symmetry of the molecule as a hint)

$$(CH_3)_2CHCH_2CH_2Br + NaC{\equiv}CH \longrightarrow (CH_3)_2CHCH_2CH_2C{\equiv}CH \xrightarrow{NaNH_2} \xrightarrow{BrCH_2CH_2CH(CH_3)_2} \;\downarrow H_2O$$

$$(CH_3)_2CH(CH_2)_6CH(CH_3)_2 \xleftarrow[Pt]{H_2} (CH_3)_2CHCH_2CH_2C{\equiv}CCH_2CH_2CH(CH_3)_2 \longleftarrow$$

d. Don't forget the aldol condensation!

$$2\ (CH_3)_2CHCH_2\overset{O}{\overset{\|}{C}}H \xrightarrow[\Delta]{NaOH} (CH_3)_2CH\underset{O=CH}{C}{=}CHCH_2CH(CH_3)_2 \xrightarrow{Ht/PtO_2} (CH_3)_2CH\underset{O=CH}{C}HCH_2CH_2CH(CH_3)_2$$

$$(CH_3)_2CH\underset{CH_3}{C}HCH_2CH_2CH(CH_3)_2 \xleftarrow[H_2NNH_2\ /\ KOH]{Zn\ ,\ HCl\ or} (CH_3)_2CH\underset{O=CH}{C}HCH_2CH_2CH(CH_3)_2$$

e.

$$CH_2{=}CHCH_2MgBr + CH_3\overset{O}{\overset{\|}{C}}CH_2CH_2CH_3 \xrightarrow{H_2O} CH_2{=}CHCH_2\overset{CH_3}{\underset{OH}{C}}CH_2CH_2CH_3$$

$$CH_2{=}CHCH_2\overset{CH_3}{C}HCH_2CH_2CH_3 \xleftarrow[2.\ Mg\quad 3.\ H_2O]{1.\ PBr_3} CH_2{=}CHCH_2\overset{CH_3}{\underset{OH}{C}}CH_2CH_2CH_3$$

f.

$$(CH_3)_2CHCH_2CH_2MgBr \xrightarrow{H_2C=O} (CH_3)_2CHCH_2CH_2CH_2OH \xrightarrow{PCC} (CH_3)_2CHCH_2CH_2\overset{O}{\overset{\|}{C}}H$$

$$(CH_3)_2CHCH_2CH_2\overset{O}{\overset{\|}{C}}H \xrightarrow[2.\ H_2O]{1.\ (CH_3)_2CHCH_2CH_2MgBr} ((CH_3)_2CHCH_2CH_2)_2CHOH \xrightarrow[H_2SO_4]{K_2Cr_2O_7} ((CH_3)_2CHCH_2CH_2)_2C{=}O$$

$$((CH_3)_2CHCH_2CH_2)_2C{=}O \xrightarrow{(CH_3)_2CHCH_2CH_2MgBr} ((CH_3)_2CHCH_2CH_2)_3COH$$

$$((CH_3)_2CHCH_2CH_2)_3CH \xleftarrow{1.PBr_3\quad 2.\ Mg\quad 3.\ H_2O} ((CH_3)_2CHCH_2CH_2)_3COH$$

$$((CH_3)_2CHCH_2CH_2)_3CH \xleftarrow{1.\ H_2SO_4\quad 2.\ H_2\ /\ Pt} ((CH_3)_2CHCH_2CH_2)_3COH$$

In section 19.7.D of the Text, you will learn that the most efficient way to do this is:

$$3\ RMgBr + R'O\overset{O}{\overset{\|}{C}}OR' \longrightarrow \xrightarrow{H_2O} R_3COH$$

3.

a.

$$HC{\equiv}CCH_2CH_2CH_3 \xrightarrow[2.\ CH_3I]{1.\ NaNH_2} CH_3C{\equiv}CCH_2CH_2CH_3 \xrightarrow[NH_3]{Na} \underset{H_3C}{\overset{H}{}}\!\!\!C{=}C\!\!\underset{H}{\overset{CH_2CH_2CH_3}{}}$$

See comments to Problem #1(e).

b. 2-butylcyclopentanol (H, OH; $CH_2CH_2CH_2CH_3$; H) $\xleftarrow{H_2O}$ $\longleftarrow$ cyclopentene oxide + $LiCH_2CH_2CH_2CH_3$ (via: epoxide O, :R^- Li^+)

c. $CH_3CH_2CH_2$(H)C=C(H)$CH_2CH_2CH_3$ $\xleftarrow[\text{Lindlar}]{H_2}$ $CH_3CH_2CH_2C{\equiv}CCH_2CH_2CH_3$

↖ $CH_3CH_2CH_2C{\equiv}CNa$ + $CH_3CH_2CH_2Br$

See comments to Problem #1(e).

d.

HO, H; CH_3CH_2–C(H)(OH)–C–CH_2CH_3 $\xleftarrow[H_2O_2]{OsO_4}$ CH_3CH_2(H)C=C(H)CH_2CH_3 $\xleftarrow{Na / NH_3}$ $CH_3CH_2C{\equiv}CCH_2CH_3$ $\longleftarrow$ CH_3CH_2Br + $CH_3CH_2C{\equiv}CNa$

↑ H_3O^+

CH_3CH_2(H)C–C(H)CH_2CH_3 (epoxide, O) $\xleftarrow{CH_3CO_3H}$ CH_3CH_2(H)C=C(H)CH_2CH_3 $\xleftarrow[\text{Lindlar}]{H_2,}$ $CH_3CH_2C{\equiv}CCH_2CH_3$

(This is good to work through with models).

e. HO, H_3C, H; H, CH_3; $CH_2CH_2CH_3$ $\xleftarrow{H_2O}$ $\longleftarrow$ H, CH_3 epoxide (O, H, CH_3) + $LiCH_2CH_2CH_3$

(The *trans* epoxide would give the (2*RS*, 3*RS*) diastereomer).

f. cyclopentanol (OH) with (H)C=C(H)CH_2CH_3 $\xleftarrow[\text{Lindlar}]{H_2}$ cyclopentanol (OH) with $C{\equiv}CCH_2CH_3$ $\xleftarrow{H_2O}$ $\longleftarrow$ cyclopentanone (=O)

+ $NaC{\equiv}CCH_2CH_3$

g. CH_3CH_2, H; H–C–C–CH_2CH_3 (CH_2) $\xleftarrow[\text{Zn-Cu}]{CH_2I_2}$ CH_3CH_2(H)C=C(H)CH_2CH_3 $\xleftarrow[NH_3]{Na}$ $CH_3CH_2C{\equiv}CCH_2CH_3$

(from part d.)

4\. a. $HOCH_2CH_2CH_2C(H)(CH_2CH_3)(CH_3)$ $\xleftarrow{H_2O}$ $\leftarrow$ (ethylene oxide) $BrMgCH_2C(H)(CH_2CH_3)(CH_3)$ $\xleftarrow{Mg}$ $BrCH_2C(H)(CH_2CH_3)(CH_3)$

b. $CH_3CH_2CH_2C(H)(CH_3)(C{\equiv}N)$ $\xleftarrow{NaCN}$ $CH_3CH_2CH_2C(Br)(CH_3)(H)$ (S_N2 reaction, with inversion)

c. $H_3C\,C(H)(OH)CH_2C(CH_3)_3$ $\xleftarrow{H_2O}$ $\leftarrow$ $H_3C(H)C(CH_2)O$ + $LiC(CH_3)_3$

d. (D)(H) alkyne product $\xleftarrow{NaC{\equiv}CCH_2CH_2CH_3}$ Br, D, H bromide

5\. a. The aldehyde must be protected during the strongly basic elimination reaction, otherwise it will undergo aldol condensation reactions.

$CH_2{=}CH(CH_2)_2CHO$ $\xleftarrow{H_3O^+}$ $CH_2{=}CH(CH_2)_2$-dioxolane $\xleftarrow[t\text{-BuOH}]{t\text{-BuOK}}$ $Br(CH_2)_4$-dioxolane $\xleftarrow[H^+]{HOCH_2CH_2OH}$ $Br(CH_2)_4CHO$

b. The hydroxyl group in the starting material must be protected during the Grignard and oxidation reactions.

$HO(CH_2)_4CH(OH)(CH_2)_4OH$ $\xleftarrow{H_3O^+}$ $Me_3SiO(CH_2)_4CH(OH)(CH_2)_4OSiMe_3$ $\xleftarrow{Me_3SiO(CH_2)_4MgBr}$ $Me_3SiO(CH_2)_4CHO$

$\uparrow$ PCC

$HO(CH_2)_4Br$ $\xrightarrow[2.\ Mg]{1.\ Me_3SiCl,\ Et_3N}$ $Me_3SiO(CH_2)_4MgBr$ $\xrightarrow{CH_2{=}O}$ $Me_3SiO(CH_2)_5OH$

c. The aldehyde must be protected to avoid cyanohydrin formation.

$NCCH_2CH_2CHO$ $\xleftarrow{H_3O^+}$ $NCCH_2CH_2$-dioxolane $\xleftarrow{NaCN}$ $BrCH_2CH_2$-dioxolane $\xleftarrow[H^+]{HOCH_2CH_2OH}$ $BrCH_2CH_2CHO$

6. a. The logical approach to the *cis*-olefin involves partial hydrogenation of the alkyne, after assembly of the alkyne by alkylation of acetylide anions:

$$\text{(cis-alkene)} \xleftarrow[\text{Lindlar}]{H_2} \text{(alkyne)} \xleftarrow[\text{2. BuBr}]{\text{1. NaNH}_2} \text{(}t\text{-butyl)}\text{—}C\equiv C\text{—H}$$

This route will be successful from *t*-butyl acetylene on to the target, but will **not** work for the synthesis of *t*-butylacetylene itself. Attempted alkylation of a tertiary halide with the strongly basic acetylide anion will lead exclusively to elimination:

$$(CH_3)_3CBr + Na^+ \ {}^-C\equiv CH \longrightarrow CH_2{=}C(CH_3)_2 + HC\equiv CH + NaBr$$

The acetylene must therefore be formed in another way; the most reasonable is via the alkene:

$$(CH_3)_3CCH{=}CH_2 \xrightarrow{Br_2} (CH_3)_3CCHBrCH_2Br \xrightarrow{NaNH_2} (CH_3)_3CC\equiv C^-Na^+$$

The alkene in turn is available by a number of methods:

$$Ph_3P^{+}{}^{-}CH_2 + (CH_3)_3CCHO \longrightarrow (CH_3)_3CCH{=}CH_2$$

$$\uparrow \ t\text{-BuOK} \ / \ t\text{-BuOH}$$

$$\textit{or}\ (CH_3)_3CCH_2Br \xrightarrow{Mg} \xrightarrow{CH_2{=}O} (CH_3)_3CCH_2CH_2OH \xrightarrow{PBr_3} (CH_3)_3CCH_2CH_2Br$$

b. Two possibilities that involve various protecting or masking groups are shown below:

$$BrCH_2CH_2CHO + HOCH_2CH_2OH \xrightarrow{H^+} BrCH_2CH_2CH(OCH_2CH_2O) \xrightarrow{Mg} \xrightarrow{C_4H_9CHO} \xrightarrow{H_2O}$$

$$CH_3(CH_2)_3CH(OH)CH_2CH_2CH(OCH_2CH_2O) \xrightarrow{PCC} CH_3(CH_2)_3C(=O)CH_2CH_2CH(OCH_2CH_2O) \xrightarrow{H_3O^+} CH_3(CH_2)_3C(=O)(CH_2)_2CH(=O)$$

or

$$CH_3(CH_2)_3MgBr + HC(=O)CH_2CH_2CH{=}CH_2 \longrightarrow \xrightarrow{H_2O} CH_3(CH_2)_3CH(OH)(CH_2)_2CH{=}CH_2$$

$$\xrightarrow{K_2Cr_2O_7,\ H_2SO_4,\ \Delta} CH_3(CH_2)_3C(=O)(CH_2)_2CH{=}CH_2 \xrightarrow[\text{2. Zn / }H_2O]{\text{1. }O_3} CH_3(CH_2)_3C(=O)(CH_2)_2CH(=O)$$

c. The desired target is a six-carbon keto alcohol. Thus, we must add one carbon at least. Since the material is an alcohol, we may consider a Grignard synthesis. It is a secondary alcohol, so there are two possible combinations:

$$CH_3\overset{O}{\overset{\|}{C}}CH_2CH_2Br \xrightarrow{Mg} \xrightarrow{CH_3CHO} \xrightarrow{H_2O} CH_3\overset{O}{\overset{\|}{C}}CH_2CH_2\underset{OH}{\underset{|}{C}}HCH_3 \xleftarrow{H_2O} \xleftarrow{} CH_3\overset{O}{\overset{\|}{C}}CH_2CH_2\overset{O}{\overset{\|}{C}}H \xleftarrow{Mg} CH_3Br$$

In the latter route, we have a problem of selectivity. The Grignard reagent CH_3MgBr can react with either carbonyl group. Although aldehydes are more reactive than ketones, the Grignard reagent is so reactive it will probably not show much selectivity.

In the first route, we have a different problem: the Grignard reagent *can react with itself*. Polymerization will result. A simple way out of this dilemma is to **protect** the carbonyl group in 4-bromo-2-butanone so that it cannot react with a Grignard reagent. This may be done by converting it into a ketal, which does not react with Grignard reagents:

$$CH_3\overset{O}{\overset{\|}{C}}CH_2CH_2Br \xrightarrow[H^+]{HOCH_2CH_2OH} \text{(ketal, Br)} \xrightarrow{Mg} \xrightarrow[2.\ H_3O^+]{1.\ CH_3CHO} \left[\text{(ketal, OH)}\right] \xrightarrow{H_3O^+} CH_3\overset{O}{\overset{\|}{C}}(CH_2)_2\underset{OH}{\underset{|}{C}}HCH_3$$

d. The target: $(CH_3)_2C{=}CH\overset{CH_2CH_3}{\overset{|}{C}}{=}CHCH_2CH_2CH_3$ (no stereochemistry specified).

The major challenge is to introduce the double bonds in the correct positions. Routes that involve dehydration of alcohols or elimination of alkyl halides will lead to other isomers in addition to the desired one. The best way to introduce a double bond in a specific position is via the Wittig reaction. With this in mind, you can imagine two carbonyl compounds which would give the desired product:

$$2\ O{=}CHCH_2CH_2CH_3 \xrightarrow[\Delta]{NaOH} O{=}CH\overset{CH_2CH_3}{\overset{|}{C}}{=}CHCH_2CH_2CH_3 \xrightarrow{(CH_3)_2C{=}PPh_3} (CH_3)_2C{=}CH\overset{CH_2CH_3}{\overset{|}{C}}{=}CHCH_2CH_2CH_3$$

$$CH_3\overset{O}{\overset{\|}{C}}CH_3 + CH_3\overset{O}{\overset{\|}{C}}CH_2CH_3 \xrightarrow[\Delta]{NaOH} (CH_3)_2C{=}CH\overset{CH_2CH_3}{\overset{|}{C}}{=}O \xrightarrow{CH_3CH_2CH_2CH{=}PPh_3} (CH_3)_2C{=}CH\overset{CH_2CH_3}{\overset{|}{C}}{=}CHCH_2CH_2CH_3$$

You should recognize that both of these carbonyl compounds are potentially available in one step via the aldol condensation. However, the ketone is clearly a less desirable intermediate because it would require a mixed aldol condensation, leading to many products in addition to the desired one.

e.

$$(CH_3)_2C{=}CHCHO + NaC{\equiv}CCH_2CH_3 \longrightarrow \xrightarrow{H_2O} (CH_3)_2C{=}CH\overset{OH}{\overset{|}{C}}HC{\equiv}CCH_2CH_3$$

f.

$$(CH_3CH_2)_2C{=}O + NaC{\equiv}CCH{=}CH_2 \longrightarrow \xrightarrow{H_2O} (CH_3CH_2)_2\overset{OH}{\overset{|}{C}}C{\equiv}CCH{=}CH_2$$

g. The easiest way to form the carbon-carbon bond in this case would be to use a Grignard reaction:

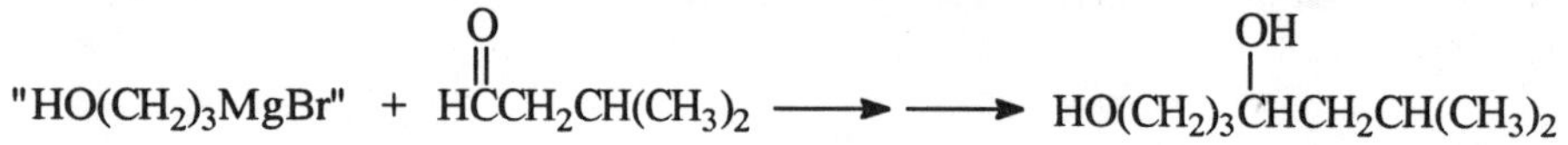

$$\text{"HO(CH}_2)_3\text{MgBr"} + \text{HC(=O)CH}_2\text{CH(CH}_3)_2 \longrightarrow \longrightarrow \text{HO(CH}_2)_3\text{CH(OH)CH}_2\text{CH(CH}_3)_2$$

However, this Grignard reagent cannot be made, because of the presence of a hydroxy group in the molecule. (A simple proton transfer reaction would destroy the reagent:

$$\text{HOCH}_2\text{CH}_2\text{CH}_2\text{MgBr} \xrightarrow{\textit{fast}} \text{BrMgOCH}_2\text{CH}_2\text{CH}_3.)$$

For this reason, we must protect the hydroxy group:

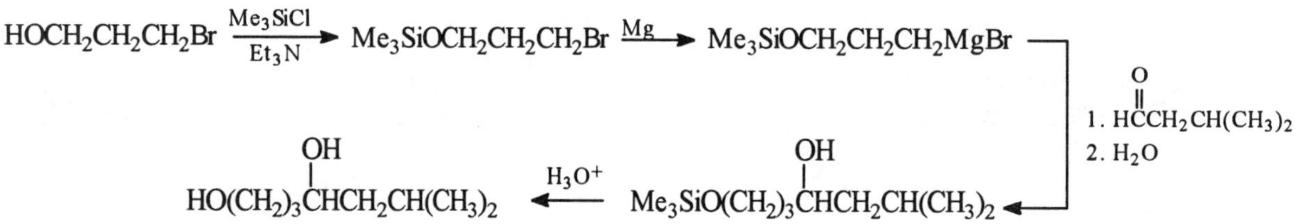

$$\text{HOCH}_2\text{CH}_2\text{CH}_2\text{Br} \xrightarrow[\text{Et}_3\text{N}]{\text{Me}_3\text{SiCl}} \text{Me}_3\text{SiOCH}_2\text{CH}_2\text{CH}_2\text{Br} \xrightarrow{\text{Mg}} \text{Me}_3\text{SiOCH}_2\text{CH}_2\text{CH}_2\text{MgBr}$$

$$\xrightarrow[\text{2. H}_2\text{O}]{\text{1. HC(=O)CH}_2\text{CH(CH}_3)_2} \text{Me}_3\text{SiO(CH}_2)_3\text{CH(OH)CH}_2\text{CH(CH}_3)_2 \xrightarrow{\text{H}_3\text{O}^+} \text{HO(CH}_2)_3\text{CH(OH)CH}_2\text{CH(CH}_3)_2$$

16.F Supplementary Problems

S1. Using monofunctional starting materials having five carbons or less, and any other reagents, outline efficient syntheses of the following compounds.

a. 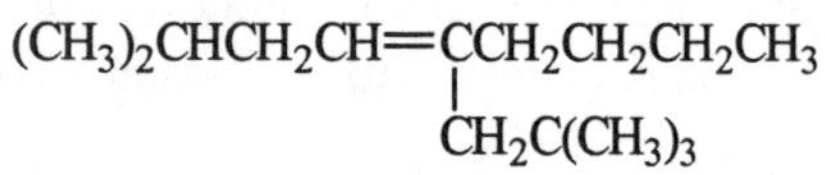 $(CH_3)_2CHCH_2CH{=}C(CH_2C(CH_3)_3)CH_2CH_2CH_2CH_3$

b. $(CH_3)_2CBr(CH_2)_6CBr(CH_3)_2$

c. 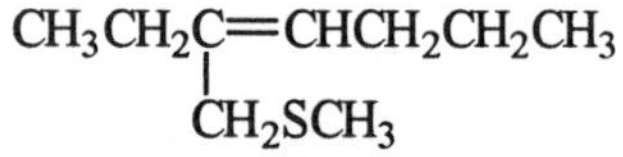 $CH_3CH_2C(CH_2SCH_3){=}CHCH_2CH_2CH_3$

d. 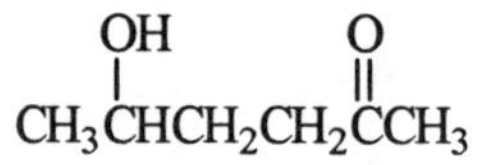 $CH_3CH(OH)CH_2CH_2C(=O)CH_3$

e. 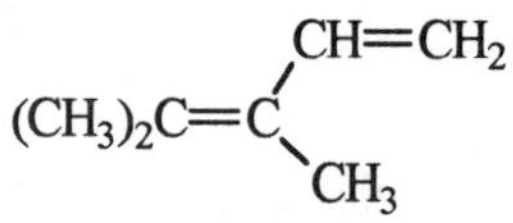 $(CH_3)_2C{=}C(CH{=}CH_2)CH_3$

f. 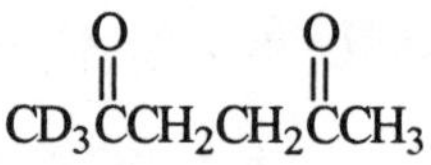 $CD_3C(=O)CH_2CH_2C(=O)CH_3$

g. 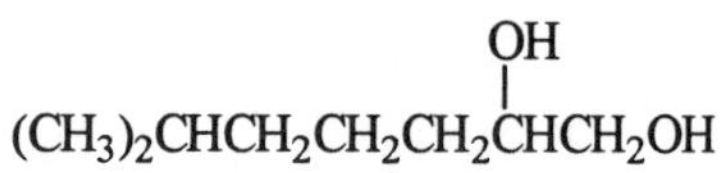 $(CH_3)_2CHCH_2CH_2CH_2CH(OH)CH_2OH$

h. 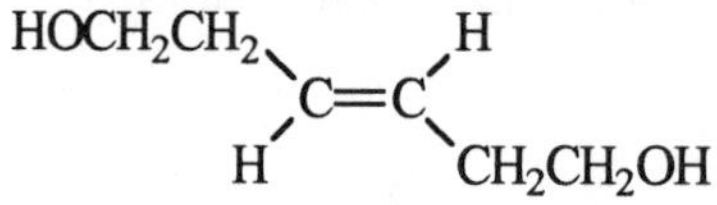 $HOCH_2CH_2(H)C{=}C(H)CH_2CH_2OH$ (trans: $HOCH_2CH_2$ and CH_2CH_2OH on opposite sides)

i. 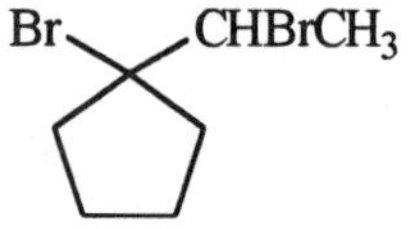 1-bromo-1-(CHBrCH$_3$)cyclopentane (Br and CHBrCH$_3$ on the same ring carbon)

j. $((CH_3)_2CHCH_2)_2CHCN$

k. $CH_3C(CH_2Br)_3$

l. 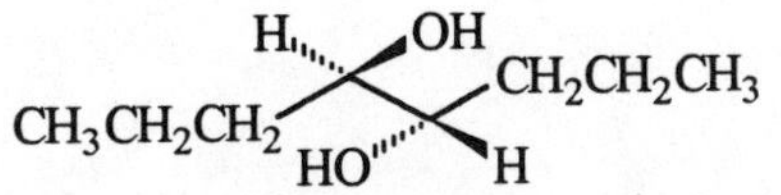 H, OH, $CH_3CH_2CH_2$, $CH_2CH_2CH_3$, HO, H

m. 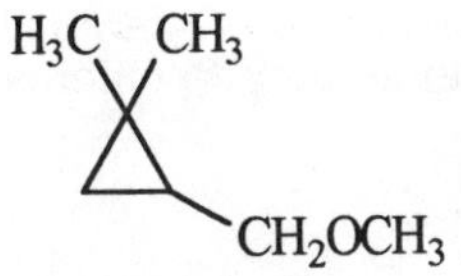 H_3C, CH_3, CH_2OCH_3

n. 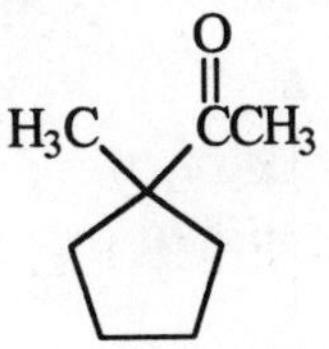 H_3C, O, CCH_3

o. CO_2H; $CH_3CH_2CCH_2CH_2CH_3$; OH

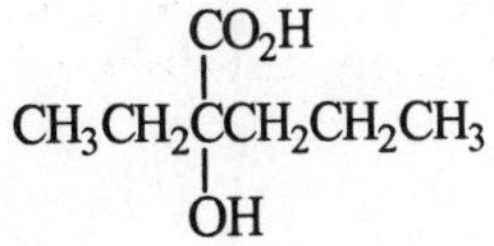

p. 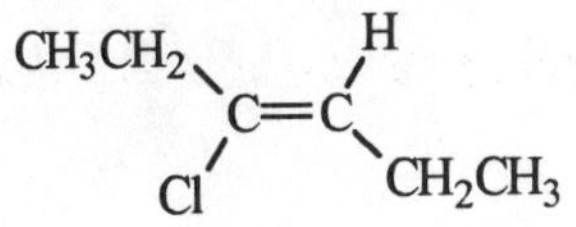 CH_3CH_2, H, $C{=}C$, Cl, CH_2CH_3

q. 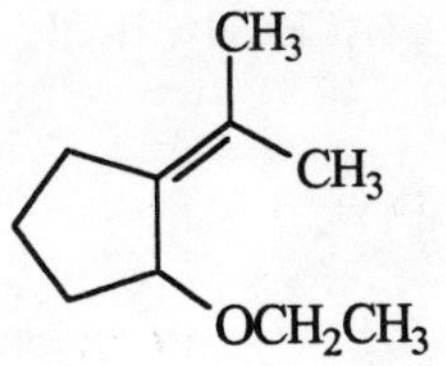 CH_3, CH_3, OCH_2CH_3

r. 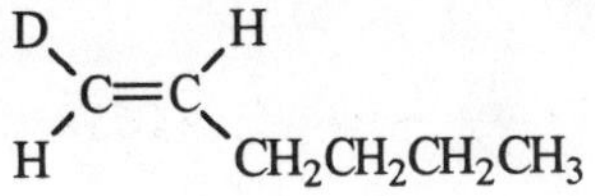 D, H, $C{=}C$, H, $CH_2CH_2CH_2CH_3$

s. 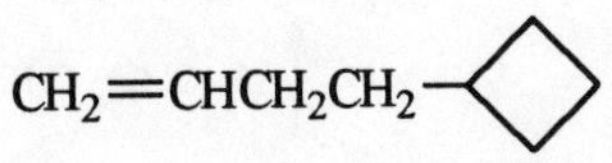 $CH_2{=}CHCH_2CH_2$–

t. 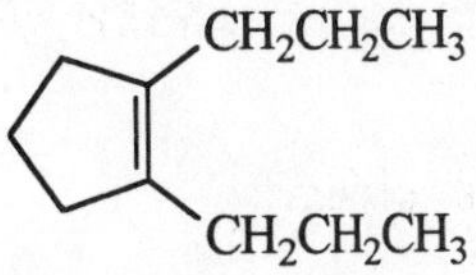 $CH_2CH_2CH_3$, $CH_2CH_2CH_3$

u. $(CH_3CH_2)_2CHC{\equiv}CCH(CH_2CH_3)_2$

v. $HON{=}CHCH_2CH_2CH_2CH{=}NOH$

w. 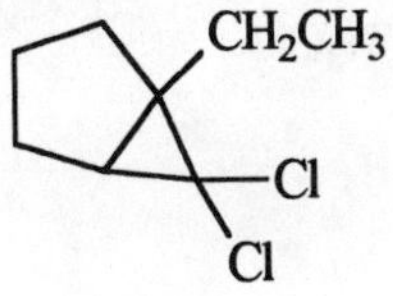 CH_2CH_3, Cl, Cl

x. 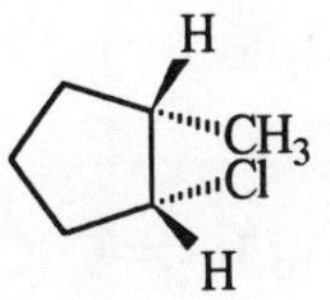 H, CH_3, Cl, H

y. 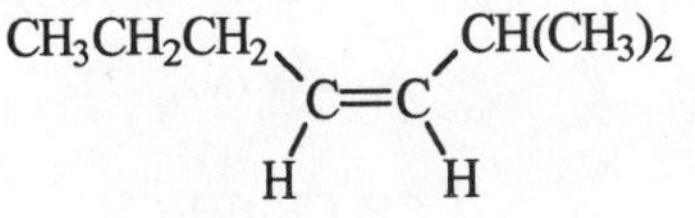 $CH_3CH_2CH_2$, $CH(CH_3)_2$, $C{=}C$, H, H

z. 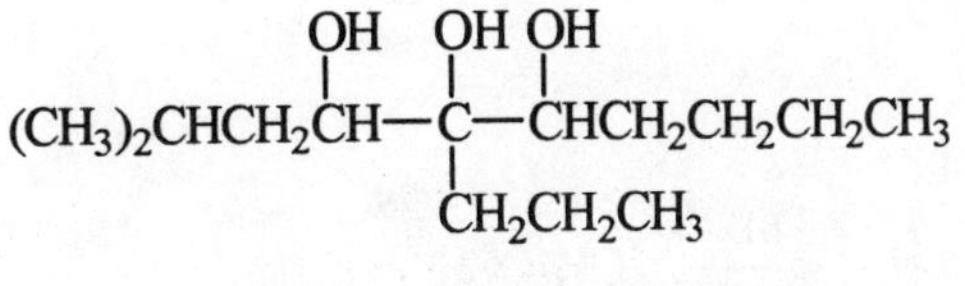 OH OH OH; $(CH_3)_2CHCH_2CH{-}C{-}CHCH_2CH_2CH_2CH_3$; $CH_2CH_2CH_3$

16.G. Answers to Supplementary Problems

S1.

a.

$(CH_3)_2CHCH_2CH{=}C(CH_2C(CH_3)_3)CH_2CH_2CH_2CH_3 \longleftarrow (CH_3)_2CHCH_2CH{=}PPh_3 + (CH_3)_3CCH_2\overset{O}{\overset{\|}{C}}CH_2CH_2CH_2CH_3$

$(CH_3)_2CHCH_2CH_2Br \xrightarrow{PPh_3} \xrightarrow{BuLi} (CH_3)_2CHCH_2CH{=}PPh_3$

$(CH_3)_3CCH_2MgBr + H\overset{O}{\overset{\|}{C}}CH_2CH_2CH_2CH_3 \longrightarrow \xrightarrow{H_2O} (CH_3)_3CCH_2\overset{OH}{\overset{|}{C}}HCH_2CH_2CH_2CH_3$

$(CH_3)_3CCH_2CH(OH)CH_2CH_2CH_2CH_3 \xrightarrow{K_2Cr_2O_7,\ H_2SO_4} (CH_3)_3CCH_2\overset{O}{\overset{\|}{C}}CH_2CH_2CH_2CH_3$

b.

$(CH_3)_2\overset{Br}{\overset{|}{C}}(CH_2)_6\overset{Br}{\overset{|}{C}}(CH_3)_2 \xleftarrow[h\nu]{2\ Br_2} (CH_3)_2CH(CH_2)_6CH(CH_3)_2 \xleftarrow[Pt]{H_2} (CH_3)_2CHCH_2CH_2C{\equiv}CCH_2CH_2CH(CH_3)_2$

$2\ NaNH_2 + HC{\equiv}CH \longrightarrow NaC{\equiv}CNa \xrightarrow{2\ (CH_3)_2CHCH_2CH_2Br} (CH_3)_2CHCH_2CH_2C{\equiv}CCH_2CH_2CH(CH_3)_2$

c.

$CH_3CH_2C(CH_2SCH_3){=}CHCH_2CH_2CH_3 \xleftarrow{NaSCH_3} CH_3CH_2C(CH_2Br){=}CHCH_2CH_2CH_3 \xleftarrow{PBr_3} CH_3CH_2C(CH_2OH){=}CHCH_2CH_2CH_3$

$2\ CH_3CH_2CH_2\overset{O}{\overset{\|}{C}}H \xrightarrow[\Delta]{NaOH} CH_3CH_2C(CH{=}O){=}CHCH_2CH_2CH_3 \xrightarrow{NaBH_4} CH_3CH_2C(CH_2OH){=}CHCH_2CH_2CH_3$

d.

$CH_3\overset{OH}{\overset{|}{C}}HCH_2CH_2\overset{O}{\overset{\|}{C}}CH_3 \longleftarrow CH_3CH\overset{O}{—}CH_2 + CH_2{=}\overset{OLi}{\overset{|}{C}}CH_3 \xleftarrow[THF]{LDA} CH_3\overset{O}{\overset{\|}{C}}CH_3$

e.

$(CH_3)_2C{=}C(CH{=}CH_2)CH_3 \xleftarrow{H_2SO_4} (CH_3)_2CH\overset{OH}{\overset{|}{C}}(CH_3)CH{=}CH_2 \longleftarrow (CH_3)_2CH\overset{O}{\overset{\|}{C}}CH_3 + BrMgCH{=}CH_2$

f.

$$CD_3\overset{O}{\overset{\|}{C}}CH_2CH_2\overset{O}{\overset{\|}{C}}CH_3 \xleftarrow[\text{pyridine}]{CrO_3} CD_3\overset{OH}{\overset{|}{C}}HCH_2CH_2\overset{OH}{\overset{|}{C}}HCH_3 \xleftarrow[Pt]{H_2} CD_3\overset{OH}{\overset{|}{C}}HC{\equiv}C\overset{OH}{\overset{|}{C}}HCH_3$$

$$HC{\equiv}CNa \xrightarrow{H\overset{O}{\overset{\|}{C}}CH_3} HC{\equiv}C\overset{ONa}{\overset{|}{C}}HCH_3 \xrightarrow{NaNH_2} NaC{\equiv}C\overset{ONa}{\overset{|}{C}}HCH_3 \xrightarrow{CD_3\overset{O}{\overset{\|}{C}}H} \xrightarrow{H_2O}$$

$$CD_3\overset{O}{\overset{\|}{C}}H \xleftarrow[D_2O]{D^+} CH_3\overset{O}{\overset{\|}{C}}H$$

g.

$$(CH_3)_2CHCH_2CH_2CH_2CHOHCH_2OH \xleftarrow[H_2O_2]{OsO_4} (CH_3)_2CHCH_2CH_2CH_2CH{=}CH_2 \xleftarrow[\text{Lindlar}]{H_2}$$

$$\xrightarrow[\Delta]{NaNH_2} (CH_3)_2CHCH_2CH_2CH_2C{\equiv}CNa \xrightarrow{H_2O} (CH_3)_2CHCH_2CH_2CH_2C{\equiv}CH$$

$$(CH_3)_2CHCH_2CH_2C{\equiv}CCH_3 \longleftarrow (CH_3)_2CHCH_2CH_2Br + NaC{\equiv}CCH_3$$

h.

$$\underset{\text{(trans; } HOCH_2CH_2 \text{ and H on one C, H and } CH_2CH_2OH \text{ on the other)}}{HOCH_2CH_2CH{=}CHCH_2CH_2OH} \xleftarrow[NH_3]{Na} HOCH_2CH_2C{\equiv}CCH_2CH_2OH \xleftarrow{2\ \text{ethylene oxide}} NaC{\equiv}CNa \xleftarrow{2\ NaNH_2} HC{\equiv}CH$$

i.

$$\text{1-bromo-1-(CHBrCH}_3\text{)cyclopentane} \xleftarrow[CCl_4]{Br_2} \text{ethylidenecyclopentane} \longleftarrow \text{cyclopentanone} + CH_3CH{=}PPh_3 \xleftarrow[\text{2. BuLi}]{\text{1. PPh}_3} CH_3CH_2Br$$

j.

$$((CH_3)_2CHCH_2)_2CHCN \xleftarrow{NaCN} ((CH_3)_2CHCH_2)_2CHBr \xleftarrow{PBr_3} ((CH_3)_2CHCH_2)_2CHOH \xleftarrow{H_2O}$$

$$(CH_3)_2CHCH_2CHO + BrMgCH_2CH(CH_3)_2$$

k.

$$CH_3C(CH_2Br)_3 \xleftarrow{3\ PBr_3} CH_3C(CH_2OH)_3 \xleftarrow{NaOH,\ CH_2O} CH_3CHO$$

l.

$$\underset{\text{(H, OH / HO, H; } CH_3CH_2CH_2 \text{ and } CH_2CH_2CH_3)}{CH_3CH_2CH_2CH(OH)CH(OH)CH_2CH_2CH_3} \xleftarrow{H_3O^+} \underset{\text{(epoxide; H and H on opposite sides)}}{CH_3CH_2CH_2CH{-}O{-}CHCH_2CH_2CH_3} \xleftarrow{CH_3CO_3H}$$

$$NaC{\equiv}CNa + 2\ CH_3CH_2CH_2Br \longrightarrow CH_3CH_2CH_2C{\equiv}CCH_2CH_2CH_3 \xrightarrow[NH_3]{Na} \underset{\text{(trans)}}{CH_3CH_2CH_2CH{=}CHCH_2CH_2CH_3}$$

h.

$HOCH_2CH_2CH{=}CHCH_2CH_2OH$ (trans; H, H) $\xleftarrow[NH_3]{Na}$ $HOCH_2CH_2C{\equiv}CCH_2CH_2OH$ $\xleftarrow{2\ \text{ethylene oxide}}$ $NaC{\equiv}CNa$ $\xleftarrow{2\ NaNH_2}$ $HC{\equiv}CH$

i.

1-bromo-1-(CHBrCH$_3$)cyclopentane $\xleftarrow[CCl_4]{Br_2}$ ethylidenecyclopentane $\longleftarrow$ cyclopentanone $+\ CH_3CH{=}PPh_3$ $\xleftarrow[2.\ BuLi]{1.\ PPh_3}$ CH_3CH_2Br

j. $((CH_3)_2CHCH_2)_2CHCN \xleftarrow{NaCN} ((CH_3)_2CHCH_2)_2CHBr \xleftarrow{PBr_3} ((CH_3)_2CHCH_2)_2CHOH \xleftarrow{H_2O}$

$(CH_3)_2CHCH_2CHO + BrMgCH_2CH(CH_3)_2$

k. $CH_3C(CH_2Br)_3 \xleftarrow{3\ PBr_3} CH_3C(CH_2OH)_3 \xleftarrow{NaOH,\ CH_2O} CH_3CHO$

l.

$CH_3CH_2CH_2CH(OH)CH(OH)CH_2CH_2CH_3$ (H, OH; HO, H) $\xleftarrow{H_3O^+}$ epoxide of $CH_3CH_2CH_2CH{=}CHCH_2CH_2CH_3$ (H, O, H) $\xleftarrow{CH_3CO_3H}$ (trans alkene below)

$NaC{\equiv}CNa + 2\ CH_3CH_2CH_2Br \longrightarrow CH_3CH_2CH_2C{\equiv}CCH_2CH_2CH_3 \xrightarrow[NH_3]{Na}$ trans-$CH_3CH_2CH_2CH{=}CHCH_2CH_2CH_3$ (H, H)

m.

1,1-dimethyl-2-(CH_2OCH_3)cyclopropane (H_3C, CH_3, CH_2OCH_3) $\xleftarrow[Zn-Cu]{CH_2I_2}$ $(CH_3)_2C{=}CHCH_2OCH_3 \longleftarrow (CH_3)_2C{=}CHCH_2Br + NaOCH_3$

n.

2-methyl-2-acetyl... (1-methylcyclopentyl methyl ketone) $\xleftarrow[\text{pyridine}]{CrO_3}$ 1-(1-methylcyclopentyl)ethanol $\longleftarrow\ \xleftarrow{CH_3CHO}$ 1-methylcyclopentylmagnesium bromide $\xleftarrow{Mg}\ \xleftarrow{HBr}$ 1-methylcyclopentanol $\xleftarrow{H_2O}\ \xleftarrow{CH_3MgI}$ cyclopentanone

o.

$CH_3CH_2C(CO_2H)(OH)CH_2CH_2CH_3 \xleftarrow[\Delta]{H_3O^+} CH_3CH_2C(CN)(OH)CH_2CH_2CH_3 \xleftarrow{HCN} CH_3CH_2C(=O)CH_2CH_2CH_3$

$\nearrow$ $K_2Cr_2O_7$, H_2SO_4

$CH_3CH_2MgBr + HC(=O)CH_2CH_2CH_3 \longrightarrow CH_3CH_2CH(OH)CH_2CH_2CH_3 \longleftarrow CH_3CH_2CH(=O) + BrMgCH_2CH_2CH_3$

p.

$(CH_3CH_2)(Cl)C{=}C(H)(CH_2CH_3) \xleftarrow{t\text{-BuOK}} CH_3CH_2CH(Cl)CH(Cl)CH_2CH_3$ (stereospecific) $\xleftarrow[CCl_4]{Cl_2}$ cis-$CH_3CH_2CH{=}CHCH_2CH_3$

$\nwarrow$ H_2, Lindlar

$NaC{\equiv}CNa + 2\ CH_3CH_2Br \longrightarrow CH_3CH_2C{\equiv}CCH_2CH_3$

q.

1-(1-methylethylidene)-2-methoxycyclopentane [$(CH_3)_2C{=}$ on ring, OCH_3] $\xleftarrow{(CH_3)_2C{=}PPh_3}$ 2-methoxycyclopentanone $\xleftarrow[H_2SO_4]{K_2Cr_2O_7}$ 2-methoxycyclopentanol $\xleftarrow[H^+]{CH_3OH}$ cyclopentene oxide

$\uparrow$ BuLi

$\xleftarrow[PPh_3]{}$ CH_3CHICH_3

r.

$(D)(H)C{=}C(H)(CH_2CH_2CH_2CH_3) \xleftarrow[\Delta]{CH_3CO_2D} (R_2B)(H)C{=}C(H)(CH_2CH_2CH_2CH_3) \xleftarrow[\Delta]{B_2H_6} HC{\equiv}CCH_2CH_2CH_2CH_3$

$\uparrow$

$HC{\equiv}CNa + BrCH_2CH_2CH_2CH_3$

s.

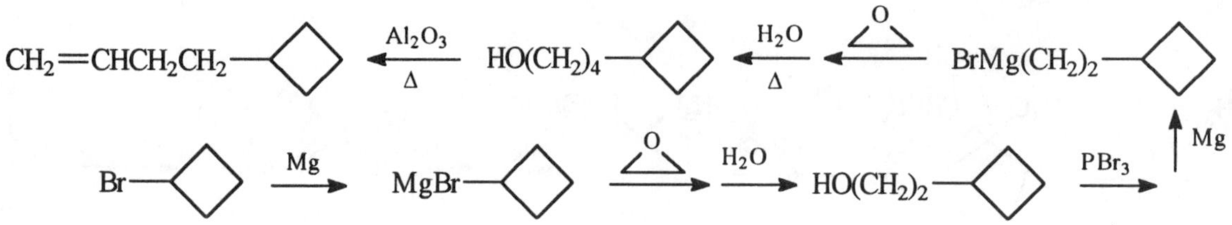

$CH_2{=}CHCH_2CH_2$–cyclobutyl $\xleftarrow[\Delta]{Al_2O_3}$ $HO(CH_2)_4$–cyclobutyl $\xleftarrow[\Delta]{H_2O}$ $\xleftarrow{\text{ethylene oxide}}$ $BrMg(CH_2)_2$–cyclobutyl

Br–cyclobutyl $\xrightarrow{Mg}$ MgBr–cyclobutyl $\xrightarrow{\text{ethylene oxide}}$ $\xrightarrow{H_2O}$ $HO(CH_2)_2$–cyclobutyl $\xrightarrow{PBr_3}$ $\xrightarrow{Mg}$

t.

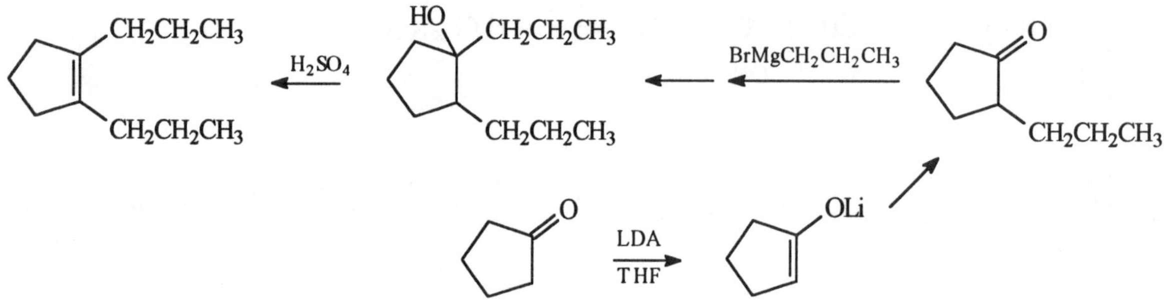

1,2-di($CH_2CH_2CH_3$)cyclopentene $\xleftarrow{H_2SO_4}$ 1-HO-1,2-di($CH_2CH_2CH_3$)cyclopentane $\longleftarrow \xleftarrow{BrMgCH_2CH_2CH_3}$ 2-($CH_2CH_2CH_3$)cyclopentanone

cyclopentanone $\xrightarrow[THF]{LDA}$ OLi (lithium enolate) $\nearrow$

u.

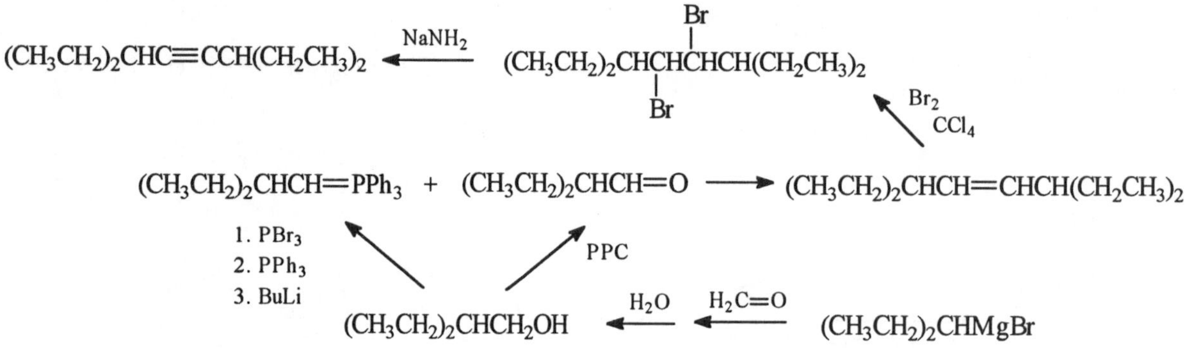

$(CH_3CH_2)_2CHC{\equiv}CCH(CH_2CH_3)_2 \xleftarrow{NaNH_2} (CH_3CH_2)_2CHCHBrCHBrCH(CH_2CH_3)_2$

$\nwarrow Br_2,\ CCl_4$

$(CH_3CH_2)_2CHCH{=}PPh_3 + (CH_3CH_2)_2CHCH{=}O \longrightarrow (CH_3CH_2)_2CHCH{=}CHCH(CH_2CH_3)_2$

1. PBr_3
2. PPh_3
3. BuLi

PPC

$(CH_3CH_2)_2CHCH_2OH \xleftarrow{H_2O} \xleftarrow{H_2C{=}O} (CH_3CH_2)_2CHMgBr$

v.

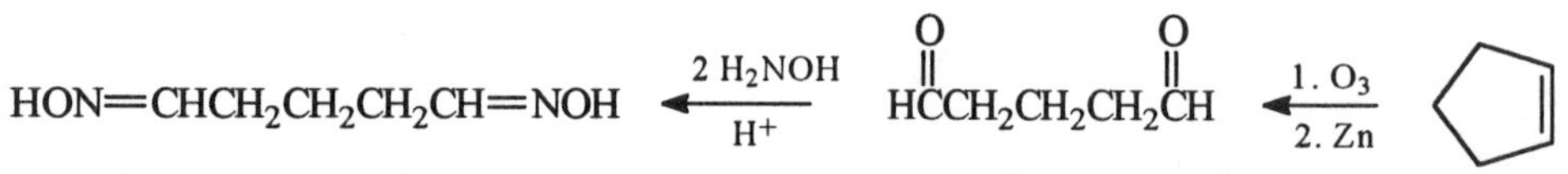

$HON{=}CHCH_2CH_2CH_2CH{=}NOH \xleftarrow[H^+]{2\ H_2NOH} HCCH_2CH_2CH_2CH$ (each terminal C with =O) $\xleftarrow[2.\ Zn]{1.\ O_3}$ cyclopentene

w.

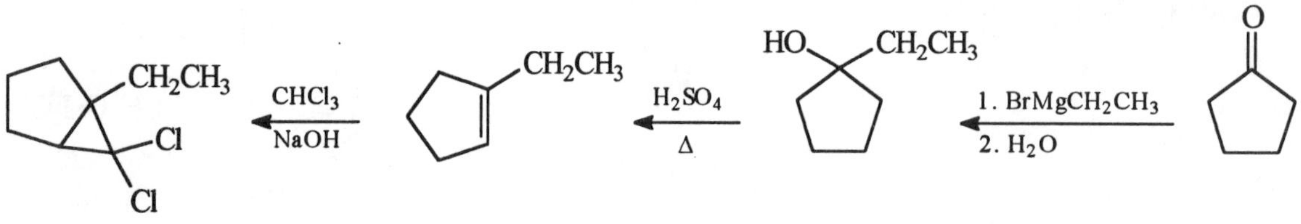

1-CH_2CH_3-6,6-di-Cl-bicyclo[3.1.0]hexane $\xleftarrow[NaOH]{CHCl_3}$ 1-CH_2CH_3-cyclopentene $\xleftarrow[\Delta]{H_2SO_4}$ 1-HO-1-CH_2CH_3-cyclopentane $\xleftarrow[2.\ H_2O]{1.\ BrMgCH_2CH_3}$ cyclopentanone

x.

cyclopentene oxide $\xrightarrow{CH_3Li}$ $\xrightarrow{H_2O}$ trans-2-methylcyclopentanol (H, CH_3; OH, H) $\xrightarrow{PCl_3}$ cis-1-chloro-2-methylcyclopentane (H, CH_3; Cl, H)

cyclopentanone $\xrightarrow{CH_3MgI}$ $\xrightarrow[\Delta]{H_2SO_4}$ 1-methylcyclopentene (CH_3) $\xrightarrow{1.\ B_2H_6;\ 2.\ H_2O_2\ /\ NaOH}$ trans-2-methylcyclopentanol

y.

$NaC{\equiv}CCH(CH_3)_2 \xrightarrow{CH_3CH_2CH_2Br} CH_3CH_2CH_2C{\equiv}CCH(CH_3)_2 \xrightarrow[\text{Lindlar}]{H_2}$ $CH_3CH_2CH_2(H)C{=}C(H)CH(CH_3)_2$ (cis)

z.

$2\ CH_3CH_2CH_2CH_2\overset{O}{\overset{\|}{C}}H \xrightarrow[\Delta]{NaOH} H\overset{O}{\overset{\|}{C}}C(CH_2CH_2CH_3){=}CHCH_2CH_2CH_2CH_3$

$\xrightarrow{(CH_3)_2CHCH_2MgBr}$ $\xrightarrow{H_2O}$ $(CH_3)_2CHCH_2CH(OH)C(CH_2CH_2CH_3){=}CH(CH_2)_3CH_3$

$\xrightarrow[H_2O_2]{OsO_4}$ $(CH_3)_2CHCH_2CH(OH){-}C(OH)(CH_2CH_2CH_3){-}CH(OH)(CH_2)_3CH_3$

17. INFRARED SPECTROSCOPY

17.A. Chapter Outline and Important Terms Introduced

17.1 The Electromagnetic Spectrum

electromagnetic radiation
spectroscopy
spectrum

$\epsilon = h\nu$, $E = Nh\nu$, $\nu = c/\lambda$, $\tilde{\nu} = 1/\lambda$
middle infrared ($\tilde{\nu}$ = 3333 - 333 cm^{-1})

17.2 Molecular Vibration

selection rule
infrared active and inactive
fundamental vibrational modes
stretching, bending
Hooke's law

degenerate vibrations
overtones
combination bands
harmonic oscillator approximation

17.3 Characteristic Group Vibrations

17.4 Alkanes

17.5 Alkenes

17.6 Alkynes and Nitriles

17.7 Alkyl Halides

17.8 Alcohols and Ethers

effects of hydrogen bonding

17.9 Aldehydes and Ketones (the most important applications of IR spectroscopy)

dependence of $\nu_{C=O}$ on structure

17.10 Use of Infrared Spectroscopy in Solving Spectral Problems

17.B. Important Reactions Introduced: (none in this chapter)

17.C. Important Concepts and Hints

What IR Spectroscopy is Good For:

IR spectroscopy is complementary to NMR because it gives you information primarily about functional groups instead of about the carbon skeleton. With this technique you can decide immediately if the compound has a hydroxy group, for example, or a carbonyl, a nitrile, and so one. Often the IR spectrum can tell you additional details about a functional group: Is the double bond *cis, trans*, or a 1,1-disubstituted olefin? Is the carbonyl group in a five- or six-membered ring, or part of an acyclic ketone? In the chapter on carboxylic acid derivatives (Chapter 19) you will learn that the position of the carbonyl stretching frequency can tell you even more.

Generally, the IR spectrum will not tell you much about the hydrocarbon backbone of a molecule. All CH_2 and CH_3 groups, for instance, have the same stretching and bending vibrations, and the fundamental vibrational modes of larger structural units are not readily assignable.

What You Should Learn:

The subject of IR spectroscopy has a logical foundation, but its routine application relies primarily on memory. The characteristic absorption frequencies of the various functional groups in organic molecules can be understood from first principles, as pointed out in the introductory sections of the chapter. However, the day-to-day interpretation of infrared spectra relies on simply knowing what IR band corresponds to what functional group. You will have to do a moderate amount of memorization in order to be able to use IR easily in solving problems. Start with Table 17.1, and then expand your knowledge until you are familiar with all the **highlighted entries** in Table 17.3. In subsequent chapters, the IR characteristics of additional functional groups will be presented, and you should add them to your memory.

What You Should Avoid:

The tendency of most students is to "overinterpret" an IR spectrum, and attempt to assign every band that appears. Although much useful information is contained in an IR spectrum, there are a lot of peaks which cannot be identified with any particular functional group and so are not useful in understanding the structure of a compound. When you look at an IR spectrum, you should first pick out the unambiguous peaks (C=O stretch, O–H stretch, aldehyde C–H, for example), and see what you can do with that information and the rest available in the problem. If necessary, you can return to the spectrum and use it to get finer details.

You should also recognize that average absorptions given for a particular vibrational mode are averages for a wide variety of compounds. As with all averages, individual cases can be very different. For example, you should not assume that all *trans* alkenes will show the C–H out-of-plane bend at exactly 970 cm^{-1}, or that a molecule with a strong band at 1720 cm^{-1} cannot be a six-membered ring ketone.

17.D. Answers to Exercises

17.1 $\tilde{\nu} = \dfrac{1}{2\pi c}\sqrt{\dfrac{f\,(m_1+m_2)}{m_1 m_2}}$ $c = 2.998 \times 10^{10}$ cm sec^{-1}

f (single bond) = 5×10^5 dynes cm^{-1}
f (double bond) = 10×10^5 dynes cm^{-1}
f (triple bond) = 15×10^5 dynes cm^{-1}

If M_1 = atomic mass, then $\dfrac{m_1+m_2}{m_1 m_2} = (\dfrac{M_1+M_2}{M_1 M_2})\ x\ (6.023\ x\ 10^{23})$

a. $\tilde{\nu} = \dfrac{1}{1.884\ x\ 10^{11}}\sqrt{6.023\ x\ 10^{23}}\sqrt{f\,(\dfrac{M_1+M_2}{M_1 M_2})} = 4.12\sqrt{5\ x\ 10^5\ x\ \dfrac{17}{16}}$

$\tilde{\nu} = 3003$ cm^{-1}

b. 2185 cm^{-1} c. 1682 cm^{-1} d. 2060 cm^{-1} e. 1113 cm^{-1}

f. 1573 cm^{-1} g. 1985 cm^{-1} h. 1074 cm^{-1}

17.2 3090 cm^{-1} indicates an alkene C−H bond:

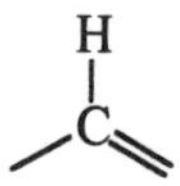

H
|
C

1780 cm^{-1} is an overtone of the strong band at 890 cm^{-1} and indicates the presence of a 1,1-disubstituted alkene: 1653 cm^{-1} is consistent with the presence of an alkene.

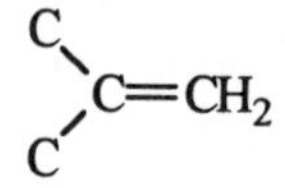

C
C=CH_2
C

17.3 Recall that absorption of infrared light can only occur if the dipole moment of the molecule is different in the two vibrational levels. For a symmetrical alkene such as *trans*-4-octene, there is no dipole moment either before or after excitation of the double bond stretching band, hence this transition is infrared inactive and the band is not observed in the IR spectrum.

17.4 Using the equation of Exercise 17.1: $\tilde{\nu} = 2225$ cm^{-1}

17.5 Because the C−C−O angle in ketene is 180°, the C=O stretching motion requires simultaneous C=C compression, a movement which will be resisted strongly by the C=C bond in comparison with the normal

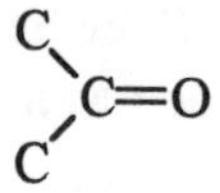

C
C=O arrangement. Therefore, the high C=O stretching frequency is to be expected.
C

17.6 The formula $C_8H_{14}O$ shows two degrees of unsaturation. The IR signal at 1710 cm^{-1} indicates a ketone. There is no evidence given for a carbon-carbon double bond; therefore, the presence of a ring is indicated. The NMR shows a sharp, 6 H singlet indicating two equivalent CH_3 groups. Therefore, a 6-membered ring with 2 methyls probably on the same carbon. There are 4 H's with a chemical shift consistent with 2 CH_2's next to the carbonyl. Since this signal is a multiplet and presumably not 2 different signals superimposed, the 2 CH_3's are not on one of the adjacent carbons. Most probable structure is 4,4-dimethylcyclohexanone.

17.E. Answers and Explanations for Problems

1. a. The intense band at ~965 cm^{-1} suggests

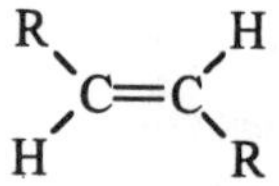

R H
C=C
H R

b. Bands at 3300, 2150, and 630 cm^{-1} conclusively indicate RC≡CH.
c. Broad band at 3350 cm^{-1} shows OH. Complex absorption peaking at 1050 cm^{-1} suggests primary.
d. Bands at 3080, 1820, 1640 and 910 cm^{-1} are conclusive evidence for $RCH=CH_2$.
e. Broad band at 3400 cm^{-1} shows OH, as does absorption at 1020 cm^{-1} (C−O stretch), probably primary. Sharp band at 2120 cm^{-1} and band at 650 cm^{-1} show terminal acetylene. Thus, there are two functional groups, OH and C≡CH.
f. Strong band at 1720 cm^{-1} (C=O stretch) could be an aldehyde or ketone (acyclic or six-membered ring), but the absence of bands at 2720 and 2820 cm^{-1} shows that it is not an aldehyde.
g. Strong band at 1725 cm^{-1} (C=O stretch) and bands at 2720 and 2820 cm^{-1} (C−H stretch) are characteristic of the aldehyde group.

2. IR: sharp bands at 3080, 1640, and 890 cm^{-1} indicate alkene; strong band at 890 cm^{-1} indicates specifically a

   1,1-disubstituted double bond: $C_2C{=}CH_2$

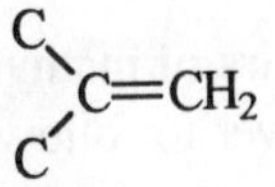

   NMR: δ 0.9, singlet, 9 hydrogens; must be a *t*-butyl group: $C(CH_3)_3$
   1.7, broad singlet (some long-range coupling), 3 hydrogens: CH_3, with some reason to be deshielded relative to an alkane,

   for example on a double bond: $CH_3{-}C{=}$

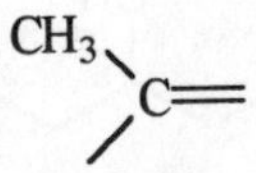

   1.9, singlet, 2 hydrogens: CH_2 with no adjacent hydrogens
   4.6 & 4.8, very little coupling, 1 hydrogen each:

   two alkene hydrogens: $C_2C{=}CH_2$

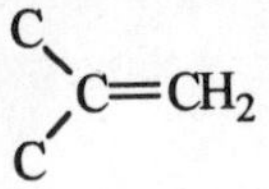

   Structure: 2,4,4-trimethyl-1-pentene:

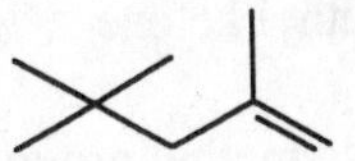

3. IR: 3400 cm^{-1}, broad: O−H
   3100, 1840 (overtone), 1640 (weak) cm^{-1}: an alkene
   1150 cm^{-1}: C−O

   NMR: δ 1.2, singlet, 6H: two equivalent, isolated methyl groups
   2.3, broad, 1H: the O−H
   5.0, multiplet, 2H, and 6.0, multiplet, 1H:
   this is the common pattern for a vinyl group: $C(H)C{=}CH_2$

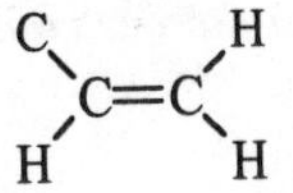

   Structure: 2-methyl-3-buten-2-ol: HO

4. IR: 3400 cm^{-1}, broad: O−H
   3300 cm^{-1} (almost hidden by O−H band), weak band at 2140 cm^{-1}, and strong band at 660 cm^{-1}:terminal acetylene, C≡C−H

   NMR: δ 1.5, singlet, 6H: two equivalent, isolated methyl groups
   2.2, singlet, 1H, sharp: consistent with C≡C−H
   2.9, singlet, 1H, broad: the O−H

   Structure: 2-methyl-3-butyn-2-ol: HO

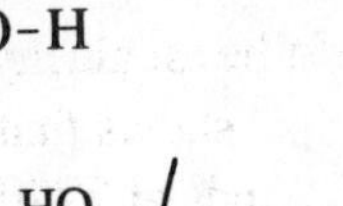

5. IR: 3350 cm^{-1}, broad: O−H

NMR: δ 0.8, two doublets, 12H: two diastereotopic pairs of methyl groups next to a CH group, probably two isopropyl groups: 2 $(CH_3)_2CH-$ (see answer to following problem)
1.0-2.0, m, 6H: CH's and CH_2's
3.6, broad, 1H: hydrogen next to an O−H group

(the OH resonance is broad and falls under the multiplet of the CH's)

Structure: 2,6-dimethyl-4-heptanol:

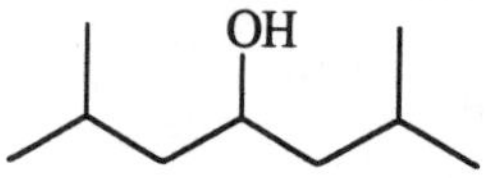

OH

6. IR: 3450 cm^{-1}, broad: O−H
1710 cm^{-1}, strong: ketone, acyclic or in six-membered ring
(1100 cm^{-1}, sharp: suggests that alcohol is 2°)

CMR: δ 212.7: confirms presence of ketone
76.8: carbon with O−H group
30.7: could be the methyl of a methyl ketone
26.4: CH
7.1 & 8.5: non-equivalent, upfield methyl groups

A reasonable structure is 4-methyl-3-hydroxy-2-pentanone:

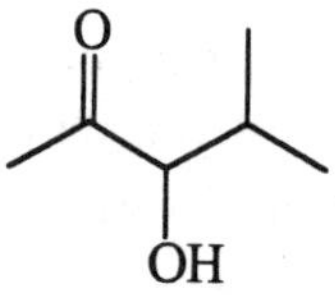

O

OH

Note that the two methyls of the isopropyl group are non-equivalent because of the stereocenter that is in the molecule. (Make two models, one with a CD_3 in place of one of the methyls, and second with it in place of the other; note that these two models represent ***diastereomers****.) The word that is used to describe the relationship between these two methyls is* ***diastereotopic****, and one of the consequences of their non-equivalence is the fact that their resonances come at different positions in the NMR spectrum.*

7. a. IR: 3400 cm^{-1}, broad: O−H
1160 cm^{-1}: tertiary O−H

CMR: 7.9: shielded methyl(s)
25.5: methyl on electron-withdrawing carbon
33.5: CH_2(s) on electron-withdrawing carbon
72.6: C−OH

Since this compound is isomeric with that in part b. and the CMR spectrum of b. shows six separate resonances, there must be two sets of equivalent carbons in the structure of a.

Structure: 3-methyl-3-pentanol

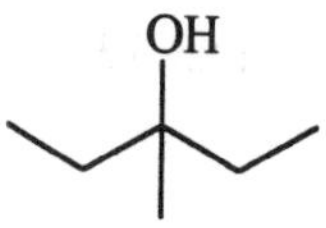

OH

b. IR: 3400 cm^{-1}, broad: O–H

There are too many bands in the 1020-1160 cm^{-1} region to tell whether the alcohol is primary, secondary, or tertiary

CMR: six resonances indicate that the alcohol cannot be derived from the 2,2-dimethylbutyl skeleton, because symmetry would result in fewer than six peaks. For the same reason, the following alcohols can also be ruled out:

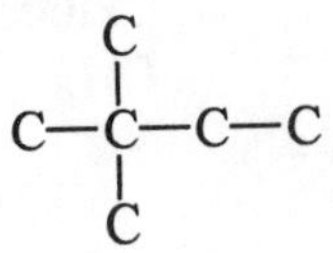

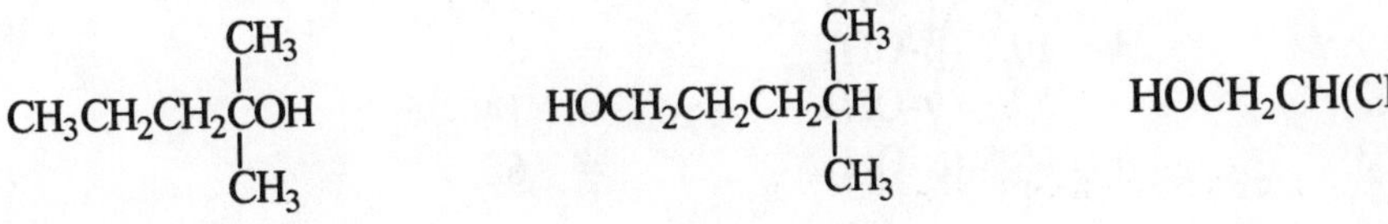

$$\begin{array}{c} C \\ | \\ C-C-C-C \\ | \\ C \end{array}$$

$$\begin{array}{r} CH_3 \\ | \\ CH_3CH_2CH_2COH \\ | \\ CH_3 \end{array} \qquad \begin{array}{r} CH_3 \\ | \\ HOCH_2CH_2CH_2CH \\ | \\ CH_3 \end{array} \qquad HOCH_2CH(CH_2CH_3)_2$$

$$\begin{array}{r} OH \\ | \\ (CH_3CH_2)_2CCH_3 \end{array} \qquad \begin{array}{r} CH_3 \\ | \\ (CH_3)_2CHCOH \\ | \\ CH_3 \end{array}$$

CMR: no resonance at $\delta < 20$ ppm indicates that there is no CH_3 at the end of a chain without a β-substituent. This rules out some more possibilities:

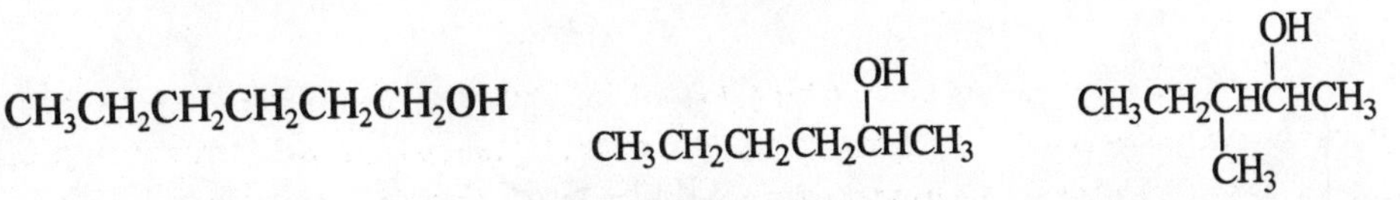

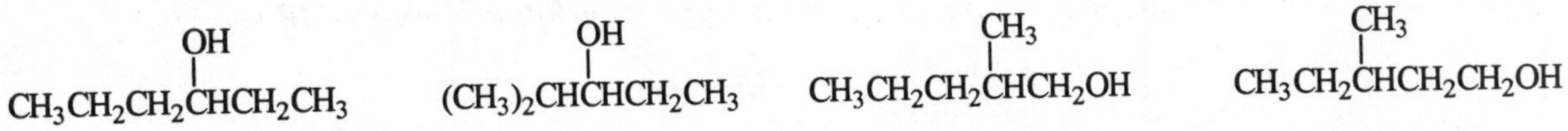

$$CH_3CH_2CH_2CH_2CH_2CH_2OH \qquad \begin{array}{r} OH \\ | \\ CH_3CH_2CH_2CH_2CHCH_3 \end{array} \qquad \begin{array}{c} OH \\ | \\ CH_3CH_2CHCHCH_3 \\ | \\ CH_3 \end{array}$$

$$\begin{array}{c} OH \\ | \\ CH_3CH_2CH_2CHCH_2CH_3 \end{array} \qquad \begin{array}{c} OH \\ | \\ (CH_3)_2CHCHCH_2CH_3 \end{array} \qquad \begin{array}{c} CH_3 \\ | \\ CH_3CH_2CH_2CHCH_2OH \end{array} \qquad \begin{array}{c} CH_3 \\ | \\ CH_3CH_2CHCH_2CH_2OH \end{array}$$

The only possibilities that remain are: $\begin{array}{c} OH \\ | \\ CH_3CHCH_2CHCH_3 \\ | \\ CH_3 \end{array}$ and $\begin{array}{c} CH_3 \\ | \\ CH_3CHCHCH_2OH \\ | \\ CH_3 \end{array}$

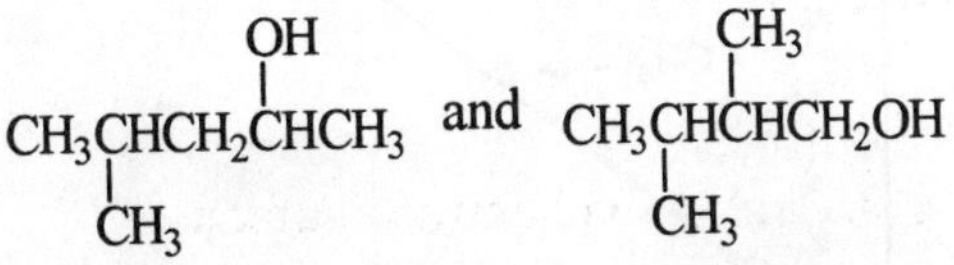

(<u>NOTE</u>: *in both of these compounds, the two methyls of the isopropyl group are diastereotopic (see answer to problem #6) and therefore are non-equivalent.)*

To distinguish between these two possibilities, estimate the CMR chemical shifts using Table 13.3 and the α-, β-, and γ-substituent effects of OH (see problem #20, Chapter 13):

	δ (2-methylpentane)	+ substituent effect	= predicted δ	(actual δ)
$(CH_3)_2$	22.5	+ 0.7 (δ-OH)	= 23 (2 resonances)	23.5, 24.3
CH	27.8	- 5.6 (γ-OH)	= 22	22.7
CH_2	41.8	+ 10.3 (β-OH)	= 52	49.2
CHOH	20.7	+ 48.5 (α-OH)	= 69	65.5
CH_3	14.1	+ 10.3 (β-OH)	= 24	25.1

	δ (2,3-dimethylbutane)	+ substituent effect	= predicted δ
$(CH_3)_2$	19.3	+ 0.7 (δ-OH)	= 20 (2 resonances)
CH	34.1	- 5.6 (γ-OH)	= 28.5
CH	34.1	+ 10.3 (β-OH)	= 44
side chain CH_3	19.3	- 5.6 (γ-OH)	= 14
CH_2OH	19.3	+ 48.5 (α-OH)	= 68

The chemical shifts predicted for 4-methyl-2-pentanol clearly fit the data best.

Notice that the only piece of information gathered from the IR spectrum was that the compound is an alcohol. Nevertheless, that was crucial, because it eliminated from consideration all fifteen ethers that have the formula $C_6H_{14}O$.

8. a. CH_3Cl, $f = 2.83 \times 10^5$, DH° = 84 kcal mole^{-1} $\quad f/$DH° = 3.37×10^3 dyne mole/kcal cm
CH_3Br, $f = 2.29 \times 10^5$, DH° = 70 kcal mole^{-1} $\quad f/$DH° = 3.27×10^3 dyne mole/kcal cm
CH_3I, $f = 1.83 \times 10^5$, DH° = 56 kcal mole^{-1} $\quad f/$DH° = 3.27×10^3 dyne mole/kcal cm

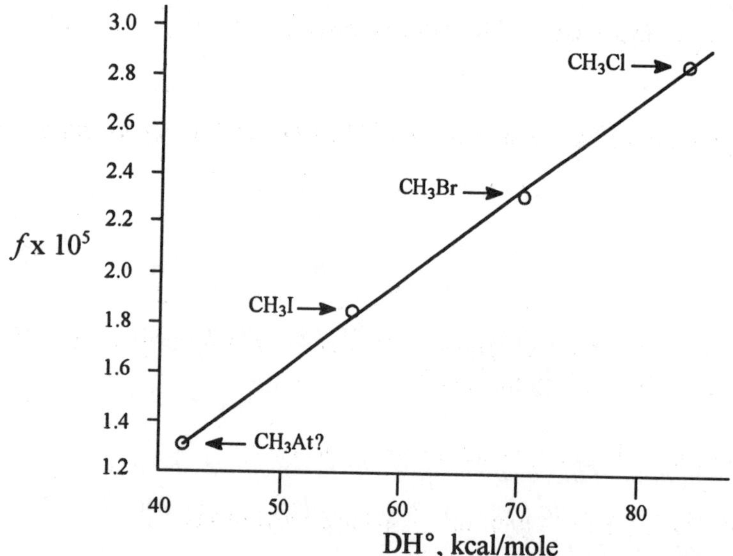

b. Notice that DH° goes down by 14 kcal mole^{-1} from CH_3Cl to CH_3Br and from CH_3Br to CH_3I. If we project this incremental change to CH_3At, then DH° ≈ 42 kcal mole^{-1}, and f from the above graph is ≈ 1.3 x 10^5 dynes cm^{-1}. Or using an average value of 3.30×10^3 dyne mole/kcal cm x 42 kcal/mole = 1.4×10^5 dyne/cm for f.

$$\tilde{\nu} = 4.120\sqrt{1.3 \times 10^5 \frac{12+210}{12 \cdot 210}} = 441 cm^{-1}$$

c. For CH_3SH, the Hooke's Law model gives $f = 2.56 \times 10^5$ dynes cm^{-1}. From the graph in part a, this corresponds to DH° = 76. kcal mol^{-1}.

9. IR: 3350, 1080 cm^{-1}: O–H
   1175 cm^{-1}, strong: C–F

   CMR: two carbons, each split into a quartet; each carbon coupled to three fluorines

   Structure: 2,2,2-trifluoroethanol, CF_3CH_2OH

10. The O–O stretch is not infrared active if the molecule exists in a conformation where there is no dipole moment change.

    i.e.

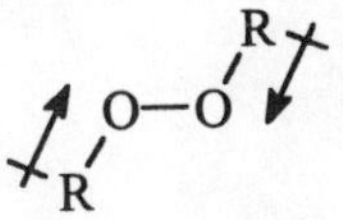

    R
    O—O
    R

    For a gauche conformation, there is a change in dipole moment and the band is active.

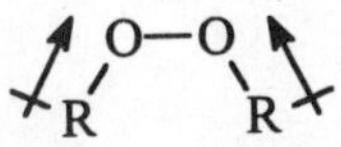

    O—O
    R R

    For a 900 cm^{-1} stretch, $f = 3.82 \times 10^5$ dynes cm^{-1}. The value is lower due to the low DH° of O–O bond.

11. The two signals arise from the presence of both unassociated $(CH_3)_3COH$ and the hydrogen-bonded dimer of the alcohol.

17.F. Supplementary Problems

S1. IR spectral bands and some nmr characteristics are given for four isomers of formula C_6H_{12}. What can you deduce about the structures from the information for each?

   a. IR: 1170, 1374, 1450, 2900 cm^{-1} (all strong); nmr: single resonance.

   b. IR: 960 (s), 1374 (w), 1450, 2940 (s) cm^{-1}; nmr: only ethyl and vinyl protons.

   c. IR: 885, 1370 (w), 1640, 2900, 3000 (w) cm^{-1}; nmr: only ethyl and vinyl protons.

   d. IR: 1450 (s), 2950 (s) cm^{-1}; nmr: single resonance.

S2. What are the structures of the compounds in the sequence below? Assign the resonances

$$A \xrightarrow{KMnO_4} B + CO_2 \xrightarrow{NaBH_4} C \xrightarrow[\Delta]{H_2SO_4} D \xrightarrow[Pt]{H_2} E$$

	A	B	C	D	E
IR:	886(s)	1450	1067(s)	1440	nmr: singlet
	1445	1714(s)	1450	1645(vw)	
	1618	2875	2860	2865	
	2860	2950(s)	2940	3000	
	2940		3300(s)		
	3082				

S3. How could you use IR spectroscopy to distinguish between the following pairs of isomers?

a. $HOCH_2CH_2CH_2CH_2CH$(=O) and 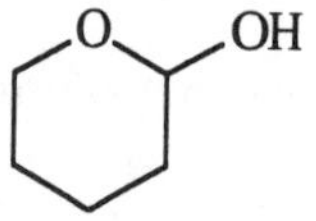

b. 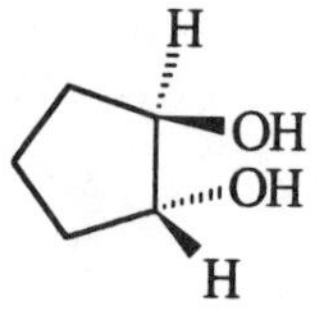 and 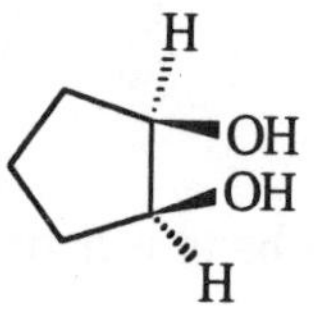

c.  and 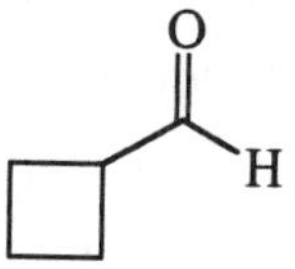

S4. Some IR and nmr spectral properties of five isomers of C_4H_8O are given below: Assign the structure for each isomer.

a. IR: 1176, 1370, 1718, 3000 cm^{-1}; nmr: singlet (3H) at 2.0 ppm.
b. IR: 962(s), 1000, 1075, 1666, 2860, 2940, 3300(br) cm^{-1}; nmr: 2 resonances (1H each) between 5 and 6 ppm.
c. IR: 1724, 2718, 2825, 2878, 2970 cm^{-1}; nmr: triplet (1H) at 9.5 ppm.
d. IR: 917(s), 1450, 1640(w), 2860, 2940, 3300 (br) cm^{-1}; nmr: 3 resonances (1H each) between 5 and 6 ppm, doublet (3H) at 1.3 ppm.
e. IR: 806, 962, 1030, 1200, 1312, 1612(s), 2950, 3000 cm^{-1}; nmr: indicates presence of ethyl group.

17.G. Answers to Supplementary Problems

S1. a. and d.: nmr: single resonance. The only possibilities are [cyclohexane] and $(H_3C)_2C{=}C(CH_3)_2$.

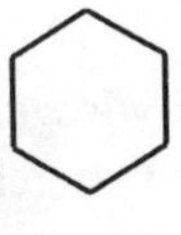

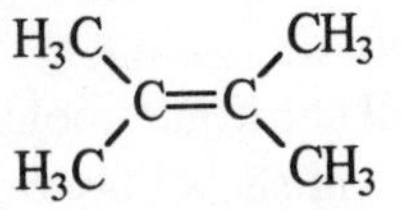

These are hard to assign by IR: the symmetrically substituted double bond and the lack of vinyl hydrogens mean that no characteristic olefinic absorbances will appear in the spectrum of 2,3-dimethyl-2-butene. However, spectrum a. has a band at 1374 cm^{-1}, characteristic of $-CH_3$ rocking, indicating that a. is the olefin. Spectrum d. therefore corresponds to cyclohexane.

b. and c.: nmr indicates only ethyl and vinyl hydrogens present. There are three possibilities: $(CH_3CH_2)_2C{=}CH_2$ and *cis-* and *trans*-3-hexene. Spectrum b. has a strong band at 960 cm^{-1}, consistent with the C–H out-of-plane bending absorption expected for *trans*-3-hexene; c. has bands at 885 and 1640 cm^{-1}, which are what you would expect to find in the spectrum of 2-ethyl-1-butene (C–H out-of-plane bending and C=C stretching respectively).

S2. It's not hard to figure out that A is $R_2C{=}CH_2$, B a ketone, C the corresponding alcohol, D an alkene and E a cyclic alkane. The 6-membered ring is indicated by the value of 1714 cm^{-1} for the C=O stretch.

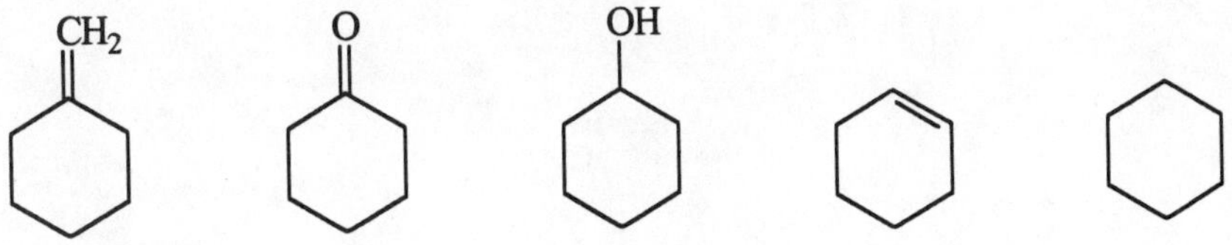

A (CH_2)	B (O)	C (OH)	D	E
886 (C–H bend)	1450 (CH_2 scissor)	1067 (C–O stretch)	1440 (CH_2 scissor)	
1445 (CH_2 scissor)	1714 (C=O stretch)	1450 (CH_2 scissor)	1645 (C=C stretch)	
1618 (C=C stretch)	2875,2950 (C–H stretch)	2860,2940 (C–H stretch)	2865 (alkane C–H)	
2860,2940 (alkane C–H stretch)		3300 (O–H stretch)	3000 (alkene C–H)	
3082 (alkene C–H stretch)				

S3. a. Open chain form will show strong band due to C=O stretch at 1725 cm^{-1}; cyclic hemiacetal form will not.

b. *cis* isomer will show intramolecular hydrogen-bonded O−H stretch (320-3450 cm^{-1}) at all concentrations; *trans* isomer will show only free O−H stretch (3620-3640 cm^{-1}) at low concentration.

c. Ketone: C=O stretch at 1745 cm^{-1}, no bands in region 2700-2850 cm^{-1}.
Aldehyde: C=O stretch at 1725 cm^{-1}, two medium bands for H−CO stretch at 2720, 2820 cm^{-1}.

S4. a. $CH_3CH_2C(=O)CH_3$

b. H_3C–CH=CH–CH_2OH (trans; H above and H below the double bond)

c. $CH_3CH_2CH_2CH(=O)$

d. $CH_3CH(OH)CH{=}CH_2$

e. $CH_3CH_2OCH{=}CH_2$

18. CARBOXYLIC ACIDS

18.A. Chapter Outline and Important Terms Introduced

18.1 Structure

carboxy group

18.2 Nomenclature

-oic acid

1,2,3... vs. α, β, γ,...

18.3 Physical Properties

b.p. increased by hydrogen bonding

18.4 Acidity (pK_a of acetic acid = 4.7)

A. Ionization
 conjugated system
 resonance stabilization

B. Inductive Effects

C. Salt Formation
 -ate

D. Soaps
 micelle
 biodegradable
 alkanesulfonates
 amphiphilic/amphipathic
 hydrophobic effect

18.5 Spectroscopy

A. Nuclear Magnetic Resonance
 $C\underline{H}_3CO_2H$ 2-2.5 ppm
 $\underline{C}H_3CO_2H$ 21 ppm
 $CH_3CO_2\underline{H}$ 10-13 ppm
 $CH_3\underline{C}O_2H$ 170-185 ppm

B. Infrared
 C=O 1710-1760 cm^{-1}
 O–H 2400-3400 cm^{-1}

18.6 Synthesis

A. Hydrolysis of Nitriles
 acid- or base-catalyzed

B. Carbonation of Organometallic Reagents

C. Oxidation of Primary Alcohols or Aldehydes

18.7 Reactions

A. Reactions Involving the O-H bond
 alkylation of salts
 esterification with diazomethane

B. Reactions Involving the Hydrocarbon Side Chain
   bromination of α-position
C. Formation of Esters
   esterification (← **important mechanism)**
   tetrahedral intermediate
D. Formation of Acyl Halides
E. Reaction with Ammonia: Formation of Amides
F. Reduction of the Carboxy Group
G. One-Carbon Degradation of Carboxylic Acids
   Hunsdiecker reaction Kochi reaction

18.8 Natural Occurrence of Carboxylic Acids

18.B. Important Reactions Introduced

Hydrolysis of nitriles (18.6.A)

Equation: $RC{\equiv}N + H_2O \xrightarrow{\text{catalysis}} RC(=O)NH_2 \longrightarrow RCO_2H + NH_3$

Generality: R = alkyl, aryl

Key features: acid- or base-catalyzed, heat required
reaction can be stopped at amide intermediate

Reaction of organometallic reagents with carbon dioxide (18.6.B)

Equation: $RM + CO_2 \longrightarrow RCO_2^-M^+ \xrightarrow{H^+} RCO_2H$

Generality: R = alkyl, aryl; M = Li or MgX

Key features: useful synthesis of carboxylic acids; increments carbon chain by one.

Oxidation of primary alcohols (or aldehydes) to carboxylic acids (18.6.C)

Equation: $RCH_2OH \xrightarrow{[Ox]} [RCH{=}O] \xrightarrow{[Ox]} RCO_2H$

Generality: [Ox] = $KMnO_4$, aq. Cr(VI), HNO_3; Ag_2O specific for aldehydes.

Key features: often a side reaction in oxidation of 1° alcohols to aldehydes; Ag_2O too mild for alcohols.

Alkylation of carboxylate anions (18.7.A)

Equation: $RCO_2^-M^+ + R'X \longrightarrow RCO_2R' + M^+X^-$

Generality: R′ = alkyl

Key features: typical S_N2 displacement reaction (R′ = 1° > 2° >> 3° alkyl)
useful method for replacing R′–X bond with O–R′ bond with inversion

Esterification of carboxylic acids with diazomethane (18.7.A)

Equation: $RCO_2H + CH_2N_2 \longrightarrow RCO_2CH_3 + N_2$

Generality: R = alkyl, aryl, H

Key features: useful and convenient on small scale
dangerous on large scale (diazomethane is toxic and explosive)

α-Bromination of carboxylic acids (18.7.B)

Equation: $RCH_2CO_2H + Br_2 \xrightarrow{PBr_3} RCHBrCO_2H$

Generality: Usually bromination, less commonly chlorination.

Key features: reaction occurs via enol of intermediate acyl halide

Esterification of carboxylic acids (18.7.C)

Equation: $RCO_2H + HOR' \overset{H^+}{\rightleftharpoons} RCO_2R' + H_2O$

Generality: R′ = alkyl

Key features: mechanism is important to know (addition/elimination reaction)
acid-catalyzed equilibrium process, often driven to product by removal of water or use of large excess of less expensive reactant.
oxygen in H_2O product comes from the carboxylic acid OH

Formation of acyl halides (18.7.D)

Equation: $RCO_2H \xrightarrow{[reagent]} RCOX$

Generality: R = alkyl, aryl; [reagent] = $SOCl_2$, PCl_3, PBr_3, PCl_5

Amide formation from ammonium carboxylate salts (18.7.E)

Equation: $RCO_2H + H_3N \xrightarrow{\Delta} RCO_2^-NH_4^+ \longrightarrow RCONH_2 + H_2O$

Generality: carboxylic acid must be stable to heat

Key features: not commonly used; employed only for simple cases

Reaction of carboxylic acids and esters with lithium aluminum hydride ($LiAlH_4$) (18.7.F)

Equation: $RCO_2R' + LiAlH_4 \longrightarrow RCH_2OH + R'OH$

Generality: R = alkyl, aryl; R′ = alkyl, aryl, or H

Degradation of carboxylic acids to alkyl halides (18.7.G)

Equation: $RCO_2H \longrightarrow RX + CO_2$

Generality: Br_2 with Ag^+ or Hg^{2+} salt of acid (Hunsdiecker reaction)
$Pb(OAc)_4$ + LiCl (Kochi reaction)

Key features: proceeds via $RCO_2\cdot$ and R· radicals; Hunsdiecker is best for 1 alkyl° acids; Kochi for 2° and 3° alkyl acids. Degrades carbon chain by one.

18.C. Important Concepts and Hints

An important new type of reaction mechanism is introduced in this chapter: the nucleophilic displacement of a substituent on a carbonyl group. The acid-catalyzed formation of an ester from a carboxylic acid and an alcohol is the most important example of this type. The mechanism involves first **addition** of the nucleophile to the carbon-oxygen double bond, and then **elimination** of the leaving group to regenerate the carbonyl group. The full scope of this substitution mechanism will become apparent in the following chapter, when it will be discussed at length. For the moment, you should examine closely the comparison of the esterification reaction and the formation, then decomposition of a hemiketal. It's a good analogy, and it helps tie this new mechanism in with one you've seen before.

Aside from that, no other new concepts are introduced in this chapter. The effects of electron-withdrawing substituents on acidity was presented in the chapter on alcohols (Section 10.4), and the synthetic reactions which lead to carboxylic acids have been introduced previously. Although the reactions of carboxylic acids are for the most part new, their mechanisms involve the nucleophilic displacement, enolization, or carbonyl-addition mechanisms you have seen before. Consequently there is little else for us to comment on in this section of the Study Guide. This situation will become more common in subsequent chapters of this book, although we will still make what we hope are helpful comments where appropriate.

18.D. Answers to Exercises

18.1 a. 5-chloropentanoic acid
b. 5,5-dimethylhexanoic acid
c. 2-iodo-4-methylhexanoic acid
d. 2,4-pentadienoic acid (or penta-2,4-dienoic acid)

18.2 Graphing exercise - no answer.

18.3 $pK_a = -\log K_a = 3.75$

$$\frac{[H^+][HCO_2^-]}{[HCO_2H]} = 1.77 \times 10^{-4}$$

$[H^+] = [HCO_2^-] = x$; $[HCO_2H] = 0.1 - x$

$$\frac{x^2}{0.1-x} = 1.77 \times 10^{-4}$$

$x = [HCO_2^-] = 4.12 \times 10^{-3}\,M$

18.4

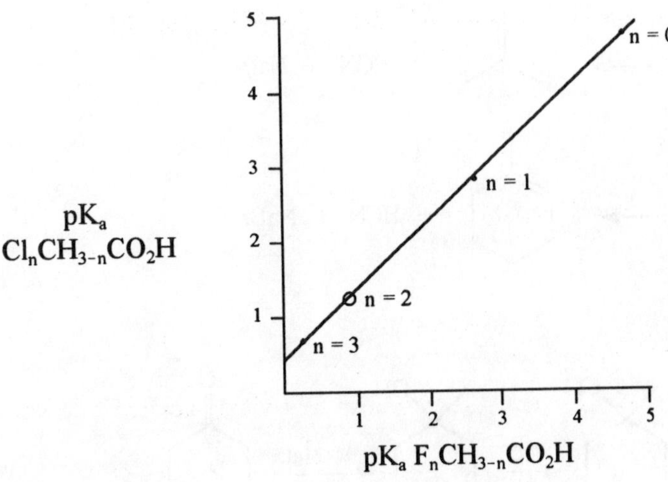

pK_a predicted for $F_2CHCO_2H = 0.9$

18.5 $K = \dfrac{[RCO_2^-]}{[RCO_2H][OH^-]} = 1.3 \times 10^9\ M^{-1}$

pH	$[OH^-]$	Ratio: $[RCO_2^-]/[RCO_2H]$
2	10^{-12}	1.3×10^{-3}
4	10^{-10}	0.13
6	10^{-8}	13
8	10^{-6}	1300

18.6 2,2-Dimethylbutanoic acid: δ_H given with δ_C in parentheses for C's with H's.

$CH_3-CH_2-C(CH_3)_2-CO_2H$

0.95 (9.5) → CH_3; 1.50 (33.7) → CH_2; 1.25 (24.8) → CH_3; (43.0) → C; 12.5 → CO_2H (H); 182.2 → CO_2H (C)

18.7 b. and d. 2-Methylpentanoic acid and 7-methyloctanoic acid can be prepared by this route:

$\text{Br} \xrightarrow{\text{NaCN}} \xrightarrow{H_3O^+} CO_2H$

$\text{Br} \xrightarrow{\text{NaCN}} \xrightarrow{H_3O^+} CO_2H$

a. and c. 2,2-Dimethylpropanoic acid and 1-methylcyclohexanecarboxylic acid <u>cannot</u> be prepared via the alkylation of cyanide ion with a tertiary halide; cyanide ion is too basic, and elimination is the major reaction:

(tert-butyl bromide) + NaCN ⟶ (2-methylpropene) + HCN + NaBr

(1-bromo-1-methylcyclohexane) + NaCN ⟶ (1-methylcyclohexene) + HCN + NaBr

18.8 a. (1-methylcyclohexanecarboxylic acid, CO_2H) ⟵ 1. CO_2, 2. H_3O^+ ⟵ (MgCl) ⟵ Mg ⟵ (Cl) ⟵ HCl ⟵ (OH) ⟵ MeMgI ⟵ (cyclohexanone, O)

b. $CH_3CH_2C(CH_3)_2CO_2H \xleftarrow{H^+} \xleftarrow{CO_2} \xleftarrow{Mg} CH_3CH_2CBr(CH_3)_2 \xleftarrow[h\nu]{Br_2} CH_3CH_2CH(CH_3)_2$

c. (3-methylbutanoic acid, CO_2H) ⟵ H_3O^+ ⟵ CO_2 ⟵ Mg ⟵ (1-chloro-2-methylpropane, Cl) ⟵ PCl_3 ⟵ (2-methyl-1-propanol, OH)

18.9 $3\ CH_3CH_2OH + 4\ KMnO_4 \longrightarrow 3\ CH_3CO_2H + 4\ MnO_2 + 4\ K^+ + H_2O + 4\ OH^-$

18.10 (sodium propanoate, $O^-\ Na^+$) + (1-deuterio-1-iodopropane: H, D, I) ⟶ (ester: O, D, H)

(*S*)-1-deuteriopropyl propanoate

18.11 (2-heptenoic acid, CO_2H) ⟵ 1. $NaNH_2$, 2. H_3O^+ ⟵ (2-bromoheptanoic acid, CO_2H, Br) ⟵ P , Br_2 ⟵ (heptanoic acid, CO_2H) ⟵ 1. NaCN, 2. H_3O^+ ⟵ (1-chlorohexane, Cl)

18.12 $RCO_2H + H^+ \overset{K}{\rightleftharpoons} \underset{[1]}{RC(=\overset{+}{O}H)OH} + CH_3OH \overset{k}{\rightleftharpoons} RC(OH)(OH)\overset{+}{O}HCH_3$ The last step is the slow one.

Rate = k x $[CH_3OH]$ x [1] [1] = K x $[RCO_2H]$ x $[H^+]$

Rate = k x K x $[CH_3OH]$ x $[RCO_2H]$ x $[H^+]$

18.13

$$CH_3CH_2CH_2CO_2^- + NH_4^+ \overset{K}{\rightleftharpoons} CH_3CH_2CH_2CO_2H + NH_3$$

pK_a: 9.24 4.82

$$K_a(NH_4^+) = \frac{[NH_3][H^+]}{[NH_4^+]} = 10^{-9.24}$$

$$K_a(RCO_2H) = \frac{[RCO_2^-][H^+]}{[RCO_2H]} = 10^{-4.82}$$

$$K = \frac{[RCO_2H][NH_3]}{[RCO_2^-][NH_4^+]} = \frac{K_a(NH_4^+)}{K_a(RCO_2H)} = 10^{-4.42} = 3.80 \times 10^{-5}$$

18.14 (pentyl)CN $\xrightarrow[\Delta]{H_3O^+}$ (pentyl)CO_2H $\xrightarrow{LiAlH_4}$ (hexyl)OH

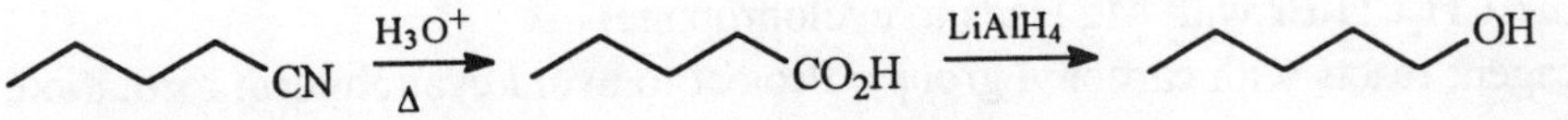

18.15 (2,2-dimethylpentyl)CO_2H $\xrightarrow[LiCl]{Pb(OAc)_4}$ (2-methylpentyl)Cl $\xrightarrow[t\text{-BuOH}]{t\text{-BuOK}}$ (2-methyl-1-pentene)

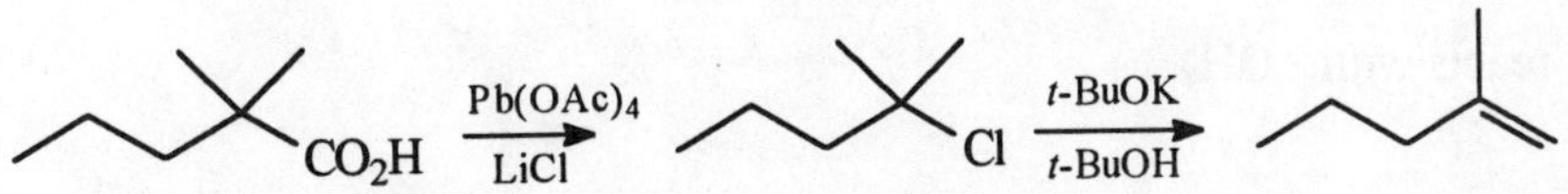

18.E Answers and Explanations for Problems

IUPAC

1. a. 3-methylpentanoic acid
   b. 2,2-dimethylpropanoic acid (NOTE: trivial name is pivalic acid)
   c. 4-bromobutanoic acid
   d. iodoacetic acid (NOTE: iodoethanoic acid is a systematic name that is never used in practice)
   e. 2-hydroxybutanoic acid
   f. decanoic acid
   g. cyclobutanecarboxylic acid
   h. 3-methoxybutanoic acid
   i. 3,4-dimethylpentanoic acid
   j. cyanoacetic acid

2. a. $CH_3CHClCH_2CO_2H$
   b. $CH_3(CH_2)_4CO_2H$
   c. $CH_3CH(OCH_3)CH_2CH_2CO_2H$
   d. (cyclopentyl)$-CO_2H$
   e. $BrCH_2CH_2CO_2H$
   f. 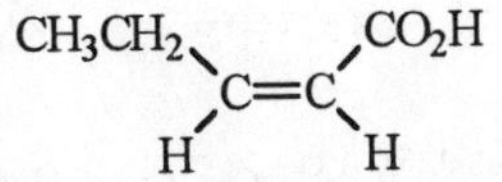 $CH_3CH_2(H)C{=}C(H)CO_2H$
   g. 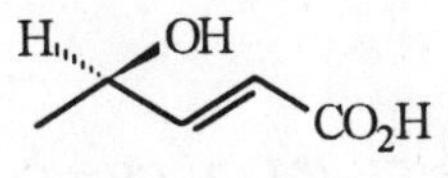 (H, OH stereocenter)$-CH{=}CH-CO_2H$
   h. $BrCH_2CHClCO_2H$

3\. a. CH_2OH b. Br CO_2H c. CO_2CH_3 d. $CO_2CH(CH_3)_2$

e. O NH_2 f. O $NHCH_3$ g. O CCl h. O CBr

i. No reaction j. O CO^- Na^+ k. Br l. Cl

4\. a. Grignard (displacement reaction is poor with a tertiary halide).
b. cyanide (reaction of $BrCH_2CH_2CH_2Br$ with Mg leads to cyclopropane).
c. cyanide (the Grignard reagent reacts with carbonyl group). Protect to avoid cyanohydrin formation.
d. Grignard (displacement reaction is slow with neopentyl halides).
e. Both methods work well.
f. cyanide (Grignard reagent reacts with -OH).

5\.

a. $(CH_3)_3CCH_3 \xrightarrow[h\nu]{Br_2} (CH_3)_3CCH_2Br \xrightarrow[ether]{Mg} (CH_3)_3CCH_2MgBr \xrightarrow{CO_2} \xrightarrow{H^+} (CH_3)_3CCH_2CO_2H$

b. $(CH_3)_3CCH_2COOH$ (from a.) $\xrightarrow[P]{Br_2} \xrightarrow{H_2O} (CH_3)_3CCHBrCOOH$

c. $(CH_3)_3CCH_2MgBr$ (from a.) $\xrightarrow{CH_2O} \xrightarrow{H^+} (CH_3)_3CCH_2CH_2OH$

or $(CH_3)_3CCH_2COOH$ (from a.) $\xrightarrow{LiAlH_4} (CH_3)_3CCH_2CH_2OH$

d. $(CH_3)_3CCH_2CO_2H$ (from a.) $\xrightarrow[\Delta]{SOCl_2} (CH_3)_3CCH_2COCl$

6\.

a. $CH_3CH_2CH_2CHO \xrightarrow{Ag_2O \text{ or } HNO_3} CH_3CH_2CH_2CO_2H$

b. $CH_3CH_2CH_2\overset{O}{\overset{\|}{C}}H \xrightarrow{\text{dil. } OH^-} CH_3CH_2CH_2CH{=}\overset{CHO}{\overset{|}{C}}CH_2CH_3 \xrightarrow{Ag_2O} CH_3CH_2CH_2CH{=}\overset{CO_2H}{\overset{|}{C}}CH_2CH_3$

c. $CH_3CH_2CH_2CH(=O) \xrightarrow{NaCN} CH_3CH_2CH_2CH(OH)CN \xrightarrow[\Delta]{H_3O^+} CH_3CH_2CH_2CH(OH)CO_2H$

d. $CH_3CH_2CH_2CHO \xrightarrow[LiAlH_4]{NaBH_4\text{ or}} CH_3CH_2CH_2CH_2OH \xrightarrow[ether]{PBr_3} CH_3CH_2CH_2CH_2Br \xrightarrow{Mg}$

$\xrightarrow[2.\ H^+]{1.\ CO_2} CH_3CH_2CH_2CH_2CO_2H$

7. a. $(CH_3)_2CHCH_2CO_2H \xrightarrow{LiAlH_4} (CH_3)_2CHCH_2CH_2OH$

b. $(CH_3)_2CHCH_2CH_2OH$ (from a.) $\xrightarrow{PBr_3} \xrightarrow{CN^-} \xrightarrow[\Delta]{H^+\text{ or }OH^-} (CH_3)_2CHCH_2CH_2CO_2H$

c. $(CH_3)_2CH(CH_2)_2CO_2H$ (from b.) $\xrightarrow{LiAlH_4} (CH_3)_2CH(CH_2)_3OH \xrightarrow{PBr_3} \xrightarrow[ether]{Mg} \xrightarrow{CH_2=O}$

$(CH_3)_2CH(CH_2)_4OH \xrightarrow[\Delta]{Al_2O_3} (CH_3)_2CH(CH_2)_2CH=CH_2 \xrightarrow{Br_2} (CH_3)_2CH(CH_2)_2CHBrCH_2Br$

d. $(CH_3)_2CHCH_2CO_2H \xrightarrow[Br_2]{P} (CH_3)_2CHCHBrCO_2H \xrightarrow[H^+,\ \Delta]{CH_3OH} (CH_3)_2CHCHBrCO_2CH_3$

$\xrightarrow[CH_3OH]{CH_3ONa} (CH_3)_2C=CHCO_2CH_3$

e. 1. CH_3Li, 2. H_2O f. NH_3, Δ g. Ag_2O, Br_2

8.

a. (cyclohexylidene)$=CH_2 \xrightarrow{B_2H_6} \xrightarrow[OH^-]{H_2O_2}$ (cyclohexyl)$-CH_2OH \xrightarrow{PBr_3} \xrightarrow[ether]{Mg} \xrightarrow[2.\ H_3O^+]{1.\ CO_2}$ (cyclohexyl)$-CH_2CO_2H$

b. $(CH_3)_3CCH=CH_2 \xrightarrow{KMnO_4} (CH_3)_3CCO_2H$

or $\xrightarrow{O_3} \xrightarrow[CH_3CO_2H]{Zn} (CH_3)_3CCHO \xrightarrow{Ag_2O} (CH_3)_3CCO_2H$

c.

$$CH_3\overset{O}{\overset{\|}{C}}(CH_2)_3\underset{CH_3}{\underset{|}{C}}BrCH_3 \xrightarrow[H^+]{HOCH_2CH_2OH} CH_3\overset{O\quad O}{C}(CH_2)_3\underset{CH_3}{\underset{|}{C}}BrCH_3 \xrightarrow[ether]{Mg} \xrightarrow[2.\ H_3O^+]{1.\ CO_2} CH_3\overset{O\quad O}{C}(CH_2)_3\underset{CH_3}{\underset{|}{\overset{CH_3}{\overset{|}{C}}}}CO_2H$$

$$\xrightarrow{H_3O^+} CH_3\overset{O}{\overset{\|}{C}}(CH_2)_3\underset{CH_3}{\underset{|}{\overset{CH_3}{\overset{|}{C}}}}CO_2H$$

d. $CH_3CH_2CO_2H \xrightarrow{LiAlH_4} CH_3CH_2CH_2OH \xrightarrow{PBr_3} CH_3CH_2CH_2Br \xrightarrow{CN^-} \xrightarrow[\Delta]{H^+ \text{ or } OH^-} CH_3CH_2CH_2CO_2H$

e. $CH_3CH_2CH_2CO_2H \xrightarrow[Br_2]{Ag_2O} CH_3CH_2CH_2Br \xrightarrow{OH^-} CH_3CH_2CH_2OH \xrightarrow{HNO_3} CH_3CH_2CO_2H$

f. $CH_3CH_2CH_2Br$ (from e.) $\xrightarrow{N_3^-} CH_3CH_2CH_2N_3$

9. a. The dissociation equilibrium can be written as: $AcOH \rightleftharpoons H^+ + AcO^-$

with $K = \dfrac{[H^+][AcO^-]}{[AcOH]_{eq}} = 1.8 \times 10^{-5}$

If we make the simplification that each H comes from ionization of AcOH, then $[H^+] = [AcO^-]$. Also, $[AcOH]_{total} = [AcOH]_{eq} + [AcO^-]$.

The equation then becomes:

$$K = \frac{[AcO^-]^2}{[AcOH]_{total} - [AcO^-]} = 1.8 \times 10^{-5}$$

$$[AcO^-]^2 + (1.8 \times 10^{-5})[AcO^-] - (1.8 \times 10^{-5})[AcOH]_{total} = 0$$

This can be solved using the solution to the quadratic equation:

$$x = \frac{-b \pm \sqrt{b^2 - 4ac}}{2a} \quad \text{where } x = [AcO^-]$$

$a = 1$
$b = 1.8 \times 10^{-5}$
and $c = -(1.8 \times 10^{-5})\,[AcOH]_{total}$

a. $[AcOH]_{total} = 0.1$ M: $[AcO^-]/[AcOH] = 0.0133 = 1.33\%$

b. $[AcOH]_{total} = 0.01$ M: $[AcO^-]/[AcOH] = 0.0415 = 4.15\%$

c. $[AcOH]_{total} = 0.001$ M: $[AcO^-]/[AcOH] = 0.1255 = 12.55\%$

A simpler way to solve the problem is to recognize that only a small fraction of AcOH dissociates, so that $[AcOH]_{eq} \approx [AcOH]_{total}$:

$$\frac{[AcO^-]^2}{[AcOH]_{total}} = 1.8 \times 10^{-5} \quad [AcO^-] = \sqrt{(1.8 \times 10^{-5})[AcOH]_{total}}$$

Using this equation, the following ratios are obtained:

a. $[AcOH]_{total} = 0.1$ M: $[AcO^-]/[AcOH] = 0.0134 = 1.34\%$
b. $[AcOH]_{total} = 0.01$ M: $[AcO^-]/[AcOH] = 0.0424 = 4.24\%$
c. $[AcOH]_{total} = 0.001$ M: $[AcO^-]/[AcOH] = 0.134 = 13.4\%$

Note that the percent dissociation increases as the concentration decreases. Because dissociation is a unimolecular reaction, its rate will be unaffected by concentration; recombination on the other hand is a bimolecular reaction and is therefore slower in more dilute solution. The position of equilibrium therefore shifts toward dissociation in dilute solution.

10. $pK_a = -\log K_a$

a. 4.85 b. 4.20 c. 1.26 d. 0.64 e. 3.68

11. a. $CH_3CH_2O^-$
In acetate ion the negative charge is shared between two oxygens, whereas in ethoxide ion the charge is essentially localized on a single oxygen. Ethoxide ion is therefore relatively less stable and has a greater tendency to react with a proton.

b. $CH_3CH_2CH_2CO_2^-$
β-Chloropropionate is stabilized by the electron-attracting inductive effect of chlorine.

c. $ClCH_2CH_2CO_2^-$
In α-chloropropionate ion, the chlorine is closer to the center of negative charge and provides greater stabilization.

d. $FCH_2CO_2^-$
Two electron-attracting fluorines provide greater stabilization of the anion than one does.

e. $CH_3CH_2CH_2CO_2^-$
The ethynyl group has a somewhat electron-attracting inductive effect compared to an alkyl group, and provides greater stabilization of an anion.

f. $CH_3CO_2^-$
HCl is a stronger acid than acetic acid, hence acetate ion is a stronger base than chloride ion. The reason that HCl is the stronger acid is, however, more difficult to explain. HCl is a stronger acid than CH_3CO_2H even in the gas phase, but the difference, ΔpK = 4-5, is smaller than in aqueous solution, ΔpK = 11. This difference means that the solvation energy of Cl^- is greater than that for $CH_3CO_2^-$. The greater acidity of HCl in the gas phase appears to be due in part to the greater bond strength of $CH_3CO_2\text{-}H$ compared to H-Cl, and in part to the higher electron affinity of Cl· compared to CH_3CO_2· .

12. $HO_2CCH_2CH_2CHClCH_2\underline{COOH}$ more acidic since the Cl is closer to this $-CO_2H$

13. Compare the effect of chlorine substitution on pK_a for the various isomers:

pK_a (butanoic acid) - pK_a (2-chlorobutanoic acid) = 1.96

> ratio = 2.55

pK_a (butanoic acid) - pK_a (3-chlorobutanoic acid) = 0.77

> ratio = 2.57

pK_a (butanoic acid) - pK_a (4-chlorobutanoic acid) = 0.30

pK_a (butanoic acid) - pK_a (3-cyanobutanoic acid) = 0.38, therefore

pK_a (butanoic acid) - pK_a (2-cyanobutanoic acid) is expected to be 2.56 x 0.38 = 0.97:

4.82 - 0.97 = 3.85 = calculated pK_a of 2-cyanobutanoic acid

Note that the pK_a of 2-cyanobutanoic acid has yet to be determined experimentally.

14.

$$CH_3CH_2C(=O)OH + H^+ \rightleftharpoons CH_3CH_2C(={}^+OH)OH$$

$$CH_3CH_2C(={}^+OH)OH + H_2{}^{18}O \rightleftharpoons CH_3CH_2C(OH)_2{}^{18}\overset{+}{O}H_2 \underset{}{\overset{-H^+}{\rightleftharpoons}} CH_3CH_2C(OH)_2{}^{18}OH \overset{H^+}{\rightleftharpoons} CH_3CH_2C(OH)({}^+OH_2){}^{18}OH$$

$$\overset{-H_2O}{\rightleftharpoons}$$

$$CH_3CH_2C(={}^{18}O)OH \overset{-H^+}{\rightleftharpoons} CH_3CH_2C(OH)={}^{18}OH^+ \text{ or } CH_3CH_2C(={}^+OH){}^{18}OH \overset{-H^+}{\rightleftharpoons} CH_3CH_2C(=O){}^{18}OH$$

15. 5-hydroxyhexanoic acid (OH, O, OH) + H^+ ⇌ protonated acid (HO, HO^+, OH; curved arrow from HO to C) ⇌ cyclic tetrahedral intermediate (CH_3, $\overset{+}{O}$–H, OH, OH)

⇌

lactone (CH_3, O, O) + H^+ ⇌ protonated lactone (CH_3, O, O^+H) + H_2O ⇌ (CH_3, O, $^+OH_2$, OH) ⇌ (CH_3, O, OH, OH) + H^+

In normal esterification, two reactant molecules (alcohol and acid) give two product molecules (ester and water), with a consequent $\Delta S°$ of about zero. In the present cyclization, one molecule gives two. The additional freedom of motion of the products corresponds to a positive $\Delta S°$ and a larger equilibrium constant.

16.

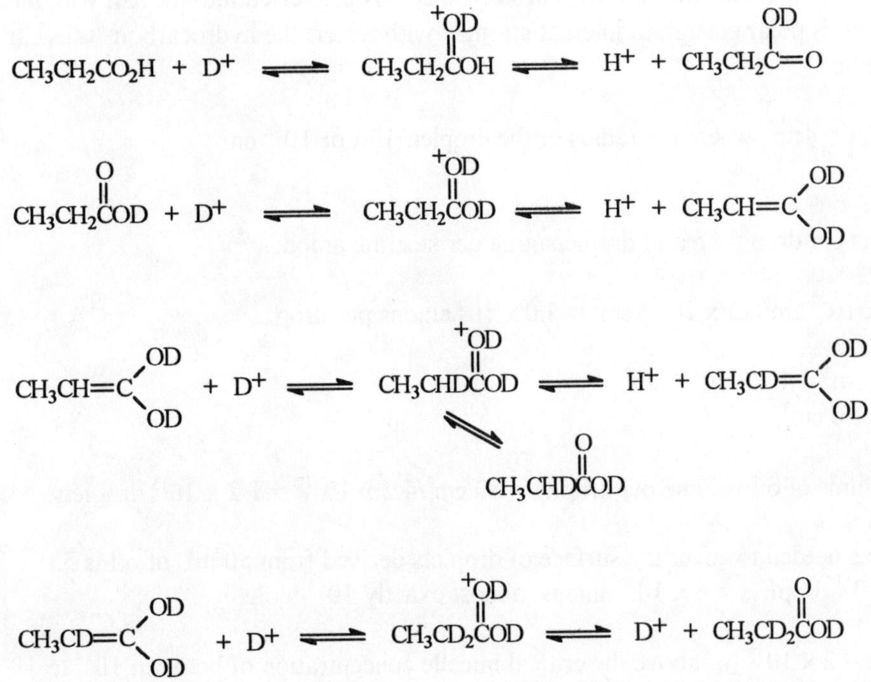

$$CH_3CH_2CO_2H + D^+ \rightleftharpoons CH_3CH_2C(=\overset{+}{O}D)OH \rightleftharpoons H^+ + CH_3CH_2C(OD)=O$$

$$CH_3CH_2C(=O)OD + D^+ \rightleftharpoons CH_3CH_2C(=\overset{+}{O}D)OD \rightleftharpoons H^+ + CH_3CH=C(OD)_2$$

$$CH_3CH=C(OD)_2 + D^+ \rightleftharpoons CH_3CHDC(=\overset{+}{O}D)OD \rightleftharpoons H^+ + CH_3CD=C(OD)_2$$

$$CH_3CHDC(=\overset{+}{O}D)OD \rightleftharpoons CH_3CHDC(=O)OD$$

$$CH_3CD=C(OD)_2 + D^+ \rightleftharpoons CH_3CD_2C(=\overset{+}{O}D)OD \rightleftharpoons D^+ + CH_3CD_2C(=O)OD$$

17. $\Delta H°(gas) = -4.6$ kcal mole^{-1}. The liquid phase involves solvation energies not present in the ideal gas state. The reactants are solvated more strongly than the products, probably because of hydrogen-bonding, and the esterification reaction is less exothermic.

18. a. The C-Cl dipole stabilizes the carboxylate ion by electrostatic attraction.

b. Trigonometry gives the following geometric results:

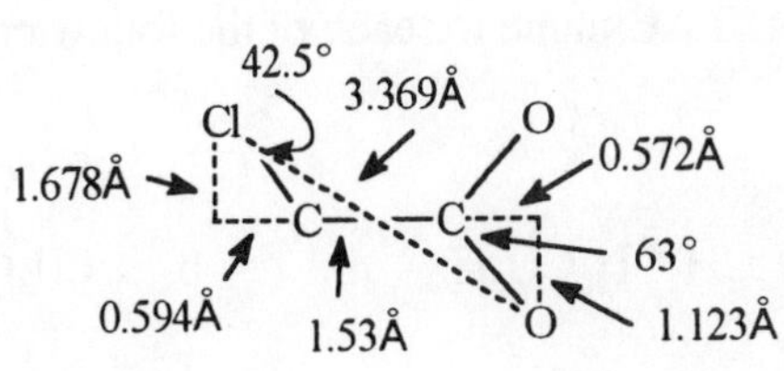

42.5°
3.369Å
Cl
O
0.572Å
1.678Å
C
C
63°
0.594Å
1.53Å
O
1.123Å

$$E = \frac{(q=0.5)(\mu=1.9)(\cos 42.5°)}{(3.369)^2}(69\ kcal\ mole^{-1}) = 4.3\ kcal\ mole^{-1}$$

This result is per oxygen. Total stabilization of both oxygens is 8.5 kcal mole^{-1}. The acidity difference between $ClCH_2CO_2H$ and CH_3CO_2H in the gas phase corresponds to an energy difference of about 13 kcal mole^{-1}; hence, even this crude electrostatic calculation on one conformer gives a result of the right order of magnitude.

19. Micelle formation is the mechanism by which hydrocarbons are solubilized by amphiphilic agents. The hydrocarbon is captured in the center of the micelle by the long hydrocarbon tails of the amphiphilic agent. The polar ionic heads interact strongly with the water and enable the hydrocarbon to be removed. The acid does not form micelles because the carboxylic acid is not polar enough to overcome the exclusion of the hydrocarbon "tails" from the structured water. (See Sec. 18.4.D) The carboxylic acid is converted into the salt with base. The resulting carboxylate anion is polar enough to interact strongly with water; the hydrocarbon "tails" are excluded. The result is a micelle.

1. Surface area of the droplet = $4\pi r^2$, where r = radius of the droplet, 10μ or 10^{-4} cm. Area = 1.3×10^{-7} cm^2.

2. Number of stearate anions per drop = area of droplets/area per stearate anion.

   Number per drop = 1.3×10^{-7} cm^2/25×10^{-16} cm^2 = 5.0×10^7 anions per drop.

3. volume of drop = $4/3\pi r^3$ cm^3 = 4.2×10^{-12} cm^2
   volume of oil in 50 mL = 50 cm^3.

   Number of droplets = volume of oil/volume per droplet = 50 cm^3/4.2×10^{-12} = 1.2×10^{13} droplets.

4. Total number of molecules needed to cover the surface of droplets derived from 50 mL of oil is 5.0×10^7 anions per drop x 1.2×10^{13} droplets = 6×10^{20} anions, almost exactly 10^{-3} mole.

5. Concentration of stearate = 2×10^{-3} M, above the critical micelle concentration of between 10^{-4} to 10^{-3}. If the same amount of agent were dissolved in the water of an olympic swimming pool with a volume of 1500 m^3, the concentration of 10^{-3} mole agent in the swimming pool would be about 10^{-9} M. One hundred thousand to one million times as much agent would be needed to reach the critical micelle concentration. The calculation shows why controlling oil spills in the ocean is not possible with detergent, the amounts of detergent required for dispersal are too large.

18.F. Supplementary Problems

S1. Give the IUPAC name for each of the following compounds:

a. $DCH_2CH_2CH_2CO_2H$

b. $CH_3C(=O)$–CH=CH–CO_2H (H, H)

c. $(CH_3)_2CHCH_2$–C(H)(OH)–CO_2H

d. H, H; C=C; $CH_3CH_2CH_2$, CO_2^- K+

e. HO_2C–CH(OH)–CH(OH)–CO_2H

f. H–C(=O)–cyclohexane(H)–CO_2H (H)

g. CH_2=C=$CHCO_2H$

S2. Write the structure for each of the following compounds:

a. γ,γ-dibromobutyric acid
b. 3-chloro-2-hydroxypropanoic acid
c. sodium 4-hydroxybutanoate
d. 2-oxopropanoic acid
e. ammonium valerate
f. *cis*-cyclopentane-1,2-dicarboxylic acid

S3. What is the major organic product of each of the following reaction sequences?

a. $(CH_3)_2CHCO_2H \xrightarrow[Cl_2]{PCl_3} \xrightarrow[\Delta]{H_2O}$

b. 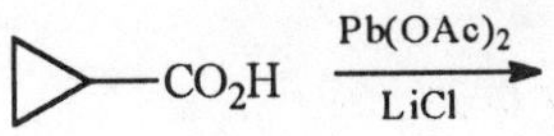 cyclopropyl$-CO_2H \xrightarrow[LiCl]{Pb(OAc)_2}$

c. $(CH_3)_3CCH_2C{\equiv}N \xrightarrow[\Delta]{H_3O^+} \xrightarrow{Ag_2O} \xrightarrow{I_2}$

d. $CH_2{=}CHCH_2CH_2CH_2CO_2H \xrightarrow{LiAlH_4} \xrightarrow[H^+]{CH_3CO_2H}$

e. $HO_2CC{\equiv}CCO_2H \xrightarrow{CH_2N_2} \xrightarrow{1\ mole\ Br_2}$

f. 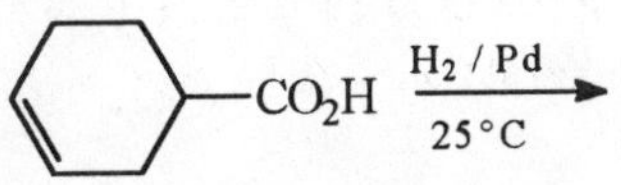 cyclohexenyl$-CO_2H \xrightarrow[25°C]{H_2\ /\ Pd}$

g. 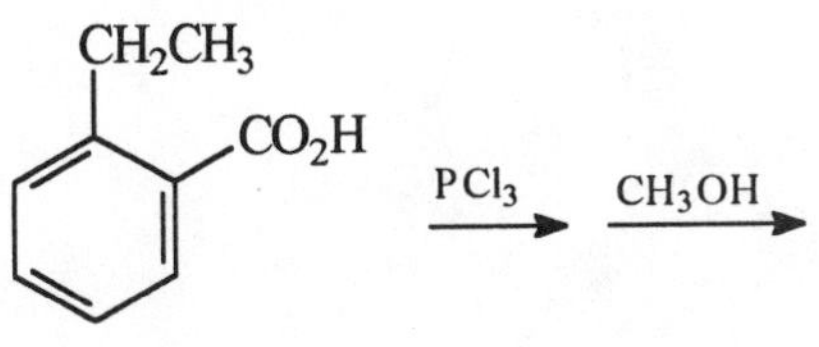 2-ethylbenzoic acid (CH_2CH_3, CO_2H) $\xrightarrow{PCl_3} \xrightarrow{CH_3OH}$

h. cyclobutyl$-\underset{H}{C}{=}O \xrightarrow{KMnO_4} \xrightarrow{NaOH}$

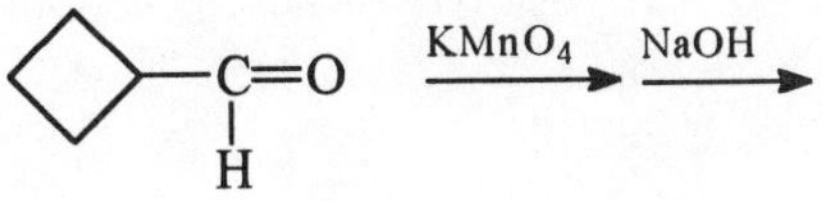

S4. Show how to carry out the following conversions:

a. $CH_3CH_2CH_2CO_2H \longrightarrow CH_3CH_2CH_2CH_2CH_2OH$

b. $CH_3CH_2CH_2CO_2H \longrightarrow CH_3CH{=}CHCO_2H$

c. (cyclopentyl)$-CH_2OH$ ⟶ (cyclopentyl)$-Br$

d. $(CH_3)_2CHCH_2CH{=}CH_2$ ⟶ $(CH_3)_2CHCH_2CHOHCO_2H$

e. (cyclohexene) ⟶ (cyclohexyl)$-CH_2CO_2H$

f. $CH_3CH_2CH_2CH_2Cl$ ⟶ $CH_3CH_2CH_2CHClCO_2CH_3$

g. $CH_3CH_2CH_2CO_2H$ ⟶ $CH_3CHBrCHBrCO_2CH_3$

h. (cyclopentanone) $=O$ ⟶ (cyclopentane with $COCl$ and CH_3)

i. $(CH_3)_3CCH_2\overset{O}{\overset{\|}{C}}CH_3$ ⟶ $(CH_3)_3CCH_2O\overset{O}{\overset{\|}{C}}CH_3$

S5. Show how to synthesize the following compounds using starting materials of five carbons or less and any other reagents:

a. $(CH_3)_3CCH_2CO_2CH(CH_3)_2$

b. $(CH_3)_2CHCH_2CH_2CH_2CH_2CO_2H$

c. $CH_3CH_2CH_2\underset{}{CH}(CO_2CH_3)CH_2CH_2CH_2CH_2CH_3$

d. (cyclopentene)$-\overset{O}{\overset{\|}{C}}Cl$

e. $CH_3CH_2CH_2(H)C{=}C(H)CO_2H$

f. (six-membered lactone, C=O, O, with CH_3)

g. 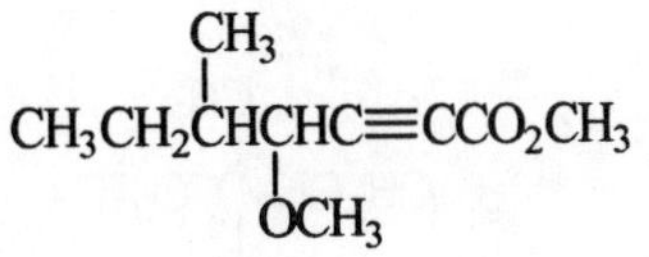 $CH_3CH_2CH(CH_3)CH(OCH_3)C{\equiv}CCO_2CH_3$

h. 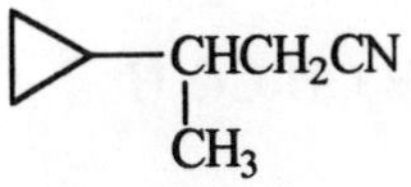 (cyclopropyl)$-CH(CH_3)CH_2CN$

i. 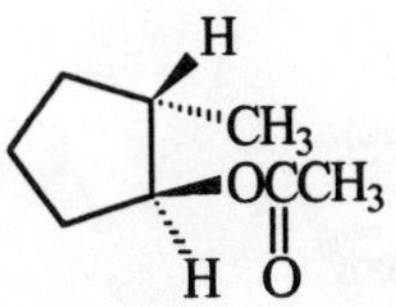 (cyclopentane ring bearing H, CH_3, $OCCH_3$ (with C=O), H)

j. $CH_3CH_2CO_2CH_2CH_2CH_2CH_2O_2CCH_2CH_3$

S6. Calculate the pH when chloroacetic acid is dissolved in water at 1.0 *M* concentration; at 0.1 *M*; at 0.001 *M*.

S7. In a solution that is 1.0 *M* in dichloroacetic acid and 1.0 *M* in acetic acid, what are the concentrations of dichloroacetate and acetate ions?

S8. Determine the structure of each of the compounds in the sequence below:

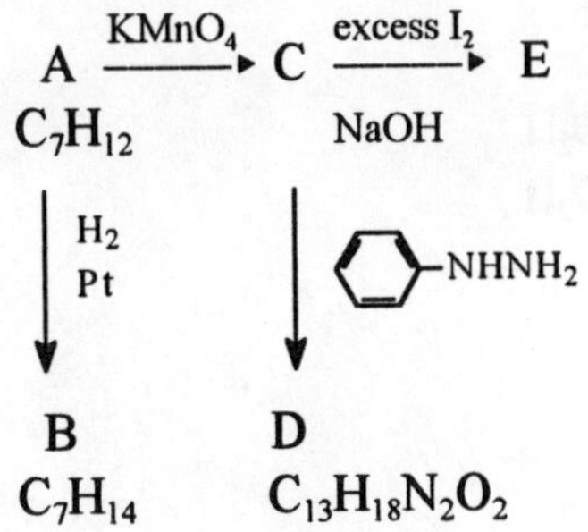

$$\underset{C_7H_{12}}{A} \xrightarrow{KMnO_4} C \xrightarrow[NaOH]{\text{excess } I_2} E \quad \text{nmr of E: } \delta\ 1.3\ (t, 4),\ 2.4\ (t, 4),\ 13\ (s, 2)$$

A $\xrightarrow{H_2,\ Pt}$ B (C_7H_{14})

C $\xrightarrow{C_6H_5NHNH_2}$ D ($C_{13}H_{18}N_2O_2$)

S9. Write a mechanism for the following transformation:

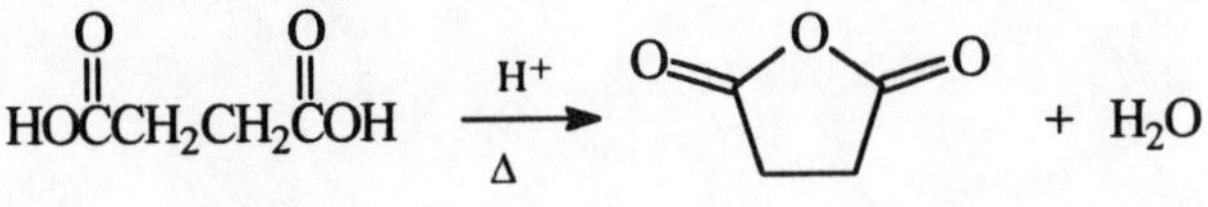

$$HO\overset{O}{\overset{\|}{C}}CH_2CH_2\overset{O}{\overset{\|}{C}}OH \xrightarrow[\Delta]{H^+} \text{(succinic anhydride)} + H_2O$$

S10. Predict the major product of the following reaction and justify your answer:

$$CH_3CH{=}CHCO_2H + Br_2 \xrightarrow[\text{solvent}]{CH_3OH}$$

S11. Rank the following compounds in order of increasing acidity:

F_2CHCO_2H $HOCH_2CH_2CO_2H$ $CH_3CH_2CH_2CH_2OH$ $CH_3CH_2C(CH_3)_2CO_2H$

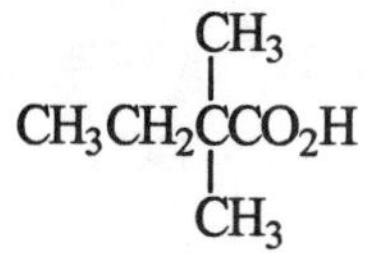

$ClCH_2CO_2H$ $CF_3CO_2CH_3$ CF_3CH_2OH $CH_3CH_2CO_2H$

18.G. Answers to Supplementary Problems

S1. a. 4-deuteriobutanoic acid

b. (*E*)-4-oxo-2-pentenoic acid

c. (*S*)-2-hydroxy-4-methylpentanoic acid

d. potassium (*Z*)-2-hexenoate

e. (*R,S*)-2,3-dihydroxybutanedioic acid (or *meso*-)

f. *trans*-4-formylcyclohexanecarboxylic acid

g. 2,3-butadienoic acid

S2. a. $Br_2CHCH_2CH_2CO_2H$

b. 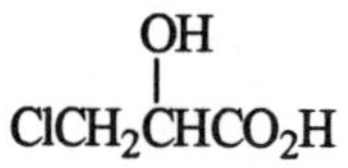 $ClCH_2CH(OH)CO_2H$

c. $HOCH_2CH_2CH_2CO_2^-\ Na^+$

d. 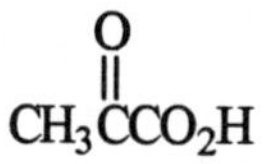 $CH_3C(=O)CO_2H$

e. $CH_3CH_2CH_2CH_2CO_2^-\ NH_4^+$

f. 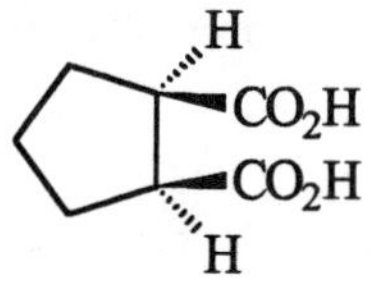 cyclopentane-1,2-dicarboxylic acid (H, CO_2H, CO_2H, H; both CO_2H wedged)

S3. a. 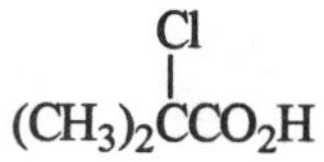 $(CH_3)_2C(Cl)CO_2H$

b.  cyclopropyl–Cl

c. $(CH_3)_3CCH_2I$

d. $CH_2{=}CHCH_2CH_2CH_2CH_2O_2CCH_3$

e. 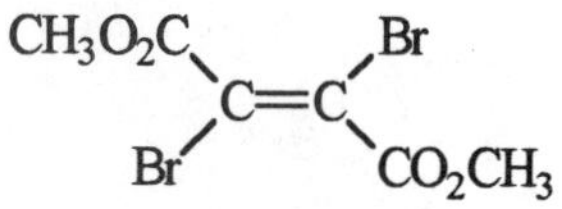 $(CH_3O_2C)(Br)C{=}C(Br)(CO_2CH_3)$

f. (cyclohexane)$-CO_2H$

g. (benzene ring with CH_2CH_3 and CO_2CH_3 ortho substituents)

h. (cyclobutane)$-CO_2^-\ Na^+$

S4. a. $CH_3CH_2CH_2CO_2H \xrightarrow{LiAlH_4} \xrightarrow{H_2O} CH_3CH_2CH_2CH_2OH \xrightarrow{PBr_3} \xrightarrow{Mg} \xrightarrow{CH_2=O} \xrightarrow{H_2O} CH_3CH_2CH_2CH_2CH_2OH$

b. $CH_3CH_2CH_2CO_2H \xrightarrow{P + Br_2} \xrightarrow[\Delta]{H_2O} CH_3CH_2CHBrCO_2H \xrightarrow[HOC_2H_5]{NaOC_2H_5} \xrightarrow{H^+} CH_3CH{=}CHCO_2H$

c. (cyclopentane)$-CH_2OH \xrightarrow{KMnO_4}$ (cyclopentane)$-CO_2H \xrightarrow[2.\ Br_2]{1.\ HgO}$ (cyclopentane)$-Br$

d.

$(CH_3)_2CHCH_2CH{=}CH_2 \xrightarrow[2.\ Zn]{1.\ O_3} (CH_3)_2CHCH_2CH(=O) \xrightarrow{HCN} (CH_3)_2CHCH_2CH(OH)CN \xrightarrow[\Delta]{H_3O^+} (CH_3)_2CHCH_2CH(OH)CO_2H$

e. (cyclohexene) $\xrightarrow{HBr} \xrightarrow[ether]{Mg} \xrightarrow{\text{(ethylene oxide)}} \xrightarrow{H_2O}$ (cyclohexane)$-CH_2CH_2OH \xrightarrow{HNO_3}$ (cyclohexane)$-CH_2CO_2H$

f. $CH_3CH_2CH_2CH_2Cl \xrightarrow{NaCN} \xrightarrow{H_3O^+} CH_3(CH_2)_3CO_2H \xrightarrow{P + Cl_2} \xrightarrow{H_2O} CH_3CH_2CH_2CHClCO_2H$

$\xrightarrow[CH_3OH]{H^+} CH_3CH_2CH_2CHClCO_2CH_3$

g. $\underset{\text{from S4.b}}{CH_3CH{=}CHCO_2H} \xrightarrow{Br_2} CH_3CHBrCHBrCO_2H \xrightarrow[CH_3OH]{H^+} CH_3CHBrCHBrCO_2CH_3$

h. (cyclopentanone) $\xrightarrow{CH_3MgI} \xrightarrow{HCl}$ (1-chloro-1-methylcyclopentane; H_3C, Cl) $\xrightarrow[ether]{Mg} \xrightarrow[2.\ H_3O^+]{1.\ CO_2}$ (1-methylcyclopentanecarboxylic acid; H_3C, CO_2H) $\xrightarrow{SOCl_2}$ (acid chloride; H_3C, COCl)

i. $(CH_3)_3CCH_2\overset{O}{\overset{\|}{C}}CH_3 \xrightarrow{CH_3CO_3H} (CH_3)_3CCH_2O\overset{O}{\overset{\|}{C}}CH_3$

S5. a. $(CH_3)_3CCH_2CO_2CH(CH_3)_2 \xleftarrow[H^+]{(CH_3)_2CHOH} (CH_3)_3CCH_2CO_2H \xleftarrow{H^+} \xleftarrow{CO_2} \xleftarrow{Mg} (CH_3)_3CCH_2Br$

(Note: $(CH_3)_3CCH_2Br \xrightarrow{NaCN}$ (×) very slow, because of hindrance from β-branching)

b. $(CH_3)_2CH(CH_2)_4CO_2H \xleftarrow[\Delta]{H_3O^+} \xleftarrow{NaCN} \xleftarrow{PBr_3} (CH_3)_2CH(CH_2)_4OH \xleftarrow{\text{ethylene oxide}} (CH_3)_2CHCH_2CH_2Br$

c. $CH_3(CH_2)_2\underset{|}{\overset{CO_2CH_3}{CH}}(CH_2)_4CH_3 \xleftarrow[H^+\ \Delta]{CH_3OH} CH_3(CH_2)_2\overset{CO_2H}{CH}(CH_2)_4CH_3 \xleftarrow[Pt]{H_2} \xleftarrow{Ag_2O} CH_3(CH_2)_2\overset{CHO}{C}{=}CH(CH_2)_3CH_3$

$2\ CH_3(CH_2)_3\overset{O}{\overset{\|}{C}}H \xrightarrow{NaOH} CH_3(CH_2)_2\overset{CHO}{C}{=}CH(CH_2)_3CH_3$

d. 1-cyclopentenyl–COCl $\xleftarrow{SOCl_2}$ 1-cyclopentenyl–CO_2H $\xleftarrow{H_3O^+} \xleftarrow{H_2SO_4}$ 1-hydroxycyclopentanecarbonitrile (OH, CN) $\xleftarrow{NaCN}$ cyclopentanone

e. (Z)-$CH_3CH_2CH_2CH{=}CHCO_2H$ (H, H cis) $\xleftarrow[Lindlar]{H_2} CH_3CH_2CH_2C{\equiv}CCO_2H \xleftarrow{H_2O} \xleftarrow{CO_2} \xleftarrow[\Delta]{CH_3MgBr} CH_3CH_2CH_2C{\equiv}CH$

f. 6-methyl-tetrahydropyran-2-one (lactone, O, CH_3) $\xleftarrow{CH_3CO_3H}$ 2-methylcyclopentanone (CH_3) $\xleftarrow{CH_3I}$ lithium cyclopentenolate (OLi) $\xleftarrow[THF,\ -78°]{LDA}$ cyclopentanone

g. $CH_3CH_2\overset{CH_3}{CH}\underset{OCH_3}{CH}C{\equiv}CCO_2CH_3 \xleftarrow{CH_2N_2} CH_3CH_2\overset{CH_3}{CH}\underset{OCH_3}{CH}C{\equiv}CCO_2H \xleftarrow{H^+} \xleftarrow{CO_2} \xleftarrow{BuLi}$

$CH_3CH_2\underset{CH_3}{CH}\overset{O}{\overset{\|}{C}}H \xrightarrow{NaC{\equiv}CH} CH_3CH_2\overset{CH_3}{CH}\underset{OH}{CH}C{\equiv}CH \xrightarrow[(CH_3)_2SO_4]{NaOH} CH_3CH_2\overset{CH_3}{CH}\underset{OCH_3}{CH}C{\equiv}CH$

h. $\text{cyclopropyl}-\underset{\underset{CH_3}{|}}{C}HCH_2CN \xleftarrow{NaCN} \xleftarrow{PBr_3} \text{cyclopropyl}-\underset{\underset{CH_3}{|}}{C}HCH_2OH \xleftarrow{H_2O} \xleftarrow{CH_2=O} \xleftarrow{Mg} \text{cyclopropyl}-\underset{\underset{Br}{|}}{C}HCH_3$

i. (cyclopentane with H, CH_3, O_2CCH_3, H) $\xleftarrow[H^+]{CH_3CO_2H}$ (cyclopentane with H, CH_3, OH, H) $\xleftarrow{H_2O} \xleftarrow{CH_3Li}$ (cyclopentene oxide)

j. $CH_3CH_2\overset{O}{\overset{||}{C}}OCH_2CH_2CH_2CH_2O\overset{O}{\overset{||}{C}}CH_2CH_3 \xleftarrow{CH_3CH_2COCl} HOCH_2CH_2CH_2CH_2OH$

S6. K_a of chloroacetic acid is 1.4×10^{-3} *M* (Table 18.3).

at 1.0 *M* $\dfrac{[H^+][ClCH_2CO_2^-]}{[ClCH_2CO_2H]} = 1.4 \times 10^{-3}$ $[H^+] = [ClCH_2CO_2^-] = X$

$[ClCH_2CO_2H] = 1.0 - X$

$\dfrac{X^2}{1-X} = 1.4 \times 10^{-3}$ $X^2 + 1.4 \times 10^{-3} X - 1.4 \times 10^{-3} = 0;$

$$X = \frac{-1.4 \times 10^{-3} \pm \sqrt{(1.4 \times 10^{-3})^2 + 5.6 \times 10^{-3}}}{2} = 3.67 \times 10^{-2};\ pH = 1.44$$

at 0.1 *M* $\dfrac{X^2}{0.1-X} = 1.4 \times 10^{-3}$ $X^2 + 1.4 \times 10^{-3} X - 1.4 \times 10^{-4} = 0;$

$X = 1.12 \times 10^{-2}$; pH = 1.95

at 0.001 *M* $\dfrac{X^2}{1 \times 10^{-3}-X} = 1.4 \times 10^{-3}$ $X^2 + 1.4 \times 10^{-3} X - 1.4 \times 10^{-6} = 0;$

$X = 6.75 \times 10^{-3}$; pH = 3.17

S7. Because dichloroacetic acid is a much stronger acid than acetic acid, only a very small amount of the acetic acid will be ionized; that is, essentially all of the protons will come from the dichloroacetic acid:

$\dfrac{[H^+][Cl_2CHCO_2^-]}{[Cl_2CHCO_2H]} = 5.5 \times 10^{-2}$ $[H^+] \cong [Cl_2CHCO_2^-] = X$, $[Cl_2CHCO_2H] = 1 - X$

$$\frac{X^2}{1 - X} = 5.5 \times 10^{-2}$$

$$X = \frac{-5.5 \times 10^{-2} \pm \sqrt{(5.5 \times 10^{-2})^2 + 0.22}}{2} = 0.21\,M = [Cl_2CHCO_2^-] = [H^+];\ pH = 0.68.$$

$$\frac{[H^+][CH_3CO_2^-]}{[CH_3CO_2H]} = 1.8 \times 10^{-5} = \frac{(0.21)\,y}{1 - y}\ ; \quad y = 8.6 \times 10^{-5} = [CH_3CO_2^-]$$

$$\frac{[Cl_2CH_2CO_2^-]}{[CH_3CO_2^-]} = \frac{0.21}{8.6 \times 10^{-5}} = 2.4 \times 10^3$$

S8.

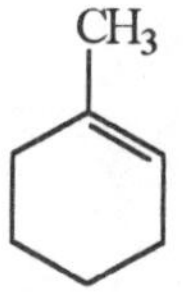

A: 1-methylcyclohexene (CH_3 on ring)

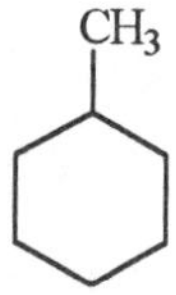

B: methylcyclohexane (CH_3 on ring)

C: $CH_3\overset{O}{\overset{\|}{C}}(CH_2)_4CO_2H$

D: $CH_3\overset{NNHC_6H_5}{\overset{\|}{C}}(CH_2)_4CO_2H$

E: $HO_2CCH_2CH_2CH_2CH_2CO_2H$

S9.

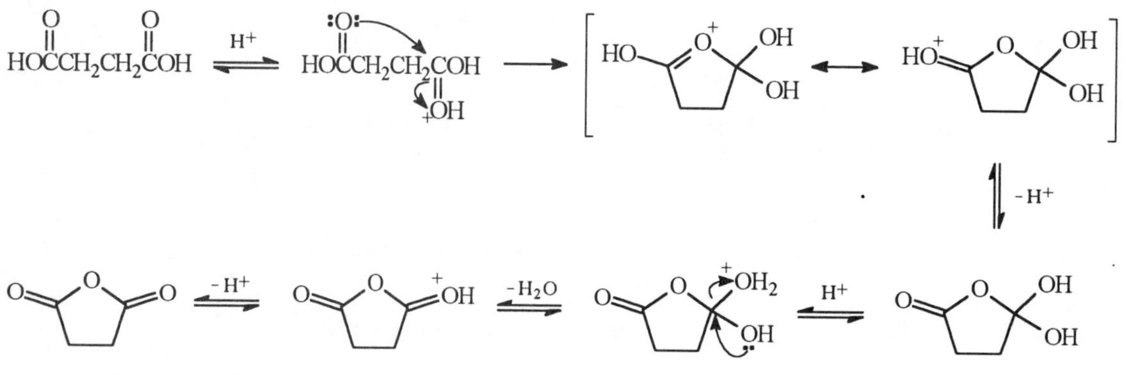

$HO\overset{O}{\overset{\|}{C}}CH_2CH_2\overset{O}{\overset{\|}{C}}OH \overset{H^+}{\rightleftharpoons} HO\overset{:O:}{\overset{\|}{C}}CH_2CH_2\overset{\overset{+}{O}H}{\overset{\|}{C}}OH \longrightarrow$ [HO–(ring, $\overset{+}{O}$)–OH, OH $\longleftrightarrow$ $H\overset{+}{O}$=(ring)–OH, OH] $\overset{-H^+}{\rightleftharpoons}$ O=(ring)–OH, OH $\overset{H^+}{\rightleftharpoons}$ O=(ring)–$\overset{+}{O}H_2$, OH $\overset{-H_2O}{\rightleftharpoons}$ O=(ring)=$\overset{+}{O}H$ $\overset{-H^+}{\rightleftharpoons}$ O=(ring)=O

S10.

$CH_3CH{=}CHCO_2H$ + Br–Br $\longrightarrow$ [$CH_3\overset{+}{C}H-\overset{Br}{C}HCO_2H$ ($\underline{1}$) $\longleftrightarrow$ $CH_3CH-CHCO_2H$ with $\overset{+}{Br}$ bridging ($\underline{2}$) $\longleftrightarrow$ $CH_3\overset{Br}{C}H-\overset{+}{C}HCO_2H$ ($\underline{3}$)]

$\xrightarrow{H^+,\ CH_3OH}$ $CH_3\underset{OCH_3}{C}H\overset{Br}{C}HCO_2H$

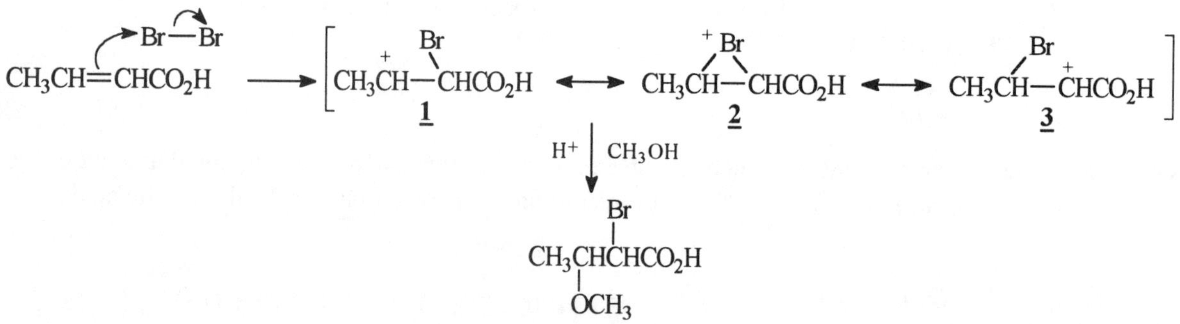

Resonance structure 3 is the *least* favored, because it puts a positive charge next to an already partially positive center: the carbonyl group. Therefore, "Markovnikov orientation" in this case favors nucleophilic attack at the β- and not the α-position.

S11. *least* acidic $CF_3CO_2CH_3$ (no ionizable hydrogens)

$CH_3CH_2CH_2CH_2OH$

CF_3CH_2OH

$CH_3CH_2C(CH_3)_2CO_2H$

$CH_3CH_2CO_2H$

$HOCH_2CH_2CO_2H$

$ClCH_2CO_2H$

most acidic F_2CHCO_2H

19. DERIVATIVES OF CARBOXYLIC ACIDS

19.A. Chapter Outline and Important Terms Introduced

19.1 Structure

esters
amides
acyl halides
acid anhydrides
[nitriles]
double bond character

19.2 Nomenclature

alkyl alkanoate
alkanamide
alkanoyl halide
alkanoic anhydride
-carboxamide
-carbonyl halide

19.3 Physical Properties

19.4 Spectroscopy

A. Nuclear Magnetic Resonance

CMR C=O 168-178 ppm

B. Infrared

$-CO_2R$ 1735 cm^{-1}
$-COCl$ 1800 cm^{-1}
$-C(=O)OC(=O)-$ 1820, 1760 cm^{-1}
lactones
lactams
$-CONR_2$, 1650-1690 cm^{-1}

19.5 Basicity of the Carbonyl Oxygen

contribution of resonance structures
O-protonation of amides

19.6 Hydrolysis; Nucleophilic Addition-Elimination

acid catalysis vs. base catalysis
sequence of bond-making vs. bond-breaking steps
nucleophilic addition-elimination mechanism
acylium ion

19.7 Other Nucleophilic Substitution Reactions

A. Reaction with Alcohols
ester formation from acyl halides or acid anhydrides
transesterification

B. Reaction with Amines and Ammonia
amide formation from acyl halides, acid anhydrides or esters

C. Reaction of Acyl Halides and Anhydrides with Carboxylic Acids and Carboxylate salts. Synthesis of Anhydrides

D. Reaction with Organometallic Compounds
 ketone formation from acyl halides
 tertiary alcohol formation from esters

19.8 Reduction

acid chloride to aldehyde (Rosenmund reduction)
ester to alcohol (Bouveault-Blanc reaction)
amide to amine
secondary amide to aldehyde
nitrile to amine (Chap 12.6.A)

19.9 Acidity of the α-Protons

$CH_3C(=O)CH_3$ pK_a 19 $CH_3C(=O)OCH_3$, CH_3CN pK_a 25 $CH_3C(=O)N(CH_3)_2$ pK_a 30

enolate formation
Reformatsky reaction

Claisen condensation
acetoacetic ester condensation

19.10 Reactions of Amides that Occur on Nitrogen

amide ionization:

$$RC(=O)NH_2 \rightleftharpoons RC(=O)NH^- + H^+ \quad pK_a\ 15$$

nitrile formation

19.11 Complex Derivatives

A. Waxes (spermaceti, bee's wax, carnauba wax)
 pharmacologically inactive

B. Fats
 fatty acids
 triglycerides
 saponification

 hardened fats
 cross-linked

C. Eicosanoids - biologically active
 prostaglandins
 thromboxanes
 leukotriens

D. Phospholipids
 phosphatidic acids
 phosphatidyl amines
 sphingomyellin
 lipid bilayer

19.B. Important Reactions Introduced

Hydrolysis of carboxylic acid derivatives (19.6)

Equation:

$$\mathrm{R\overset{O}{\overset{\|}{C}}Y + H_2O \longrightarrow R\overset{O}{\overset{\|}{C}}OH + HY}$$

Generality: Y = halogen, OR′, NR'_2, O_2CR'
Key features: for Y = OR′, NR'_2: acid-or base-catalyzed NR'_2 requires vigorous conditions)
base is also a reactant, requires stoichiometric quantities.

Reaction of carboxylic acid derivatives with alcohols (19.7.A)

Equation:

$$\mathrm{R\overset{O}{\overset{\|}{C}}Y + HOR' \longrightarrow R\overset{O}{\overset{\|}{C}}OR' + HY}$$

Generality: R = H, alkyl, aryl, OR; R′ = alkyl
Y = Cl or O_2CR (often with pyridine or Et_3N as reagent)
Y = OR" (transesterification, acid- or base-catalyzed)

Key features: equilibrium process for Y = OR"

Reactions of carboxylic acid derivatives with amines (19.7.B)

Equation:

$$\mathrm{R\overset{O}{\overset{\|}{C}}Y + 2\,HNR'_2 \longrightarrow R\overset{O}{\overset{\|}{C}}NR'_2 + HY}$$

Generality: R = alkyl, aryl, H, or OR; R′ = alkyl, aryl, or H
Y = Cl, O_2CR, OR"

Key features: **not** acid-catalyzed

Reactions of acyl halides and anhydrides with carboxylic acids (19.7.C)

Equation:

$$\mathrm{R\overset{O}{\overset{\|}{C}}Y + R'CO_2H\ (or\ R'CO_2^-\ Na^+) \longrightarrow R\overset{O}{\overset{\|}{C}}O\overset{O}{\overset{\|}{C}}R' + HY\ (or\ Na^+\ Y^-)}$$

Generality: Y = Cl, O_2CR (anhydride exchange)

Reaction of carboxylic acid derivatives with organometallic reagents (19.7.D)

To give alcohols:

Equation:

$$\mathrm{R\overset{O}{\overset{\|}{C}}Y + 2\,R'M \longrightarrow R\overset{OH}{\underset{R'}{C}}R' + MY}$$

Generality: Y = Cl, O_2CR, OR"; R′ = alkyl, aryl; M = Li, MgX

Key features: reaction proceeds via addition/elimination to give ketone, then another addition reaction to give alcohol

To give ketones:

Equation:

$$\mathrm{RCCl}(\text{=O}) + \mathrm{R'M} \longrightarrow \mathrm{RCR'}(\text{=O}) + \mathrm{MCl}$$

Generality: R′M = Grignard (RMgX) or organocuprate (R'_2CuLi)

Key features: reaction with Grignard reagent must be carried out at −78 °C to avoid formation of 3° alcohol

Reduction of carboxylic acid derivatives (19.8)

To aldehyde:

Equation:

$$\mathrm{RCY}(\text{=O}) \xrightarrow{[H]} \mathrm{RCH}(\text{=O})$$

Generality: Y = Cl, [H] = H_2/Pd-$BaSO_4$-poison (quinoline): Rosenmund reduction
Y = Cl, [H] = LiAlH(O-*t*-Bu)$_3$
Y = NR'_2, [H] = $LiAlH(OEt)_3$

Key features: these special reagents are required to avoid over-reduction to alcohol

To alcohol:

Equation:

$$\mathrm{RCY}(\text{=O}) \xrightarrow{[H]} \mathrm{RCH_2OH}$$

Generality: Y = Cl, O_2CR, OR′
[H] = $LiAlH_4$, $LiBH_4$
Y = OR', [H] = Na: Bouveault-Blanc reaction

Reduction of amides to amines (19.8)

Equation:

$$\mathrm{RCNR'_2}(\text{=O}) \xrightarrow{\mathrm{LiAlH_4}} \mathrm{RCH_2NR'_2}$$

Generality: R, R′ = alkyl, aryl, or H

Alkylation of ester enolates (19.9)

Equation:

$$\mathrm{RCH_2CO_2R'} + \mathrm{LDA} \longrightarrow \mathrm{RCH{=}C(OLi)OR'} \xrightarrow{\mathrm{R''X}} \mathrm{RR''CHCO_2R'}$$

Generality: R = H, alkyl, aryl; R′ = alkyl; R" = methyl or 1° alkyl

Key features: similar to alkylation of ketone enolate, but ester enolate is more reactive

Reaction of ester enolates with ketones and aldehydes (19.9)

Equation: $RCH_2CO_2R' \xrightarrow[\text{THF, -78°}]{\text{LDA}} \left[RCH{=}\overset{O^- Li^+}{\overset{|}{C}}OR' \right] \xrightarrow{R''\overset{O}{\overset{\|}{C}}R''} R''{-}\overset{OH}{\overset{|}{\underset{R''}{\underset{|}{C}}}}{-}\underset{R}{\underset{|}{C}}HCO_2R'$

$$R''\overset{O}{\overset{\|}{C}}R'' + R\underset{Br}{\underset{|}{C}}HCO_2R' + Zn \xrightarrow{\text{toluene}} \xrightarrow{H_3O^+} R''{-}\overset{OH}{\overset{|}{\underset{R''}{\underset{|}{C}}}}{-}\underset{R}{\underset{|}{C}}HCO_2R'$$

Generality: R, R′, R" = various alkyl groups; reaction of α-haloester with Zn is the Reformatsky reaction.

Key features: similar to aldol addition reaction

Claisen condensation = acetoacetic ester condensation (19.9)

Equation: $2\ RCH_2CO_2R' \xrightarrow{R'O^-} RCH_2\overset{O}{\overset{\|}{C}}\underset{R}{\underset{|}{C}}HO_2R'$

Generality: R = H, alkyl, aryl, R′ = alkyl

Key features: mixed condensations are only practical if just one ester can enolize

Formation of nitriles from amides (19.10)

Equation: $R\overset{O}{\overset{\|}{C}}NH_2 \xrightarrow{[-H_2O]} RC{\equiv}N$

Generality: R = alkyl, aryl
reagent = $SOCl_2$, $POCl_3$, P_2O_5, Ac_2O

19.C Important Concepts and Hints

Addition-Elimination Mechanism. The addition-elimination mechanism for substitution reactions of carboxylic acids was introduced in Chapter 18. Its importance is emphasized by the number of reactions discussed in this chapter that proceed by this mechanism. It holds the same importance in the chemistry of carboxylic acid derivatives as the S_N2 mechanism holds for substitution reactions of alkyl systems.

The addition-elimination mechanism applies to both acidic and basic reaction conditions, and examples can be found in which the same overall transformation is catalyzed by either one. Ester and amide hydrolysis are two examples; for both of them, you should compare closely the acid- and base-catalyzed mechanisms (see Section 19.6, and the answer to the Exercise at the end of that section), and be familiar with their similarities and differences (see Section 18.B of this Study Guide for a similar comparison in another reaction sequence).

You should also be completely familiar with the general aspects of the mechanisms which hold under acidic and basic conditions, as depicted below.

Acidic Conditions:

$$RC(=O)Y \xrightleftharpoons{H^+} RC(=\overset{+}{O}H)Y \;(:NuH) \rightleftharpoons RC(OH)(Y)\overset{+}{N}uH \xrightleftharpoons{-H^+} RC(OH)(Y)Nu \xrightleftharpoons{H^+} RC(:OH)(\overset{+}{Y}H)Nu \xrightleftharpoons{-YH} RC(=\overset{+}{O}H)Nu \xrightleftharpoons{-H^+} RC(=O)Nu$$

Basic Conditions:

$$RC(=O)Y \;(:Nu^-) \rightleftharpoons RC(:\ddot{O}:^-)(Y)Nu \rightleftharpoons RC(=O)Nu + Y^-$$

Notice the symmetry of these sequences: the steps leading to the formation of the tetrahedral intermediate are simply reversed in going from there to the product.

There are some reactions in which the distinction is somewhat blurred; for example, in the addition of ammonia to an acid chloride. In this case the nucleophile attacks before it loses its proton (as in the "acidic mechanism") and the leaving group departs before it gains one (as in the "basic mechanism"):

$$RC(=O)Cl \;(:NH_3) \rightleftharpoons RC(:\ddot{O}:^-)(Cl)\overset{+}{N}H_3 \xrightleftharpoons{-H^+} RC(:\ddot{O}:^-)(Cl)NH_2 \rightleftharpoons RC(=O)NH_2 + Cl^-$$

Because of the symmetrical nature of these sequences, the reactions could conceivably go in either direction. In fact, the reaction:

$$\text{ester + water} \rightleftharpoons \text{acid + alcohol}$$

is such an equilibrium process (see Section 18.7.C). There is a convenient way to predict which way the equilibrium lies for reactions that proceed by the **basic** mechanism: compare the pK_a's of "HNu" and "HL"; the reaction will go in the direction that generates the least basic (most stable) leaving group:

$$CH_3C(=O)Cl \xrightarrow[-Cl^-]{CH_3CO_2^-} CH_3C(=O)OC(=O)CH_3 \xrightarrow[-CH_3CO_2^-]{CH_3O^-} CH_3C(=O)OCH_3 \xrightarrow[-CH_3O^-]{\text{"}CH_3^-\text{" } (CH_3MgBr)} \left[CH_3C(=O)CH_3\right] \longrightarrow (CH_3)_2COMgBr$$

$$CH_3C(=O)OCH_3 \xrightarrow[-CH_3OH]{NH_3} CH_3C(=O)NH_2 \xrightarrow[-NH_3]{OH^-} CH_3C(=O)O^- \xrightarrow[-\text{"}O^{2-}\text{" } (LiOAlH_3)]{\text{"}H^-\text{" } (LiAlH_4)} \left[CH_3C(=O)H\right] \longrightarrow CH_3CH_2OAlH_3$$

For acid-catalyzed amide hydrolysis, and for reactions with hydroxide ion, additional reactions of the products serve to remove one of the components of the equilibrium, thereby driving it completely to one side:

reaction	equilibrium	additional reaction
acid-catalyzed hydrolysis of amides: $RC(=O)NH_2 + H_2O$	$\overset{H^+}{\rightleftharpoons}$ $RC(=O)OH + NH_3$	$\overset{H^+}{\rightleftharpoons}$ $RC(=O)OH + NH_4^+$
base-catalyzed hydrolysis of esters, amides $RC(=O)L + H_2O$	$\overset{OH^-}{\rightleftharpoons}$ $RC(=O)OH + HL$ (L = OR′ or NR′$_2$)	$\overset{OH^-}{\rightleftharpoons}$ $RC(=O)O^- + HL$

19.D Answers to Exercises

19.1 The dipolar resonance structure has a double bond between the carbonyl C and the Y group. $R-C(-O^-)=Y^+$

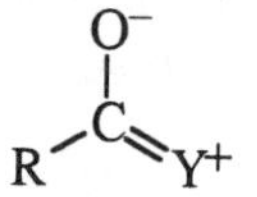

Therefore, the more important this resonance structure is, the greater the degree of shortening of the carbonyl-to-Y bond relative to the single bond model, CH_3-Y.

CH_3NH_2 vs. $CH_3C(=O)N(CH_3)_2$ — $C(=O)-N$ bond shorter by 0.11 Å

CH_3OCH_3 vs. $CH_3C(=O)OCH_3$ — $C(=O)-O$ bond shorter by 0.06 Å

CH_3F vs. $CH_3C(=O)F$ — $C(=O)-F$ bond shorter by 0.01 Å

There is a direct correlation between the basicity of Y^- and the degree of bond shortening.

19.2
a. methyl hexanoate, hexanamide, hexanoyl chloride, hexanoic anhydride
b. methyl 4-methylpentanoate, 4-methylpentanamide, 4-methylpentanoyl chloride, 4-methylpentanoic anhydride
c. methyl 4-pentenoate, 4-pentenamide, 4-pentenoyl chloride, 4-pentenoic anhydride
d. methyl 3-bromopropanoate, 3-bromopropanamide, 3-bromopropanoyl chloride, 3-bromopropanoic anhydride
e. 2-methylpropyl 4-methylpentanoate
f. N,N-diethyl-4,4-dimethylpentanamide

19.3 $CH_3-CH_2-CH_2-CH_2-CH_2-C(=O)-O-CH(CH_3)_2$

IR, 1740 cm^{-1}

^{1}H NMR: 0.9 (CH_3); 1.2 - 1.9 (CH_2—CH_2—CH_2); 2.3 (CH_2—C=O); 5.0 (O—CH); CH_3 ← 1.25 ppm

^{13}C NMR: 14.3, 23.4, 32.9, 25.5, 33.9, 172 (C=O), 65.6 (O—CH); CH_3 ← 20.2 ppm

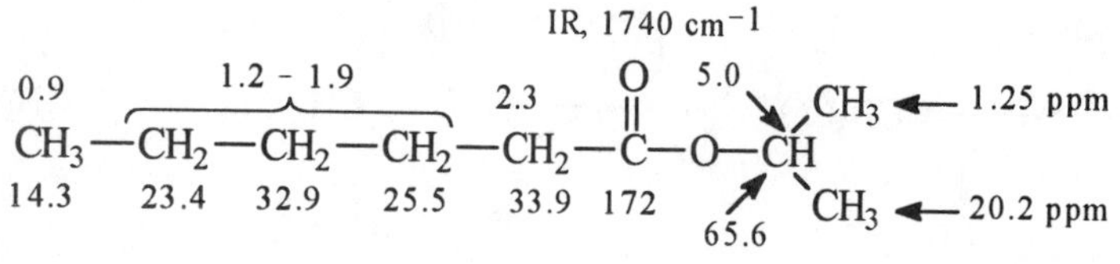

19.4 There is a good correlation:

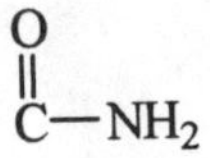

$C(=O)-NH_2$ bond shorter than CH_3-NH_2 by 0.11Å; pK_a of conjugate acid: 0.0

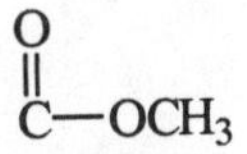

$C(=O)-OCH_3$ bond shorter than CH_3-OCH_3 by 0.06Å; pK_a of conjugate acid: -6.5

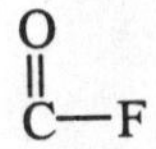

$C(=O)-F$ bond shorter than CH_3-F by 0.01Å; pK_a of conjugate acid: ~ -9

19.5 *Acid-Promoted*:

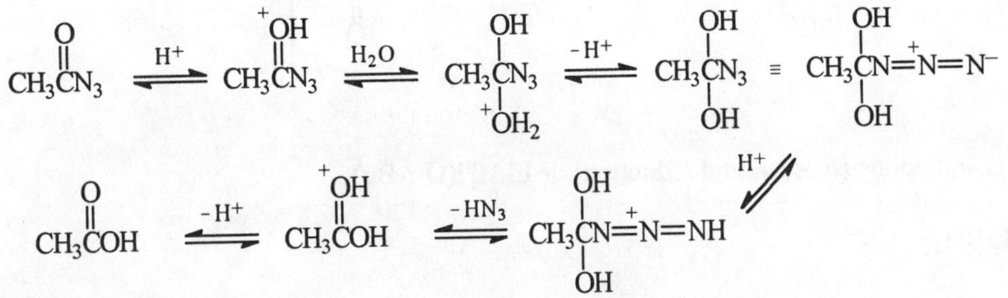

$$CH_3C(=O)N_3 \overset{H^+}{\rightleftharpoons} CH_3C(=\overset{+}{O}H)N_3 \overset{H_2O}{\rightleftharpoons} CH_3C(OH)(\overset{+}{O}H_2)N_3 \overset{-H^+}{\rightleftharpoons} CH_3C(OH)_2N_3 \equiv CH_3C(OH)_2N{=}\overset{+}{N}{=}N^-$$

$$\overset{H^+}{\rightleftharpoons} CH_3C(OH)_2N{=}\overset{+}{N}{=}NH \overset{-HN_3}{\rightleftharpoons} CH_3C(=\overset{+}{O}H)OH \overset{-H^+}{\rightleftharpoons} CH_3C(=O)OH$$

Base-Promoted:

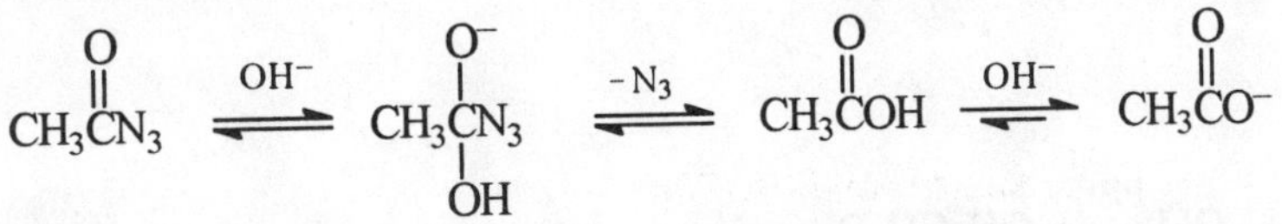

$$CH_3C(=O)N_3 \overset{OH^-}{\rightleftharpoons} CH_3C(O^-)(OH)N_3 \overset{-N_3}{\rightleftharpoons} CH_3C(=O)OH \overset{OH^-}{\rightleftharpoons} CH_3C(=O)O^-$$

19.6

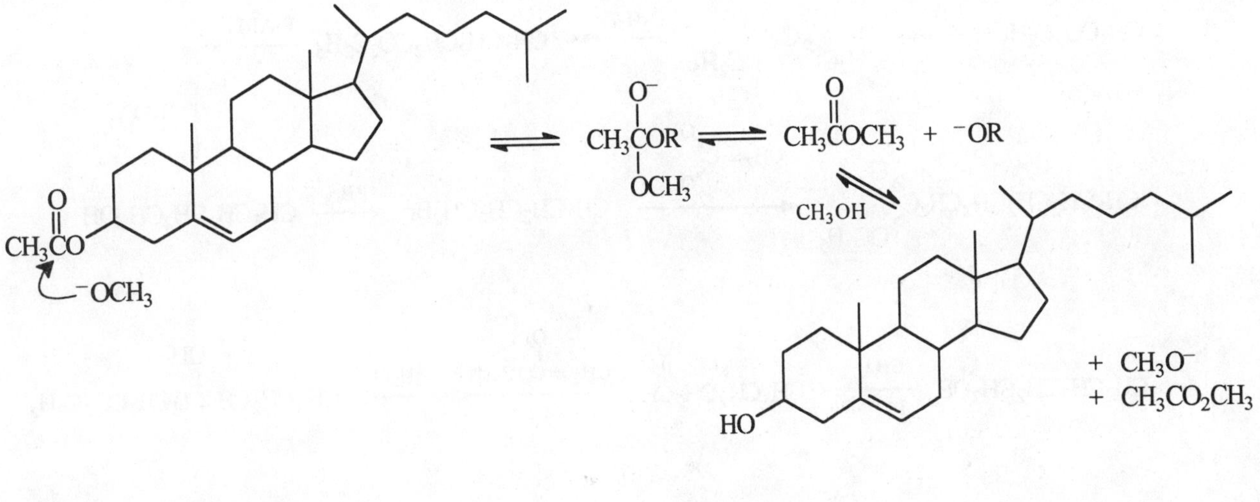

Cholesteryl acetate ($CH_3C(=O)O$–R) + $^-OCH_3$ $\rightleftharpoons$ $CH_3C(O^-)(OR)OCH_3$ $\rightleftharpoons$ $CH_3C(=O)OCH_3$ + ^-OR $\overset{CH_3OH}{\rightleftharpoons}$ cholesterol (HO–R) + CH_3O^- + $CH_3CO_2CH_3$

19.7

(cyclic dimer: 3,6-dimethyl-1,4-dioxane-2,5-dione, lactide)

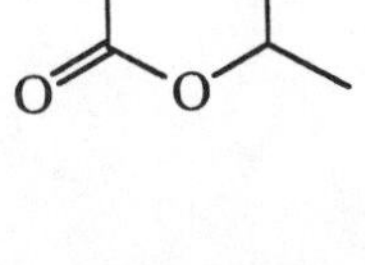

19.8 a. 2 (2-chlorobutane) $\xrightarrow{Li}$ $\xrightarrow{CuI}$ $(\text{sec-butyl})_2CuLi$ $\xrightarrow{CH_3COCl}$ 3-methyl-2-pentanone

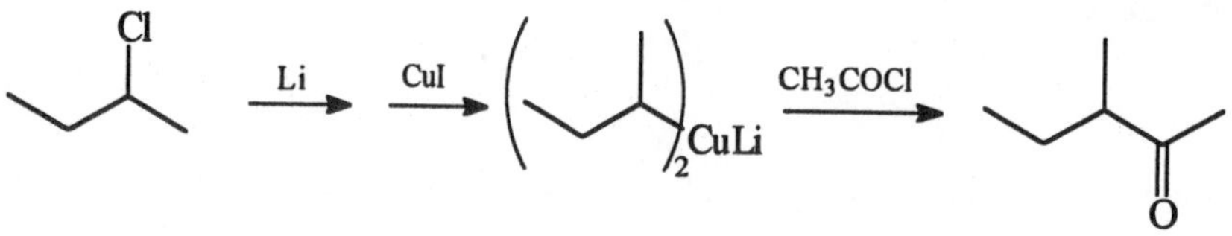

b. R = CH_3

c. R = H

2 (2-chlorobutane) $\xrightarrow{Mg}$ 2 (sec-butyl)MgCl $\xrightarrow{RCO_2Et}$ [tetrahedral intermediate: sec-butyl–C(R)(OEt)–$^-$O $ClMg^+$] → [sec-butyl–C(=O)–R] $\xrightarrow{\text{further reaction}}$ (sec-butyl)$_2$C(R)OH

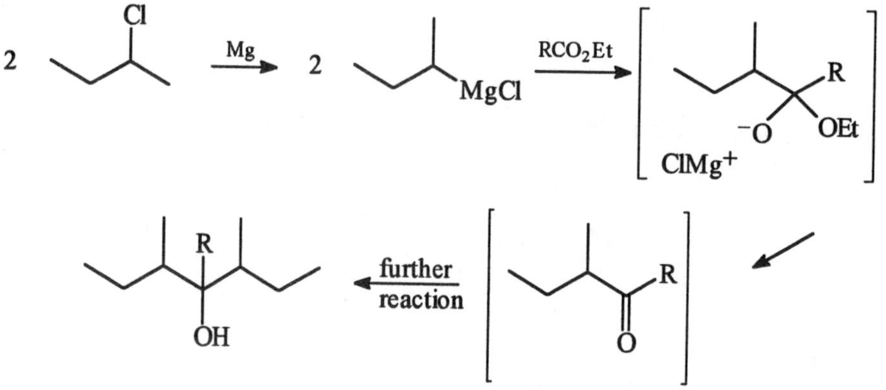

19.9 a. H_2/Pd-$BaSO_4$-quinoline (Rosenmund reduction) or $LiAlH(O\text{-}t\text{-}Bu)_3$

b. $LiAlH_4$ or $LiBH_4$

c. $LiAlH_4$

d. $LiAlH(OC_2H_5)_3$

19.10 a. $CH_3CO_2C_2H_5 \xrightarrow{LiAlH_4} \xrightarrow{H_2O} C_2H_5OH \xrightarrow{PBr_3} CH_3CH_2Br$

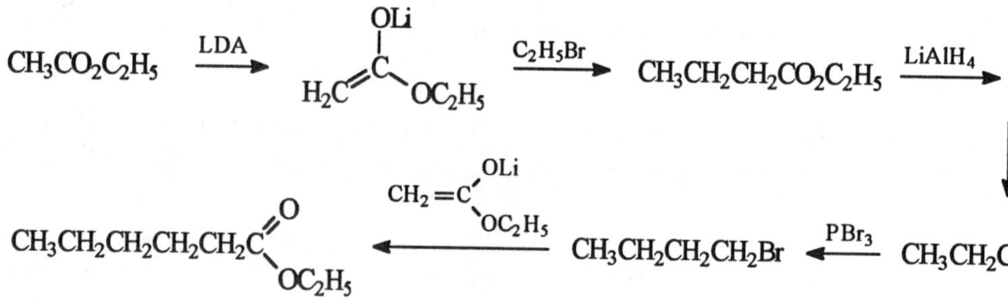

$CH_3CO_2C_2H_5 \xrightarrow{LDA} H_2C{=}C(OLi)OC_2H_5 \xrightarrow{C_2H_5Br} CH_3CH_2CH_2CO_2C_2H_5 \xrightarrow{LiAlH_4}$

$\xrightarrow{H_2O} CH_3CH_2CH_2CH_2OH \xrightarrow{PBr_3} CH_3CH_2CH_2CH_2Br \xrightarrow{CH_2=C(OLi)OC_2H_5} CH_3CH_2CH_2CH_2CH_2C(=O)OC_2H_5$

b. $CH_3CH_2CH_2CH_2OH \xrightarrow[\text{pyridine}]{CrO_3} CH_3CH_2CH_2CH{=}O \xrightarrow{CH_2=C(OLi)OC_2H_5} \xrightarrow{H_2O} CH_3CH_2CH_2CH(OH)CH_2CO_2C_2H_5$

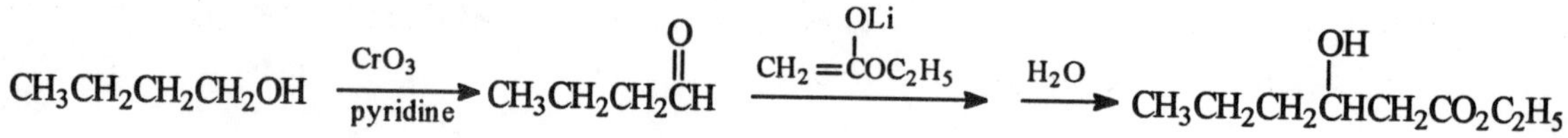

c. $$CH_3CO_2C_2H_5 \xrightarrow[\text{THF}, -78°]{\text{LDA}} \xrightarrow{CH_3\overset{O}{\overset{\|}{C}}CH_3} (CH_3)_2\underset{OH}{C}CH_2CO_2C_2H_5$$

d. $$CH_3CO_2C_2H_5 \xrightarrow[\text{EtOD}]{\text{EtO}^-} CD_3CO_2C_2H_5$$

e. $$CH_3\overset{O}{\overset{\|}{C}}OCH_2CH_3 \xrightarrow{H_3O^+} \xrightarrow{SOCl_2} \xrightarrow[-15°]{CH_3CH_2MgBr} CH_3\overset{O}{\overset{\|}{C}}CH_2CH_3$$

$$CH_3CO_2C_2H_5 \xrightarrow[\text{THF}, -78°]{\text{LDA}} \xrightarrow{CH_3\overset{O}{\overset{\|}{C}}CH_2CH_3} CH_3CH_2\underset{OH}{\overset{CH_3}{C}}CH_2CO_2C_2H_5 \xrightarrow{LiAlH_4} CH_3CH_2\underset{OH}{\overset{CH_3}{C}}CH_2CH_2OH$$

f. $$CH_3\overset{O}{\overset{\|}{C}}OC_2H_5 + H\overset{O}{\overset{\|}{C}}OC_2H_5 \xrightarrow{^-OC_2H_5} H\overset{O}{\overset{\|}{C}}CH_2\overset{O}{\overset{\|}{C}}OC_2H_5$$

19.11

Starting Material	C_2H_5Br	$CH_3CH_2CO_2H$	$CH_3CH_2CONH_2$	$CH_3CH_2C{\equiv}N$
C_2H_5Br		18.6.B		9.2
$CH_3CH_2CO_2H$	18.7.G		18.7.E	
$CH_3CH_2CONH_2$		19.6		19.10
$CH_3CH_2C{\equiv}N$		18.6.A	12.6.B	

19.12 The polarity of the carbonyl is offset by the large non-polar hydrocarbon side chain. The large surface area allows for the maximum van der Waal's attraction between the side chains leading to the high melting points.

19.13 The naturally occuring fatty acids are characterized by unbranched chains. When a fat is hardened, the saturated fatty acid portion of the fat is a long regular hydrocarbon. The fats can stack easily with maximum surface contact. In the unsaturated fats, the presence of the *cis*-double bond(s) puts a kink in the chain. The surface contact is not as complete as with the saturated fats, the van der Waal's interactions are diminished and the melting point is higher.

19.14 Besides the carboxylic acid function, the only other functional groups in arachadonic acid are the double bonds. Therefore, the C_{21} acid could be synthesized by reducing arachidonic acid to the C_{20} alcohol with a reducing agent that will not react with the double bonds such as $LiAlH_4$, conversion of the alcohol to the bromide with PBr_3, formation of the Grignard with Mg in ether, reaction of the Grignard with CO_2 and workup with cold aqueous acid.

Answers and Explanation for Problems

1. a. propyl 3-ethylpentanoate
   b. ethyl cyclohexanecarboxylate
   c. ethyl cyclopentanecarboxylate
   d. propanoyl chloride
   e. butanoic anhydride
   f. N-methylpropanamide
   g. N-cyclohexyl-2-methylpropanamide
   h. methyl 2-chlorobutanoate
   i. 3-methylpentanoyl bromide
   j. cyclohexyl ethanoate *(cyclohexyl acetate is used more commonly)*

2. a. $(CH_3CH_2CH_2)(CH_3CH_2)CHCH_2C(=O)NHCH_2CH_3$

   b. $HC(=O)N(CH_3)_2$

   c. $CH_3CH_2CH_2C(=O)OC_2H_5$

   d. $ClCH_2CH_2CO_2CH_3$

   e. $CH_3CH_2C(=O)OC(=O)CH_2CH_3$

   f. $CH_3C(=O)OC(=O)H$

   g. cyclohexyl–$C(=O)NH_2$

   h. $CH_3CH_2CH_2C(=O)Br$

   i. cyclobutyl–$OC(=O)H$

   j. $CH_3CH_2CH_2CH(CHO)CH_2CO_2H$

   k. $CH_3C(=O)CH_2CO_2C_2H_5$

   l. $CH_3C(=O)NHBr$

3. Acetic acid is a weak acid and a weak base as shown by:

$$CH_3CO_2H \rightleftharpoons H^+ + CH_3CO_2^-$$

$$K_2 = \frac{[CH_3CO_2^-][H^+]}{[CH_3CO_2H]} = 1.8 \times 10^{-5}$$

$$CH_3C(OH)_2^+ \rightleftharpoons H^+ + CH_3CO_2H \text{ (see Appendix IV)}$$

$$K_1 = \frac{[H^+][CH_3CO_2H]}{[CH_3C(OH)_2^+]} = 1.3 \times 10^6$$

Make the assumption that in 0.1 *M* HCl, most of the acetic acid is unionized; that is, present as CH_3CO_2H.

a. $$\frac{[CH_3CO_2^-]\cdot 10^{-1}}{[CH_3CO_2H]} = 1.8 \times 10^{-5}$$

b. $$\frac{[CH_3CO_2H]\cdot 10^{-1}}{[CH_3C(OH)_2^+]} = 1.3 \times 10^6$$

$$\frac{[CH_3CO_2^-]}{[CH_3CO_2H]} = 1.8 \times 10^{-4}$$

% as $CH_3CO_2^-$ = 0.018%

$$\frac{[CH_3CO_2H]}{[CH_3C(OH)_2^+]} = 1.3 \times 10^{7}$$

$$\frac{[CH_3C(OH)_2^+]}{[CH_3CO_2H]} = 7.9 \times 10^{-8}$$

% as $CH_3C(OH)_2^+$ = 0.00001%
(% as $CH_3CO_2H \cong$ 99.98%)

4.

$$CH_3C(=O)OH_2^+ \rightleftharpoons H^+ + CH_3CO_2H \qquad K' = \frac{[H^+][CH_3CO_2H]}{[CH_3C(=O)OH_2^+]} = 10^{12} \quad \text{(eq. 1)}$$

$$CH_3C(=OH^+)OH \rightleftharpoons H^+ + CH_3CO_2H \qquad K = \frac{[H^+][CH_3CO_2H]}{[CH_3C(=OH^+)OH]} = 10^{6} \quad \text{(eq. 2)}$$

Divide eq. 1 by eq. 2:

$$\frac{[CH_3C(=OH^+)OH]}{[CH_3C(=O)OH_2^+]} = 10^{6}$$

5. The formula $C_7H_{13}O_2Br$ shows that the molecule has one ring or double bond. The infrared band at 1740 cm^{-1} and the CMR signal at 168 ppm strongly suggest an ester, which would account for both oxygens and the double bond:
We make the hypothesis that the compound is a bromine-containing ester.
Some of the NMR bands may be: δ 1.0 (3H) triplet: $C\underline{H}_3-CH_2-$
1.3 (6H) doublet: $(C\underline{H}_3)_2CH-O$

A possible structure that fits this information is:

$$CH_3CH_2CH(Br)-C(=O)-O-CH(CH_3)_2$$

- δ 2.1 (2H) multiplet (CH_2)
- δ 4.2 (1H) triplet ($CHBr$)
- δ 4.6 (1H) multiplet ($O-CH$)

6. Formula, $C_6H_{11}BrO_2$, indicates one double bond or one ring.

IR: no OH
1740 cm^{-1} (strong) suggests ester

NMR: δ 1.3 (t, 3H) and δ 4.2 (q, 2H) must be CH_3CH_2; furthermore, the downfield position of the CH_2 resonance (δ 4.2 ppm) means that it must be attached to an electronegative atom.

δ 1.9 (s, 6H): two equivalent, isolated methyl groups; downfield position of resonance (δ 1.8 ppm) indicates that the methyl groups are attached to an electron-withdrawing carbon.

From this information, we conclude that the structural pieces are:

CH_3CH_2-X $\overset{O}{\overset{\|}{C}}O$ $H_3C-\underset{Z}{\overset{Y}{C}}-CH_3$ –Br

Structure: $CH_3CH_2O\overset{O}{\overset{\|}{C}}\underset{CH_3}{C}BrCH_3$

7. a. $CH_3CH{=}CH_2 + CH_3CO_2H$ b. (cyclohexene) + CH_3SH c. (3-methylcyclopentene)–CH_3 + CH_3SH

d. $CH_3CH_2CH_2CO_2^-Na^+$ e. (benzene ring)–CH_2NH_2 f. (benzene ring)–CN

8.

a. $(CH_3)_3C\overset{O}{\overset{\|}{C}}ND_2$

b. (steroid structure with HO and side chain CO_2CH_3) + $CH_3\overset{O}{\overset{\|}{C}}OCH_3$

c. (bicyclic structure)–CH_2OH

d. $CH_3CH_2\overset{O}{\overset{\|}{C}}\underset{CH_3}{C}HCO_2C_2H_5$

e. $CH_3CH_2\overset{OH}{C}HCH_2CO_2CH_3$

f. $CH_3CH_2\overset{OH}{C}H\underset{CH_3}{C}HCO_2C_2H_5$

9. a. $CH_3CH_2CH_2\overset{O}{\overset{\|}{C}}OH \xrightarrow[\text{or } PCl_3]{SOCl_2} CH_3CH_2CH_2\overset{O}{\overset{\|}{C}}Cl$

b. $CH_3CH_2CH_2\overset{O}{\overset{\|}{C}}OH \xrightarrow{NaOH} CH_3CH_2CH_2\overset{O}{\overset{\|}{C}}O^-\ Na^+ \xrightarrow{CH_3CH_2CH_2\overset{O}{\overset{\|}{C}}Cl} CH_3(CH_2)_2\overset{O}{\overset{\|}{C}}O\overset{O}{\overset{\|}{C}}(CH_2)_2CH_3$

c. $CH_3CH_2CH_2\overset{O}{\overset{\|}{C}}OH \xrightarrow{LiAlH_4} CH_3CH_2CH_2CH_2OH$

$$CH_3CH_2CH_2\overset{O}{\overset{\|}{C}}OH \xrightarrow{SOCl_2} \xrightarrow{CH_3CH_2CH_2CH_2OH} CH_3CH_2CH_2\overset{O}{\overset{\|}{C}}OCH_2CH_2CH_2CH_3$$ or

$$CH_3CH_2CH_2\overset{O}{\overset{\|}{C}}OH \xrightarrow[H^+]{CH_3CH_2CH_2CH_2OH} CH_3CH_2CH_2\overset{O}{\overset{\|}{C}}OCH_2CH_2CH_2CH_3$$

d. $$CH_3CH_2CH_2\overset{O}{\overset{\|}{C}}OH + (CH_3)_2NH \xrightarrow[170°]{} CH_3CH_2CH_2\overset{O}{\overset{\|}{C}}N(CH_3)_2$$

or $$CH_3CH_2CH_2\overset{O}{\overset{\|}{C}}OH \xrightarrow{SOCl_2} \xrightarrow[pyridine]{(CH_3)_2NH} CH_3CH_2CH_2\overset{O}{\overset{\|}{C}}N(CH_3)_2$$

e. $$CH_3CH_2CH_2\overset{O}{\overset{\|}{C}}OH \xrightarrow{SOCl_2} \xrightarrow{NH_3} CH_3CH_2CH_2\overset{O}{\overset{\|}{C}}NH_2 \xrightarrow[\Delta]{P_2O_5} CH_3CH_2CH_2CN$$

f. $CH_3CH_2CH_2CN$ (from e.) $\xrightarrow{H_2/catalyst} CH_3CH_2CH_2CH_2NH_2$

or $$CH_3CH_2CH_2\overset{O}{\overset{\|}{C}}NH_2 \xrightarrow[ether]{LiAlH_4} CH_3CH_2CH_2CH_2NH_2$$

g. $$CH_3CH_2CH_2CO_2H \xrightarrow[Br_2]{Ag_2O} CH_3CH_2CH_2Br$$

h. $$CH_3CH_2CH_2CO_2H \xrightarrow[\text{2. } HN(C_2H_5)_2 \text{ 3. } LiAlH_4]{\text{1. } SOCl_2} CH_3CH_2CH_2CH_2N(C_2H_5)_2$$

10. a. $$CH_3CH_2CH_2CO_2H \xrightarrow[\text{or } PCl_3]{SOCl_2} CH_3CH_2CH_2\overset{O}{\overset{\|}{C}}Cl \xrightarrow{\left(CH_3\overset{CH_3}{\overset{|}{C}H}CH_2\right)_2CuLi} CH_3CH_2CH_2\overset{O}{\overset{\|}{C}}CH_2CH(CH_3)_2$$

b. $$CH_3CH_2CH_2CO_2H \xrightarrow[H^+]{CH_3OH} CH_3CH_2CH_2CO_2CH_3 \xrightarrow{CH_3MgBr} \xrightarrow{H_3O^+} CH_3CH_2CH_2\underset{CH_3}{\underset{|}{\overset{CH_3}{\overset{|}{C}}}}OH$$

c. $$CH_3CH_2CH_2CO_2H \xrightarrow{SOCl_2} \xrightarrow{LiAlH(OtBu)_3} CH_3CH_2CH_2CHO$$

d. $$CH_3CH_2CH_2CO_2CH_3 \xrightarrow{LiAlH_4} CH_3CH_2CH_2CH_2OH$$

e.

$$CH_3(CH_2)_2CO_2H \xrightarrow[H^+]{C_2H_5OH} CH_3(CH_2)_2CO_2C_2H_5 \xrightarrow{NaOC_2H_5} CH_3(CH_2)_2\overset{O}{\overset{\|}{C}}\underset{CH_2CH_3}{\underset{|}{C}H}CO_2C_2H_5$$

$$\xrightarrow{LiAlH_4} CH_3(CH_2)_2\overset{OH}{\overset{|}{C}H}\underset{CH_2CH_3}{\underset{|}{C}H}CH_2OH$$

f. $$CH_3CH_2CH_2CO_2C_2H_5 \xrightarrow[THF,\ -78°]{LDA} \xrightarrow{CH_3CH_2CH_2\overset{O}{\overset{\|}{C}}H} \xrightarrow{H_2O} CH_3CH_2CH_2\overset{OH}{\overset{|}{C}H}\underset{CH_2CH_3}{\underset{|}{C}H}CO_2C_2H_5$$

g. See answer to e.

11. a.

$$H_2N\overset{O}{\overset{\|}{C}}Cl + CH_3O^- \longrightarrow H_2N\overset{O}{\overset{\|}{C}}OCH_3 + Cl^-$$

After addition of CH_3O^-, the intermediate is: $H_2N\overset{O^-}{\overset{|}{C}}\underset{OCH_3}{\underset{|}{}}Cl$

This may decompose in three ways. The three leaving groups are NH_2^-, CH_3O^-, Cl^-. The best leaving group is Cl^-, so the product is:

$$H_2N\overset{O^-}{\overset{|}{C}}\underset{OCH_3}{\underset{|}{}}\!-\!Cl \longrightarrow H_2N\overset{O}{\overset{\|}{C}}OCH_3$$

b. $$H_3CO\overset{O}{\overset{\|}{C}}Cl + NH_2^- \longrightarrow H_3CO\overset{O}{\overset{\|}{C}}NH_2 + Cl^-$$; same reasoning as in (a).

12. a. $$CH_3CO_2CH_3 \xrightarrow[THF,\ -78°]{LDA} H_2C{=}\overset{OLi}{\overset{|}{C}}OCH_3 \xrightarrow{(CH_3)_2CHCH_2CH_2Br} (CH_3)_2CHCH_2CH_2CH_2CO_2CH_3$$

b. $$H_2C{=}\overset{OLi}{\overset{|}{C}}OC_2H_5 + CH_3CH_2CH_2\overset{O}{\overset{\|}{C}}H \longrightarrow \longrightarrow CH_3CH_2CH_2\overset{OH}{\overset{|}{C}H}CH_2CO_2C_2H_5$$

c. $$(CH_3)_3C\overset{O}{\overset{\|}{C}}H + BrCH_2CO_2C_2H_5 \xrightarrow[\text{benzene } \Delta]{Zn} (CH_3)_3C\overset{OH}{\overset{|}{C}H}CH_2CO_2C_2H_5 \xrightarrow[H_2SO_4]{K_2Cr_2O_7} (CH_3)_3C\overset{O}{\overset{\|}{C}}CH_2CO_2C_2H_5$$

(NOTE: although one ester would be enolizable, a mixed Claisen condensation would not be a good choice, because the ethyl acetate enolate would condense faster with ethyl acetate itself than with the sterically more congested ethyl 2,2-dimethylpropanoate.)

d. cyclopentanone (=O) + $BrCH_2CO_2C_2H_5$ $\xrightarrow{Zn}$ $\xrightarrow{H_3O^+}$ 1-hydroxycyclopentyl (OH) $CH_2CO_2C_2H_5$

(or via lithium enolate)

13. $CH_3\overset{\ddot{O}:}{\overset{||}{C}}NHCH_3$ + $CH_3CH_2-\overset{+}{O}(CH_2CH_3)_2$ $\longrightarrow$ $\left[CH_3\overset{+OCH_2CH_3}{\overset{||}{C}}NHCH_3 \longleftrightarrow CH_3\overset{OCH_2CH_3}{\overset{|}{C}}=\overset{+}{N}HCH_3 \right]$

$\xrightarrow{-H^+}$ $CH_3\overset{OCH_2CH_3}{\overset{|}{C}}=NCH_3$ $\xrightarrow{H_3O^+}$ $CH_3\overset{O}{\overset{||}{C}}OCH_2CH_3$

14. $(CH_3)_3COH + H_2SO_4 \rightleftharpoons (CH_3)_3C\overset{+}{O}H_2 \rightleftharpoons (CH_3)_3C^+ + H_2O$; $(CH_3)_3C^+$ + $:N{\equiv}CCH_3$ $\longrightarrow (CH_3)_3C\overset{+}{N}{\equiv}CCH_3 \xrightleftharpoons{H_2O}$

$(CH_3)_3CN=\overset{\overset{+}{O}H_2}{\overset{|}{C}}CH_3 \xrightleftharpoons{-H^+} (CH_3)_3CN=\overset{OH}{\overset{|}{C}}CH_3 \xrightleftharpoons{\text{tautomerism}} (CH_3)_3CNH-\overset{O}{\overset{||}{C}}CH_3$

With 2-methyl-2,4-pentanediol, the tertiary carbocation forms more easily than the secondary carbocation. After addition of acetonitrile, the resulting cation reacts *intramolecularly* with the other hydroxyl group. The product is a heterocyclic compound, called *tetrahydrooxazine*.

2-methyl-2,4-pentanediol (OH, OH) $\xrightarrow[-H_2O]{H^+}$ tertiary cation (OH, +) $\xrightarrow{CH_3CN}$ nitrilium ion (HO:, $+N{\equiv}C-CH_3$) $\longrightarrow$ cyclic oxonium (HO+, N) $\xrightarrow{-H^+}$ tetrahydrooxazine (O, N)

15. The eicosanoid compounds are biologically active 20 carbon carboxylic acids such as the following:

Prostaglandin G_2, PGG_2 (O–O bridge, CO_2H, OOH)

Prostaglandin H_2, PGH_2 (O–O bridge, CO_2H, OH)

Thromboxane A_2, TXA_2 (O, O bridge, CO_2H, OH)

16.

$$CH_3C(OC_2H_5)_3 + H^+ \rightleftharpoons CH_3C(\overset{+}{H}OC_2H_5)(OC_2H_5)_2 \underset{}{\overset{-EtOH}{\rightleftharpoons}} CH_3C(=\overset{+}{O}C_2H_5)-OC_2H_5 \overset{H_2O}{\rightleftharpoons} CH_3C(OC_2H_5)_2(\overset{+}{O}H_2)$$

$$\overset{-H^+}{\rightleftharpoons} CH_3C(OC_2H_5)_2(OH) \overset{H^+}{\rightleftharpoons} CH_3C(\overset{+}{H}OC_2H_5)(OC_2H_5)(OH) \overset{-EtOH}{\rightleftharpoons} CH_3C(=\overset{+}{O}H)OC_2H_5 \overset{-H^+}{\rightleftharpoons} CH_3\overset{O}{\overset{\|}{C}}OC_2H_5$$

17.

$$C_2H_5O\overset{O}{\overset{\|}{C}}(CH_2)_4\overset{O}{\overset{\|}{C}}OC_2H_5 \xrightarrow{NaOC_2H_5} C_2H_5O-C(O^-)=CH(CH_2)_3\overset{O}{\overset{\|}{C}}OC_2H_5 \longrightarrow$$

(cyclic tetrahedral intermediate: $C_2H_5O\overset{O}{\overset{\|}{C}}$–cyclopentane bearing H, with ^-O and OC_2H_5 on the ring carbon) $\rightleftharpoons$ 2-carbethoxycyclopentanone (H) $+ {}^-OC_2H_5$ $\rightleftharpoons$ enolate of 2-carbethoxycyclopentanone $+ HOC_2H_5$

$C_8H_{12}O_3$ *(isolated on acidification)*

18.

Cyclohexyl methyl ketone $\xrightarrow{C_6H_5CO_3H}$ tetrahedral intermediate (H_3C–C(OH)(cyclohexyl)–O–O–C(=O)–C_6H_5) $\longrightarrow$ $CH_3\overset{O}{\overset{\|}{C}}$–O–cyclohexyl $+ C_6H_5CO_2H$

Note that the Baeyer-Villiger oxidation occurs exclusively with migration of the cyclohexyl group.

19. a. S_N2 reaction with inversion of configuration.

b. Convert to a sulfonate ester and displace with sodium methoxide.

20. This is an example of an ester exchange catalyzed by isopropoxide. The initial product, A, is the mixed ester. As its concentration becomes appreciable compared to that of the dimethyl carbonate, it will react further to give di-isopropyl carbonate:

$$CH_3O\overset{O}{\overset{\|}{C}}OCH_3 + {}^-OCH(CH_3)_2 \rightleftharpoons CH_3O-\overset{O^-}{\overset{|}{C}}(OCH_3)-OCH(CH_3)_2 \rightleftharpoons CH_3O\overset{O}{\overset{\|}{C}}OCH(CH_3)_2 + CH_3O^- \quad (A)$$

$$CH_3O^- + (CH_3)_2CHOH \rightleftharpoons CH_3OH + (CH_3)_2CHO^-$$

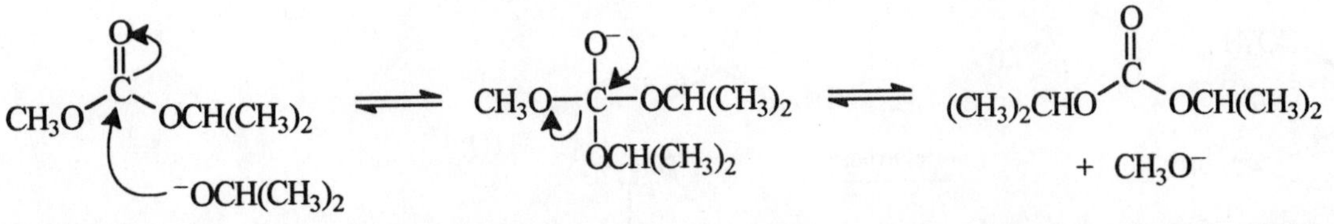

$$CH_3O{-}C(=O){-}OCH(CH_3)_2 + {}^-OCH(CH_3)_2 \rightleftharpoons CH_3O{-}C(O^-)(OCH(CH_3)_2){-}OCH(CH_3)_2 \rightleftharpoons (CH_3)_2CHO{-}C(=O){-}OCH(CH_3)_2 + CH_3O^- \quad \text{(B)}$$

The reaction is pushed towards the formation of the diisopropyl ester by the large excess of isopropyl alcohol in the reaction.

21. (arachidonic acid, CO_2H) $\xrightarrow[\text{cyclooxygenase}]{O_2}$ (cyclic endoperoxide, CO_2H) $\xrightarrow{O_2}$ (endoperoxide hydroperoxide, OOH, CO_2H) $\xrightarrow{\text{peroxidase}}$ (PGH$_2$, OH, CO_2H)

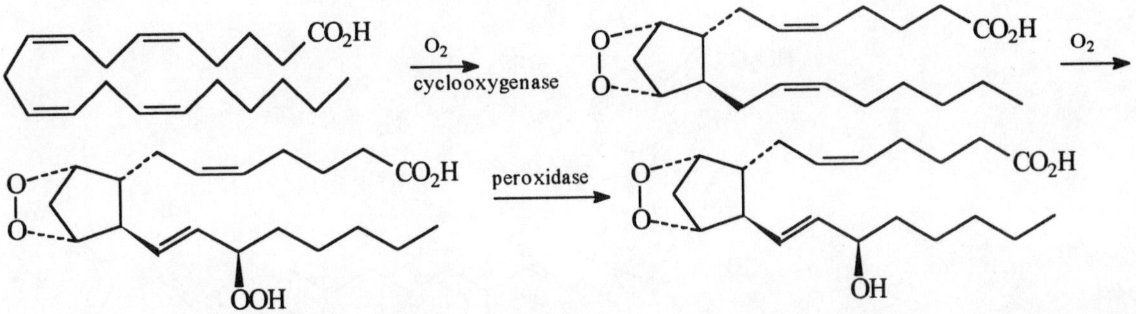

If the original arachidonic acid is labelled with deuterium at the carbons adjacent to the double bonds:

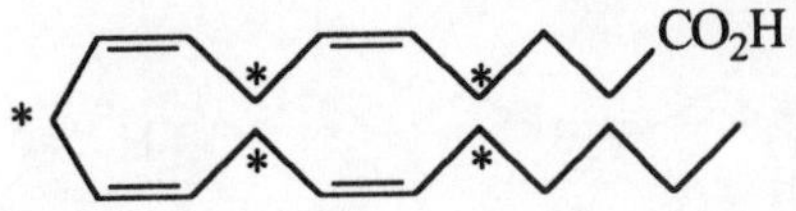

(labelled arachidonic acid, CO_2H, * labels), the final product, PGH_2, will be labelled as indicated:

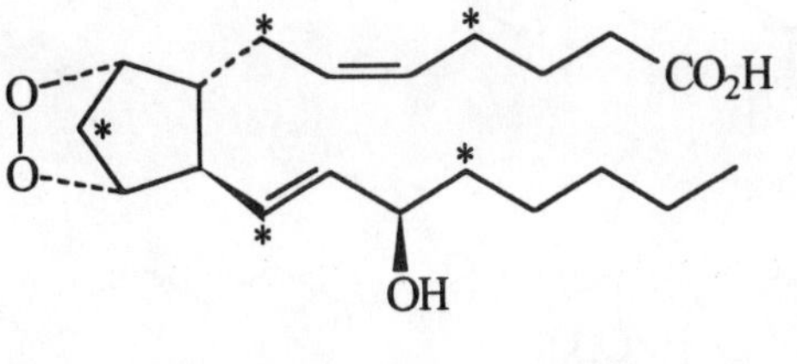

(labelled PGH$_2$: O–O, CO_2H, OH, * labels)

If the label is at the carbons in the double bond, the final product will be labelled:

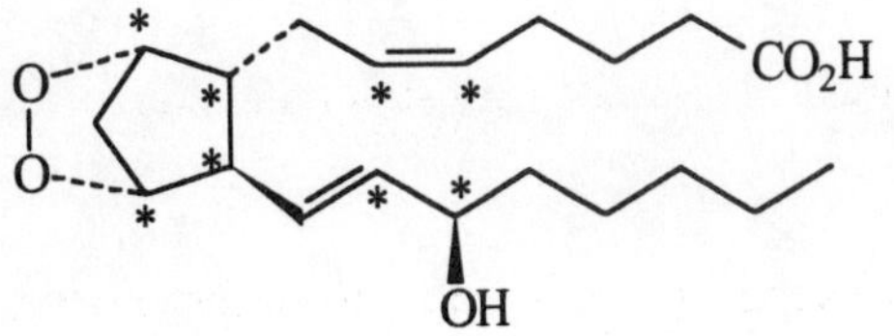

(labelled PGH$_2$: O–O, CO_2H, OH, * labels)

22. PGH_2 to PGI_2

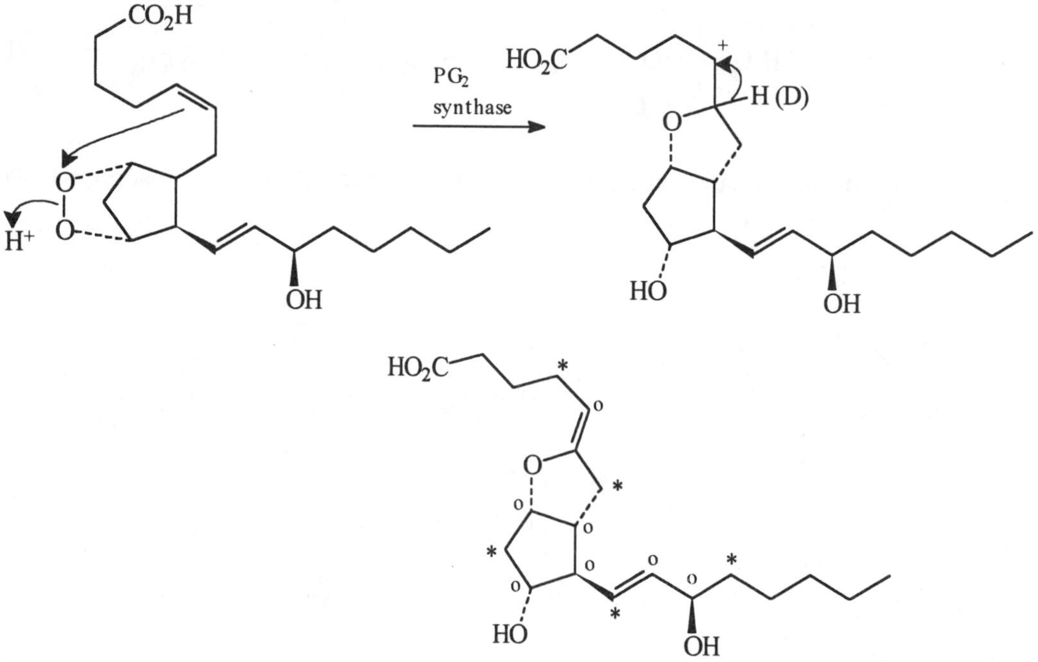

CO_2H

PG₂ synthase →

HO_2C H (D)

HO OH

OH

HO_2C

HO OH

The *'s indicate the position of the deuterium label when the initial deuterium in the arachidonic acid was next to the double bond; the °'s indicate the position of the deuterium label that was on the double bond.

PGH_2 to TXA_2:

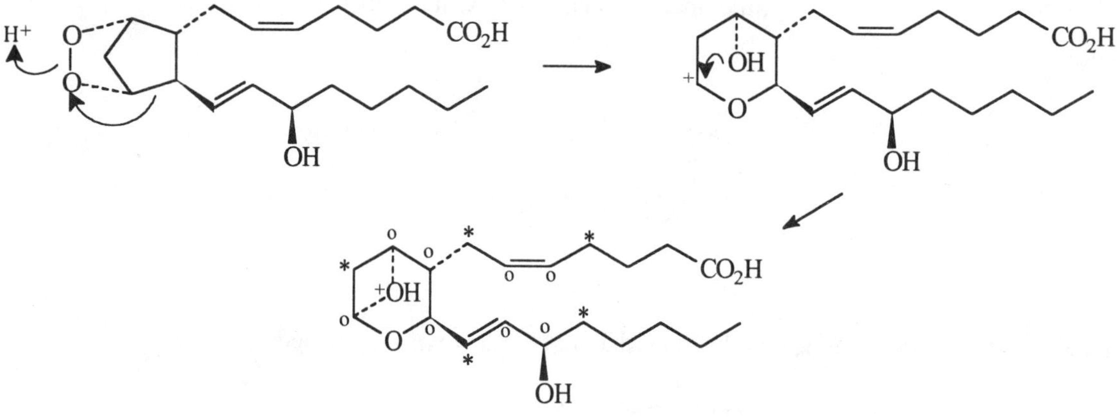

H^+ CO_2H → OH CO_2H

OH OH

CO_2H

+OH

OH

23. Formation of leukotriene A:

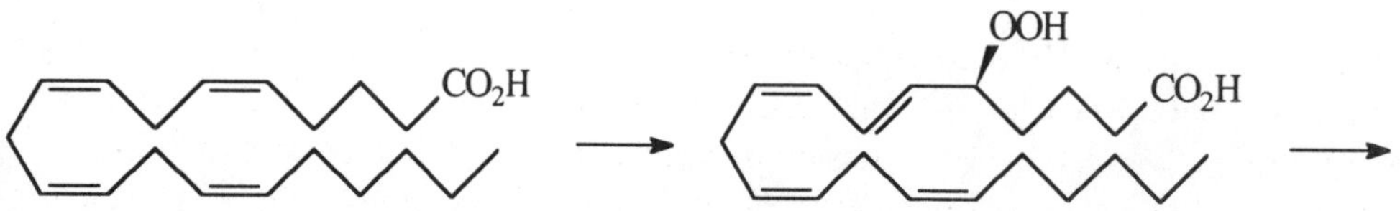

CO_2H → OOH CO_2H →

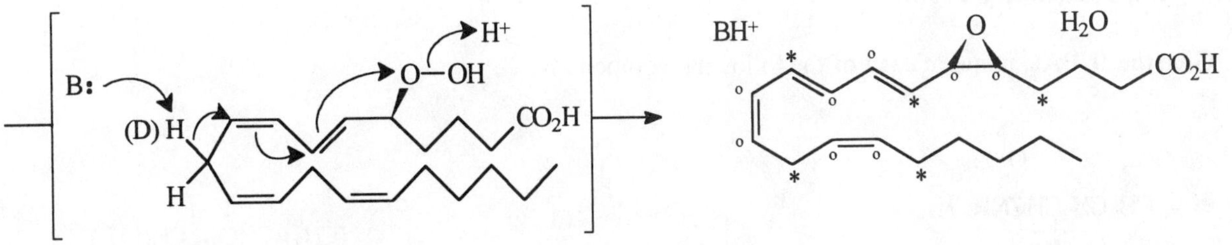

B: (D) H H O–OH H+ CO2H BH+ O H2O CO2H

24. Formation of Leukotriene D:

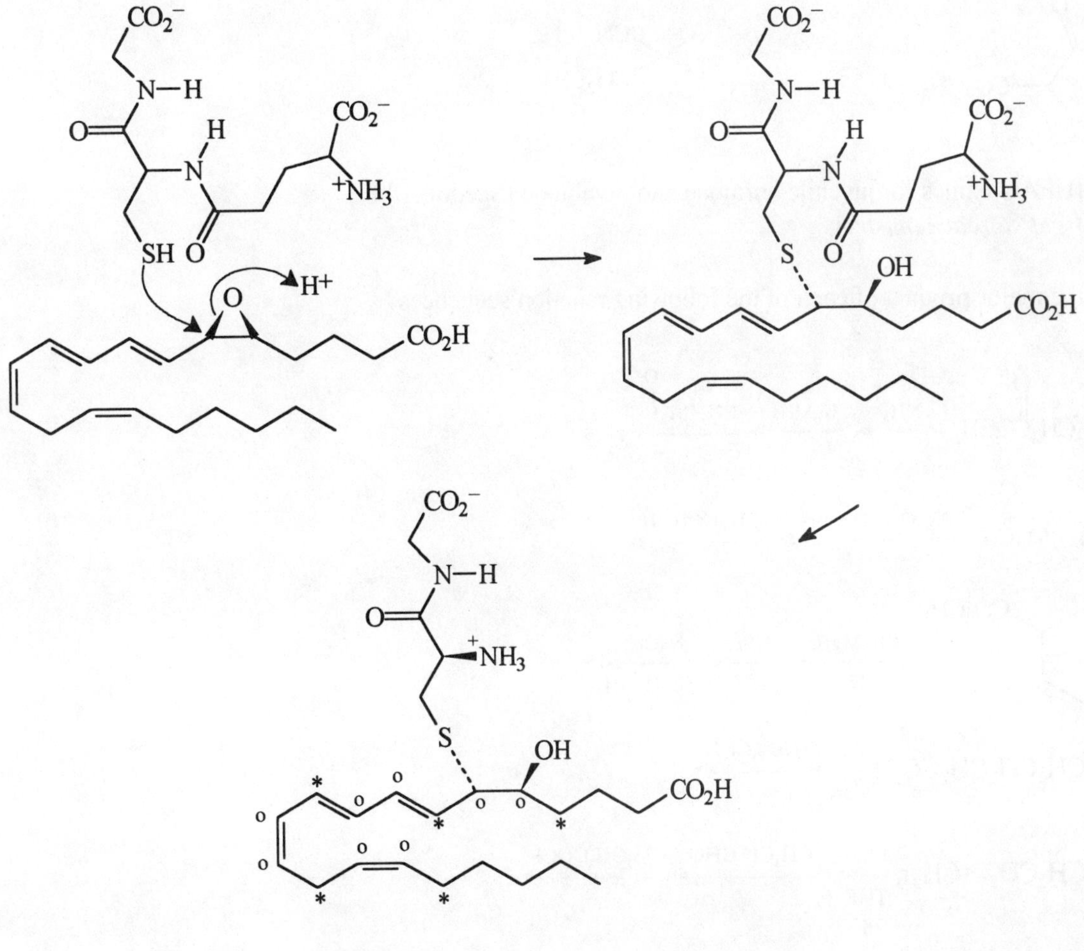

25. The phospholipid bilayer forms by the attraction of the polar "heads" for the aqueous medium of the cell. The hydrocarbon "tails" are not attracted to the water and congregate at the interior of the bilayer. The saturated fats stack in an orderly fashion while the *cis*-unsaturated fats have a kink in the chain that disrupts this stacking. This introduces an element of disorder in the membrane. This increases the membrane fluidity and allows for the increased biological activity of the proteins embedded in the bilayer. The elevated temperatures in hot springs can increase the fluidity.

19.F. Supplementary Problems

S1. Give the IUPAC name of each of the following compounds.

a. $CH_3CH_2CH(CH_3)C(=O)NHCH_3$

b. 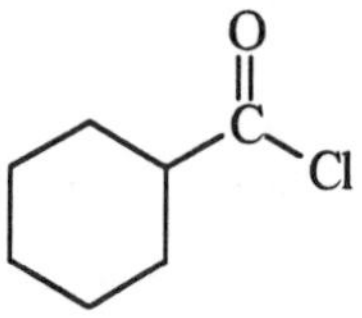 Cyclohexyl–C(=O)Cl

c. $(CH_3)_2CHC(=O)OC(=O)CH(CH_3)_2$

d. $CH_2{=}CHCH{=}CHOC(=O)CH_3$

e. $CH_3CH_2CH_2C(=O)C{\equiv}N$

f. $CH_3CH_2C(=O)NHBr$

g. 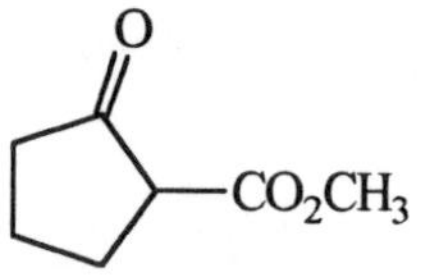 2-oxocyclopentyl–CO_2CH_3

h. 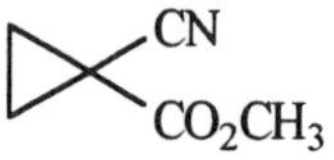 cyclopropane bearing CN and CO_2CH_3 on the same carbon

S2. Provide IUPAC names for juvenile hormone and juvabione (Section 18.8.) *(Don't forget stereochemistry)*.

S3. What is the major product of each of the following reaction sequences?

a. $CH_3CH_2C(=O)OCH_3 \xrightarrow{NH_3} \xrightarrow{LiAlH_4} \xrightarrow{CH_3CH_2C(=O)Cl}$

b. $(CH_3)_2CHC(=O)H \xrightarrow{Ag_2O} \xrightarrow[\Delta]{P + Br_2} \xrightarrow{(CH_3)_2CHOH}$

c. 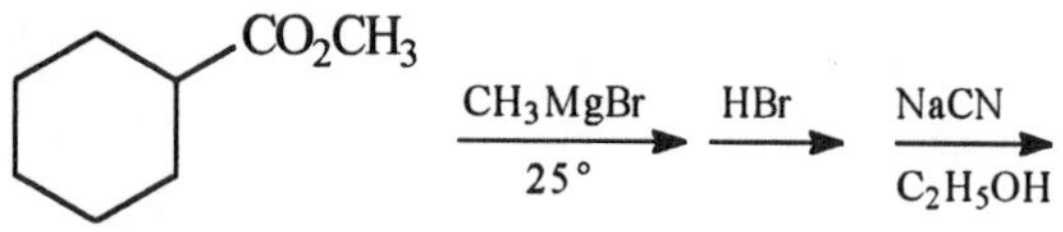 Cyclohexyl–CO_2CH_3 $\xrightarrow[25°]{CH_3MgBr} \xrightarrow{HBr} \xrightarrow[C_2H_5OH]{NaCN}$

d. $CH_3CH_2CH_2CH_2CO_2H \xrightarrow[Br_2]{Ag_2O} \xrightarrow{NaO_2CCH_3}$

e. $CH_3CH_2CO_2C(CH_3)_3 \xrightarrow[THF, -78°C]{LDA} \xrightarrow{CH_3CH_2CHO} \xrightarrow{CH_3CH_2COCl}$

f. 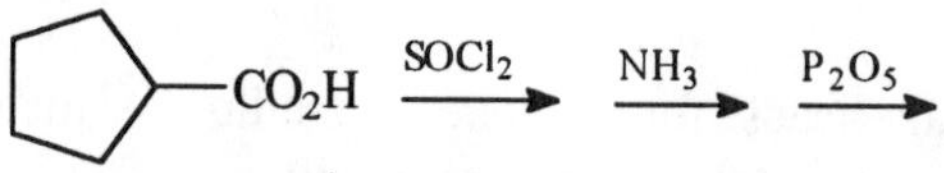 Cyclopentyl–CO_2H $\xrightarrow{SOCl_2} \xrightarrow{NH_3} \xrightarrow{P_2O_5}$

g. $(CH_3)_2CHCH_2CO_2H \xrightarrow{SOCl_2} \xrightarrow{LiAlH(O\text{-}t\text{-}Bu)_3} \xrightarrow{H_2O}$

h. 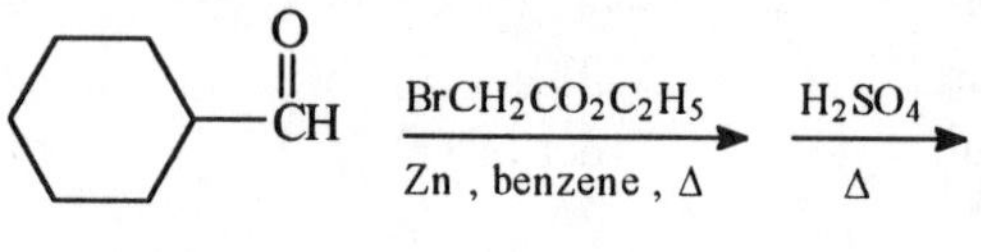 Cyclohexyl–C(=O)H $\xrightarrow[Zn,\ benzene,\ \Delta]{BrCH_2CO_2C_2H_5} \xrightarrow[\Delta]{H_2SO_4}$

S4. Show how to carry out the following transformations:

a. 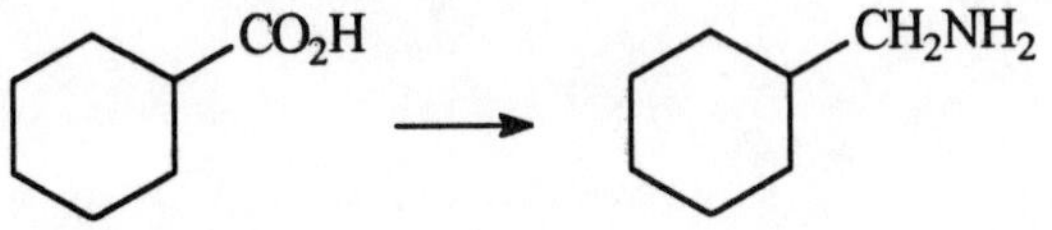 cyclohexyl–CO_2H ⟶ cyclohexyl–CH_2NH_2

b. $(CH_3)_2CHCH_2CH_2OH$ ⟶ $(CH_3)_2CHCH_2CH_2CH(CH_3)CO_2CH(CH_3)_2$

c. 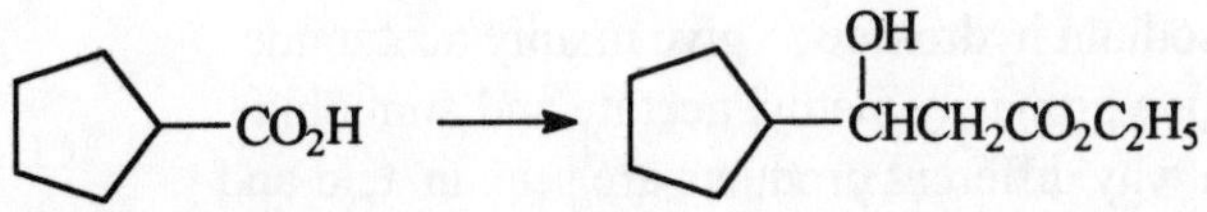 cyclopentyl–CO_2H ⟶ cyclopentyl–$CH(OH)CH_2CO_2C_2H_5$

d. 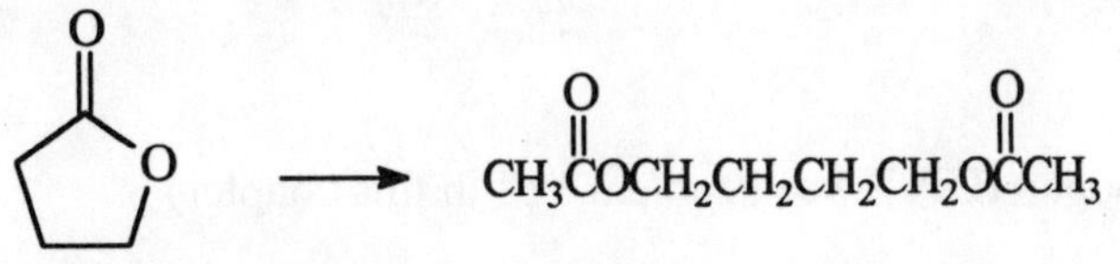 γ-butyrolactone ⟶ $CH_3\overset{O}{\overset{\|}{C}}OCH_2CH_2CH_2CH_2O\overset{O}{\overset{\|}{C}}CH_3$

e. $(CH_3)_2CHCH_2OH$ ⟶ $((CH_3)_2CHCH_2)_3CBr$

f. 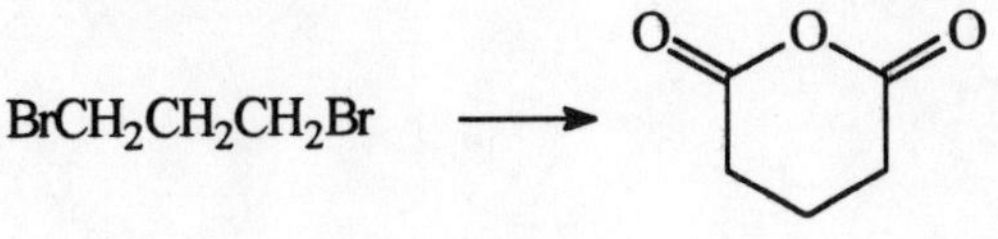 $BrCH_2CH_2CH_2Br$ ⟶ glutaric anhydride (cyclic anhydride, O=C–O–C=O in six-membered ring)

g. $CH_3CH_2CH_2CO_2C_2H_5$ ⟶ $CH_3CH_2CH_2CH(OH)CH(CH_2OH)CH_2CH_3$

S5. a. Provide structures for A-D below, and assign all the spectral information.

$$A \xrightarrow[\text{2. } H^+]{\text{1. NaOCH}_3} B \xrightarrow{SOCl_2} C \xrightarrow[\text{quinoline}]{H_2\text{, Pd-BaSO}_4} D$$

A	B	C	D
IR: 1725,1820 cm^{-1}	IR: 1740,1710, 2500-3000 (br) cm^{-1}	IR: 1735, 1785 cm^{-1}	IR:1725,1740 cm^{-1}
NMR: δ 2.0(qnt,2H) 2.8(t,4H)	NMR: δ 3.8(3H), 13(s, 1H) and resonances for 6 other H's		

b. When compound **D** is treated with HCN and a trace of base, compound **E** is formed. The NMR spectrum of **E** reveals the presence of only seven hydrogens: the IR spectrum shows bands at 1735 and 2130 cm^{-1}. Propose a structure for **E** and write a reasonable mechanism for its formation.

S6. Reaction of acetonitrile in methanol with dry HCl gives initially methyl acetimidate hydrochloride (1) and then methyl orthoacetate (2). Write a mechanism for this reaction.

$$\overset{\overset{+}{N}H_2}{\overset{\|}{CH_3COCH_3}}\ Cl^- \qquad CH_3C(OCH_3)_3$$

1   2

S7. Methyl acetimidate (3) is hydrolyzed in aqueous sodium hydroxide to give mainly acetamide and methanol. In aqueous acid, it hydrolyzes to give primarily methyl acetate and ammonia. Write mechanisms for these reactions and explain why different products are seen in acid and base.

$$\overset{NH}{\overset{\|}{CH_3COCH_3}}$$

3

S8. Write a mechanism for the following transformation. (**HINT**: see Problem #18 in this chapter)

2-methyl-2-(CO_2CH_3)cyclopentanone $\xrightarrow[\Delta]{NaOCH_3}$ $\xrightarrow[\text{(work up)}]{H^+}$ H_3C–(cyclopentanone)–CO_2CH_3

S9. Starting with benzoic acid (4), propanoic acid, dimethylamine $(CH_3)_2NH$), and any needed reagents, outline a synthesis of propoxyphene (the active ingredient in Darvon®).

C_6H_5–CO_2H

4

$(CH_3)_2NCH_2CH(CH_3)C(C_6H_5)(OC(=O)CH_2CH_3)CH_2$–$C_6H_5$

S10. A useful preparation of deuterioethanol (C_2H_5OD) involves refluxing diethyl carbonate with D_2O and a small amount of strong acid. Why is the preparation so convenient?

S11. Trimyristin is a white crystalline fat, mp 54-55°, obtainable from nutmeg, and is the principal constituent of nutmeg butter. Hydrolysis of trimyristin with hot aqueous sodium hydroxide gives an excellent yield of myristic acid, mp 52-53°, as the only fatty acid. What is the structure of trimyristin?

19.G. Answers to Supplementary Problems

S1. a. N-2-dimethylbutanamide
b. cyclohexanecarbonyl chloride
c. 2-methylpropanoic anhydride
d. 1,3-butadienyl acetate
e. 2-oxopentanenitrile
f. N-bromopropanamide
g. methyl 2-oxocyclopentanecarboxylate
h. methyl 1-cyanocyclopropanecarboxylate

S2. Juvenile hormane: methyl (*Z*)-10,11-epoxy-7-ethyl-3,11-dimethyl-(*E*,*E*)-2,6-tridecadienoate

Juvabione: methyl (*R*)-4((*R*)-1,5-dimethyl-3-oxohexyl)-1-cyclohexenecarboxylate

S3. a. $CH_3CH_2C(=O)NHCH_2CH_2CH_3$

b. $(CH_3)_2C(Br)C(=O)OCH(CH_3)_2$

c. cyclohexyl–CO_2CH_3 $\xrightarrow{CH_3MgBr}$ $\xrightarrow{HBr}$ cyclohexyl–$C(Br)(CH_3)_2$ $\xrightarrow{NaCN}$ cyclohexyl–$C(=CH_2)CH_3$ + cyclohexylidene=$C(CH_3)CH_3$

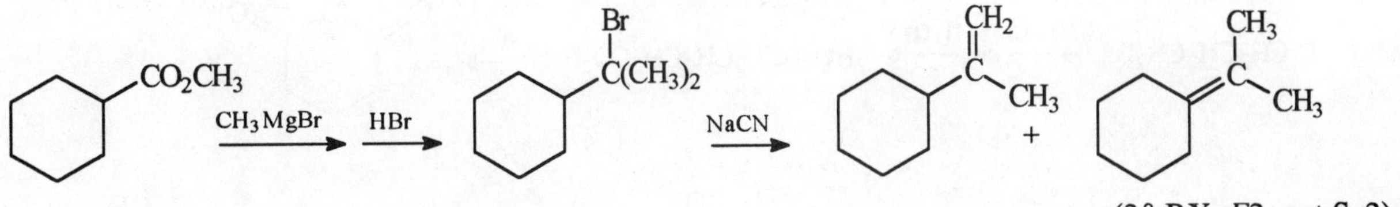

(3° RX, *E*2, not S_N2)

d. $CH_3CH_2CH_2CH_2O_2CCH_3$

e. $CH_3CH_2CH(C(=O)CH_2CH_3)CH(CH_3)C(=O)OC(CH_3)_3$

f. cyclopentyl–CN

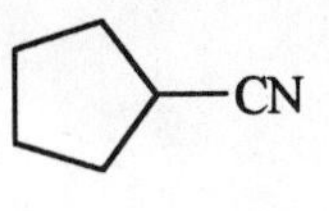

g. $(CH_3)_2CHCH_2C(=O)H$

h. cyclohexyl–$CH{=}CHCO_2C_2H_5$

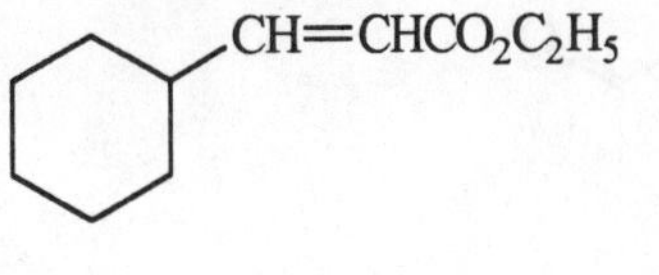

S4.

a. cyclohexyl–CO_2H $\xrightarrow{1.\ SOCl_2;\ 2.\ NH_3}$ (or $\xrightarrow{1.\ NH_3;\ 2.\ 135°}$) cyclohexyl–$C(=O)NH_2$ $\xrightarrow{LiAlH_4}$ cyclohexyl–CH_2NH_2

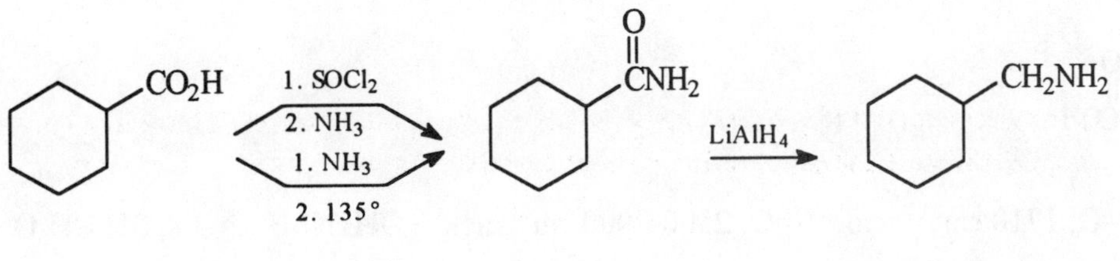

b. $(CH_3)_2CHCH_2CH_2OH$ $\xrightarrow{PBr_3}$ $(CH_3)_2CHCH_2CH_2Br$ + $CH(CH_3){=}C(OLi)OCH(CH_3)_2$ $\xleftarrow{LDA,\ THF,\ -78°}$ $CH_3CH_2C(=O)OCH(CH_3)_2$

$\rightarrow (CH_3)_2CHCH_2CH_2CH(CH_3)C(=O)OCH(CH_3)_2$

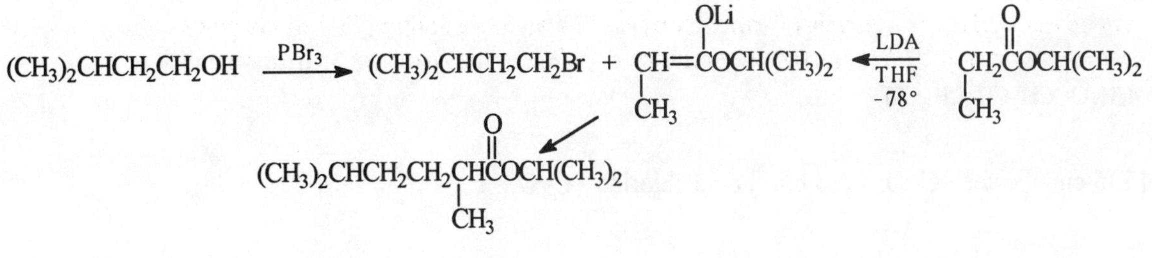

c. Cyclopentyl–CO_2H $\xrightarrow{SOCl_2}$ $\xrightarrow[\text{Lindlar}]{H_2}$ Cyclopentyl–CHO $\xrightarrow[\text{Zn}]{BrCH_2CO_2C_2H_5}$ Cyclopentyl–$CH(OH)CH_2CO_2C_2H_5$

d. γ-Butyrolactone $\xrightarrow{LiAlH_4}$ $\xrightarrow{H_2O}$ $HOCH_2CH_2CH_2CH_2OH$ $\xrightarrow[\text{pyridine}]{2\,(CH_3C(=O))_2O}$ $CH_3C(=O)OCH_2CH_2CH_2CH_2OC(=O)CH_3$

e. $(CH_3)_2CHCH_2OH \xrightarrow{PBr_3} \xrightarrow{Mg} \xrightarrow{(CH_3O)_2C=O} \xrightarrow{HBr} ((CH_3)_2CHCH_2)_3CBr$

f. $BrCH_2CH_2CH_2Br \xrightarrow{2\,NaCN} \xrightarrow[\Delta]{H_3O^+} HO_2CCH_2CH_2CH_2CO_2H \xrightarrow{P_2O_5}$ glutaric anhydride (six-membered cyclic anhydride)

g. $CH_3CH_2CH_2CO_2C_2H_5 \xrightarrow{NaOC_2H_5} CH_3CH_2CH_2C(=O)CH(CO_2C_2H_5)CH_2CH_3 \xrightarrow{LiAlH_4} \xrightarrow{H_2O} CH_3CH_2CH_2CH(OH)CH(CH_2OH)CH_2CH_3$

S5.

a. A: glutaric anhydride (six-membered cyclic anhydride)

The 2 IR signals at 1755 and 1820 cm^{-1} arise from the acid anhydride. The 4H triplet at 2.8 ppm arises from the two $-CH_2$ groups next to the $-C=O$; the 2H quintet at 2.0 ppm arises from the remaining ring $-CH_2$ group.

B: $CH_3OC(=O)CH_2CH_2CH_2C(=O)OH$

IR: 1740 cm^{-1}, ester $-C=O$; 1710 cm^{-1}, acid $-C=O$; 2500-3000 cm^{-1}, acid $-OH$. NMR: δ 3.8, 3H CH_3O-; δ 13, carboxylic acid $-OH$.

C: $CH_3OC(=O)CH_2CH_2CH_2C(=O)Cl$

IR: 1735 cm^{-1}, ester $-C=O$; 1785 cm^{-1}, acid chloride $-C=O$.

D: $CH_3OC(=O)CH_2CH_2CH_2CH(=O)$

IR: 1740 cm^{-1}, ester –C=O; 1725 cm^{-1}, aldehyde –C=O.

b. $CH_3OC(=O)CH_2CH_2CH_2CH(=O) + HCN \longrightarrow CH_3OC(=O)CH_2CH_2CH_2CH(OH)CN \rightleftharpoons CH_3OC(=O)CH_2CH_2CH_2CH(O^-)CN$

$\rightleftharpoons$ cyclic tetrahedral intermediate (CH_3O, O^- on ring carbon; H, CN) $\rightleftharpoons$ lactone (6-membered ring, C=O, CN)

S6.

$CH_3CN \underset{}{\overset{H^+}{\rightleftharpoons}} CH_3C\equiv\overset{+}{N}H$ (+ $CH_3\ddot{O}H$) $\rightleftharpoons CH_3C(=NH)\overset{+}{H}OCH_3 \overset{-H^+}{\rightleftharpoons} CH_3C(=NH)OCH_3 \overset{H^+}{\rightleftharpoons} CH_3C(=\overset{+}{N}H_2)OCH_3$ (+ $CH_3\ddot{O}H$) $\rightleftharpoons CH_3C(NH_2)(OCH_3)\overset{+}{H}OCH_3$

$\overset{-H^+}{\rightleftharpoons}$ $CH_3C(NH_2)(OCH_3)OCH_3 \overset{H^+}{\rightleftharpoons} CH_3C(\overset{+}{N}H_3)(OCH_3)\ddot{O}CH_3 \overset{-NH_3}{\rightleftharpoons} [CH_3\overset{+}{C}(OCH_3)OCH_3 \leftrightarrow CH_3C(OCH_3)=\overset{+}{O}CH_3] \overset{CH_3OH}{\rightleftharpoons} CH_3C(OCH_3)(OCH_3)\overset{+}{H}OCH_3 \overset{-H^+}{\rightleftharpoons} CH_3C(OCH_3)_3$

S7:

With OH$^-$:

$CH_3C(=NH)OCH_3$ + HO$^-$ (HOH) $\rightleftharpoons$ $CH_3C(NH_2)(OCH_3)OH$ (HO$^-$) $\rightleftharpoons CH_3C(NH_2)(OCH_3)O^-$ (HOH) $\rightleftharpoons CH_3C(=O)NH_2 + CH_3O^-$

(best leaving group lost: pK_a (CH_3OH) = 16; pK_a (NH_3) = 35)

With H_3O^+:

$$CH_3C(=NH)OCH_3 \overset{H^+}{\rightleftharpoons} CH_3C(=\overset{+}{N}H_2)OCH_3 \;(H_2\ddot{O}:) \rightleftharpoons CH_3C(NH_2)(\overset{+}{O}H_2)OCH_3 \overset{-H^+}{\rightleftharpoons} CH_3C(NH_2)(OH)OCH_3$$

$$\overset{H^+}{\rightleftharpoons} CH_3C(\overset{+}{N}H_3)(:OH)OCH_3 \overset{-NH_3}{\rightleftharpoons} \left[CH_3C(=\overset{+}{O}CH_3)OH \longleftrightarrow CH_3C(OCH_3)=\overset{+}{O}H \right] \overset{-H^+}{\rightleftharpoons} CH_3\overset{O}{\overset{\|}{C}}OCH_3$$

(in acid, the *amine* is protonated (pK_a RN^+H_3 ≈ 9), and NH_3 is now a better leaving group than $^-OCH_3$.)

S8.

$$CH_3O^- + \text{2-methyl-2-(methoxycarbonyl)cyclopentanone} \rightleftharpoons \text{tetrahedral alkoxide (}CH_3O, O^-\text{; }CH_3, COCH_3(=O)\text{)} \rightleftharpoons CH_3O\overset{O}{\overset{\|}{C}}(CH_2)_3\underset{CH_3}{\underset{|}{C}}=\overset{O^-}{\overset{|}{C}}OCH_3$$

$$\overset{CH_3OH}{\rightleftharpoons} CH_3O\overset{O}{\overset{\|}{C}}(CH_2)_3\underset{CH_3}{\underset{|}{C}H}\overset{O}{\overset{\|}{C}}OCH_3 + CH_3O^- \rightleftharpoons CH_3O\overset{O^-}{\overset{|}{C}}=CH(CH_2)_2\underset{CH_3}{\underset{|}{C}H}\overset{O}{\overset{\|}{C}}OCH_3 + CH_3OH$$

$$\rightleftharpoons CH_3O\overset{O}{\overset{\|}{C}}\text{–cyclopentane(}O^-, OCH_3\text{)–}CH_3 \rightleftharpoons CH_3O^- + \text{2-methyl-5-(methoxycarbonyl)cyclopentanone } (CH_3CO_2, H) \rightleftharpoons \text{enolate (}CH_3O\text{–C(=O)–C}^-\text{–C=O, ring, }CH_3\text{)}$$

S9. This is clearly a case where you have to work backward!

1. What part of the target came from which starting material?

   The 2 phenyl (Ph) rings and their adjacent carbon atoms come from benzoic acid; the two 3-carbon fragments come from the propanoic acid and the nitrogen with two methyl groups comes from the dimethylamine.

2. The CH_2 next to the nitrogen could come from $LiAlH_4$ reduction of the amide, and the propanoate ester from the 3° alcohol:

$$(CH_3)_2NCH_2\underset{CH_3}{\underset{|}{C}H}\text{—}\underset{OCOCH_2CH_3}{\underset{|}{\overset{Ph}{\overset{|}{C}}}}\text{—}CH_2Ph \xleftarrow[H^+]{CH_3CH_2CO_2H} (CH_3)_2NCH_2\underset{CH_3}{\underset{|}{C}H}\text{—}\underset{OH}{\underset{|}{\overset{Ph}{\overset{|}{C}}}}CH_2Ph \xleftarrow[LiAlH_4]{excess} (CH_3)_2N\overset{O}{\overset{\|}{C}}\underset{CH_3}{\underset{|}{C}H}\text{—}\underset{OH}{\underset{|}{\overset{Ph}{\overset{|}{C}}}}CH_2Ph$$

3. The hydroxyl β to the carbonyl suggests that an ester-enolate or Reformatsky reaction could be used to form the carbon-carbon bond.

$$(CH_3)_2NC(=O)CH(CH_3)-C(OH)(Ph)CH_2Ph \xleftarrow[\Delta]{(CH_3)_2NH} CH_3CH_2OC(=O)CH(CH_3)-C(OH)(Ph)CH_2Ph \xleftarrow[\text{Zn, benzene } \Delta]{CH_3CH_2OC(=O)CHBrCH_3} PhC(=O)CH_2Ph$$

4. And so on:

$$CH_3CH_2CO_2H \xrightarrow{P,\ Br_2} \xrightarrow{C_2H_5OH} CH_3CHBrCO_2C_2H_5$$

$$C_6H_5CO_2H \xrightarrow{SOCl_2} C_6H_5C(=O)Cl + BrMgCH_2C_6H_5 \xleftarrow{Mg} \xleftarrow{PBr_3} \xleftarrow{LiAlH_4} HO_2CC_6H_5$$

$$\xrightarrow{\text{low temp.}} C_6H_5C(=O)CH_2C_6H_5$$

S10.

$$C_2H_5OC(=O)OC_2H_5 + D_2O \longrightarrow C_2H_5OD + \left[C_2H_5OC(=O)OD\right] \longrightarrow C_2H_5OD + CO_2$$

The only other product is gaseous CO_2.

S11.

$$\begin{array}{l} CH_3(CH_2)_{12}COOCH_2 \\ CH_3(CH_2)_{12}COOCH \\ CH_3(CH_2)_{12}COOCH_2 \end{array}$$

20. CONJUGATION

20.A. Chapter Outline and Important Terms Introduced

*(***NOTE: except for the Diels-Alder reaction, most of the reactions discussed in this chapter are not "new". What is different about them is how they proceed in conjugated systems.)*

20.1 Allylic Systems

A. Allylic Cations (resonance structures)

$$\left.\begin{matrix} CH_3CH{=}CHCH_2OH \\ \textit{and} \\ CH_3CHOHCH{=}CH_2 \end{matrix}\right\} \underset{Ag^+,\ H_2O}{\overset{HX}{\rightleftharpoons}} \left\{\begin{matrix} CH_3CH{=}CHCH_2X \\ \textit{and} \\ CH_3CHXCH{=}CH_2 \end{matrix}\right.$$

allylic rearrangements

B. S_N2 Reactions
with or without rearrangement

C. Allylic Anions

"E^{+}" = electrophile

$$\left.\begin{matrix} CH_3CH{=}CHCH_2Br \\ \textit{and} \\ CH_3CHBrCH{=}CH_2 \end{matrix}\right\} \xrightarrow{Mg} \xrightarrow{"E^{+}"} \left\{\begin{matrix} CH_3CH{=}CHCH_2E \\ \textit{and} \\ CH_3CHECH{=}CH_2 \end{matrix}\right.$$

major product

dilution principle
conjugated carbons

D. Allylic Radicals

allylic bromination

cyclohexene + N-bromosuccinimide (N—Br, with two C=O) $\xrightarrow{h\nu}$ 3-bromocyclohexene (Br) + succinimide (NH, with two C=O)

20.2 Dienes

A. Structure and Stability

conjugated vs. unconjugated dienes
isolated double bonds

B. Addition Reactions

$$CH_2{=}CHCH{=}CH_2 + Br_2 \xrightarrow{-15^\circ} BrCH_2CH{=}CHCH_2Br\ (46\%) + CH_2{=}CHCHBrCH_2Br\ (54\%) \xrightarrow{60^\circ} (E)\text{-}BrCH_2CH{=}CHCH_2Br\ (90\%)$$

kinetic vs. thermodynamic control

C. 1,2-Dienes: Allenes

sp-hybridization of central carbon
stereoaxis
cumulated double bonds

D. Preparation of Dienes

(dehydration, Grignard coupling, Wittig reaction, etc.)

20.3 Unsaturated Carbonyl Compounds

A. Unsaturated Aldehydes and Ketones

α,β- vs. β,γ-unsaturation

$$\overset{\gamma}{CH_2}{=}\overset{\beta}{CH}{-}CH_2{-}\overset{O}{\overset{\|}{CH}} \rightleftharpoons CH_2{=}CH{-}CH{=}\overset{OH}{\overset{|}{CH}} \rightleftharpoons CH_3{-}\overset{\beta}{CH}{=}\overset{\alpha}{CH}{-}\overset{O}{\overset{\|}{CH}}$$

(unconjugated)
("move into conjugation")

(conjugated)
facilitated by acid/base catalysis → enol

formation via aldol condensation:

$$R^1\overset{O}{\overset{\|}{C}}R^2 + \underset{R^3}{\underset{|}{CH_2}}\overset{O}{\overset{\|}{C}}R^4 \xrightarrow[\text{or } OH^-]{H^+} R^1R^2C{=}\underset{R^3}{\underset{|}{C}}\overset{O}{\overset{\|}{C}}R^4$$ acid- or base-catalyzed

ease of oxidation of allylic alcohols:

$$>C{=}\underset{|}{C}{-}\overset{OH}{\overset{|}{CH}}{-} \xrightarrow{MnO_2} >C{=}\underset{|}{C}{-}\overset{O}{\overset{\|}{C}}{-}$$

1,2-additions (normal additions) vs. 1,4-additions (conjugate additions):

$$>C{=}\underset{|}{C}{-}\overset{O}{\overset{\|}{C}}{-} + HCN \longrightarrow NC{-}\underset{|}{\overset{|}{C}}{-}\underset{|}{CH}{-}\overset{O}{\overset{\|}{C}}{-}$$

cuprates (1,4) vs. organolithium (1,2) (and Grignard - but it depends) reagents:

$$R{-}\overset{|}{\underset{|}{C}}{-}\underset{|}{CH}{-}\overset{O}{\overset{\|}{C}}{-} \xleftarrow{H_2O} \xleftarrow{R_2CuLi} {>}C{=}\underset{|}{C}{-}\overset{O}{\overset{\|}{C}}{-} \xrightarrow[\text{(or RMgBr)}]{RLi} \xrightarrow{H_2O} {>}C{=}\underset{|}{C}{-}\overset{OH}{\underset{|}{C}}{-}R$$

reduction methods

$$\text{2-cyclohexen-1-ol} \xleftarrow{H_2O} \xleftarrow{LiAlH_4} \text{2-cyclohexen-1-one} \xrightarrow[Na\ /\ NH_3]{H_2\ /\ Pt\ \text{or}} \text{cyclohexanone}$$

B. Unsaturated Carboxylic Acids and Derivatives

cross-conjugated
Perkin reaction

$$RCH{=}O + (CH_3\overset{O}{\overset{\|}{C}})_2O \xrightarrow{CH_3CO_2Na} RCH{=}CHCO_2H$$

C. Ketenes

$$CH_2{=}C{=}O + HNu: \longrightarrow CH_3\overset{O}{\overset{\|}{C}}Nu$$

20.4 **Higher Conjugated Systems (trienes)**

20.5 **The Diels-Alder Reaction**

cycloaddition reaction
[4+2] vs. [2+2] or [4+4] cycloaddition

exo vs. *endo* stereochemistry
bicyclic products available

20.B. Important Reactions Introduced

NOTE: Except for the Diels-Alder reaction, most of the reactions discussed in this chapter are not "new". What is different about them is how they proceed in conjugated systems.

Diels-Alder reactions (20.5)

Equation:

$$\text{R-diene} + \text{CH}_2{=}\text{CH–Y} \longrightarrow \text{R,Y-substituted cyclohexene}$$

Generality: Y = usually an electron-withdrawing group
R = a variety of substituents, cyclic diene, etc.

Key features: cycloaddition reaction
often proceeds with specific stereochemistry, i.e. *endo* or *exo*
important reaction for forming cyclic and bicyclic compounds

20.C. Important Concepts and Hints

Conjugation, or the Double Bond Relay

You are familiar by now with many of the reactions of carbon-carbon double bonds and of carbonyl groups. When both functional groups are present in one molecule, the same reactions can usually be observed. However, when the *p*-orbitals of a double bond overlap with those of an adjacent double bond or carbonyl group, special chemical behavior is often seen. Unusual reactivity is also observed if an intermediate or transition state involves the formation of an sp^2-hybridized carbon adjacent to a double bond (allylic system). All of this comes under the heading of conjugation. One of the ways to understand conjugation intuitively is to think of it in the following way: any chemical behavior which involves an sp^2-hybridized carbon can be relayed two carbons away by an adjacent double bond. The following summary illustrates this point:

S_N1 Substitution (carbocation intermediate)

alkyl system: C–X —($-X^-$)→ [C^+] —(Nu^-)→ C–Nu

allylic system: C=C–C(X) —($-X^-$)→ [C=C–C^+ ↔ C^+–C=C] —(Nu^-)→ C=C–C(Nu) [S_N1] ; (Nu)C–C=C [S_N1']

S_N2 Substitution

alkyl system: Nu^- + C–X → [$\delta-$ Nu-----C-----X $\delta-$] → Nu–C X^- [S_N2]

allylic system: Nu^- + C=C–C(X) → Nu–C–C=C + X^- [S_N2']

(also S_N2 without allylic rearrangement)

Grignard Reaction (carbanion intermediate)

alkyl system: —C—X →(Mg) —C—MgX →(E^+) —C—E "E^{+}" = electrophile

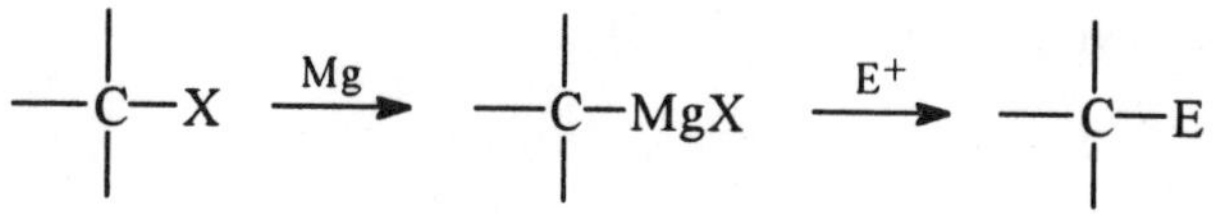

allylic system:

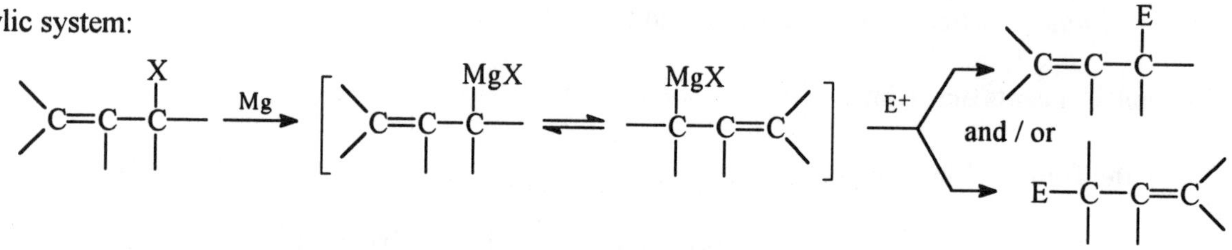

Free-Radical Halogenation(radical intermediate)

alkyl system: —C—H →(X·) —C· + HX →(X_2) —C—X

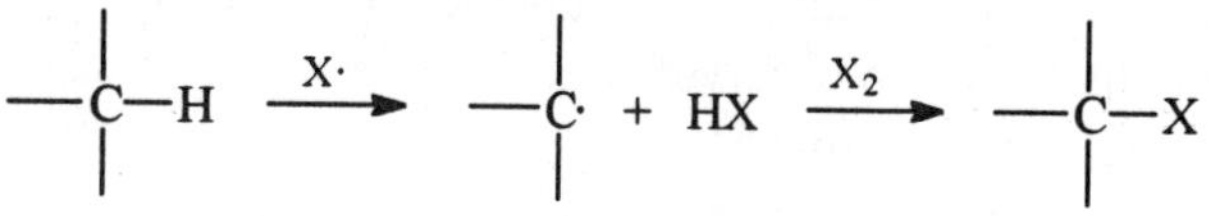

allylic system:

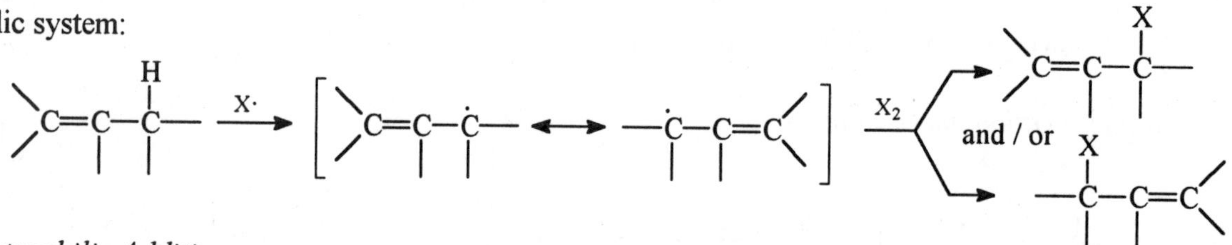

Electrophilic Addition

isolated double bond: C=C →(X_2, H_2O) HO—C—C—X

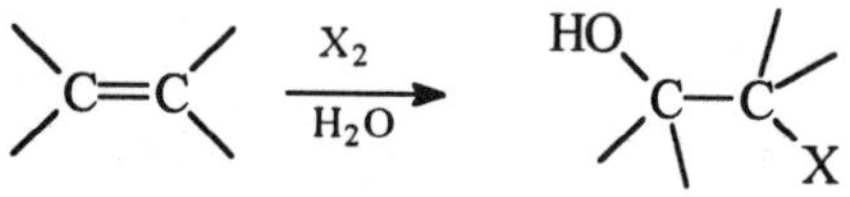

conjugated diene: C=C—C=C →(X_2, H_2O) HO—C—C=C—C—X (normal addition can also occur)

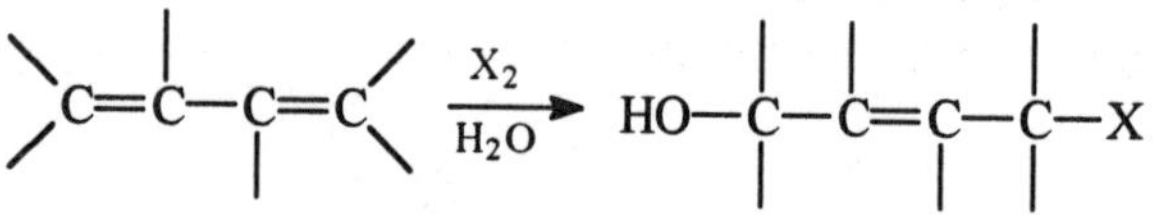

Ketone Enolization

isolated carbonyl group:

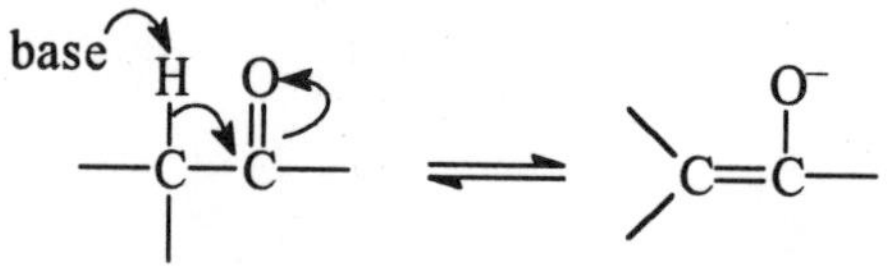

α,β-unsaturated carbonyl group:

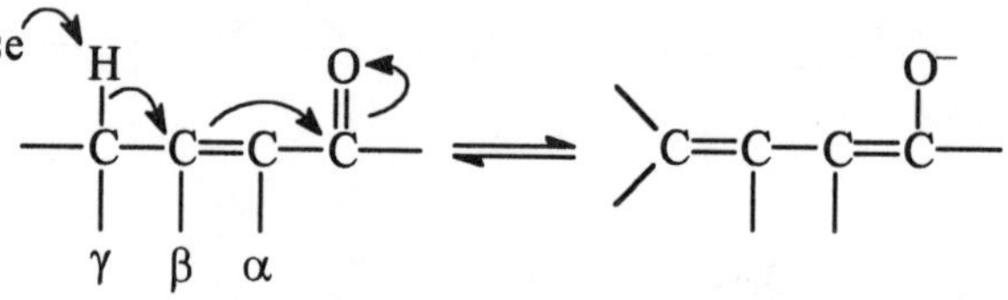

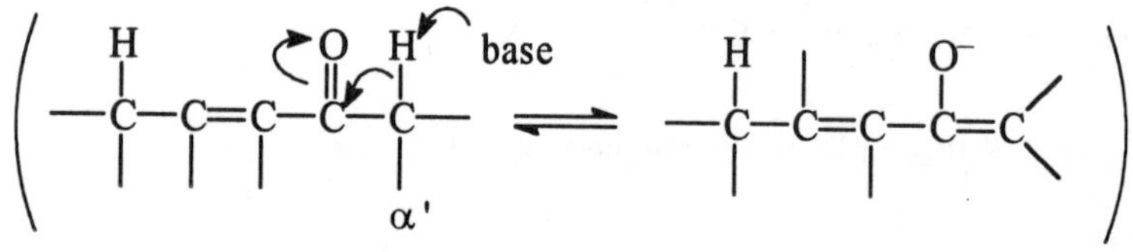

γ β α

(... α' ...) can still occur

Addition to a Carbonyl

isolated carbonyl group: (1,2-addition)

α,β-unsaturated carbonyl group:

(1,4-addition; 1,2-addition can still occur)

Notice how in each case a reaction that can occur at one carbon atom can take place at the other end of a double bond that is conjugated to it. This does not mean that **<u>all</u>** reactions of allylic systems involve rearrangement, or that all additions to α,β-unsaturated carbonyl compounds are 1,4-; the "normal" modes of reaction are observed as well. Often by choosing specific reaction conditions or reagents, you can favor one over the other.

The Diels-Alder Reaction, or Electrons-going-around-in-a-circle

In contrast to the Diels-Alder reaction, reactions that look similar but involve four or eight electrons in a circle (rather than six) occur only in exceptional circumstances:

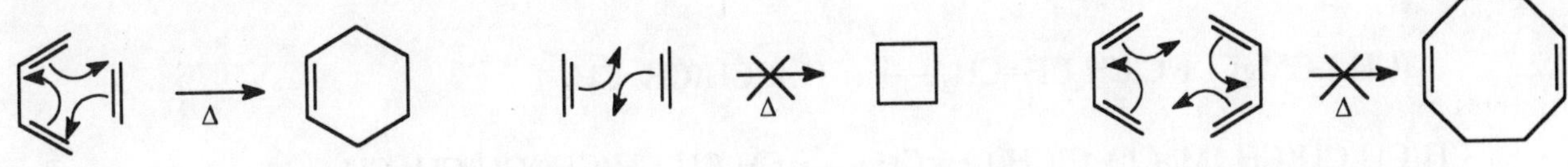

You will encounter additional cases such as this one, in which systems involving six electrons in a cyclic arrangement (in π-bonds (for example, benzene) or in transition states (the pyrolysis illustrated below is an example)) are favored relative to the analogous four-or eight-electron systems. A unifying explanation for this phenomenon is given in Chapter 21.

The six-electron pyrolysis shown below occurs readily; the analogous four-electron transformation does not occur.

ROH +

20.D. Answers to Exercises

20.1

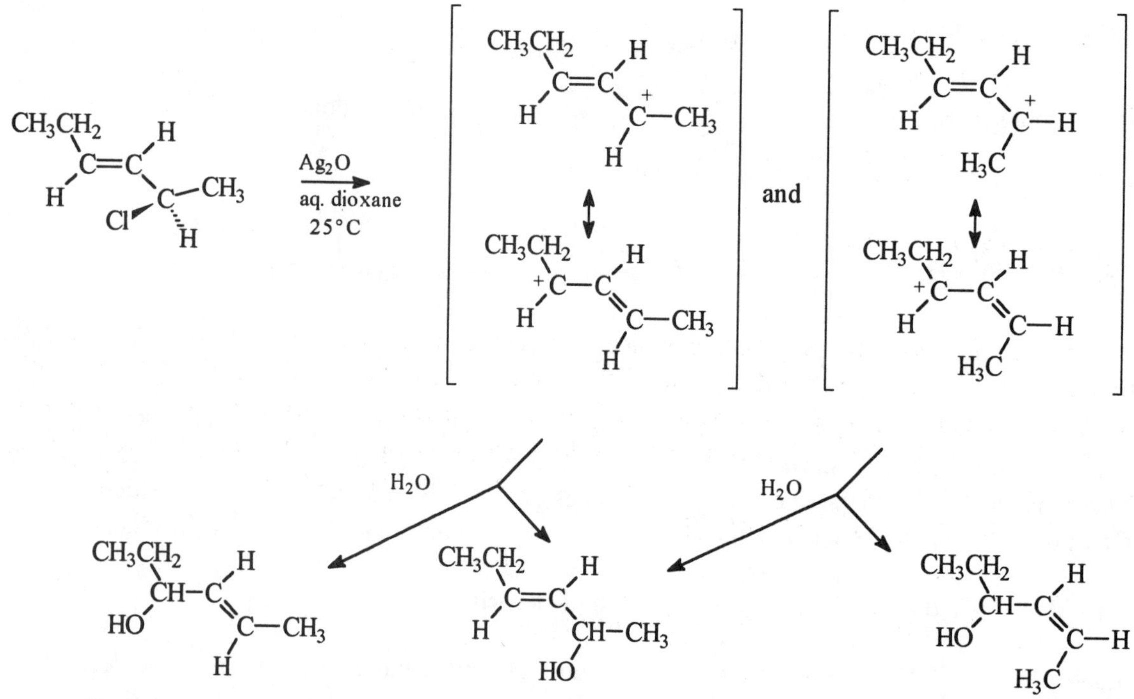

$CH_3CH_2(H)C{=}C(H)\text{–}CH(Cl)CH_3$ $\xrightarrow[\text{aq. dioxane, 25°C}]{Ag_2O}$ [$CH_3CH_2(H)C{=}C(H)\text{–}\overset{+}{C}H\text{–}CH_3$ ↔ $CH_3CH_2(H)\overset{+}{C}\text{–}C(H){=}CH\text{–}CH_3$] and [$CH_3CH_2(H)C{=}C(H)\text{–}\overset{+}{C}H\text{–}H$ (H_3C) ↔ $CH_3CH_2(H)\overset{+}{C}\text{–}C(H){=}C(H)(H_3C)$]

H_2O → $CH_3CH_2CH(HO)\text{–}C(H){=}C(H)\text{–}CH_3$ + $CH_3CH_2(H)C{=}C(H)\text{–}CH(HO)\text{–}CH_3$ ← H_2O → $CH_3CH_2CH(HO)\text{–}C(H){=}C(H)(H_3C)$

Each isomer will result in a racemic mixture.

20.2 $(CH_3)_3CMgCl + ClCH_2CH{=}CH_2 \longrightarrow (CH_3)CCH_2CH{=}CH_2$

$CH_3CH_2CH(CH_3)MgCl + ClCH_2CH{=}CH_2 \longrightarrow CH_3CH_2CH(CH_3)CH_2CH{=}CH_2$

20.3 $CH_3CH_2CH(CO_2H)CH{=}CH_2$ $\quad$ $CH_3CH_2(H)C{=}C(H)CH_2CO_2H$ $\quad$ $CH_3CH_2(H)C{=}C(H)CH_2CO_2H$

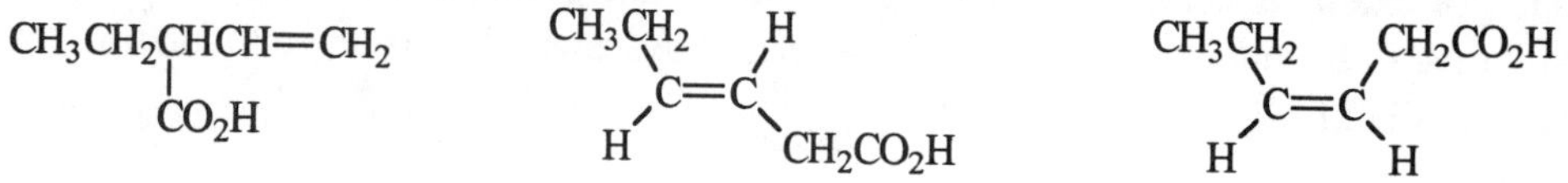

20.4 $R\text{–}C({=}\ddot{N}H)\text{–}\ddot{N}H\text{–}H$ + H^+ ⇌ $R\text{–}C({=}\overset{+}{N}H_2)\text{–}\ddot{N}H_2$ ⟷ $R\text{–}\overset{+}{C}(\ddot{N}H_2)(\ddot{N}H_2)$ ⟷ $R\text{–}C(\ddot{N}H_2){=}\overset{+}{N}H_2$

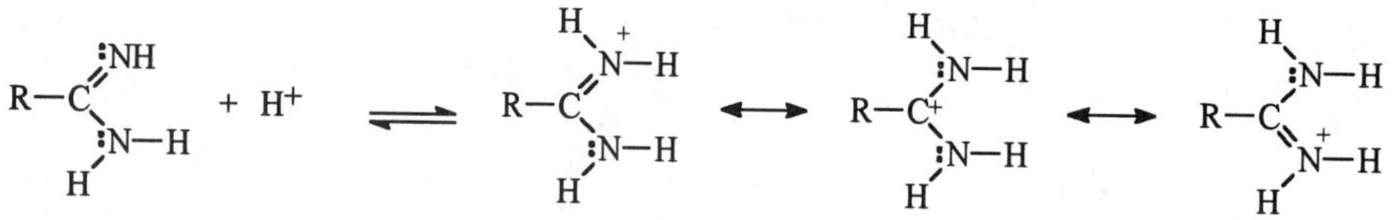

Since imines are more basic than ketones (pK_a of immonium ion ≈ 9, pK_a of protonated acetone ≈ -7), amidines are expected to be more basic than amides. In fact they are: pK_a of protonated acetamidine = 12.5; pK_a of protonated acetamide = 0.

20.5 From the theory of absolute rates (Section 4.4 in the Text): $k = \nu^{\ddagger}e^{-\Delta G^{\ddagger}/RT}$

Assuming that the proportionality constant, $\nu^{\ddagger}$, is the same for both reactions, the ratio of the two rate constants will be given by:

$$\frac{k_A}{k_B} = \frac{\nu^{\ddagger} e^{-\Delta G_A^{\ddagger}/RT}}{\nu^{\ddagger} e^{-\Delta G_B^{\ddagger}/RT}} = e^{-(\Delta G_A^{\ddagger} - \Delta G_B^{\ddagger})/RT} = e^{-\Delta\Delta G^{\ddagger}/RT} = e^{-(\Delta\Delta H^{\ddagger} - T\Delta\Delta S^{\ddagger})/RT}$$

For $\Delta\Delta H^{\ddagger} = 10$ kcal mole^{-1} and $\Delta\Delta S^{\ddagger} = 20$ e.u.,

at 300K = $k_A/k_B = 1.2 \times 10^{-3}$
at 700K = $k_A/k_B = 18$

This exercise serves to emphasize the importance of entropy at higher temperatures. At 300K, the **enthalpy** of activation, $\Delta H^{\ddagger}$, is the major contribution to the free energy of activation ($\Delta G^{\ddagger} = \Delta H^{\ddagger} - T\Delta S^{\ddagger}$). Since $\Delta H^{\ddagger}$ is more unfavorable for A ($\Delta\Delta H^{\ddagger} = +10$ kcal mole^{-1}), the reaction of A proceeds more slowly than that of B at the lower temperature. At 700K, the **entropy** of activation, which is multiplied by the temperature ($T\Delta S^{\ddagger}$) becomes dominant. Since the entropy of activation is more favorable for A ($\Delta\Delta S^{\ddagger} = +20$ e.u.), the reaction of A proceeds faster at the higher temperatures.

20.6 $BrCH_2CH(OH)CH{=}CH_2 \xrightarrow{NaOH} CH_2{-}CHCH{=}CH_2$ (epoxide, O bridging CH_2 and CH)

(cis) $BrCH_2CH{=}CHCH_2OH \xrightarrow{NaOH}$ 2,5-dihydrofuran ; (trans) $BrCH_2CH{=}CHCH_2OH \xrightarrow{NaOH} HOCH_2CH{=}CHCH_2OH$ (trans)

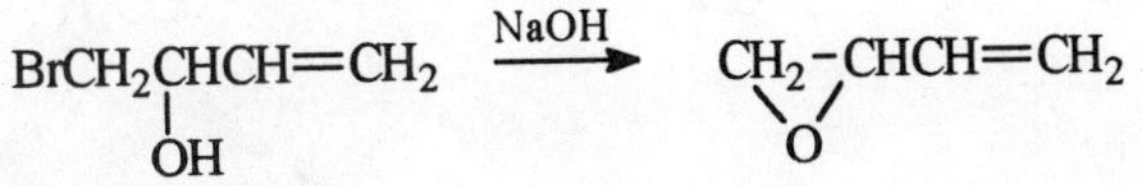

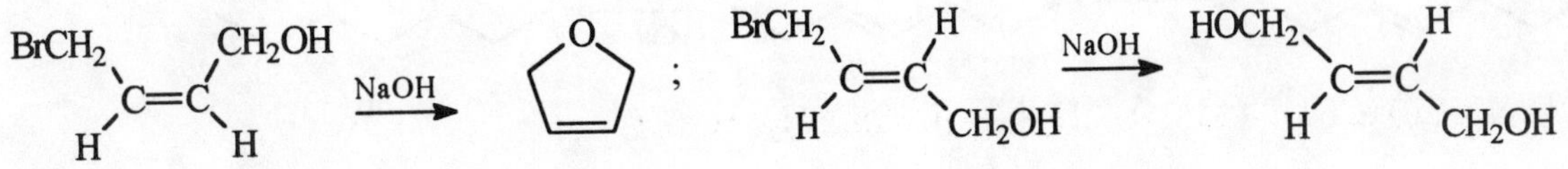

20.7 and 20.8 These are model-building exercises

20.9 a. 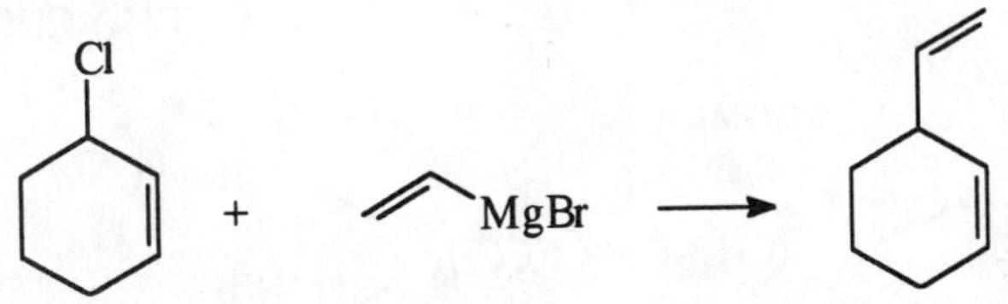 3-chlorocyclohexene + $CH_2{=}CHMgBr$ ⟶ 3-vinylcyclohexene

b. 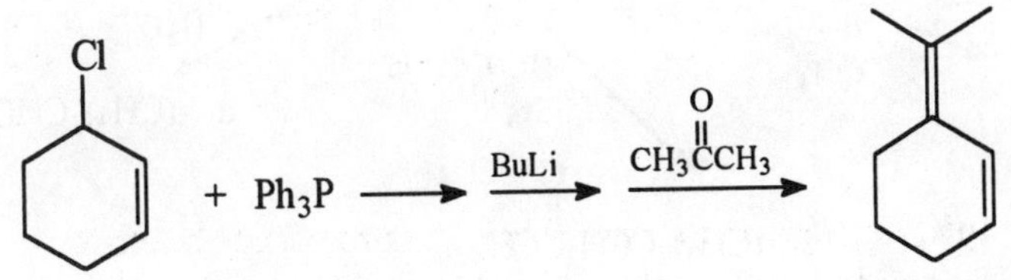 3-chlorocyclohexene + Ph_3P ⟶ $\xrightarrow{BuLi}$ $\xrightarrow{CH_3\overset{O}{\overset{\|}{C}}CH_3}$ 3-isopropylidenecyclohexene

20.10 (structures of deuterated cyclohexenones: D at C-2, C-4 (D_2) and C-6 (D_2); D at C-2, C-4 (D_2), C-6 (D, CH_3); H_3C at C-2, D at C-4 (D_2) and C-6 (D_2); D at C-2, D_3C at C-3, D at C-4 (D_2) and C-6 (D_2))

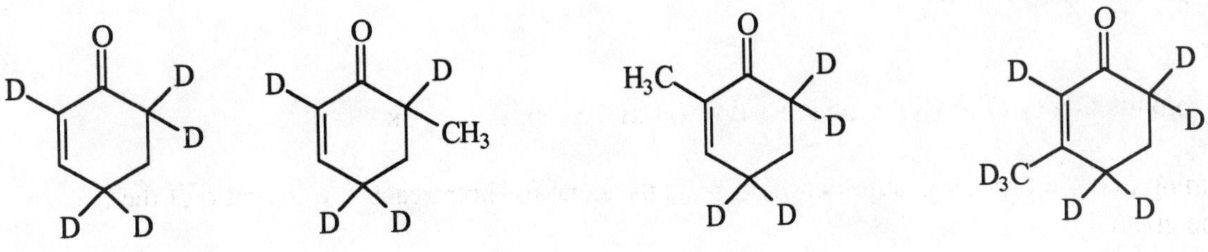

20.11

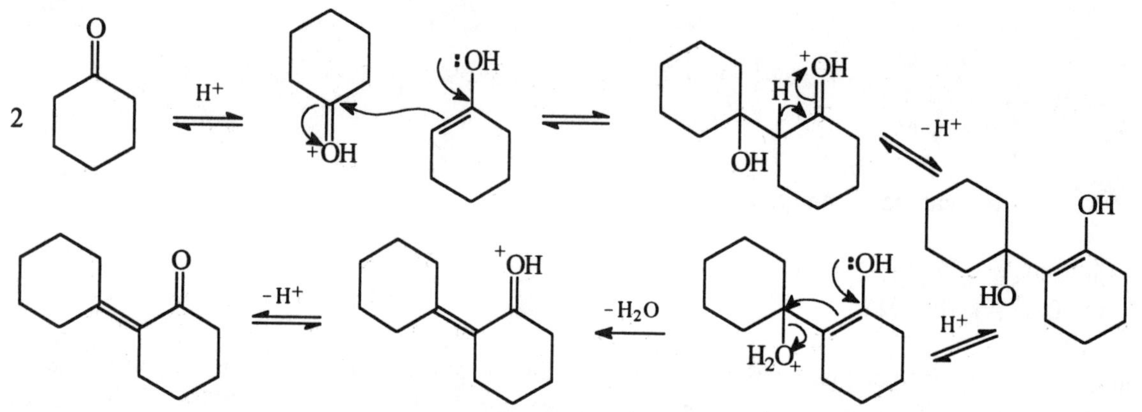

2 cyclohexanone $\xrightarrow{H^+}$ ... $\xrightarrow{-H^+}$ ... $\xrightarrow{H^+}$ ... $\xrightarrow{-H_2O}$ ... $\xrightarrow{-H^+}$

20.12

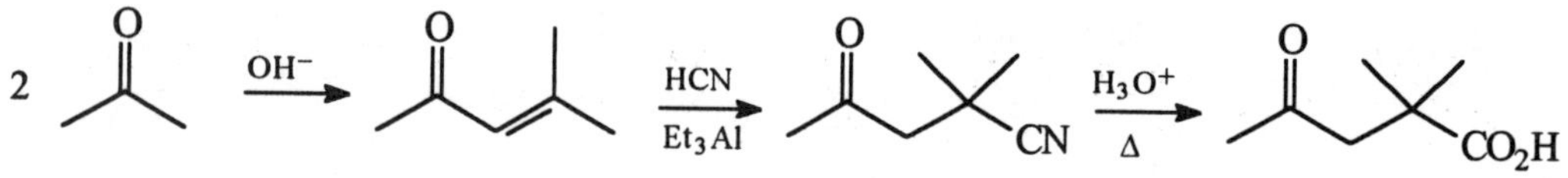

2 acetone $\xrightarrow{OH^-}$ $\xrightarrow[Et_3Al]{HCN}$ CN $\xrightarrow[\Delta]{H_3O^+}$ CO_2H

20.13

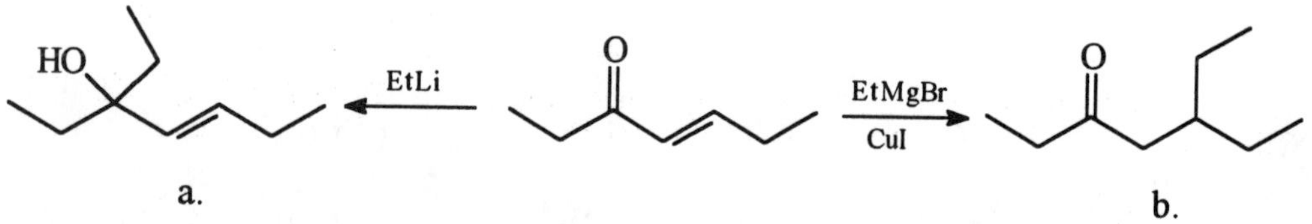

a. $\xleftarrow{EtLi}$ $\xrightarrow[CuI]{EtMgBr}$ b.

20.14

$(CH_3)_2C{=}CHCCH_3$ (C=O)

a. $(CH_3)_2C{=}CHC(OH)(CH_3)CH_2CH_2CH_2CH_3$ — BuLi; H_2O

b. $CH_3CH_2CH_2CH_2C(CH_3)_2CH_2CCH_3$ (C=O) — C_4H_9MgBr, CuBr; H_2O

c. $(CH_3)_2CHCH_2CCH_3$ (C=O) — H_2 / Pt

d. $(CH_3)_2CHCH_2CCH_3$ (C=O) — Li / NH_3; H_2O

e. $(CH_3)_2C(CN)CH_2CCH_3$ (C=O) — HCN, Et_3Al

f. $(CH_3)_2C(Br)CH(Br)CCH_3$ (C=O) — Br_2, CCl_4

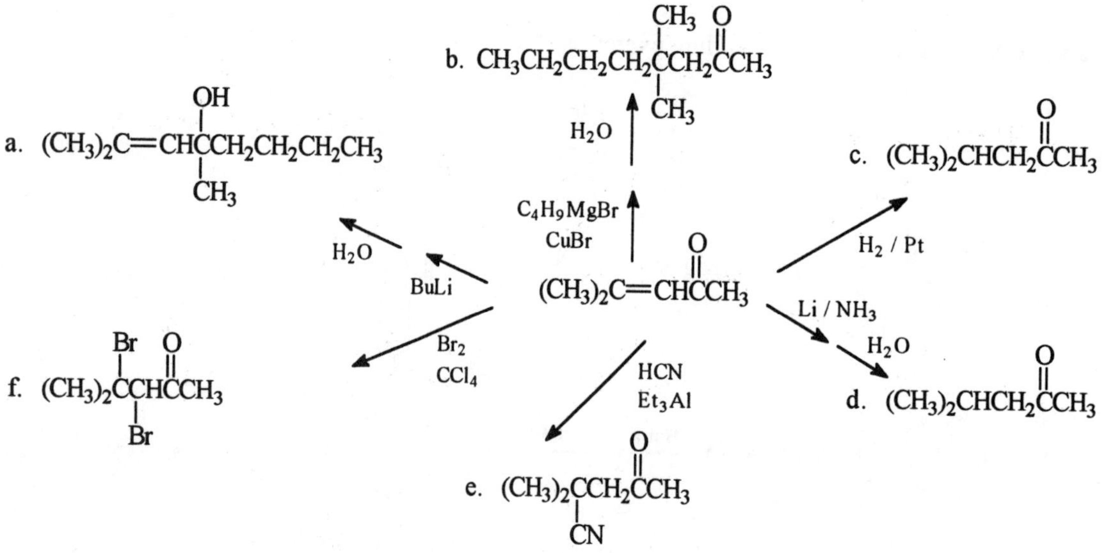

20.15 Ethanol:

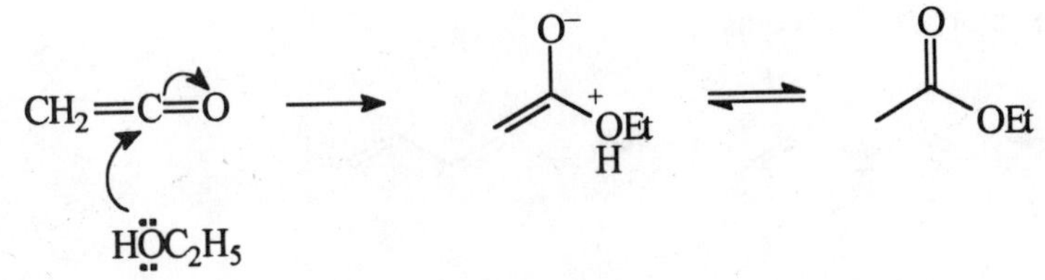

$CH_2{=}C{=}O$ + $H\ddot{O}C_2H_5$ → (O⁻, OEt⁺H) ⇌ (O, OEt)

Acetic Acid:

$CH_2{=}C{=}O + HO_2CCH_3 \rightleftharpoons CH_2{=}C{=}\overset{+}{O}H$ + $^{-}O_2CCH_3$ → (OH, O, O) ⇌ (O, O, O)

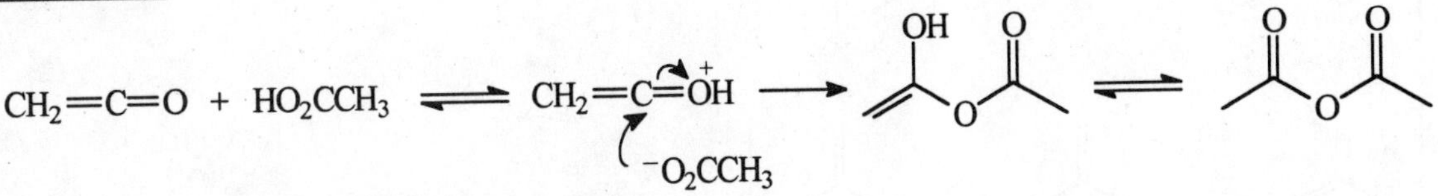

20.16 Not including *cis* and *trans* isomers:

Br^+ $CH_2{=}CH{-}CH{=}CH{-}CH{=}CH_2$ ⟶ $BrCH_2{-}\overset{+\delta}{C}H{-}CH{-}\overset{+\delta}{C}H{-}CH{-}\overset{+\delta}{C}H_2$

$BrCH_2CH{=}CHCH{=}CHCH_2Br$ + $BrCH_2CH{=}CHCH(Br)CH{=}CH_2$ + $BrCH_2CH(Br)CH{=}CHCH{=}CH_2$

↑

This isomer is expected to be the major product at equilibrium, because it is conjugated and has the more highly substituted double bonds.

20.17 a. bicyclo[2.2.2]oct-5-en-2-one
b. 7,7-dichlorobicyclo[3.2.0]hept-3-en-6-one

20.18

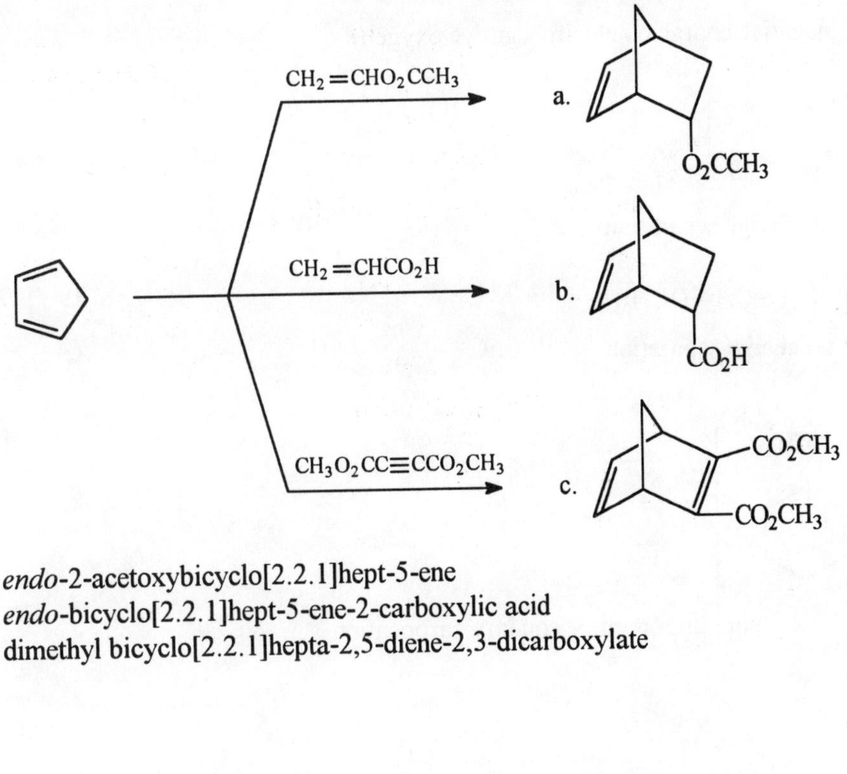

Cyclopentadiene + $CH_2{=}CHO_2CCH_3$ → a. (O_2CCH_3)

Cyclopentadiene + $CH_2{=}CHCO_2H$ → b. (CO_2H)

Cyclopentadiene + $CH_3O_2CC{\equiv}CCO_2CH_3$ → c. (CO_2CH_3, CO_2CH_3)

a. *endo*-2-acetoxybicyclo[2.2.1]hept-5-ene
b. *endo*-bicyclo[2.2.1]hept-5-ene-2-carboxylic acid
c. dimethyl bicyclo[2.2.1]hepta-2,5-diene-2,3-dicarboxylate

20.E Answers and Explanations for Problems

1. a. 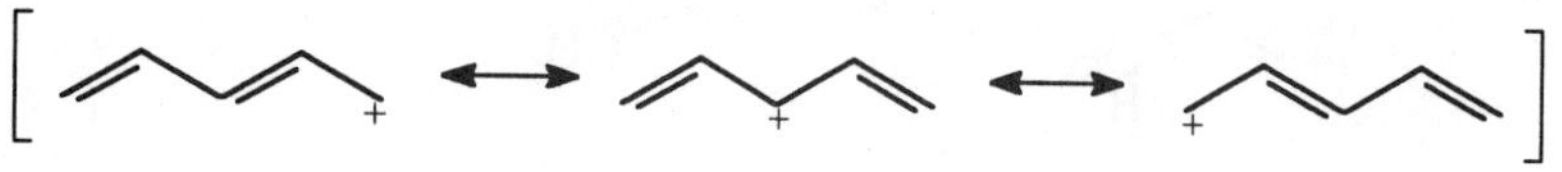 [structures with + charges ↔ ↔]

b.  [structures ↔]

c. 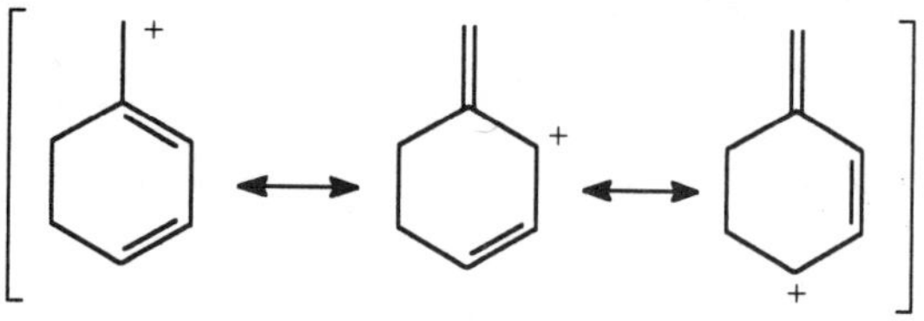 [structures ↔ ↔]

2. a. [$^+CH_2-CH=CH-CH_3$ ↔ $CH_2=CH-{}^+CH-CH_3$]
↑ The secondary carbocation is more important.

b. [$CF_3{}^-CH-CH=CH-CH_3$ ↔ $CH_3CH=CH-{}^-CHCH_3$]
↑ more important; inductive effect of CF_3

c. [$CF_3{}^+CH-CH=CHCH_3$ ↔ $CF_3CH=CH-{}^+CHCH_3$]
↑ more important (CF_3 destabilizes cation)

d. [$CH_2=CH-O^-$ ↔ $^-CH_2-CH=O$]
↑ more important; negative charge on electronegative oxygen

e. [$CH_3C(=O)NH_2$ ↔ $CH_3C(O^-)=N^+H_2$]
↑ more important; no charge separation

f. [$CH_2=CH-OCH_3$ ↔ $^-CH_2-CH={}^+OCH_3$]
↑ more important; no charge separation

g. 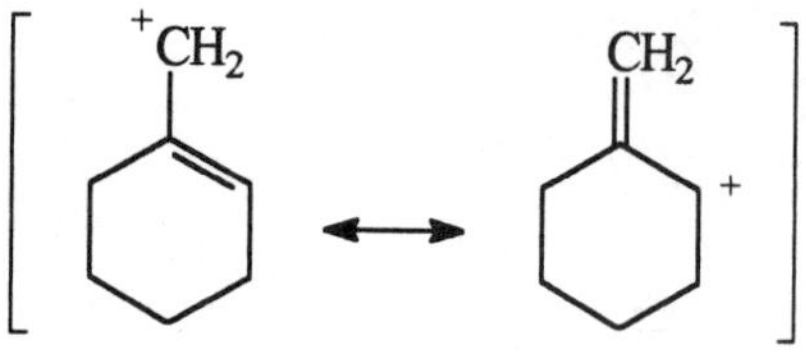 [$^+CH_2$ (on cyclohexene) ↔ CH_2 (= on cyclohexane ring, +)]
↑ more important; secondary carbocation

3. a. $CH_3CH_2MgBr + BrCH_2CH{=}CH_2 \longrightarrow CH_3CH_2CH_2CH{=}CH_2$

b. 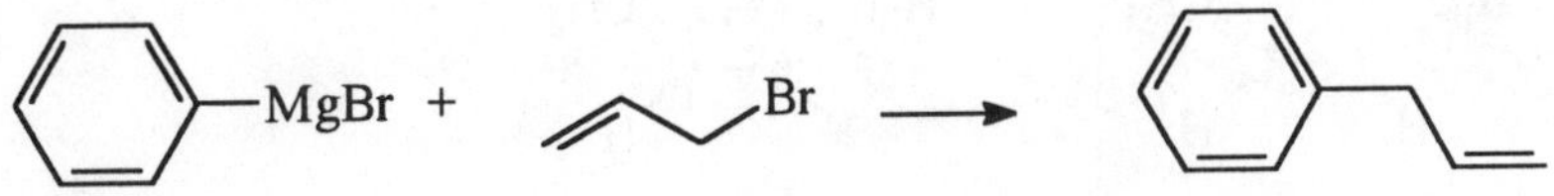 —MgBr + Br ⟶

c. $CH_2{=}CH{-}MgBr + BrCH_2CH{=}CH_2 \longrightarrow CH_2{=}CH{-}CH_2{-}CH{=}CH_2$

d. $(CH_3)_3CMgBr + BrCH_2CH{=}CH_2 \longrightarrow (CH_3)_3CCH_2CH{=}CH_2$

4. a. $CH_3CH{=}CHCH_2OH \xrightarrow{MnO_2} CH_3CH{=}CHCHO$

b. $CH_3CH_2CH_2CH_2OH \xrightarrow{MnO_2}$ no reaction

c. $HOCH_2CH_2CH{=}CHCH_2OH \xrightarrow{MnO_2} HOCH_2CH_2CH{=}CHCHO$

d. 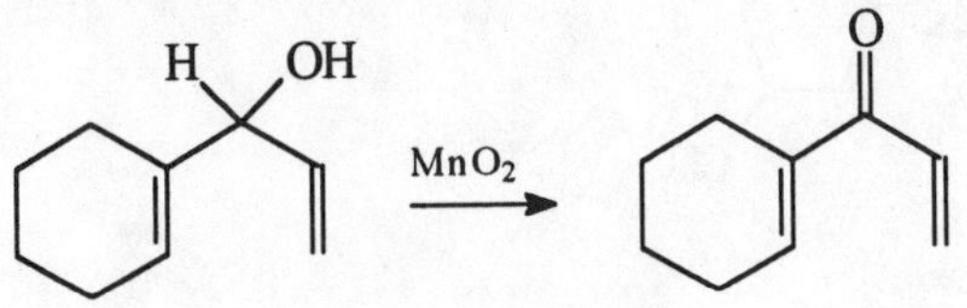 H OH $\xrightarrow{MnO_2}$ O

e. 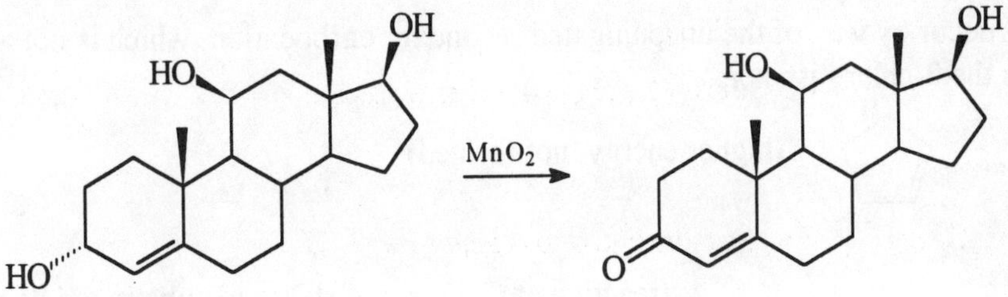 HO OH HO $\xrightarrow{MnO_2}$ HO OH O

f. $CH_3C{\equiv}CCH(OH)CH_3 \xrightarrow{MnO_2} CH_3C{\equiv}CC({=}O)CH_3$

5. Abstraction of the allylic H gives the allyl free radical

$$C_5H_{11}CH_2CH{=}CH_2 + Br \longrightarrow C_5H_{11}\dot{C}HCH{=}CH_2 + HBr$$

There are two isomeric allylic radicals:

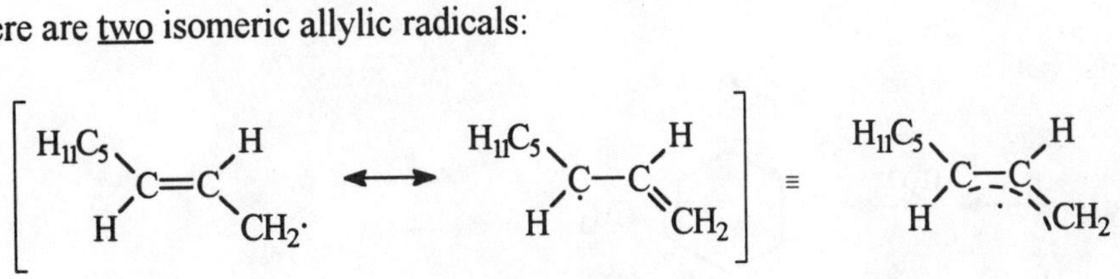

$$\left[\begin{matrix} H_{11}C_5 & & H \\ & C{=}C & \\ H & & CH_2\cdot \end{matrix} \longleftrightarrow \begin{matrix} H_{11}C_5 & & H \\ & \dot{C}{-}C & \\ H & & CH_2 \end{matrix} \right] \equiv \begin{matrix} H_{11}C_5 & & H \\ & C{\cdots}C & \\ H & & CH_2 \end{matrix}$$

allylic radical A, *transoid*

$$\left[\begin{matrix} H_{11}C_5 & & CH_2\cdot \\ & C{=}C & \\ H & & H \end{matrix} \longleftrightarrow \begin{matrix} H_{11}C_5 & & CH_2 \\ & \dot{C}{-}C & \\ H & & H \end{matrix} \right] \equiv \begin{matrix} H_{11}C_5 & & CH_2 \\ & C{\cdots}C & \\ H & & H \end{matrix}$$

allylic radical B, *cisoid*

If either A or B reacts with Br_2 at C-3, the same product is produced:

$$\text{A or B} + Br_2 \longrightarrow C_5H_{11}\underset{}{\overset{Br}{\overset{|}{C}}}HCH{=}CH_2 + Br\cdot$$

However, if reaction with Br_2 occurs at C-1, then the **transoid** allylic radical A gives *trans*-1-bromo-2-octene, whereas the **cisoid** allylic radical B gives *cis*-1-bromo-2-octene. The barrier to interconversion of these two isomeric radicals is about 10 kcal $mole^{-1}$, much higher than the activation energy for reaction of either with Br_2 (see Chapter 6).

6. (cyclopentadiene) + HCl ⟶ [(cyclopentenyl cation, + at C-3) ⟷ (cyclopentenyl cation, + at C-1)] $\xrightarrow{Cl^-}$ 3-chlorocyclopentene

allylic cation

The 4-chloro isomer can only occur by way of the unconjugated secondary carbocation, which is not as stable as the allylic cation that gives the 3-chloro isomer.

i.e., (cyclopentadiene) $\overset{H^+}{\not\longrightarrow}$ (cyclopent-3-enyl cation) (higher energy, not formed)

7.

$$CH_2{=}CH{-}CH{=}CH_2 \xrightarrow{H^+} \left[CH_2{=}CH\overset{+}{C}HCH_3 \longleftrightarrow {}^{+}CH_2CH{=}CHCH_3 \right] \xrightarrow{Cl^-}$$

(reaction at C-1) more stable (disubstituted double bond) $ClCH_2CH{=}CHCH_3$, 22%

(reaction at C-3) less stable (monosubstituted double bond) $CH_2{=}CHCHClCH_3$, 78%

Kinetic control: reaction occurs faster at site of the greater positive charge (secondary)

If equilibration is allowed to occur, the *thermodynamic mixture* is produced. Equilibration occurs by ionization back to the carbocation.

8. a. (3-methyl-2-cyclohexenone, CH_3) + $(CH_3)_2CuLi$ ⟶ $\xrightarrow{H_3O^+}$ (3,3-dimethylcyclohexanone, CH_3, CH_3)

b. 3-methylcyclohex-2-enone + CH_3Li ⟶ $\xrightarrow{H_3O^+}$ 1,3-dimethylcyclohex-2-en-1-ol (HO, CH_3; CH_3)

c. 5,5-dimethylcyclohex-2-enone (H_3C, H_3C) $\xrightarrow[\text{or Li / NH}_3]{H_2\ \ Pd/C}$ 3,3-dimethylcyclohexanone (H_3C, H_3C)

d. 5,5-dimethylcyclohex-2-enone (H_3C, H_3C) $\xrightarrow{LiAlH_4}$ $\xrightarrow{H_2O}$ 5,5-dimethylcyclohex-2-en-1-ol (H, OH; H_3C, H_3C)

e. 3-methylcyclohex-2-enone (CH_3) + HCN $\xrightarrow{Et_3Al}$ 3-cyano-3-methylcyclohexanone (CH_3, CN) $\xrightarrow[\Delta]{H_3O^+}$ 3-methyl-3-carboxycyclohexanone (CH_3, CO_2H)

9. a. $Ba^{14}CO_3 + H_2SO_4 \longrightarrow BaSO_4 + H_2O + {}^{14}CO_2$

$CH_2{=}CHBr + Mg \xrightarrow{ether} CH_2{=}CHMgBr \xrightarrow{^{14}CO_2} \xrightarrow{H_3O^+} CH_2{=}CH^{14}CO_2H \xrightarrow{LiAlH_4} \xrightarrow{H_2O} CH_2{=}CH^{14}CH_2OH$

$CH_2{=}CH^{14}CH_2OH \xrightarrow{SOCl_2}$ $\left.\begin{array}{c} CH_2{=}CH^{14}CH_2Cl \\ \textbf{or} \\ {}^{14}CH_2{=}CHCH_2Cl \end{array}\right\}$ $\xrightarrow{KMnO_4} CO_2 + ClCH_2CO_2H$

Collect and count the CO_2. If rearrangement is complete, the CO_2 will have 100% of the ^{14}C. If there is no rearrangement, it will have 0% of the ^{14}C.

b. $CH_2{=}CHCO_2CH_3 + LiAlD_4 \longrightarrow \xrightarrow{H_2O} CH_2{=}CHCD_2OH$

NMR will show only the vinyl H's and the OH. Recall that δ(OH) is concentration-dependent.

$CH_2{=}CHCD_2OH \xrightarrow{SOCl_2} CH_2{=}CHCD_2Cl + CD_2{=}CHCH_2Cl$
not rearranged rearranged

The NMR spectrum will show a mixture of two chlorides. One component shows only vinyl H's as a complex multiplet, while the other shows a vinyl H (triplet) and $-CH_2Cl$ (doublet) with an area ratio of 1:2. The $-CH_2Cl$ group will appear at about δ=4 (3 for $-CH_2Cl$ plus 1 for allylic)

10.

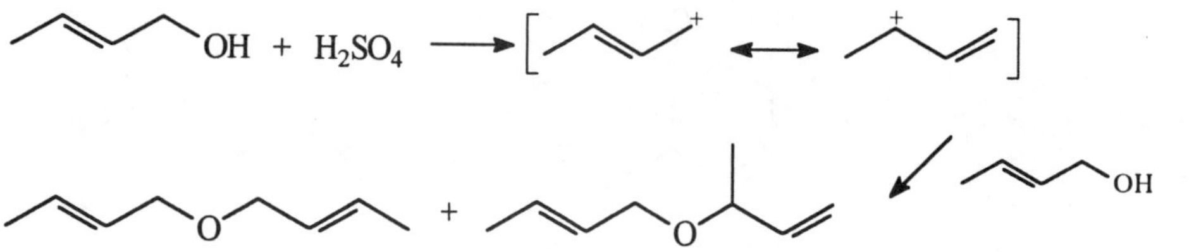

OH + H_2SO_4 ⟶ [+ ⟷ +]

OH

O + O

Of course the allylic cation can also react with water to give the isomeric alcohol:

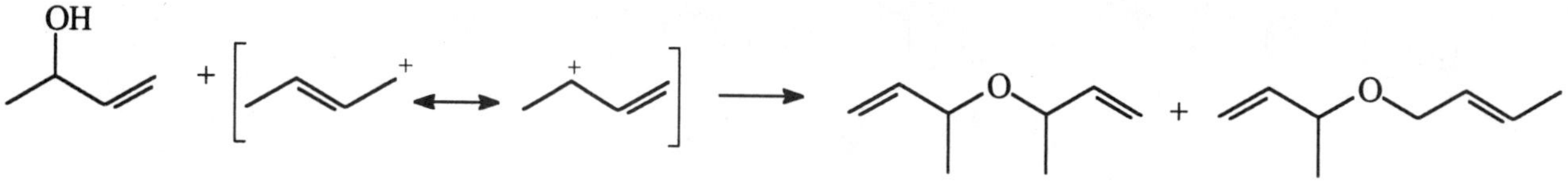

OH

+ [+ ⟷ +] ⟶ O + O

11. a. $CH_3CH_2CH_2C{\equiv}CH \rightleftharpoons CH_3CH_2C{\equiv}CCH_3 \rightleftharpoons CH_3CH_2CH{=}C{=}CH_2$ (175°C; 448K)

A, 1.3% B, 95.2% C, 3.5%

Take B, the most stable, as the point of reference. The equilibrium constant for the equilibrium between B and A is:

$$K = [A]/[B] = 1.3/95.2 = 0.0137$$
$$\Delta G° = -RT \ln K = -1.987 \text{ x } 448 \text{ x } (-4.29) \text{ cal mole}^{-1}$$
$$\Delta G° = +3.82 \text{ kcal mole}^{-1}$$

Similarly, the equilibrium constant for $B \rightleftharpoons C$ is:

$$K = [C]/[B] = 3.5/95.2 = 0.0368$$
$$\Delta G° = -1.987 \text{ x } 448 \text{ x } (-3.30) \text{ cal mole}^{-1}$$
$$\Delta G° = +2.94 \text{ kcal mole}^{-1}$$

Note that allene C is more stable than the 1-alkyne and less stable than the internal alkyne.

Mechanism:

$RCH_2C{\equiv}CH + OH^- \rightleftharpoons [R\ddot{C}^-HC{\equiv}CH \longleftrightarrow RCH{=}C{=}\ddot{C}^-H]$

$RCH{=}C{=}CH^- + H_2O \rightleftharpoons RCH{=}C{=}CH_2 + OH^-$

$RCH{=}C{=}CH_2 + OH^- \rightleftharpoons [R\ddot{C}^-{=}C{=}CH_2 \longleftrightarrow RC{\equiv}C\ddot{C}^-H_2]$

$RC{\equiv}CCH_2^- + H_2O \rightleftharpoons RC{\equiv}CCH_3 + OH^-$

b. In the case of $Na^+NH_2^-$, the amide ion is so basic that it converts the terminal alkyne completely to the carbanion, thus shifting the equilibrium quantitatively in this direction.

$$RCH_2C{\equiv}CH + NH_2^- \rightleftarrows RCH_2C{\equiv}C^- + NH_3 \quad K \approx 10^{10}$$

$pK_a = 25$ $pK_a = 34$

12.

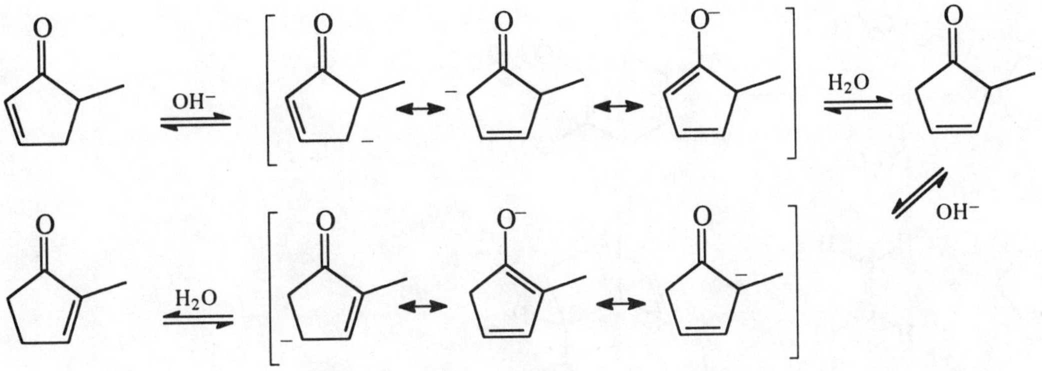

OH^- ⇌ [↔ ↔] $\overset{H_2O}{\rightleftharpoons}$

OH^-

$\overset{H_2O}{\rightleftharpoons}$ [↔ ↔]

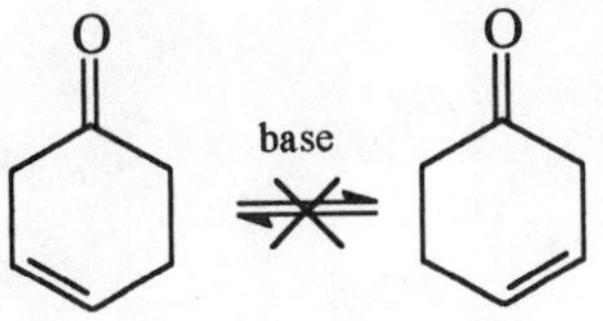

base

The β,γ- form is common to two α,β- forms in this case. Note that the analogous transformation is not possible for a cyclohexenone:

13. $H_3C(H)C{=}C(H)CH_2Cl \longrightarrow Cl^- + [H_3C(H)C{=}C(H)\overset{+}{C}H_2 \longleftrightarrow H_3C(H)\overset{+}{C}{-}C(H){=}CH_2]$

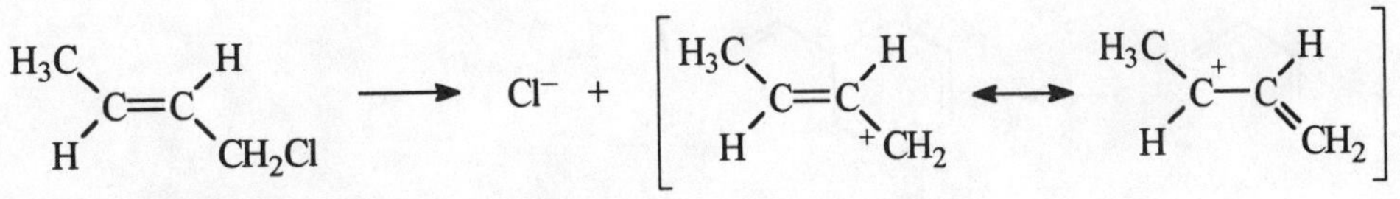

$\Delta H° = +161$ kcal mole^{-1}

$CH_2{=}C(CH_3)CH_2Cl \longrightarrow Cl^- + [CH_2{=}C(CH_3)\overset{+}{C}H_2 \longleftrightarrow \overset{+}{C}H_2C(CH_3){=}CH_2]$

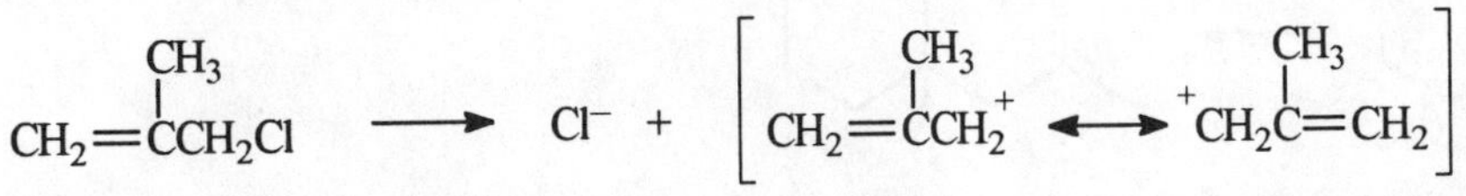

$\Delta H° = +169$ kcal mole^{-1}

The allylic cation from 1-chloro-2-butene has the dual character of primary and secondary carbocations. That from 3-chloro-2-methyl-1-propene is primary-primary.

14. a. + $CH_3O_2C{-}C{\equiv}C{-}CO_2CH_3 \longrightarrow$ CO_2CH_3, CO_2CH_3

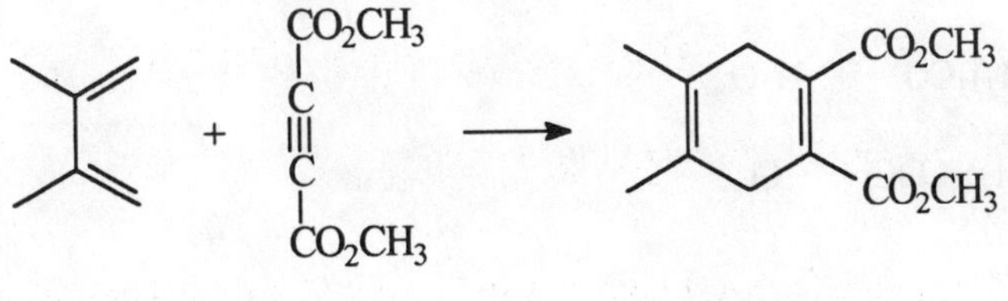

b. + $CH_2{=}CH{-}CN \longrightarrow$ CN

c. 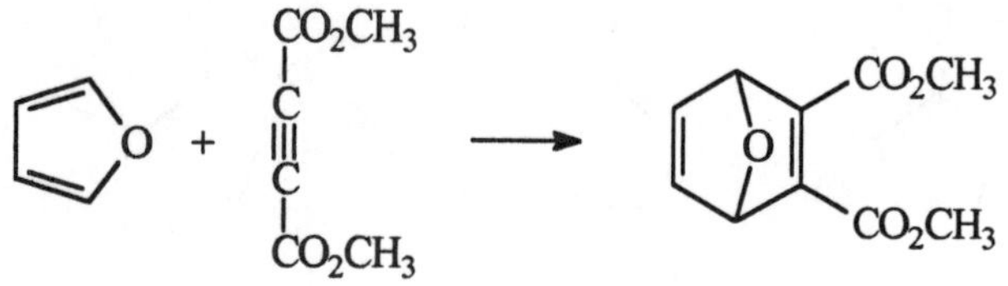 furan + CO_2CH_3–C≡C–CO_2CH_3 ⟶ 7-oxabicyclo adduct with two CO_2CH_3 groups

d. 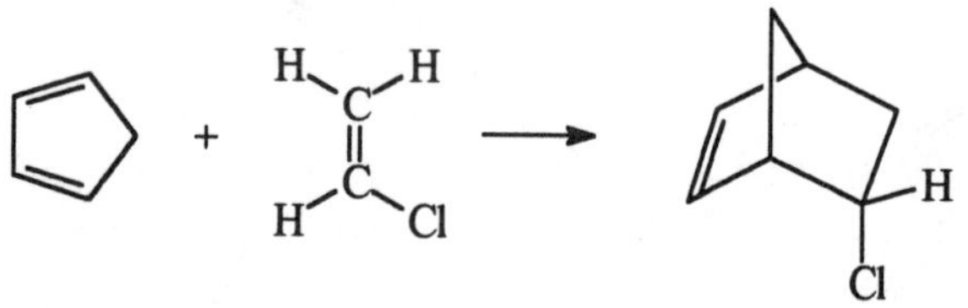 cyclopentadiene + H(H)C=C(H)Cl ⟶ bicyclic adduct with H and Cl

e.

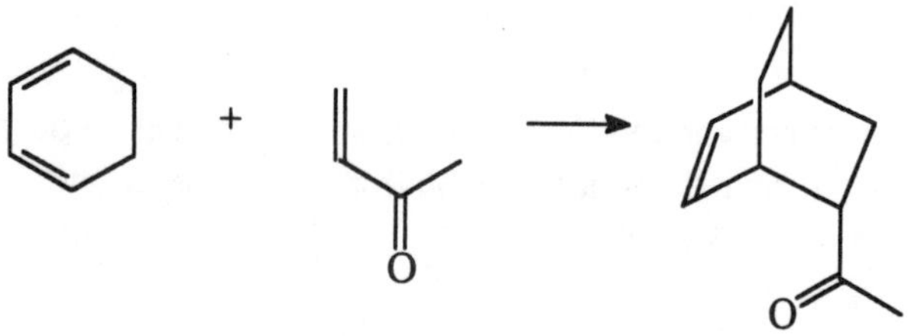

f.

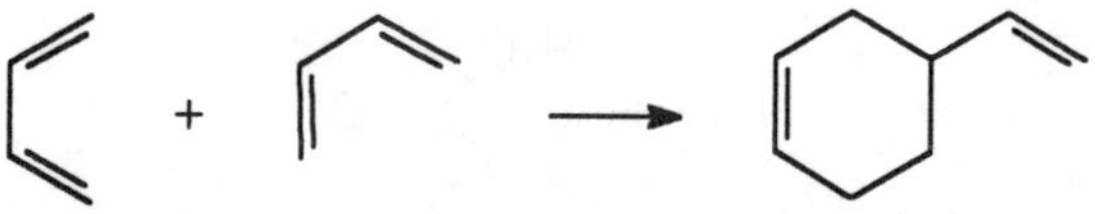

g.

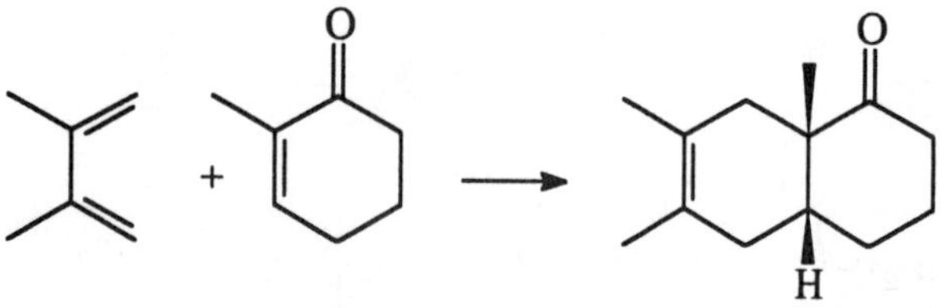

h.

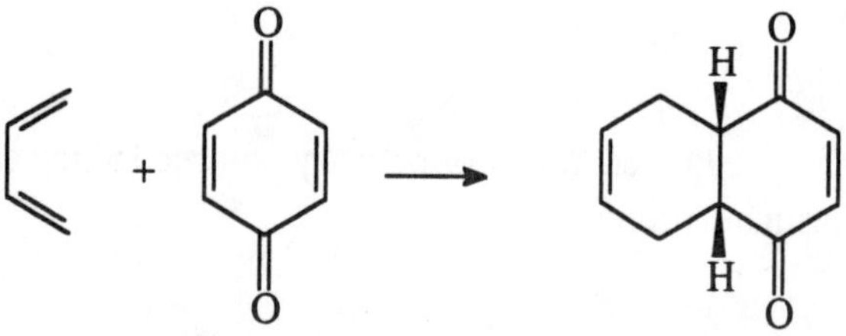

15. a. $(CH_3)_3COH + OH^- \rightleftharpoons (CH_3)_3CO^- + H_2O$

$(CH_3)_3CO^- + Cl{-}Cl \rightleftharpoons (CH_3)_3COCl + Cl^-$

b. 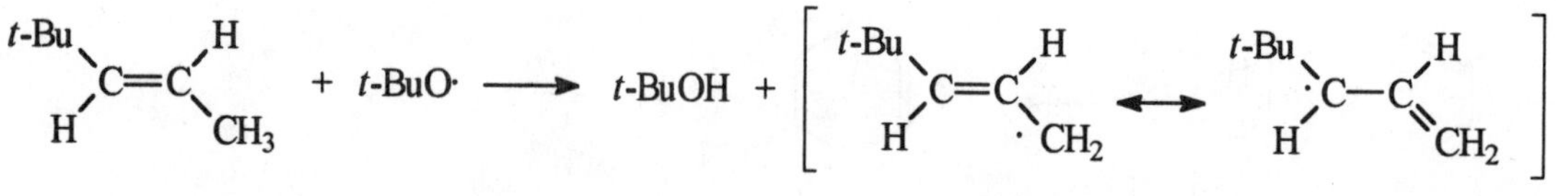 t-Bu(H)C=C(H)CH_3 + t-BuO· ⟶ t-BuOH + [t-Bu(H)C=C(H)·CH_2 ⟷ t-Bu(H)·C–C(H)=CH_2]

$$t\text{-Bu(H)C}\overset{\cdot}{=}\text{C(H)CH}_2 + t\text{-BuOCl} \longrightarrow t\text{-BuO}\cdot + \;(t\text{-Bu})(H)C{=}C(H)(CH_2Cl) \;\text{ or }\; (t\text{-Bu})(Cl)(H)C{-}C(H){=}CH_2$$

c. This experiment shows that the *transoid* ⇌ *cisoid* equilibration at −78 °C must be slower than reaction of the radical with *t*-BuOCl.

d. The first experiment shows either that the two types of allylic radical (see problem #5) do not interconvert to a significant extent at −78 °C, or that the *transoid* radical is more stable than the *cisoid*. The second experiment establishes that the rates of interconversion of the radicals are slow.

e. The resonance structures show that there is "double-bond character" between C-2 and C-3:

$$\left[\;(R)(H)C{=}C(H){-}CH_2\cdot \;\longleftrightarrow\; (R)(H)\dot{C}{-}C(H){=}CH_2 \;\right]$$

16. $$CH_2{=}C(CH_3){-}C(CH_3){=}CH_2 + Cl_2 \xrightarrow{CCl_4} (ClCH_2)(H_3C)C{=}C(CH_3)(CH_2Cl) + A + B$$

45% 54% 1%

$$A = ClCH_2C(=CH_2){-}C(CH_3){=}CH_2$$

NMR: δ 4.20 (2H) $-CH_2Cl$
δ 1.90 (3H) $-CH_3$
The rest are vinyl H's

$$B = ClCH{=}C(CH_3){-}C(CH_3){=}CH_2$$

NMR: δ 6.20 (1H) =CHCl
δ 5.08 & 5.00 (1H each) $=CH_2$
δ 1.78 & 1.85 (3H singlets), 2 $-CH_3$'s

Mechanism (A + B)

$$CH_2{=}C(CH_3){-}C(CH_3){=}CH_2 + Cl_2 \xrightarrow{CCl_4} ClCH_2{-}\overset{+}{C}(CH_3){-}C(CH_3){=}CH_2 + Cl^- \xrightarrow{-H^+}$$

$$\longrightarrow ClCH_2{-}C(=CH_2){-}C(CH_3){=}CH_2 + HCl \quad \textbf{A}$$

$$\longrightarrow ClCH{=}C(CH_3){-}C(CH_3){=}CH_2 + HCl \quad \textbf{B}$$

17. 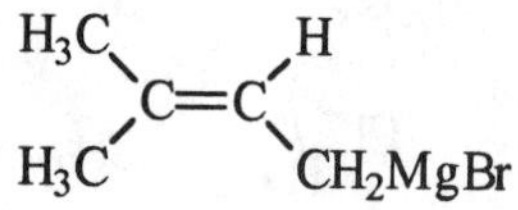 $(H_3C)_2C{=}CH{-}CH_2MgBr$

The two peaks of the doublet at δ 1.60 are due to the two methyls, which are not equivalent. One is *cis* to CH_2MgBr and one is *trans* to it. The signal at δ 5.6 (triplet 1H) arises from the vinyl H. The signal at δ 0.6 (doublet 2H) arises from the CH_2MgBr and is upfield because of negative charge in the carbanion.

At room temperature, the two methyl groups become equivalent on the NMR time scale by the following mechanism.

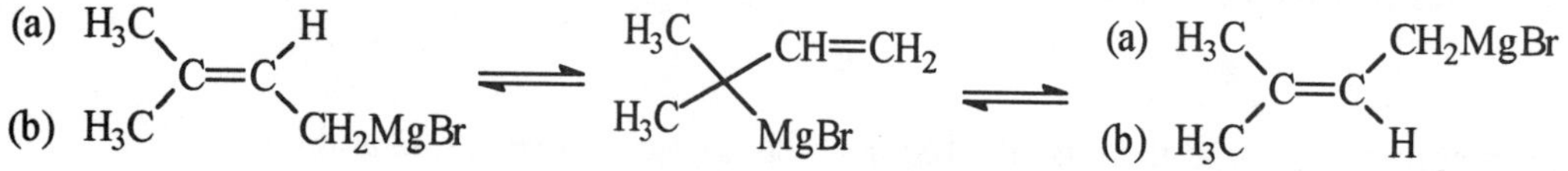

$$\text{(a) }H_3C,\ \text{(b) }H_3C{-}C{=}CH{-}CH_2MgBr \rightleftharpoons (H_3C)_2C(MgBr){-}CH{=}CH_2 \rightleftharpoons \text{(a) }H_3C,\ \text{(b) }H_3C{-}C{=}CH{-}CH_2MgBr$$

(present only in small amount)

The equilibria are rapid at room temperature.

18. Radical chain mechanism. The propagation steps are:

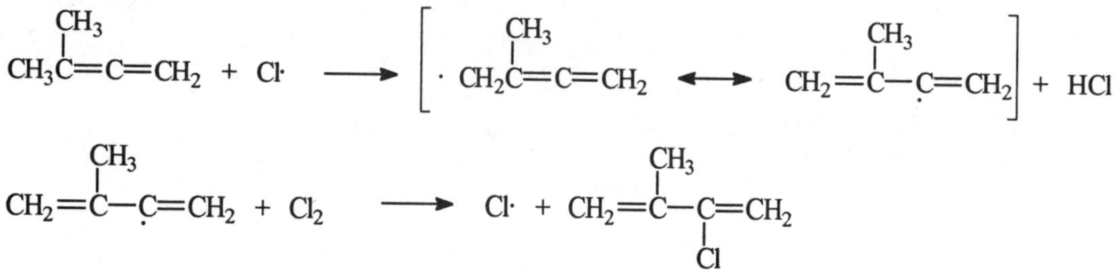

$$CH_3C(CH_3){=}C{=}CH_2 + Cl\cdot \longrightarrow \left[\cdot CH_2C(CH_3){=}C{=}CH_2 \longleftrightarrow CH_2{=}C(CH_3){-}\dot{C}{=}CH_2\right] + HCl$$

$$CH_2{=}C(CH_3){-}\dot{C}{=}CH_2 + Cl_2 \longrightarrow Cl\cdot + CH_2{=}C(CH_3){-}C(Cl){=}CH_2$$

19.

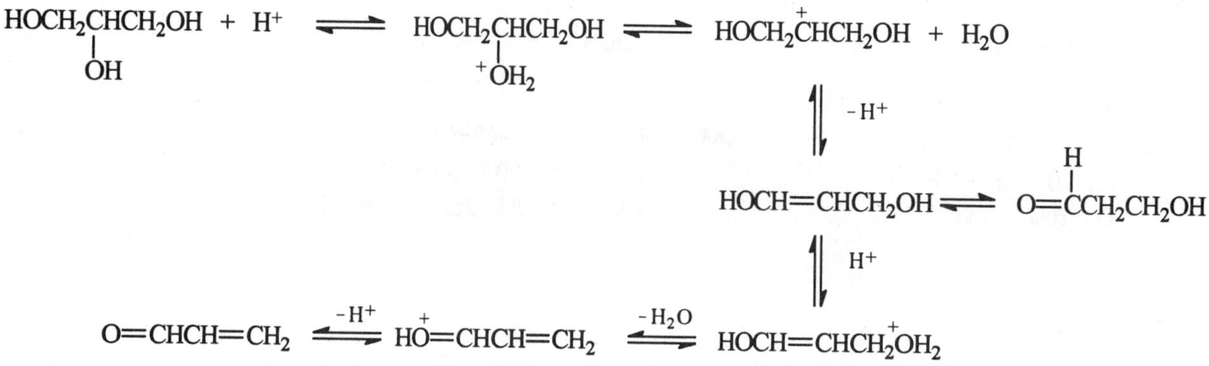

$$HOCH_2CH(OH)CH_2OH + H^+ \rightleftharpoons HOCH_2CH(\overset{+}{O}H_2)CH_2OH \rightleftharpoons HOCH_2\overset{+}{C}HCH_2OH + H_2O$$

$$\xrightarrow{-H^+} HOCH{=}CHCH_2OH \rightleftharpoons O{=}C(H)CH_2CH_2OH$$

$$HOCH{=}CHCH_2OH \xrightarrow{H^+} HOCH{=}CHCH_2\overset{+}{O}H_2 \xrightleftharpoons{-H_2O} H\overset{+}{O}{=}CHCH{=}CH_2 \xrightleftharpoons{-H^+} O{=}CHCH{=}CH_2$$

20.

$$CH_3CH_2CH(CH_3){-}CH{=}C(CH_3)CO_2Et \rightleftharpoons CH_3CH_2(CH_3)C{=}CH{-}CH(CH_3)CO_2Et$$

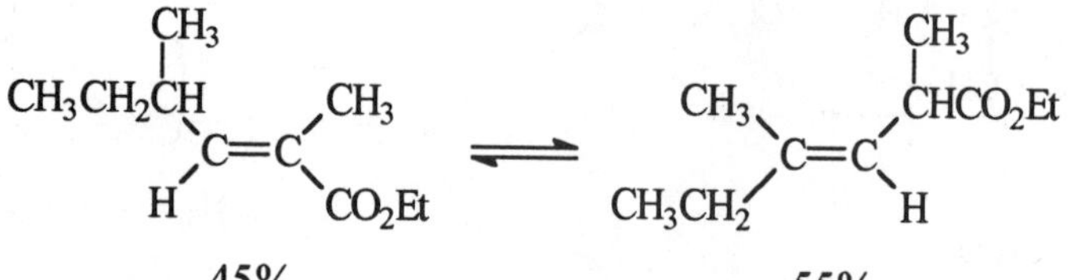

45% 55%

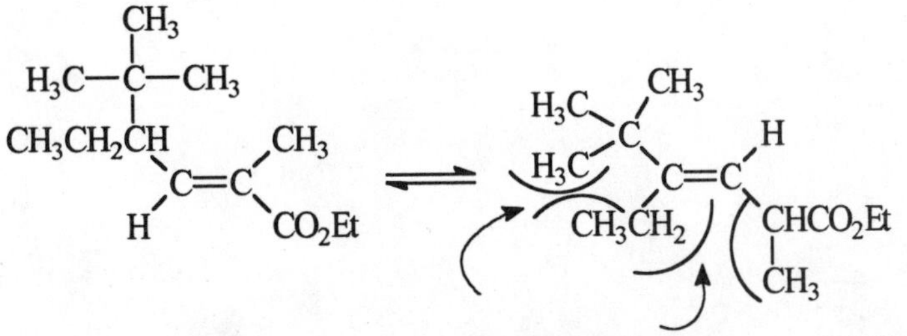

$$\underset{86\%}{(CH_3)_3C(CH_3CH_2)CH\text{–}CH{=}C(CH_3)CO_2Et} \rightleftharpoons \underset{14\%}{(CH_3)_3C(CH_3CH_2)C{=}CH\text{–}CH(CH_3)CO_2Et}$$

Steric hindrance

Steric hindrance is greater in the R = *t*-Bu case and is clearly evident using molecular models.

21. 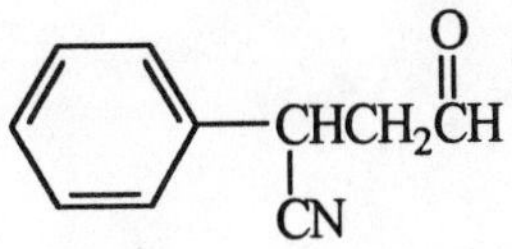 $C_6H_5\text{–}CH(CN)CH_2CHO$

CMR: aldehyde C, δ ~200; benzene C's, δ ~ 128; –CN δ ~ 117; CH_2 δ ~65; CH δ ~43 ppm.

NMR: aldehyde H, δ ~ 10; benzene H's, δ ~7.2; CH δ ~3.5 - 4; CH_2 δ ~3 ppm.
The aldehyde will appear as a triplet with a small value of J (1 - 3 Hz). The CH_2 H's will appear as a doublet of doublets with a J of about 6 - 7 from coupling to the CH and a small, J = 1-3, coupling from the aldehyde superimposed. The CH signal will be a doublet.

1,2-addition is a possible side reaction.

20.F. Supplementary Problems

S1. What is the major product to result from each of the following reaction sequences?

a. 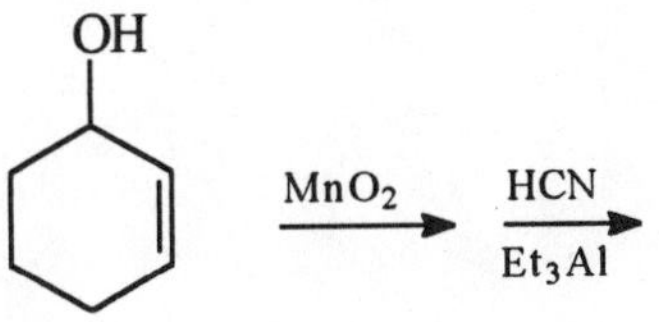 2-cyclohexen-1-ol $\xrightarrow{MnO_2}$ $\xrightarrow[Et_3Al]{HCN}$

b. 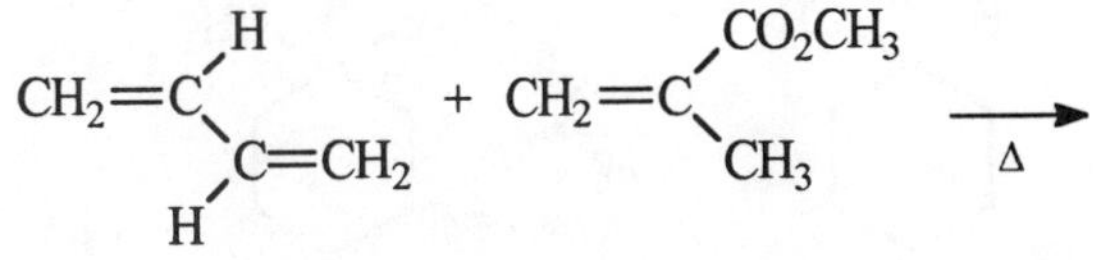 $CH_2{=}CH\text{–}CH{=}CH_2$ + $CH_2{=}C(CH_3)CO_2CH_3$ $\xrightarrow{\Delta}$

c. $(CH_3)_3CCHO$ + $(CH_3CO)_2O$ $\xrightarrow[180°]{CH_3CO_2Na}$ $\xrightarrow{Br_2}$

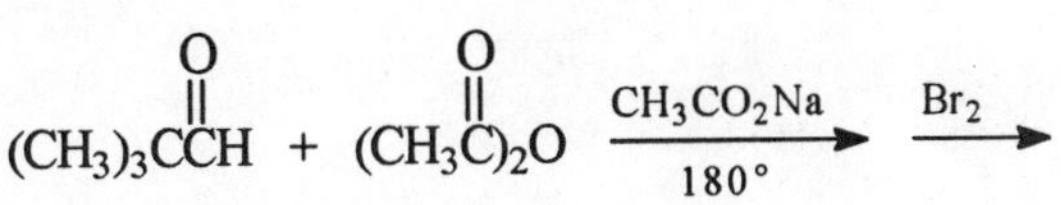

d. 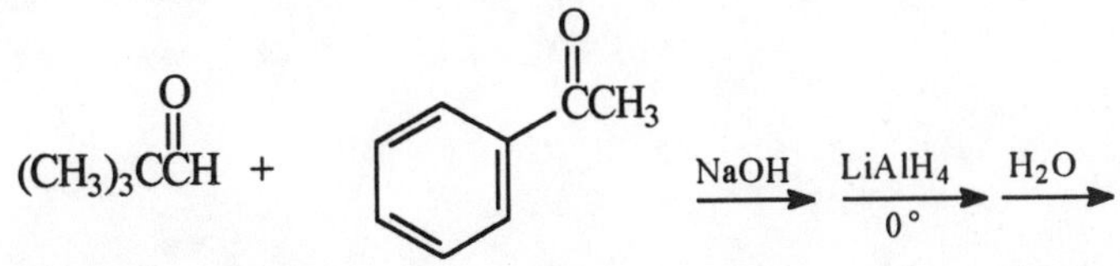 $(CH_3)_3CCHO$ + $C_6H_5COCH_3$ $\xrightarrow{NaOH}$ $\xrightarrow[0°]{LiAlH_4}$ $\xrightarrow{H_2O}$

e. $CH_3CH{=}CHCH_2Br \xrightarrow[ether]{Mg} \xrightarrow{CO_2} \xrightarrow[H^+]{CH_3OH} \xrightarrow{CH_3O^-}$

f. $(CH_3)_2C{=}CHCH{=}CH_2 \xrightarrow[0°C]{HCl} \xrightarrow[50°C]{}$

g. cyclopentadiene + HOOC(H)C=C(H)COOH (trans) $\xrightarrow[\Delta]{}$

S2. Give a series of steps to show how the following transformation can be carried out:

cyclohexane–CO_2H ⟶ cyclohexene–CO_2CH_3

S3. Provide an explanation for the following differences in chemical behavior.

a. $ClCH{=}CHCH_3 + NaCN \longrightarrow$ no reaction

$ClCH{=}CHCCH_3$ (C=O) + NaCN ⟶ $NCCH{=}CHCCH_3$ (C=O)

b. $CH_3OCH_2CH_2CH_3 \xrightarrow{KO\textit{t}\text{-Bu}}$ no reaction

$CH_3OCH_2CH_2CO_2CH_3 \xrightarrow{KO\textit{t}\text{-Bu}} CH_2{=}CHCO_2CH_3 + CH_3OH$

c. 1-methylcyclohexane (H, CH_3) $\xrightarrow[h\nu]{NBS}$ 1-bromo-1-methylcyclohexane (Br, CH_3)   NBS = N-bromosuccinimide (N—Br)

3-methylcyclohexene (H, CH_3) $\xrightarrow[h\nu]{NBS}$ 1-methyl-3-bromocyclohexene (CH_3, Br)

d. 1-methylcyclopentanol (H_3C, OH) $\xrightarrow[H_2SO_4,\ 0°]{K_2Cr_2O_7}$ no reaction

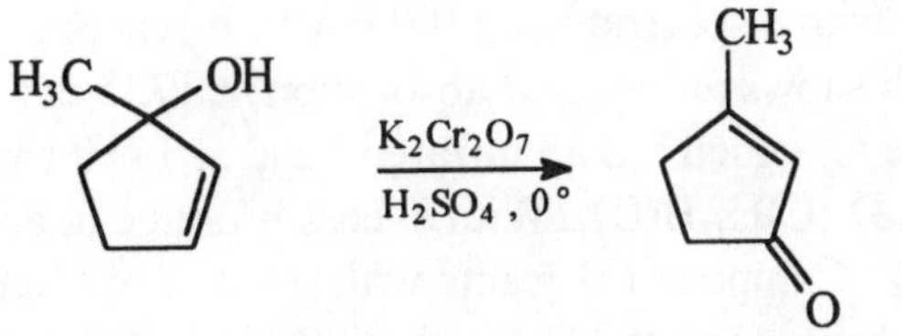

H_3C OH → $K_2Cr_2O_7$ / H_2SO_4, 0° → CH_3 ... O

S4. The C=O stretch in the infrared spectrum of conjugated ketones comes at lower frequency than that for the analogous saturated systems. Explain why.

$CH_3\overset{O}{\overset{\|}{C}}CH_2CH(CH_3)_2$ $\nu_{C=O} = 1710\ cm^{-1}$

$CH_3\overset{O}{\overset{\|}{C}}CH{=}C(CH_3)_2$ $\nu_{C=O} = 1695\ cm^{-1}$, $\nu_{C=C} = 1625\ cm^{-1}$

S5. Using Appendices I and II, calculate the change in enthalpy expected for the propagation steps of the free-radical chlorination of ethane, ethylene, and propene to give ethyl chloride, vinyl chloride, and allyl chloride, respectively.

S6. The prostaglandins are a class of compounds whose occurrence, structures, and potent biological effects have been studied and elucidated only within recent years. They are found throughout the body, and are implicated in many diverse biological processes, often at nanomolar (10^{-9} M) concentrations. PGE_2, the most potent of the prostaglandins, is unstable in the presence of base: it loses water to give PGA_2, which then isomerizes to the physiologically inactive PGB_2 isomer. Write a reasonable structure for PGA_2, as well as step-by-step mechanisms for these two transformations.

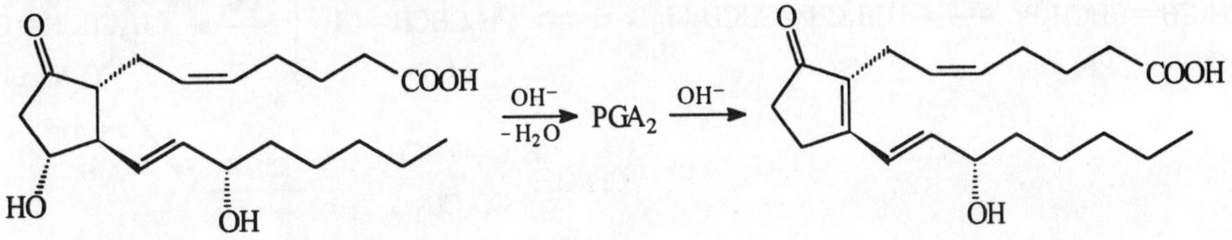

O COOH HO OH → OH^- / $-H_2O$ → PGA_2 → OH^- → O COOH OH

S7. When 2-methyl-3-cyclohexenone is treated with base, it readily isomerizes to 2-methyl-2-cyclohexenone. Similar treatment of bicyclo[2.2.2]oct-5-en-2-one does not produce any reaction, however. Why do these two compounds differ so much in reactivity?

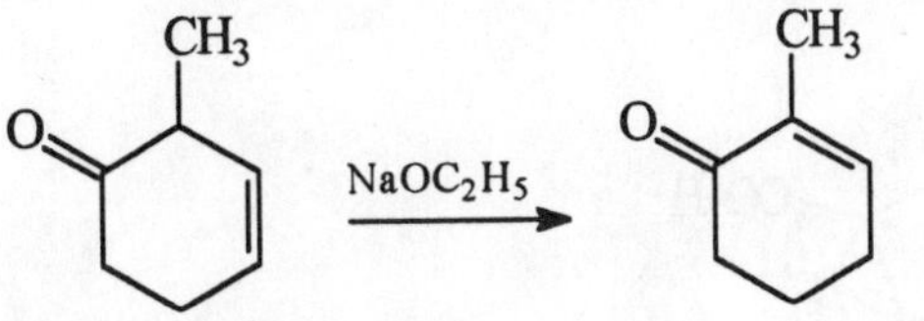

CH_3 O → $NaOC_2H_5$ → CH_3 O

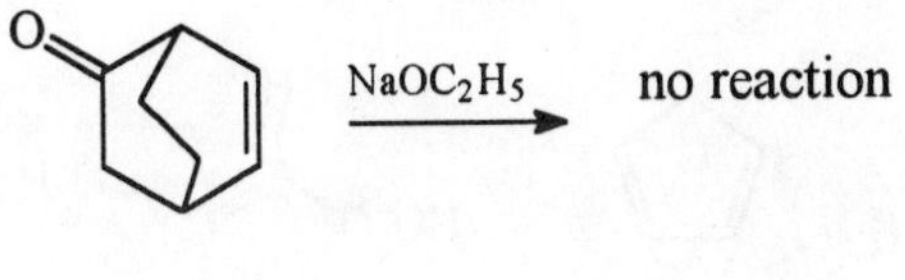

O → $NaOC_2H_5$ → no reaction

S8. Write a mechanism for the following transformation.

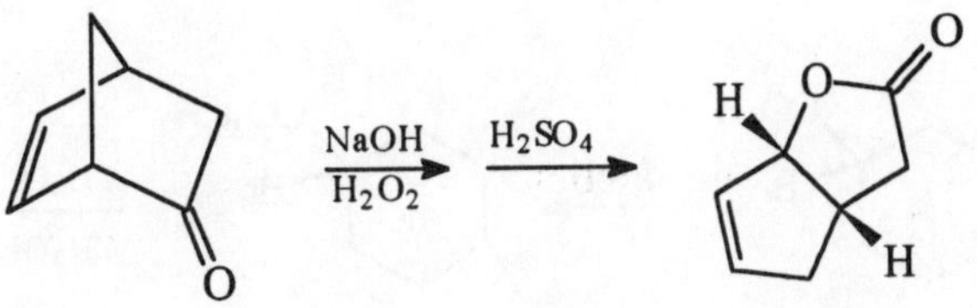

O → NaOH / H_2O_2 → H_2SO_4 → H O O H

S10. Compound A ($C_7H_{14}O$) has a strong absorption in its infrared spectrum at 3400 cm^{-1}. It reacts with acetic anhydride to give a new compound (B, $C_9H_{16}O_2$), which shows an infrared absorption at 1735 cm^{-1}. Compound A reacts with Na_2CrO_4 in acetic acid to give C, which has an infrared band at 1710 cm^{-1}. Compound C reacts with bromine in acetic acid to give D ($C_7H_{11}BrO$). With excess bromine in aqueous NaOH, C gives a tetrabromo compound (E, $C_7H_8Br_4O$). Compound D reacts with potassium *t*-butoxide in refluxing *t*-butyl alcohol to give F, which has infrared absorptions at 1685 and 1670 cm^{-1}. When either C or F is treated with $NaOCH_3$ in CH_3OD, it is found to exchange four of its protons for deuterium. What are compounds A-F?

20.G. Answers to Supplementary Problems

S1. a. 3-cyanocyclohexanone (O, CN)

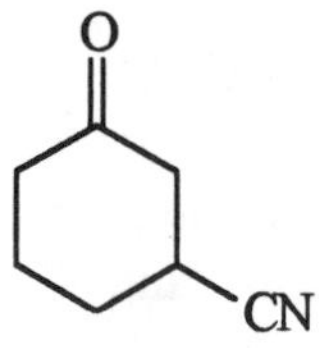

b. cyclohexene bearing CH_3 and CO_2CH_3 on the same carbon

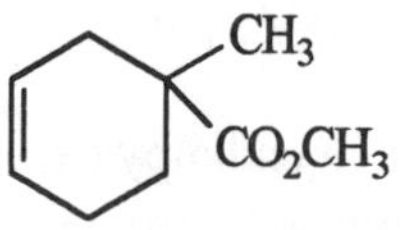

c. $(CH_3)_3CCHBrCHBrCO_2H$

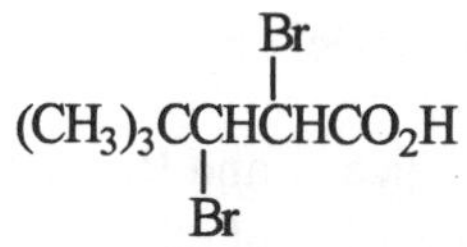

d. $(CH_3)_3CCH{=}CHCH(OH)$–C_6H_5

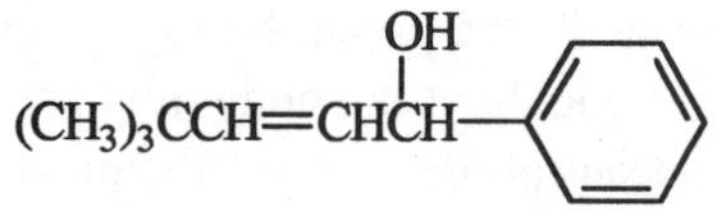

e.

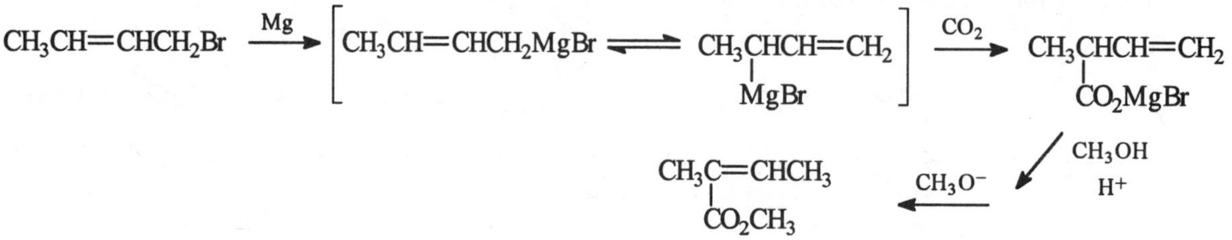

$CH_3CH{=}CHCH_2Br \xrightarrow{Mg} [CH_3CH{=}CHCH_2MgBr \rightleftharpoons CH_3CH(MgBr)CH{=}CH_2] \xrightarrow{CO_2} CH_3CH(CO_2MgBr)CH{=}CH_2 \xrightarrow[H^+]{CH_3OH}$ (ester) $\xrightarrow{CH_3O^-} CH_3C(CO_2CH_3){=}CHCH_3$

f. $(CH_3)_2C{=}CHCH{=}CH_2 \xrightarrow[0^\circ]{HCl} (CH_3)_2CClCH{=}CHCH_3 \xrightarrow{50^\circ} (CH_3)_2C{=}CHCHClCH_3$

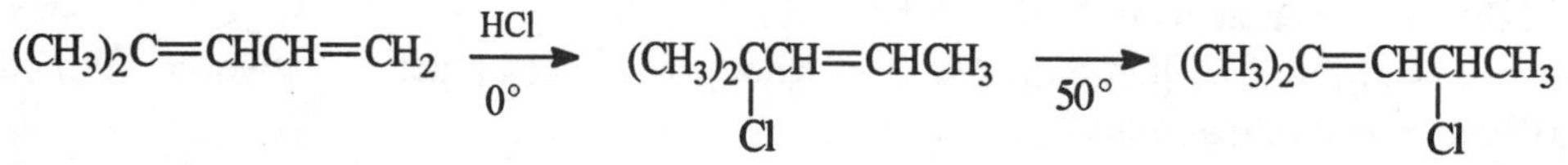

g. cyclopentadiene + HO_2C–CH=CH–CO_2H (trans) → bicyclic adduct (norbornene) with CO_2H, H, H, CO_2H

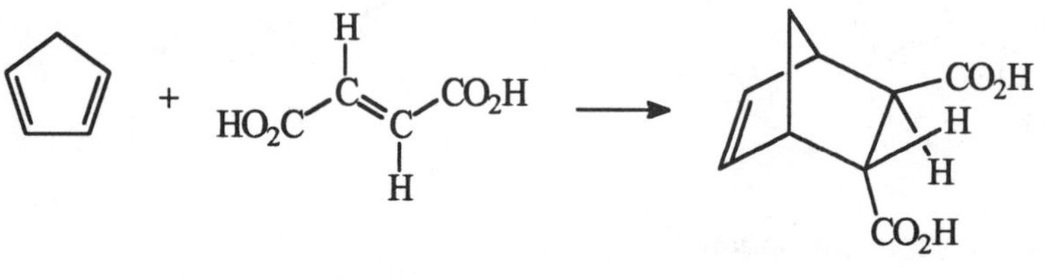

S2. cyclohexyl–$CO_2H \xrightarrow{Br_2}$ 1-bromocyclohexane-1-carboxylic acid (CO_2H, Br) $\xrightarrow[H^+]{CH_3OH}$ (CO_2CH_3, Br) $\xrightarrow[CH_3OH,\ \Delta]{CH_3O^-}$ cyclohexene–CO_2CH_3

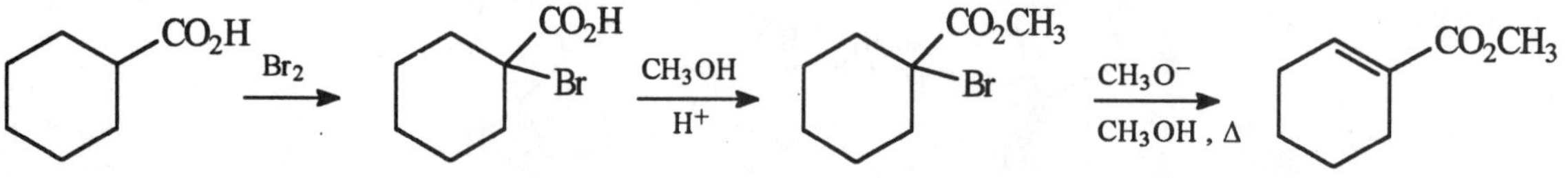

S3. a. The chloroketone can undergo substitution by a conjugate addition-elimination sequence. Such a mechanism is not possible for 1-chloropropene.

$$ClCH{=}CH{-}\overset{O}{\overset{\|}{C}}CH_3 + CN^- \rightleftharpoons NC{-}CH(Cl){-}CH{=}\overset{O^-}{\overset{|}{C}}CH_3 \rightleftharpoons NCCH{=}CH\overset{O}{\overset{\|}{C}}CH_3$$

b. The β-methoxyester can undergo elimination via an enolate, as in the second step of an aldol condensation. Again, the simple alkyl system cannot react in this manner.

$$CH_3OCH_2CH(H){-}\overset{O}{\overset{\|}{C}}OCH_3 \xrightarrow{\text{base}} CH_3O{-}CH_2{-}CH{=}\overset{O^-}{\overset{|}{C}}OCH_3 \rightleftharpoons CH_2{=}CH\overset{O}{\overset{\|}{C}}OCH_3$$

c. In both free radical reactions, the hydrogen atom-abstraction is selective for the tertiary hydrogen. In the case of 3-methylhexene, this produces an allylic radical, which can subsequently react at the other end of the original double bond.

(Scheme: 3-methylcyclohexene (H, CH_3) $\xrightarrow{Br\cdot}$ [allylic radical ↔ allylic radical] $\xrightarrow{NBS}$ 3-bromo-1-methylcyclohexene (Br))

d. Tertiary alcohols are not oxidized under mild conditions, but the tertiary allylic alcohols can undergo ionization, allylic isomerization, and then oxidation:

(Scheme: 1-methylcyclopent-2-en-1-ol (HO, CH_3) $\xrightarrow[-H_2O]{H^+}$ [allylic cation ↔ allylic cation] $\xrightarrow[-H^+]{H_2O}$ 3-methylcyclopent-2-en-1-ol (OH) $\xrightarrow{[O]}$ 3-methylcyclopent-2-en-1-one (O))

S4. Because of the contribution of resonance structures that have a C–O single bond, there is slightly less double-bond character in the carbonyl bond of an enone:

$$\left[CH_3\overset{O}{\overset{\|}{C}}CH{=}C(CH_3)_2 \longleftrightarrow CH_3\overset{O^-}{\overset{|}{\underset{+}{C}}}CH{=}C(CH_3)_2 \longleftrightarrow CH_3\overset{O^-}{\overset{|}{C}}{=}CH\underset{+}{C}(CH_3)_2 \right]$$

S5.

	Reaction				$\Delta H°$ (kcal mole^{-1})
	CH_3CH_3 + $Cl\cdot$ → $CH_3CH_2\cdot$ + HCl				
ΔH_f° =	−20.2	28.9	29	−22.1	−2
	$CH_3CH_2\cdot$ + Cl_2 → CH_3CH_2Cl + $Cl\cdot$				
ΔH_f° =	29	0	−26.1	28.9	−26
	$CH_2{=}CH_2$ + $Cl\cdot$ → $CH_2{=}CH\cdot$ + HCl				
ΔH_f° =	12.5	28.9	72	−22.1	+9
	$CH_2{=}CH\cdot$ + Cl_2 → $CH_2{=}CHCl$ + $Cl\cdot$				
ΔH_f° =	72	0	8.6	28.9	−35
	$CH_2{=}CHCH_3$ + $Cl\cdot$ → $CH_2{=}CHCH_2\cdot$ + HCl				
ΔH_f° =	4.9	28.9	39	−22.1	−17
	$CH_2{=}CHCH_2\cdot$ + Cl_2 → $CH_2{=}CHCH_2Cl$ + $Cl\cdot$				
ΔH_f° =	39	0	0*	28.9	−10

***Calculate** $\Delta H_f^\circ(CH_2{=}CHCH_2Cl) = \Delta H_f^\circ(CH_2{=}CHCH_2\cdot) + \Delta H_f^\circ(Cl\cdot) - DH°(\text{allyl-Cl})$

$$= 39 + 28.9 - 68 = 0$$

S6.

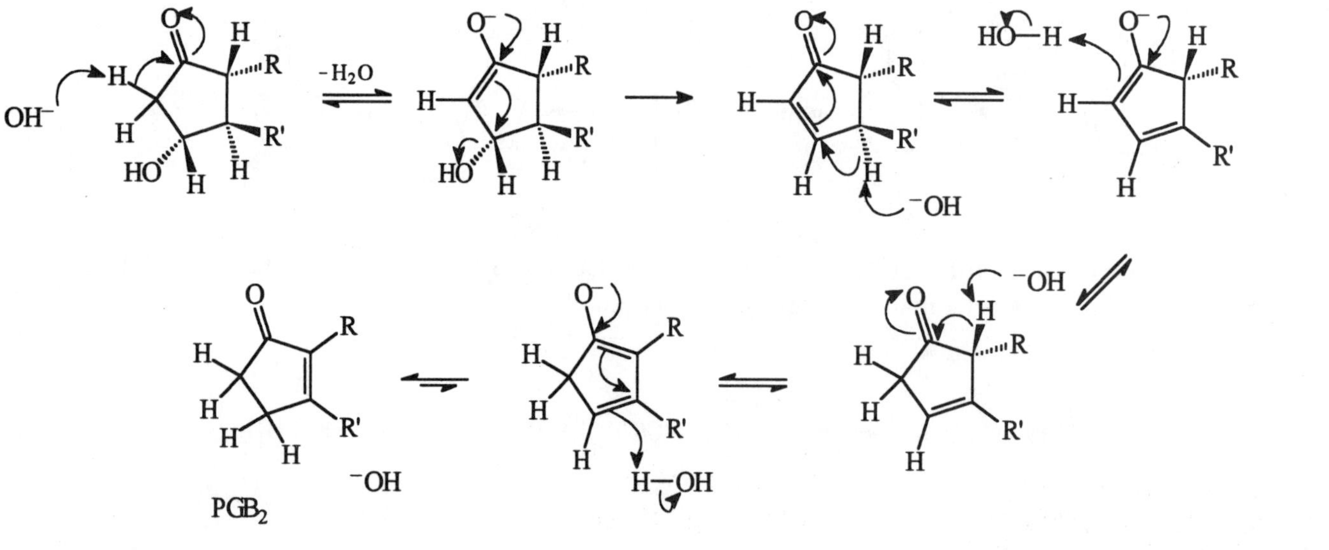

OH⁻   −H₂O   −OH   H–OH   PGB₂

S7. Enolization and subsequent conjugation of the enone system both require that all the atoms involved are able to line up their p-orbitals:

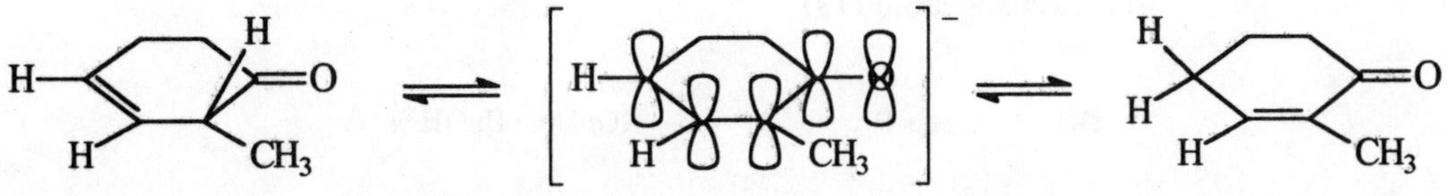

This sort of configuration is not possible for the bicyclic β,γ-enone:

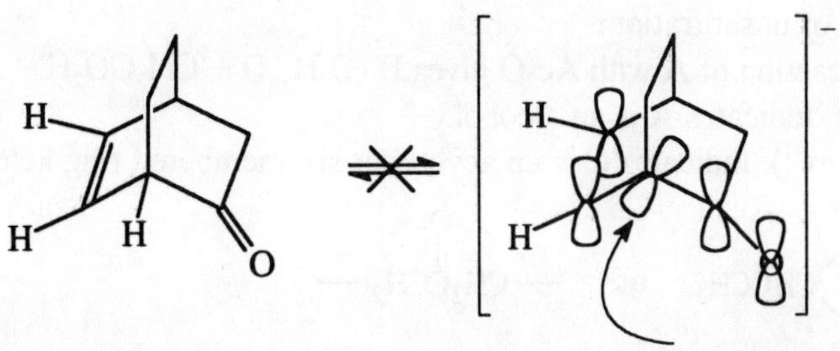

This orbital is perpendicular to the other p-orbitals and cannot overlap with them.

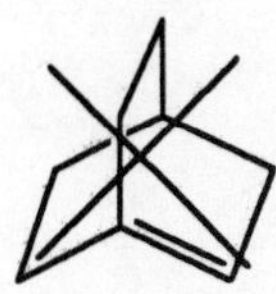

(The generalization that bicyclic systems cannot have a double bond at the bridge-head carbon is known as Bredt's rule).

S8.

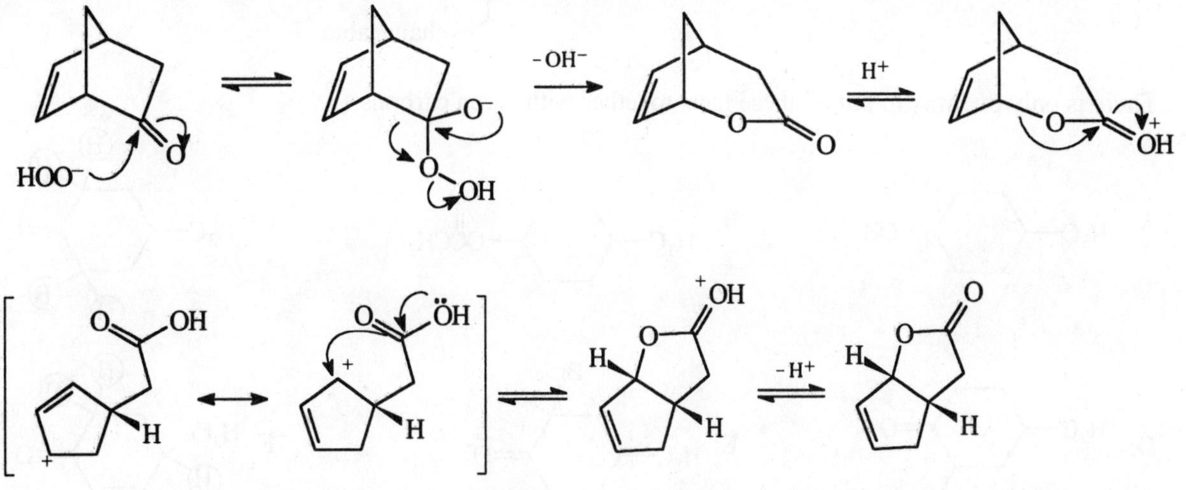

This alkaline Baeyer-Villiger reaction is possible because of the strain in the bicyclo[2.2.1]heptane skeleton.

S9. A $\xrightarrow[H^+]{Cr(VI)}$ C $\xrightarrow[CH_3CO_2H]{Br_2}$ D

A: $C_7H_{14}O$ (IR: 3400cm^{-1})

C: (IR: 1710 cm^{-1}, four exchangeable H's)

D: $C_7H_{11}BrO$

A ↓ Ac_2O → B: $C_9H_{16}O_2$ (IR: 1735 cm^{-1})

C: NaOH ↓ excess Br_2 → E: $C_7H_8Br_4O$

D: KO*t*-Bu, *t*-BuOH ↓ Δ → F: (IR: 1685, 1670 cm^{-1}, four exchangeable H's)

1. Formula for A ($C_7H_{14}O$): indicates one degree of unsaturation.
2. IR of A (3400 cm^{-1}), as well as the fact that reaction of A with Ac_2O gives B ($C_7H_{14}O + CH_3CO_2H - H_2O = C_9H_{16}O_2$ with IR 1735 cm^{-1} (indicates ester): indicates A is an alcohol.
3. A reacts with Cr(VI), H^+ to give C (IR 1710 cm^{-1}); indicates C is an acyclic or six-membered ring ketone.
4. C has four exchangeable H's: indicates C is $>CHC(=O)CH_3$ or $-CH_2C(=O)CH_2-$
5. C reacts with excess Br_2 and NaOH to give E ($C_7H_8Br_4O$); indicates that C is <u>not</u> a methyl ketone (which would give $-CO_2H + HCBr_3$. See Chap. 15, problem S8, p.265).
6. C $\xrightarrow[H^+]{Br_2}$ $\xrightarrow[t\text{-BuOH}]{KOt\text{-Bu}}$ F (IR 1685, 1670 cm^{-1}): indicates that F is a conjugated ketone.
7. F has four exchangeable H's: indicates F must be $>CH-C(H)=CH-C(=O)-CH_2-$ (exchangeable)
8. There is only one way to put all these facts together with seven carbons:

A: H_3C–(cyclohexane)–OH

B: H_3C–(cyclohexane)–$OCCH_3$ (O above C)

C: H_3C–(cyclohexanone) with Ⓗ Ⓗ on each α-carbon

D: H_3C–(2-bromocyclohexanone), Br

E: H_3C–(2,2,6,6-tetrabromocyclohexanone), Br, Br, Br, Br

F: H_3C, Ⓗ, Ⓗ, Ⓗ, Ⓗ (cyclohexenone)

Ⓗ = exchangeable H's

21. BENZENE AND THE AROMATIC RING

21.A. Chapter Outline and Important Terms Introduced

21.1 Benzene

A. The Benzene Enigma
- phenyl
- aromatic

B. Resonance Energy of Benzene
- delocalization energy
- empirical resonance energy
- aromatic stability
- cyclic system of six π-electrons

C. Symbols for the Benzene Ring

D. Formation of Benzene

dehydrogenation of cyclohexane

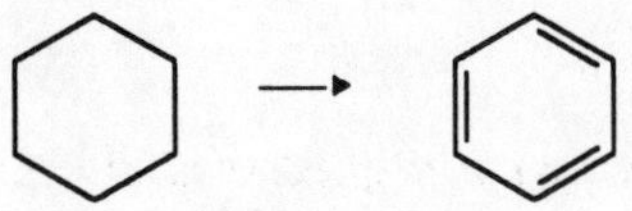

cyclohexane ⟶ benzene (using Pd or Pt/Δ; or S, Δ)

hydroforming process

hexane, heptane ⟶ 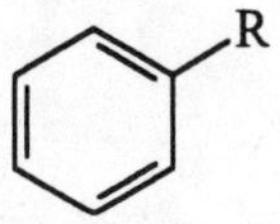 R-benzene (using Cr_2O_3 or Pt as catalysts; industrial process)

21.2 Substituted Benzenes

Nomenclature

- *ortho-*, *meta*, and *para-* (*o-*, *m-*, *p-)*
- phenyl, tolyl, xylyl, mesityl
- arene
- benzyl

21.3 NMR and CMR Spectra

ring current

δ 7.3 (NMR)

δ 128.5 (CMR)

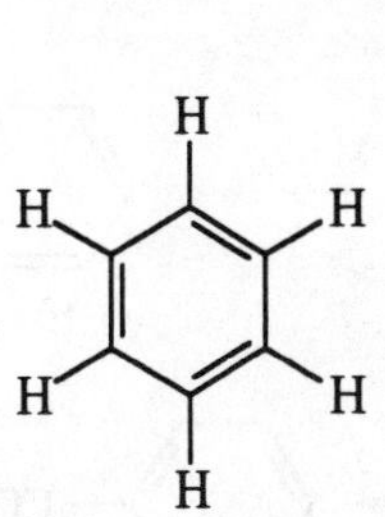

21.4 **Dipole Moments in Benzene Derivatives**

21.5 **Side -Chain Reactions**

A. Free Radical Halogenation

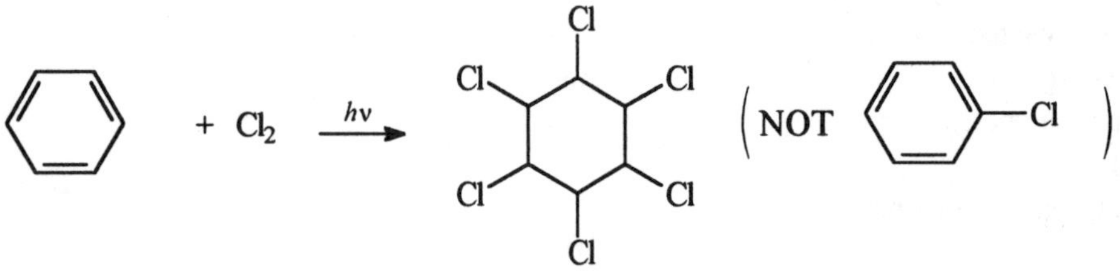

benzene + Cl_2 $\xrightarrow{h\nu}$ 1,2,3,4,5,6-hexachlorocyclohexane (NOT chlorobenzene)

benzene–CH_3 + X_2 $\longrightarrow$ benzene–CH_2X + HX (very easily at benzylic positions)

benzyl radical

B. Benzylic Displacement and Carbocation Reactions

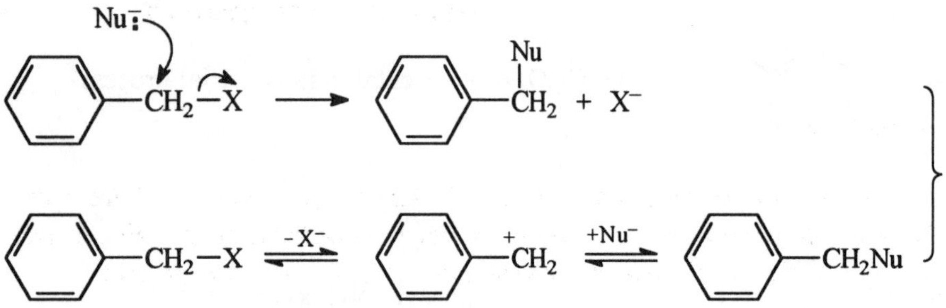

Nu^- + benzene–CH_2–X $\longrightarrow$ benzene–CH_2–Nu + X^-

benzene–CH_2–X $\underset{}{\overset{-X^-}{\rightleftharpoons}}$ benzene–$\overset{+}{C}H_2$ $\overset{+Nu^-}{\rightleftharpoons}$ benzene–CH_2Nu

both S_N1 and S_N2 are fast

C. Oxidation

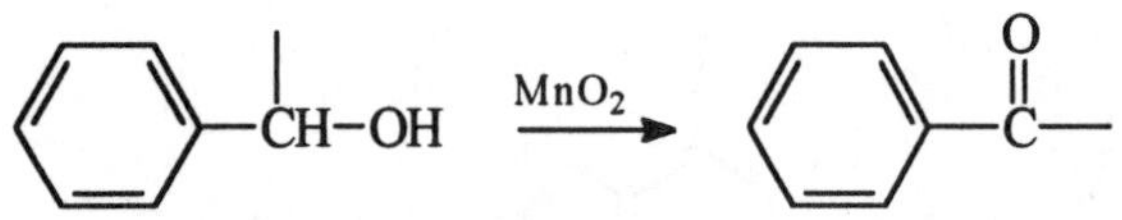

benzene–CH(–)–OH $\xrightarrow{MnO_2}$ benzene–C(=O)– (as easy as allylic alcohol)

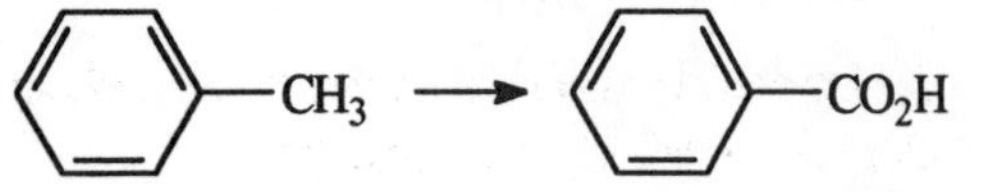

benzene–CH_3 $\longrightarrow$ benzene–CO_2H (using $Na_2Cr_2O_7/H_2SO_4$ or $KMnO_4$; at least one benzylic H is needed)

D. Acidity of Alkylbenzenes

benzylic carbanion

pK_a of toluene ~41

$$C_6H_5-CH_3 \rightleftharpoons C_6H_5-CH_2^- + H^+$$

21.6 Reduction

A. Catalytic Hydrogenation

$$C_6H_6 \xrightarrow[1000\ psi,\ 100°\ C]{H_2/Rh} C_6H_{12}$$ (can't stop short of complete hydrogenation)

B. Hydrogenolysis of Benzylic Groups

C. Birch Reduction (dissolving metal reduction)

21.7. Aromatic Transition States

pericyclic reactions
Cope rearrangement
Claisen rearrangement
sigmatropic rearrangements

21.B. Important Reactions Introduced

NOTE: Many of the reactions that occur at benzylic positions [21.5, Side-Chain Reactions] have been discussed earlier. They are different in the case of benzylic systems only because of the benzene π-system and the increased stability of benzylic carbocations, radicals, and carbanions.

Hydrogenolysis of benzylic groups (21.6.B)

Equation: $$Ar-\overset{|}{\underset{|}{C}}-Y \xrightarrow[Pd,\ H^+]{H_2} Ar-\overset{|}{\underset{|}{C}}-H$$

Generality: Ar = aryl (i.e. phenyl, tolyl, xylyl, etc.)
Y = halogen, OH, OR, O_2CR

Key features: useful way to make alkyl benzenes
benzylic esters and ethers can be used as protected versions of carboxylic acids and alcohols, since the benzyl group can be cleaved under neutral conditions

Birch reduction (21.6 C)

alkyl-substituted benzenes:

R

Na/NH_3 / ROH →

R

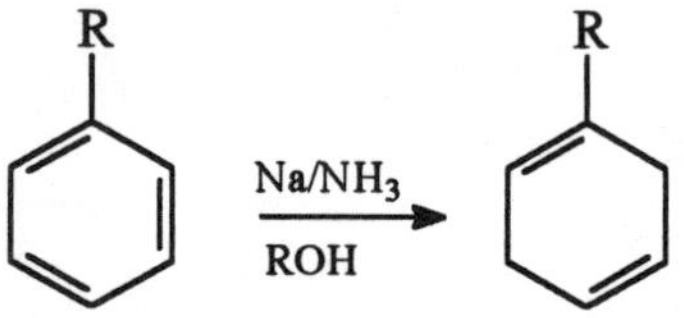

(via cyclohexadienyl anion)

alkoxy-substituted benzenes:

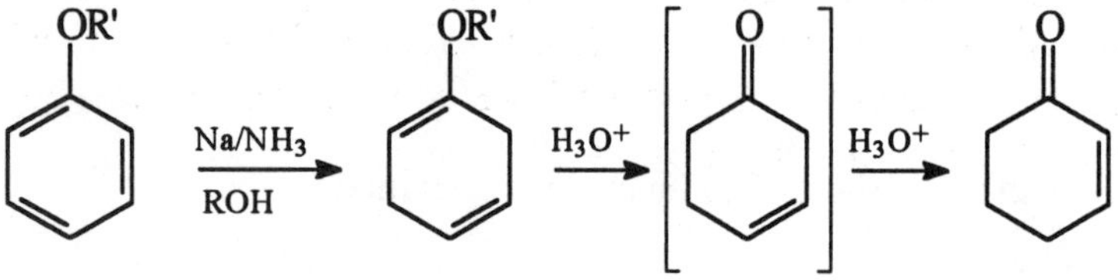

OR'

Na/NH_3 / ROH →

OR'

H_3O^+ →

[O]

H_3O^+ →

O

(2-cyclohexenone synthesis)

carbonyl-substituted benzenes:

O=C–R'

Na/NH_3 / ROH →

O=C–R'

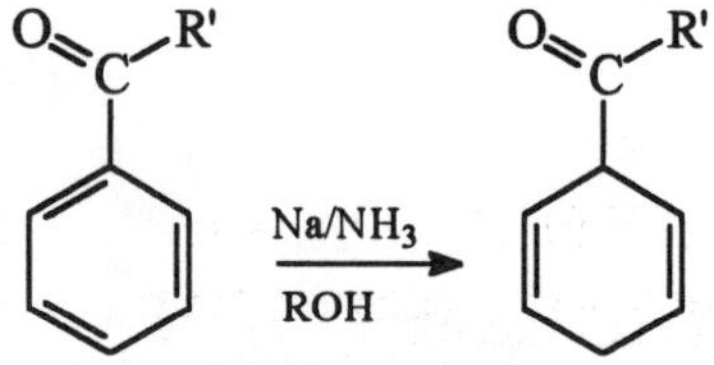

(R' = O^-, alkyl)

alkenyl-substituted benzenes (conjugatd):

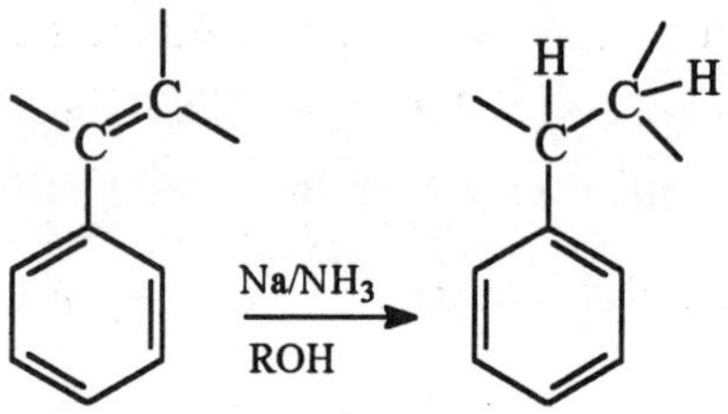

C=C

Na/NH_3 / ROH →

H–C–C–H

Generality: if substitutent is an alkenyl group (e.g. vinyl), it will be reduced too
substituent must not be any other easily reduced group (such as X, NO_2)

Key features: dissolving metal reduction, using Li or Na metal in liquid ammonia as solvent gives 1,4-cyclohexadiene derivatives for Y = OR′, product can be hydrolyzed with aq. H^+ to give α,β-unsaturated ketone

[3,3] Sigmatropic rearrangements (21.7)

Equation:

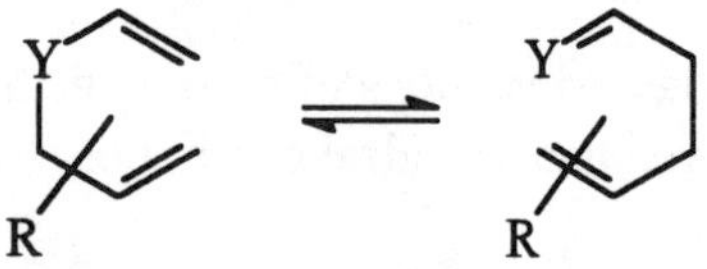

Y, R ⇌ Y, R

Generality: R = various substituents
Y = CH_2 = Cope rearrangement
Y = O = Claisen rearrangement

Key features: pericyclic reaction with aromatic transition state (4n + 2 electrons)

21.C. Important Concepts and Hints

This chapter introduces you to the benzene ring and some of its chemistry. You will notice in this and subsequent chapters that derivatives of benzene (aryl compounds) react quite differently from the alkyl compounds discussed in earlier chapters. To give you an overview of these differences, and a brief justification as to why they occur, we can divide the reactions into three groups: reactions of aryl σ-bonds, reactions on the ring that involve the π-electrons, and reactions on carbons directly attached to the ring (benzylic positions).

I. Reactions of Aryl σ-Bonds:

Because such reactions involve an sp^2- instead of an sp^3-hybrid orbital, they are usually more difficult; for example, free radical halogenation is not successful with benzene.

II. Reactions on the Ring which involve the π-Electrons:

Hydrogenation of the π-system of the benzene ring is substantially more difficult than it is for the π-bond of an alkene. Furthermore, electrophilic addition, which is so important for alkenes, does not occur in aryl compounds except under unusual conditions. This behavior reflects the extra stability, known as aromatic stabilization, of the benzene ring's π-system. In Chapter 23, you will see that **substitution** is the most important reaction of electrophiles with aromatic systems, rather than addition.

The Birch reduction is an important reaction of aryl compounds (and acetylenes) that simple alkenes do not undergo. At first glance, the fact that aryl compounds undergo the Birch reduction while alkenes are stable to such conditions would seem to conflict with the idea that the benzene π-system is more stable than that of an alkene. However, the first, and hardest, step in these reductions is the addition of an electron to the lowest-energy antibonding molecular orbital. For benzene, this orbital is lower in energy than that of an alkene, and benzene can accept the electron more readily. It is not until the next step that aromatic stabilization of the π-system is lost by protonation.

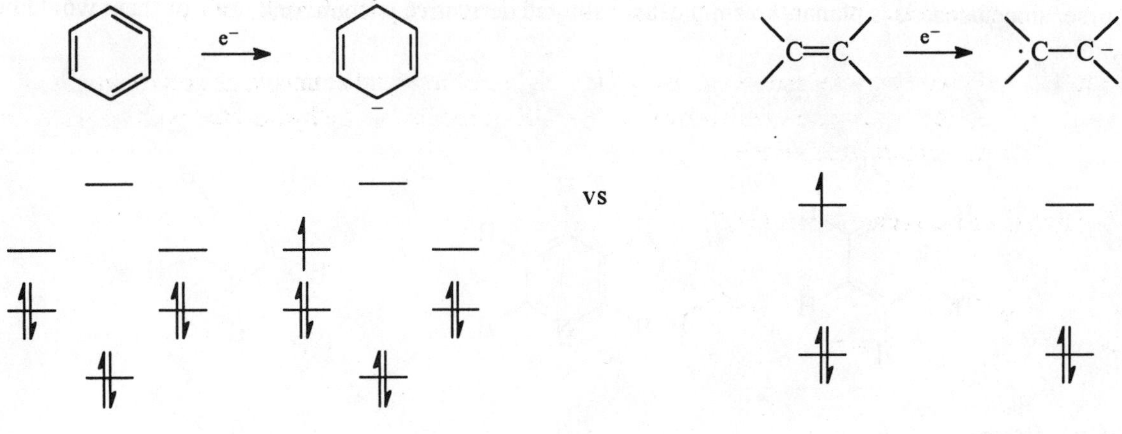

e^- vs   C=C $\xrightarrow{e^-}$ ·C–C⁻

III. Reactions Occurring at Benzylic Positions:

This class of reactions receives the most attention in Chapter 21 and it serves to tie in chemistry that you have learned from previous chapters with aromatic compounds. Because the π-system of the benzene ring can stabilize a *p*-orbital at the benzylic position via conjugation, it makes virtually every reaction at such a position easier (i.e., faster) than for the alkyl counterpart. Reactions which involve cations (S_N1 reactions; oxidation of alkyl side chains), radicals (free radical halogenation; MnO_2 oxidation of benzylic alcohols; hydrogenolysis of benzylic groups), anions (acidity), or even sp^2-hybridized transition states (S_N2 reactions) are all accelerated by the overlap of the π-system of the ring with the *p*-orbital on the benzylic carbon. This overlap is depicted in Figure 21.10 for the benzyl radical.

The subject of aromaticity extends beyond derivatives of benzene alone, as pointed out in the last part of this chapter. Hückel's 4n+2 rule, and our more sophisticated understanding of the molecular orbital interactions which underlie it, help to describe many cyclic transition states (Cope and Claisen rearrangements) as well as other cyclic conjugated systems (elaborated in Section 22.3 in the Text).

21.D. Answers to Exercises

21.1

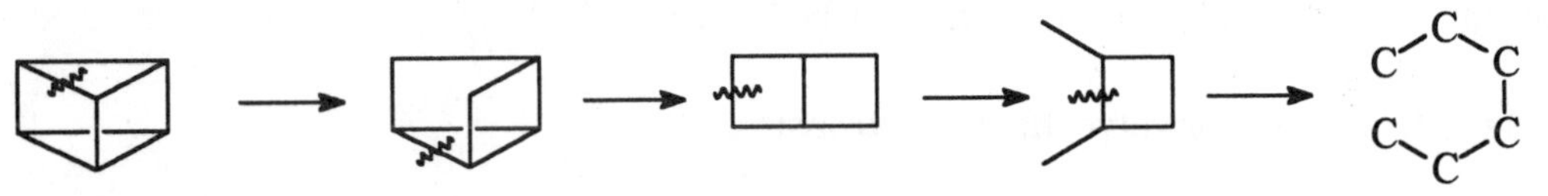

21.2 There are three structural isomers of prismane substituted with two different groups:

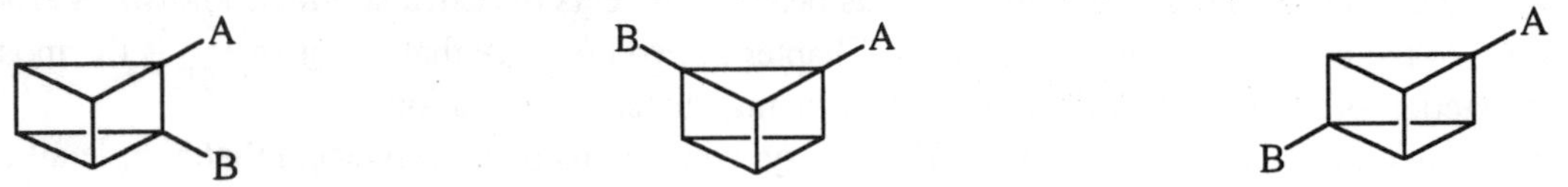

Two of these are chiral and would be capable of being resolved into enantiomers:

Of course, since benzene is planar, no simple disubstituted derivative is chiral and none of them would be resolvable.

21.3

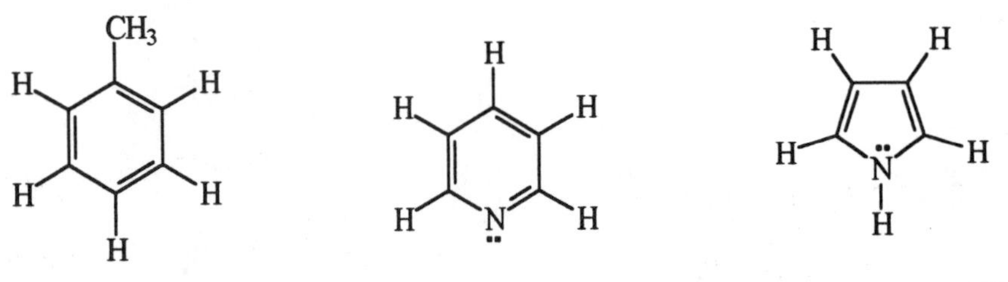

No. of π electrons: 6 6 6

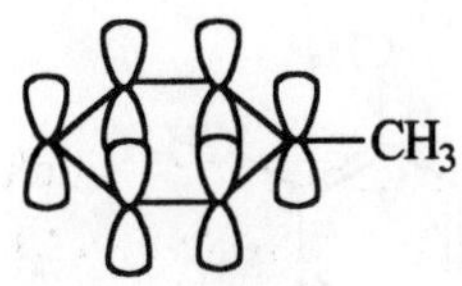

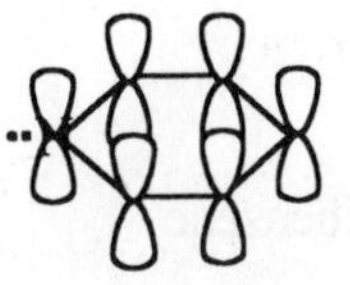

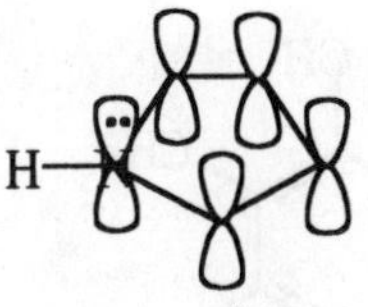

No. of p orbitals: 6 *Toluene* — 6 *Pyridine* — 5 *Pyrrole*

21.4

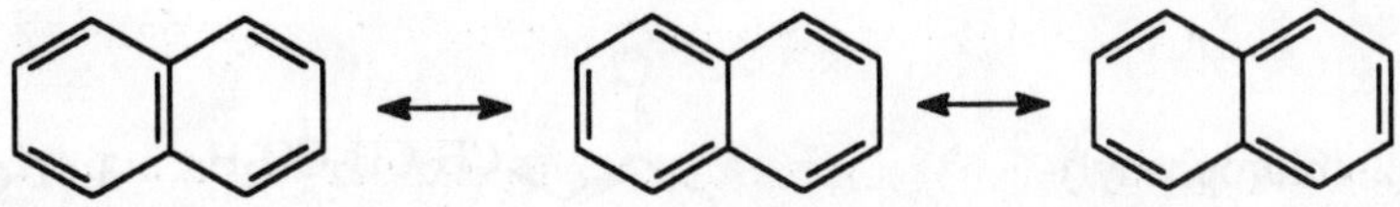

each resonance structure has six π electrons in each of its two rings

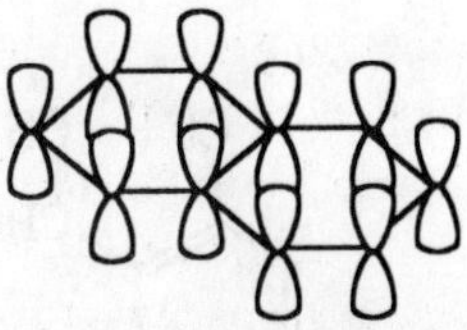

10 π electrons total
10 p orbitals

21.5

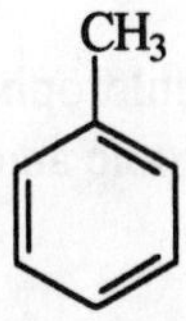

toluene (methylbenzene)

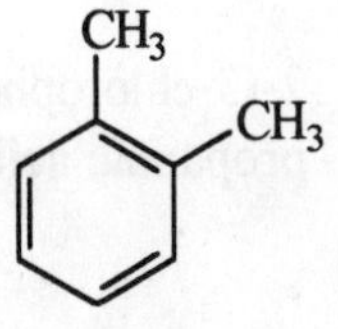

o-xylene (1,2-dimethylbenzene)

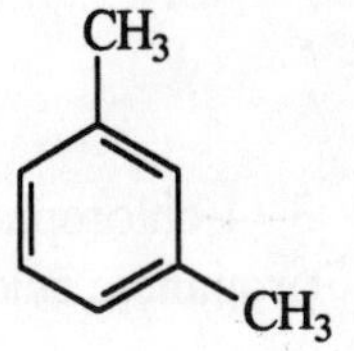

m-xylene (1,3-dimethylbenzene)

p-xylene (1,4-dimethylbenzene)

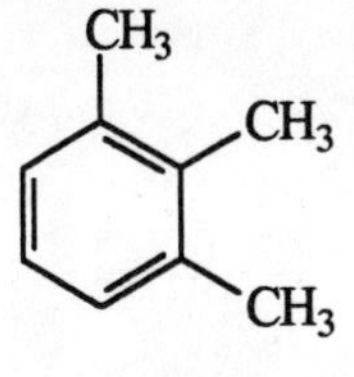

1,2,3-trimethylbenzene (hemimellitene)

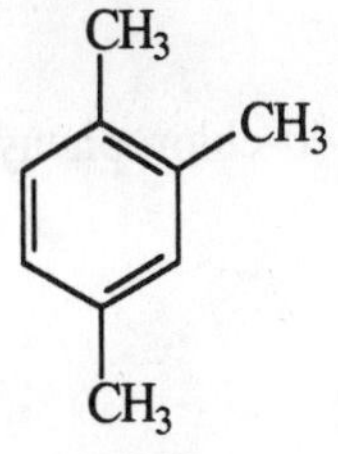

1,2,4-trimethylbenzene

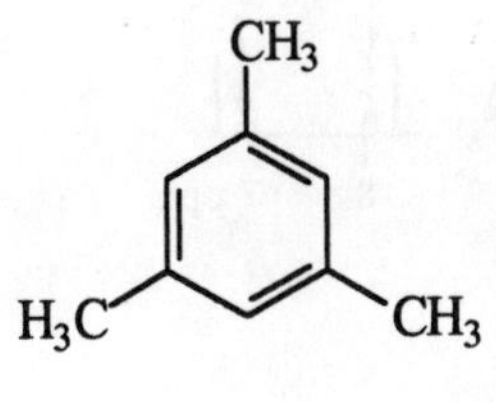

1,3,5-trimethylbenzene (mesitylene)

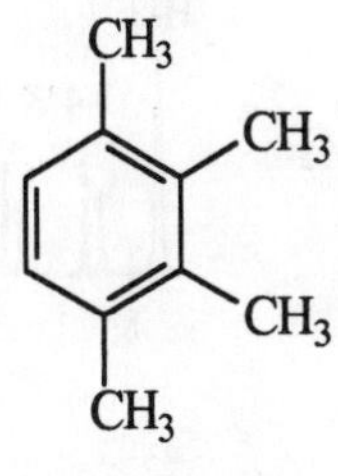

1,2,3,4-tetramethylbenzene (prehitnene)

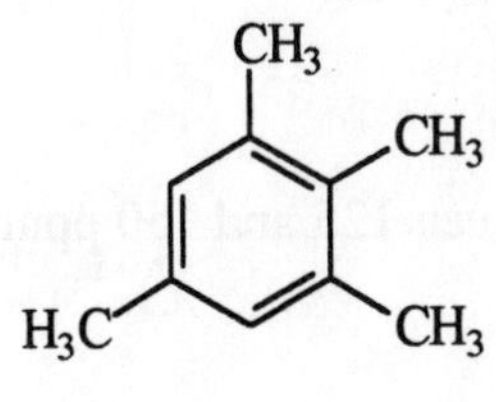

1,2,3,5-tetramethyl-benzene(isodurene)

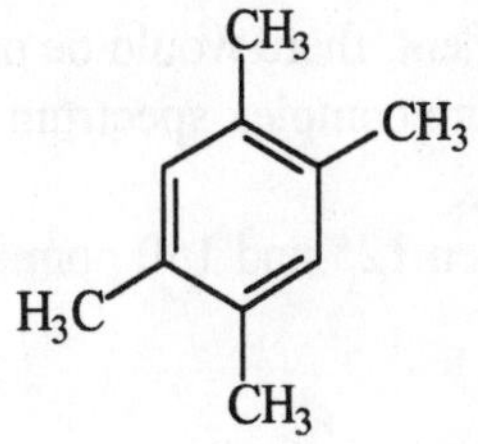

1,2,4,5-tetramethylbenzene (durene)

pentamethylbenzene

hexamethylbenzene

2-(2-chlorophenyl)-propanoic acid

3-(2-chlorophenyl)-propanoic acid

2-(3-chlorophenyl)-propanoic acid

3-(3-chlorophenyl)-propanoic acid

2-(4-chlorophenyl)propanoic acid

3-(4-chlorophenyl)-propanoic acid

(common names: α-(*o*-chlorophenyl)propionic acid, etc.)

21.6

a. NMR: H-2, H-4 & 6, H-5; 8, 7 ppm

On a 60-MHz instrument, there would be overlap, and a significantly more complex spectrum

CMR: six lines between 125 and 150 ppm.

b. NMR: 8, 7 ppm

CMR: four lines between 125 and 150 ppm.

21.7 F—(C$_6$H$_4$)—CH_3

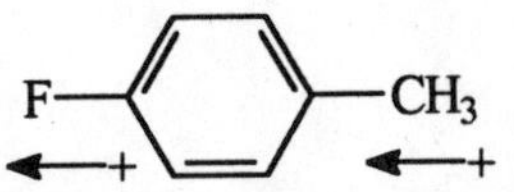

1.63D + 0.37D = 2.00D (actual value = 2.01D)

21.8 $C_6H_5CH(CH_3)_2$ $\xrightarrow[h\nu]{Br_2}$ $C_6H_5CBr(CH_3)_2$ $\xrightarrow[C_2H_5OH]{NaOC_2H_5}$ $C_6H_5C(CH_3){=}CH_2$

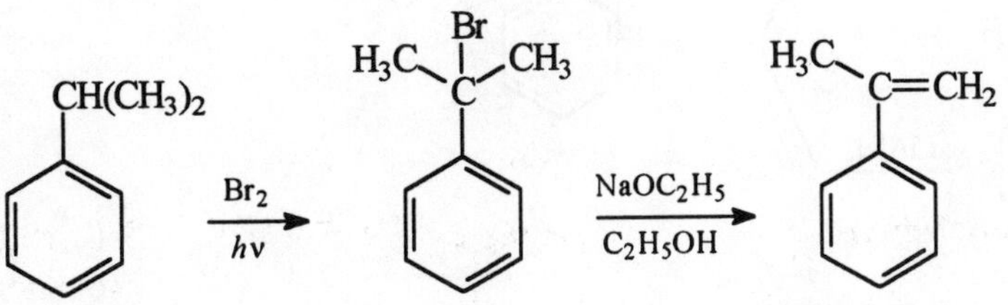

21.9

initiation Br_2 $\xrightarrow{h\nu \text{ or } \Delta}$ Br·

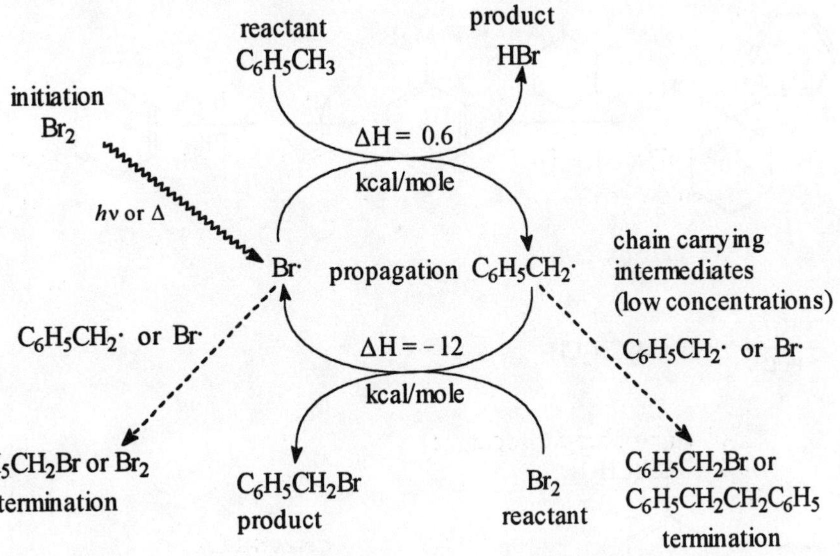

reactant $C_6H_5CH_3$ → product HBr; ΔH = 0.6 kcal/mole

Br· propagation $C_6H_5CH_2$·

chain carrying intermediates (low concentrations)

Br$_2$ reactant → $C_6H_5CH_2Br$ product; ΔH = − 12 kcal/mole

$C_6H_5CH_2$· or Br· → $C_6H_5CH_2Br$ or Br_2 termination

$C_6H_5CH_2$· or Br· → $C_6H_5CH_2Br$ or $C_6H_5CH_2CH_2C_6H_5$ termination

21.10

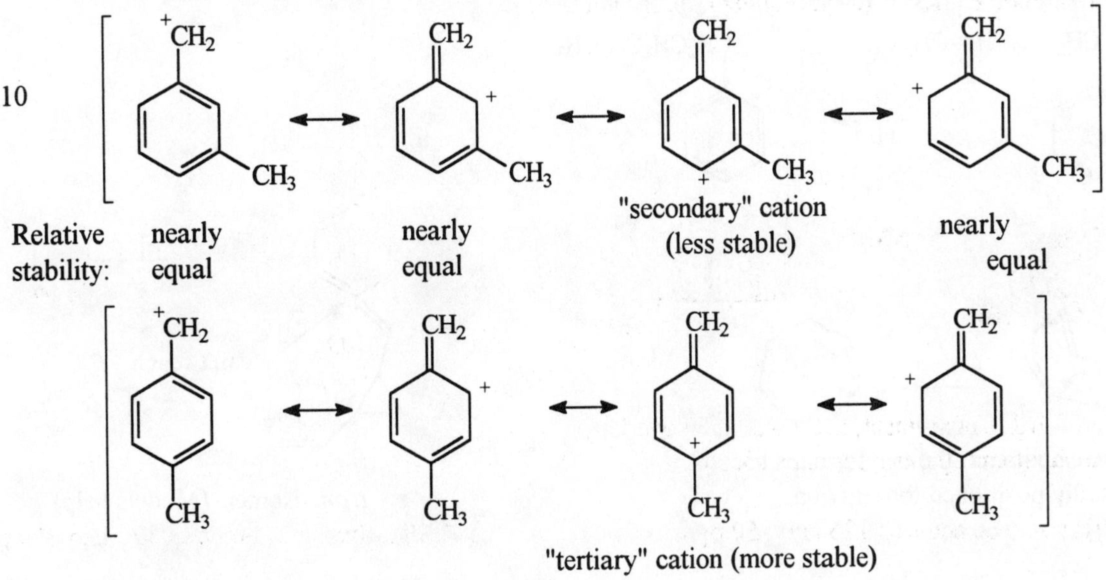

[m-methylbenzyl cation ↔ ↔ ↔ resonance structures]

Relative stability: nearly equal; nearly equal; "secondary" cation (less stable); nearly equal

[p-methylbenzyl cation ↔ ↔ ↔ resonance structures]

"tertiary" cation (more stable)

The *p*-methylbenzyl cation is more stable

21.11

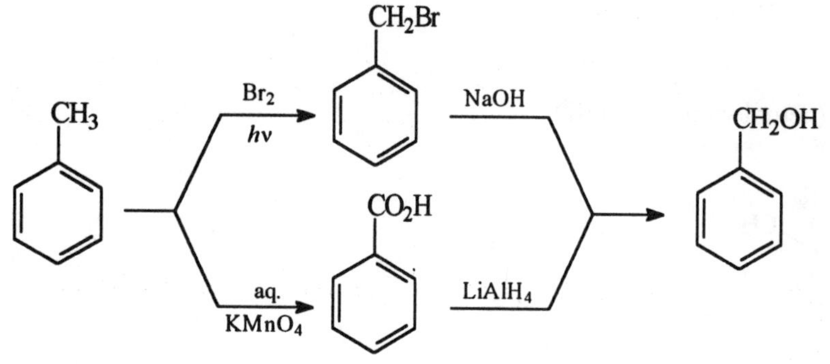

Toluene (CH_3) $\xrightarrow[h\nu]{Br_2}$ CH_2Br (benzyl bromide) $\xrightarrow{NaOH}$ CH_2OH (benzyl alcohol)

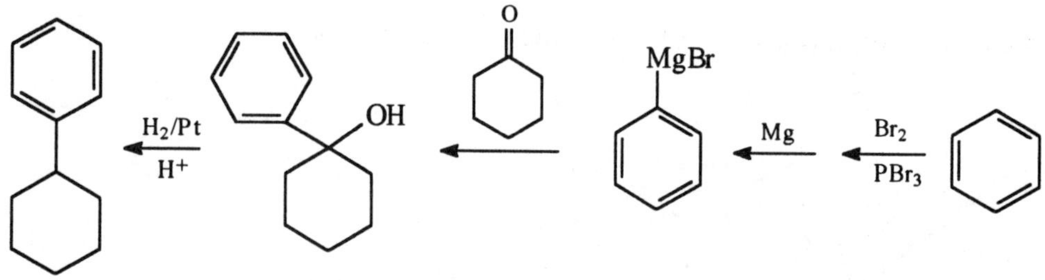

Toluene $\xrightarrow[KMnO_4]{aq.}$ CO_2H (benzoic acid) $\xrightarrow{LiAlH_4}$ CH_2OH (benzyl alcohol)

21.12 $\xleftarrow[H^+]{H_2/Pt}$ OH $\longleftarrow$ (cyclohexanone, O) MgBr $\xleftarrow{Mg}$ $\xleftarrow[PBr_3]{Br_2}$

21.13 H_3C, CH_3, C=C, H, H $\xrightarrow[Pt/C]{H_2}$ $CH_3CH_2CH_2CH_3$

H_3C, C=CH_2 $\xrightarrow[Pt/C]{H_2}$ $CH(CH_3)_2$

$CH_2CH_2CH{=}CH_2$ $\xrightarrow[Pt/C]{H_2}$ $CH_2CH_2CH_2CH_3$

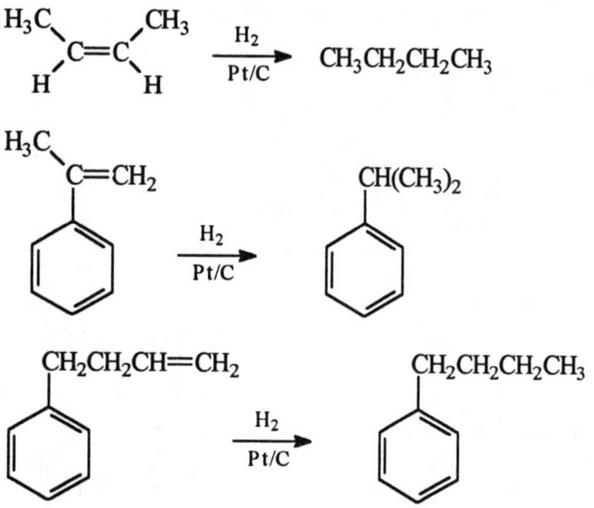

21.14 $\longrightarrow$

In the *cis* isomer, the vinyl groups are sterically positioned for reaction

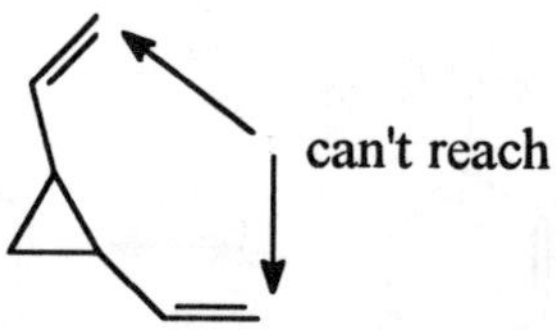

can't reach

trans isomer (Models help)

21.15 OH, OEt, O, ≡, O, ≡, O, H

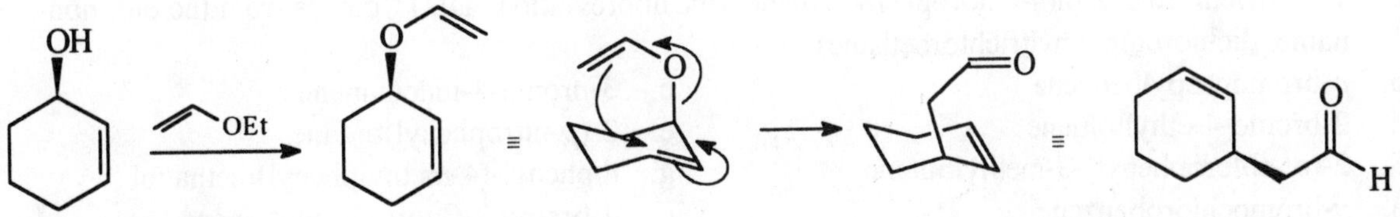

21.16 a.  H

more stable, because double bonds are more substituted.

The transition states for the forward and reverse reactions are the same. H

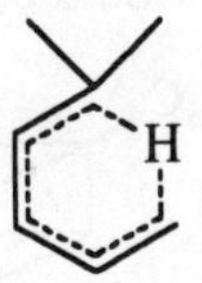

b. 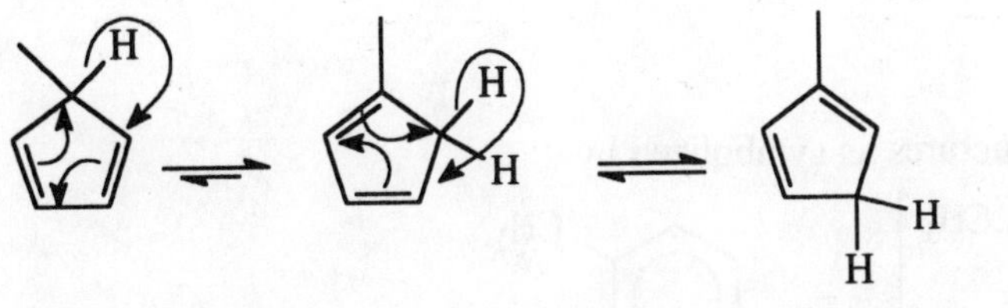 H, H, H, H, H

All of the rearrangements are [1.5]sigmatropic rearrangements

21.E Answers and Explanations for Problems

1. a. 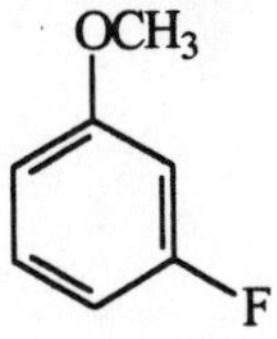 OCH_3, F b. 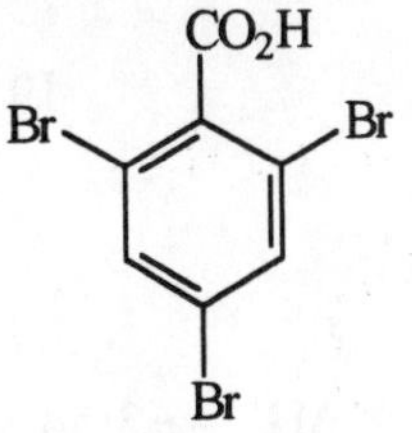 CO_2H, Br, Br, Br c. 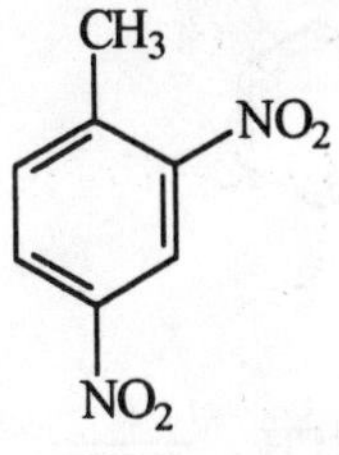 CH_3, NO_2, NO_2

d. 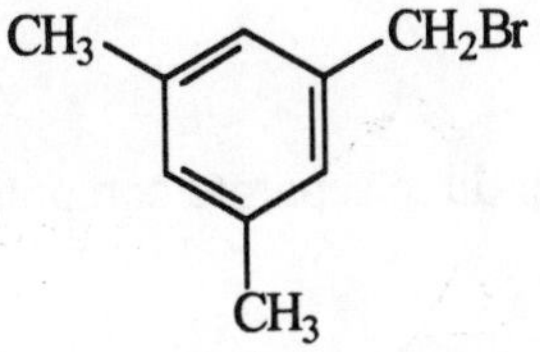 CH_3, CH_2Br, CH_3 e. 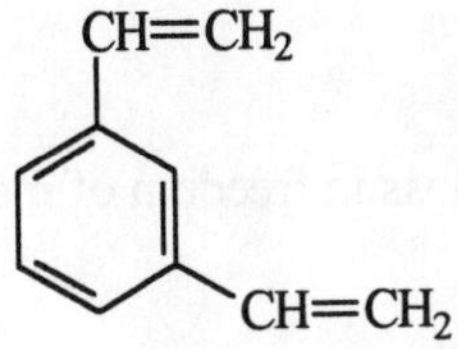 $CH{=}CH_2$, $CH{=}CH_2$ f. NC—, —C≡CH

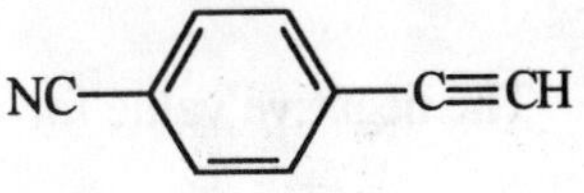

g. 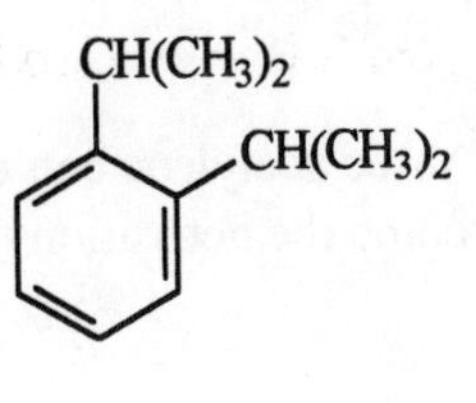 $CH(CH_3)_2$, $CH(CH_3)_2$ h. 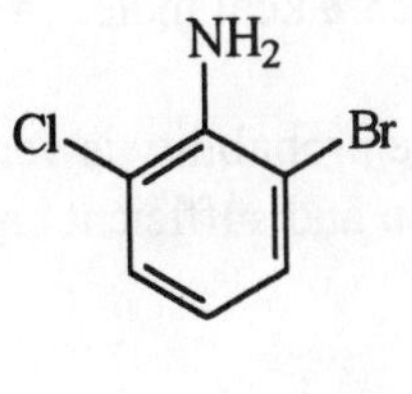 NH_2, Cl, Br

2. a. 1,1,1-trichloro-2,2-di(4-chlorophenyl)ethane (the abbreviation "DDT" comes from the old, non-systematic name: dichlorodiphenyltrichloroethane)
   b. *p*-bromopropylbenzene
   c. 3-bromo-4-iodocumene
   d. 2-bromo-4-ethyltoluene
   e. 2-(*p*-nitrophenyl)butane
   f. 2-(*m*-chlorophenyl)-3-methylbutane
   g. diphenyl-(4-methylphenyl)methanol
   h. *p*-bromochlorobenzene
   i. 4-bromo-3-fluoro-2-iodotoluene
   j. *m*-methoxybenzaldehyde or *m*-anisaldehyde
   k. 4-bromo-2,6-dimethylbenzoic acid
   l. 1,2,4-trimethylbenzene

3. a. No. There would be two isomers of the form

H_3C CH_3 and CH_3 CH_3

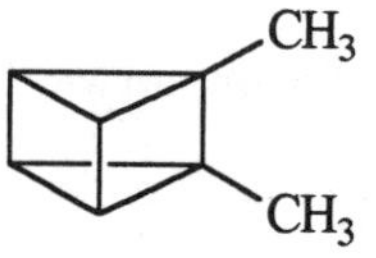

   b. The two structures are resonance structures as symbolized by

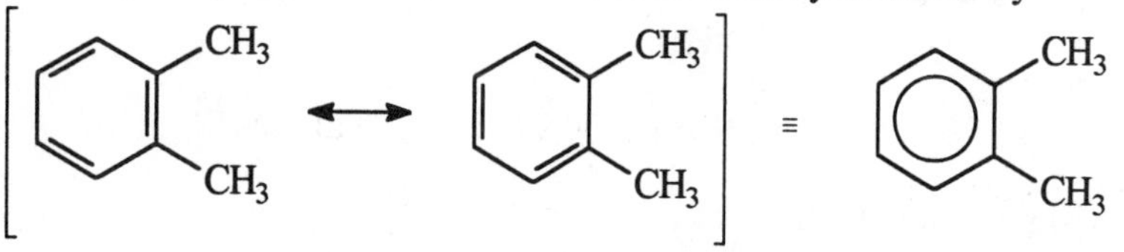

CH_3 CH_3 ⟷ CH_3 CH_3 ≡ CH_3 CH_3

4.

	ΔH_f°
CH≡CH	54.3 kcal $mole^{-1}$
 (benzene)	19.8 kcal $mole^{-1}$

3 HC≡CH ⇌ (benzene) $\Delta H^\circ = -3(54.3) + 19.8 = -143.1$ kcal $mole^{-1}$

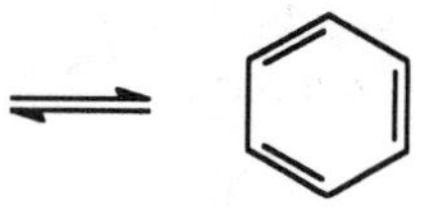

The negative value for ΔS° reflects the loss in freedom of motion when three separate compounds form one.

$$\Delta G^\circ = \Delta H^\circ - T\Delta S^\circ$$

At 298K: $\Delta G^\circ = -143.1 \text{ kcal mole}^{-1} - (298\text{K} \times -79.7 \text{ cal K}^{-1}\text{ mole}^{-1})$

$\Delta G^\circ = -143.1 \text{ kcal mole}^{-1} + 23.8 \text{ kcal mole}^{-1} = -119.3 \text{ kcal mole}^{-1}$

The equilibrium lies far to the right, but the probability is very small that three acetylene can collide at the same time with the proper orientation for reaction and sufficient energy to overcome the activation energy barrier.

5. a. cyclohexene $+ H_2 \longrightarrow$ cyclohexane $\Delta H = -28.4$ kcal mole^{-1}

∴ benzene $+ 3\ H_2 \longrightarrow$ cyclohexane $\Delta H_{calc} = 3 \times (-28.4) = -85.2$ kcal mole^{-1}

$-85.2 - (-49.3) = -35.9$ kcal mole^{-1}

Empirical resonance energy = 35.9 kcal mole^{-1}

b. cyclooctene $+ H_2 \longrightarrow$ cyclooctane $\Delta H° = -23.3$ kcal mole^{-1}

cyclooctatetraene $+ 4\ H_2 \longrightarrow$ cyclooctane $\Delta H° = -100.9$ kcal mole^{-1}

$-93.2 - (-100.9) = +7.7$ kcal mole^{-1}

Empirical resonance energy = -7.7 kcal mole^{-1}

The negative value implies that the four double bonds in the tetraene are less stable than 4 x 1 double bonds; i.e., this value represents a destabilization energy or negative resonance energy. This means that not only is there no stabilization energy (resonance), but that cyclooctatetraene is probably more strained because of the four double bonds.

c. benzene (C_6H_6, H on each carbon)

$6 \times E(C{-}H) = 6 \times 99 = 594$ kcal mole^{-1}
$3 \times E(C{-}C) = 3 \times 83 = 249$ kcal mole^{-1}
$3 \times E(C{=}C) = 3 \times 146 = \underline{438 \text{ kcal mole}^{-1}}$
calc. $\Delta H°_{atom}$ 1281 kcal mole^{-1}

Empirical resonance energy = 1318 - 1281 = 37 kcal mole^{-1}

6. $C_6H_6 + Cl\cdot \longrightarrow C_6H_5\cdot + HCl$

$\Delta H°_f =$ 19.8 28.9 79 -22.1 $\Delta H° = +8$ kcal mole^{-1}

$C_6H_5\cdot + Cl_2 \longrightarrow C_6H_5Cl + Cl\cdot$

$\Delta H°_f =$ 79 0 12.2 28.9 $\Delta H° = -38$ kcal mole^{-1}

Although the overall reaction is highly exothermic, the first step is endothermic and occurs quite slowly in comparison to other chlorination reactions. The overall reaction proceeds cleanly and in high yield at 300-400°C.

7\. a. three aromatic resonances

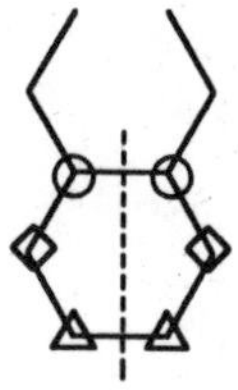

b\. four aromatic resonances

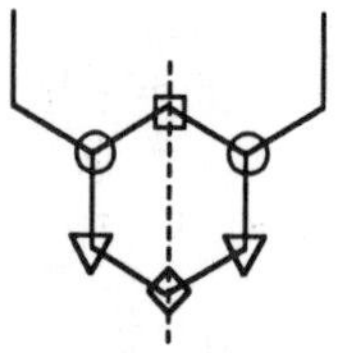

c\. two different aromatic resonances

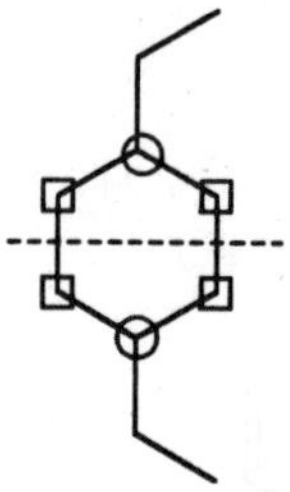

8\. a. $CH_3CHCH_2CH_2OH$ (phenyl on C-2)

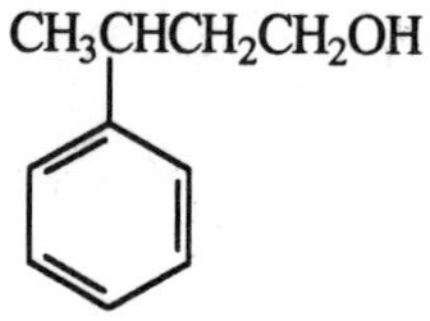

b\. CH_3 / CH_3 (1,4-dimethyl-1,4-cyclohexadiene)

c\. $(CH_3)_2COCH_3$ (phenyl)

d\. $(CH_3)_2COH$ (phenyl)

e\. CH_2CH_3 / CH_2CH_2OH (para-disubstituted benzene)

f\. O=CCH$_3$ / CH_2CH_2OH (para-disubstituted benzene)

g\. H_3C (4-methylcyclohex-2-enone) and H_3C (4-methylcyclohex-3-enone)

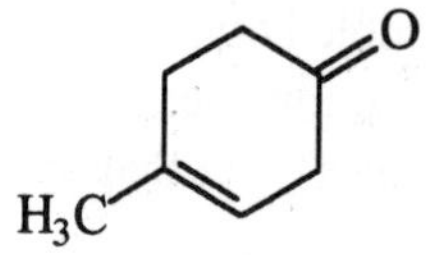

9\. a. benzene $\xrightarrow[AlBr_3]{Br_2}$ $\xrightarrow{Mg}$ $\xrightarrow{CH_3CH_2COCH_2CH_3}$ $CH_3CH_2C(OH)CH_2CH_3$ (phenyl) $\xrightarrow[HClO_4]{H_2/Pd}$ $CH(CH_2CH_3)_2$ (phenyl)

b\. CH_3 (toluene) $\xrightarrow[liq.\ NH_3]{Na,\ C_2H_5OH}$ CH_3 (1-methyl-1,4-cyclohexadiene)

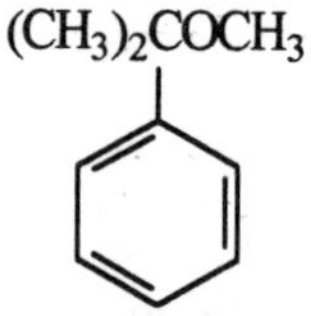

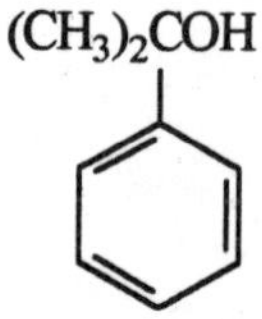

c\. CH_2CH_3 (ethylbenzene) $\xrightarrow[h\nu]{Br_2}$ CBr_2CH_3 (phenyl) $\xrightarrow[\Delta]{NaOH}$ $C\equiv CH$ (phenyl)

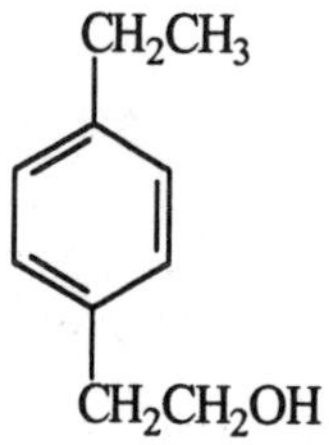

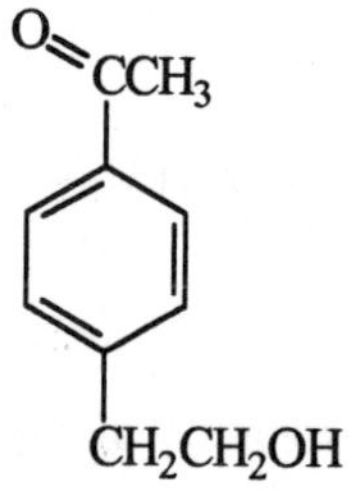

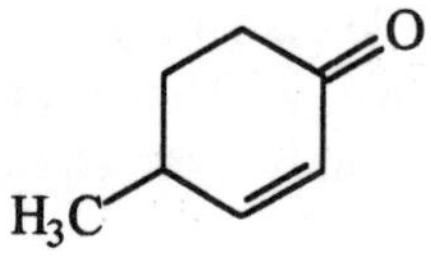

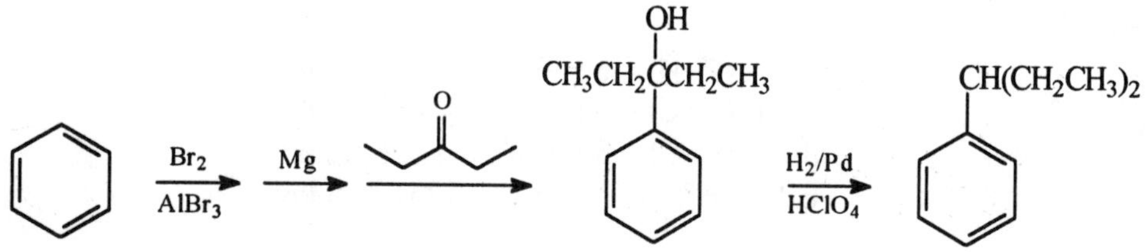

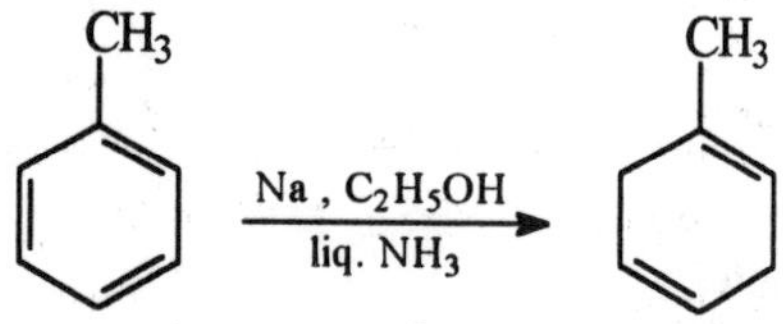

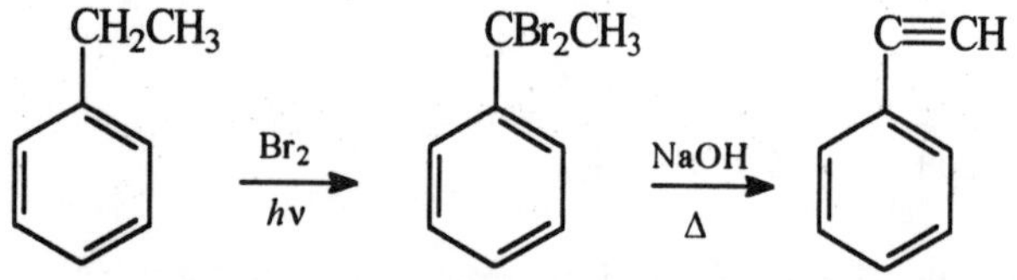

d. Toluene ($C_6H_5CH_3$) $\xrightarrow[h\nu]{Cl_2}$ $C_6H_5CH_2Cl$ $\xrightarrow[ether]{Mg}$ $C_6H_5CH_2MgCl$ $\xrightarrow{C_6H_5CH_2Cl}$ C_6H_5–CH_2CH_2–C_6H_5

e. p-Xylene (CH_3–C_6H_4–CH_3) $\xrightarrow[h\nu]{B_2}$ $BrCH_2$–C_6H_4–CH_3 $\xrightarrow[\text{or } LiAlD_4]{D_2/Pt}$ DCH_2–C_6H_4–CH_3

f. $C_6H_5CH_2Cl$ (from d.) $\xrightarrow{NaC{\equiv}CH}$ $C_6H_5CH_2C{\equiv}CH$

10. a. and b. $Br_2 \longrightarrow 2\ Br\cdot$ *(initiation)*

$C_6H_5CH_3$ + Br· $\longrightarrow$ $C_6H_5CH_2\cdot$ + HBr

$\Delta H_f^\circ = 12.0$ 26.7 48 -8.7 $\Delta H^\circ = 1$ kcal mole^{-1}

$C_6H_5CH_2\cdot$ + Br_2 $\longrightarrow$ $C_6H_5CH_2Br$ + Br·

$\Delta H_f^\circ = 48$ 7.4 17 26.7 $\Delta H^\circ = -12$* kcal mole^{-1}

*Since Appendix I does not have ΔH_f° for benzyl bromide, the ΔH_f° can be calculated as follows:

$\Delta H_f^\circ (C_6H_5CH_2Br) = \Delta H_f^\circ (C_6H_5CH_2\cdot) + \Delta H_f^\circ (Br\cdot) - DH^\circ (C_6H_5CH_2Br)$

48 26.7 -58 = 17 kcal mole^{-1}

The first step is barely endothermic and occurs readily; the second step is quite exothermic.

c. For ethane: the first step is very endothermic and occurs slowly:

$CH_3CH_3 + Br\cdot \longrightarrow CH_3CH_2\cdot + HBr$ $\quad \Delta H° = 14$ kcal mole^{-1}

-20.2 26.7 29 -8.7

For 2-methylpropane: this step is still endothermic, but less so:

$(CH_3)_3CH + Br\cdot \longrightarrow (CH_3)_3C\cdot + HBr$ $\quad \Delta H° = 9$ kcal mole^{-1}

-32.4 26.7 12 -8.7

11. [*Note*: in the drawings below, only Cl bonds are shown. Each position also has a hydrogen.]

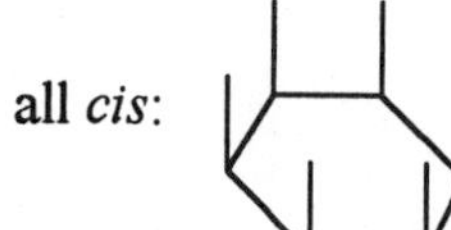

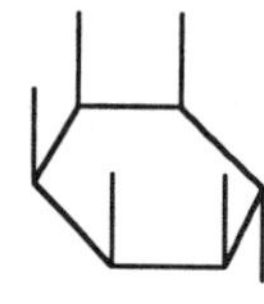
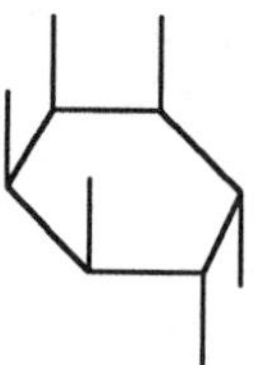
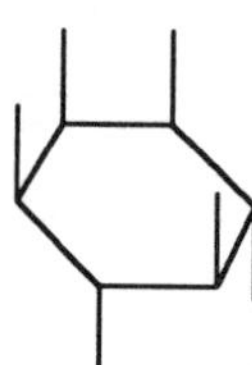

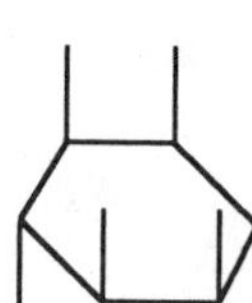
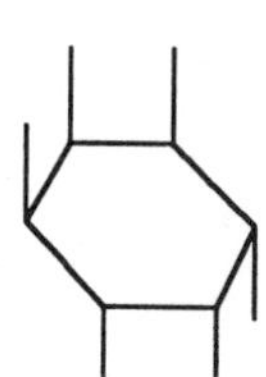
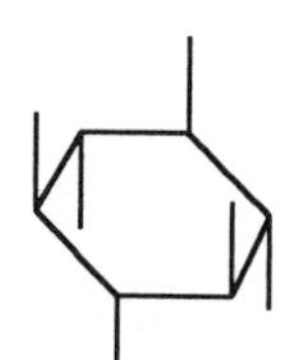
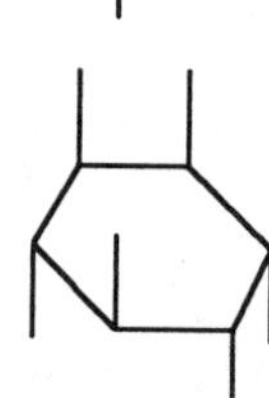

slowest in E_2 elimination since there are no H's anti to a Cl

chiral: all others have a plane of symmetry.

12. $C_6H_4(CH_2OH)_2 + H^+ \rightleftharpoons C_6H_4(CH_2\overset{+}{O}H_2)(CH_2\ddot{O}H) \xrightarrow{-H_2O}$ (cyclic $^+$OH) $\xrightleftharpoons{-H^+}$ (cyclic ether, O)

13. $CH_3\overset{O}{\overset{\|}{C}}OCH_2CH{=}CH_2 \longrightarrow \left[{}^-CH_2\overset{O}{\overset{\|}{C}}OCH_2CH{=}CH_2 \longleftrightarrow \right.$ (cyclic, O^-, O) $\left. \right] \longrightarrow$ (O$^-$, O) $\xrightarrow{H^+}$ $CH_2{=}CHCH_2CH_2CO_2H$

14. (D, H) $\xrightarrow[{[3.3]}]{\text{suprafacial}}$ (D, H) $\equiv$ (D, H)

This is a product from an allowed [3.3]sigmatropic rearrangement. The other isomer cannot arise by an allowed pathway.

15. 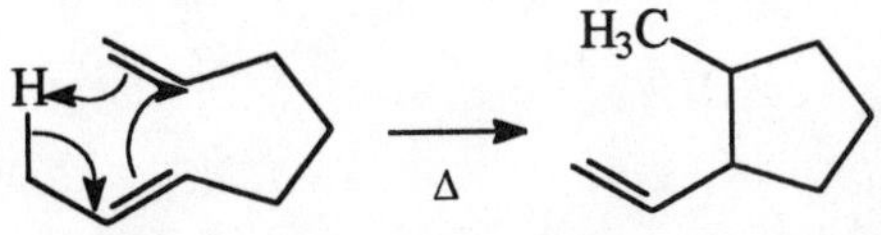 H H_3C Δ

six electrons moving in a ring = aromatic transition state

16. a. 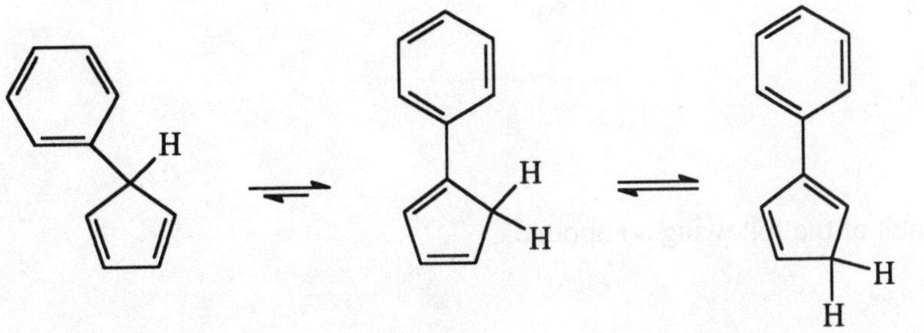 H H H H H

b. 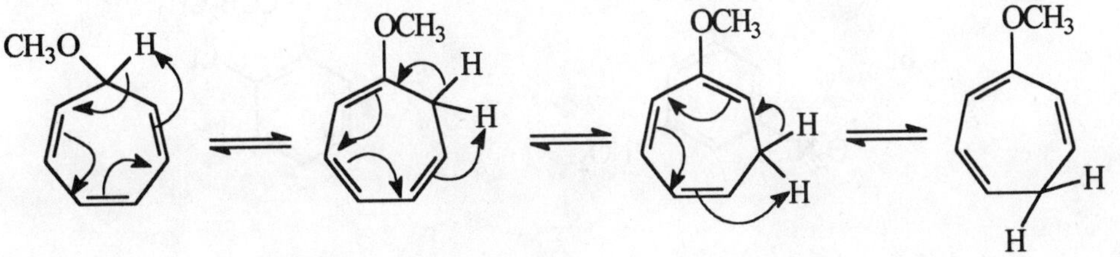 CH_3O H OCH_3 H H OCH_3 H H OCH_3 H H

c. 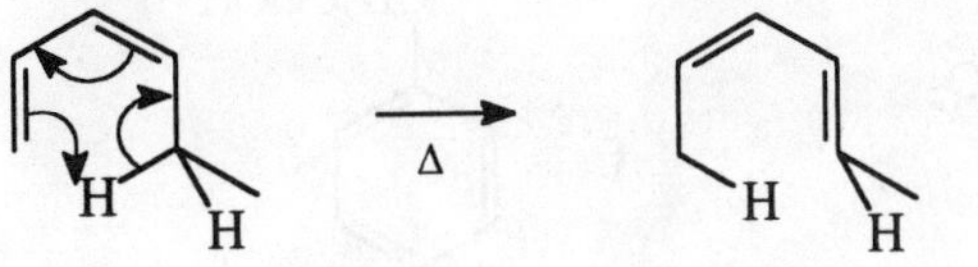 Δ H H H H

17. 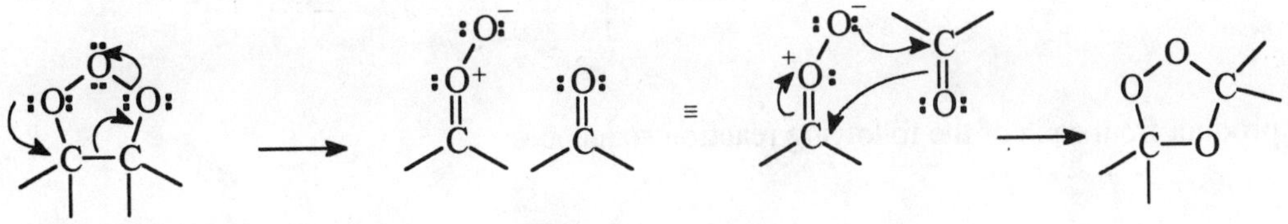 :O: :O: :O: :O: :O: C C C O C O O C O

18. *m*-Xylene will give three different monobromoisomers:

CH_3 Br CH_3 CH_3 CH_3 Br CH_3 Br CH_3

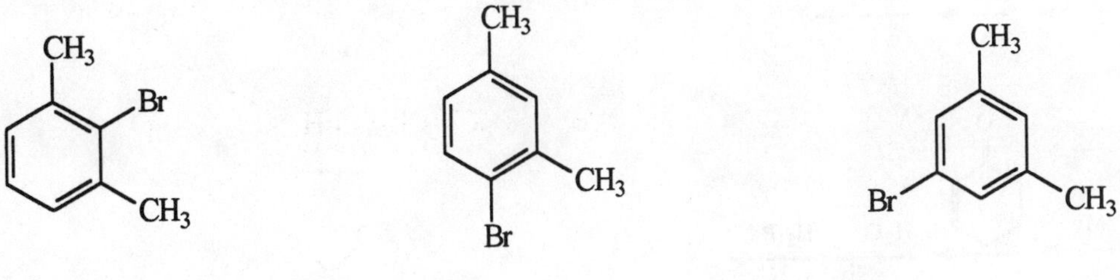

o-Xylene gives two different monobromoisomers:

CH_3 CH_3 Br CH_3 CH_3 Br

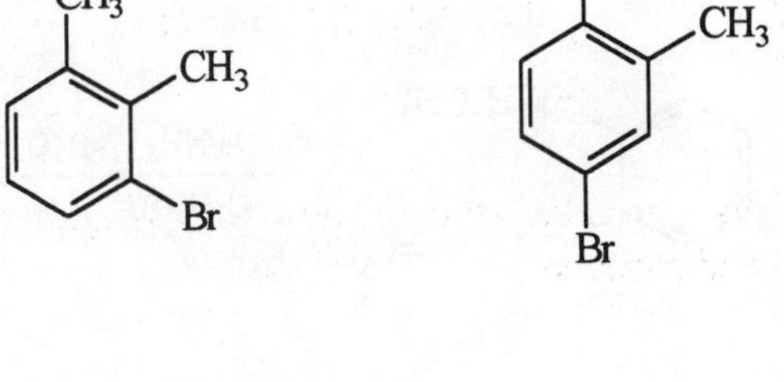

p-Xylene gives only one monobromoisomer:

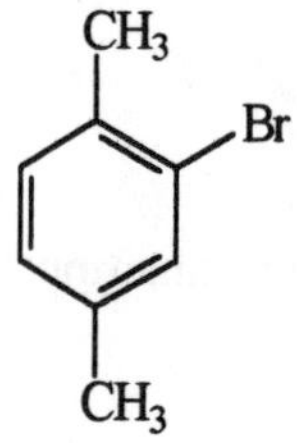

CH_3 / Br / CH_3

21.F. Supplementary Problems

S1. Provide an acceptable name for each of the following compounds.

a. $C(CH_3)_3$ / $C(CH_3)_3$

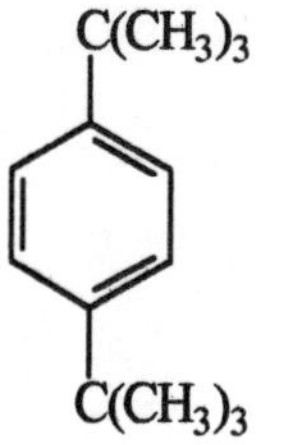

b. COCl / O_2N / NO_2

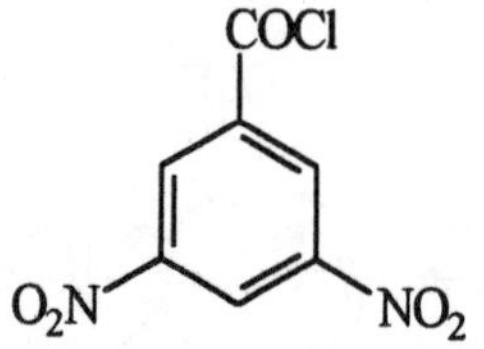

c. $O_2CCH_2CH_3$ / Cl

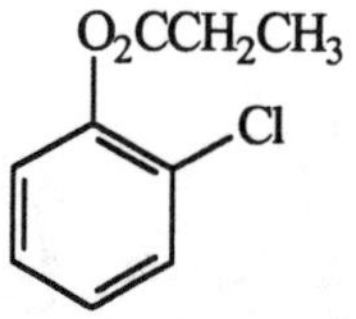

d. O / $CCH_2CH_2CH(CH_3)_2$

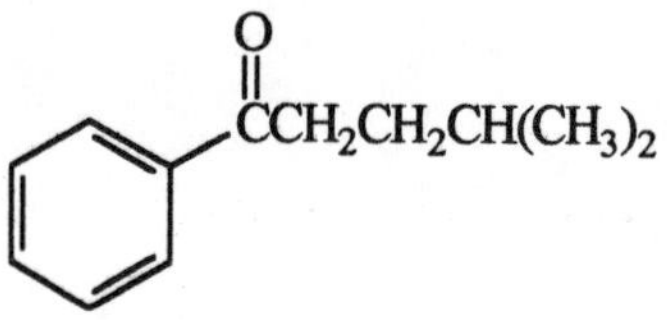

e. Cl / NO_2 / NO_2

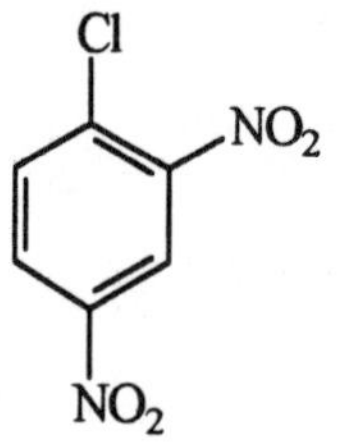

f. CH_2COOH / CH_3

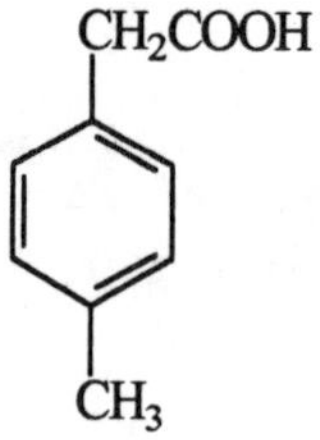

S2. Predict the major product from each of the following reaction sequences:

a. CH_2CH_3 $\xrightarrow[h\nu]{Br_2}$ $\xrightarrow[\Delta]{H_2O}$ $\xrightarrow{MnO_2}$

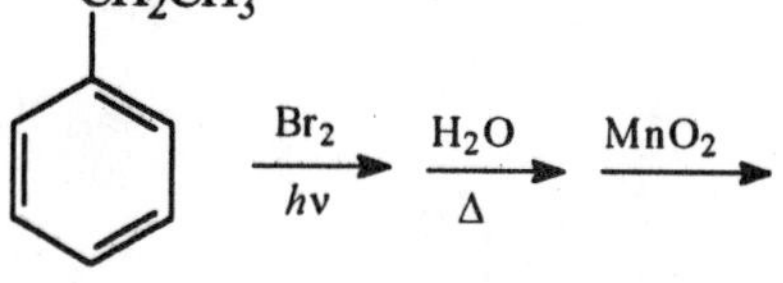

b. Br $\xrightarrow{Mg}$ (O, cyclohexenone) $\xrightarrow{}$ $\xrightarrow{H_2O}$ $\xrightarrow[H^+]{H_2/Pd}$

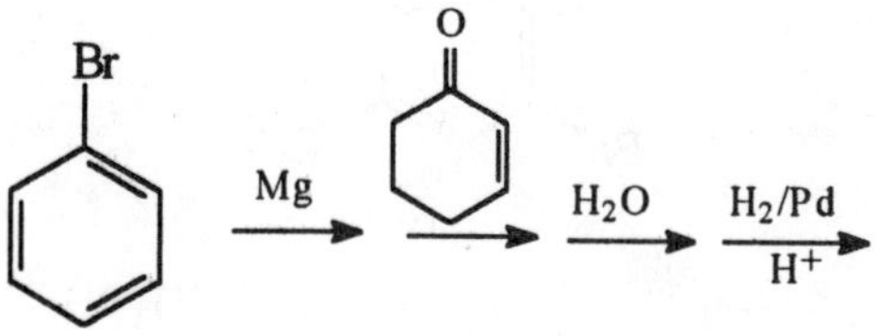

c. OCH_2CH_3 / $CH_2CH{=}C(CH_3)_2$ $\xrightarrow[C_2H_5OH]{Na/NH_3}$ $\xrightarrow{H_3O^+}$

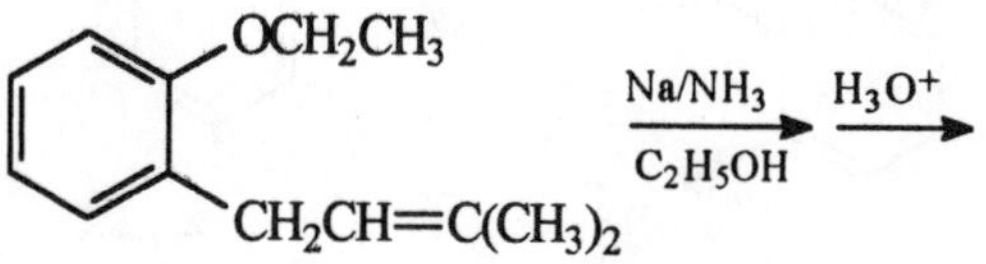

d. 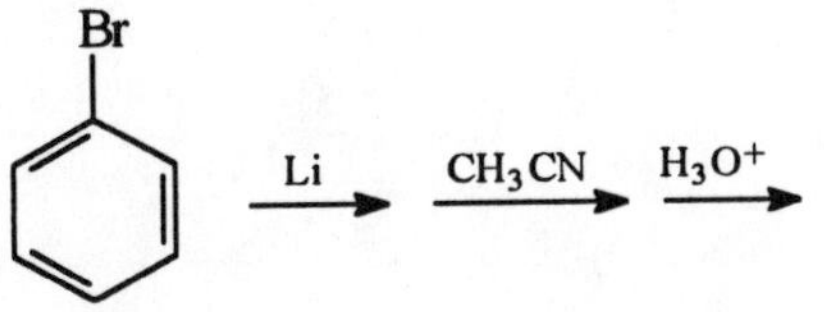 Br-benzene $\xrightarrow{Li}$ $\xrightarrow{CH_3CN}$ $\xrightarrow{H_3O^+}$

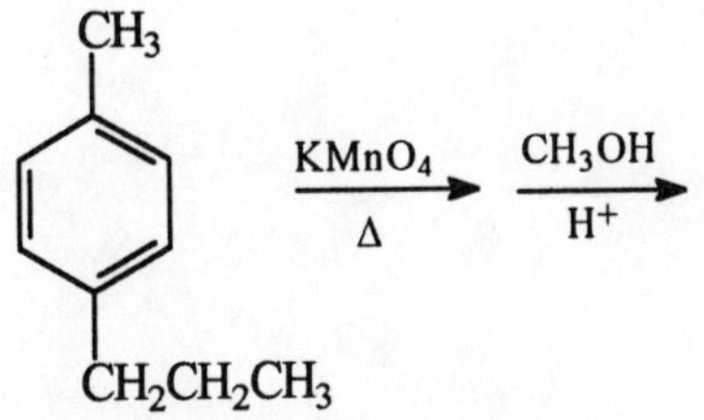

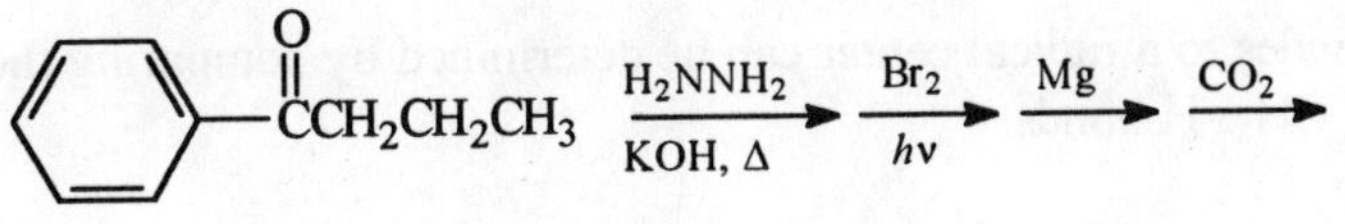

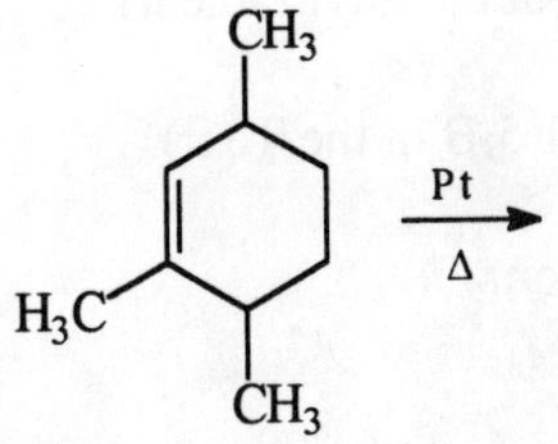

e. CH_3 / $CH_2CH_2CH_3$ (p-substituted benzene) $\xrightarrow[\Delta]{KMnO_4}$ $\xrightarrow[H^+]{CH_3OH}$

f. $C_6H_5CCH_2CH_2CH_3$ (C=O) $\xrightarrow[KOH,\ \Delta]{H_2NNH_2}$ $\xrightarrow[h\nu]{Br_2}$ $\xrightarrow{Mg}$ $\xrightarrow{CO_2}$

g. CH_3, H_3C, CH_3 (trimethylcyclohexene) $\xrightarrow[\Delta]{Pt}$

h. CH_2OCH_3 (benzene) $\xrightarrow[D^+]{D_2/Pd}$

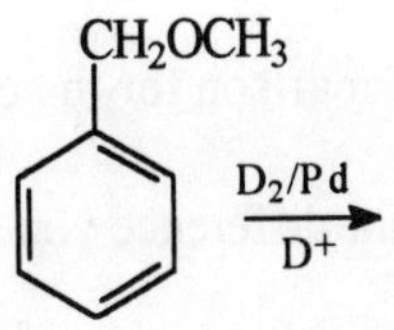

S3. Predict the major product from reaction of *cis*-1-phenylpropene with each of the following reagents, and explain your choice.

a. HCl
b. Br_2/CH_3OH
c. HBr, peroxides
d. $Hg(OAc)_2/CH_3OH$; then $NaBH_4$

S4. Which of the two compounds illustrated below is more acidic? Why?

CH_3 (toluene) CH_2, H, H (methylenecyclohexadiene)

S5. Which of the following retro-Diels-Alder reactions will take place most easily? Which one will be the most difficult? Why?

a. 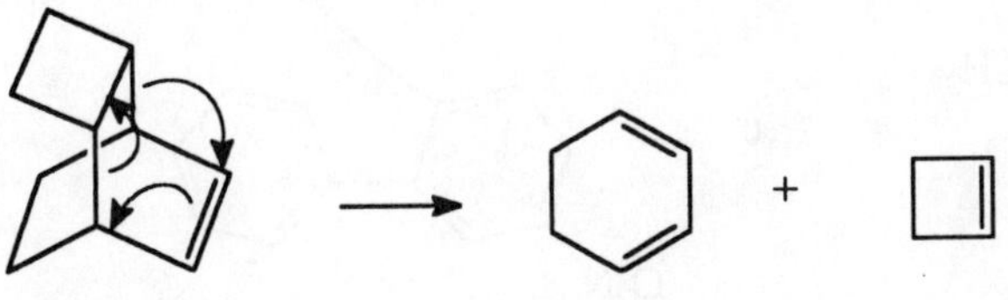 → (cyclohexadiene) + (cyclobutene)

b.

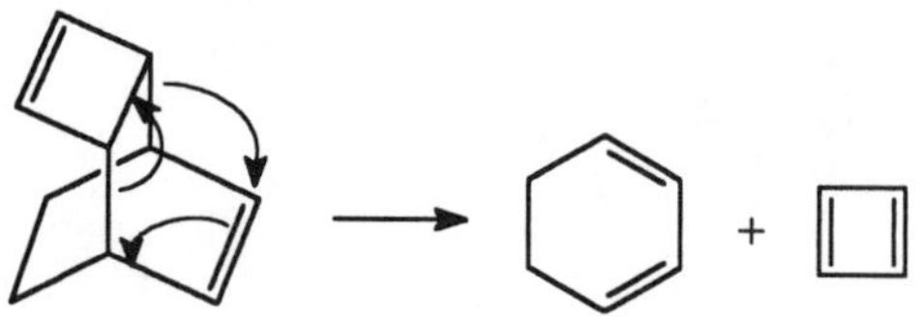

$\longrightarrow$ +

c.

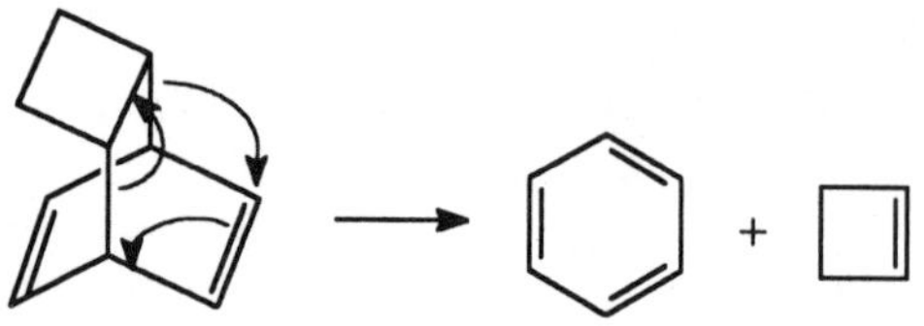

$\longrightarrow$ +

S6. The stabilization that a phenyl group provides to a radical center can be determined by comparing the bond dissociation energies of $C_6H_5CH_2{-}H$ and $CH_3{-}H$ bonds.

a. How does this stabilization compare with that provided by a vinyl group in the allyl radical?

b. Perform the same comparison for the cations (see Sections 20.1.A and 21.5.B in the Text).

c. Explain any significant difference you see in radical vs. cation stabilization.

S7. Benzylic alcohols and ethers are cleaved under the conditions of the Birch reduction:

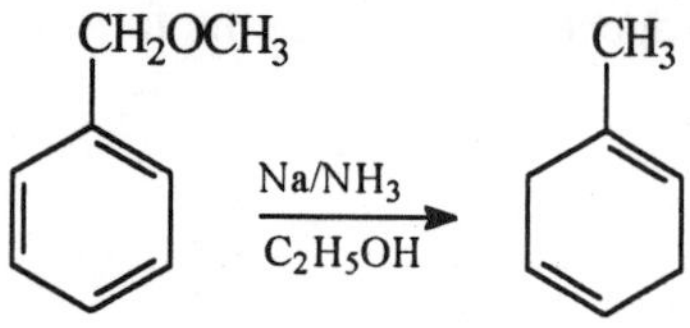

CH_2OCH_3 $\xrightarrow[C_2H_5OH]{Na/NH_3}$ CH_3

a. Write a mechanism for this reaction.

b. Explain why the following ether undergoes Birch reduction without cleavage:

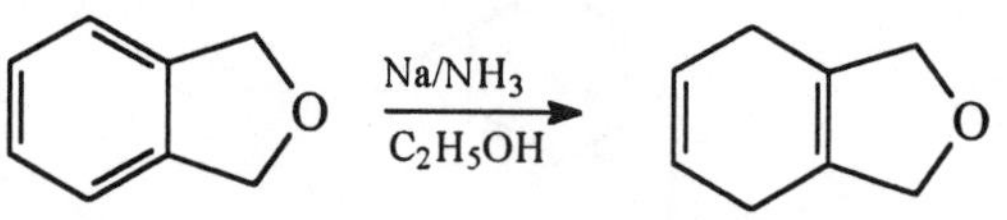

O $\xrightarrow[C_2H_5OH]{Na/NH_3}$ O

S8. Rank the following compounds in order of acidity.

a. H H

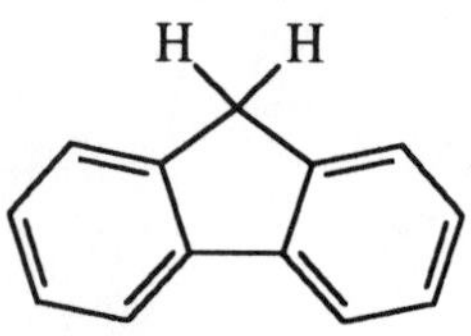

b. CH_3

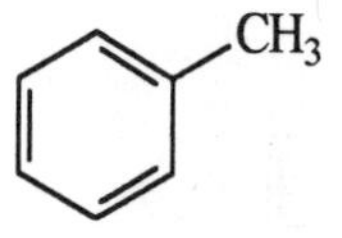

c. H H O_2N

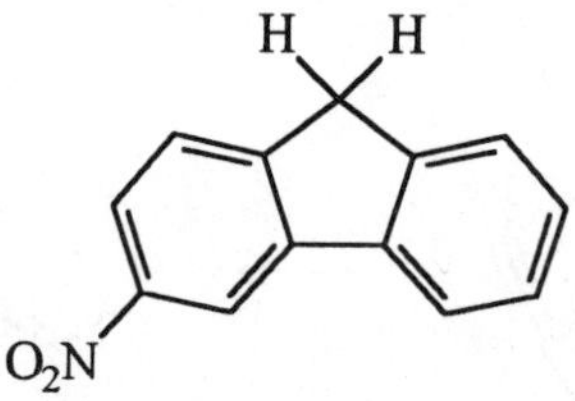

d. 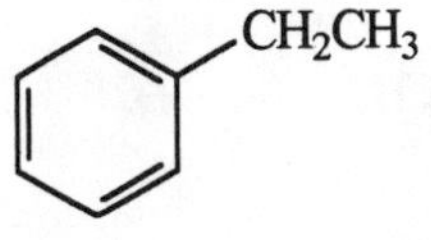 $C_6H_5CH_2CH_3$ e. 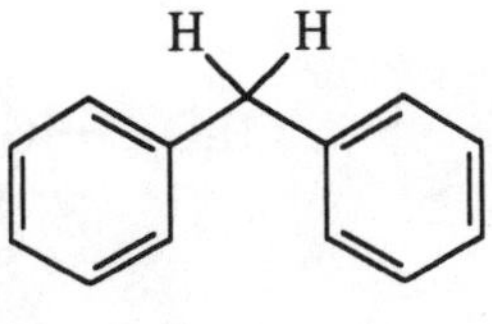 $C_6H_5CH_2C_6H_5$

S9. The aliphatic Claisen rearrangement involves a "chair-like" transition state, as depicted below.

a. If a secondary allylic vinyl ether is the substrate for the rearrangement, a *trans*-olefin is produced selectively. Show how this observation is consistent with the transition state illustrated above.

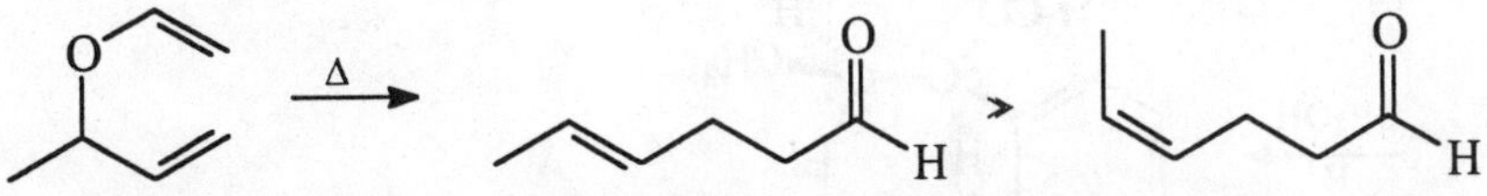

Δ

b. What products would you expect from the Claisen rearrangement of the following enol ethers?

1.

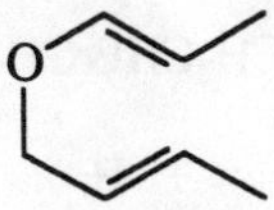

2.

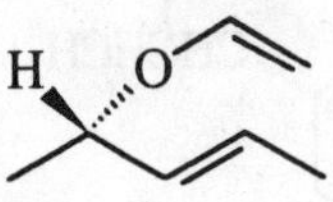

S10. *Para*-methoxybenzyl bromide, illustrated below, hydrolyzes in water many times faster than the *meta* isomer. Explain why this is so.

$$CH_3O-C_6H_4-CH_2Br + H_2O \longrightarrow CH_3O-C_6H_4-CH_2OH + HBr$$

21.G. Answers to Supplementary Problems

S1. a. *p*-di-*t*-butylbenzene
b. 3,5-dinitrobenzoyl chloride
c. *o*-chlorophenyl propionate, or 2-chlorophenyl propanoate
d. 4-methyl-1-phenyl-1-pentanone
e. 1-chloro-2,4-dinitrobenzene
f. *p*-methylphenylacetic acid

S2. a. 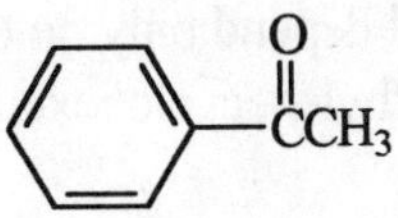 $C_6H_5\overset{O}{\overset{\|}{C}}CH_3$ b. $C_6H_5-C_6H_{11}$ c. $CH_2CH{=}C(CH_3)_2$

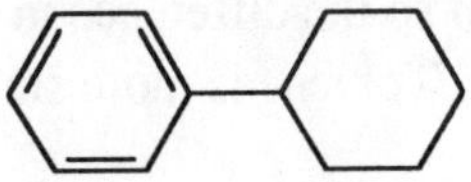

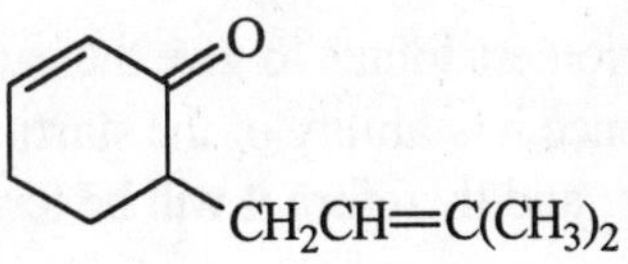

d. 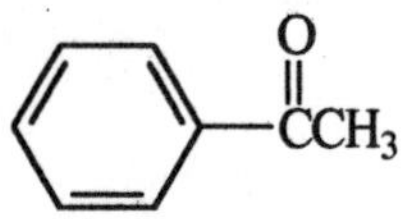 $C_6H_5CCH_3$ (with C=O)

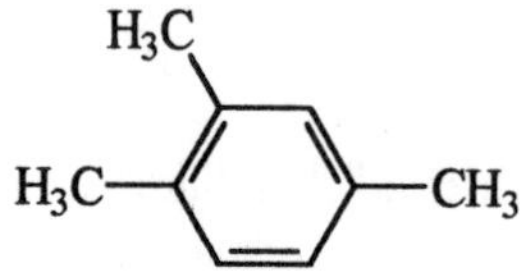

e. 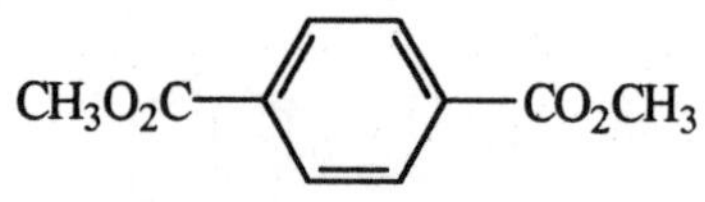 $CH_3O_2C-C_6H_4-CO_2CH_3$

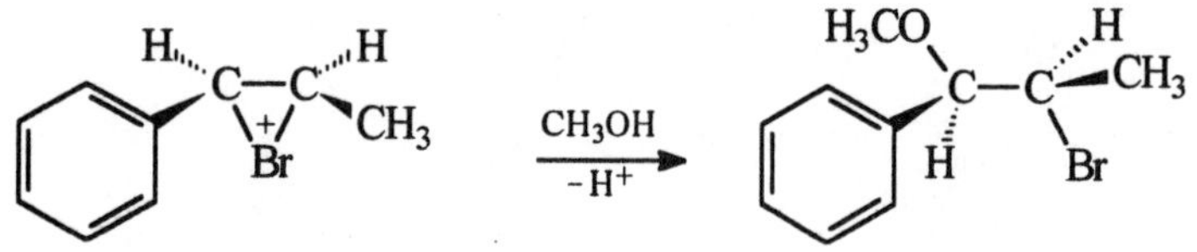

f. 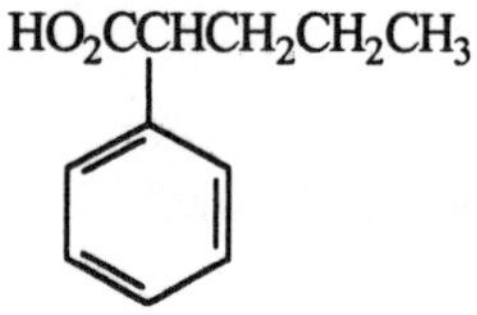 $HO_2CCHCH_2CH_2CH_3$ (with C_6H_5 on C-2)

g. H_3C, H_3C-, $-CH_3$ (trimethylbenzene)

h. $C_6H_5-CH_2D$

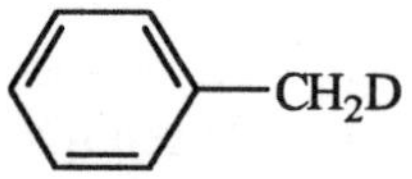

S3. a. 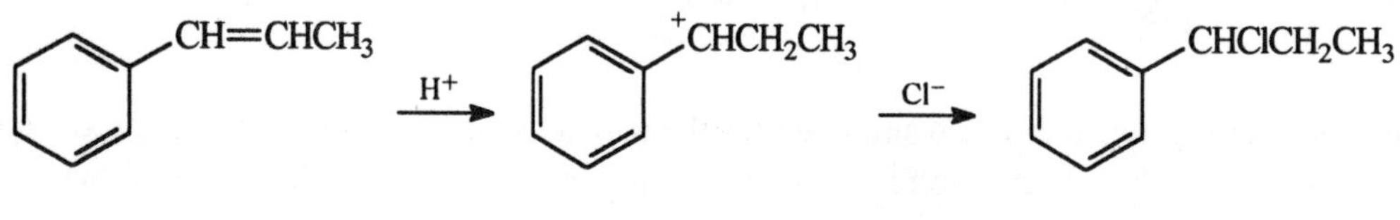 $C_6H_5CH{=}CHCH_3 \xrightarrow{H^+} C_6H_5\overset{+}{C}HCH_2CH_3 \xrightarrow{Cl^-} C_6H_5CHClCH_2CH_3$

b. (bromonium ion: H, C—C, H, CH_3, Br$^+$) $\xrightarrow[-H^+]{CH_3OH}$ H_3CO, H, C—C, CH_3, H, Br

c. $C_6H_5CH{=}CHCH_3 \xrightarrow{Br\cdot} C_6H_5\dot{C}HCHCH_3$ (Br) $\xrightarrow{HBr} C_6H_5CH_2CHBrCH_3$ + Br·

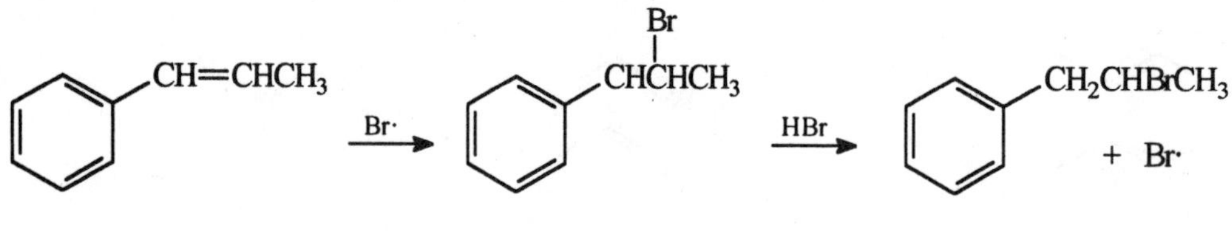

d. 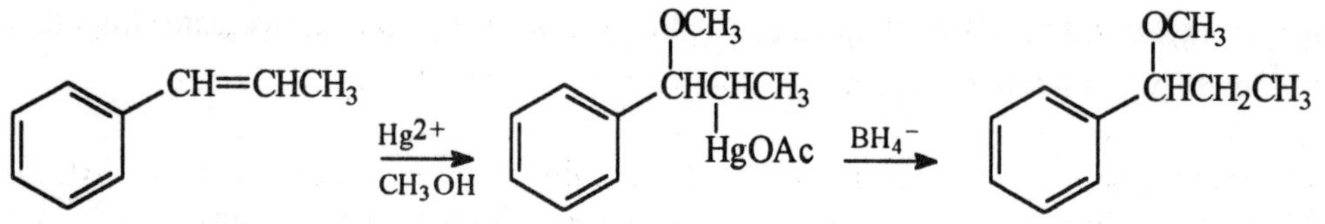 $C_6H_5CH{=}CHCH_3 \xrightarrow[CH_3OH]{Hg^{2+}} C_6H_5CHCHCH_3$ (OCH_3, HgOAc) $\xrightarrow{BH_4^-} C_6H_5CHCH_2CH_3$ (OCH_3)

In a., b., and d., the additions proceed in the Markovnikov sense (the most stable carbocation is formed); in c., Br· adds so as to afford the most stable radical. Phenyl stabilizes radicals and carbocations better than a simple alkyl group does.

S4. 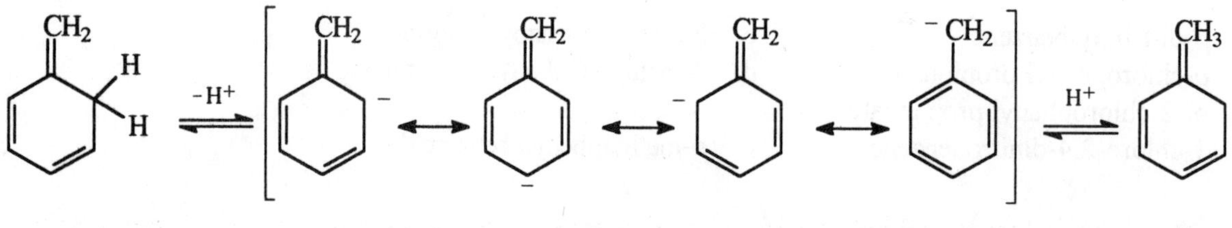 (CH_2, H, H) $\underset{}{\overset{-H^+}{\rightleftharpoons}}$ [CH_2 $^-$ ⟷ CH_2 $^-$ ⟷ CH_2 $^-$ ⟷ $^-CH_2$] $\overset{H^+}{\rightleftharpoons}$ CH_3

Both isomers ionize to give the same carbanion so the difference in equilibrium will depend only on the difference in stability of the starting materials. Toluene is more stable than the methylenecyclohexadiene isomer, and therefore it will be less acidic.

S5. Reaction c. will occur most rapidly because the aromatic stabilization of benzene is gained. Reaction b. will be the most difficult because the very unstable (high-energy) cyclobutadiene is being formed. In reaction a., no aromatic or antiaromatic compounds are being produced.

S6. a. $C_6H_5CH_2-H$ $\quad$ CH_3-H $\quad$ $CH_2=CH-CH_2-H$

	$C_6H_5CH_2-H$	CH_3-H	$CH_2=CH-CH_2-H$	
DH° =	88	105	86	kcal mole^{-1}
Stabilization:	17		19	kcal mole^{-1}

Stabilization of a radical is about the same for phenyl and vinyl.

b.

<u>ΔH° (kcal mole^{-1})</u>

$$C_6H_5CH_2Cl \longrightarrow C_6H_5CH_2^+ + Cl^- \quad 155$$
(4.5) (214) (−54.5)

$$CH_3Cl \longrightarrow CH_3^+ + Cl^- \quad 234$$
(−20.6) (262) (−54.5)

$$CH_2{=}CHCH_2Cl \longrightarrow CH_2{=}CHCH_2^+ + Cl^- \quad 171$$
(0) (225) (−54.5)

Stabilization: 79 kcal mole^{-1} (benzyl); 63 kcal mole^{-1} (allyl)

The ΔH_f° for $CH_2{=}CHCH_2Cl$ can be calculated as usual for the ΔH_f° for the allyl radical and the chlorine atom minus the bond dissociation energy for the allyl chloride bond: 39 + 28.9 − 68 = 0.

c. The phenyl and vinyl groups provide essentially the same stabilization to a radical center. (Although more carbons can share the radical in the benzylic case, those resonance structures lack the cyclic valence bond structure of an aromatic ring:

$\cdot CH_2$–C_6H_5 ⟷ CH_2= (radical ortho) ⟷ CH_2= (radical para) ⟷ CH_2= (radical ortho)

"*non-aromatic*"

In the case of a cation, the ability to distribute the positive charge over several atoms becomes more important, and the benzylic cation therefore becomes more stabilized than the allyl cation.

S7. a.

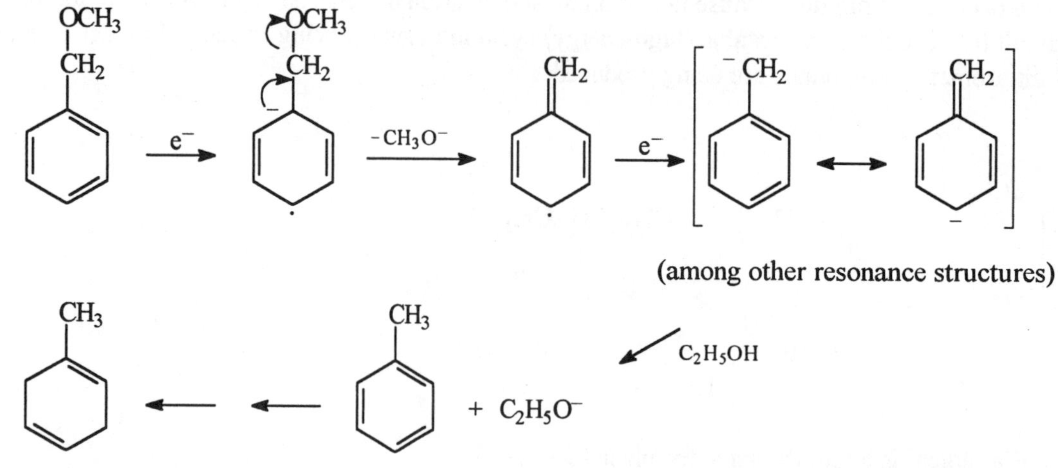

OCH_3 CH_2 → (e^-) → OCH_3 CH_2 → ($-CH_3O^-$) → CH_2 → (e^-) → [$^-CH_2$ ↔ CH_2]

(among other resonance structures)

C_2H_5OH

CH_3 ← ← CH_3 + $C_2H_5O^-$

b.

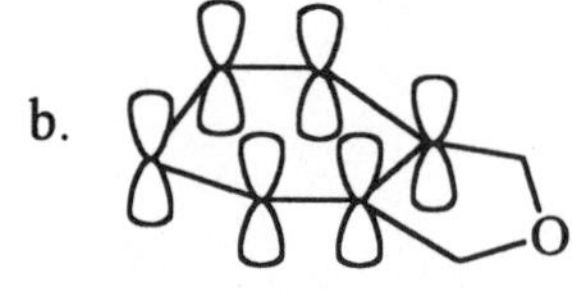

The C-O bonds are perpendicular to the p-orbitals of the π-system. Therefore the orbitals of the C-O bond cannot overlap with the aromatic p-orbitals, and bond cleavage cannot take place.

S8. *Most acidic*: c., because of stabilization of the carbanion by both coplanar rings and electron-withdrawing nitro group.

a. is more acidic than e. because the tricyclic structure holds both rings coplanar for maximum overlap with the p-orbital of the benzylic carbanion.

e. is more acidic than b. and d. because it has two phenyl rings which stabilize the carbanion.

b. is more acidic than d. because it does not have the electron-releasing methyl substitutent on the carbanionic carbon.

Least acidic: d.

S9. a. Since there is a chiral center in the molecule there are two diastereomeric chair-like transition states possible:

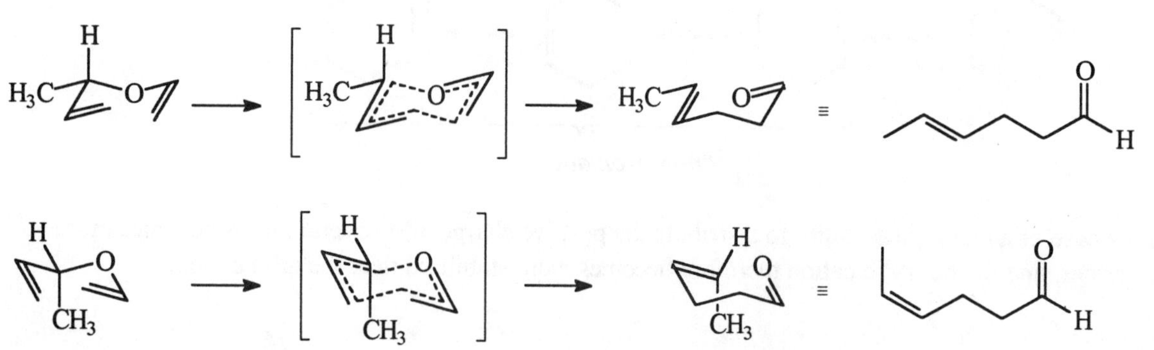

H H_3C O → [H H_3C O] → H_3C O ≡ O H

H O CH_3 → [H O CH_3] → H O CH_3 ≡ O H

Rearrangement by the first transition state is preferred because the methyl group is in an equatorial-like position, rather than the axial-like orientation in the transition state which leads to the *cis*-product.

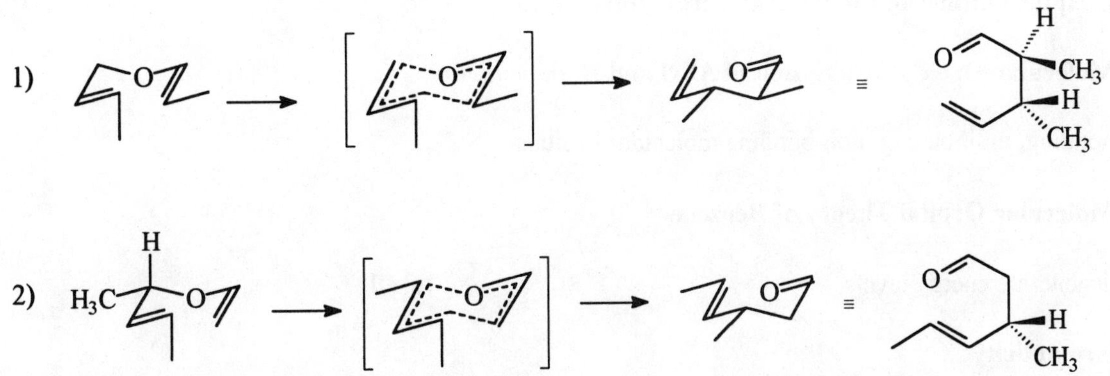

1)

2) H
H_3C

S10. The benzylic cation from S_N1-type cleavage of the *para*-isomer can be stabilized by the oxygen lone pair electrons by resonance:

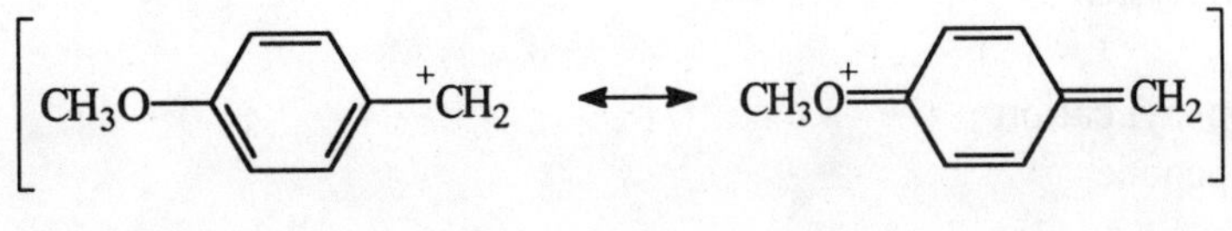

$$\left[CH_3O-C_6H_4-\overset{+}{C}H_2 \longleftrightarrow CH_3\overset{+}{O}=C_6H_4=CH_2 \right]$$

No such resonance structure is possible for the *meta*-isomer.

22. MOLECULAR ORBITAL THEORY

22.A. Chapter Outline and Important Terms Introduced

22.1 Molecular Orbital Description of Allyl and Butadiene

bonding, antibonding, non-bonding molecular orbitals

22.2 Molecular Orbital Theory of Benzene

degenerate energy levels shell

22.3 Aromaticity

A. Cyclooctatetraene: The Hückel 4n + 2 Rule

aromatic vs. non-aromatic or antiaromatic

B. Two-Electron Systems

cyclopropenyl cation
cyclopropenone

C. Six-Electron Systems

<u>not</u> cyclobutadiene (antiaromatic)
cyclopentadienyl anion
cycloheptatrienyl cation

D. Ten-Electron Systems

cyclononatetraenyl anion

E. Larger Cyclic π-Systems

annulenes

22.4 Hückel Transition States

Stabilization of carbocation rearrangements

22.5 Möbius Transition States

A. Electrocyclic Reactions
conrotatory vs. disrotatory motion
Woodward-Hoffmann Rules
Hückel vs. Möbius molecular orbital systems
B. Cycloaddition Reactions
suprafacial vs. antarafacial

22.6. Ultraviolet Spectroscopy

A. Electronic Transitions
   excited states — ground state
B. $\pi \rightarrow \pi^*$ Transitions
   longer conjugation → longer wavelength
C. $n \rightarrow \pi^*$ Transitions
   extinction coefficient
D. Charge-Transfer Transitions
   electron donor — electron acceptor
   empirical parameters of solvent polarity
E. Alkyl Substituents
   hyperconjugation
F. Benzene
   symmetry forbidden
G. Other Functional Groups
H. Photochemical Reactions

22.7. Perturbational MO Approach to Reactivity

Highest **O**ccupied **M**olecular **O**rbital (HOMO)
Lowest vacant (= **U**noccupied) **M**olecular **O**rbital (LUMO)
frontier orbitals
donor-acceptor

22.B. Important Reactions Introduced

Electrocyclic reactions (22.5.A)

Equation:

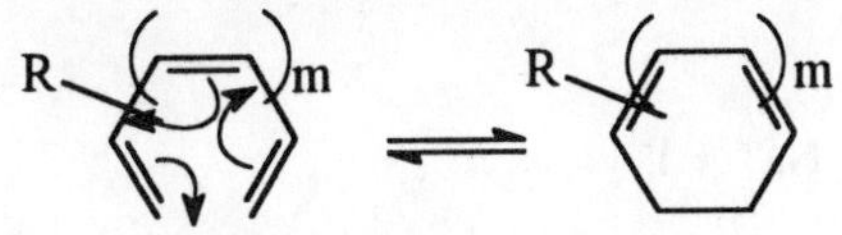

Generality: R = various substituents, rings sytems, etc.
m = 0, 1, 2, ...

Key features: m = even, i.e. 4n electron systems: Möbius transition state; conrotatory motion
m = odd, i.e. 4n + 2 systems: Hückel transition state; disrotatory motion

Cycloaddition reactions (22.5.B)

Equation:

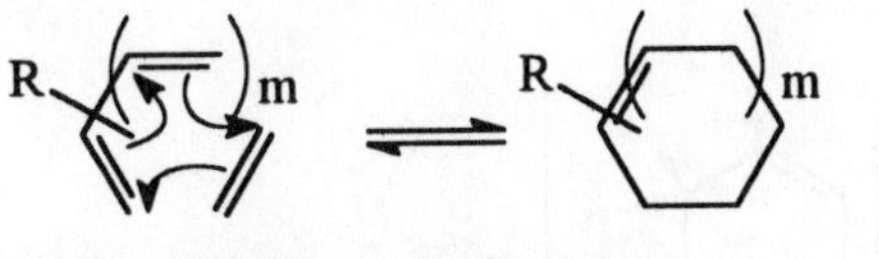

Generality: R = various substituents
m = 0, 1, 2, ...

Key features: p + q cycloaddition reaction: p π-electrons in one part, q in the other
if p + q = 4n: Möbius transition state, antarafacial stereochemistry
if p + q = 4n + 2: Hückel transition state, suprafacial stereochemistry

Photochemical *cis/trans* isomerization of alkenes (22.5.H)

Equation: (R)(H)C=C(H)(R) $\overset{h\nu}{\rightleftharpoons}$ (R)(H)C=C(R)(H)

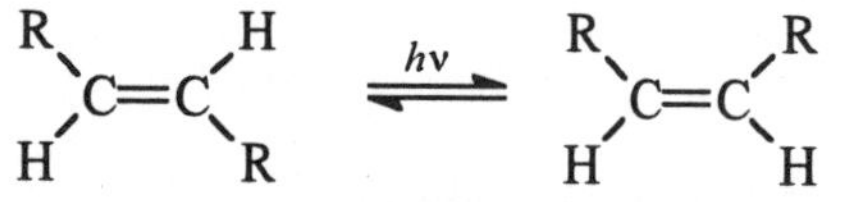

Generality: R = alkyl, aryl, etc.

Key features: reactions proceeds via $\pi \rightarrow \pi^*$ electronic excitation
equilibrium process

22.C Important Concepts and Hints

For many years, the concept of a "reaction mechanism" in organic chemistry consisted of simply showing what fragments of the starting materials moved around and became attached to each other in the products. As our understanding and sophistication increased, so did the finesse with which we displayed reaction mechanisms. The "lasso" diagrams (Eq. 1) gave way to the convention of using curved arrows to represent the movement of electron pairs that are involved in bond cleavage and formation. This style is now universally used because it helps us keep track of the crucial features in a reaction mechanism: the valence electrons. In the process of writing the reaction of sodium cyanide with methyl iodide as in Eq. 2, we can easily see why it is the cyanide ion that bonds to the methyl group, and why it becomes attached via the carbon and not the nitrogen.

1. Na(CN + CH_3)I → CH_3CN + NaI

2. :N≡C:⁻ Na^+ + CH_3—I ⟶ :N≡C—CH_3 + Na^+ + I^-

Nevertheless, for many reactions, our detailed understanding of the mechanism has now advanced beyond the level that can be described by curved arrows. For example, the cyclization of *E,Z,E*-2,4,6-octatriene to *cis*-5,6-dimethyl-1,3-cyclohexadiene can be written with curved arrows (Eq. 3), showing us the cyclic movement of electrons, but this diagram cannot let us see why the *cis* product is formed and not the *trans*. To describe this type of reaction adequately, an understanding of molecular orbital theory has become indispensable. Interestingly, the types of reactions for which MO theory is most applicable are in turn becoming increasingly important in modern organic chemistry. Which of these parallel developments has the greatest importance is a present-day version of the chicken-and-the-egg question.

3. (octatriene) ⟶ (cis product, H, H) **not** [(trans product, H, H)]

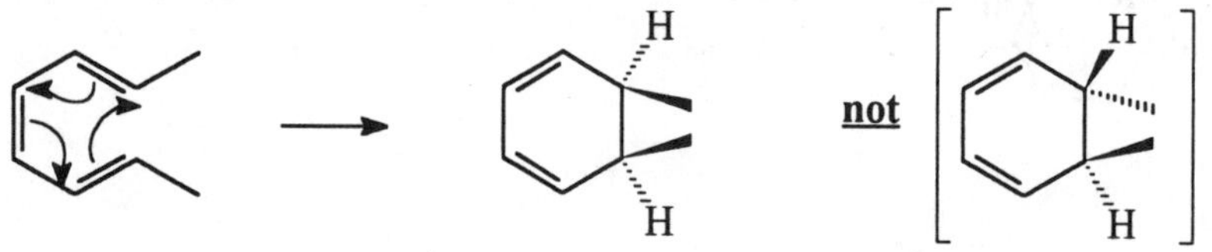

For many types of reactions, molecular orbital theory can be distilled down to some straight-forward rules. It is of course a good idea to understand the derivation of these rules, but at the least you should learn what they are:

1. For cyclic π-systems:

   4n + 2 electrons involved in a ring = Hückel = aromatic = stabilized
   4n electrons in a ring = antiaromatic = destabilized

2. For electrocyclic reactions

   4n + 2 electrons involved in a ring requires:

      a Hückel system in the transition state
      an even number of nodes
      disrotatory ring closure/opening

   4n electrons in a ring requires:

      a Möbius transition state
      an odd number of nodes
      conrotatory ring closure/opening

3. For cycloaddition reactions

   4n + 2 = Hückel = suprafacial addition
   4n = Möbius = antarafacial addition

One other point can be made which may help to clear up a common confusion about molecular orbitals, their energies, and what effect their occupancy by electrons has on the energy of the molecule. Think of an orbital as a landlord thinks of an apartment: A vacant apartment (orbital) exists and has a defined price (energy level) even when it is unoccupied, but this rent (energy) only counts when a paying tenant (an electron) moves in. Of course, in any MO system there are an equal number of bonding and anti-bonding orbitals, so the apartment analogy isn't perfect. Whereas there is an unfavorable energy change when an electron moves into an antibonding orbital, it is unlikely that any apartments can be found where the landlord pays money to the tenants...

UV Spectroscopy

As a useful spectroscopic method, UV spectroscopy is limited to conjugated systems. Therefore, it is not as generally applicable as NMR or IR. Nevertheless it can provide important information on structure and conformation in such systems.

The process of electronic excitation, on which UV spectroscopy is based, is also important in photochemistry. Many reactions that do not occur thermally with a molecule in its ground state can be made to go when the molecule is in an excited state. For example, the cyclization of two olefins to give a cyclobutane can be made to occur under irradiation, but not thermally, as pointed out in Chapter 21.

22.D. Answers to Exercises

22.1. $[CH_2{=}CH{-}CH{=}CH{-}{}^{-}CH_2 \leftrightarrow CH_2{=}CH{-}{}^{-}CH{-}CH{=}CH_2 \leftrightarrow {}^{-}CH_2{-}CH{=}CH{-}CH{=}CH_2]$

pentadienyl anion

π_5 HIGHEST

(least stable)

ENERGY

22.2 1,3,5-hexatriene anion

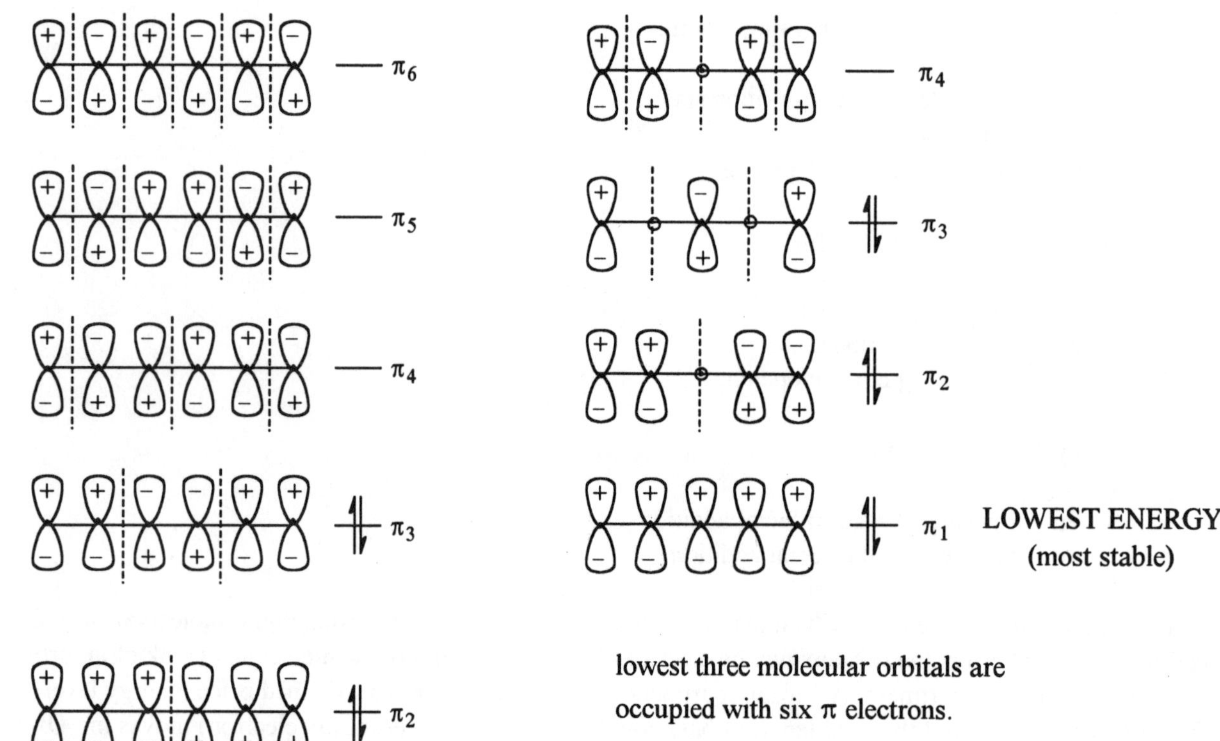

lowest three molecular orbitals are occupied with six π electrons.

π_1

lowest three molecular orbitals occupied with six π electrons

Lewis structure: CH_2::CH:CH::CH:CH::CH_2

The bond between carbons 1 and 2 is bonding in all three of the occupied molecular orbitals. The bond between carbons 2 and 3 is bonding in the lowest 2 occupied molecular orbitals, but antibonding in the the third. The bond between carbons 3 and 4 is bonding in the first and third molecular orbitals but antibonding in the second. Therfore, the bond between carbons 1 and 2 is always bonding in the occupied orbitals; for the other two bonds, there are two bonding and one antibonding in the occupied molecular orbitals.

22.3 Model-building exercise.

22.4

4π-electron, "antiaromatic" system; very reactive 6π-electron, aromatic system; stabilized

22.5 b. (22 π electrons) d. (6 π electrons) e. (26 π electrons) f. (10 π electrons)

22.6 a. 2+2: 4+4:

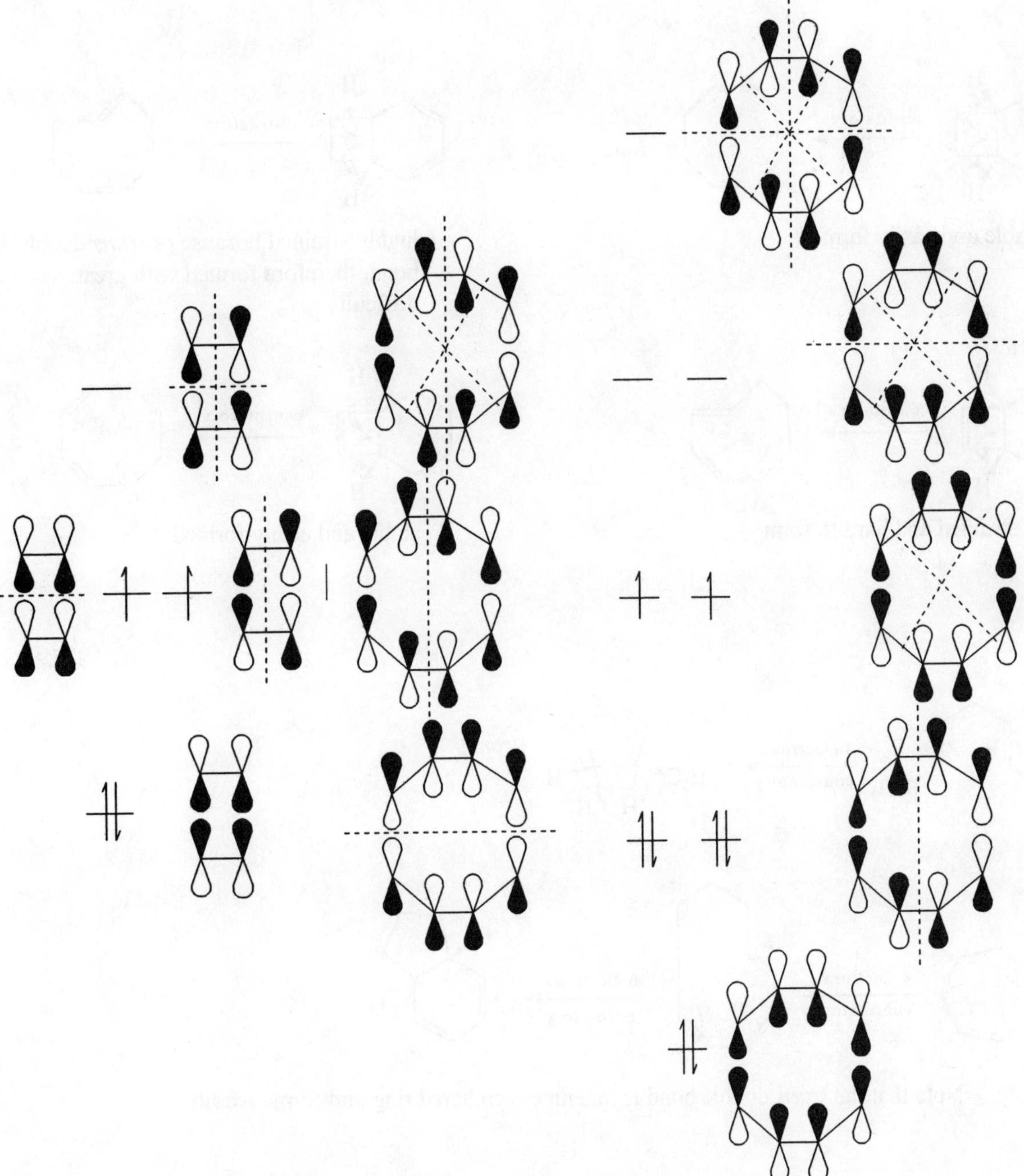

(NOTE: dashed lines indicate nodes in the molecular orbitals)

22.7 The transition states for the forward and reverse reactions are the same.

22.8

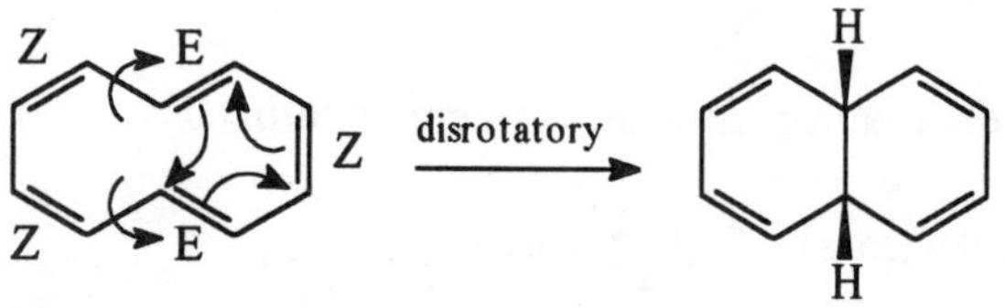

Z E Z Z E disrotatory H H (cis)

six = 4n + 2 electrons involved in the electrocyclization; therefore disrotatory

22.9

a. 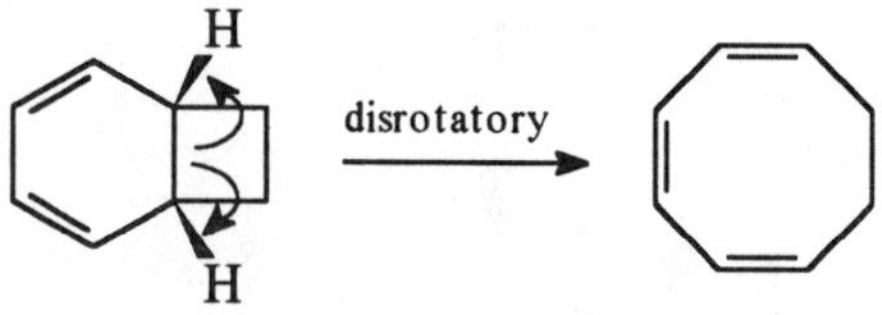 H H disrotatory

stable and easily formed

H H disrotatory

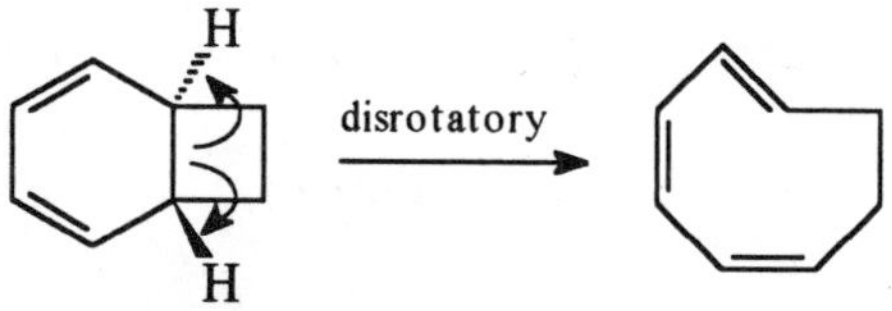

highly strained because of *trans* double bond, therefore formed with greater difficulty

b. 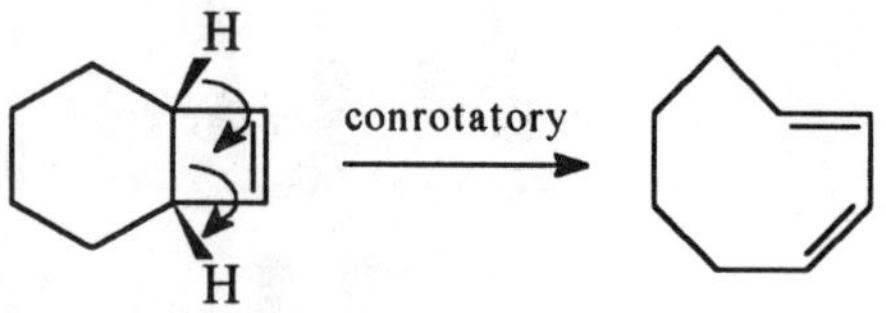 H H conrotatory

highly strained and hard to form

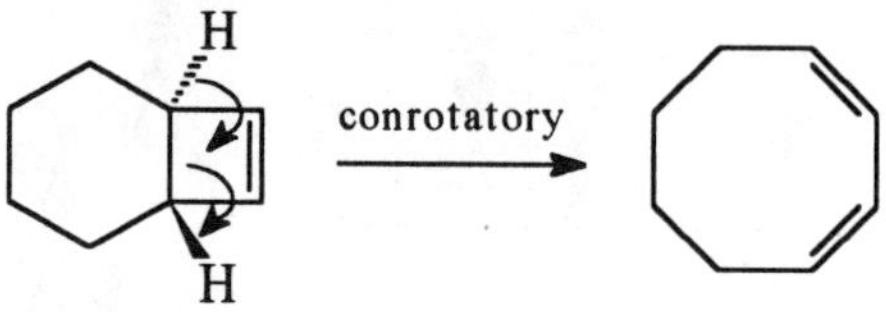

H H conrotatory

stable and easily formed

22.10

a. 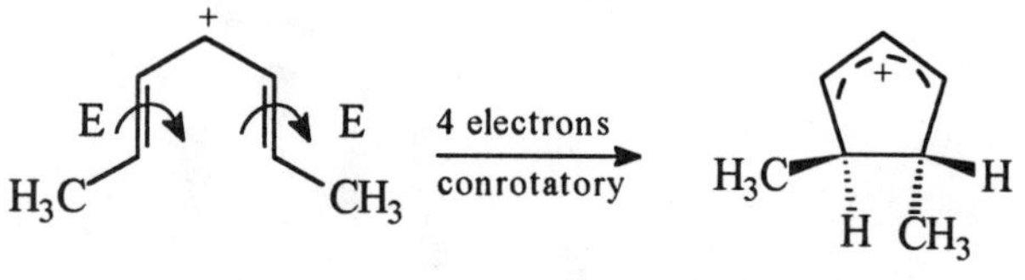 E E H_3C CH_3 + 4 electrons conrotatory H_3C H H CH_3 + (*trans*)

b. 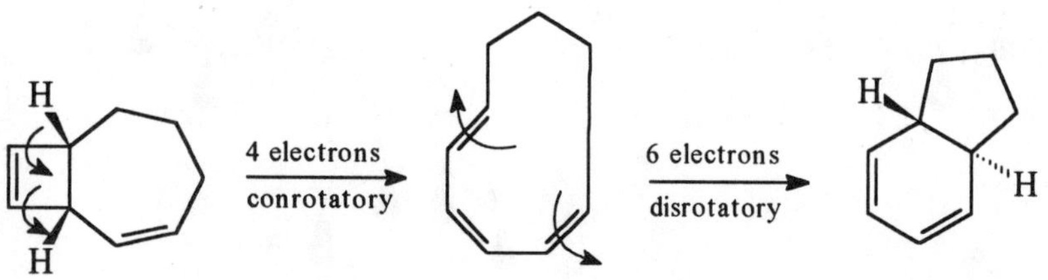 H H 4 electrons conrotatory 6 electrons disrotatory H H

Note that the *trans* double bond is in a nine-membered ring and forms readily

c. C_6H_5, E, E, E, CH_3, CH_3 $\xrightarrow[\text{conrotatory}]{\text{8 electrons}}$ C_6H_5, H_3C, CH_3, H H (*trans*)

22.11 i. ${}_4\pi_s + {}_4\pi_s$: disallowed (the allowed ${}_4\pi_s + {}_4\pi_a$ is geometrically unlikely)

ii. ${}_8\pi_s + {}_2\pi_s$: allowed

iii. ${}_2\pi_s + {}_2\pi_s + {}_2\pi_s$: allowed

iv. ${}_{12}\pi_s + {}_2\pi_s$: allowed

22.12 Allyl anion:

π_3, π_2, π_1 $\xrightarrow{h\nu}$

Pentadienyl anion:

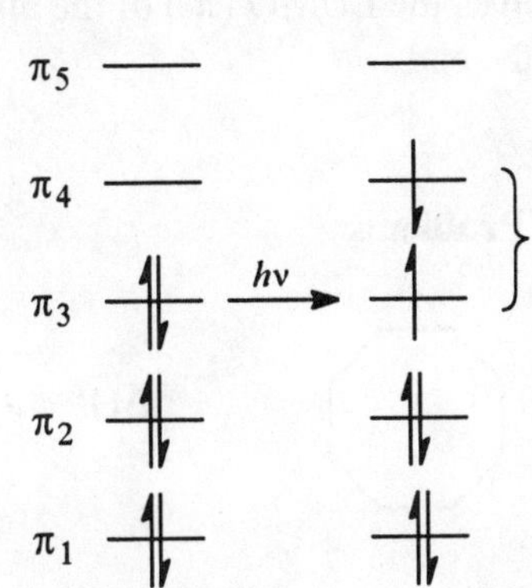

smaller energy difference, therefore longer wavelength absorption

22.13 n → π*: 0.00731 g crotonic acid/10 mL = 8.5 x $10^{-3}M$

$$\epsilon_{250} = \frac{0.77}{(8.5 \times 10^{-3})(1)} = 91$$

π → π*: 8.5 x $10^{-3}M$ diluted 100-fold = 8.5 x 10^{-5} M

$$\epsilon_{200} = \frac{0.86}{(8.5 \times 10^{-5})(1)} = 10{,}120$$

22.14 $E = \dfrac{hc}{\lambda}$; E(kcal mole^{-1}) = $\dfrac{2.857 \times 10^{-5}}{500 \times 10^{-9}} = 57\ kcal\ mole^{-1}$

22.15 There is a large lobe available for bonding behind the carbon of the C-F bond. Overlap there with an incoming nucleophile is involved in S_N2 displacement.

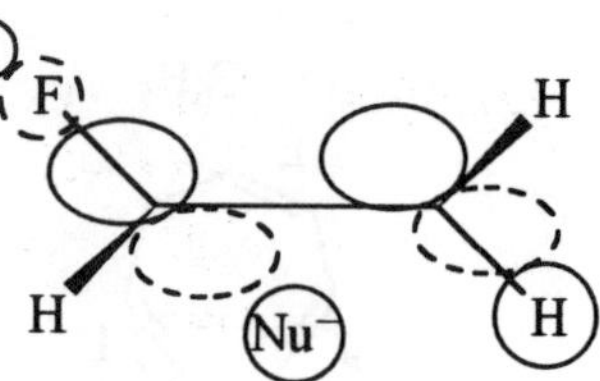

There is another lobe accessible for bonding at the *anti*-hydrogen. Overlap there leads to E2 elimination, via the transition state:

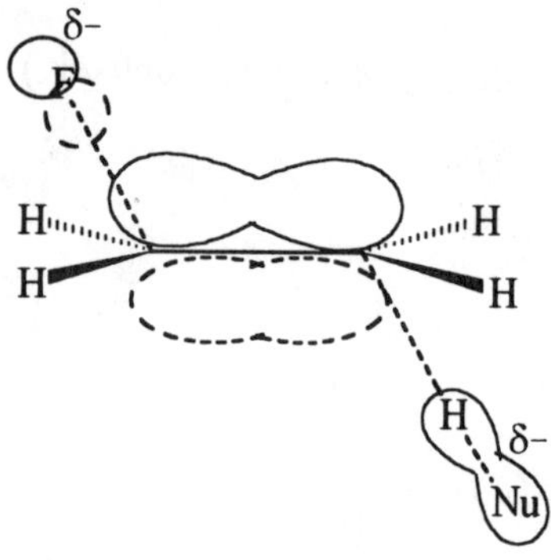

22.16 Reaction with a nucleophile involves the LUMO (π_2) of the allyl cation. Since this orbital has a node at the central carbon, attack occurs at the end.

22.E Answers and Explanations for Problems

1. a. cyclooctene + H_2 ⟶ cyclooctane $\Delta H° = -23.3$ kcal mole^{-1}

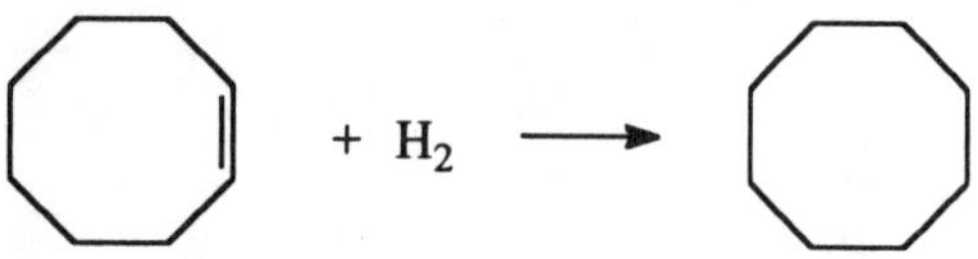

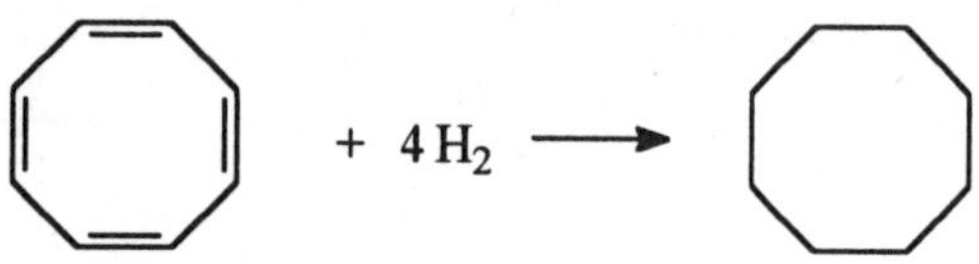

cyclooctatetraene + $4\,H_2$ ⟶ cyclooctane $\Delta H° = -100.9$ kcal mole^{-1}

$4 \times (-23.3) - (-100.9) = +7.7$ kcal mole^{-1}

Empirical resonance energy: -7.7 kcal mole^{-1}.

The negative value implies that the four double bonds in cyclooctatetraene are less stable than 4 x 1 double bonds; i.e., this value represents a *destabilization* energy or negative resonance energy. This means that not only is there no stabilization energy (resonance), but that cyclooctatetraene is probably more strained because of the four double bonds.

b. cyclooctatetraene ⟶ $8 \cdot \dot{\underset{\cdot}{C}} \cdot$ + $8\ H\cdot$

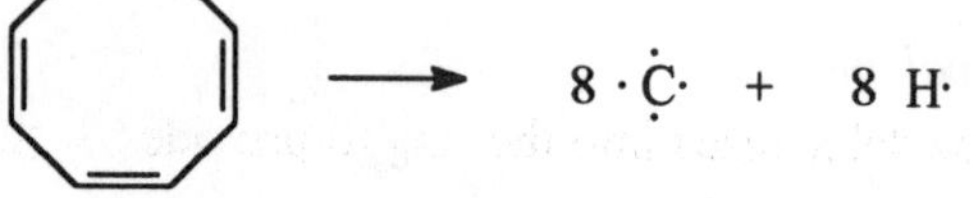

<u>Breaking</u>	four C–C: $4 \times 83 =$	332
	four C=C: $4 \times 146 =$	584
	eight C–H: $8 \times 99 =$	<u>792</u>
	Sum:	1708 kcal mole^{-1}

Empirical resonance energy: $1713 - 1708 = 5$ kcal mole^{-1}

c.

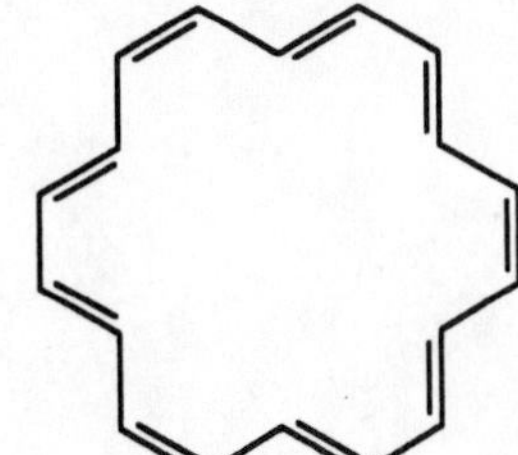

Calculated empirical resonance energy:

$9 \times 83 = 747$
$9 \times 146 = 1314$
$18 \times 99 = \underline{1782}$
3843

[18] annulene $3890 - 3843 = 47$ kcal mole^{-1}

This resonance energy implies that [18]annulene is aromatic, as the Hückel 4n + 2 rule would suggest.

2. The anion that results from proton removal from the first isomer is the most stable.

CH_2^- ⟷ $-CH{=}CH_2$

aromatic, 6π-electron cyclopentadienyl anion

[H_2C^- CH_2 ⟷ H_2C CH_2 ⟷ H_2C $^-CH_2$]

$^-CH_2$ CH_2 ⟷ CH_2 CH_2 ⟷ CH_2 CH_2 ⟷ CH_2 $^-CH_2$

3. a. Nonaromatic: the boron has only six electrons in its valence shell and cannot contribute any electrons to the π system (total of 4 π electrons in the ring).

b. Nonaromatic: each nitrogen has two lone pair electrons, which, combined with those from the double bonds, makes eight.

c. Aromatic: one of the oxygen lone pairs (not both!) can be delocalized into the ring to provide the third pair of π electrons (total = 6).

d. Aromatic: cyclo-$C_7H_7^{3-}$ has seven π orbitals, but the 3-charge indicates that there are three extra electrons, for a total of 10 π electrons in the ring.

e. Aromatic: cyclo-$C_8H_8^{2+}$ has eight π orbitals and 2+ charge, therefore six π electrons.

f. Aromatic: eight π electrons from the double bonds and two from the nitrogen lone pair = 10 π electrons.

4. b (12) and c (8) have have 4n π electrons, not 4n + 2, and are non-aromatic.

5.

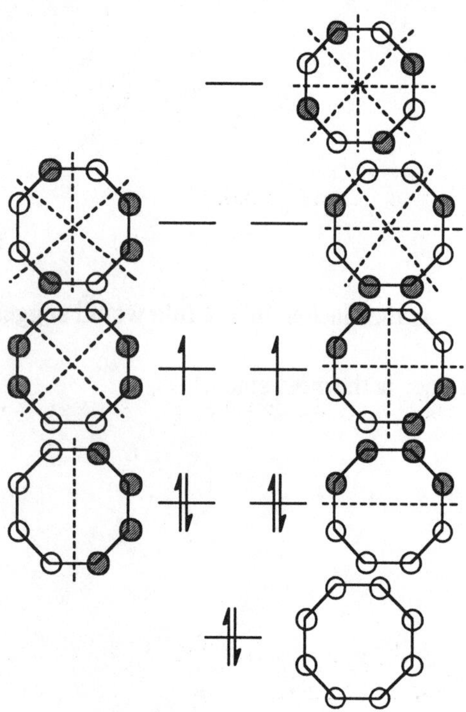

6. The pentadienyl cation is a 4π electron system, therefore the highest occupied molecular orbital is the second, represented by figure d. (See answer to Exercise 22.1)

7. 4n+2: b. c.

Not 4n+2: a. d.

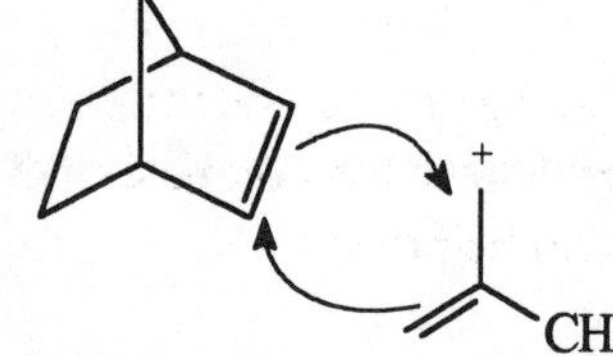

(4 electrons)

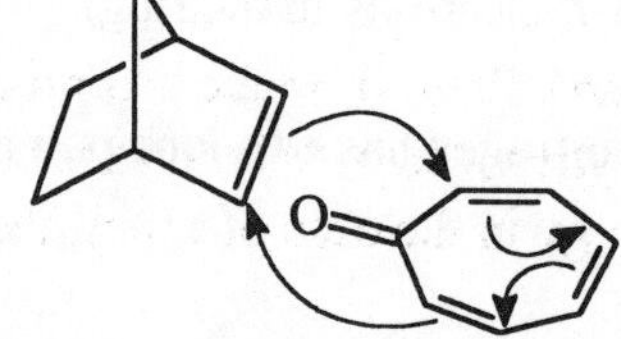

(8 electrons)

8. The anionic product is stabilized by aromatic resonance in the case of the five-membered ring product (6 π electrons), whereas the corresponding seven-membered ring anion (8 π electrons) is not aromatic.

9. The reaction with butadiene is favored according to HOMO-LUMO interactions; it also has a 4n+2 (Hückel aromatic) transition state:

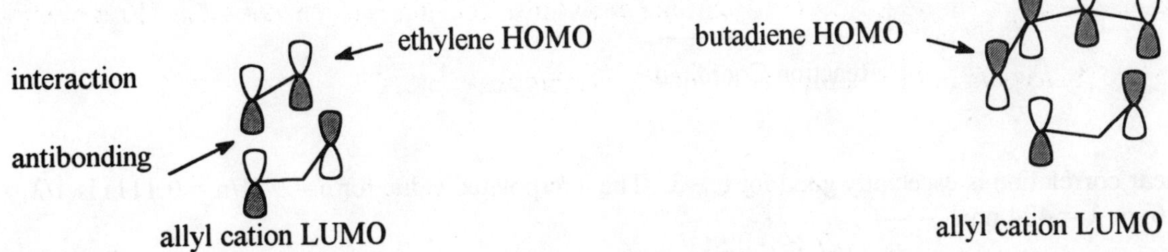

10. The first step is the photoinitiated reverse of the addition of HX to a double bond:

$$RCH_2CHXR' \xrightarrow{h\nu} RCH_2\dot{C}HR' + X^{\cdot}$$

$$X^{\cdot} + RCH_2CHXR' \longrightarrow RCH_2\dot{C}HR' + X_2$$

The color arises from the I_2 or Br_2 produced. Eventually the alkyl radicals combine or disproportionate:

$$2\ RCH_2\dot{C}HR' \longrightarrow RCH{=}CHR' + RCH_2CH_2R'$$

$$2\ RCH_2\dot{C}HR' \longrightarrow RCH_2CHR'CHR'CH_2R$$

11. Suitable solvents are those that do not have any significant UV absorption above 220 nm: methanol, perfluoropropane, 1-chlorobutane, ethyl ether, cyclohexane, and acetonitrile.

The alkyl bromides and iodides absorb in the region 250-260 nm ($n \rightarrow \sigma^*$ transition); sulfides at 210 and 230 nm (n-d transitions); the benzene π-systems absorbs strongly below 280 nm.

12. The polyene in a. has the longest ($n = 5$) conjugated polyene system, and will have the longest wavelength absorption; d. will have the next longest ($n = 4$); c. has $n = 3$ and b. only $n = 2$.

13. a. Both the dielectric constant D and its microscopic counterpart Z are a measure of the ability of the solvent to develop an induced charge. With its small, highly polarized structure and two hydrogen-bonding OH groups, water has a greater ability to do this than methanol.

b. Since *t*-butyl chloride will go from the neutral compound in the ground state to the ion pair in the transition state, the ionic transition state will be stabilized by the water more than by methanol. Since the transition state is lower in energy in the case of water, the reaction will be faster.

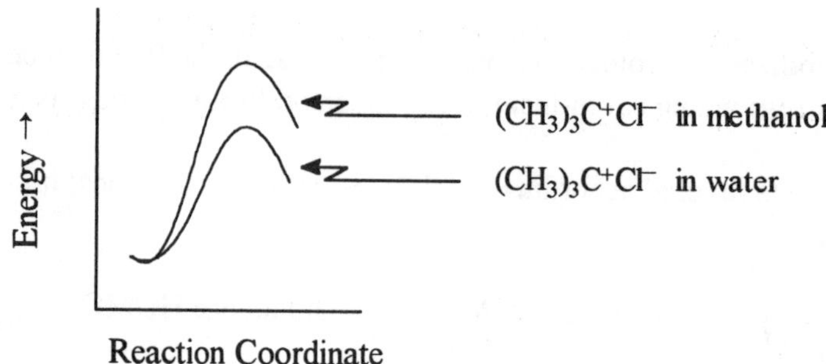

14. The linear correlation is especially good for n ≥ 3. The interpolated value for n = 9 (1/n = 0.111) is 1/λ = 0.00236, or λ = 424 nm.

15. Molecular weight = 138; 1.486×10^{-5} g mL^{-1} = 1.08×10^{-4} M

$$\epsilon = \frac{1}{1.08 \times 10^{-4}} = 9260 \qquad \lambda_{max} = 232 \text{ nm}$$

16. The UV spectrum of butadiene depends on the energy difference between the highest occupied (HOMO) and lowest unoccupied (LUMO) molecular orbitals. These orbitals are the π_2 and π_3 molecular orbitals, as depicted in Figure 22.5. As you can see, the magnitude of the wave function is largest at carbons 1 and 4 in these two orbitals (the π lobes are biggest at the ends), hence a substituent at these positions will have more of an influence than at the 2 and 3 positions.

17. The I_2 forms an electron donor-electron acceptor complex with toluene. The toluene acts as the electron donor and the I_2 as the electron acceptor. Light causes the transfer of an electron to give the excited state $[\text{toluene}]^+[I_2]^-$. This transfer of an electron from the π_3 level in toluene to the σ* in I_2 is an example of a charge-transfer transition. The electron donating ability of the CH_3 group on toluene helps to stabilize the complex.

22.F. Supplementary Problems

S1. A popular mnemonic for determining the relative ordering of MO's in an [n]-annulene is to inscribe the appropriate polygon inside a circle, with one of the vertices down. Each vertex then corresponds to the energy level of a molecular orbital, with those below the center of the circle representing bonding orbitals and those above it, antibonding orbitals. When enough electrons are present to fill the bonding orbitals and give a closed shell, the molecule is aromatic. The mnemonic is illustrated below for benzene:

To convince yourself of its validity, apply this concept to cyclooctatetraene, cyclopentadienyl anion, cyclopropenyl anion, and cycloheptatrienyl cation.

S2. Figure 22.18 in the Text shows the energy level pattern for the transition states of alternative conrotation and disrotation in the closure of octatetraene. According to the perturbation approach to reactivity, the transition state is dominated by the HOMO of the reactant. Sketch the occupied molecular orbitals of octatetraene, and determine which mode of cyclization is predicted by the form of the HOMO.

S3. Show that each of the following reactions is allowed in terms of HOMO-LUMO interactions:

a. 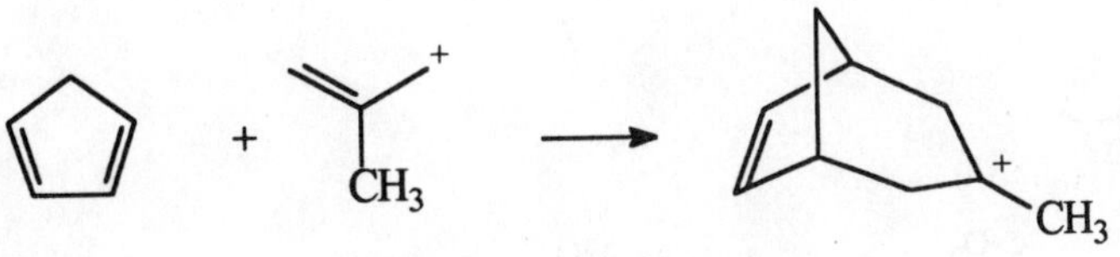 + →

CH$_3$ CH$_3$

b. 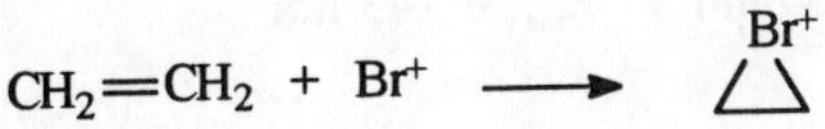 $CH_2{=}CH_2$ + Br^+ → Br$^+$

c. 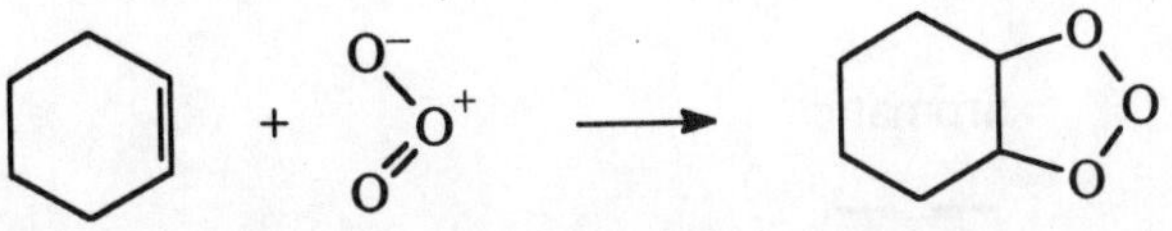 + O$^-$ O$^+$ O → O O O

S4. During the course of a research project in organic synthesis, we attempted to hydrolyze the complex acetate ester **1** with NaOH. Instead of the desired product **2**, we isolated the tricyclic compound **3**. The mechanism of this unanticipated reaction involves enolate formation, followed by loss of methoxide ion to form intermediate **4**.

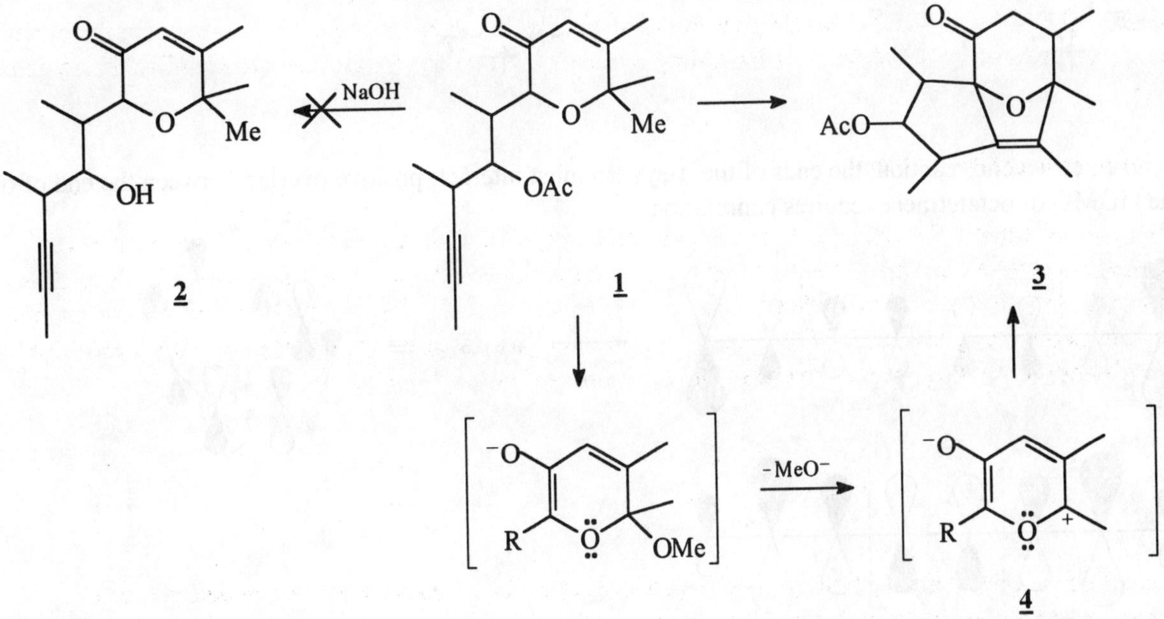

O O Me OH **2** NaOH O O Me OAc **1** O AcO O **3**

$^-$O R O OMe –MeO$^-$ $^-$O R O + **4**

a. Do you expect intermediate **4** to be aromatic or antiaromatic, or neither?
b. Write a mechanism for the transformation of **4** to the observed product **3**. Is this a Hückel-allowed reaction?

S5. A phenyl substituent normally raises the energy of a HOMO and lowers the energy of a LUMO. What effect would you predict that a phenyl group has on an electronic transition corresponding to HOMO ⟶ LUMO?

S6. What are plausible structures for intermediate **A** and product **B**?

HO

cholesterol, $C_{27}H_{46}O$ $\xrightarrow[\text{pyridine}]{CrO_3}$ **A** $C_{27}H_{44}O$ λ_{max} = < 200 nm $\xrightarrow[\text{MeOH}]{\text{MeO}^-}$ **B** $C_{27}H_{44}O$ λ_{max} = 243 nm

22.G Answers to Supplementary Problems

S1.

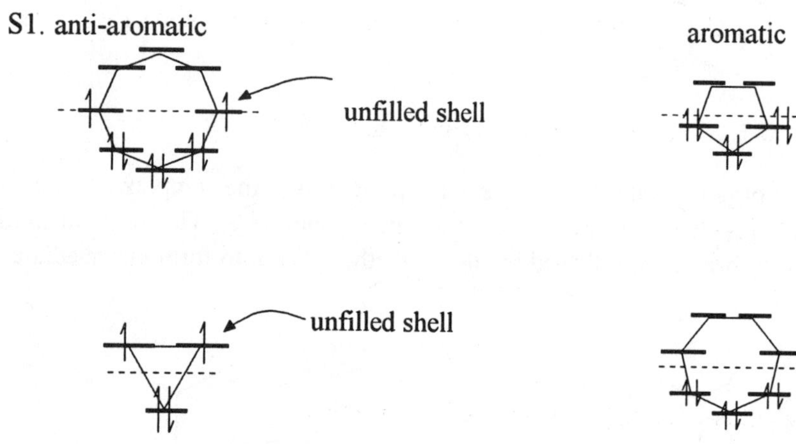

S2. In an electrocyclic reaction, the ends of the π-system must interact; positive overlap between the end of orbitals the HOMO of octatetraene requires conrotation:

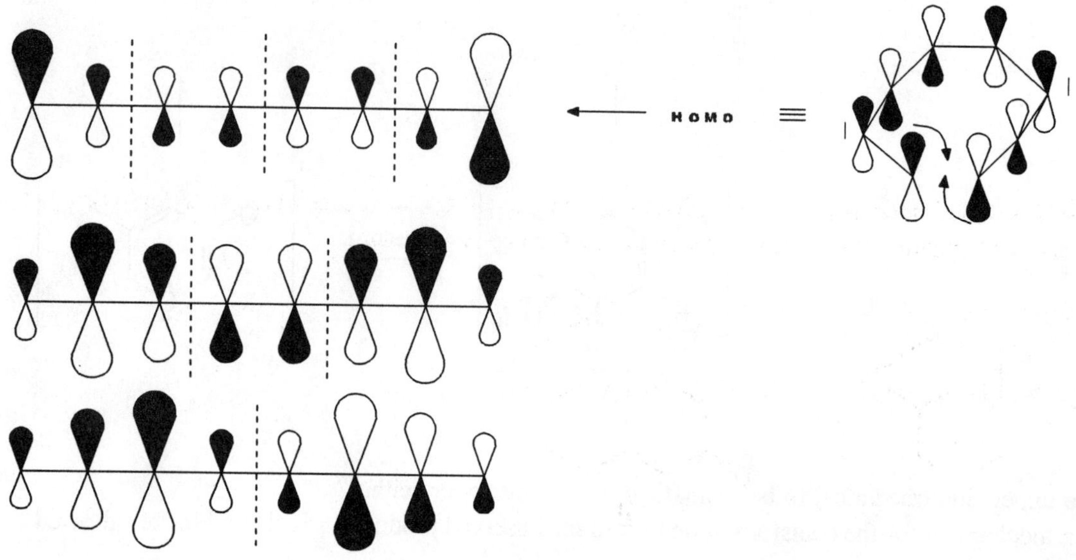

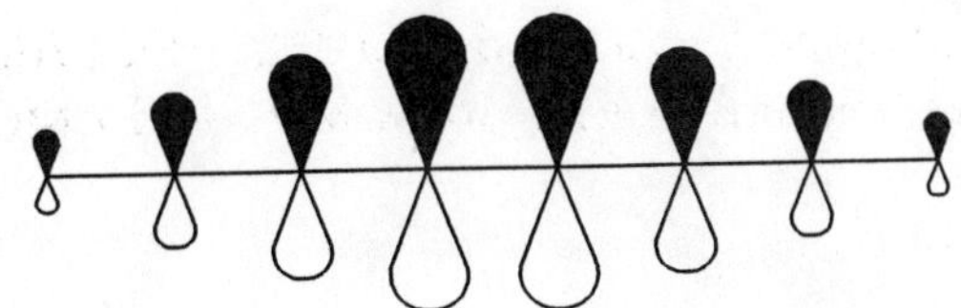

S3. a. 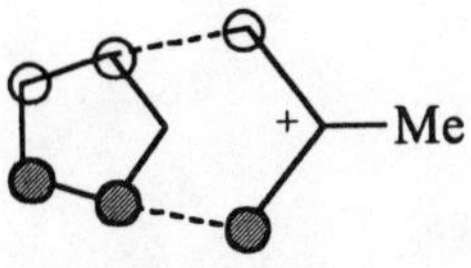 HOMO LUMO

b.  HOMO LUMO

c. 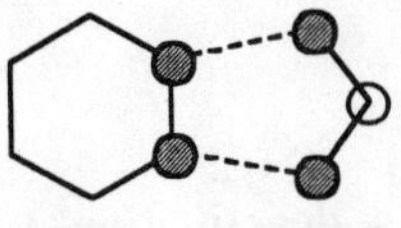 HOMO LUMO

O_3 has four π-electrons in the "allyl MO's" with an electronic configuration similar to allyl anion. Hence the LUMO is ψ_3, as shown.

S4 a. Intermediate 4 is a cyclic system of six π-electrons and hence is Hückel aromatic.

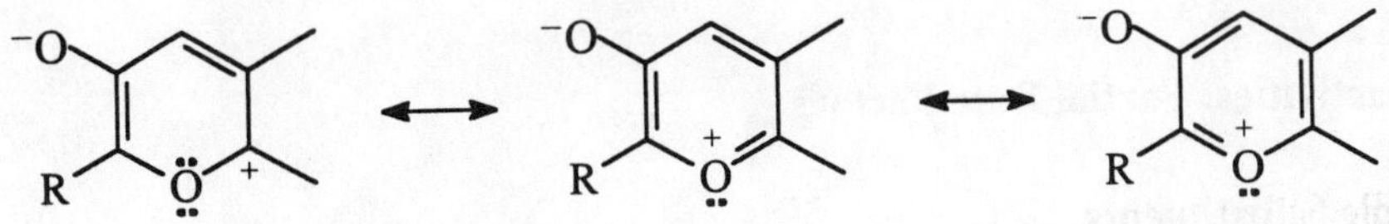

b. Formation of product 3 involves a Hückel-allowed [4+2] cycloaddition reaction:

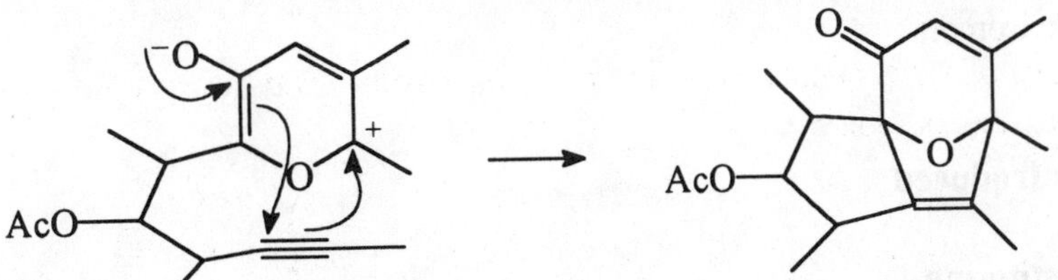

4 3

S5. If the HOMO is raised in energy and the LUMO lowered, the energy difference between them becomes smaller and the electronic transition will occur at lower energy = longer wavelength.

S6.

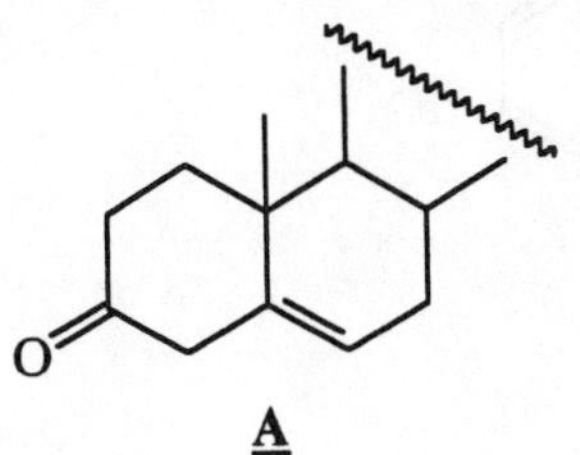

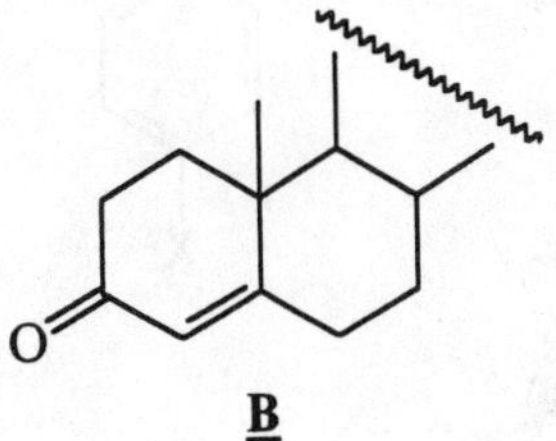

A B

23. ELECTROPHILIC AROMATIC SUBSTITUTION

23.A. Chapter Outline and Important Terms Introduced

23.1 Halogenation

23.2 Protonation
tracer isotope vs. macroscopic isotope
liquid scintillation counter

23.3 Nitration

23.4 Friedel-Crafts Reactions (will not work on strongly deactivated rings)
A. Acylations — acylium ion
B. Alkylations — carbocations
with alkyl halides and olefins, chloromethylation

23.5 Orientation in Electrophilic Aromatic Substitution
ortho, para, vs. *meta*-directors — activating vs. deactivating
1. *o,p*- with activation
2. *o,p*- with deactivation
3. *m*- with deactivation

23.6 Theory of Orientation in Electrophilic Aromatic Substitution
stabilization of resonance structures

23.7 Quantitative Reactivities: Partial Rate Factors

23.8 Effects of Multiple Substituents

23.9 Synthetic Utility of Electrophilic Aromatic Substitution
acylation + deoxygenation = "alkylation"
limitations of orientation and reactivity
avoidance of inseparable isomers

23.B. Important Reactions Introduced

Electrophilic aromatic substitution

Halogenation (23.1):

Equation:

o, p-directors — *m*-directors

Y-benzene $\xrightarrow[\text{Lewis acid}]{X_2}$ *o*-X-Y-benzene and *p*-X-Y-benzene or *m*-X-Y-benzene

Generality: $X = Cl, Br$
Lewis acid = FeX_3 or AlX_3
(for X = I, use I_2/HNO_3 or H_3AsO_4)
Y = various substituents

Key features: X is a deactivating substituent, so stepwise formation of mono-, di-, or tri-halogenated products is possible

Nitration (23.3):

Equation:

o, p-directors — m-directors

Y-benzene $\xrightarrow[H_2SO_4]{HNO_3}$ o-nitro (Y, NO_2) and p-nitro (Y, NO_2) or m-nitro (Y, NO_2)

Generality: Y = various substituents

Key features: NO_2 is a deactivating substituent, so stepwise formation of mono-, di-, or tri-halogenated products is possible

Friedel-Crafts acylation (23.4.A):

Equation: Y-benzene + R–C(=O)–Cl $\xrightarrow{AlCl_3}$ o-(Y)(C(=O)R) and p-(Y)(C(=O)R)

Generality: Y **cannot** be a *meta*-directing substituent

Key features: coupled with deoxygenation, acylation is a useful method for introduction of 1° alkyl groups in aromatic rings (gives only mono-substitution, and is not susceptible to carbocation rearrangements)

intramolecular reaction useful for making cyclic aromatic compounds

Friedel-Crafts alkylation (23.4.B):

Equation: Y-benzene + R—Cl $\xrightarrow{AlCl_3}$ o-(Y)(R) and p-(Y)(R)

Generality: Y **cannot** be *meta*-directing substituent
R = 2° or 3° alkyl
other reagents: alkene or alcohol + HF + BF_3
chloromethylation: CH_2O + HCl

Key features: an alkyl group is an activating substituent, therefore polyalkylation is a common side reaction

carbocation rearrangements also common
R $\neq 1°$ alkyl

Other things to keep in mind:

Y = OR′, NR'_2: strong *ortho-*, *para*-directing, strongly activating
Y = alkyl: weak *ortho-*, *para*-directing, activating
Y = halide: *ortho-*, *para*-directing, deactivating
Y = COR, CO_2R', CN, NO_2: strongly *meta*-directing, strongly deactivating

23.C Important Concepts and Hints

This entire chapter is devoted to the reaction depicted below: electrophilic aromatic substitution. We have drawn it in such a way as to show how the p-orbitals and electron distribution of the ring change throughout the process:

$6\pi\ e^-$ E^+ H *sp*²-hybridized → [$4\pi\ e^-$ $2\sigma\ e^-$ E H *sp*³-hybridized]$^+$ → $6\pi\ e^-$ H^+ E *sp*²-hybridized

In the course of forming the intermediate, the π-system goes from neutral (six electrons/six orbitals) to positively charged (four electrons/five orbitals). Two electrons and one orbital are taken up in the new bond to the electrophile E^+. As pointed out in the text, instead of a nucleophile **adding** to the cationic intermediate, a proton is lost. This is simply the reverse of the initial attack, with the two electrons and the carbon orbital of the C−H bond going to reform the aromatic, 6π electron system.

A substituent (Z) that is attached to the ring can affect this process in two ways: by adding or withdrawing electron density through the σ-bond:

Z→(ring) Z←(ring) (this is usually called the "inductive effect")

or by adding or withdrawing electron density via a π-type interaction:

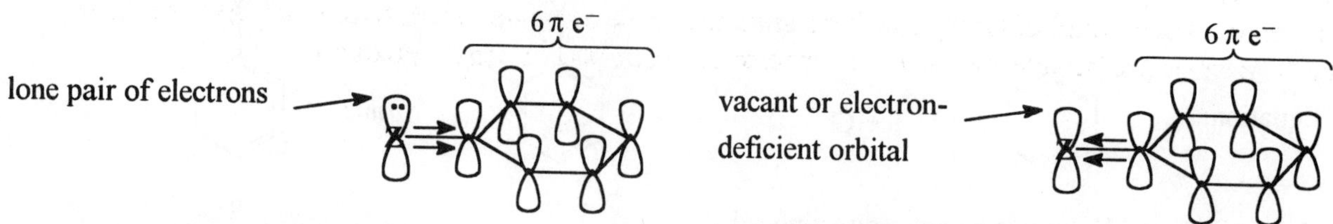

(This is usually called the "resonance" or "mesomeric" effect)

The effect through the σ-framework decreases in the order $o > m > p$, but it does not greatly influence the position of electrophilic attack. On the other hand, the effect through the π-system is only felt in the *ortho* and *para* positions, so it is the determining factor in orientation. The table below illustrates the influence of σ- and π-effects for a variety of substituent types: (Substitution rate relative to benzene)

Type of Substituent	Group	π-effect	σ-effect	*o*-,*p*-positions	*m*-position
$R\ddot{\underset{\cdot\cdot}{O}}$, $R_2\ddot{N}$, $R\overset{O}{\overset{\|}{C}}\underset{\cdot\cdot}{N}H$	I	strongly electron-releasing	electron-withdrawing	+++	–
Alkyl	II	electron-releasing via hyperconjugation	electron-releasing	++	+
Halogen	II	electron-releasing	electron-withdrawing	–	– –
NO_2, $R\overset{O}{\overset{\|}{C}}$, $R\overset{O}{\overset{\|}{\underset{\underset{O}{\|}}{S}}}$	III	strongly electron-withdrawing	electron-withdrawing	– – –	– –

Notice how the *meta* reactivity (relative to benzene) parallels the σ-effect, because the resonance (π) effect can only influence the *ortho* and *para* positions. The π-effect combines with the σ-effect in influencing the *ortho* and *para* rates. For strongly activating substituents, the π-effect dominates the σ-effect and the *ortho* and *para* positions are strongly activated. For alkyl substituents, the two effects act together, and *ortho, para* substitution gets an extra boost relative to *meta.* For halogens, the σ- and π-effects work in opposite directions again, but this time the π-effect is weaker: substitution at all positions is deactivated, but the *ortho, para* ones are less so. Finally, for strongly deactivating groups, the σ- and π-effects combine to deactivate *ortho* and *para* the most, resulting in *meta*-direction.

In predicting the results of competition between the directive effects of two substituents on the same ring, it is best to remember **three** groups of substituents (instead of the four above). Group I includes the strongly activating *ortho,para*-directors such as RO- and R_2N-; their influence dominates that of the other groups. Group II includes the "moderate" *ortho,para*-directors such as alkyl and halogen; these substituents will yield to Group I effects, and will win out only over those of Group III. Group III includes all the *meta*-directing substituents; they will control orientation only in the absence of Group I or II substituents. If the competition is between members of the same group, no "winner" is predictable, and you can expect to see mixtures of isomeric products.

23.D Answers to Exercises

23.1

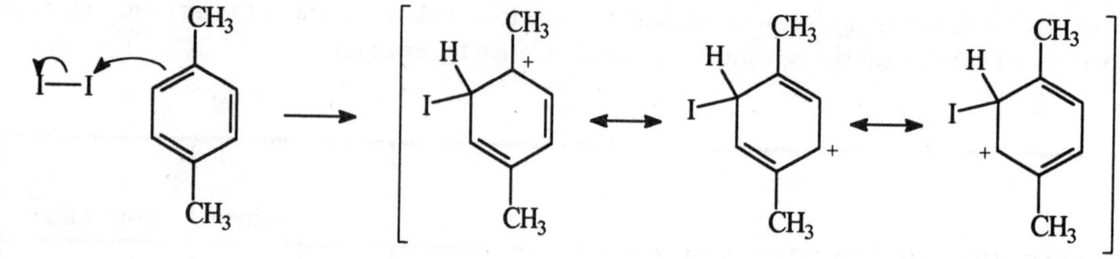

I—I + p-xylene (CH_3, CH_3) → [sigma complex resonance structures: H, I, CH_3, + ↔ H, I, CH_3, + ↔ H, I, CH_3, +]

23.2

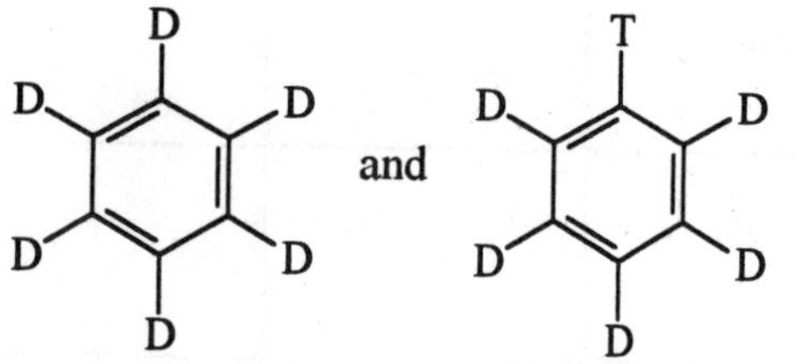

Hexadeuteriobenzene (D, D, D, D, D, D) and pentadeuterio-tritiobenzene (T, D, D, D, D, D)

23.3

$HO-\overset{+}{N}(=O)O^-$ $\underset{}{\overset{H^+}{\rightleftharpoons}}$ $H_2\overset{+}{O}-\overset{+}{N}(=O)O^-$ $\overset{-H_2O}{\rightleftharpoons}$ $O=\overset{+}{N}=O$ + p-xylene (CH_3, CH_3) →

[O_2N, H, CH_3, + ↔ O_2N, H, CH_3, + ↔ O_2N, H, CH_3, +] $\xrightarrow{-H^+}$ 2-nitro-p-xylene (O_2N, CH_3, CH_3)

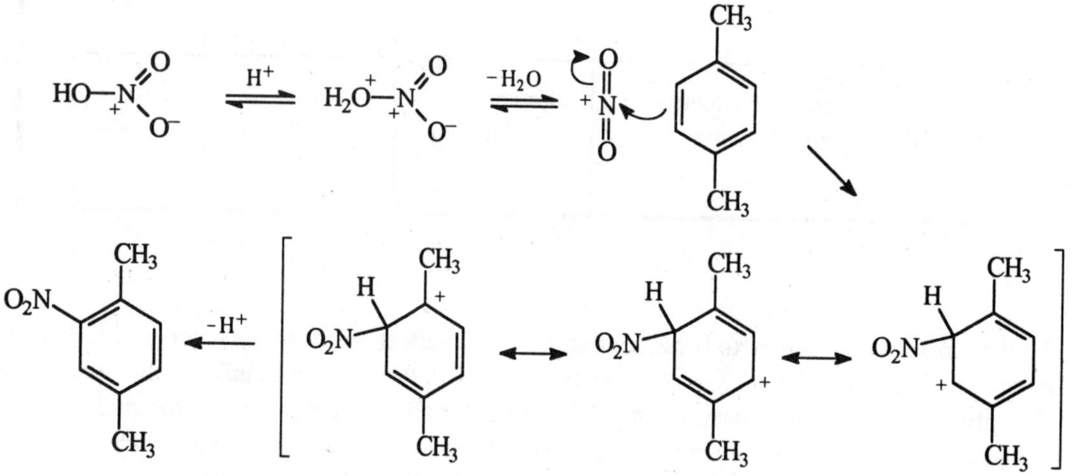

(more stabilized than those from reaction with benzene)

23.4 a. Benzene $\xrightarrow[AlCl_3]{ClCCH_2CH_3 \text{ (C=O)}}$ propiophenone $\xrightarrow{Zn,\ HCl}$ propylbenzene

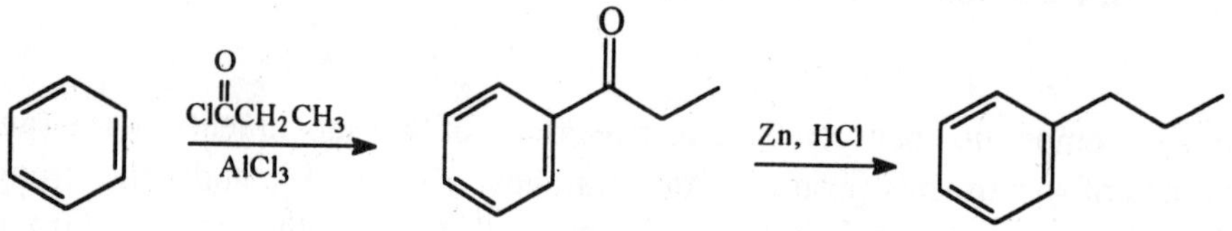

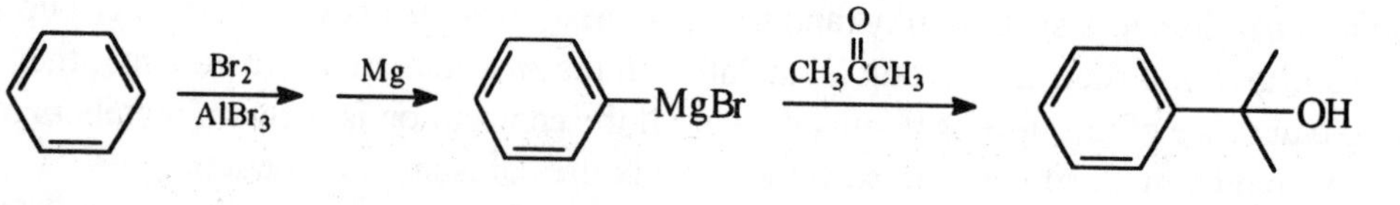

b. Benzene $\xrightarrow[AlBr_3]{Br_2}$ $\xrightarrow{Mg}$ PhMgBr $\xrightarrow{CH_3CCH_3 \text{ (C=O)}}$ PhC(CH_3)$_2$OH

c. Benzene $\xrightarrow[AlCl_3]{ClCCH_3 \text{ (C=O)}}$ $\xrightarrow{NaBH_4}$ PhCH(OH)CH_3 $\xrightarrow{HBr}$ $\xrightarrow{KOH}$ styrene

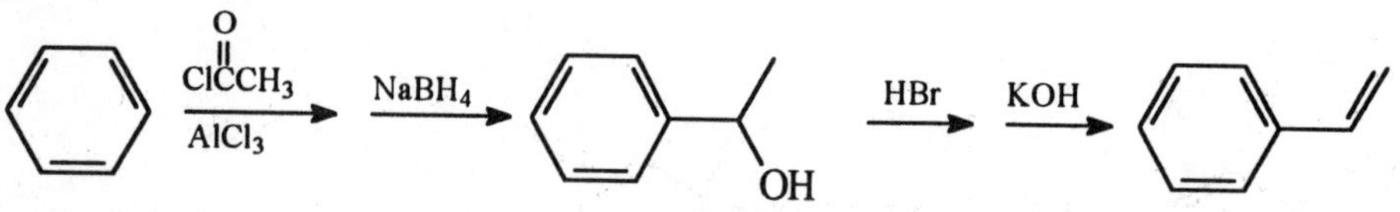

23.5 The carbocation that would be produced from 3-methyl-2-butanol rearranges to the tertiary cation faster than it undergoes Friedel-Crafts reaction:

BF$_3$ → [H + → +] benzene →

H OH

The desired product can be obtained from the following sequence:

$\xrightarrow[AlCl_3]{ClCCH(CH_3)_2 \text{ (O)}}$ $\xrightarrow{CH_3MgI}$ HO $\xrightarrow[Pt, H^+]{H_2}$

23.6 a.

$CH_3CH_2CH_2CCl$ (O) + $AlCl_3$ ⇌ $AlCl_4^-$ + $CH_3CH_2CH_2C{\equiv}O^+$

$CH_3CH_2CH_2C$(O)–Ph $\xleftarrow{-H^+}$ [$CH_3CH_2CH_2C$(O) H + + other resonance structures]

b.

HF + BF_3 ⇌ BF_4^- H^+ $\xrightarrow{CH_2{=}C(CH_3)_2}$ $(CH_3)_3C^+$

$(CH_3)_3C$–Ph $\xleftarrow{-H^+}$ [H $(CH_3)_3C$ + ↔ etc]

c. $CH_3CH_2CHCH_3$ (OH) $\overset{H_2SO_4}{\rightleftharpoons}$ $CH_3CH_2CHCH_3$ ($^+OH_2$) $\overset{-H_2O}{\rightleftharpoons}$ $CH_3CH_2\overset{+}{C}HCH_3$

$CH_3CH_2CH(CH_3)$–Ph $\xleftarrow{-H^+}$ [H $CH_3CH_2CH(CH_3)$ + ↔ etc]

d.

$$H_2C{=}O + ZnCl_2 \rightleftharpoons Cl_2\bar{Z}n{-}\overset{+}{O}{=}CH_2 \xrightarrow{C_6H_6} \left[Cl_2\bar{Z}nOCH_2{-}C_6H_6^{+} \longleftrightarrow \text{etc} \right] \xrightarrow{-H^+,\ H_3O^+} HOCH_2C_6H_5$$

$$HOCH_2C_6H_5 \underset{}{\overset{H^+}{\rightleftharpoons}} H_2\overset{+}{O}CH_2C_6H_5 \overset{-H_2O}{\rightleftharpoons} {}^+CH_2C_6H_5 \overset{Cl^-}{\rightleftharpoons} ClCH_2C_6H_5$$

23.7

	$Br_2 + FeBr_3$	$HNO_3 + H_2SO_4$	$(CH_3)_3CCl + AlCl_3$	CH_3CCl (C=O) $+ AlCl_3$
$C_6H_5CH_3$	o-$BrC_6H_4CH_3$ and p-isomer; *faster*	o-$O_2NC_6H_4CH_3$ and p-isomer; *faster*	o-$(CH_3)_3CC_6H_4CH_3$ and p -isomer; *faster*	o-$CH_3CC_6H_4CH_3$ (C=O) and p-isomer; *faster*
$C_6H_5CCH_3$ (C=O)	($C_6H_5CCH_2Br$ (C=O))	m-$O_2NC_6H_4CCH_3$ (C=O) *slower*	*no reaction*	*no reaction*
C_6H_5Br	o-$C_6H_4Br_2$ and p-isomer; *slower*	o-$O_2NC_6H_4Br$ and p -isomer; *slower*	o-$(CH_3)_3CC_6H_4Br$ and p-isomer; *slower*	o-$CH_3CC_6H_4Br$ (C=O) and p-isomer; *slower*
$C_6H_5OCH_3$	o-$BrC_6H_4OCH_3$ and p-isomer; *faster*	o-$O_2NC_6H_4OCH_3$ and p-isomer; *faster*	o-$(CH_3)_3CC_6H_4OCH_3$ and p-isomer; *faster*	o-$CH_3CC_6H_4OCH_3$ (C=O) and p-isomer; *faster*
$C_6H_5NO_2$	m-$BrC_6H_4NO_2$; *slower*	m-$C_6H_4(NO_2)_2$; *slower*	*no reaction*	*no reaction*
$C_6H_5NHCCH_3$ (C=O)	o-$BrC_6H_4NHCCH_3$ (C=O) and p-isomer; *faster*	o-$O_2NC_6H_4NHCCH_3$ (C=O) and p-isomer; *faster*	o-$(CH_3)_3CC_6H_4NHCCH_3$ (C=O) and p-isomer; *faster*	o-$CH_3CC_6H_4NHCCH_3$ (both C=O) and p-isomer; *faster*

"*faster*" = reaction occurs faster than with benzene; "*slower*" = reaction occurs slower than with benzene

23.8

a.

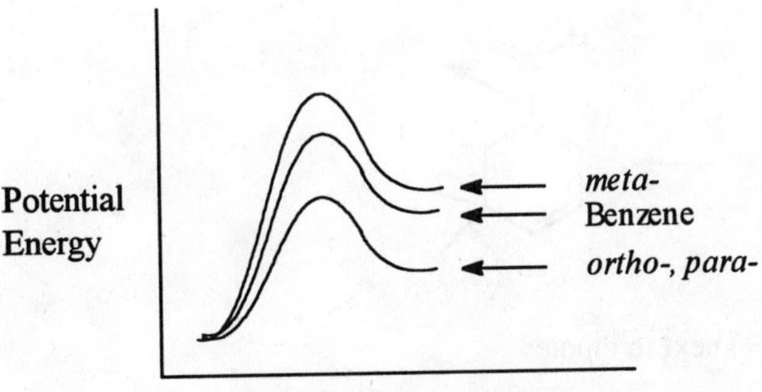

b.

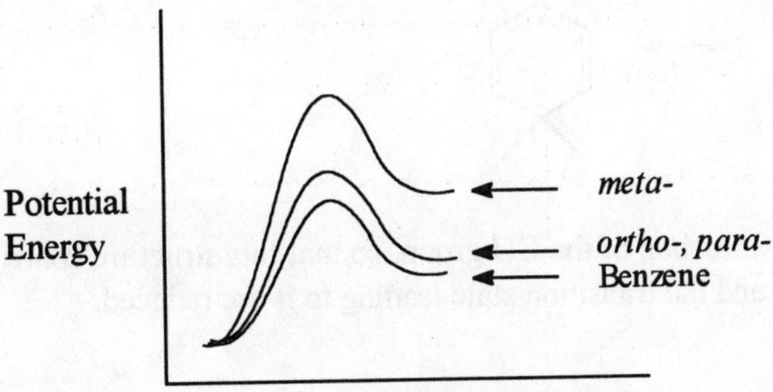

c.

In the *o*- and *p*-positions, the vinyl group helps to distribute positive charge; thus, it is *o,p*-directing and activating. In the *m*-position, the vinyl group does not conjugate with the charge:

large dipole directed away from ring:

m-Substitution keeps charge away from dipole in all three structures:

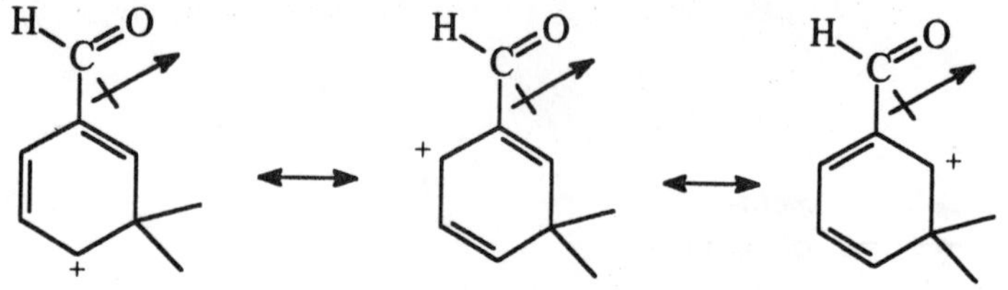

In *o,p*-substitution, one structure has (+) next to dipole:

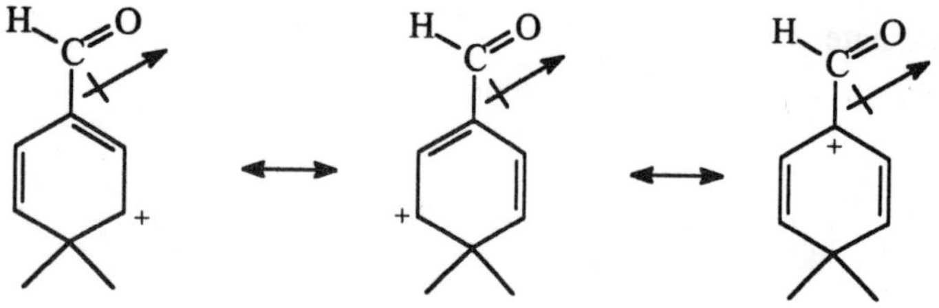

the positive charge is next to the positive carbon of the CO group, so that this structure contributes less; the resonance stabilization of intermediate and the transition state leading to it are reduced.

23.9 Partial rate factor x number of positions = relative amounts

ortho	0.03	x	2	= 0.06 = 30%
meta	0.0009	x	2	= 0.0018 = 1%
para	0.14	x	1	= 0.14 = 69%

23.10 a.

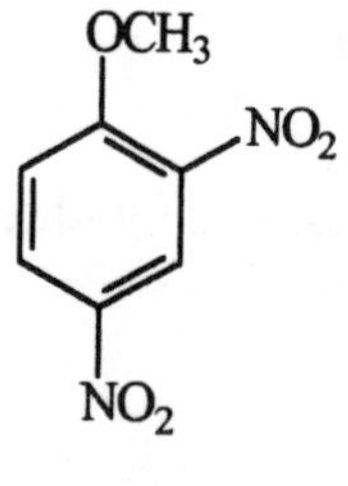

b.

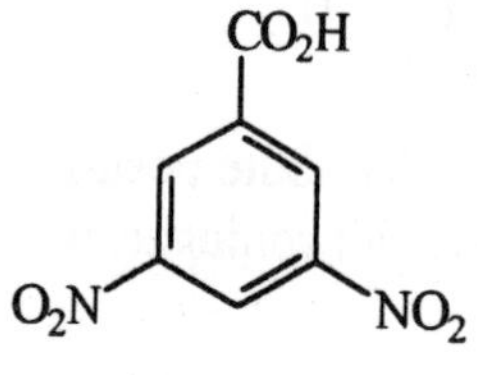

c.

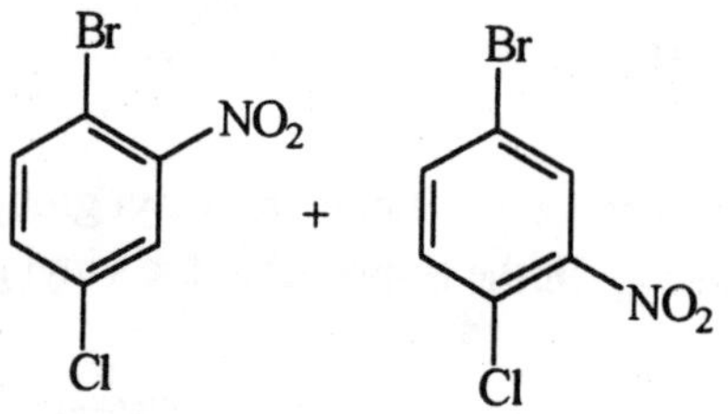

d.

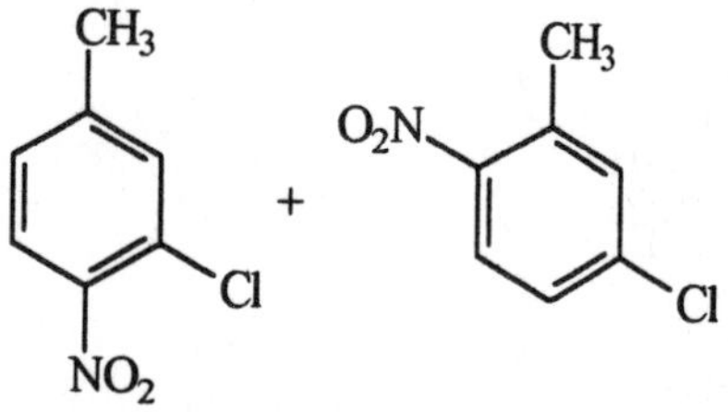

23.11 All of these transformations of Ar–Br require formation of a Grignard reagent, which cannot be generated in the presence of the $-CH{=}O$, $-CO_2H$, or $-NO_2$ functional groups. The mono-Grignard reagents can be prepared from *m*- and *p*-dibromobenzene, however.

23.12 Partial rate factor x number of positions = relative amounts

nitrobenzene	*ortho*	1.8×10^{-6}	x 2	= 3.6	= 6%
	meta	28×10^{-6}	x 2	= 56	= 94%
	para	0.2×10^{-6}	x 1	= 0.2	= 0.3%
ethyl benzoate	*ortho*	1×10^{-4}	x 2	= 2	= 28%
	meta	2.5×10^{-4}	x 2	= 5	= 69%
	para	0.24×10^{-4}	x 1	= 0.24	= 3.3%

23.E Answers and Explanations for Problems

1. $I_2 + H_2O_2 \longrightarrow 2\ HOI \rightleftharpoons H_2^+OI + {}^-OI$

Benzene + $I{-}OH_2^+$ ⟶ (cation with H and I) + H_2O

(cation with I and H) + ^-OI ⟶ iodobenzene + HOI

2. a. Mechanism A:

1. $CH_2{=}O + ZnCl_2 \rightleftharpoons CH_2{=}\overset{+}{O}{-}\overline{Z}nCl_2$

2. Benzene + $CH_2{=}\overset{+}{O}{-}\overline{Z}nCl_2 \rightleftharpoons$ (cation with H and $CH_2\overline{O}ZnCl_2$)

3. (cation with H and $CH_2\overline{O}ZnCl_2$) $\rightleftharpoons$ $C_6H_5CH_2\overline{O}ZnCl_2$ + H^+

4. $C_6H_5CH_2\overline{O}ZnCl_2 + H^+ \rightleftharpoons C_6H_5CH_2OH + ZnCl_2$

5. $C_6H_5CH_2OH + HCl \xrightleftharpoons{ZnCl} C_6H_5CH_2Cl + H_2O$

Mechanism B:

1. $CH_2{=}O + ZnCl_2 \rightleftharpoons CH_2{=}\overset{+}{O}{-}\overset{-}{Zn}Cl_2$

2. $CH_2{=}\overset{+}{O}{-}\overset{-}{Zn}Cl_2 + Cl^- \rightleftharpoons ClCH_2O\overset{-}{Zn}Cl_2$

3. $ClCH_2O\overset{-}{Zn}Cl_2 + H^+ \rightleftharpoons ClCH_2\overset{H}{\underset{+}{O}}\overset{-}{Zn}Cl_2$

4. $ClCH_2\overset{H}{\underset{+}{O}}\overset{-}{Zn}Cl_2 \rightleftharpoons Cl\overset{+}{C}H_2 + HO\overset{-}{Zn}Cl_2$

5. $C_6H_6 + {}^+CH_2Cl \rightleftharpoons \left[C_6H_6^+(H)(CH_2Cl) \longleftrightarrow \text{etc} \right]$

6. $C_6H_6^+(H)(CH_2Cl) \rightleftharpoons C_6H_5CH_2Cl + H^+$

b. 1. $CH_2{=}\ddot{O}\colon + {}^+CH_2Cl \rightleftharpoons CH_2{=}\overset{+}{O}{-}CH_2Cl$

2. $Cl^- + CH_2{=}\overset{+}{O}{-}CH_2Cl \rightleftharpoons ClCH_2OCH_2Cl$

or

1. $CH_2{=}\ddot{O}\colon + CH_2{=}\overset{+}{O}{-}\overset{-}{Zn}Cl_2 \rightleftharpoons CH_2{=}\overset{+}{O}{-}CH_2O\overset{-}{Zn}Cl_2$

2. $Cl^- + CH_2{=}\overset{+}{O}{-}CH_2O\overset{-}{Zn}Cl_2 \rightleftharpoons ClCH_2OCH_2O\overset{-}{Zn}Cl_2$

3. $ClCH_2OCH_2O\overset{-}{Zn}Cl_2 \xrightleftharpoons{H^+} ClCH_2OCH_2\overset{H}{\underset{+}{O}}\overset{-}{Zn}Cl_2$

4. $ClCH_2\ddot{O}{-}CH_2{-}\overset{H}{\underset{+}{O}}\overset{-}{Zn}Cl_2 \rightleftharpoons ClCH_2\overset{+}{O}{=}CH_2 + HO\overset{-}{Zn}Cl_2$

5. $ClCH_2\overset{+}{O}{=}CH_2 + Cl^- \longrightarrow ClCH_2OCH_2Cl$

3. $CH_3C(=O)OHgOC(=O)CH_3 + H^+ \rightleftharpoons CH_3C(=\overset{+}{O}H)OHgOC(=O)CH_3$

$CH_3C(=\overset{+}{O}H)OHgOC(=O)CH_3 \rightleftharpoons CH_3C(=O)OH + {}^+HgOAc$

benzene + ${}^+HgOAc \rightleftharpoons$ [arenium ion with H and HgOAc, + ↔ etc]

arenium ion (H, HgOAc, +) $\rightleftharpoons$ $C_6H_5HgOAc + H^+$

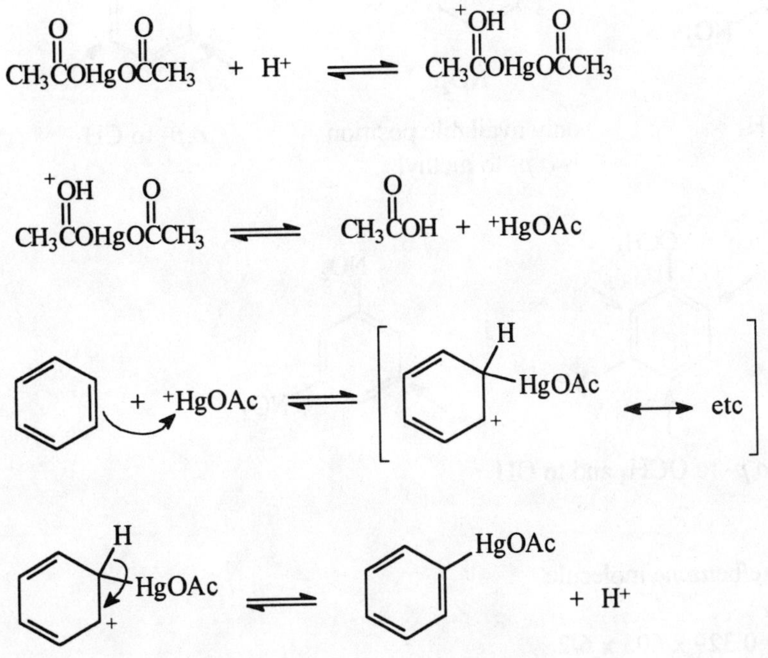

4. [Resonance structures of the arenium ions from electrophilic attack (E, H) on biphenyl]

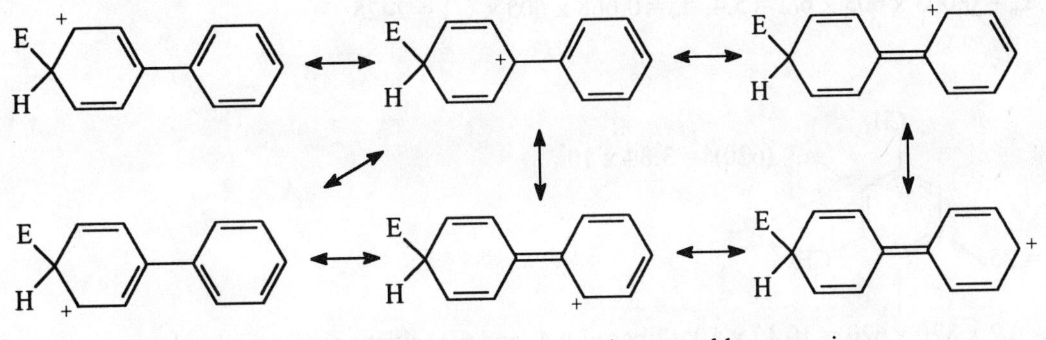

In the *o*- and *p*-positions, the charge is distributed to the second benzene ring.

5. The carboxy group is *meta*-directing and deactivating for the same reason that the formyl group is (see answer to Exercise 23.8).

6. a. [1-chloro-2-fluorobenzene with arrows at positions of attack]

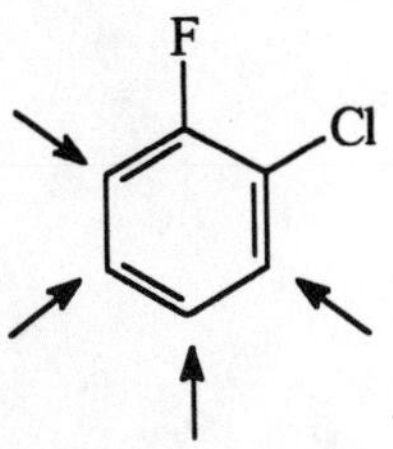

serious mixture!
o,p to F and Cl

b. [4-bromotoluene with arrows]

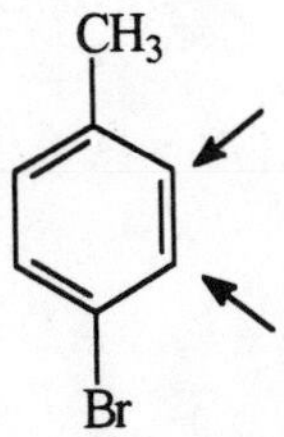

o- to both Br and CH_3

c. [1,3-dibromobenzene with arrows]

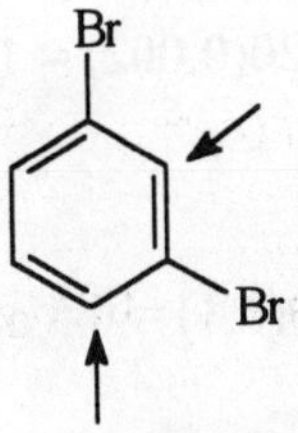

some *o*- to both Br

mostly *o*- to one Br, *p*- to the other

d. NH_2 / CH_3

NH_2 dominates

e. CH_3 / NO_2

o,p to CH_3

f. CH_3 / NO_2 / NO_2

only available position is *o,p*- to methyl

g. CO_2CH_3 / CH_3

o,p- to CH_3

h. $N(CH_3)_2$ / NO_2

i. OCH_3 / OH

o,p- to OCH_3 and to OH

j. 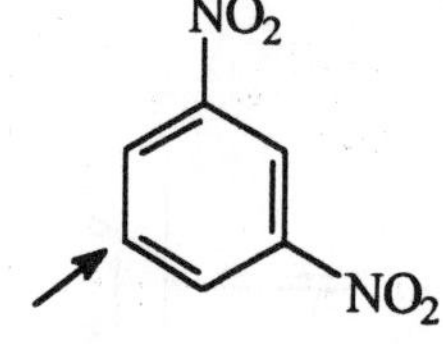

7. a. Rate/toluene molecule = 605 x rate/benzene molecule

Rate/toluene *ortho* position = $f_o = 0.329 \times 605 \times 6/2$

$f_o = 597$, $f_m = 0.003 \times 605 \times 6/2 = 5.4$, $f_p = 0.668 \times 605 \times 6/1 = 2425$

b. CH_3 / CH_3

$(620)^2 = 3.84 \times 10^5$

$(5.0)^2 = 25$

$2 \times 820 \times 620 = 10.17 \times 10^5$ [2 because 4- and 6-positions are equivalent]

2-position: $(3.84 \times 10^5)/[(3.84 \times 10^5) + (10.17 \times 10^5) + 25] = 27.4\%$
4-position: $(10.17 \times 10^5)/[3.84 \times 10^5) + (10.17 \times 10^5) + 25] = 72.6\%$
5-position: $25/[3.84 \times 10^5) + (10.17 \times 10^5) + 25] = 0.0018\%$

c. CH_3 / Cl

$620(0.002) = 1.24$ or 71%

$5.0(0.1) = 0.5$ or 29%

8. a. No; both Cl and Br are *o,p*-directors
   b. No; Friedel-Crafts acylation cannot be applied to a ketone.

The sequence: benzene → $C_6H_5COCH_3$ ⟶✕ m-$C_6H_4(COCH_3)_2$ will not work

   c. OK; benzene $\xrightarrow[AlCl_3]{CH_3COCl}$ $C_6H_5COCH_3$ $\xrightarrow[H_2SO_4]{HNO_3}$ m-$NO_2C_6H_4COCH_3$

   Note that nitration first, followed by Friedel-Crafts acylation, will not work. Acylation cannot be applied to $ArNO_2$ compounds.

   d. No; the sequence benzene → C_6H_5Cl → o-$ClC_6H_4COCH_2CH_3$

   gives mostly *para*. The *ortho*-isomer is difficult to isolate pure. This method is okay for the *para*-isomer, since it is higher-melting and less soluble and can be separated pure from the *ortho*-isomer by crystallization.

   e. OK; benzene $\xrightarrow[FeCl_3]{Cl_2}$ C_6H_5Cl $\xrightarrow[H_2SO_4]{HNO_3}$ p-$ClC_6H_4NO_2$

   The *para*-chloronitrobenzene can be separated from the *ortho*-isomer by crystallization (see Table 23.4)

   f. OK; nitration of *t*-butylbenzene strongly favors *para*-substitution.

9. a. OK; p-$CH_3OC_6H_4NO_2$ $\xrightarrow[H_2SO_4]{HNO_3}$ 2,4-$(NO_2)_2C_6H_3OCH_3$ (no other isomers possible)

b. OK; $\xrightarrow[H_2SO_4]{HNO_3}$ (no other isomers possible)

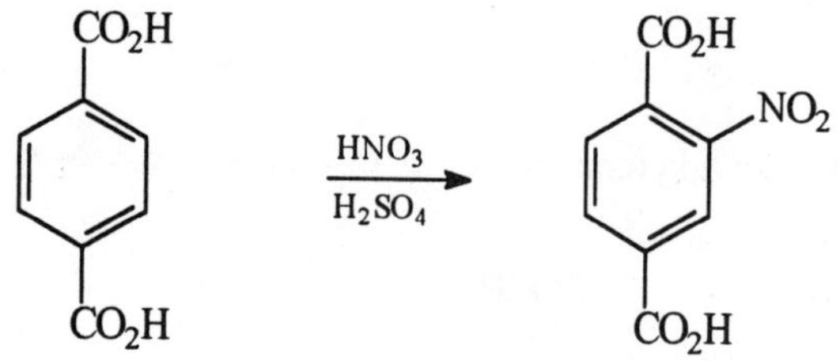

c. OK; $\xrightarrow[\text{fuming } H_2SO_4]{\text{fuming } HNO_3}$ (both NO_2 groups direct *meta*-)

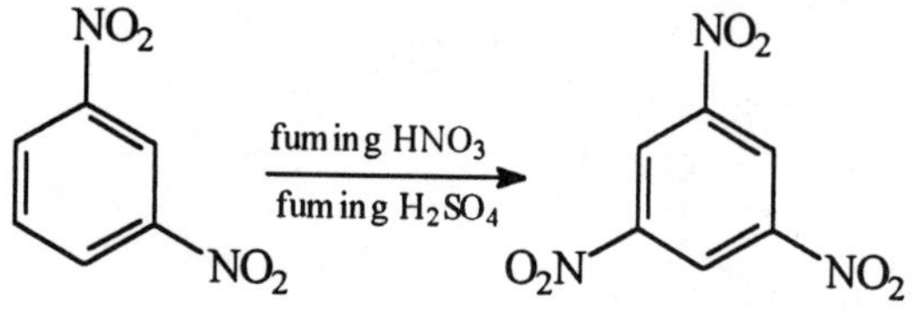

d. OK; $\xrightarrow[\text{fuming } H_2SO_4]{\text{fuming } HNO_3}$ (no other isomers possible)

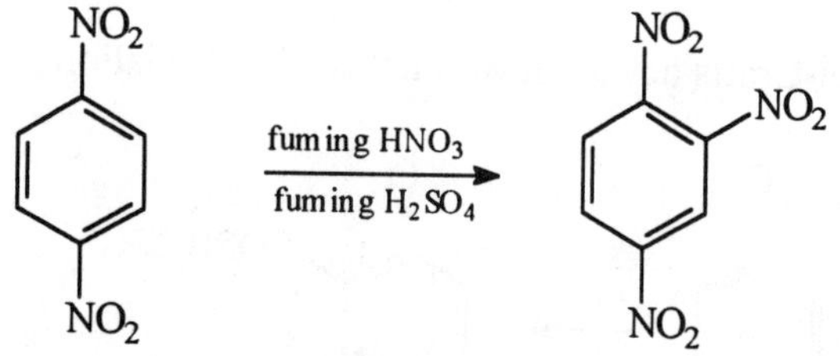

e. No; nitration of *o*-dichlorobenzene gives mostly 4-nitro; chlorination of *o*- or *m*- chloronitrobenzene would give a mixture of isomers.

f. OK; $\xrightarrow{Br_2}$ (OH dominates)

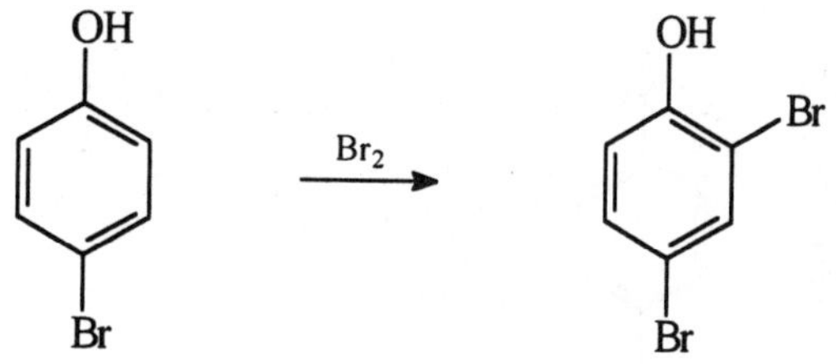

g. No. Chlorination of [p-ethyltoluene] goes *ortho* to both CH_3 and CH_2CH_3 and gives a mixture.

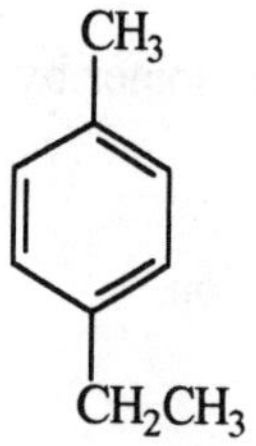

h. No. $\xrightarrow{HNO_3}$ or

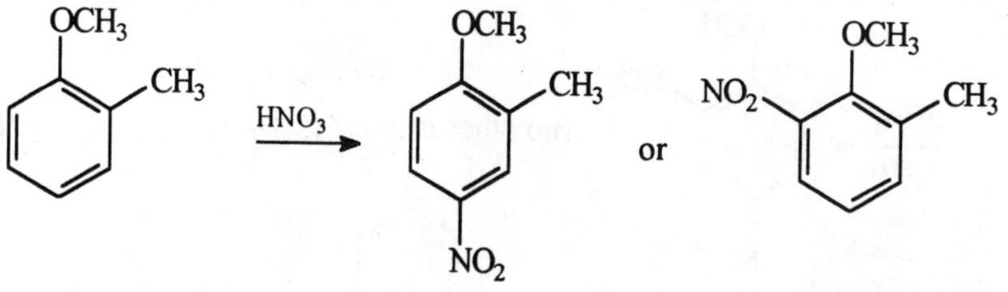

These isomers are difficult to separate (compare $O_2NC_6H_5OCH_3$ isomers in Table 23.5).

i. OK; (NO_2, CO_2H benzene) $\xrightarrow[FeBr_3\ \Delta]{Br_2}$ (NO_2, Br, CO_2H benzene) both groups orient *meta-*

j. No. Nitration of (OCH_3, CHO benzene) gives (OCH_3, NO_2, CHO benzene)

Friedel-Crafts acylation reactions cannot be applied to nitro compounds, so *m*-nitroanisole cannot be formylated.

10. The trifluoromethyl group is electron-withdrawing because of the C-F bond dipole $\left(\overset{+\!\!\rightarrow}{—CF_3}\right)$ The explanation is analogous to that given for the COOH group in the answer to problem #5.

11. $HCl + AlCl_3 \rightleftharpoons AlCl_4^- \ H^+ \ :C{=}\ddot{O} \rightleftharpoons AlCl_4^- + H{-}C{\equiv}\overset{+}{O}:$

CHO / CH_3 $\xleftarrow{-H^+}$ [OHC H + / CH_3 $\longleftrightarrow$ OHC H + / CH_3 $\longleftrightarrow$ O=C(H) H + / CH_3] $\longleftarrow$ CH_3

12. a. $H_3C(H)C(CH_3){-}CH_2{-}\overset{+}{O}(H)BF_3 \longrightarrow (CH_3)_3C^+ \ HOBF_3^-$ *rearrangement of primary carbocation to tertiary*

b. $CH_3\overset{OH}{C}DCH_2CH_3 + BF_3 \rightleftharpoons CH_3\overset{H^+O-BF_3}{C}DCH_2CH_3 \xrightleftharpoons{-BF_3OH^-} CH_3\overset{+}{C}DCH_2CH_3 \rightleftharpoons CH_3CHD\overset{+}{C}HCH_3$

$CH_3\overset{+}{C}DCH_2CH_3 \xrightarrow{C_6H_6} CH_3CDCH_2CH_3$ (phenyl)

$CH_3CHD\overset{+}{C}HCH_3 \xrightarrow{C_6H_6} CH_3CHDCHCH_3$ (phenyl)

c. Scrambling would require a primary carbocation:

$$CD_3\overset{+}{C}HCH_3 \rightleftharpoons {}^{+}CD_2CHDCH_3 \rightleftharpoons CHD_2\overset{+}{C}DCH_3$$

Primary carbocations are so unstable relative to secondary ones that the reaction does not occur.

d. Carbocations have a planar central carbon which is achiral. Consequently, a racemic reaction product is anticipated. In practice, the reaction product is 99% racemized.

13. Aluminum chloride, unless specially purified and handled on a vacuum line, always has traces of H_2O and HCl.

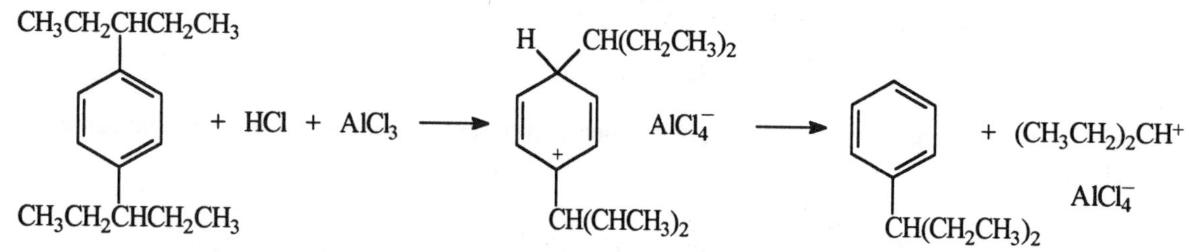

1,4-bis($CH_3CH_2CHCH_2CH_3$)benzene + HCl + $AlCl_3$ ⟶ [arenium ion with H and $CH(CH_2CH_3)_2$ on the same carbon, + in ring, $CH(CHCH_3)_2$ para] $AlCl_4^-$ ⟶ $C_6H_5CH(CH_2CH_3)_2$ + $(CH_3CH_2)_2CH^+$ $AlCl_4^-$

Rapid rearrangements and slower alkylation of benzene:

$$\underset{AlCl_4^-}{CH_3CH_2CH_2{}^{+}CHCH_3} \overset{fast}{\rightleftharpoons} \underset{AlCl_4^-}{CH_3CH_2{}^{+}CHCH_2CH_3} \overset{fast}{\rightleftharpoons} \underset{AlCl_4^-}{CH_3{}^{+}CHCH_2CH_2CH_3}$$

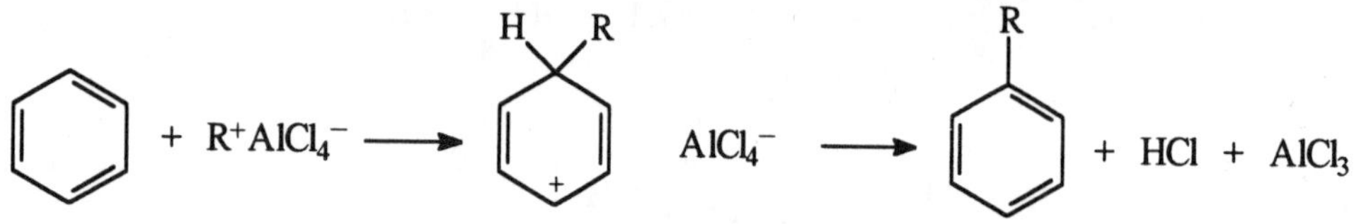

Benzene + $R^+AlCl_4^-$ ⟶ [arenium ion with H and R, + in ring] $AlCl_4^-$ ⟶ C_6H_5R + HCl + $AlCl_3$

$CH_3{}^{+}CHCH_2CH_2CH_3$ and $CH_3CH_2{}^{+}CHCH_2CH_3$ are so similar in structure, we may confidently expect that their rates of alkylating benzene will be closely similar as well. Thus, the relative rates of formation of 2-phenylpentane and 3-phenylpentane will be the same as the relative populations of the two carbocations, or 2:1.

Consequently, one mole of 1,4-di(3-pentyl)benzene gives one mole of 3-phenylpentane (after cleavage of one pentyl group), and the cleaved pentyl group gives: 0.33 moles 3-phenylpentane, 0.67 moles 2-phenylpentane

Total: 1.33 moles of 3-phenylpentane and 0.67 moles of 2-phenylpentane
or a ratio of 2:1 for 3-isomer/2-isomer.

14. 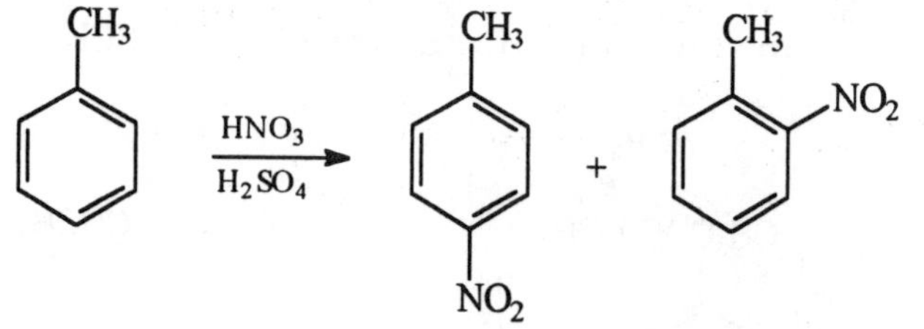 Toluene ($C_6H_5CH_3$) $\xrightarrow[H_2SO_4]{HNO_3}$ p-nitrotoluene (CH₃, NO₂ para) + o-nitrotoluene (CH₃, NO₂ ortho)

separate by crystallization

33% (para) 62% (ortho)

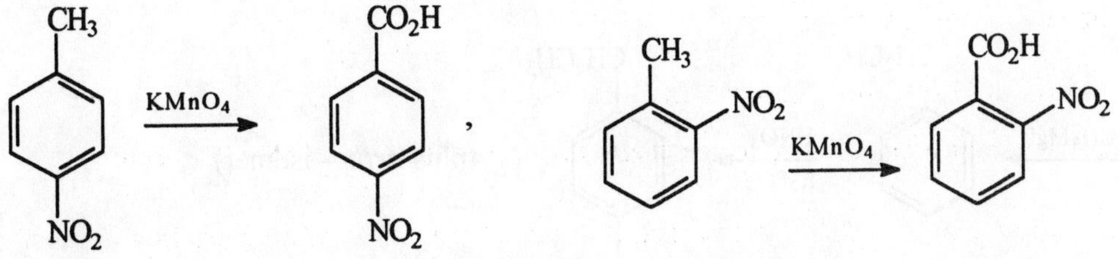

CH_3 / NO_2 $\xrightarrow{KMnO_4}$ CO_2H / NO_2 , CH_3 / NO_2 $\xrightarrow{KMnO_4}$ CO_2H / NO_2

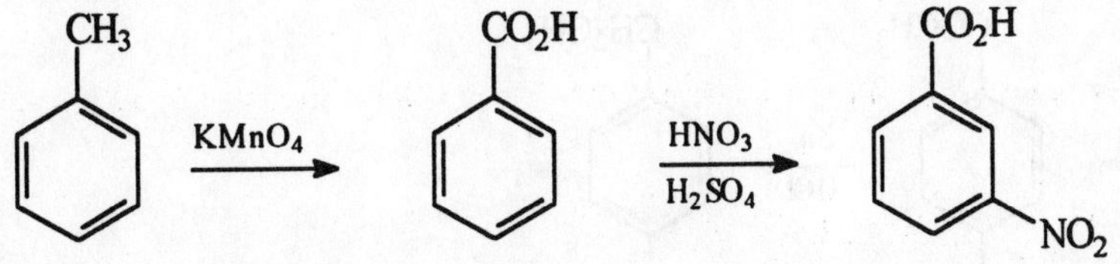

CH_3 $\xrightarrow{KMnO_4}$ CO_2H $\xrightarrow[H_2SO_4]{HNO_3}$ CO_2H / NO_2

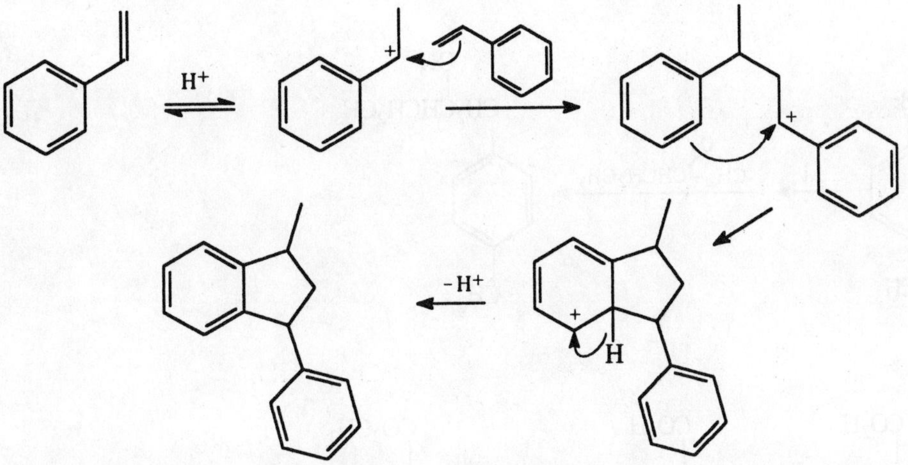

15.

H^+ ; $-H^+$; + ; H

16.

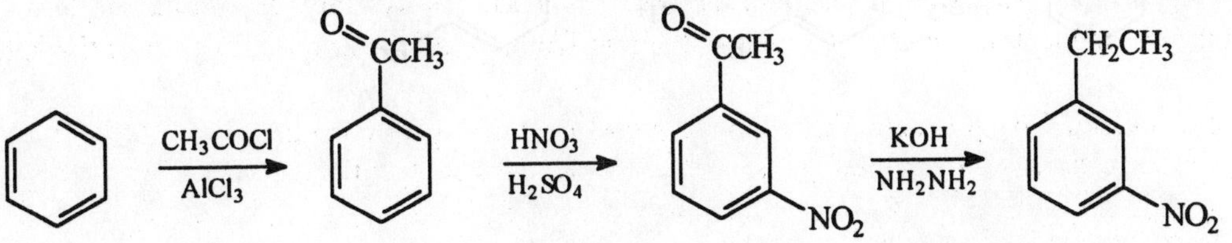

a.

$\xrightarrow[AlCl_3]{CH_3COCl}$ O=CCH$_3$ $\xrightarrow[H_2SO_4]{HNO_3}$ O=CCH$_3$ / NO_2 $\xrightarrow[NH_2NH_2]{KOH}$ CH_2CH_3 / NO_2

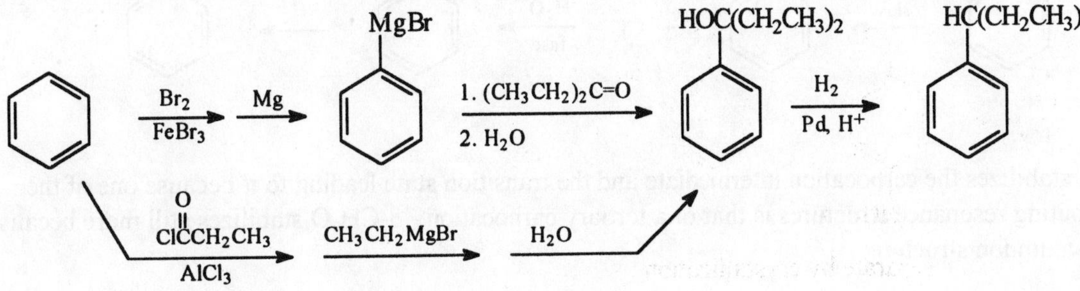

b.

$\xrightarrow[FeBr_3]{Br_2}$ $\xrightarrow{Mg}$ MgBr $\xrightarrow[2.\ H_2O]{1.\ (CH_3CH_2)_2C{=}O}$ HOC(CH_2CH_3)$_2$ $\xrightarrow[Pd,\ H^+]{H_2}$ HC(CH_2CH_3)$_2$

$\xrightarrow[AlCl_3]{ClCCH_2CH_3\ (\text{O})}$ $\xrightarrow{CH_3CH_2MgBr}$ $\xrightarrow{H_2O}$

c. $C_6H_5CH_3$ $\xrightarrow[h\nu]{Br_2}$ $\xrightarrow{CH_3MgI}$ $C_6H_5CH_2CH_3$ $\xrightarrow[H_2SO_4]{HNO_3}$ $p\text{-}O_2NC_6H_4CH_2CH_3$ (plus *ortho-* isomer)

d. C_6H_6 $\xrightarrow[FeBr_3]{Br_2}$ C_6H_5Br $\xrightarrow[AlCl_3]{CH_3COCl}$ $p\text{-}BrC_6H_4COCH_3$ $\xrightarrow[HCl]{Zn}$ $p\text{-}BrC_6H_4CH_2CH_3$

e. $C_6H_5CH_3$ $\xrightarrow[FeBr_3]{Br_2}$ $p\text{-}BrC_6H_4CH_3$ $\xrightarrow{Li}$ $\xrightarrow{\overset{O}{CH_2-CHCH_2CH_3}}$ $p\text{-}CH_3C_6H_4CH_2\overset{OH}{C}HCH_2CH_3$

f. $C_6H_5CH_3$ $\xrightarrow{KMnO_4}$ $C_6H_5CO_2H$ $\xrightarrow[FeBr_3]{Br_2}$ $m\text{-}BrC_6H_4CO_2H$ $\xrightarrow[H^+]{CH_3OH}$ $m\text{-}BrC_6H_4CO_2CH_3$

17. a. $H_3C-\overset{Cl}{C}(C_6H_5)-CH_3$ $\xrightarrow{slow}$ $H_3C-\overset{+}{C}(C_6H_5)-CH_3$ + Cl^- $\xrightarrow[fast]{H_2O}$ $H_3C-\overset{^+OH_2}{C}(C_6H_5)-CH_3$ $\xrightarrow{-H^+}$ $H_3C-\overset{OH}{C}(C_6H_5)-CH_3$

b. p-CH_3 stabilizes the carbocation intermediate and the transition state leading to it because one of the contributing resonance structures is that of a tertiary carbocation. p-CH_3O stabilizes still more because of an oxonium ion structure:

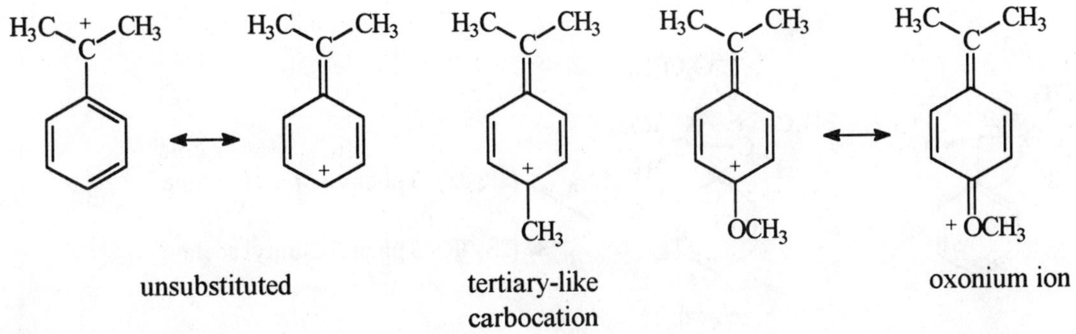

unsubstituted

tertiary-like carbocation

oxonium ion

The *p*-NO_2 group destabilizes the carbocation by electrostatic repulsion:

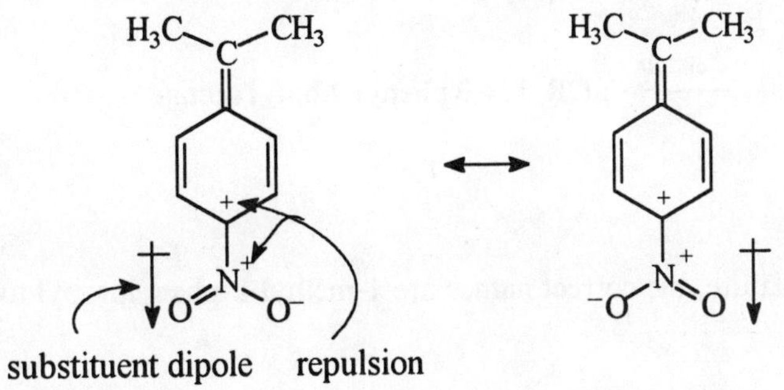

substituent dipole repulsion

18. *Lowest dissociation constant*: $(p\text{-}O_2NC_6H_4)_3CCl$
$(m\text{-}ClC_6H_4)_3CCl$
$(C_6H_5)_3CCl$
$(m\text{-}CH_3C_6H_4)_3CCl$
$(p\text{-}CH_3C_6H_4)_3CCl$
Highest dissociation constant: $(p\text{-}CH_3OC_6H_4)_3CCl$

Substituents that activate the ring toward electrophilic substitution will also stabilize the carbocation. The influence of these substituents is greatest in the *para* position because of resonance.

19. a.

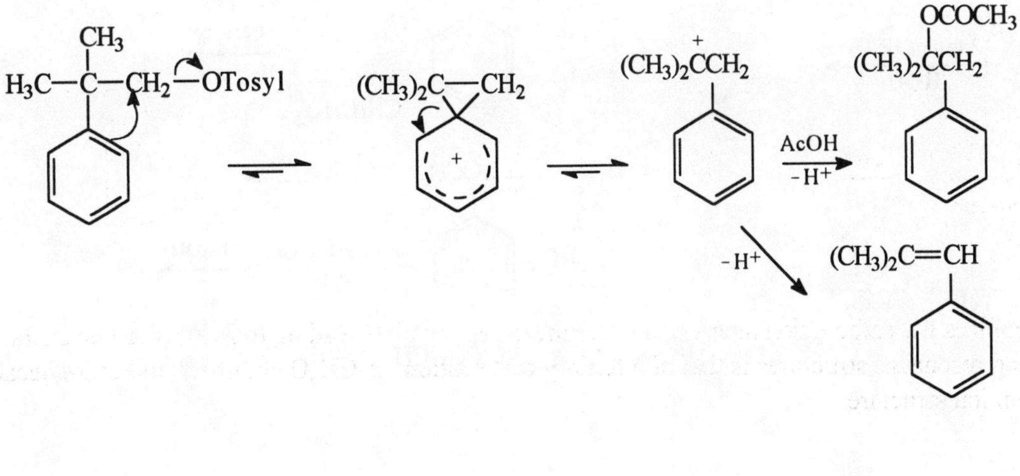

AcOH, -H^+

-H^+

$(CH_3)_2C{=}CH$

b.

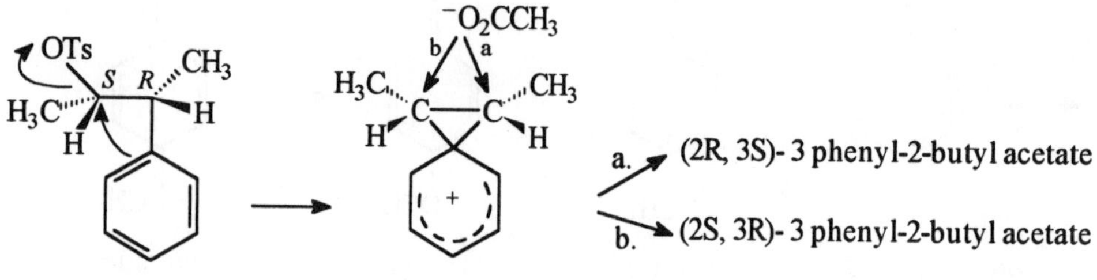

a. (2R, 3S)- 3 phenyl-2-butyl acetate

b. (2S, 3R)- 3 phenyl-2-butyl acetate

achiral

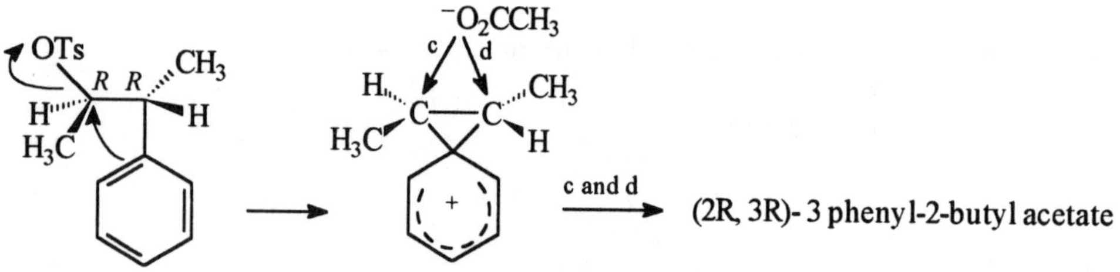

c and d

(2R, 3R)- 3 phenyl-2-butyl acetate

chiral

NOTE: By the current rules of nomenclature, the correct names are 1-methyl-2-phenylpropyl tosylate, etc.

23.F Supplementary Problems

S1. Predict the products from the following reactions. If more than one product is anticipated, indicate which (if any) will predominate.

a. Toluene (excess) $\xrightarrow[BF_3\ +\ HF]{CH_2=C(CH_3)CH_2CH_3}$

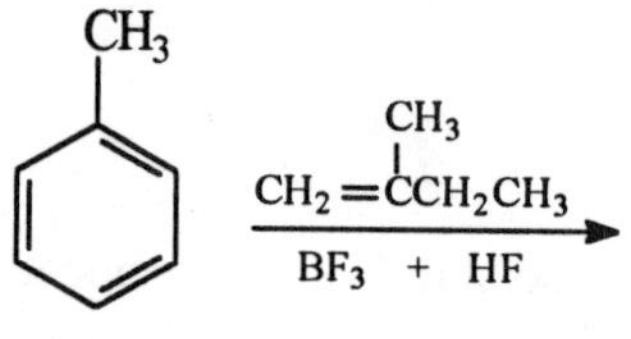

b. Benzoic acid (CO_2H) $\xrightarrow[FeCl_3]{Cl_2}$ $\xrightarrow[H^+]{CH_3OH}$ $\xrightarrow{CH_3MgI}$

c. Chlorobenzene $\xrightarrow[H_2SO_4]{2\ moles\ HNO_3}$

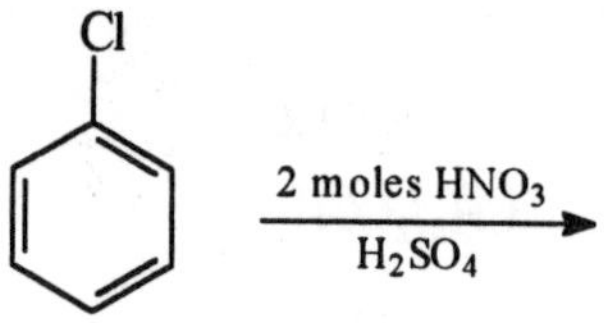

d. 3-Isopropylanisole (OCH_3, $CH(CH_3)_2$) $\xrightarrow[H_2SO_4]{HNO_3}$

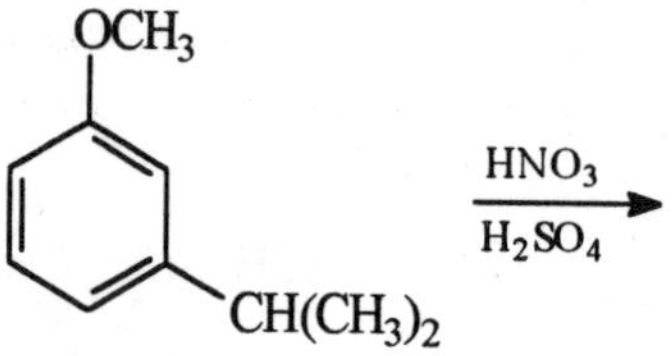

e. Nitrobenzene (NO_2) $\xrightarrow[FeBr_3]{Br_2}$ $\xrightarrow[FeCl_3]{Cl_2}$

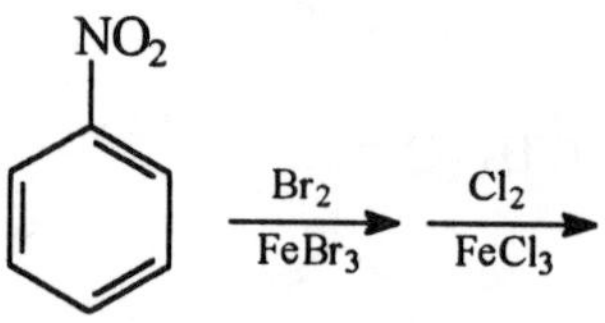

f. Methyl 4-methoxybenzoate (OCH_3, $COOCH_3$) $\xrightarrow[FeBr_3]{2\ moles\ Br_2}$ $\xrightarrow{NaOH}$

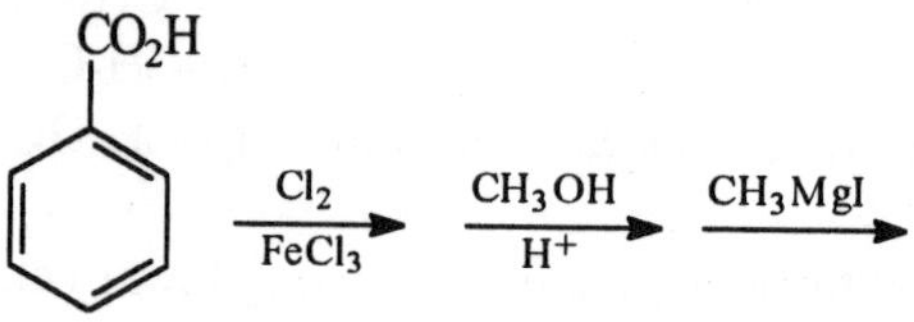

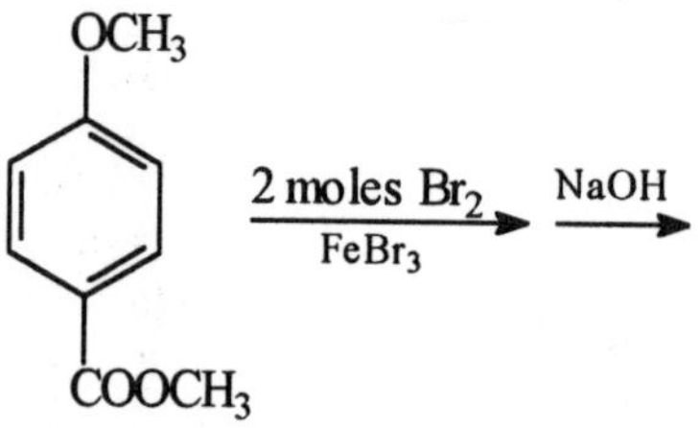

g. 4-Bromo-N-methylacetanilide ($CH_3N(COCH_3)$ and Br on the ring) $\xrightarrow[AlCl_3]{CH_3COCl}$

h. Toluene $\xrightarrow[AlCl_3]{CH_3CH_2CH(CH_3)CH_2Cl}$

i. Anisole (OCH_3 on the ring) $\xrightarrow[AlCl_3]{CH_3COCl}$ $\xrightarrow[HCl]{Zn}$

j. Benzaldehyde (CHO on the ring) $\xrightarrow[AlCl_3]{ClCOCH_2CH_3}$

k. $HNCOCH_3$, $CH_3OC(O)$ and Cl on the ring $\xrightarrow[H_2SO_4]{HNO_3}$

l. CO_2CH_3 and NO_2 para on the ring $\xrightarrow[FeBr_3]{Br_2}$

S2. Show how to make each of the following compounds (as free of isomers as possible), starting with benzene or toluene and any other reagents.

a. Br and $CH(CH_3)_2$ meta on the ring

b. CH_3 and CH_2CH_3 para on the ring

c. CH_3, NO_2 (ortho to CH_3) and CH_2CH_3 (para to CH_3) on the ring

d. Br, Cl (ortho to Br) and NO_2 (para to Br) on the ring

e. Two CH_3-substituted rings (para) joined by C=O: H_3C–C_6H_4–C(O)–C_6H_4–CH_3

f. CO_2CH_3 and CO_2CH_3 para on the ring

g. CO_2H, Cl (ortho to CO_2H) and NO_2 (para to CO_2H) on the ring

h. CH_2Cl and NO_2 para on the ring

i. $CH_2CH_2CH_2OH$ and $C(CH_3)_3$ para on the ring

j. CH_2Cl and Cl meta on the ring

S3. Write a reasonable mechanism for the cyclization illustrated below, showing all of the intermediates involved. Do you expect to see any other compound(s) as products of this reaction?

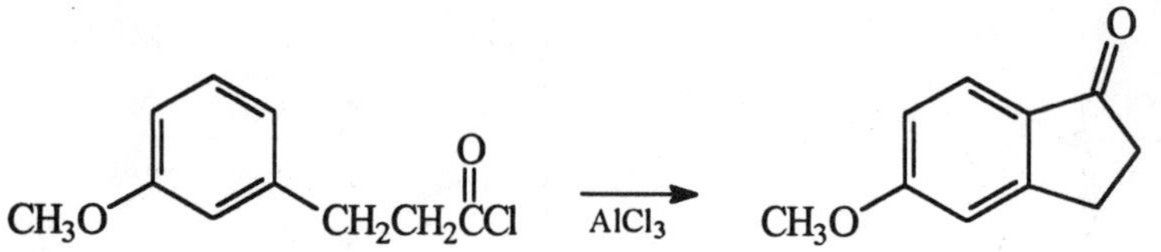

CH_3O $CH_2CH_2\overset{O}{\overset{\|}{C}}Cl$ $\xrightarrow{AlCl_3}$ CH_3O O

S4. Treatment of phenol with sodium nitrite and HCl results in the formation of *p*-nitrosophenol. Write a reasonable mechanism for the formation of the active electrophilic species and its reaction with phenol.

S5. None of the reaction sequences outlined below will lead to the compounds that are shown as the major products. For each case, show what the main product(s) actually would be, and provide an alternative sequence that will produce the desired compound as the major isomer, **using the same starting materials**.

a. 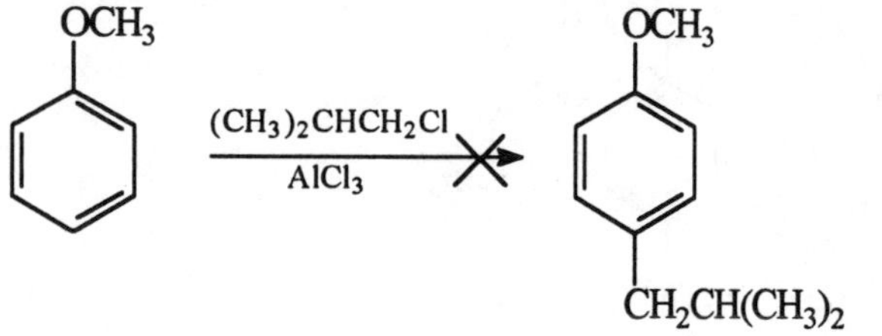 OCH_3 $\xrightarrow[AlCl_3]{(CH_3)_2CHCH_2Cl}$ ✗ OCH_3, $CH_2CH(CH_3)_2$

b. 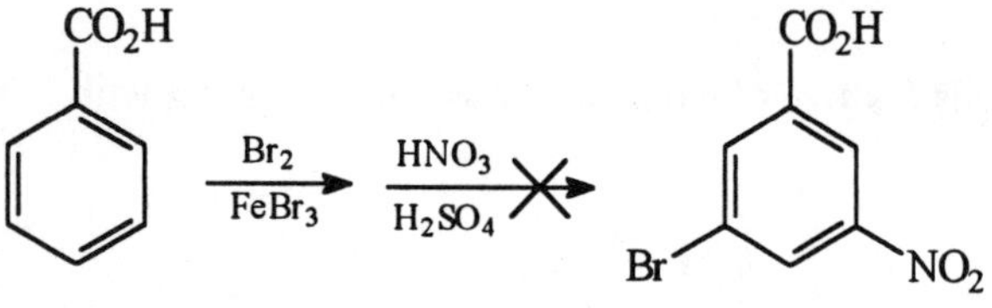 CO_2H $\xrightarrow[FeBr_3]{Br_2}$ $\xrightarrow[H_2SO_4]{HNO_3}$ ✗ CO_2H, Br, NO_2

c. 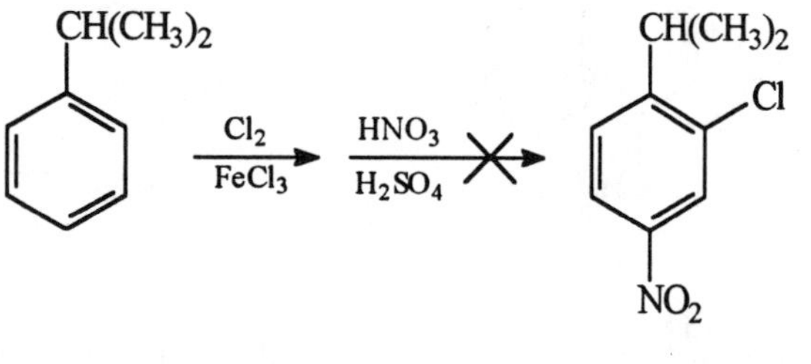 $CH(CH_3)_2$ $\xrightarrow[FeCl_3]{Cl_2}$ $\xrightarrow[H_2SO_4]{HNO_3}$ ✗ $CH(CH_3)_2$, Cl, NO_2

d. 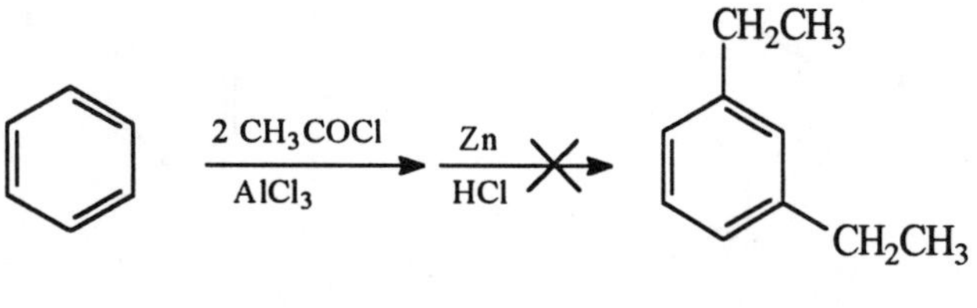 $\xrightarrow[AlCl_3]{2\ CH_3COCl}$ $\xrightarrow[HCl]{Zn}$ ✗ CH_2CH_3, CH_2CH_3

S6. Predict the favored position of electrophilic aromatic substitution of the following compounds, and justify your answer.

a. 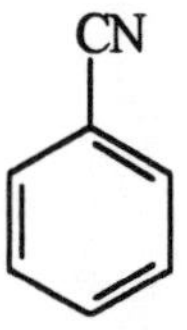 CN

b. 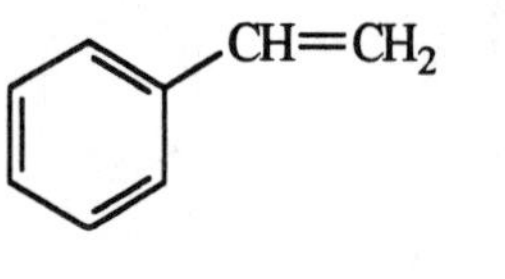 $CH{=}CH_2$

c. 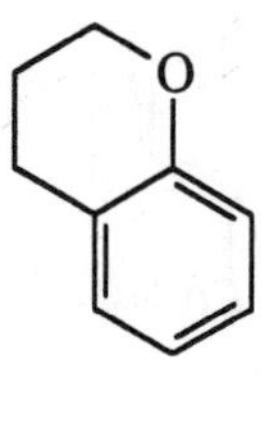 O

d. 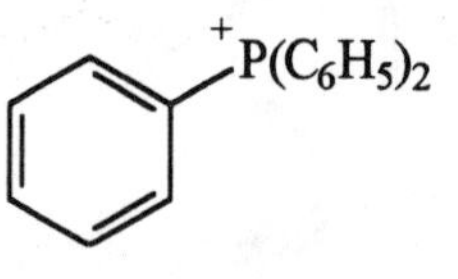 $\overset{+}{P}(C_6H_5)_2$

S7. If sodium triethylphenylborate is treated with D_2SO_4/D_2O, it undergoes cleavage to give deuteriobenzene and a triethylboron compound:

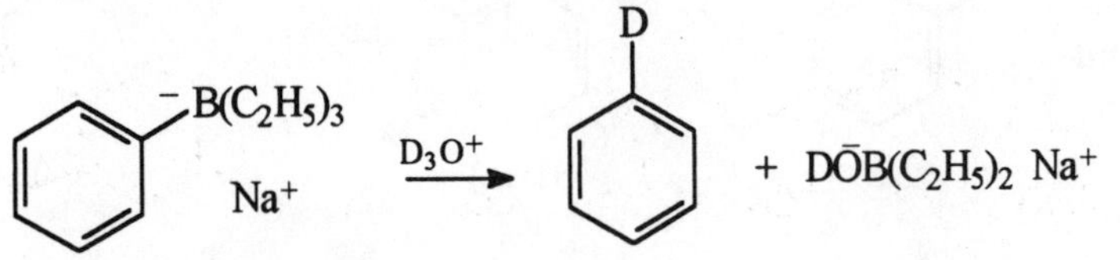

$C_6H_5\overset{-}{B}(C_2H_5)_3\ Na^+ \xrightarrow{D_3O^+} C_6H_5D + D\bar{O}B(C_2H_5)_2\ Na^+$

a. Write a mechanism for this transformation.

b. What product would you expect if you treated the same compound with Br_2? Write a mechanism for the reaction you predict.

S8. Write a mechanism for the following transformation:

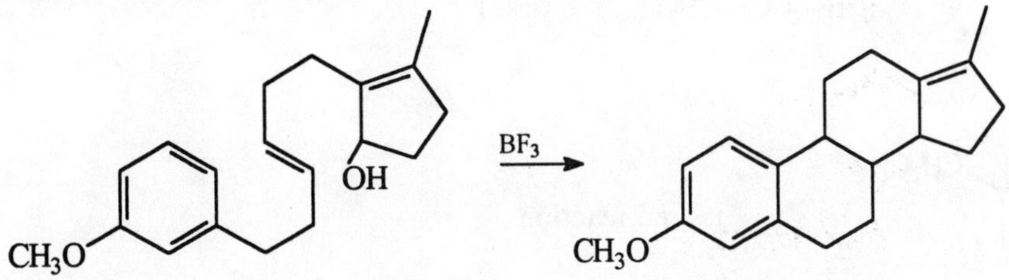

$\xrightarrow{BF_3}$

(Structures: CH_3O-substituted aryl alkenyl cyclopentenol (OH) → CH_3O-substituted tetracyclic product)

23.G. Answers to Supplementary Problems

a. CH_3 / $(CH_3)_2CCH_2CH_2$ (para-substituted ring); CH_3, CH_2CH_3 / $C(CH_3)_2$ (ortho-substituted ring) (less)

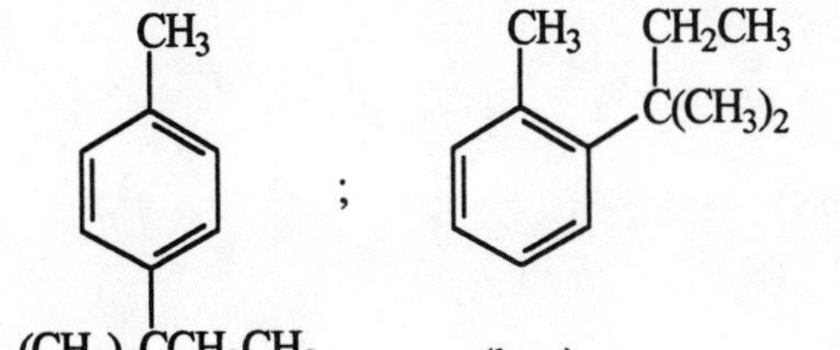

b. $(CH_3)_2COH$ / Cl (meta-substituted ring)

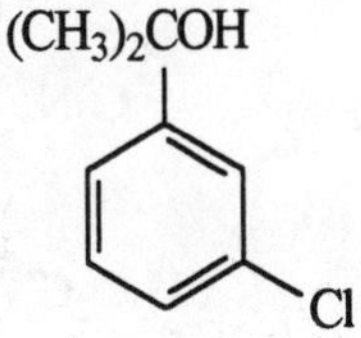

c. Cl, NO_2, NO_2 (2,4-dinitrochlorobenzene); O_2N, Cl, NO_2 (2,6-dinitrochlorobenzene) (less)

d. OCH_3, O_2N, $CH(CH_3)_2$; OCH_3, NO_2, $CH(CH_3)_2$ (less)

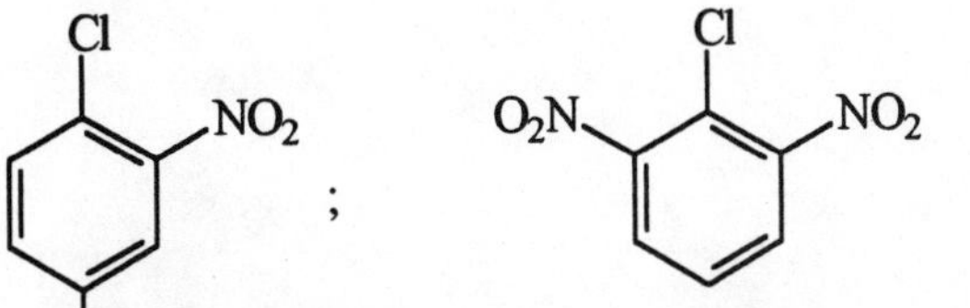

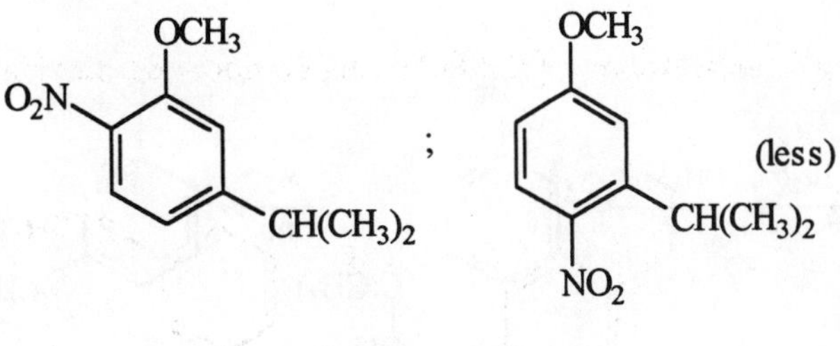

e. 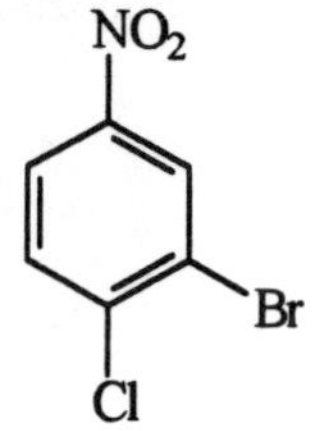 and 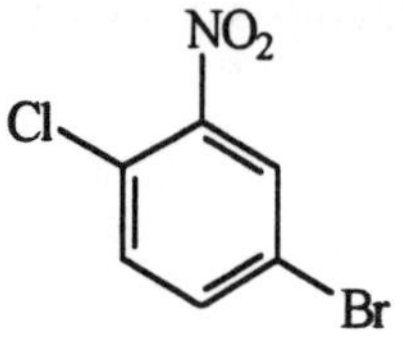 ; 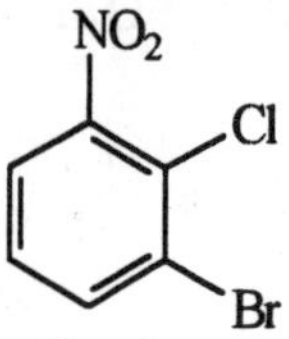 (less)

f. 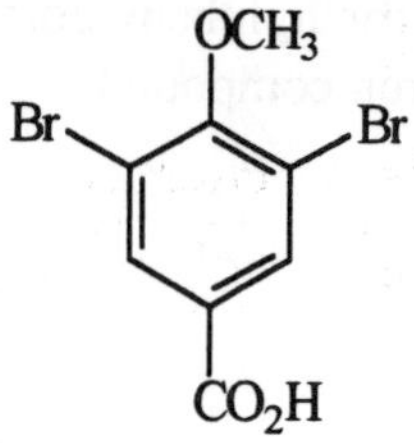

g. 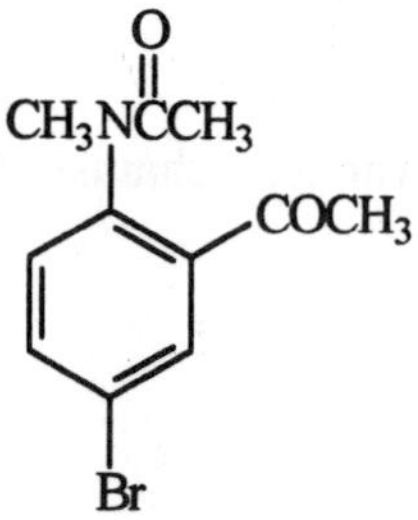

h. 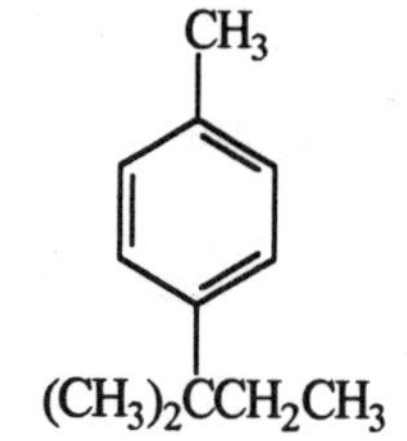 ; 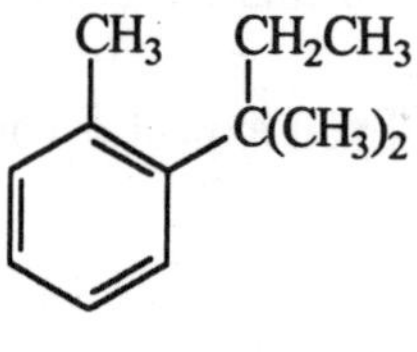 (less)

i. 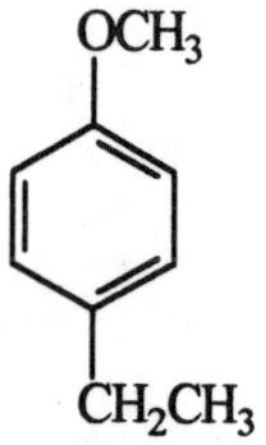 and 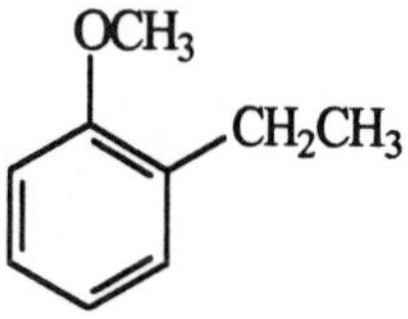

j. no reaction

k. 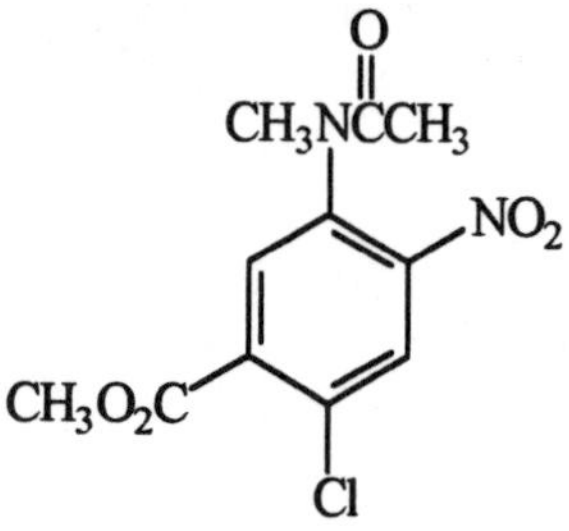 ; 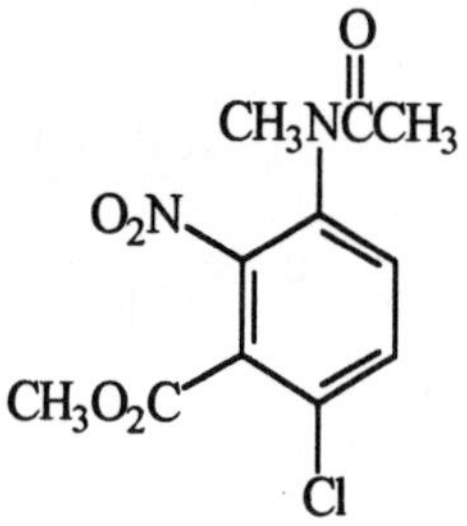 (less)

l. 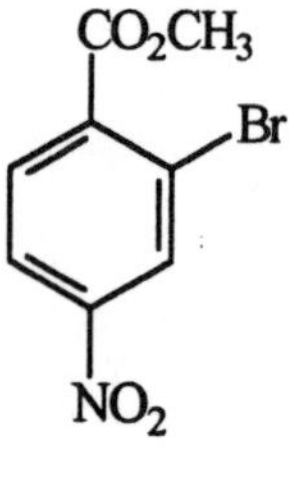 and 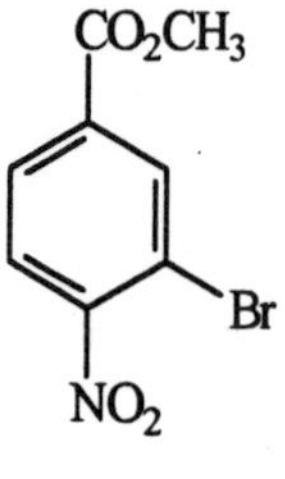

S2. a.

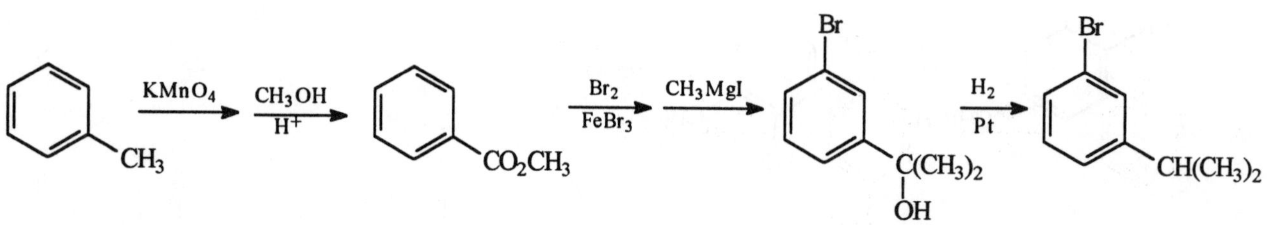
KMnO4
CH3OH
H+
Br2
FeBr3
CH3MgI
H2
Pt

b. Toluene $\xrightarrow[AlCl_3]{CH_3COCl}$ p-$CH_3C_6H_4COCH_3$ $\xrightarrow[HCl]{Zn}$ p-$CH_3C_6H_4CH_2CH_3$

c. Toluene $\xrightarrow[AlCl_3]{CH_3COCl}$ p-$CH_3C_6H_4COCH_3$ $\xrightarrow[H_2SO_4]{HNO_3}$ (CH_3, NO_2, $COCH_3$) $\xrightarrow[H_2NNH_2]{KOH}$ (CH_3, NO_2, CH_2CH_3)

d. Benzene $\xrightarrow[FeBr_3]{Br_2}$ $\xrightarrow[H_2SO_4]{HNO_3}$ p-$BrC_6H_4NO_2$ $\xrightarrow[FeCl_3]{Cl_2}$ (Br, Cl, NO_2)

e. H_3C–C_6H_5 (excess) $\xrightarrow[HCl,\ ZnCl_2]{CH_2=O}$ H_3C–C_6H_4–CH_2–C_6H_4–CH_3 $\xrightarrow{CrO_3}$ H_3C–C_6H_4–C(=O)–C_6H_4–CH_3

f. p-$CH_3C_6H_4COCH_3$ (from c.) $\xrightarrow{KMnO_4}$ $\xrightarrow[H^+]{CH_3OH}$ p-CH_3O_2C–C_6H_4–CO_2CH_3

g. Toluene $\xrightarrow[H_2SO_4]{HNO_3}$ $\xrightarrow[FeCl_3]{Cl_2}$ (CH_3, Cl, NO_2) $\xrightarrow{KMnO_4}$ (CO_2H, Cl, NO_2)

h. Toluene $\xrightarrow[H_2SO_4]{HNO_3}$ p-$CH_3C_6H_4NO_2$ (separate from *ortho*) $\xrightarrow[h\nu]{Cl_2}$ p-$ClCH_2C_6H_4NO_2$

i.

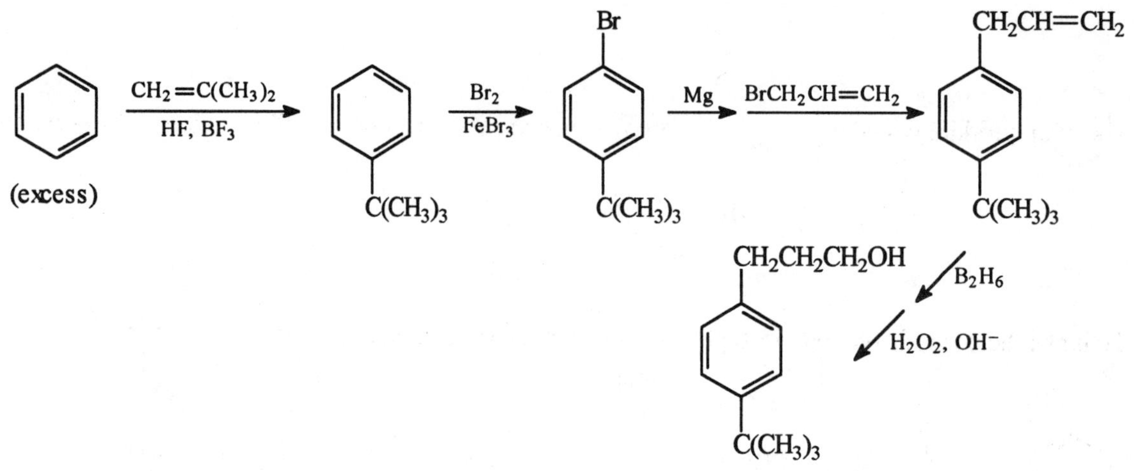

Benzene (excess) $\xrightarrow[\text{HF, BF}_3]{CH_2=C(CH_3)_2}$ $C_6H_5C(CH_3)_3$ $\xrightarrow[\text{FeBr}_3]{Br_2}$ p-$BrC_6H_4C(CH_3)_3$ $\xrightarrow{Mg}$ $\xrightarrow{BrCH_2CH=CH_2}$ p-$CH_2=CHCH_2C_6H_4C(CH_3)_3$ $\xrightarrow{B_2H_6}$ $\xrightarrow{H_2O_2,\ OH^-}$ p-$HOCH_2CH_2CH_2C_6H_4C(CH_3)_3$

j.

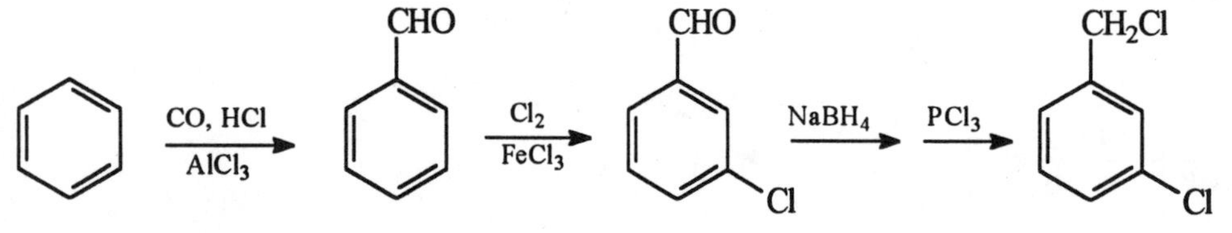

Benzene $\xrightarrow[\text{AlCl}_3]{\text{CO, HCl}}$ C_6H_5CHO $\xrightarrow[\text{FeCl}_3]{Cl_2}$ m-ClC_6H_4CHO $\xrightarrow{NaBH_4}$ $\xrightarrow{PCl_3}$ m-$ClC_6H_4CH_2Cl$

(see Problem #11)

S3.

CH_3O–C_6H_4–$CH_2CH_2C(=O)$–Cl---$AlCl_3$ ⟶ CH_3O–C_6H_4–$CH_2CH_2C{\equiv}O^+$ $AlCl_4^-$ ⟶ (CH_3O^+ cation intermediate with H) $\xrightarrow{-H^+}$ 5-methoxy-1-indanone (CH_3O, O)

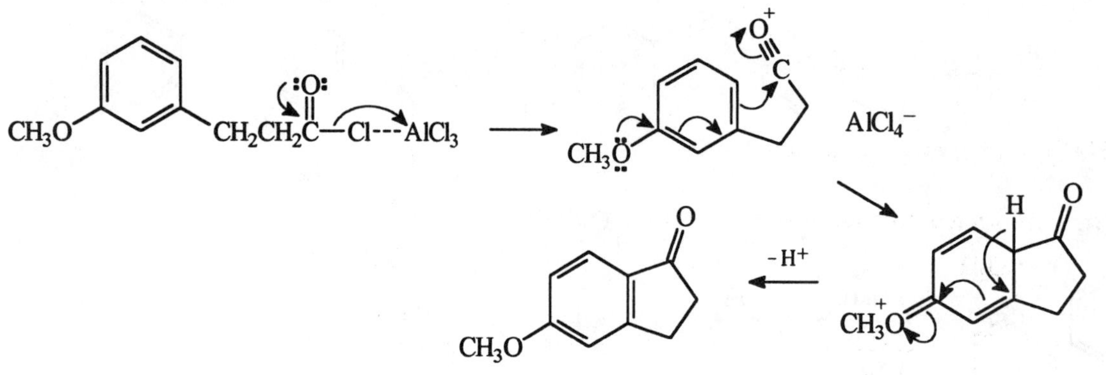

Some product from *ortho* attack will also be seen: 7-methoxy-1-indanone (CH_3O, O)

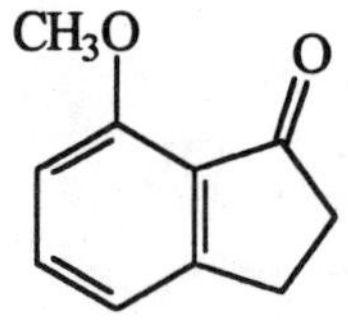

S4.

$O=N-O^- + 2H^+ \rightleftharpoons$:Ö=N̈–$\overset{+}{O}H_2$ $\rightleftharpoons$:$\overset{+}{O}$≡N: + phenol (:OH) ⟶ ($\overset{+}{O}H$ intermediate with :O=N and H) ⟶ p-nitrosophenol (OH, O=N:)

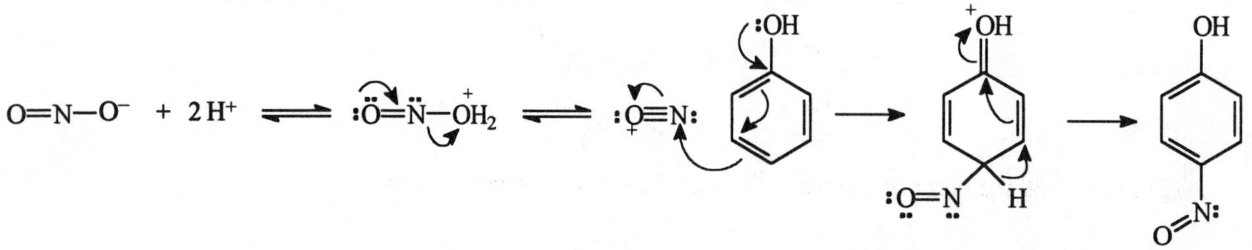

(plus other resonance structures)

S5. a. The main product would be [structure: 4-tert-butyltoluene, CH_3 para to $C(CH_3)_3$] from arrangement of the primary alkyl halide before substitution.

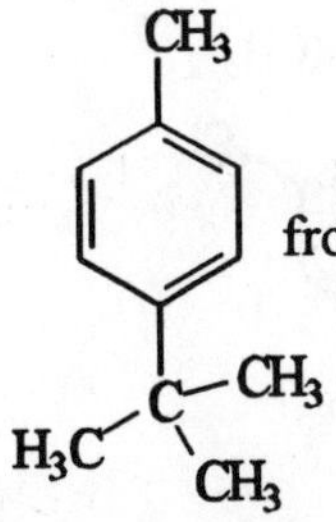

To make the desired compound, employ an acylation/deoxygenation sequence:

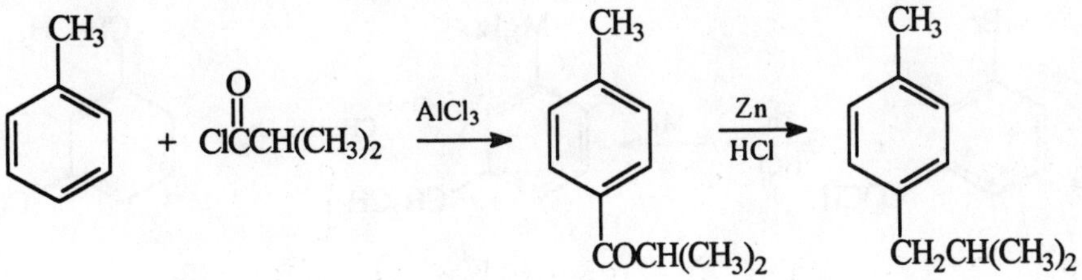

$$C_6H_5CH_3 + ClCOCH(CH_3)_2 \xrightarrow{AlCl_3} p\text{-}CH_3C_6H_4COCH(CH_3)_2 \xrightarrow[HCl]{Zn} p\text{-}CH_3C_6H_4CH_2CH(CH_3)_2$$

b. Main products would be [structure: CO_2H with NO_2 ortho and Br para to NO_2] and [structure: CO_2H with Br meta and NO_2 para, ortho to Br] because Br (Group II) dominates COOH (Group III) in directing power.

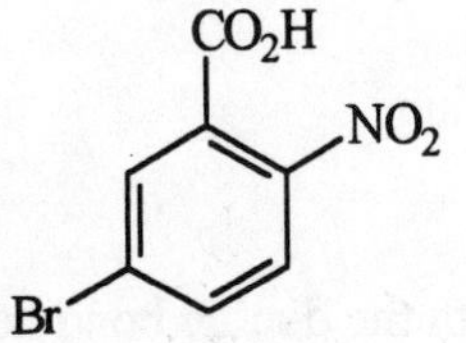

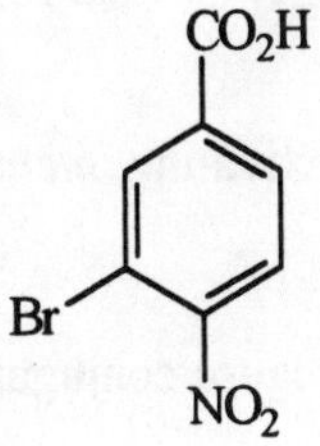

To obtain the desired product, reverse the sequence of steps:

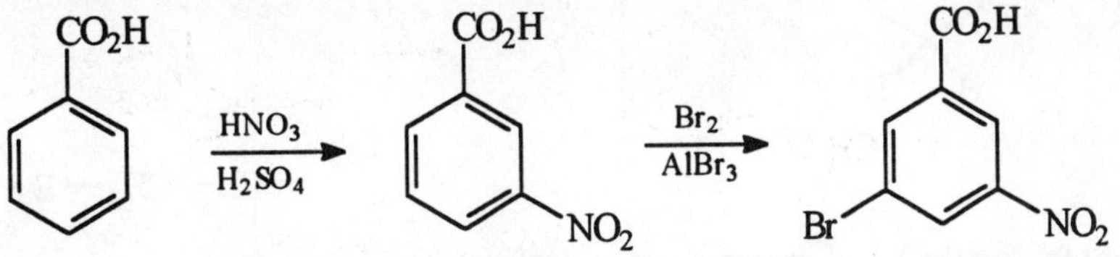

$$C_6H_5CO_2H \xrightarrow[H_2SO_4]{HNO_3} m\text{-}NO_2C_6H_4CO_2H \xrightarrow[AlBr_3]{Br_2} 3\text{-}Br\text{-}5\text{-}NO_2C_6H_3CO_2H$$

c. The chlorine will go preferentially *para* (with some *ortho*), and then will compete with the isopropyl group to give a mixture of isomers:

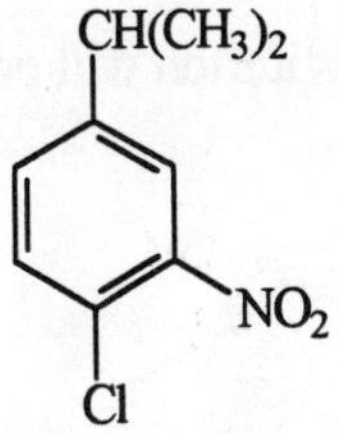

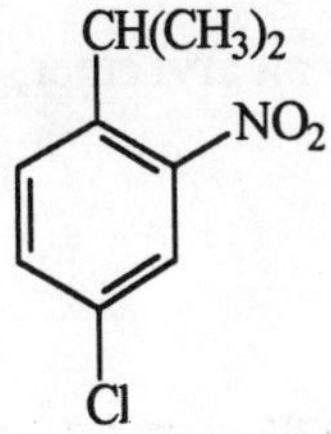

[structure: $CH(CH_3)_2$, Cl para, NO_2 ortho to Cl] > [structure: $CH(CH_3)_2$, Cl para, NO_2 ortho to $CH(CH_3)_2$] (for steric reasons)

Again, reversing the order of the reactions will furnish the correct product

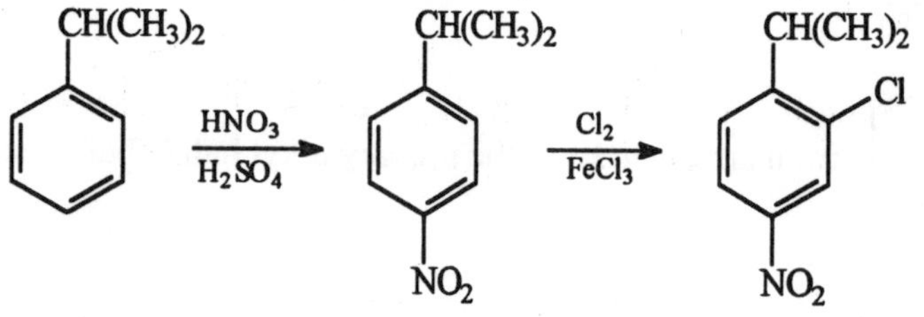

$CH(CH_3)_2$ benzene $\xrightarrow[H_2SO_4]{HNO_3}$ p-nitro-$CH(CH_3)_2$-benzene $\xrightarrow[FeCl_3]{Cl_2}$ 2-chloro-4-nitro-$CH(CH_3)_2$-benzene

d. Only one acyl group can be attached in a Friedel-Crafts acylation reaction, so ethylbenzene will be the overall product. To get around this involves a somewhat more involved sequence.

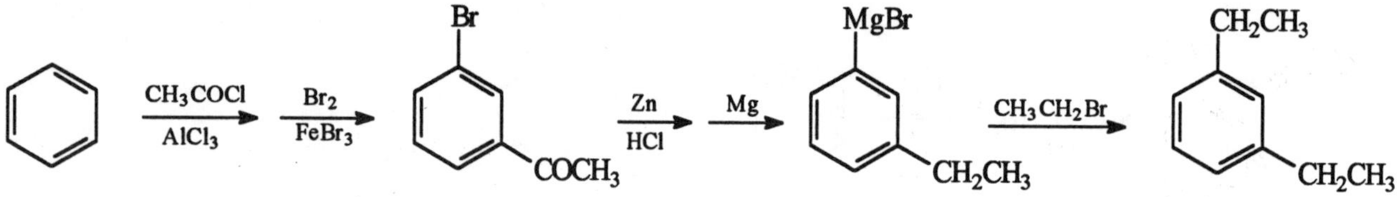

benzene $\xrightarrow[AlCl_3]{CH_3COCl}$ $\xrightarrow[FeBr_3]{Br_2}$ (Br, $COCH_3$ meta) $\xrightarrow[HCl]{Zn}$ $\xrightarrow{Mg}$ (MgBr, CH_2CH_3 meta) $\xrightarrow{CH_3CH_2Br}$ (CH_2CH_3, CH_2CH_3 meta)

S6. a. Dipole moment (benzene–C≡N, dipole arrow) and resonance $(-^{+}C{=}N^{-})$ both disfavor adjacent positive charge, so this group will be deactivating, *meta*-directing.

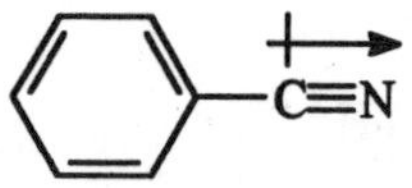

b. A vinyl group is *o,p*-directing because conjugation with the double bond helps to stabilize the positive charge:

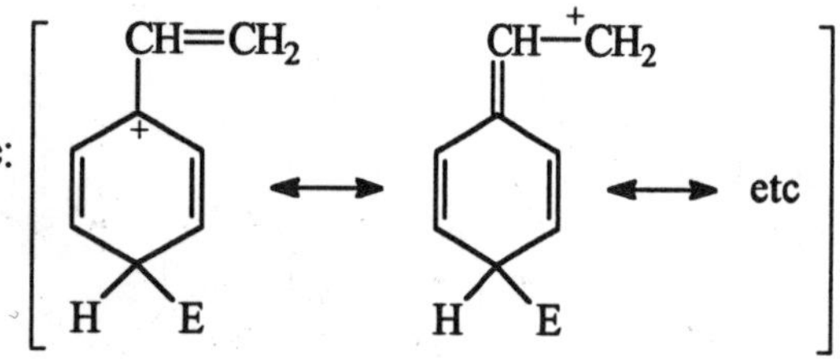

[$CH{=}CH_2$ (ring +, H E) ⟷ $CH{-}\overset{+}{C}H_2$ (H E) ⟷ etc]

(NOTE: electrophilic attack on the vinyl group itself is preferred:) $^{+}CH{-}CH_2{-}E$ on benzene

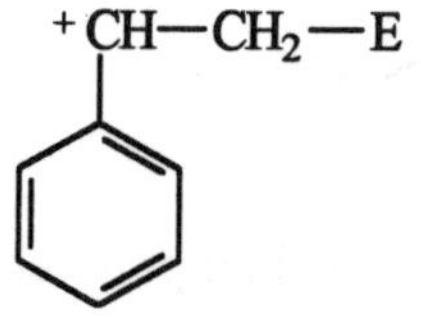

c. 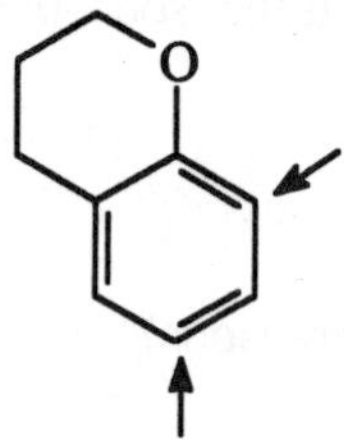 This is simply an aryl ether, and substitution will occur *ortho* and *para* to the oxygen.

d. 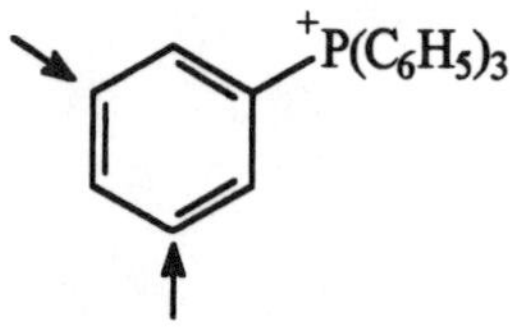 $^{+}P(C_6H_5)_3$ — Because of the cationic phosphorus, the ring is deactivated, with the greatest effect at the *ortho* and *para* positions; this substituent is deactivating, *meta*-directing.

S7. a. D^+ $^-B(C_2H_5)_3$ ⇌ D $^-B(C_2H_5)_3$ ⟶ D + $B(C_2H_5)_3$ $\xrightarrow{-D^+}$ D + $D\bar{O}B(C_2H_5)_3$

D_2O:

(This is simply the reverse of the electrophilic aromatic substitution reactions which we have been focusing on.)

b. Br—Br $^-B(C_2H_5)_3$ ⇌ Br $^-B(C_2H_5)_3$ Br^- ⟶ Br + $B(C_2H_5)_3$ + Br^-

S8. CH_3O OH BF_3 ⇌ [CH_3O + ⟷ R +] ⟶ CH_3O + ⟶ CH_3O H + $\xrightarrow{-H^+}$ CH_3O

24. AMINES

24.A. Chapter Outline and Important Terms Introduced

24.1 **Structure**

primary, secondary, and tertiary amines — nonbonding electron pair
quaternary ammonium compounds — nitrogen inversion

24.2 **Nomenclature**

-amine — aniline, benzenamine
amino-, alkylamino- — N-alkyl...

24.3 **Physical Properties and Spectra**

A. Physical Properties
B. Infrared Spectra (weak N-H stretch)
C. Nuclear Magnetic Resonance Spectra
  C<u>H</u>$_3$NR$_2$: 2.2 ppm — N<u>H</u>: 0.6-3.0 ppm
  <u>C</u>NR$_2$: 30 - 45 ppm

24.4 **Basicity**

$R_3N: + H^+ \rightleftharpoons R_3N^+H$ — pK_a of alkylammonium compounds: 9-11
pK_a of anilinium ion: 4.6

resolution of racemic carboxylic acids via formation of diastereomeric salts

24.5 **Quaternary Ammonium Compounds**

A. Tertiary Amines as Nucleophiles
B. Phase-Transfer Catalysis
  salts soluble in organic synthesis
  greater reactivity because of reduced solvation

24.6 **Synthesis**

A. Direct Alkylation of Ammonia or other Amines
  (polyalkylation problems)
B. Indirect Alkylation: The Gabriel Synthesis
C. Reduction of Nitro Compounds
D. Reduction of Nitriles
E. Reduction of Oximes
F. Reduction of Imines: Reductive Amination
  Eschweiler-Clarke reaction — immonium ion
  Leukart reaction
G. Reduction of Amides
H. Reduction of Azides
  Staudinger Reaction
I. Preparation of Amines from Carboxylic Acids:
  The Hofmann, Curtius, and Schmidt Rearrangements
  acyl nitrene — isocyanate
  carbamic acid

24.7. **Reactions** (see Sections 24.4, 24.5, and 24.6, as well)

A. Formation of Amides (see Section 19.7.B)
B. Reaction with Nitrous Acid
nitrosation
nitrosamines
nitrosonium ion
diazonium compound
diazotization

C. Oxidation
hydroxylamines
amine oxides

D. Electrophilic Aromatic Substitution
(strong *ortho, para*-directing)
Vilsmeier reaction

E. Elimination of the Amino Group: the Cope and Hofmann elimination reactions)

24.8 **Enamines, Immonium Ions**
Mannich reaction

24.B Important Reactions Introduced

Direct alkylation of ammonia and other amines (24.6.A)

Equation:

$$H_3N: + RX \longrightarrow H_3N^+R\,X^- \overset{:NH_3}{\rightleftharpoons} H_2NR \xrightarrow{RX} H_2N^+R_2, \text{ etc.}$$

Generality: R ≠ 3° alkyl (E2 occurs instead)

Key features: because of equilibria between ammonium species, polyalkylation is a problem

Gabriel synthesis of amines (24.6.B)

Equation:

phthalimide anion + R–X ⟶ N–R (phthalimide) $\xrightarrow[\text{or } H_2NNH_2]{H_3O^+ \text{ or } OH^-}$ RNH_2

phthalimide anion

Generality: R ≠ 3° alkyl, (E2 occurs instead)

Key features: method for the synthesis of 1° amines without problems of overalkylation substitution reaction is an S_N2 displacement

Reduction of nitro compounds (24.6.C)

Equation:

$$ArNO_2 \xrightarrow{[H]} ArNH_2$$

Generality: Ar = aryl group
[H] = H_2/Ni, Fe/HCl, $SnCl_2$/HCl, Zn/HCl, Sn/HCl, or NaSH

Key features: this is the best method for the formation of aryl amines
with polynitro compounds, selective reduction is sometimes possible

Reduction of nitriles (24.6.D)

Equation:

$$RC{\equiv}N \xrightarrow{[H]} RCH_2NH_2$$

Generality: R = alkyl or aryl
[H] = H_2/cat. or $LiAlH_4$

Key features: in combination with displacement of RX with $^-$CN, useful method for synthesis of 1° amines
hydrogenation method sometimes leads to secondary amines; this can be minimized by conducting the reaction in presence of excess NH_3

Reduction of oximes (24.6.E)

Equation:

$$R{-}\overset{NOH}{\overset{\|}{C}}{-}R' \xrightarrow{[H]} R\overset{NH_2}{\overset{|}{C}}HR'$$

Generality: R, R′ = H, alkyl, or aryl
[H] = H_2/Ni or $LiAlH_4$

Reductive amination (24.6.F)

Equation:

$$R{-}\overset{O}{\overset{\|}{C}}{-}R' + H_2NR'' \rightleftharpoons R{-}\overset{NR''}{\overset{\|}{C}}{-}R' \xrightarrow{[H]} R\overset{HNR''}{\overset{|}{C}}HR'$$

Generality: R, R′, R″ = H, alkyl, or aryl
[H] = H_2/Ni; CH_2O/HCO_2H (Eschweiler-Clarke reaction); $HCONR_2$, heat, (Leukart reaction)

Key features: general method for the synthesis of a wide range of 1°, 2°, and 3° amines
proceeds via immonium ion intermediate

Reduction of amides (24.6.G)

Equation:

$$R{-}\overset{O}{\overset{\|}{C}}{-}NR'_2 \xrightarrow{[H]} RCH_2NR'_2$$

Generality: R, R′ = H, alkyl, or aryl
[H] = $LiAlH_4$ or B_2H_6

Key features: useful for the preparation of 1°, 2°, or 3° amines

Reduction of Azides (24.6.H)

Equation:

$$RX + N_3^- \longrightarrow RN_3 + X^- \xrightarrow[Pd/CH_3OH]{H_2NNH_2} RNH_2$$

Generality: R = alkyl, aryl

Key Features: azides can be explosive.

Staudinger Reaction:

Equation:

$$RN_3 + Ph_3P \xrightarrow[THF]{H_2O} RNH_2 + Ph_3PO + N_2$$

Generality: R = aryl or alkyl

Key Features: Useful for preparation of amine with other functional groups such as nitro, epoxy, hydroxy, cyano, ethoxycarbonyl.

Degradation of carboxylic acid derivatives to primary amines (24.6.I)

Equation:

Hofmann: $R{-}\overset{O}{\overset{\|}{C}}{-}NH_2 + X_2 + OH^- \longrightarrow R\overset{O}{\overset{\|}{C}}NHX$ (X = Br, Cl)

$R\overset{O}{\overset{\|}{C}}NHX \xrightarrow[-HX]{OH^-} R\overset{O}{\overset{\|}{C}}N{:}$

Curtius: $R{-}\overset{O}{\overset{\|}{C}}{-}Cl + NaN_3 \longrightarrow R\overset{O}{\overset{\|}{C}}N_3 \xrightarrow{\Delta} R\overset{O}{\overset{\|}{C}}N{:}$ (acyl nitrene) $\longrightarrow RN{=}C{=}O$ isocyanate

Schmidt: $R{-}\overset{O}{\overset{\|}{C}}{-}OH + NaN_3 \xrightarrow{H_2SO_4} R\overset{O}{\overset{\|}{C}}N_3$

$RN{=}C{=}O \xrightarrow{R'OH} RNH\overset{O}{\overset{\|}{C}}OR'$ carbamate

$RN{=}C{=}O \xrightarrow{H_2O} \left[RNH\overset{O}{\overset{\|}{C}}OH\right]$ carbamic acid $\xrightarrow{-CO_2} RNH_2$

Generality: R = alkyl, aryl

Key features: mechanism involves rearrangement of acyl nitrene intermediate

Reaction of amines with nitrous acid (24.7.B)

Secondary amines:

Equation: $R_2NH + HONO \longrightarrow R_2N{-}N{=}O + H_2O$

Generality: R = alkyl, aryl

Key features: compounds with R = methyl can be carcinogenic

Primary amines; diazotization:

Equation: $RNH_2 + HONO \longrightarrow \underset{\text{diazonium ion}}{R^+N{\equiv}N} \longrightarrow \text{"}R^+\text{"} \longrightarrow$ carbocation reactions

Generality: if R = alkyl, diazonium ion loses N_2 and products arising from R^+ are obtained
if R = aryl, diazonium ion can be isolated

Key features: diazonium ions are important intermediates in chemistry of aromatic compounds
3° amines generally don't react with nitrosating agents, unless one of the substituents is an aryl group (ring nitrosation then occurs)

Oxidation of amines (24.7.C)

Equation:

2° amines: $R_2NH \xrightarrow{[Ox]} R_2N{-}OH$ (hydroxylamines)

3° amines: $R_3N{:} \xrightarrow{[Ox]} R_3N{-}O^-$ (amine oxides)

Generality: R = alkyl, aryl
[Ox] = H_2O_2 or $R'CO_3H$

Key features: oxidation to hydroxylamines usually occurs in poor yield

Vilsmeier reaction (24.7.D)

Equation: $HC(=O)N(CH_3)_2 \xrightarrow{POCl_3} \underset{\text{Vilsmeier reagent}}{HC(Cl){=}\overset{+}{N}(CH_3)_2} \xrightarrow{C_6H_5{-}Y} \xrightarrow{H_2O} HC(=O){-}C_6H_4{-}Y$

Generality: Y = strongly activating, *o*-,*p*-directing substituent

Elimination of amines (24.7.E)

Hofmann degradation:

Equation:

$$\mathrm{HO^-} + \mathrm{H{-}C{-}C{-}\overset{+}{N}R'_3} \longrightarrow \mathrm{>C{=}C<} + \mathrm{:NR_3} + \mathrm{H_2O}$$

Generality: R′ ≠ H, usually CH_3
elimination usually involves pyrolysis of quaternary ammonium hydroxide salt

Key features: mechanism involves *anti*-elimination
Hofmann rule: least-substituted alkene is the major product
S_N2 displacement of CH_3 group side reaction when R′ = CH_3 and no β-H.

Cope elimination:

Equation:

$$\mathrm{R_2\overset{O^-}{N}{-}C{-}C{-}H} \longrightarrow \mathrm{>C{=}C<} + \mathrm{R_2NOH}$$

Generality: R ≠ H
elimination proceeds at 150-200°

Key features: mechanism involves *syn*-elimination

Enamine formation and alkylation (24.8)

Equation:

$$\mathrm{{-}\overset{O}{\overset{\|}{C}}{-}CH<} + \mathrm{R_2NH} \rightleftharpoons \mathrm{{-}\overset{NR_2}{\overset{|}{C}}{=}C<} \xrightarrow{\mathrm{R'X}} \mathrm{{-}\overset{\overset{+}{N}R_2}{\overset{\|}{C}}{-}CR'<} \xrightarrow{\mathrm{H_2O}} \mathrm{{-}\overset{O}{\overset{\|}{C}}{-}CR'<}$$

Generality: R′ must be reactive toward S_N2 displacement

Key features: useful method for mono-alkylation of ketones

Mannich reaction (24.8)

Equation:

$$\mathrm{{-}\overset{OH}{\overset{|}{C}}{=}C<} + \mathrm{>C{=}\overset{+}{N}R_2} \xrightarrow{-\mathrm{H^+}} \mathrm{{-}\overset{O}{\overset{\|}{C}}{-}\overset{|}{\underset{|}{C}}{-}\overset{|}{\underset{|}{C}}{-}NR_2}$$

Generality: acid-catalyzed reaction
immonium ion usually derived from an aldehyde and 2° amine

Key features: useful carbon-carbon bond-forming reaction

24.C Important Concepts and Hints

The amino group is different from the other functional groups you have encountered because, in its neutral form, it is basic and appreciably nucleophilic. Other functional groups require strong acid to be completely protonated, and usually it is only when they are in a deprotonated, anionic form that they are good nucleophiles (for instance: acetylide anions from acetylenes, enolates from ketones, etc.). As you know, acidity and basicity are commonly indicated by referring to the "pK_a" of a compound. The pK_a value indicates the position of the following equilibrium:

$$H{-}Y \overset{K_a}{\rightleftharpoons} Y^- + H^+; \qquad K_a = \frac{[H^+][Y^-]}{[HY]}; \qquad pK_a = -\log K_a$$

It is clearly convenient to use the same term to indicate both how acidic H−Y is and how basic Y^- is, since they are related. But confusion can arise if you forget that pK_a literally refers to a compound functioning as an acid; i.e., losing a proton. This has not been a problem when discussing the functional groups presented in the text previously, but it can arise in the chemistry of amines. In Chapter 4 of this Study Guide, we pointed out what was wrong with the common statement: "The pK_a of ammonia is 9." Two *correct* statements are: "The pK_a of ammonium ion is 9", or "The pK_a of ammonia is 34." Because acid-base chemistry plays such an important role in the reactions of amines, it would be a good idea to review Chapter 4 of the text and of this Study Guide.

Much of the chemistry of amines involves the formation and reactions of imines and immonium ions. These reactions are analogous to those of carbonyl compounds to a great extent, as you can see from the list below. In general, imines (the neutral forms) are less reactive toward nucleophilic attack than are their carbonyl counterparts, and immonium ions (the cationic forms) are more reactive than their carbonyl analogs.

Starting Material	Reaction	Product
$C{=}O$ (ketone)	reduction (H_2/catalyst; $LiAlH_4$; $NaBH_4$; $(i\text{-}PrO)_3Al$; etc.	$H{-}C{-}OH$
$C{=}\overset{+}{N}HR$ or $C{=}NR$	reduction (H_2/catalyst; $LiAlH_4$; $NaBH_4$ or $NaBH_3CN$; Eschweiler-Clarke and Leuckart)	$H{-}C{-}NHR$
$C{=}O$ (ketone)	aldol condensation ($C{=}C{-}O^-$)	$-C(OH){-}C{-}C(=O){-}$

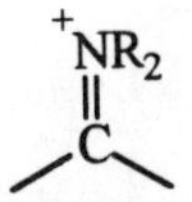

$\overset{+}{NR_2}=C<$ Mannich reaction $\left(>C=C<^{OH}\right)$ $-C(NR_2)-C-C(=O)-$

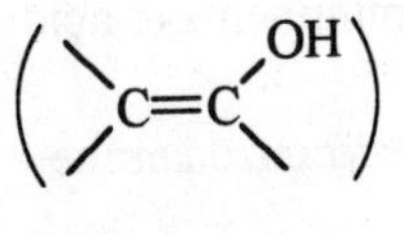

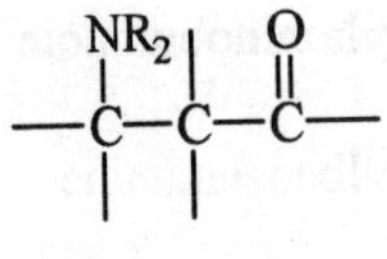

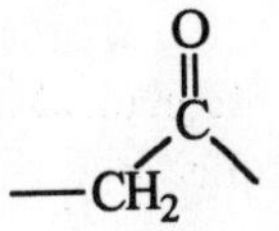

$-CH_2-C(=O)-$ enol formation (acid- or base-catalyzed) $-CH=C(OH)-$

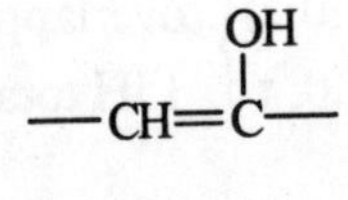

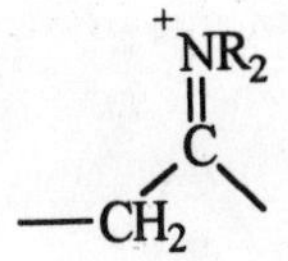

$-CH_2-C(=\overset{+}{N}R_2)-$ enamine formation $-CH=C(:NR_2)-$

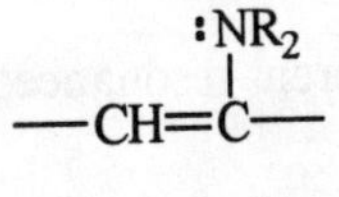

24.D. Answers to Exercises

24.1 $R_1R_2R_3R_4N^+ + X^+ \rightleftharpoons R_1X + R_2R_3R_4N$ (inverts)

In this specific case, the most reactive group for S_N2 reaction by X^- is allyl:

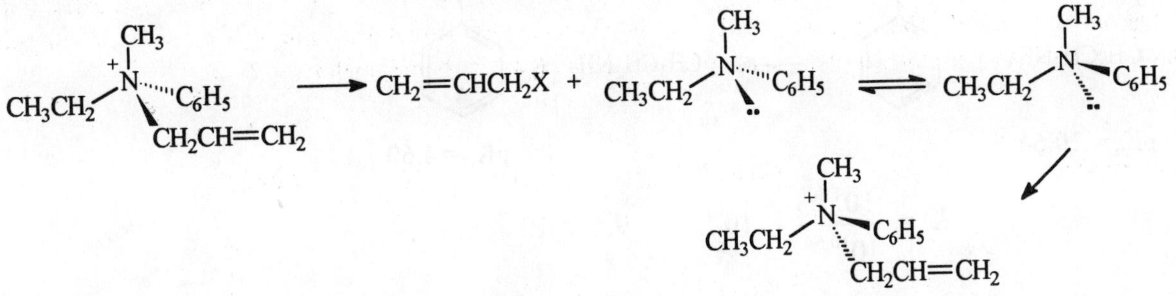

$CH_3CH_2\overset{+}{N}(CH_3)(C_6H_5)(CH_2CH{=}CH_2) \longrightarrow CH_2{=}CHCH_2X + CH_3CH_2\ddot{N}(CH_3)(C_6H_5) \rightleftharpoons CH_3CH_2\ddot{N}(CH_3)(C_6H_5)$ (inverted) $\rightarrow CH_3CH_2\overset{+}{N}(CH_3)(C_6H_5)(CH_2CH{=}CH_2)$

I^- is generally more nucleophilic and faster in S_N2 reactions than Br^-.

24.2 methanamine N-ethylethanamine N,N-dipropylpropanamine

N-methylethanamine N-ethyl-N-methylethanamine N-ethyl-N-methylcyclopropanamine

1-methylpropanamine 1-ethyl-3-methylbutanamine

N,N-dimethylethanamine N,N-diethyl-1,3-dimethylbutanamine

N-ethyl-N,1-dimethylpropanamine N,3-dimethylpentanamine

benzenamine 3-bromobenzenamine 4-nitrobenzenamine N,N-dimethylbenzenamine

4-methylaminobutanoic acid 4-aminobenzoic acid 4-aminoazobenzene

4-methylbenzenamine 3-methoxybenzenamine 2-ethoxybenzenamine

24.3 IR: no N-H

NMR: δ 1.0, d; 1.0, t — overlapped, 15H total — {4 CHC**H**$_3$; 1 -CH$_2$C**H**$_3$}

2.5, q, 2H: -C**H**$_2$CH$_3$
3.0, septet, 2H: -C**H**(CH$_3$)$_2$

CMR: four different resonances

<u>Structure</u>: ethyldiisopropylamine, $CH_3CH_2N(CH(CH_3)_2)_2$

24.4 $K_a = \dfrac{[H^+][RNH_2]}{[RNH_3^+]}$ $\qquad K_a \times K_b = \dfrac{[H^+][RNH_2]}{[RNH_3^+]} \times \dfrac{[RNH_3^+][OH^-]}{[RNH_2]} = [H^+][OH^-] = K_w = 10^{-14}\ M^2$

$$-\log(K_a \times K_b) = pK_a + pK_b = -\log(10^{-14}) = 14$$

$pK_a(NH_2^-) = 14 - pK_a(NH_3)$ $\qquad 14 - 34 = -20$

24.5

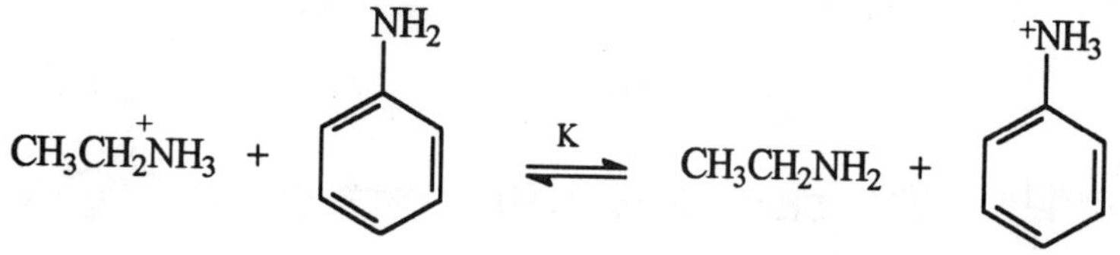

$$CH_3CH_2\overset{+}{N}H_3 + C_6H_5NH_2 \overset{K}{\rightleftharpoons} CH_3CH_2NH_2 + C_6H_5NH_3^+$$

$pK_a = 10.64$ $\qquad pK_a = 4.60$

$$K = \frac{10^{4.60}}{10^{10.64}} = 10^{-6.04} = 9.1 \times 10^{-7}$$

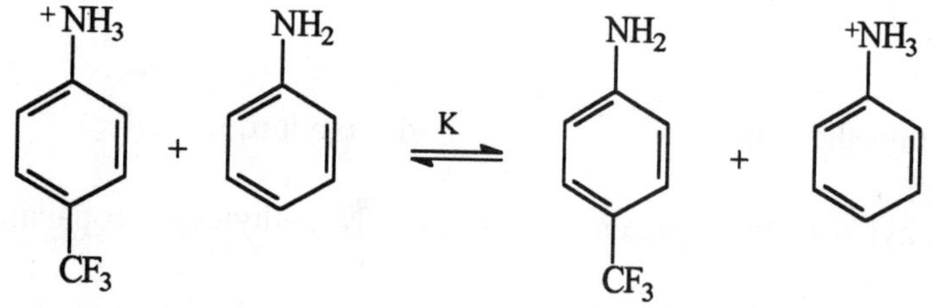

$$4\text{-}CF_3C_6H_4NH_3^+ + C_6H_5NH_2 \overset{K}{\rightleftharpoons} 4\text{-}CF_3C_6H_4NH_2 + C_6H_5NH_3^+$$

$pK_a = 2.75$ $\qquad pK_a = 4.60$

$$K = \frac{10^{4.60}}{10^{2.75}} = 10^{1.85} = 71$$

24.6 If you make the ester of racemic 2-octanol with (*S*)-1-methoxy-1-phenylacetic acid, two diastereomers will be obtained:

(*R*)-1-methylheptyl-(*S*)-1-methoxy-1-phenylacetate (*S*)-1-methylheptyl-(*S*)-1-methoxy-1-phenylacetate

These isomers have different physical properties and can be separated by crystallization, chromatography, etc. After separation of the isomers, hydrolysis of each ester separately will release the optically active alcohols.

24.7 a. reaction of $CH_3(CH_2)_7CH{=}CH_2$:
In the organic phase: $CH_3(CH_2)_7CH{=}CH_2$, reactant; $CH_3(CH_2)_7CO_2H$ product.
In the aqueous phase: K^+, Cl^-
In both: $(CH_3(CH_2)_6CH_2)_3N^+CH_3$
Goes from the aqueous to the organic: MnO_4^-
Formed in the organic: MnO_2 precipitates; CO_2 gas.

b. reaction of $C_6H_5-CH_2\overset{O}{\overset{\|}{C}}CH_3$:

In the organic phase: $C_6H_5-CH_2\overset{O}{\overset{\|}{C}}CH_3$, $CH_3(CH_2)_3Br$ reactants;

$C_6H_5-\underset{CH_2CH_2CH_2CH_3}{\underset{|}{CH}}\overset{O}{\overset{\|}{C}}CH_3$ product.

In the aqueous phase: Na^+
In both: $C_6H_5CH_2N^+Et_3$
Goes from the aqueous to the organic: OH^-
Formed in the organic, goes to the aqueous: Br^-

c. reaction of $CH_3(CH_2)_9Br$:
In the organic phase: $CH_3(CH_2)_9Br$ reactant; $CH_3(CH_2)_9SCN$ product.
In the aqueous phase: Na^+, Cl^-
In both: $(n\text{-}C_6H_{13})_3N^+CH_3$
Goes from the aqueous to the organic: SCN^-
Formed in the organic, goes to the aqueous: Br^-

24.8 Alkylation of the phthalimide ion involves S_N2 displacement, which will not occur with a neopentyl halide (too sterically hindered) or a *t*-butyl halide (E2 elimination instead). In addition, the sequence can only give primary amines, and cannot provide di-*n*-propylamine.459

24.9 m-xylene (1,3-dimethylbenzene) $\xrightarrow[H_2SO_4]{HNO_3}$ 2,4-dimethylnitrobenzene $\xrightarrow[\text{or other [H]}]{H_2/Ni}$ 2,4-dimethylaniline

anisole ($C_6H_5OCH_3$) $\xrightarrow[H_2SO_4]{HNO_3}$ $\xrightarrow{[H]}$ o-methoxyaniline and p-methoxyaniline, not m-methoxyaniline

benzaldehyde ($C_6H_5CH{=}O$) $\xrightarrow[H_2SO_4]{HNO_3}$ $\xrightarrow[HCl]{SnCl_2}$ m-aminobenzaldehyde (not *ortho*)

24.10 a. $(CH_3)_2CHCH_2CH_2Br \xrightarrow{NaCN} (CH_3)_2CHCH_2CH_2C{\equiv}N \xrightarrow{LiAlH_4} (CH_3)_2CHCH_2CH_2CH_2NH_2$

b. $(CH_3)_2CHCH{=}O \xrightarrow{HCN} (CH_3)_2CHCH(OH)C{\equiv}N \xrightarrow{H_2/Ni} (CH_3)_2CHCH(OH)CH_2NH_2$

24.11 a. $p\text{-}CH_3C_6H_4CH_2NH_2$ b. $C_5H_9NHCH_2CH_2CH_3$ (N-propylcyclopentylamine) c. $C_6H_5CH_2CH(CH_3)N(CH_3)_2$ d. 1,2-dimethylpyrrolidine (N–CH_3, 2-CH_3)

24.12 a. $CH_3(CH_2)_3NH_2 \xrightarrow[Et_3N]{(CH_3)_2CHCCl(=O)} CH_3(CH_2)_3NHC(=O)CH(CH_3)_2 \xrightarrow{LiAlH_4} CH_3(CH_2)_3NHCH_2CH(CH_3)_2$

b. $(CH_3)_2CHNH_2 \xrightarrow{CH_3CCl(=O)} \xrightarrow{LiAlH_4} (CH_3)_2CHNHCH_2CH_3 \xrightarrow[\text{2. }LiAlH_4]{\text{1. }CH_3CH_2CH_2CCl(=O)} (CH_3)_2CHN(CH_2CH_3)(CH_2)_3CH_3$

24.13

$CH_3CH_2CH(CH_3)CH_2CH_2CO_2H$ or (2-bromobenzoic acid: Br, CO_2H)

$\xrightarrow{SOCl_2}$ $RCCl$ (O) $\xrightarrow{NH_3}$ $RCNH_2$ (O) $\xrightarrow{NaOH,\ Cl_2}$ RNH_2

$RCCl$ (O) $\downarrow NaN_3$ RCN_3 (O)

$\xrightarrow[NaN_3,\ H_2SO_4]{}$ RCN_3 (O) $\xrightarrow{\Delta}$ $\xrightarrow{H_2O}$ RNH_2

24.14 a. $CH_3CH_2CH_2NH_2 \xrightarrow[HCl]{NaNO_2} CH_3CH_2CH_2Cl + CH_3CHClCH_3 + CH_3CH_2CH_2OH + CH_3CHOHCH_3$

b. $(CH_3CH_2CH_2)_2NH \xrightarrow[HCl]{NaNO_2} (CH_3CH_2CH_2)_2NN{=}O$

c. $(CH_3CH_2CH_2)_3N \xrightarrow[HCl]{NaNO_2} (CH_3CH_2CH_2)_3N^+H\ Cl^-$ *only product*

d. $C_6H_5NH_2 \xrightarrow[HCl]{NaNO_2} C_6H_5N_2^+\ Cl^-$

e. $C_6H_5NHCH_2CH_3$ (CH_3CH_2NH–) $\xrightarrow[HCl]{NaNO_2}$ $C_6H_5N(N{=}O)CH_2CH_3$ ($CH_3CH_2NN{=}O$–)

f. $C_6H_5N(CH_2CH_3)_2 \xrightarrow[HCl]{NaNO_2}$ $N(CH_2CH_3)_2$ with para $N{=}O$ (p-nitroso-N,N-diethylaniline)

24.15 a. $C_6H_5NH_2 + 3\ Br_2 \xrightarrow{H_2O}$ 2,4,6-tribromoaniline (NH_2; Br, Br, Br)

3-nitroaniline (NH_2, NO_2) $+ Br_2 \xrightarrow{H_2O}$ 4-bromo-3-nitroaniline (NH_2, NO_2, Br), 2-bromo-5-nitroaniline (NH_2, Br, NO_2)

4-methoxyaniline (NH_2, OCH_3) + 2 Br_2 $\xrightarrow{H_2O}$ 2,6-dibromo-4-methoxyaniline (Br, Br, NH_2, OCH_3)

b. aniline (NH_2) $\xrightarrow[\text{pyridine}]{Ac_2O}$ acetanilide ($HNCCH_3$, C=O) $\xrightarrow[CH_3CO_2H]{HNO_3}$ p-nitroacetanilide ($HNCCH_3$, NO_2) + o-nitroacetanilide ($HNCCH_3$, NO_2) $\xrightarrow[\text{2. } OH^-]{\text{1. } H^+}$ nitroaniline (NH_2, NO_2)

mixture of *ortho* and *para*

3-nitroaniline (NH_2, NO_2) $\xrightarrow{\text{same sequence}}$ 2,5-dinitroaniline (NH_2, O_2N, NO_2) + 3,4-dinitroaniline (NH_2, NO_2, NO_2)

4-methoxyaniline (NH_2, OCH_3) $\xrightarrow{\text{same sequence}}$ 4-methoxy-2-nitroaniline (NH_2, NO_2, OCH_3)

24.16 a. $CH_3(CH_2)_2CH(CH_3)N(CH_3)_2 \xrightarrow{CH_3I} CH_3(CH_2)_2CH(CH_3)N^+(CH_3)_3\ I^- \xrightarrow{AgOH}$

$CH_3(CH_2)_2CH(CH_3)N^+(CH_3)_3\ OH^- \xrightarrow[\Delta]{} CH_3CH_2CH_2CH{=}CH_2 + (CH_3)_3N$

b. $CH_3(CH_2)_7N(CH_3)_2 \xrightarrow[\text{2. AgOH}]{\text{1. } CH_3I} \xrightarrow{\Delta} CH_3(CH_2)_5CH{=}CH_2 + N(CH_3)_3$

c. $(CH_3CH_2)_3N \xrightarrow{H_2O_2} (CH_3CH_2)_3N\text{-}O^- \xrightarrow{\Delta} (CH_3CH_2)_2NOH + CH_2{=}CH_2$

d.

CH_3CH_2, H_3C, D, H, H, *S*, *R*, $N(CH_3)_2$ $\xrightarrow{H_2O_2}$ CH_3CH_2, H_3C, D, H, H, $\overset{+}{N}(CH_3)_2$, ^-O $\xrightarrow{\Delta}$ CH_3CH_2, H_3C C=C H, D + $(CH_3)_2NOH$

24.17 $CH_3CH_2C(=O)CH_2CH_3 + CH_2{=}O + (CH_3)_2\overset{+}{N}H_2\ Cl^- \longrightarrow CH_3CH_2C(=O)CH(CH_3)CH_2N(CH_3)_2$

24.E Answers and Explanations for Problems

1. a. 2-methylbutanamine
   b. trimethylamine (N,N-dimethylmethanamine)
   c. N-ethyl-N-methyl-2-propenamine
   d. *p*-bromophenyltrimethylammonium chloride
   e. N-ethyl-N-nitrosobenzenamine
   f. N-ethyl-4-nitrosobenzenamine
   g. ethyltrimethylammonium iodide
   h. N,N-dimethylpropanamine oxide
   i. 1-ethyl-N,N-dimethylpropanamine
   j. N-ethyl-3-methylbenzenamine
   k. N,4-diisopropyl-N-methylbenzenamine
   l. 2,4,6-trichloroaniline

2. IR: doublet at 3290 and 3370 cm^{-1} and bands at 1600, 1160, and 850 cm^{-1} suggest $R\text{-}NH_2$

   NMR: δ 2.8, sextuplet, 1H: hydrogen with 5 adjacent H's, perhaps: $RCH_2{-}CH(N){-}CH_3$

   The 10:1 ratio of other hydrogens suggests: $CH_3CH_2{-}CH(NH_2){-}CH_3$

   CMR: four different carbons, and chemical shifts, are consistent with this structure

3. IR: 3300 cm^{-1}, weak: R_2NH
   700 cm^{-1}: NH wag

   NMR: δ 0.4, very broad, 1: NH
   1.0, d, 12: four equivalent CH_3's next to one hydrogen
   2.8, septet, 2: $(CH_3)_2CH$

   Structure: diisopropylamine, $(CH_3)_2CHNHCH(CH_3)_2$

4. In a planar amine, the bonding orbitals from nitrogen are sp^2 hybrids which have a natural angle of 120°. The ring strain compared to a three-membered ring is greater than the difference between the angle in the aziridine ring and the 107° angle in the open amine.

   An alternative but equivalent explanation focuses on the lone pair. In an ordinary amine, this lone pair is in an orbital having some s-character; in the planar amine the lone pair is in a pure p-orbital. Removing s-character from the lone pair electrons takes some energy. In aziridine, the narrow bond angle of the three-membered ring requires more p-character and leaves more s-character for the lone pair. This greater degree of s-character requires more energy to form the planar system with a p-lone pair.

5. a.

$$K_a = \frac{[RNH_2][H^+]}{[RNH_3^+]} \qquad -\log K_a = -\log [H^+] - \log \frac{[RNH_2]}{[RNH_3^+]}$$

At equal $[RNH_2]$ and $[RNH_3^+]$, the last term = 0, so that $pK_a = pH$
for CH_3NH_2, $pK_a = pH = 10.62$

b. $\log [RNH_2]/[RNH_3^+] = pH - pK_a$
for CH_3NH_2:

pH	$\log [RNH_2]/[RNH_3^+]$	$[RNH_2]/[RNH_3^+]$
6	-4.62	2.4×10^{-5}
8	-2.62	2.4×10^{-3}
10	-0.62	0.24
12	1.38	24

6. a.

$$CH_3NH_2 + H^+ \rightleftharpoons CH_3NH_3^+$$

$$CH_3COOH \rightleftharpoons CH_3CO_2^- + H^+$$

$$CH_3NH_2 + CH_3COOH \rightleftharpoons CH_3NH_3^+ + CH_3CO_2^-$$

$$K = \frac{[CH_3NH_3^+][CH_3CO_2^-]}{[CH_3NH_2][CH_3COOH]} = \frac{[CH_3NH_3^+]}{[CH_3NH_2][H^+]} \cdot \frac{[H^+][CH_3CO_2^-]}{[CH_3COOH]}$$

$$= \frac{K_a(CH_3COOH)}{K_a(CH_3NH_3^+)} = \frac{1.75 \times 10^{-5}}{2.40 \times 10^{-11}} = 7.29 \times 10^5$$

b. Assume equal concentration of amine and acid to start, then $7.29 \times 10^5 = \dfrac{x^2}{(a-x)^2}$

$$\frac{x}{a-x} = 854 = \frac{[CH_3NH_3^+]}{[CH_3NH_2]} \; ; \qquad \log\frac{[CH_3NH_2]}{[CH_3NH_3^+]} = \frac{1}{854}$$

(from Problem #5b:) $-\log 854 = -pK_a + pH$
$pH = 10.62 - \log 854 = 7.69$

7. a. Extract with dilute HCl to remove amine; extract with dilute NaOH or Na_2CO_3 to remove the carboxylic acid. Hydrocarbon remains.

b. A mixture of both enantiomers of the hydrogen phthalate results. Reaction with an optically pure amine such as naturally-occurring brucine or strychnine gives two diastereomeric salts, e.g.,

(−)brucine-H$^+$ (+) [$C_6H_4(CO_2R)(CO_2^-)$] and (−)brucine-H$^+$ (−) [$C_6H_4(CO_2R)(CO_2^-)$]

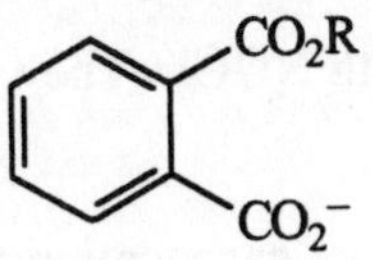

CH_3O OCH_3

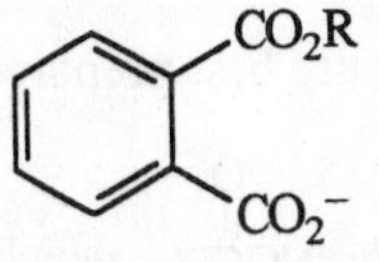

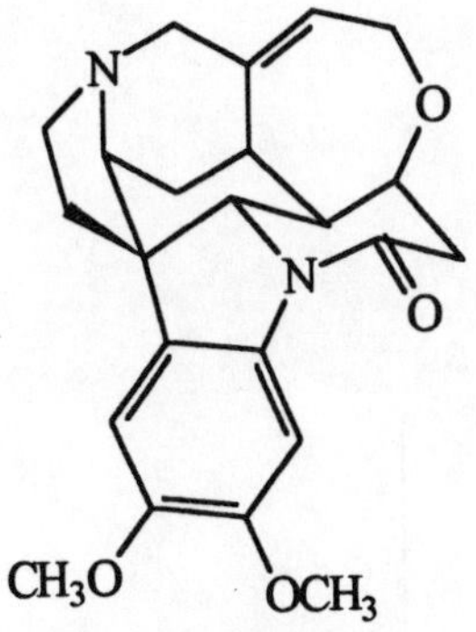

(−) brucine

These salts are separated by crystallization. Actually, the brucine (−)(+) salt is usually less soluble in acetone. The individual salts are treated with dilute HCl (which removes the alkaloid as the soluble hydrochloride). Heating each hydrogen phthalate with aqueous NaOH hydrolyzes the ester and gives each enantiomer of the alcohol.

8. a. 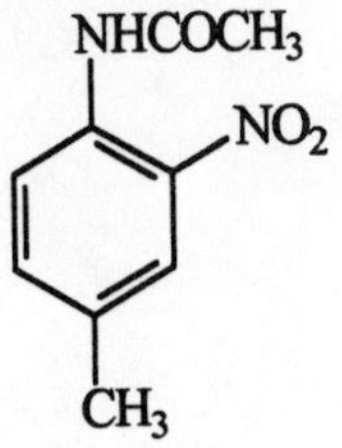 $NHCOCH_3$, NO_2, CH_3

b. 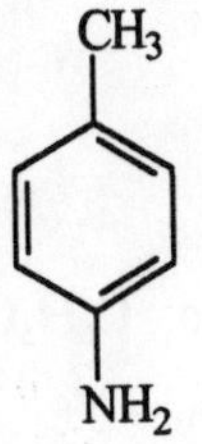 CH_3, NH_2

c. 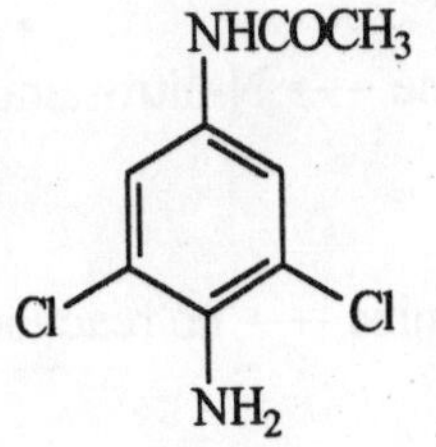 $NHCOCH_3$, Cl, Cl, NH_2

d. 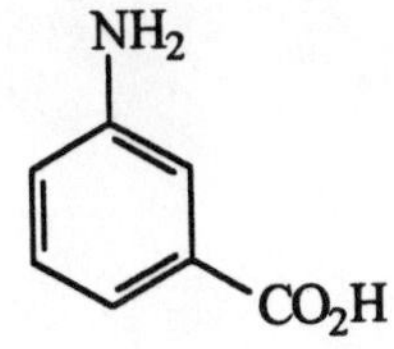 NH_2, CO_2H

e. 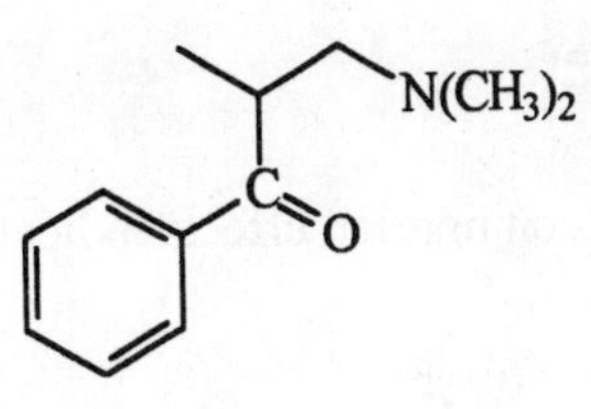 $N(CH_3)_2$, C=O

f. 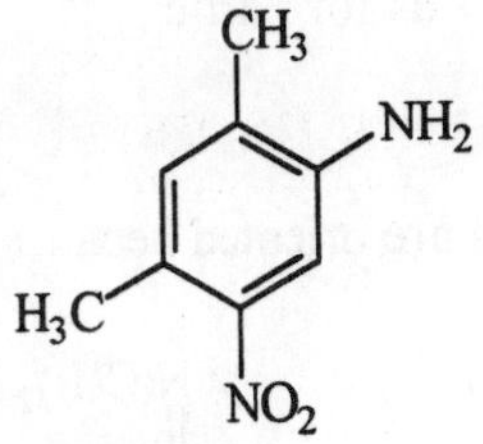 CH_3, NH_2, H_3C, NO_2

9.

$$CH_3\overset{O}{\overset{\|}{C}}CH_3 + CH_2{=}O + (CH_3)_2\overset{+}{N}H_2Cl^- \longrightarrow CH_3\overset{O}{\overset{\|}{C}}CH_2CH_2N(CH_3)_2$$

$$\downarrow CH_3I$$

$$CH_3\overset{O}{\overset{\|}{C}}CH{=}CH_2 + (CH_3)_3N \xleftarrow[\Delta]{Ag_2O} CH_3\overset{O}{\overset{\|}{C}}CH_2CH_2\overset{+}{N}(CH_3)_3\ I^-$$

10. a. primary arylamine ⟶ diazonium ion: $C_6H_5N_2^+$

b. secondary amine ⟶ N-nitrosamine: $C_6H_5N(NO)CH_3$

c. substituted amide ⟶ N-nitroso amide: $C_6H_5N(NO)(COCH_3)$

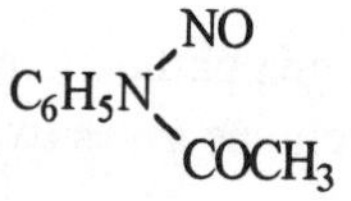

N-nitroso amides are best prepared by reaction of the amide with NOCl. They rearrange readily to provide diazoesters, $RN{=}NOC(O)R'$, which then undergo further radical or carbocation reactions involving $R^{\cdot}$ or R^+.

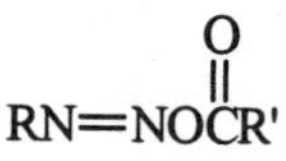

d. tertiary arylamine ⟶ electrophilic substitution: $ON{-}C_6H_4{-}N(CH_3)_2$

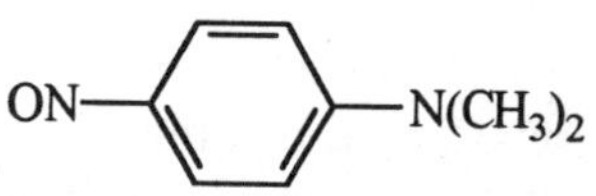

e. primary alkylamine ⟶ diazonium ion which decomposes: $C_6H_5CH_2OH$

f. 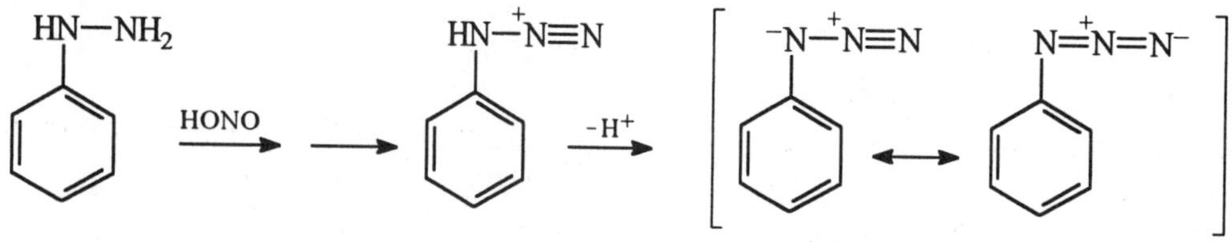 $C_6H_5NHNH_2$ $\xrightarrow{HONO}$ ⟶ $C_6H_5NH{-}\overset{+}{N}{\equiv}N$ $\xrightarrow{-H^+}$ [$C_6H_5{-}\overset{-}{N}{-}\overset{+}{N}{\equiv}N$ ⟷ $C_6H_5{-}N{=}\overset{+}{N}{=}N^-$]

(phenyl azide)

g. secondary amine ⟶ N-nitrosamine: $C_6H_5CH_2N(CH_3){-}N{=}O$

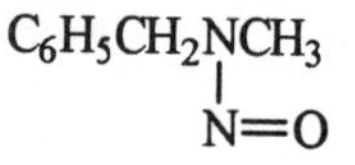

h. tertiary alkylamine ⟶ no reaction (except salt formation: $C_6H_5CH_2NH(CH_3)_2$) Cl^-

i. same answer as for part c: $C_6H_5CH_2N(COCH_3){-}N{=}O$

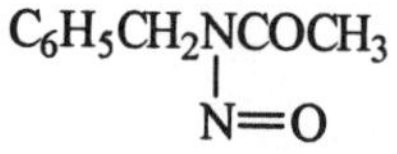

11. Dipole moments are oriented because of the conjugation effects indicated:

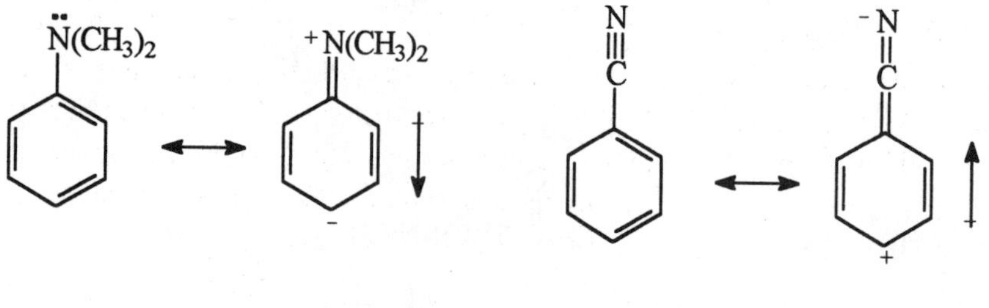

$C_6H_5\ddot{N}(CH_3)_2$ ⟷ $^+N(CH_3)_2{=}C_6H_5^-$ ↓ $C_6H_5C{\equiv}N$ ⟷ $^+C_6H_5{=}C{=}N^-$ ↑

When the functional groups are conjugated in the same molecule, the dipolar effects are enhanced by the structure drawn at the right. Since the charges are now far apart, even a small contribution by this structure has an important effect on the dipole moment.

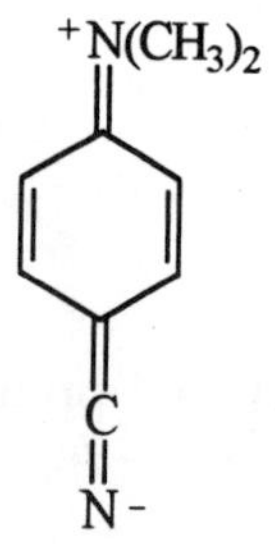

$^+N(CH_3)_2{=}C_6H_4{=}C{=}N^-$

12.

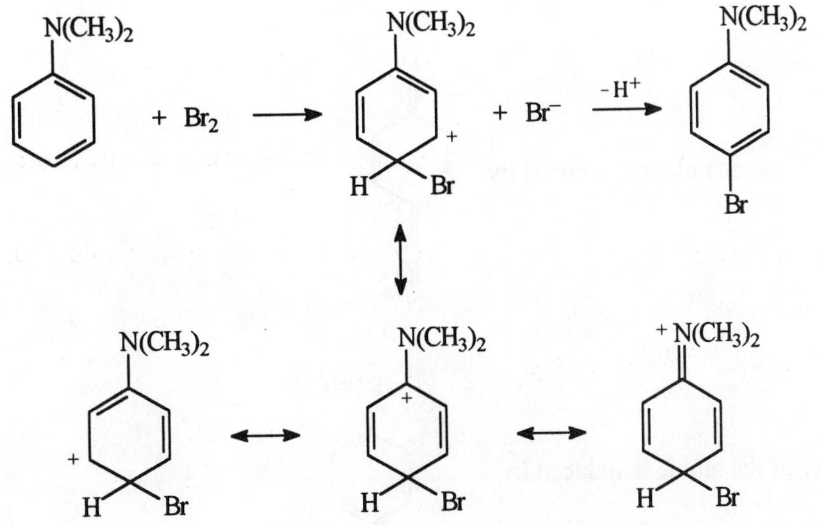

$$C_6H_5N(CH_3)_2 + Br_2 \longrightarrow [\text{arenium ion, H/Br at para carbon}] + Br^- \xrightarrow{-H^+} p\text{-}BrC_6H_4N(CH_3)_2$$

N(CH3)2 + Br2 → N(CH3)2 (H, Br, +) + Br⁻ —(-H+)→ N(CH3)2 (Br)

N(CH3)2 (+, H, Br) ↔ N(CH3)2 (+, H, Br) ↔ +N(CH3)2 (H, Br)

stable immonium ion

The transition state for reaction with benzene has no such immonium cation structure, and hence has much more energy and is much less stable.

13. Aniline is a weaker base than aliphatic amines, in part because of conjugation of the nitrogen lone pair electrons with the aromatic π-system:

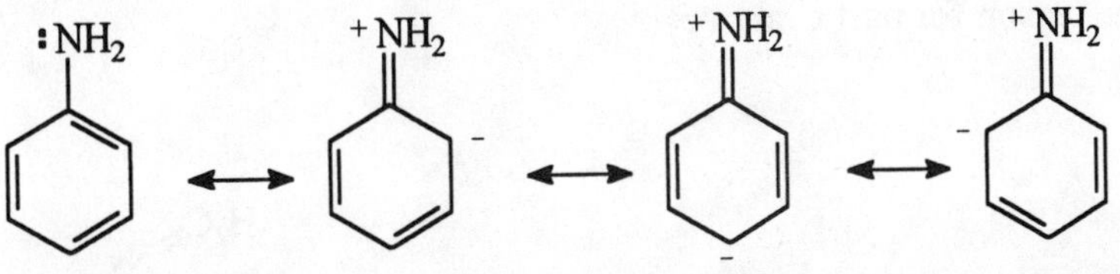

:NH2 ↔ +NH2 (-) ↔ +NH2 (-) ↔ +NH2 (-)

This conjugation is not present in the ammonium ion; hence, the amine has additional stabilization and the equilibrium, $RNH_3^+ \rightleftharpoons H^+ + RNH_2$ is displaced more to the right for R = aromatic ring compared to R = alkyl. However, in *o*-methyl-N,N-dimethylaniline, steric hindrance prevents effective conjugation (see (*I*) below). The $(CH_3)_2N$ group must twist at right angles to the ring (see (*II*) below). The nitrogen lone pair now cannot overlap with the π-system. This amine does not have additional conjugation stabilization, so it is more basic than the primary amine, in which the smaller hydrogens are not as involved in steric hindrance.

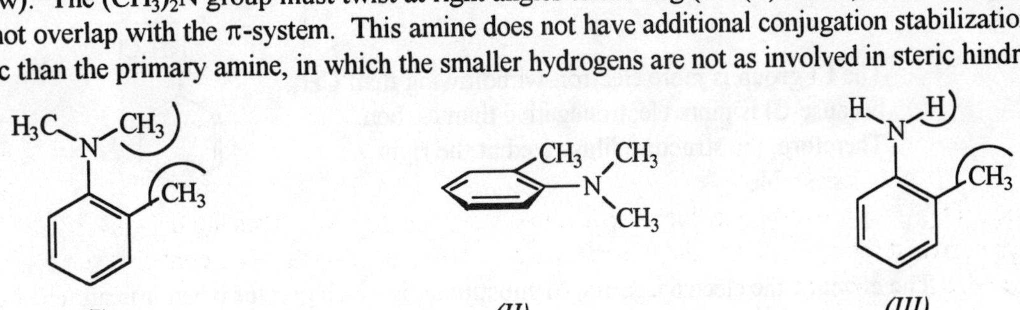

H3C N CH3, CH3 — (*I*)

CH3 CH3 N CH3 — (*II*)

H N H, CH3 — (*III*)

14\. a. 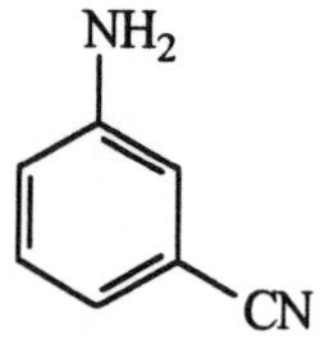 NH_2 / CN (m-cyanoaniline) The basicity of *p*-cyanoaniline is reduced by $^+NH_2$ / C / N^-

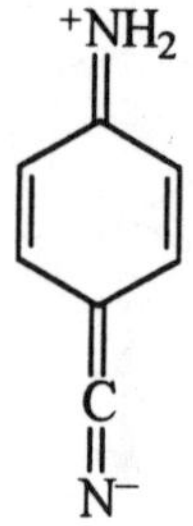

b. 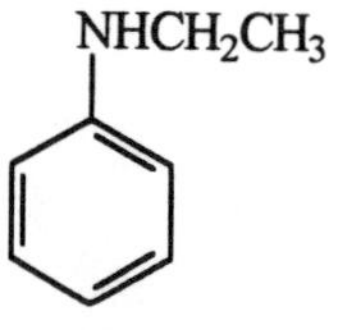 $NHCH_2CH_3$ The basicity of the amide is reduced by O^- $\overset{+}{H}N{=}CCH_3$

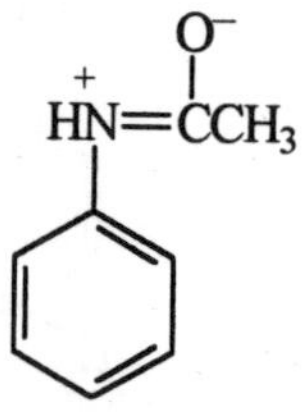

c. 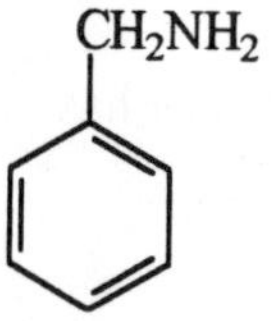 CH_2NH_2 The basicity of the aromatic amine is reduced by $^+NH_2$, etc.

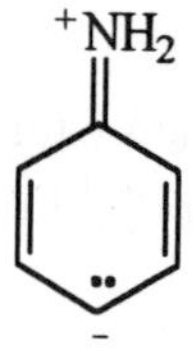

d. 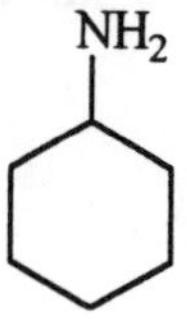 NH_2 Same reason for part c. above

e. 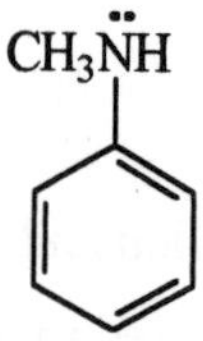 $CH_3\ddot{N}H$ The lone pair of the imine is more sp^2 in character;

more s-character = more stable electrons = less basic. $H_2C{=}N$:

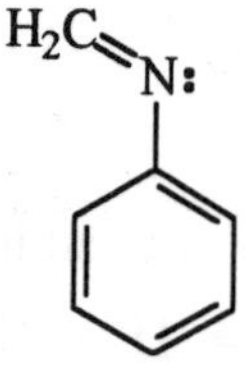

f. 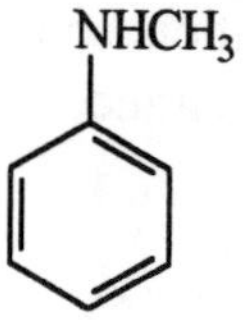 $NHCH_3$ The Cl group is more electron-withdrawing than CH_3, because Cl is more electronegative than carbon. Therefore, the structure illustrated at the right is less stable. $^+NH_2Cl$

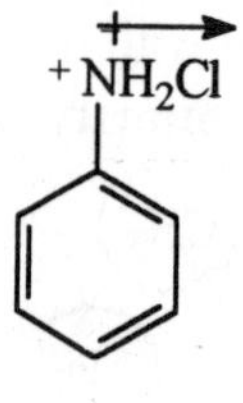

g. 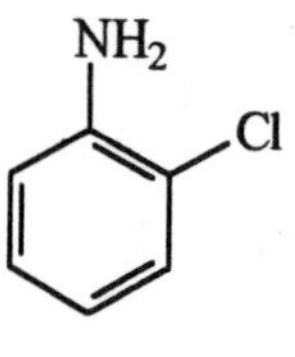 NH_2 Cl The effect of the electronegative Cl substituent is much greater when it is attached directly to the nitrogen.

15. a. $CH_3(CH_2)_3CH_2OH \xrightarrow[H_2SO_4]{HBr} CH_3(CH_2)_3CH_2Br \xrightarrow[\text{(large excess)}]{NH_3} CH_3(CH_2)_3CH_2NH_2$

<u>or</u> Gabriel synthesis

b. $CH_3(CH_2)_3CH_2NH_2$ [from a.] $\xrightarrow[HCO_2H,\ 100°]{CH_2{=}O} CH_3(CH_2)_3CH_2N(CH_3)_2$

c. $CH_3CH_2CH_2OH \xrightarrow{PCC} CH_3CH_2CHO \xrightarrow[H_2/Pt]{NH_3,\ C_2H_5OH} [CH_3CH_2CH_2NH_2] \xrightarrow[H_2/Pt]{CH_3CH_2CHO}$

$\longrightarrow (CH_3CH_2CH_2)_2NH \xrightarrow[HCOOH,\ \Delta]{CH_2{=}O} (CH_3CH_2CH_2)_2NCH_3$

d. $(CH_3)_2CHCH_2CH_2OH \xrightarrow[H_2SO_4]{HBr} (CH_3)CHCH_2CH_2Br \xrightarrow[\text{Gabriel synthesis}]{NH_3 \text{ or}} (CH_3)_2CHCH_2CH_2NH_2$

$\xrightarrow[\text{or } (CH_3CO)_2O]{CH_3COOH,\ \Delta} (CH_3)_2CHCH_2CH_2NHCOCH_3 \xrightarrow{LiAlH_4} (CH_3)_2CHCH_2CH_2NHCH_2CH_3$

16. a. Requires a single inversion of configuration:

(R)-2-octanol (OH) $\xrightarrow[\text{pyridine}]{TsCl}$ (R)-OTs $\xrightarrow{\text{phthalimide anion} = \text{"FtN}^-\text{"}}$ (S)-NFt $\xrightarrow{H_2NNH_2}$ (S)-NH_2

b. Requires two inversions, to give overall retention:

(R)-2-octanol (OH) $\xrightarrow[\text{pyridine}]{SOCl_2}$ (S)-Cl $\xrightarrow{\text{"FtN}^-\text{"}}$ (R)-NFt $\xrightarrow{H_2NNH_2}$ (R)-NH_2

17. a. 2 CH_3CH_2CHO $\xrightarrow{OH^-}$ $O{=}CHCH(CH_3)CH(OH)CH_2CH_3$ $\xrightarrow{NH_3}$ $\xrightarrow[Ni]{H_2}$ $H_2NCH_2CH(CH_3)CH(OH)CH_2CH_3$

mixture of diastereomers

b. Best method is by epoxide opening:

trans-2-butene $\xrightarrow{C_6H_5CO_3H}$ epoxide $\xrightarrow{\text{phthalimide anion} = \text{"FTN}^-\text{"}}$ FTN–CH(CH_3)–CH(OH)–CH_3 $\xrightarrow{H_2NNH_2}$ H_2N–CH(CH_3)–CH(OH)–CH_3

NOTE: what happens if phthalimide attacks the other epoxide carbon?

f. benzene $\xrightarrow[H_2SO_4]{HNO_3}$ $\xrightarrow[Ac_2O]{Zn}$ $C_6H_5NHCOCH_3$ $\xrightarrow[H_2SO_4]{HNO_3}$ p-$O_2NC_6H_4NHCOCH_3$ $\xrightarrow{1.\ H_3O^+ \quad 2.\ OH^-}$ p-$O_2NC_6H_4NH_2$

c. cyclohexanone $\xrightarrow{H^+,\ ^-CN}$ 1-hydroxycyclohexanecarbonitrile (HO, CN) $\xrightarrow[NH_3]{H_2/Pt}$ 1-(aminomethyl)cyclohexanol (HO, CH_2NH_2)

d. $CH_3CH_2COCH_2CH_3$ + $CH_2{=}O$ + $(CH_3CH_2)_2\overset{+}{N}H_2Cl^-$ $\longrightarrow$ $CH_3CH_2COCH(CH_3)CH_2N(CH_2CH_3)_2$

e. butadiene + $CH_2{=}CHCN$ $\xrightarrow{120°}$ cyclohexene–CN $\xrightarrow[Pt/C\ -\ H_2\ -\ NH_3]{Pt/H_2\ -\ NH_3\ \text{or}}$ cyclohexyl–CH_2NH_2

Catalytic hydrogenation reduces both the double bond and C≡N. (Alternatively, use $LiAlH_4$ first (to reduce CN), then Pt/H_2 (to reduce double bond)).

g. benzene $\xrightarrow[H_2SO_4]{HNO_3}$ $\xrightarrow[HCl]{Zn}$ $C_6H_5NH_2$ $\xrightarrow[H_2O]{Cl_2}$ 2,4,6-trichloroaniline

18. The basicity of aniline is lower than that of aliphatic amines because of conjugation of the nitrogen lone pair with the benzene ring. In diphenylamine, the lone pair can conjugate with two rings; hence, the basicity is still lower.

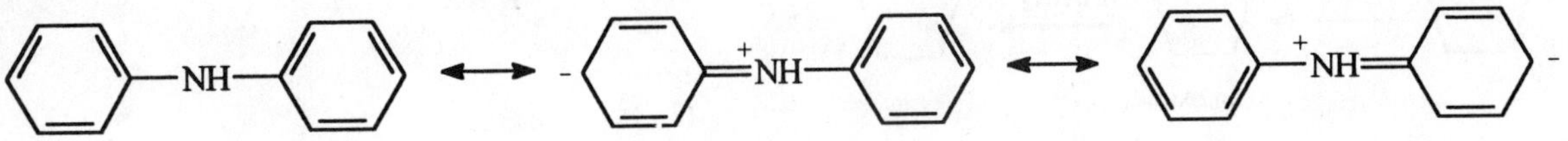

19. *ortho:*

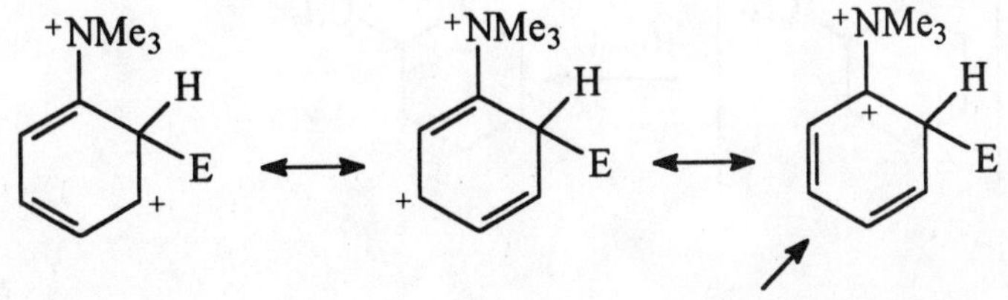

adjacent positive charges, resulting in high electrostatic repulsion. These structures are of high energy and contribute little to the resonance hybrid.

para:

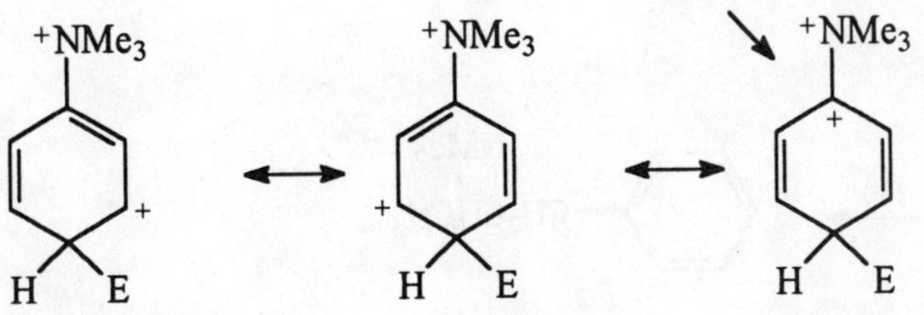

meta:

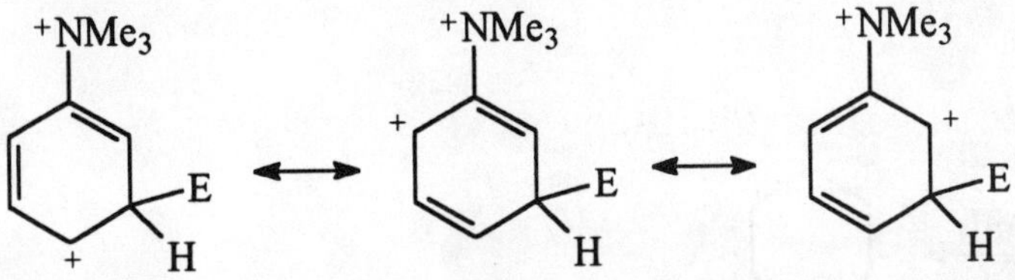

The positive charges in the *meta*-isomer are separated by at least one atom; thus, the *meta* transition state is of lowest energy. However, two positive charges in the same molecule represent substantial electrostatic repulsion. For this reason the trimethylammonium group is highly deactivating and *m*-directing.

20. a. Addition of N-C is best accomplished by a Michael reaction:

$$N{\equiv}C^- + CH_2{=}CHCOOEt \longrightarrow N{\equiv}CCH_2CH_2COOEt \xrightarrow{H^+} NC(CH_2)_2COOH$$

$$NC(CH_2)_2COOH \xrightarrow{SOCl_2} NC(CH_2)_2COCl \xrightarrow{(CH_3)_2NH} NC(CH_2)_2CON(CH_3)_2$$

$$NC(CH_2)_2CON(CH_3)_2 \xrightarrow{LiAlH_4} H_2N(CH_2)_4N(CH_3)_2$$ (LiAlH$_4$ reduces both CN and amide)

b. 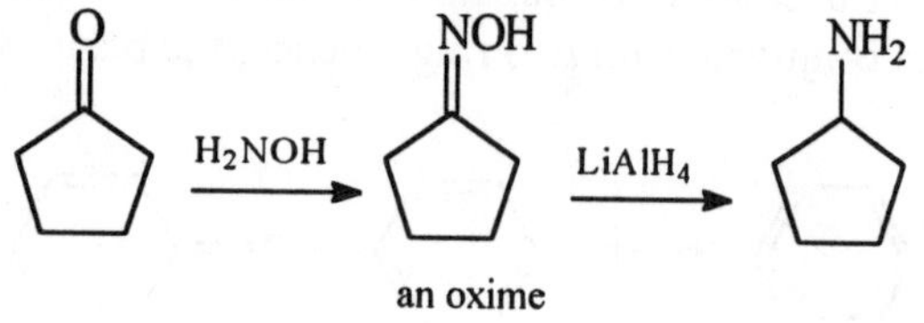 cyclopentanone $\xrightarrow{H_2NOH}$ cyclopentanone oxime (NOH) $\xrightarrow{LiAlH_4}$ cyclopentylamine (NH_2)

an oxime

c. 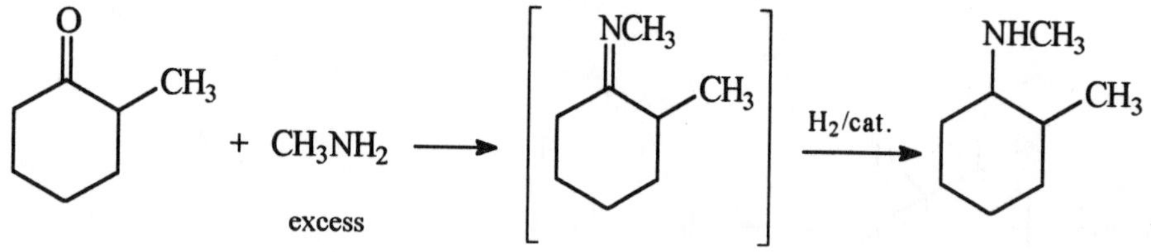 2-methylcyclohexanone (CH_3) + CH_3NH_2 (excess) $\longrightarrow$ [imine (NCH_3, CH_3)] $\xrightarrow{H_2/cat.}$ $NHCH_3$, CH_3 (N-methyl-2-methylcyclohexylamine)

d. 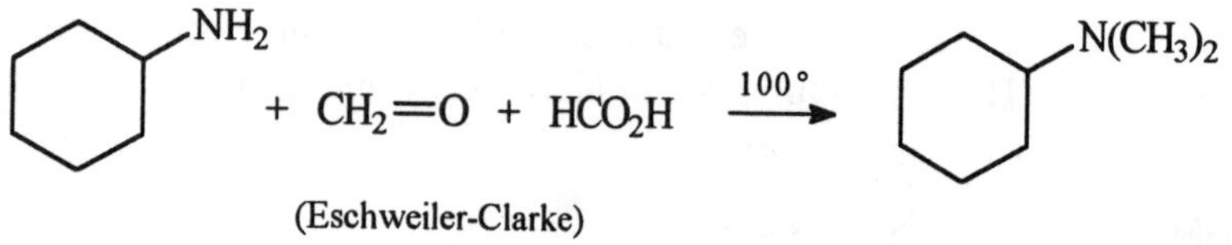 cyclohexylamine (NH_2) + $CH_2{=}O$ + HCO_2H $\xrightarrow{100°}$ cyclohexyl–$N(CH_3)_2$

(Eschweiler-Clarke)

e. 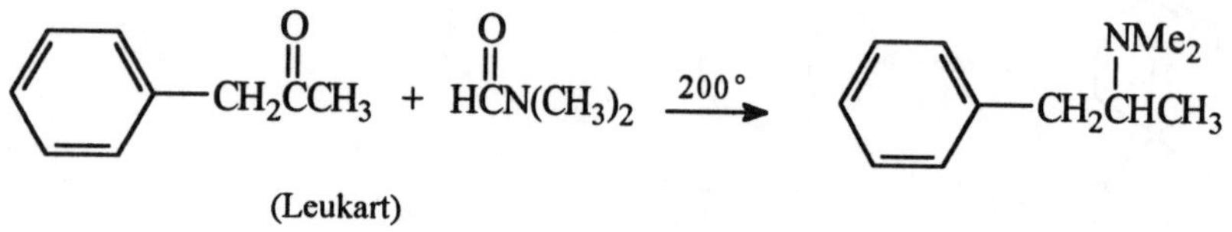 C_6H_5–$CH_2\overset{O}{\overset{\|}{C}}CH_3$ + $H\overset{O}{\overset{\|}{C}}N(CH_3)_2$ $\xrightarrow{200°}$ C_6H_5–$CH_2\overset{NMe_2}{\overset{|}{C}}HCH_3$

(Leukart)

21. a. 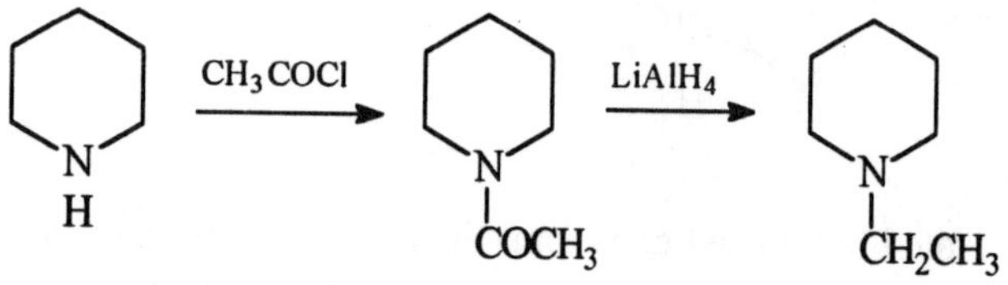 piperidine (N–H) $\xrightarrow{CH_3COCl}$ N–$COCH_3$ $\xrightarrow{LiAlH_4}$ N–CH_2CH_3

b. 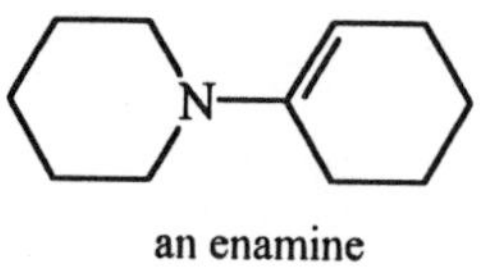 piperidine N–cyclohexenyl

an enamine

c. 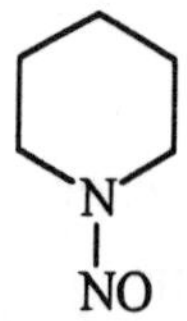 piperidine N–NO

d. 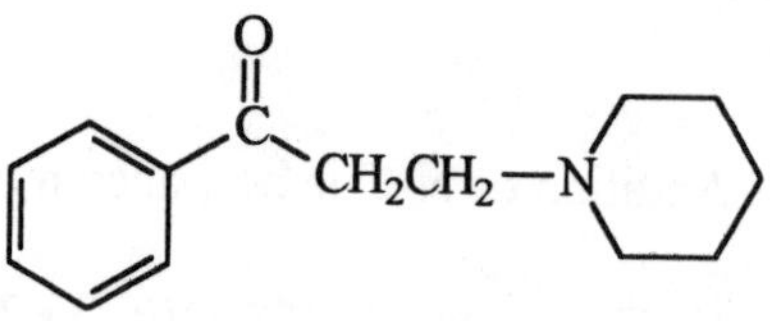 C_6H_5–$\overset{O}{\overset{\|}{C}}$–$CH_2CH_2$–N(piperidine)

e. 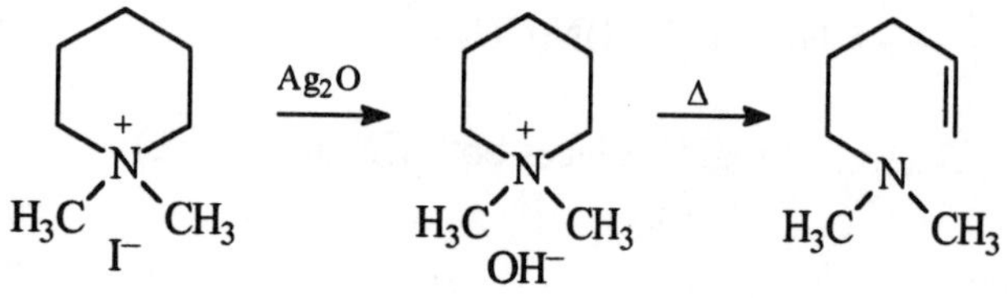 H_3C–$\overset{+}{N}$–CH_3 (piperidinium) I^- $\xrightarrow{Ag_2O}$ H_3C–$\overset{+}{N}$–CH_3 OH^- $\xrightarrow{\Delta}$ H_3C–N–CH_3 (open-chain alkenyl amine)

f. 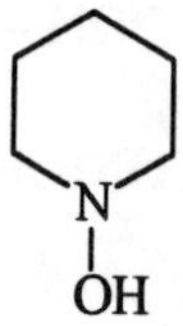 piperidine N–OH

22. 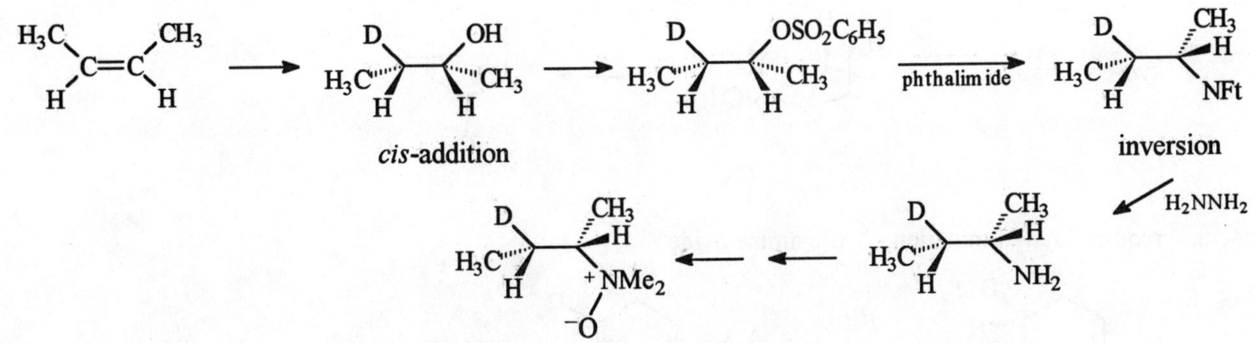 [reaction scheme: cis-2-butene → *cis*-addition (D, OH) → tosylate ($OSO_2C_6H_5$) → phthalimide, inversion → NFt → H_2NNH_2 → NH_2 → $^{+}NMe_2$ / ^{-}O]

cis-addition

inversion

There are three possible pathways for elimination. Loss of H from CH_3 gives $CH_3CHDCH{=}CH_2$ (one deuterium).

From the methylene group, the molecule may lose H or D by *syn*-elimination:

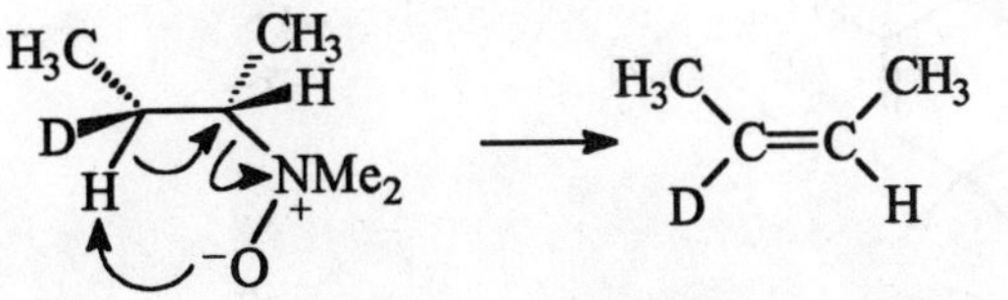

cis-2-butene has one deuterium

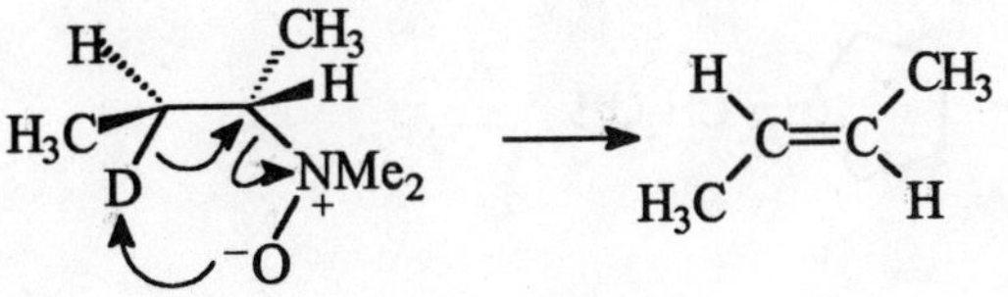

trans-2-butene has no deuterium

23. This elimination occurs with least hindered or most acidic hydrogen:

a. $CH_2{=}CH_2 + MeNCH_2CH(CH_3)_2$ (least hindered)

b. $(CH_3)_3N + (CH_3)_2CHCH{=}CH_2$ (least hindered)

c. The hydrogen α to the CO group is most acidic; it eliminates rather than the less hindered hydrogen on CH_3 of the ethyl group: $(CH_3CH_2)_2N + CH_2{=}CHCOCH_3$

d. This problem is best solved by the use of models. Recall that the Hofmann elimination proceeds by the E2 mechanism, involving the hydrogen that is *anti* to the leaving nitrogen. The preferred conformation is the one having the two methyl groups equatorial to the six-membered rings. Only the bridgehead proton has the necessary *anti*-coplanar relationship to the C-N bond which is being broken:

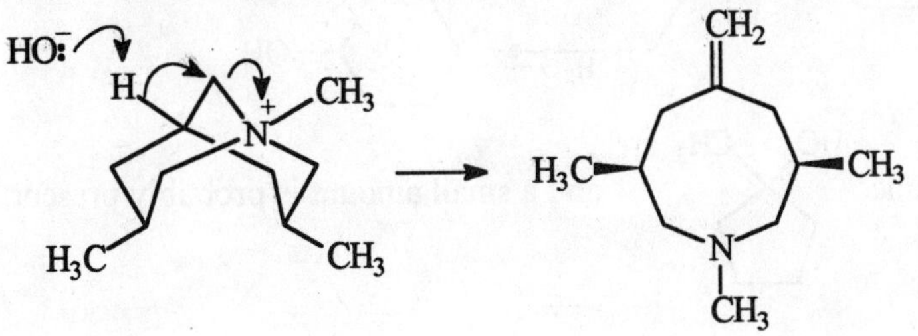

e. *syn*-elimination: H_3C, H, O^-, $\overset{+}{N}(CH_3)_2$ → CH_3

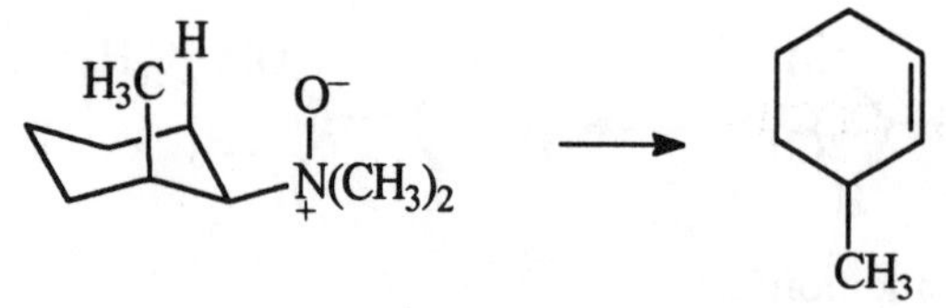

24. a. requires *syn*-elimination; ∴ use amine oxide

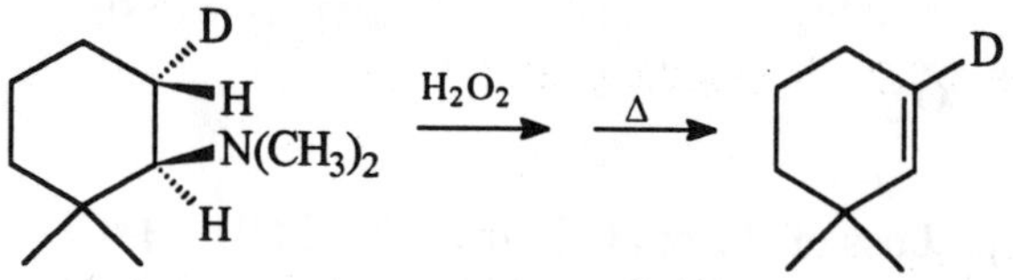

D, H, $N(CH_3)_2$, H $\xrightarrow{H_2O_2}$ $\xrightarrow{\Delta}$ D

b. requires *anti*-elimination; use Hofmann reaction

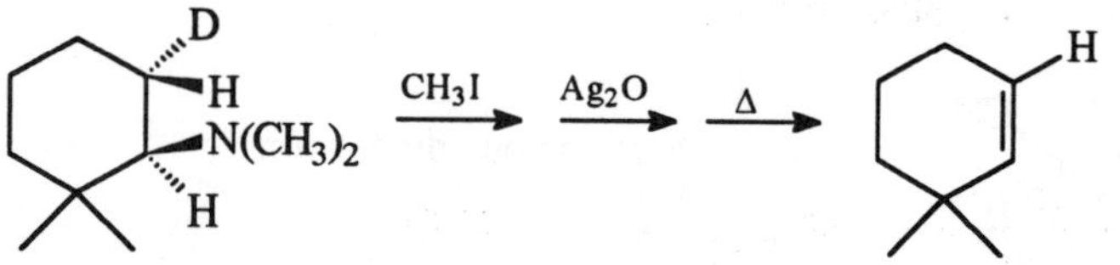

D, H, $N(CH_3)_2$, H $\xrightarrow{CH_3I}$ $\xrightarrow{Ag_2O}$ $\xrightarrow{\Delta}$ H

25. 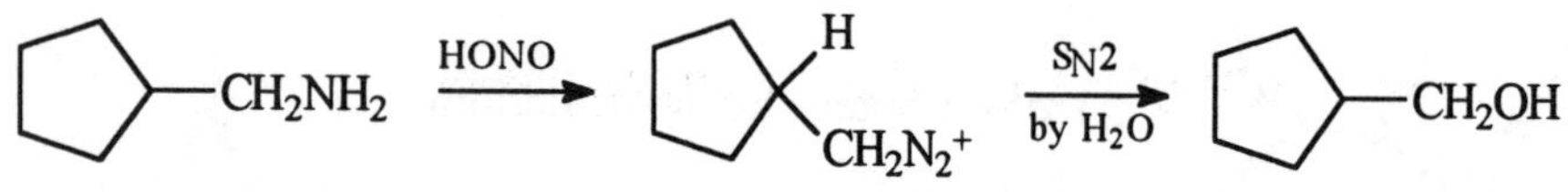 $-CH_2NH_2$ $\xrightarrow{HONO}$ H, $CH_2N_2^+$ $\xrightarrow[\text{by } H_2O]{S_N2}$ $-CH_2OH$

elimination of β-H ⟶ $=CH_2$

H, $CH_2-N_2^+$ rearrangement of β-H gives:

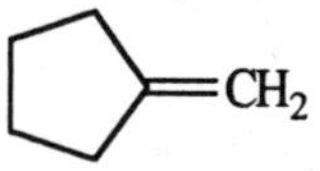

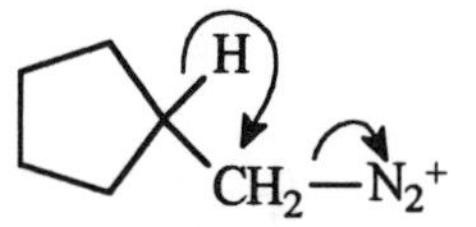

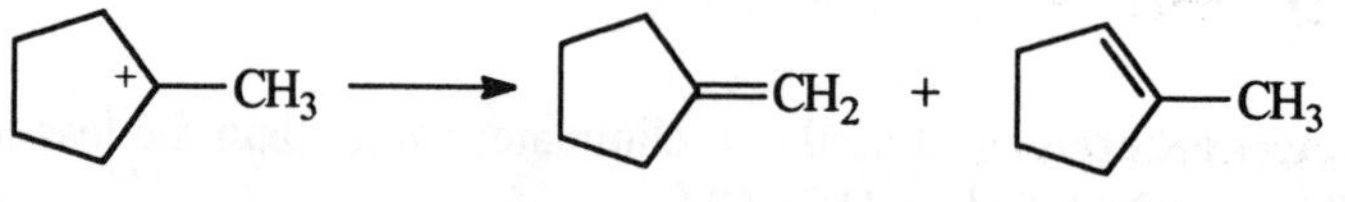

$-CH_3$ (+) ⟶ $=CH_2$ + $-CH_3$

Rearrangement of ring:

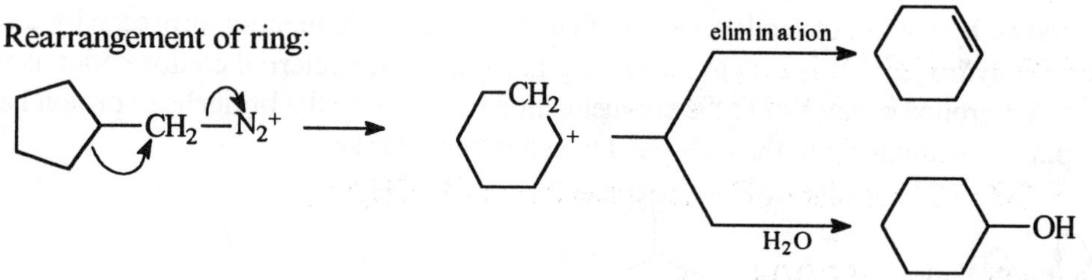

$CH_2-N_2^+$ ⟶ CH_2 (+) — elimination → ; H_2O → $-OH$

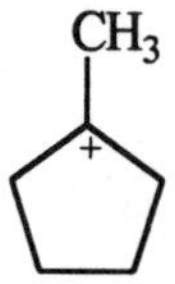

NOTE: CH_3 (+) could also give some HO CH_3 and a small amount is probably present.

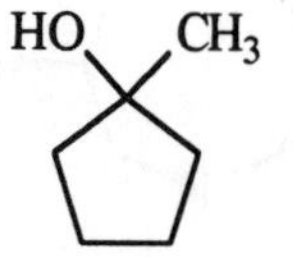

26. a. $C_{10}H_{21}Br \xrightarrow[\text{aq. } NaO_2CCH_3]{C_6H_5CH_2\overset{+}{N}Et_3Cl^-} C_{10}H_{21}O\overset{O}{\overset{\|}{C}}CH_3$

b. $3,4\text{-}(CH_3)_2C_6H_3\overset{O}{\overset{\|}{C}}CH_3 \xrightarrow[NaOD,\ D_2O]{C_6H_5CH_2\overset{+}{N}Et_3Cl^-} 3,4\text{-}(CH_3)_2C_6H_3\overset{O}{\overset{\|}{C}}CD_3$

$\left(R\overset{O}{\overset{\|}{C}}CH_3 + OD^- \rightleftharpoons R\overset{O^-}{\overset{|}{C}}{=}CH_2 + HOD \text{ in organic phase} \right)$

c. $(CH_3)_2CH(CH_2)_4OH + (CH_3)_2SO_4 \xrightarrow[50\%\ NaOH]{C_6H_5CH_2N^+Et_3\ Cl^-} (CH_3)CH(CH_2)_4OCH_3 + Na^+\ ^-O_3SOCH_3$

$(ROH + OH^- \rightleftharpoons RO^- + H_2O \text{ in organic phase})$

d. $(CH_3)_2CH\overset{O}{\overset{\|}{C}}H + C_6H_5CH_2Cl \xrightarrow[NaOD,\ D_2O]{C_6H_5CH_2\overset{+}{N}Et_3Cl^-} (CH_3)_2\underset{CH_2C_6H_5}{\underset{|}{C}}\overset{O}{\overset{\|}{C}}H + NaCl$

$\left((CH_3)_2CH\overset{O}{\overset{\|}{C}}H + OH^- \rightleftharpoons (CH_3)_2C{=}\overset{O^-}{\overset{|}{C}}H + H_2O \text{ in organic phase} \right)$

e. $C_6H_5CH{=}CH_2 + CHCl_3 \xrightarrow[50\%\ NaOH]{C_6H_5CH_2\overset{+}{N}Et_3Cl^-} C_6H_5\text{–(2,2-dichlorocyclopropyl)} + NaCl$

$(CHCl_3 + OH^- \longrightarrow :CCl_2 + H_2O + Cl^- \text{ in organic phase})$

27. a. $ClCH_2CH_2NH_3^+$; inductive effect of Cl

b. $CH_3ONH_3^+$; inductive effect of oxygen

c. $CH_3\overset{O}{\overset{\|}{C}}\overset{+}{N}H_3$ electron-withdrawing inductive effect of carbonyl; in addition, the product amide is resonance-stabilized by $R\overset{O^-}{\overset{|}{C}}{=}\overset{+}{N}H_2$

d. $CH_2{=}N^+H_2$; lone pair in $CH_2{=}\ddot{N}H$ has more s-character than that in $CH_3CH_2\ddot{N}H_2$

e. $CH_3OOCCH_2NH_3^+$; inductive effect of COOR group

f. $CH_2{=}CHNH_3^+$; amine is resonance-stabilized

28. Since the mixture of cyclooctadienes contains no 1,3-isomer, it must consist of a mixture of the 1,4- and 1,5-isomers. Hence, the starting $C_9H_{17}N$ amine must have the symmetrical bicyclic structure:

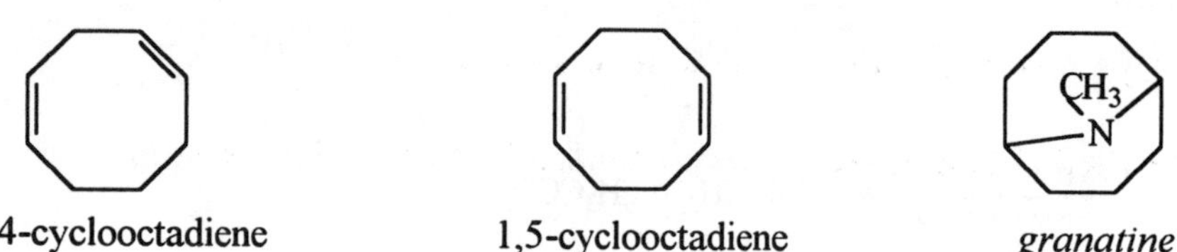

29. Cl–N(COCH$_3$)–C$_6$H$_5$ + HCl ⇌ [Cl$^-$ Cl–N(Ph)–C(CH$_3$)=$\overset{+}{O}$H] ⟶ Ph–N=C(CH$_3$)–OH + Cl$_2$ ⇌ Ph–NHCOCH$_3$

C$_6$H$_5$NHCOCH$_3$ + Cl$_2$ ⟶ mixture of *o*- and *p*-chloroacetanilide

30. 1-(aminomethyl)cyclohexanol (HO, NH$_2$) $\xrightarrow{\text{HONO}}$ 1-(diazoniomethyl)cyclohexanol (H–O, $\overset{+}{N}_2$) $\xrightarrow[-H^+]{-N_2}$ cycloheptanone

31. A: $C_{10}H_{21}N_3$ 1-azidodecane B: $C_{10}H_{21}N{=}PPh_3$ an iminophosphorane C: $C_{10}H_{21}NH_2$ 1-aminodecane

32. a. A: No, neopentyl halide too sterically hindered for S_N2
B: No, CN^- will lead primarily to E_2 with a 3° halide, $(CH_3)_3CBr$
C: Yes, $(CH_3)_3CCOCl$ and 1. NH_3, 2. $LiAlH_4$
D: Yes, $(CH_3)_3CCHO$ and 1. NH_2OH, 2. $LiAlH_4$
E. Yes, $(CH_3)_3CCH_2COCl$

b. A: No, Gabriel gives a 1° amine
B: No, same reason as A:
C: Yes, use cyclo-C_6H_{11}-$CON(CH_3)_2$
D: No, same reason as A:
E. No, same reason as A:

c. $(CH_3)_2CHCH_2CH_2NH_2$ is an unhindered 1° amine; all the methods listed are appropriate.

d. The desired amine is aryl; none of the listed methods is appropriate.

e. Methods A,B,D and E yield 1° amines and are inappropriate. Method C could be used.

24.F Supplementary Problems

S1. Draw the structure of each of the following compounds.

a. triisobutylamine
b. *cis*-1,4-diaminocyclohexane
c. 3-(aminomethyl)aniline
d. N,3-dimethylbutanamine
e. ethyltripropylammonium chloride
f. ethyldiisopropylamine oxide
g. N,O-diethylhydroxylamine
h. anilinium bromide

S2. For each pair of compounds below, explain how you would distinguish between the two without using NMR.

a. $(CH_3)_2NCH_2CH(CH_3)_2$ and $(CH_3)_2CHNHCH(CH_3)_2$

b. $(CH_3CH_2CH_2)_2NH$ and $CH_3CH_2CH(NH_2)CH_2CH_2CH_3$

c. $(CH_3)_3CCH_2NH_2$ and $CH_3CH_2CH(CH_3)CH_2NH_2$

d. 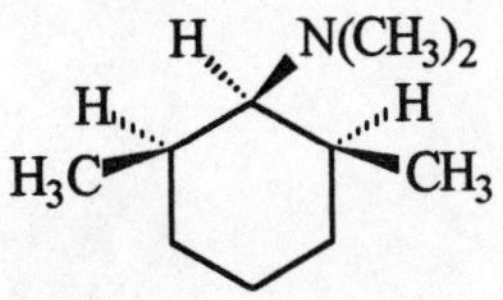

and 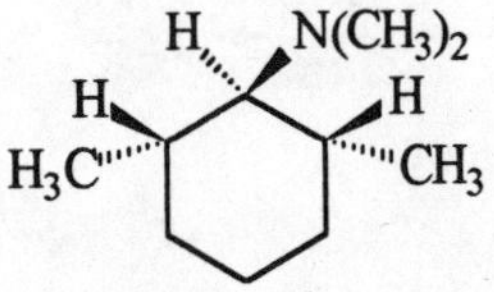

S3. Rank the following compounds in order of increasing basicity.

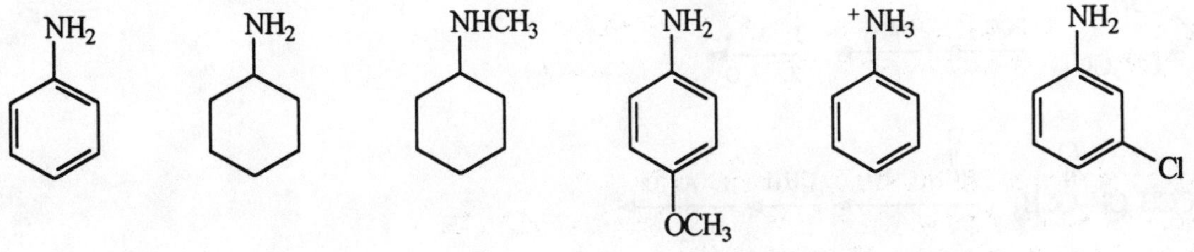

I II III IV V VI

S4. Do you expect *p*-aminoacetophenone ($H_2NC_6H_4COCH_3$) to be a stronger or weaker base than aniline itself? Justify your answer with resonance structures.

S5. What is the principal organic product to result from each of the following reaction sequences?

a. 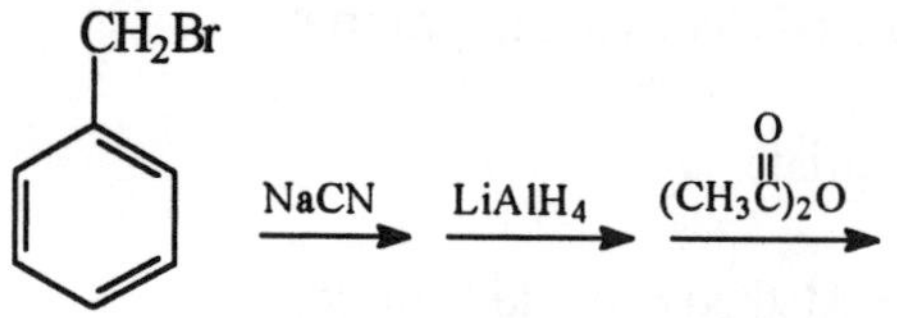

CH₂Br
NaCN
LiAlH₄
(CH₃C)₂O

b. 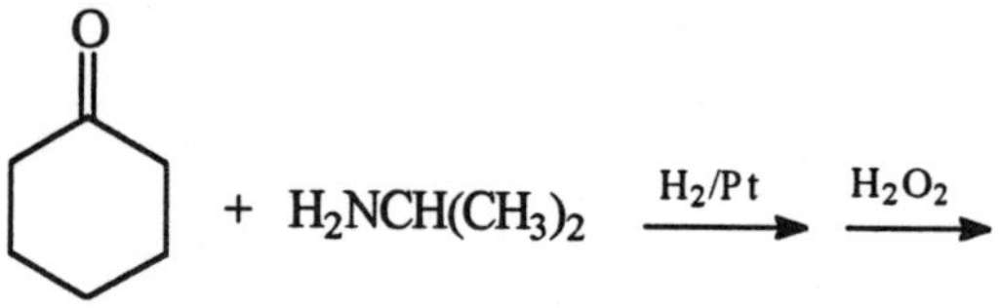

+ H₂NCH(CH₃)₂
H₂/Pt
H₂O₂

c. 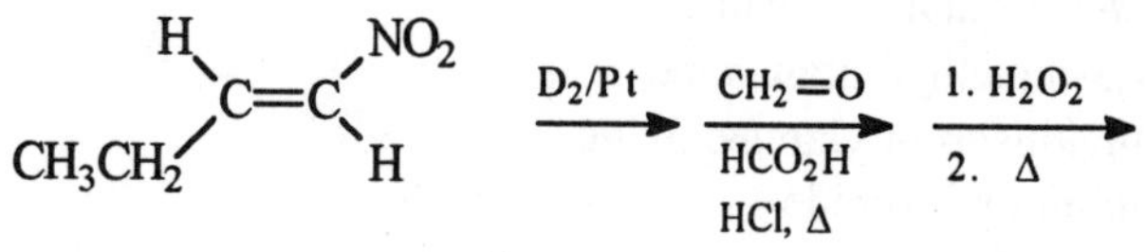

NO₂
CH₃CH₂
D₂/Pt
CH₂=O
HCO₂H
HCl, Δ
1. H₂O₂
2. Δ

d. 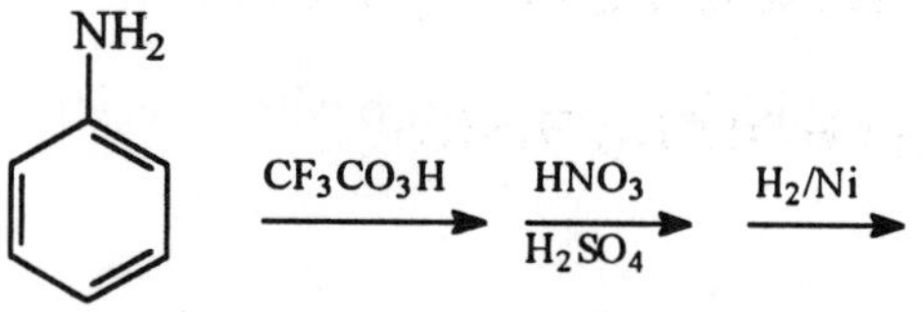

NH₂
CF₃CO₃H
HNO₃
H₂SO₄
H₂/Ni

e. 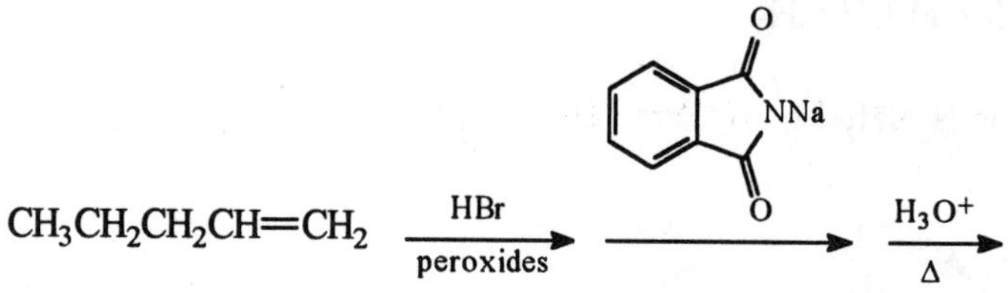

CH₃CH₂CH₂CH=CH₂
HBr
peroxides
NNa
H₃O⁺
Δ

f. $CH_3CH_2CH_2CH{=}O \xrightarrow{H_2NOH} \xrightarrow{H_2/Ni} \xrightarrow{HCl}$

g. 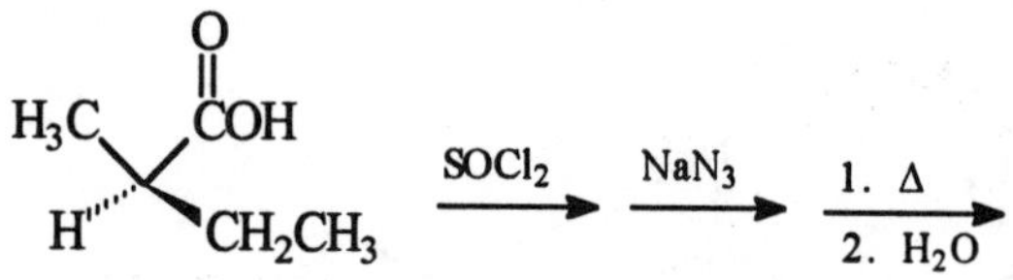

H₃C
COH
CH₂CH₃
SOCl₂
NaN₃
1. Δ
2. H₂O

h. $(CH_3)_3CCH_2CH_2\overset{O}{\overset{\|}{C}}CH_3 \xrightarrow{H\overset{O}{\overset{\|}{C}}N(CH_3)_2} \xrightarrow{CH_3I} \xrightarrow[2.\ \Delta]{1.\ Ag_2O}$

i. 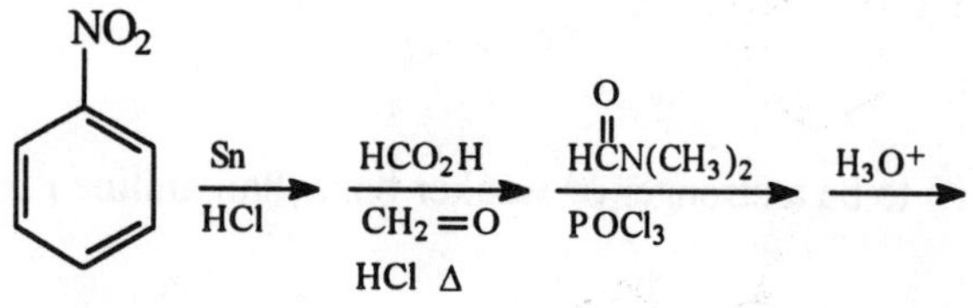

NO₂
Sn
HCl
HCO₂H
CH₂=O
HCl Δ
HCN(CH₃)₂
POCl₃
H₃O⁺

j. $(CH_3)_2CHCH_2NH_2 \xrightarrow{(CH_3)_2CH\overset{O}{\overset{\|}{C}}Cl} \xrightarrow{LiAlH_4} \xrightarrow[HCl]{NaNO_2}$

k. 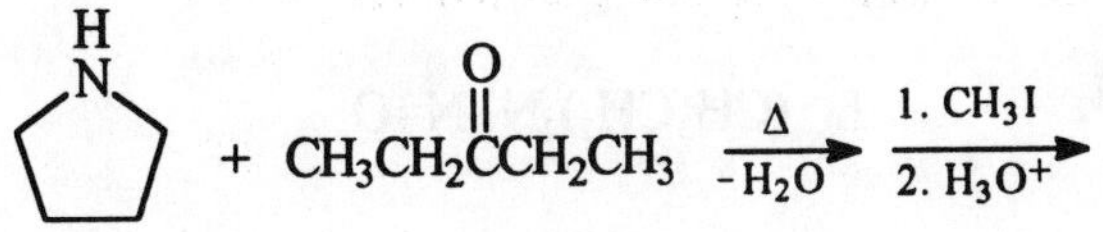 (pyrrolidine, N–H) + $CH_3CH_2\overset{O}{\overset{\|}{C}}CH_2CH_3$ $\xrightarrow[-H_2O]{\Delta}$ $\xrightarrow[2.\ H_3O^+]{1.\ CH_3I}$

l. 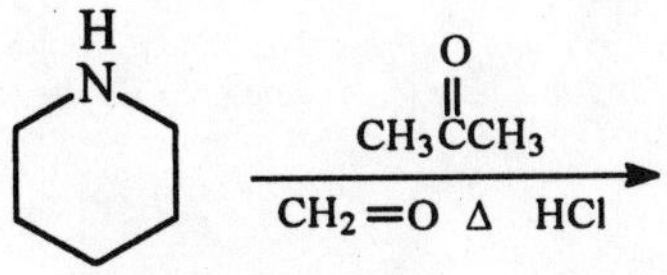 (piperidine, N–H) $\xrightarrow[CH_2{=}O\ \ \Delta\ \ HCl]{CH_3\overset{O}{\overset{\|}{C}}CH_3}$

m. 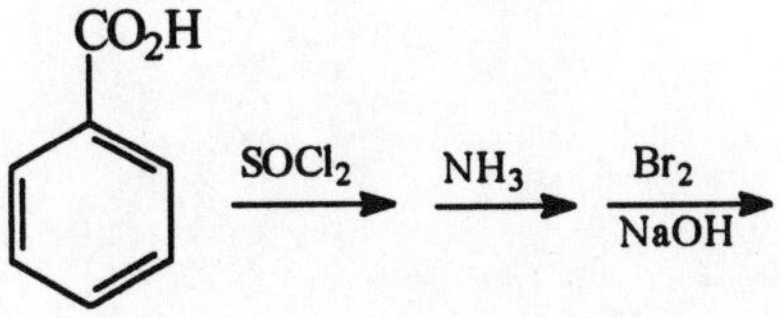 (benzoic acid, $C_6H_5CO_2H$) $\xrightarrow{SOCl_2}$ $\xrightarrow{NH_3}$ $\xrightarrow[NaOH]{Br_2}$

S6. Outline syntheses of the following compounds, using 3-methylbutanoic acid as starting material.

a. $(CH_3)_2CHCH_2NH_2$ b. $(CH_3)_2CHCH_2CH_2NH_2$ c. $(CH_3)_2CHCH_2CH_2CH_2NH_2$

S7. Write the intermediates formed during the following reaction sequence, as well as a mechanism for the last step.

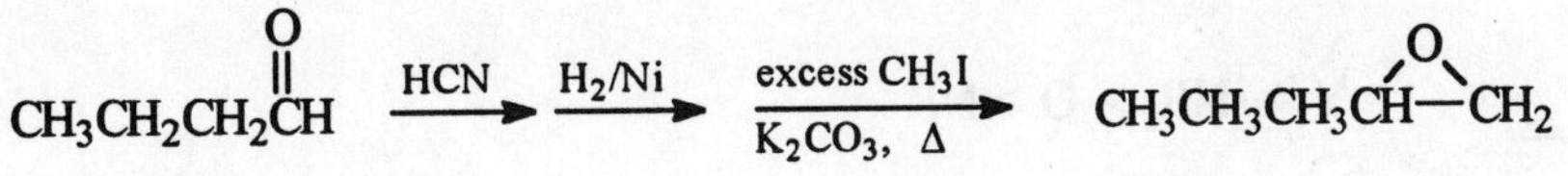

$CH_3CH_2CH_2\overset{O}{\overset{\|}{C}}H$ $\xrightarrow{HCN}$ $\xrightarrow{H_2/Ni}$ $\xrightarrow[K_2CO_3,\ \Delta]{excess\ CH_3I}$ $CH_3CH_3CH_3CH\overset{O}{—}CH_2$ (epoxide)

S8. Provide an explanation for the following behavior, and write a detailed mechanism for the conversion of B to A.

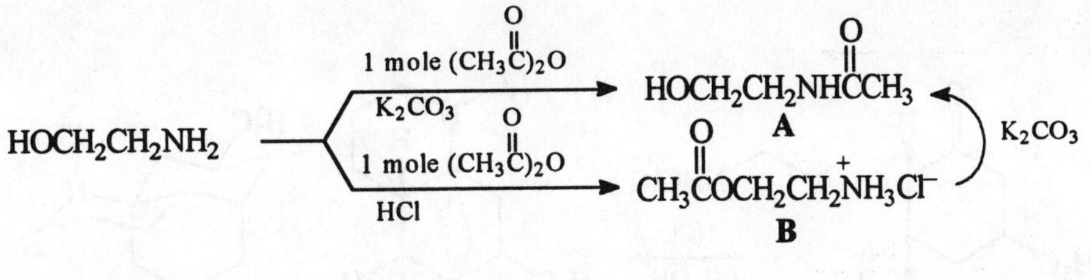

$HOCH_2CH_2NH_2$ $\xrightarrow[K_2CO_3]{1\ mole\ (CH_3\overset{O}{\overset{\|}{C}})_2O}$ $HOCH_2CH_2NH\overset{O}{\overset{\|}{C}}CH_3$ **A**

$HOCH_2CH_2NH_2$ $\xrightarrow[HCl]{1\ mole\ (CH_3\overset{O}{\overset{\|}{C}})_2O}$ $CH_3\overset{O}{\overset{\|}{C}}OCH_2CH_2\overset{+}{N}H_3Cl^-$ **B**

B $\xrightarrow{K_2CO_3}$ **A**

S9. Show how to synthesize the following compounds, using only inorganic compounds (e.g., $NaNO_2$, NH_3, NaCN, etc.) as a source of nitrogen.

a. 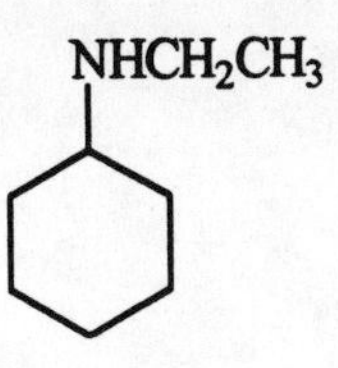 (cyclohexane ring bearing $NHCH_2CH_3$)

b. 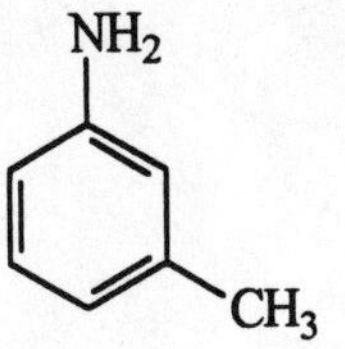 (benzene ring bearing NH_2 and, meta, CH_3)

c. 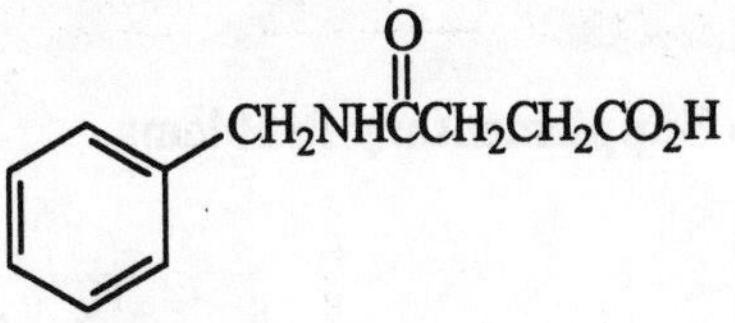 (benzene ring bearing $CH_2NH\overset{O}{\overset{\|}{C}}CH_2CH_2CO_2H$)

d. 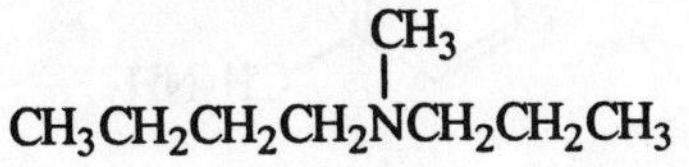 $CH_3CH_2CH_2CH_2\overset{CH_3}{\overset{|}{N}}CH_2CH_2CH_3$

e. 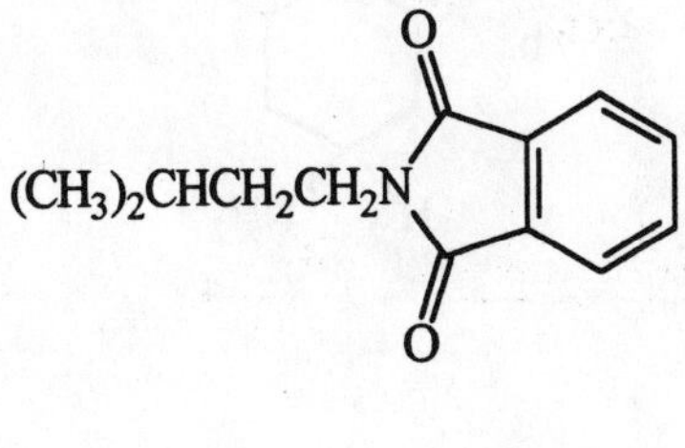 $(CH_3)_2CHCH_2CH_2N$ (phthalimide: N attached to two C=O of a fused benzene ring)

f. 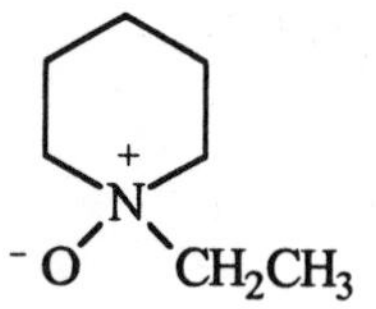 [structure: N-ethylpiperidine N-oxide; ^{-}O, N^{+}, CH_2CH_3]

g. 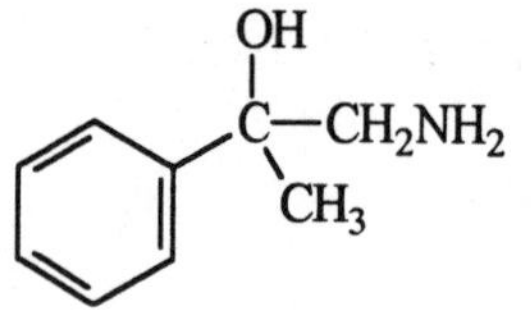 [structure: phenyl–C(OH)(CH$_3$)–CH_2NH_2]

h. $(CH_3CH_2)_2N{-}N{=}O$

S10. Write a step-by-step mechanism for the following transformations:

a. 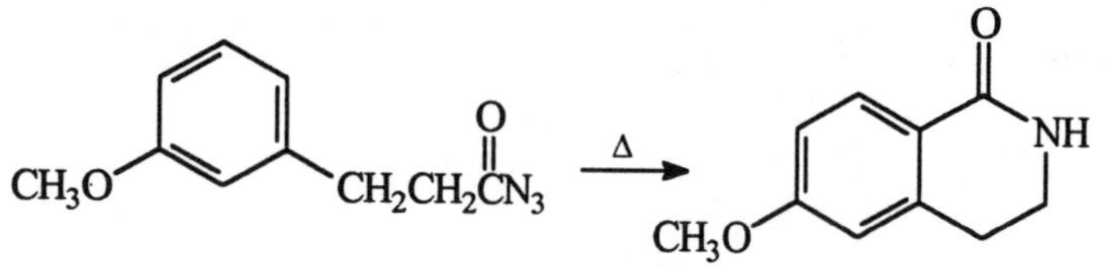 CH_3O–C$_6$H$_4$–$CH_2CH_2\overset{O}{\overset{\|}{C}}N_3$ $\xrightarrow{\Delta}$ 6-methoxy-3,4-dihydroisoquinolin-1(2H)-one (CH_3O, NH, O)

b. 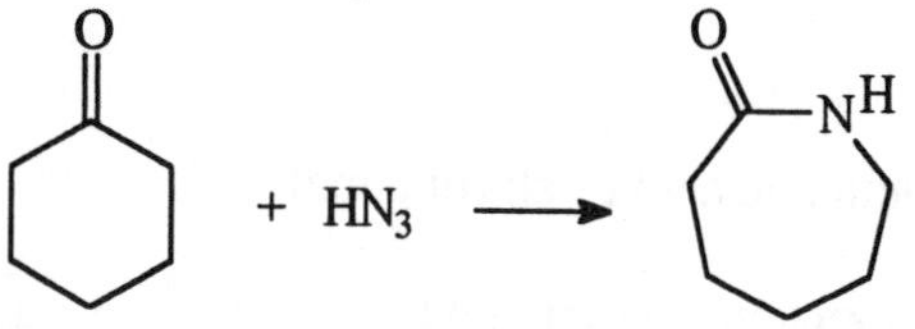 cyclohexanone + HN_3 ⟶ caprolactam (O, NH)

S11. Offer an explanation for the degree of reactivity of immonium ions, carbonyl compounds, and imines toward nucleophilic attack:

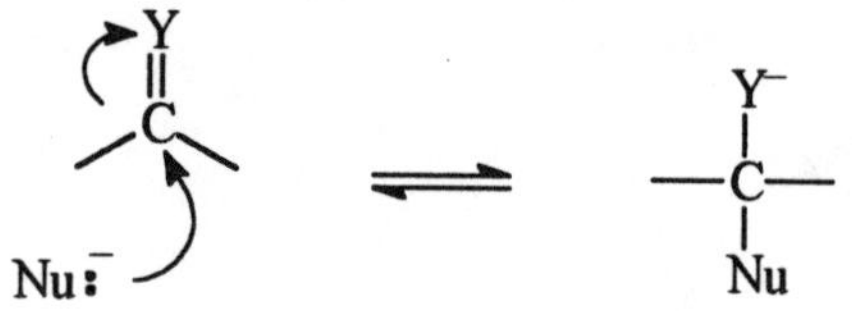

$$\mathrm{Nu}:^{-} + \;>\!C{=}Y \;\rightleftharpoons\; -\overset{Y^{-}}{\overset{|}{C}}(\mathrm{Nu})-$$

$Y = {}^{+}NR_2 > O > NR$

S12. Write a mechanism for the following transformation:

[structures: H_3C, O, CH_3, $NHCH_3$ ≡ H_3C, O, H, $NHCH_3$, H, H_3C, O $\xrightarrow[CH_3OH]{HCl}$ H_3C, N, H, H_3C, H, O ≡ H_3C, CH_3, N, O]

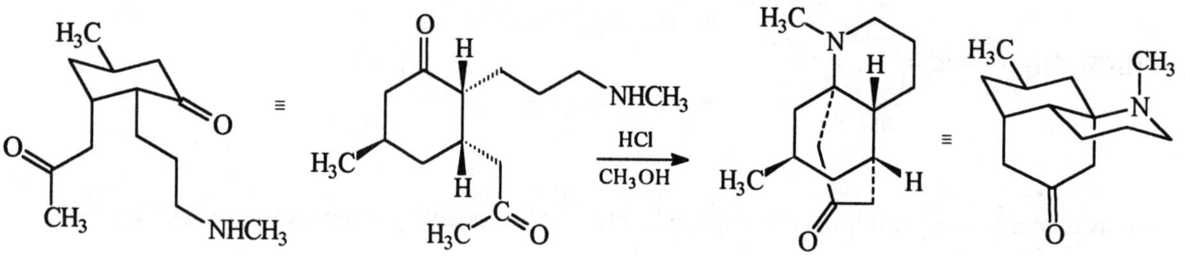

24.G. Answers to Supplementary Problems

S1. a. $((CH_3)_2CHCH_2)_3N$

b. 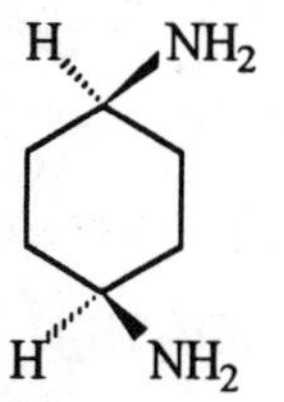 [structure: trans-1,4-diaminocyclohexane; H, NH_2, H, NH_2]

c. 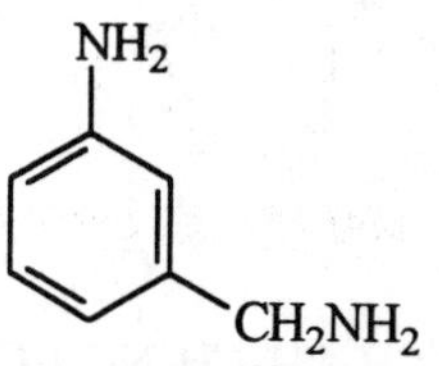 [structure: 3-aminobenzylamine; NH_2, CH_2NH_2]

d. $(CH_3)_2CHCH_2CH_2NHCH_3$ e. $(CH_3CH_2CH_2)_3{}^+NCH_2CH_3Cl^-$ f. $CH_3CH_2\overset{+}{N}(CH(CH_3)_2)_2$ with O^- on N

g. $CH_3CH_2NHOCH_2CH_3$ h. $C_6H_5\overset{+}{N}H_3\ Br^-$

S2. Only diisopropylamine will react with nitrous acid:

$$((CH_3)_2CH)_2NH \xrightarrow{HNO_2} ((CH_3)_2CH)_2N{-}N{=}O$$

b. $(CH_3CH_2CH_2)_2NH$ — very weak N-H stretch: 3310-3350 cm^{-1}

$CH_3CH_2CH(NH_2)CH_2CH_2CH_3$ — weak N-H stretch: doublet 3400, 3500 cm^{-1}; N-H band: 1580-1650 cm^{-1}

c. $$RNH_2 \xrightarrow[CH_2=O\ HCl]{HCO_2H} RN(CH_3)_2 \xrightarrow{H_2O_2} R\overset{O^-}{N}(CH_3)_2 \xrightarrow{\Delta}$$

R = $(CH_3)_3CCH_2$–: no elimination takes place

R = $CH_3CH_2CH(CH_3)CH_2$– : products are $CH_3CH_2C(CH_3){=}CH_2$ + $(CH_3)_2NOH$

d. $$RN(CH_3)_2 \xrightarrow[Ag_2O]{CH_3I} RN(CH_3)_3OH^- \xrightarrow{\Delta}$$

[cyclohexane with H and $^+N(CH_3)_3$, H, H, H_3C, CH_3 substituents] gives [H, H_3C / CH_3 cyclohexene] (resulting from *anti* elimination)

[cyclohexane with H and $^+N(CH_3)_3$, H, CH_3, H_3C, H substituents] gives [cyclohexane with H and $N(CH_3)_2$, H, CH_3, H_3C, H substituents] starting material

(Demethylation occurs because there is no hydrogen *anti* to the nitrogen)

S3. *Weakest base*: V (to act as a base, it must gain a proton, and an additional positive charge)
VI (Cl is electron-withdrawing)
I
IV (CH_3O is electron-releasing, through conjugation of the lone pair electrons)
II (alkylamines are stronger bases than arylamines)

Strongest base: III (dialkylamines are even better than alkylamines)

S4. 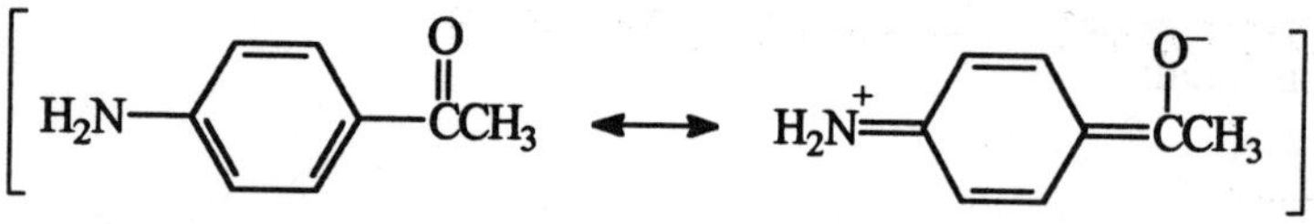 [$H_2N-C_6H_4-\overset{O}{\overset{\|}{C}}CH_3$ ⟷ $H_2\overset{+}{N}=C_6H_4=\overset{O^-}{\overset{|}{C}}CH_3$]

Because of some contribution from the second resonance structure, the nitrogen already has positive character and will be a weaker base.

S5. a. 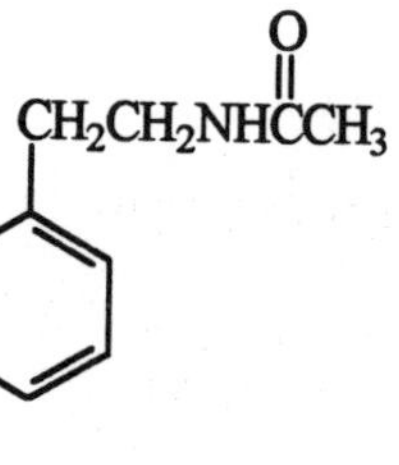 $C_6H_5CH_2CH_2NH\overset{O}{\overset{\|}{C}}CH_3$

b. 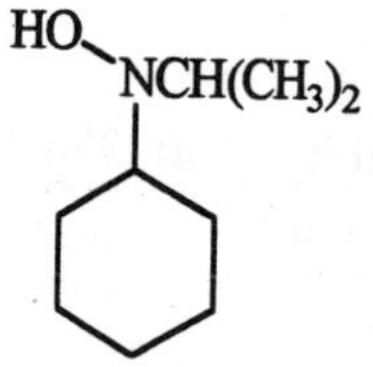 $HO-NCH(CH_3)_2$ (N attached to cyclohexane)

c. $CH_3CH_2CHD-CHD-\overset{O^-}{\overset{|}{N}}(CH_3)_2$ (H, D stereochemistry shown)

→ $CH_3CH_2(H)C=C(H)D$ + $(CH_3)_2NOD$ and $CH_3CH_2(D)C=C(D)H$ + $(CH_3)_2NOH$

d. 1,3-$C_6H_4(NH_2)_2$

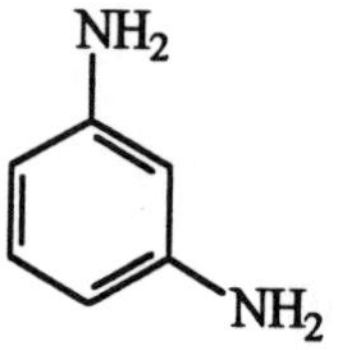

e. $CH_3CH_2CH_2CH_2CH_2NH_2$

f. $CH_3CH_2CH_2CH_2{}^+NH_3$ Cl^-

g.  CH_3, NH_2, H, CH_2CH_3 on a stereocenter

h. $(CH_3)_3CCH_2CH_2CH=CH_2 + (CH_3)_3N$:

i. 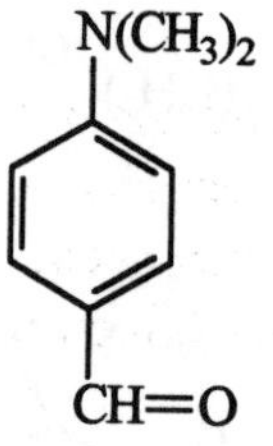 p-$(CH_3)_2N-C_6H_4-CH=O$

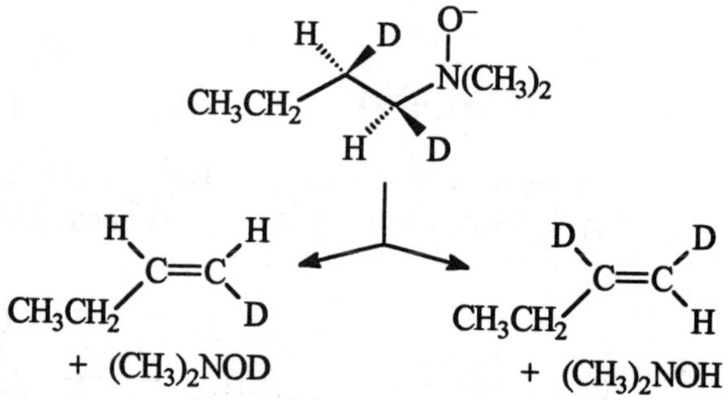

j. 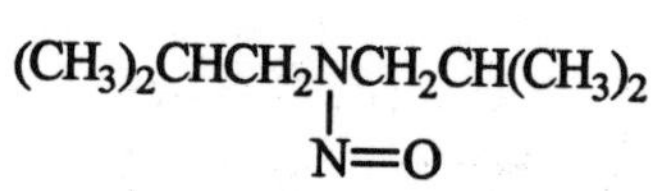 $(CH_3)_2CHCH_2NCH_2CH(CH_3)_2$ with $N=O$ on the central N

k. 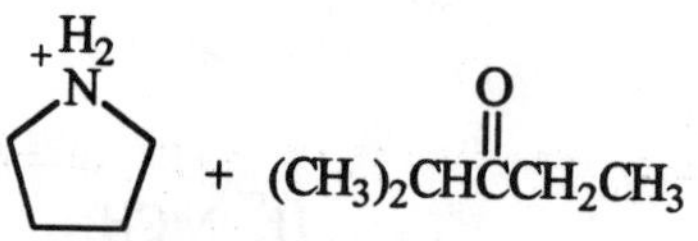 pyrrolidinium ($\overset{+}{N}H_2$) + $(CH_3)_2CH\overset{O}{\overset{\|}{C}}CH_2CH_3$

l. 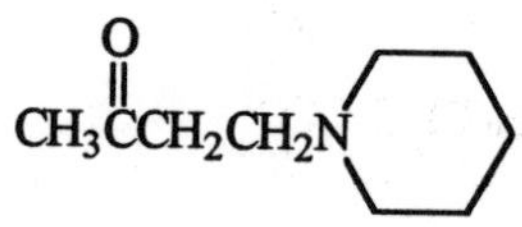 $CH_3\overset{O}{\overset{\|}{C}}CH_2CH_2N$ (piperidine ring)

m. 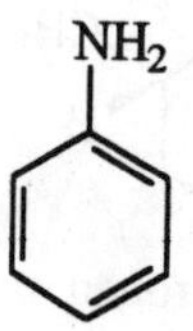 $C_6H_5NH_2$

S6. a.

$$(CH_3)_2CHCH_2CO_2H \xrightarrow{NaN_3,\ H_2SO_4} \left[(CH_3)_2CHCH_2\overset{O}{\overset{\|}{C}}N_3 \longrightarrow (CH_3)_2CHCH_2\overset{O}{\overset{\|}{C}}N{:} \longrightarrow (CH_3)_2CHCH_2N{=}C{=}O \right] \longrightarrow (CH_3)_2CHCH_2NH_2$$

b.

$$(CH_3)_2CHCH_2CO_2H \xrightarrow[\Delta]{NH_3} (CH_3)_2CHCH_2CONH_2 \xrightarrow{LiAlH_4} (CH_3)_2CHCH_2CNH_2$$

or

$$(CH_3)_2CHCH_2CO_2H \xrightarrow{SOCl_2} \xrightarrow{NH_3} (CH_3)_2CHCH_2CONH_2$$

c.

$$(CH_3)_2CHCH_2COOH \xrightarrow[2.\ PBr_3]{1.\ LiAlH_4} (CH_3)_2CHCH_2CH_2Br \xrightarrow[2.\ LiAlH_4]{1.\ NaCN} (CH_3)_2CHCH_2CH_2CH_2NH_2$$

S7.

$$CH_3CH_2CH_2\overset{O}{\overset{\|}{C}}H + HCN \longrightarrow CH_3CH_2CH_2\overset{OH}{\overset{|}{C}}HC{\equiv}N \xrightarrow{H_2/Ni} CH_3CH_2CH_2\overset{OH}{\overset{|}{C}}HCHNH_2$$

$$\xrightarrow{\text{excess } CH_3I} CH_3CH_2CH_2\overset{OH}{\overset{|}{C}}HCH_2\overset{+}{N}(CH_3)_3 \xrightarrow{K_2CO_3} CH_3CH_2CH_2\overset{O^-}{\overset{|}{C}}H{-}CH_2{-}\overset{+}{N}(CH_3)_3$$

$$\longrightarrow CH_3CH_2CH_2\overset{O}{CH{-}CH_2} + N(CH_3)_3 \xrightarrow{CH_3I} \overset{+}{N}(CH_3)_4\ I^-$$

S8. Under basic conditions, the amine is more nucleophilic than the hydroxy group, and acylation takes place preferentially on nitrogen to give the amide. In acid, the amine is unreactive because it is protonated, and acylation occurs on oxygen to furnish the ester. However, on deprotonation of the amine again, a transacylation reaction takes place to provide the more stable amide:

$$CH_3\overset{O}{\overset{\|}{C}}OCH_2CH_2\overset{+}{N}H_3 \underset{}{\overset{K_2CO_3}{\rightleftharpoons}} CH_3\overset{O}{\overset{\|}{C}}OCH_2CH_2\ddot{N}H_2 \rightleftharpoons H_3C{-}\overset{O^-}{\overset{|}{C}}{-}O\ (\text{ring with } H_2{}^{+}N) \rightleftharpoons H_3C{-}\overset{^-O}{\overset{|}{C}}{-}O\ (\text{ring with } HN)$$

$$\longrightarrow CH_3\overset{O}{\overset{\|}{C}}NHCH_2CH_2O^- \underset{}{\overset{HCO_3^-}{\rightleftharpoons}} CH_3\overset{O}{\overset{\|}{C}}NHCH_2CH_2OH$$

S9. a. cyclohexanone $\xrightarrow[H_2/Pt]{\text{excess } NH_3}$ cyclohexylamine (NH_2) $\xrightarrow{(CH_3\overset{O}{\overset{\|}{C}})_2O}$ $\xrightarrow{LiAlH_4}$ cyclohexyl–$NHCH_2CH_3$

Alternatively,

phthalic anhydride + NH_3 ⟶ phthalimide (NH) $\xrightarrow[CH_3CH_2Br]{NaOEt}$ N-ethylphthalimide (NCH_2CH_3) $\xrightarrow{H_3O^+}$ $CH_3CH_2NH_2$ $\xrightarrow[H_2/Pt]{\text{cyclohexanone}}$ cyclohexyl–$NHCH_2CH_3$

b. benzaldehyde ($CH{=}O$) $\xrightarrow[H_2SO_4]{HNO_3}$ m-nitrobenzaldehyde (NO_2, $CH{=}O$) $\xrightarrow[HCl]{Zn}$ m-toluidine (NH_2, CH_3)

(Note that both reductions can be accomplished in one step.)

c. phthalimide $\xrightarrow[C_6H_5CH_2Br]{NaOEt}$ $\xrightarrow{H_3O^+}$ $C_6H_5CH_2NH_2$ $\xrightarrow{\text{succinic anhydride}}$ $C_6H_5CH_2NH\overset{O}{\overset{\|}{C}}CH_2CH_2CO_2H$

(see part a.)

d. phthalimide $\xrightarrow[CH_3(CH_2)_3Br]{NaOEt}$ $\xrightarrow{H_3O^+}$ $CH_3(CH_2)_3NH_2$ $\xrightarrow[Et_3N]{CH_3CH_2\overset{O}{\overset{\|}{C}}Cl}$ $CH_3(CH_2)_3NH\overset{O}{\overset{\|}{C}}CH_2CH_3$ $\xrightarrow{LiAlH_4}$ $CH_3(CH_2)_3NHCH_2CH_2CH_3$ $\xrightarrow[HCO_2H,\ HCl\ \Delta]{CH_2{=}O}$ $CH_3(CH_2)_3N(CH_3)CH_2CH_2CH_3$

e. phthalimide $\xrightarrow[(CH_3)_2CHCH_2CH_2Br]{NaOEt}$ $(CH_3)_2CHCH_2CH_2N$(phthalimido)

f. $CH_3CH_2NH_2$ (see part a.) $\xrightarrow[K_2CO_3]{BrCH_2CH_2CH_2CH_2CH_2Br}$ N-ethylpiperidine (N–CH_2CH_3) $\xrightarrow{H_2O_2}$ N-ethylpiperidine N-oxide (^-O–$\overset{+}{N}$–CH_2CH_3)

g. $C_6H_5\overset{O}{\overset{\|}{C}}CH_3 \xrightarrow{HCN} C_6H_5\underset{CH_3}{\overset{OH}{C}}C{\equiv}N \xrightarrow{LiAlH_4} C_6H_5\underset{CH_3}{\overset{OH}{C}}CH_2NH_2$

h. $CH_3CH_2NH_2 \xrightarrow{(CH_3\overset{O}{\overset{\|}{C}})_2O} \xrightarrow{LiAlH_4} (CH_3CH_2)_2NH \xrightarrow[HCl]{NaNO_2} (CH_3CH_2)_2NN{=}O$

S10. a.

CH_3O–C_6H_4–$CH_2CH_2C(=O)$–$\overset{-}{N}$–$\overset{+}{N}{\equiv}N$ ⟶ CH_3O–C_6H_4–$CH_2CH_2C(=O)$–$\ddot{N}$: ⟶ $CH_3\ddot{O}$–C_6H_4–$CH_2CH_2N{=}C{=}O$ ↙

CH_3O-substituted 3,4-dihydroisoquinolin-1(2H)-one (C=O, NH) ⇌ ⇌ $CH_3\overset{+}{O}$= cyclohexadienyl intermediate (H, O^-, N:)

b.

cyclohexanone $\overset{H^+}{\rightleftharpoons}$ protonated cyclohexanone ($\overset{+}{O}H$) + $^-N_3$ ⇌ [HO, $N{=}\overset{+}{N}{=}N^-$ ⟷ HO, $\overset{-}{N}$–$\overset{+}{N}{\equiv}N$]

↙ $-N_2$ HO, $\ddot{N}$: ⟵ HO, C=N (seven-membered ring) ⇌ caprolactam (O, NH)

S11. The ability of Y to stabilize the electron pair it gains during the reaction is indicated by the pK_a of the conjugate acid:

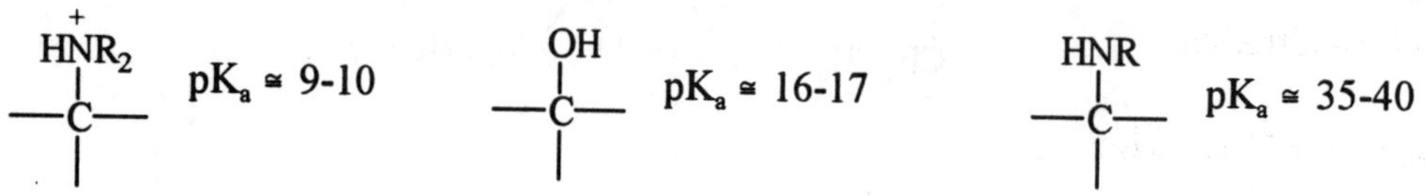

$\overset{+}{HNR_2}$—C— $pK_a \cong$ 9-10

OH—C— $pK_a \cong$ 16-17

HNR—C— $pK_a \cong$ 35-40

S12.

NHCH$_3$ ⇌ $\overset{+}{N}HCH_3$ OH ⇌ ⇌ NCH$_3$ $\overset{+}{O}H_2$

$\overset{+}{N}CH_3$ HÖ CH$_2$ ⇌(-H+) $\overset{+}{N}CH_3$ HO^+ CH$_3$ ⇌(H+) $\overset{+}{N}CH_3$ O CH$_3$

NCH$_3$ +OH ⇌(-H+) NCH$_3$ O

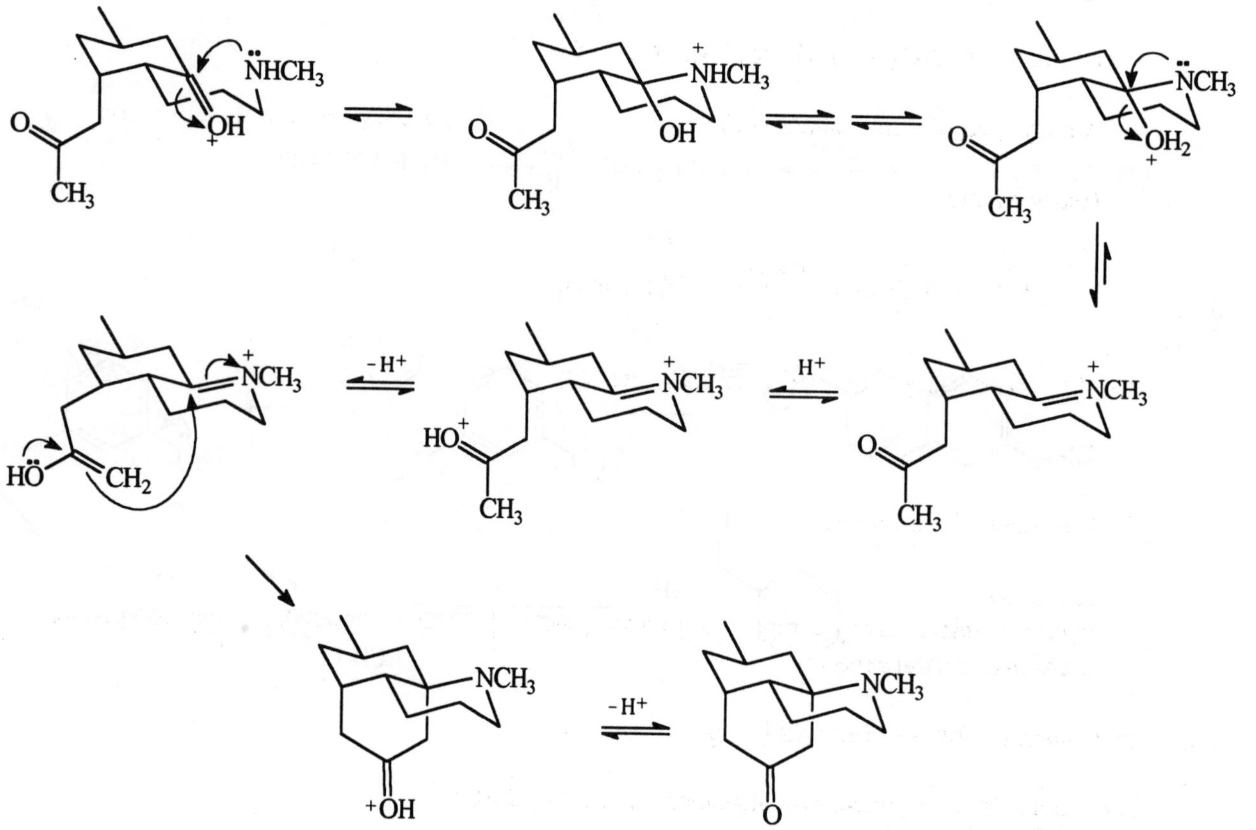

25. OTHER NITROGEN FUNCTIONS

25.A. Chapter Outline and Important Terms Introduced

25.1 Nitro Compounds

A. Nitroalkanes
from free radical nitration:

$$RH + HNO_3 \longrightarrow RNO_2 + H_2O$$

from displacement by nitrite:

$$RX + NO_2^- \longrightarrow RNO_2 + X^-$$

reduction of $-NO_2$ to $-NH_2$: see Section 24.6.C

Acidity: pK_a of nitromethane = 10.2 resonance stabilization of anion

Henry reaction

$$R'CH{=}O + RCH_2NO_2 \xrightarrow{NaOH} R'CH(OH)CH(R)NO_2$$

B. Nitroarenes (preparation covered in Chapter 23)

TNT

C. Reactions of Nitroarenes

reduction
hydroxylamines, azoxy compounds, azo compounds, hydrazo compounds, nitroso compounds
benzidine rearrangement

25.2 Isocyanates, Carbamates, and Ureas

isocyanates from Hofmann rearrangement: see Section 24.6.I
isocyanates from displacement:

$$RX + {}^-{:}N{=}C{=}O \longrightarrow RN{=}C{=}O + X^-$$

carbamates (urethanes) and ureas from isocyanates

25.3 Azides

alkyl and acyl azides by displacement:

$$RX + N_3^- \longrightarrow RN_3 + X^-$$

$$RC(=O)Cl + N_3^- \longrightarrow RC(=O)N_3 + Cl^-$$ (see Curtius rearrangement, Section 24.6.I)

reduction of alkyl azides (Section 24.6.H)

25.4 **Diazo Compounds**

preparation of esters with diazomethane: see Section 18.7.A
diazoketones and diazoesters
carbenes
ketenes

25.5 **Diazonium Salts**

preparation from aryl amines: see Section 24.7.B

A. Acid-Base Equilibria of Arenediazonium Ions

$$Ar{-}^{+}N{\equiv}N + H_2O \underset{K_1}{\overset{-H^+}{\rightleftharpoons}} Ar{-}N{=}N{-}OH \underset{K_2}{\overset{OH^-}{\rightleftharpoons}} Ar{-}N{=}N{-}O^- \quad (K_2 > K_1)$$

arenediazohydroxide *arenediazotate*

synthesis of substituted benzene derivatives via diazonium displacement

B. Thermal Decomposition of Diazonium Salts; Formation of ArOH, ArI, and ArSH
C. The Sandmeyer Reaction: Preparation of ArCl, ArBr, and ArCN
D. Preparation of Fluoro- and Nitroarenes
   Gatterman reaction Schiemann reaction
E. Replacement of the Diazonium Group by Hydrogen
F. Arylation Reactions
   Gomberg-Bachmann reaction
G. Diazonium Ions as Electrophiles: Azo Compounds
H. Synthetic Utility of Arenediazonium Salts

25.B. Important Reactions Introduced

Carbanion reactions of nitroalkanes (25.1.A)

Equation:

$$CH_3NO_2 \overset{base}{\rightleftharpoons} {}^-CH_2NO_2 \xrightarrow{RCH=O} RCH(OH)CH_2NO_2 \xrightarrow{-H_2O} RCH{=}CHNO_2$$

Key features: similar to aldol addition reaction

Reduction of nitroarenes (25.1.B and C)

Equations and generality:

to amine: $ArNO_2 \xrightarrow{[H]} ArNH_2$; [H] = Zn, Sn, or $SnCl_2$ in HCl; H_2/catalyst

to hydroxylamine: $ArNO_2 \xrightarrow{[H]} ArNHOH$; [H] = Zn/ NH_4Cl

to azoxy compound: $2\ ArNO_2 \xrightarrow{[H]} Ar{-}\overset{O^-}{\underset{+}{N}}{=}N{-}Ar$ [H] = As_2O_3, aq. NaOH

to azo compound: $2\ ArNO_2 \xrightarrow{[H]} Ar{-}N{=}N{-}Ar$; [H] = Zn, alcoholic NaOH

to hydrazo compound: $2\ ArNO_2 \xrightarrow{[H]} Ar{-}NHNH{-}Ar$; [H] = H_2NNH_2, Ru/C, alcoholic KOH

Interconversions of aromatic nitrogen compounds (25.1.C)

Equations and generality:

hydroxylamines and nitroso compounds

$ArNHOH \xrightarrow{[Ox]} Ar{-}N{=}O$; [Ox] = $Na_2Cr_2O_7$, H_2SO_4

azoxy compounds and azo compounds

$Ar{-}\overset{O^-}{\underset{+}{N}}{=}N{-}Ar \underset{[Ox]}{\overset{[H]}{\rightleftharpoons}} Ar{-}N{=}N{-}Ar$ [H] = $(EtO)_3P$; [Ox] = H_2O_2, HOAc

azo compounds and hydrazo compounds

$ArN{=}NAr \underset{[Ox]}{\overset{[H]}{\rightleftharpoons}} ArNHNHAr$; [H] = H_2NNH_2, Pd/C; [Ox] = O_2 or NaOBr

azo compounds and amines

$ArN{=}NAr \xrightarrow{[H]} 2\ ArNH_2$; [H] = $Na_2S_2O_4$, or any reagents that take $ArNO_2$ to $ArNH_2$

Benzidine rearrangement (25.1.C)

Equation:

$$Ar{-}NH{-}NH{-}Ar \xrightarrow{H^+} H_2N{-}Ar{-}Ar{-}NH_2$$

Key feature: Benzidine and related compounds are carcinogenic

Reaction of isocyanates with nucleophiles (25.2)

Equation:

$$RN{=}C{=}O + HY \longrightarrow RNH\overset{\overset{O}{\|}}{C}Y$$

Generality: Y = OH: product is a carbamic acid, which decomposes to give RNH_2 and CO_2
Y = OR′: product is a carbamate ester = a urethane
Y = NR′$_2$: product is a urea

Key features: isocyanates are key intermediates in Hofmann, Schmidt, and Curtius rearrangements

Reduction of alkyl azides (25.3)

Equation:

$$RN_3 \xrightarrow{[H]} RNH_2$$

Generality: R = alkyl or aryl
[H] = $LiAlH_4$ or H_2/catalyst

Key features: coupled with displacement of RX by N_3^-, this is an alternative to the Gabriel synthesis of 1° amines

Preparation of diazomethane from N-nitroso amides (25.4)

Equation:

$$R\overset{\overset{O}{\|}}{C}\underset{\underset{N=O}{|}}{N}CH_3 + OH^- \longrightarrow RCO_2^- + CH_2{=}N{=}N + H_2O$$

Generality: R = alkyl or aryl
applicable to preparation of other diazoalkanes

Formation and reaction of diazoketones (25.4)

Equation:

$$R\overset{\overset{O}{\|}}{C}Cl + 2\ CH_2N_2 \longrightarrow R\overset{\overset{O}{\|}}{C}CHN_2 + CH_3Cl + N_2$$

$$R\overset{\overset{O}{\|}}{C}CHN_2 \xrightarrow{HX} N_2 + R\overset{\overset{O}{\|}}{C}CH_2X$$

$$R\overset{\overset{O}{\|}}{C}CHN_2 \xrightarrow{\Delta \text{ or } h\nu} \left[R\overset{\overset{O}{\|}}{C}CH:\right] \longrightarrow RCH{=}C{=}O \xrightarrow{R'OH} RCH_2\overset{\overset{O}{\|}}{C}OR'$$

($RCH{=}C{=}O \longrightarrow$ dimer)

Carbene addition from α-diazoesters (25.4)

Equation:

$$N_2{=}CHCO_2Et \longrightarrow :CHCO_2Et \xrightarrow{\text{cyclohexene}} \text{bicyclo[4.1.0]heptane}{-}CO_2Et$$

Generality: carbene formation induced either by heat or metal catalysis

Substitution of arenediazonium salts (25.5)

Equation: $Ar-N_2^+ + M^+Nu^- \longrightarrow Ar-Nu + M^+ + N_2$

Generality: M^+Nu^- = H-OH, H-I, H-SH (better alternative for formation of Ar-SH is $EtOCS_2^-K^+$, followed by hydrolysis of xanthate product)
= CuX (X = Cl, Br, Cn; Sandmeyer reaction)
= $NaNO_2$/copper powder (Gatterman reaction)
= H_3PO_2 to give Ar-H

thermal decomposition of BF_4^- or PF_6^- salts gives Ar-F (Schiemann reaction)

Key features: substitution reactions proceed via complexed aryl cations or radical intermediates

Gomberg-Bachmann arylation reaction (25.5.F)

Equation: $Ar-N_2^+ + Ar'H \longrightarrow Ar-Ar' + H^+ + N_2$

Generality: isomer mixtures produced if Ar′H is substitutd
works best with Ar′H in excess

Key features: radical mechanism
low yield, but often the only way to form biphenyls

Formation of azo compounds (25.5.G)

Equation: $Ar-N_2^+ + Ar'-Y \longrightarrow Ar-N{=}N-Ar'-Y$

Generality: Y = strongly activating, *o*-, *p*-directing group

Key features: reaction proceeds via triazene intermediate if Y = NH_2
azo compounds are important dyes and colorings

25.C. Important Concepts and Hints

This chapter covers a true potpourri of reactions, with the common factor that each transformation concerns a nitrogen compound in which the nitrogen is not an amine. Much of this chemistry involves arene derivatives, including all of the intermediates along the way as an aryl nitro compound is reduced to the aryl amine. This topic is full of detail, and is difficult to learn in a way that doesn't ultimately rely on memorization. On the other hand, the reactions of arenediazonium salts involve the common goal of replacing the diazonium group with something else. This is an important group of reactions because it allows the introduction of a wide range of substituents onto an aromatic ring by a method different from electrophilic aromatic substitution. This can be useful when normal directive effects in electrophilic substitution make it impossible to obtain a specific isomer. Keep in mind the fact that RS, HS, RO, HO, F, Cl, Br, CN, CH=O (from CN), CO_2H (from CN), and even H can be represented by amino or nitro substituents at various stages of a synthesis.

25.D. Answers to Exercises

25.1 a. $CH_3CH_2CH_2\overset{O}{\overset{\|}{C}}CH_3 \xrightarrow{NaBH_4} \xrightarrow{PBr_3} \xrightarrow{NaCN} \xrightarrow{LiAlH_4} CH_3CH_2CH_2\overset{CH_3}{\overset{|}{C}}HCH_2NH_2$

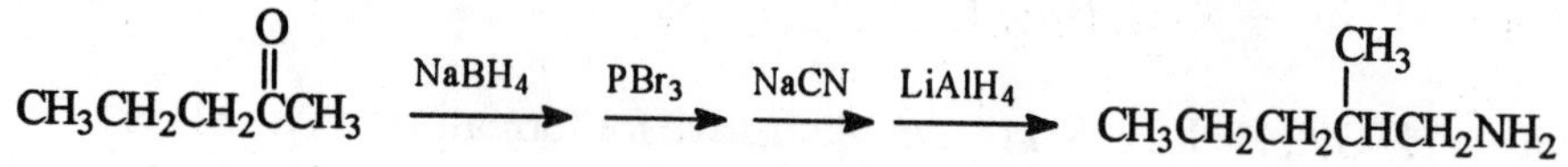

b. $CH_3CH_2CH_2\overset{O}{\overset{\|}{C}}CH_3 + CH_3NO_2 \xrightarrow{NaOH} \xrightarrow{H_2/Pt} CH_3CH_2CH_2\underset{OH}{\underset{|}{\overset{CH_3}{\overset{|}{C}}}}CH_2NH_2$

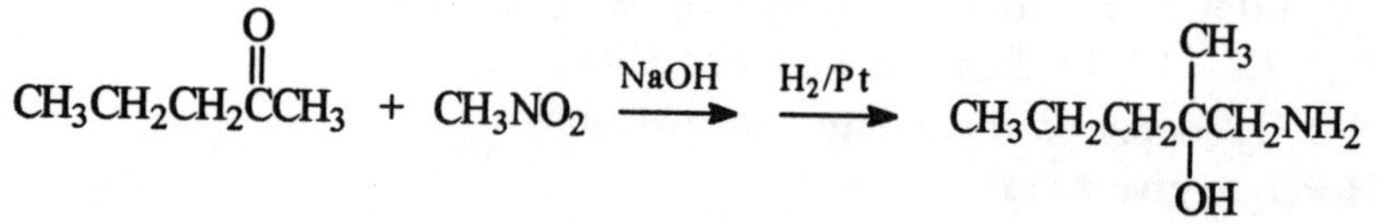

25.2 The negative charge can be delocalized via resonance onto the strongly electron-withdrawing nitro groups:

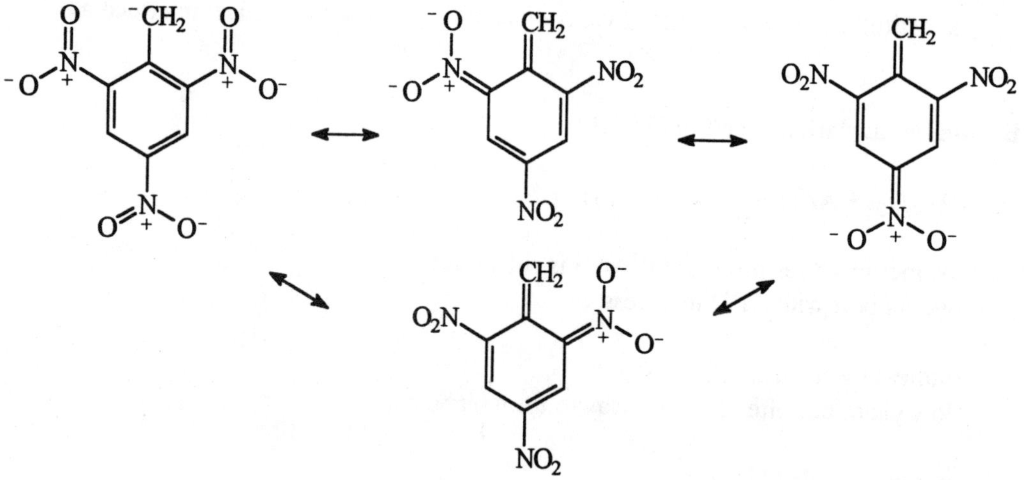

25.3

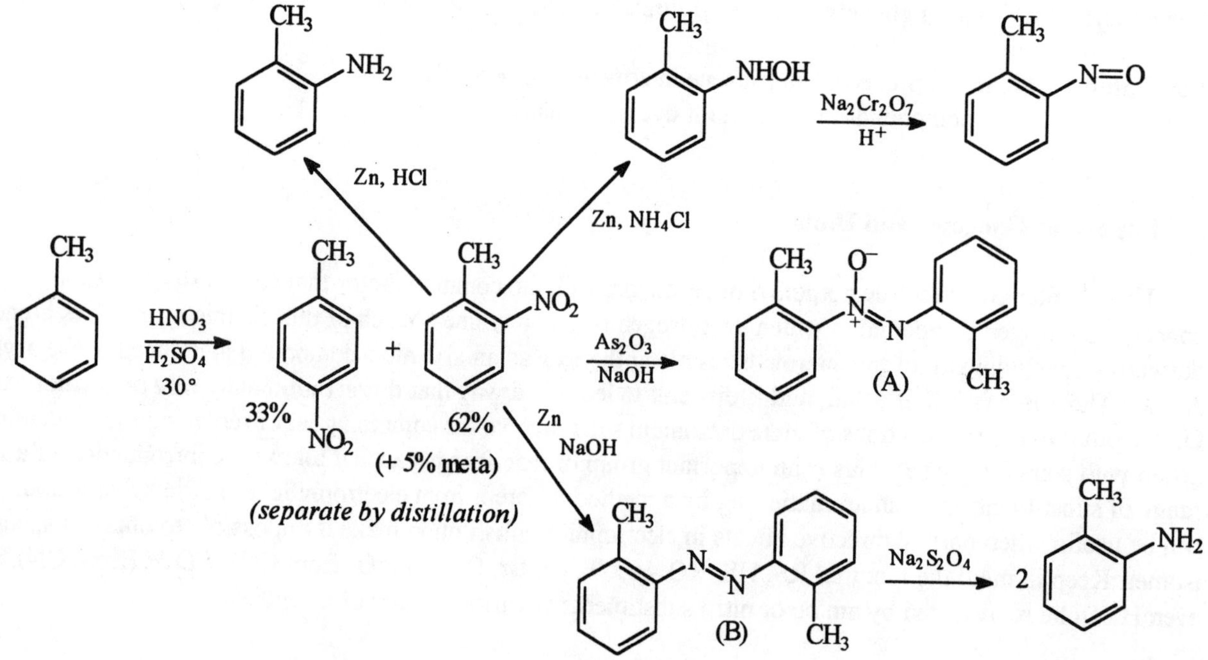

(separate by distillation)

33% 62% (+ 5% meta)

(A) (B)

$$o\text{-}CH_3C_6H_4NO_2 \xrightarrow[\text{Ru/C, KOH}]{H_2NNH_2} o\text{-}CH_3C_6H_4NHNHC_6H_4CH_3\text{-}o$$

$$\text{hydrazo compound} \underset{\text{Pd/C}}{\overset{\text{NaOBr or } O_2}{\underset{H_2NNH_2}{\rightleftarrows}}} B \underset{(EtO)_3P}{\overset{H_2O_2}{\rightleftarrows}} A$$

$$\xrightarrow{H^+,\ \Delta} H_2N\text{-}(3\text{-}CH_3)C_6H_3\text{-}C_6H_3(3'\text{-}CH_3)\text{-}NH_2$$

25.4 $R:\ddot{N}::C::\ddot{O}$ $\qquad R:\ddot{N}::\overset{+}{N}::\ddot{N}^-$

$$C_6H_5CH_2Cl + NaN{=}C{=}O \xrightarrow{CH_3OH} [C_6H_5CH_2N{=}C{=}O] \longrightarrow C_6H_5CH_2NH\overset{O}{\overset{\|}{C}}OCH_3 + NaCl$$

25.5

$$CH_3CH_2CO_2H \xrightarrow{SOCl_2} CH_3CH_2\overset{O}{\overset{\|}{C}}Cl \xrightarrow{2\ CH_2N_2} CH_3CH_2\overset{O}{\overset{\|}{C}}CH{=}N_2 \text{ (b)} + CH_3Cl + N_2 \longrightarrow \left[CH_3CH_2\overset{O}{\overset{\|}{C}}CH:\right] + N_2$$

$$CH_3CH_2CO_2H \xrightarrow{CH_2N_2} CH_3CH_2CO_2CH_3 \text{ (a)}$$

$$CH_3CH_2\overset{O}{\overset{\|}{C}}Cl \xrightarrow{CH_2N_2} CH_3CH_2\overset{O}{\overset{\|}{C}}CH_2Cl + N_2 \text{ (c)}$$

$$\left[CH_3CH_2\overset{O}{\overset{\|}{C}}CH:\right] \longrightarrow [CH_3CH_2CH{=}C{=}O] \text{ (d)} \longrightarrow \text{dimer: 3-}CH_3CH_2\text{-4-}(CH_3CH_2CH{=})\text{oxetan-2-one}$$

25.6

$$p\text{-}CH_3C_6H_4CH_3 \xrightarrow[H_2SO_4]{HNO_3} \xrightarrow[HCl]{Zn} 2,5\text{-}(CH_3)_2C_6H_3NH_2 \xrightarrow[HCl]{NaNO_2} 2,5\text{-}(CH_3)_2C_6H_3\overset{+}{N_2}$$

$$2,5\text{-}(CH_3)_2C_6H_3\overset{+}{N_2} \xrightarrow[\Delta]{KI} 2,5\text{-}(CH_3)_2C_6H_3I$$

$$2,5\text{-}(CH_3)_2C_6H_3\overset{+}{N_2} \xrightarrow[\Delta]{H_2O} 2,5\text{-}(CH_3)_2C_6H_3OH$$

25.7

Benzene $\xrightarrow[H_2SO_4]{HNO_3}$ $\xrightarrow[HCl]{Zn}$ $C_6H_5NH_2$ $\xrightarrow[HCl]{NaNO_2}$ $C_6H_5N_2^+Cl^-$ $\xrightarrow[\Delta]{CuCl_2}$ C_6H_5Cl (a)

$C_6H_5NH_2$ $\xrightarrow[HBr]{NaNO_2}$ $C_6H_5N_2^+Br^-$

(b) C_6H_5Br $\xleftarrow[\Delta]{CuBr}$ $C_6H_5N_2^+Br^-$ $\xrightarrow[0°]{CuCN}$ C_6H_5CN $\xrightarrow[\Delta]{H_3O^+}$ $C_6H_5CO_2H$ (c)

C_6H_5CN $\xrightarrow{LiAlH_4}$ $C_6H_5CH_2NH_2$ (d)

C_6H_5CN $\xrightarrow{NaOH}$ $C_6H_5CONH_2$ (e)

(d) $C_6H_5CH_2NH_2$ $\xleftarrow{LiAlH_4}$ $C_6H_5CONH_2$ (e)

(Note the use of HBr for the diazotization followed by CuBr so that no C_6H_5Cl is formed in the CuBr reaction).

25.8

Benzene $\xrightarrow[H_2SO_4]{HNO_3}$ $C_6H_5NO_2$ $\xrightarrow[H_2SO_4]{HNO_3}$ m-$C_6H_4(NO_2)_2$

$C_6H_5NO_2$ $\xrightarrow[Ac_2O]{Zn}$ C_6H_5NHAc

C_6H_5NHAc $\xrightarrow[2.\ HCl]{1.\ HNO_3}$ o-$NH_2C_6H_4NO_2$ + p-$NH_2C_6H_4NO_2$

p-$NH_2C_6H_4NO_2$ $\xrightarrow[HBF_4]{NaNO_2}$ p-$N_2^+C_6H_4NO_2$ $\xrightarrow[Cu\ powder]{NaNO_2}$ p-$C_6H_4(NO_2)_2$

o-$NH_2C_6H_4NO_2$ $\xrightarrow{NaNO_2,\ HBF_4}$ o-$N_2^+C_6H_4NO_2$ $\xrightarrow[Cu\ powder]{NaNO_2}$ o-$C_6H_4(NO_2)_2$

25.9

Benzene $\xrightarrow[H_2SO_4]{HNO_3}$ $\xrightarrow[HCl]{Zn}$ $C_6H_5NH_2$ $\xrightarrow{3\ Cl_2}$ 2,4,6-trichloroaniline $\xrightarrow[HCl]{NaNO_2}$ 2,4,6-$Cl_3C_6H_2N_2^+Cl^-$ $\xrightarrow{H_3PO_2}$ 1,3,5-trichlorobenzene

25.10

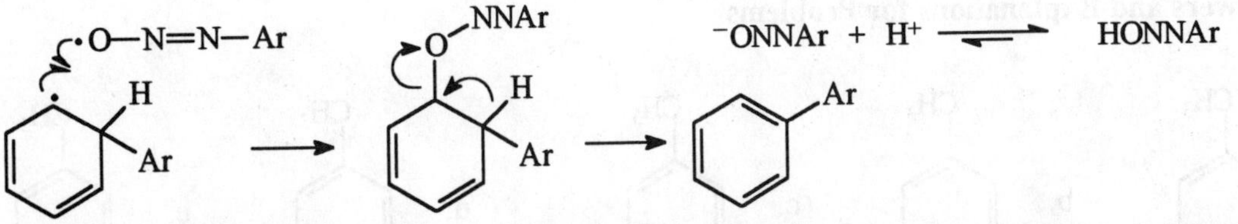

•O—N=N—Ar, H, Ar → O–NNAr, H, Ar → Ar

$^{-}$ONNAr + H$^+$ ⇌ HONNAr

This pathway involves ionic products. Since the reaction takes place in the organic phase, it is less likely than abstraction of a hydrogen atom.

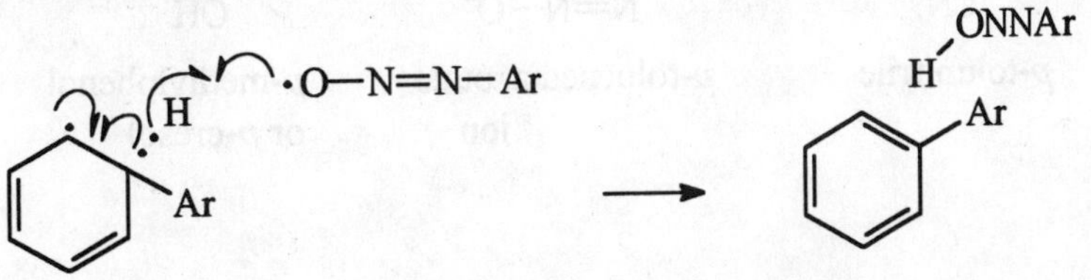

H, Ar, •O—N=N—Ar → ONNAr, H, Ar

25.11

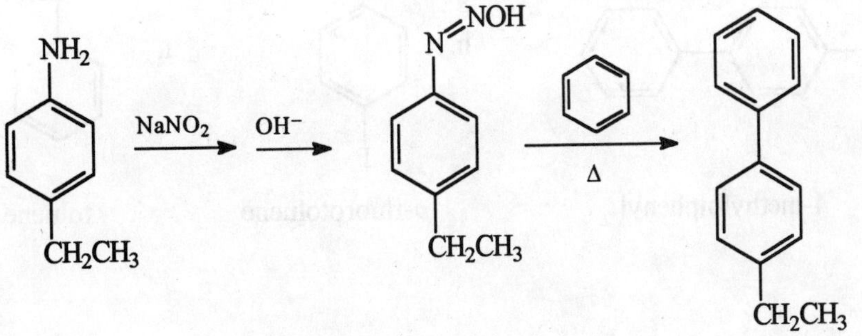

NH_2, CH_2CH_3 —$NaNO_2$→ —OH^-→ N=NOH, CH_2CH_3 —(benzene, Δ)→ CH_2CH_3 (This route can only give a single isomer.)

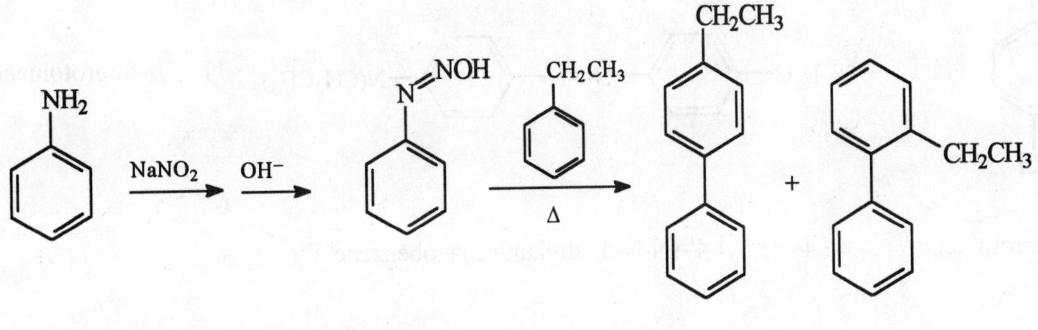

NH_2 —$NaNO_2$→ —OH^-→ N=NOH —(CH_2CH_3-benzene, Δ)→ CH_2CH_3 + CH_2CH_3

25.12

CO_2H, CH_3 —HNO_3, H_2SO_4→ CO_2H, NO_2, CH_3 —Zn, HCl, Ac_2O→ CO_2H, NHAc, CH_3 —Cl_2, $FeCl_3$→ CO_2H, Cl, NHAc, CH_3 —HCl, HNO_2→ CO_2H, Cl, N_2^+, CH_3 —H_3PO_2→ CO_2H, Cl, CH_3

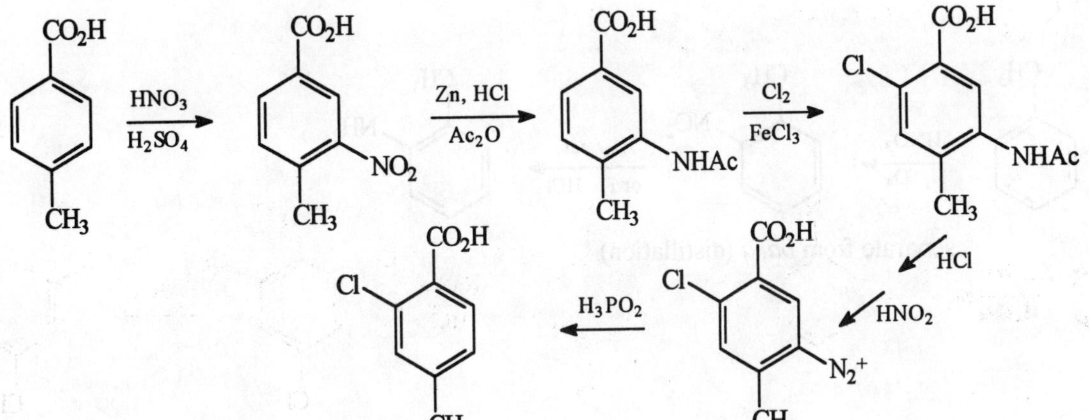

25.E. Answers and Explanations for Problems

1. a. 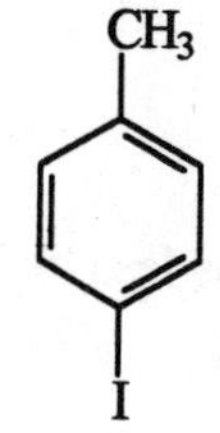

p-iodotoluene

b. 

p-tolunitrile

c. 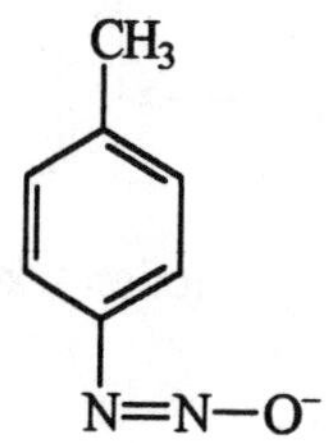

p-toluenediazotate ion

d. 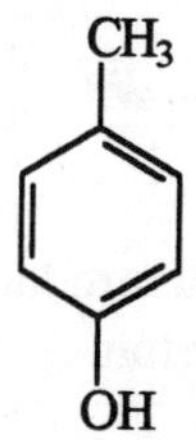

p-methylphenol or *p*-cresol

e. 

p-bromotoluene

f. 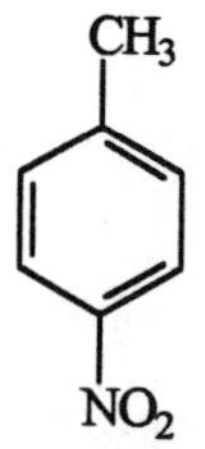

p-nitrotoluene

g. 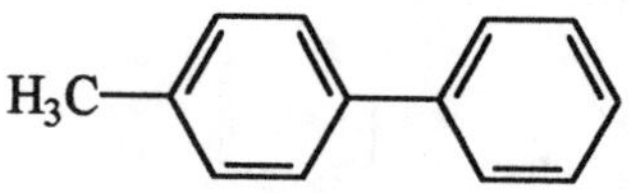

4-methylbiphenyl

h. 

p-fluorotoluene

i. 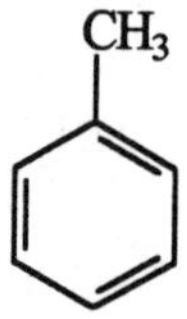

toluene

j. 

p-chlorotoluene

k. 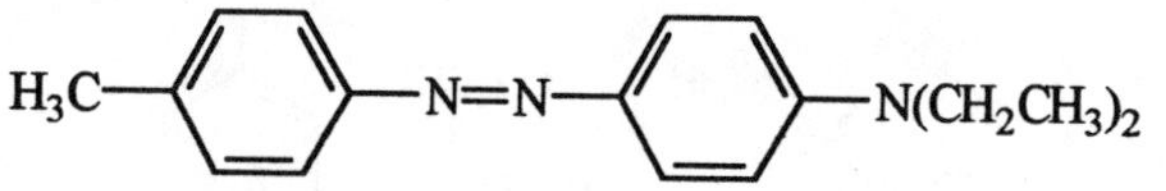

4-methyl-4'-(N,N-diethylamino)azobenzene

l. *p*-fluorotoluene (see 1 h.)

2. a. 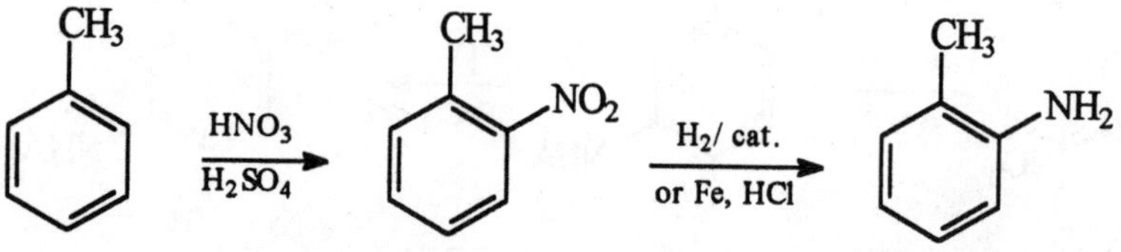

separate from *para* (distillation)

b. CH_3 → (HNO_3, H_2SO_4) → CH_3 / NO_2 → (H_2/ cat. or Fe, HCl) → CH_3 / NH_2 → (Ac_2O) → CH_3 / $NHCOCH_3$ → (HNO_3, CH_3CO_2H) → CH_3 / NO_2 / $NHCOCH_3$

separate from *ortho*

→ (HCl, Δ) → CH_3 / NO_2 / NH_2 → (HCl, $NaNO_2$, cold) → (H_3PO_2) → CH_3 / NO_2

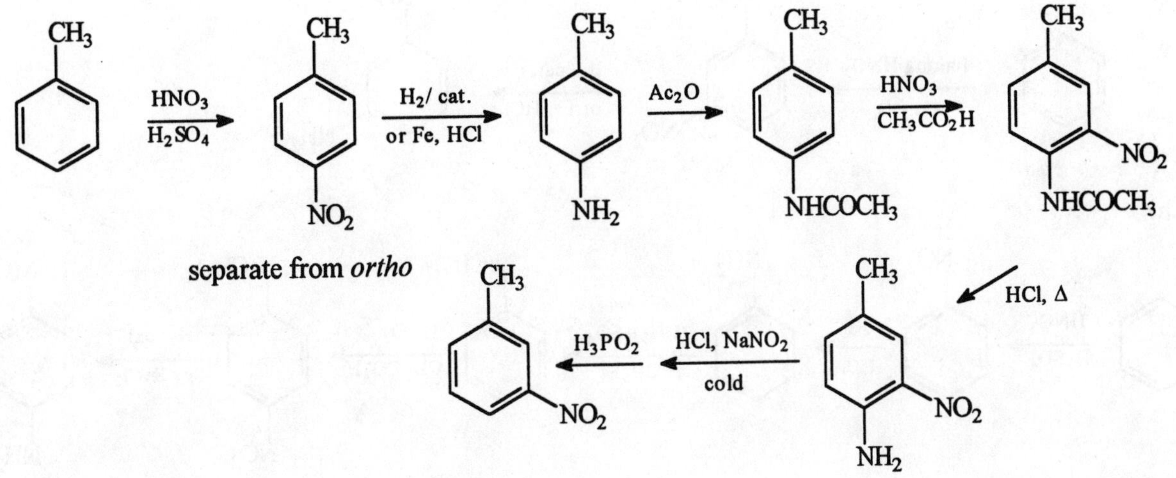

c. See b. above.

d. CH_3 → ($KMnO_4$) → CO_2H → (fuming HNO_3, H_2SO_4) → CO_2H / NO_2 → (H_2/ cat. or Fe, HCl) → CO_2^- / $\overset{+}{N}H_3$

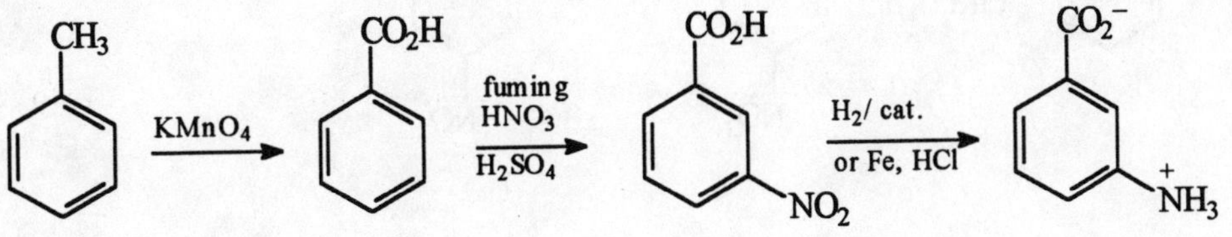

e. CH_3 / NO_2 → (Cl_2, $FeCl_3$) → CH_3 / NO_2 / Cl

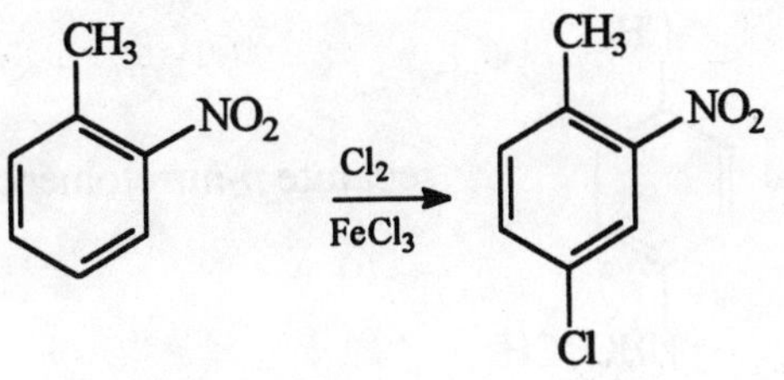

from a. above

f. CH_3 / NH_2 (from a.) → (Ac_2O) → CH_3 / $NHCOCH_3$ → (HNO_3, CH_3CO_2H) → CH_3 / $NHCOCH_3$ / O_2N

Cl_2, CH_3CO_2H *[alternate paths]*

→ (Zn, HCl) → CH_3 / $NHCOCH_3$ / H_2N → (NaNO$_2$, HCl) → (Cu_2Cl_2) → CH_3 / $NHCOCH_3$ / Cl → (HCl, Δ) → CH_3 / NH_2 / Cl

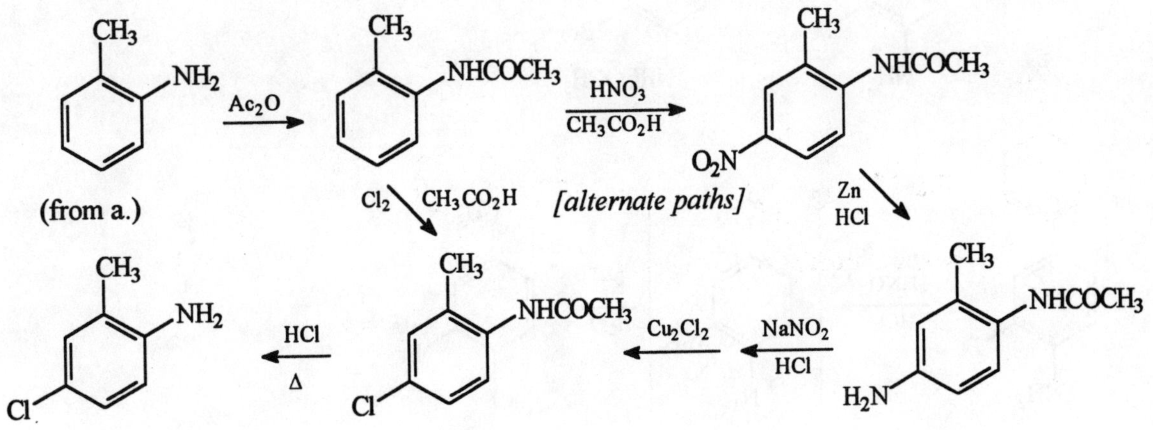

g. benzene $\xrightarrow[H_2SO_4]{\text{fuming } HNO_3}$ m-dinitrobenzene (NO_2, NO_2) $\xrightarrow[\text{or Fe, HCl}]{H_2/\text{ cat.}}$ m-phenylenediamine (NH_2, NH_2)

h. benzene $\xrightarrow[H_2SO_4]{HNO_3}$ NO_2 $\xrightarrow[HCl]{Fe}$ NH_2 $\xrightarrow[\text{or } CH_3Br]{(CH_3)_2SO_4}$ $N(CH_3)_2$ $\xrightarrow[CH_3CO_2H]{HNO_3}$ $N(CH_3)_2$ / NO_2 $\xrightarrow[\text{or } SnCl_2]{H_2/\text{ cat.}}$ $N(CH_3)_2$ / NH_2

i. NH_2 (from h.) $\xrightarrow{Ac_2O}$ $NHCOCH_3$ $\xrightarrow[CH_3CO_2H]{HNO_3}$ $NHCOCH_3$ / NO_2 $\xrightarrow[\Delta]{HCl}$ NH_2 / NO_2

3. a. CH_3 (toluene) $\xrightarrow[H_2SO_4]{HNO_3}$ CH_3 / NO_2 $\xrightarrow[\text{or Fe, HCl}]{H_2/\text{ cat.}}$ CH_3 / NH_2 $\xrightarrow{Ac_2O}$ CH_3 / $NHCOCH_3$ separate *p*-nitrotoluene from *o*-

$\xrightarrow[CH_3CO_2H]{Br_2}$ CH_3 / Br / $NHCOCH_3$ $\xrightarrow{H_3O^+}$ CH_3 / Br / NH_2

b. CH_3 / Br / NH_2 (from a.) $\xrightarrow[HCl]{NaNO_2}$ [CH_3 / Br / $^+N_2$] $\xrightarrow[H_2O]{KI}$ CH_3 / Br / I

c. CH_3 / NH_2 (from a.) $\xrightarrow[H_2O]{Br_2}$ CH_3, Br, Br, NH_2 $\xrightarrow[HCl]{NaNO_2}$ $\xrightarrow{H_3PO_2}$ CH_3, Br, Br

d. $\xrightarrow[H_2SO_4]{\text{fuming } HNO_3}$ NO_2, NO_2 $\xrightarrow[\text{alc.}]{NaSH}$ NH_2, NO_2 $\xrightarrow[HCl]{NaNO_2}$ $\xrightarrow[\text{benzene}]{NaOH}$ NO_2

e. NO_2 $\xrightarrow[FeBr_3]{Br_2}$ NO_2, Br $\xrightarrow[NaOH]{Zn}$ NHNH, Br, Br $\xrightarrow{HCl, \Delta}$ Br Br, H_2N—, —NH_2 $\xrightarrow[HCl]{NaNO_2}$ $\xrightarrow[\Delta]{CuBr}$ Br Br, Br—, —Br

f. CH_3 $\xrightarrow[H_2SO_4]{HNO_3}$ CH_3, NO_2 $\xrightarrow[HCl]{Fe}$ CH_3, NH_2 $\xrightarrow{C_6H_5CH_2Cl}$ $C_6H_5CH_2NH$—C_6H_4—CH_3

CH_3 $\xrightarrow[h\nu]{Cl_2}$ $C_6H_5CH_2Cl$

$CH_3C_6H_4NH_2$ $\xrightarrow{\text{or } C_6H_5SO_2Cl}$ $C_6H_5SO_2NH$—C_6H_4—CH_3 $\xrightarrow{NaOH}$ $\xrightarrow{C_6H_5CH_2Cl}$ CH_3, $C_6H_5CH_2NSO_2C_6H_5$ $\xrightarrow[\Delta]{NaOH}$ $C_6H_5CH_2NH$—C_6H_4—CH_3

(separate *p*-nitrotoluene from *ortho*)

g. [aniline] $\xrightarrow[\text{aq. HCl}]{Cl_2}$ [2,4,6-trichloroaniline] $\xrightarrow{CF_3CO_3H}$ [1,3,5-trichloro-2-nitrobenzene]

h. [toluene] $\xrightarrow[H_2SO_4]{HNO_3}$ [o-nitrotoluene] $\xrightarrow[\text{or Fe, HCl}]{H_2\text{/ cat.}}$ [o-toluidine] $\xrightarrow{Ac_2O}$ [o-methylacetanilide] $\xrightarrow[CH_3CO_2H]{HNO_3}$ [NHCOCH$_3$, CH$_3$, NO$_2$] $\xrightarrow[\Delta]{HCl}$ [NH$_2$, CH$_3$, NO$_2$] $\xrightarrow[HCl]{NaNO_2}$ $\xrightarrow[\Delta]{CuCN}$ [CN, CH$_3$, NO$_2$]

separate from *para*

4. From $pK_a = -\log \dfrac{[H^+][AcO^-]}{[AcOH]}$, for $[AcO^-] = [AcOH]$, $pK_a = pH$

Hence, the pH of a solution with equimolar amounts of AcO^- and AcOH is 4.74. This value is greater than the pK_a of methyl orange, 3.5 (see Sect. 25.5.G); hence, methyl orange is in the yellow, unprotonated, form.

5. On reaction at the *p*-position, the odd electron can conjugate with the nitro group:

[resonance structures: NO$_2$ / H R ↔ NO$_2$ / H R ↔ NO$_2$ / H R ↔ •O–N$^+$–O$^-$ / H R]

The nitro group helps to stabilize the odd electron and directs *o, p*-, just as in *nucleophilic* aromatic substitution. The nitro group is *m*-directing for *electrophilic* substitution.

6. a. [H$_3$C–C$_6$H$_4$–N=N–C$_6$H$_4$–CH$_3$]

b. [NH$_2$, CH$_3$]

c. [NHOH, CH$_3$]

d. (3-methylaniline: NH_2, CH_3)

e. H_3C–C$_6$H$_4$–NHNH–C$_6$H$_4$–CH_3

f. H_3C–C$_6$H$_4$–N=N$^+$(O$^-$)–C$_6$H$_4$–CH_3

7. a. Loss of N_2 gives a carbene, which rearranges to a ketene:

$$RC(=O)-CHN_2 \xrightarrow{-N_2} RC(=O)-\ddot{C}H \longrightarrow RCH{=}C{=}O$$

This rearrangement is similar to that which leads to a nitrene in the Curtius reaction. The ketene rapidly adds water to give a carboxylic acid.

$$RCH{=}C{=}O + H_2O \longrightarrow RCH{=}C(-O^-)-\overset{+}{O}H_2 \longrightarrow RCH_2CO_2H$$

b. $CH_3COCHN_2 \xrightarrow[-N_2]{h\nu} CH_3CO\ddot{C}H \longrightarrow CH_3CH{=}C{=}O \xrightarrow{CH_3OH} CH_3CH_2CO_2CH_3$

8. (cyclohexanone) $+ \ {}^-CH_2\overset{+}{N}{\equiv}N \longrightarrow$ (1-oxido-1-($CH_2N_2^+$)cyclohexane: ^-O, $CH_2N_2^+$)

Two modes of further reaction:

Displacement of -N_2 by -O^-

^-O, CH_2–N_2^+ (cyclohexane) ⟶ (spiro-epoxide)

Rearrangement:

^-O, CH_2–N_2^+ (cyclohexane) ⟶ (cycloheptanone, O)

9. This is an example of the Hofmann rearrangement. When one gets to the isocyanate stage, the nucleophilic species now present is methanol. The product, a carbamic acid ester, is stable.

$$C_6H_{11}-N{=}C{=}O + CH_3OH \longrightarrow C_6H_{11}-NHC(=O)OCH_3$$

10. The predominant base in methanolic potassium hydroxide is actually methoxide ion, because of the equilibrium:

$$HO^- + CH_3OH \rightleftharpoons CH_3O^- + H_2O$$

The large excess of methanol drives the equilibrium to the right.

The reaction is slow, since the ethyl carbamate is also an amide. The main reaction observed is **transesterification**:

$$C_6H_{11}NHCOC_2H_5\,(\text{C=O}) + CH_3O^- \rightleftharpoons C_6H_{11}NHC(O^-)(OCH_3)OC_2H_5\ (\mathbf{A}) \rightleftharpoons C_6H_{11}NHCOCH_3\,(\text{C=O}) + C_2H_5O^-$$

The alternative path (elimination of amide from A) is not observed, since $C_6H_{11}\ddot{N}H^-$ is such a poor leaving group. After a long time, of course, cyclohexylamine will be produced, since occasionally a hydroxide ion will attack to give an irreversible hydrolysis:

$$C_6H_{11}NHCOC_2H_5\,(\text{C=O}) + OH^- \rightleftharpoons C_6H_{11}NHC(O^-)(OH)OC_2H_5\ (\mathbf{A}) \rightleftharpoons C_6H_{11}NHCOH\,(\text{C=O}) + C_2H_5O^- \xrightarrow{-CO_2} C_6H_{11}NH_2$$

11. $CuBr + Cl^- \rightleftharpoons CuCl + Br^-$

The use of CuBr + HCl gives a mixture of bromocumene and chlorocumene.

12. The diazonium group is strongly electron-withdrawing and facilitates the nucleophilic aromatic substitution of groups in *o*- and *p*-positions:

$$p\text{-Br}C_6H_4N^+\equiv N \xrightarrow{Cl^-} \left[\text{(Cl)(Br)C}_6H_4^-\text{-}N^+\equiv N \longleftrightarrow \text{(Cl)(Br)C}_6H_4\text{=}N^+\text{=}N^- \right] \xrightarrow{-Br^-} p\text{-Cl}C_6H_4N^+\equiv N \xrightarrow{H_2O} p\text{-Cl}C_6H_4OH$$

13. a. $\xrightarrow[H_2O]{KMnO_4}$ $\xrightarrow{SOCl_2}$ CCl(=O) $\xrightarrow{NaN_3}$ $\xrightarrow[\Delta]{H_2O}$ NH_2

b. OH $\xrightarrow{PBr_3}$ Br $\xrightarrow{NH_3 \text{ excess}}$ or $\xrightarrow{1.\ NaN_3;\ 2.\ LiAlH_4}$ NH_2

c. OH (C=O) $\xrightarrow{SOCl_2}$ $\xrightarrow{CH_2N_2}$ CH_2N_2 (C=O) + HCl ⟶ CH_2Cl (C=O) + N_2

d. NO_2 $\xrightarrow[OH^-]{CH_3CH_2CHO}$ NO_2, OH $\xrightarrow{LiAlH_4}$ $\xrightarrow{H_3O^+}$ NH_2, OH

14. OH, NH_2 $\xrightarrow[HCl]{NaNO_2}$:OH, N_2^+ $\xrightarrow{-N_2}$ $\xrightarrow{-H^+}$ O

15. This method is suitable when the substrate ketone is symmetric. Otherwise, since the CH_2 group can be inserted on either side of the carbonyl, a mix of products would result.

a. Suitable: a symmetric substrate.

b. Unsuitable: asymmetric substrate. This would give a mixture of 2-methyl-4-octanone and 7-methyl-4-octanone.

c. Suitable: a symmetric substrate. The same compound is produced regardless of the CH_2 insertion.

O ≡ O

d. Unsuitable: asymmetric substrate. This would give a mixture of 3,3-dimethylcyclohexanone and 4,4-dimethylcyclohexanone.

25.F Supplementary Problems

S1. a. Nitroethylene can be prepared from 2-nitroethanol by distillation from phthalic anhydride. Write a mechanism for this reaction.

$$\text{phthalic anhydride} + HOCH_2CH_2NO_2 \longrightarrow \text{phthalic acid } (C_6H_4(CO_2H)_2) + CH_2{=}CHNO_2$$

b. Nitroethylene is a powerful electrophile and reacts rapidly with dimethylamine to give N,N-dimethyl-2-nitroethanamine, for example. Write a reasonable mechanism for this reaction.

S2. a. If nitrocyclohexane is treated with a strong base to form the anion, and then protonated with acid and isolated rapidly, the so-called *aci* tautomer is isolated. Propose a structure for this compound.

b. If the *aci* form of nitrocyclohexane is kept in the presence of aqueous acid, it undergoes hydrolysis to give cyclohexanone (the Nef reaction). Write a reasonable mechanism for this transformation.

$$\text{nitrocyclohexane } (NO_2) \xrightarrow{NaOH} \xrightarrow{H^+} \textit{aci}\text{ tautomer} \xrightarrow{H_3O^+} \text{cyclohexanone} + \text{"HNO"} \rightarrow 1/2\,(N_2O + H_2O)$$

S3. What side product would you expect if the reaction depicted in Supplementary Problem #S10a in Chapter 24 of this Study Guide were carried out in ethanol?

S4. Phenyl azide can be prepared by treatment of benzenediazonium ion with sodium azide or by reaction of phenylhydrazine with nitrous acid. What intermediates are involved in this latter method?

S5. Show how to synthesize the following compounds from benzene or any monosubstituted derivative.

a. Benzene ring with Br, HO_2C, Cl (1,3,5-substituted)

b. Benzene ring with SH, Cl, Cl (ortho to SH), Cl (para to SH)

c. Benzene ring with OCH_3, Cl, Cl (ortho to OCH_3), D, D

d. Benzene ring with F, Cl, CHO (para to F)

e. HO_2C–C_6H_4–C_6H_4–CO_2H

f. $(CH_3)_3C$–C_6H_4–C_6H_4–Cl

S6. Write a reasonable mechanism for the formation of diazomethane on treatment of N-methyl-N-nitrosoacetamide with sodium hydroxide.

S7. In contrast to ethyl diazoacetate, the diazoacetate anion is very unstable, decomposing in aqueous solution to give diazomethane and CO_2. Propose a mechanism for this transformation. (NOTE: the diazomethane anion N_2CH^- is not involved.)

$$N_2CHCO_2^- \xrightarrow{H_2O} N_2CH_2 + CO_2 + OH^-$$

S8. What is the major product from each of the following reactions.

a. 3-nitrotoluene (ring with NO_2, CH_3) $\xrightarrow[HCl]{Zn}$ $\xrightarrow[HBF_4]{NaNO_2}$ $\xrightarrow{\Delta}$

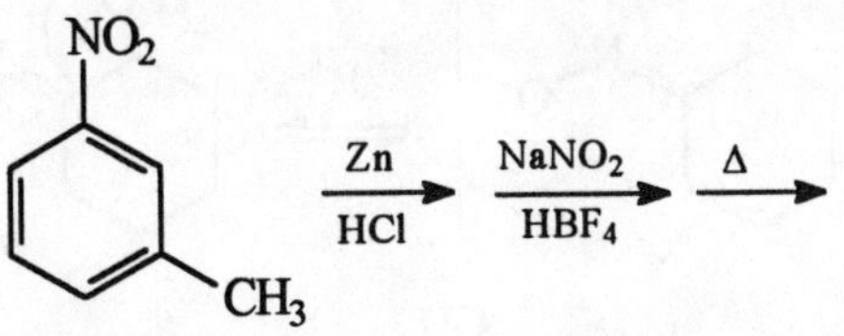

b. 3,5-dinitrotoluene (ring with NO_2, O_2N, CH_3) $\xrightarrow[CH_3OH]{NaSH}$ $\xrightarrow{Ac_2O}$

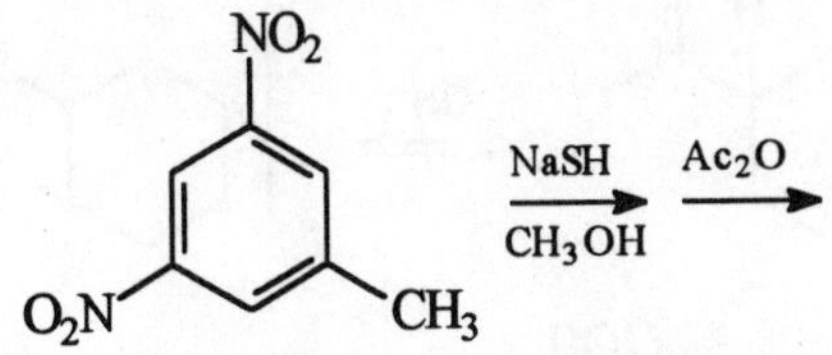

c. 4-chloronitrobenzene (ring with NO_2, Cl) $\xrightarrow[\text{alc. NaOH}]{Zn}$ $\xrightarrow{H_2O_2}$

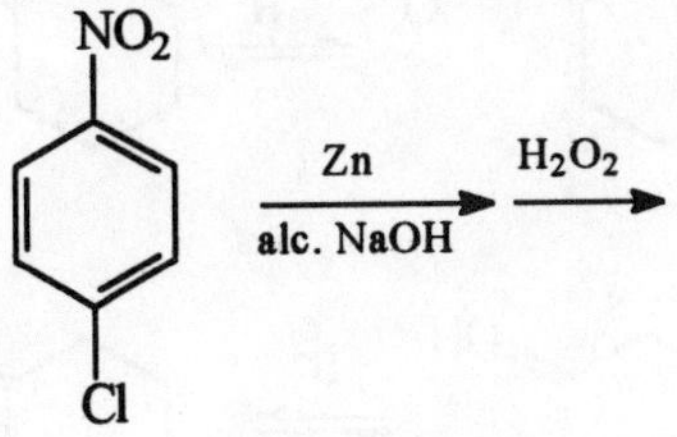

d. $H_3C-C_6H_4-\overset{+}{N}(O^-)=N-C_6H_5$ $\xrightarrow{(EtO)_3P:}$

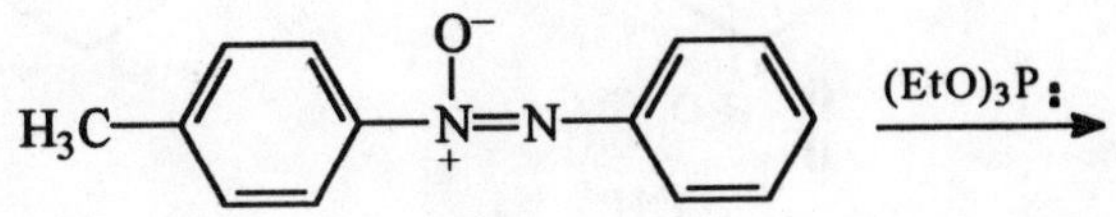

e. 3-nitrobenzoic acid (ring with NO_2, HO_2C) $\xrightarrow[\text{aq. } NH_4Cl]{Zn}$ $\xrightarrow[H^+]{K_2Cr_2O_7}$

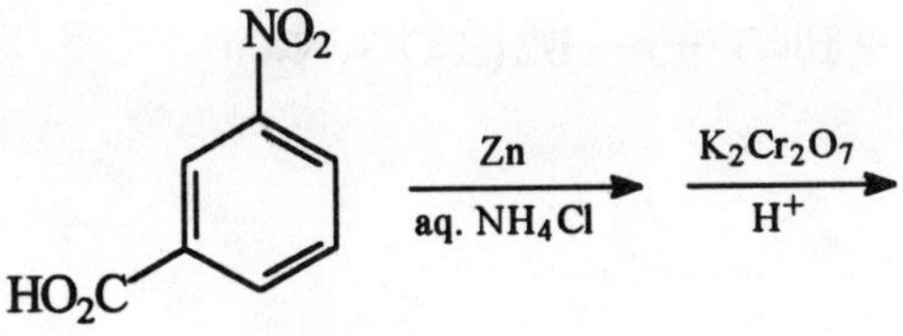

f. acetanilide (ring with $NHCOCH_3$) $\xrightarrow{Br_2}$ $\xrightarrow{H_3O^+}$ $\xrightarrow{CF_3CO_3H}$

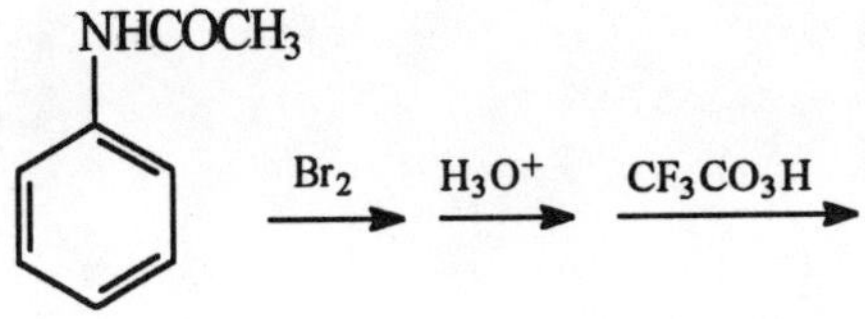

25.G Answers to Supplementary Problems

S1. a.

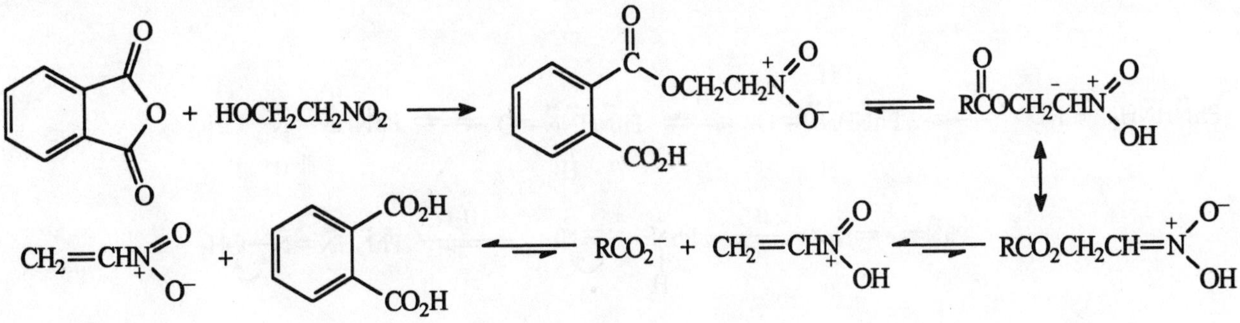

phthalic anhydride + $HOCH_2CH_2NO_2$ ⟶ $C_6H_4(CO_2H)C(=O)OCH_2CH_2\overset{+}{N}(=O)O^-$ ⇌ $RCOCH_2\overset{-}{C}H\overset{+}{N}(=O)OH$ (RC(=O)OCH₂CHN(=O)OH)

↕

$RCO_2CH_2CH=\overset{+}{N}(O^-)OH$ ⇌ $RCO_2^- + CH_2=CH\overset{+}{N}(=O)OH$ ⇌ $CH_2=CH\overset{+}{N}(=O)O^-$ + $C_6H_4(CO_2H)_2$

b.

$$(CH_3)_2\ddot{N}H + CH_2{=}CHNO_2 \rightleftharpoons \left[(CH_3)_2\overset{+}{N}HCH_2\overset{-}{C}H\overset{+}{N}(=O)O^- \longleftrightarrow (CH_3)_2\overset{+}{N}HCH_2CH{=}\overset{+}{N}(O^-)O^-\right]$$

$$\xrightleftharpoons{-H^+} (CH_3)_2NCH_2\overset{-}{C}H\overset{+}{N}(=O)O^- \xrightleftharpoons{H^+} (CH_3)_2NCH_2CH_2NO_2$$

S2.

a. Cyclohexyl–$\overset{+}{N}(=O)O^-$ (with H on the ring carbon) $\xrightleftharpoons{OH^-}$ [carbanion $\longleftrightarrow$ nitronate] $\xrightleftharpoons{H^+}$ cyclohexylidene–$\overset{+}{N}(OH)O^-$ (H_2O: attacks)

b. $\rightleftharpoons$ $H_2\overset{+}{O}$–C(cyclohexyl)–N(OH)O^- $\xrightleftharpoons{-H^+}$ HO–C–$\ddot{N}$(OH)O^- $\xrightleftharpoons{H^+}$ HO–C–$\ddot{N}$(OH)OH $\xrightleftharpoons{H^+}$ HO–C–$\ddot{N}$(:OH)$\overset{+}{O}H_2$

$\xrightleftharpoons{-H_2O}$ OH–C–N$=\overset{+}{O}H$ $\rightleftharpoons$ OH–C–N=O (H^+) $\longrightarrow$ cyclohexanone=$\overset{+}{O}H$ $\xrightleftharpoons{H^+}$ cyclohexanone

$$+ HNO \longrightarrow 1/2\,(H_2O + N_2O)$$

S3.

$$CH_3O\text{-}C_6H_4\text{-}CH_2CH_2\overset{O}{\overset{\|}{C}}N_3 \xrightarrow[EtOH]{\Delta} [ArCH_2CH_2N{=}C{=}O] \xrightarrow{EtOH} ArCH_2CH_2NH\overset{O}{\overset{\|}{C}}OEt$$

S4.

$$Ph\ddot{N}H\ddot{N}H_2 + NO^+ \longrightarrow Ph\ddot{N}H\overset{+}{N}H_2\ddot{N}{=}O \xrightleftharpoons{-H^+} Ph\ddot{N}H\ddot{N}H\ddot{N}{=}O \rightleftharpoons Ph\ddot{N}H\ddot{N}{=}\ddot{N}{-}OH$$

$$\xrightleftharpoons{H^+} Ph\ddot{N}H\ddot{N}{=}\ddot{N}{-}\overset{+}{O}H_2 \xrightleftharpoons{-H_2O} Ph\ddot{N}(H){-}\overset{+}{N}{\equiv}N: \xrightleftharpoons{-H^+} Ph\ddot{N}{=}\overset{+}{N}{=}\ddot{N}:$$

S5. a.

HO_2C–C$_6H_5$ $\xrightarrow[H_2SO_4]{HNO_3}$ $\xrightarrow[FeCl_3]{Cl_2}$ 3-NO_2-5-Cl-C$_6H_3$-CO_2H $\xrightarrow[HCl]{Zn}$ $\xrightarrow[HBr]{NaNO_2}$ 3-$^{+}N_2\,Br^-$-5-Cl-C$_6H_3$-CO_2H $\xrightarrow{CuBr}$ 3-Br-5-Cl-C$_6H_3$-CO_2H

b. NH_2–C$_6H_5$ $\xrightarrow{Cl_2}$ $\xrightarrow[HBF_4]{NaNO_2}$ 2,4,6-trichloro-C$_6H_2$-$^{+}N_2\,BF_4^-$ $\xrightarrow{KSC(=S)OC_2H_5}$ $\xrightarrow{NaOH}$ 2,4,6-trichloro-C$_6H_2$-SH

c. OCH_3–C$_6H_5$ $\xrightarrow[H_2SO_4]{HNO_3}$ 4-NO_2-C$_6H_4$-OCH_3 $\xrightarrow[\Delta]{D_2SO_4}$ 2,3,5,6-tetradeutero-4-NO_2-C$_6D_4$-OCH_3 $\xrightarrow[FeCl_3]{Cl_2}$ 2,6-Cl$_2$-3,5-D$_2$-4-NO_2-C$_6$-OCH_3 $\xrightarrow[DCl]{Zn}$ $\xrightarrow[DCl]{NaNO_2}$ $\xrightarrow{H_3PO_2}$ 2,6-Cl$_2$-3,5-D$_2$-4-H-C$_6$-OCH_3

d. NO_2–C$_6H_5$ $\xrightarrow[FeCl_3]{Cl_2}$ 3-Cl-C$_6H_4$-NO_2 $\xrightarrow[Ac_2O]{Zn}$ 3-Cl-C$_6H_4$-$NHCOCH_3$ $\xrightarrow[CO,\ HCl]{AlCl_3}$ 3-Cl-4-CHO-C$_6H_3$-$NHCOCH_3$ $\xrightarrow{H_3O^+}$ $\xrightarrow[HBF_4]{NaNO_2}$ 3-Cl-4-CHO-C$_6H_3$-$^{+}N_2\,BF_4^-$ $\xrightarrow{\Delta}$ 4-F-2-Cl-C$_6H_3$-CHO

e.

$$2\ C_6H_5NO_2 \xrightarrow[\text{Ru/C KOH}]{H_2NNH_2} C_6H_5NHNHC_6H_5 \xrightarrow[\Delta]{H^+} H_2N{-}C_6H_4{-}C_6H_4{-}NH_2 \xrightarrow[HBF_4]{NaNO_2} \xrightarrow{2\ CuCN} NC{-}C_6H_4{-}C_6H_4{-}CN \xrightarrow{H_3O^+} HO_2C{-}C_6H_4{-}C_6H_4{-}CO_2H$$

f.

$$C_6H_5Cl \xrightarrow[H_2SO_4]{HNO_3} O_2N{-}C_6H_4{-}Cl \xrightarrow[HCl]{Zn} H_2N{-}C_6H_4{-}Cl \xrightarrow[HCl]{NaNO_2} \xrightarrow{NaOH} \xrightarrow{C_6H_5C(CH_3)_3} Cl{-}C_6H_4{-}C_6H_4{-}C(CH_3)_3$$

S6.

$$CH_3C(=O)N(CH_3)N{=}O + HO^- \rightleftharpoons CH_3C(O^-)(OH){-}N(CH_3)N{=}O \rightleftharpoons CH_3COOH + CH_3N{=}NO^-$$

$$CH_3N{=}NO^- \rightleftharpoons CH_3N{=}NOH + CH_3CO^-$$

$$CH_3N{=}NOH \underset{OH^-}{\rightleftharpoons} HO{-}N{=}N{-}CH_2^- \underset{-OH^-}{\rightleftharpoons} [\,^-N{=}\overset{+}{N}{=}CH_2 \longleftrightarrow :N{\equiv}\overset{+}{N}{-}CH_2^-\,]$$

S7.

$$^-O{-}C(=O){-}CH{=}\overset{+}{N}{=}N^- + H{-}OH \rightleftharpoons {}^-O{-}C(=O){-}CH_2{-}\overset{+}{N}{\equiv}N \longrightarrow CO_2 + [\,^-CH_2{-}\overset{+}{N}{\equiv}N \longleftrightarrow CH_2{=}\overset{+}{N}{=}N^-\,]$$

S8. a. [structure: 3-fluorotoluene; F, CH_3]

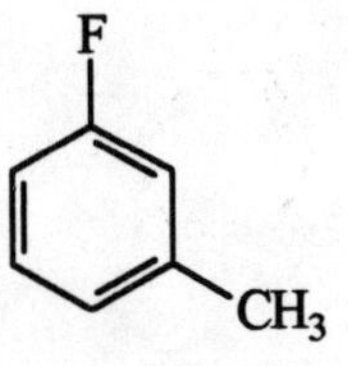

b. [structure: NHCOCH$_3$, O_2N, CH_3]

c. [structure: azoxy compound with two 4-chlorophenyl groups; Cl, O^-, N^+, N, Cl]

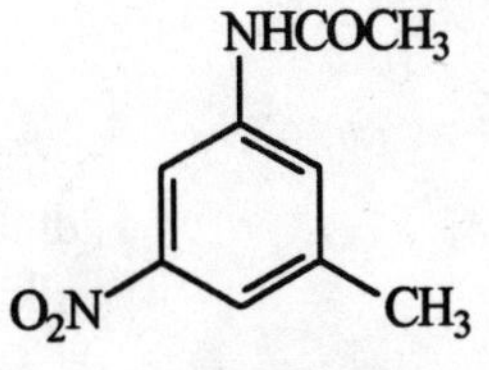

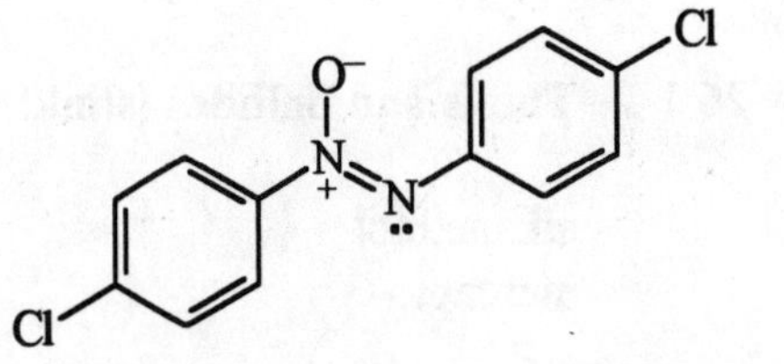

d. [structure: azo compound; H_3C, N=N, phenyl]

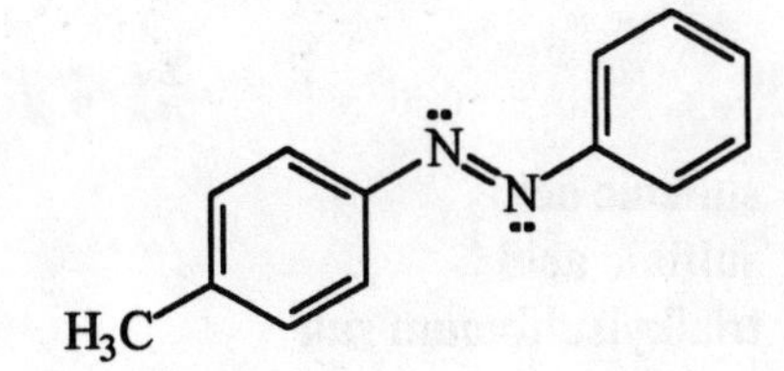

e. [structure: :N=O, HO_2C]

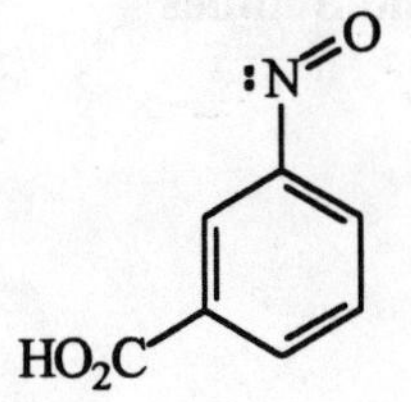

f. [structure: NO_2, Br]

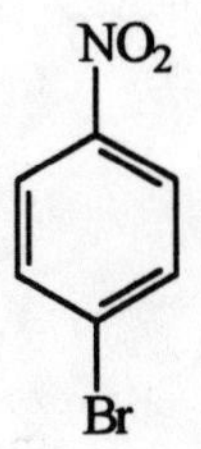

26. SULFUR, PHOSPHORUS AND SILICON COMPOUNDS

26.A. Chapter Outline and Important Terms Introduced

26.1 Thiols and Sulfides (stink!)

alkanethiol
mercapto-
dialkyl sulfide
alkylthio-

26.2 Preparation of Thiols and Sulfides

26.3 Reactions of Thiols and Sulfides

pK_a of RSH = 10.5
sulfide
sulfoxide
sulfone
sulfenic acid
sulfinic acid
trialkylsulfonium salt

26.4 Sulfate Esters

alkylsulfuric acid, $ROSO_3H$
dialkyl sulfate, $ROS(=O)_2OR$

26.5 Sulfonic Acids

A. Alkanesulfonic Acids
B. Arenesulfonic Acids
nucleophilic aromatic substitution
sulfonyl halides

26.6 Phosphines and Phosphonium Salts

alkyl-, dialkyl-, trialkylphosphines

26.7 Phosphates and Phosphonate Esters

phosphoric acid and esters
pyrophosphoric acid and esters
phosphorous acid and esters
alkylphosphonic acids and esters
pk_a's of H_3PO_4: 2.15, 7.2, 12.4
Arbuzov-Michaelis reaction

26.8 Sulfur- and Phosphorus-Stabilized Carbanions

phosphorus ylids, the Wittig reagent (phosphonium ion pK_a = 15-18)
phosphonate carbanions, Horner-Emmons reaction
sulfur-stabilized carbanions
sulfonium ylids, epoxide formation
anions of sulfones (pK_a= 31) and sulfoxides (pK_a= 35)

26.9 Organosilicon Compounds: Structure and Properties

26.10 Organosilicon Compounds: Preparation

26.11 Organosilicon Compounds: Reactions

A. Nucleophilic Substitution at Silicon
siliconate ions pseudorotation
B. Electrophilic Cleavage of the Carbon-Silicon Bond
electrophilic substitution
C. Silyl Ethers as Protecting Groups

26.B. Important Reactions Introduced

Formation of thiols and sulfides by nucleophilic substitution (26.2)

Equation: $RX + R'S^-M^+ \longrightarrow RSR' + MX$

Generality: R = 1° or 2° alkyl; R′ = H, alkyl, or aryl

Key features: S_N2 displacement reaction (not for 3° RSH)
with R′ = H, sulfide formation (RSR) is often observed; excess SH^- necessary
thiourea, $(NH_2)_2C{=}S$, followed by hydrolysis of product salt, used in place of NaSH to avoid sulfide formation

Reaction of Grignard reagents with sulfur (26.2)

Equation: $RMgX + S_8 \longrightarrow RSMgx \longrightarrow RSH$

Key features: useful method for making 3° RSH or ArSH

Oxidation of thiols (26.3)

Equation:

$$2\ RSH \xrightarrow{\text{mild oxidation}} RSSR \xrightarrow[\text{oxidation}]{\text{strong}} 2\ RSO_3H$$

Generality: R = alkyl or aryl
mild oxidation = I_2 or air
strong oxidation = $KMnO_4$ or HNO_3 (Cl_2/HNO_3 gives RSO_2Cl)

Reactions of sulfides as electrophiles (26.3)

Reduction:

Equation: $RSSR \xrightarrow{[H]} 2\ RSH$

Generality: [H] = Na or Li in liquid NH_3

Thiolation of enolates

Equation:

$$RS{-}SR + \;\; >C{=}C{<}^{O^-} \longrightarrow RS{-}\overset{|}{\underset{|}{C}}{-}\overset{O}{\overset{||}{C}}{-} \;+\; RS^-$$

Generality: enolate = ester or ketone enolate

Oxidation of sulfides (26.3)

Equation:

$$R{-}S{-}R \xrightarrow{[Ox]} R{-}\overset{O}{\overset{||}{S}}{-}R \xrightarrow{[Ox]} R{-}\overset{O}{\overset{||}{\underset{\underset{O}{||}}{S}}}{-}R$$

sulfoxide *sulfone*

Generality: R = alkyl or aryl; [Ox] = H_2O_2, RCO_3H, $NaIO_4$

Key feature: $NaIO_4$ best reagent for stopping oxidation at sulfoxide level

Thermal elimination of sulfoxides (26.3)

Equation:

$$R{-}\overset{O}{\overset{||}{S}}{-}\overset{|}{C}{-}\overset{H}{\overset{|}{C}}{-} \longrightarrow R{-}S{-}OH \;+\; >C{=}C<$$

Key features: mechanism is syn elimination
initial product is sulfenic acid, which disproportionates to sulfinic acid and sulfide

Formation of sulfonium salts (26.3)

Equation: $RSR + R'X \longrightarrow R_2^+SR'X^-$

Generality: R, R′ = alkyl: X^- = good leaving group (e.g. I^-)

Formation and reductive cleavage of dithioacetals (26.3)

Equation:

$$R{-}\overset{O}{\overset{||}{C}}{-}R' \;+\; HS(CH_2)_3SH \xrightarrow{BF_3} \text{(1,3-dithiane, } R,R'\text{ on C2)} \xrightarrow{H_2/Ni} R{-}\overset{H\;\;H}{C}{-}R'$$

Generality: R, R′ = H, alkyl, or aryl

Key features: useful method for deoxygenation of ketones and aldehydes under neutral conditions

Formation of alkanesulfonic acids (26.5)

Equation: $RX + Na^{+}HSO_3^{-} \longrightarrow RSO_3H + NaX$

Generality: R = 1° or 2° alkyl; X = good leaving group

Formation of α-hydroxysulfonic acids (26.5)

Equation: $R{-}C({=}O){-}R' \xrightarrow{NaHSO_3} R{-}C(OH)(R'){-}SO_3^{-}\,Na^{+}$

Generality: R, R′ = H or alkyl

Key features: seldom works for ketones
reaction is reversible with acid or base

Formation of sulfonate esters (26.5)

Equation: $RSO_3H + PCl_5 \longrightarrow RSO_2Cl \xrightarrow[\text{pyridine}]{R'OH} RSO_3R'$

Generality: R = alkyl or aryl
$SOCl_2$ can be used in place of PCl_5 as well

Electrophilic aromatic sulfonation (26.5)

Equation: $ArH + SO_3 \longrightarrow ArSO_3H$ *arenesulfonic acid*

$ArH + ClSO_3H \longrightarrow ArSO_2Cl$ *arenesulfonyl chloride*

Generality: reagent is "fuming sulfuric acid" (SO_3 in conc. H_2SO_4) or chlorosulfonic acid (HSO_3Cl)

Key features: mechanism is electrophilic aromatic substitution
reaction can be reversed: heating $ArSO_3H$ in dilute H_2SO_4 gives ArH

Nucleophilic aromatic substitution of arenesulfonic acid (26.5)

Equation: $ArSO_3H + Na^{+}X^{-} \xrightarrow[\Delta]{} ArX + NaHSO_3$

Generality: $X^{-} = OH^{-}$ or CN^{-}

Key features: requires vigorous temperatures (300°C)
sulfonate group useful as a blocking group in synthesis of substituted aromatic compounds

Reactions of arenesulfonyl halides (26.5)

Equation: $ArSO_2Cl + Nu: \longrightarrow ArSO_2Nu$

Generality: Nu: = Ar′H: Friedel-Crafts reaction, $AlCl_3$ as catalyst, to give sulfone as product
= HNR_2: amide formation, to give sulfonamide as product
= HOR′: ester formation, to give sulfonate ester as product (see above)
Reduction (Zn, H_2O) gives $ArSO_2H$, an arenesulfinic acid

Alkylation of sulfinic acid salts (26.5)

Equation: $RSO_2^- \ Na^+ + R'X \longrightarrow R{-}S(=O)_2{-}R' + NaX$

Generality: R = alkyl or aryl; R′ = 1° or 2° alkyl

Key features: byproduct is sulfinate ester, $R{-}S(=O){-}OR'$

Formation of phosphines from Grignard reaction (26.6)

Equation: $3\ RMgX + PX_3 \longrightarrow R_3P: + 3\ MgX_2$

Generality: R = alkyl or aryl; X = halogen
other organometallic reagents can be used (eg. RLi)

Formation of phosphate esters (26.7)

Equation: $POCl_3 + 3\ ROH \longrightarrow RO{-}P(=O)(OR){-}OR$

Generality: R = H, alkyl, or aryl

with less than 3 equivalents of ROH, can stop at $ROP(=O)Cl_2$ or $(RO)_2P(=O)Cl$

hydrolysis of phosphate esters occurs with C-O bond cleavage in acid, with P-O bond cleavage in base (mostly)

Arbuzov-Michaelis reaction (26.7)

Equation: $RX + (R'O)_3P: \longrightarrow R{-}P(=O)(OR'){-}OR' + R'X$

Generality: R, R′ = alkyl; X = halogen

Key features: reactions proceeds via S_N2 displacement and trialkoxyphosphonium salt intermediate

Formation and reactions of sulfur- and phosphorus-stabilized carbanions (26.8)

Equation: $RCH_2{-}Y \xrightarrow{[base]} {}^{-}\underset{R}{\underset{|}{C}}H{-}Y \xrightarrow{["E"]} E{-}\underset{R}{\underset{|}{C}}H{-}Y \longrightarrow$

Generality: [base] = LDA or *n*-BuLi
R = H, alkyl, or aryl
Y = $^{+}PR'_3$: phosphorus ylide = Wittig reagent; E = ketone or aldehyde; product = alkene
Y = $PO_3R'_2$: phosphonate anions = Horner-Emmons reagent; E = ketone or aldehyde; product = alkene
Y = SR′, SOR′, SO_2R' (or dithiane): sulfide, sulfoxide, or sulfone α-anions; can be alkylated with R′X, carbonyl compounds, etc.
Y = $^{+}SR'_2$: sulfonium ylide; E = aldehyde or ketone; product is epoxide

Nucleophilic substitution reactions at silicon (26.11.A)

Equation: $R_3SiX + Nu{:} \longrightarrow NuSiR_3 + X^-$

Generality: X = good leaving group; Nu: = good nucleophile

Key features: reaction involves addition/elimination process, with intermediacy of pentavalent "siliconate" ion
reaction usually proceeds with inversion of configuration, but retention via "pseudorotation" is also possible
S_N1-type mechanism **not** observed

Electrophilic cleavage of carbon-silicon bonds (26.11.B)

Equation: $RSiR_3 + E^+X^- \longrightarrow R{-}E + R_3SiX$

Generality: R = aryl, vinyl, or allyl group; $E^+ = H^+$ or other electrophile

Key features: for R = allyl, allylic rearrangement occurs during substitution reaction
useful method for formation of a wide variety of carbon-carbon bonds

Formation and hydrolysis of silyl ethers (26.11.C)

Equation: $ROH + R'_3SiX \longrightarrow ROSiR'_3 + HX$

$ROSiR'_3 + H_2O \longrightarrow ROH + HOSiR'_3$

Generality: R, R′ = alkyl (usually); X = Cl or O_3SCF_3
formation of silyl ether by reaction of RO^- or with 3° amine catalysts
hydrolysis of silyl ether catalyzed with H^+ or F^-

Key feature: because of stability of silyl ethers to basic reagents, they are useful in alcohol protecting groups

26.C. Important Concepts and Hints

The chemistry of sulfur, phosphorus, and silicon compounds is becoming increasingly important because of the invention of many functional group interconversions and carbon-carbon bond-forming reactions that involve these intermediates. In many respects, analogies can be drawn between thiols/sulfides/disulfides and alcohols/ethers/peroxides, and between tetravalent silicon compounds and tetrahedral carbon compounds. However, there are no oxygen counterparts to the more highly oxidized sulfur compounds: sulfoxides, sulfones, and sulfinic and sulfonic acids. In a similar manner, the organic chemistry of phosphorus ranges from the reduced derivatives (phosphines) to the higher oxidation states (phosphonic and phosphoric acids). Silicon derivatives have a less varied chemical behavior, undergoing primarily substitution reactions.

The major utility of sulfur- and phosphorus-containing compounds as carbon-carbon bond-forming reagents lies in the diverse ways that these functional groups can stabilize adjacent carbanions. With strong base (*n*-butyllithium or LDA) a proton can be removed from carbons that are adjacent to sulfide, sulfoxide, sulfone, sulfonium ion, phosphonium ion, and phosphonate groups (among others!). These carbanionic reagents react with alkyl halides or carbonyl derivatives to form new carbon-carbon bonds. Particularly important examples are the Wittig reaction, and the use of dithioacetal anions as synthetic equivalents to acyl anions. In contrast, for carbon-carbon bond formation, silane derivatives derive their reactivity from silicon's ability to stabilize adjacent positive charge.

More "synthetic methods" that use sulfur, phosphorus, and silicon functional groups are being invented all the time – perhaps you can think of some new ones yourself!

26.D. Answers to Exercises

26.1 a. $CH_3CH_2SCH(CH_3)_2$ b. $CH_3CH_2CH_2CH_2SH$ c. $CH_3(CH_2)_4CH(SCH_3)CH_2CH_3$

d. $C_6H_5SC_6H_5$ e. $CH_3CH_2CH(SH)CH_2CH_3$ f. $CH_3CH{=}S$ g. $CH_3C(=O)SCH_2CH_3$

26.2 a. $(CH_3)_3CCH_2Br \xrightarrow{Mg} \xrightarrow{S_8} \xrightarrow{H^+} (CH_3)_3CCH_2SH$ *(neopentyl bromide is too hindered to undergo displacement reaction)*

b. $2\ CH_3CH_2CH_2OH \xrightarrow{PBr_3} 2\ CH_3CH_2CH_2Br \xrightarrow{Na_2S} (CH_3CH_2CH_2)_2S$

26.3 cyclohexanecarbonitrile (CN) $\xrightarrow{LDA}$ carbanion (NC, −) $\xrightarrow{(CH_3S)_2}$ 1-(methylthio)cyclohexanecarbonitrile (NC, SCH_3) $\xrightarrow{NaIO_4}$ sulfoxide (NC, $S(=O)CH_3$) $\xrightarrow{\Delta}$ 1-cyclohexenecarbonitrile (CN)

26.4

$$CH_3CH{=}\overset{+}{O}{-}\bar{B}F_3 + HS(CH_2)_3SH \rightleftharpoons CH_3C(H)(O\bar{B}F_3){-}\overset{+}{S}H(CH_2)_3SH \underset{}{\overset{-H^+}{\rightleftharpoons}} CH_3CH(O\bar{B}F_3){-}S(CH_2)_3SH$$

$$\overset{H^+}{\rightleftharpoons} CH_3CH(\overset{+}{O}H\bar{B}F_3){-}S(CH_2)_3SH \rightleftharpoons CH_3CH{=}\overset{+}{S}(CH_2)_3SH + HO\bar{B}F_3 \rightleftharpoons \text{(cyclic sulfonium, 2-methyl-1,3-dithiane-SH}^+) \overset{-H^+}{\rightleftharpoons} \text{2-methyl-1,3-dithiane}$$

26.5 a.

$$CH_3(CH_2)_3SH + CH_3Br \xrightarrow{NaOH} CH_3(CH_2)_3SCH_3 \xrightarrow{H_2O_2} CH_3(CH_2)_3\overset{O}{\overset{\|}{S}}CH_3 \xrightarrow{150^\circ} CH_3CH_2CH{=}CH_2 + [CH_3SOH]$$

b. $$2\ CH_3(CH_2)_3SH \xrightarrow{I_2,\ KI} CH_3(CH_2)_3SS(CH_2)_3CH_3 \xrightarrow{Li,\ NH_3} 2\ CH_3(CH_2)_3SH$$

c. $$CH_3(CH_2)_3SH + CH_3CH_2I \xrightarrow{NaOH} CH_3(CH_2)_3SCH_2CH_3 \xrightarrow{CH_3I} CH_3(CH_2)_3\overset{+}{S}(CH_3)CH_2CH_3\ I^-$$

d. $$CH_3(CH_2)_3SH + CH_3CH_2Br \xrightarrow{NaOH} CH_3(CH_2)_3SCH_2CH_3 \xrightarrow{NaIO_4} CH_3(CH_2)_3\overset{O}{\overset{\|}{S}}CH_2CH_3$$

26.6 $$HO{-}SO_2{-}OH \rightleftharpoons HO{-}SO_2{-}O^-\ H^+ \ ; \ F{-}SO_2{-}OH \rightleftharpoons F{-}SO_2{-}O^-\ H^+$$

The OH group in sulfuric acid is replaced by the more strongly electron-withdrawing F substituent in fluorosulfonic acid.

26.7 $$(CH_3)_2CHCH_2CH_2Cl + NaHSO_3 \longrightarrow (CH_3)_2CHCH_2CH_2SO_3H \xrightarrow{PCl_5}$$

$$(CH_3)_2CHCH_2CH_2SO_2Cl + CH_3OH \xrightarrow{pyridine} (CH_3)_2CHCH_2CH_2SO_3CH_3$$

26.8 In the toluenesulfonic acids, you would expect that the *o*-isomer would be less stable than the others because of the steric interaction between the methyl and the sulfonate groups. On the other hand, it is unlikely that there would be much difference between the *m*- and *p*-isomers, neither of which has such a steric interaction. The fact that there is a significantly greater amount of *p*-isomer than *m*-isomer in the product of reaction at 100°C suggests that the mixture still does not reflect the thermodynamic ratio.

26.9 SO_3H, Cl (para) —fuming HNO_3→ SO_3H, O_2N, NO_2, Cl —H_3O^+, Δ→ O_2N, NO_2, Cl

26.10 CH_3 —SO_3, H_2SO_4→ CH_3, SO_3H —HNO_3, H_2SO_4→ CH_3, NO_2, SO_3H —KOH, Δ→ CH_3, NO_2, OH —Zn, HCl→ —HNO_2→ —H_2O, Δ→ CH_3, OH, OH

26.11 Br–C₆H₄–SO_3Na —PCl_5→ Br–C₆H₄–SO_2Cl

a. Br–C₆H₄–$SO_2OCH_2CH_2CH_2CH_3$ b. Br–C₆H₄–SO_2–C₆H₄–CH_3

c. Br–C₆H₄–SO_2–CH_2–C₆H₅ d. Br–C₆H₄–SH

26.12 (b) and (e) are chiral: (c) and (d) could be chiral if the oxygens were isotopically different:

Ph, CH_3, S=O (lone pair); Ph, CH_3O, S=O (lone pair); Ph, CH_3, S(=O)=O*; Ph, CH_3O, S(=O)=O*

26.13 $C_6H_5MgCl + PCl_3 \longrightarrow (C_6H_5)_3P + ClCH_2C_6H_5 \longrightarrow (C_6H_5)_3{}^+PCH_2C_6H_5\ Cl^-$

$\xrightarrow{n\text{-BuLi}} (C_6H_5)_3PCHC_6H_5 \xrightarrow{CH_3CHO} (C_6H_5)_3P{=}O + CH_3CH{=}CHC_6H_5$

26.14

$$(CH_3CH_2O)_3P: + \underset{(trace)}{CH_3I} \longrightarrow CH_3\overset{O}{\overset{\|}{P}}(OCH_2CH_3)_2 + CH_3CH_2I$$

$$\xrightarrow{(CH_3CH_2O)_3P:} \underset{(major\ product)}{CH_3CH_2\overset{O}{\overset{\|}{P}}(OCH_2CH_3)_2} + CH_3CH_2I$$

26.15 a. $2\,C_2H_5OH + POCl_3 \xrightarrow{\text{pyridine}} (C_2H_5O)_2\overset{O}{\overset{\|}{P}}Cl \xrightarrow{CH_3ONa} (C_2H_5O)_2\overset{O}{\overset{\|}{P}}OCH_3$

b. $3\ i\text{-}C_4H_9OH + POCl_3 \xrightarrow{\text{pyridine}} (i\text{-}C_4H_9O)_3P{=}O$

c. $3\,C_2H_5OH + PCl_3 \xrightarrow{\text{pyridine}} (C_2H_5O)_3P: \xrightarrow{CH_3CH_2CH_2I} (C_2H_5O)_2\overset{O}{\overset{\|}{P}}CH_2CH_2CH_3 + CH_3CH_2I$

d. $\underset{from\ c.}{(C_2H_5O)_3P:} \xrightarrow[\Delta]{C_2H_5I\ (catalytic)} (C_2H_5O)_2\overset{O}{\overset{\|}{P}}CH_2CH_3 \xrightarrow[2.\ H^+]{1.\ NaOH} C_2H_5O\underset{OH}{\overset{O}{\overset{\|}{P}}}CH_2CH_3$

26.16 a. $(CH_3O)_3P: \xrightarrow{CH_3I} (CH_3O)_2\overset{O}{\overset{\|}{P}}CH_3 \xrightarrow{BuLi} (CH_3O)_2\overset{O}{\overset{\|}{P}}CH_2Li \xrightarrow{CH_3\overset{O}{\overset{\|}{C}}CH_3} (CH_3)_2\overset{OH}{\overset{|}{C}}CH_2\overset{O}{\overset{\|}{P}}(OCH_3)_2$

b. $(CH_3O)_2\overset{O}{\overset{\|}{P}}CH_2Li \xrightarrow{(CH_3O)_2C{=}O} CH_3O\overset{O}{\overset{\|}{C}}CH_2\overset{O}{\overset{\|}{P}}(OCH_3)_2 \xrightarrow[2.\ CH_3CH{=}O]{1.\ NaOCH_3} CH_3CH{=}CH\overset{O}{\overset{\|}{C}}OCH_3$

26.17

a.

1,3-dithiane (S, S ring) $\xrightarrow[2.\ C_2H_5I]{1.\ BuLi}$ 2-ethyl-1,3-dithiane (S, S) $\xrightarrow[2.\ CH_3(CH_2)_2CH{=}O]{1.\ BuLi}$ 2-ethyl-2-(1-hydroxybutyl)-1,3-dithiane (S, S, OH) $\xrightarrow[H_2O]{HgCl_2}$ 4-hydroxy-3-heptanone (O, OH)

b.

S S — 1. BuLi; 2. CH_3I → S S — 1. BuLi; 2. O → S S OH — $HgCl_2$, H_2O → O OH

26.18 a. $CH_3\overset{O}{\overset{\|}{S}}CH_2\overset{OH}{\overset{|}{C}}HC_6H_5$ b. O c. S S OH

26.19 a. $C_6H_5MgBr + (CH_3)_3SiCl \longrightarrow C_6H_5Si(CH_3)_3 + MgX_2$

b. $2\ C_6H_5MgBr + SiCl_4 \longrightarrow (C_6H_5)_2SiCl_2 + MgX_2$

c. $CH_2{=}CHCH_2MgCl + (CH_3)_3SiCl \longrightarrow CH_2{=}CHCH_2Si(CH_3)_3 + MgCl_2$

d. $CH_2{=}CHMgCl + (CH_3)_3SiCl \longrightarrow CH_2{=}CHSi(CH_3)_3 + MgCl_2$

26.20 Two:

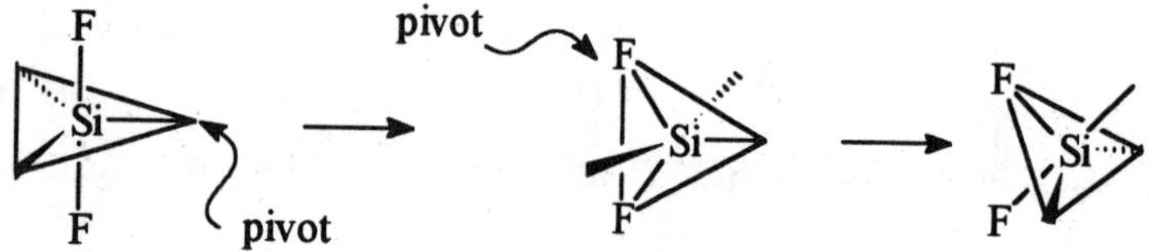

26.21 No pseudorotation:

Nu: a b Si–Cl c → [Nu–Si(a, b, c)–Cl] → Nu–Si(b, a, c) *inversion*

One pseudorotation:

a b Si–Cl c Nu: → [b Nu Si Cl a c] → (pivot on a) → [b–Si(a, Nu, c)–Cl] → b–Si(Nu, a, c)

retention

Two pseudorotations:

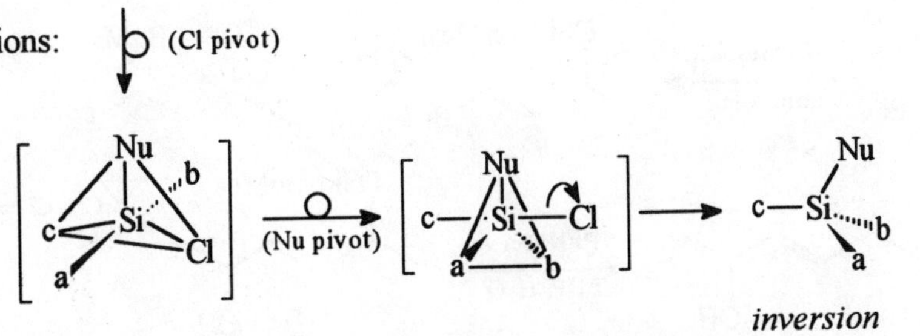
(Cl pivot)
Nu
b
Si
c
Cl
a
(Nu pivot)
inversion

26.22 a.

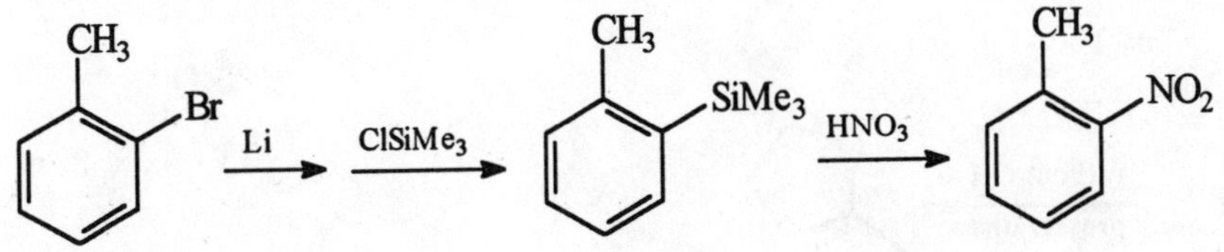
CH3
Br
Li
ClSiMe3
SiMe3
HNO3
NO2

b.

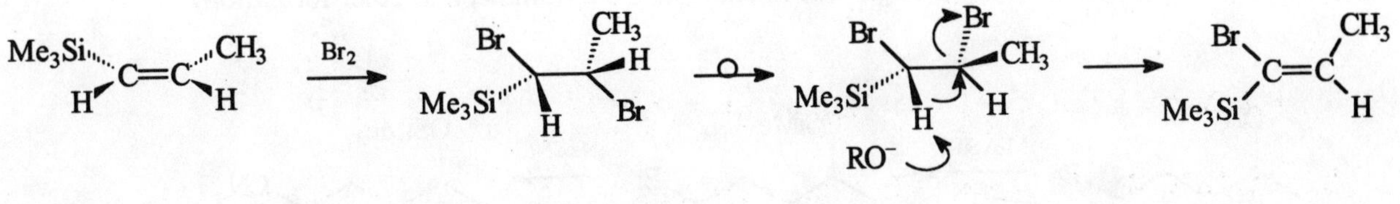
Me3Si
CH3
H
Br2
Br
RO⁻

26.23 a.

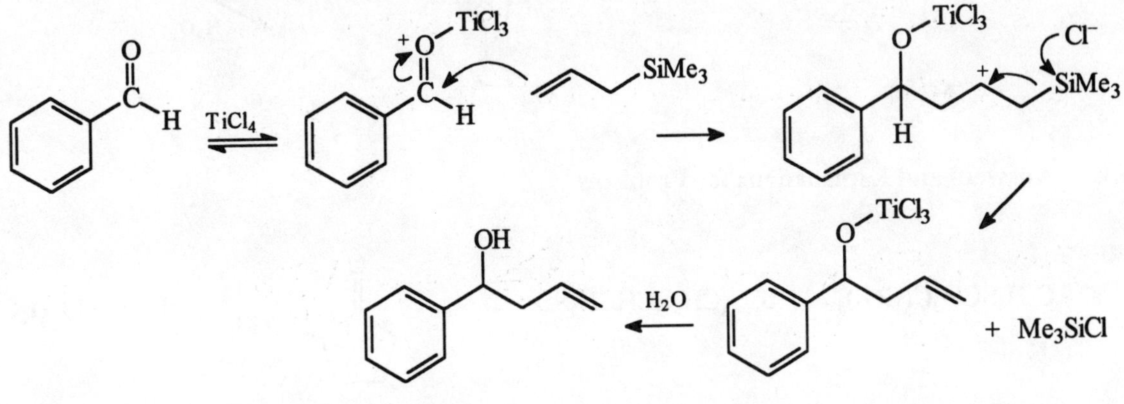
TiCl4
TiCl3
SiMe3
Cl⁻
OH
H2O
+ Me3SiCl

b.

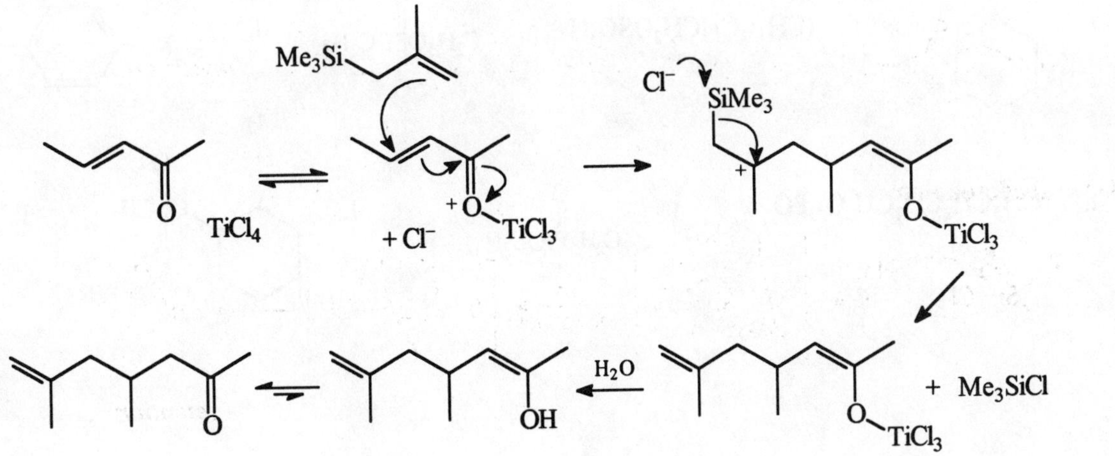
Me3Si
TiCl4
+ Cl⁻
TiCl3
Cl⁻
SiMe3
H2O
OH
+ Me3SiCl

26.24 a.

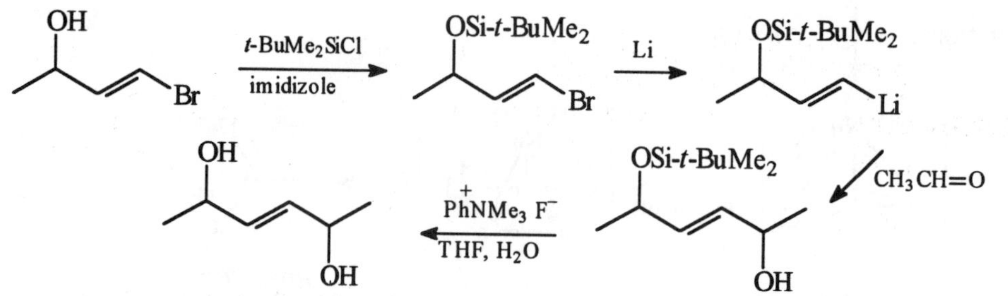

OH / Br (allylic alcohol) $\xrightarrow[\text{imidizole}]{t\text{-BuMe}_2\text{SiCl}}$ OSi-*t*-BuMe$_2$ / Br $\xrightarrow{\text{Li}}$ OSi-*t*-BuMe$_2$ / Li $\xrightarrow{CH_3CH{=}O}$ OSi-*t*-BuMe$_2$ / OH $\xrightarrow[\text{THF, }H_2O]{Ph\overset{+}{N}Me_3\ F^-}$ OH / OH

(any suitable reagent, like Li or Mg/ether, would react with the -OH to give an insoluble salt)

b.

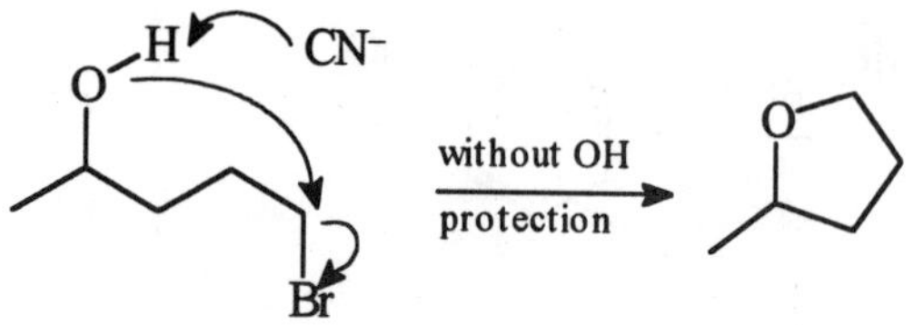

O–H ← CN⁻ ; Br $\xrightarrow[\text{protection}]{\text{without OH}}$ (2-methyltetrahydrofuran)

(cyanide ion is a strong enough base to catalyze the intramolecular ether formation)

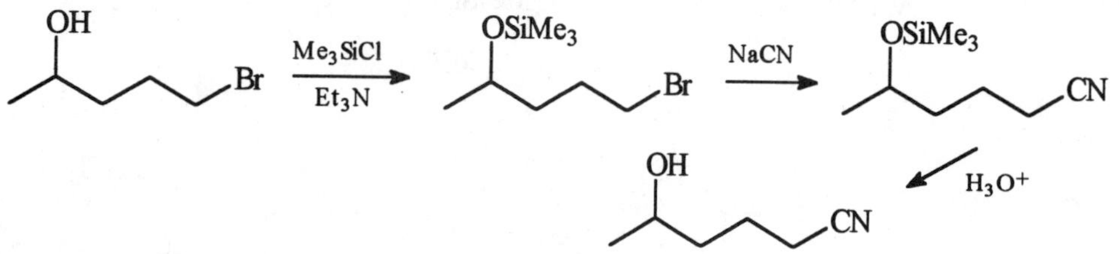

OH / Br $\xrightarrow[Et_3N]{Me_3SiCl}$ OSiMe$_3$ / Br $\xrightarrow{\text{NaCN}}$ OSiMe$_3$ / CN $\xrightarrow{H_3O^+}$ OH / CN

26.E. Answers and Explanations for Problems

1. a. $C_2H_5SCH_2C(CH_3)_3$ b. $(CH_3)_2CHCH_2SH$ c. 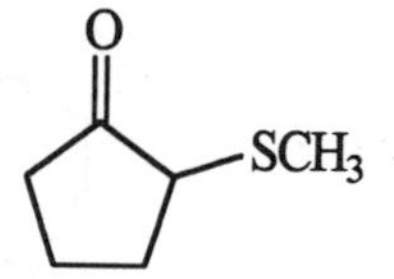 (2-(methylthio)cyclopentanone: O, SCH$_3$) d. $(CH_3CH_2CH_2CH_2)_2S_2$

e. 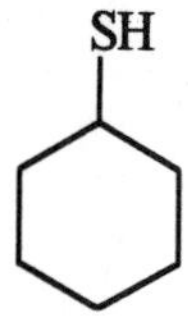 (cyclohexanethiol: SH) f. $(CH_3)_2CHCH_2OSO_3H$ g. $C_2H_5OSOC_2H_5$ (S with two =O) h. 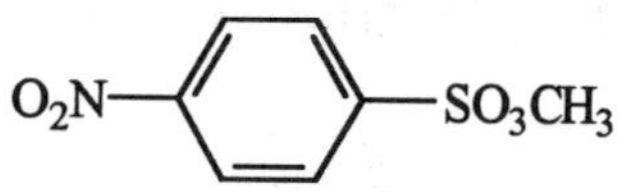 O_2N–C$_6$H$_4$–SO_3CH_3

i. $(CH_3CH_2CH_2CH_2O)_3PO$ j. $C_2H_5\overset{O}{\overset{\|}{P}}(OC_2H_5)_2$ k. 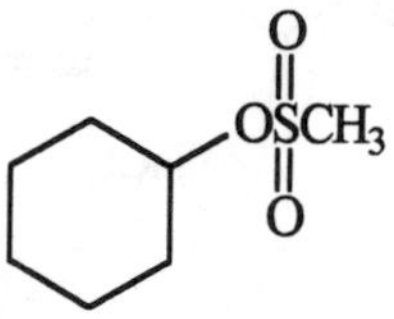 (cyclohexyl–OSCH$_3$, S with two =O) l. $(C_6H_5O)_3P$

m. 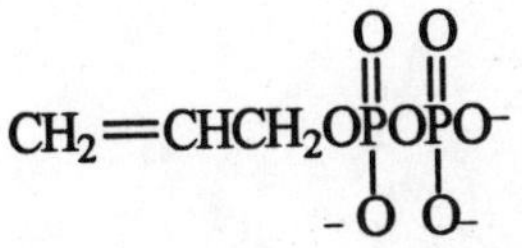 $CH_2{=}CHCH_2OP(=O)(O^-)OP(=O)(O^-)O^-$ n. $Me_3SiCH_2CH_2CH_3$ o. 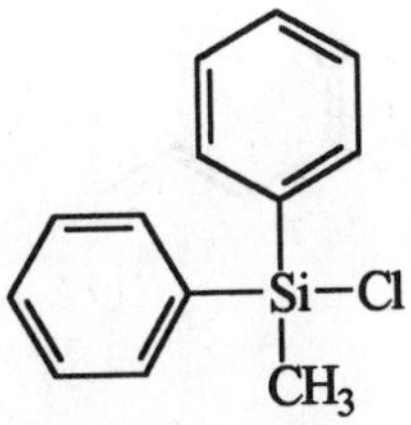 $(C_6H_5)_2Si(CH_3)-Cl$

2. a. 4-methyl-2-pentanethiol
   b. 2-methyl-3-methylthiopentane
   c. tetracyclopentylphosphonium bromide
   d. N-methyl-4-isopropylbenzenesulfonamide
   e. 3-(2,2-dimethylpropyl)oxycarbonyl-4-bromobenzenesulfonic acid
   (In a complex case such as this, the sulfonic acid group can be expressed as a prefix, sulfo; hence neopentyl 2-bromo-5-sulfobenzoate)
   f. *p*-nitrobenzenesulfonyl chloride
   g. methyl-*p*-bromophenyl sulfone
   h. dicyclopropyldisulfide
   i. diethyl ethylphosphonate
   j. cyclohexyltrimethylsilane

3. a. *t*-$BuOCH_3 + CH_3OSO_3^-K^+$ b. $CH_3CH_2SCH_3 + CH_3I$ c. $(CH_3)_3CCH_2SCH_2CH_3 + CH_3CH_2Br$
   d. 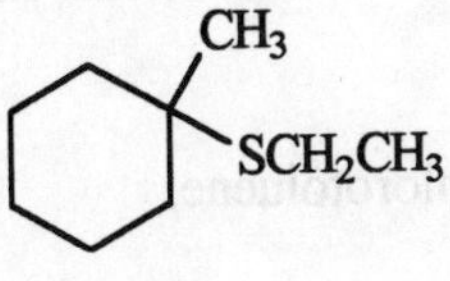 1-methyl-1-(SCH_2CH_3)cyclohexane $+ CH_3CH_2Cl$ e. 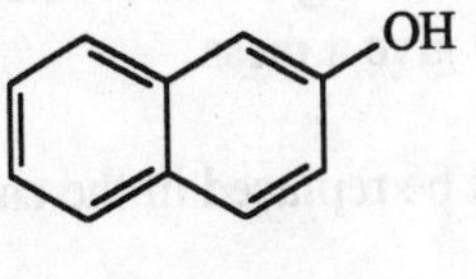 2-naphthol (OH) f. 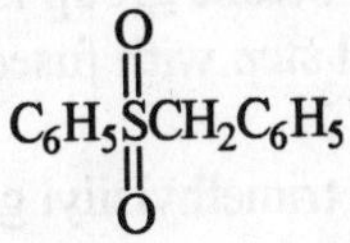 $C_6H_5S(=O)_2CH_2C_6H_5$
   g. 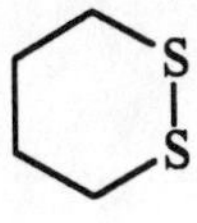 1,2-dithiane (S–S ring) h. 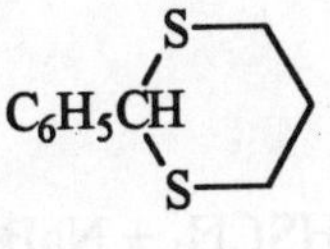 C_6H_5CH (1,3-dithiane) i. 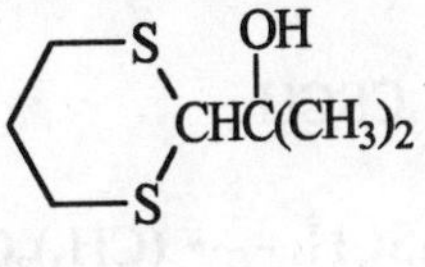 1,3-dithiane–$CH(OH)C(CH_3)_2$ j. 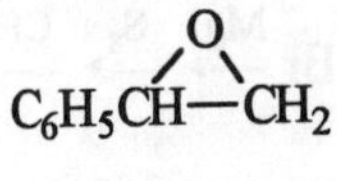 $C_6H_5CH{-}CH_2$ (epoxide, O) $+ CH_3SCH_3$

4. a. 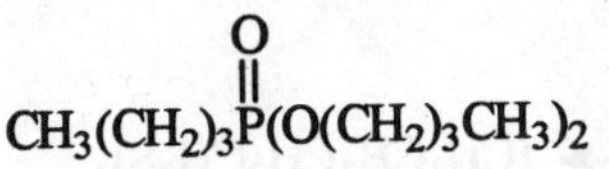 $CH_3(CH_2)_3P(=O)(O(CH_2)_3CH_3)_2$ b. 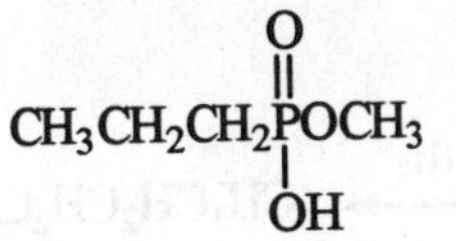 $CH_3CH_2CH_2P(=O)(OH)OCH_3$ c. 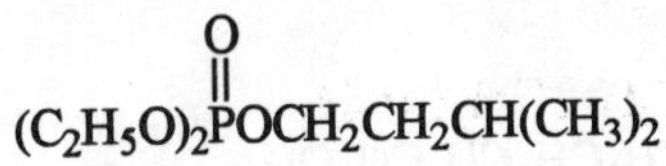 $(C_2H_5O)_2P(=O)OCH_2CH_2CH(CH_3)_2$
   d. $(CH_3O)_3P{=}O$ e. $(C_6H_5)_3{}^+PCH_2CH{=}CH_2\ Br^-$ f. 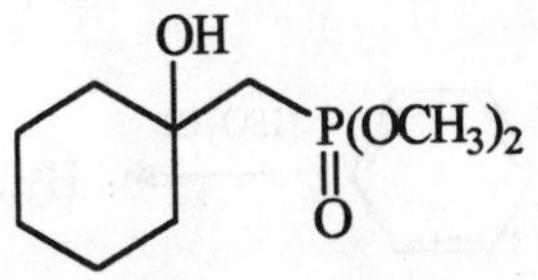 1-hydroxycyclohexyl–$CH_2P(=O)(OCH_3)_2$ (OH)
   g. 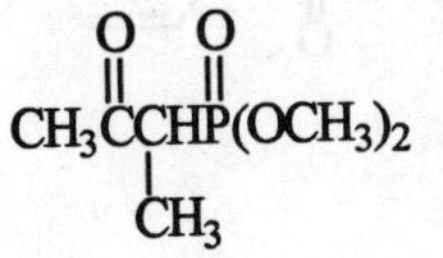 $CH_3C(=O)CH(CH_3)P(=O)(OCH_3)_2$

5. a. 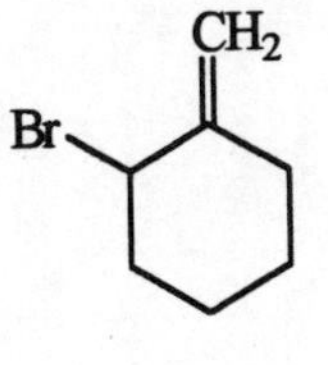 2-bromo-1-methylenecyclohexane (CH_2, Br) b. 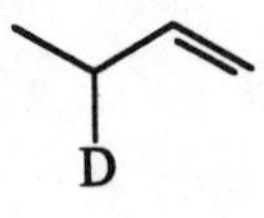 (D-substituted alkene) c. 3,5-dimethylchlorobenzene (CH_3, H_3C, Cl) d. $C_6H_5Si(CH_3)_2-Cl$

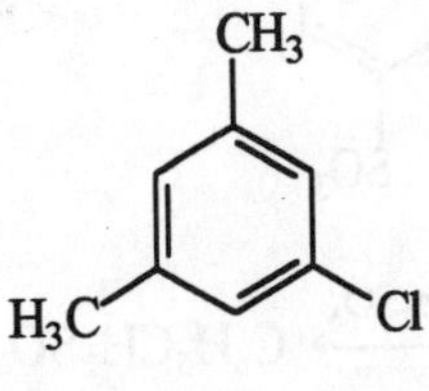

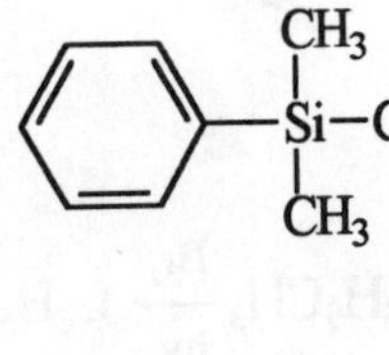

e. [structure: phenyl ketone with C=O attached to CH=CH–CH₃]

6. a. $CH_3SH + 2\ KMnO_4 \longrightarrow CH_3SO_3K + KOH + 2\ MnO_2$

b. $3\ (CH_3)_2S + 4\ KMnO_4 + 2\ H_2O \longrightarrow 3\ (CH_3)_2SO_2 + 4\ MnO_2 + 4\ KOH$

c. $(CH_3)_2S + 4\ HNO_3 \longrightarrow (CH_3)_2SO_2 + 4\ NO_2 + 2\ H_2O$

7. a. E2 elimination would occur during either step to give isobutylene rather than the desired phosphite triester or alkylphosphonate.

b. Friedel-Crafts reactions do not apply with *meta*-directing groups such as $-SO_3H$.

c. The ketone group is sensitive to strong base. Hydroxide promotes aldol condensation reactions; hence the final step with fused KOH will give a mess.

d. The trimethylsilyl group would be replaced in the chlorination step to give 4-chlorotoluene.

8. a. $(CH_3)_3CBr \xrightarrow{Mg} \xrightarrow{S_8} \xrightarrow{CH_3I} (CH_3)_3CSCH_3$

(<u>NOTE</u>: $(CH_3)_3CBr + NaSCH_3 \longrightarrow (CH_3)_2C{=}CH_2 + HSCH_3 + NaBr$)

b. $CH_3CH_2CH_2CH_2OH \xrightarrow{PBr_3} \xrightarrow[2.\ OH^-]{1.\ H_2N-C(=S)-NH_2} CH_3CH_2CH_2CH_2SH \xrightarrow{I_2} (CH_3CH_2CH_2CH_2S)_2$

c. $H_3C-C_6H_5 \xrightarrow{HSO_3Cl} H_3C-C_6H_4-SO_2Cl \xrightarrow[AlCl_3]{toluene} H_3C-C_6H_4-SO_2-C_6H_4-CH_3$

d. $C_6H_5NH_2 \xrightarrow[180-190°C]{H_2SO_4} {}^+NH_3-C_6H_4-SO_3^- \xrightarrow{CF_3CO_3H} NO_2-C_6H_4-SO_3H$

e. $C_6H_5CH_3 \xrightarrow[hv]{Br_2} C_6H_5CH_2Br \xrightarrow{Na_2SO_3} C_6H_5CH_2SO_3^-Na^+ \xrightarrow{H^+} C_6H_5CH_2SO_3H$

f. $p\text{-}CH_3C_6H_4SO_3H \xrightarrow[350°C]{NaCN} p\text{-}CH_3C_6H_4CN \xrightarrow[\Delta]{H_3O^+} p\text{-}CH_3C_6H_4CO_2H$

g. cyclohexanone $\xrightarrow{(CH_3)_2S^{+}\text{–}CH_2}$ spiro-epoxide $\xrightarrow[\text{or } OH^-]{H_3O^+}$ 1-(hydroxy)cyclohexyl–CH_2OH

h. sulfolane (SO_2 ring) $\xrightarrow{BuLi}\xrightarrow{ICH_2CH_2CH_3}$ 2-$CH_2CH_2CH_3$-sulfolane

9. a. 3 cyclopentanol (OH) $\xrightarrow{PBr_3}$ 3 bromocyclopentane (Br) $\xrightarrow{Mg}\xrightarrow[\text{pyridine}]{PCl_3}$ $(C_5H_9)_3P$

b. $C_6H_5CH_2OH + PCl_3 \longrightarrow (C_6H_5CH_2O)_3P \xrightarrow{\Delta} C_6H_5CH_2\overset{O}{\overset{\|}{P}}(OCH_2C_6H_5)_2$

$C_6H_5CH_2OH \xrightarrow{HBr} C_6H_5CH_2Br \xrightarrow{\text{catalytic amount}}$ (to the Δ step)

c. $C_6H_5\overset{O}{\overset{\|}{C}}H + HS(CH_2)_3SH \xrightarrow{HCl}$ 2-C_6H_5-1,3-dithiane (C_6H_5CH with S, S) $\xrightarrow[\text{then}]{BuLi}$ $\overset{O}{CH_2\text{–}CHCH_2CH_2CH_3}$ → dithiane (S, S) with phenyl and $CH_2CH(OH)CH_2CH_2CH_3$ (OH) $\xrightarrow[H_2O]{HgCl_2}$ $C_6H_5COCH_2CH(OH)CH_2CH_2CH_3$ (O, OH)

d. cyclopentanone (=O) $\xrightarrow[\Delta]{(C_6H_5O)_2\overset{O}{\overset{\|}{P}}\bar{C}HCN}$ cyclopentylidene=CHCN

10. $ClCH_2CH_2\ddot{S}(CH_2CH_2\text{–}Cl) \xrightarrow{-Cl^-} ClCH_2CH_2\overset{+}{S}(CH_2CH_2)$ (episulfonium ring) $\xrightarrow{:Nu^-} ClCH_2CH_2SCH_2CH_2Nu$

11. $CH_3S\cdot + H\cdot = CH_3SH$ $\Delta H° = -91$

$CH_4 = CH_3\cdot + H\cdot$ $\Delta H° = DH° = 105$

$CH_4 + CH_3S\cdot = CH_3SH + CH_3\cdot$ $\Delta H° = +14$ kcal mole^{-1}

$\Delta S°$ is probably about zero, therefore $\Delta G° \approx \Delta H°$, and the equilibrium lies far to the left. Hence, CH_3SH works as an inhibitor by reacting with alkyl radicals to stop propagation of the radical chains. The $CH_3S\cdot$ formed cannot abstract hydrogen atoms from carbon, so nothing happens until two $CH_3S\cdot$ radicals come together to form the disulfide.

12. RSO_2OH are much more acidic than RCO_2H; the ΔpK_a is about 10. RSO_2NH_2 are more acidic than $RCONH_2$; the ΔpK_a is about 5. With these analogies, we would expect RSO_2CH_3 to be more acidic than $RCOCH_3$; actually, RSO_2CH_3 are less acidic than $RCOCH_3$ in aqueous solution, but the difference is only a few pK_a units.

13. The shorter bond results from the increased coulombic attraction of the dipolar dative bond:

$>P\rightarrow O \;=\; >\overset{+}{P}-O^-$. This attraction also results in a stronger bond.

(bond strengths: dative P→O bond ≈ 130 kcal mole^{-1}, single P–O bond ≈ 90 kcal mole^{-1}).

The conversion of a P–O single bond into a P→O dative bond provides most of the 47 kcal mole^{-1} which is the driving force of the rearrangement.

14. Because of the strain in the smaller five-membered ring, the P–O bonds in the ring do not both want to occupy equatorial positions simultaneously (O–P–O angle of 120°); instead, one of them is equatorial and one is apical (O–P–O) angle of 90°). A ring P–O bond (apical) is therefore cleaved in preference to the methoxy bond in the five-membered ring ester. For the six-membered ring ester, the larger ring can accommodate the pentacoordinate structure with apical methoxy, so both types of cleavage are seen.

15.

Br–Br SO_3^- Br Br :NH_2 → Br^- Br SO_3^- Br Br $+NH_2$ → Br Br Br NH_2

Ipso is the Latin word for "the same".

16.

$$C_6H_5\overset{O}{\overset{\|}{C}}CH_2\overset{O}{\overset{\|}{S}}CH_3 \rightleftharpoons C_6H_5\overset{OH}{\overset{|}{C}}=CH\overset{O}{\overset{\|}{S}}CH_3 \overset{H^+}{\rightleftharpoons} C_6H_5C=CH-\overset{OH}{\overset{|}{S^+}}-CH_3 \overset{-H^+}{\rightleftharpoons} C_6H_5\overset{O}{\overset{\|}{C}}CH=\overset{OH}{\overset{|}{S}}CH_3$$

$$\overset{H^+}{\rightleftharpoons} C_6H_5\overset{O}{\overset{\|}{C}}CH=\overset{^+OH_2}{\overset{|}{S}}CH_3 \rightleftharpoons C_6H_5\overset{O}{\overset{\|}{C}}CH=\overset{+}{S}CH_3 \;(H_2O:) \rightleftharpoons C_6H_5\overset{O}{\overset{\|}{C}}\underset{^+OH_2}{\underset{|}{C}}HSCH_3 \overset{-H^+}{\rightleftharpoons} C_6H_5\overset{O}{\overset{\|}{C}}\underset{OH}{\underset{|}{C}}HSCH_3$$

$$C_6H_5\overset{O}{\overset{\|}{C}}CH_2\overset{O}{\overset{\|}{S}}CH_3 \xrightarrow[\text{pyridine}]{Ac_2O} C_6H_5\overset{O}{\overset{\|}{C}}\underset{OCOCH_3}{\underset{|}{C}}HSCH_3$$

17. cyclohexanone $\xrightarrow{LDA}$ $\xrightarrow{C_6H_5SSC_6H_5}$ 2-(SC_6H_5)cyclohexanone $\xrightarrow{H_2O_2}$ 2-($S(=O)C_6H_5$)cyclohexanone $\xrightarrow{\Delta}$ 2-cyclohexenone + C_6H_5SOH

18. $CH_3SCH_3 + \frac{1}{2}O_2 \longrightarrow CH_3\overset{O}{\overset{\|}{S}}CH_3 \qquad \Delta H° = -27.2 \text{ kcal mole}^{-1}$

$$CH_3\overset{O}{\overset{\|}{S}}CH_3 + \tfrac{1}{2}O_2 \longrightarrow CH_3\overset{O}{\overset{\|}{\underset{O}{\underset{\|}{S}}}}CH_3 \qquad \Delta H° = -53.0 \text{ kcal mole}^{-1}$$

The second oxidation step is much more exothermic than the first. Assuming that the analogy holds true for the disulfide systems, one would expect the thiosulfonic ester to be thermodynamically favored over the disulfoxide.

19. 1,3,5-trithiane $\xrightarrow{[O]}$ **A** + **B** (isomeric 1,3,5-trithiane 1,3-dioxides); **A** $\xrightarrow{H_2O_2}$ 1,3,5-trioxide, and **B** $\xrightarrow{H_2O_2}$ the same trioxide and **D**; the two trioxide structures, labelled **C**, interconvert ($\rightleftharpoons$)

C

D

equivalent by chair $\rightleftharpoons$ chair interconversion

20. a.

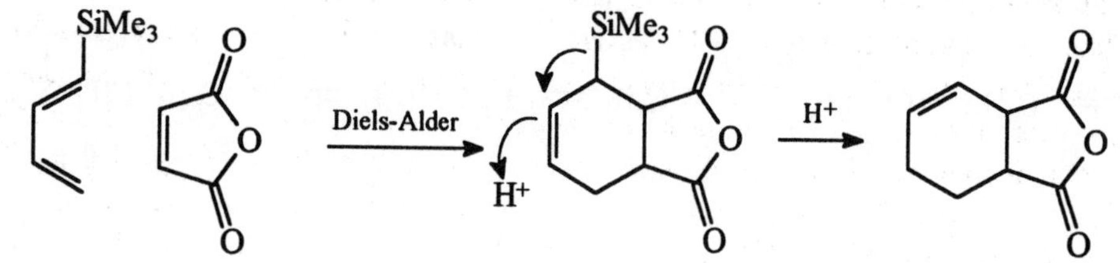

SiMe3 (diene) + maleic anhydride —Diels-Alder→ adduct; H+ → product

b.

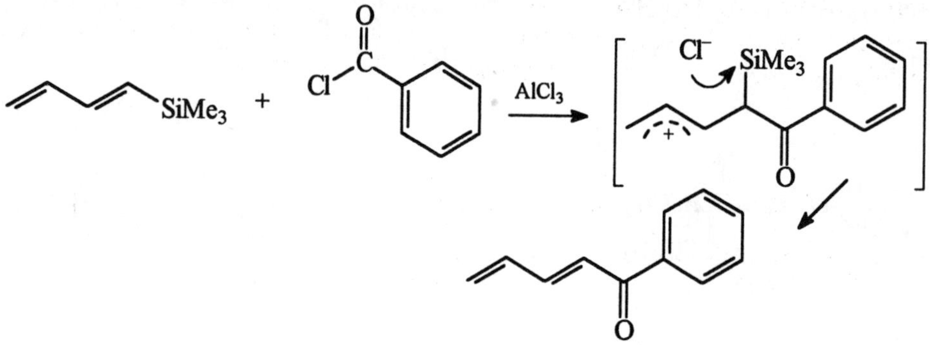

$\ce{SiMe3}$ diene + benzoyl chloride —$AlCl_3$→ [Cl^- SiMe3 intermediate] → dienyl phenyl ketone

21.

OH, SiMe3 — KOH — [O⁻ SiMe3 ⇌ Si ⁻ → ⁻CHOSiMe3] — EtOH — CH_2OSiMe_3

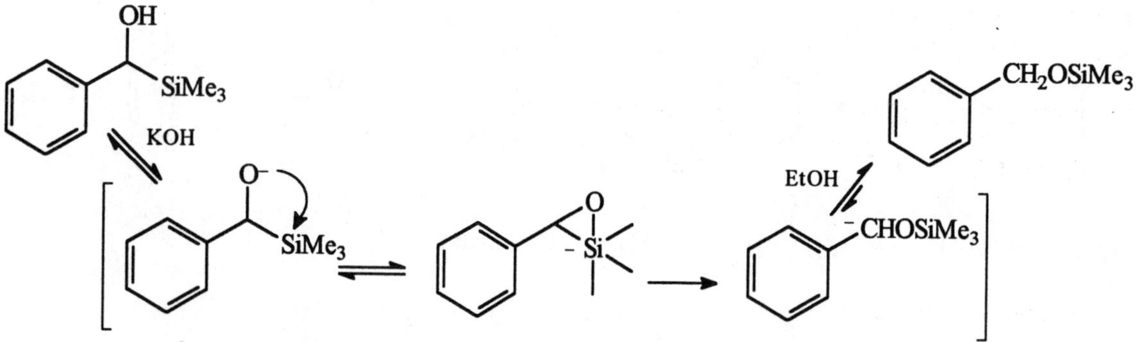

front-side attack, results in retention of configuration

22.

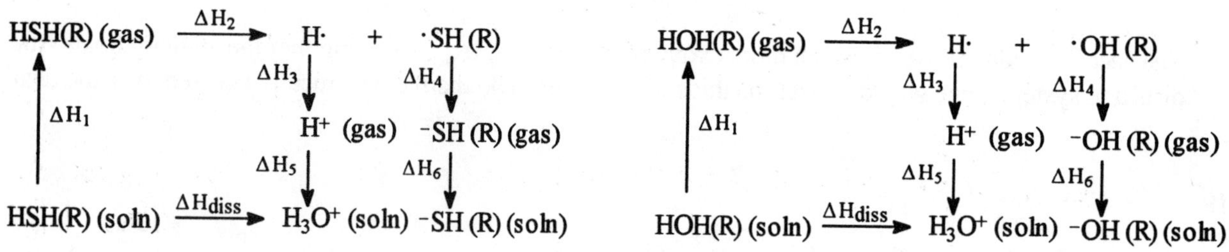

HSH(R) (gas) —ΔH_2→ H· + ·SH (R); ΔH_3, ΔH_4 ↓ H+ (gas) ⁻SH (R) (gas); ΔH_5, ΔH_6 ↓; ΔH_1 ↑ from HSH(R) (soln) —ΔH_{diss}→ H_3O^+ (soln) ⁻SH (R) (soln)

HOH(R) (gas) —ΔH_2→ H· + ·OH (R); ΔH_3, ΔH_4 ↓ H+ (gas) ⁻OH (R) (gas); ΔH_5, ΔH_6 ↓; ΔH_1 ↑ from HOH(R) (soln) —ΔH_{diss}→ H_3O^+ (soln) ⁻OH (R) (soln)

$\Delta H_{diss} = \Delta H_1 + \Delta H_2 + \Delta H_3 + \Delta H_4 + \Delta H_5 + \Delta H_6$ $\Delta H_{diss} = \Delta H_1 + \Delta H_2 + \Delta H_3 + \Delta H_4 + \Delta H_5 + \Delta H_6$

$\Delta\Delta H_{diss} = \Delta\Delta H_1 + \Delta\Delta H_2 + \Delta\Delta H_3 + \Delta\Delta H_4 + \Delta\Delta H_5 + \Delta\Delta H_6$; $\Delta\Delta H_3 = \Delta\Delta H_5 = 0$

$\Delta\Delta H_{diss}(\text{HSH} - \text{HOH}) = \Delta\Delta H_1 + \Delta\Delta H_2 + \Delta\Delta H_4 + \Delta\Delta H_6$

$= (1-10) + (89-119) + (-52 -(-40)) + (-341 -(-371))$

$\Delta\Delta H_{diss}(\text{HSH} - \text{HOH}) = (-9) + (-30) + (-12) + (+30) = -21$ kcal/mole

The difference in dissociation constants is related to $\Delta\Delta G_{diss}(HSH-HOH) = \Delta\Delta H_{diss} - T\Delta\Delta S_{diss}$. It seems likely that $\Delta\Delta S_{diss}$ will be small. Thus, the conclusion is that HSH is more acidic than HOH. The experimental difference corresponds to 9.6 kcal/mole. Consider now the effect of R on the ΔH of each step. (1) Thiols are considerably more volatile than alcohols (see text, p.807). (2) The dissociation energy for the S–H bond should not be affected much differently than the dissociation energy for the O–H bond by the substitution of R for H in HSH or HOH. (3) The nature of R does not affect step 3. (4) The electron affinity of ·SR might be different than that of ·SH but the difference should be similar to that between ·OR and ·OH. (5) The nature of R does not affect step 5. (6) R will decrease the solvation energies of both $^-$SR and $^-$OR in comparison to $^-$SH and $^-$OH. It is likely that the oxygen anion solvation will be decreased more than that of the sulfur anion. We conclude that thiols will be more acidic than alcohols but that the difference will be smaller than that between HSH and HOH.

23. Glutathione $^-O_2C-CH(NH_3^+)-CH_2CH_2-C(=O)-NH-CH(CH_2SH)-C(=O)-NH-CH_2-CO_2^-$ = GSH

$GSH \rightleftharpoons GS^- + H^+$

$GS^- + I_2 \rightleftharpoons GSI + I^-$ sulfenyl iodide

$GS^- + GSI \rightleftharpoons GSSG + I^-$ disulfide

26.F. Supplementary Problems

S1. Name the following compounds:

a. $(CH_3O)_3P$

b. $Br-C_6H_4-S(=O)OCH_3$ (para)

c. $C_6H_5-\overset{+}{S}(CH_2CH_3)_2$ I^-

d. $CH_3(CH_2)_{10}CH_2OSO_2O^-$ Na^+

e. $(CH_3CH_2CH_2CH_2)_3{}^+PCH_3$ Cl^-

f. $C_6H_5-P(=O)(OCH_2CH_3)(OCH_3)$

g. $H_2NCH_2PO_3H_2$

h. $Cl_3CSO_2C_6H_5$

i. $(CH_3CH_2CH_2)_2S{=}O$

j. $(C_6H_5)_2Si(Br)C(CH_3)_3$

k. cyclohexyl–O–Si(isopropyl)$_3$

l. $(CH_3CH_2O)_2P(=O)OP(=O)(OCH_2CH_3)_2$

S2. What is the major product to result from each of the following reaction sequences?

a. $(C_6H_5)_3P \xrightarrow{CH_3CH_2I} \xrightarrow{C_6H_5Li} \xrightarrow{\text{cyclohexanone}}$

b. $(CH_3)_2CHCH{=}O \xrightarrow[BF_3]{HS(CH_2)_2SH} \xrightarrow{BuLi} \xrightarrow{CH_3CH_2OSO_2C_6H_5} \xrightarrow[HgCl_2]{H_2O}$

c. $(CH_3O)_3P \xrightarrow[\Delta]{BrCH_2CO_2CH_3} \xrightarrow[\Delta]{NaH} \xrightarrow{\text{cyclohexanone}}$

d. toluene ($C_6H_5CH_3$) $\xrightarrow{ClSO_3H} \xrightarrow{Zn} \xrightarrow{NaOH} \xrightarrow{BrCH_2CH{=}CH_2}$

e. $BrCH_2CH_2CH_2CH_2Br \xrightarrow[C_2H_5OH]{NaSH}$

f. $BrCH_2CH_2CH_2CH_2Br \xrightarrow{2\ (NH_2)_2C{=}S} \xrightarrow{NaOH}$

g. $(CH_3O)_2\overset{O}{\overset{\|}{P}}O\overset{O}{\overset{\|}{P}}(OCH_3)_2 \xrightarrow[CH_3CH_2OH]{CH_3CH_2ONa}$

h. $C_6H_5SCH_3 \xrightarrow{n\text{-BuLi}} \xrightarrow[2.\ 2\ CH_3CO_3H]{1.\ (CH_3)_2C{=}O}$

i. $(CH_3)_2SiCl_2 + t\text{-BuLi} \longrightarrow$

j. $OHC{-}CH_2CH_2CH_2CH_2CH{=}CHCH_2SiMe_3 \xrightarrow{TiCl_4} \xrightarrow{H_2O}$

S3. Show how to accomplish the following transformations in a practical manner.

a. $CH_3CH_2CH_2\overset{O}{\overset{\|}{C}}CH_2CH_3 \longrightarrow CH_3CH_2CH_2\overset{O-CH_2}{C}CH_2CH_3$ (epoxide)

b. $CH_3CH_2CH_2\overset{O}{\overset{\|}{C}}H \longrightarrow CH_3CH_2CH_2\overset{O}{\overset{\|}{C}}D$

c. $(CH_3)_2CHCH_2\overset{O}{\overset{\|}{C}}CH_3 \longrightarrow (CH_3)_2CHCH_2\overset{CH_3}{\overset{|}{C}}{=}CHCO_2CH_2CH_3$

d. $C_6H_5CH_3 \longrightarrow C_6H_5CH_2SO_2{-}C_6H_4{-}CH_3$ (p-tolyl sulfone)

e. benzene $\longrightarrow (C_6H_5O)_3P{=}O$

f. (structure: 4-(hydroxymethyl)cyclohexanone, O= and OH) ⟶ (structure: 4-hydroxycyclohexanecarbaldehyde, HO and O)

S4. The Arbuzov-Michaelis reaction is a very important one for the preparation of dialkyl alkyl-phosphonates from trialkylphosphites, but it fails if applied to the synthesis of diphenyl alkylphosphonates from triphenylphosphite. Why? What products do you expect to see instead?

S5. A chemist tried to synthesize 1-methyl-2-(phenylthio)ethanamine from propylene oxide by sequential treatment with sodium benzothiolate, *p*-toluenesulfonyl chloride, and ammonia. However, 2-(phenylthio)propanamine was the actual product. Draw the intermediates involved in this transformation and explain why the unanticipated isomer was produced.

$$CH_3CH\overset{O}{-}CH_2 \xrightarrow{C_6H_5SNa} \xrightarrow{CH_3C_6H_4SO_2Cl} \xrightarrow{NH_3} CH_3\overset{SC_6H_5}{\overset{|}{C}H}CH_2NH_2$$

S6. If a carboxylic ester is to be converted to a β-ketosulfone by reaction with the carbanion derived from dimethyl sulfone, two equivalents of the carbanion are required; when one equivalent is used, only 50% of the ester reacts.

a. Why are two equivalents necessary for complete conversion?

b. β-Ketosulfones are "desulfonylated", to give the ketone itself, using aluminum amalgam in wet THF:

$$-\overset{O}{\overset{\|}{C}}-\overset{|}{\underset{|}{C}}-SO_2R \xrightarrow[H_2O,\ THF]{AlHg} -\overset{O}{\overset{\|}{C}}-CH\!\!< \; + \; RSO_2^-$$

Using this reaction, devise a method for the conversion of a carboxylic ester into a ketone using dimethyl sulfone and an alkyl halide:

$$RCO_2CH_3 + CH_3\overset{O}{\underset{O}{\overset{\|}{\underset{\|}{S}}}}CH_3 + BrR' \xrightarrow{?} R\overset{O}{\overset{\|}{C}}CH_2R'$$

S7. Dithioketals are often difficult to hydrolyze, even using mercuric chloride as catalyst, particularly if there is an acid-sensitive functional group in the molecule. A method for accomplishing this hydrolysis under neutral conditions using methyl iodide in aqueous acetone with sodium bicarbonate has been devised. Write a step-by-step mechanism to illustrate how this reaction occurs.

(structure: 2,2-dialkyl-1,3-dithiane, S, S, C, R, R) $\xrightarrow[NaHCO_3,\ H_2O,\ acetone]{excess\ CH_3I}$ $R\overset{O}{\overset{\|}{C}}R$

S8. Depending on the order in which substituents are introduced onto the bicyclic lactone illustrated below, the final sulfoxide elimination reaction products one or the other double bond isomer. How do you account for this difference in behavior?

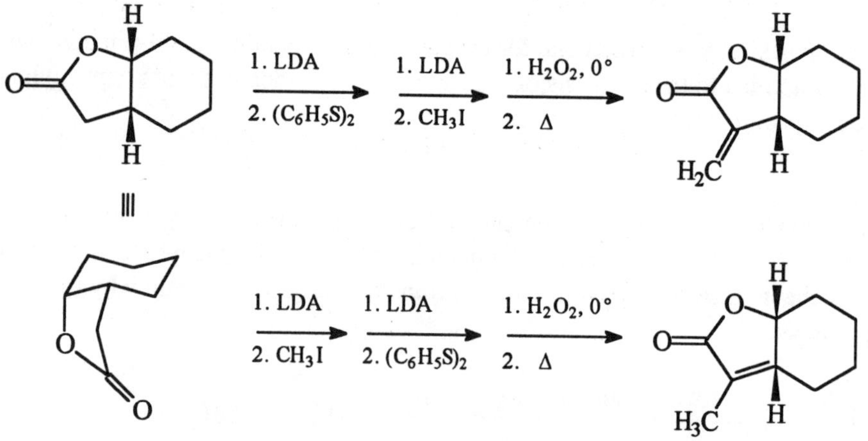

1. LDA 2. $(C_6H_5S)_2$ → 1. LDA 2. CH_3I → 1. H_2O_2, 0° 2. Δ → (product with H_2C=, H, H, O)

1. LDA 2. CH_3I → 1. LDA 2. $(C_6H_5S)_2$ → 1. H_2O_2, 0° 2. Δ → (product with H_3C, H, O)

S9. β-Hydroxysilanes undergo elimination under either acidic or basic conditions to give olefins and a silanol. The elimination reactions are stereospecific, and opposite products are formed, depending on the conditions employed:

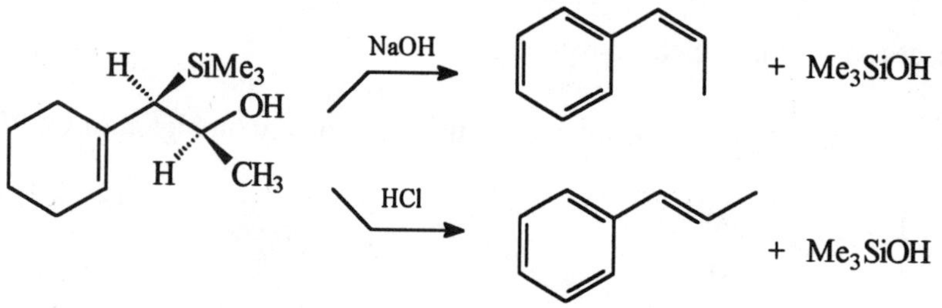

(H, $SiMe_3$, OH, H, CH_3) NaOH → + Me_3SiOH

HCl → + Me_3SiOH

Write mechanisms that rationalize the stereochemical results of these two reactions.

26.G. Answers to Supplementary Problems

S1. a. trimethyl phosphite
b. methyl *p*-bromobenzenesulfinate
c. diethylphenylsulfonium iodide
d. sodium dodecyl sulfate (an important detergent)
e. tributylmethylphosphonium chloride
f. ethyl methyl phenylphosphonate
g. aminomethylphosphonic acid
h. trichloromethyl phenyl sulfone
i. dipropyl sulfoxide
j. *t*-butyldiphenylsilyl bromide
k. cyclohexyl triisopropylsilyl ether
l. tetraethyl pyrophosphate

S2. a. 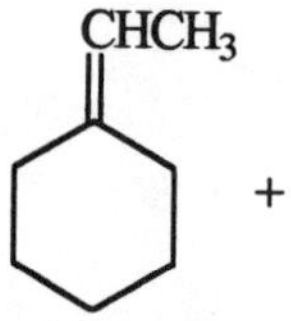 (cyclohexylidene $CHCH_3$) + $(C_6H_5)_3P{=}O$

b. $(CH_3)_2CHCCH_2CH_3$ (C=O)

c. (cyclohexylidene $CHCO_2CH_3$)

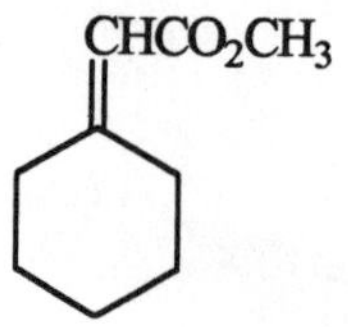

d. $H_3C-C_6H_4-S(=O)_2-CH_2CH=CH_2$

e. (tetrahydrothiophene, ring with S)

($HSCH_2CH_2CH_2CH_2Br + HS^- \rightleftharpoons H_2S + {}^-SCH_2CH_2CH_2CH_2Br \rightarrow C_4H_8S$ is faster than intermolecular displacement by a second ^-SH group)

f. $HSCH_2CH_2CH_2CH_2SH$

g. $(CH_3O)_2PO_2^- + CH_3CH_2OP(=O)(OCH_3)_2$

h. $C_6H_5S(=O)_2CH_2C(OH)(CH_3)_2$

i. $(CH_3)_3CSi(CH_3)_2Cl$

j. 2-vinylcyclopentanol (OH on ring carbon, adjacent carbon bearing $CH=CH_2$)

S3. a. $RC(=O)R' + (CH_3)_2\overset{+}{S}-\overset{-}{C}H_2 \longrightarrow$ $RC(O-CH_2)R'$ (epoxide) $+ (CH_3)_2S$

b. $CH_3CH_2CH_2CH(=O) \xrightarrow[BF_3]{HS(CH_2)_3SH}$ 1,3-dithiane with H and $CH_2CH_2CH_3$ at C-2 $\xrightarrow{BuLi}$ 1,3-dithiane with Li and $CH_2CH_2CH_3$ at C-2 $\xrightarrow{D_2O}$ 1,3-dithiane with D and $CH_2CH_2CH_3$ at C-2 $\xrightarrow[HgCl_2]{H_2O}$ $CH_3CH_2CH_2C(=O)D$

c. $RC(=O)R' + (EtO)_2\overset{O}{\overset{\|}{P}}\overset{-}{C}HCO_2C_2H_5 \xrightarrow{\Delta} (R')(R)C=CHCO_2C_2H_5 + (EtO)_2PO_2^-$

d. $C_6H_5CH_3 \xrightarrow{ClSO_3H}$ $p-CH_3C_6H_4SO_2Cl$ $\xrightarrow[2.\ NaHCO_3]{1.\ Zn,\ H_2O}$ $p-CH_3C_6H_4SO_2^-\ Na^+$ + $C_6H_5CH_2Br$ $\xleftarrow[h\nu]{Br_2}$ $C_6H_5CH_3$

$\longrightarrow H_3C-C_6H_4-S(=O)_2CH_2-C_6H_5$

e. Benzene $\xrightarrow[H_2SO_4]{SO_3}$ $C_6H_5SO_3H$ $\xrightarrow[\Delta\ \Delta]{KOH}$ C_6H_5OH $\xrightarrow[pyridine]{1/3\ POCl_3}$ $(C_6H_5O)_3P{=}O$

f. 4-(hydroxymethyl)cyclohexanone + t-$BuMe_2SiCl$ $\longrightarrow$ 4-(CH_2OSiR_3)cyclohexanone $\xrightarrow{NaBH_4}$ $\xrightarrow{Ac_2O}$ 4-(CH_2OSiR_3)cyclohexyl OAc $\xrightarrow{Ph\overset{+}{N}Me_3\ F^-}$ 4-(CH_2OH)cyclohexyl OAc $\xrightarrow{PCC}$ 4-(CHO)cyclohexyl OAc $\xrightarrow{NaOH}$ 4-(CHO)cyclohexanol

S4.

$(C_6H_5O)_3P\!: + RX \longrightarrow (C_6H_5O)_2\overset{+}{P}(OC_6H_5)\text{—}R$ (attack by X^- on phenyl carbon ✗) $\not\longrightarrow$ $(C_6H_5O)_2\overset{O}{\overset{\|}{P}}R + C_6H_5X$

Alkylation of phosphorus occurs readily, but cleavage of a phenyl-oxygen bond to form the phosphonate is not possible, due to the difficulty of nucleophilic substitution on the benzene ring.

S5.

$CH_3CH\overset{O}{—}CH_2$ + $^-SC_6H_5$ $\longrightarrow$ $CH_3\overset{O^-}{\overset{|}{C}}HCH_2SC_6H_5$ $\xrightarrow{TsCl}$ $CH_3\overset{OTs}{\overset{|}{C}}HCH_2\text{—}\!:\!\ddot{S}C_6H_5$ $\not\xrightarrow{NH_3}$ $CH_3\overset{NH_2}{\overset{|}{C}}HCH_2SC_6H_5$

$\longrightarrow$ $CH_3CH\text{—}CH_2$ (episulfonium, $\overset{+}{S}$—C_6H_5) + $:NH_3$ $\longrightarrow$ $CH_3\underset{SC_6H_5}{\underset{|}{C}}HCH_2NH_2$

S_N2 attack on less substituted end of epoxide. S_N2 attack at the secondary position is slower than the competitive intramolecular cyclization. Attack occurs the at less substituted end of the sulfonium intermediate.

S6.

$$RCOOCH_3 + {}^-CH_2SO_2CH_3 \longrightarrow RC(O^-)(OCH_3)CH_2SO_2CH_3 \longrightarrow RCOCH_2SO_2CH_3 + CH_3O^- \rightleftharpoons RCO\bar{C}HSO_2CH_3 + CH_3OH$$

$$RCOCH_2SO_2CH_3 + {}^-CH_2SO_2CH_3 \longrightarrow CH_3SO_2CH_3 + RCO\bar{C}HSO_2CH_3 + CH_3O^-$$

The dimethyl sulfone carbanion is more basic than both methanol and the β-keto sulfone product, and one mole will be protonated for each mole of β-keto sulfone that is formed. It is commonly necessary to use two equivalents of a carbanion whenever the product is more acidic than the acid of the carbanion. Formation of the enolate of the product also prevents addition of another equivalent of carbanion to the ketone carbonyl.

b. $$RCOOCH_3 + 2\,NaCH_2SO_2CH_3 \longrightarrow RCO\bar{C}HSO_2CH_3 \xrightarrow{R'Br} RCOCH(R')SO_2CH_3 \xrightarrow{Al(Hg)} RCOCH_2R'$$

(<u>NOTE</u>: if alkylation of the sulfonyl carbanion is carried out first, it will be impossible to obtain the desired carbanion for the acylation reaction:

$$RCH_2SO_2CH_3 \xrightarrow{NaH} RCH_2SO_2\bar{C}H_2\ Na^+ \text{, not } Na^+R\bar{C}HSO_2CH_3$$

S7.

$$\text{2,2-dialkyl-1,3-dithiane (R}_2\text{C) } \xrightarrow{CH_3I} \text{S-methylsulfonium iodide } (CH_3\overset{+}{S},\ I^-) \longrightarrow \left[CH_3SCH_2CH_2CH_2\ddot{S}{-}\overset{+}{C}R_2 \longleftrightarrow CH_3SCH_2CH_2CH_2\overset{+}{S}{=}CR_2 \right]$$

$$\xrightarrow{H_2O} CH_3SCH_2CH_2CH_2\ddot{S}{-}CR_2{-}\overset{+}{O}H_2 \underset{}{\overset{NaHCO_3}{\rightleftharpoons}} CH_3SCH_2CH_2CH_2\ddot{S}{-}CR_2{-}O^- \longrightarrow RCOR + CH_3SCH_2CH_2CH_2SH$$

S8. Because of the folded structure of the bicyclic lactone, all reactions of the enolates (alkylations and sulfenylations) occur from the top side of the molecule (*exo* face), as illustrated below, for steric reasons. If alkylation follows sulfenylation, compound **A** will result. The subsequent sulfoxide elimination can occur only toward the methyl group, because the bridgehead hydrogen is *trans* to it. If sulfenylation follow alkylation, diastereomer **B** will be produced. In this case, the sulfoxide can eliminate in either direction because the bridgehead hydrogen is *cis* to the sulfur group; as it turns out, elimination to give the more highly substituted double bond is preferred.

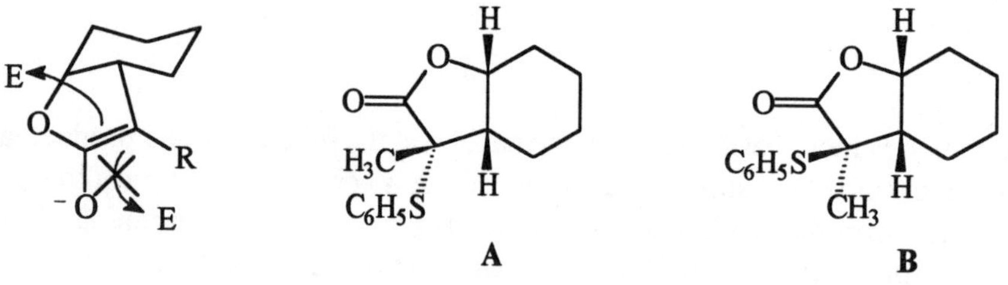

E R O $^-$O E H O O H_3C C_6H_5S H **A** H O O C_6H_5S CH_3 H **B**

S9.

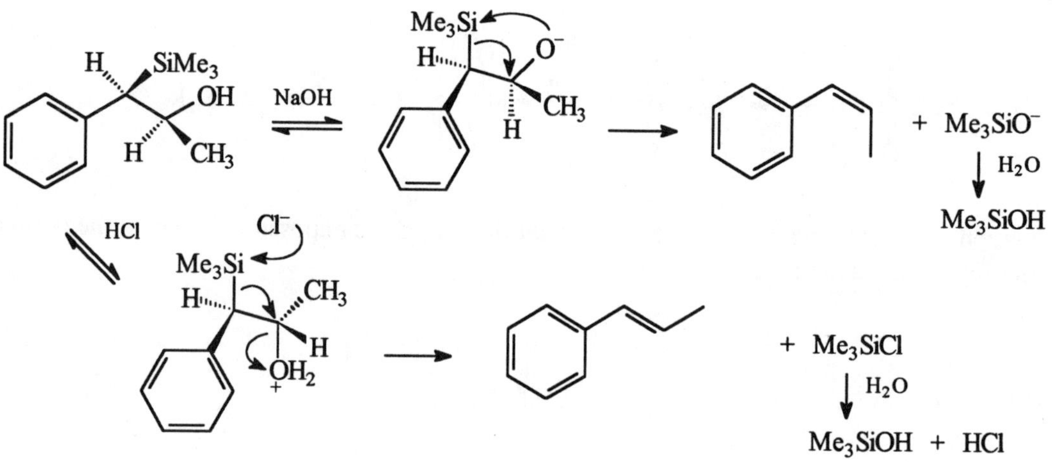

H $SiMe_3$ OH H CH_3 NaOH Me_3Si H O^- CH_3 H + Me_3SiO^- H_2O Me_3SiOH

HCl Cl^- Me_3Si H CH_3 H $\overset{+}{O}H_2$ + Me_3SiCl H_2O Me_3SiOH + HCl

27. DIFUNCTIONAL COMPOUNDS

27.A. Chapter Outline and Important Terms Introduced

27.1 Introduction

Scope of chapter: diols, dicarbonyl compounds, hydroxy carbonyl compounds

27.2 Nomenclature of Difunctional Compounds

diene, diyne, diol, dione, dicarboxylic acid
glycols
hydroxyalkanoic acid
hydroxyalkanone

27.3 Diols

A. Preparation (1,2-diols)
- hydroxylation (see Section 11.6.E and 10.11)
- *syn* ($KMnO_4$ or OsO_4) or *anti* addition (via epoxide and hydrolysis)
- *erythro* and *threo*
- asymmetric synthesis: OsO_4 + chiral amine, e.g. MEQ, Fe(III) cooxidant
- reductive dimerization (pinacol reaction)
  - Pederson for coupling of 2 different aldehydes
  - Sharpless asymmetric epoxidation of allylic alcohols
- ketyl radical anion
- higher diols from same reactions as in monofunctional systems

B. Reactions of Diols
- pinacol rearrangement
- periodate cleavage
- cyclic ethers from 1,4- and 1,5-diols
- Lemieux-Johnson reaction

27.4 Hydroxy Aldehydes and Ketones

A. Synthesis
- acyloin condensation
- aldol condensation (see Section 15.2)
- benzoin condensation

B. Reactions
- dehydration of β-hydroxycarbonyl compounds - acid or base catalyzed
  - α-hydroxy carbonyls dehydrated with difficulty
- cyclic hemiacetals and hemiketals
- periodate cleavage of α-hydroxyketones and α-aldehydes

27.5 Hydroxy Acids

A. Natural Occurrence (lactic, citric, malic, tartaric acids, etc.)
- Citric acid cycle

B. Synthesis
- hydrolysis of α-halo acids
- hydrolysis of cyanohydrins
- Baeyer-Villiger reaction/lactone cleavage
- aldol-like additions
- Evans asymmetric aldol reaction

C. Reactions
   - lactonization — lactones and lactides
   - polymerization — dehydration

27.6 Dicarboxylic Acids

A. Synthesis
   - chemistry of carbonic acid

B. Acidity
   - electrostatic effects

C. Reactions of Dicarboxylic Acids and Their Derivatives
   - decarboxylation of malonic acids — Dieckmann condensation
   - anhydride and imide formation from succinic and glutaric acids

27.7 Diketones, Keto Aldehydes, Keto Acids, and Keto Esters

A. Synthesis
   - 1,2-diketones (α-diketones): oxidation of acyloins; SeO_2 oxidation of ketones
   - 1,3-dicarbonyl compounds: Claisen condensation

B. Keto-Enol Equilibria in Dicarbonyl Compounds
   - vinylogy, vinylogs

C. Decarboxylation of β-Keto Acids

D. 1,3-Dicarbonyl Compounds as Carbon Acids
   - 2,4-pentanedione $pK_a = 9$

E. The Malonic Ester and Acetoacetic Ester Synthesis

F. The Knoevenagel Condensation

G. The Michael Addition Reaction
   - Robinson annulation

27.B. Important Reactions Introduced

Hydroxylation of alkenes to form 1,2-diols (27.3.A; see also Section 11.6.E and 10.11.A)

Equation:

$$R_2C{=}CR_2 \xrightarrow{[Ox]} R{-}\underset{R}{\overset{OH}{C}}{-}\underset{R}{\overset{OH}{C}}{-}R$$

Generality: [Ox] = $KMnO_4$ or OsO_4 (*syn*) or 1. $R'CO_3H$ 2. H_3O^+ (*anti*)
Addition of chiral amine, e.g. MEQ, cooxidant Fe(III) yields enantiomeric excess of one diol.

Key features: see Sections 11.6.E and 10.11.A

Pinacol reaction (reductive dimerization) (27.3.A)

Equation: $2\ \mathrm{RC(=O)R'} + \mathrm{M} \longrightarrow \mathrm{R{-}C(OH)(R'){-}C(OH)(R'){-}R}$

Generality: R, R' = H, alkyl, aryl; M = 2 Na or Mg

Key features: reaction proceeds via dimerization of ketyl radical anion

Pederson reaction (unsymmetric reductive dimerization) (27.3.A)

Equation: $\mathrm{RCHO} + \mathrm{HC(=O)R'} \xrightarrow{\text{cat.}} \mathrm{RCH(OH){-}CH(OH)R'}$

Generality: R, R' = alkyl, aryl; catalyst = V(II)

Key features: one of the aldehydes should have a coordinating group elsewhere in the molecule.

Sharpless Asymmetric Epoxidation (27.3.A)

Equation: $\mathrm{RCH{=}CHCH_2OH} \xrightarrow{\text{cat.}} \text{R-epoxide-}\mathrm{CH_2OH} \xrightarrow{\mathrm{LiAlH_4}} \mathrm{RCH(OH)CH_2CH_2OH}$

Generality: Specific for allyl alcohols; catalyst = $(i\text{-}C_3H_7O)_4Ti$, $t\text{-}C_4H_9OOH$ and enantiomeric DET, diethyl tartrate.

Key features: Yields 1,2 or 1,3 diols depending on workup. Choice of enantiomer of DET determines enantiomer of product.

Pinacol rearrangement (27.3.B)

Equation: $\mathrm{R{-}C(OH)(R'){-}C(OH)(R'){-}R} \xrightarrow{\mathrm{H^+}} \mathrm{R{-}C(=O){-}C(R')_2{-}R}$

Generality: R, R' = H, alkyl, aryl

Key features: carbocation rearrangement
with an unsymmetrical diol, it is the more substituted carbon that loses the OH and gains the migrating group

Dehydration of 1,4- and 1,5-diols to cyclic ethers (27.3.B)

Equation: HO—( )—OH $\xrightarrow{H^+}$ (cyclic ether, O in ring) + H_2O

Generality: formation of 5- and 6-membered rings

Key features: intramolecular version of dehydration of alcohols to ethers: see Section 10.6.D

Periodate cleavage of 1,2-diols (27.3.B)

Equation: $R-C(OH)(R)-C(OH)(R')-R' + NaIO_4 \longrightarrow R_2C{=}O + O{=}CR'_2$

Generality: R, R' = H, alkyl, aryl
reaction can also be applied to α-hydroxycarbonyl compounds
(product is a ketone or aldehyde and a carboxylic acid [section 27.4.B])

Key features: useful for structure elucidation in carbohydrate chemistry
hydroxylation of an alkene (OsO_4) coupled with periodate cleavage (together called the Lemieux-Johnson reaction) is an alternative to ozonolysis

Acyloin condensation (27.4.A)

Equation: $2\ R-C(=O)-OR' \xrightarrow[\text{ether}]{4\ Na} \left[R-C(O^-){=}C(O^-)-R'\right] \xrightarrow{H_2O} R-C(=O)-CH(OH)-R'$

Generality: R, R' = alkyl

Key feature: useful for synthesizing large ring compounds from α,ω-diesters

Benzoin condensation (27.4.A)

Equation: $2\ Ar-C(=O)-H \xrightarrow{NaCN} Ar-C(=O)-CH(OH)-Ar$

Generality: limited to aromatic aldehydes and cyanide catalysis

Key features: proceeds via benzylic anion of aldehyde cyanohydrin

Synthesis and dehydration of β-hydroxy carbonyl compounds by aldol and related addition-condensation reactions (27.4.A and B; see also 15.2, 19.9, and 20.3.A)

Equation:

$$2\ R_2C{=}O + R{-}CH_2{-}C({=}O){-}R' \xrightarrow{\text{base}} R{-}C(OH)(R){-}CH(R){-}C({=}O){-}R' \longrightarrow R_2C{=}C(R){-}C({=}O){-}R'$$

Generality: R = various combinations of H, alkyl, aryl; R′ = alkyl, aryl, or OR
dehydration step acid- or base-catalyzed
Evans asymmetric aldol synthesis forms the enolate from a chiral carbonyl compound.

Key features: see Sections 15.2, 19.9, 20.3.A

Cyclic hemiacetals, hemiketals, acetals and ketals from hydroxy carbonyl compounds (27.4.B; see also 14.6.B)

Equation:

$$HO{\frown}C({=}O)R \rightleftharpoons \text{cyclic } O{-}C(R)(OH) \underset{H^+}{\overset{R'OH}{\rightleftharpoons}} \text{cyclic } O{-}C(R)(OR') + H_2O$$

Generality: R = H or alkyl
formation of 5- and 6-membered rings

Key features: simply the intramolecular version of hemiacetal and acetal formation (see Section 14.6.B)

Synthesis of α-hydroxyacids from α-haloacids (27.5.B)

Equation:

$$R{-}CH(X){-}CO_2H \xrightarrow{\text{NaOH}} R{-}CH(OH){-}CO_2H$$

Generality: X = Br (also Cl or I)

Key features: S_N2 process
useful combination with α-bromination of carboxylic acids (Section 18.7.B)

Synthesis of α-hydroxyacids from cyanohydrins (27.5.B)

Equation:

$$R{-}C(OH)(R){-}CN \xrightarrow{H_3O^+} R{-}C(OH)(R){-}CO_2H$$

Key features: useful combination with formation of cyanohydrin (Section 14.7.B)
process must be acid catalyzed; in presence of base, cyanohydrins decompose to NaCN and carbonyl compound

Lactonization (27.5.C)

Equation: HO—( )—CO_2R ⇌ (lactone, ring O and C=O) + ROH

Generality: R = H: acid catalyzed equilibrium; R = alkyl: acid or base catalyzed equilibrium for formation of five- and six-membered rings;
longer-chain hydroxy acids tend to polymerize instead of lactonize

Key features: simply an intramolecular esterification or transesterification reaction

Lactide formation (27.5.C)

Equation: 2 R—CH(OH)—CO_2R' ⇌ (lactide ring with two R groups, O=, =O) + 2 R'OH

Generality: limited to α-hydroxy acids and esters

Synthesis of dicarboxylic acids by oxidative cleavage of cycloalkenes and cycloalkanones (27.6.A)

Equation: (cycloalkene) $\xrightarrow{KMnO_4}$ HO_2C—( )—CO_2H $\xleftarrow[\Delta]{HNO_3}$ (cycloalkanone, C=O)

Decarboxylation of β-carbonyl carboxylic acid derivatives (27.6.C and 27.7.C)

Equation: R—C(=O)—CH(R')—CO_2H $\xrightarrow{\Delta}$ R—C(=O)—CHR' + CO_2

Generality: R = H, alkyl, aryl, OR"; RC=O group can be replaced by CN, too

Key features: reaction proceeds via cyclic transition state and enol form of carbonyl product
key step in malonic ester and acetoacetic ester syntheses

Cyclic anhydride and imide formation (27.6.C)

Equation: HOC(=O)—( )—C(=O)YH $\xrightarrow{KMnO_4}$ (cyclic O=C—Y—C=O ring) + H_2O

Generality: ring size = 5 or 6
Y = O (anhydride) or NR' (imide)
reaction is caused by heating or with a dehydrating agent such as PCl_5, $SOCl_2$, P_2O_5, $POCl_3$, etc.

Dieckmann condensation (27.6.C)

Equation:

$$RO_2CCH_2\text{—(ring)—}CO_2R \xrightarrow{RO^-} \text{2-(}RO_2C\text{)-cycloalkanone} + H_2O$$

Generality: ring size = 5 or 6

Key features: reaction is simply an intramolecular Claisen condensation

Oxidation of α-hydroxy ketones to α-diketones (27.7.A)

Equation:

$$R\text{—CH(OH)—C(=O)—}R' \xrightarrow{Cu(OAc)_2} R\text{—C(=O)—C(=O)—}R'$$

Key features: α-hydroxy ketones are sensitive to oxidative cleavage, so cupric acetate is specific reagent

Oxidation of ketones and aldehydes to α-dicarbonyl compounds (27.7.A)

Equation:

$$R\text{—}CH_2\text{—C(=O)—}R' \xrightarrow{SeO_2} R\text{—C(=O)—C(=O)—}R'$$

Claisen condensation (27.7.A.; see also 19.9)

Equation:

$$R\text{—C(=O)—OR''} + H_2C(R')\text{—C(=O)—Y} \xrightarrow{R''O^-} R\text{—C(=O)—CH(R')—C(=O)—Y}$$

Generality: R, R' = H, alkyl, aryl; Y = alkyl, aryl, O-alkyl; R" = alkyl

Key features: see section 19.9

Malonic ester and acetoacetic ester syntheses (27.7.E)

Equation:

$$R\text{—C(=O)—CH(R')—C(=O)—OR'''} + R''X \xrightarrow{\text{base}} R\text{—C(=O)—C(R'')(R')—C(=O)—OR'''} \xrightarrow[2.\ H^+]{1.\ OH^-} R\text{—C(=O)—CHR''(R')} + CO_2$$

Generality: R = R'''O: malonic ester synthesis; R = H, alkyl, or aryl: acetoacetic ester synthesis
R' = alkyl, aryl; R"X = good alkylating agent

Key features: important method for the synthesis of substituted ketones and carboxylic acids

Knoevenagel condensation (22.7.F)

Equation: $CH_2(CO_2R)_2$ + R'CH=O ⟶ R'CH=$CHCO_2R$

Generality: R, R' = H or alkyl; can be applied to ketones as well, although the yields are poor catalyzed by a weak base such as an amine

Key features: related to the aldol condensation (section 15.2, 19.9, 20.3)

Michael addition reaction (27.7.G)

Equation:

$$R{-}\overset{O}{\overset{\|}{C}}{-}CH_2{-}\overset{O}{\overset{\|}{C}}{-}R + CH_2{=}CH{-}\overset{O}{\overset{\|}{C}}{-}R \xrightarrow{RO^-} (R\overset{O}{\overset{\|}{C}})_2CHCH_2CH_2\overset{O}{\overset{\|}{C}}R$$

Generality: R = various combinations of H, alkyl, aryl, OR, etc.
very general process
RO^- used only in catalytic amounts

Key features: important method for the synthesis of 1,5-dicarbonyl compounds
coupled with an intramolecular aldol condensation, this is an important component of the **Robinson annulation**

27.C. Important Concepts and Hints

For the most part, previous chapters have focused on the chemistry of one functional group at a time. When two are present in a molecule, the reactions the compound undergoes often involve each group independently. However, in many cases the two functional groups interact, and together undergo reactions that are not typical of either group alone. In a sense, such a combination can be considered to be a new functional group, with its own set of reactions and ways to be synthesized. Some examples of this sort were presented in the chapter on conjugation (Chapter 20).

This chapter is organized in a logical fashion, presenting the various pairwise combinations of alcohol, aldehyde and ketone, and carboxylic acid functional groups, and outlining the special behavior thatresults when they are in 1,2- (adjacent), 1,3-, 1,4- or more remote relationships.

You may have seen many of the reactions in this chapter before; for instance, hydroxylation of olefins to give 1,2-diols and the cleavage of olefins to give dicarboxylic acids were both discussed in the chapter on alkenes (Chapter 11). Similarly, the synthesis of β-hydroxy ketones and their dehydration were discussed in Chapter 15 in connection with the aldol condensation. The reactions of difunctional compounds frequently involve the formation or cleavage of cyclic compounds, by reactions that you have previously learned for the intermolecular cases. For example, the hydrolysis or formation of a lactone is simply the hydrolysis or formation of an ester in which the hydroxy and carboxylic acid groups are part of the same molecule.

From the synthetic standpoint, the synthesis of β-keto esters by the Claisen condensation, its intramolecular counterpart, the Dieckmann condensation, the alkylation of β-keto esters and malonic esters and their subsequent decarboxylation (the "acetoacetic ester synthesis" and the "malonic ester synthesis") are among the most important reactions in organic synthesis. (The decarboxylation of β-keto acids and malonic acids is another example of six electrons going around a circle – see Section 20.B of this Study Guide). The Robinson annulation is also a noteworthy sequence that involves a Michael addition and subsequent intramolecular aldol condensation, and it has been used extensively in the synthesis of natural products.

27.D. Answers to Exercises

27.1 For example:

$HO(CH_2)_6OH$

1,6-hexanediol

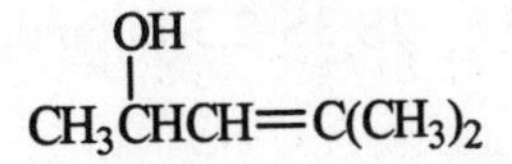

$CH_3CH(OH)CH{=}C(CH_3)_2$

4-methyl-3-penten-2-ol

$(CH_3CH_2)_2C(OH)CH{=}O$

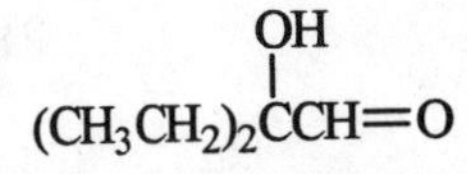

2-ethyl-2-hydroxybutanal

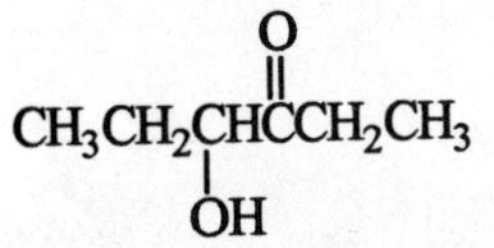

$CH_3CH_2CH(OH)C(=O)CH_2CH_3$

4-hydroxy-3-hexanone

$(CH_3)_2CHCH(OH)CH_2CO_2H$

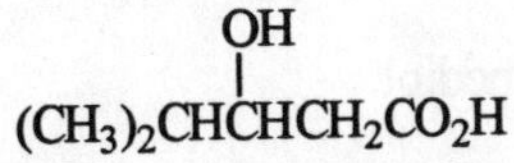

3-hydroxy-4-methylpentanoic acid

$(CH_3)_2C{=}CHCH{=}CH_2$

4-methyl-1,3-pentadiene

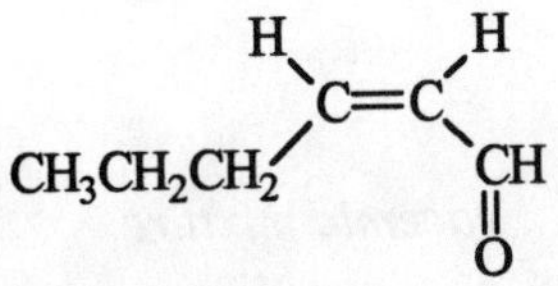

$CH_3CH_2CH_2CH{=}CHCH{=}O$ (H and H cis)

(*Z*)-2-hexenal

$CH_2{=}CHCH_2CH_2CH_2COOH$

5-hexenoic acid

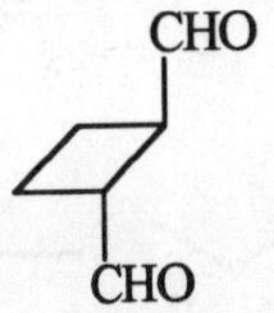

CHO / CHO (cyclobutane)

trans-1,2-cyclobutane dicarbaldehyde (or -dicarboxaldehyde)

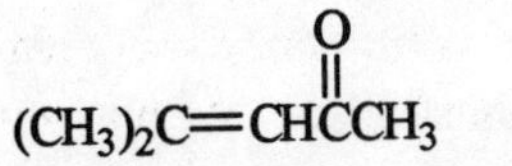

$(CH_3)_2C{=}CHC(=O)CH_3$

4-methyl-3-penten-2-one

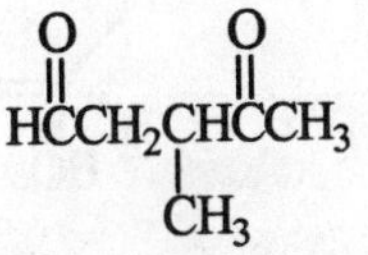

$HC(=O)CH_2CH(CH_3)C(=O)CH_3$

3-methyl-4-oxopentanal

$HC(=O)CH_2C(CH_3)_2CO_2H$

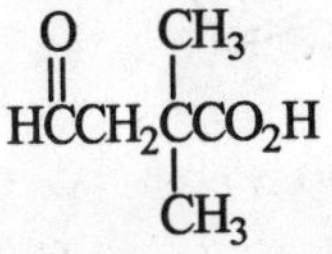

2,2-dimethyl-4-oxobutanoic acid

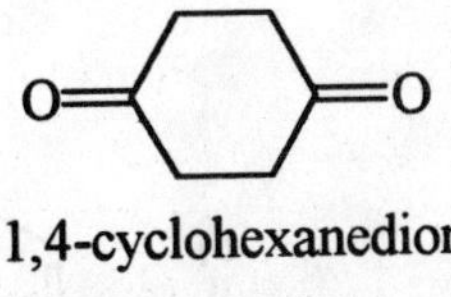

O=⟨cyclohexane⟩=O

1,4-cyclohexanedione

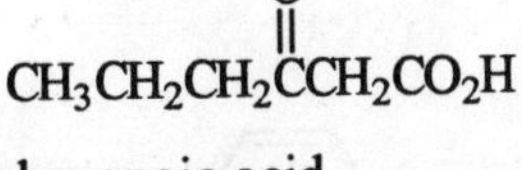

$CH_3CH_2CH_2C(=O)CH_2CO_2H$

3-oxohexanoic acid

$HO_2CCH(CH_2CH_3)CH_2CO_2H$

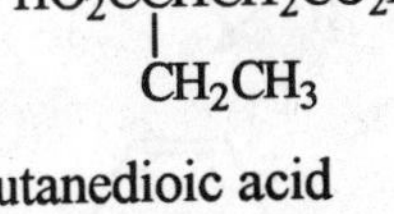

2-ethylbutanedioic acid

27.2 cyclooctene $\xrightarrow{KMnO_4}$ (H, OH / HO, H) ⟶ (H, H / HO OH) *trans*-1,2-cyclooctanediol

27.3 a. → 2R,3R-2,3-butandiol + 2S,3S-2,3-butandiol *racemic mixture*

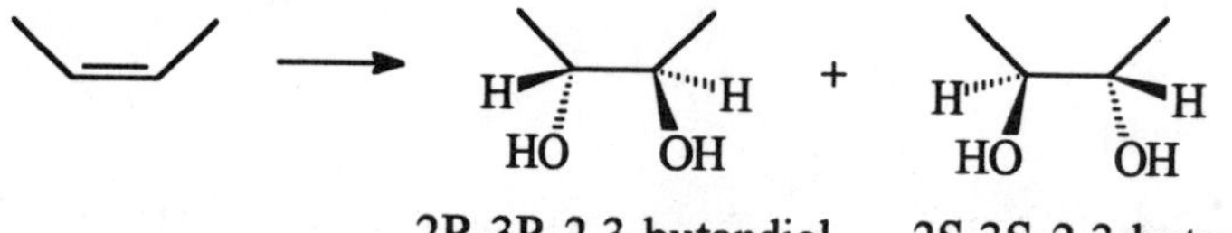

b. → 2R,3S-2,3-butanediol *meso compound*

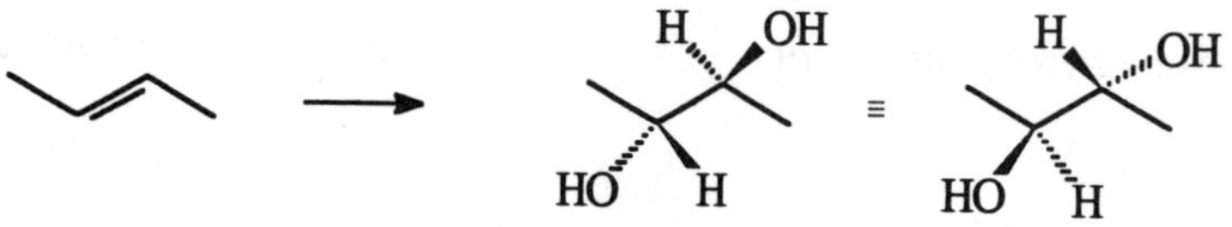

c. → 2R,3R-2,3-pentanediol + 2S,3S-2,3-pentanediol *racemic mixture*

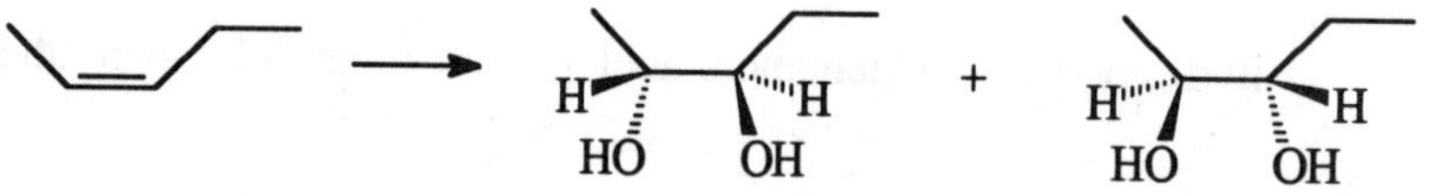

d. → 2R,3S-2,3-pentanediol + 2S,3R-2,3-pentanediol *racemic mixture*

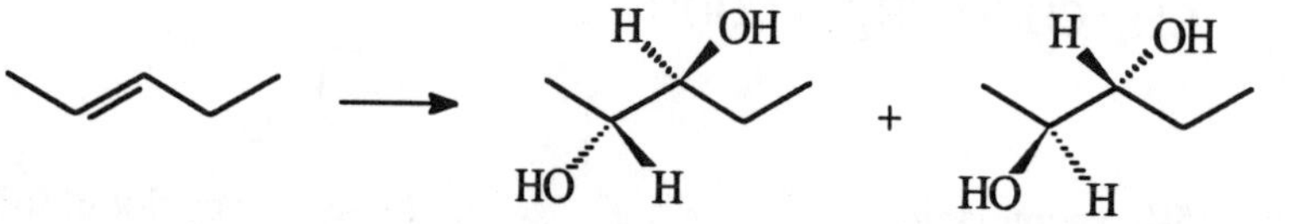

e. → *cis*-1,2-cyclo-octanediol

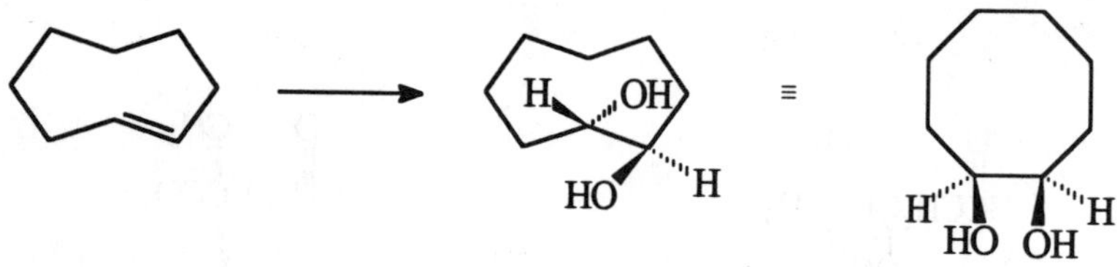

27.4 $Ar_2C(OH)-C(OH)Ar_2$ $\xrightarrow{H^+, -H_2O}$ → etc. $\xrightarrow{-H^+}$

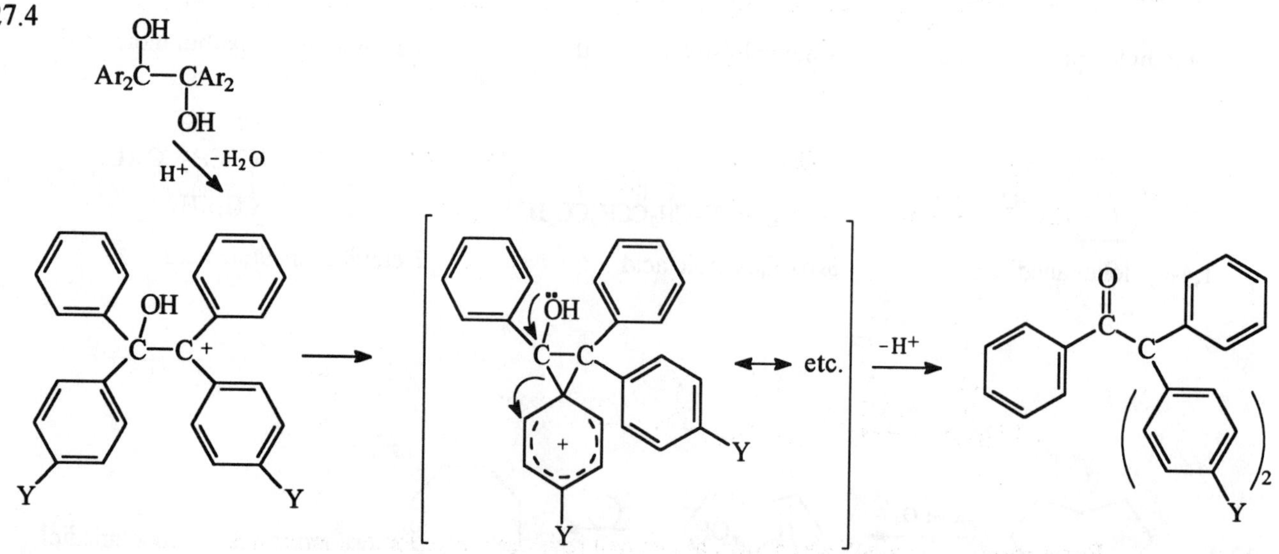

favored relative to phenyl migration when Y = CH_3, disfavored for Y = NO_2

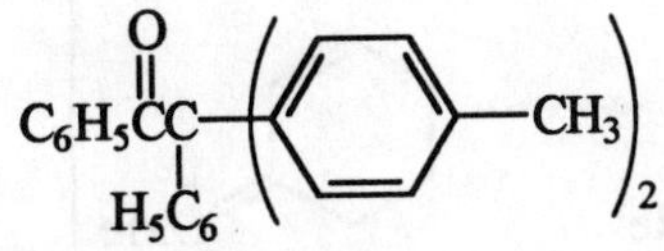

$C_6H_5C(=O)C(C_6H_5)(C_6H_4\text{-}CH_3)_2$ and $O_2N\text{-}C_6H_4\text{-}C(=O)C(C_6H_5)_2(C_6H_4\text{-}NO_2)$ are the major products.

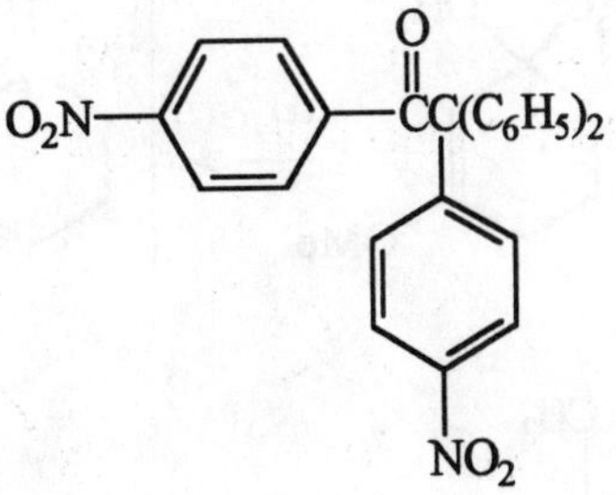

27.5 2 cyclohexanone $\xrightarrow[\text{benzene}]{\text{Mg}}$ 1,1'-bi(cyclohexyl)-1,1'-diol (OH, HO) $\xrightarrow{H_2SO_4}$ spiro ketone (cycloheptanone spiro-fused to cyclohexane)

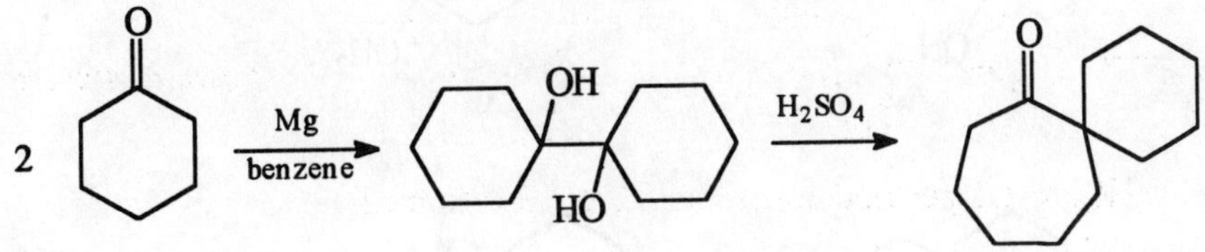

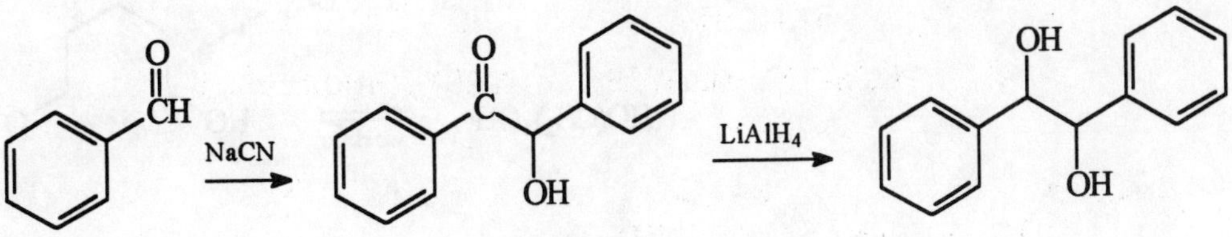

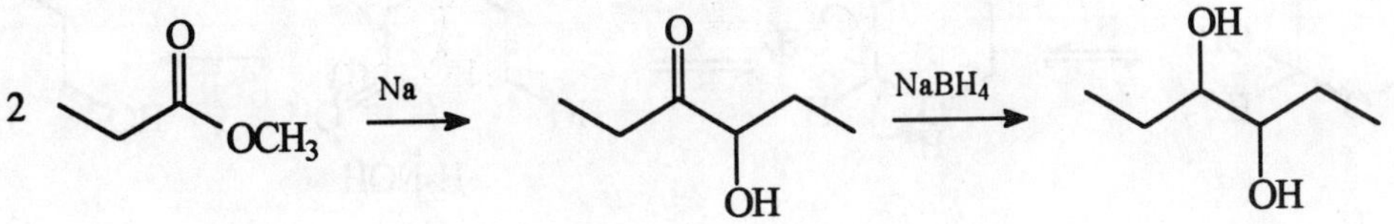

27.6 a. benzaldehyde ($C_6H_5CH=O$) $\xrightarrow{\text{NaCN}}$ $C_6H_5C(=O)CH(OH)C_6H_5$ $\xrightarrow{\text{LiAlH}_4}$ $C_6H_5CH(OH)CH(OH)C_6H_5$

b. 2 $CH_3CH_2C(=O)OCH_3$ $\xrightarrow{\text{Na}}$ $CH_3CH_2C(=O)CH(OH)CH_2CH_3$ $\xrightarrow{\text{NaBH}_4}$ $CH_3CH_2CH(OH)CH(OH)CH_2CH_3$

c. 2 $CH_3CH=O$ $\xrightarrow{\text{NaOH}}$ $CH_3CH(OH)CH_2CH=O$ $\xrightarrow{\text{NaBH}_4}$ $CH_3CH(OH)CH_2CH_2OH$

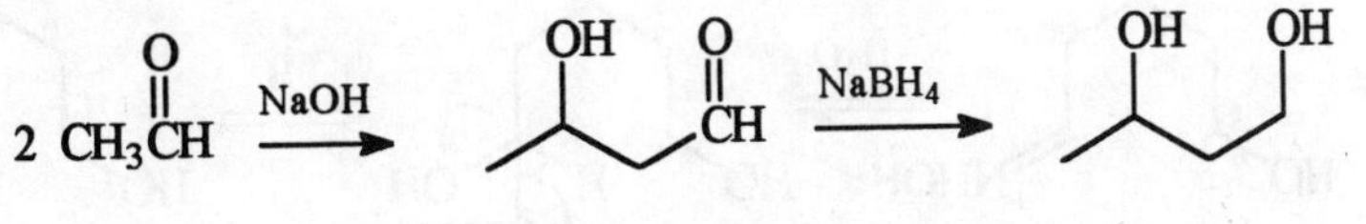

d. 2 $CH_3CH=O$ $\xrightarrow{\text{NaOH}}$ $CH_3CH(OH)CH_2CH=O$ $\xrightarrow{CH_3CH_2MgBr}$ $CH_3CH(OH)CH_2CH(OH)CH_2CH_3$

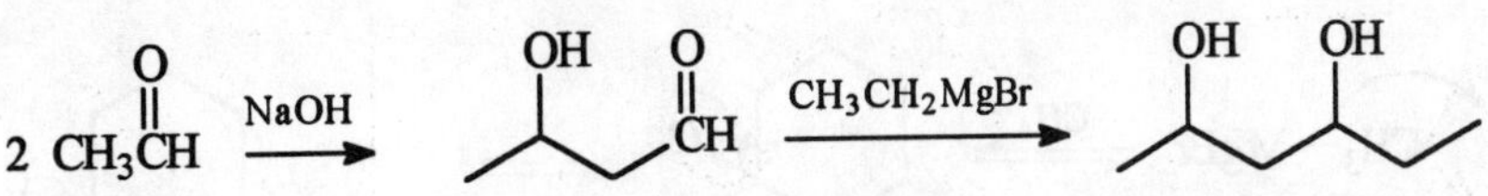

Except for c., the syntheses shown above will result in diasteriomeric mixtures.

27.7 The intermediate cation is stabilized by resonance with the lone pair electrons on the ether oxygen:

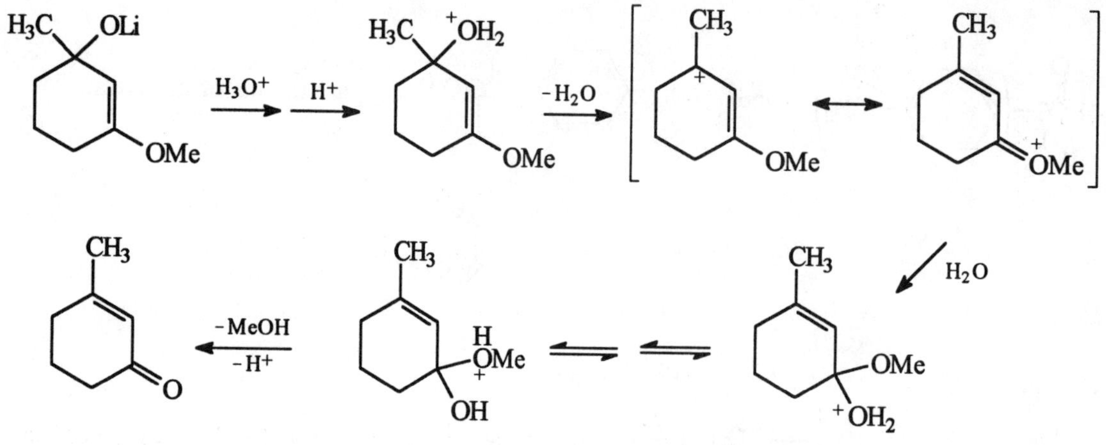

H_3C OLi OMe $\xrightarrow{H_3O^+}$ $\xrightarrow{H^+}$ H_3C $^+OH_2$ OMe $\xrightarrow{-H_2O}$ [CH_3 OMe ⟷ CH_3 $\overset{+}{O}Me$]

$\xrightarrow{H_2O}$ CH_3 OMe $^+OH_2$ ⇌ ⇌ CH_3 H $\overset{+}{O}Me$ OH $\xrightarrow[-H^+]{-MeOH}$ CH_3 O

27.8

OH H O ⇌ O–H H O ⇌ HO H O H ; H–BH_3 → HO O^- $\underset{workup}{\overset{H_2O}{\rightleftharpoons}}$ $HO(CH_2)_5OH$

OH H O $\overset{H^+}{\rightleftharpoons}$ $\ddot{O}H$ H O H^+ ⇌ HO $\overset{+}{O}H$ $H_2\ddot{N}OH$ → HO H OH $H_2\overset{+}{N}OH$ $\overset{-H^+}{\rightleftharpoons}$ HO H OH HNOH $\overset{H^+}{\rightleftharpoons}$ HO H $\overset{+}{O}H_2$:NHOH $\overset{-H_2O}{\rightleftharpoons}$ HO $\overset{+}{N}HOH$ $\overset{-H^+}{\rightleftharpoons}$ $HO(CH_2)_4CH{=}NOH$

O–H H O CH_3–MgBr $\overset{-CH_4}{\rightleftharpoons}$ O^- H O Mg^{2+} Br^- ⇌ BrMgO O CH_3–MgBr → BrMgO H_3C OMgBr $\overset{H_2O}{\rightleftharpoons}$ $HO(CH_2)_4CH(OH)CH_3$

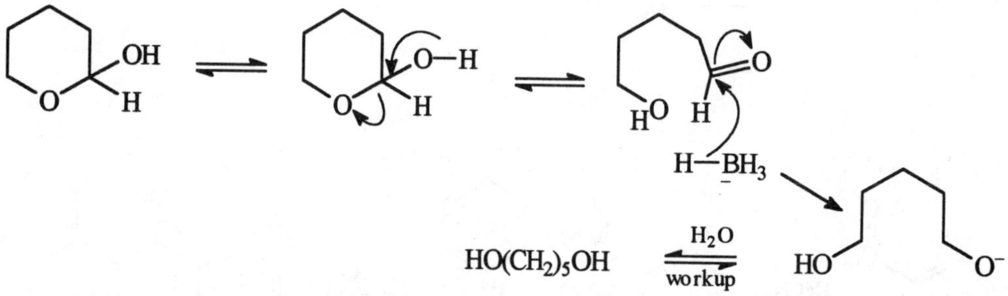

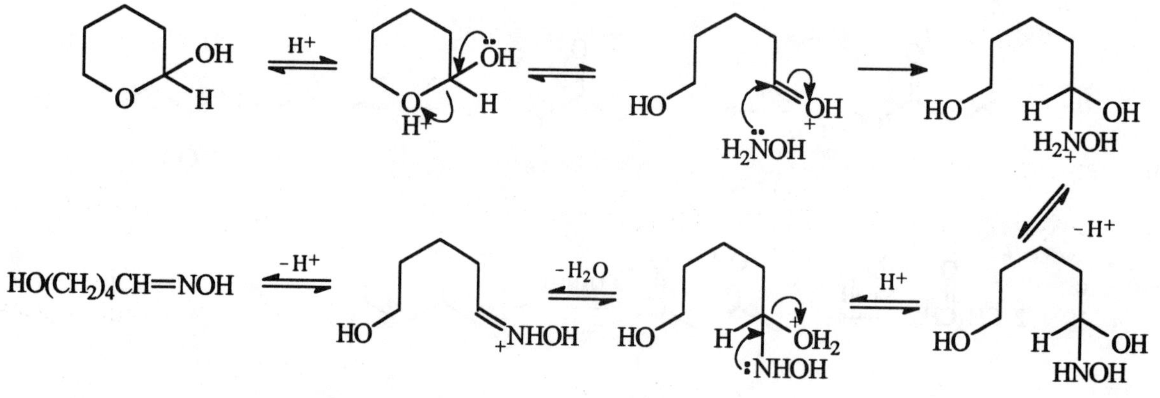

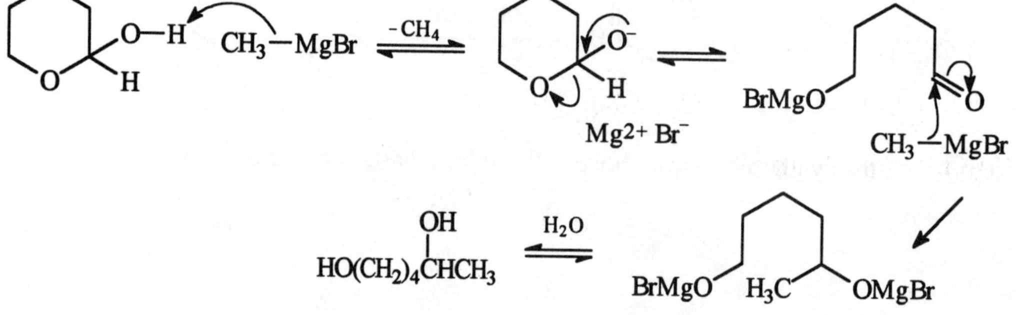

$$\text{(2-hydroxytetrahydropyran)} + CH_3C(=O)Cl \longrightarrow \text{ring-}\overset{+}{O}H\text{-}C(O^-)(Cl)\text{-}CH_3 \xrightleftharpoons{-H^+} \text{ring-}O\text{-}C(O^-)(Cl)\text{-}CH_3 \xrightarrow{-Cl^-} \text{ring-}O\text{-}C(=O)\text{-}CH_3$$

$$\text{(2-hydroxytetrahydropyran)} \xrightleftharpoons{H_2Cr_2O_7} \text{ring-}O\text{-}\overset{VI}{Cr}(=O)_2\text{-}OH \longrightarrow \text{(lactone)} + (\text{Cr(IV)} \rightarrow \text{disproportionates})$$

27.9 a. $CH_3(CH_2)_2CH_2CO_2H \xrightarrow{P, Br_2} \xrightarrow{H_2O} CH_3(CH_2)_2CHBrCO_2H \xrightarrow{NaOH} CH_3(CH_2)_2CHOHCO_2H$

b. $CH_3CH_2CH_2CH{=}O \xrightarrow{HCN} CH_3CH_2CH_2CH(OH)CN \xrightarrow{H_2SO_4} CH_3CH_2CH_2CH(OH)CO_2H$

$CH_3C(=O)OC_2H_5 \xrightarrow[THF\ -70°]{LDA} \xrightarrow{CH_3CH_2CH=O} \xrightarrow{OH^-} CH_3CH_2CH(OH)CH_2CO_2H$

27.10 $(CH_3)_2C{=}CHCH_2Br \xrightarrow{^-CH_2CO_2Et} (CH_3)_2C{=}CHCH_2CH_2CH_2CO_2Et \xrightarrow{H_3O^+} (CH_3)_2C{=}CHCH_2CH_2CH_2CO_2H \xrightarrow{\text{aq. } H_2SO_4 \text{ or } Hg(OAc)_2 : NaBH_4} (CH_3)_2C(OH)CH_2CH_2CH_2CH_2CO_2H \xrightarrow[-H_2O]{H^+}$ (6,6-dimethyl lactone)

27.11 a. $2\ CH_3CH_2CH(OH)CO_2H \xrightarrow{H^+}$ (cyclic dilactide bearing CH_3CH_2 and CH_2CH_3 substituents)

b.

$$CH_3CH(OH)CH_2CO_2H \xrightarrow{H^+} CH_3CH{=}CHCO_2H$$

c.

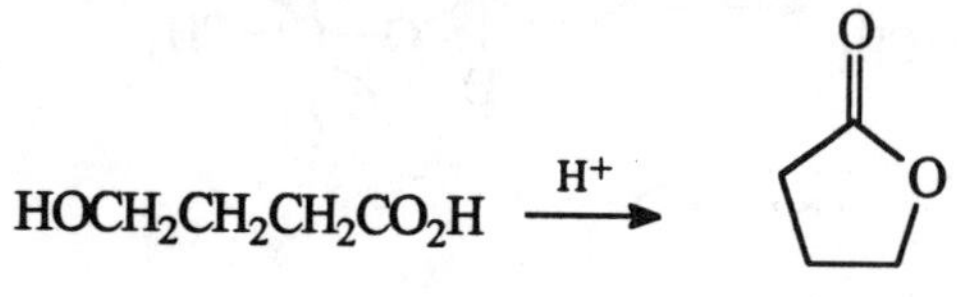

$$HOCH_2CH_2CH_2CO_2H \xrightarrow{H^+}$$ (γ-butyrolactone)

27.12

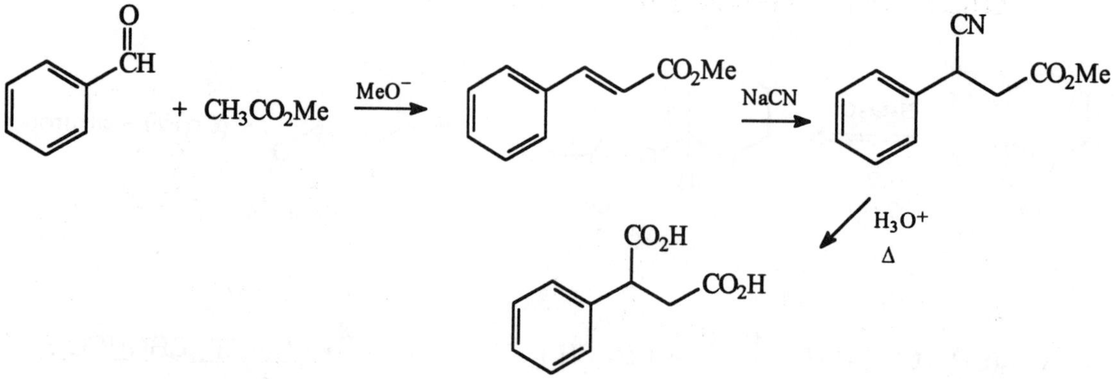

Benzaldehyde + CH_3CO_2Me $\xrightarrow{MeO^-}$ $PhCH{=}CHCO_2Me$ $\xrightarrow{NaCN}$ $PhCH(CN)CH_2CO_2Me$ $\xrightarrow{H_3O^+, \Delta}$ $PhCH(CO_2H)CH_2CO_2H$

27.13

$$HO_2CCH_2CH_2CH_2CO_2H \xrightarrow{\Delta} \text{(glutaric anhydride)} + H_2O$$

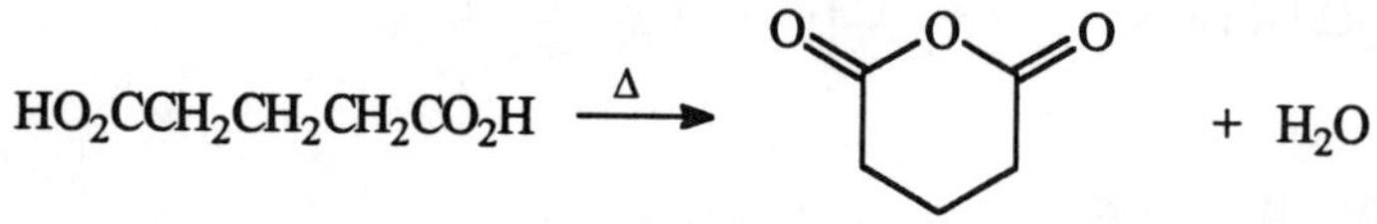

$$CH_3CH_2CH(CO_2H)_2 \xrightarrow{\Delta} CH_3CH_2CH_2CO_2H + CO_2$$

$$(CH_3)_2C(CO_2H)_2 \xrightarrow{\Delta} (CH_3)_2CHCO_2H + CO_2$$

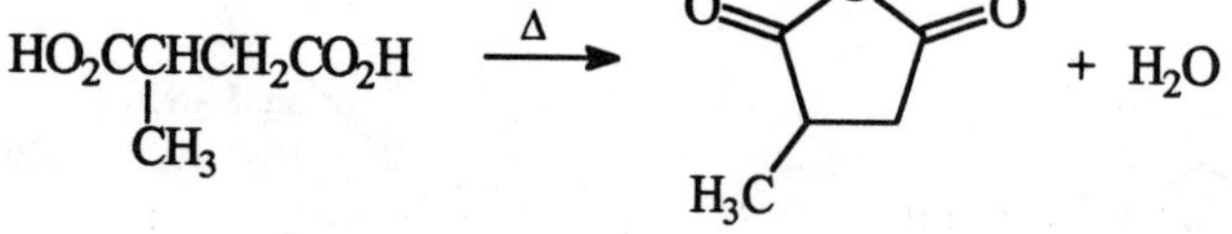

$$HO_2CCH(CH_3)CH_2CO_2H \xrightarrow{\Delta} \text{(methylsuccinic anhydride, } H_3C\text{)} + H_2O$$

27.14

$$EtO_2CCH_2CH_2CH_2CH_2CO_2Et \overset{EtO^-}{\rightleftharpoons} EtO_2C\bar{C}HCH_2CH_2CH_2COEt \rightleftharpoons$$

(cyclic tetrahedral intermediate with EtO_2C, H, ^-O, OEt) $\rightleftharpoons$ (2-carbethoxycyclopentanone; EtO⁻ removes H) $\rightleftharpoons$ (enolate, EtO_2C, O^-) $\xrightarrow[\text{(work-up)}]{H^+}$ (2-carbethoxycyclopentanone, EtO_2C, H)

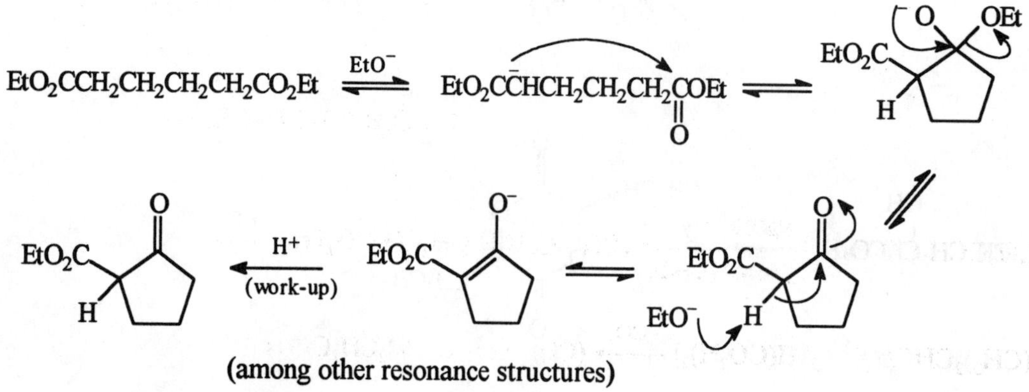

(among other resonance structures)

27.15 a. $2\ CH_3CH_2CO_2Et \xrightarrow[\text{ether}]{\text{Na}} \xrightarrow{H_2O} CH_3CH_2C(=O)CH(OH)CH_2CH_3 \xrightarrow{Cu(OAc)_2} CH_3CH_2C(=O)C(=O)CH_2CH_3$

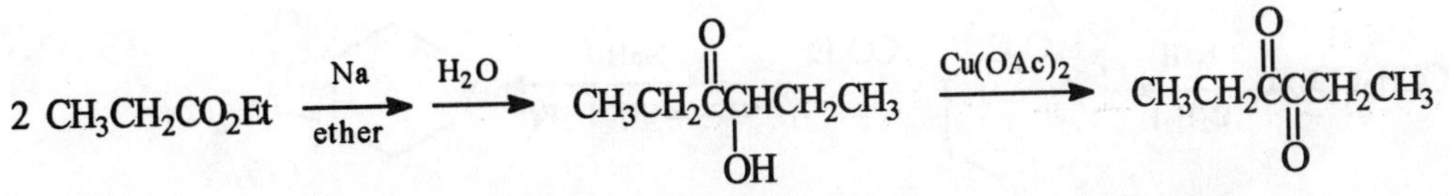

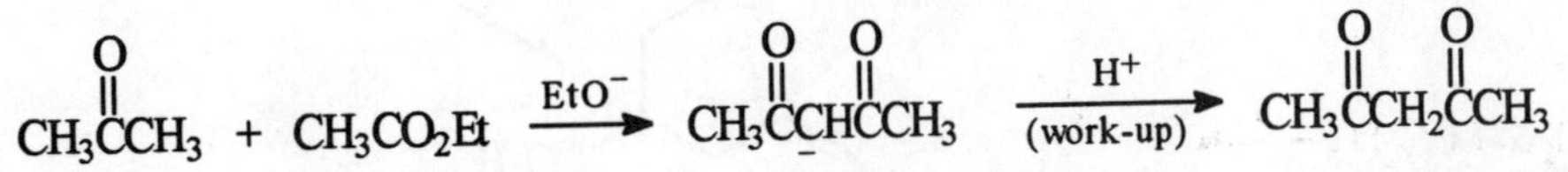

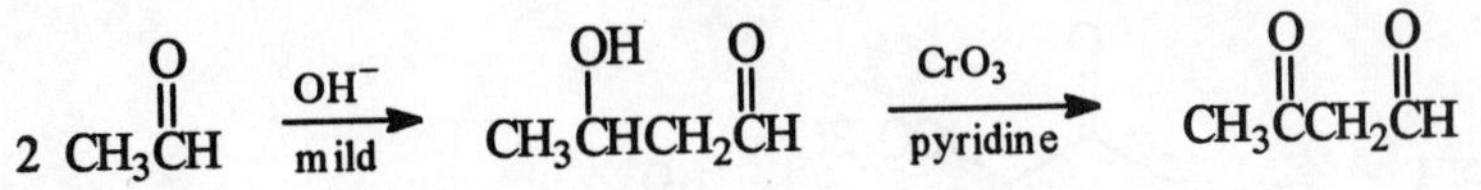

b. $CH_3C(=O)CH_3 + CH_3CO_2Et \xrightarrow{EtO^-} CH_3C(=O)\underline{C}HC(=O)CH_3 \xrightarrow[\text{(work-up)}]{H^+} CH_3C(=O)CH_2C(=O)CH_3$

c. $2\ CH_3CH(=O) \xrightarrow[\text{mild}]{OH^-} CH_3CH(OH)CH_2CH(=O) \xrightarrow[\text{pyridine}]{CrO_3} CH_3C(=O)CH_2CH(=O)$

27.16 (H-bonded enol forms of 2,4-pentanedione, H_3C, CH_3, H, interconverting: fast)

from small amount of diketo form: CH_2, CH_3; area 1; area 6; $-CH_3$

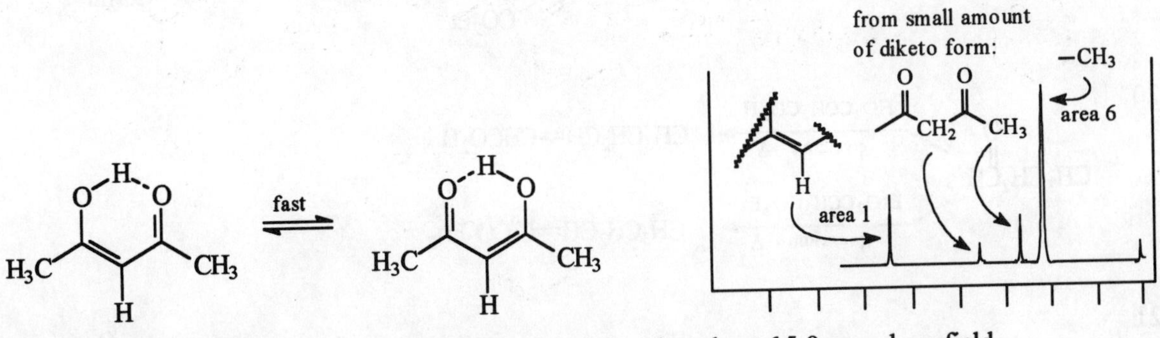

The H-bonded H from the OH is strongly shifted and found at about 15.8 ppm downfield.

27.17 Model-building, no answer required.

27.18 2,4-pentanedione + MeO^- ⇌ enolate (O^-, O) + MeOH $K = 10^{6.5} = 3.2 \times 10^6$

pK$_a$: 9 ; 15.5 (Table 10.3)

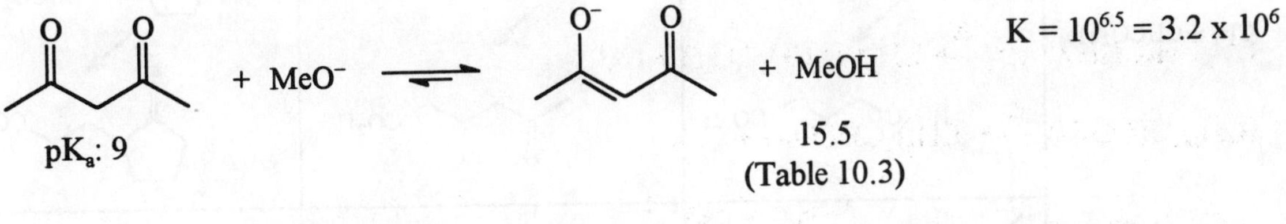

27.19 a. $CH_3C(=O)CH_2CO_2Et \xrightarrow[CH_2=CHCH_2Br]{EtO^-} CH_3C(=O)CH(CH_2CH=CH_2)CO_2Et \xrightarrow{H_3O^+} [CH_3C(=O)CH(CH_2CH=CH_2)CO_2H] \xrightarrow{-CO_2} CH_3C(=O)CH_2CH_2CH=CH_2$

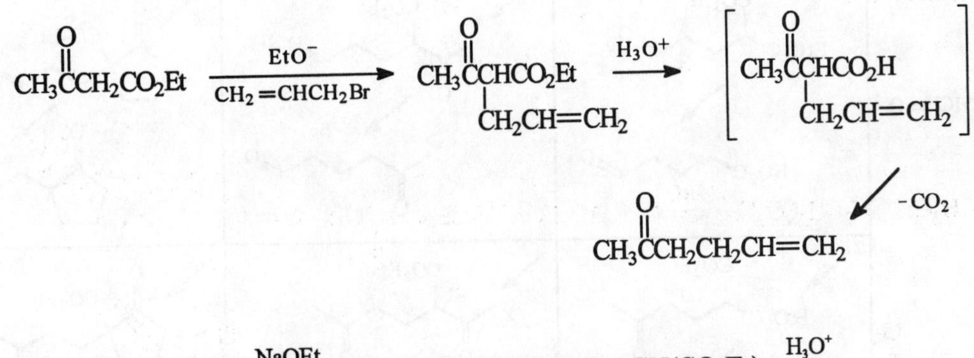

b. $EtO_2CCH_2CO_2Et \xrightarrow[(CH_3)_2CHCH_2CH_2Br]{NaOEt} (CH_3)_2CHCH_2CH_2CH(CO_2Et)_2 \xrightarrow[\Delta]{H_3O^+}$

$[(CH_3)_2CHCH_2CH_2CH(CO_2H)_2] \xrightarrow{-CO_2} (CH_3)_2CHCH_2CH_2CH_2CO_2H$

c.

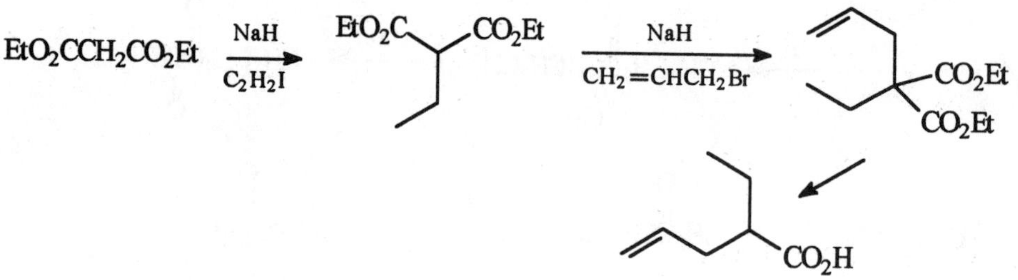

$EtO_2CCH_2CO_2Et$ $\xrightarrow[C_2H_5I]{NaH}$ EtO_2C–CH(Et)–CO_2Et $\xrightarrow[CH_2=CHCH_2Br]{NaH}$ (allyl)(ethyl)C(CO_2Et)$_2$ → $CH_2=CHCH_2CH(CH_2CH_3)CO_2H$

d. $CH_3COCH_2CO_2Et$ $\xrightarrow[BrCH_2CO_2Et]{NaH}$ $CH_3COCH(CO_2Et)CH_2CO_2Et$ $\xrightarrow[\Delta]{H_3O^+}$ $CH_3COCH_2CH_2CO_2Et$

27.20

CH_3CH_2CHO $\xrightarrow[\text{pyridine } \Delta]{HO_2CCH_2CO_2H}$ $CH_3CH_2CH{=}CHCO_2H$

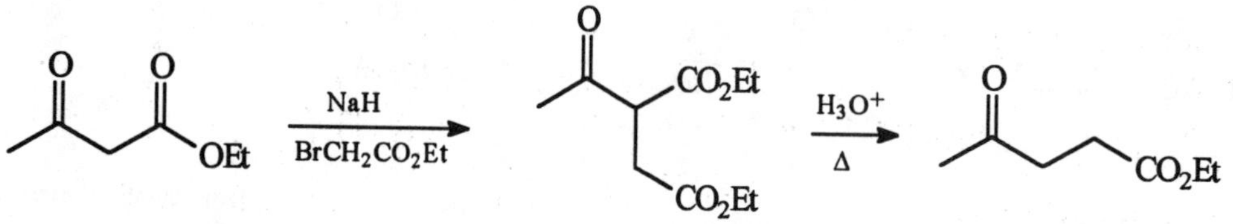

CH_3CH_2CHO $\xrightarrow[\text{piperidine } \Delta]{EtO_2CCH_2CO_2Et}$ $CH_3CH_2CH{=}C(CO_2Et)_2$

27.21

	$EtO_2CCH_2CO_2Et$	$CH_3COCH_2CO_2Et$	2-(CO_2Et)cyclopentanone
$CH_2=CHCO_2Me$	EtO_2C–CH(CO_2Et)$CH_2CH_2CO_2Me$ → $HO_2CCH_2CH_2CH_2CO_2H$	CH_3CO–CH(CO_2Et)$CH_2CH_2CO_2Me$ → $CH_3COCH_2CH_2CH_2CO_2H$	2-(CO_2Et)-2-($CH_2CH_2CO_2Me$)cyclopentanone → 2-($CH_2CH_2CO_2H$)cyclopentanone
$CH_2=CHCH=O$	EtO_2C–CH(CO_2Et)CH_2CH_2CHO → $HO_2CCH_2CH_2CH_2CHO$	CH_3CO–CH(CO_2Et)CH_2CH_2CHO → $CH_3COCH_2CH_2CH_2CHO$	2-(CO_2Et)-2-(CH_2CH_2CHO)cyclopentanone → 2-(CH_2CH_2CHO)cyclopentanone
$CH_2=CHCOCH_3$	EtO_2C–CH(CO_2Et)$CH_2CH_2COCH_3$ → $HO_2CCH_2CH_2CH_2COCH_3$	CH_3CO–CH(CO_2Et)$CH_2CH_2COCH_3$ → $CH_3COCH_2CH_2CH_2COCH_3$	2-(CO_2Et)-2-($CH_2CH_2COCH_3$)cyclopentanone → 2-($CH_2CH_2COCH_3$)cyclopentanone

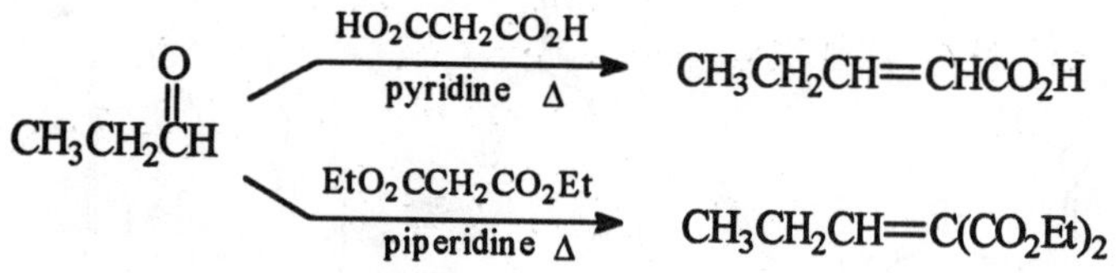

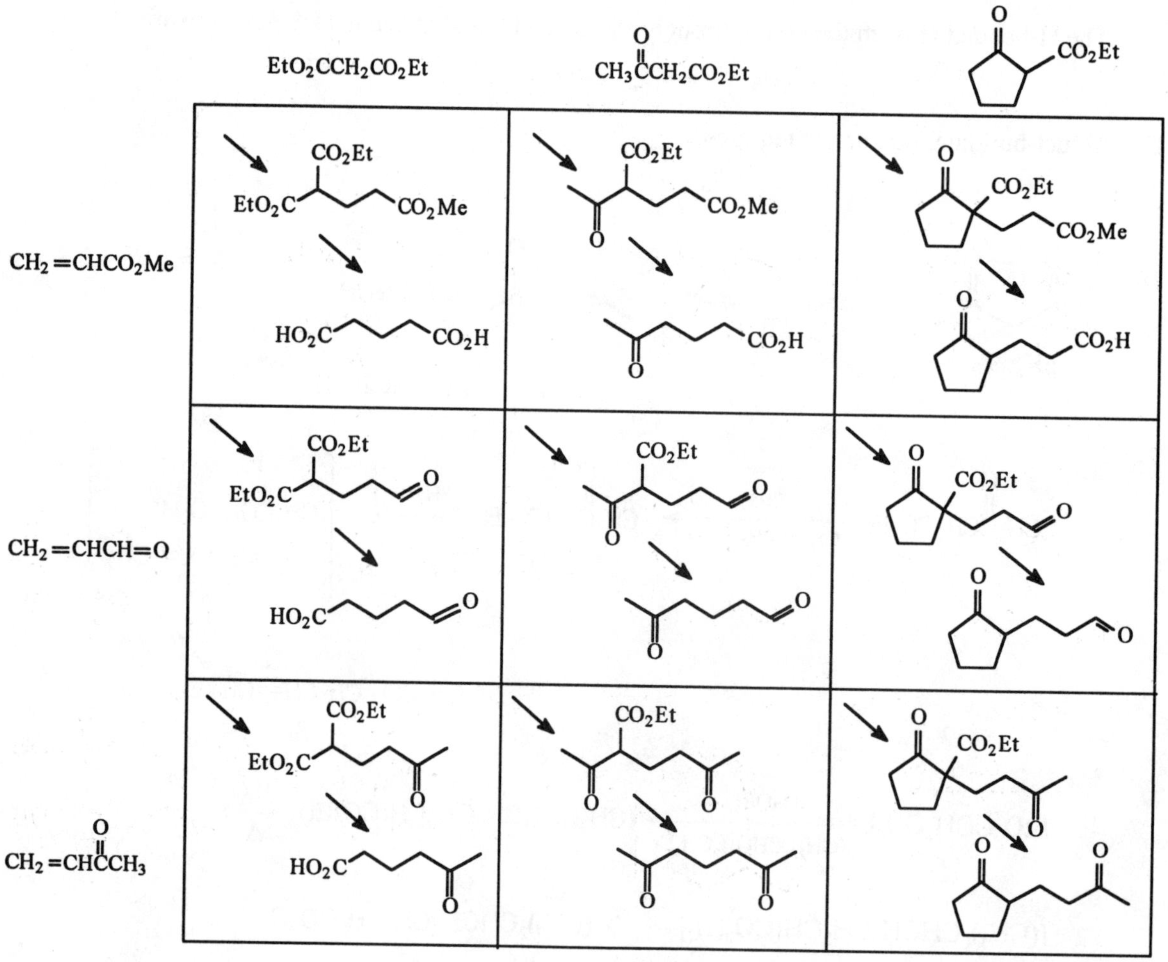

27.22 (cyclopentanecarbaldehyde) + $CH_2{=}CHCCH_3$ (O) $\xrightarrow{KOH}$ (spiro keto aldehyde: $CH_2{-}CH_2$, H, C=O, H_3C, O) $\xrightarrow{KOH}$ (spiro cyclohexenone)

27.E. Answers and Explanations for Problems

1. a. methyl 4-methyl-3-oxopentanoate
   b. 2,2-dimethyl-1,3-propanediol
   c. 4-hydroxybutanoic acid
   d. 4-oxopentanenitrile
   e. β-methylglutaric acid; 3-methylpentanedioic acid
   f. α-methyladipic acid; 2-methylhexanedioic acid
   g. dimethyl methylpropylmalonate; dimethyl 2-methyl-2-propylpropanedioate
   h. 3-methyl-5-oxopentanoic acid
   i. 4-hydroxybutanamide
   j. hex-5-en-2-one
   k. 2,2-dimethylcyclohexane-1,4-dione
   l. α-methyl-γ-ketovaleric acid; 2-methyl-4-oxopentanoic acid
   m. glutaronitrile; pentanedinitrile
   n. ϵ-methyl-γ-ketoenanthaldehyde; 6-methyl-4-oxoheptanal

2.

(cis-cyclododecene) $\xrightarrow[\text{or 1. } Br_2,\ H_2O\ \ 2.\ NaOH]{CH_3CO_3H}$ (epoxide, H, O, H) $\xrightarrow[H_2O]{NaOH}$ (diol: H, OH, OH, H) *trans*

$\xrightarrow[H_2O_2]{OsO_4}$ [osmate ester: O, O, Os, O, O] $\longrightarrow$ (diol: H, OH, OH, H) *cis*

(trans-cyclododecene) $\xrightarrow[\text{or 1. } Br_2,\ H_2O\ \ 2.\ NaOH]{CH_3CO_3H}$ (epoxide: O, H, H) $\xrightarrow[H_2O]{NaOH}$ (H, OH, HO, H) ≡ (H, OH, OH, H) *cis*

$\xrightarrow[H_2O_2]{OsO_4}$ [osmate ester: O=Os=O, O, O, H, H] $\longrightarrow$ (HO, H, HO, H) ≡ (H, OH, OH, H) *trans*

3. Loss of one of the two hydroxy groups can give a primary carbocation (unfavored); loss of the other gives a tertiary carbocation that can be further stabilized by delocalization into the two benzene rings.

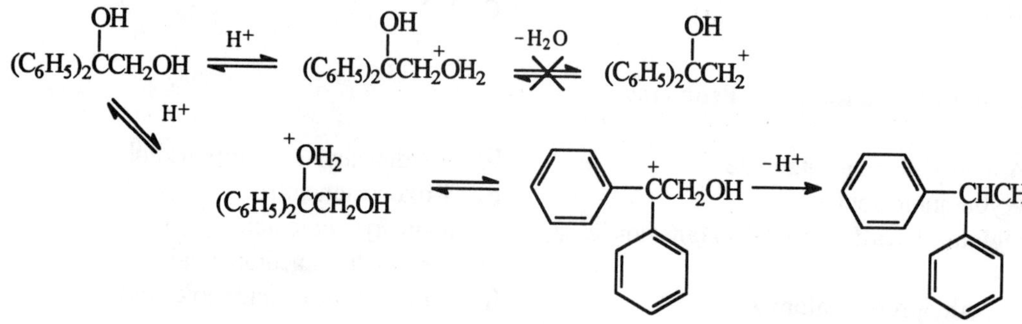

$(C_6H_5)_2C(OH)CH_2OH \overset{H^+}{\rightleftharpoons} (C_6H_5)_2C(OH)CH_2\overset{+}{O}H_2 \overset{-H_2O}{\not\rightleftharpoons} (C_6H_5)_2C(OH)\overset{+}{C}H_2$

$(C_6H_5)_2C(OH)CH_2OH \overset{H^+}{\rightleftharpoons} (C_6H_5)_2C(\overset{+}{O}H_2)CH_2OH \rightleftharpoons (C_6H_5)_2\overset{+}{C}CH_2OH \xrightarrow{-H^+} (C_6H_5)_2CHCHO$

resonance-stabilized tertiary carbocation

4. a. 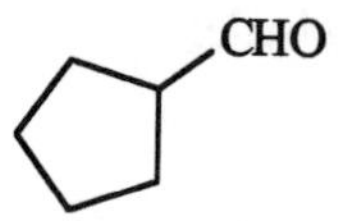 cyclopentane–CHO

b. $HOH_2CC(CH_3)_2CHO + H_2C{=}O$

c. $\left(O(CH_2)_6\overset{O}{\overset{\|}{C}} \right)_n$

polymer

d. 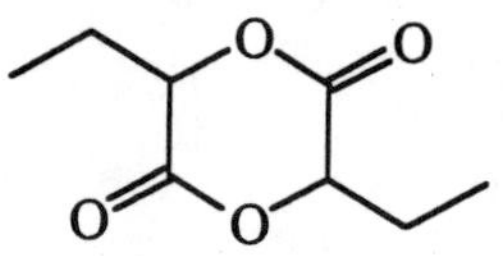 (cyclic diester: 3,6-diethyl-1,4-dioxane-2,5-dione)

e. 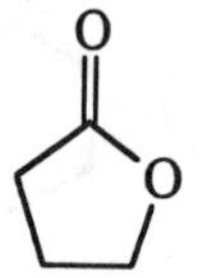 (γ-butyrolactone)

f. 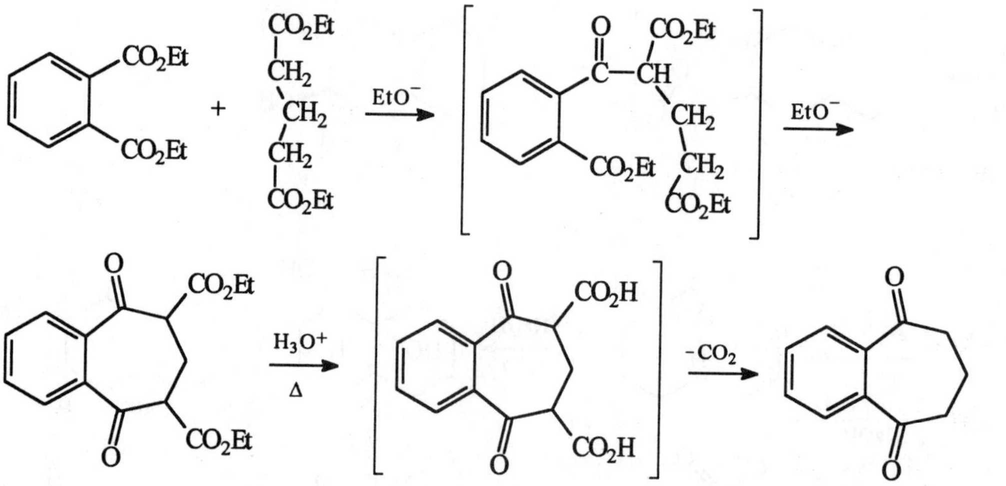 (benzene-1,2-di-CO_2Et) + EtO_2C–CH_2–CH_2–CH_2–CO_2Et $\xrightarrow{EtO^-}$ [$C_6H_4(CO_2Et)$–C(=O)–CH(CO_2Et)–CH_2–CH_2–CO_2Et] $\xrightarrow{EtO^-}$

(benzocycloheptanedione with two CO_2Et) $\xrightarrow[\Delta]{H_3O^+}$ [(benzocycloheptanedione with two CO_2H)] $\xrightarrow{-CO_2}$ (benzocycloheptane-1,5-dione)

Note that both CO_2H groups are β-keto acids

g.

$$H\!-\!C(D)(CH_3)\!-\!OH\ (S) \xrightarrow[\text{pyridine}]{C_6H_5SO_2Cl} H\!-\!C(D)(CH_3)\!-\!OSO_2C_6H_5 \xrightarrow[EtO^-]{CH_2(CO_2Et)_2} (EtO_2C)_2CH\!-\!C(D)(CH_3)\!-\!H \xrightarrow[\Delta]{H^+} HO_2CCH_2\!-\!C(D)(CH_3)\!-\!H\ (R)$$

note inversion of configuration

h.

$$(EtO_2C)_2CH_2 + EtO^- \rightleftharpoons (EtO_2C)_2CH^-$$

$$(EtO_2C)_2CH^- + C_6H_5CHO \rightleftharpoons (EtO_2C)_2CHCH(O^-)C_6H_5$$

$$(EtO_2C)_2CHCH(O^-)C_6H_5 + EtOH \rightleftharpoons (EtO_2C)_2CHCH(OH)C_6H_5 + EtO^-$$

$$\rightleftharpoons (EtO_2C)_2\bar{C}\!-\!CH(OH)C_6H_5 \xrightarrow{-OH^-} (EtO_2C)_2C{=}CHC_6H_5 + (EtO_2C)_2CH^-$$

$$\longrightarrow (EtO_2C)_2\bar{C}CHCH(CO_2Et)_2\ (C_6H_5) \xrightarrow[\Delta]{H^+} (HO_2C)_2CHCHCH(CO_2H)_2\ (C_6H_5) \xrightarrow{-CO_2} HO_2CCH_2CHCH_2CO_2H\ (C_6H_5)$$

i.

$$C_6H_5CCH_2CO_2Et\ (C{=}O) \xrightarrow[2.\ CH_3CH_2I]{1.\ EtO^-/EtOH} C_6H_5C(O)CHCO_2Et\ (CH_2CH_3) \xrightarrow[CH_3I]{EtO^-} C_6H_5COC(CH_3)(CH_2CH_3)CO_2Et \xrightarrow[\Delta]{H_3O^+} C_6H_5C(O)CH(CH_3)CH_2CH_3$$

j.

$$CH_3CO_2Et + EtO_2CCO_2Et \xrightarrow{EtO^-} EtO_2CCH_2COCO_2Et$$

5. a.

$$H\!-\!\ddot{O}\!-\!I^{+3}(\!-\!\ddot{O}{:}^-)_3 \longleftrightarrow H\!-\!\ddot{O}\!-\!I({=}O)_3 \qquad H\!-\!\ddot{O}\!-\!I^{+2}{:}(\!-\!\ddot{O}{:}^-)_2 \longleftrightarrow H\!-\!\ddot{O}\!-\!I{:}({=}O)_2$$

$$>C(\!-\!O\!-\!)\!-\!C(\!-\!O\!-\!)<\ \text{cyclic periodate ester}\ (I({=}O)(=O)OH) \longrightarrow >C{=}O + >C{=}O + {:}I({=}O)_2OH$$

b. The cyclic iodate intermediate forms more easily with the *cis* isomer. This is easier to see using molecular models.

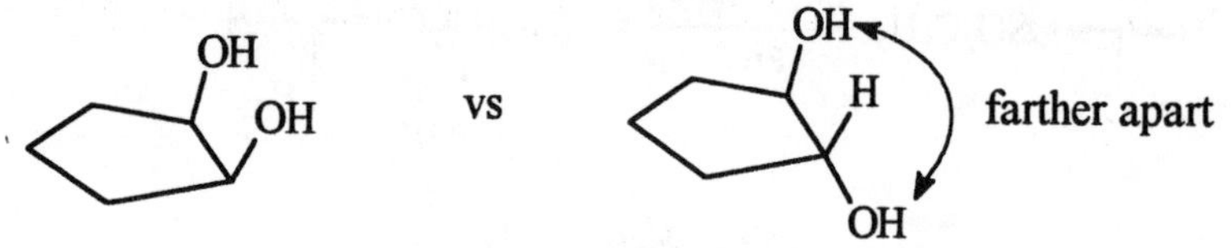

OH OH vs OH H OH farther apart

c. 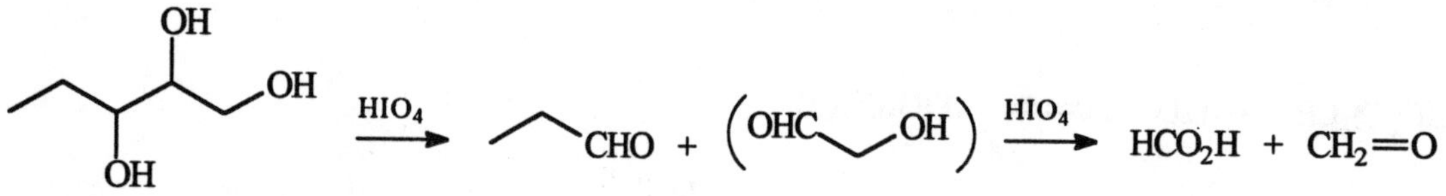 OH OH OH $\xrightarrow{HIO_4}$ CHO + (OHC OH) $\xrightarrow{HIO_4}$ $HCO_2H + CH_2{=}O$

The 1,2,3-isomer will consume two equivalents of HIO_4.

The 1,2,4-isomer only reacts once. OH OH OH $\xrightarrow{HIO_4}$ OH CHO + $CH_2{=}O$

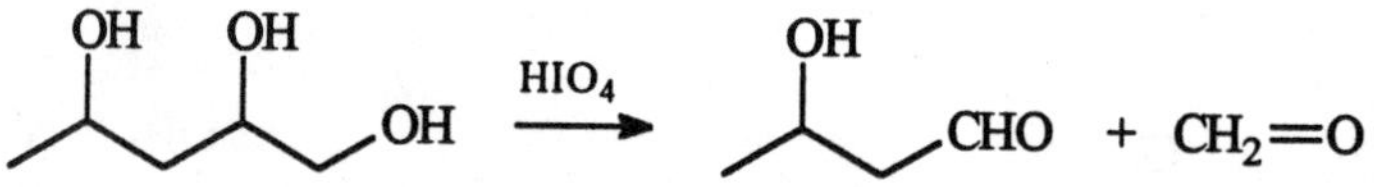

6. a. 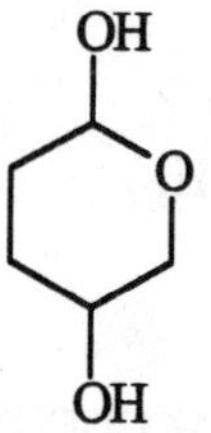 OH O OH

major isomer; six membered saturated ring is preferred

b. 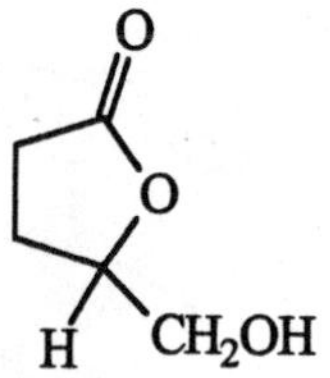 O O H CH_2OH

five-membered lactone ring is preferred.

7. a. 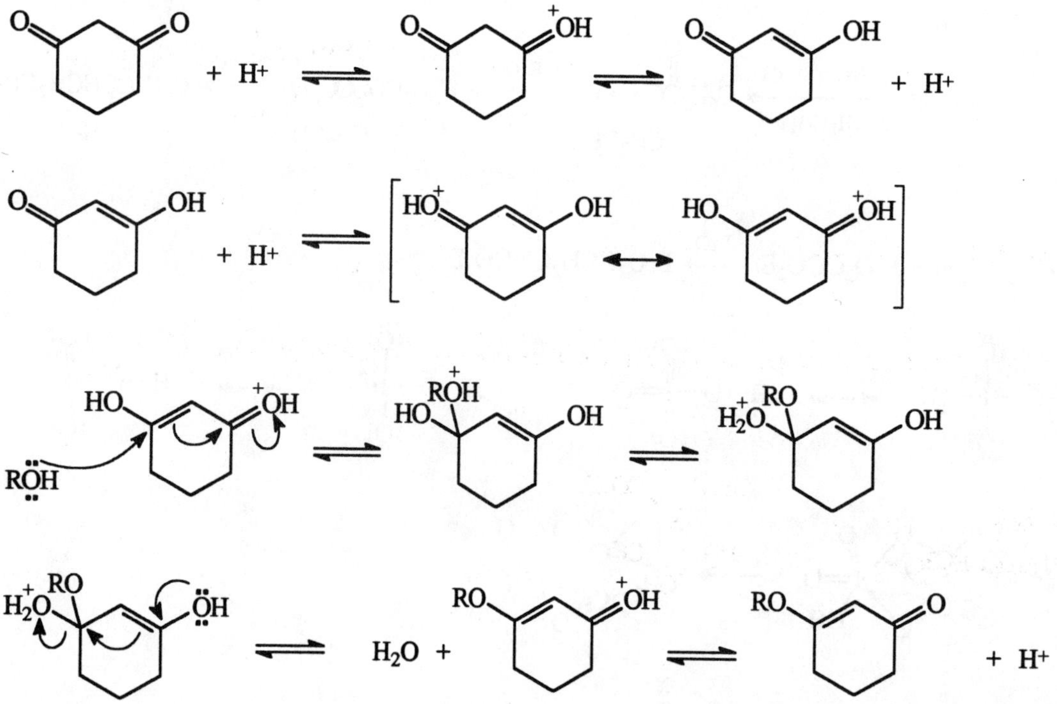 O O + H^+ ⇌ O $\overset{+}{O}H$ ⇌ O OH + H^+

O OH + H^+ ⇌ [$H\overset{+}{O}$ OH ↔ HO $\overset{+}{O}H$]

HO $\overset{+}{O}H$ ROH ⇌ $R\overset{+}{O}H$ HO OH ⇌ RO $H_2\overset{+}{O}$ OH

RO $H_2\overset{+}{O}$ OH ⇌ H_2O + RO $\overset{+}{O}H$ ⇌ RO O + H^+

b. $(MeO_2C)_2\overset{-}{C}H$ + ethylene oxide (H_2C–CH_2–O) → $MeO_2CC(CO_2Me)HCH_2CH_2O^-$ ⇌ tetrahedral intermediate (MeO–C(O^-)–O) ⇌ MeO_2C-substituted γ-lactone

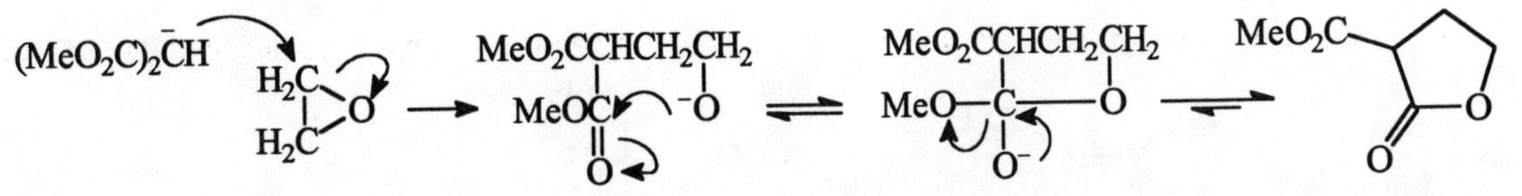

c. $CH_2{=}CHCH_2CH_2CH(CO_2Me)_2$ $\xrightarrow{C_6H_5CO_3H}$ epoxide–$CH_2CH_2CH(CO_2Me)_2$ $\xrightarrow{MeO^-}$ epoxide–$CH_2CH_2\overset{-}{C}(CO_2Me)_2$ → ^-O–cyclopentane–$C(CO_2Me)_2$ $\xrightarrow{MeOH}$ HO–cyclopentane–$C(CO_2Me)_2$

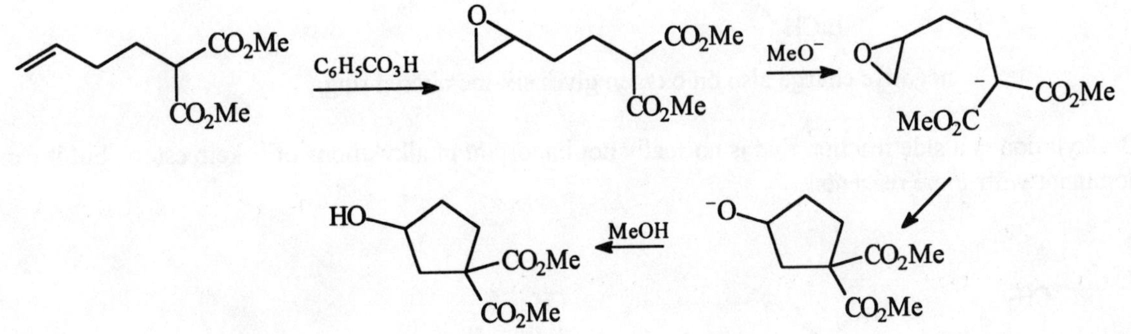

d. $(C_6H_5)_2C(C_6H_4OCH_3\text{-}p)CH_2$–O–$SO_2$–Cl → spiro cyclopropane intermediate (${}^+OCH_3$) → $(C_6H_5)_2\overset{+}{C}$–$CH_2C_6H_4OCH_3$ → $(C_6H_5)_2C{=}CH$–$C_6H_4OCH_3$

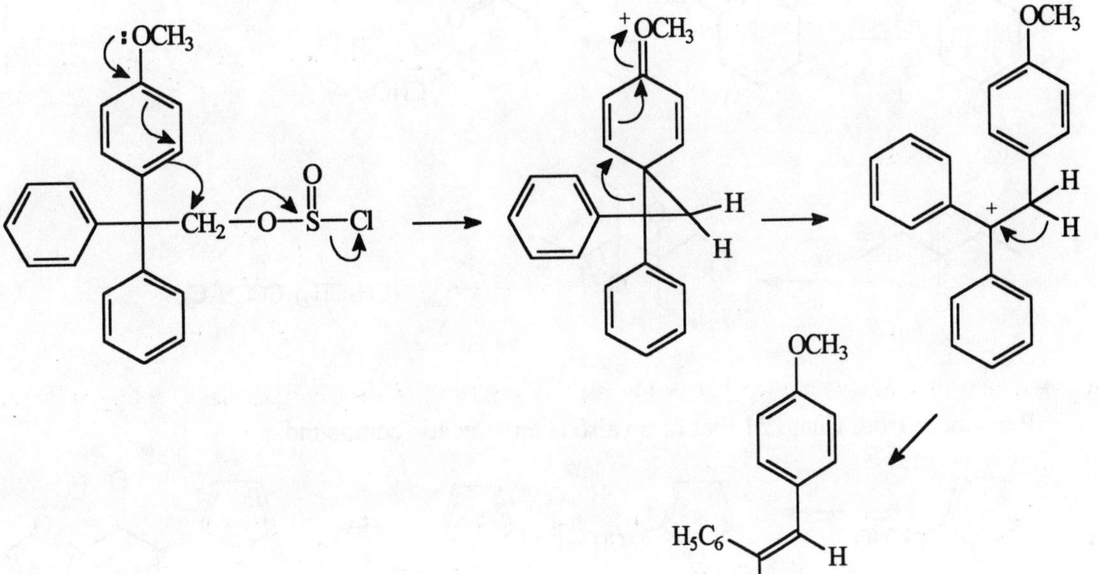

8. $CH_3COCH_2CO_2Et + BrCH_2CH_2CH_2Br \xrightarrow{EtO^-} CH_3COCH(CO_2Et)CH_2CH_2CH_2Br$

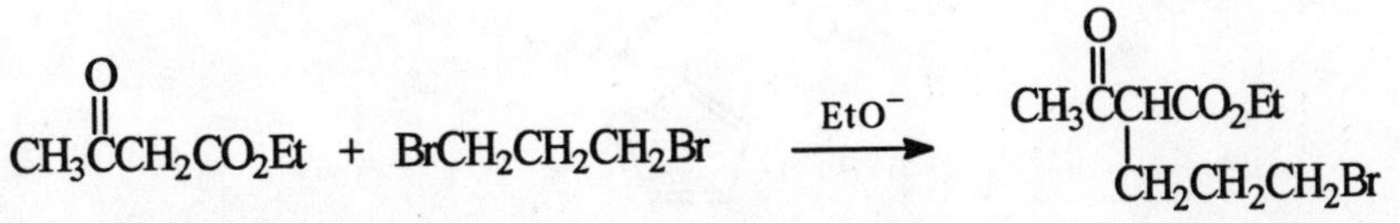

While the first alkylation is normal, the second would give a strained four-membered ring:

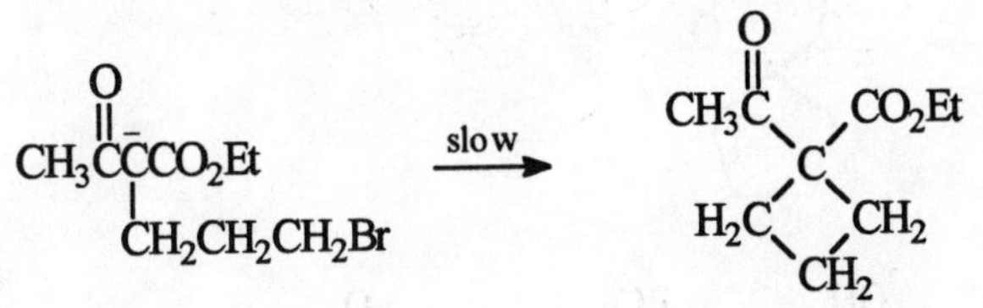

$CH_3CO\overset{-}{C}(CO_2Et)CH_2CH_2CH_2Br \xrightarrow{slow}$ 1-acetyl-1-(ethoxycarbonyl)cyclobutane (CH_3CO, CO_2Et on C; ring H_2C, CH_2, CH_2)

but note: [structure: $H_3C(^-O)C{=}C(CO_2Et)CH_2CH_2CH_2Br$] $\xrightarrow{\text{faster}}$ [cyclic enol ether with H_3C and CO_2Et]

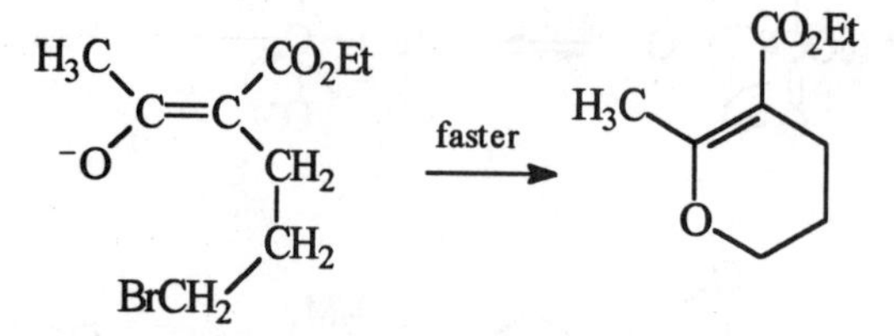

negative charge also on oxygen gives six-membered ring

O-alkylation is a side reaction that is normally not important in alkylations of β-keto esters, but it can become dominant with some reagents.

9. a. [1-methylcyclohexene] → [cis-diol with H_3C, OH, OH] → [keto aldehyde with CH_3, O, CHO]

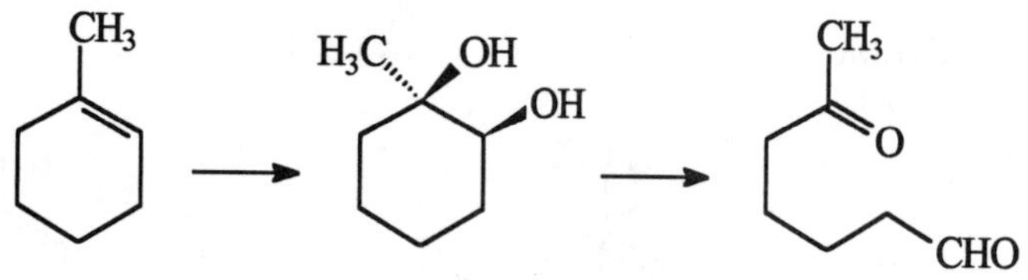

b. [3-pentanone] → [diol with OH, OH] → $(CH_3CH_2)_3C\overset{O}{\overset{\|}{C}}CH_2CH_3$

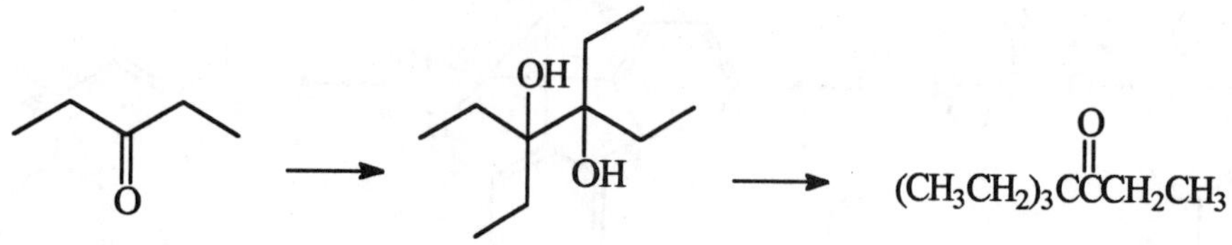

c. Remember from Chap. 21 that furan also is an aromatic compound.

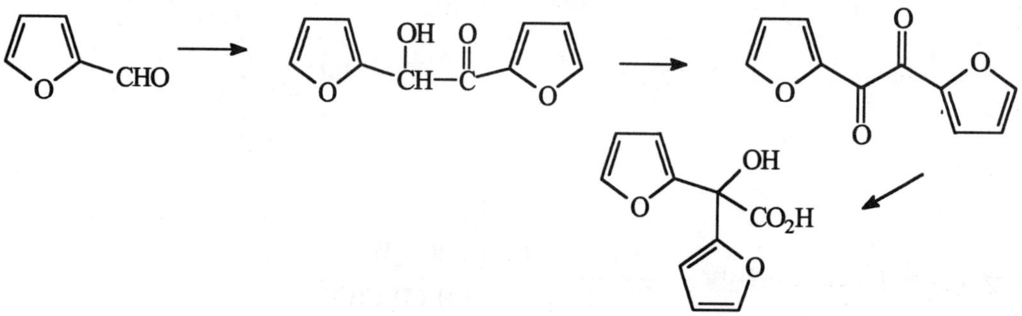

[furfural, CHO] → [furoin: OH O, CH–C] → [furil] → [furilic acid: OH, CO_2H]

d. [2-butene] → [diol: HO, H, CH_3, H_3C, H, OH] → [cyclopentylidene ketal: H_3C, CH_3, H, H, O, O]

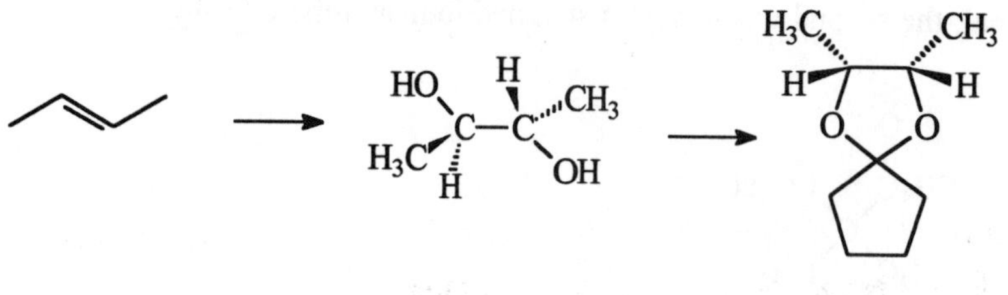

10. a.

$$CH_3CH_2CO_2H \xrightarrow{SOCl_2} CH_3CH_2\overset{O}{\overset{\|}{C}}Cl \xrightarrow[Pd/BaSO_4]{H_2} CH_3CH_2\overset{O}{\overset{\|}{C}}H \xrightarrow{\text{dil. }OH^-} CH_3CH_2\overset{OH}{\overset{|}{C}}H\underset{CH_3}{\underset{|}{C}}H\overset{O}{\overset{\|}{C}}H \xrightarrow{Ag_2O} CH_3CH_2\overset{OH}{\overset{|}{C}}H\underset{CH_3}{\underset{|}{C}}H\overset{O}{\overset{\|}{C}}OH$$

or:

$$CH_3CH_2CO_2H \xrightarrow[Br_2]{PBr_3} CH_3CHBr\overset{O}{\overset{\|}{C}}Br \xrightarrow{CH_3OH} CH_3CHBr\overset{O}{\overset{\|}{C}}OCH_3 \xrightarrow[CH_3CH_2CHO]{Zn} CH_3CH_2\overset{OH}{\overset{|}{C}}H\underset{CH_3}{\underset{|}{C}}H\overset{O}{\overset{\|}{C}}OCH_3 \xrightarrow[H_2O]{OH^-} CH_3CH_2\overset{OH}{\overset{|}{C}}H\underset{CH_3}{\underset{|}{C}}H\overset{O}{\overset{\|}{C}}OH$$

b.

$$H_3C\text{-}C_6H_4\text{-}\overset{O}{\overset{\|}{C}}H \xrightarrow{CN^-} H_3C\text{-}C_6H_4\text{-}\underset{OH}{\underset{|}{C}}H\text{-}\overset{O}{\overset{\|}{C}}\text{-}C_6H_4\text{-}CH_3 \xrightarrow{LiAlH_4} H_3C\text{-}C_6H_4\text{-}\underset{OH}{\underset{|}{C}}H\text{-}\overset{OH}{\overset{|}{C}}H\text{-}C_6H_4\text{-}CH_3$$

c.

$$\text{cyclopentanone} \xrightarrow{HCN} \text{1-hydroxycyclopentanecarbonitrile (HO, CN)} \xrightarrow[H_2O,\ \Delta]{H^+} \text{1-hydroxycyclopentanecarboxylic acid (HO, }CO_2H) \xrightarrow[\Delta]{H^+} \text{dispiro cyclic diester (lactide: O–C(=O) ring)}$$

d.

$$CH_3CH_2CO_2H \xrightarrow[H^+]{CH_3OH} CH_3CH_2CO_2CH_3 \xrightarrow[2.\ H^+]{1.\ Na,\ ether} CH_3CH_2\underset{OH}{\underset{|}{C}}H\overset{O}{\overset{\|}{C}}CH_2CH_3 \xrightarrow{LiAlH_4} CH_3CH_2\overset{OH}{\overset{|}{C}}H\underset{OH}{\underset{|}{C}}HCH_2CH_3$$

e.

$$\text{cyclohexanone} \xrightarrow[NaOEt]{CO(OEt)_2} \text{2-}(CO_2Et)\text{cyclohexanone anion} \xrightarrow[\text{dil.}]{H_3O^+} \text{2-}(CO_2Et)\text{cyclohexanone}$$

f.

$$C_6H_5CH_2CO_2Et \xrightarrow[NaOEt]{CO(OEt)_2} C_6H_5CH(CO_2Et)_2 \xleftarrow{H^+} C_6H_5\bar{C}(CO_2Et)_2$$

$$C_6H_5CH_2CO_2Et \rightleftharpoons C_6H_5\bar{C}HCO_2Et$$

$$C_6H_5\bar{C}HCO_2Et \xrightarrow{EtOC(=O)OEt} C_6H_5CH(CO_2Et)\text{—}C(OEt)_2O^- \rightleftharpoons C_6H_5CH(CO_2Et)CO_2Et + EtO^-$$

$$C_6H_5CH(CO_2Et)_2 \rightleftharpoons C_6H_5\bar{C}(CO_2Et)_2$$

g.

$$C_6H_5CH(CO_2Et)_2 + EtO^- \rightleftharpoons C_6H_5\bar{C}(CO_2Et)_2 \xrightarrow{ClCH_2CO_2Et} C_6H_5C(CO_2Et)_2(CH_2CO_2Et) + Cl^-$$

$$\xrightarrow[2.\ H^+]{1.\ OH^-} C_6H_5C(CO_2H)_2(CH_2CO_2H) \xrightarrow[-CO_2]{\Delta} C_6H_5CH(CO_2H)CH_2CO_2H$$

11. a. $(CH_3)_2CHCH_2Br + CH_2(CO_2Et)_2 \xrightarrow{NaOEt} \xrightarrow{OH^-} \xrightarrow[\Delta]{H^+} (CH_3)_2CHCH_2CH_2CO_2H$

b. methylallylacetic acid:

$$CH_2{=}CHCH_2Br + CH_2(CO_2Et)_2 \xrightarrow{NaOEt} \xrightarrow[CH_3I]{NaOEt} CH_2{=}CHCH_2C(CO_2Et)_2CH_3 \xrightarrow[2.\ H^+\ \Delta]{1.\ OH^-} CH_2{=}CHCH_2CH(CO_2H)CH_3$$

Note that the allyl group is added first. Some dialkylation occurs as a side reaction. Separation of mono- from dialkylated ester is easier when the first group is larger.

c.

$$CH_2(CO_2Et)_2 + CH_2{=}C(CH_3)CH_2Cl \xrightarrow{NaOEt} \xrightarrow{OH^-} \xrightarrow[\Delta]{H^+} CH_2{=}C(CH_3)CH_2CH_2CO_2H$$

d. 1,5-keto acid; Michael addition: 2-cyclohexenone + $CH_2(CO_2Et)_2 \xrightarrow{NaOEt}$ etc.

e. $CH_2(CO_2Et)_2 + BrCH_2CO_2Et \xrightarrow{NaOEt} (EtO_2C)_2CHCH_2CO_2Et \xrightarrow[\Delta]{H_3O^+} HO_2CCH_2CH_2CO_2H$

f. $CH_2(COOEt)_2 \xrightarrow[Br(CH_2)_5Br]{NaOEt\ (2\ moles)} [(EtOOC)_2CH(CH_2)_5Br]$

$$\rightarrow \text{cyclohexane}{=}C(CO_2Et)_2 \xrightarrow{OH^-} \xrightarrow[\Delta]{H^+} \text{cyclohexane}\text{—}CO_2H$$

12. a. "isopentylacetone":

$$CH_3\overset{O}{\overset{\|}{C}}CH_2CO_2Et + BrCH_2CH_2CH(CH_3)_2 \xrightarrow{NaOEt} CH_3\overset{O}{\overset{\|}{C}}\underset{CO_2Et}{\underset{|}{C}}H(CH_2)_2CH(CH_3)_2 \xrightarrow[2.\ H^+\ \Delta]{1.\ OH^-} CH_3\overset{O}{\overset{\|}{C}}(CH_2)_3CH(CH_3)_2$$

b. a 1,5-keto acid: Michael addition:

$$CH_3\overset{O}{\overset{\|}{C}}CH_2CO_2Et + CH_2{=}CHCO_2Et \xrightarrow{NaOEt} CH_3\overset{O}{\overset{\|}{C}}\underset{CO_2Et}{\underset{|}{C}}HCH_2CH_2CO_2Et \longrightarrow \text{etc.}$$

c. "propylethylacetone":

$$CH_3\overset{O}{\overset{\|}{C}}CH_2CO_2Et \xrightarrow[CH_3CH_2CH_2Br]{NaOEt} \xrightarrow[CH_3CH_2Br]{NaOEt} CH_3CO\overset{CH_2CH_3}{\underset{CH_2CH_2CH_3}{C}}CO_2Et \xrightarrow[2.\ H^+\ \Delta]{1.\ OH^-} CH_3\overset{O}{\overset{\|}{C}}\underset{CH_2CH_3}{\underset{|}{C}}HCH_2CH_2CH_3$$

d. Note symmetrical nature; both ends can be attached by an acetoacetic ester alkylation:

$$CH_3\overset{O}{\overset{\|}{C}}CH_2CO_2Et \xrightarrow[CH_3I]{NaOEt} CH_3\overset{O}{\overset{\|}{C}}\underset{CH_3}{\underset{|}{C}}HCOEt_2 \xrightarrow[\substack{BrCH_2CH_2Br\\(0.5\ mole)}]{NaOEt} \left(CH_3CO\overset{CO_2Et}{\underset{CH_3}{C}}CH_2-\right)_2 \xrightarrow[2.\ H^+\ \Delta]{1.\ OH^-}$$

13. (Ethyl esters that contain 5 or fewer carbons in the acid portion are considered to be eqyuivalent to a 5 carbon compound since the esters are easily formed.)

a. This product is a substituted succinic acid. Two possible routes:

$$CH_3CH_2CH{=}CHCO_2Et \xrightarrow{NaCN} CH_3CH_2\overset{CN}{\overset{|}{C}}HCH_2CO_2Et \xrightarrow[\Delta]{H^+} CH_3CH_2\overset{CO_2H}{\overset{|}{C}}HCH_2CO_2H$$

or

$$CH_2(CO_2Et)_2 \xrightarrow{NaOEt} \xrightarrow{C_2H_5Br} CH_3CH_2CH(CO_2Et)_2 \xrightarrow{NaOEt} \xrightarrow{ClCH_2CO_2Et} CH_3CH_2\overset{CH_2CO_2Et}{\overset{|}{C}}(CO_2Et)_2$$

$$\xrightarrow{H_3O^+} CH_3CH_2\overset{CH_2CO_2H}{\overset{|}{C}}(CO_2H)_2 \xrightarrow[-CO_2]{\Delta} CH_3CH_2\overset{CH_2CO_2H}{\overset{|}{C}}HCO_2H$$

b. $$\underset{\text{(from a.)}}{CH_3CH_2CH(CO_2Et)_2} \xrightarrow{NaOEt} \xrightarrow[CH_3CHBrCO_2Et]{CH_3CHClCO_2Et\ or} CH_3CH_2\underset{\underset{CH_3}{\underset{|}{C}HCO_2Et}}{\underset{|}{C}}(CO_2Et)_2 \xrightarrow[2.\ H^+\ \Delta]{1.\ OH^-} \begin{matrix} CH_3CH_2CHCO_2H \\ | \\ CH_3CHCO_2H \end{matrix}$$

Note that a mixture of two diastereomers results.

c. (glutaric anhydride) $\xrightarrow{(CH_3)_2NH}$ $CON(CH_3)_2$ / CO_2H

d. (1,3-cyclopentanedione) $\xrightarrow[CH_3I]{NaOMe}$ (2,2-dimethyl-1,3-cyclopentanedione; H_3C, CH_3)

e. This compound obviously can be prepared from

$(CH_3)_2CHCOCH_2CO_2Et + ClCH_2CO_2Et \xrightarrow{EtO^-}$ etc.

The problem is to make the β-keto ester. It cannot be made by the Claisen condensation, but it is available by a sequence starting with:

$CH_3COCH(CH_3)_2 \xrightarrow[\substack{EtOH \\ HCO_2Et}]{NaOEt} (CH_3)_2CHCOCH_2CHO \xrightarrow{Ag_2O} RCO_2H \xrightarrow[\text{or } CH_2N_2]{EtOH/H^+}$ ester

Why must the condensation go in the methyl group?

f. (cyclopentanone) + $BrCH_2\overset{O}{\overset{\|}{C}}OMe \xrightarrow{Zn} \xrightarrow{H_3O^+}$ (1-hydroxycyclopentyl: HO, CH_2CO_2Me) $\xrightarrow[2.\ H_3O^+]{1.\ OH^-}$ (HO, CH_2CO_2H)

g. EtO_2C–(β-keto ester) $\xrightarrow[C_2H_5I]{NaOEt} \xrightarrow{H_3O^+}$ (ketone) $\xrightarrow{SeO_2}$ (diketone)

[from e. above]

h. This is a 1,6-dicarbonyl compound for which none of the special methods applies. Try treating it as an ethyl ketone: $CH_3CH_2\overset{O}{\overset{\|}{C}}CH_2 \mid (CH_2)_3CO_2H$

$CH_3CH_2\overset{O}{\overset{\|}{C}}\overset{-}{C}HCO_2Et + Br(CH_2)_3CO_2Et \longrightarrow CH_3CH_2\overset{O}{\overset{\|}{C}}\underset{CO_2Et}{CH}(CH_2)_3CO_2Et \xrightarrow[\Delta]{H^+} CH_3CH_2\overset{O}{\overset{\|}{C}}CH_2(CH_2)_3CO_2H$

The ketoester can be synthesized by a Reformatsky reaction with bromoacetic ester and propanal followed by oxidation of the hydroxy ester.

$$CH_3CH_2\overset{O}{\overset{\|}{C}}H + BrCH_2CO_2Et \xrightarrow{Zn} \xrightarrow{H_3O^+} CH_3CH_2\overset{OH}{\overset{|}{C}}HCH_2CO_2Et \xrightarrow{PCC} CH_3CH_2\overset{O}{\overset{\|}{C}}CH_2CO_2Et$$

The bromoester can be made from commercially available γ-butyrolactone or by:

$$H_2C\overset{O}{—}CH_2 + CH_2(CO_2Et)_2 \xrightarrow{NaOEt} \xrightarrow[\Delta]{H^+} \text{(γ-butyrolactone)} \xrightarrow[MeOH,\ \Delta]{HBr} Br(CH_2)_3CO_2Me$$

i. $CH_3CH_2CO_2Et \xrightarrow{NaOEt} CH_3CH_2COCH(CH_3)CO_2Et \xrightarrow[C_2H_5I]{NaOEt} CH_3CH_2COC(CH_3)(C_2H_5)CO_2Et \xrightarrow{H_3O^+} CH_3CH_2COCH(CH_3)CH_2CH_3$

j. $(CH_3)_2C{=}O \xrightarrow{Mg} (CH_3)_2C(OH)C(OH)(CH_3)_2 \xrightarrow{H^+} (CH_3)_3CCOCH_3$

k. α-Keto esters are available by Claisen condensations with diethyl oxalate:

$$CH_3(CH_2)_3CO_2Et + EtO_2CCO_2Et \xrightarrow{NaOEt} CH_3CH_2CH_2\overset{CO_2Et}{\overset{|}{C}}H\underset{\underset{O}{\|}}{C}CO_2Et \xrightarrow[2.\ H^+\ \Delta]{1.\ OH^-} \left[CH_3CH_2CH_2\overset{CO_2H}{\overset{|}{C}}H\underset{\underset{O}{\|}}{C}CO_2H\right]$$

$$\xrightarrow[-CO_2]{\Delta} CH_3CH_2CH_2CH_2\overset{O}{\overset{\|}{C}}CO_2H$$

(an α-keto acid and a β-keto acid; β-COOH decarboxylates)

l. Substituted glutaric acid available by Michael addition:

$$CH_3CH_2CH(CO_2Et)_2 \xrightarrow[2.\ CH_2{=}CHCO_2Et]{1.\ NaOEt} EtO_2CCH_2CH_2\overset{CO_2Et}{\underset{CO_2Et}{C}}CH_2CH_3 \xrightarrow[2.\ H^+\ \Delta]{1.\ OH^-} HO_2CCH_2CH_2\overset{CO_2H}{\overset{|}{C}}HCH_2CH_3$$

[from a.]

m.

$$(EtO_2C)_2CH_2 + H_2C\overset{O}{—}CHCH_2CH_3 \xrightarrow{NaOEt} (EtO_2C)_2CHCH_2\overset{O^-}{\overset{|}{C}}HCH_2CH_3$$

[See problem 7b.]

$$\rightleftharpoons \left[\text{cyclic } {}^-O\text{–}C(OEt)\text{: } EtO_2CCH,\ O,\ H_2C—CHCH_2CH_3\right] \rightleftharpoons EtO_2CCH\text{–}C(=O)\text{–}O\text{–}CH(CH_2CH_3)\text{–}CH_2 \xrightarrow[\Delta]{H^+} \text{5-ethyl-γ-butyrolactone}$$

n.  cyclopentanone + HCN → 1-hydroxycyclopentanecarbonitrile (OH, CN) $\xrightarrow[\Delta]{H_3O^+}$ 1-hydroxycyclopentanecarboxylic acid (OH, CO_2H)

14. $EtO_2CCH_2CO_2Et + 2\ C_6H_5CH_2Cl \xrightarrow[50\%\ KOH]{C_6H_5CH_2{}^+NEt_3Cl^-} (C_6H_5CH_2)_2C(CO_2Et) \xrightarrow[EtOH]{NaOH} \xrightarrow[\Delta]{H^+} (C_6H_5CH_2)_2CHCO_2H$

In the organic phase, the following equilibrium occurs faster than ester hydrolysis:

$(EtO_2CCH_2CO_2Et + OH^- \rightleftharpoons EtO_2C^-CHCO_2Et + H_2O)$

15. a. 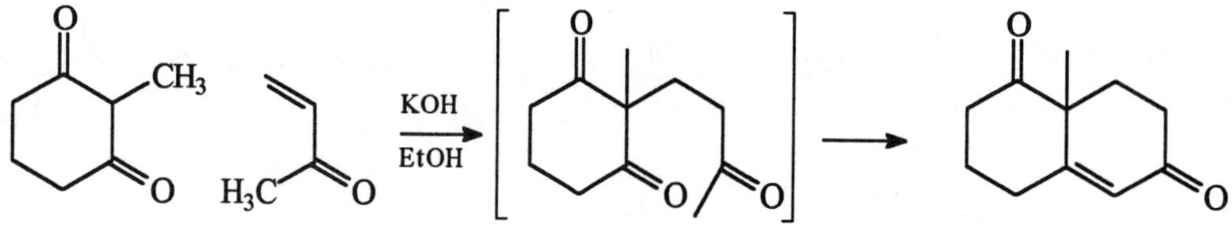 2-methylcyclohexane-1,3-dione (CH_3) + H_3C-CO-CH=CH_2 $\xrightarrow[EtOH]{KOH}$ [Michael adduct] → bicyclic enone (O, O)

b. 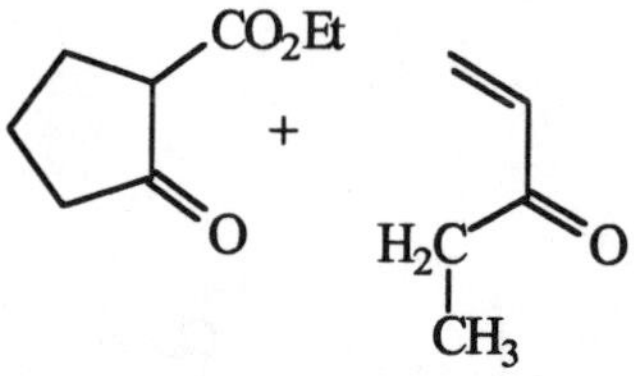 2-(CO_2Et)cyclopentanone + H_2C=CH-CO-CH_2CH_3 (H_2C, CH_3, O)

c. 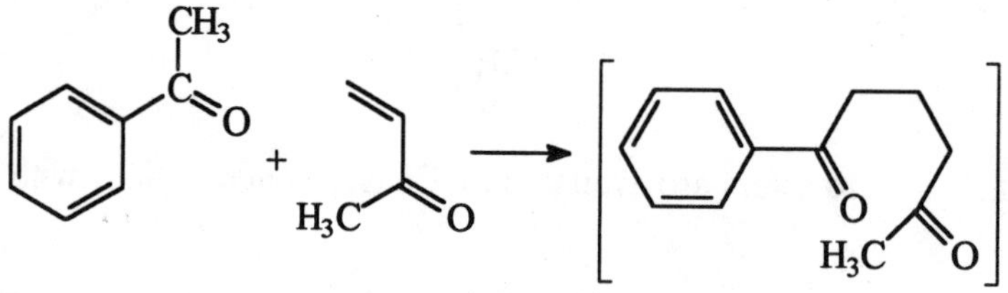 C_6H_5-C(=O)-CH_3 + H_3C-CO-CH=CH_2 → [C_6H_5-CO-$CH_2CH_2CH_2$-CO-CH_3] (H_3C, O)

d. 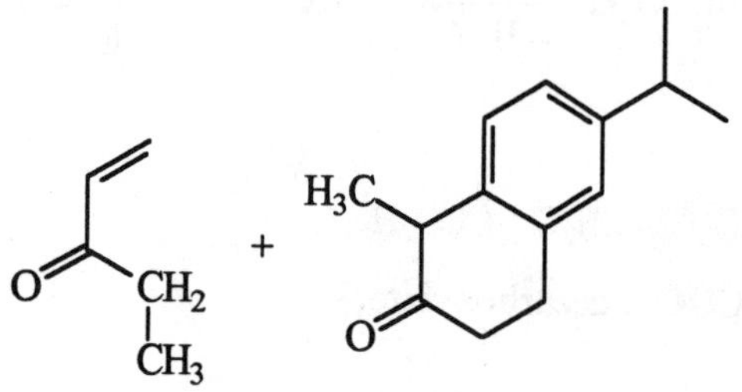 CH_2=CH-CO-CH_2CH_3 (CH_2, CH_3) + bicyclic ketone bearing H_3C and isopropyl groups

16.

CO_2H H–C–OH C_6H_5 (-)-mandelic acid *(R)*	→	CO_2Et H–C–OEt C_6H_5 (-)-C *(R)*	→	CH_2OH H–C–OEt C_6H_5 (-)-D *(R)*	→ →	CH_3 H–C–OEt C_6H_5 (-)-E *(S)*

CH_3 D–C–H C_6H_5 (-)-B *(R)*	←	CH_3 H–C–OH C_6H_5 (-)-F *(S)*	↙ (from (-)-E)

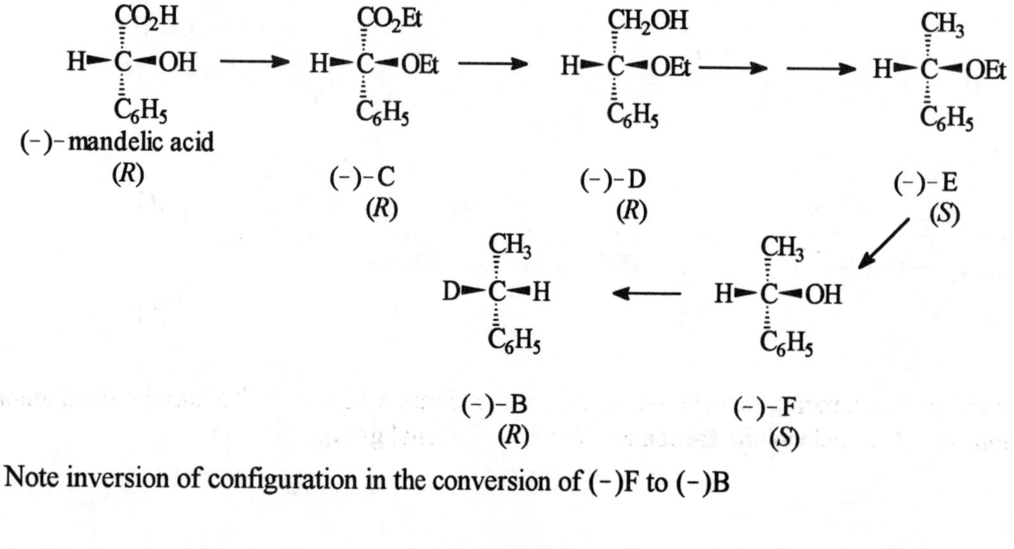

Note inversion of configuration in the conversion of (−)F to (−)B

17. a.

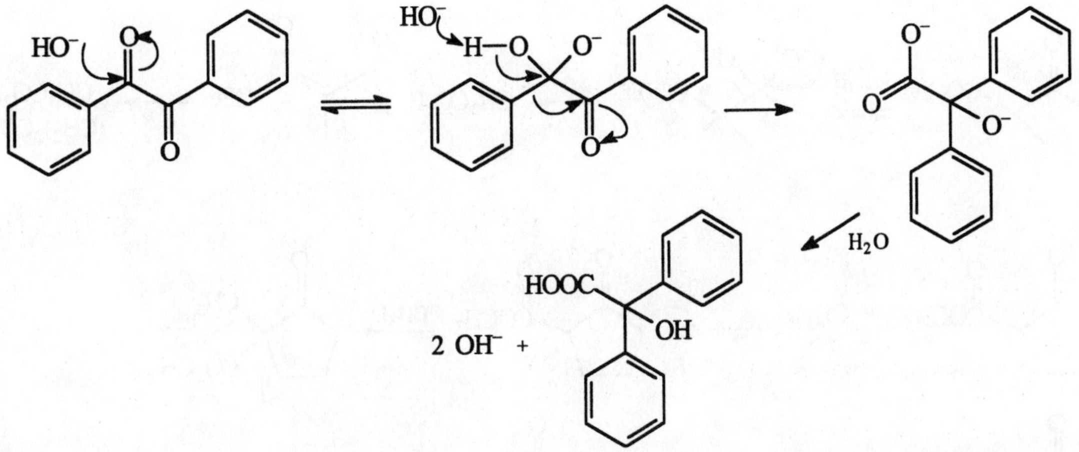

HO^- ⇌ HO^- H–O, O^- → O^-, O, O^-

H_2O

2 OH^- + HOOC, OH

b. Cl, Cl ⇌ (OH^-) HO^- H–O, O^-, Cl, Cl → O^-, Cl → (H^+) CO_2H, Cl

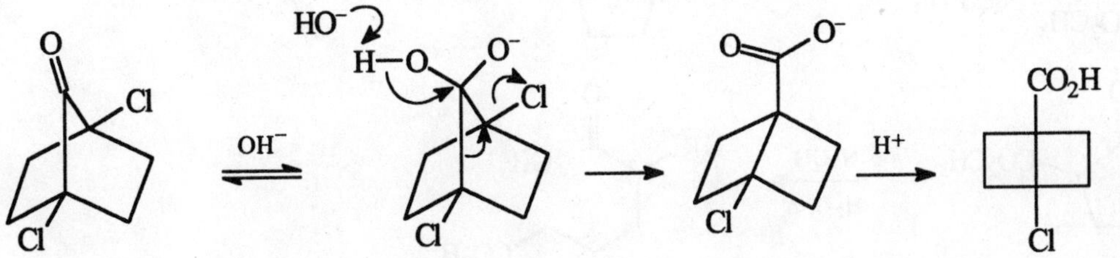

c. → (SeO_2) → (KOH) HO, CO_2H

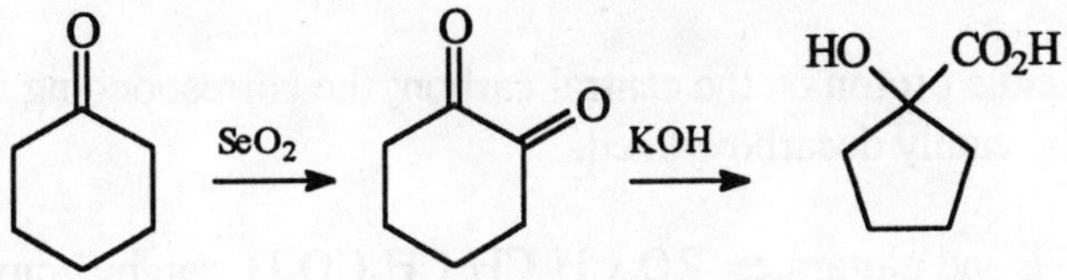

18.

CH_2–C≡CH + Li → [CH_2, C, CH, O] → O^-, C–H

Li

OH, CH_2 ← (H_2O (work-up)) NH_2^- + O^-, CH_2 ← (NH_3) O^-, CH^-

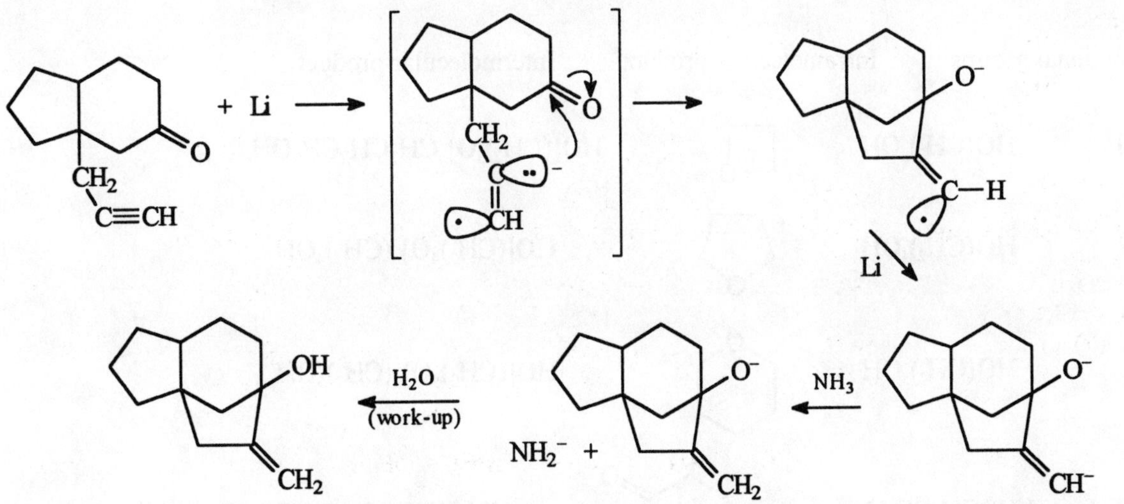

The first step is like the metal-ammonia reduction of an alkyne (see Section 12.6.A). The initial radical anion undergoes cyclization, with the nucleophilic carbon attacking the carbonyl group.

19. a. [reaction scheme: 2,2-dimethyl-3-oxobutanoate ethyl ester + NaOH ⇌ tetrahedral alkoxide intermediate → CH_3CO_2H + enolate of ethyl isobutyrate (O⁻, OEt) → CH_3CO_2H + $(CH_3)_2CHCO_2H$]

CO_2Et NaOH ⇌ ^-O OH CO_2Et → CH_3CO_2H + O^- OEt → CH_3CO_2H + $(CH_3)_2CHCO_2H$

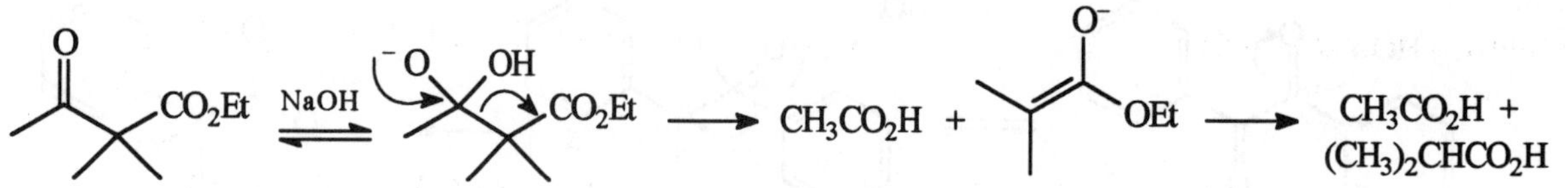

b. [reaction schemes]

O O $\overset{\|}{C}OCH_3$ + CH_3O^- ⇌ O O $^-$ $\overset{\|}{C}OCH_3$ $\xrightarrow{CH_3I}$ O CH_3 CO_2CH_3

O CH_3 CO_2CH_3 $\xrightarrow[CH_3OH]{NaOCH_3}$ $\xrightarrow{H_3O^+}$ H_3C O CO_2CH_3

H_3C O CO_2CH_3 $\xrightarrow[H_2O,\ \Delta]{NaOH}$ H_3C O OH CO_2H

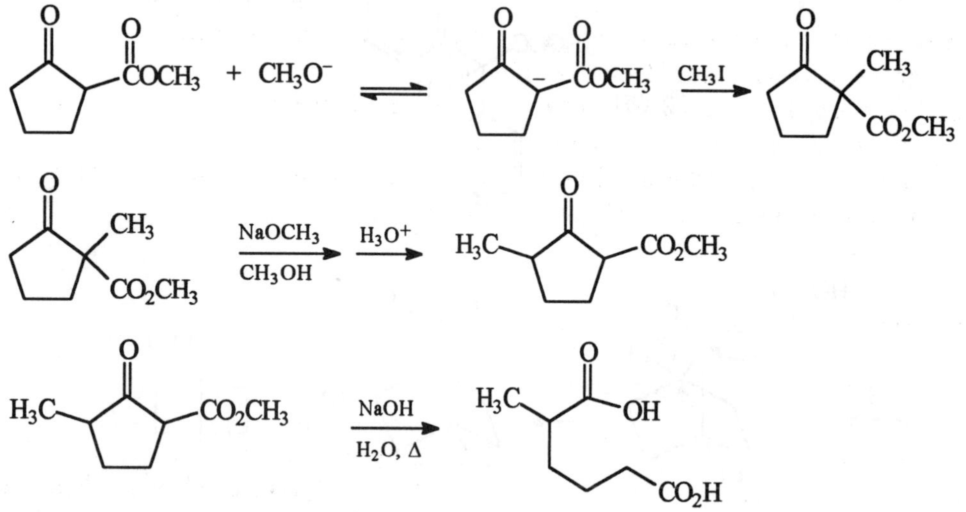

20. a. Alkyl malonates, $RO_2CCH_2CO_2H$, have an acidic proton on the central carbon; the corresponding base is a useful nucleophile. The subsequent product is readily decarboxylated.

b. and c. alkyl succinates, $RO_2CCH_2CH_2CO_2H$, and glutarates, $RO_2CH_2CH_2CH_2CO_2H$, can be converted to cyclic anhydrides and imides.

21.

Functional groups		Intramolecular product	Intermolecular product
1,3-diol	$HO(CH_2)_3OH$	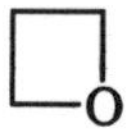 [oxetane] O	$HO[(CH_2)_3O]_nCH_2CH_2CH_2OH$
1,4-diol	$HO(CH_2)_4OH$	 [tetrahydrofuran] O	$HO[(CH_2)_4O]_n(CH_2)_4OH$
1,5-diol	$HO(CH_2)_5OH$	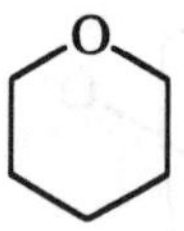 [tetrahydropyran] O	$HO[(CH_2)_5O]_n(CH_2)_5OH$
α-mercaptoacid	$HSCH_2CO_2H$	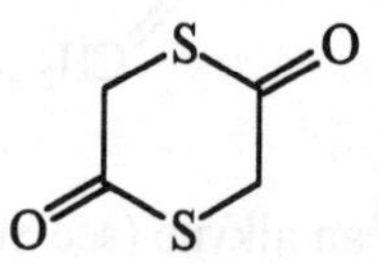 [1,4-dithiane-2,5-dione] S O O S	$HS[CH_2COS]_nCH_2CO_2H$

β-mercaptoacid $HSCH_2CH_2CO_2H$ → HS[CH_2CH_2COS]$_n CH_2CH_2CO_2H$

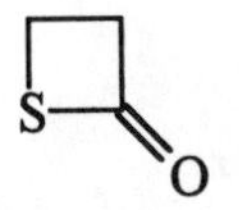

γ-mercaptoacid $HS(CH_2)_3CO_2H$ → HS[$(CH_2)_3COS$]$_n(CH_2)_3CO_2H$

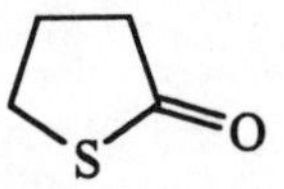

22. Knoevenagel: malonic acid or ester and aldehyde with amine catalyst; fairly easy and inexpensive.
Wittig and Horner-Emmons reactions with aldehydes and ketones.
Perkin reaction: aromatic aldehydes with acetic anhydride and sodium acetate.
dehydrohalogenation of β-haloacids or dehydration of β-hydroxyacids
mild oxidation of α,β-unsaturated aldehydes, available from aldol condensation. This route is probably the most expensive, since these oxidants are frequently metals like Ag^+ or Cu^{2+}.

27.F. Supplementary Problems

S1. Write the structure of each of the following compounds:

a. *trans*-4-cyclopenten-1,3-diol
b. 4-hydroxy-3-hexanone
c. 4-oxocyclohexanecarboxaldehyde
d. (2*R*,4*S*)-2,4-dimethylpentanedioic acid
e. (*R*)-2,3-dihydroxypropanal
f. methyl 2-oxocyclopentanecarboxylate
g. diethyl methylmalonate
h. 1,1,2,2-tetraphenyl-1,2-ethanediol

S2. Give the major product of each of the following reaction sequences:

a. 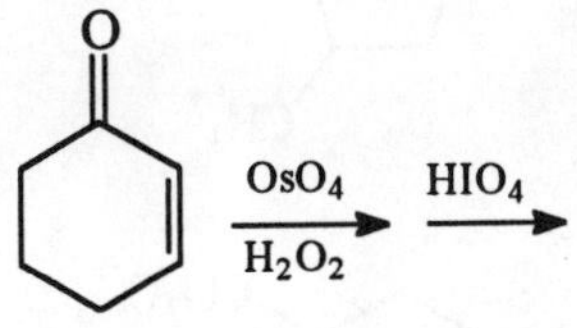 2-cyclohexenone $\xrightarrow[H_2O_2]{OsO_4}$ $\xrightarrow{HIO_4}$

b. $(CH_3)_2CHCO_2Et$ $\xrightarrow[toluene]{Na}$ $\xrightarrow{Cu(OAc)_2}$

c. $EtO_2CCH_2CO_2Et$ $\xrightarrow[NaOEt]{CH_2=CHCO_2Et}$ $\xrightarrow{H_3O^+}$ $\xrightarrow{\Delta}$

d. $CH_3O_2CCH_2CH_2CO_2CH_3$ $\xrightarrow{Na}$ $\xrightarrow{H_2O}$

e. 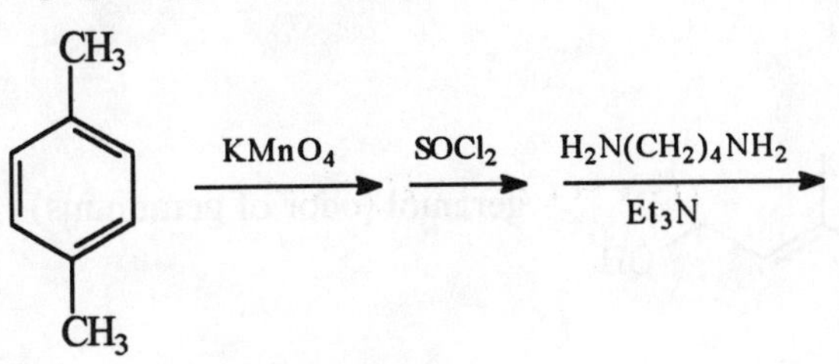 p-xylene (CH_3, CH_3) $\xrightarrow{KMnO_4}$ $\xrightarrow{SOCl_2}$ $\xrightarrow[Et_3N]{H_2N(CH_2)_4NH_2}$

f. 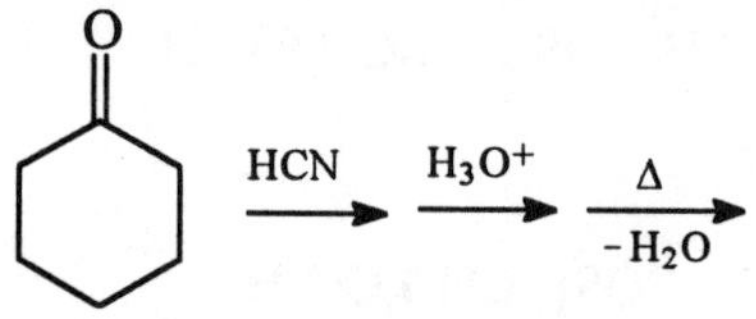 (cyclohexanone) $\xrightarrow{HCN}$ $\xrightarrow{H_3O^+}$ $\xrightarrow[-H_2O]{\Delta}$

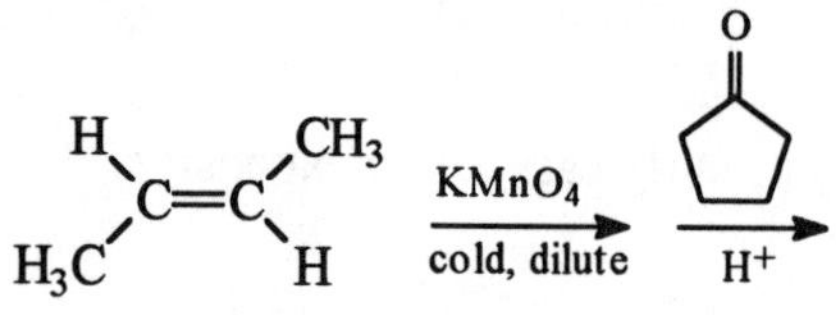

g. H, CH_3 / H_3C, H: C=C $\xrightarrow[\text{cold, dilute}]{KMnO_4}$ $\xrightarrow[H^+]{\text{cyclopentanone}}$

h. (1,2-dimethylcyclopentane-1,2-diol: CH_3, OH, OH, CH_3) $\xrightarrow{H^+}$

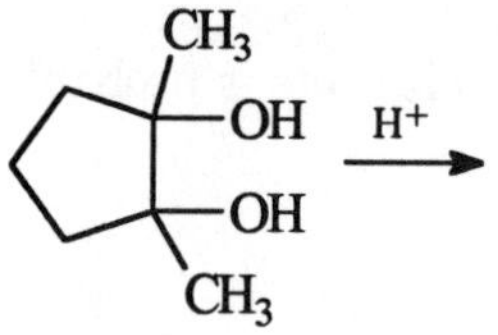

i. 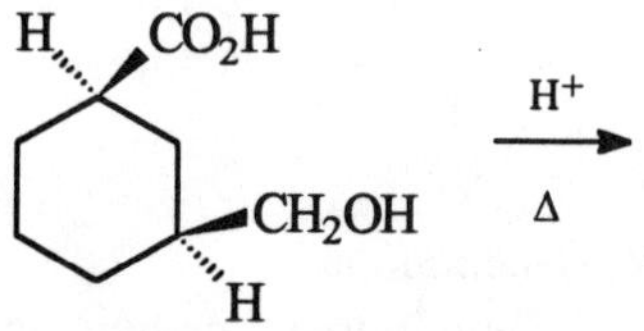 (cyclohexane with H, CO_2H and CH_2OH, H) $\xrightarrow[\Delta]{H^+}$

S3. Propose efficient syntheses of the following compounds, using any starting material containing five carbons or less.

a. 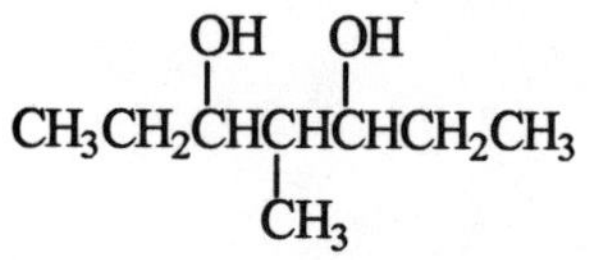 $CH_3CH_2\overset{OH}{C}H\overset{OH}{C}H\underset{CH_3}{C}HCH_2CH_3$

b. (cyclohexanone)

c. $CH_3\overset{O}{\overset{\|}{C}}$– (lactone ring with O, CH_3)

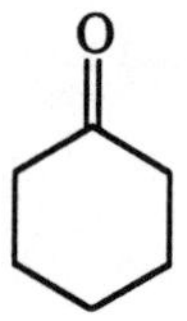

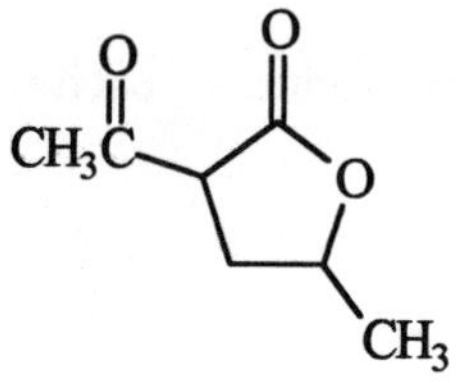

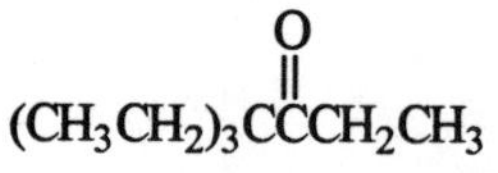

d. $(CH_3CH_2)_3C\overset{O}{\overset{\|}{C}}CH_2CH_3$

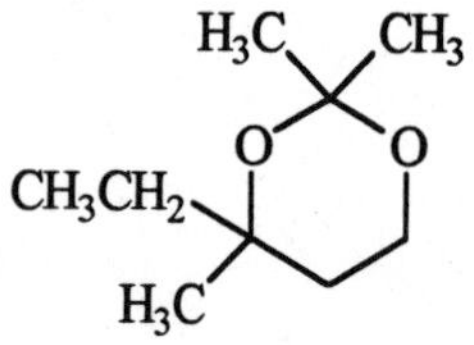

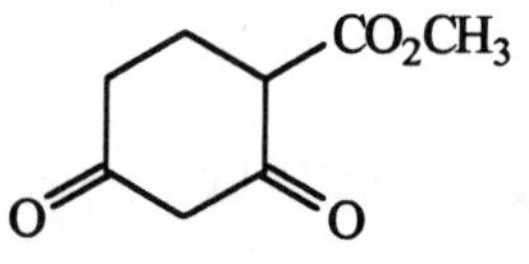

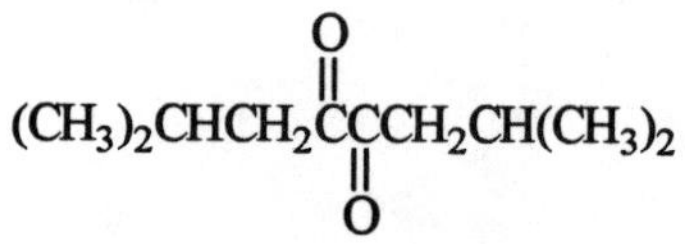

e. (1,3-dioxane with H_3C, CH_3, CH_3CH_2, H_3C)

f. (cyclohexane-1,3-dione with CO_2CH_3)

g. $(CH_3)_2CHCH_2\overset{O}{\overset{\|}{C}}\underset{O}{\underset{\|}{C}}CH_2CH(CH_3)_2$

h. (tetralin) (You may use benzene as a starting material for this one.)

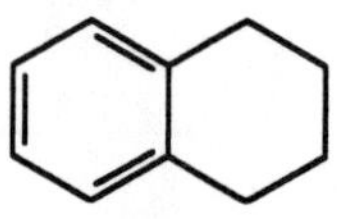

i. 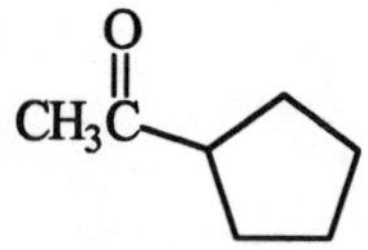 $CH_3\overset{O}{\overset{\|}{C}}$–(cyclopentyl)

j. 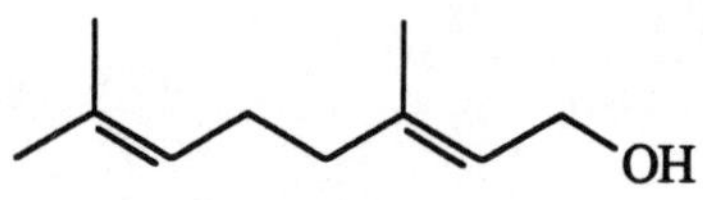 (structure) OH geraniol (odor of geraniums)

S4. Carbamates can be synthesized from an alcohol, NaN=C=O, and acid (see below). Given this reaction, outline a synthesis of the tranquilizer meprobamate.

$$ROH + NaN{=}C{=}O \xrightarrow{H^+} RO\overset{O}{\overset{\|}{C}}NH_2$$

$$H_2N\overset{O}{\overset{\|}{C}}OCH_2\underset{CH_2CH_3}{\overset{CH_3}{C}}CH_2O\overset{O}{\overset{\|}{C}}NH_2 \quad \textit{meprobamate}$$

S5. Provide structures that are consistent with the information given below.

$$B \xleftarrow[I_2]{NaOH} A \xrightarrow[CH_3OH]{H^+} C \xrightarrow{LiAlH_4} D \xrightarrow{H^+ \text{ (catalytic)}} E$$

B	A	C	D	E
$C_4H_6O_4$	$C_5H_8O_3$	$C_8H_{16}O_4$	$C_7H_{16}O_3$	$C_6H_{12}O_2$
NMR: δ 2.3 (s, area 2) δ 12 (s, area 1)	IR (dilute solution): 1710, 1760 2400-3400 cm^{-1}		IR: 1050, 1100, 3400 cm^{-1}	IR: 1070, 1120 cm^{-1}

S6. The benzilic acid rearrangement (see Problem 17) is a useful way to contract the ring of a cyclic α-diketone. Show how this rearrangement could be applied in a synthesis of bicyclo[2.1.1]-5-hexanone from bicyclo[2.2.1]-2-heptanone.

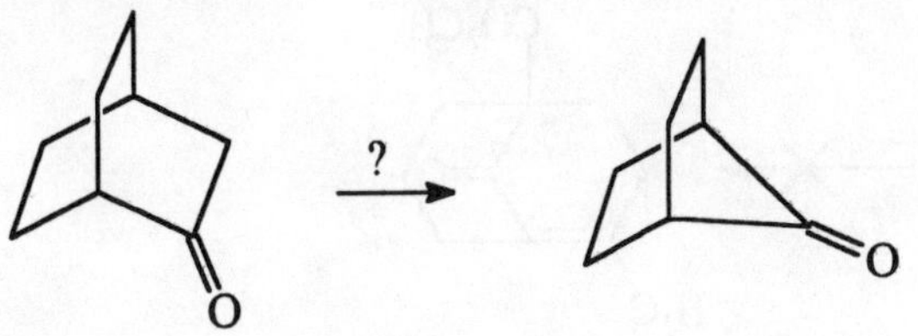

?

S7. α-Cyanocarboxylic acids also lose carbon dioxide on heating, although less readily than malonic or β-keto acids do. Write a mechanism for this decarboxylation, showing all of the intermediates involved.

$$N{\equiv}CCH_2CO_2H \xrightarrow{\Delta} N{\equiv}CCH_3 + CO_2$$

S8. Write mechanisms for each of the following transformations, showing all of the steps involved.

a. $2\ (EtO_2C)_2CH_2 + CH_2{=}O \xrightarrow[EtOH]{NaOEt} (EtO_2C)_2CHCH_2CH(CO_2Et)_2$

b. $HOCH_2CH_2CH_2\overset{O}{\overset{\|}{C}}CH_2CO_2CH_3 \xrightarrow{H^+}$ (tetrahydrofuran-2-ylidene)$=CHCO_2CH_3 + H_2O$

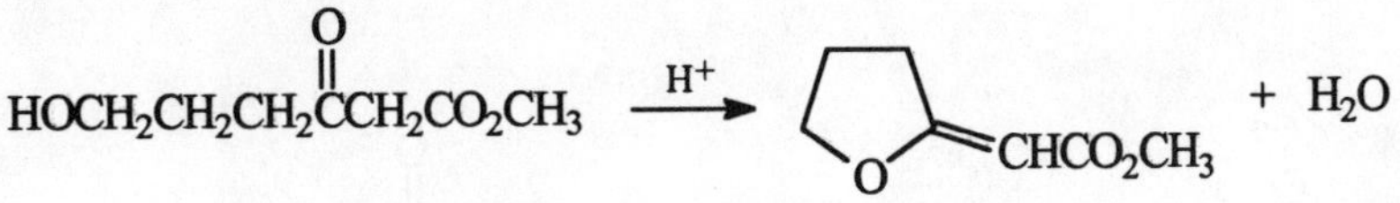

S9. The molecule illustrated at the right is very unstable, rapidly decomposing with loss of CO_2. What is the product and how is it formed?

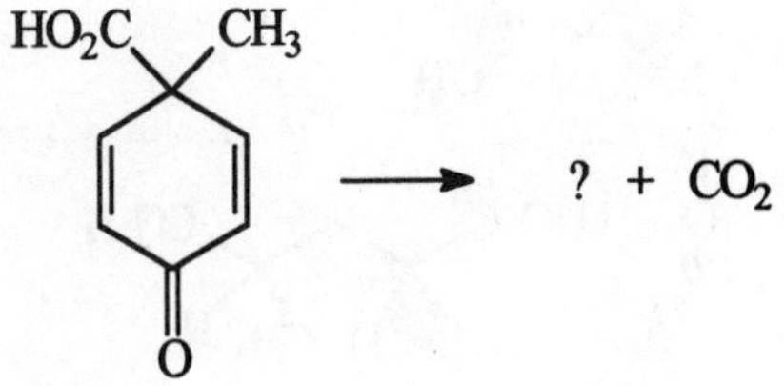

HO_2C, CH_3 ... O → ? + CO_2

S10. A byproduct that can sometimes be isolated during the course of a Robinson annulation is the β-hydroxy ketone drawn below.

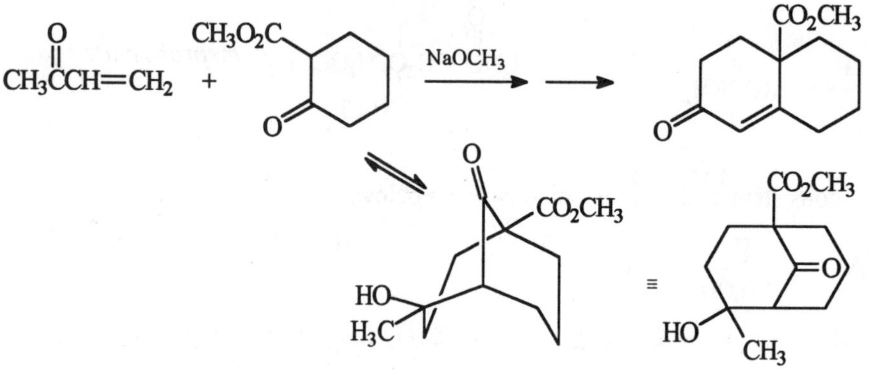

$CH_3\overset{O}{\overset{\|}{C}}CH{=}CH_2$ + CH_3O_2C NaOCH$_3$ → → CO_2CH_3 HO H_3C CO_2CH_3 ≡ CO_2CH_3 =O HO CH_3

a. Write a mechanism for the formation of this compound.

b. The formation of this β-hydroxy ketone is reversible, and on continued treatment with base it is converted to the normal annulation product. Why doesn't it undergo the normal dehydration reaction of β-hydroxy ketones and lead to an unsaturated ketone with the same carbon skeleton?

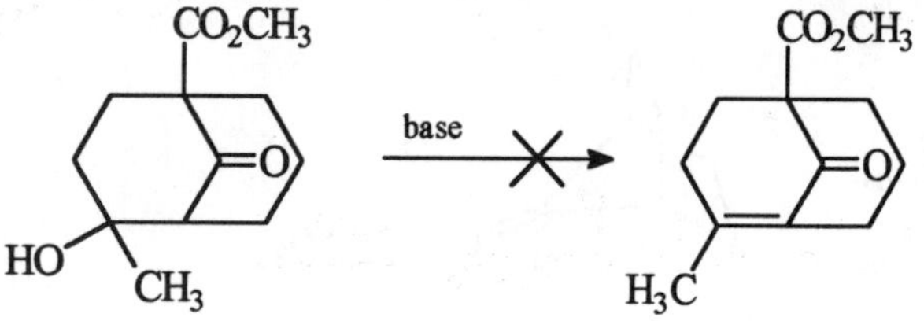

CO_2CH_3 =O HO CH_3 base (crossed arrow) CO_2CH_3 =O H_3C

S11. Write a reasonable mechanism for the following transformation.

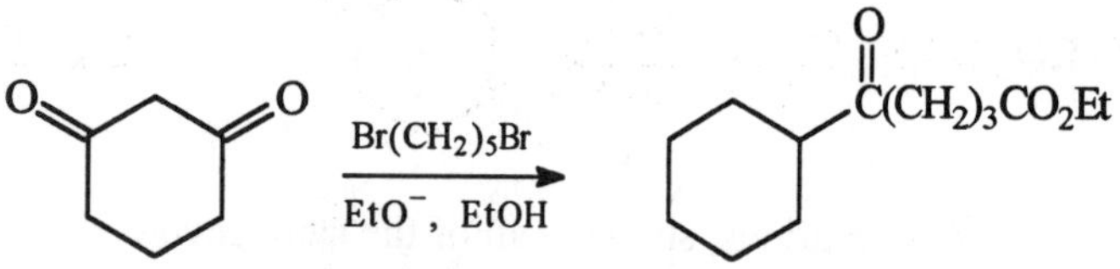

O O Br(CH$_2$)$_5$Br / EtO$^-$, EtOH → $\overset{O}{\overset{\|}{C}}(CH_2)_3CO_2Et$

27.G. Answers to Supplementary Problems

S1. a. OH OH

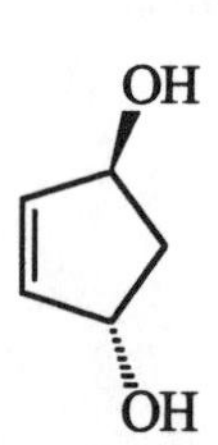

b. $CH_3CH_2\overset{O}{\overset{\|}{C}}\underset{OH}{\underset{|}{C}}HCH_2CH_3$

c. O $\overset{O}{\overset{\|}{C}}H$

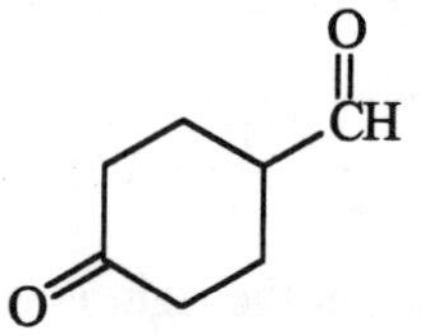

d. HO_2C CO_2H H_3C H CH_3 H

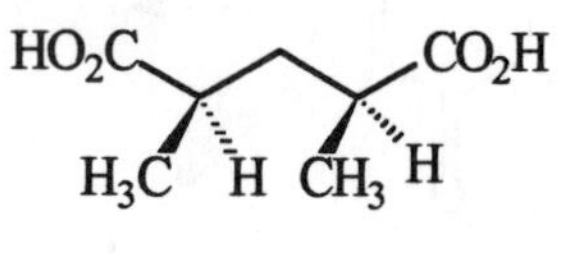

e. CHO H OH CH_2OH

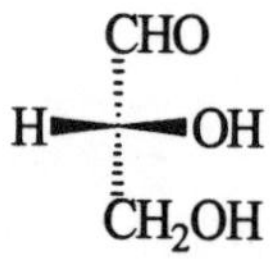

f. O CO_2CH_3

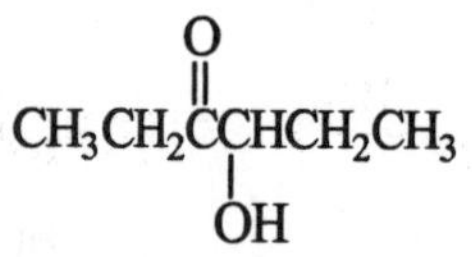

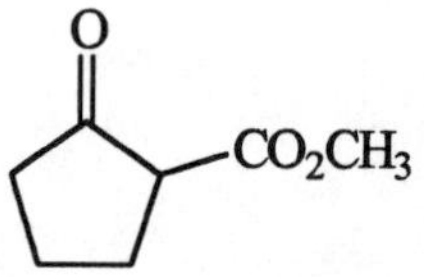

g. 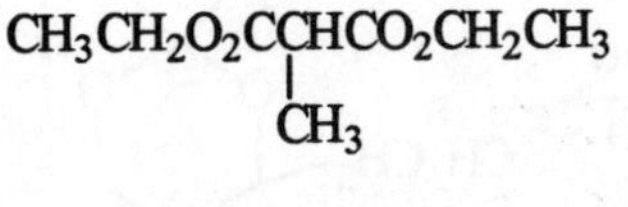 $CH_3CH_2O_2CCH(CH_3)CO_2CH_2CH_3$

h. 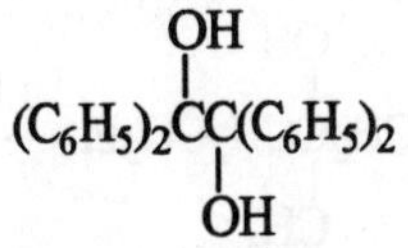 $(C_6H_5)_2C(OH)C(OH)(C_6H_5)_2$

S2. a. $HO_2CCH_2CH_2CH_2CH{=}O + HCO_2H$

b. 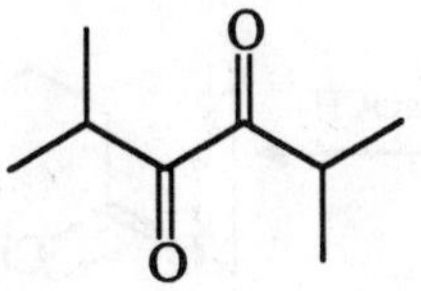 c.

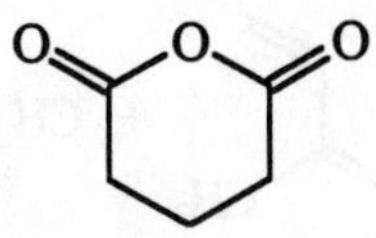

d. 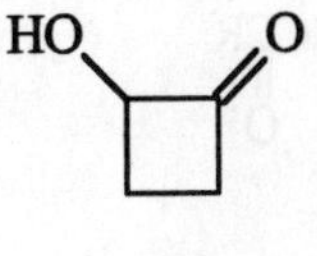 HO, O

e. 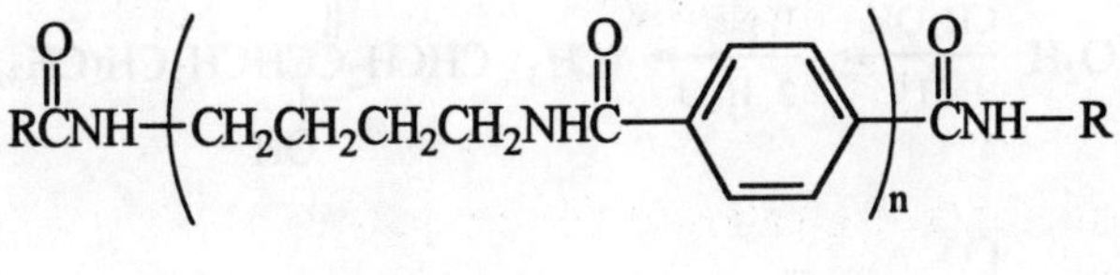 $RCNH\text{—}(CH_2CH_2CH_2CH_2CH_2NHC(=O)\text{—}C_6H_4\text{—}C(=O))_n\text{—}NH\text{—}R$ (with RC=O)

f. 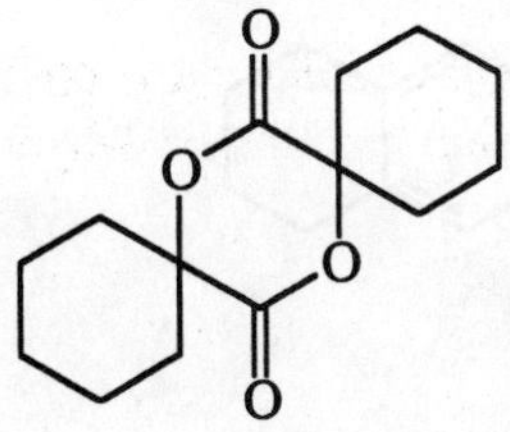 O, O, O, O

g. 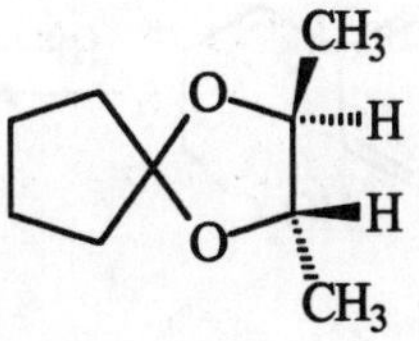 CH_3, O, H, H, O, CH_3

h. O, CH_3, CH_3

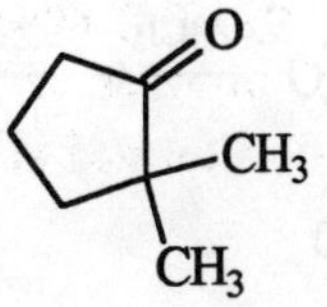

i. 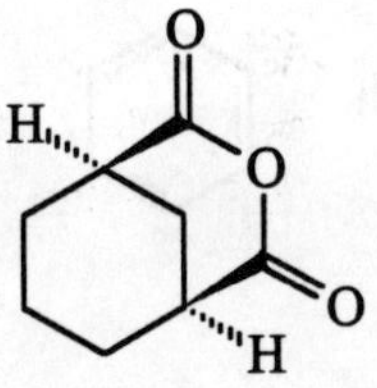 H, O, O, O, O, H

S3. a. 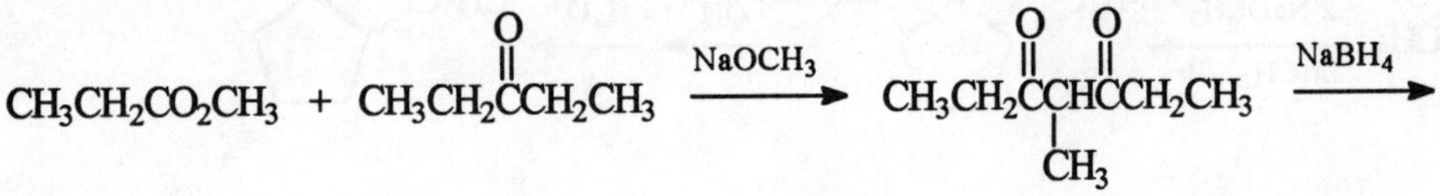 $CH_3CH_2CO_2CH_3 + CH_3CH_2CCH_2CH_3$ (C=O) $\xrightarrow{NaOCH_3}$ $CH_3CH_2C(=O)CH(CH_3)C(=O)CH_2CH_3$ $\xrightarrow{NaBH_4}$

b. $Br(CH_2)_5Br \xrightarrow[\Delta]{2\ NaCN} \xrightarrow{H_3O^+} HO_2C(CH_2)_5CO_2H \xrightarrow[H^+]{EtOH} \xrightarrow[EtOH]{NaOEt}$ (2-oxocyclohexane with CO_2Et) $\xrightarrow{OH^-} \xrightarrow[\Delta]{H_3O^+}$

c. $CH_3CCH_2CO_2CH_3$ (C=O) $\xrightarrow[CH_3CH\text{—}CH_2\ (epoxide)]{NaOCH_3}$ [H_3C, O, CO_2CH_3, O^-, CH_3] $\longrightarrow$ H_3C, O, O, O, CH_3 $+ CH_3O^-$

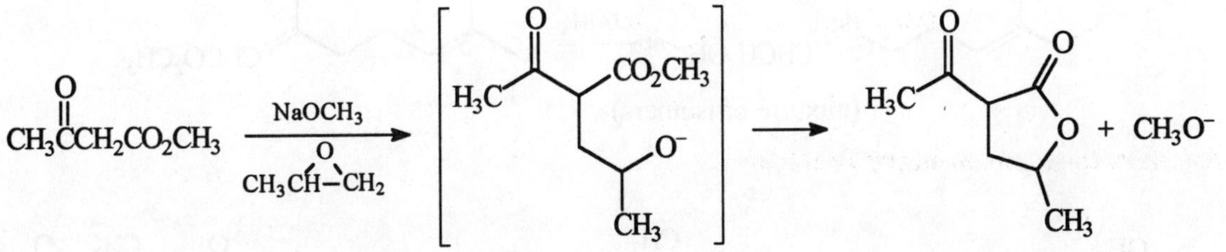

d. 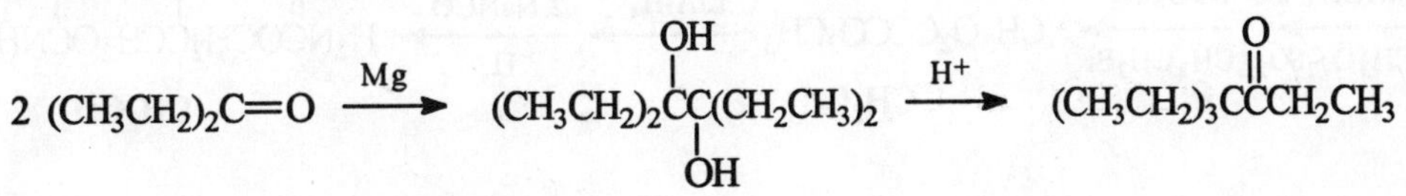 $2\ (CH_3CH_2)_2C{=}O \xrightarrow{Mg} (CH_3CH_2)_2C(OH)C(OH)(CH_2CH_3)_2 \xrightarrow{H^+} (CH_3CH_2)_3CCCH_2CH_3$ (C=O)

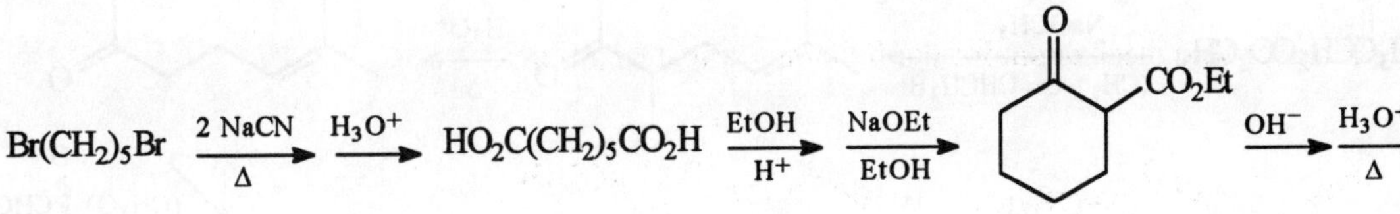

e. $CH_3CH_2CCH_3$ (O) + $BrCH_2CO_2Et$ $\xrightarrow[\text{benzene}, \Delta]{\text{Zn}}$ $CH_3CH_2C(OH)(CH_3)CH_2CO_2Et$ $\xrightarrow{\text{LiAlH}_4}$ $\xrightarrow[\text{H}^+, -\text{H}_2\text{O}]{(\text{CH}_3)_2\text{C=O}}$ (cyclic acetal: H_3C, CH_3, O, O, CH_3CH_2, H_3C)

f. (methyl vinyl ketone, O, CH_3) + $CH_2(CO_2CH_3)_2$ $\xrightarrow{\text{NaOCH}_3}$ [CO_2CH_3, CO_2CH_3, O, CH_3] $\longrightarrow$ (cyclohexanedione, CO_2CH_3, O, O)

g. $(CH_3)_2CHCH_2CO_2H$ $\xrightarrow[\text{H}^+]{\text{CH}_3\text{OH}}$ $\xrightarrow[\text{2. H}_2\text{O}]{\text{1. Na}}$ $(CH_3)_2CHCH_2CCHCH_2CH(CH_3)_2$ (O, OH) $\xrightarrow{\text{Cu(OAc)}_2}$ RCCR (O, O)

h. (benzene) + (succinic anhydride, O, O, O) $\xrightarrow{\text{AlCl}_3}$ $\xrightarrow[\text{HCl}]{\text{Zn}}$ (4-phenylbutanoic acid, HO_2C) $\xrightarrow{\text{HF}}$ $\xrightarrow[\text{HCl}]{\text{Zn}}$ (tetralin)

i. $CH_3CCH_2CO_2CH_3$ (O) $\xrightarrow[\text{Br(CH}_2)_4\text{Br}]{\text{2 NaOCH}_3}$ (H_3C, O, CO_2CH_3, cyclopentane) $\xrightarrow{\text{OH}^-}$ $\xrightarrow[\Delta]{\text{H}_3\text{O}^+}$ CH_3C(O)-cyclopentyl

j. $CH_3CCH_2CO_2CH_3$ (O) $\xrightarrow[(\text{CH}_3)_2\text{C=CHCH}_2\text{Br}]{\text{NaOCH}_3}$ (O, CO_2CH_3) $\xrightarrow[\Delta]{\text{H}_3\text{O}^+}$ (O) $\xrightarrow[\Delta]{(\text{CH}_3\text{O})_2\overset{\text{O}}{\text{P}}\bar{\text{C}}\text{HCO}_2\text{CH}_3}$ $CHCO_2CH_3$ $\xrightarrow{\text{LiAlH}_4}$ $CHCH_2OH$

(mixture of isomers)

S4. $CH_2(CO_2CH_3)_2$ $\xrightarrow[(\text{CH}_3\text{O})_2\text{SO}_2]{\text{NaOCH}_3}$ $\xrightarrow[\text{CH}_3\text{CH}_2\text{Br}]{\text{NaOCH}_3}$ $CH_3O_2CC(CH_3)(CH_2CH_3)CO_2CH_3$ $\xrightarrow{\text{LiAlH}_4}$ $\xrightarrow[\text{H}^+]{\text{2 NaNCO}}$ $H_2NCOCH_2C(CH_3)(CH_2CH_3)CH_2OCNH_2$ (O, O)

S5. $CH_3CCH_2CH_2CO_2H$ (O) — A

$HO_2CCH_2CH_2CO_2H$ — B

$CH_3C(OCH_3)_2CH_2CH_2CO_2CH_3$ — C

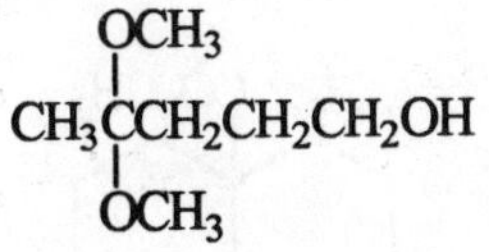

$CH_3C(OCH_3)_2CH_2CH_2CH_2OH$

D

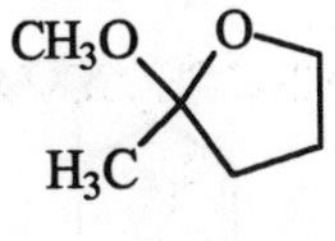

E

S6. 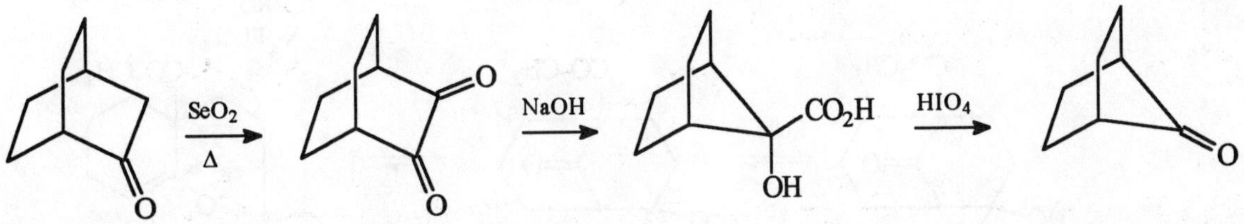 $\xrightarrow[\Delta]{SeO_2}$ $\xrightarrow{NaOH}$ CO_2H, OH $\xrightarrow{HIO_4}$

S7. 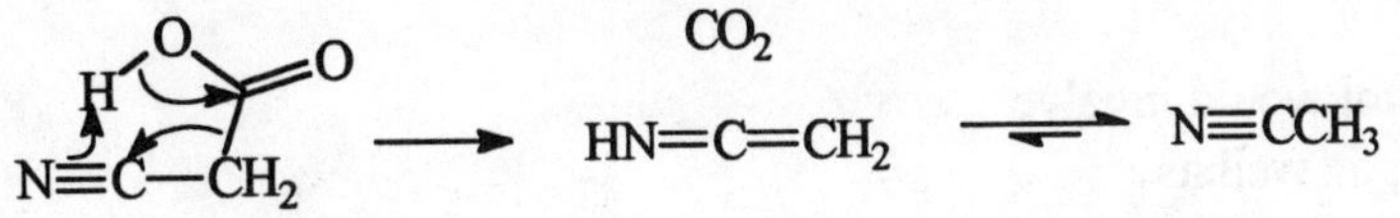 CO_2 $\longrightarrow$ $HN{=}C{=}CH_2 \rightleftharpoons N{\equiv}CCH_3$

S8. a. 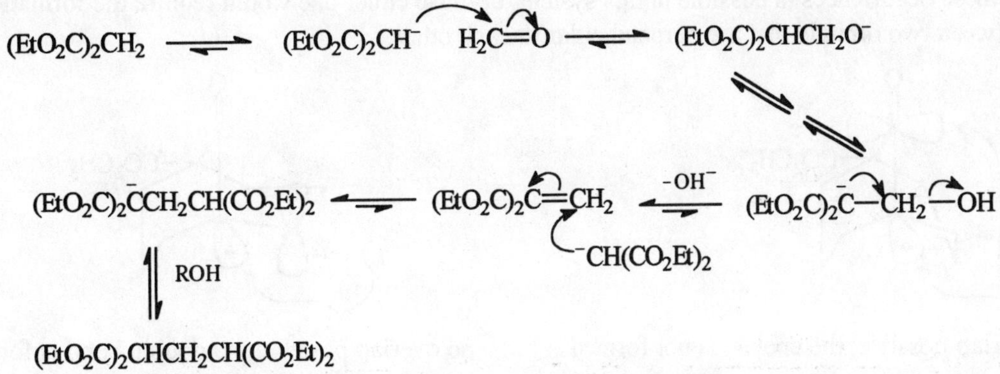 $(EtO_2C)_2CH_2 \rightleftharpoons (EtO_2C)_2CH^- \; H_2C{=}O \rightleftharpoons (EtO_2C)_2CHCH_2O^-$

$(EtO_2C)_2\bar{C}CH_2CH(CO_2Et)_2 \rightleftharpoons (EtO_2C)_2C{=}CH_2 \;\; {}^-CH(CO_2Et)_2 \xleftarrow{-OH^-} (EtO_2C)_2\bar{C}{-}CH_2{-}OH$

ROH

$(EtO_2C)_2CHCH_2CH(CO_2Et)_2$

b. 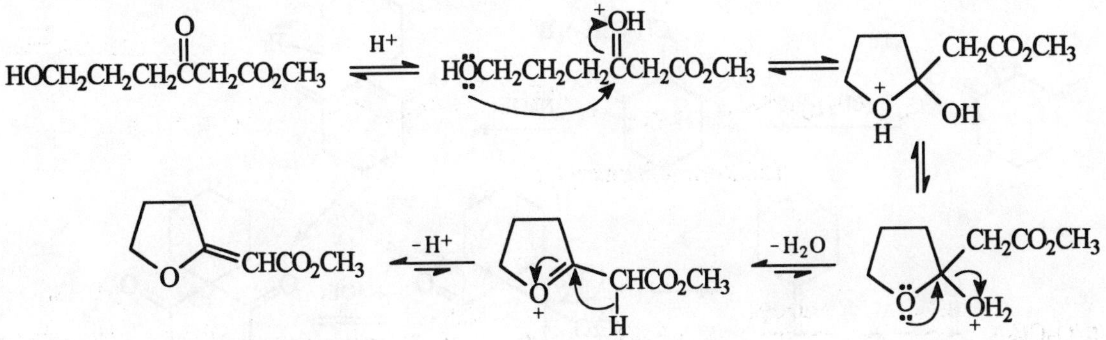 $HOCH_2CH_2CH_2\overset{O}{\overset{\|}{C}}CH_2CO_2CH_3 \overset{H^+}{\rightleftharpoons} H\ddot{O}CH_2CH_2CH_2\overset{^+OH}{\overset{\|}{C}}CH_2CO_2CH_3 \rightleftharpoons$ $CH_2CO_2CH_3$, OH, H

$CHCO_2CH_3 \overset{-H^+}{\rightleftharpoons}$ $CHCO_2CH_3$, H $\overset{-H_2O}{\rightleftharpoons}$ $CH_2CO_2CH_3$, $\overset{+}{O}H_2$

S9. 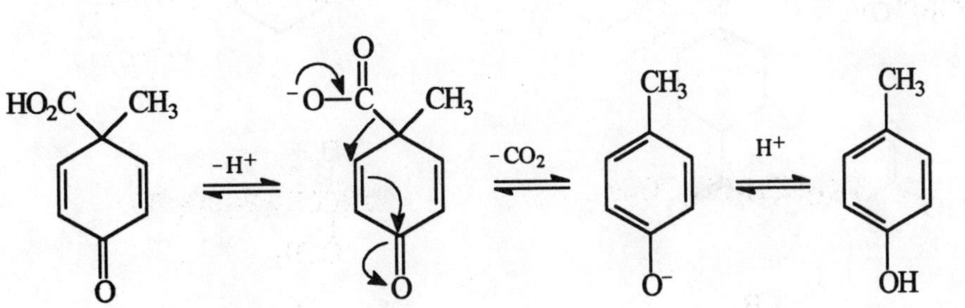 HO_2C, CH_3 $\overset{-H^+}{\rightleftharpoons}$ $^-O{-}\overset{O}{\overset{\|}{C}}$, CH_3 $\overset{-CO_2}{\rightleftharpoons}$ CH_3, O^- $\overset{H^+}{\rightleftharpoons}$ CH_3, OH

S10.

a.

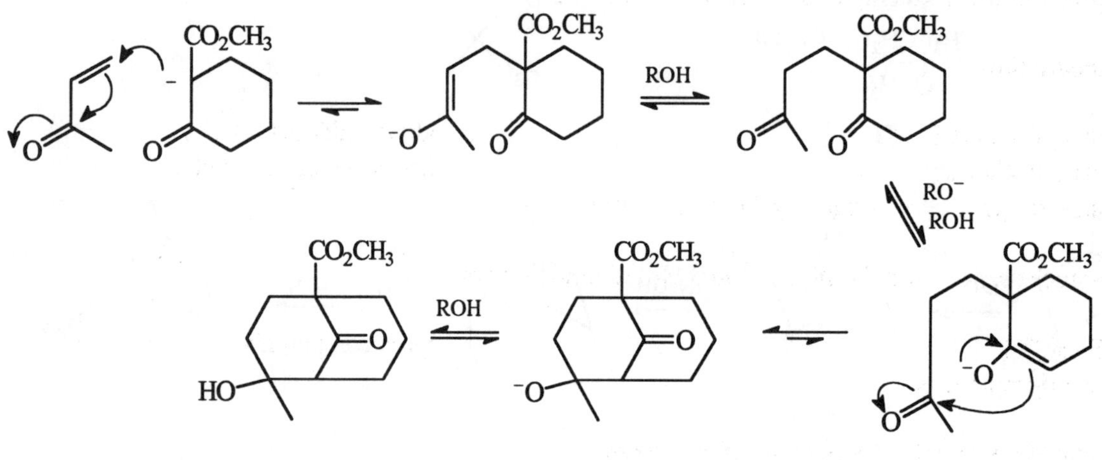

ROH

RO^-
ROH

ROH

b. Elimination of water from this bicyclic ketol would involve
1) formation of the enolate of the ketone, as well as
2) loss of hydroxide to generate the double bond.
Neither of these occurrences is possible in this system, because either one would require the formation of a π-bond between two orbitals that are perpendicular to each other:

no overlap possible; the enolate is not formed

no overlap possible; no double bond is formed

S11.

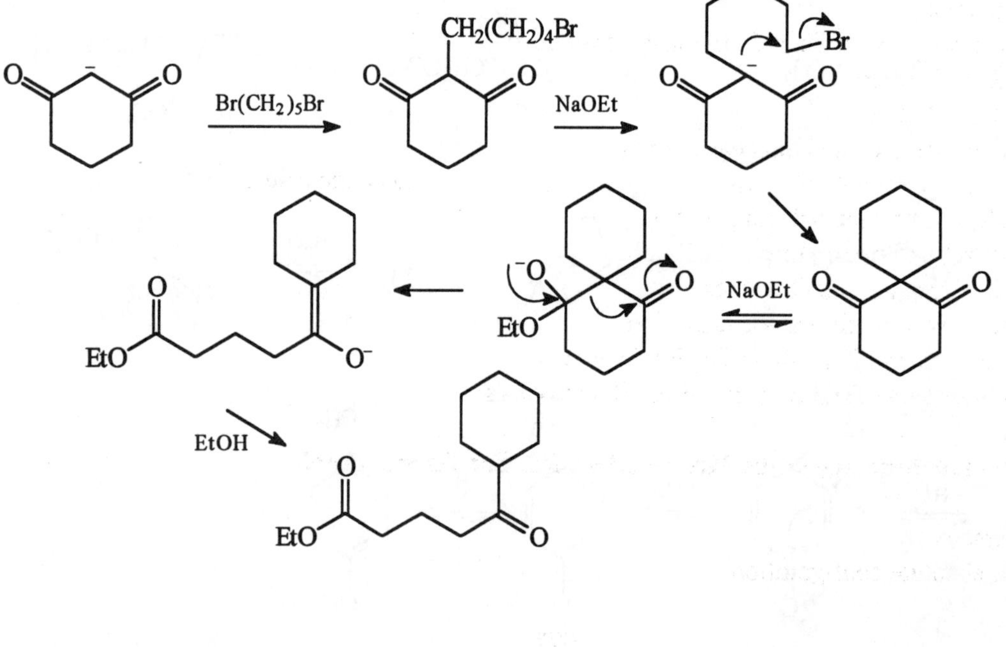

$CH_2(CH_2)_4Br$

$Br(CH_2)_5Br$

NaOEt

NaOEt

EtOH

28. CARBOHYDRATES

28.A. Chapter Outline and Important Terms Introduced

28.1 Introduction

carbohydrate: $(CH_2O)_n$
sugar (saccharide)
mono-, di-, tri-, tetra-, oligo-, and polysaccharides

aldose, aldopentose, etc.
ketose, ketohexose, etc.

28.2 Stereochemistry and Configurational Notation of Sugars

R,S vs. D,L
Fischer projections

meso compounds

28.3 Cyclic Heimacetals: Anomerism; Glycosides

α- and β-anomers
Haworth projections
pyranose vs. furanose
mutarotation

glycoside
glucoside, mannoside, etc.
enzymatic hydrolysis

28.4 Conformations of the Pyranoses

chair conformations

axial vs. equatorial substituents

28.5 Reactions of Monosaccharides

A. Ether Formation
   - glycoside formation (see Section 28.3)
   - protection of anomeric carbon — methylation

B. Formation of Cyclic Acetals and Ketals

C. Esterification
   - equatorial anomeric hydroxyl faster than axial
   - anomeric effect

D. Reduction: Alditols

E. Oxidation: Aldonic and Saccharic Acids
   - Tollens and Fehling's tests — saccharic acid
   - reducing vs. non-reducing sugars

F. Oxidation by Periodic Acid

G. Phenylhydrazones and Osazones
   - for derivatization and identification

H. Chain Extension: The Kiliani-Fischer Synthesis

I. Chain Shortening: The Ruff and Wohl Degradations

28.6 Relative Stereochemistry of the Monosaccharides: The Fischer Proof

use of symmetry
relative vs. absolute configuration

28.7 Oligosaccharides

structure proof via methylation/cleavage
invertase enzymes

28.8 Polysaccharides

starch — amylose, amylopectin
cellulose, cellulase

28.9 Sugar Phosphates

ribonucleic acid (RNA) — deoxyribonucleic acid (DNA)

28.10 Natural Glycosides

glycosyl residue (glycon) vs. aglycon
amygdalin (laetrile) vs. erythromycin

28.11 Aminosaccharides and Poly(aminosaccharides)

chiton

28.B. Important Reactions Introduced

NOTE: There are essentially no **new** reactions introduced in this Chapter; it is simply the context – the fact that the substrates are sugars – that makes the reactions different from what you have already learned.

Mutarotation

Equation: [structure: glucopyranose with anomeric OH equatorial, H axial] ⇌ [structure: glucopyranose with anomeric H equatorial, OH axial]

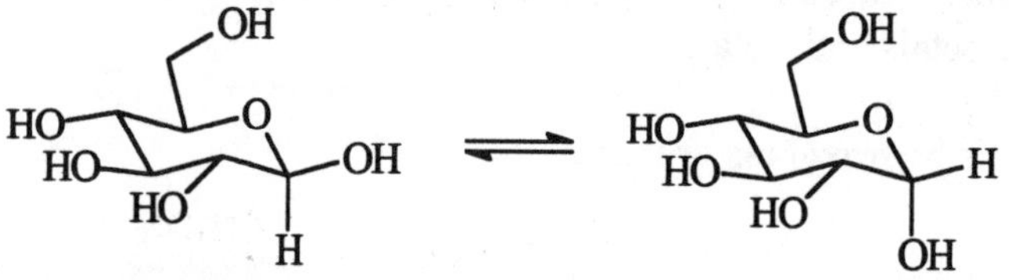

Generality: acid- or base-catalyzed
saccharide must be a free hemiacetal (or hemiketal)
observed for a wide variety of saccharides

Key features: mechanism involves equilibration between hemiacetal and hydroxy aldehyde (or hemiketal and hydroxyketone)

Glycoside formation and hydrolysis

Equation: [structure: glucopyranose]–OH + CH_3OH ⇌ [structure: glucopyranoside]–OCH_3

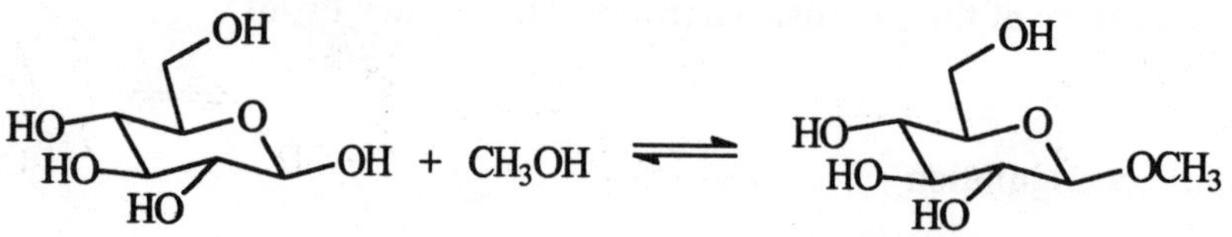

Generality: catalyzed by acid or glycosidase enzyme
mixture of anomers often formed

Key features: mechanism is simply a formation of a cyclic acetal
useful for protection of anomeric (= carbonyl) carbon toward basic reagents

Methylation (28.5.A)

Equation:

$$\text{methyl glycoside (HO, HO, HO, }CH_2OH\text{, }OCH_3\text{)} \underset{}{\overset{Me_2SO_4}{\rightleftharpoons}} \text{permethylated glycoside (}CH_3O\text{, }CH_3O\text{, }CH_3O\text{, }CH_2OCH_3\text{, }OCH_3\text{)}$$

Generality: strongly basic conditions (= Williamson ether synthesis) require anomeric carbon to be protected as glycoside (acetal)

Key features: used in structure elucidation to determine acetal ring size

Formation of cyclic acetals and ketals (28.5.B)

Equation:

$$\text{sugar (HO, HO, HO, OH, }CH_2OH\text{)} + CH_3\overset{O}{\overset{\|}{C}}CH_3 \overset{H^+}{\rightleftharpoons} \text{di-isopropylidene derivative (}H_3C\text{, }H_3C\text{, }H_3C\text{, }CH_3\text{, OH)} + 2\ H_2O$$

Generality: many possibilities, exact product depends on specific saccharide

Key features: useful for selective protection of hydroxyls

Reduction to alditols (28.5.D)

Equation:

$$\text{cyclic hemiacetal (O, OH)} \rightleftharpoons \begin{matrix} CHO \\ \wr \\ CH_2OH \end{matrix} \xrightarrow[H_2/Ni]{NaBH_4\text{ or}} \begin{matrix} CH_2OH \\ \wr \\ CH_2OH \end{matrix}$$

Key features: requires hemiacetal form of saccharide
reaction proceeds via open chain (hydroxy aldehyde) form
useful in symmetrization for structure elucidation

Oxidation to aldonic acids (28.5.E)

Equation:

$$\text{cyclic hemiacetal (O, OH)} \rightleftharpoons \begin{matrix} CHO \\ \wr \\ CH_2OH \end{matrix} \xrightarrow{[Ox]} \begin{matrix} CO_2H \\ \wr \\ CH_2OH \end{matrix} \rightleftharpoons \text{lactone (C=O, O, }CH_2OH\text{)} + H_2O$$

(a reducing sugar) (an aldonic acid)

Generality: [Ox] = $Ag^+(NH_3)_2$ (Tollens reagent), $Cu(OH)_2$ (Fehling's reagent), or Br_2 in H_2O (bromine water)

Key features: requires hemiacetal form of saccharide
product often lactonizes

Oxidation to saccharic acids (28.5.E)

Equation: cyclic hemiacetal (O, OH) ⇌ CHO / ~~~ / CH_2OH $\xrightarrow[\Delta]{HNO_3}$ CO_2H / ~~~ / CO_2H

Key features: useful in symmetrization for structure elucidation

Oxidation by periodic acid (28.5.F; see Section 27.3.B)

Formation of osazones (28.5.G)

Equation: CH=O / CHOH / ~~~ $\xrightarrow[HOAc]{C_6H_5NHNH_2}$ $CH{=}NNHC_6H_5$ / CHOH / ~~~ $\xrightarrow{2\ C_6H_5NHNH_2}$ $CH{=}NNHC_6H_5$ / $C{=}NNHC_6H_5$ / ~~~ + NH_3 + $C_6H_5NH_2$

Key features: important as crystalline derivatives of saccharides, with characteristic mp as well as rate of formation for each saccharide
stereocenter at C-2 destroyed, hence C-2 epimeric saccharides give the same osazone (albeit at different rates)

Kiliani-Fischer synthesis (28.5.H)

Equation: CH=O / ~~~ $\xrightarrow{HCN}$ CN / CHOH / ~~~ $\xrightarrow[2.\ \Delta]{1.\ Ba(OH)_2 \text{ or } H_3O^+}$ O=C / CHOH / ~~~ (lactone, O) $\xrightarrow[pH\ 3.5]{Na(Hg) \text{ or } NaBH_4}$ OH / HC / CHOH / ~~~ (cyclic, O)

Key features: leads to formation of both C-2 epimers of the chain-extended saccharide

Chain degradations (28.5.I)

Equations:

Ruff degradation CHO / CHOH / ~~~ $\xrightarrow[2.\ Ca(OH)_2]{1.\ Br_2}$ $(CO_2^-$ / CHOH / ~~~$)_2$ Ca^{2+} $\xrightarrow[Fe^{3+}]{H_2O_2}$ CHO / ~~~ + CO_2

Wohl degradation

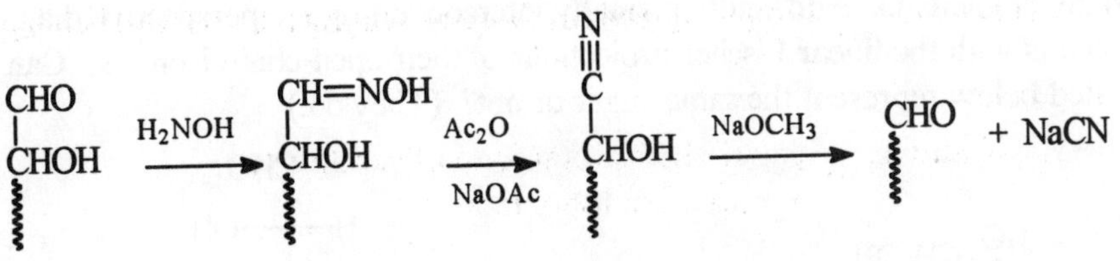

$$\text{CHO}-\text{CHOH}\;\xrightarrow{H_2NOH}\;\text{CH=NOH}-\text{CHOH}\;\xrightarrow[\text{NaOAc}]{Ac_2O}\;\text{N}\equiv\text{C}-\text{CHOH}\;\xrightarrow{NaOCH_3}\;\text{CHO} + \text{NaCN}$$

Key features: saccharides that are epimeric at C-2 give the same degradation product, since the stereocenter at that position is destroyed

28.C. Important Concepts and Hints

Appropriately, the chemistry of carbohydrates follows the discussion of difunctional compounds. With only a couple of exceptions, all of the reactions presented in this chapter are ones you have seen before. What makes them different is the fact that the molecules involved have a functional group on every carbon! Nonetheless, these molecules behave according to the principles outlined in the preceding chapter, and only a few transformations that are specific to carbohydrate chemistry have to be learned. In essence, this topic requires you to apply the knowledge you've already gained to more complicated systems.

The topic of sugar chemistry also requires you to review the subject of stereochemistry. You have to be able to manipulate Fischer projections, determine R and S configurations, and recognize the presence or absence of planes of symmetry and *meso* compounds. The presence or absence of optical activity has been the single most important observation in schemes for the determination of the relative stereochemistry of sugars. Many of the techniques devised for probing the stereochemistry of a sugar have involved bringing the two ends of the carbon chain to the same oxidation state ($-CH_2OH$ or $-CO_2H$) and then looking for optical activity. An optically active compound cannot be *meso*, and an optically inactive compound is assumed to be *meso*.

Fischer projections are by convention drawn with the carbon chain extending vertically and the most oxidized end of the molecule at the top, if there is a difference. To determine whether a Fischer projection represents a *meso* compound, draw (or imagine) a line across the picture exactly half-way down. If there is an even number of carbons, this line will cross one of the carbon-carbon bonds; if there is an odd number of carbons, the line will cut across the middle carbon atom and its substituents. With this line in place (or in mind), compare the pattern above the line with that below: if they are mirror images, the compound is *meso*; if they differ in any way the compound is chiral, and thus capable of being optically active. Some examples are illustrated below:

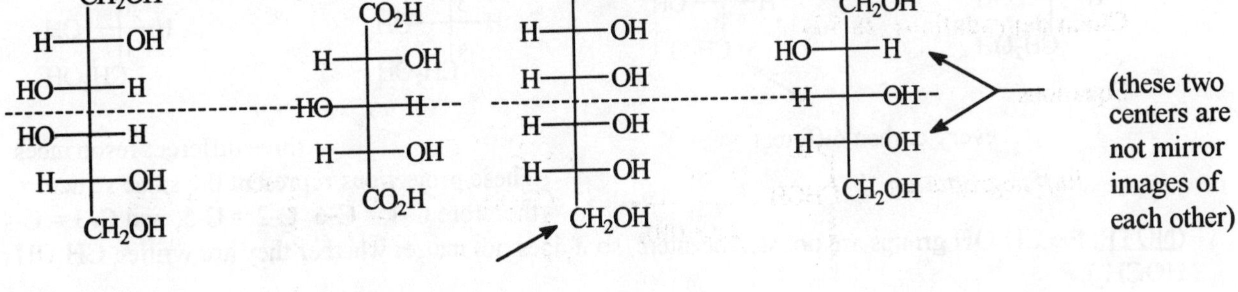

meso: CH_2OH / H–OH / HO–H / - - - / HO–H / H–OH / CH_2OH; CO_2H / H–OH / HO–H (on line) / H–OH / CO_2H

chiral: CH=O / H–OH / H–OH / - - - / H–OH / H–OH / CH_2OH (the two ends of this molecule are different; CH_2OH / HO–H / H–OH (on line) / H–OH / CH_2OH (these two centers are not mirror images of each other)

Many students have difficulty mentally interconverting the perspective diagrams of the chair forms of sugar hemiacetals with the linear Fischer projections of their open-chain isomers. Can you tell whether the two structures illustrated below represent the same sugar or not? (They do.)

One hint that we can give you (in addition to **using models!!**) is to become familiar with Haworth projections. These are intermediate between the perspective formulas and Fischer projections, and serve to relate the two in a logical way. A Haworth projection is essentially a flattened form of the cyclic structure; the distinction between axial and equatorial is ignored and all that matters is whether substituents point up or down. To go from the perspective formula to the Haworth projection, you simply look for up or down orientations. The Haworth projection of the sugar above is is given at the right.

The substituents that stick **up** in a Haworth formula are on the **left** in a Fischer projection, with two exceptions:

1. the next-to-last carbon gets mixed up because the substituent in the Haworth formula is part of the carbon chain itself (if $-CH_2OH$ sticks up (D-sugar), the hydroxy group is on the right);
2. the configuration of the anomeric carbon (α or β) is not represented in the open-chain Fischer projection.

As usual, these correlations can be most easily visualized and understood with the help of models.

28.D. Answers to Exercises

28.1 One of the epimeric hexaols has an axis of symmetry, such that carbons 1-3 are equivalent to carbons 6-4, respectively. The other hexaol is not symmetrical, and every carbon is different in the CMR spectrum.

rotate 180° in plane of paper

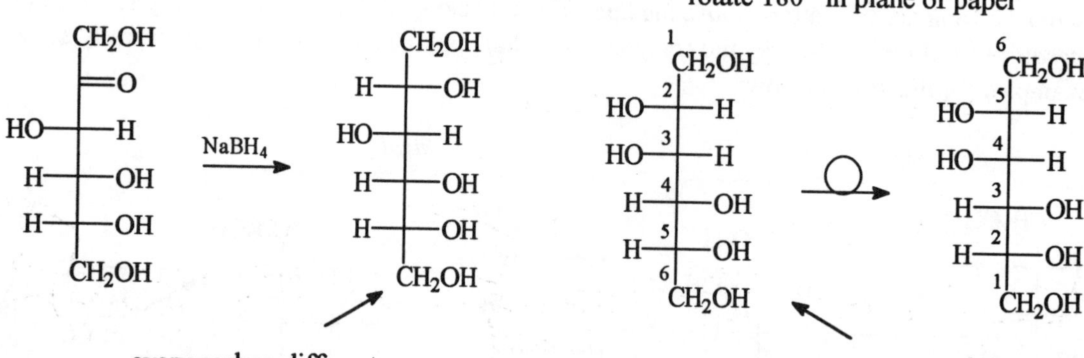

These projections represent the same structure, therefore C-1 = C-6, C-2 = C-5, and C-3 = C-4.

(NOTE: the CH_2OH groups are not stereocenters, so it does not matter whether they are written CH_2OH or $HOCH_2$).

28.2 Model-building exercise.

28.3

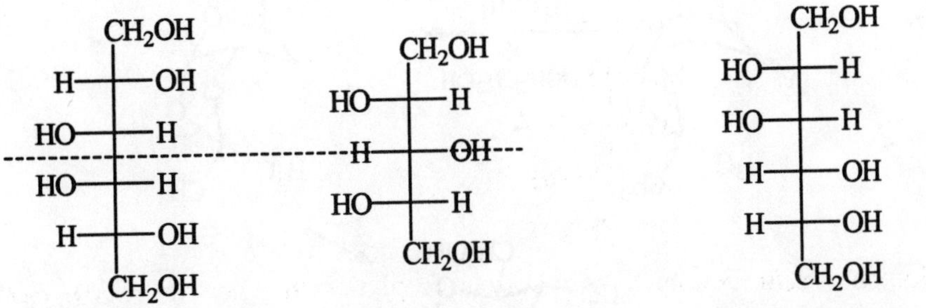

from D-galactose (achiral)

from L-xylose (achiral)

from D-mannose (chiral)

28.4

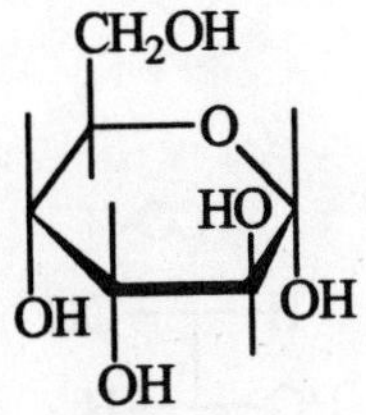

α-D-altrose

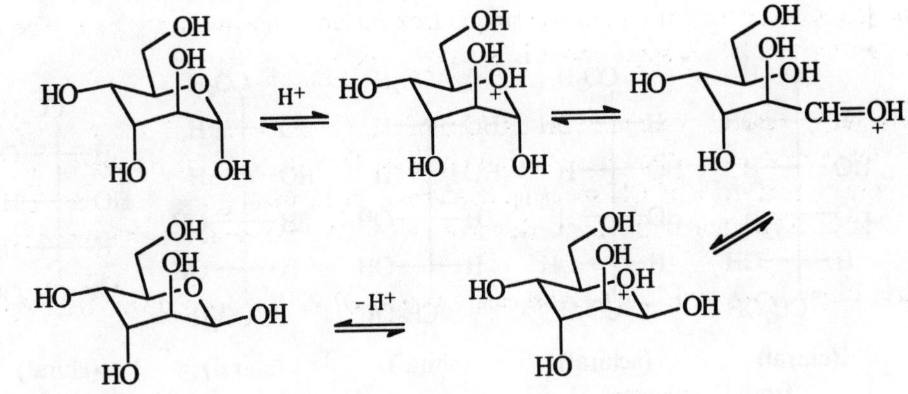

β-D-altrose

28.5

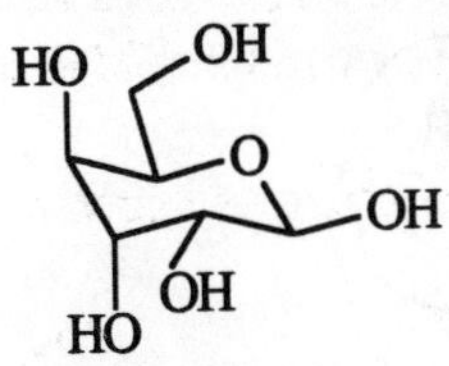

β-D-gulose

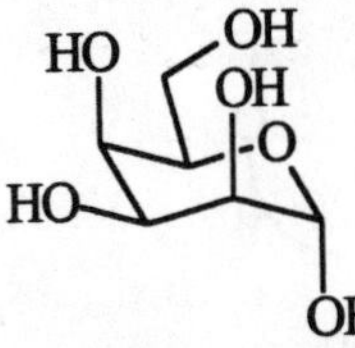

α-D-talose

28.6

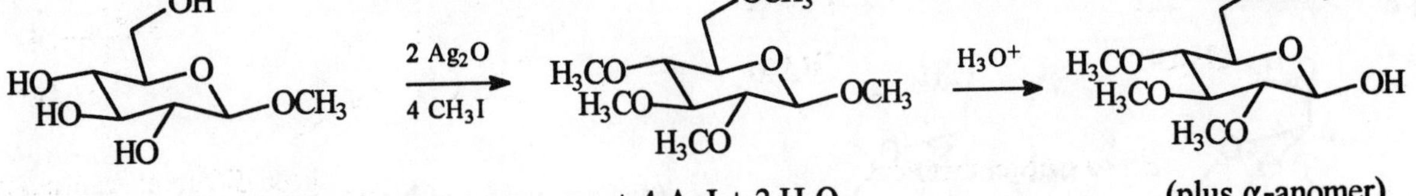

$+ 4\ AgI + 2\ H_2O$

(plus α-anomer)

28.7

RuO_4

H_2NNH_2, KOH, $HOCH_2CH_2OH$, Δ

(plus α-anomer)

28.8

(chiral) (achiral)
from D-galactose

(chiral) (chiral)
from D-mannose

(chiral) (achiral)
from D-xylose

28.9

a.

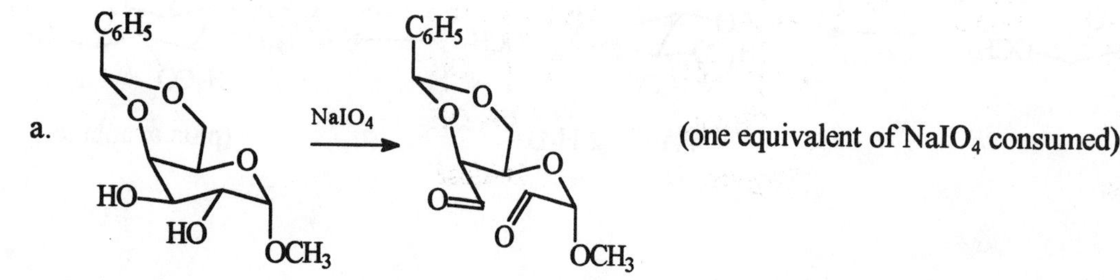

b. (methyl pyranoside: HO, HO, HO, O, OCH_3) $\xrightarrow{NaIO_4}$ (dialdehyde: O=, O, OCH_3, O) + $HCOH$ (two equivalents of $NaIO_4$ consumed)

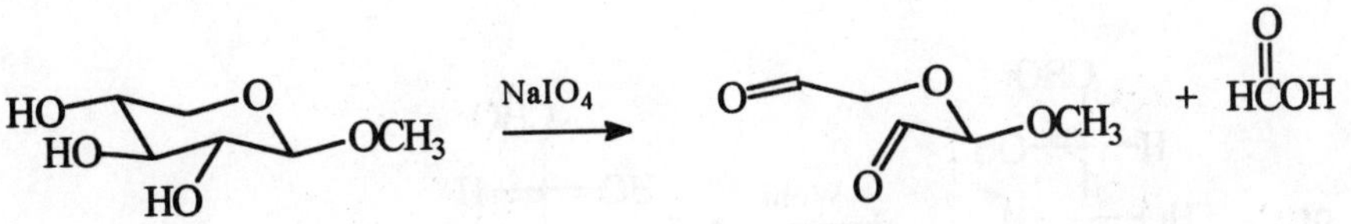

c. CH_2OH / H—OH / HO—H / CH_2OH $\xrightarrow{NaIO_4}$ 2HCOOH + 2 CH_2=O (three equivalents of $NaIO_4$ consumed)

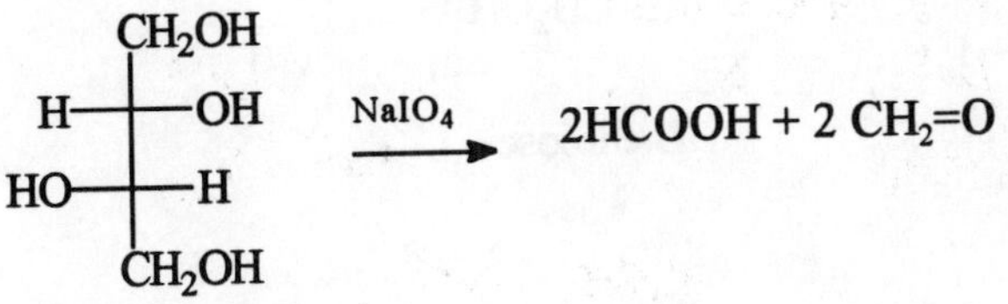

28.10

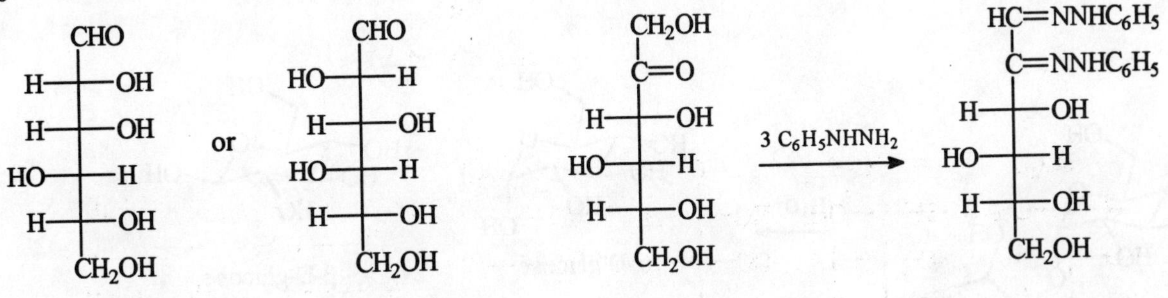

CHO / H—OH / H—OH / HO—H / H—OH / CH_2OH — D-(−)-gulose

or

CHO / HO—H / H—OH / HO—H / H—OH / CH_2OH — D-(−)-idose

CH_2OH / C=O / H—OH / HO—H / H—OH / CH_2OH — D-(+)-sorbose $\xrightarrow{3\ C_6H_5NHNH_2}$ $HC{=}NNHC_6H_5$ / $C{=}NNHC_6H_5$ / H—OH / HO—H / H—OH / CH_2OH

28.11 CHO / H—OH / H—OH / H—OH / CH_2OH — D-ribose $\xrightarrow{\text{Kiliani Fischer}}$ CHO / H—OH / H—OH / H—OH / H—OH / CH_2OH — D-allose + CHO / HO—H / H—OH / H—OH / H—OH / CH_2OH — D-altrose

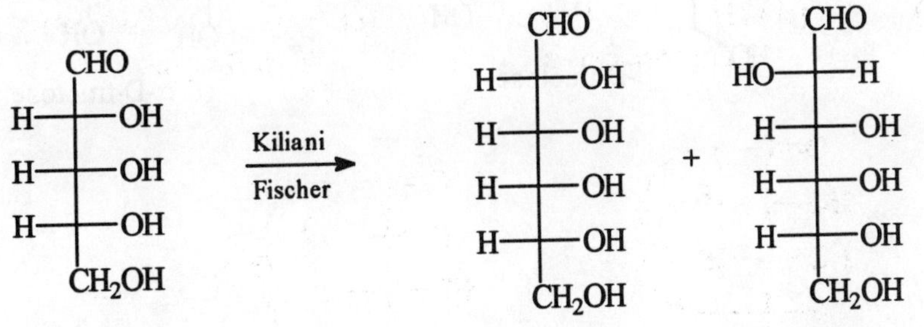

CHO / H—OH / HO—H / H—OH / CH_2OH — D-xylose $\xrightarrow{\text{Kiliani Fischer}}$ CHO / H—OH / H—OH / HO—H / H—OH / CH_2OH — D-gulose + CHO / HO—H / H—OH / HO—H / H—OH / CH_2OH — D-idose

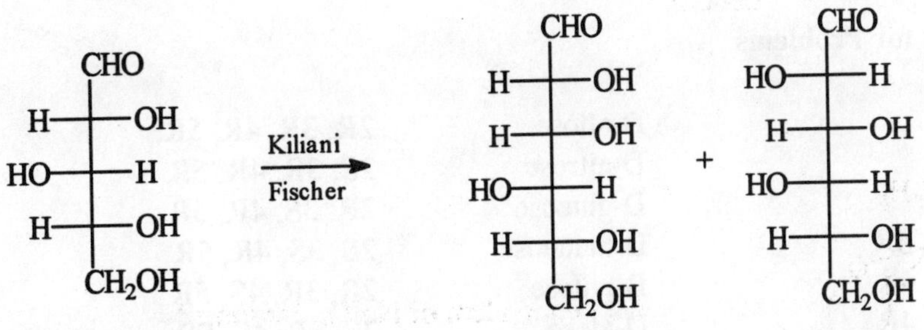

28.12

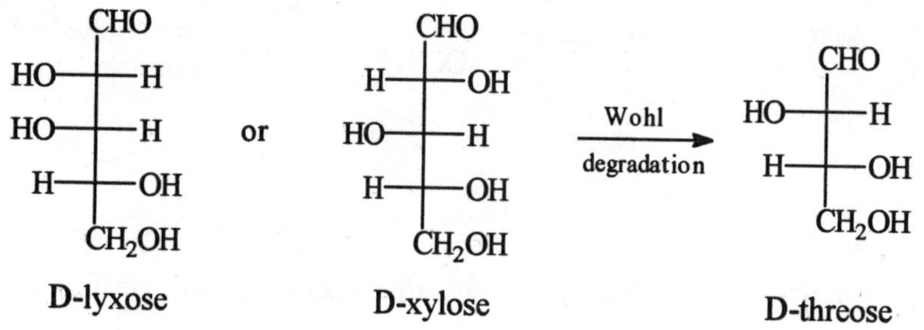

D-lyxose or D-xylose → (Wohl degradation) → D-threose

28.14

sucrose → (H_3O^+) → α-D-glucose, β-D-glucose, β-D-fructose, α-D-fructose

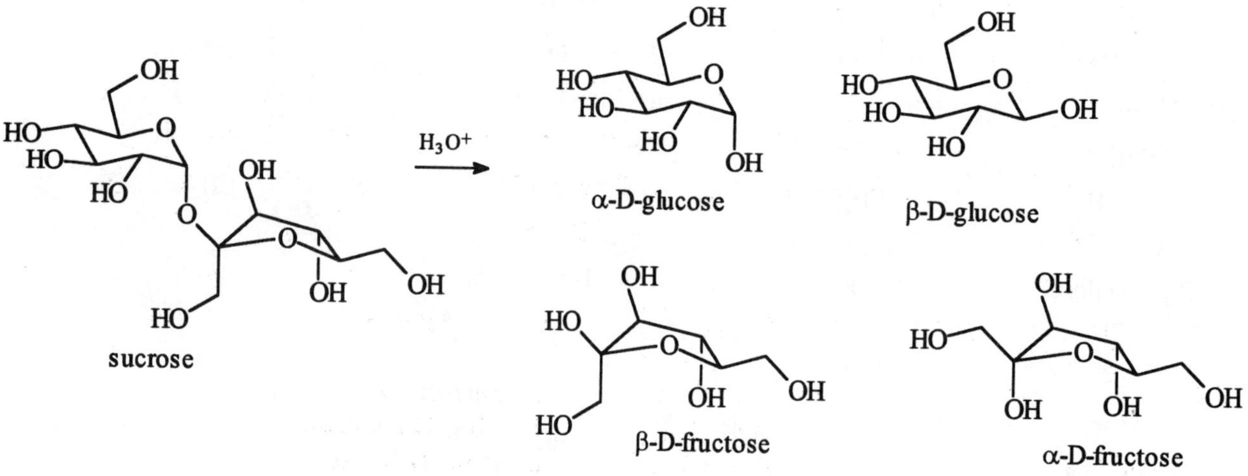

28.E. Answers and Explanations for Problems

1.

D-glyceraldehyde	2R	D-allose	2R, 3R, 4R, 5R
D-erythrose	2R, 3R	D-altrose	2S, 3R, 4R, 5R
D-threose	2S, 3R	D-glucose	2R, 3S, 4R, 5R
D-ribose	2R, 3R, 4R	D-mannose	2S, 3S, 4R, 5R
D-arabinose	2S, 3R, 4R	D-gulose	2R, 3R, 4S, 5R
D-xylose	2R, 3S, 4R	D-idose	2S, 3R, 4S, 5R
D-lyxose	2S, 3S, 4R	D-galactose	2R, 3S, 4S, 5R
		D-talose	2S, 3S, 4S, 5R

2.

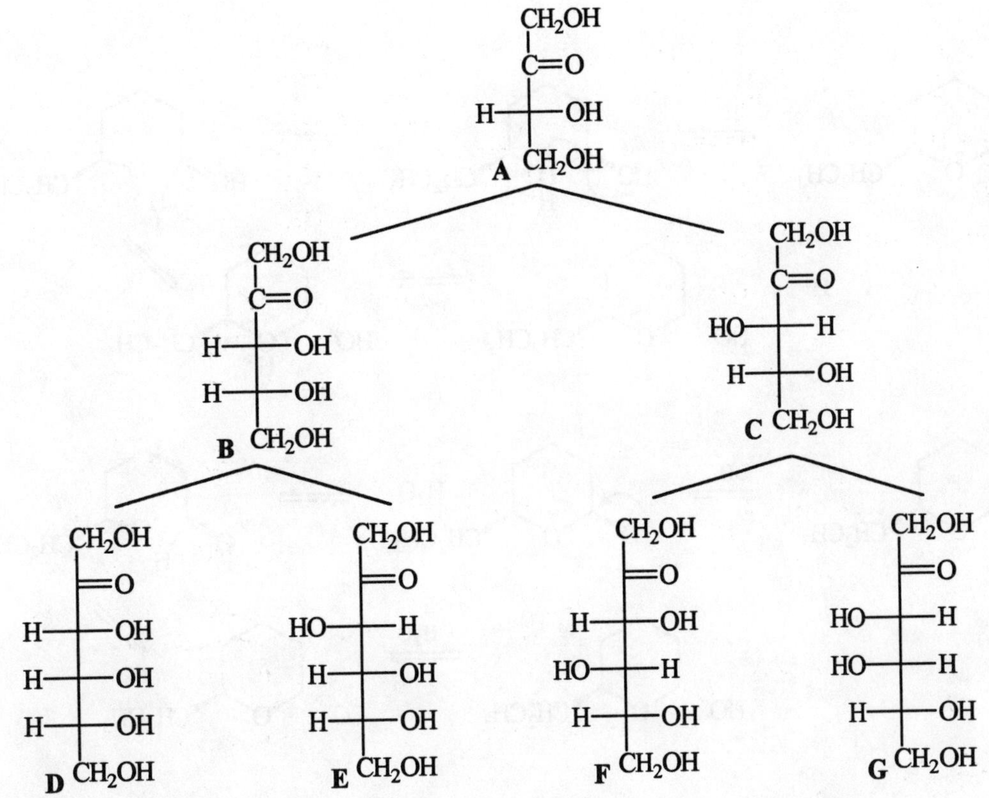

A: CH_2OH / C=O / H—C—OH / CH_2OH

B: CH_2OH / C=O / H—C—OH / H—C—OH / CH_2OH

C: CH_2OH / C=O / HO—C—H / H—C—OH / CH_2OH

D: CH_2OH / C=O / H—C—OH / H—C—OH / H—C—OH / CH_2OH

E: CH_2OH / C=O / HO—C—H / H—C—OH / H—C—OH / CH_2OH

F: CH_2OH / C=O / H—C—OH / HO—C—H / H—C—OH / CH_2OH

G: CH_2OH / C=O / HO—C—H / HO—C—H / H—C—OH / CH_2OH

IDENTICAL OSAZONES

	Ketose	**Aldoses**
A	D-erythrulose	D-erythrose, D-threose
B	D-ribulose	D-ribose, D-arabinose
C	D-xylulose	D-xylose, D-lyxose
D	D-psicose	D-allose, D-altrose
E	D-fructose	D-glucose, D-mannose
F	D-sorbose	D-gulose, D-idose
G	D-tagatose	D-galactose, D-talose

3. D-erythrose, D-ribose, D-xylose, D-allose, D-galactose

4.

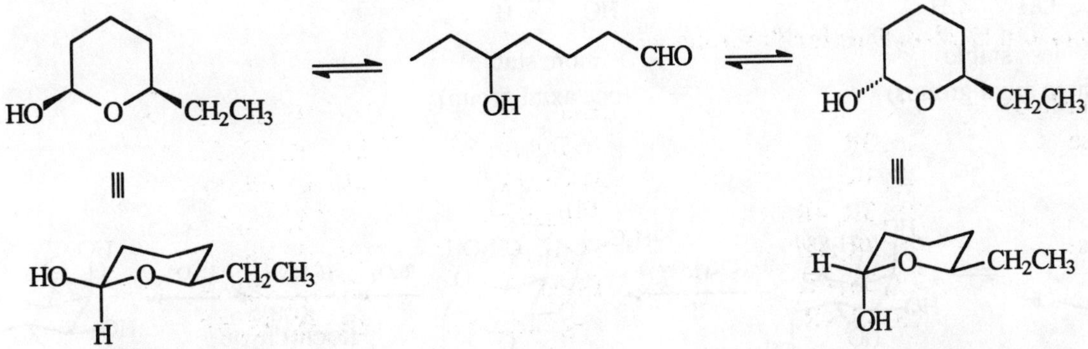

HO ... O ... CH_2CH_3 ⇌ (OH) ... CHO ⇌ HO ... O ... CH_2CH_3

HO ... O ... CH_2CH_3, H — diequatorial (more stable)

H ... O ... CH_2CH_3, OH — axial-equatorial (less stable)

Mechanism:

Acid:

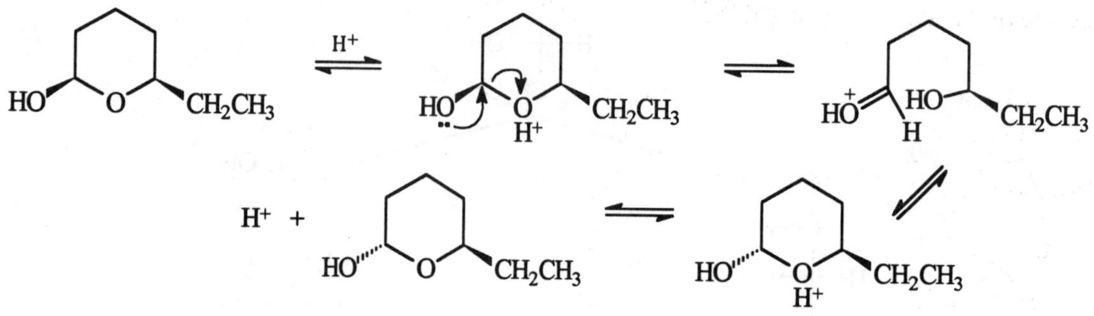

HO O CH_2CH_3 $\overset{H^+}{\rightleftharpoons}$ HO O H^+ CH_2CH_3 $\rightleftharpoons$ $H\overset{+}{O}$ H HO CH_2CH_3

H^+ + HO O CH_2CH_3 $\rightleftharpoons$ HO O H^+ CH_2CH_3

Base:

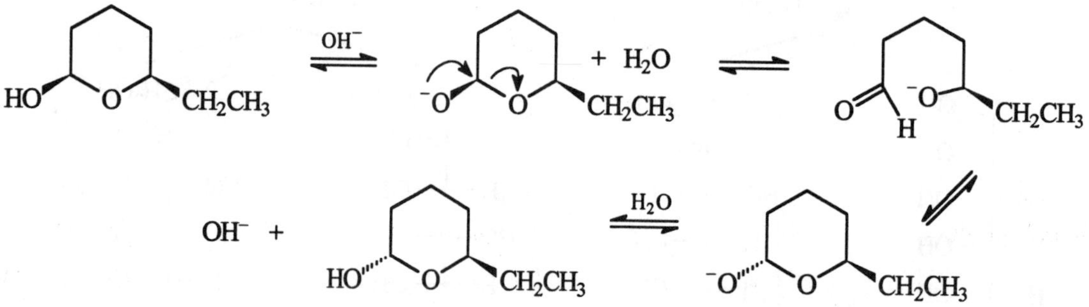

HO O CH_2CH_3 $\overset{OH^-}{\rightleftharpoons}$ ^-O O CH_2CH_3 + H_2O $\rightleftharpoons$ O H ^-O CH_2CH_3

OH^- + HO O CH_2CH_3 $\overset{H_2O}{\rightleftharpoons}$ ^-O O CH_2CH_3

5. a.

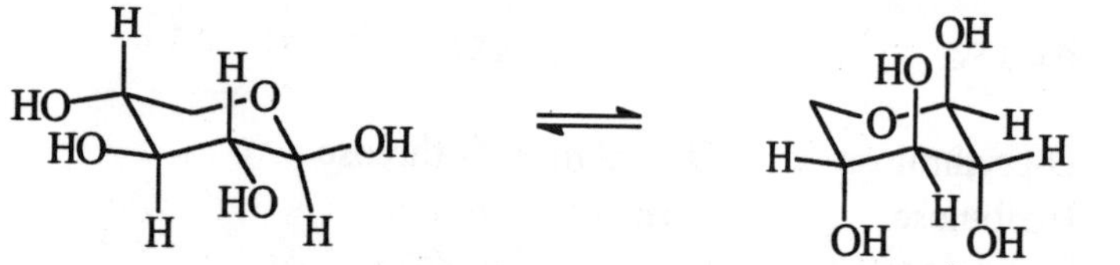

more stable $\rightleftharpoons$ less stable (four axial groups)

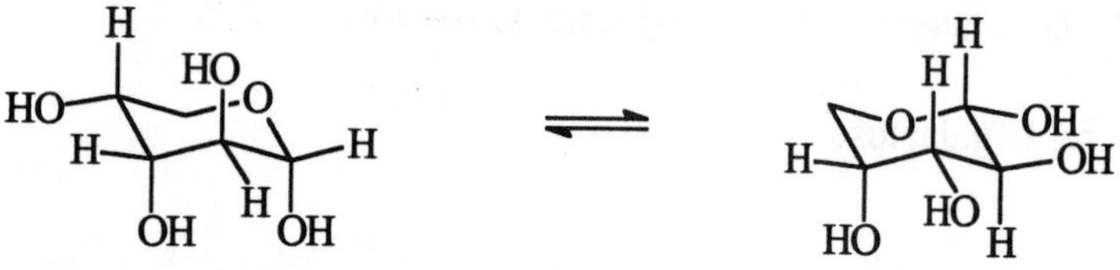

less stable (three axial groups) $\rightleftharpoons$ more stable (one axial group)

6.

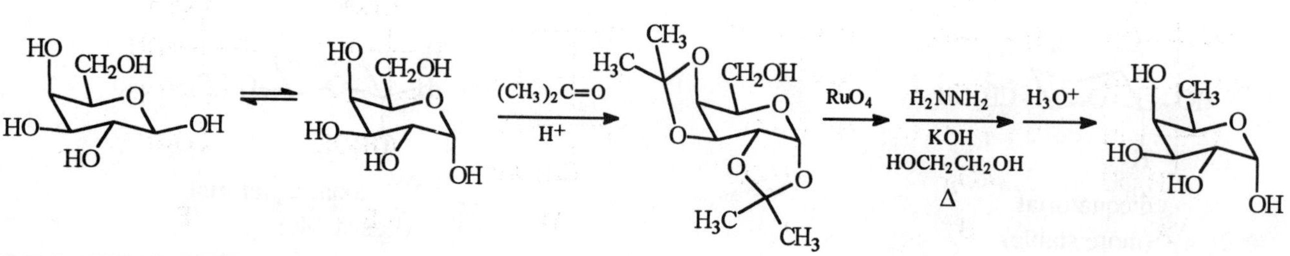

$\overset{(CH_3)_2C=O}{\underset{H^+}{\longrightarrow}}$ $\xrightarrow{RuO_4}$ $\xrightarrow[\substack{KOH \\ HOCH_2CH_2OH \\ \Delta}]{H_2NNH_2}$ $\xrightarrow{H_3O^+}$

7. Kiliani-Fischer chain extension follows the aldose tree in Table 28.1.

8. Reverse of the aldose tree in Table 28.1

9.

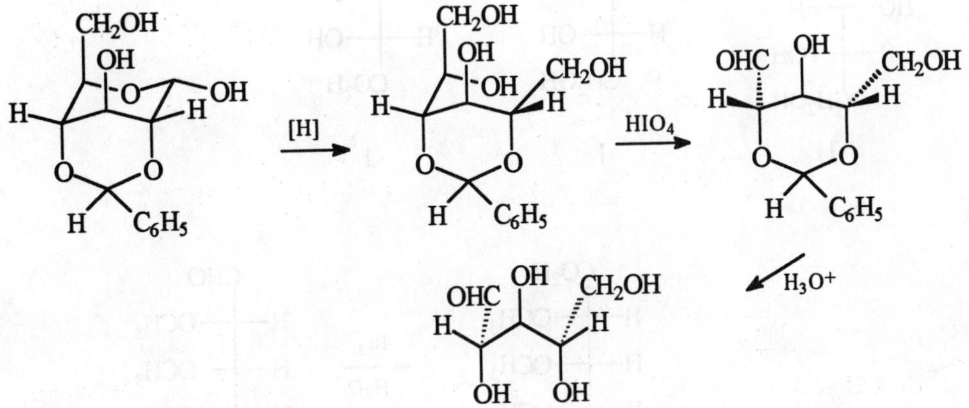

[H]

HIO_4

H_3O^+

Overall process:

[H]

HIO_4

H_3O^+

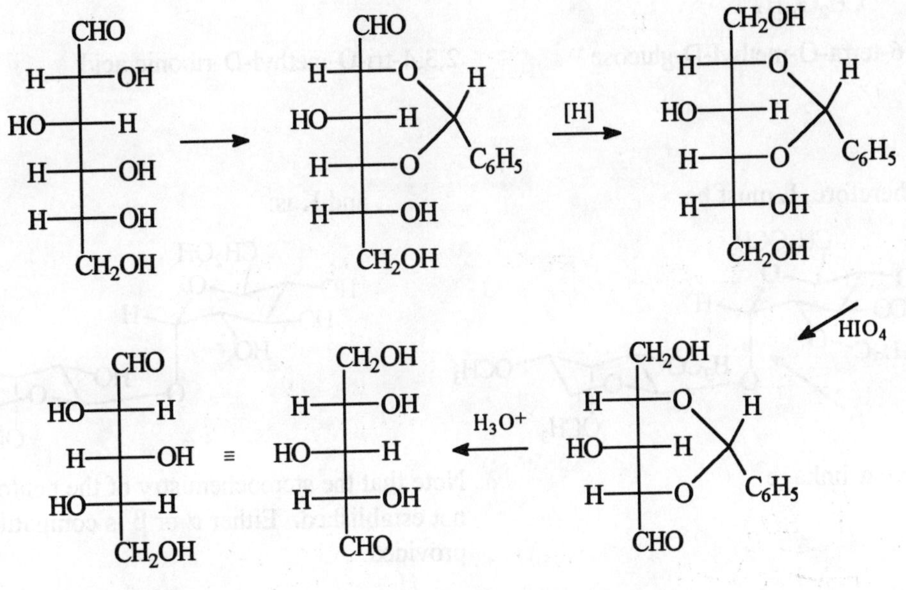

L-xylose

10.

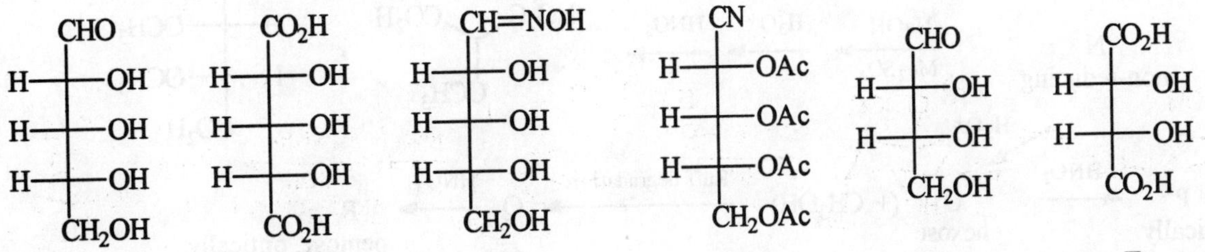

A B C D E F

11.

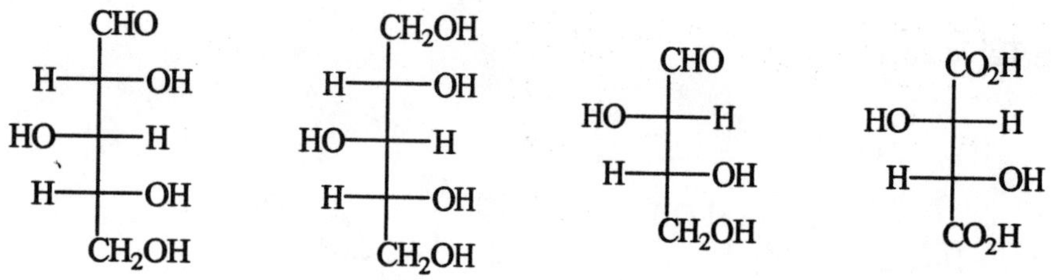

CHO, H—OH, HO—H, H—OH, CH_2OH

G

CH_2OH, H—OH, HO—H, H—OH, CH_2OH

H

CHO, HO—H, H—OH, CH_2OH

I

CO_2H, HO—H, H—OH, CO_2H

J

12.

CHO, H—OCH_3, H_3CO—H, H—OCH_3, H—OH, CH_2OCH_3

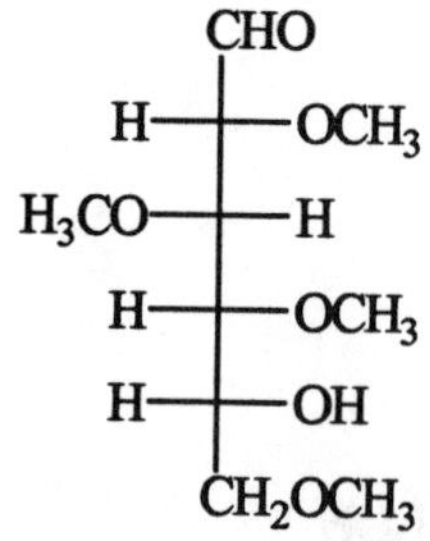

2,3,4,6-tetra-O-methyl-D-glucose

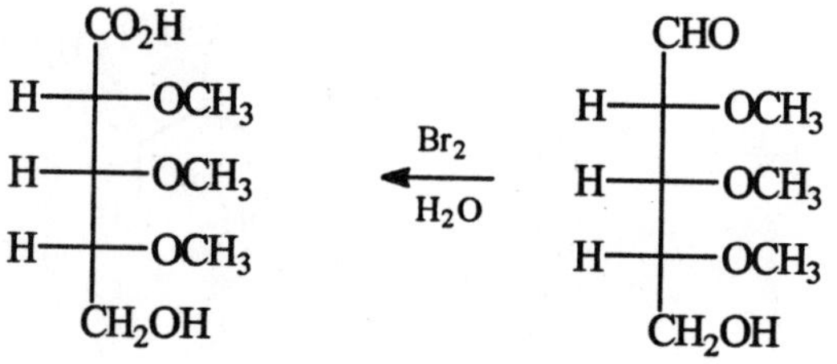

CO_2H, H—OCH_3, H—OCH_3, H—OCH_3, CH_2OH ← (Br_2, H_2O) ← CHO, H—OCH_3, H—OCH_3, H—OCH_3, CH_2OH (M)

2,3,4-tri-O-methyl-D-ribonic acid

Therefore, L must be:

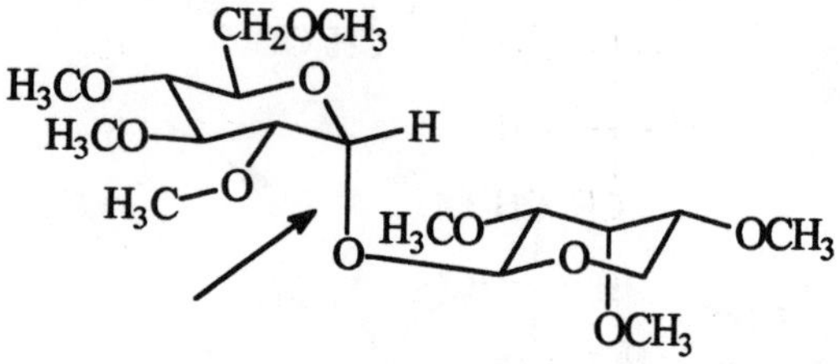

note α linkage

and K is:

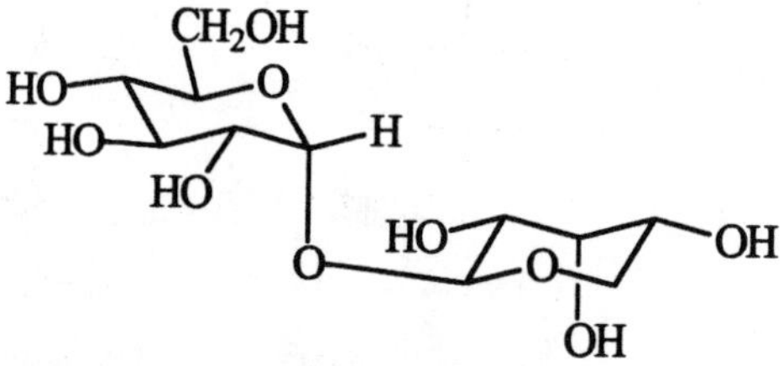

Note that the stereochemistry of the pentose anomeric carbon is not established. Either α or β is compatible with the data provided.

13.

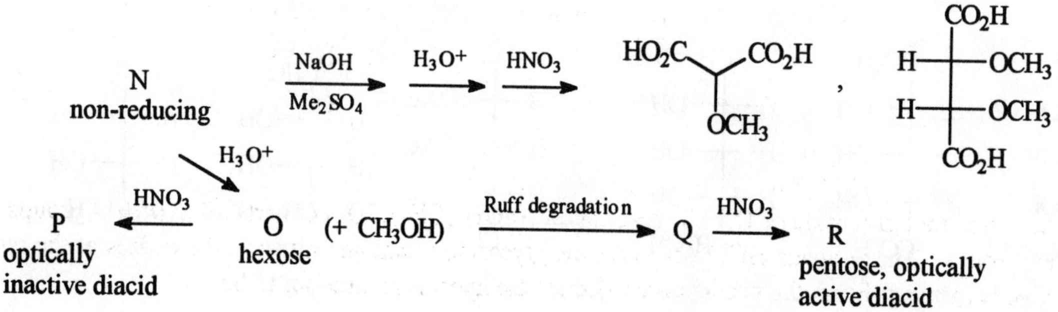

N (non-reducing) → NaOH, Me_2SO_4 → H_3O^+ → HNO_3 → HO_2C–CH(OCH_3)–CO_2H , CO_2H, H—OCH_3, H—OCH_3, CO_2H

N → H_3O^+ → O (+ CH_3OH) hexose

O → HNO_3 → P optically inactive diacid

O → Ruff degradation → Q → HNO_3 → R

pentose, optically active diacid

1. **R is optically active:**

```
     CO2H                 CO2H
 HO ─┼─ H             HO ─┼─ H
 HO ─┼─ H      ≡       H ─┼─ OH
  H ─┼─ OH             H ─┼─ OH
     CO2H                 CO2H
                 R
```

2. $Q \xrightarrow{HNO_3} R$:

Therefore, Q can be

```
     CHO                  CHO
 HO ─┼─ H             HO ─┼─ H
 HO ─┼─ H      or      H ─┼─ OH
  H ─┼─ OH             H ─┼─ OH
     CH2OH                CH2OH
```

3. $O \xrightarrow{\text{Ruff}} Q$:

Therefore **O** can be

```
     CHO               CHO               CHO               CHO
 HO ─┼─ H           H ─┼─ OH         HO ─┼─ H           H ─┼─ OH
 HO ─┼─ H     or   HO ─┼─ H     or   HO ─┼─ H     or   HO ─┼─ H
 HO ─┼─ H          HO ─┼─ H           H ─┼─ OH          H ─┼─ OH
  H ─┼─ OH          H ─┼─ OH          H ─┼─ OH          H ─┼─ OH
     CH2OH             CH2OH             CH2OH             CH2OH
```

4. $O \xrightarrow{HNO_3} P$ (optically inactive):

```
     CO2H                 CHO
  H ─┼─ OH             H ─┼─ OH
 HO ─┼─ H             HO ─┼─ H
 HO ─┼─ H             HO ─┼─ H
  H ─┼─ OH             H ─┼─ OH
     CO2H                 CH2OH
      P                    O
```

N is evidently a methyl glycoside of O. The reaction of $NaOH/(CH_3)_2SO_4$ converts all hydroxyl groups to methyl ether groups. The aqueous HCl hydrolyzes the glycoside acetal and nitric acid oxidizes at the resulting CHO and C-OH groups. From the products, we deduce the hydrolysis product to be:

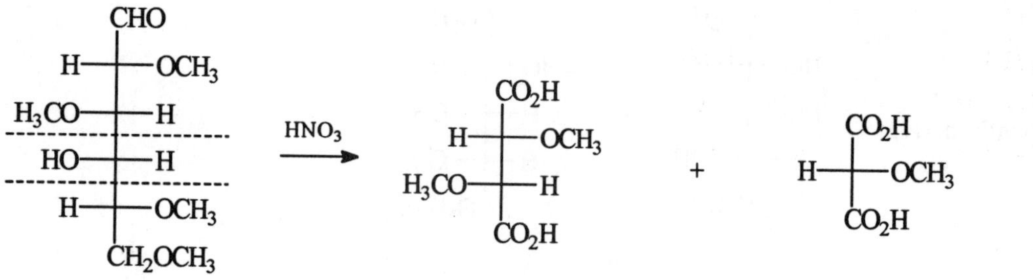

CHO
H—OCH$_3$
H$_3$CO—H
HO—H
H—OCH$_3$
CH$_2$OCH$_3$

$\xrightarrow{HNO_3}$

CO$_2$H
H—OCH$_3$
H$_3$CO—H
CO$_2$H

+

CO$_2$H
H—OCH$_3$
CO$_2$H

α-β-dimethoxysuccinic acid   α-methoxymalonic acid

Thus, N must be: (The configuration at the top carbon in N is not determined.)

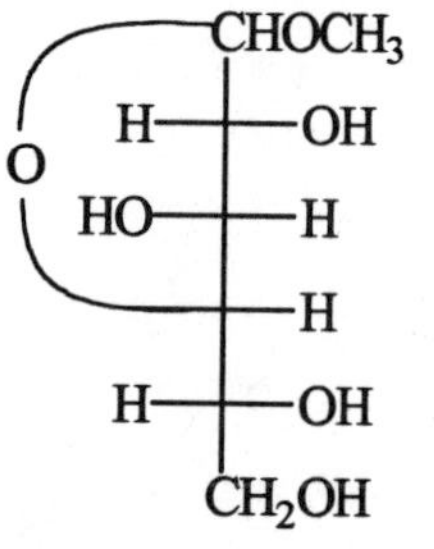

CHOCH$_3$
H—OH
O
HO—H
—H
H—OH
CH$_2$OH

Note that

CHO
H—OCH$_3$
HO—H
H$_3$CO—H
H—OCH$_3$
CH$_2$OCH$_3$

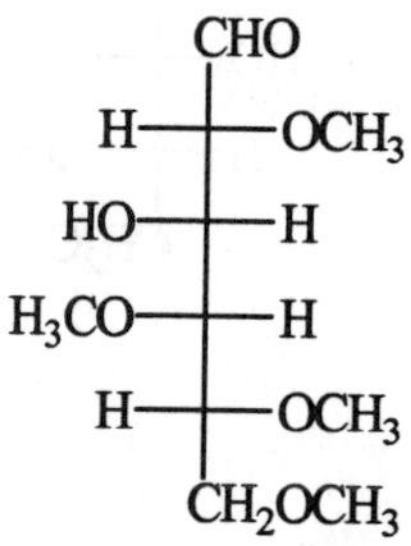

could give α,β-dimethoxysuccinic acid and α-methoxymalonic acid on oxidation, but the corresponding cyclic acetal has a four-membered ring and is not a reasonble structure.

NOTE also that the α-methoxymalonic acid could derive from further oxidation of the dimethoxysuccinic acid, rather than from oxidation of the 6-methoxy ether.

b. Not only has the configuration (α or β) of the anomeric carbon not been established, but we have assumed D-configurations in the above structures. The available data do not allow a distinction between D and L.

14.

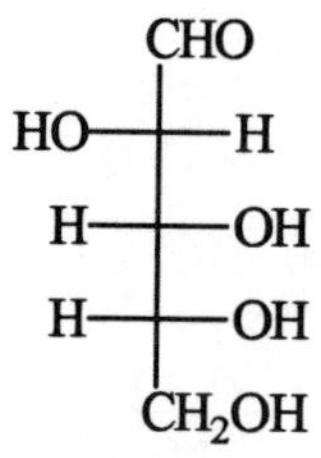

CHO
HO—H
H—OH
H—OH
CH$_2$OH

S

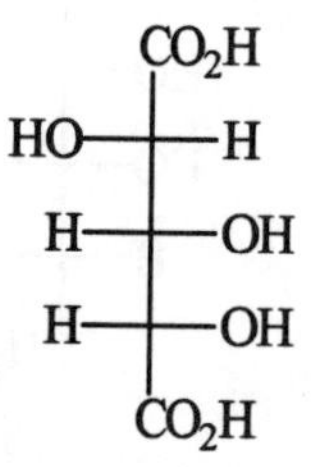

CO$_2$H
HO—H
H—OH
H—OH
CO$_2$H

T

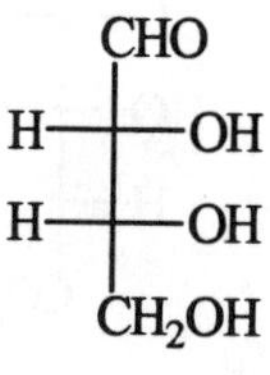

CHO
H—OH
H—OH
CH$_2$OH

U

CO$_2$H
H—OH
H—OH
CO$_2$H

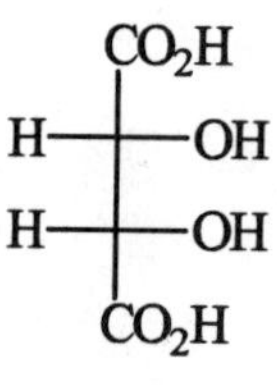

V

15.

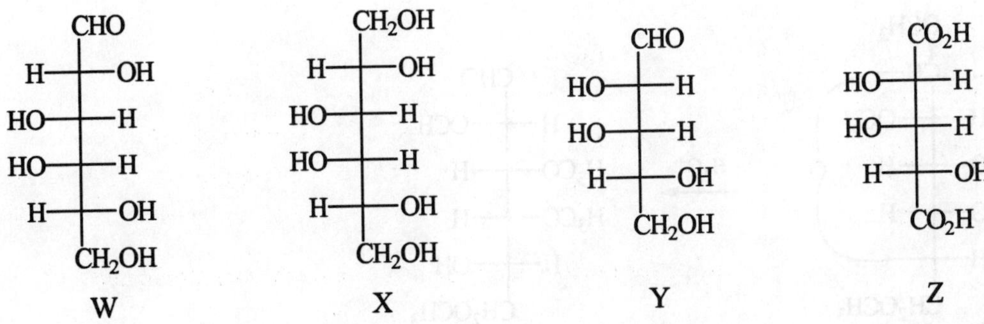

16.

A': CHO; HO—H; H—OH; HO—H; H—OH; CH_2OH

B': CO_2H; HO—H; H—OH; HO—H; H—OH; CO_2H

C': CHO; H—OH; HO—H; H—OH; CH_2OH

D': CO_2H; H—OH; HO—H; H—OH; CO_2H

17. Let n_α = fraction α, and $1 - n_\alpha$ = fraction β

$$n_\alpha(29.3) + (1 - n_\alpha)(-17.0) = 14.2$$

$n_\alpha = 0.674$, or 67.4% $\quad$ $n_\beta = 0.326$, or 32.6%

18.

tetra-O-methylglucaric acid: CO_2H; H—OCH_3; H_3CO—H; H—OCH_3; H—OCH_3; CO_2H

←

E', tetra-O-methylgluconic acid: CO_2H; H—OCH_3; H_3CO—H; H—OCH_3; H—OCH_3; CH_2OH

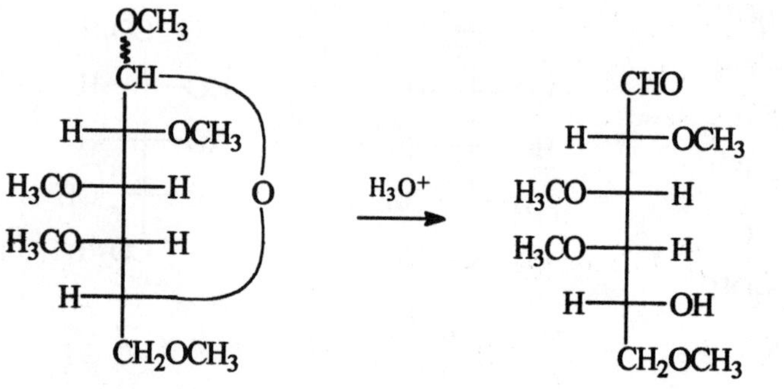

OCH_3
CH
H — OCH_3
H_3CO — H
H_3CO — H
H
CH_2OCH_3

$\xrightarrow{H_3O^+}$

CHO
H — OCH_3
H_3CO — H
H_3CO — H
H — OH
CH_2OCH_3

methyl-2,3,4,6-tetra-O-methyl galactopyranoside

F', tetra-O-methylgalactose

Melibionic acid must join C-6 of gluconic acid to C-1 of galactose:

melibionic acid:

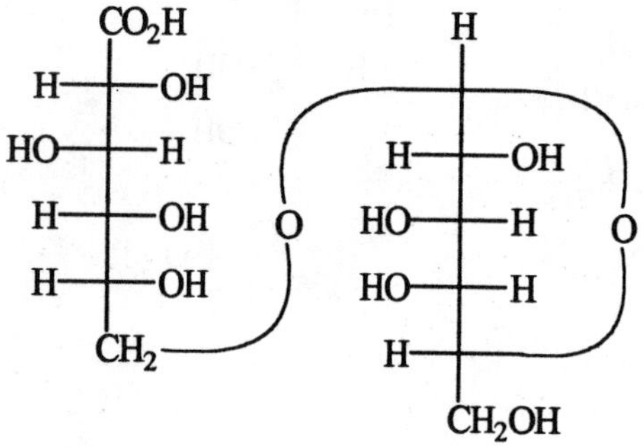

CO_2H
H — OH
HO — H
H — OH
H — OH
CH_2
O

H
H — OH
HO — H
HO — H
H
CH_2OH
O

note α-linkage

or:

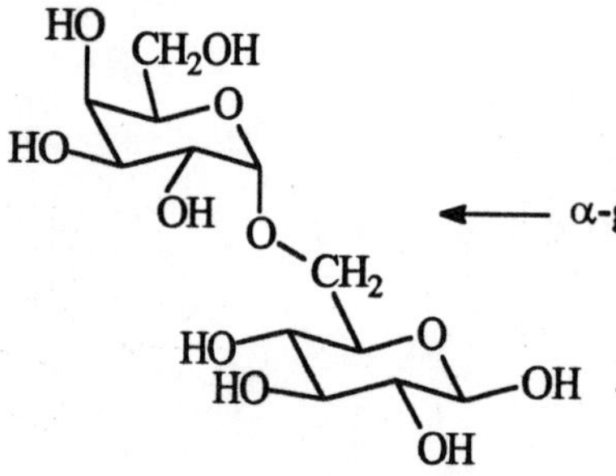

HO CH_2OH O HO OH O CH_2

← α-glycoside

HO HO O OH OH ← hemiacetal, reducing sugar

melibiose, β-anomer shown

19.

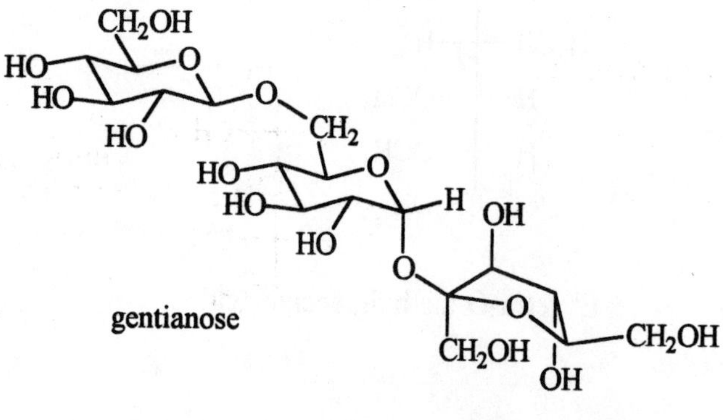

CH_2OH HO HO O HO O CH_2 HO HO O H HO O OH O CH_2OH OH CH_2OH

gentianose

20.

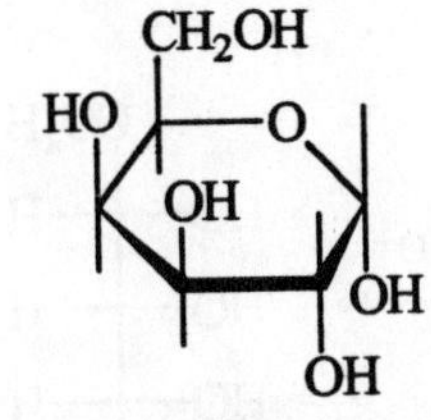

α-D-galactopyranose

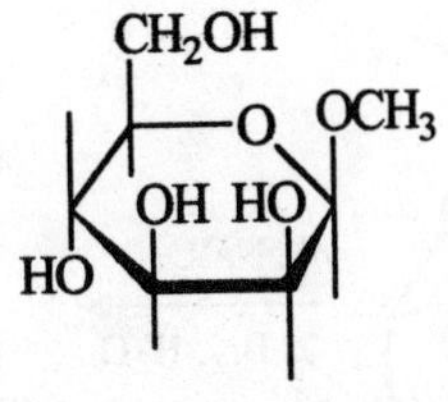

methyl β-D-mannoside

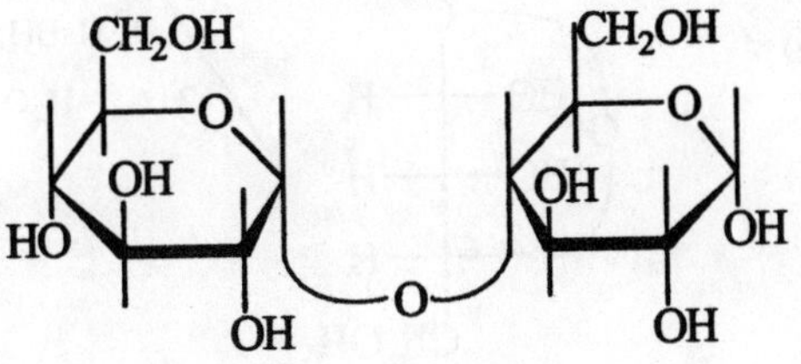

α-maltose

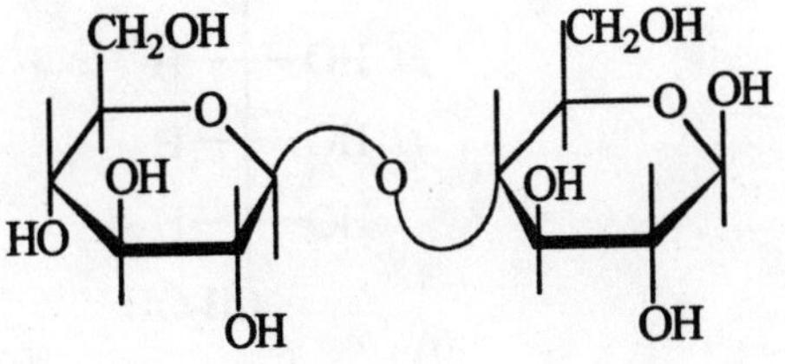

β-cellobiose

21. a.

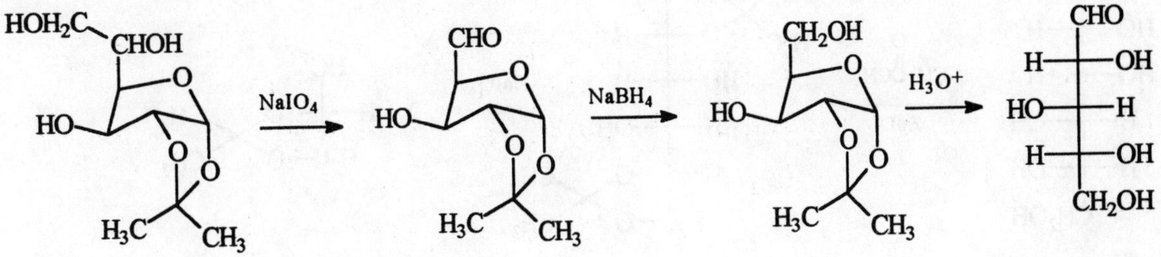

$NaIO_4$ → $NaBH_4$ → H_3O^+ →

1,2-O-isopropylidene-α-D-glucofuranose

D-xylose

HOAc
H_2O

b.

1. RuO_4
2. H_2NNH_2
KOH, Δ

HOAc
H_2O

$NaIO_4$

H_3O^+

HNO_3

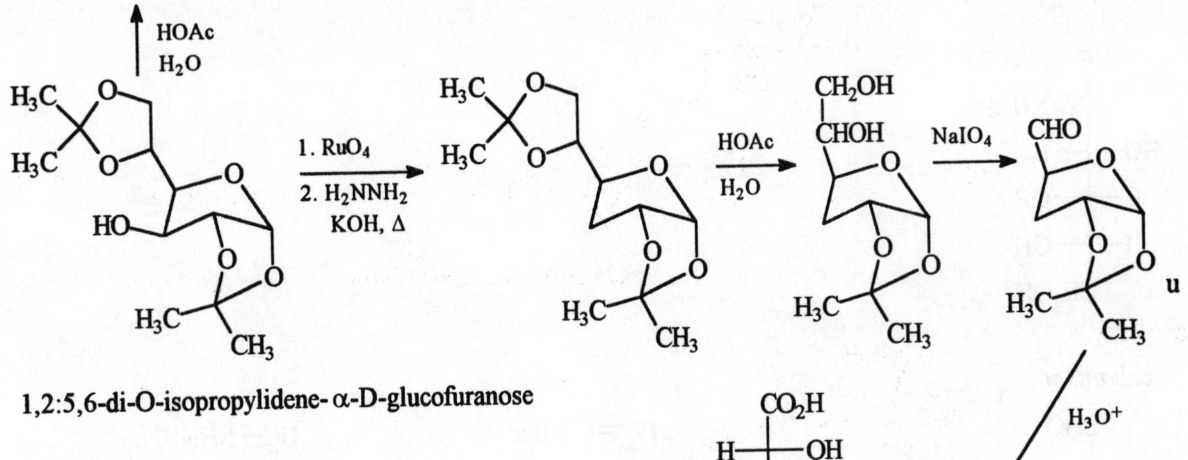

1,2:5,6-di-O-isopropylidene- α-D-glucofuranose

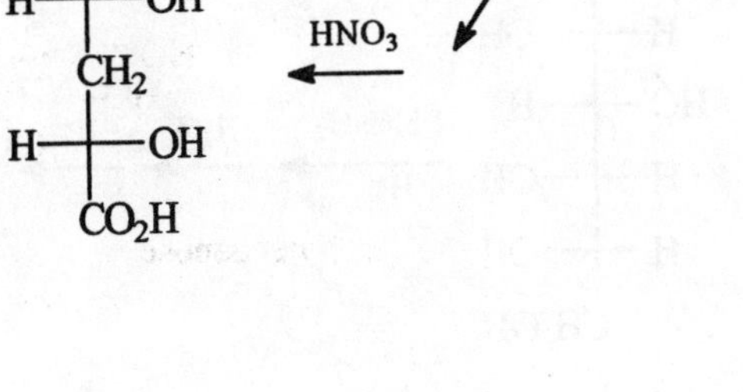

(see Exercise 28.7)

c. 1. RuO_4 2. $LiAlH_4$ → HOAc / H_2O → 1. $NaIO_4$ 2. Br_2, H_2O → H_3O^+ → 1. $NaBH_4$ 2. Δ, $-H_2O$ → Na (Hg) or $NaBH_4$ →

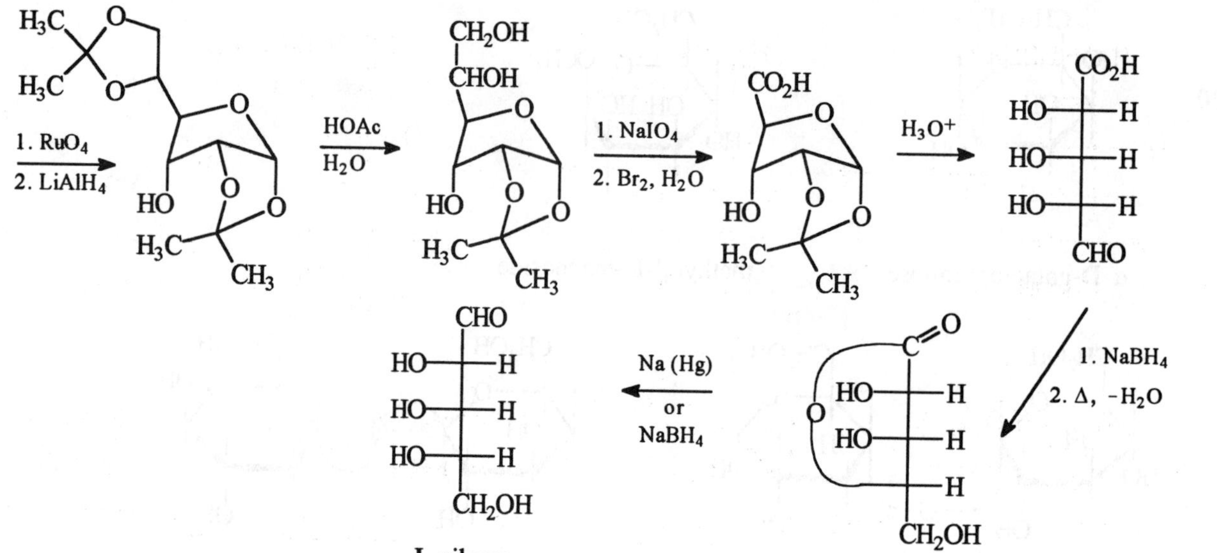

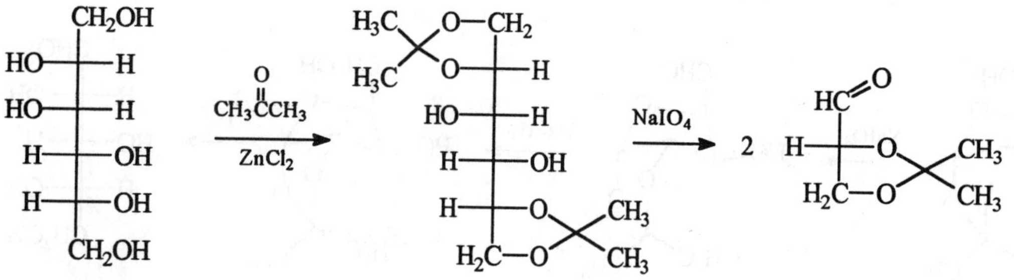

L-ribose

22. D-mannitol → $CH_3\overset{O}{\overset{\|}{C}}CH_3$ / $ZnCl_2$ → $NaIO_4$ → 2

23. HC=NNHPh, C=NNHPh, HO—H, H—OH, H—OH, CH_2OH

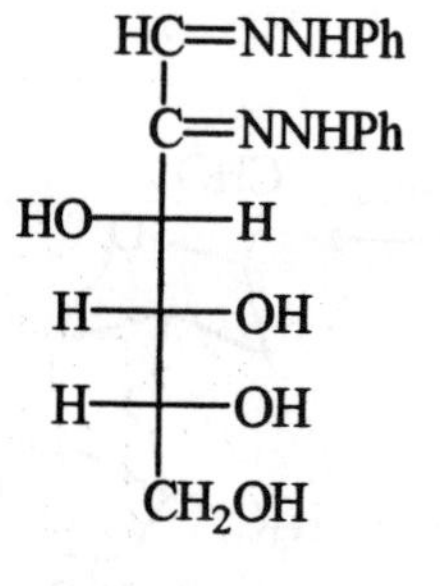

Mechanism:

PhNNH$_2$ / H^+ → $-H_2O$ → $-H^+$ → H^+ → $-H^+$ → H^+ → $-H^+$ →

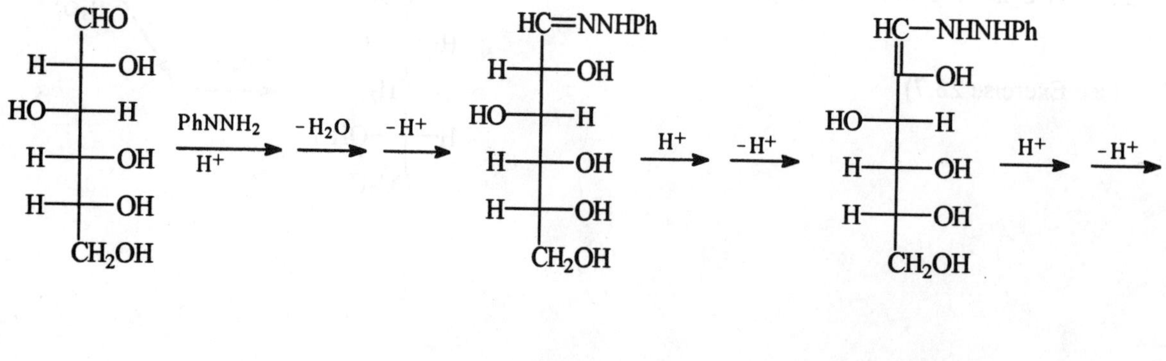

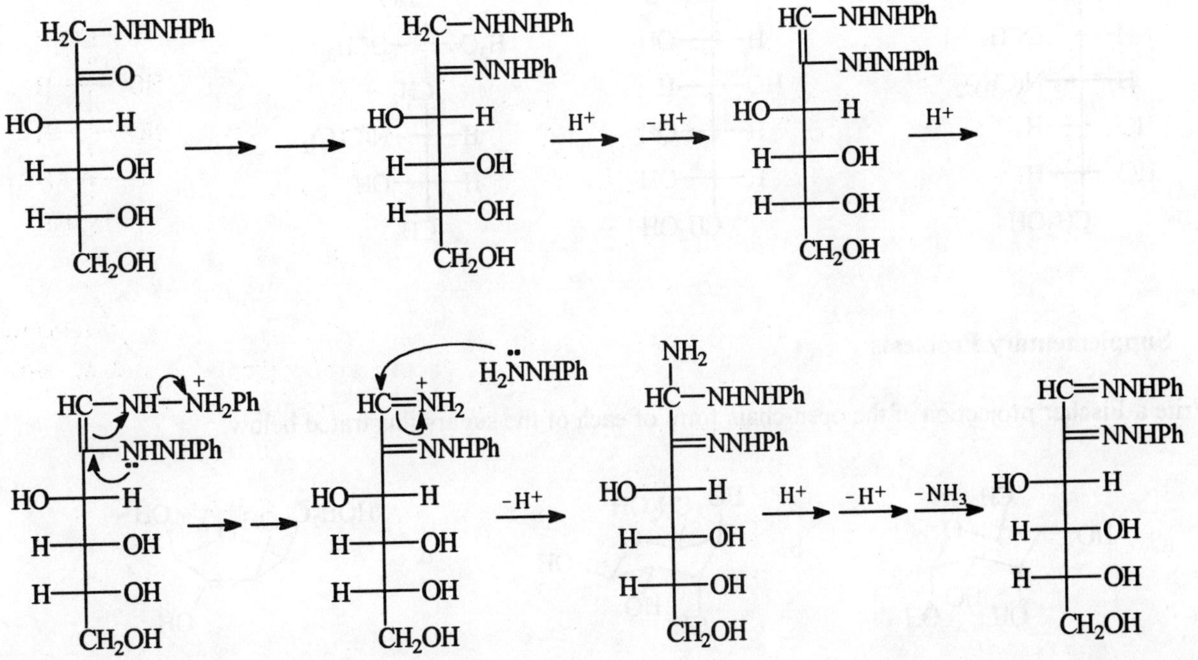

24.

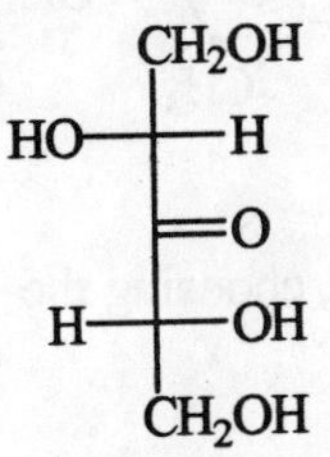

G'

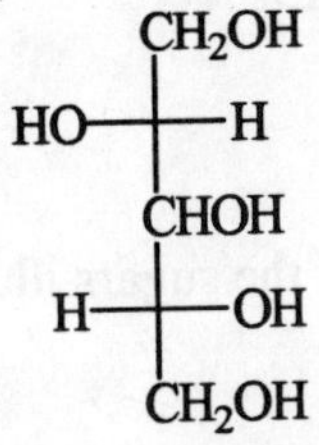

H'

The central C is not a chiral center.

25.

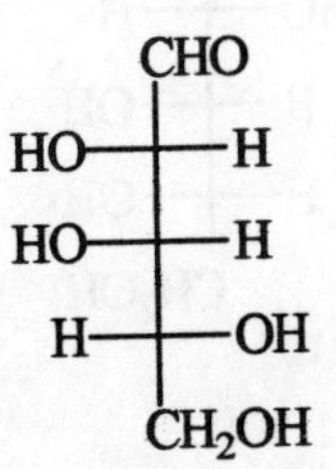

I' (lyxose)

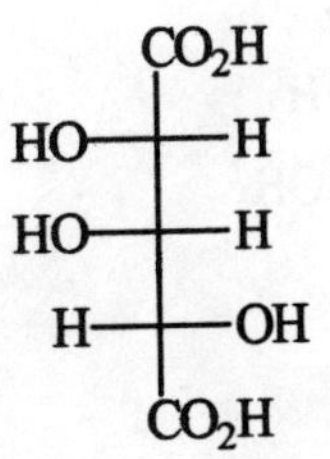

J'

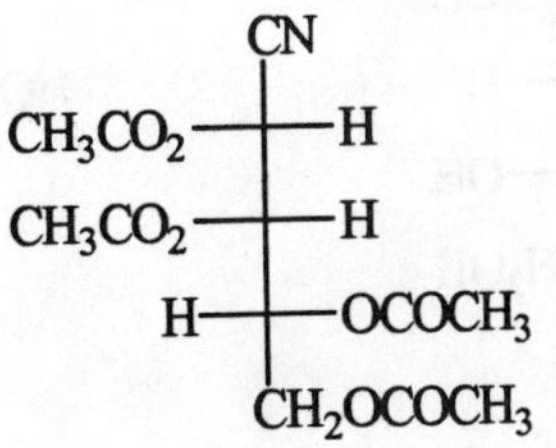

K'

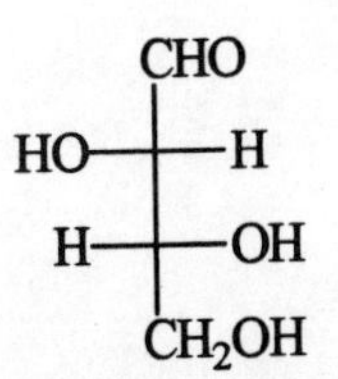

L' (threose)

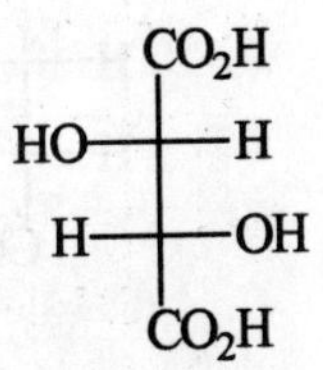

M'

26. c and d.

27.

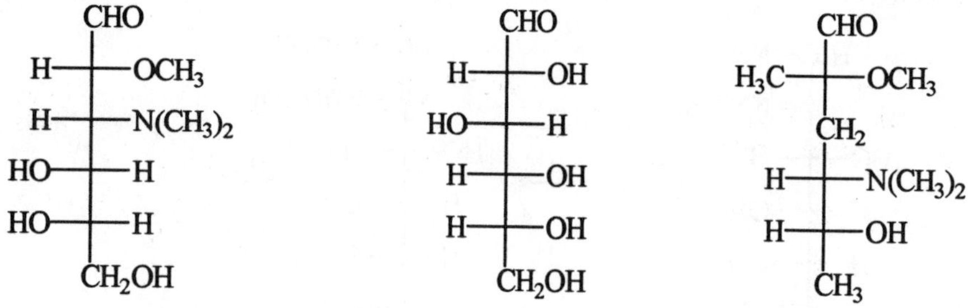

	CHO	
H		OCH_3
H		$N(CH_3)_2$
HO		H
HO		H
	CH_2OH	

	CHO	
H		OH
HO		H
H		OH
H		OH
	CH_2OH	

	CHO	
H_3C		OCH_3
	CH_2	
H		$N(CH_3)_2$
H		OH
	CH_3	

28.F. Supplementary Problems

S1. Write a Fischer projection of the open-chain form of each of the sugars illustrated below.

a. 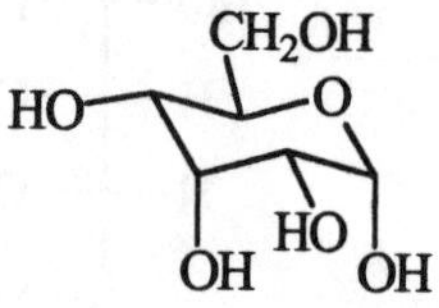 b. 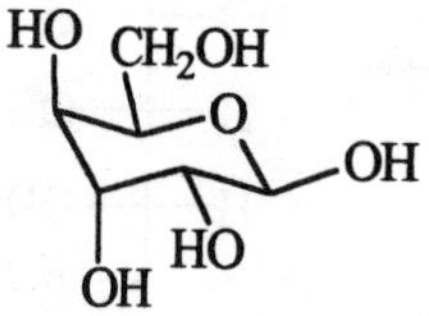 c. 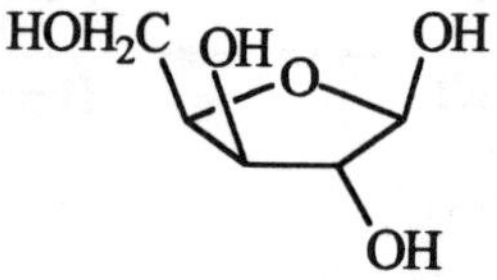 d. 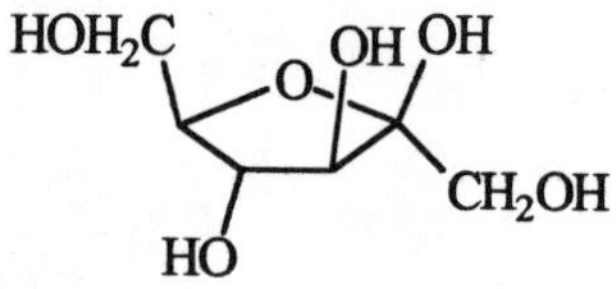 e. 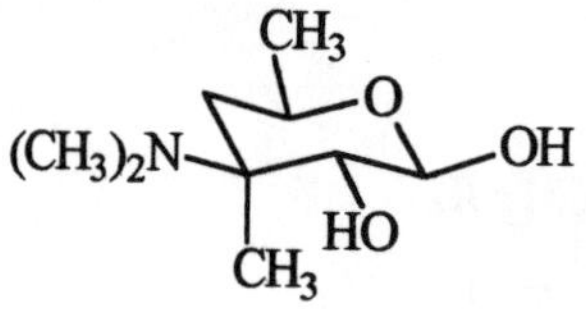 f.

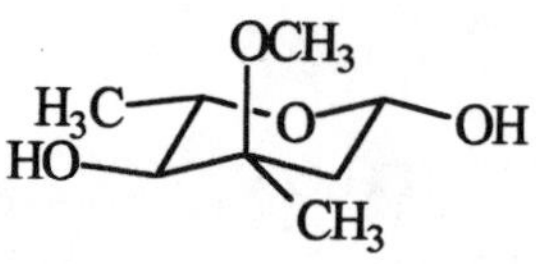

S2. Write perspective formulas of the pyranose β-anomers of each of the sugars illustrated below, choosing the most stable conformation where appropriate.

a.

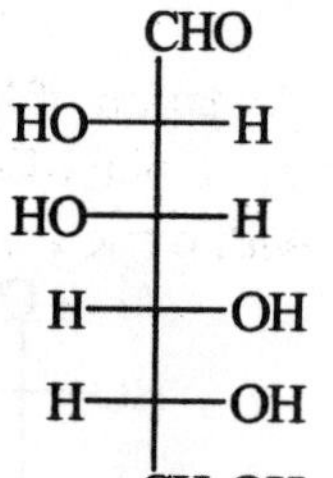

	CHO	
HO		H
HO		H
H		OH
H		OH
	CH_2OH	

b.

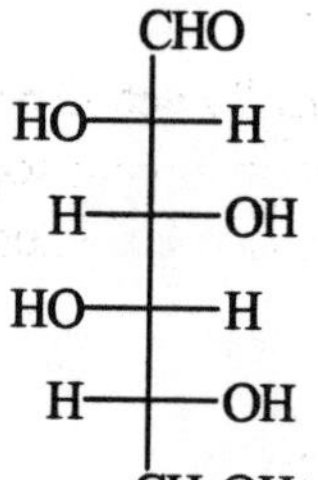

	CHO	
HO		H
H		OH
HO		H
H		OH
	CH_2OH	

c.

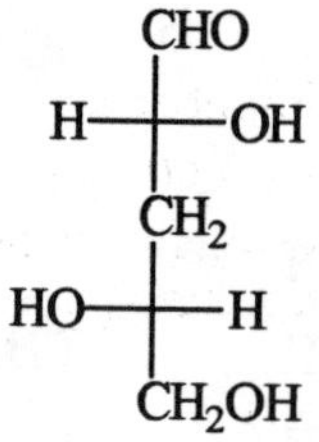

	CHO	
H		OH
	CH_2	
HO		H
	CH_2OH	

d.

	CHO	
H		$NHCCH_3$ (C=O)
HO		H
H		OH
H		OH
	CH_2OH	

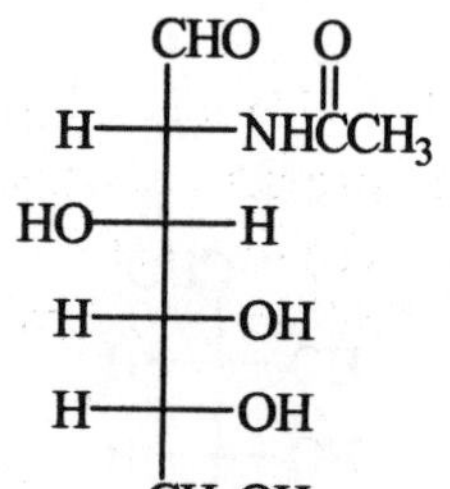

S3. Write a step-by-step mechanism for the base-catalyzed interconversion of the furanose and pyranose forms of fructose:

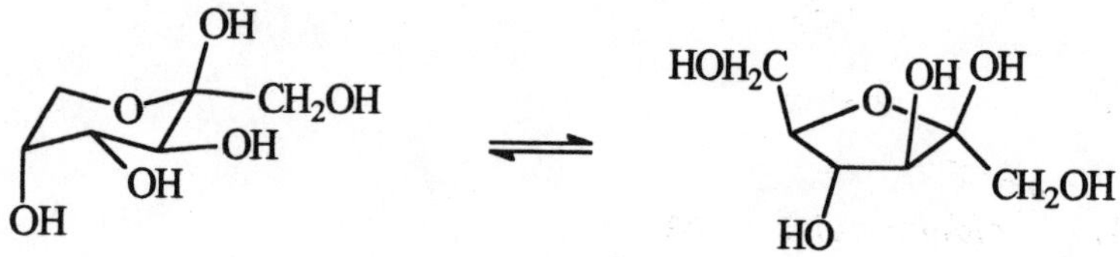

S4. Predict the major product(s) from each of the following reaction sequences:

a. 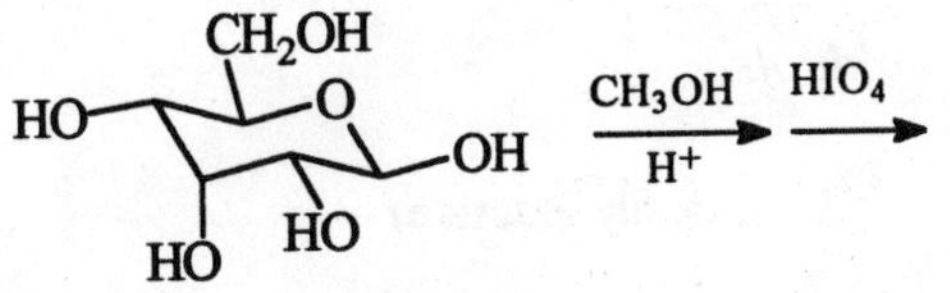 (pyranose structure: CH_2OH, HO, O, OH, HO, HO) $\xrightarrow[H^+]{CH_3OH}$ $\xrightarrow{HIO_4}$

b. 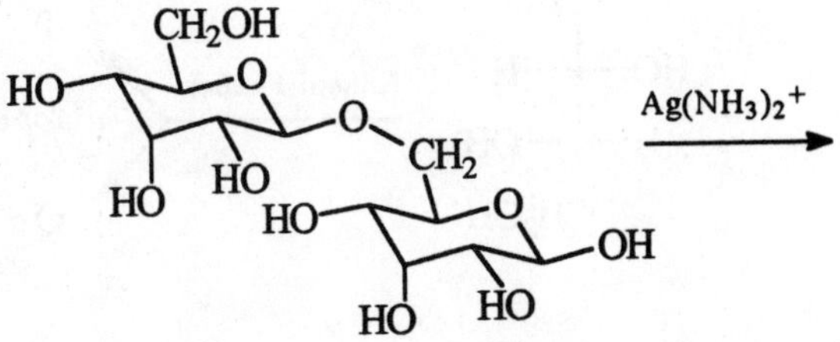 (disaccharide structure: CH_2OH, HO, O, HO, HO, O, CH_2, HO, O, OH, HO, HO) $\xrightarrow{Ag(NH_3)_2^+}$

c. 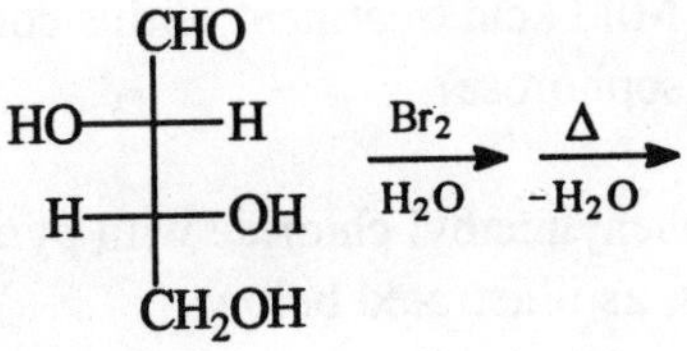 CHO / HO–H / H–OH / CH_2OH $\xrightarrow[H_2O]{Br_2}$ $\xrightarrow[-H_2O]{\Delta}$

d. 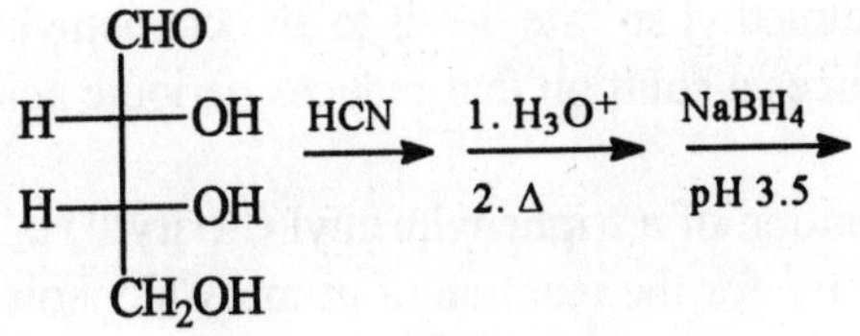 CHO / H–OH / H–OH / CH_2OH $\xrightarrow{HCN}$ $\xrightarrow[2.\ \Delta]{1.\ H_3O^+}$ $\xrightarrow[pH\ 3.5]{NaBH_4}$

e. 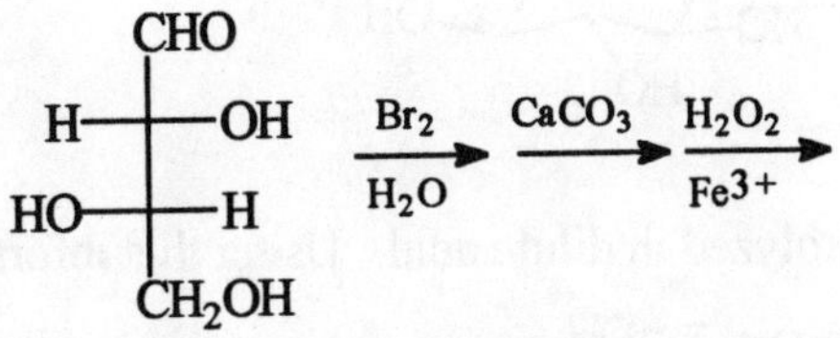 CHO / H–OH / HO–H / CH_2OH $\xrightarrow[H_2O]{Br_2}$ $\xrightarrow{CaCO_3}$ $\xrightarrow[Fe^{3+}]{H_2O_2}$

f. 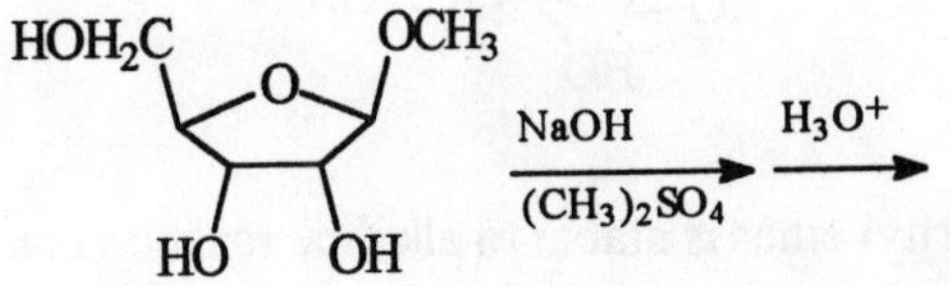 (furanose structure: HOH_2C, O, OCH_3, HO, OH) $\xrightarrow[(CH_3)_2SO_4]{NaOH}$ $\xrightarrow{H_3O^+}$

S5. Treatment of D-aldopentose A with sodium borohydride gives an optically inactive alditol, B. Reaction of A with HCN, followed by acid-catalyzed hydrolysis, produces two aldonic acids, C and D, **both** of which afford optically active saccharic acids after treatment with nitric acid. What are A, B, C, and D?

S6. Oxidation of L-aldohexose E with nitric acid leads to an optically active product. Ruff degradation of E provides aldopentose F, which loses all optical acitvity on reaction with sodium borohydride. Kiliani-Fischer chain extension of F gives E back again, along with an isometric aldohexose G. G reacts with nitric acid to afford an optically inactive diacid. What are E, F, and G?

S7. J $\xleftarrow[2.\ CaCO_3,\ H_2O_2,\ Fe^{+3}]{1.\ Br_2,\ H_2O}$ H $\xrightarrow{HNO_3}$ I (optically active)

aldopentose   *D-aldohexose*   *saccharic acid*

J ↓ $NaBH_4$

K (optically active) *alditol*

H → 1. HCN; 2. H_3O^+ → L and M (*aldonic acids*)

L ↓ HNO_3 M ↓ HNO_3

(optically inactive) N   *saccharic acids*   O (optically active)

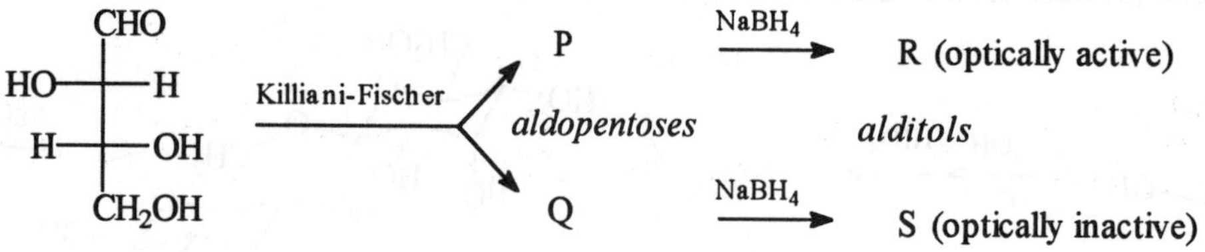
CHO
HO—H
H—OH
CH2OH
Killiani-Fischer
P
Q
aldopentoses
NaBH4
NaBH4
R (optically active)
S (optically inactive)
alditols

S8. Hydrolysis of the disaccharide sophorose furnishes two moles of D-glucose; it is not hydrolyzed by α-glucosidase. Oxidation of sophorose with bromine water, followed by permethylation with sodium hydroxide and dimethyl sulfate, leads to an octamethylsophoronic acid derivative. Mild acid treatment of this compound produces a solution that reduces periodic acid. What is the structure of sophorose?

S9. Formation of a triphenylmethyl ("trityl") ether from an alcohol and triphenylmethyl chloride with pyridine is selective for the reaction of primary alcohols in carbohydrate derivatives, as illustrated below.

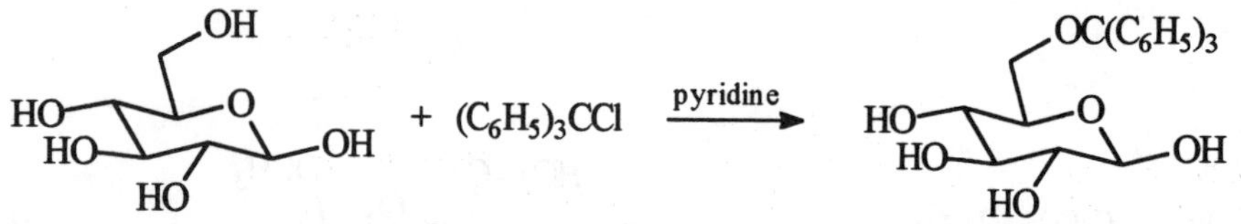
OH
HO
HO
HO
OH
O
+ (C6H5)3CCl
pyridine
OC(C6H5)3
OH

A trityl ether is stable to alkaline reaction conditions, but is hydrolyzed in dilute acid. Using this information, show how to synthesize L-fucose from D-galactose.

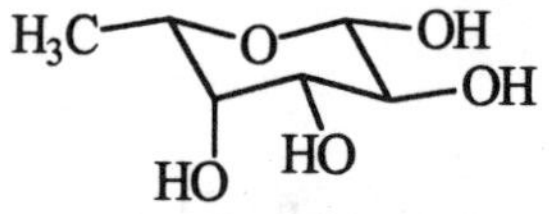
H3C
O
OH
OH
HO
HO

L-fucose

28.G. Answers to Supplementary Problems

S1. a.

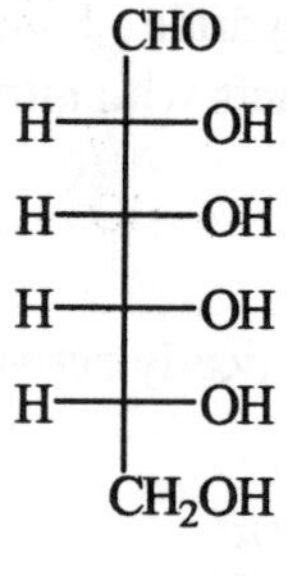

b.

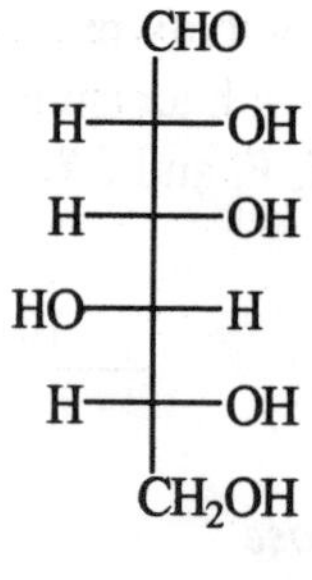

c.

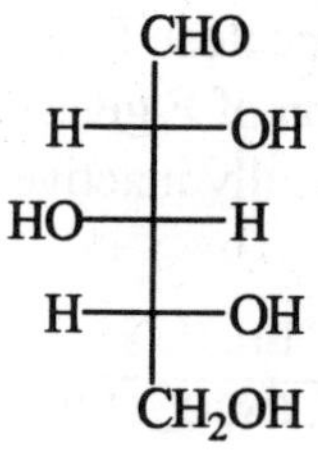

d.

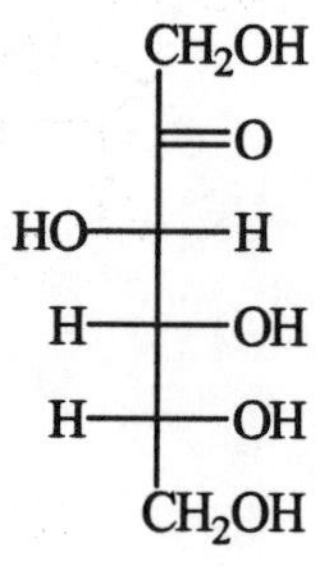

e.

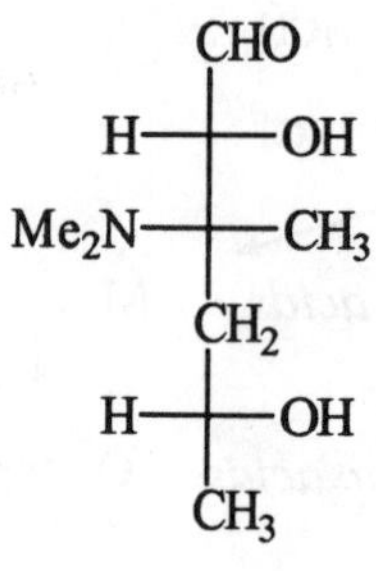

f.

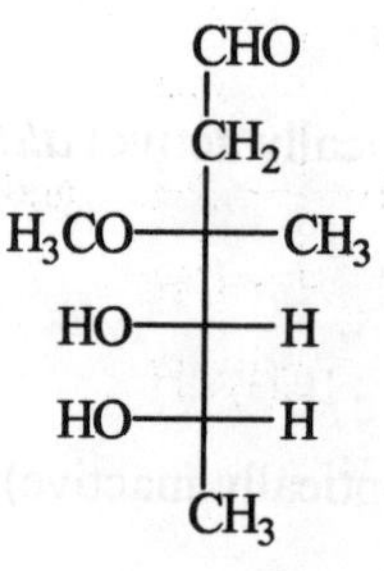

S2. a. 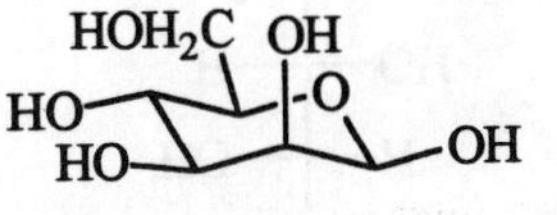 HOH$_2$C OH HO O HO OH b. HOH$_2$C OH HO O OH HO ⇌ OH OH OH O OH OH

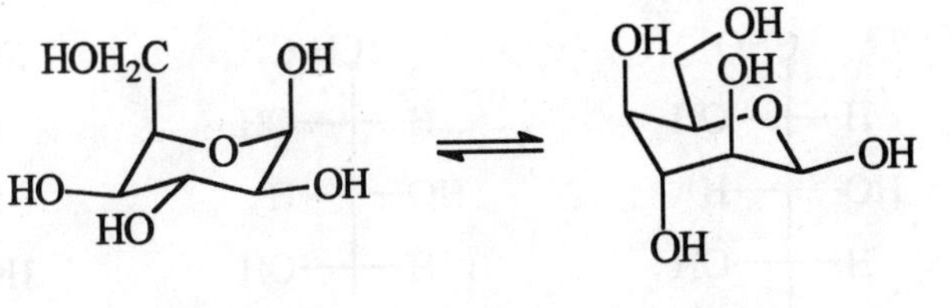

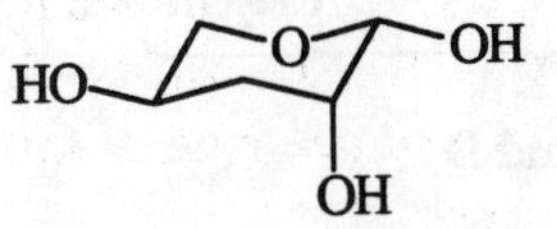

c. HO O OH OH d. 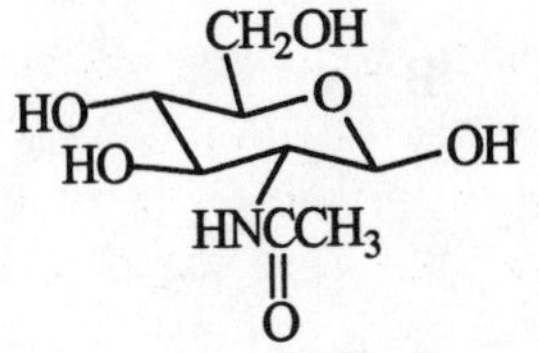 CH$_2$OH HO O HO OH HNCCH$_3$ O

NOTE: for an L-sugar, this is the β-anomer

S3. OH O CH$_2$OH OH HO OH OH^- ⇌ O$^-$ O CH$_2$OH OH HO OH ⇌ CH$_2$OH =O HO—H H—OH H—OH CH$_2$O$^-$ ⇌ CH$_2$OH O HO—H H—OH H—O$^-$ CH$_2$OH

HOH$_2$C O OH OH CH$_2$OH HO H_2O ⇌ HOH$_2$C O OH O$^-$ CH$_2$OH HO ⇌

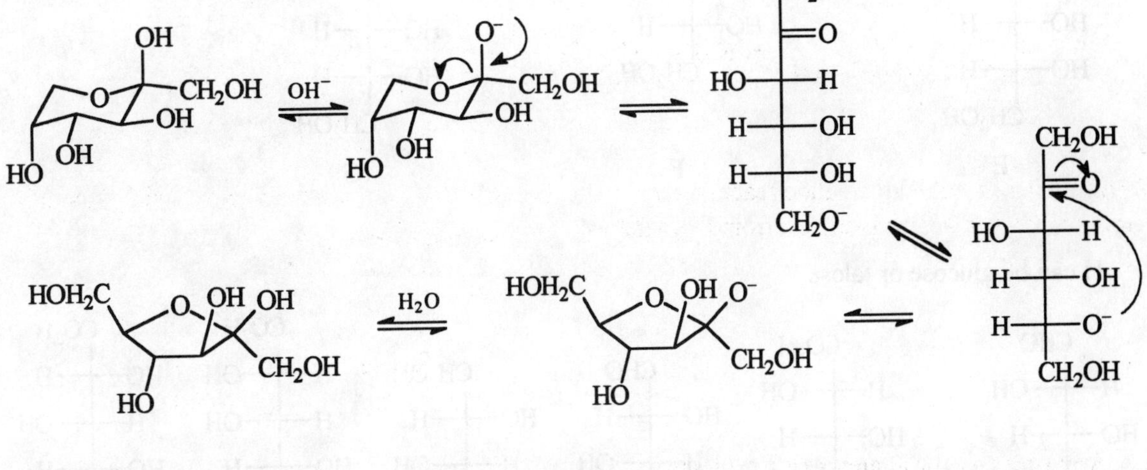

S4. a. 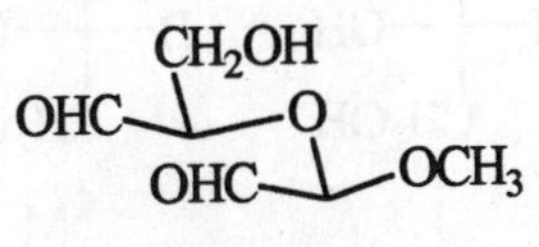 CH$_2$OH OHC O OHC OCH$_3$ + HCOOH b. 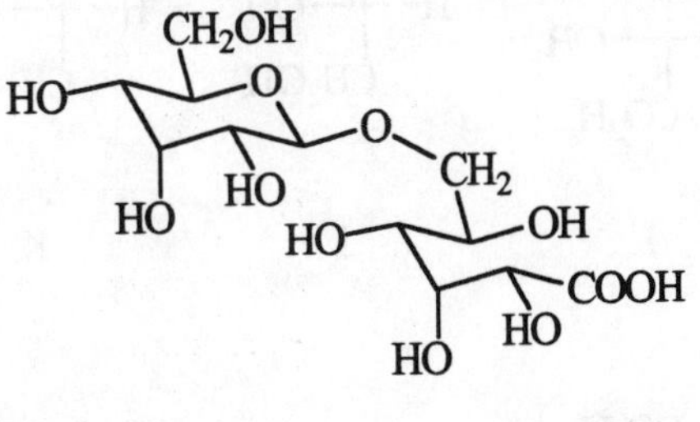 CH$_2$OH HO O HO HO O CH$_2$ HO OH COOH HO HO c. 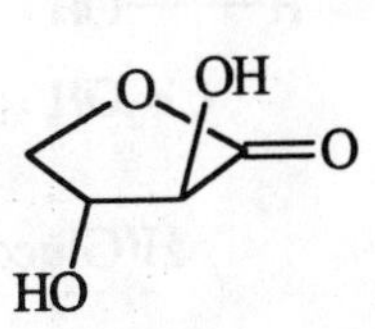 O OH =O HO

d. 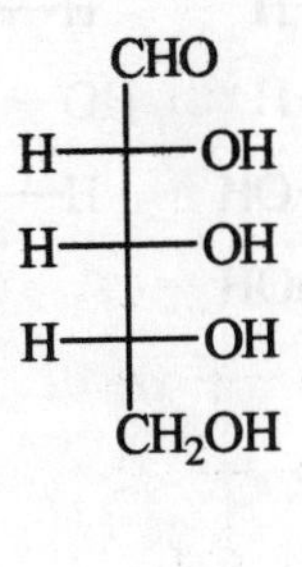 CHO H—OH H—OH H—OH CH$_2$OH and CHO HO—H H—OH H—OH CH$_2$OH e. 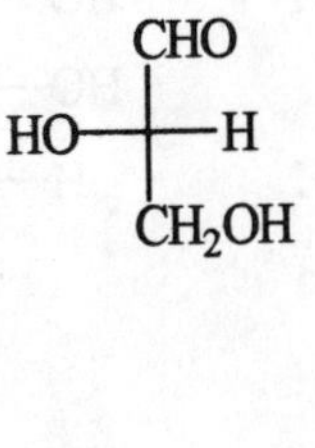 CHO HO—H CH$_2$OH f. 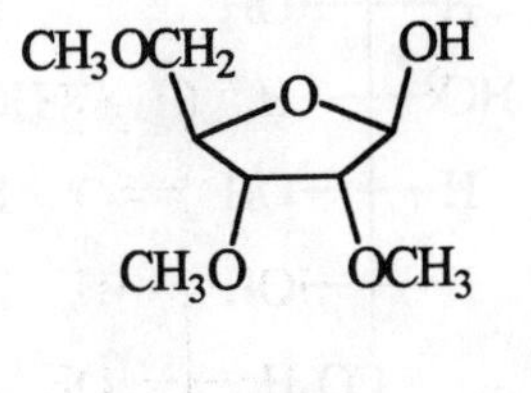 CH$_3$OCH$_2$ O OH CH$_3$O OCH$_3$

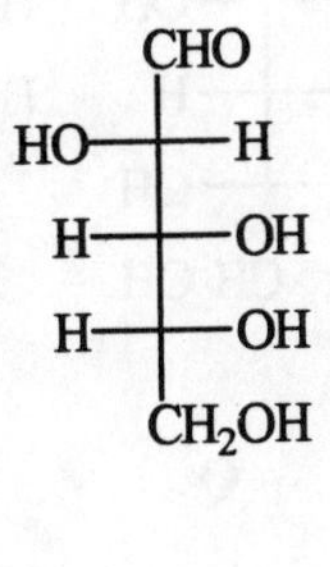

S5.

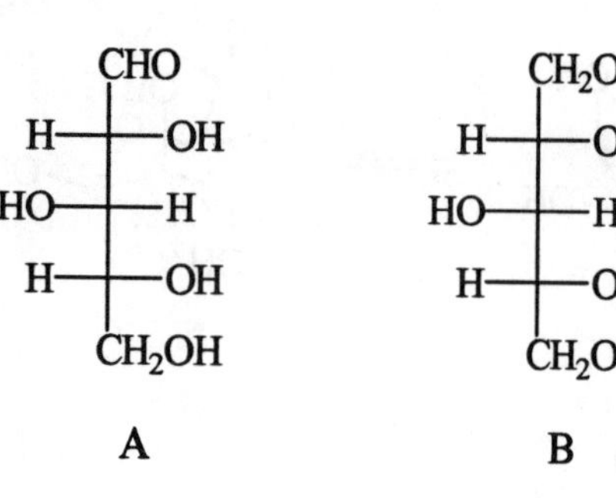

CO₂H / H—OH / H—OH / HO—H / H—OH / CH₂OH

CO₂H / HO—H / H—OH / HO—H / H—OH / CH₂OH

C and D

S6.

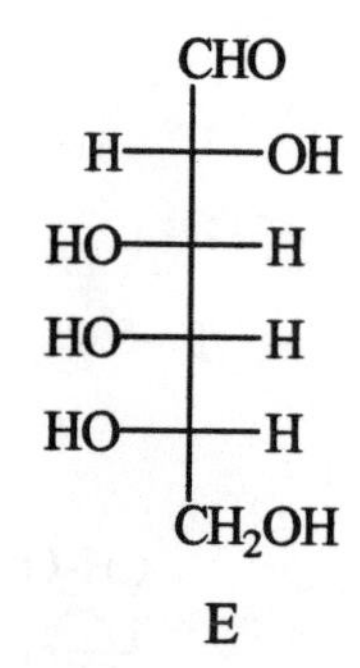

CHO / HO—H / HO—H / HO—H / CH₂OH

F

CHO / HO—H / HO—H / HO—H / HO—H / CH₂OH

G

S7. H can be glucose or talose.

CHO / H—OH / HO—H / H—OH / H—OH / CH₂OH — H(Glucose)

CO₂H / H—OH / HO—H / H—OH / H—OH / CO₂H — I

CHO / HO—H / H—OH / H—OH / CH₂OH — J

CH₂OH / HO—H / H—OH / H—OH / CH₂OH — K

CO₂H / H—OH / H—OH / HO—H / H—OH / H—OH / CH₂OH — L

CO₂H / HO—H / H—OH / HO—H / H—OH / H—OH / CH₂OH — M

CO₂H / H—OH / H—OH / HO—H / H—OH / H—OH / CO₂H — N

CO₂H / HO—H / H—OH / HO—H / H—OH / H—OH / CO₂H — O

CHO / HO—H / HO—H / H—OH / CH₂OH — P

CHO / H—OH / HO—H / H—OH / CH₂OH — Q

CH₂OH / HO—H / HO—H / H—OH / CH₂OH — R

CH₂OH / H—OH / HO—H / H—OH / CH₂OH — S

S8.

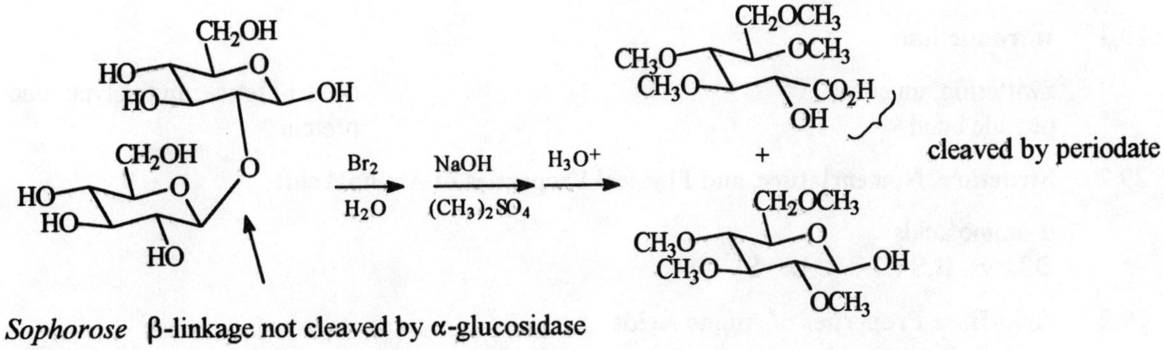

Sophorose β-linkage not cleaved by α-glucosidase

$\xrightarrow[H_2O]{Br_2}$ $\xrightarrow[(CH_3)_2SO_4]{NaOH}$ $\xrightarrow{H_3O^+}$

cleaved by periodate

S9.

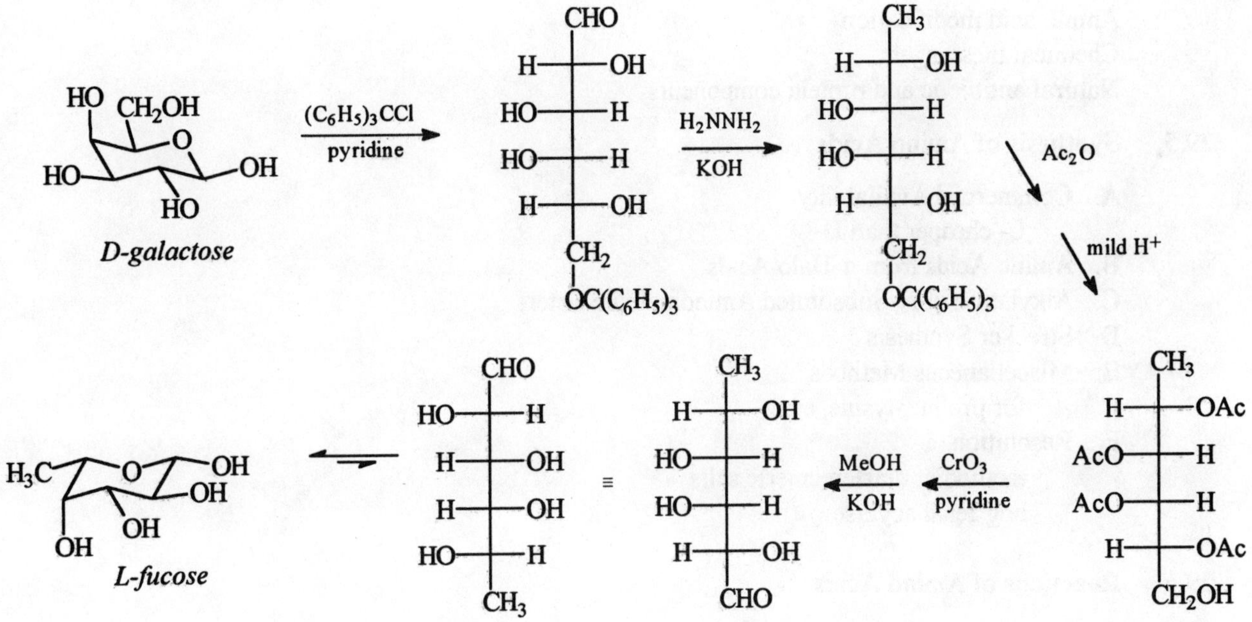

D-galactose $\xrightarrow[\text{pyridine}]{(C_6H_5)_3CCl}$ $\xrightarrow[KOH]{H_2NNH_2}$ Ac₂O, mild H⁺

$\xleftarrow[\text{pyridine}]{CrO_3}$ $\xleftarrow[KOH]{MeOH}$ ≡ ⇌ L-fucose

29. AMINO ACIDS, PEPTIDES AND PROTEINS

29.A. Chapter Outline and Important Terms Introduced

29.1 Introduction

zwitterion, inner salt
peptide bond
di-, tri-, tetra-, and polypeptides
protein

29.2 Structure, Nomenclature, and Physical Properties of Amino Acids

α-amino acids
D,L. vs. R,S (D = R, L = S)

29.3 Acid-Base Properties of Amino Acids

amphoterism
pK_a's = 2.4, 9.8 for α-amino acids (plus side chain groups)
isoelectric point

29.4 Occurrence of Amino Acids

Naturally occurring "rare" amino acids
Amino acid modification
Chemical messengers
Natural antibiotic and protein components

29.5 Synthesis of Amino Acids

A. Commercial Availability
 L- cheaper than D-
B. Amino Acids from α-Halo Acids
C. Alkylation of N-Substituted Aminomalonic Esters
D. Strecker Synthesis
E. Miscellaneous Methods
 for proline, lysine, etc.
F. Resolution
 method of diastereomeric salts
 hog renal acylase

29.6 Reactions of Amino Acids

A. Esterification

$$H_3\overset{+}{N}\underset{\displaystyle R}{\underset{|}{C}}HCO_2H + R'OH \xrightarrow{H^+} H_3\overset{+}{N}\underset{\displaystyle R}{\underset{|}{C}}HCO_2R'$$

B. Amide Formation

$$H_3\overset{+}{N}\underset{\displaystyle R}{\underset{|}{C}}HCO_2^- + R'\overset{O}{\overset{\|}{C}}X \xrightarrow{\text{base}} R'\overset{O}{\overset{\|}{C}}NH\underset{\displaystyle R}{\underset{|}{C}}HCO_2R' \qquad (X = Cl,\ O_2CR')$$

C. Ninhydrin Reaction
 amino acid analysis

29.7 Peptides

A. Structure and Nomenclature
- amino acid sequence
- disulfide bond
- N-terminal vs. C-terminal amino acids
- partial hydrolysis of proteins

B. Synthesis of Peptides
- homopolymer
- 2,5-diketopiperazine formation
- protecting groups: carbobenzoxy ("Cbz") and *t*-butoxycarbonyl ("Boc") groups
- coupling with dicyclohexylcarbodiimide ("DCC")
- Merrifield solid-phase technique: polymer-bound peptides

C. Structure Determination
- amino acid analyzer
- identification of the N-terminal amino acid
  - Sanger method
  - Edman degradation
- identification of the C-terminal amino acid
  - sequential removal of C-terminal amino acids with carboxypeptidase
- fragmentation of the peptide chain
  - proteases: serine protease; trypsin, chymotrypsin; carboxy protease; pepsin, thiol protease; Papain
  - cyanogen bromide (cleaves at methionine carbonyl)
  - partial degradation and peptide mapping
  - labelling

29.8 Proteins

- protein folding
- protein function

A. Molecular Shape
- fibrous vs. globular
- enzymes - usually globular
- prosthetic group - non-protein molecule

B. Factors that Influence Molecular Shape
- primary structure
  - amino acid sequence
- secondary structure
  - hydrogen bonds
  - Ramachandran maps
  - extended chain
  - disulfide bridges
  - reverse turn
  - α-helix, 3_{10}-helix
- tertiary structure
  - electronic and steric properties of the side chain groups
  - hydrophobic vs. hydrophilic
  - supersecondary structure
  - domains
- quaternary structure

C. Structure of the Fibrous Proteins
- α-helix
- random coil
- super helix
- β-pleated sheet

D. Structure of the Globular Proteins
- denaturation/renaturation

E. Biological Function of Protein and Polypeptides
- enzymes
- transport
- defensive substances
- hormones

storage
Cytochrome P-450
monoxygenase

(amphiphilic character)

29.B. Important Reactions Introduced

<u>NOTE</u>: Just as in the last Chapter, there are very few **new** reactions introduced; it is the combination of an acidic and a basic functional group in the same molecule that makes the chemistry look different.

Synthesis of α-amino acids by amination of α-haloacids (29.5.B)

Equation:

$$\underset{\displaystyle X}{RCHCO_2H} \xrightarrow{NH_3} \underset{\displaystyle {}^{+}NH_3}{RCHCO_2^-}$$

Generality: X is usually Br

Key features: reaction proceeds better than most direct aminations with NH_3 because NH_2 group in product is less reactive than normal alkyl amine (see Section 24.6.A)

Synthesis of α-amino acids by alkylation of N-substituted aminomalonic esters (29.5.C)

Equation:

$$\underset{\displaystyle RCONH}{EtO_2CCHCO_2Et} + R'X \xrightarrow{EtO^-} \underset{\displaystyle RCONH}{\overset{\displaystyle R'}{EtO_2CCCO_2Et}} \xrightarrow{H_3O^+} \underset{\displaystyle R'}{H_3\overset{+}{N}CHCO_2^-}$$

Generality: RCONH = AcNH, CbzNH, Phthalimidyl, *t*-BocNH, etc.

Key features: version of the classic malonic ester synthesis adapted to preparation of amino acids (see Section 27.7.E)

Strecker amino acid synthesis (29.5.D)

Equation:

$$NH_3 + RCH{=}O + HCN \longrightarrow \underset{\displaystyle R}{H_2NCHC{\equiv}N} \xrightarrow{H_3O^+} \underset{\displaystyle R}{H_3\overset{+}{N}CHCO_2^-}$$

Key features: amino version of cyanohydrin formation (Section 14.7.B)
reaction proceeds via ^-CN addition to immonium intermediate

Resolution of racemic amino acids with hog renal acylase (29.5.F)

Equation: $\overset{+}{H_3N}CH(R)CO_2^-$ (*racemic*) $\xrightarrow{Ac_2O}$ $CH_3C(=O)NHCH(R)CO_2H$ $\xrightarrow[\text{acylase}]{\text{hog renal}}$ $\overset{+}{H_3N}CH(R)CO_2^-$ (L-enantiomer) + $CH_3CONHCH(R)CO_2H$ (D-enantiomer)

Ninhydrin reaction (29.6.C)

Equation: $RCH(\overset{+}{NH_3})CO_2^-$ $\xrightarrow{\text{ninhydrin}}$ $RCH(N{=}nin)CO_2^-$ $\xrightarrow{H_2O}$ $RCH{=}N{-}nin^-$ + CO_2 $\xrightarrow[\text{-RCHO}]{\text{ninhydrin}}$ *purple color*

Key features: used as analytical reaction to detect presence of α-amino acids; not used preparatively
gives a different reaction with proline

Diketopiperazine formation (29.7.B)

Equation: $2\ \overset{+}{H_3N}CH(R)CO_2^-$ $\xrightarrow{-H_2O}$ (2,5-diketopiperazine ring bearing two R groups)

Key feature: generally a reaction to be avoided in the synthesis of peptides

Introduction and removal of N-protecting groups for peptide synthesis (29.7.B)

Equation: $R'OC(=O)X$ + $\overset{+}{H_3N}CH(R)CO_2^-$ $\xrightarrow{\text{base}}$ $R'OC(=O)NHCH(R)CO_2H$

$R'OC(=O)NHCH(R)CO_2R''$ $\xrightarrow{[\text{"}-R'\text{"}]}$ CO_2 + $H_2NCH(R)CO_2R''$

Generality: "R'OCOX" is usually $C_6H_5CH_2OCOCl$ ("CbzCl") or t-$BuO_2CON{=}C(CN)C_6H_5$ ("Boc-ON")
["–R' "] = H_2/Pt or HBr for Cbz protecting group
= HCl or CF_3CO_2H for Boc protecting group

Peptide synthesis with dicyclohexylcarbodiimide (DCC) (29.7.B)

Equation: RCO_2H + H_2NR' + $C_6H_{11}N{=}C{=}NC_6H_{11}$ (*DCC*) $\longrightarrow$ $RC(=O)NHR'$ + $C_6H_{11}NHC(=O)NHC_6H_{11}$ (*dicyclohexylurea (DCU)*)

Generality: general method for formation of amide and ester bonds

Key features: reaction proceeds via O-acylisourea, an activated carboxylic acid derivative
in combination with polymer-bound amino acid derivative (R' part above), used with Merrifield automated technique for peptide synthesis

Sanger method for N-terminal amino acid identification (29.7.C)

Equation:

$$2,4\text{-}(O_2N)_2C_6H_3F + H_2NCH(R)CNH(pep)_n \ (\text{C=O}) \longrightarrow 2,4\text{-}(O_2N)_2C_6H_3\text{-}NHCH(R)C(=O)NH(pep)_n \xrightarrow[\Delta]{H_3O^+} 2,4\text{-}(O_2N)_2C_6H_3\text{-}NHCH(R)CO_2H$$

+ amino acids

Key features: hydrolysis of polypeptide gives free amino acids, with the N-terminal one derivatized with the dinitrophenyl group (lysine ϵ-amino group derivatized too)
reaction is example of nucleophilic aromatic substitution (Section 26.5.B and 30.3.B)

Edman method for sequential degradation of polypeptide chains (29.7.C)

Equation: $C_6H_5N{=}C{=}S + H_2NCH(R)C(=O)NH(pep)_n \longrightarrow C_6H_5NHC(=S)NHCH(R)C(=O)NH(pep)_n \xrightarrow{HCl}$ (cyclic thiohydantoin: C_6H_5N, C=S, NH, CH–R, C=O)

Key features: can be repeated sequentially to remove one amino acid at a time from N-terminus of peptide chain
cyclic product (thiohydantoin) is isolated and identified

Cyanogen bromide cleavage of methionine peptide bonds (29.7.C)

Equation:

A–NH–CH(CH$_2$CH$_2$SMe)–C(=O)–HN–B $\xrightarrow{BrCN}$ A–NH–(homoserine lactone) + H_2N–B + MeSCN

Key features: cleaves peptide chain **only** at methionine carbonyl position
useful in selective peptide degradation and sequence analysis

Cytochrome P-450 Oxidation (29.8.E)

Equation:

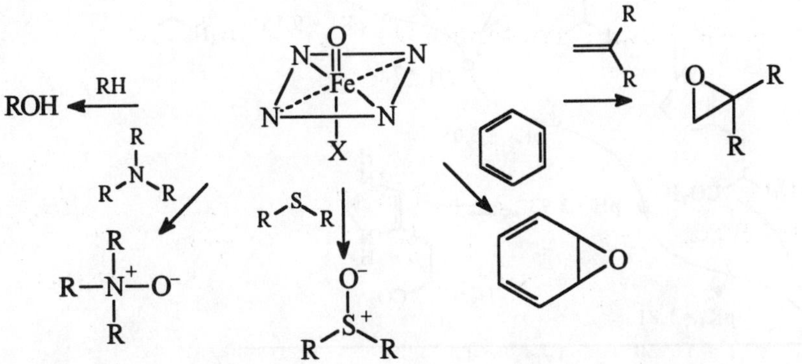

Generality: A monoxygenase - inserts 1 O atom in a C-H or C=C bond.

Key features: Used biologically as a detoxifying agent.

29.C. Important Concepts and Hints

Like the last chapter, this one discusses the special chemistry that arises when two familiar functional groups are present in the same molecule. Although most of the reactions of amino acids are ones you've seen before, complications can arise because of the juxtaposition of the acidic and basic groups. For example, a sequence of protection and deprotection steps is necessary for controlled formation of the amide linkage in a polypeptide, whereas it is a very straightforward process in monofunctional molecules.

The topic of this chapter is important not only for the chemistry it presents, but also because of the biological significance of amino acids, oligopeptides, and proteins. The frontiers of biology have reached the molecular level, and it is necessary to understand the chemistry of one of its most important groups of building blocks. It really is useful to know the structures, names, and abbreviations of the amino acids, and we urge you to learn them if you have any interest in the life sciences.

29.D. Answers to Exercises

29.1 This is especially important for biology and biochemistry students.

29.2 (D) = (R) and (L) = (S) for the α-position of amino acids (except for the amino acid cysteine).

29.3 $H_3N^+CH_2CO_2H$ $H_3N^+CH_2CO_2^-$ $H_2NCH_2CO_2^-$

pH 2 pH 4 pH 8 pH 11

29.4

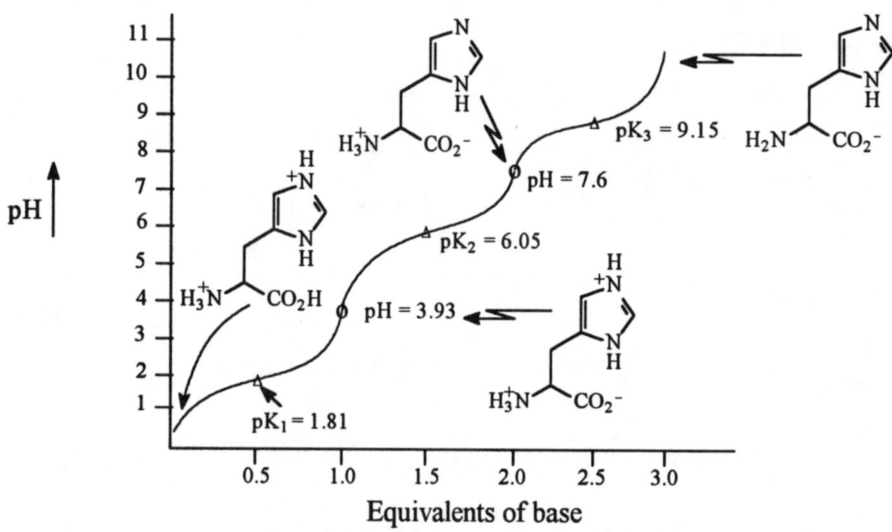

29.5 $C_6H_5CH_2CH_2CO_2H \xrightarrow[\Delta]{P,\ Br_2} \xrightarrow{H_2O} C_6H_5CH_2CH(Br)CO_2H \xrightarrow{NH_3} C_6H_5CH_2CH(^{+}NH_3)CO_2^{-} + NH_4Br$

$(CH_3)_2CHCH_2CO_2H \xrightarrow[\Delta]{P,\ Br_2} \xrightarrow{H_2O} (CH_3)_2CHCH(Br)CO_2H \xrightarrow{NH_3} (CH_3)_2CHCH(^{+}NH_3)CO_2^{-} + NH_4Br$

$(CH_3)_2CHCH_2CH_2CO_2H \xrightarrow[\Delta]{P,\ Br_2} \xrightarrow{H_2O} (CH_3)_2CHCH_2CH(Br)CO_2H \xrightarrow{NH_3} (CH_3)_2CHCH_2CH(^{+}NH_3)CO_2^{-} + NH_4Br$

for serine: $HOCH_2CH_2CO_2H \xrightarrow[\Delta]{P,\ Br_2} \xrightarrow{H_2O} BrCH_2CH(Br)CO_2H$

for tyrosine: $HO\text{-}C_6H_4\text{-}CH_2CH_2CO_2H \xrightarrow[\Delta]{P,\ Br_2} \xrightarrow{H_2O}$ HO–(3,5-dibromophenyl, Br, Br)–$CH_2CH(Br)CO_2H$

activated aromatic ring

29.6

$EtO_2C\text{-}CH(CO_2Et)\text{-}N$(phthalimide) $\xrightarrow[RBr]{NaOEt}$ $EtO_2C\text{-}CR(CO_2Et)\text{-}N$(phthalimide) $\xrightarrow{H_3O^+}$

$HO_2C\text{-}CH(^{+}NH_3)\text{-}CH_2CO_2H$ or $HO_2C\text{-}CH(^{+}NH_3)\text{-}CH_2C_6H_5$ or $HO_2C\text{-}CH(^{+}NH_3)\text{-}CH(CH_3)_2$

aspartic acid (if R=CH_2CO_2Et) *phenylalanine* (if R=$CH_2C_6H_5$) *valine* (if R=$CH(CH_3)_2$)

29.7 (Acetamidomalonic acid) $\xrightarrow[\text{RBr or } H_2C{=}O]{\text{NaOEt}}$ $\xrightarrow{H_3O^+}$ *serine* or *tyrosine* or *valine*

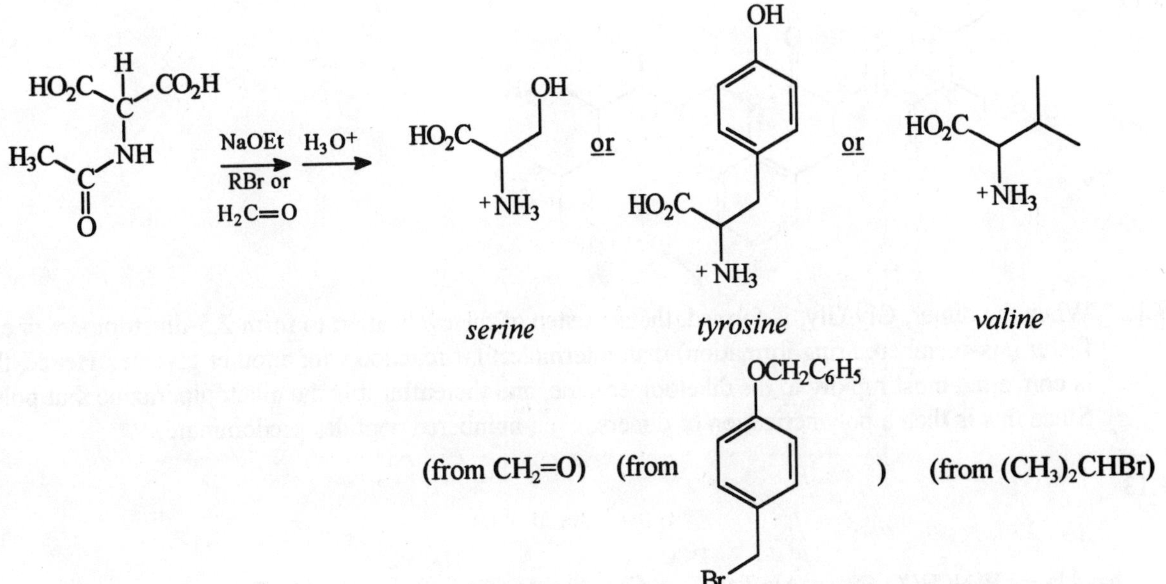

(from $CH_2{=}O$) (from p-benzyloxybenzyl bromide ($OCH_2C_6H_5$ / Br) followed by hydrogenation to remove protecting group) (from $(CH_3)_2CHBr$)

29.8 For tyrosine:

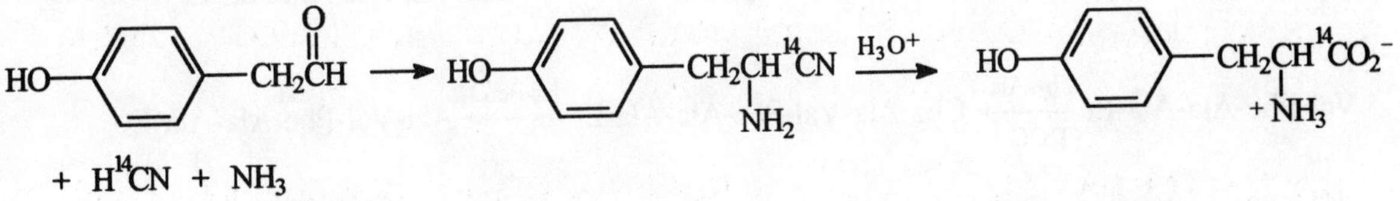

$HO{-}C_6H_4{-}CH_2CHO + H^{14}CN + NH_3 \rightarrow HO{-}C_6H_4{-}CH_2CH(NH_2)^{14}CN \xrightarrow{H_3O^+} HO{-}C_6H_4{-}CH_2CH(^+NH_3)^{14}CO_2^-$

for lysine:

$H_2NCH_2CH_2CH_2CH_2CHO \xrightleftharpoons{-H_2O}$ (cyclic imine) $+ {}^-CN \rightarrow$ (2-cyanopiperidine)

$\downarrow\uparrow NH_3$

$H_2NCH_2CH_2CH_2CH_2CH{=}\overset{+}{N}H_2 + {}^-CN \rightarrow H_2NCH_2CH_2CH_2CH_2CH(NH_2)CN \dashrightarrow$

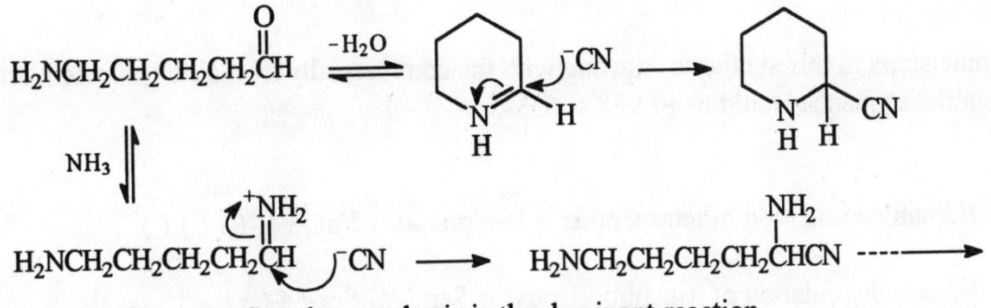

The intramolecular Strecker synthesis is the dominant reaction.

29.9 (Ethyl 2-oxocyclopentanecarboxylate) $\xrightarrow[H_2SO_4]{HN_3}$ (6-oxopiperidine-2-carboxylic acid) $\xrightarrow{H_3O^+}$ $HO_2C(CH_2)_3CH(^+NH_3)CO_2H$

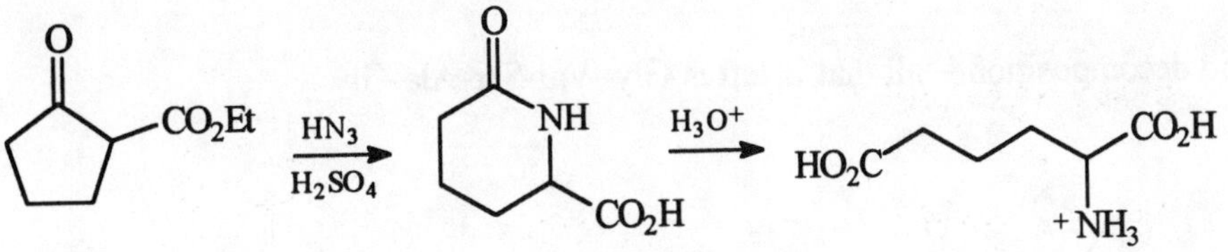

29.10 $C_6H_5CH_2CH(^+NH_3)CO_2^- \xrightarrow[CH_3OH,\ \Delta]{HCl} \xrightarrow{\text{neutralize}} C_6H_5CH_2CH(NH_2)CO_2CH_3 \xrightarrow[Et_3N]{Ac_2O} C_6H_5CH_2CH(NHCOCH_3)CO_2CH_3$

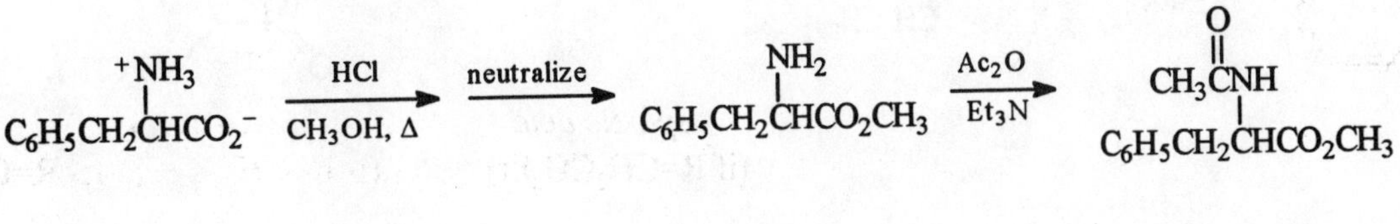

29.11

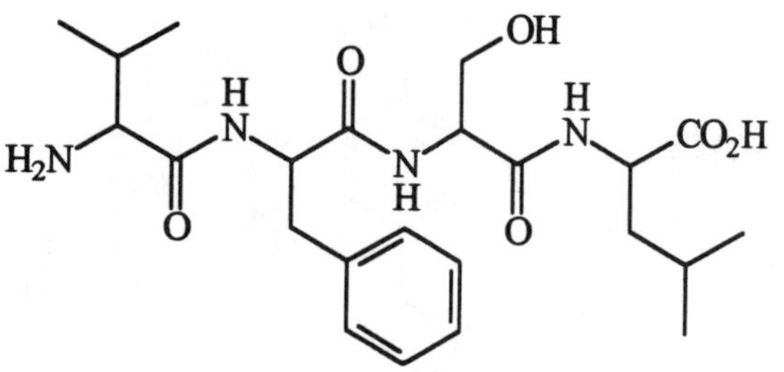

H_2N–Val–Phe–Ser–Leu–CO_2H (structure: H₂N, H–N, O, OH, N–H, CO_2H)

29.12 When the dimer, Gly-Gly, is formed, the next step of the cyclization to form 2,5-diketopiperazine is much faster (six-membered ring formation) than intermolecular reaction with another glycine. Hence, the glycine is converted most rapidly to the diketopiperazine, and thereafter it is the diketopiperazine that polymerizes. Since this is then a polymerization of dimers, even-numbered peptides predominate.

29.13

Cbz-Ala + $H_2NCH(CH_3)CO_2Et$ (≡ Ala-Et) $\xrightarrow{\text{DCC}}$ Cbz-Ala-Ala-Et $\xrightarrow{H_2\text{-Pd/C}}$ Ala-Ala-Et

$\xrightarrow[\text{DCC}]{\text{Cbz-Phe}}$ Cbz-Phe-Ala-Ala-Et $\xrightarrow{H_2\text{-Pd/C}}$ Phe-Ala-Ala-Et $\xrightarrow[\text{DCC}]{\text{Cbz-Val}}$ Cbz-Val-Phe-Ala-Ala-Et $\xrightarrow{H_2\text{-Pd/C}}$

Val-Phe-Ala-Ala-Et $\xrightarrow[\text{DCC}]{\text{Cbz-Ala}}$ Cbz-Ala-Val-Phe-Ala-Ala-Et $\xrightarrow{H_2\text{-Pd/C}}$ Ala-Val-Phe-Ala-Ala-Et

$\xrightarrow[\text{hydrolysis}]{\text{mild}}$ Ala-Val-Phe-Ala-Ala

Since there are nine steps in this synthesis (starting with the commercially available protected amino acids), the overall yield from the sequence would be $(0.95)^9 = 63\%$.

29.14

1. Edman degradation of tetrapeptide – N-terminal is Val: Val-( )-( )-( )
2. Edman degradation of tripeptide – next is Ser: Val-Ser-( )-( )
3. Sanger method on dipeptide – then Ala: Val-Ser-Ala-( )
4. Amino acid decomposition – all that is left is Gly: Val-Ser-Ala-Gly

First Edman: $C_6H_5N{=}C{=}S$ + $H_3\overset{+}{N}CH(CH(CH_3)_2)C(=O)NH$-Ser-Ala-Gly $\longrightarrow$ $\xrightarrow{\text{HCl}}$ C_6H_5N(thiohydantoin: C=S, NH, C=O, $CH(CH_3)_2$) + Ser-Ala-Gly

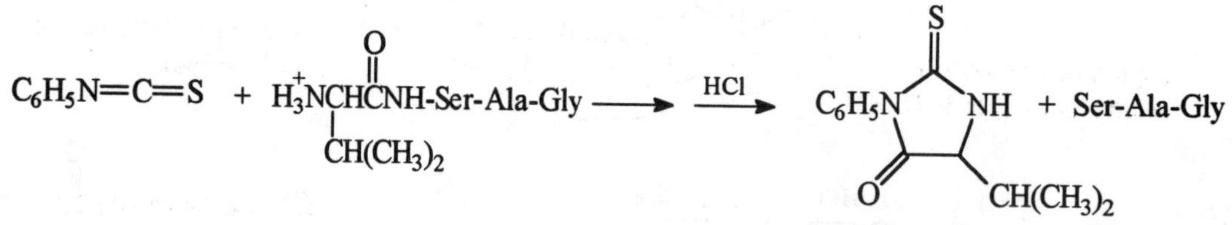

Second Edman: $C_6H_5N{=}C{=}S + H_3\overset{+}{N}CH(CH_2OH)C(O)NH\text{-Ala-Gly} \xrightarrow{HCl}$ [C_6H_5N–C(=S)–NH–CH(CH_2OH)–C(=O)– ring] + Ala-Gly

Sanger: O_2N–$C_6H_3(NO_2)$–F + $H_3\overset{+}{N}CH(CH_3)C(O)NHCH_2CO_2^-$ $\xrightarrow[\Delta]{H_3O^+}$ O_2N–$C_6H_3(NO_2)$–$NHCH(CH_3)CO_2H$ + $H_3\overset{+}{N}CH_2CO_2^-$

29.15 Gly-Ser-Phe

29.16 Gly-Ala-Leu–Leu–Phe
↑ ↑ *pepsin cleavage*

29.17

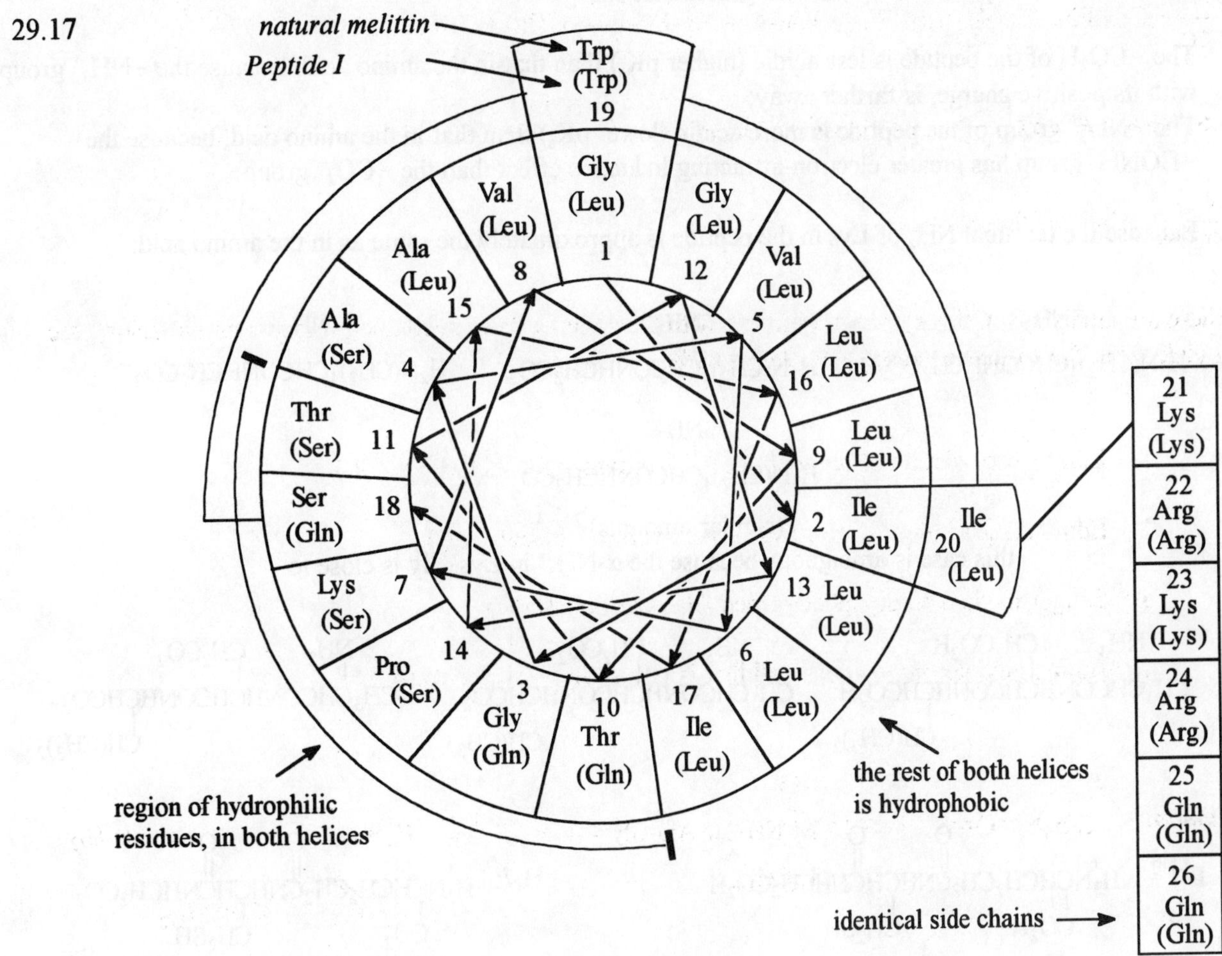

Peptide I shows the same lytic activity that melittin does.

29.E. Answers and Explanations for Problems

1. This question requires Tables 29.2 and 29.4

	pH = 2	pH = 7	pH = 12
a.	$CH_3CH_2CH(CH_3)CH(^+NH_3)CO_2H$	$CH_3CH_2CH(CH_3)CH(^+NH_3)CO_2^-$	$CH_3CH_2CH(CH_3)CH(NH_2)CO_2^-$
b.	$HO_2CCH_2CH(^+NH_3)CO_2^-$	$^-O_2CCH_2CH(^+NH_3)CO_2^-$	$^-O_2CCH_2CH(NH_2)CO_2^-$
c.	$H_3\overset{+}{N}(CH_2)_4CH(^+NH_3)CO_2H$	$H_3\overset{+}{N}(CH_2)_4CH(^+NH_3)CO_2^-$	$H_2N(CH_2)_4CH(NH_2)CO_2^-$
d.	$H_3^+NCH_2CONHCH_2CO_2H$	$H_3^+NCH_2CONHCH_2CO_2^-$	$H_2NCH_2CONHCH_2CO_2^-$

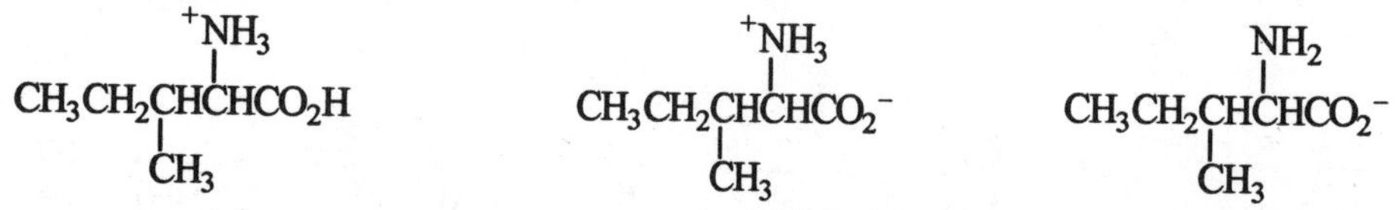

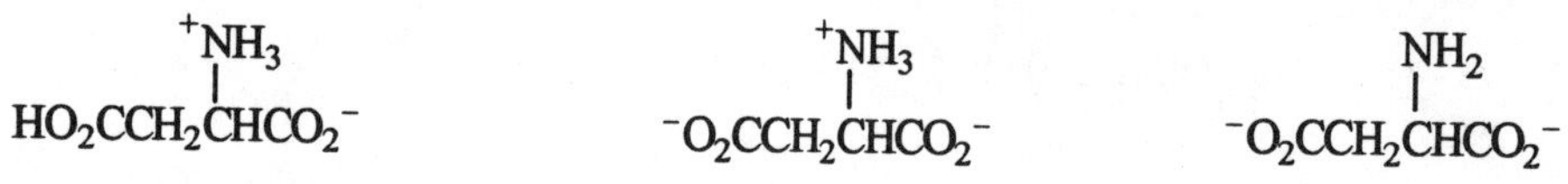

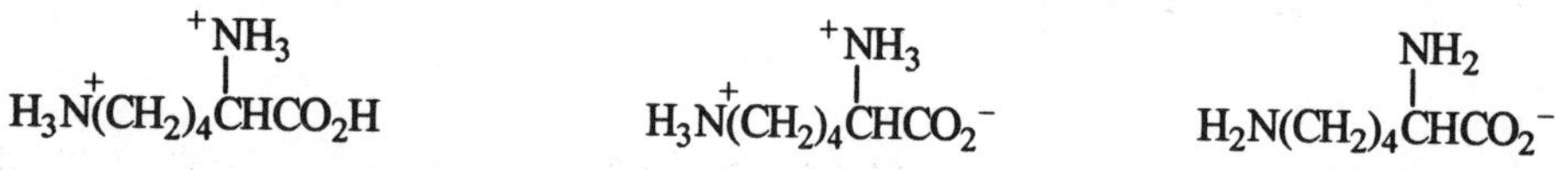

The remainder of the question requires two generalizations:

1. The $-CO_2H$ of the peptide is less acidic (higher pK_a) than that in the amino acid, because the $-NH_3^+$ group, with its positive charge, is farther away;
2. The $-NH_3^+$ group of the peptide is more acidic (lower pK_a) than that in the amino acid, because the $-CONH-$ group has greater electron-attracting inductive effect than the $-CO_2^-$ group.

e. Because the terminal NH_2 of Lys in the peptide is approximately the same as in the amino acid:

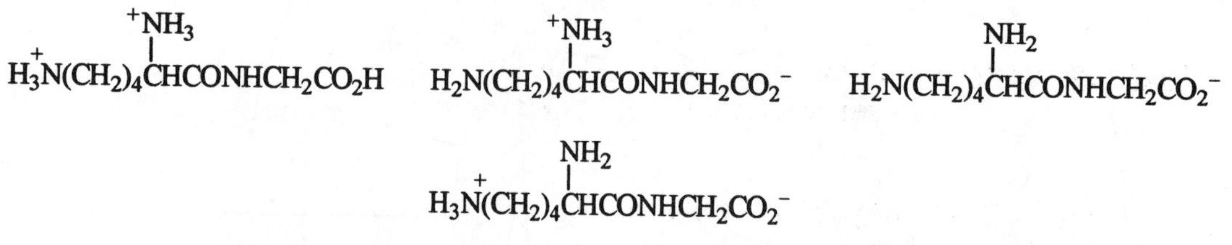

$$H_3\overset{+}{N}(CH_2)_4CH(^+NH_3)CONHCH_2CO_2H \qquad H_2N(CH_2)_4CH(^+NH_3)CONHCH_2CO_2^- \qquad H_2N(CH_2)_4CH(NH_2)CONHCH_2CO_2^-$$

$$H_3\overset{+}{N}(CH_2)_4CH(NH_2)CONHCH_2CO_2^-$$

(similar amounts)

this case is ambiguous because the α-NH_3^+ in Lys-Gly is close to 7

f. 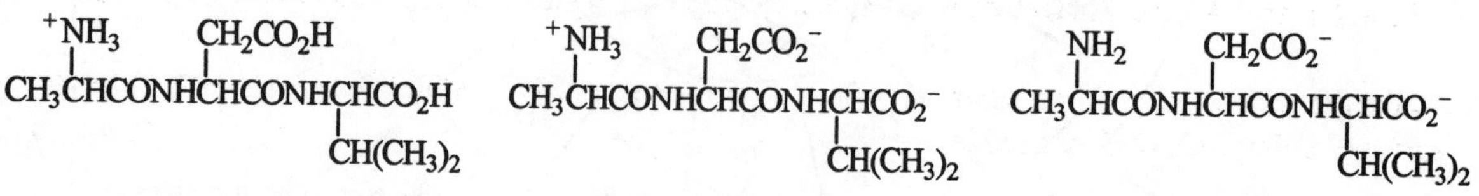 $CH_3CH(^+NH_3)CONHCH(CH_2CO_2H)CONHCH(CH(CH_3)_2)CO_2H \qquad CH_3CH(^+NH_3)CONHCH(CH_2CO_2^-)CONHCH(CH(CH_3)_2)CO_2^- \qquad CH_3CH(NH_2)CONHCH(CH_2CO_2^-)CONHCH(CH(CH_3)_2)CO_2^-$

g. pH 2: $H_3\overset{+}{N}CH(CO_2H)CH_2CH_2\overset{O}{\overset{\|}{C}}NHCH(CH_2SH)\overset{O}{\overset{\|}{C}}NHCH_2CO_2H$

pH 7: $H_3\overset{+}{N}CH(CO_2^-)CH_2CH_2\overset{O}{\overset{\|}{C}}NHCH(CH_2SH)\overset{O}{\overset{\|}{C}}NHCH_2CO_2^-$

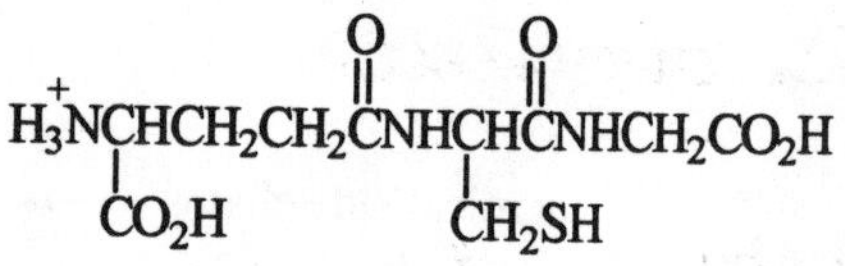

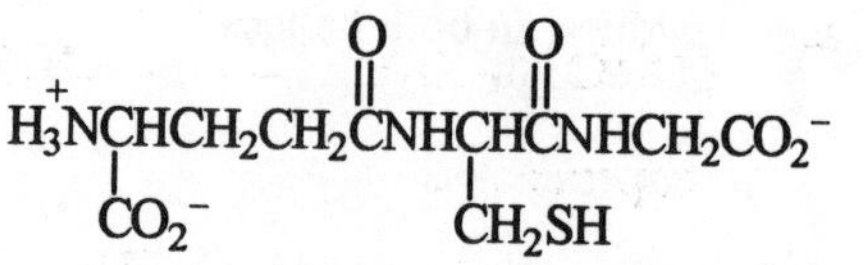

pH 10: $H_2NCH(CO_2^-)CH_2CH_2\overset{O}{\overset{\|}{C}}NHCH(CH_2S^-)\overset{O}{\overset{\|}{C}}NHCH_2CO_2^-$

2.

$$K_1 = \frac{[H_3N^+\sim\sim\sim CO_2^-][H^+]}{[H_3N^+\sim\sim\sim CO_2H]} \quad K_2 = \frac{[H_2N\sim\sim\sim CO_2^-][H^+]}{[H_3N^+\sim\sim\sim CO_2^-]} \quad K_1K_2 = \frac{[H_2N\sim\sim\sim CO_2^-]}{[H_3N^+\sim\sim\sim CO_2H]} \cdot [H^+]^2$$

At the isoelectric point, $[H_2N\sim\sim\sim CO_2^-] = [H_3N^+\sim\sim\sim CO_2H]$

At this point, $[H^+] = (K_1K_2)^{1/2}$; or $pH = \frac{1}{2}[pK_1 + pK_2]$

3. The $-CO_2H$ of Ala is more acidic than that of β-alanine; i.e., the distance to the $-NH_3^+$ group is greater in β-alanine. Thus there is lower electrostatic attraction in the conjugate base, $^+H_3N\sim\sim\sim CO_2^-$. The NH_3^+ group in $H_3N^+CH(CH_3)CO_2^-$ is more acidic than that in $H_3N^+CH_2CH_2CO_2^-$. For 4-aminobutanoic acid, the NH_3^+ is even farther from the acid function and the pK_a is even closer to the value for butanoic acid itself, 4.82.

As suggested in Section 29.3, the difference is probably associated with solvation. When (+) and (−) are far apart they can be separately solvated, but when they are close together the solvation is less efficient. As the distance from the COO^- increases, the pK_2 increases until it is almost the same as that for $CH_3NH_3^+$, 10.62.

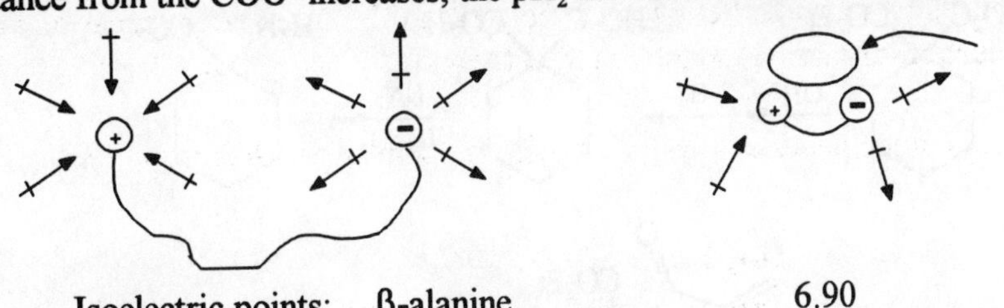
solvent molecules in this region don't know which way to point!

Isoelectric points: β-alanine 6.90
4-aminobutanoic acid 7.30

4. 8.60 ⟶ $H_3\overset{+}{N}CH_2CONHCH(CH_2CO_2H)CO_2H$ ⟵ 2.8; CH_2CO_2H ⟵ 4.4

The value of 4.45 is close to that for simple aliphatic carboxylic acids; 2.81 is typical of peptide $-CO_2H$.

Make the diethyl ester of Asp; convert Gly to Cbz-Gly; react the 2 with DCC; cleave with catalytic hydrogenation followed by mild hydrolysis

5. a. $CH_3(CH_2)_4CO_2H \xrightarrow{P,\ Br_2} CH_3(CH_2)_3CHBrCO_2H \xrightarrow{NH_3} CH_3(CH_2)_3CH(\overset{+}{N}H_3)CO_2^-$

caproic acid

or $CH_3(CH_2)_3Br + {}^-C(NHCOCH_3)(CO_2Et)_2 \longrightarrow CH_3(CH_2)_3C(NHCOCH_3)(CO_2Et)_2 \xrightarrow{OH^-}$ ↑ H^+, Δ

b. $C_6H_5CHO + NH_3 + HCN \longrightarrow C_6H_5CH(NH_2)CN \xrightarrow[\Delta]{\text{aq. NaOH}} \xrightarrow{H^+} C_6H_5CH(NH_3)CO_2H$

c. $(CH_3)_3CCHO + NH_3 + HCN \longrightarrow (CH_3)_3CCH(NH_2)CN \xrightarrow{\text{aq. NaOH}} \xrightarrow{H^+} (CH_3)_3CCH(^+NH_3)CO_2^-$

d. $CH_2(CO_2Et)_2 \xrightarrow[\text{EtBr}]{\text{NaOEt}} EtCH(CO_2Et)_2 \xrightarrow[CH_3I]{\text{NaOEt}} CH_3CH_2C(CH_3)(CO_2Et)_2 \xrightarrow{OH^-} \xrightarrow{H^+} CH_3CH_2C(CH_3)(CO_2H)CO_2H \xrightarrow[H_2SO_4]{HN_3} CH_3CH_2C(CH_3)(^+NH_3)CO_2^-$

(Schmidt reaction)

e. cyclohexanone + NH_3 + HCN $\longrightarrow$ 1-amino-1-cyanocyclohexane (H_2N, CN) $\xrightarrow[\Delta]{OH^-} \xrightarrow{H^+}$ 1-aminocyclohexanecarboxylic acid (H_3N^+, CO_2^-)

or $CH_2(CO_2Et)_2 \xrightarrow[Br(CH_2)_5Br]{\text{NaOEt, 2 moles}}$ cyclohexane-1,1-(CO_2Et)$_2$ (EtO_2C, CO_2Et) $\xrightarrow{OH^-} \xrightarrow{H^+}$ cyclohexane-1,1-(CO_2H)$_2$ (HO_2C, CO_2H) $\xrightarrow[H_2SO_4]{HN_3}$ 1-aminocyclohexanecarboxylate (H_3N^+, CO_2^-)

f. phthalimido-$NCH(CO_2Et)_2 \xrightarrow[Br(CH_2)_4Br]{\text{NaOEt}}$ phthalimido-$NC(CO_2Et)_2(CH_2)_4Br \longrightarrow$ piperidine-2,2-(CO_2^-)$_2$ (N–H) $\xrightarrow[\Delta]{H^+}$ piperidinium-2-carboxylate ($\overset{+}{N}H_2$, CO_2^-)

6. a. $CH_3CHO + NH_3 + H{*}CN \longrightarrow$ etc.

b. $CH_3MgX + {*}CO_2 \longrightarrow CH_3{*}CO_2H \xrightarrow{LiAlH_4} CH_3{*}CH_2OH \xrightarrow{PCC} CH_3{*}CHO \xrightarrow[HCN]{NH_3}$ etc

c. $ROCO_2R \xrightarrow{LiAlD_4} CD_3OD \xrightarrow{HI} CD_3I \xrightarrow[\text{EtOH, NaOEt}]{CH_3CONHCH(CO_2Et)_2}$ etc.
a carbonate ester

<u>or</u> $CO + D_2 \xrightarrow[\text{catalyst}]{\Delta} CD_3OD$ <u>NOTE</u>: would you use dimethyl carbonate for the first step?

d.

$$C_6H_5CH_2Cl + CH_3\overset{O}{\overset{\|}{C}}NHCH(CO_2Et)_2 \xrightarrow[EtOH]{NaOEt} C_6H_5CH_2C(NHCOCH_3)(CO_2Et)_2 \xrightarrow{OH^-} \xrightarrow{H^+} C_6H_5CH_2C(\overset{+}{N}H_3)(CO_2H)CO_2H$$

$$\xrightarrow{D_2O} C_6H_5CH_2C(\overset{+}{N}D_3)(CO_2D)CO_2D \xrightarrow[\Delta]{D^+} C_6H_5CH_2CD(\overset{+}{N}D_3)CO_2D \xrightarrow[wash]{H_2O} C_6H_5CH_2CD(\overset{+}{N}H_3)CO_2^-$$

e. $$*CO_2 \xrightarrow{LiAlH_4} *CH_3OH \xrightarrow{HI} *CH_3I \xrightarrow{SH^-} *CH_3SH \xrightarrow[\text{ethylene oxide}]{OH^-} *CH_3SCH_2CH_2OH$$

$$\xrightarrow{HBr} *CH_3SCH_2CH_2Br \xrightarrow[\text{phthalimido-}NCH(CO_2Et)_2]{NaOEt} \text{etc.}$$

f. $$CH_3COCH_3 \xrightarrow[Na_2CO_3]{D_2O} CD_3COCD_3 \xrightarrow{LiAlH_4} CD_3CH(OH)CD_3 \xrightarrow{HBr} (CD_3)_2CHBr \xrightarrow[\text{of e. above}]{\text{see last step}} \text{etc.}$$

g. $$CH_3MgX + *CO_2 \longrightarrow CH_3{*}CO_2H \xrightarrow[P]{Br_2} BrCH_2{*}CO_2H \xrightarrow[H^+]{EtOH} BrCH_2{*}CO_2Et \xrightarrow[\text{of e.}]{\text{last step}} \text{etc.}$$

h.

$$\text{PhthN}C(CO_2Et)_2CH_2CH_2CH_2Br + CN^- \longrightarrow \text{PhthN}C(CO_2Et)_2CH_2CH_2CH_2CN \xrightarrow[\text{catalyst}]{D_2} \text{PhthN}C(CO_2Et)_2CH_2CH_2CH_2CD_2ND_2$$

$$\xrightarrow[EtOH]{NaOEt} \xrightarrow{H^+} H_2NCH(CO_2^-)(CH_2)_3CD_2\overset{+}{N}H_3$$

(Section 29.5.E)

$LiAlD_4$ clearly cannot be used in the reduction step because of the other reducible groups present; however, the catalytic reduction is not straightforward either. If excess NH_3 is present to cut down secondary amine formation (Section 24.6.F), some catalysts will promote exchange:

$$NH_3 + D_2 \xrightarrow{\text{catalyst}} ND_3 + H_2, \text{ etc.}$$

and the deuterium will become diluted. In practice, experiments would be required to find the best conditions for the hydrogenation with D_2.

7. Acid-catalyzed esterification starts with a protonated carbonyl: $R{-}C(={}^{+}OH)OH$

Protonation of an amino acid is more difficult because of electrostatic repulsion with the $-NH_3^+$ group:

$-CH(NH_3^+)-C(={}^{+}OH)OH$ ← repulsion (between $^{+}NH_3$ and ^{+}OH)

8. Hydrolysis of the benzoyl amide group requires conditions that also cause hydrolysis of the peptide bond; both are normal amide functions.

9. a. $BrCH_2COBr + CH_3CH(NH_2)CO_2^- \longrightarrow BrCH_2CONHCH(CH_3)CO_2^- \xrightarrow{NH_3} H_3\overset{+}{N}CH_2CONHCH(CH_3)CO_2^-$

$BrCH_2COBr + H_2NCH_2CONHCH(CH_3)CO_2^- \longrightarrow BrCH_2CONHCH_2CONHCH(CH_3)CO_2^- \xrightarrow{NH_3}$

b. $CH_3CHBrCOBr$; $(CH_3)_2CHCHBrCOBr$

c. Val-Ala or Ala-Ala Note that the amino acid used forms the CO_2H end of the peptide.

d. Stereochemistry: would require optically active α-bromo acyl bromides, which are difficult to prepare and which readily racemize with mild base:

$RCHBrCOBr + base \rightleftharpoons R^-CBrCOBr + base\text{-}H^+$ Configuration is lost at carbanion.

Thus a mixture of diastereomers would result. As an example, in c. above, the use of racemic bromides will produce L-Ala-L-Ala, D-Ala-L-Ala, L-Ala-D-Ala, D-Ala-D-Ala, etc.

10. $C_6H_5CH_2OC(=O)Cl + H_3\overset{+}{N}CH(R)CO_2^- \longrightarrow C_6H_5CH_2OC(=O)NHCH(R)CO_2H$ ≡ Cbz-amino acid

$H_2NCHCO_2CH_2CH_3$ (with $CH(CH_3)_2$ on the α-carbon) *(Val-Et)* $\xrightarrow[DCC]{Cbz\text{-}Ala}$ Cbz-Ala-Val-Et $\xrightarrow{H_2/Pt}$ Ala-Val-Et $\xrightarrow[DCC]{Cbz\text{-}Ala}$ Cbz-Ala-Ala-Val-Et

$\xrightarrow{H_2/Pt}$ Ala-Ala-Val-Et $\xrightarrow[DCC]{Cbz\text{-}Pro}$ Cbz-Pro-Ala-Ala-Val-Et $\xrightarrow{H_2/Pt}$ Pro-Ala-Ala-Val-Et

$\xrightarrow[DCC]{Cbz\text{-}Ala}$ Cbz-Ala-Pro-Ala-Ala-Val-Et $\xrightarrow{H_2/Pt}$ Ala-Pro-Ala-Ala-Val-Et

$\xrightarrow[DCC]{Cbz\text{-}Gly}$ Cbz-Gly-Ala-Pro-Ala-Ala-Val-Et $\xrightarrow[2.\ H^+]{1.\ H_2/Pd}$ Gly-Ala-Pro-Ala-Ala-Val

11. Boc-Gly + $ClCH_2$–C_6H_4–Polymer ⟶ Boc-Gly-CH_2-Ⓟ $\xrightarrow{H^+}$ Gly-CH_2-Ⓟ

$\xrightarrow[DCC]{Boc\text{-}Gly}$ Boc-Gly-Gly-CH_2-Ⓟ $\xrightarrow{H^+}$ $\xrightarrow{BocNHCH(CH_2CH_2CO_2CH_2C_6H_5)CO_2H}$ Boc-Glu(δ-Bz)-Gly-Gly-CH_2-Ⓟ

$\xrightarrow{H^+}$ $\xrightarrow[DCC]{Boc\text{-}Lys(\epsilon\text{-}Cbz)}$ Boc-Lys(ε-Cbz)-Glu(δ-Bz)-Gly-Gly-CH_2-Ⓟ $\xrightarrow{H^+}$ $\xrightarrow[DCC]{Boc\text{-}Ala}$

Boc-Ala-Lys(ε-Cbz)-Glu(δ-Bz)-Gly-Gly-CH_2-Ⓟ $\xrightarrow{H_2/Pt}$ $\xrightarrow{HF}$ $\xrightarrow{H^+}$ Ala-Lys-Glu-Gly-Gly

12. Either of the methods depicted in answers #10 or #11 can be used by continuing to add groups in the sequence Ala, Ala, Phe, Val, Ala. However, the first and second halves of this decapeptide are the same. Hence, one could simply couple two of the pentapeptides:

Cbz-Ala-Val-Phe-Ala-Ala-Et $\xrightarrow{H_2\text{-}Pd/C}$ Ala-Val-Phe-Ala-Ala-Et

Cbz-Ala-Val-Phe-Ala-Ala-Et $\xrightarrow[hydrolysis]{mild}$ Cbz-Ala-Val-Phe-Ala-Ala

$\xrightarrow[DCC]{}$ Cbz-Ala-Val-Phe-Ala-Ala-Ala-Val-Phe-Ala-Ala-Et $\xrightarrow{H_2\text{-}Pd/C}$ $\xrightarrow[hydrolysis]{mild}$

Ala-Val-Phe-Ala-Ala-Ala-Val-Phe-Ala-Ala

13. TRH ≡ pyroGlu-His-Pro-NH_2 or

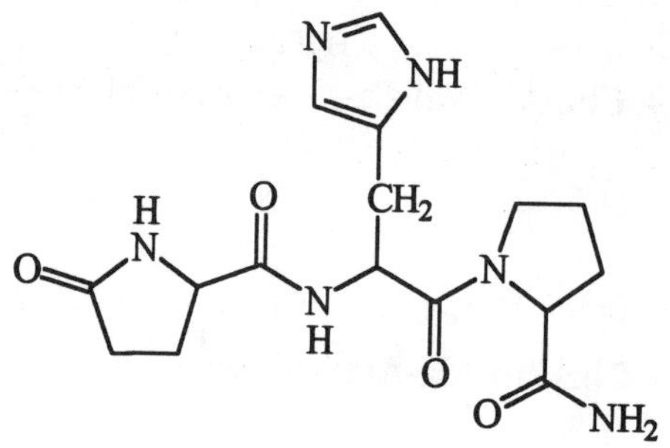

Synthesis:

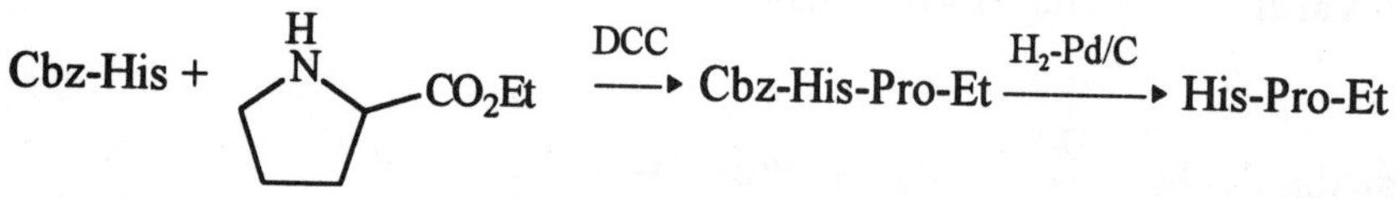

Cbz-His + (proline ethyl ester, CO_2Et) $\xrightarrow{DCC}$ Cbz-His-Pro-Et $\xrightarrow{H_2\text{-}Pd/C}$ His-Pro-Et

$\xrightarrow[DCC]{pyroGlu}$ pyroGlu-His-Pro-Et $\xrightarrow[\text{1. mild hydrolysis; 2. } NH_3\text{, DCC}]{NH_3 \text{ or}}$ pyroGlu-His-Pro-NH_2

NOTE: conversion to the acid chloride with $SOCl_2$, followed by treatment with NH_3 to form the final amide is an alternative, but will probably cause some racemization at the proline α-carbon to give a mixture of diastereomers.

14.

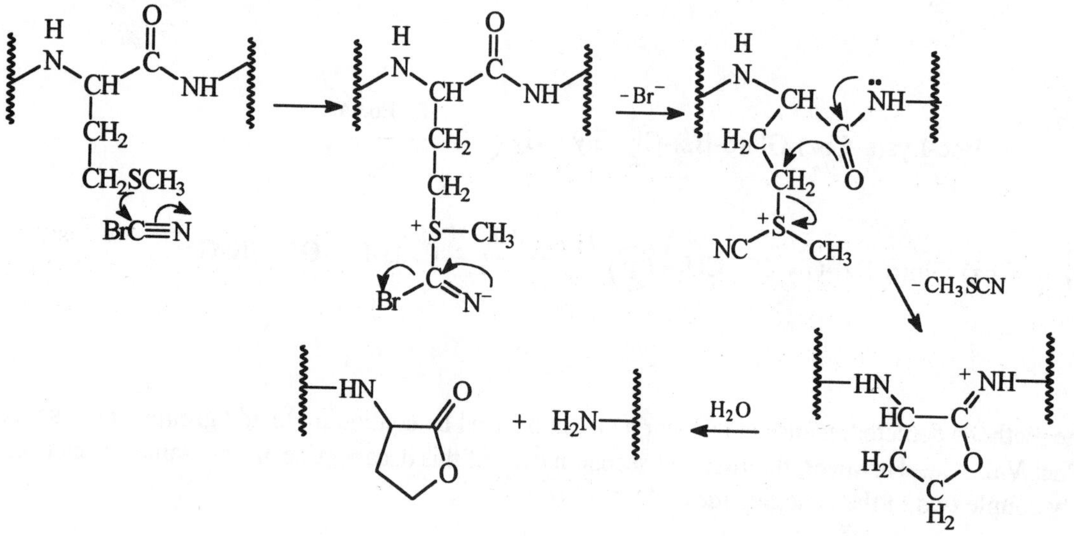

$-Br^-$

$-CH_3SCN$

H_2O

15. a. Lys; Glu-Thr-Ala-Ala-Ala-Lys; Phe-Glu-Arg; Glu-His~ 15 20 25 30 ~Met-Lys; Ser-Arg;

Asn-Leu-Thr-Lys; Asp-Arg; Cys-Lys; Pro-Val~ 45 50 55 60 ~Glu-Lys; Asn-Val-Ala-Cys-Lys;

Asn-Gly~ 70 75 80 85 ~Arg; Glu-Thr-Gly-Ser-Ser-Lys; Try-Pro-Asn-Cys-Ala-Tyr-Lys;

105 110 115 120 124
Thr-Thr-Glu-Ala-Asn-Lys; His~~~~~~~~~~~~~~~~~Val

b. Ten polypeptides: three ending with Phe, six with Tyr, and one with Val (terminal piece). No Trp is present.

c. BrCN cleaves only at Met, of which four are present (two of which are joined). Five pieces will be formed, one of which is cyclized Met itself: H_2N–(γ-lactone ring, C=O, O)

16. I-II-IV-III or I-IV-II-III. II must be the amino end because it is the only fragment that starts with Glu; similarly, III must be the carboxy end. Positions of II and IV are not established.

17. Consider the following logic:

a. T-1 through T-8 contain all 56 amino acids. T-9 is clearly (T-7) - (T-6). T-7 must be a product of abnormal cleavage, since trypsin does not normally hydrolyze at Tyr. Thus, T-9 presumably is the primary hydrolysis product that partially hydrolyzed to T-6 and T-7.

b. Since the protein N-end starts with Thr and the C-end is Cys, the sequence can initially be represented as (T-4)~~~~~~~~~(T-3).

c. Furthermore, this also tells us that the chains from chymotrypsin digestion are in the sequence: (Ch-1)—(Ch-2)—(Ch-3).

d. Comparing the amino acid compositions, we find that Ch-2 is the same as T-7 + Tyr + Ile. Similarly, T-8 is part of Ch-1 (note that there is only one Ala in the protein). Since T-6 follows T-7, we can write the protein as:

(T-9) = (T7)–(T-6)

(T-4)—(T-8)—(T-7)—(T-6)—(T-5,1,2)—(T-3)

e. The only question left is where the single units, T-1 and T-2, fit with relation to T-5. For this answer we turn to the T* series. T*-2 contains the Cys and comes at the end; in fact, the composition of T*-2 is that of T-5 and T-3. Thus, T-1 and T-2 come between (T-6) and (T-5), but with CH_3NCS, Lys has been modified so that it will not cleave; hence, T*-3 must terminate with Asp.

f. We conclude that the complete sequence is:

(T-4)—(T-8)—(T-7)—(T-6)—(T-1)—(T-2)—(T-5)—(T-3)

18.

$^-O_2CCH_2$–NH–C(=O)–CHCH$_2$SH–NH–C(=O)–CH(CO$_2^-$)CH$_2$CH$_2$–$\overset{+}{H_3N}$ ... $\Psi = 180°$, $\Phi = 180°$

$^-O_2CCH_2$–HN–C(=O)–CHCH$_2$SH–NH–C(=O)–CH$_2$CH$_2$CH(CO$_2^-$)$\overset{+}{N}H_3$ ... $\Psi = 0°$, $\Phi = 0°$

19. The most common are α-helix and β-pleated sheet, see figures 29.12 and 29.15 in the text; also extended chains, reverse turns and 3_{10} helices.

20. Naturally-occurring L-cysteine is available and inexpensive. Acylate with acetic anhydride. The remaining acid functions are the $-CO_2H$ and the -SH. Cysteine can be synthesized from serine (Exercise 29.7) reacted with PBr_3 and then NaSH, but this will produce a racemic mixture that must be resolved.

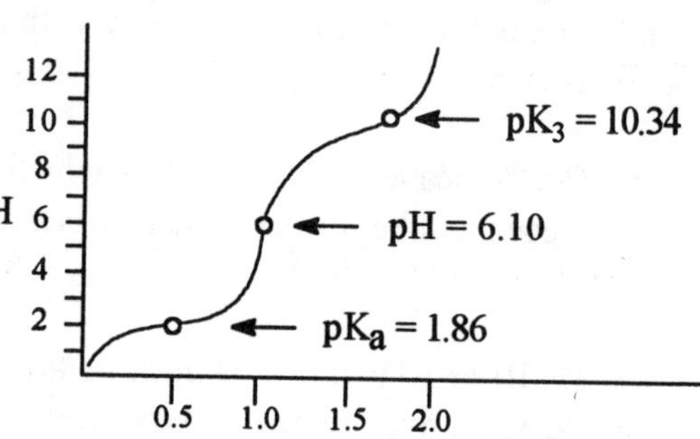

21. a. hexanol

b. tri-*n*-propylamine oxide

$R_3N \longrightarrow R_3N^+-O^-$, $R = CH_3CH_2CH_2-$

c. cyclohexene oxide

d. di-*n*-butyl sulfoxide

$R-S-R \longrightarrow R-S^+(O^-)-R$, $R = CH_3CH_2CH_2CH_2-$

23.

$C_6H_5CH_2CN \rightleftharpoons C_6H_5\bar{C}HCN$ + O=N–OCH$_3$ $\xrightarrow{-OCH_3}$ $C_6H_5C(CN)(H)-N=O$ (B:) + $Cl-C(=O)-Cl$ $\xrightarrow{-BH,\ -Cl}$ $C_6H_5-C(CN)=N-O-C(=O)-Cl$ $\xrightarrow{(CH_3)_3CO^-}$ $C_6H_5-C(CN)=N-O-C(=O)-O-C(CH_3)_3$

The first step will require a base like methoxide to form the α-carbanion. The second step will require a base that will not compete for the phosgene; possibly a tertiary amine. Methyl nitrite is a gas at room temperature, b.p. $-12°$. It can be formed by an S_N2 reaction of NO_2^- and CH_3I.

25. a. Phe-Leu-Gly-Val-Tyr-Ala-Lys-Pro; FLGVYAKP

b. Valine has the R configuration.

c. Proline is the C-terminal amino acid.

d. Phenylalanine is the first amino acid released in an Edmann degradation.

e. The net charge will be +1 because of the lysine residue. At pH 7, the ionized groups will be the $-NH_3^+$ of the N-terminal phenylalanine, the $-CO_2^-$ of the C-terminal proline and the ε-NH_3^+ of the lysine.

29.F. Supplementary Problems

S1. What is the principle ionic form of the dipeptide histidyltyrosine (His-Tyr) at pH 2, 5, 8, and 11?

S2. What is the approximate isoelectric point of the following amino acids?

a. Valine b. Glutamic Acid c. Lysine

S3. Show how to synthesize the following amino acids.

a. 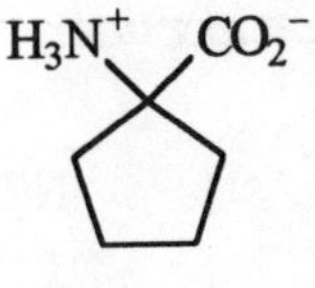 H_3N^+, CO_2^- (cyclopentane ring bearing both groups on one carbon)

b. 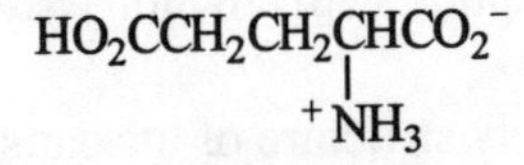 $HO_2CCH_2CH_2CHCO_2^-$ with $^+NH_3$ on the CH

c. 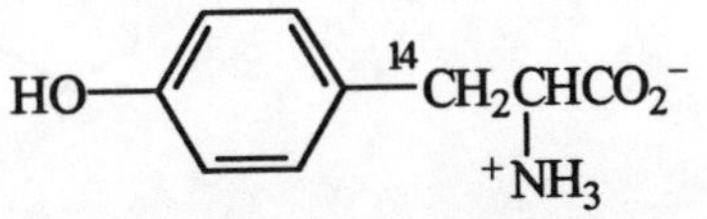 HO–(para-phenylene)–$^{14}CH_2CHCO_2^-$ with $^+NH_3$ on the CH

d. 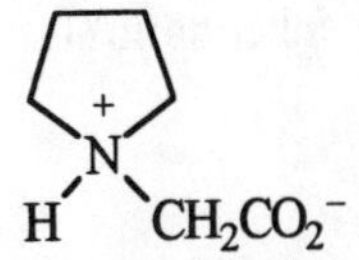 pyrrolidinium ring, $\overset{+}{N}$ bearing H and $CH_2CO_2^-$

(^{14}C available as $Ba^{14}CO_3$ or $Na^{14}CN$)

S4. Outline syntheses of the following tripeptides, using the N-protecting group indicated and appropriate coupling reagents and other protecting groups as necessary.

a. Ser-Leu-Tyr (*t*-Boc)
b. Phe-Ile-Asp (Cbz)
c. Gly-Val-Met (*t*-Boc)

S5. How many isomers are produced in the Strecker synthesis of isoleucine? What methods may be used to separate them from each other?

S6. Conversion an of N-acyl amino acid to the acid chloride often leads to formation of an azlactone and racemization. Write reasonable mechanisms for these reactions.

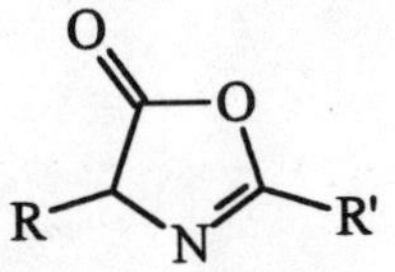

(structure: five-membered ring with C=O, O, C(R'), N, C(R)) an azlactone

S7. Another method for peptide bond formation is the "mixed-anhydride" coupling method, which is illustrated in the example below:

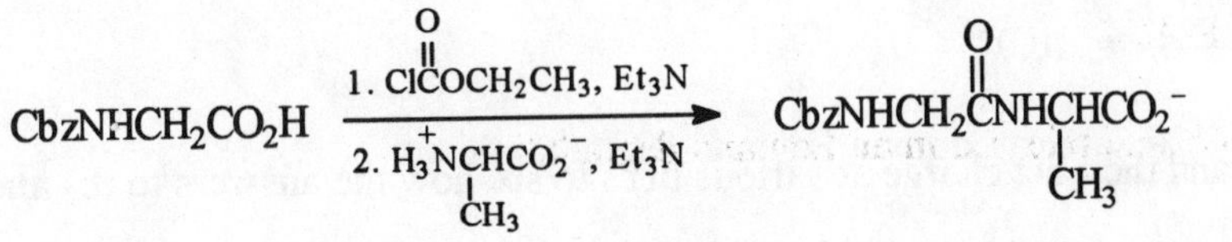

$$CbzNHCH_2CO_2H \xrightarrow[2.\ H_3\overset{+}{N}CH(CH_3)CO_2^-,\ Et_3N]{1.\ ClC(=O)OCH_2CH_3,\ Et_3N} CbzNHCH_2C(=O)NHCH(CH_3)CO_2^-$$

a. Write a balanced equation for this reaction sequence and indiate what intermediates are involved.
b. What would happen if an unprotected amino acid were used as the starting material?
c. What do you expect is the major side product to result from this sequence?

S8. The enkephalins are believed to be the natural compounds in the central nervous system whose activity is imitated by morphine. The amino acid composition of one of the enkephalins is Gly-Gly-Met-Phe-Tyr. Reaction of this pentapeptide with 8-dimethylaminonaphthalenesulfonyl chloride (dansyl chloride) and subsequent acid-catalyzed hydrolysis affords the dansyl derivative of tyrosine as the only modified amino acid. When the enkephaline is treated with cyanogen bromide, no cleavage of the chain takes place. Partial hydrolysis of the pentapeptide with chymotrypsin gives only tryosine, methionine, and a tripeptide.

a. What is the most likely structure of the dansyl derivative of tyrosine?

b. What is the sequence of the enkephalin?

$N(CH_3)_2$

SO_2Cl

Dansyl chloride

29.G. Answers to Supplementary Problems

S1. pH 2:

HN, $\overset{+}{N}$H — $CH_2CHCNHCHCH_2$ (C=O), $^{+}NH_3$, CO_2H — OH

pH 5:

HN, $\overset{+}{N}$H — $CH_2CHCNHCHCH_2$ (C=O), $^{+}NH_3$, CO_2^- — OH

pH 8:

HN, N — $CH_2CHCNHCHCH_2$ (C=O), $^{+}NH_3$, CO_2^- — OH

pH 11:

HN, N — $CH_2CHCNHCHCH_2$ (C=O), NH_2, CO_2^- — O^-

S2. a.
$$\frac{2.29(CO_2H) + 9.72(^{+}NH_3)}{2} = 6.0$$

b.
$$\frac{2.13(\alpha\text{-}CO_2H) + 4.32(\gamma\text{-}CO_2H)}{2} = 3.23$$

c.
$$\frac{9.20(\alpha\text{-}NH_2) + 10.8(\epsilon\text{-}NH_2)}{2} = 10.0$$

Write out the major ionic forms and their net charge at various pH's to see how the answers to (b) and (c) were obtained.

S3. a. Cyclopentanone $\xrightarrow[NH_3]{HCN}$ $\xrightarrow{aq.\ NaOH}$ 1-aminocyclopentanecarboxylate (H_3N^+, CO_2^-)

b. $(EtO_2C)_2CH(NHAc)$ + $CH_2{=}CHCO_2Et$ $\xrightarrow[EtOH]{NaOEt}$ $(EtO_2C)_2C(NHAc)CH_2CH_2CO_2Et$ $\xrightarrow[\Delta]{H_3O^+}$ $^-O_2CCH(^+NH_3)CH_2CH_2CO_2^-$

c. Me_3SiO–C_6H_4–$MgBr$ + $^{14}CO_2$ (from $Ba^{14}CO_3$, H_2SO_4) $\longrightarrow$ $\xrightarrow[H^+]{CH_3OH}$ $\xrightarrow{LiAlH_4}$ Me_3SiO–C_6H_4–$^{14}CH_2OH$ $\xrightarrow{HBr}$ Me_3SiO–C_6H_4–$^{14}CH_2Br$ $\xrightarrow{AcNH\bar{C}(CO_2Et)_2}$ $\xrightarrow[\Delta]{HBr}$ Me_3SiO–C_6H_4–$^{14}CH_2CH(^+NH_3)CO_2^-$ $\xrightarrow{H_3O^+}$

d. Pyrrolidine (NH) + $BrCH_2CO_2CH_3$ $\longrightarrow$ $\xrightarrow{aq.\ NaOH}$ pyrrolidinium (N^+–H, $CH_2CO_2^-$)

S4.
a. (Tyr(OBz)Bz): BzO–C_6H_4–$CH_2CH(NH_2)CO_2Bz$ + Boc-Leu $\xrightarrow{DCC}$ Boc-Leu-Tyr(OBz)Bz $\xrightarrow{TFA}$ Leu-Tyr(OBz)Bz

$\xrightarrow{BocHN\text{–}CH(CO_2H)CH_2OCH_2Ph,\ DCC}$ BzSer(OBz)Leu-Tyr(OBz)Bz $\xrightarrow{H_2/Pt}$ Ser-Leu-Tyr

b. (Asp(Et)$_2$): H_2N–$CH(CO_2Et)CH_2CO_2Et$ + Cbz-Ile $\xrightarrow{DCC}$ Cbz-Ile-Asp(Et)$_2$ $\xrightarrow{H_2/Pt}$ Ile-Asp(Et)$_2$ $\xrightarrow[DCC]{Cbz\text{-}Phe}$

$$\text{Cbz-Phe-Ile-Asp(Et)}_2 \xrightarrow[\text{2. OH}^-]{\text{1. H}_2\text{/Pt}} \text{Phe-Ile-Asp}$$

c. $$\text{Met-Et} + \text{Boc-Val} \xrightarrow{\text{DCC}} \text{Boc-Val-Met-Et} \xrightarrow[(=\text{TFA})]{\text{CF}_3\text{CO}_2\text{H}} \text{Val-Met-Et} \xrightarrow[\text{DCC}]{\text{Boc-Gly}}$$

$$\text{Boc-Gly-Val-Met-Et} \xrightarrow[\text{2. OH}^-]{\text{1. TFA}} \text{Gly-Val-Met}$$

S5.

$$CH_3CH_2CH(CH_3)CHO \xrightarrow[NH_3]{HCN} \xrightarrow{\text{aq. NaOH}}$$

chiral center

Fischer projections (top CO_2^-, bottom CH_2CH_3): (R,R) H / $\overset{+}{N}H_3$, H / CH_3; (S,R) $H_3\overset{+}{N}$ / H, H / CH_3; (S,S) $H_3\overset{+}{N}$ / H, H_3C / H; (R,S) H / $\overset{+}{N}H_3$, H_3C / H

The racemic mixture of RS and SR isomers can be separated from the RR,SS racemate by crystallization or chromotography because they are diastereomeric. Separation of the RS enantiomer from the SR enantiomer, and separation of the RR from the SS enantiomer, require resolution procedures like those discussed in Section 29.4.F.

S6.

R–CH(NHC(=O)R')–COOH $\xrightarrow[-HCl,\ -SO_2]{SOCl_2}$ R–CH(NHC(=O)R')–C(=O)Cl ⟶ tetrahedral oxazoline intermediate ($^-$O, Cl) $\xrightarrow[-H^+]{-Cl^-}$ oxazolone (R, H, R')

oxazolone enolate (O^-) $\overset{\text{base}}{\rightleftharpoons}$ oxazolone (H, R, R') $\overset{H^+}{\rightleftharpoons}$ hydroxy-oxazole (OH)

both planar and achiral

S7. a.

$$\text{CbzNHCH}_2\text{CO}_2^-\ \text{Et}_3\overset{+}{\text{N}}\text{H} \xrightarrow{\text{ClC(=O)OEt}} \text{CbzNHCH}_2\text{C(=O)OC(=O)OEt} + \text{Et}_3\overset{+}{\text{N}}\text{H Cl}^-$$

Major | Minor

$H_2\ddot{N}CH(CH_3)CO_2^-\ Et_3\overset{+}{N}H$

$$\text{CbzNHCH}_2\text{C(=O)NHCH(CH}_3\text{)CO}_2^-\ \text{Et}_3\overset{+}{\text{N}}\text{H} + CO_2 + EtOH$$

b. $H_2NCH_2CO_2^-\ Et_3\overset{+}{N}H \xrightarrow{ClCOEt\ (C{=}O)} EtOC(=O)NHCH_2CO_2H + Et_3\overset{+}{N}H\ Cl^-$

c. $EtOC(=O)NHCH(CH_3)CO_2^-$, from attack at the wrong carbonyl of the mixed anhydride.

S8. a. $H\overset{+}{N}(CH_3)_2$ (naphthalene ring) $O{=}S{=}O$ $NHCH(CO_2^-)CH_2$—(benzene ring)—OH

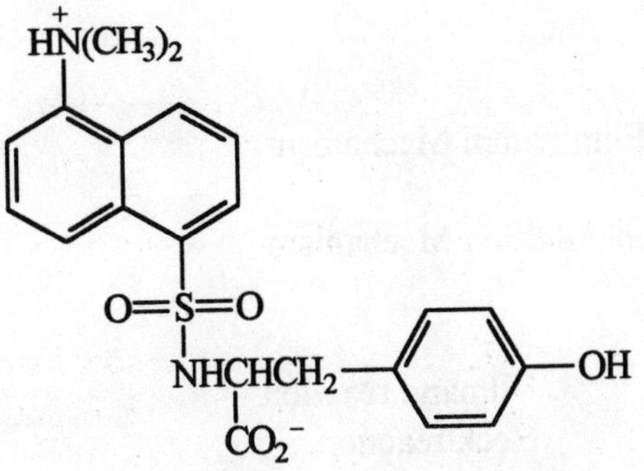

b. 1. Enkephalin $\xrightarrow[\text{chloride}]{\text{dansyl}}\ \xrightarrow{H_3O^+}$ modified Tyr: Tyr is N-terminal residue.

2. Enkephalin $\xrightarrow{\text{BrCN}}$ no cleavage: Met must be C-terminal, (otherwise cleavage would have occurred).

3. Enkephalin $\xrightarrow{\text{chymotrypsin}}$ Tyr, tripeptide, Met: C-terminal amino acid of tripeptide must be Phe.

4. Therefore the sequence is: Tyr-Gly-Gly-Phe-Met

30. AROMATIC HALIDES, PHENOLS, PHENYL ETHERS AND QUINONES

30.A. Chapter Outline and Important Terms Introduced

30.1 Introduction

benzyl vs. aryl halides

30.2 Preparation of Halobenzenes

electrophilic aromatic substitution (see Section 23.1)
substitution of arenediazonium salts (see Section 25.5)

30.3 Reactions of Halobenzenes

A. Nucleophilic Aromatic Substitution: The Addition-Elimination Mechanism
   addition-elimination vs. elimination-addition
B. Nucleophilic Aromatic Substitution: The Elimination-Addition Mechanism
   benzyne
C. Metallation
   transmetallation
   Wurtz-Fittig reaction
   Ullmann reaction
   Heck reaction

30.4 Nomenclature of Phenols and Phenyl Ethers

phenol
(benzenol)

phenyl ethers
(alkoxyarenes)

30.5 Preparation and Properties of Phenols and Phenyl Ethers

A. Preparation of Phenols
B. Acidity of Phenols ($pK_a = 10$)
   effects of substitution
C. Preparation of Ethers (Williamson ether synthesis)

30.6 Reactions of Phenolate Ions

A. Halogenation
B. Addition to Aldehydes
   Bakelite
C. Kolbe synthesis
D. Reimer-Tiemann Reaction
E. Diazonium coupling

30.7 Reactions of Phenols and Phenyl Ethers

A. Esterification
   requires acid chloride or anhydride

B. Electrophilic Substitutions on Phenols and Phenyl Ethers
   highly activated system — phthaleins
   chelation effects in orientation — Fries rearrangement

C. Reactions of Ethers
   Claisen rearrangement — Cope rearrangement

30.8 Quinones

A. Nomenclature
   benzoquinone, 1,2-naphthoquinone, etc.

B. Preparation

C. Reduction-Oxidation Equilibria
   reversibility of electron transfer
   dependence of reduction potentials on substituents
   radical anions, antioxidants

D. Charge-Transfer Complexes
   quinhydrone — donor-acceptor complexes

E. Reactions of Quinones
   addition — Diels-Alder reaction

30.B. Important Reactions Introduced

Nucleophilic aromatic substitution

Addition-elimination mechanism (30.3.A)

Equation: Nu^- + (X, Y-disubstituted benzene) ⇌ (Nu, X addition anion intermediate, Y) ⇌ (Nu, Y-disubstituted benzene) + X^-

Generality: X = halogen, SO_3^-, or other good leaving group, Y = electron-withdrawing group in *o*- or *p*-positions, such as NO_2, COR, CO_2R, etc.
$Nu:^-$ = RO^-, other strong nucleophile

Key features: the more Y groups in *o*- and *p*-positions, the easier the reaction is

Elimination-addition mechanism (30.3.B)

Equation: (X, R-disubstituted benzene) + Nu^- ⟶ [benzyne intermediate, R] ⇌ (Nu, R para product) + (Nu, R meta product)

benzyne intermediate

Generality: X = halogen; $Nu:^-$ = RO^-, $^-:NH_2$, or other strong base

Key features: strongly basic conditions
mechanism involves benzyne intermediate, therefore mixtures of isomers possible with substituted aromatic compounds

Metallation of aryl halides (30.3.C)

Equation: $ArX + M \longrightarrow ArMX$ or ArM

Generality: Ar must contain no reactive groups, such as NO_2, COR, SO_3R, CN, OH, NH_2, etc.
X = Br, I (Cl possible in THF solvent)
M = 2 Li, 2 Na, Mg, etc.

Key features: ordinary formation of Grignard reagent, aryllithium, etc.

Transmetallation of aryl halides (30.3.C)

Equation: $ArX + RM \longrightarrow ArM + RX$

Generality: Ar must contain no reactive groups, such as NO_2, COR, SO_3R, CN, OH, NH_2, etc.
X = Br, I; RM = alkyllithium

Key features: reaction is poor with X = Cl

Aryl coupling reactions (30.3.C)

Equation:

$$ArX + RX' \xrightarrow{[M]} ArR + MXX'$$

$$2\ ArLi \xrightarrow{CuX} Ar_2CuLi \xrightarrow{O_2} ArAr$$

Generality: Wurtz-Fittig reaction: M = 2 Na; Ar contains no reactive groups; RX = 1° or 2° alkyl bromide or iodide
Ullmann reaction: M = Cu powder; RX = **aryl**-Cl, -Br or -I; reaction is better if Ar contains electron withdrawing groups (e.g. NO_2 or CN); enhanced by ultrasound
cuprate coupling: Ar contains no reactive groups

Key features: Ullmann coupling is the best way to make functionalized biaryls
Ullmann coupling is probably a radical process

Heck Reaction (30.3.C)

Equation: $$ArX + CH_2{=}CHCO_2R \xrightarrow[R_3N,\ 100°]{Pd(OAc)_2} ArCH{=}CHCO_2R$$

Generality: Substituents like NO_2 do not interfere
Heterocyclic, benzyl and vinyl halides can be used as well as aryl
X = I, Br.

Oxidation of cumene (30.5.A)

Equation:

$$C_6H_5CH(CH_3)_2 \xrightarrow{O_2} \xrightarrow{H^+} C_6H_5OH + CH_3CCH_3 \text{ (C=O)}$$

Key features: important industrial process for the preparation of phenol and acetone
free radical reaction

Halogenation of phenols (30.6.A)

Equation:

$$C_6H_5O^- \xrightarrow[\text{fast}]{4\,X_2} \text{2,4,4,6-tetrahalocyclohexa-2,5-dienone} \xrightarrow{NaHSO_3} \text{2,4,6-}X_3C_6H_2OH$$

Generality: X = Cl, Br, or I

Key features: under basic conditions, form tetrahalocyclohexadienone
stepwise halogenation is only possible under acidic conditions (30.7.B)

Condensation of phenolate ions with aldehydes (30.6.B)

Equation:

$$C_6H_5OH + H_2C{=}O \xrightarrow{\text{base}} p\text{-}HOC_6H_4CH_2OH \longrightarrow \textit{Bakelite plastic}$$

Key features: polymeric compound obtained is Bakelite, the first commercial plastic

Kolbe synthesis (30.6.C)

Equation:

$$C_6H_5OH + CO_2 \longrightarrow o\text{-}HOC_6H_4CO_2^- \xrightarrow[\Delta]{K_2CO_3} p\text{-}HOC_6H_4CO_2^-$$

Key features: the product of kinetic control (first formed product) is the *o*-isomer
the product of thermodynamic control (more stable product) is the *p*-isomer
o-hydroxybenzoic acid (salicylic acid) is starting material for aspirin

Reimer-Tiemann reaction (30.6.D)

Equation:

OH + $CHCl_3$ $\xrightarrow{OH^-}$ (OH, CH=O)

Key features: mechanism involves reaction of $:CCl_2$ to give dichloromethyl intermediate

Diazonium coupling (30.6.E)

Equation:

O^- + $^+N_2$ ⟶ HO, N=N

Key features: see also Section 25.5.G
method for synthesis of azo dyes

Synthesis of phthaleins (30.7.B)

Equation:

2 (OH) + (O, O, O) $\xrightarrow{H_2SO_4}$ HO, OH, O, O

Key features: phthaleins are important dyes, pH indicators, and laxatives....

Synthesis of fluoresceins (30.7.B)

Equation:

2 (OH, OH) + (O, O, O) $\xrightarrow[\Delta]{H_2SO_4}$ HO, O, O, CO_2H

Fries rearrangement (30.7.B)

Equation:

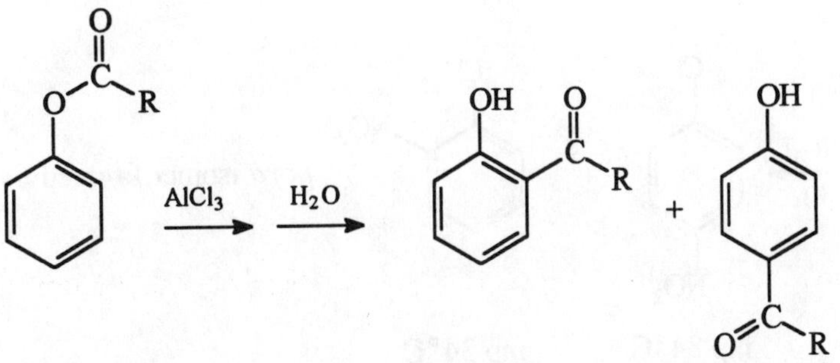

Key features: best method for Friedel-Crafts acylation of phenols

Aromatic Claisen rearrangement (30.7.C)

Equation:

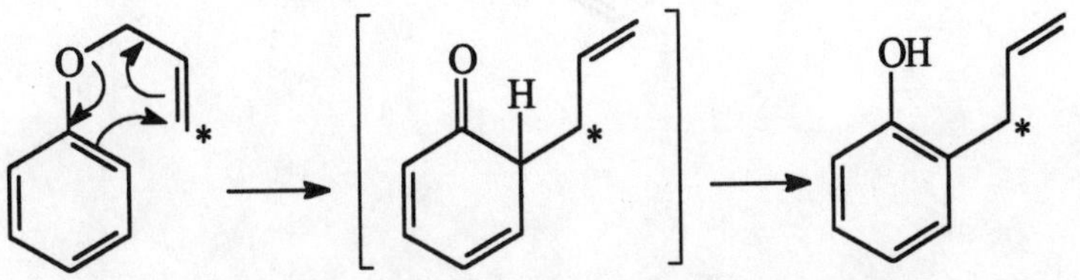

Key features: reaction proceeds with allylic rearrangement of allylic group
if both *o*-positions are blocked, *p*-substituted products from double rearrangement (Claisen plus Cope) are seen

Preparation of quinones by oxidation of hydroquinone derivatives (30.8.B)

Equation:

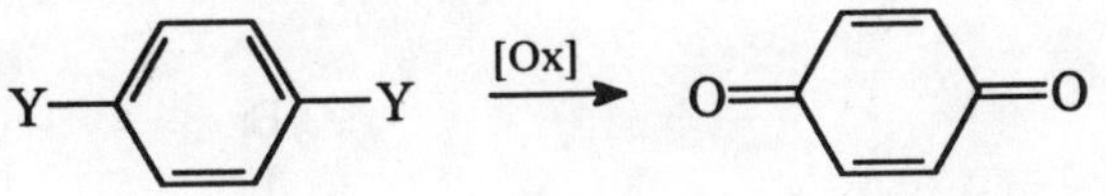

Generality: Y = combinations of OH, OR, NR_2, either *o*- or *p*-substituted
[Ox] = air, HNO_3, Cr(VI) salts, Fe^{+3}, N_2O_4, $NaClO_3/VO_5$, etc.

Key features: redox couples involving quinone hydroquinone are important in biological systems and in photography
quinones and hydroquinones also are important in forming charge-transfer complexes

Quinone addition reactions (30.8.E)

Equation:

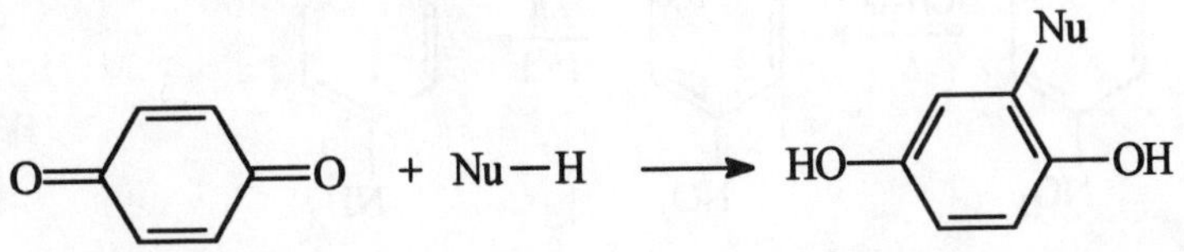

Key features: if Nu is electron releasing, hydroquinone product is oxidized by quinone starting material

30.D. Answers to Exercises

30.1 Benzene $\xrightarrow[FeCl_3]{Cl_2}$ $\xrightarrow[H_2SO_4]{HNO_3}$ p-chloronitrobenzene (mp 84°C) + o-chloronitrobenzene (mp 34°C)

para isomer favored 69%; see Exercise 23.9

p-chloronitrobenzene $\xrightarrow[FeBr_3]{Br_2}$ $\xrightarrow[HCl]{Zn}$ $\xrightarrow[H_2SO_4]{NaNO_2}$ $\xrightarrow{H_3PO_2}$ o-bromochlorobenzene

p-chloronitrobenzene $\xrightarrow[HCl]{Zn}$ $\xrightarrow[HBr]{NaNO_2}$ $\xrightarrow{CuBr}$ p-bromochlorobenzene

Benzene $\xrightarrow[H_2SO_4]{HNO_3}$ $\xrightarrow[FeBr_3]{Br_2}$ m-bromonitrobenzene $\xrightarrow[HCl]{Zn}$ $\xrightarrow{HNO_2}$ $\xrightarrow[\Delta]{CuCl}$ m-bromochlorobenzene

30.2 Chlorobenzene $\xrightarrow[H_2SO_4]{HNO_3}$ p-chloronitrobenzene $\xrightarrow[\Delta]{CH_3O^-}$ p-nitroanisole (OCH_3, NO_2) $\xrightarrow[HCl]{Zn}$ p-anisidine (OCH_3, NH_2)

30.3

CH3, Br → KNH2 / NH3 → [CH3 benzyne] → CH3, NH2 + CH3, NH2

CH3, Br → KNH2 / NH3 → [CH3 benzyne] [CH3 benzyne] → CH3, NH2 + CH3, NH2 + CH3, NH2

CH3, Br → KNH2 / NH3 → [CH3 benzyne] → CH3, NH2 + CH3, H2N

30.4

CH3, Br → Mg → CO2 → H+ → CH3, CO2H

CH3, Br → 2 Li → CuBr → O2 → H3C–C6H4–C6H4–CH3

CH3, Br → Na / BrCH2CH2CH3 → CH3, CH2CH2CH3

30.5

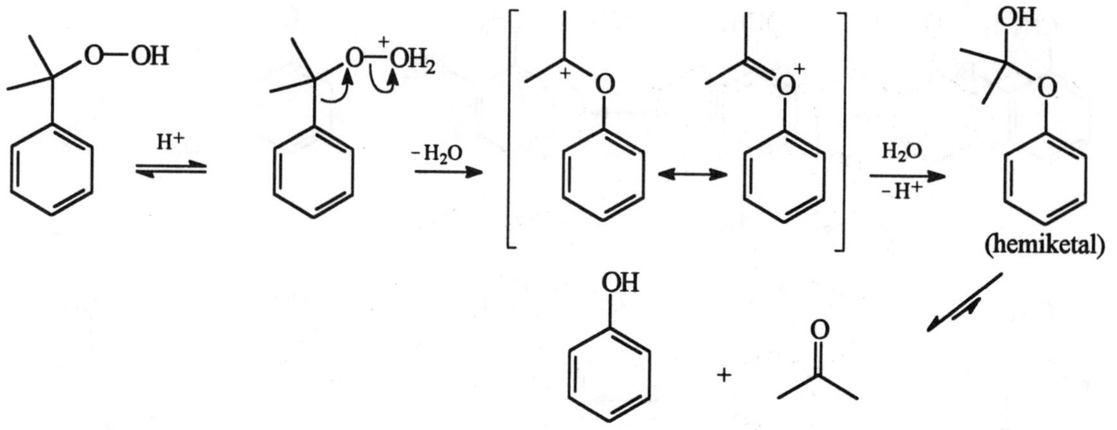

(hemiketal)

30.6

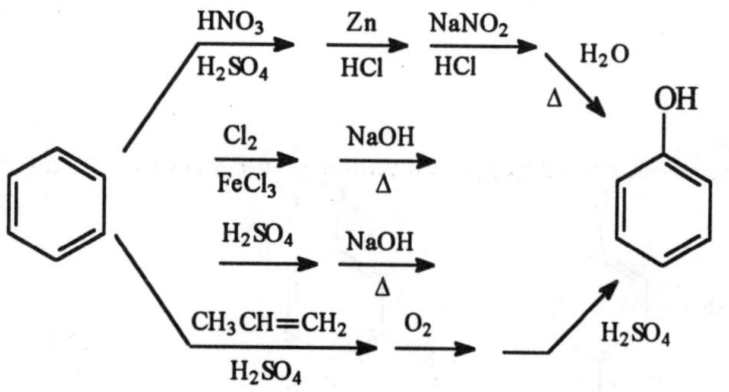

30.7. Because the acidity of phenols and anilinium ions is concerned with the loss of a proton from a hetero atom attached to an aromatic ring, similar substituent effects are seen in both cases.

30.8 pK_a of phenol = 10.0; so that $K = \frac{[C_6H_5O^-][H^+]}{[C_6H_5OH]} = 10^{-10}$

Assuming that a negligible amount of the phenol ionizes, and that all protons arise from the ionization of phenol:

$[C_6H_5O^-] = [H^+]$, and $[C_6H_5OH] = 0.1$, so that $K = \frac{[H^+]^2}{0.1} = 10^{-10}$

Therefore $[H^+] = 3.16 \times 10^{-6}\,M$; pH = 5.5

When [HA] and [A⁻] are equal, the pH is the same as the pK_a of HA. Therefore, the pH is 10.0.

30.9 $C_6H_5OH + BrCH_2CH{=}CH_2 + NaOH \longrightarrow C_6H_5OCH_2CH{=}CH_2 + NaBr + H_2O$

30.10

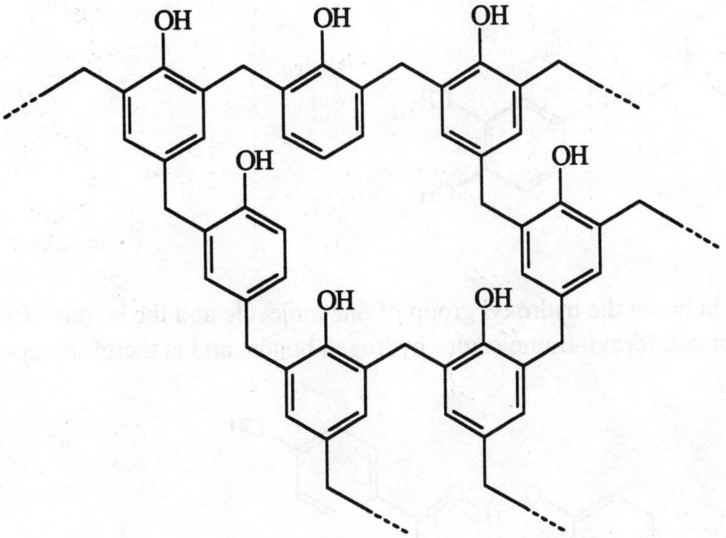

30.11 The hydrolysis reaction involves an elimination-addition mechanism which is not possible for the non-aromatic *p*-product:

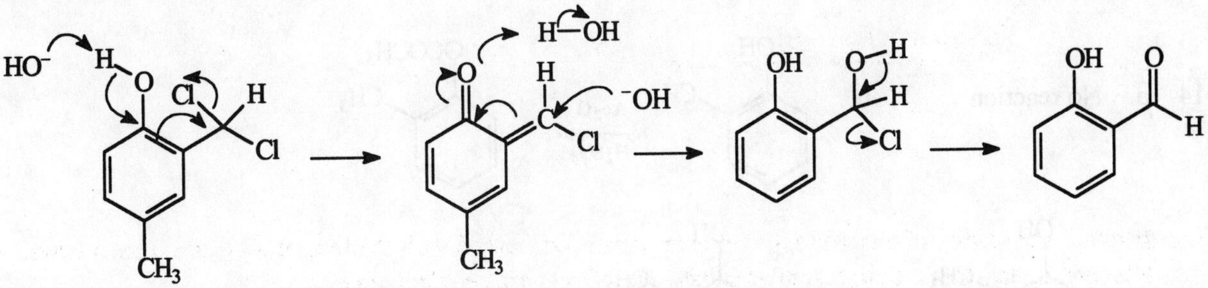

30.12 a. No reaction

b. 1. Br_2, H_2O 2. $NaHSO_3$

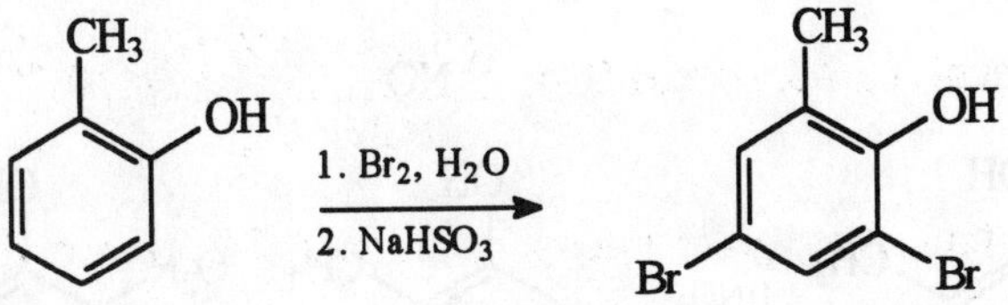

c. KOH, $CHCl_3$

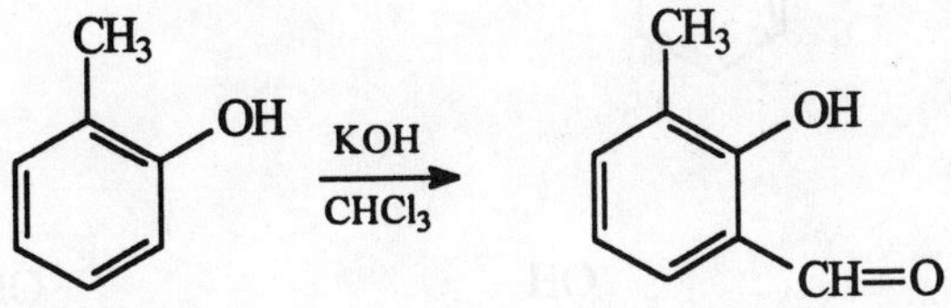

d. NaOH, CO_2, 150°

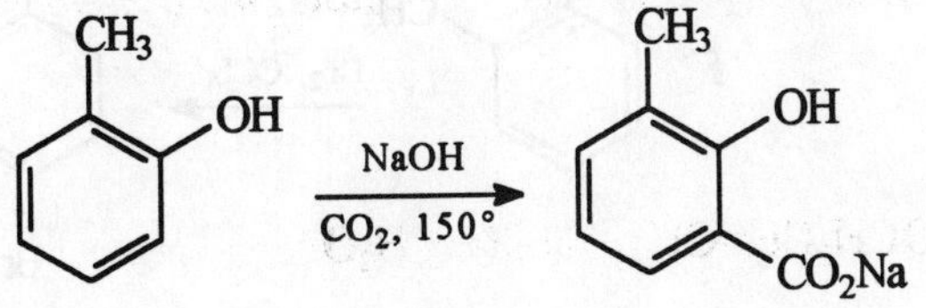

e. 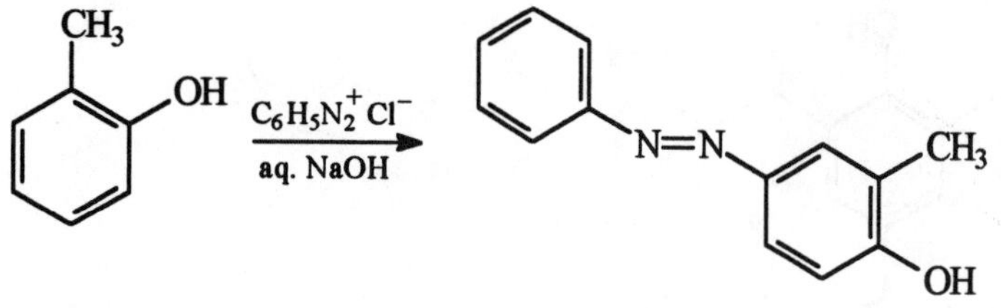 [reaction scheme: 2-methylphenol with $C_6H_5N_2^+Cl^-$, aq. NaOH → 4-(phenylazo)-2-methylphenol; labels: CH_3, OH, N=N]

30.13 The *p*-isomer has hydrogen bonds between the hydroxyl group of one molecule and the ketone of another molecule. In contrast, the *o*-isomer can form intramolecular hydrogen bonds, and is therefore more volatile.

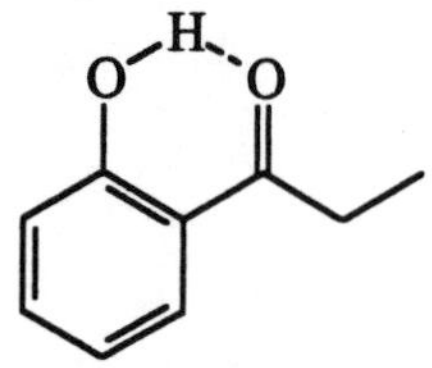

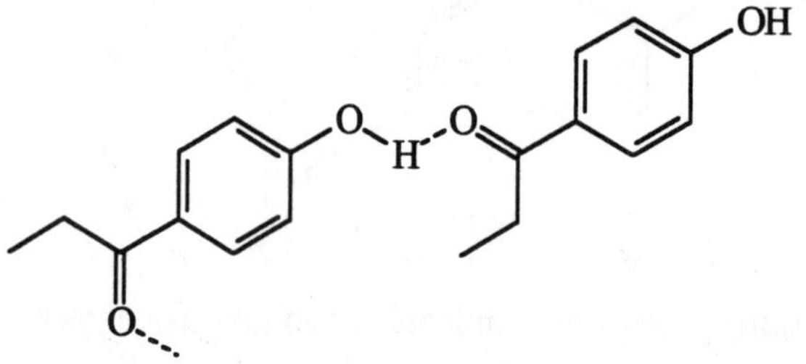

[structures: intramolecular O–H···O hydrogen bond in the *o*-isomer; intermolecular O–H···O hydrogen bonds between *p*-isomer molecules]

30.14 a. No reaction

b. 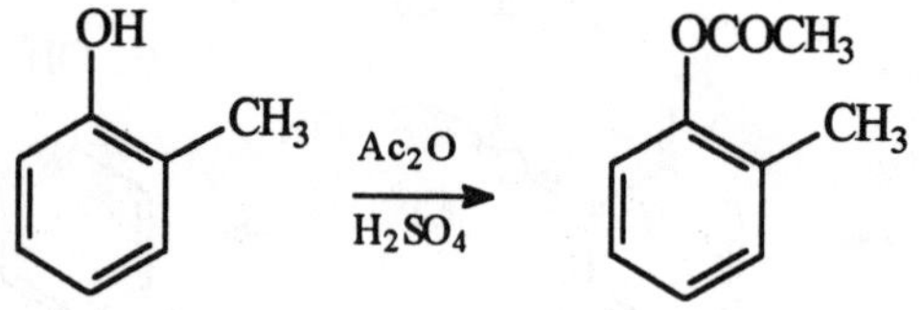 [2-methylphenol, Ac_2O, H_2SO_4 → $OCOCH_3$ / CH_3 product]

c. 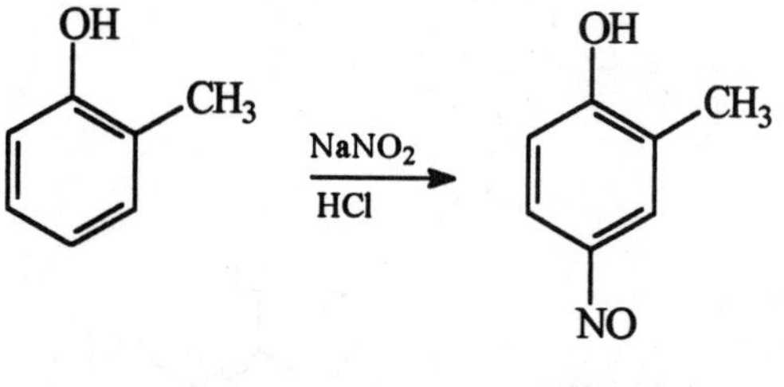 [2-methylphenol, $NaNO_2$, HCl → 2-methyl-4-nitrosophenol (OH, CH_3, NO)]

d. 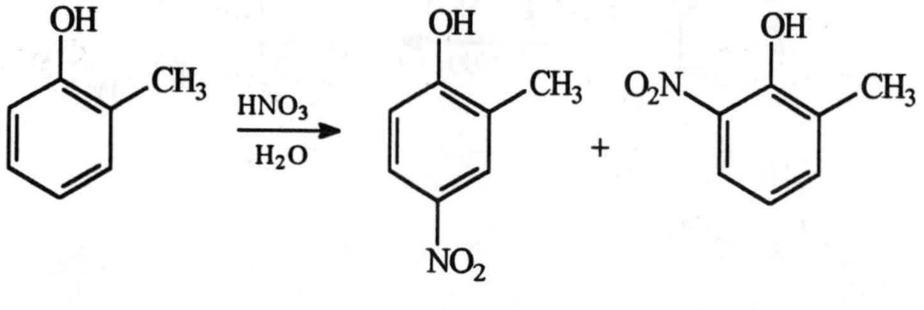 [2-methylphenol, HNO_3, H_2O → 2-methyl-4-nitrophenol (OH, CH_3, NO_2) + 2-methyl-6-nitrophenol (O_2N, OH, CH_3)]

e. 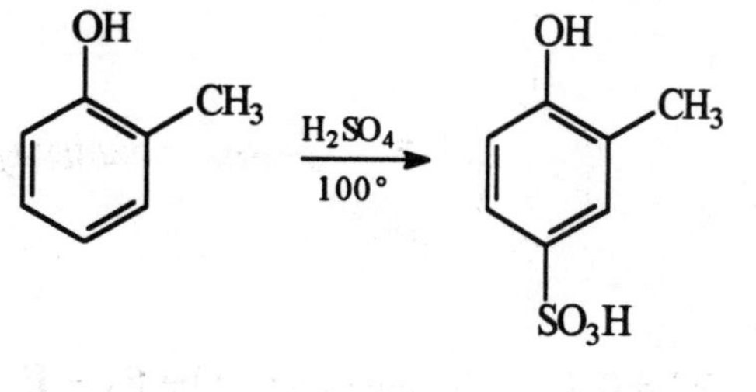 [2-methylphenol, H_2SO_4, 100° → OH, CH_3, SO_3H product]

f. 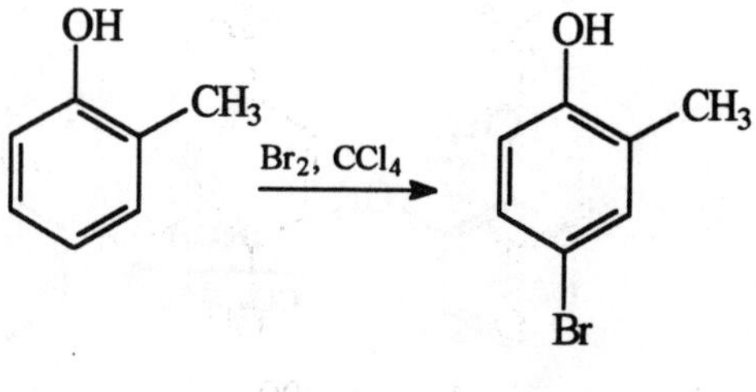 [2-methylphenol, Br_2, CCl_4 → 4-bromo-2-methylphenol (OH, CH_3, Br)]

g. o-cresol (OH, CH_3) $\xrightarrow[BF_3]{Ac_2O}$ 4-hydroxy-3-methylacetophenone (OH, CH_3, $COCH_3$)

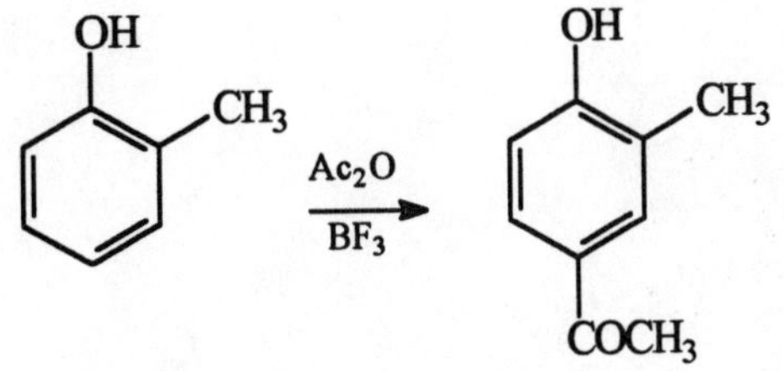

h. o-cresol (OH, CH_3) $\xrightarrow[2.\ AlCl_3,\ \Delta]{1.\ Ac_2O}$ 4-hydroxy-3-methylacetophenone (OH, CH_3, $COCH_3$) + 2-hydroxy-3-methylacetophenone ($CH_3C(=O)$, OH, CH_3)

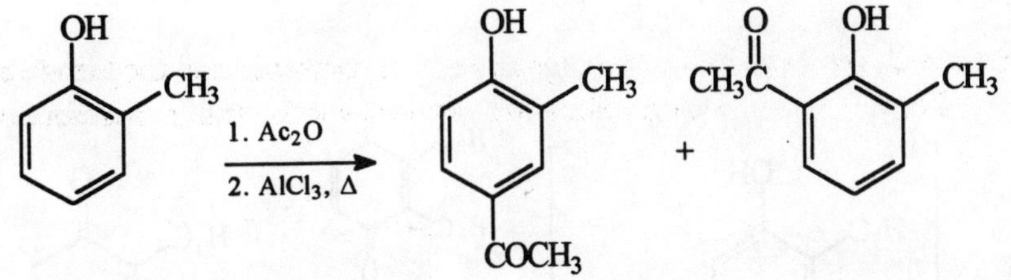

i. o-cresol (OH, CH_3) + phthalic anhydride $\xrightarrow{H_2SO_4,\ \Delta}$ o-cresolphthalein (HO, CH_3, CH_3, OH; lactone O, O)

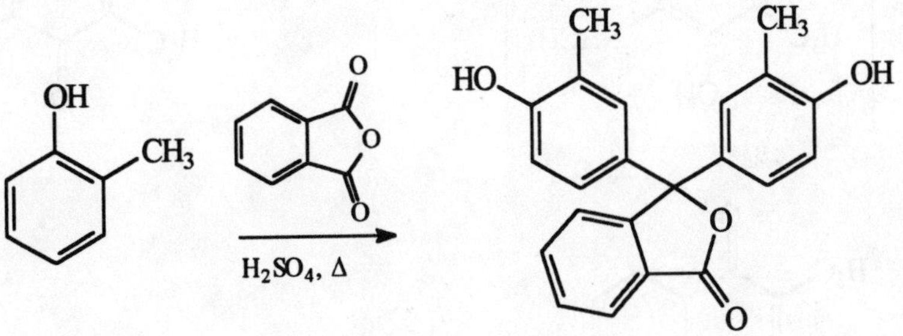

30.15

a. p-cresol (OH, CH_3) $\xrightarrow{CH_3CH{=}CHCH_2Br}$ p-tolyl crotyl ether (O–$CH_2CH{=}CHCH_3$, CH_3) $\xrightarrow[200°]{\Delta}$ 2-(1-methylallyl)-4-methylphenol (OH, $CH(CH_3)CH{=}CH_2$, CH_3)

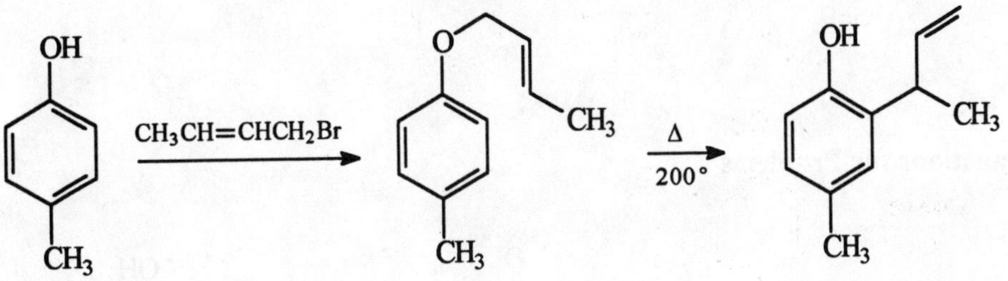

b. p-cresol (OH, CH_3) $\xrightarrow[Me_2SO_4]{NaOH}$ p-methylanisole (OCH_3, CH_3) $\xrightarrow[Ac_2O,\ 10°]{HNO_3}$ 4-methyl-2-nitroanisole (OCH_3, NO_2, CH_3) $\xrightarrow[100°]{HBr}$ 4-methyl-2-nitrophenol (OH, NO_2, CH_3) + CH_3Br

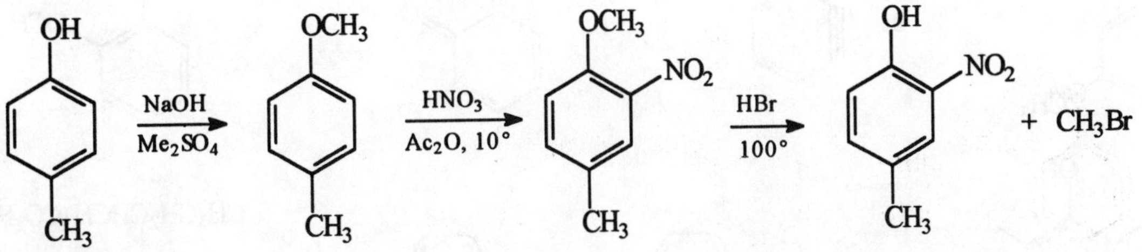

30.16

p-benzoquinone (O, O) + chlorohydroquinone (OH, OH, Cl) $\rightleftharpoons$ hydroquinone (OH, OH) + chloro-p-benzoquinone (O, O, Cl) ← (stronger oxidizing agent)

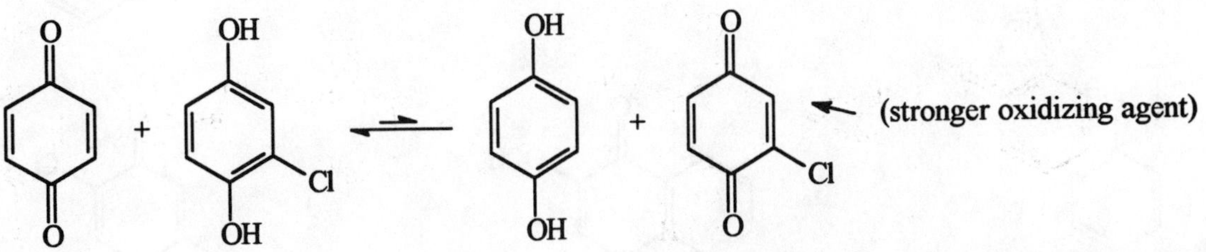

Reduction potential: $E_1^\circ = 0.699$

$E_2^\circ = 0.713$ cell potential will be $E_2^\circ - E_1^\circ = 0.014$ volt

30.17

a. 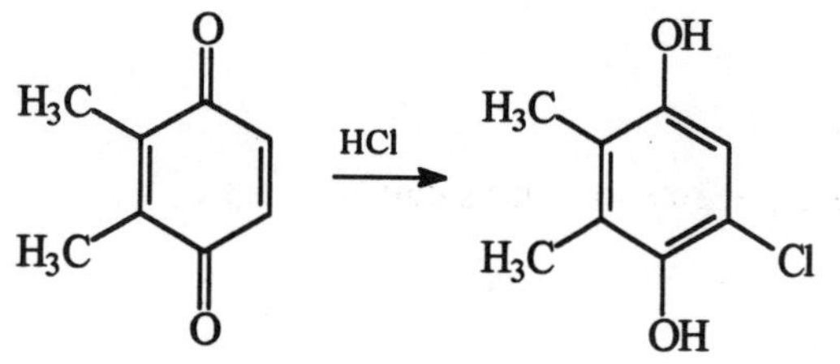 [structure: 2,3-dimethyl-1,4-benzoquinone] $\xrightarrow{HCl}$ [structure: 5-chloro-2,3-dimethylhydroquinone]

b. 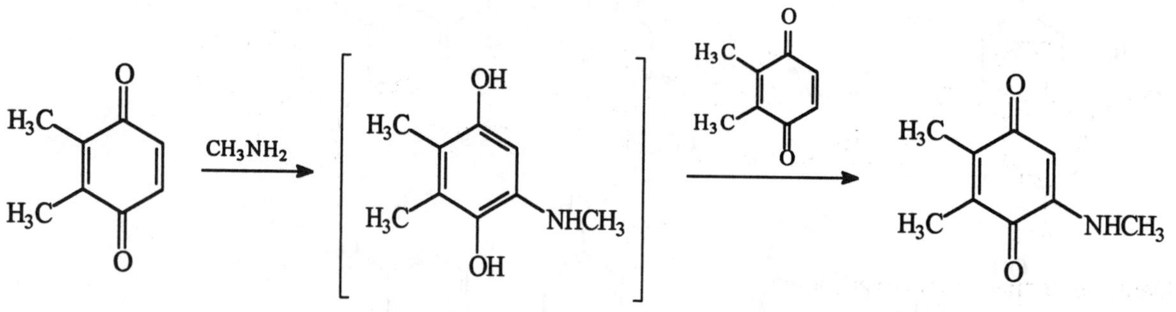 [structure: 2,3-dimethyl-1,4-benzoquinone] $\xrightarrow{CH_3NH_2}$ [[structure: 2,3-dimethyl-5-(methylamino)hydroquinone]] $\xrightarrow{\text{2,3-dimethyl-1,4-benzoquinone}}$ [structure: 2,3-dimethyl-5-(methylamino)-1,4-benzoquinone]

c. 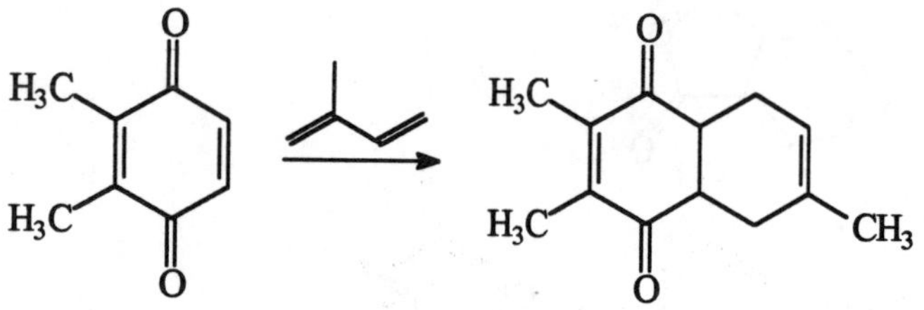 [structure: 2,3-dimethyl-1,4-benzoquinone] $\xrightarrow{\text{isoprene}}$ [structure: Diels–Alder adduct, dimethyl tetrahydronaphthoquinone with CH_3]

30.E. Answers and Explanations for Problems

1. a. 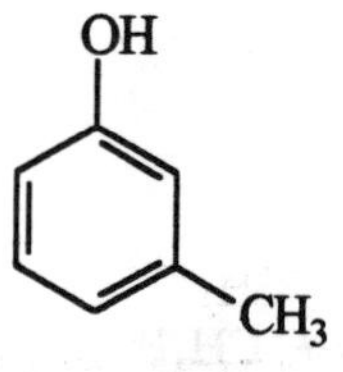 [structure: 3-methylphenol (OH, CH_3)]

b. 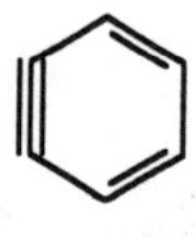 [structure: benzyne]

c. 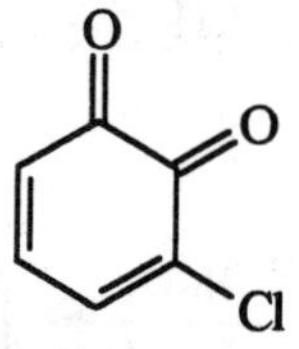 [structure: 3-chloro-1,2-benzoquinone]

d. 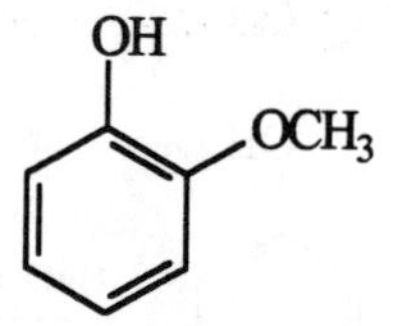 [structure: 2-methoxyphenol (OH, OCH_3)]

e. 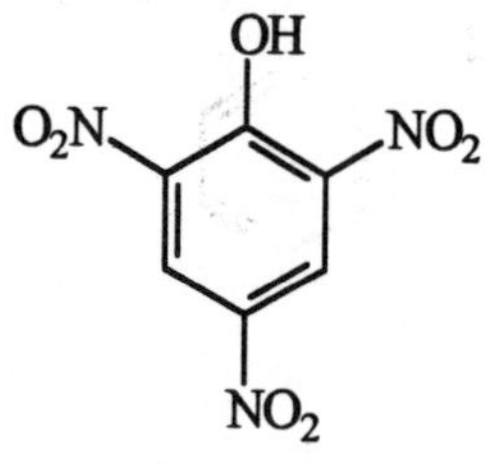 [structure: 2,4,6-trinitrophenol (OH, O_2N, NO_2, NO_2)]

f. 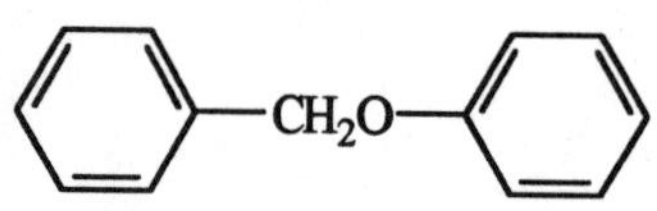 [structure: C_6H_5–CH_2O–C_6H_5]

g. 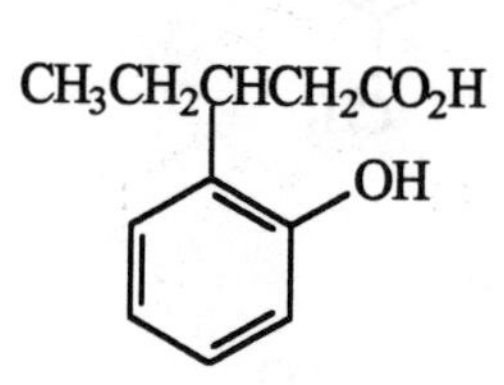 $CH_3CH_2CHCH_2CO_2H$ with 2-hydroxyphenyl on C-3 (OH)

h. 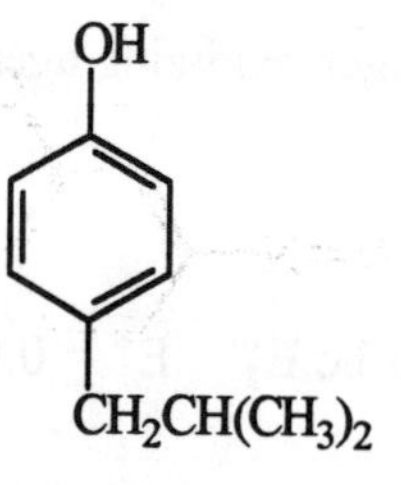 [structure: 4-isobutylphenol (OH, $CH_2CH(CH_3)_2$)]

i. 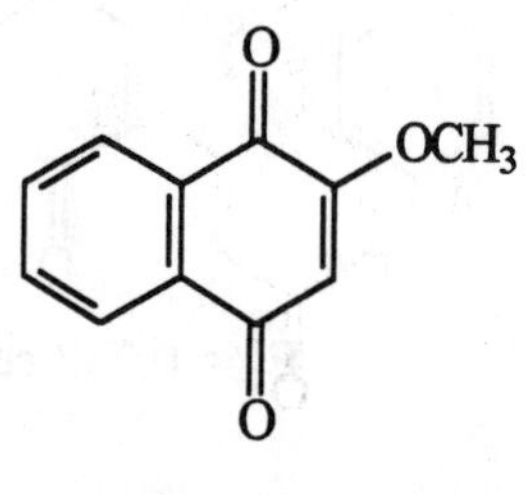 [structure: 2-methoxy-1,4-naphthoquinone (OCH_3)]

j. 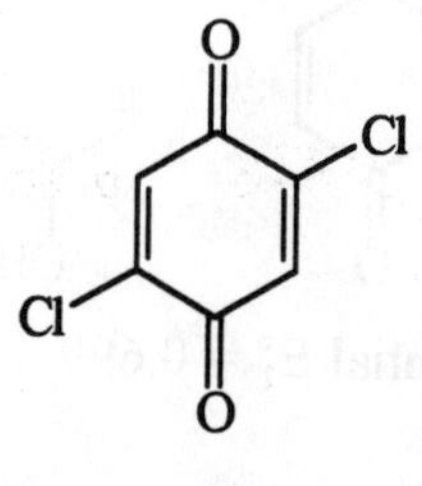 [structure: 2,5-dichloro-1,4-benzoquinone (Cl, Cl)]

2.

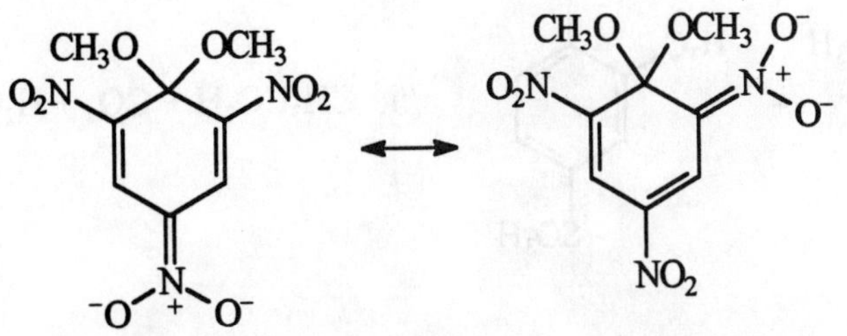

CH_3O OCH_3; O_2N, NO_2; $=N^+(O^-)_2$ ⟷ CH_3O OCH_3 O^-; O_2N, $=N^+ O^-$; NO_2 , etc.; many structures can be written

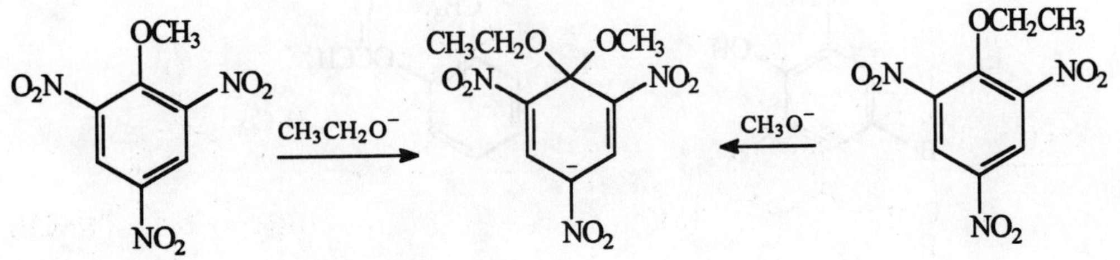

OCH_3; O_2N, NO_2; NO_2 $\xrightarrow{CH_3CH_2O^-}$ CH_3CH_2O OCH_3; O_2N, NO_2; NO_2 $\xleftarrow{CH_3O^-}$ OCH_2CH_3; O_2N, NO_2; NO_2

(same compound from either route)

3. OH * $\xleftarrow{1-X}$ Cl * $\xrightarrow{X}$ * (benzyne) ⟶ OH * + OH *

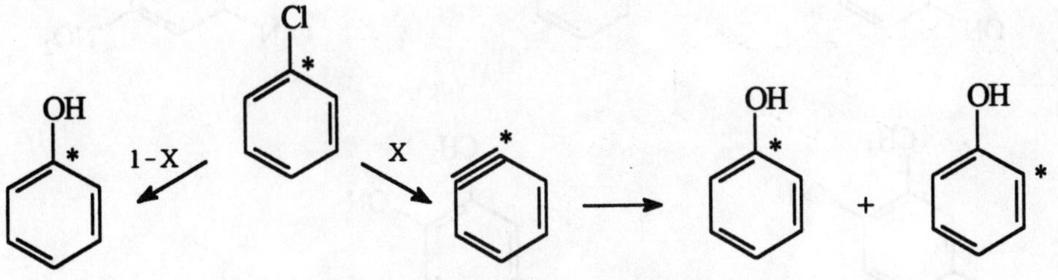

Let x = fraction that proceeds via the benzyne intermediate
Percent of label in the 2-position = 100(x/2) = 42%; therefore x = 0.84
Only 16% proceeds via normal nucleophilic substitution

4. a. CH_3, Cl, D b. NH_2, NO_2, NO_2 c. NO_2, Cl, OCH_3 d. CH_3 $\overset{|}{C}HCH_2CH_2OH$

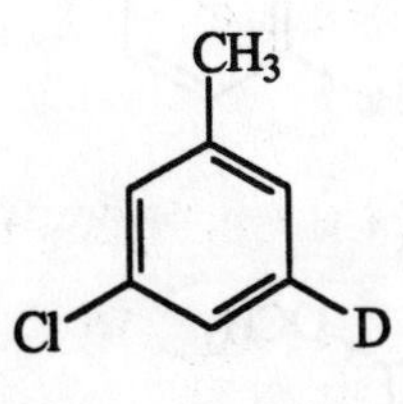

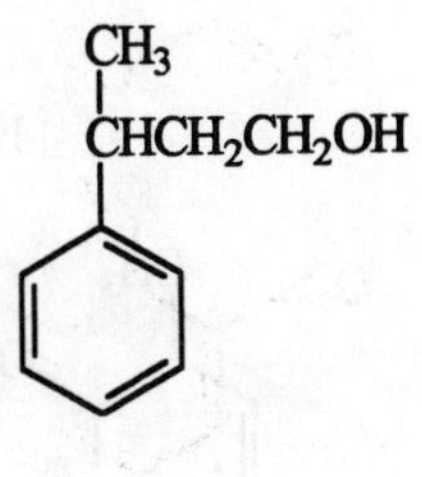

5. a. CH_3, OCH_3 b. and c. no reaction d. O, CH_3, O e. CH_3, –N=N–, OH

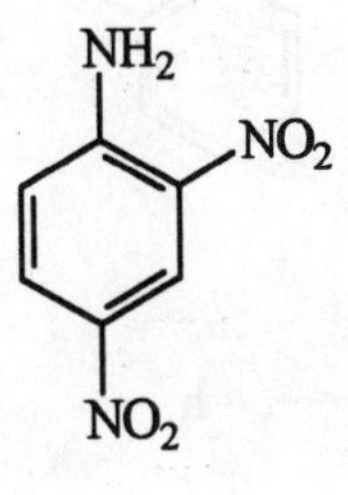

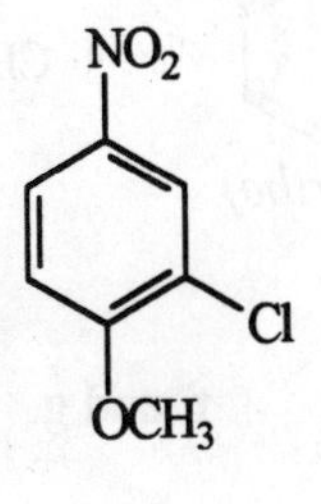

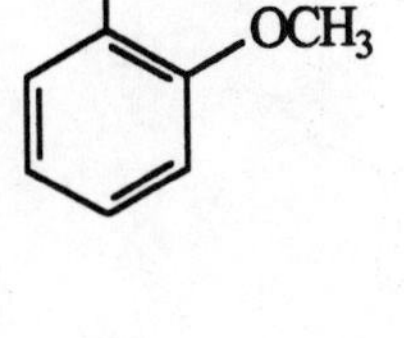

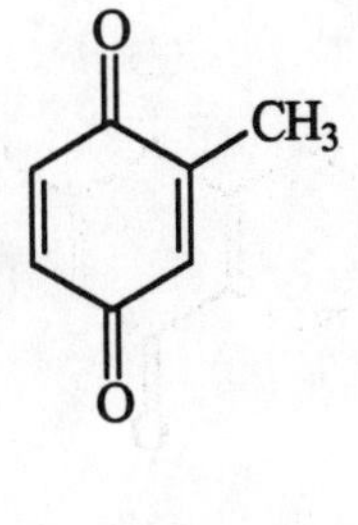

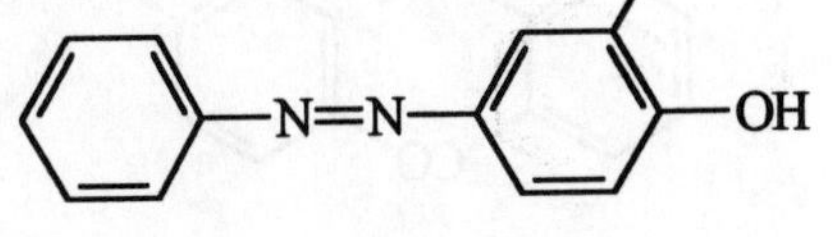

f. and g. no reaction

h. (structures: 2-hydroxy-3-methylbenzenesulfonic acid with OH, H_3C, SO_3H labels) + (4-hydroxy-3-methylbenzenesulfonic acid with OH, H_3C, SO_3H labels)

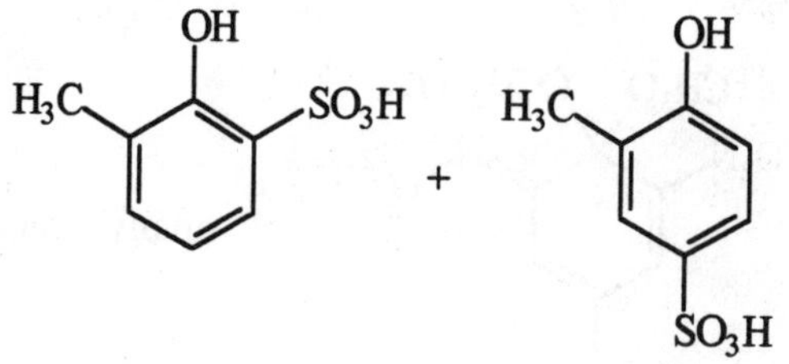

i. $CH_3CO_2H + CO_2 + H_2O$

j. (structure: O, O, CH_3)

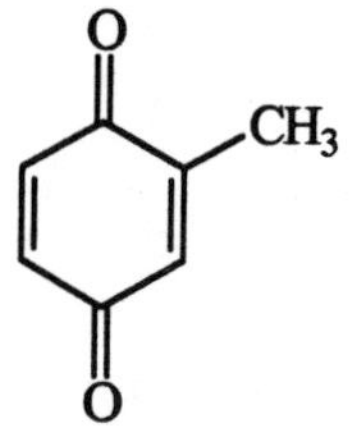

k. (structure: CH_3, OH, Br, Br)

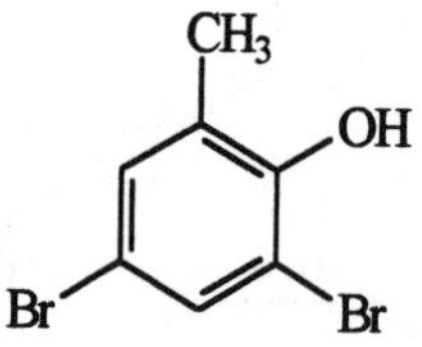

l. (structure: CH_3, $OCCH_3$ with C=O)

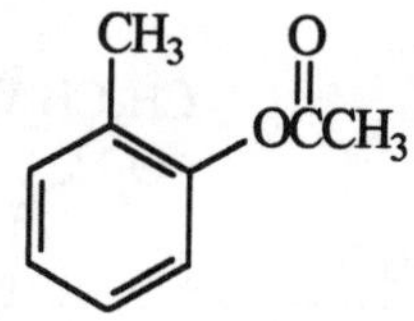

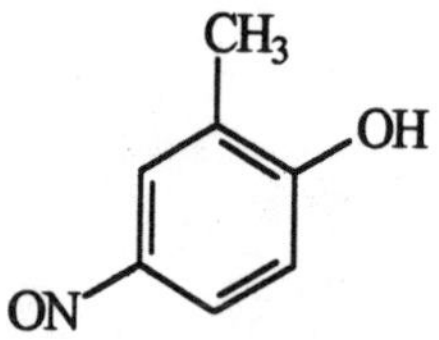

m. (structure: CH_3, OH, CHO)

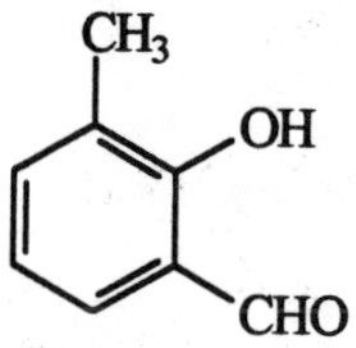

n. (structure: CH_3, OH, ON)

o. (structure: CH_3, O^-)

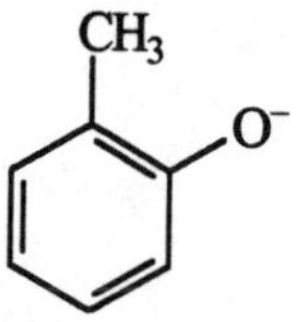

p. (structure: CH_3, OH, O_2N, NO_2)

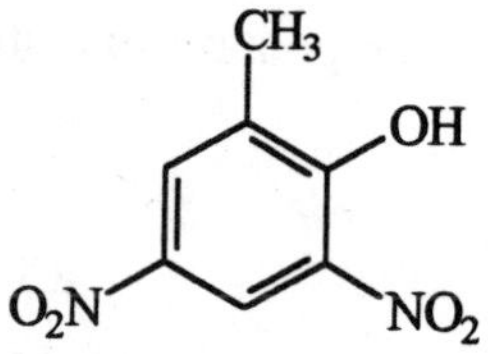

q. no reaction

r. (structure: CH_3, OH, HO_2C)

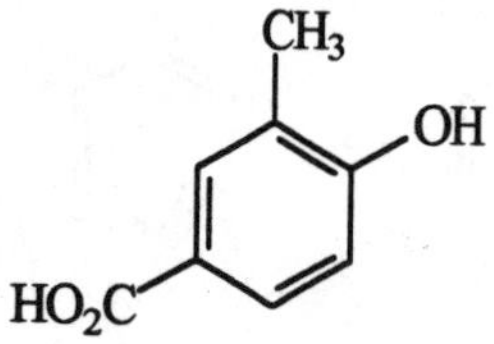

s. (structure: CH_3, OH, Br)

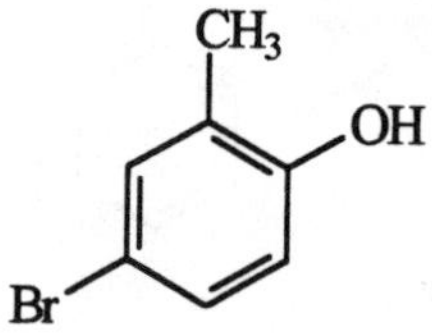

6. a. no reaction

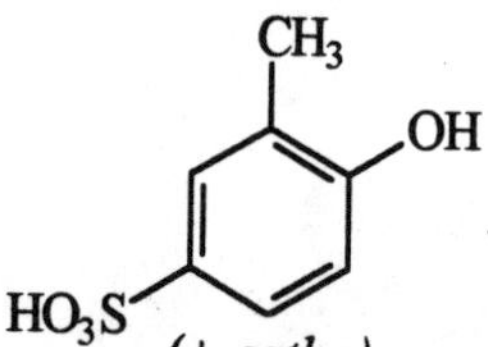

b. (structure: CH_3, OH, HO_3S) + CH_3Br (+ *ortho*)

c. (structure: CH_3, OH)

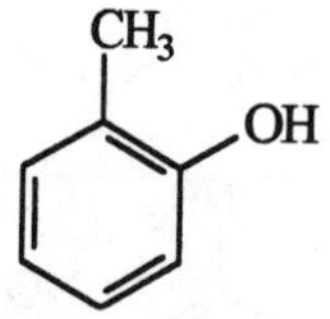

d. (structure: CH_3, OCH_3, Br)

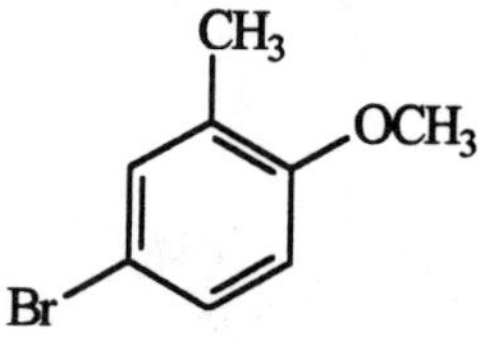

e. (structure: CH_3, OCH_3, O_2N, NO_2)

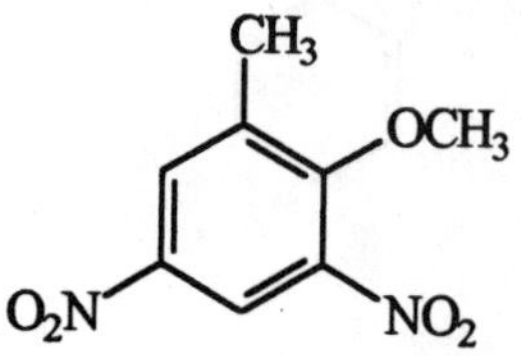

f. and g. no reaction

h. (structure: CH_3, OCH_3, CH_3CO)

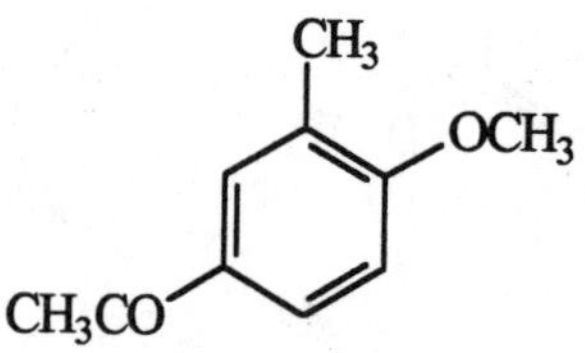

i. (structure: CO_2H, CO, CH_3, OCH_3)

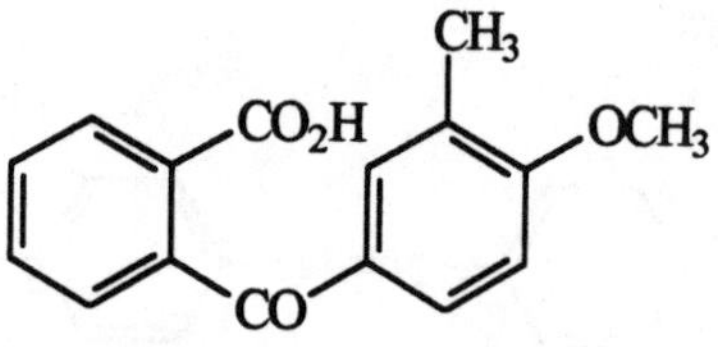

j. From the following analogies:

CH_3 (toluene) → CH_3 (1-methylcyclohexa-1,4-diene); OCH_3 (anisole) → OCH_3 (1-methoxycyclohexa-1,4-diene)

the expected reaction would be: CH_3, OCH_3 (2-methylanisole) → CH_3, OCH_3 (cyclohexadiene)

7. a.

gallic acid (CO_2H; HO, OH, OH) $\xrightarrow[NaOH]{(CH_3O)_2SO_2}$ CO_2CH_3 (CH_3O, OCH_3, OCH_3) $\xrightarrow[ether]{LiAlH_4}$ CH_2OH (CH_3O, OCH_3, OCH_3) $\xrightarrow{HCl\ or\ SOCl_2}$ CH_2Cl (CH_3O, OCH_3, OCH_3) $\xrightarrow{CN^-}$ CH_2CN (CH_3O, OCH_3, OCH_3) $\xrightarrow{LiAlH_4\ or\ H_2\ cat./NH_3}$ $CH_2CH_2NH_2$ (CH_3O, OCH_3, OCH_3)

gallic acid

2-(3,4,5-trimethoxyphenyl)ethanamine

b.

OH (phenol) $\xrightarrow{NaOH}$ $\xrightarrow{CH_3CH{=}CHCH_2Cl}$ $OCH_2CH{=}CHCH_3$ $\xrightarrow{\Delta}$ OH, CH_3 $\overset{|}{C}HCH{=}CH_2$ $\xrightarrow{H_2/\ cat.}$ OH, CH_3 $\overset{|}{C}HCH_2CH_3$ $\xrightarrow{HNO_3}$ OH, CH_3 $\overset{|}{C}HCH_2CH_3$, O_2N, NO_2

2-(1-methylpropyl)-4,6-dinitrophenol

c.

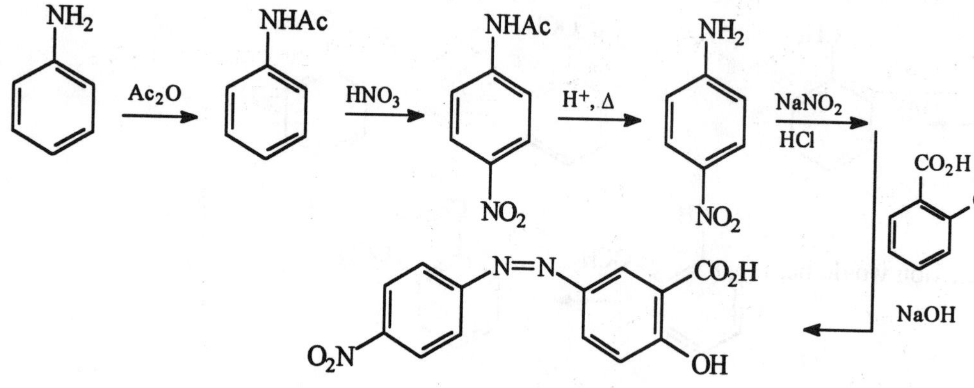

NH_2 (aniline) $\xrightarrow{Ac_2O}$ NHAc $\xrightarrow{HNO_3}$ NHAc, NO_2 $\xrightarrow{H^+, \Delta}$ NH_2, NO_2 $\xrightarrow[HCl]{NaNO_2}$

CO_2H, OH (salicylic acid)

NaOH

N=N, CO_2H, O_2N, OH

5-(*p*-nitrophenyl)azo-2-hydroxybenzoic acid

d.

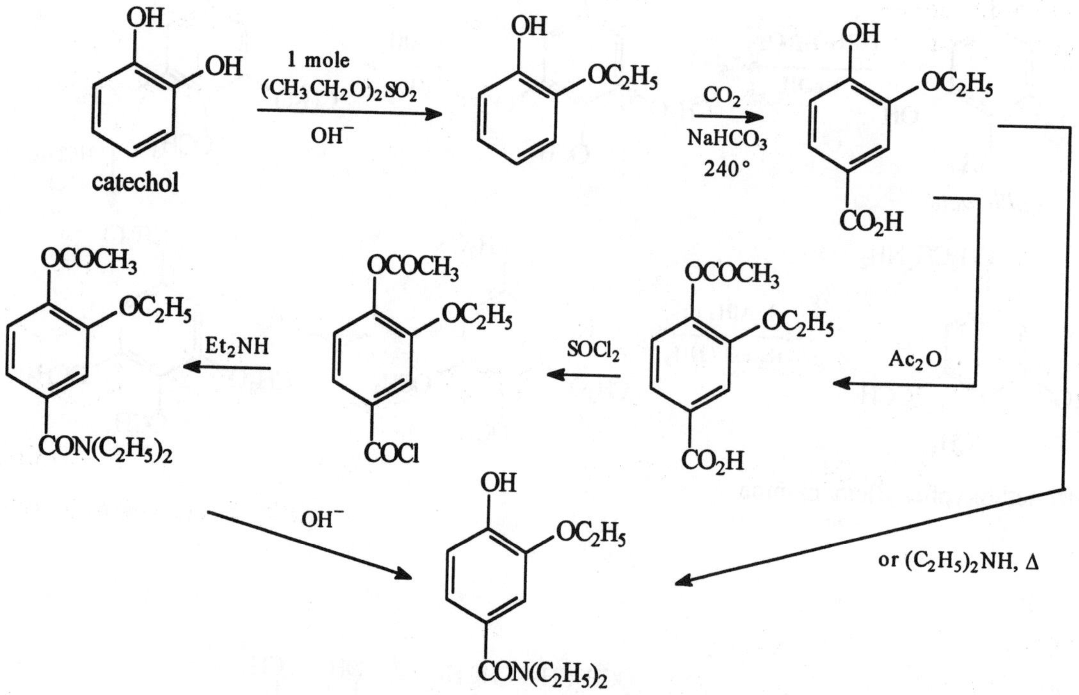

OH, OH (catechol) $\xrightarrow[OH^-]{\text{1 mole } (CH_3CH_2O)_2SO_2}$ OH, OC_2H_5 $\xrightarrow[NaHCO_3, 240°]{CO_2}$ OH, OC_2H_5, CO_2H

catechol

$\xrightarrow{Ac_2O}$ $OCOCH_3$, OC_2H_5, CO_2H $\xrightarrow{SOCl_2}$ $OCOCH_3$, OC_2H_5, COCl $\xrightarrow{Et_2NH}$ $OCOCH_3$, OC_2H_5, $CON(C_2H_5)_2$

$\xrightarrow{OH^-}$ OH, OC_2H_5, $CON(C_2H_5)_2$

or $(C_2H_5)_2NH$, Δ

3-ethoxy-N,N-diethyl-4-hydroxybenzamide

e.

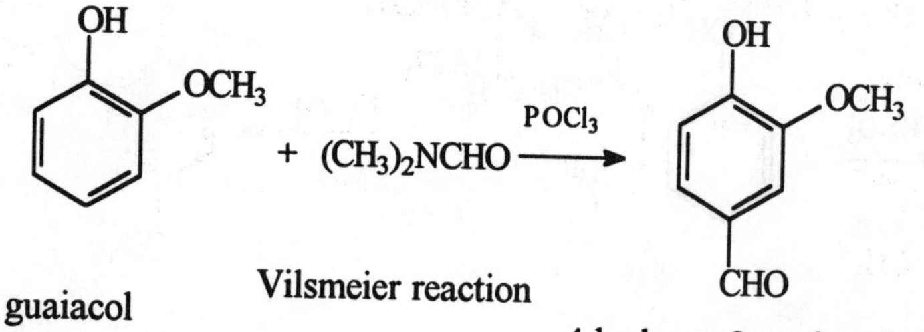

OH, OCH_3 (guaiacol) + $(CH_3)_2NCHO$ $\xrightarrow{POCl_3}$ OH, OCH_3, CHO

guaiacol

Vilsmeier reaction

4-hydroxy-3-methoxybenzaldehyde

f.

OH, OH (catechol) —Me_2SO_4 / OH^-→ OCH_3, OCH_3 —CH_2O, Zn, HCl→ OCH_3, OCH_3, CH_2Cl —$CH_3CONHCH(CO_2Et)_2$ / EtONa→

or NaHCO$_3$ / Δ → OH, OH, CO_2H —Me_2SO_4 / OH^-→ OCH_3, OCH_3, CO_2CH_3 —$LiAlH_4$→ —HCl or $SOCl_2$→ OCH_3, OCH_3, CH_2Cl

OCH_3, OCH_3, $CH_2C(CO_2Et)_2NHCOCH_3$ —H^+ / Δ→ OCH_3, OCH_3, $CH_2CH(NH_2)CO_2H$ —HBr / Δ→ OH, OH, $CH_2CH(NH_2)CO_2H$

2-amino-3-(3,4-dihydroxy-phenyl)propanoic acid

g.

OH (phenol) —$NaHCO_3$ / Δ→ OH, CO_2H —HNO_3→ OH, NO_2, CO_2H —H_2/ cat. or Fe, HCl→ OH, NH_2, CO_2H —MeOH / H^+, Δ→ OH, NH_2, CO_2CH_3

methyl 3-amino-4-hydroxybenzoate

8. a.

OCH_3, OCH_3 —HBr / Δ→ OH, OH —Ag_2O, ether, Na_2SO_4→ O, O (o-benzoquinone)

b.

OH, CH_3 —NaOH→ —$CH_3CH{=}CHCH_2Cl$→ $OCH_2CH{=}CHCH_3$, CH_3 —Δ→ OH, $CH(CH_3)CH{=}CH_2$, CH_3 —H_2/ cat.→ OH, $CH(CH_3)CH_2CH_3$, CH_3

c. CH_3 / NO_2 (o-nitrotoluene) $\xrightarrow[\text{or Fe, HCl}]{H_2/\text{ cat.}}$ CH_3 / NH_2 $\xrightarrow[H_2SO_4\ \text{cold}]{NaNO_2}$ CH_3 / $\overset{+}{N_2}$ $\xrightarrow[\Delta]{H_2O}$ CH_3 / OH

CH_3 / $\overset{+}{N_2}$ $\xrightarrow{OH^-}$ CH_3–C$_6$H$_4$–N=N–C$_6$H$_3$(CH_3)(OH)

d. CH_3 / CH_3 (m-xylene) $\xrightarrow{H_2SO_4}$ CH_3 / CH_3 / SO_3H $\xrightarrow[\Delta]{NaOH}$ CH_3 / CH_3 / OH $\xrightarrow[OH^-]{CHCl_3}$ CH_3 / OHC / CH_3 / OH

e. OH $\xrightarrow[OH^-]{Me_2SO_4}$ OCH_3 $\xrightarrow[AlCl_3\ \Delta]{(CH_3CH_2CO)_2O}$ OCH_3 / $COCH_2CH_3$ $\xrightarrow[\text{glycol, }\Delta]{H_2NNH_2,\ KOH}$ OCH_3 / $CH_2CH_2CH_3$

9. a. $C_6H_5CH{=}CHCH_2Cl$ ⟶ $C_6H_5CH{=}CHCH_2O$–C$_6$H$_5$ ⟶ OH / CH(CH=CH$_2$)–C$_6$H$_5$

b. OH / NH_2 (naphthalene) ⟶ O / O (naphthoquinone) ⟶ $OCOCH_3$ / $OCOCH_3$ / $OCOCH_3$

c. Claisen rearrang. → $(CH_3)_2C(CH_2CHO)(CH{=}CH_2)$ Wittig → $(CH_3)_2C(CH_2CH{=}CH_2)(CH{=}CH_2)$ Cope rearrang. → $(CH_3)_2C{=}CHCH_2CH_2CH{=}CH_2$

d. NH_2 → $^+N_2$ → N=N, OCH_3, OH

e. OH, CH_3 → $OCOCH_3$, CH_3 → OH, $COCH_3$, CH_3

10. OH $\xrightarrow{HNO_3}$ OH, NO_2 $\xrightarrow{\text{aq. } Cl_2}$ OH, Cl, Cl, NO_2 $\xrightarrow{\text{Fe, HCl}}$ OH, Cl, Cl, NH_2 $\xrightarrow{NaNO_2,\ HCl}$ OH, Cl, Cl, $^+N_2$ $\xrightarrow{H_3PO_2}$ OH, Cl, Cl

11. OH $\xrightarrow{HNO_3}$ OH, NO_2 (plus *para*-isomer, separate by steam distillation) $\xrightarrow{Me_2SO_4,\ OH^-}$ OCH_3, NO_2 $\xrightarrow{H_2\text{/ cat. or Fe, HCl}}$ OCH_3, NH_2 $\xrightarrow{NaNO_2,\ HBr;\ CuBr}$ OCH_3, Br $\xrightarrow{CH_3CH_2CH_2CH_2Li,\ \text{ether, cold}}$ OCH_3, Li $\xrightarrow{D_2O}$ OCH_3, D

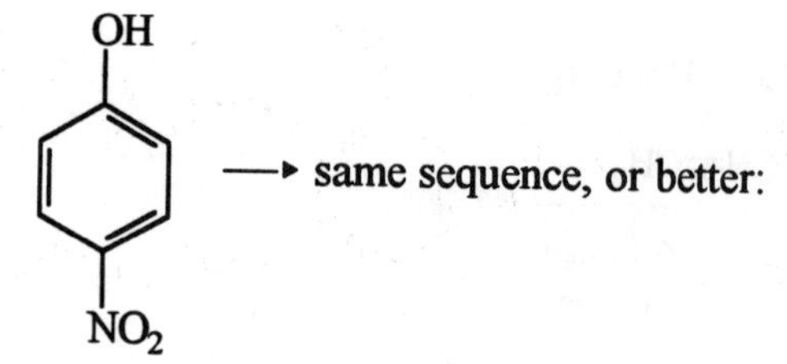

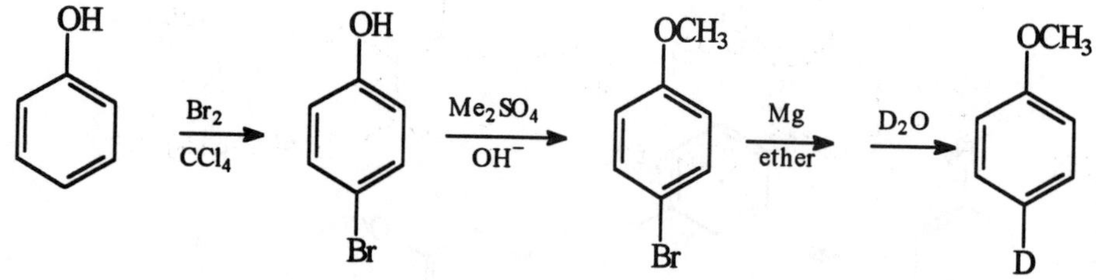

OH / NO$_2$ (p-nitrophenol) → same sequence, or better:

OH (phenol) $\xrightarrow[CCl_4]{Br_2}$ p-bromophenol $\xrightarrow[OH^-]{Me_2SO_4}$ p-bromoanisole $\xrightarrow[ether]{Mg}$ $\xrightarrow{D_2O}$ OCH$_3$ / D

Actually, anisole itself can be metallated to give a far simpler sequence for obtaining the *ortho*-deuterated compound:

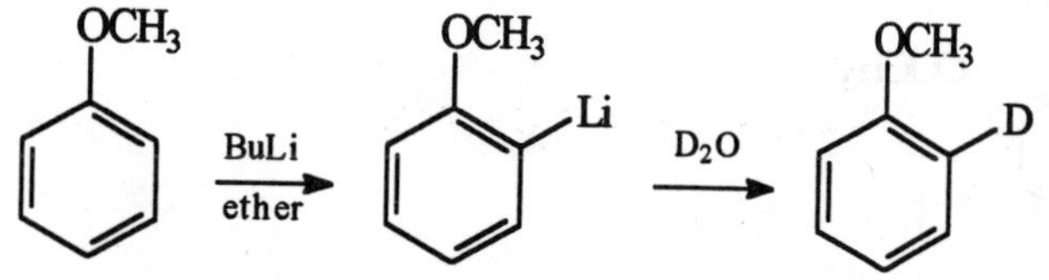

OCH$_3$ $\xrightarrow[ether]{BuLi}$ OCH$_3$ / Li $\xrightarrow{D_2O}$ OCH$_3$ / D

The only problem with this sequence is that it is difficult to accomplish complete metallation. Thus, the final deuterioanisole is accompanied by undeuterated anisole which cannot be separated. For many purposes, a partially deuterated compound will suffice. This procedure can be applied without problems for tracer-labeled anisole-2-*t*.

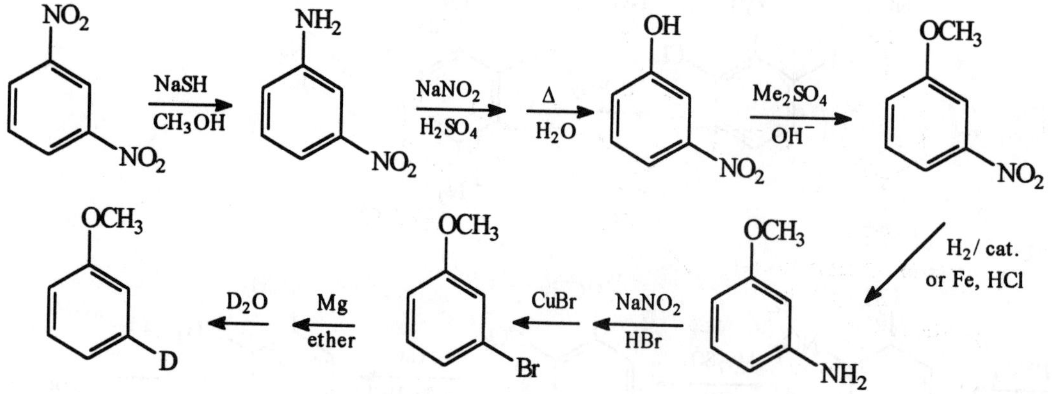

NO$_2$ / NO$_2$ $\xrightarrow[CH_3OH]{NaSH}$ NH$_2$ / NO$_2$ $\xrightarrow[H_2SO_4]{NaNO_2}$ $\xrightarrow[H_2O]{\Delta}$ OH / NO$_2$ $\xrightarrow[OH^-]{Me_2SO_4}$ OCH$_3$ / NO$_2$

$\xrightarrow{H_2/\text{ cat. or Fe, HCl}}$ OCH$_3$ / NH$_2$ $\xrightarrow[HBr]{NaNO_2}$ $\xrightarrow{CuBr}$ OCH$_3$ / Br $\xrightarrow[ether]{Mg}$ $\xrightarrow{D_2O}$ OCH$_3$ / D

12.

OH, OH (catechol) —$CHCl_3$, OH^-→ OH, OH, CHO —Me_2SO_4, OH^-→ OCH_3, OCH_3, CHO —$C_{14}H_{29}MgBr$→ OCH_3, OCH_3, $CH(CH_2)_{13}CH_3$ / OH —H^+, Δ or $AlCl_3$, Δ→ OCH_3, OCH_3, $CH{=}CH(CH_2)_{12}CH_3$ —H_2/ cat.→ OCH_3, OCH_3, $(CH_2)_{14}CH_3$ —HBr, Δ→ OH, OH, $(CH_2)_{14}CH_3$

13. Benzyl alcohol is only slightly more acidic than ethanol; hence, a solution of potassium benzyloxide in ethanol contains substantial amounts of potassium ethoxide. Both can undergo S_N2 reactions to give a mixture of ethers:

$$C_6H_5CH_2O^-\ K^+ + C_2H_5OH \rightleftharpoons C_6H_5CH_2OH + C_2H_5O^-\ K^+$$

p-Cresol, however, is much more acidic than ethanol. A solution of potassium *p*-methylphenolate in ethanol contains very little potassium ethoxide.

$$H_3C{-}C_6H_4{-}O^-\ K^+ + C_2H_5OH \rightleftharpoons H_3C{-}C_6H_4{-}OH + C_2H_5O^-\ K^+$$

14.

OH (phenol) —HNO_3→ OH, NO_2 (separate from *ortho*-) —H_2/ cat. or Fe, HCl→ OH, NH_2 —Ac_2O→ OH, $NHCOCH_3$

or —$C_6H_5N_2^+$, OH^-→ $C_6H_5N{=}N{-}C_6H_4{-}OH$ —$Na_2S_2O_4$→ OH, NH_2

15. OH + $CH_3\overset{+}{C}CH_3$ (OH) ⇌ [... ↔ ... ↔ etc.] $\xrightarrow{-H^+}$ (CH_3CCH_3, OH)

$\xrightarrow{H^+}$ ... $\underset{}{\overset{-H_2O}{\rightleftharpoons}}$ [H_3C—CH_3 ↔ $CH_3\overset{+}{C}CH_3$] ; C_6H_5OH

HO—C(CH_3)(H)—$\overset{+}{=}$OH $\xrightarrow{-H^+}$ HO—C(CH_3)(CH_3)—OH

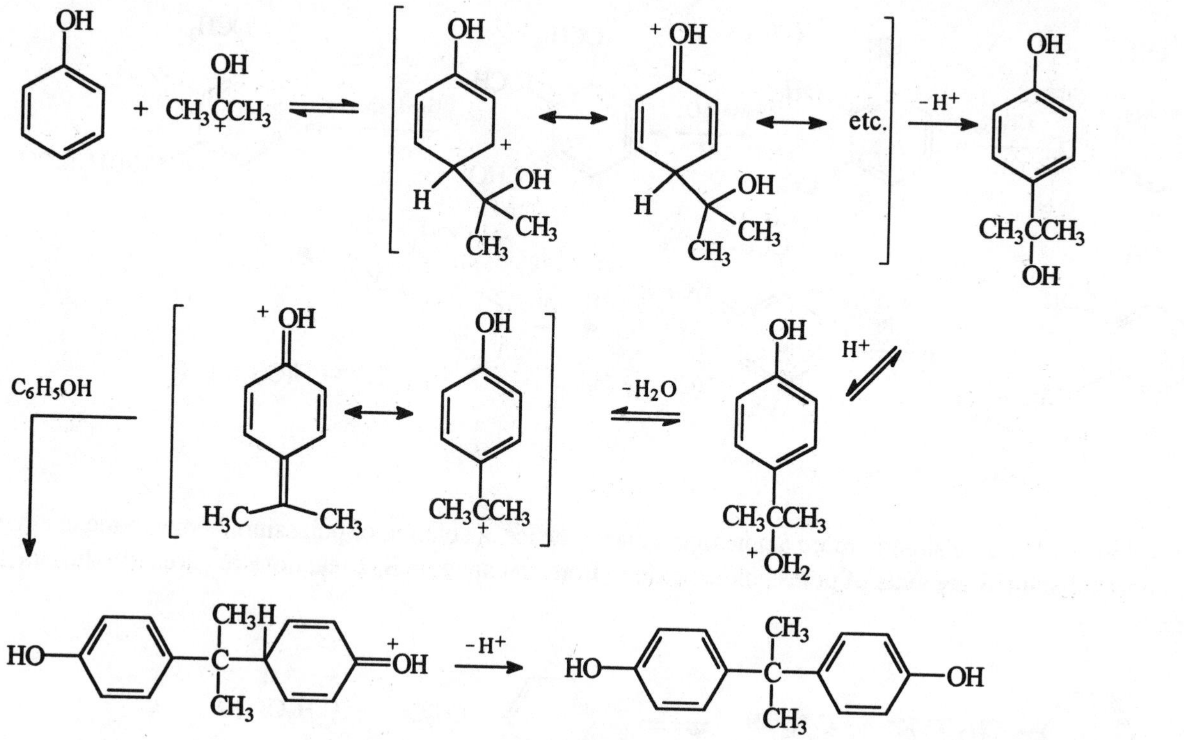

16. Yes. SO_3H is a bulky group and goes *ortho* to the smaller CH_3 group rather than to the larger $(CH_3)_2CH$ group.

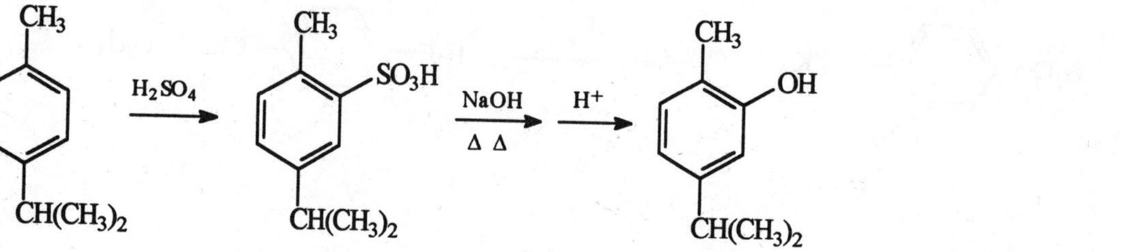

CH_3 / $CH(CH_3)_2$ $\xrightarrow{H_2SO_4}$ CH_3, SO_3H / $CH(CH_3)_2$ $\xrightarrow[\Delta\ \Delta]{NaOH}$ $\xrightarrow{H^+}$ CH_3, OH / $CH(CH_3)_2$

carvacrol

17. OH, CH_3 + $POCl_3$ $\xrightarrow{\text{pyridine}}$ (CH_3—C$_6$H$_4$—O)$_3$PO

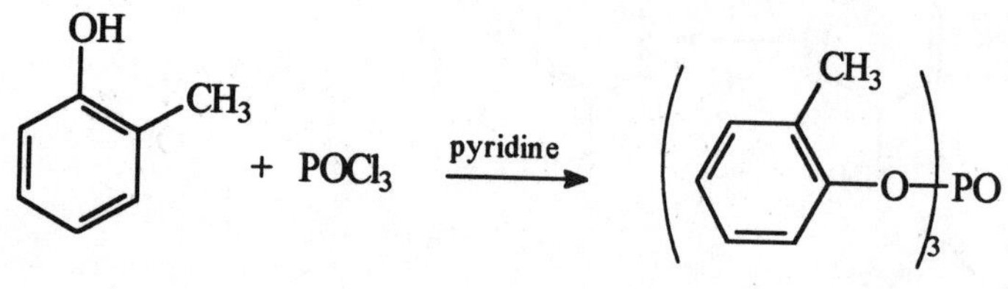

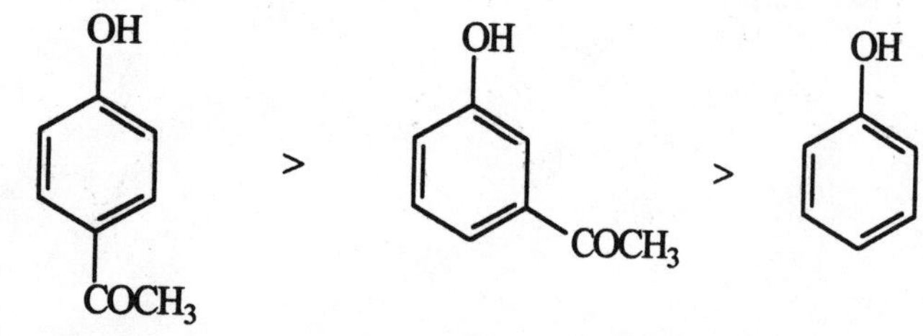

18. a. OH / $COCH_3$ (conjugation) > OH / $COCH_3$ (inductive effect) > OH

b.

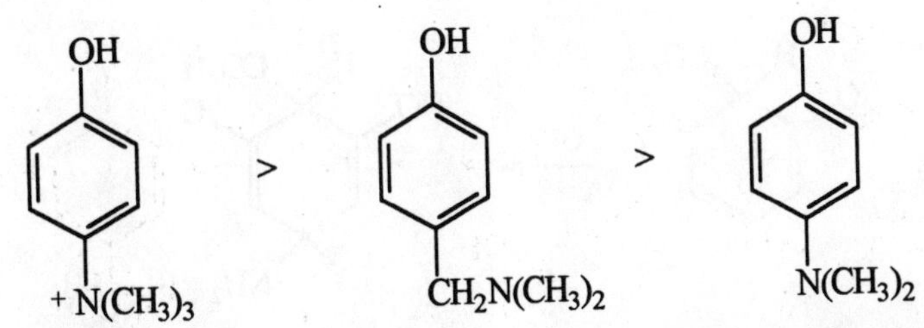

OH, +N(CH$_3$)$_3$ > OH, CH$_2$N(CH$_3$)$_2$ > OH, N(CH$_3$)$_2$

(inductive effect) (electron-donating group)

c.

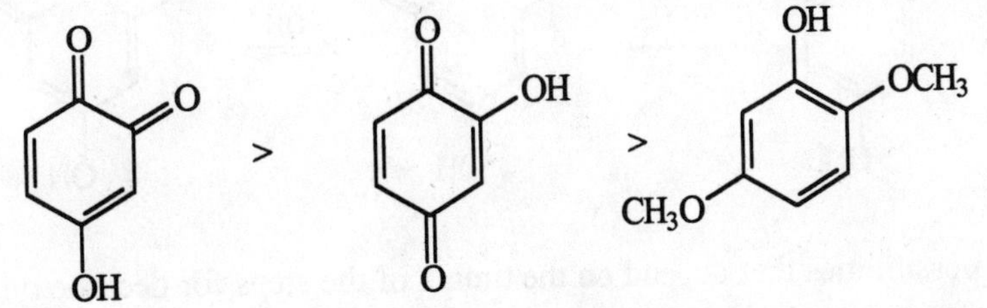

O, O, OH > O, OH, O > OH, OCH$_3$, CH$_3$O

(NOTE: the two quinones give the same anion, but the *ortho*-quinone is less stable; thus, it is more acidic.)

19. Oxidation of a phenol gives the quinone.

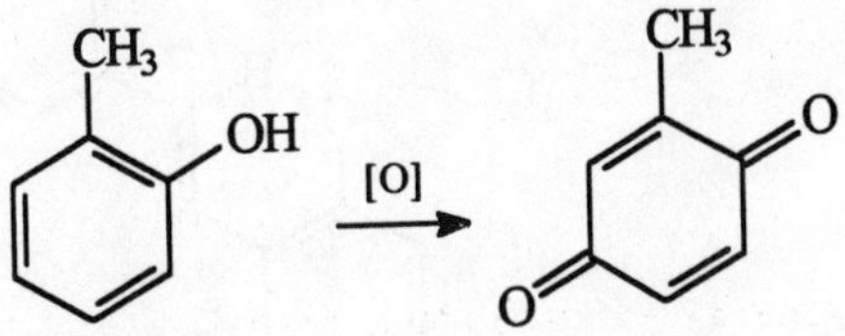

CH$_3$, OH —[O]→ CH$_3$, O, O

20.

CO$_2^-$, H$_2$N, NH$_2$, NH$_2$ ⇌ [CO$_2$H, H$_2$N, $^-$NH, NH$_2$ ↔ CO$_2$H, H$_2$N, NH, NH$_2$] ⇌ H, CO$_2$H, H$_2$N, NH, NH$_2$

OH$^-$

H, CO$_2$H, H$_2$N, OH, NH$^-$, NH$_2$ ⇌ H, CO$_2$H, H$_2$N, O$^-$, NH$_2$, NH$_2$ —(-NH$_2^-$)→ H, CO$_2$H, H$_2$N, O, NH$_2$

OH$^-$

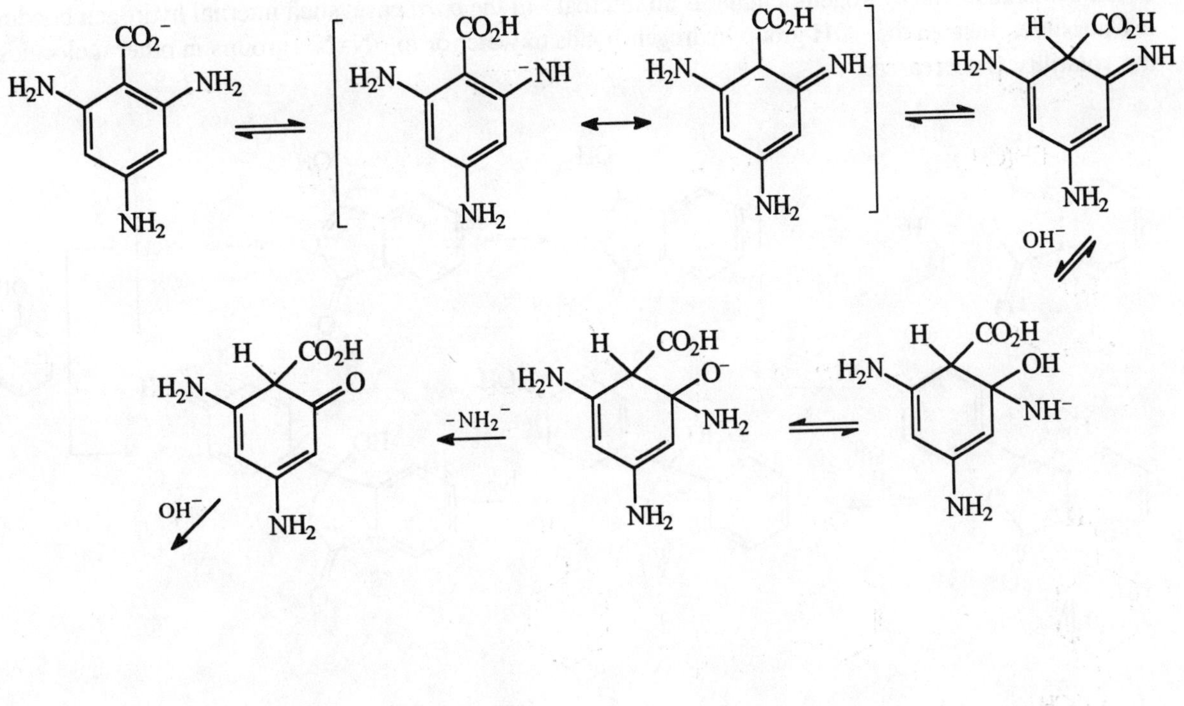

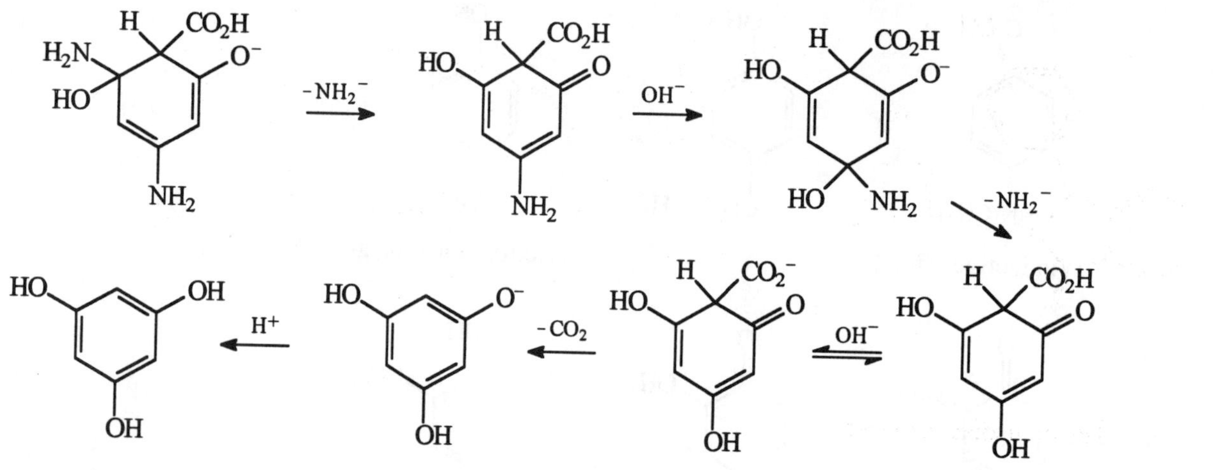

NOTE: this mechanism has many possibilities that depend on the timing of the steps for decarboxylation and hydrolysis of ammonia groups.

21. Internal hydrogen-bonding:

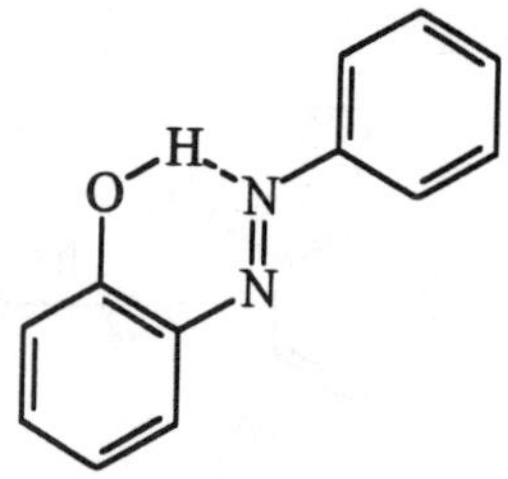

In the *ortho* case, the hydrogen-bonding is all internal. In the *para* case, such internal hydrogen bonding is not possible. Instead the - OH group hydrogen-bonds to water or to - N=N- groups in other molecules and the volatility is decreased.

22. a.

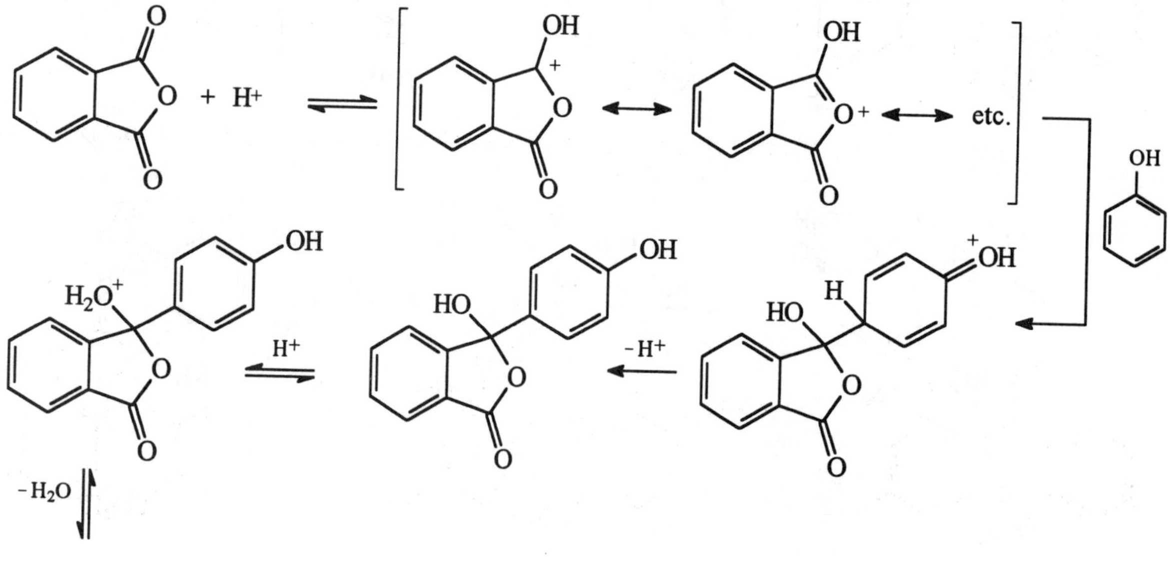

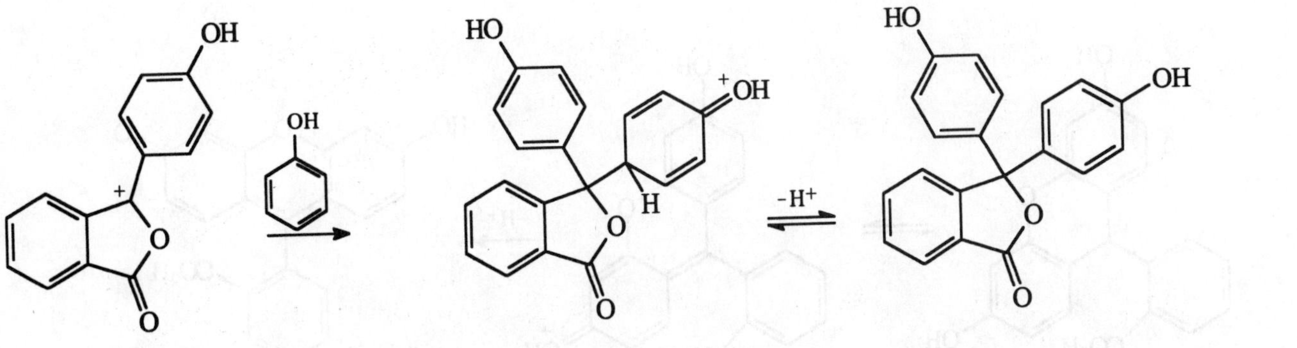

OH OH HO +OH H -H+ HO OH O O O O O O

b. The immediate product is:

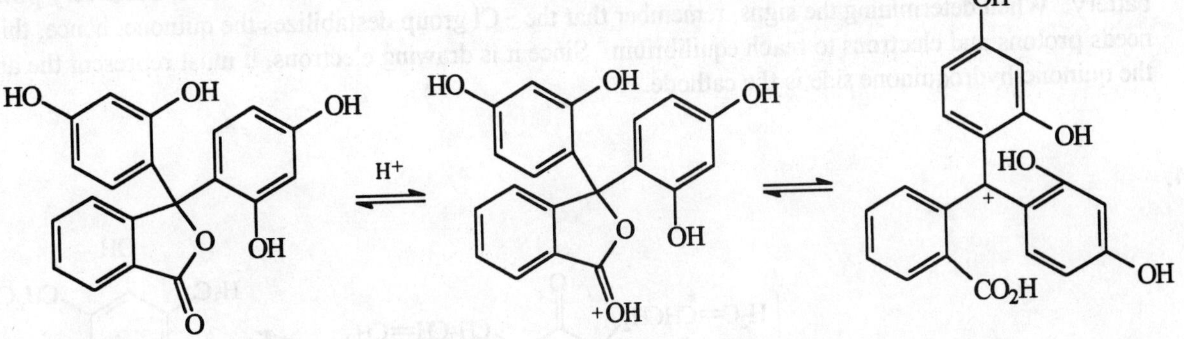

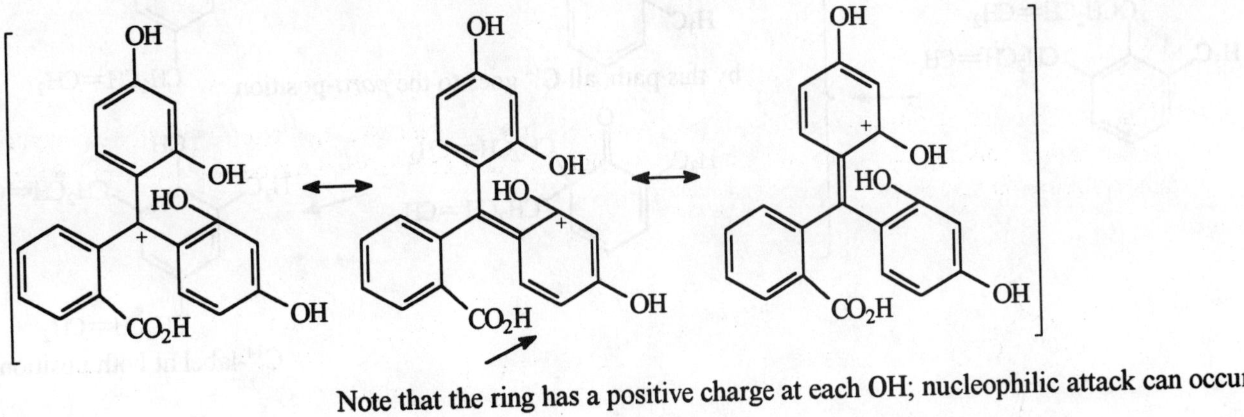

HO OH OH OH H+ HO OH OH OH +OH OH OH HO OH CO_2H OH

OH OH HO OH CO_2H OH OH HO OH CO_2H OH OH HO OH CO_2H OH

Note that the ring has a positive charge at each OH; nucleophilic attack can occur:

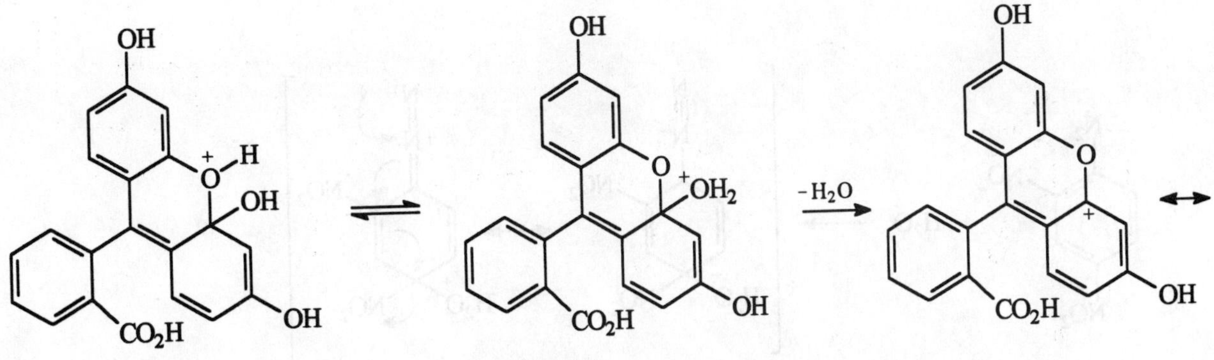

OH +O H OH CO_2H OH OH O $^+OH_2$ CO_2H OH $-H_2O$ OH O + CO_2H OH

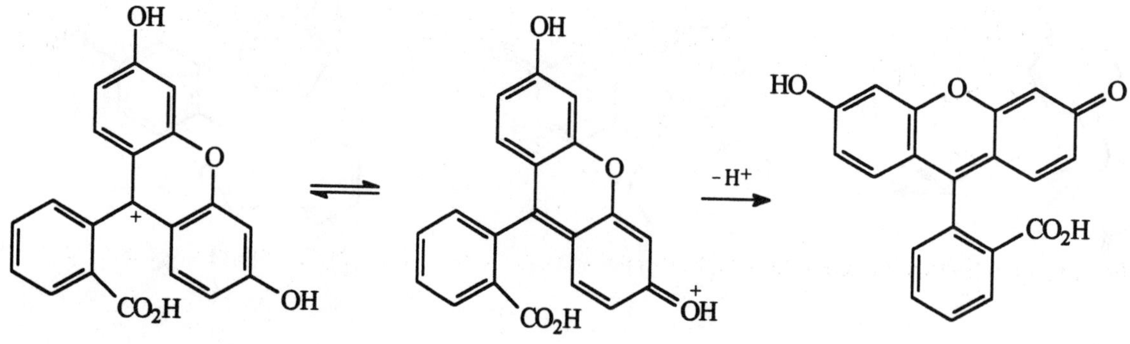

OH OH HO O O + O O -H+ CO_2H CO_2H OH CO_2H $\overset{+}{O}H$

23. The potential difference is given directly by Table 30.2 as 0.713 - 0.699 = 0.014 volts, not a very powerful battery. When determining the signs, remember that the -Cl group destabilizes the quinone; hence, this side needs protons and electrons to reach equilibrium. Since it is drawing electrons, it must represent the anode, and the quinone-hydroquinone side is the cathode.

24.

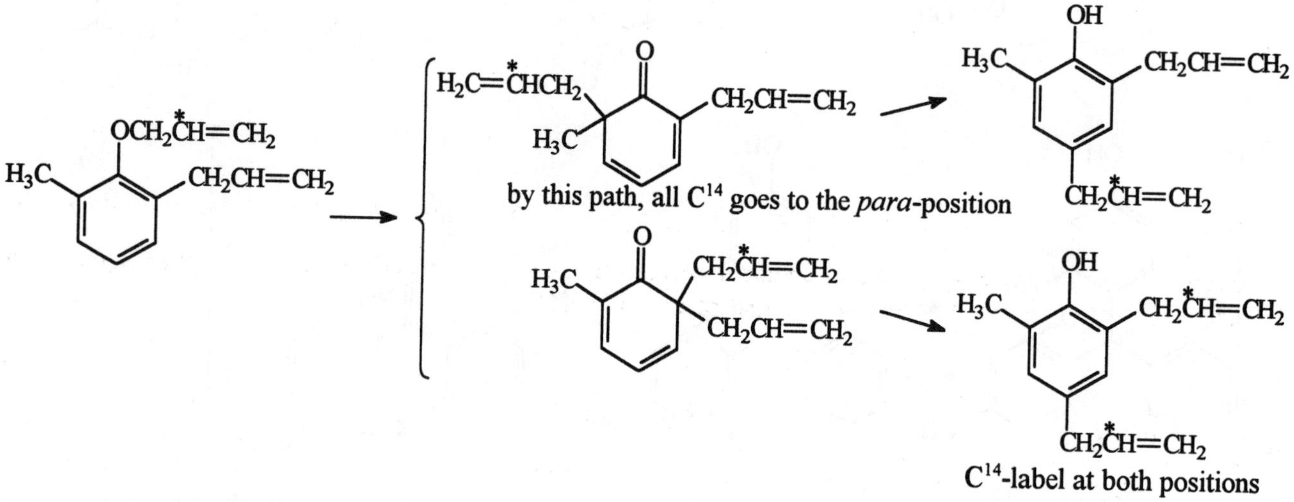

by this path, all C^{14} goes to the *para*-position

C^{14}-label at both positions

25.

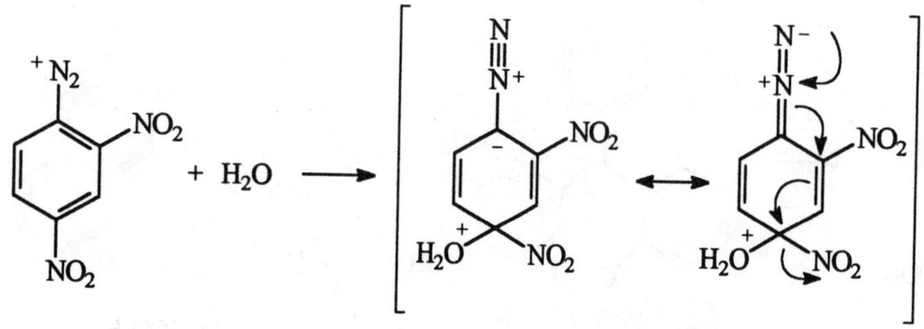

$\longrightarrow$ 4-($^{+}OH_2$)-2-nitrobenzenediazonium + NO_2^- $\rightleftharpoons$ 4-hydroxy-2-nitrobenzenediazonium + HNO_2

The $-N_2^+$ group activates nucleophilic aromatic substitution, much as the $-N{=}O$ and $-NO_2$ groups do.

26. Aniline ($:NH_2$) + Cl–Cl $\rightleftharpoons$ $^{+}NH_2$=C (H, Cl) cyclohexadiene, attacked by $H_2O:$ $\rightleftharpoons$ H_2O^{+}/NH_2 adduct (H, Cl) $\rightleftharpoons$ $H\ddot{O}$/$^{+}NH_3$ adduct (H, Cl) $\rightleftharpoons$ ^{+}OH (H, Cl) $\xrightarrow{-H^+}$ 4-chlorophenol (OH, Cl)

27. $:\ddot{O}-\ddot{Cl}-\dot{O}: \longleftrightarrow :\ddot{O}-\dot{Cl}-\ddot{O}: \longleftrightarrow :\dot{O}-\ddot{Cl}-\ddot{O}:$ Since there is an uneven number (19) of valence electrons in ClO_2, there is necessarily an unpaired electron. Since O has an electronegativy slightly higher than that of Cl, the unpaired electron will be found on O as well as Cl.

28. a. $C_6H_5-NH_2 \xrightarrow[HCl]{NaNO_2} \xrightarrow{KI} C_6H_5-I \xrightarrow[H_2SO_4]{HNO_3} O_2N-C_6H_4-I$

$O_2N-C_6H_4-I + CH_2{=}CHCO_2CH_3 \xrightarrow[R_3N,\ 100°]{Pd(OAc)_2} \xrightarrow{H_3O^+} O_2N-C_6H_4-CH{=}CHCO_2H$

b. $C_6H_5-OH \xrightarrow{HNO_3}$ 2-nitrophenol (NO_2, OH) $\xrightarrow{SnCl_2} \xrightarrow[HCl]{NaNO_2} \xrightarrow{KI}$ 2-iodophenol (I, OH)

2-iodophenol (OH, I) $+ CH_2{=}CHCO_2CH_3 \xrightarrow[R_3N,\ 100°]{Pd(OAc)_2} \xrightarrow{H_3O^+}$ (OH) $-CH{=}CHCO_2H$

c. 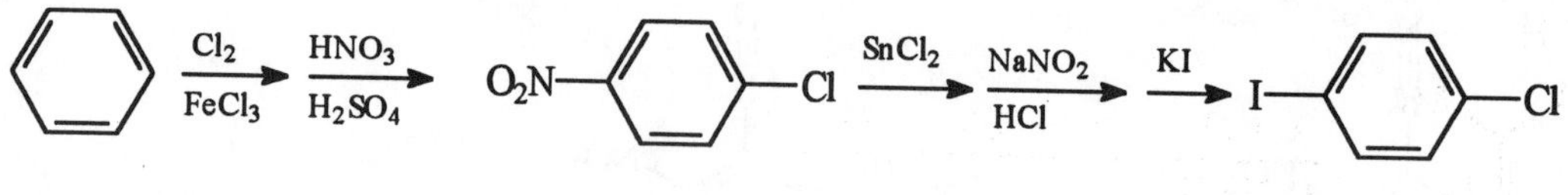 Benzene $\xrightarrow[FeCl_3]{Cl_2}$ $\xrightarrow[H_2SO_4]{HNO_3}$ $O_2N-C_6H_4-Cl$ $\xrightarrow{SnCl_2}$ $\xrightarrow[HCl]{NaNO_2}$ $\xrightarrow{KI}$ $I-C_6H_4-Cl$

$Cl-C_6H_4-I$ + $CH_2{=}CHCO_2CH_3$ $\xrightarrow[R_3N,\ 100°]{Pd(OAc)_2}$ $\xrightarrow{H_3O^+}$ $Cl-C_6H_4-CH{=}CHCO_2H$

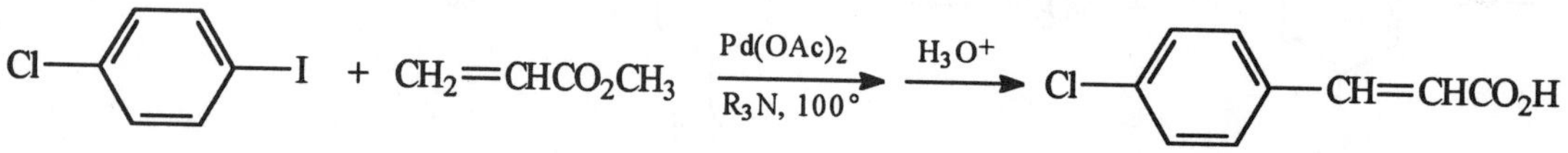

29. Sodium dithionate is a reducing agent that will gradually reduce the benzoquinone to hydroquinone. As the concentrations of these two become nearly equal the solution will take on the dark green color of the charge transfer complex as described on p. 1039 of the Text.

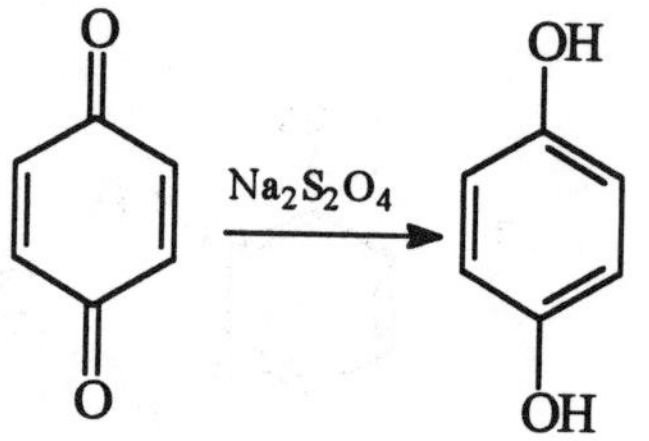

Benzoquinone $\xrightarrow{Na_2S_2O_4}$ hydroquinone (OH, OH)

30.F. Supplementary Problems

S1. Name the following compounds:

a. 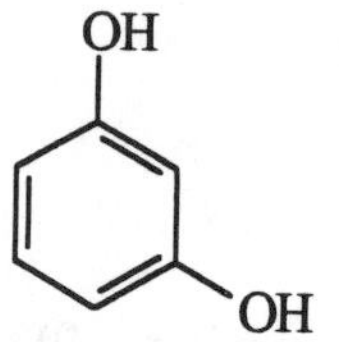 (benzene ring with OH, OH)

b. 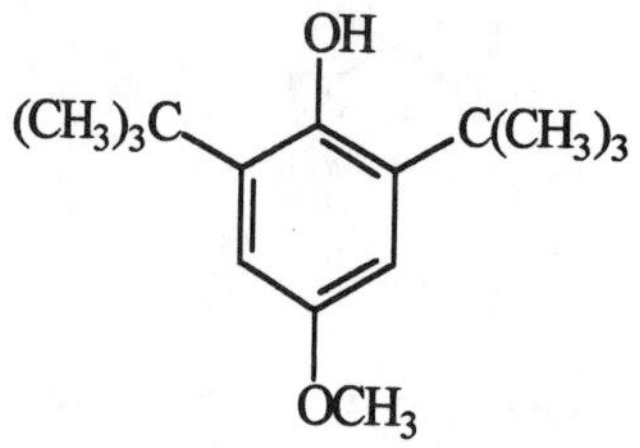 (benzene ring with OH, $(CH_3)_3C$, $C(CH_3)_3$, OCH_3)

c. (cyclohexane ring with HO_2C, H, H, O–phenyl)

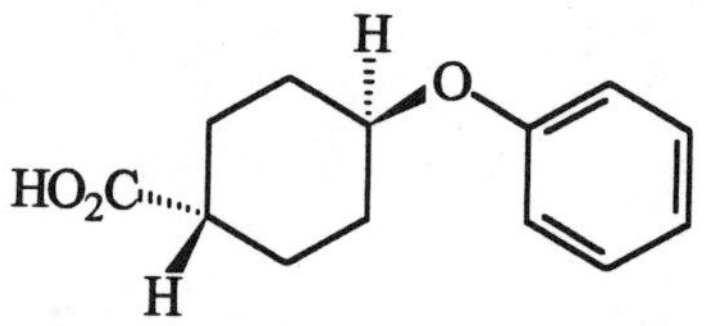

d. 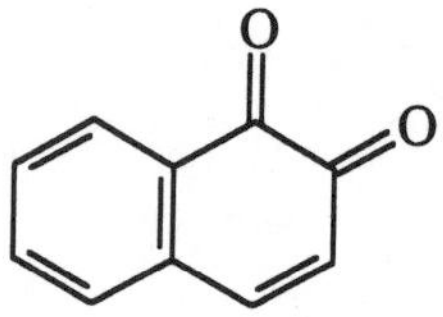 (naphthalene-dione, O, O)

e. 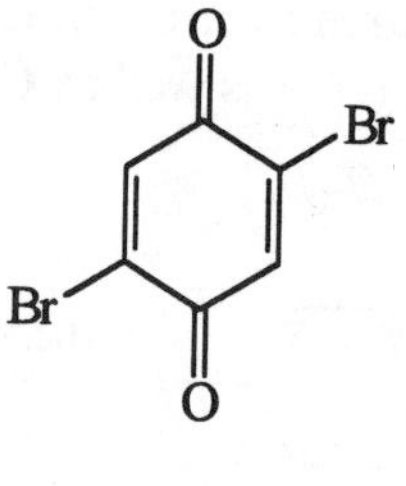 (benzoquinone with O, O, Br, Br)

f. 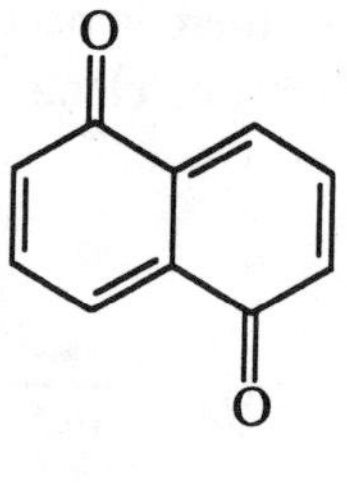 (naphthoquinone with O, O)

S2. Predict the major product to result from each of the following reaction sequences.

a. 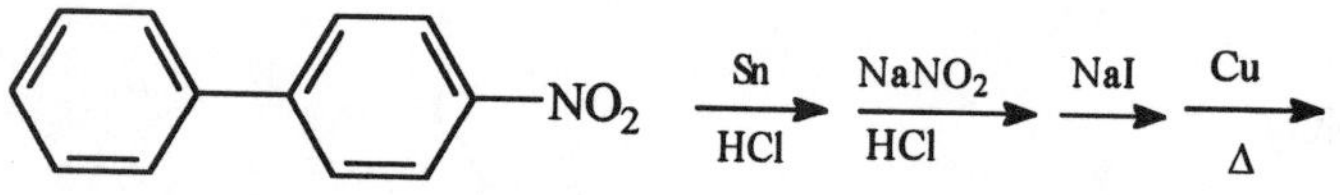 $C_6H_5-C_6H_4-NO_2$ $\xrightarrow[HCl]{Sn}$ $\xrightarrow[HCl]{NaNO_2}$ $\xrightarrow{NaI}$ $\xrightarrow[\Delta]{Cu}$

b. 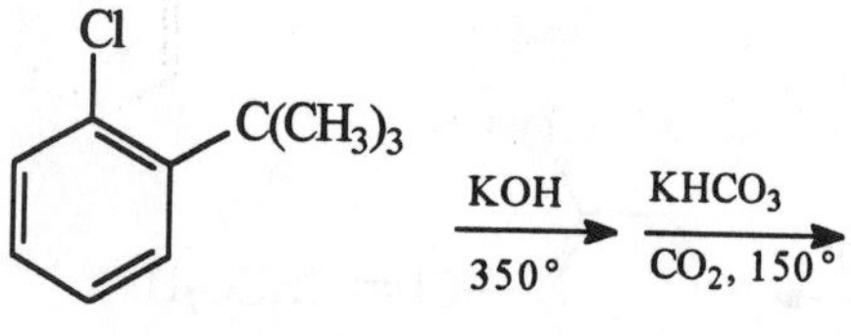 (benzene ring with Cl and $C(CH_3)_3$) $\xrightarrow[350°]{KOH}$ $\xrightarrow[CO_2,\ 150°]{KHCO_3}$

c. 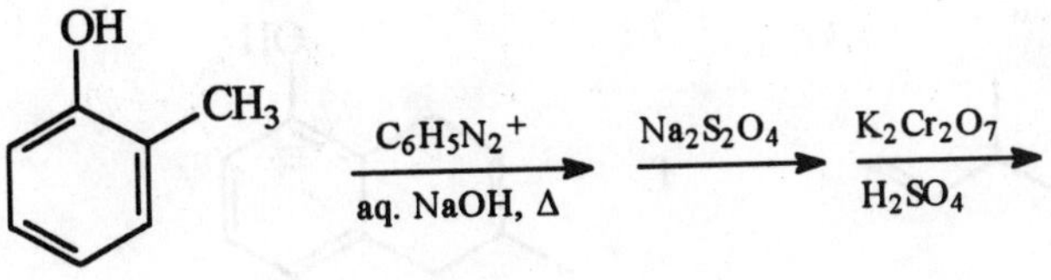 [2-methylphenol] $\xrightarrow[\text{aq. NaOH, }\Delta]{C_6H_5N_2^+}$ $\xrightarrow{Na_2S_2O_4}$ $\xrightarrow[H_2SO_4]{K_2Cr_2O_7}$

d. benzene (excess) $\xrightarrow[BF_3]{(CH_3)_3COH}$ $\xrightarrow[FeBr_3]{Br_2}$ $\xrightarrow[\text{ether}]{n\text{-BuLi}}$ $\xrightarrow{(CH_3)_2C{=}O}$ $\xrightarrow{H_2O}$

e. 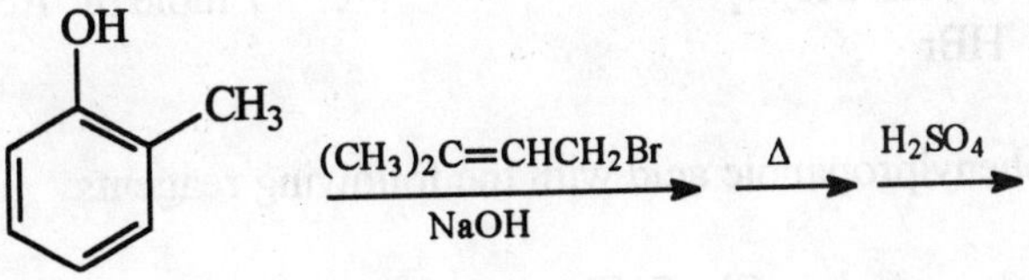 [2-methylphenol] $\xrightarrow[NaOH]{(CH_3)_2C{=}CHCH_2Br}$ $\xrightarrow{\Delta}$ $\xrightarrow{H_2SO_4}$

f. 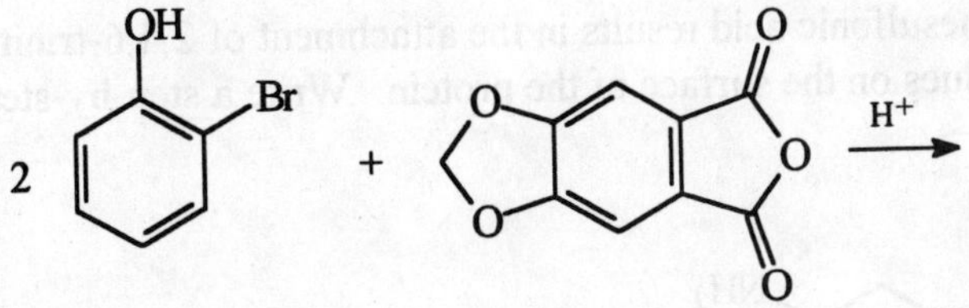 2 [2-bromophenol] + [methylenedioxy phthalic anhydride] $\xrightarrow{H^+}$

g. 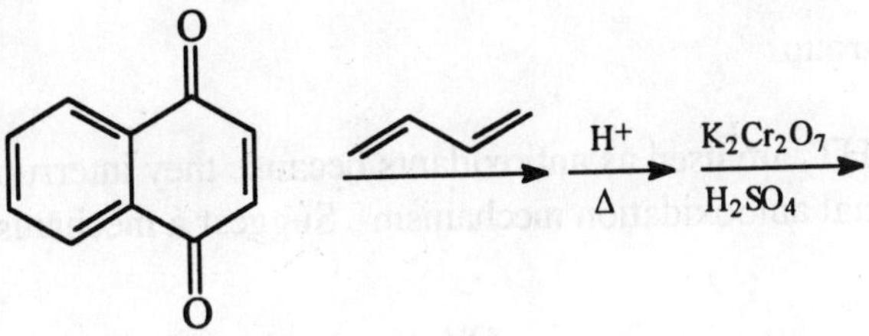 [1,4-naphthoquinone] $\xrightarrow{\text{butadiene}}$ $\xrightarrow[\Delta]{H^+}$ $\xrightarrow[H_2SO_4]{K_2Cr_2O_7}$

S3. Show how to synthesize the following compounds, starting with any monosubstituted benzene derivatives and any other non-aromatic compounds.

a. 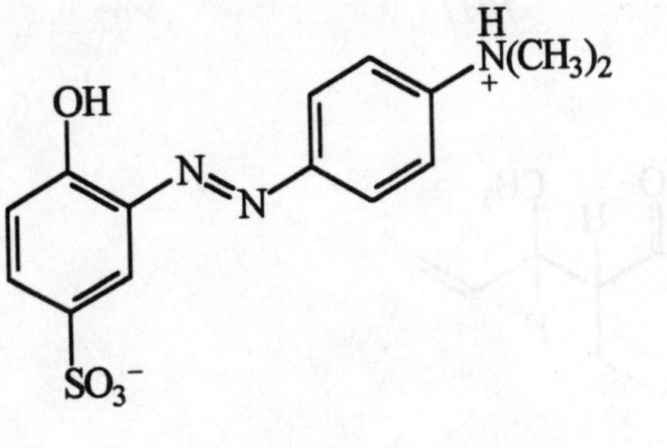 [OH, N=N azo to 4-($\overset{+}{N}H(CH_3)_2$)phenyl, SO_3^-]

b. 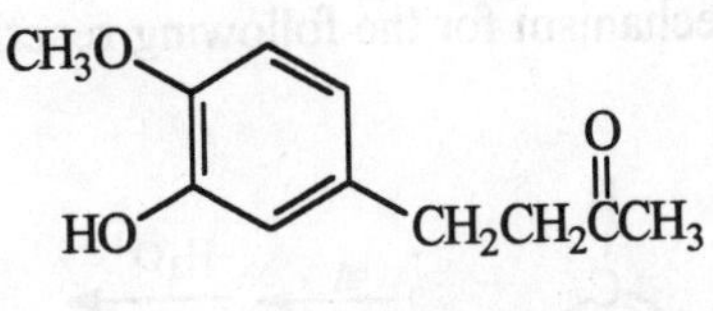 [CH_3O, HO, $CH_2CH_2\overset{O}{\overset{\|}{C}}CH_3$]

Zingerone (a constituent of ginger)

c. 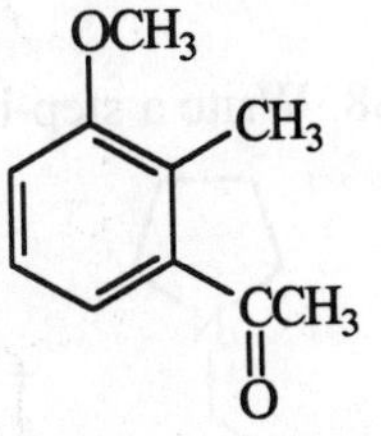 [OCH_3, CH_3, $\overset{}{C}CH_3$ with C=O]

d. 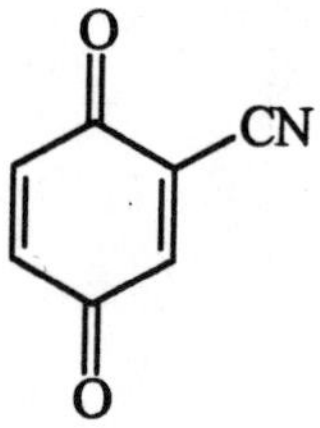 e. 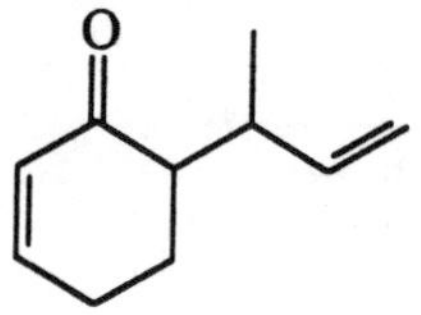 f.

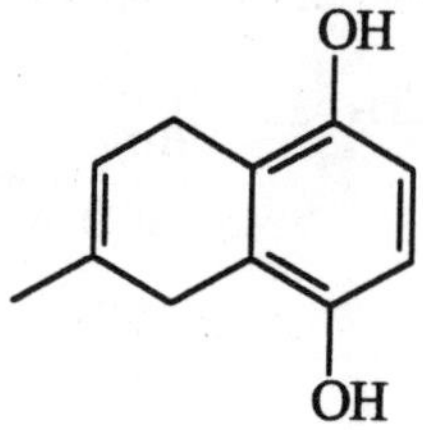

S4. Write the expected products from reaction of 2-hydroxy-5-(hydroxymethyl)benzoic acid with the following reagents:

a. NaOH, CO_2, Δ
b. excess CH_2N_2
c. 1 mole of Ac_2O
d. NaOH, $CHCl_3$
e. HBr

S5. Write the expected products from reaction of 3-phenylpropanoic acid with the following reagents:

a. Cl_2, catalytic amount of PCl_3
b. Cl_2, $FeCl_3$
c. Cl_2, $h\nu$
d. $SOCl_2$

S6. Treatment of a protein with 2,4,6-trinitrobenzenesulfonic acid results in the attachment of 2,4,6-trinitrophenyl groups to the ε-amino groups of the lysine residues on the surface of the protein. Write a step-by-step mechanism for this reaction.

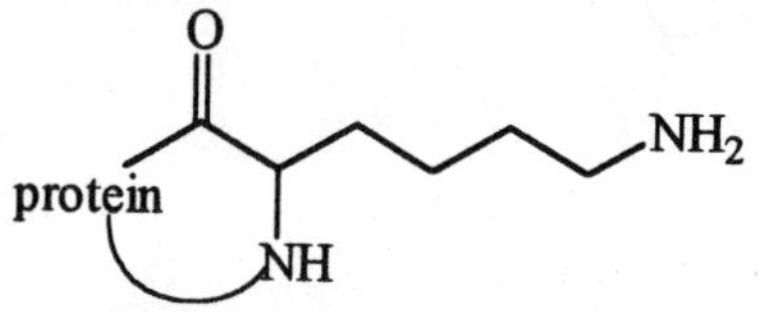

Lysine ε-amino group

S7. Hydroquinone, and molecules such as BHA and BHT, are used as antioxidants because they interrupt propagation of the radical chain involved in the usual autooxidation mechanism. Suggest a mechanism for this interruption.

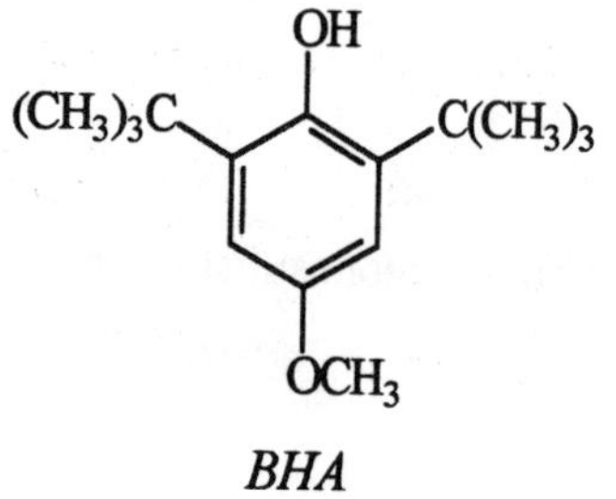

BHA 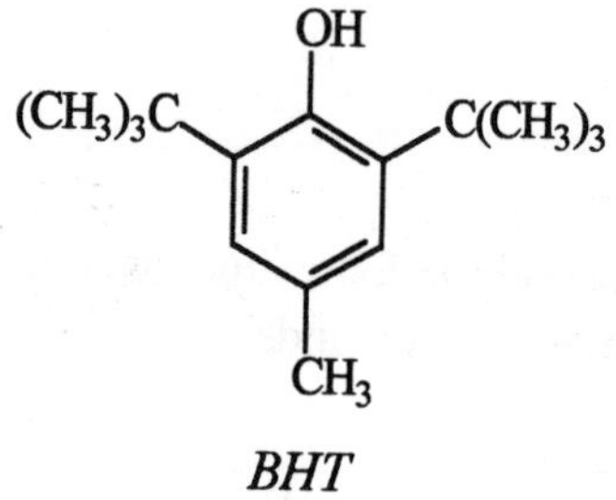 *BHT*

S8. Write a step-by-step mechanism for the following reaction:

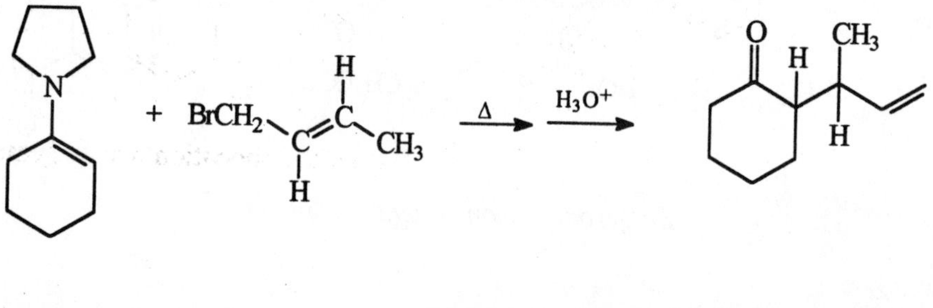

S9. a. How do you account for the difference in the following reactions?

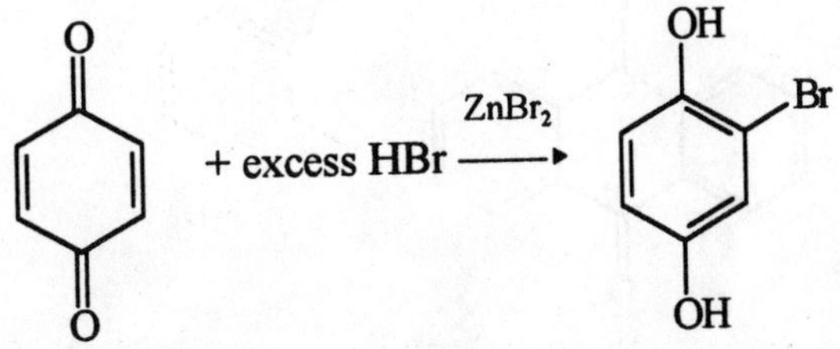

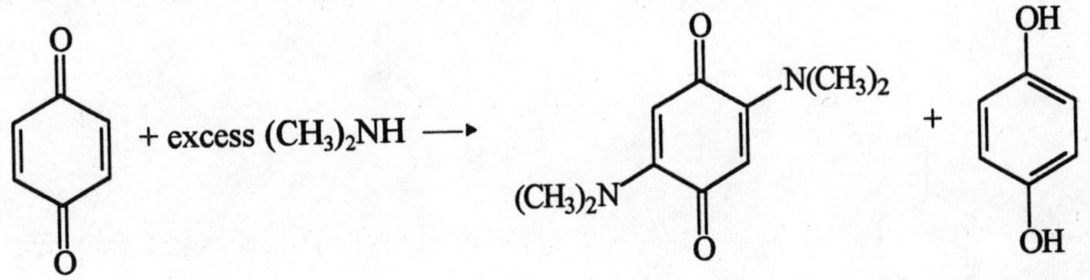

1,4-benzoquinone + excess HBr $\xrightarrow{ZnBr_2}$ 2-bromohydroquinone (OH, Br, OH)

1,4-benzoquinone + excess $(CH_3)_2NH$ ⟶ 2,5-di(dimethylamino)-1,4-benzoquinone (O, $N(CH_3)_2$, $(CH_3)_2N$, O) + hydroquinone (OH, OH)

one-third mole   two-thirds mole

b. Why is the 2,5-di(dimethylamino) isomer produced in the second reaction above, rather than the 2,3- or 2,6-isomer?

30.G. Answers to Supplementary Problems

S1. a. 1,3-dihydroxybenzene (*m*-dihydroxybenzene, resorcinol)
b. 2,6-di(*t*-butyl)-4-methoxyphenol ("butylated hydroxyanisole", BHA)
c. *trans*-4-phenoxycyclohexanecarboxylic acid
d. 1,2-naphthoquinone
e. 2,5-dibromo-1,4-benzoquinone
f. 1,5-naphthoquinone

S2.

a. 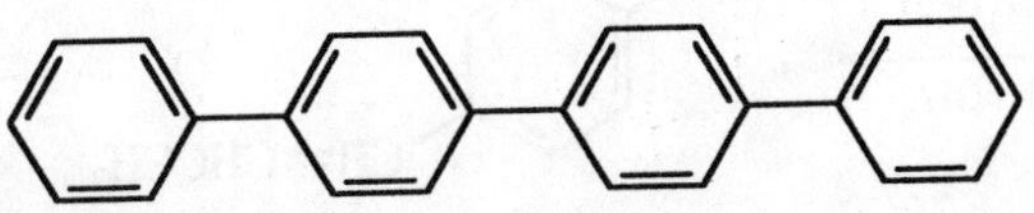 (p-quaterphenyl: four linked benzene rings)

b. HO, $C(CH_3)_3$, HO_2C — no reaction occurs here because of steric hindrance

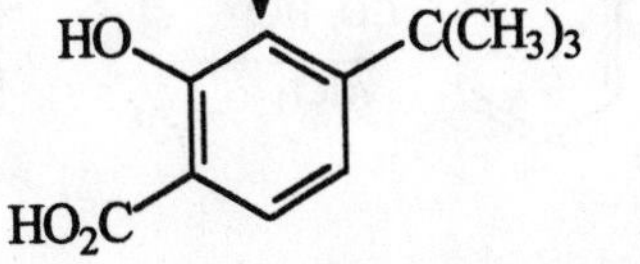

c. 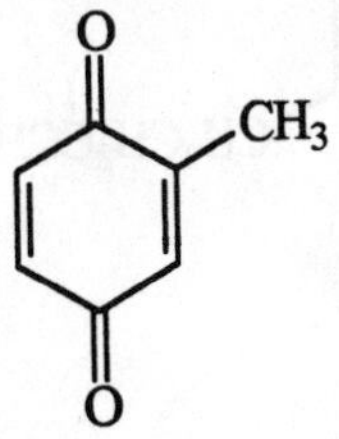 (2-methyl-1,4-benzoquinone: O, CH_3, O)

d. 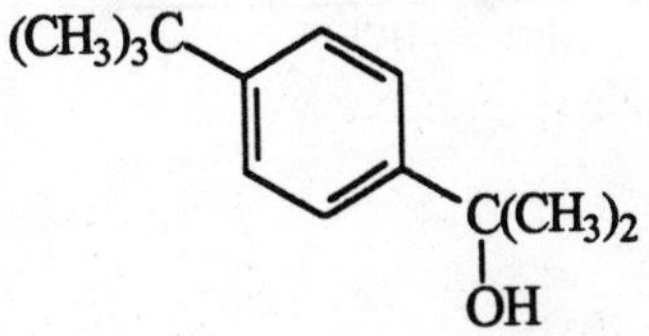 $(CH_3)_3C$–C₆H₄–$C(CH_3)_2$OH

e. 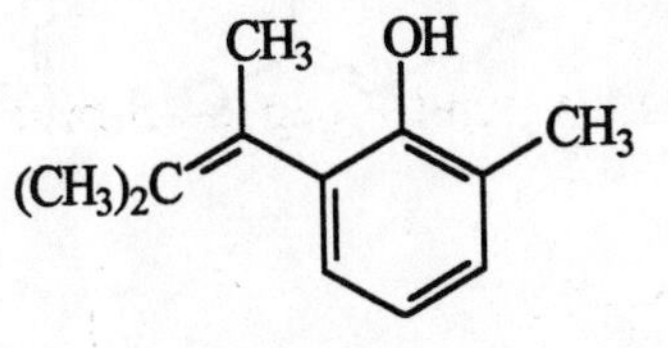 CH_3, OH, CH_3, $(CH_3)_2C=$

(from a carbocation rearrangement)

f.

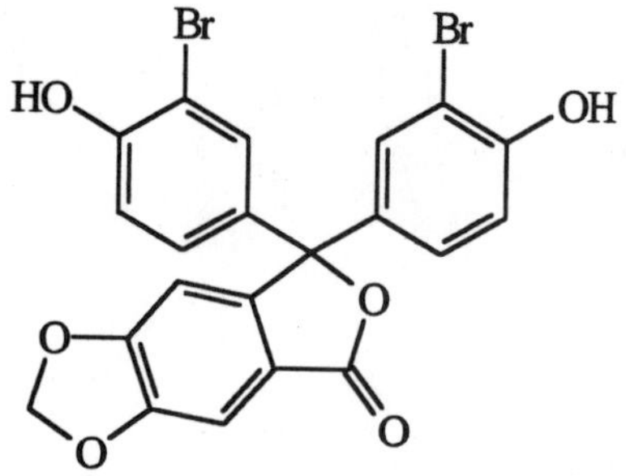

g.

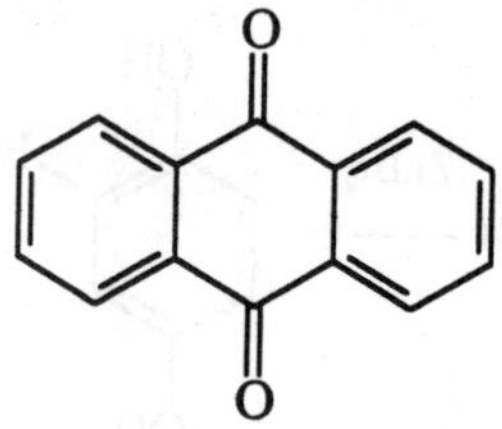

S3. a.

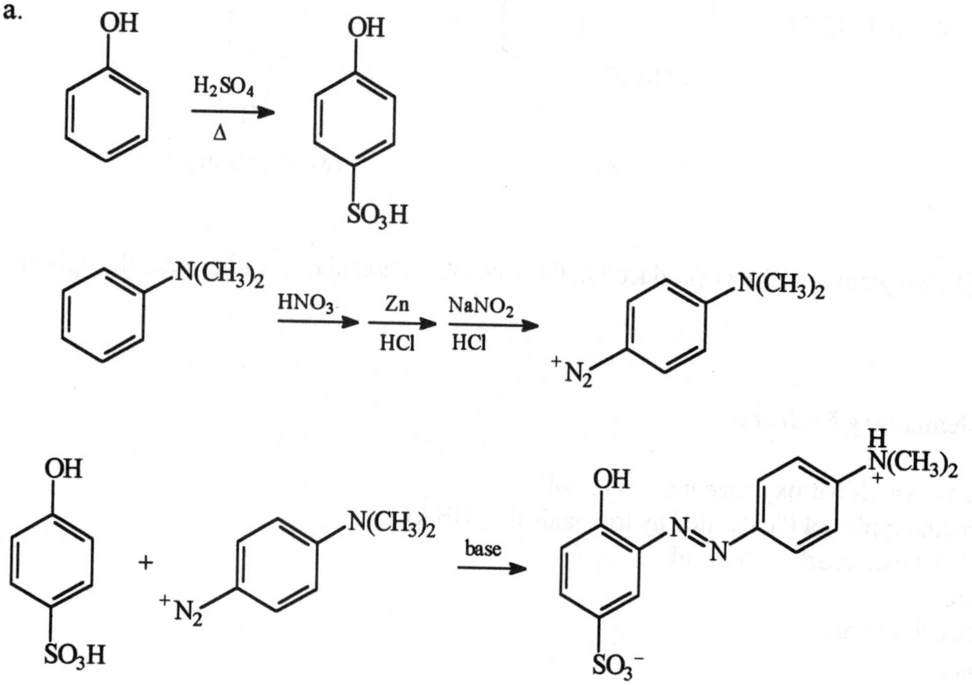

H_2SO_4, Δ

SO_3H

$N(CH_3)_2$ $\xrightarrow{HNO_3}$ $\xrightarrow[HCl]{Zn}$ $\xrightarrow[HCl]{NaNO_2}$ $N(CH_3)_2$, $^+N_2$

OH, SO_3H + $^+N_2$, $N(CH_3)_2$ $\xrightarrow{base}$ OH, SO_3^-, N=N, $\overset{+}{N}H(CH_3)_2$

b.

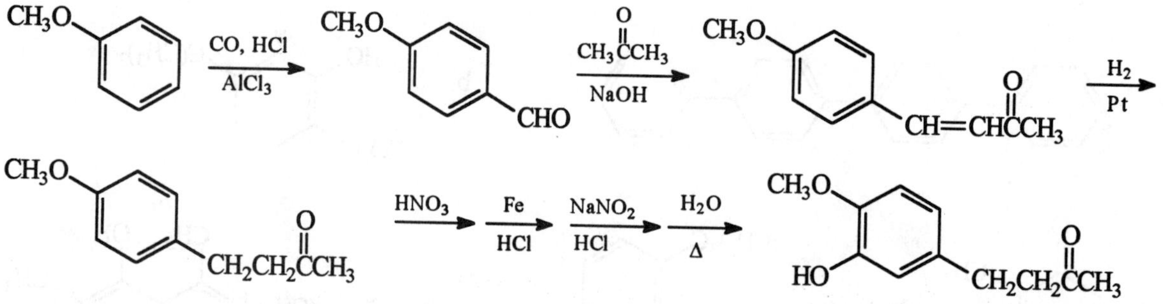

CH_3O $\xrightarrow[AlCl_3]{CO,\ HCl}$ CH_3O, CHO $\xrightarrow[NaOH]{CH_3\overset{O}{\overset{\|}{C}}CH_3}$ CH_3O, $CH{=}CH\overset{O}{\overset{\|}{C}}CH_3$ $\xrightarrow[Pt]{H_2}$

CH_3O, $CH_2CH_2\overset{O}{\overset{\|}{C}}CH_3$ $\xrightarrow{HNO_3}$ $\xrightarrow[HCl]{Fe}$ $\xrightarrow[HCl]{NaNO_2}$ $\xrightarrow[\Delta]{H_2O}$ CH_3O, HO, $CH_2CH_2\overset{O}{\overset{\|}{C}}CH_3$

c.

CH_3 $\xrightarrow[\Delta]{H_2SO_4}$ HO_3S, CH_3 $\xrightarrow[H_2SO_4]{HNO_3}$ $\xrightarrow[FeBr_3]{Br_2}$ HO_3S, NO_2, CH_3, Br $\xrightarrow{H_3O^+}$ $\xrightarrow[HCl]{Zn}$

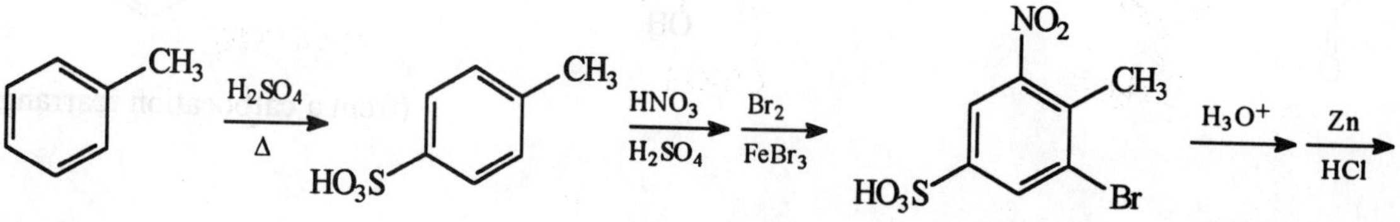

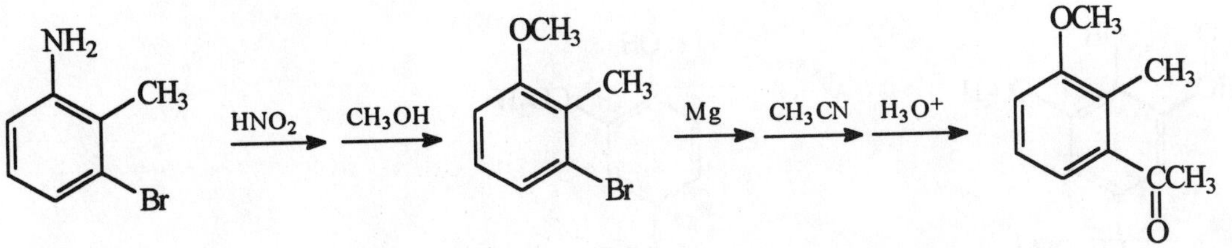

NH$_2$, CH$_3$, Br $\xrightarrow{HNO_2}$ $\xrightarrow{CH_3OH}$ OCH$_3$, CH$_3$, Br $\xrightarrow{Mg}$ $\xrightarrow{CH_3CN}$ $\xrightarrow{H_3O^+}$ OCH$_3$, CH$_3$, CCH$_3$ (=O)

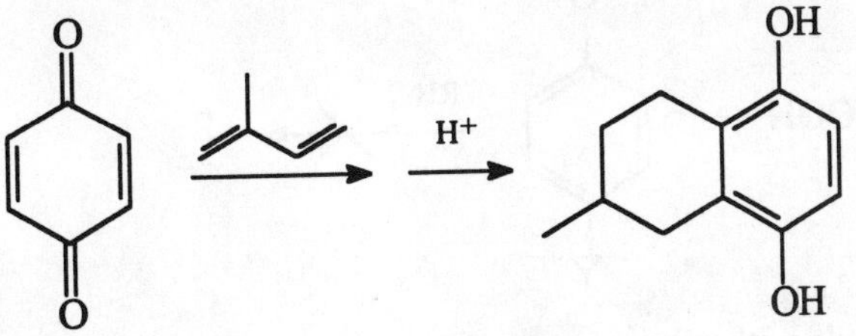

d. OH $\xrightarrow[H_2SO_4]{HNO_3}$ $\xrightarrow[HCl]{Zn}$ $\xrightarrow[H_2SO_4]{K_2Cr_2O_7}$ O, O $\xrightarrow[NaCN]{HCN}$ OH, CN, OH $\xrightarrow{KMnO_4}$ O, CN, O

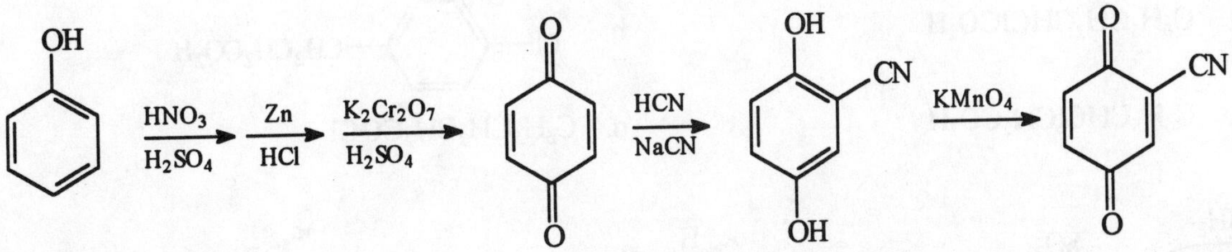

e. OH $\xrightarrow[CH_3CH{=}CHCH_2Br]{NaOH}$ $\xrightarrow{\Delta}$ OH $\xrightarrow[OH^-]{Me_2SO_4}$ OCH$_3$ $\xrightarrow{Na, NH_3}$ OCH$_3$ $\xrightarrow{H_3O^+}$ O

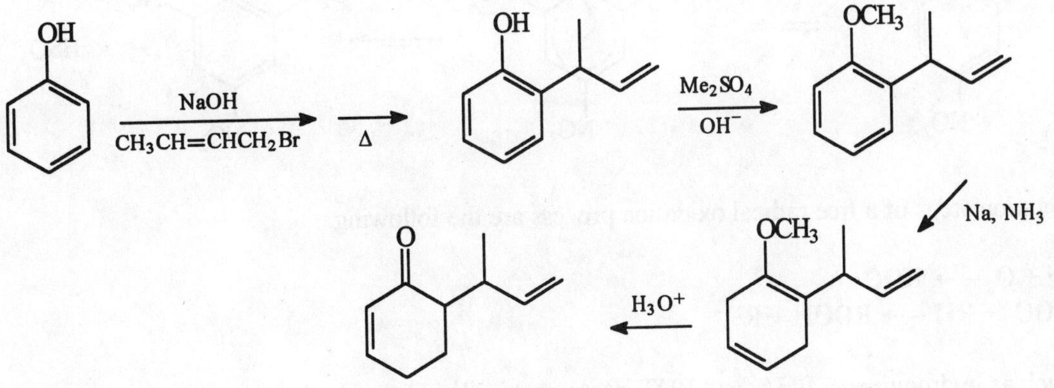

f. O, O (from (d) above) $\longrightarrow$ $\xrightarrow{H^+}$ OH, OH

S4. a. OH, HO$_2$C, CO$_2$H, CH$_2$OH b. OCH$_3$, CO$_2$CH$_3$, CH$_2$OH c. OH, CO$_2$H, CH$_2$OCCH$_3$ (=O)

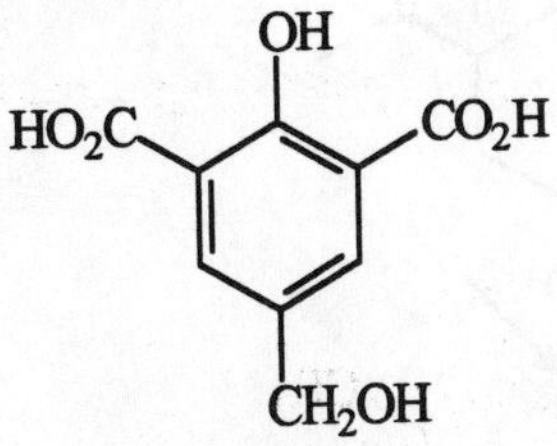

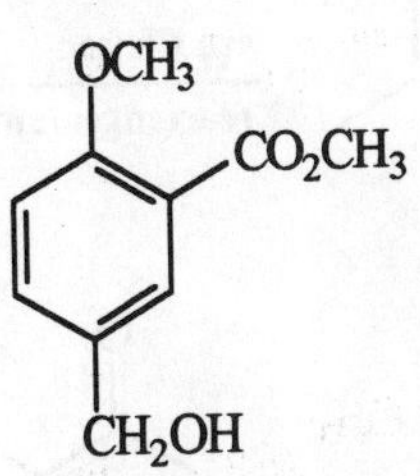

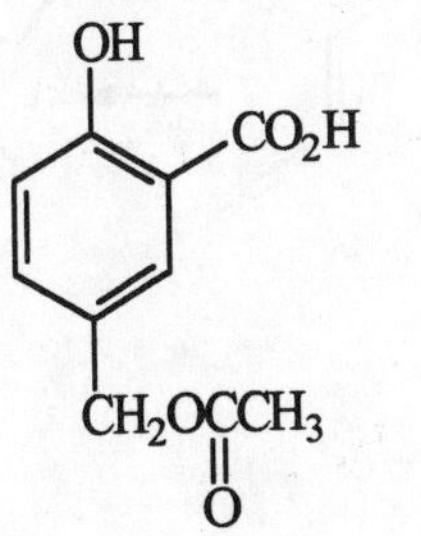

d. 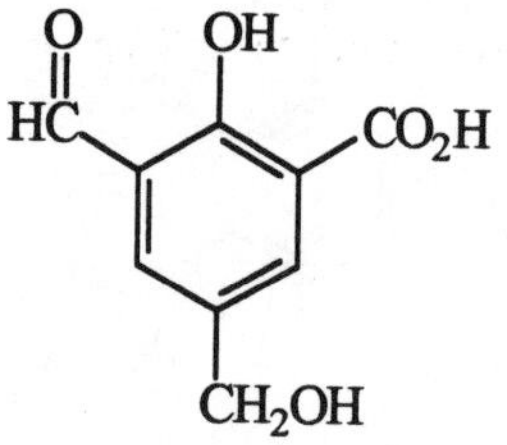 (structure: benzene ring with HC(=O), OH, CO_2H, CH_2OH substituents)

e. 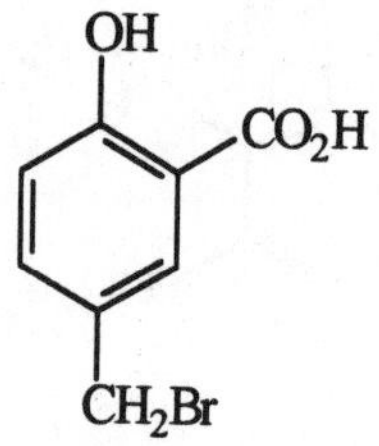 (structure: benzene ring with OH, CO_2H, CH_2Br substituents)

S5. a. $C_6H_5CH_2CHClCO_2H$

b. 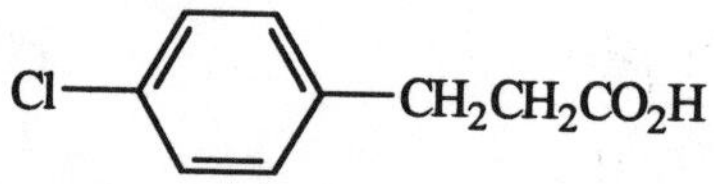 Cl–C_6H_4–$CH_2CH_2CO_2H$

c. $C_6H_5CHClCH_2CO_2H$

d. $C_6H_5CH_2CH_2COCl$

S6. 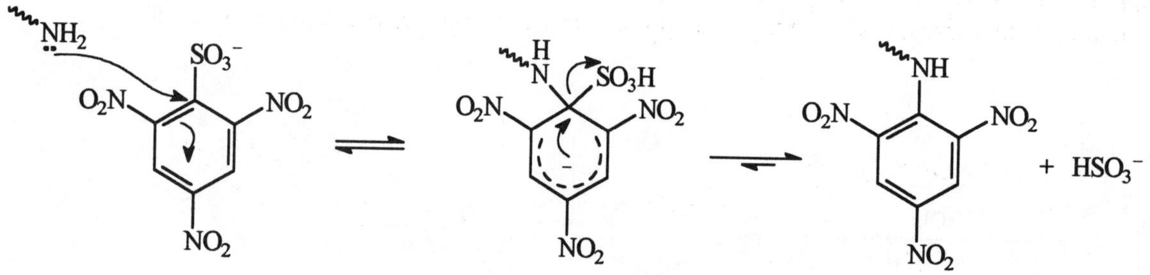 NH_2 SO_3^- O_2N NO_2 NO_2 ⇌ H N SO_3H O_2N NO_2 NO_2 ⇌ NH O_2N NO_2 NO_2 + HSO_3^-

S7. The propagation steps of a free radical oxidation process are the following:

$$R\cdot + O_2 \longrightarrow ROO\cdot$$
$$ROO\cdot + RH \longrightarrow ROOH + R\cdot$$

Phenols such as hydroquinone, BHA, and BHT interrupt this chain because they lose a hydrogen atom more readily than RH. By doing so, they form a stable radical which is unable to continue the propagation chain.

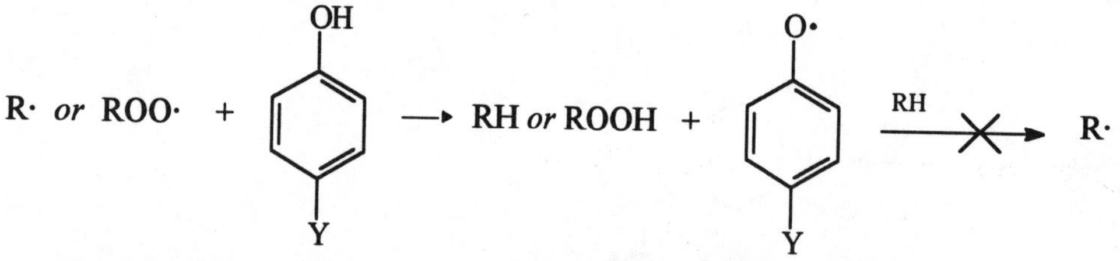

R· *or* ROO· + (OH, Y-substituted benzene) ⟶ RH *or* ROOH + (O·, Y-substituted benzene) $\xrightarrow{RH}$ ✗ R·

S8. The clue is the fact that the enamine alkylation proceeds with allylic rearrangement:

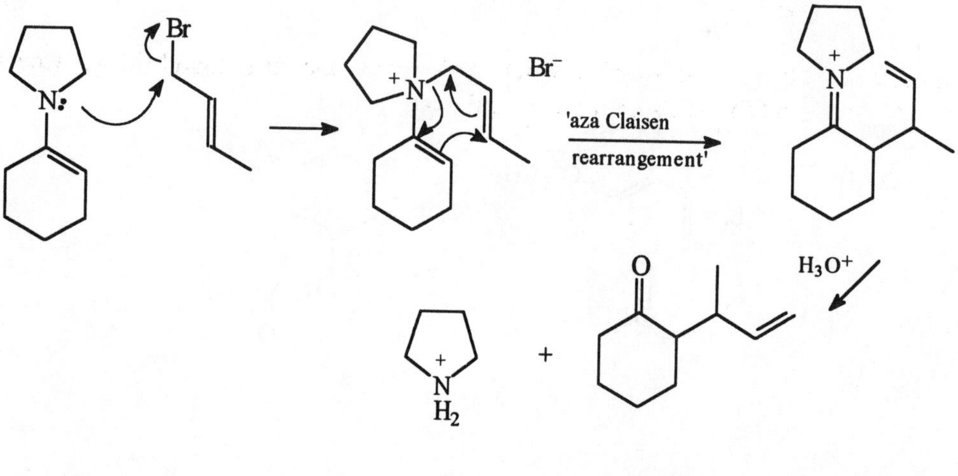

N: Br ⟶ +N Br⁻ 'aza Claisen rearrangement' ⟶ +N

H_3O^+

+N H_2 + O

S9. The product of HBr addition has a higher oxidation potential than hydroquinone itself, therefore only addition occurs:

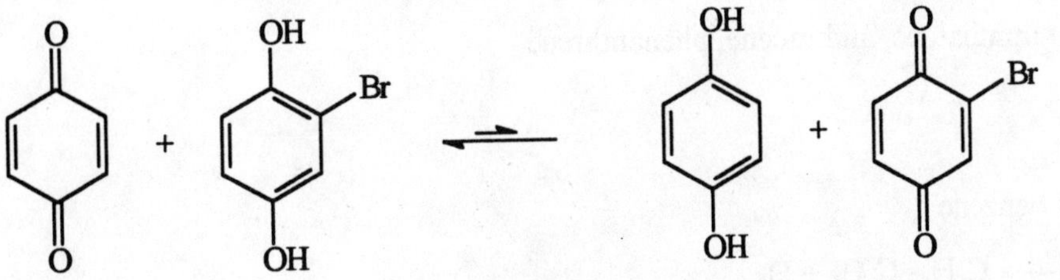

The opposite is true for the dimethylamine adduct. As soon as it is formed, it is oxidized by the starting material and another amine adds. This second product in turn undergoes oxidation as well:

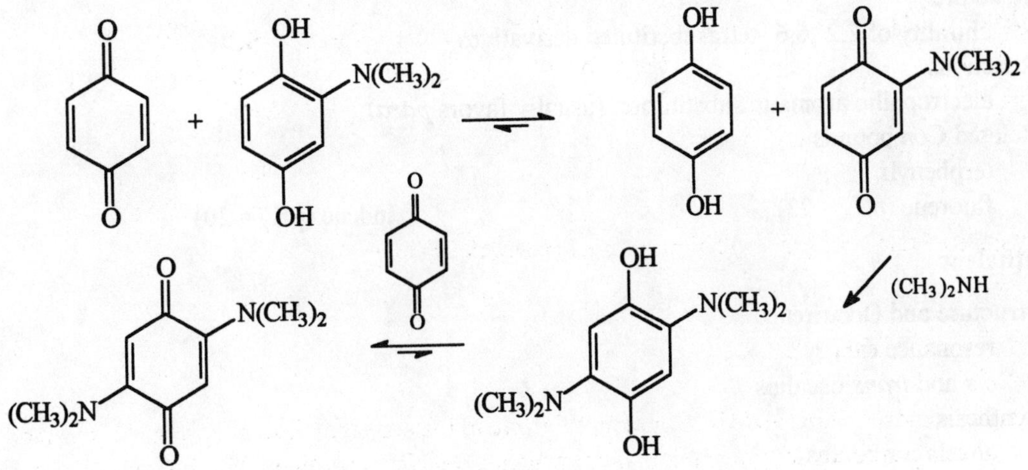

$(CH_3)_2NH$

b.

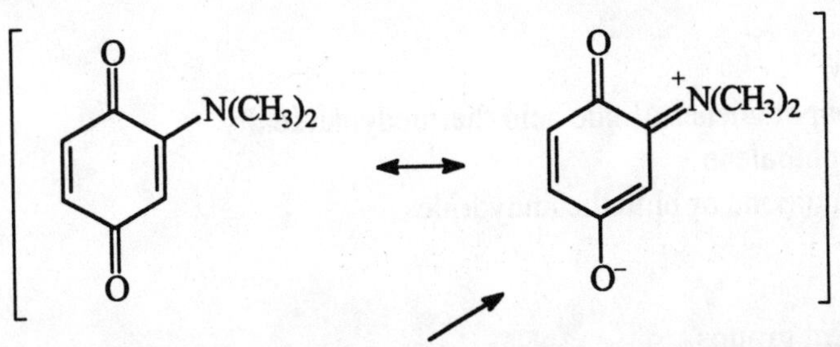

Because of this resonance contribution, 4-carbonyl is much less reactive, and the second Michael occurs at the β-position of the other carbonyl (β to α,β-unsaturated carbonyl)

31. POLYCYCLIC AROMATIC HYDROCARBONS

31.A. Chapter Outline and Important Terms Introduced

31.1 Nomenclature

biaryls
fused-ring systems: naphthalene, anthracene, phenanthrene

31.2 Biphenyl

A. Synthesis
   pyrolysis of benzene

$$2\ C_6H_6 \longrightarrow C_6H_5\text{-}C_6H_5 + H_2$$

   benzidine rearrangement (see Section 25.1.C)
   Ullmann reaction (see Section 30.3.C)
   Gomberg-Bachmann reaction (see Section 25.5.F)

B. Structure
   chirality of 2,2′,6,6′-tetrasubstituted derivatives

C. Reactions
   electrophilic aromatic substitution (usually favors *para*)

D. Related Compounds
   terphenyls
   fluorene ($pK_a = 23$)   indene ($pK_a = 20$)

31.3 Naphthalene

A. Structure and Occurrence
   resonance energy
   cis and *trans* decalins

B. Synthesis
   annelation routes
   aromatization and hydroaromatics
   Diels-Alder reactions of *p*-benzoquinone

C. Electrophilic Substitution
   favors 1-position kinetically
   sulfonation can provide 2-naphthalenesulfonic acid thermodynamically

D. Oxidation and Reduction of Naphthalene
   naphthalene to 1,4-naphthoquinone or phthalic anhydride
   Birch reduction

E. Substituted Naphthalenes
   transformations of substituent groups
   directive effects in electrophilic aromatic substitution reactions
   Bucherer reaction

31.4 Anthracene and Phenanthrene

A. Structure and Stability

B. Preparation of Anthracenes and Phenanthrenes
   annelation methods

C. Reactions
   oxidation to quinones

reduction to dihydro compounds
Diels-Alder reactions of anthracene
electrophilic aromatic substitution (occurs on central ring)

31.5 Higher Polybenzenoid Hydrocarbons

acenes
graphite
pyrene
C_{60} - Buckminsterfullerene
benz-, benzo- derivatives
carcinogens
hexahelicene

31.B. Important Reactions Introduced

Synthesis of polycyclic aromatic hydrocarbons by annelation:

Friedel-Crafts acylation (31.3.B)

Equation:
e.g.

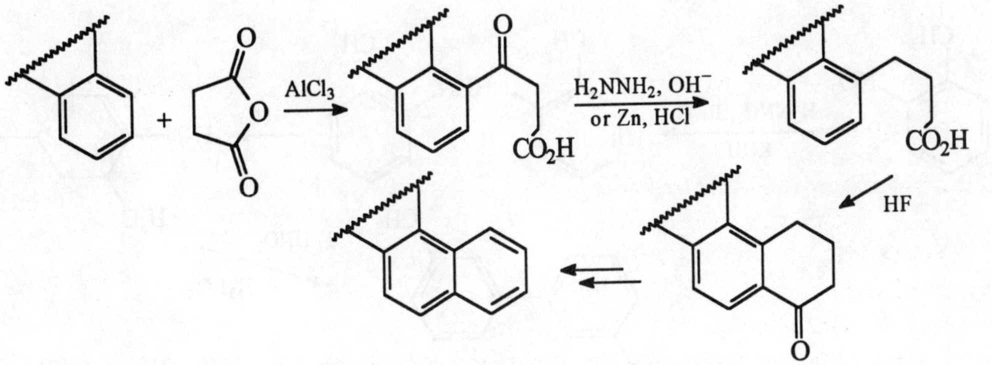

AlCl3; H_2NNH_2, OH^- or Zn, HCl; CO_2H; HF

Diels-Alder reaction (31.3.B)

Equation:

e.g.

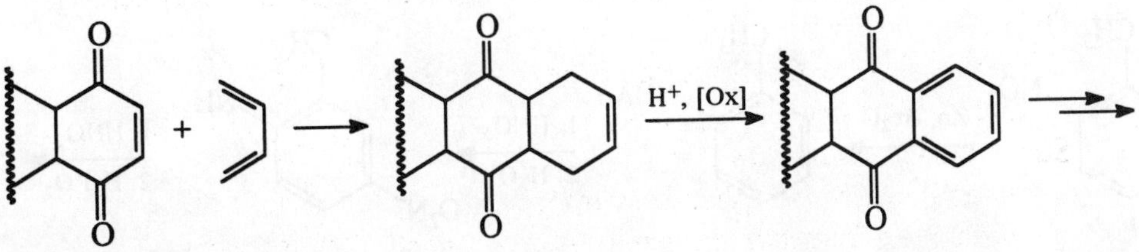

H^+, [Ox]

Bucherer reaction (31.3.E)

Equation:

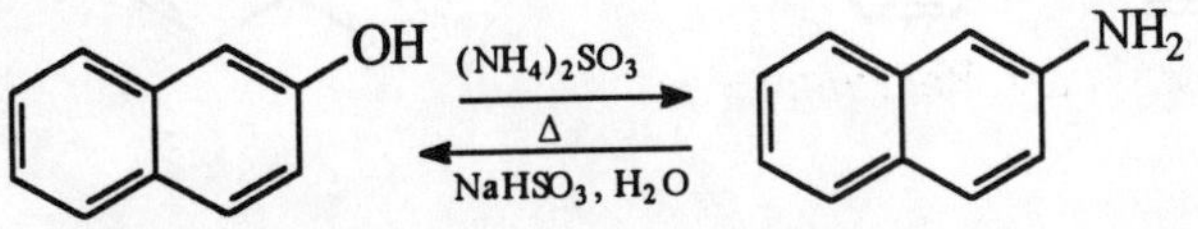

OH; $(NH_4)_2SO_3$, Δ; $NaHSO_3$, H_2O; NH_2

Generality: Reversible, required catalysis by sulfite
Specific for naphthalenes and higher fused polycyclics

Diels-Alder reaction of anthracene (31.4.C)
Equation:

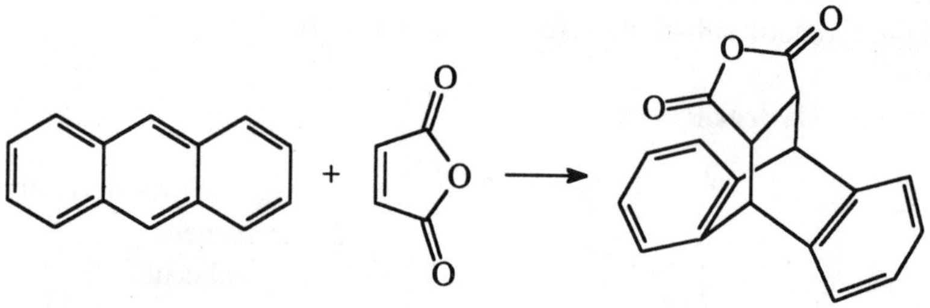

31.D. Answers to Exercises

31.1 a. 6-bromonaphthalene-2-carboxylic acid
b. 1-bromo-2,5-dimethylanthracene
c. 3-chloro-9,10-dihydrophenanthrene

31.2
a.

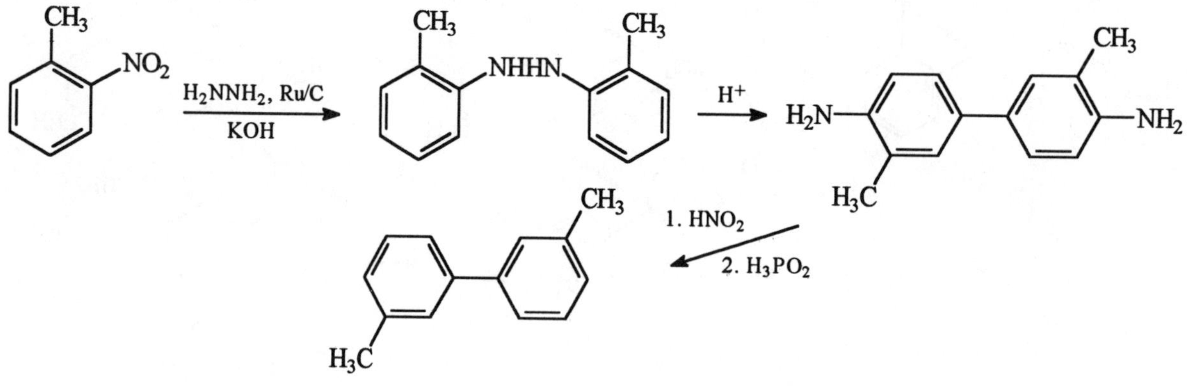

CH_3, NO_2; H_2NNH_2, Ru/C, KOH; CH_3 NHHN CH_3; H^+; H_2N, CH_3, NH_2, H_3C; 1. HNO_2 2. H_3PO_2; CH_3, H_3C

b.

CH_3, NO_2; Zn, Ac_2O; CH_3, NHAc; 1. HNO_3 2. H_3O^+; CH_3, NH_2, O_2N; 1. HNO_2 2. H_3PO_2; CH_3, O_2N; Zn, HCl; HNO_2; CH_3, $^+N_2$; benzene, KOH, Δ; CH_3

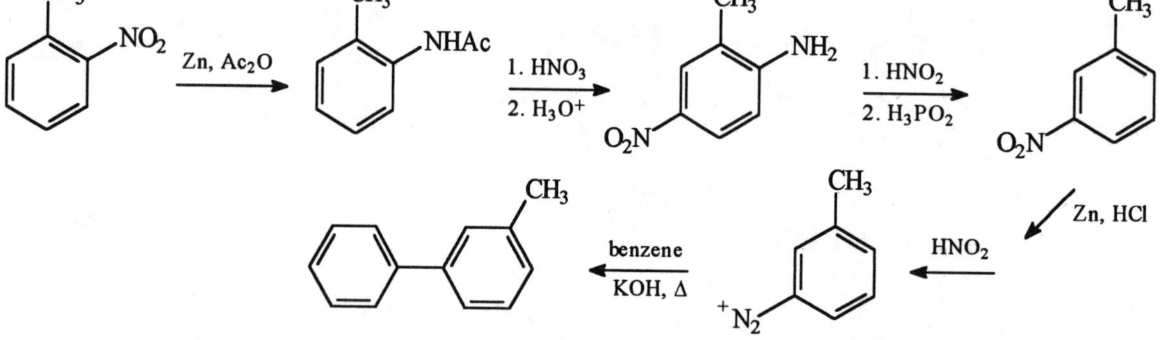

31.3 [there is no answer for a.] b.

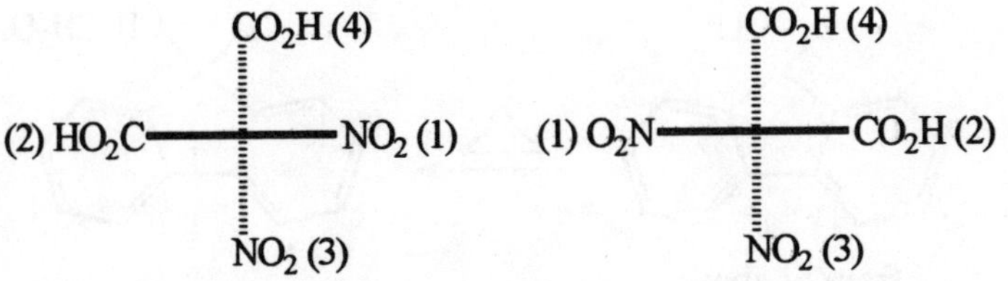

CO_2H (4), (2) HO_2C, NO_2 (1), NO_2 (3) — *S*

CO_2H (4), (1) O_2N, CO_2H (2), NO_2 (3) — *R*

31.4 *para* attack:

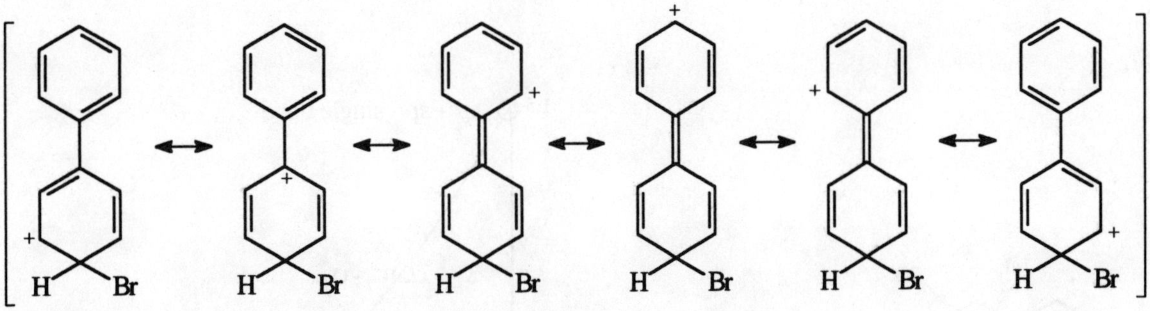

H Br ↔ H Br ↔ H Br ↔ H Br ↔ H Br ↔ H Br

meta attack:

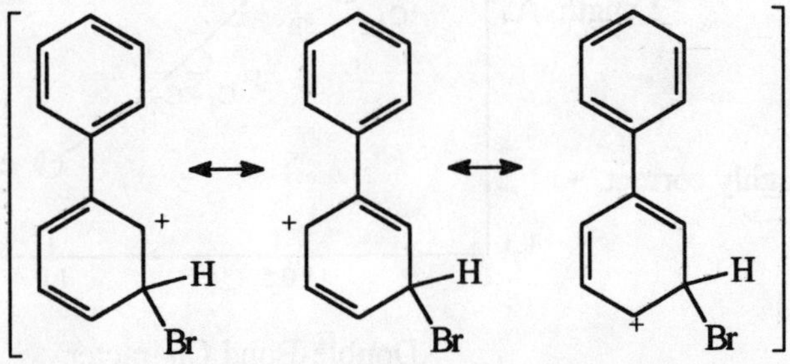

H Br ↔ H Br ↔ H Br

The phenyl substituent itself can stabilize the positive charge via resonance when it is in the *para* position.

31.5 a.

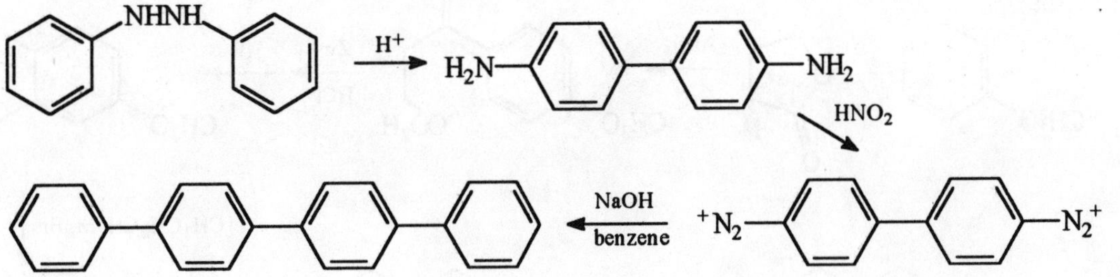

NHNH → H^+ → H_2N–C₆H₄–C₆H₄–NH_2 → HNO_2 → $^+N_2$–C₆H₄–C₆H₄–N_2^+ → NaOH, benzene →

b.

H H → BuLi → H Li^+ → CO_2 → H CO_2H

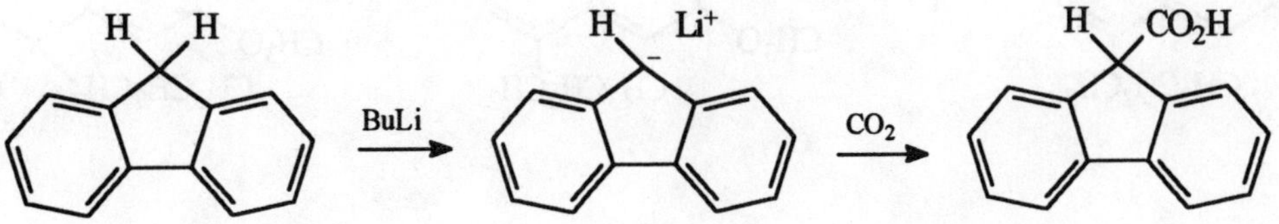

c.

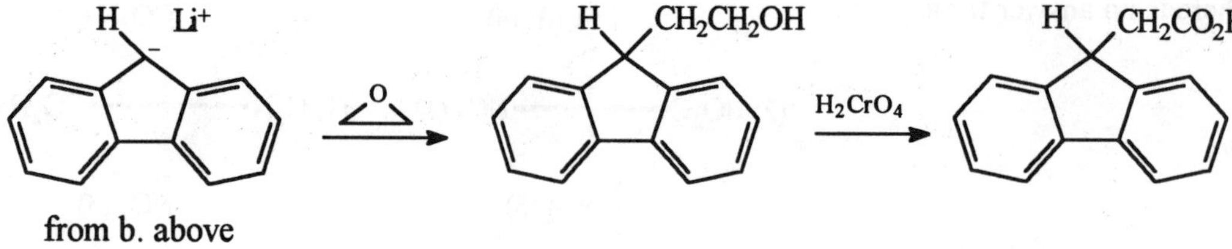

H Li+ → (ethylene oxide) → H CH_2CH_2OH → H_2CrO_4 → H CH_2CO_2H

from b. above

31.6 Fractional double bond character = the number of resonance structures with a double bond divided by the total number of resonance structures: (Benzene = 1.40Å, CH_2=CH_2 = 1.33Å, see Text p.606)

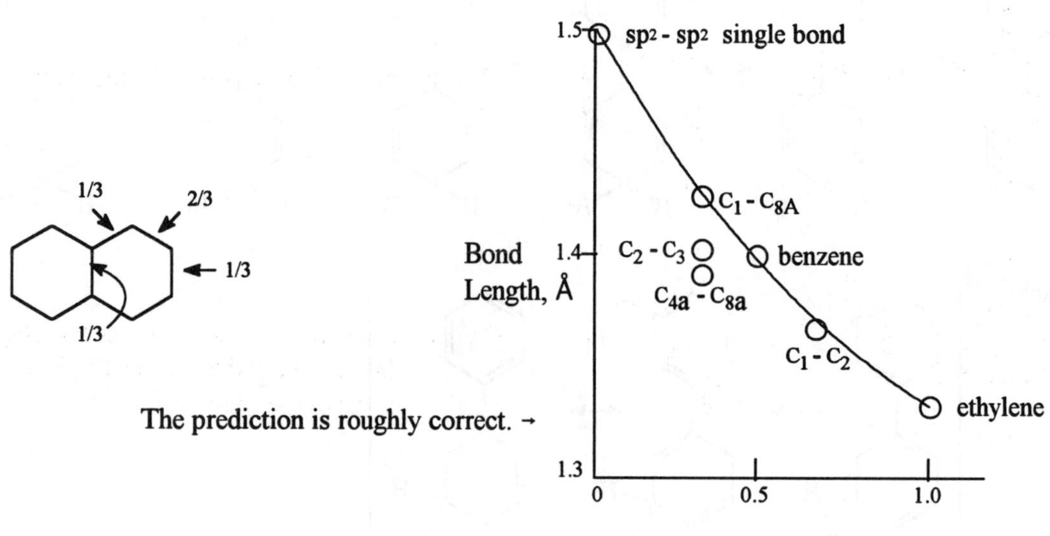

31.7

a.

CH_3O-benzene + succinic anhydride → $AlCl_3$ → CH_3O-aryl ketone with CO_2H → Zn, HCl → HF → CH_3O-tetralone (O)

↓ $CH_3CH_2CH_2MgBr$

CH_3O, $CH_3CH_2CH_2$, OH ← H_2SO_4, Δ ← CH_3O, $CH_3CH_2CH_2$ ← S, Δ ← CH_3O, $CH_3CH_2CH_2$ (naphthalene)

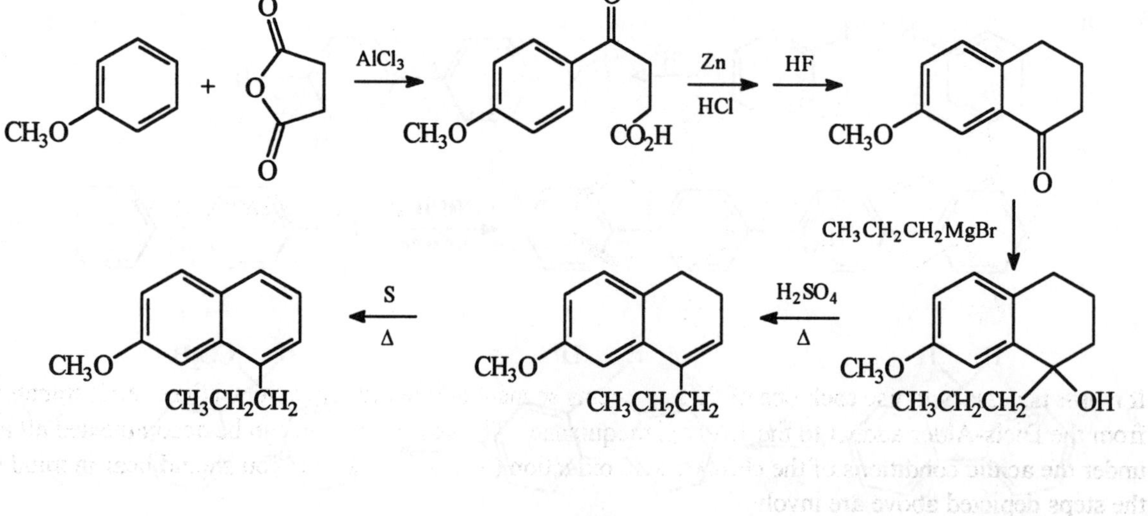

b. [2,3-dimethyl-1,4-benzoquinone] + [1,3-butadiene] → [Diels-Alder adduct] $\xrightarrow[H_2SO_4]{K_2Cr_2O_7}$ [2,3-dimethyl-1,4-naphthoquinone]

NOTE: The reaction sequence outlined below, which is depicted in the text, often confuses students. It is intended to illustrate that, with proper choice of reagents, it is possible to carry out the sequence in a stepwise fashion.

$\xrightarrow[\Delta]{HCl}$ (This reaction is simply a double enolization.)

$\xrightarrow{HNO_2}$ (This is an oxidation, which stops at this stage when HNO_2 is the oxidant. Note that HNO_2 is a mild oxidizing agent and a weak acid.)

$\xrightarrow{H^+}$ (This reaction is another double enolization. It occurs under the acidic conditions of $K_2Cr_2O_7/H_2SO_4$ oxidation).

$\xrightarrow[H_2SO_4]{K_2Cr_2O_7}$ (This is the final oxidation step.)

It is **not** necessary to use each one of these reagents sequentially in order to achieve the overall transformation from the Diels-Alder adduct to the 1,4-naphthoquinone. The transformation can be accomplished all at once under the acidic conditions of the chromic acid oxidation (below). However, you should bear in mind that all of the steps depicted above are involved.

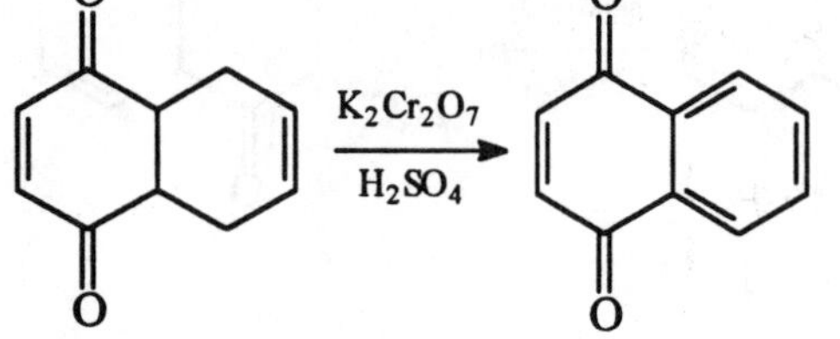

$K_2Cr_2O_7$ / H_2SO_4

31.8 a.

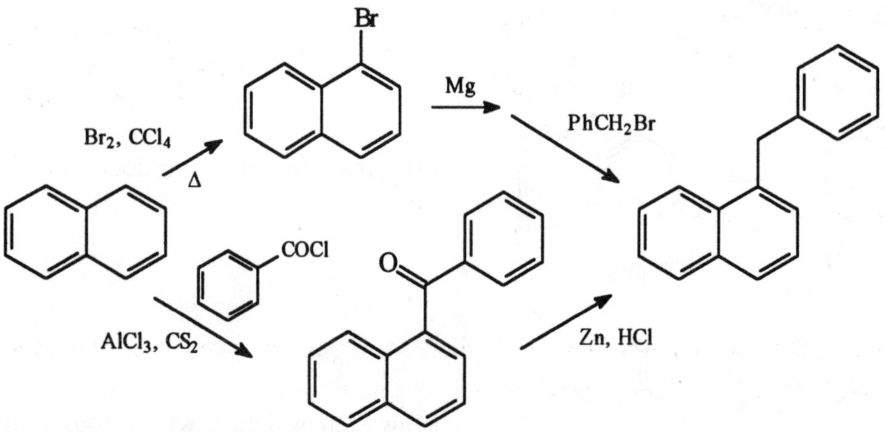

Br_2, CCl_4, Δ

Br

Mg

$PhCH_2Br$

COCl

$AlCl_3$, CS_2

O

Zn, HCl

b.

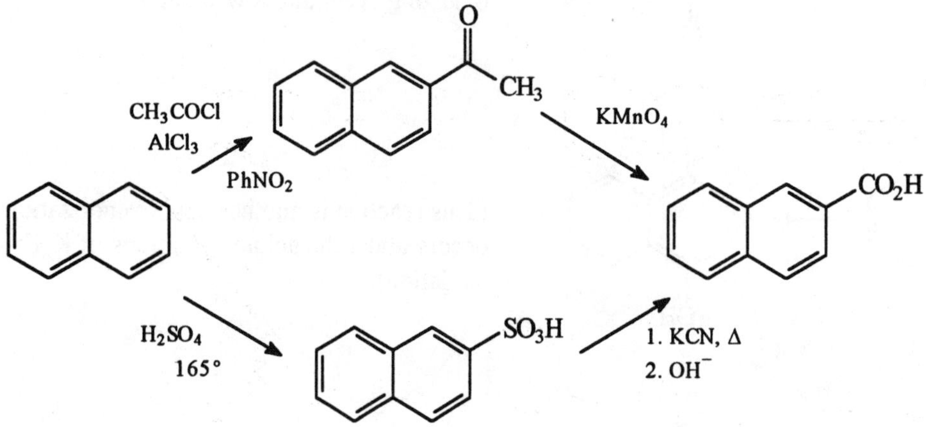

CH_3COCl / $AlCl_3$ / $PhNO_2$

O

CH_3

$KMnO_4$

CO_2H

H_2SO_4 / 165°

SO_3H

1. KCN, Δ
2. OH^-

31.9 a. air, V_2O_5, Δ

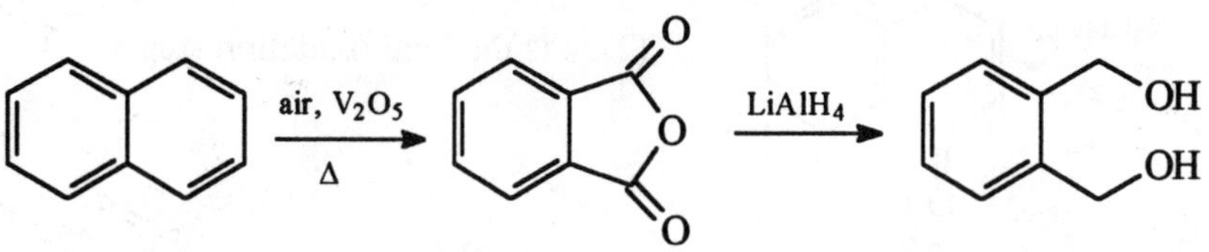

O

O

O

$LiAlH_4$

OH

OH

b. Na / NH_3(l)

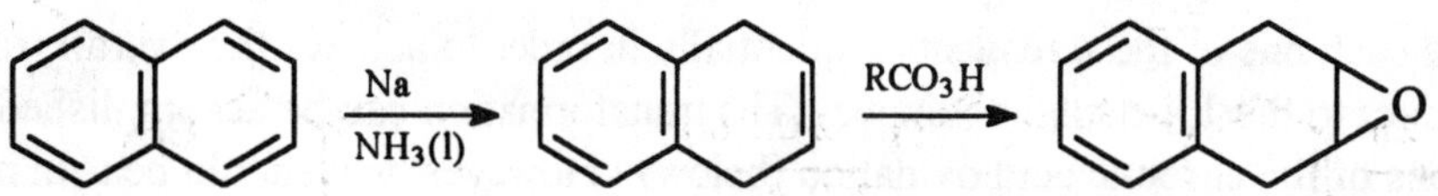

RCO_3H

O

31.10 1: 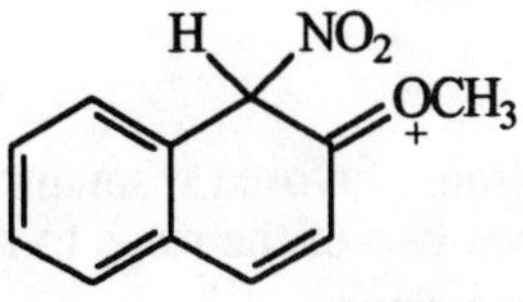 3:

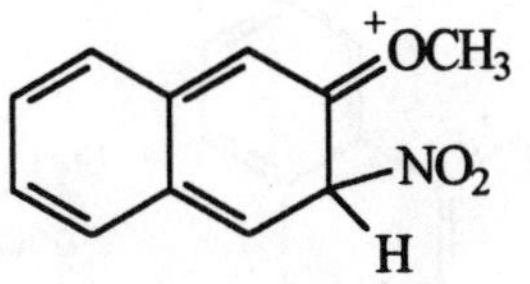

4:

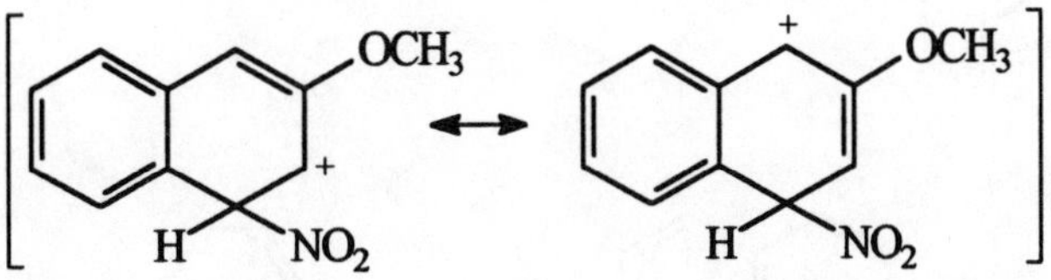

5: 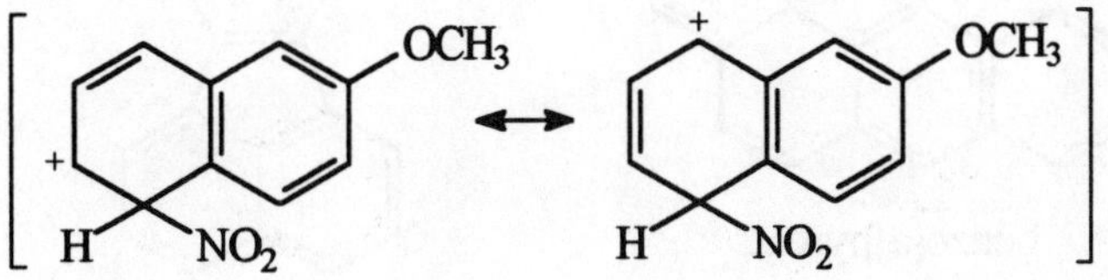 6:

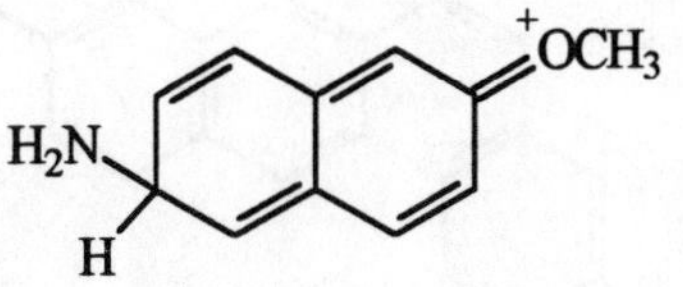

7: 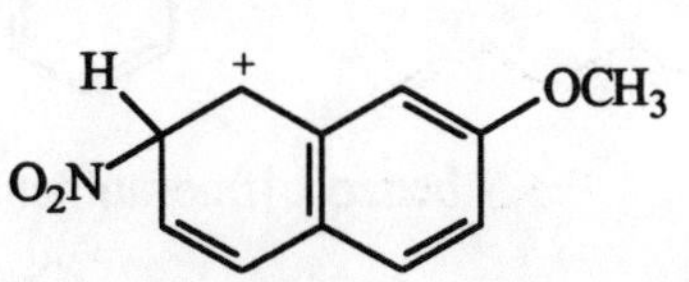 8:

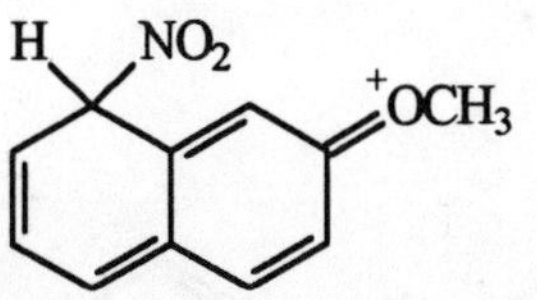

Only the 1-substitution intermediate can utilize the oxygen lone pair electrons to stabilize the positive charge and maintain an aromatic ring.

31.11

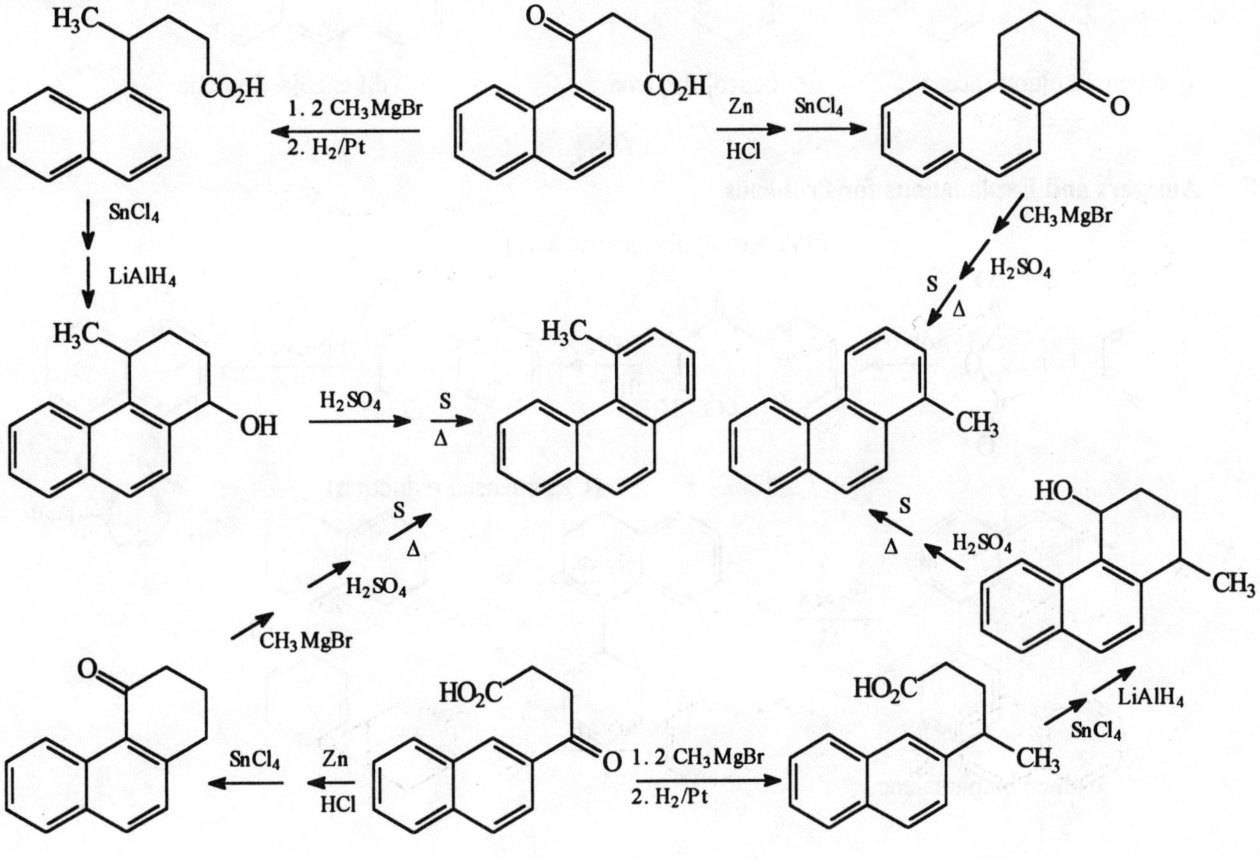

1. 2 CH_3MgBr
2. H_2/Pt

Zn, HCl; $SnCl_4$

$SnCl_4$; $LiAlH_4$

CH_3MgBr; H_2SO_4; S, Δ

H_2SO_4; S, Δ

CH_3MgBr; H_2SO_4; S, Δ

H_2SO_4; S, Δ

$LiAlH_4$; $SnCl_4$

$SnCl_4$; Zn, HCl

1. 2 CH_3MgBr
2. H_2/Pt

31.12

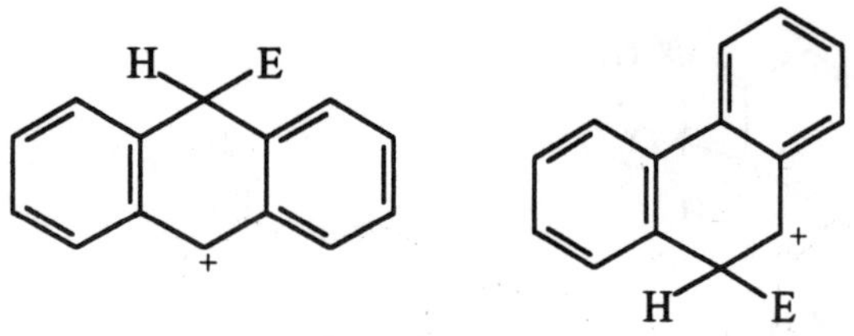

H E

+

+

H E

These are the two most favored resonance structures, because each allows two of the rings to remain fully aromatic as benzene rings.

31.13

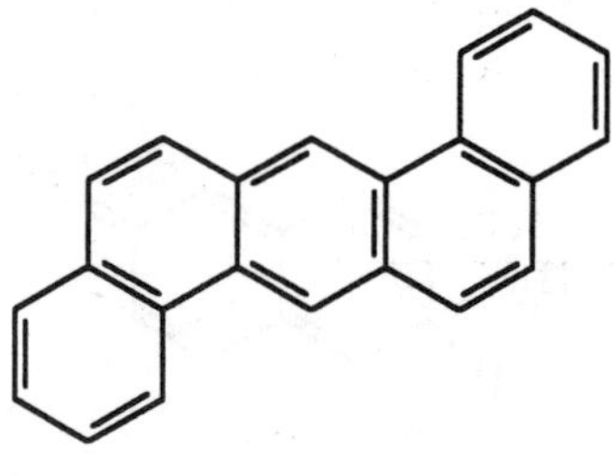

dibenz[a,h]-anthracene

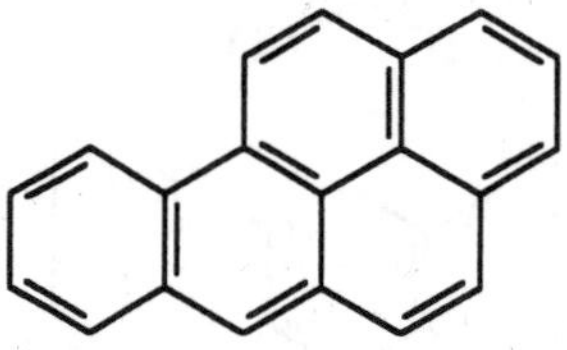

benzo[a]pyrene

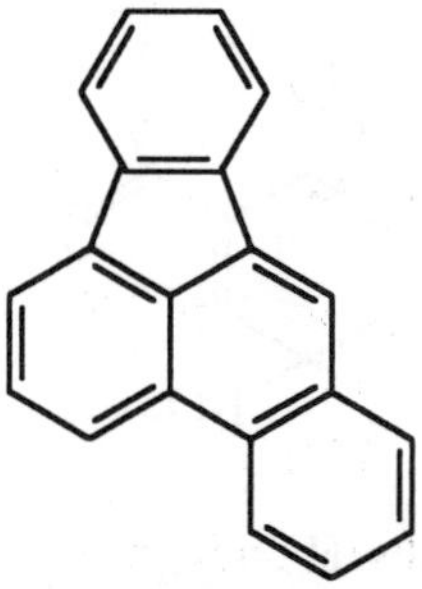

benzo[b]fluoranthene

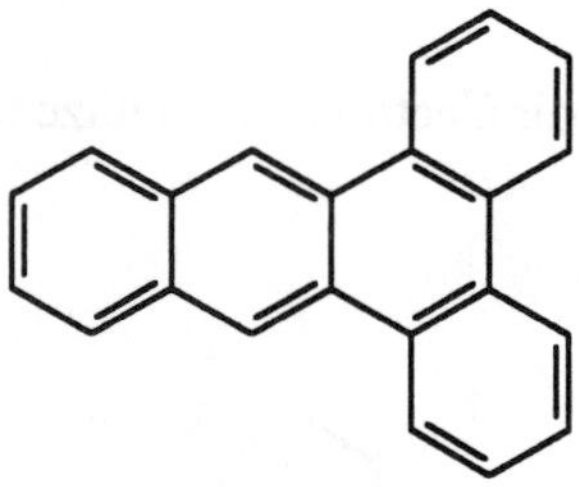

dibenz[a,c]anthracene

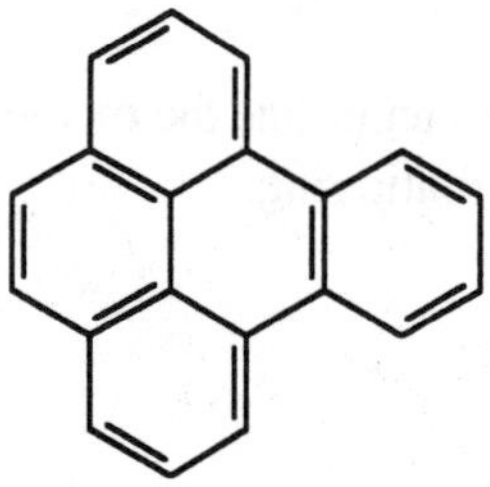

benzo[e]pyrene

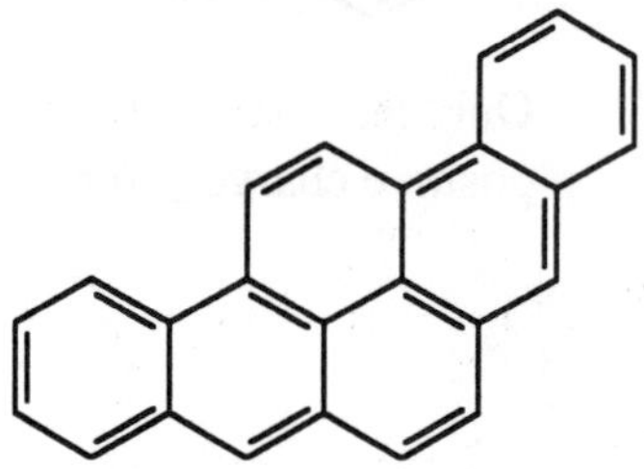

dibenzo[a,i]pyrene

31.E. Answers and Explanations for Problems

1. a.

PPA = (polyphosphoric acid)

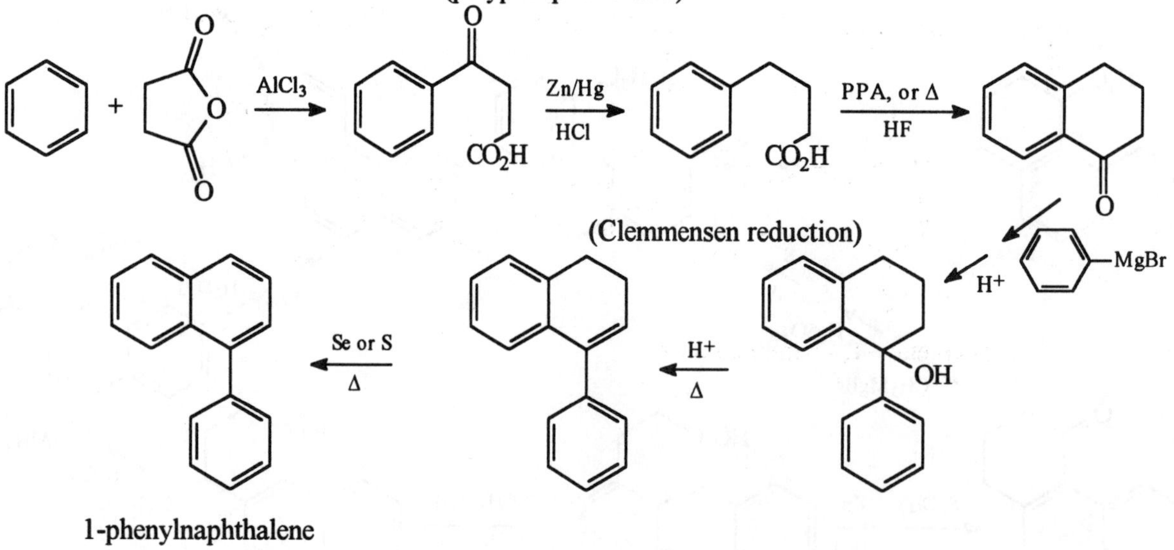

$AlCl_3$

Zn/Hg, HCl

CO_2H

CO_2H

PPA, or Δ, HF

O

(Clemmensen reduction)

H+

MgBr

OH

H+, Δ

Se or S, Δ

1-phenylnaphthalene

b.

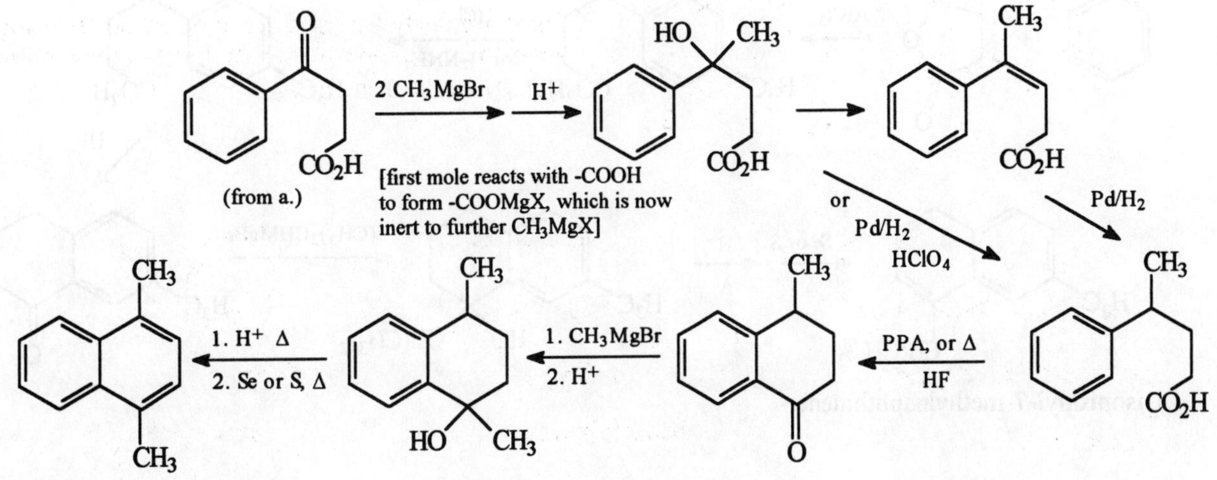

2 CH_3MgBr → H^+ →

CO_2H (from a.)

[first mole reacts with -COOH to form -COOMgX, which is now inert to further CH_3MgX]

HO CH_3 CO_2H → CH_3 CO_2H

or Pd/H_2 $HClO_4$

Pd/H_2

CH_3 CO_2H

PPA, or Δ HF

1. CH_3MgBr 2. H^+

1. H^+ Δ 2. Se or S, Δ

1,4-dimethylnaphthalene

[NOTE: if not hydrogenated first, we would obtain a naphthol at this point]

Alternative: CH_3 CH_3 + (succinic anhydride) → CH_3 O CO_2H CH_3 → etc.

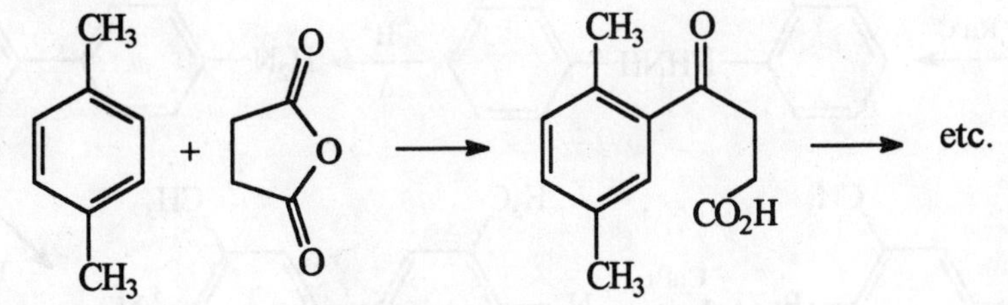

c.

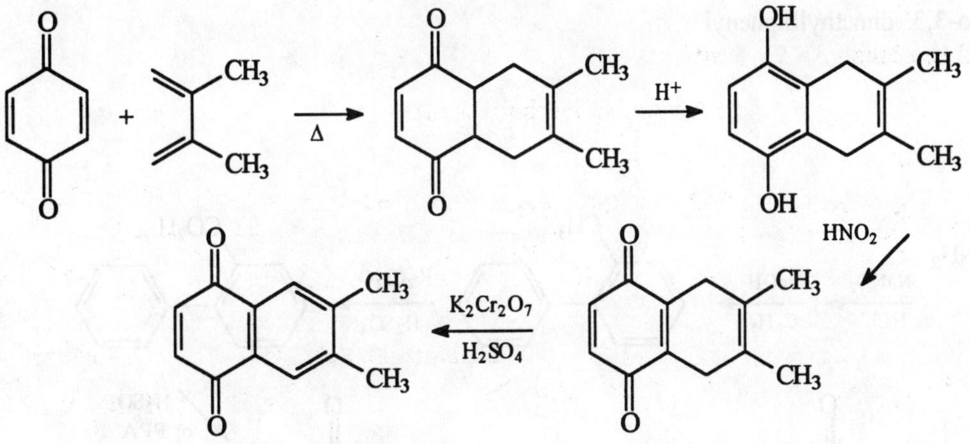

O O + CH_3 CH_3 → Δ → O O CH_3 CH_3 → H^+ → OH OH CH_3 CH_3

HNO_2

O O CH_3 CH_3 ← $K_2Cr_2O_7$ H_2SO_4 ← O O CH_3 CH_3

6,7-dimethyl-1,4-naphthoquinone

d.

CH_3 + (succinic anhydride) $\xrightarrow{AlCl_3}$ H_3C–C_6H_4–$CO(CH_2)_2CO_2H$ $\xrightarrow[\text{or } H_2NNH_2 \text{ glycerol, KOH}]{Zn/Hg,\ HCl}$ H_3C–C_6H_4–$(CH_2)_3CO_2H$ $\xrightarrow{HF}$ (tetralone, H_3C, O)

$\xrightarrow{(CH_3)_2CHMgBr}$ (H_3C, HO, $CH(CH_3)_2$) $\xrightarrow[\Delta]{H^+}$ $\xrightarrow[\Delta]{\text{Se or S}}$ (H_3C, $CH(CH_3)_2$)

1-isopropyl-7-methylnaphthalene

e.

(NO_2, CH_3) $\xrightarrow[\text{alc. KOH}]{H_2NNH_2\ \ Ru/C}$ (CH_3, H_3C, –NHNH–) $\xrightarrow[\Delta]{H^+}$ (H_3C, CH_3, H_2N–, –NH_2) $\xrightarrow[HBr]{NaNO_2}$ (H_3C, CH_3, $^+N_2$–, –$\overset{+}{N_2}$) $\xrightarrow{CuBr}$ (H_3C, CH_3, Br–, –Br)

4,4′-dibromo-3,3′-dimethylbiphenyl

f.

(CH_3, NH_2) $\xrightarrow[HCl]{NaNO_2}$ $\xrightarrow[C_6H_6]{NaOH}$ (CH_3) $\xrightarrow[H_2SO_4]{K_2Cr_2O_7}$ (CO_2H) $\xrightarrow[\text{or PPA}]{H_2SO_4}$ (O) $\xrightarrow[H_2SO_4]{HNO_3}$ (O, NO_2)

2-nitrofluorenone

g. CH_3 + O O O $\xrightarrow{AlCl_3}$ O H_3C HO_2C $\xrightarrow[\Delta]{H_2SO_4}$ O O H_3C

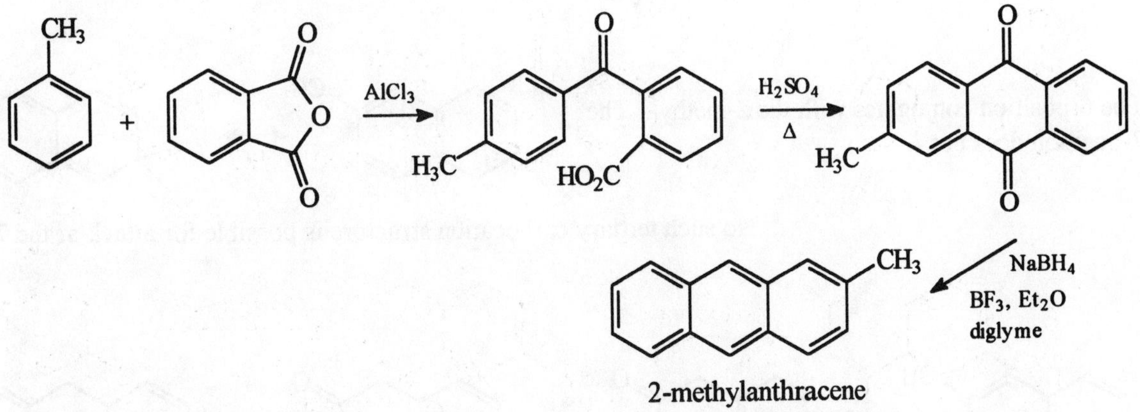

CH_3 NaBH$_4$ BF$_3$, Et$_2$O diglyme

2-methylanthracene

h. CHO + $(CH_3CO)_2O$ $\xrightarrow{NaOCOCH_3}$ CO_2H cinnamic acid $\xrightarrow[Pt]{H_2}$ CO_2H

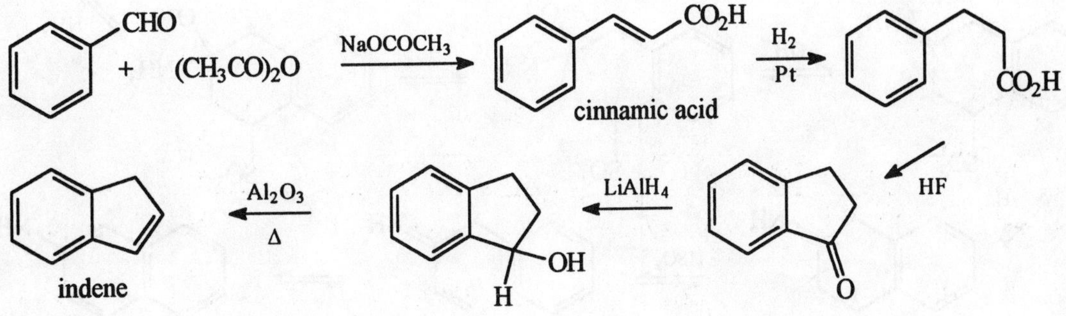

HF

O

LiAlH$_4$

OH H

Al_2O_3 Δ

indene

2. CO_2H NH_2 anthranilic acid $\xrightarrow{(CH_3CO)_2O}$ CO_2H $NHCOCH_3$ $\xrightarrow[(CH_3CO)_2O]{HNO_3}$ CO_2H $NHCOCH_3$ NO_2

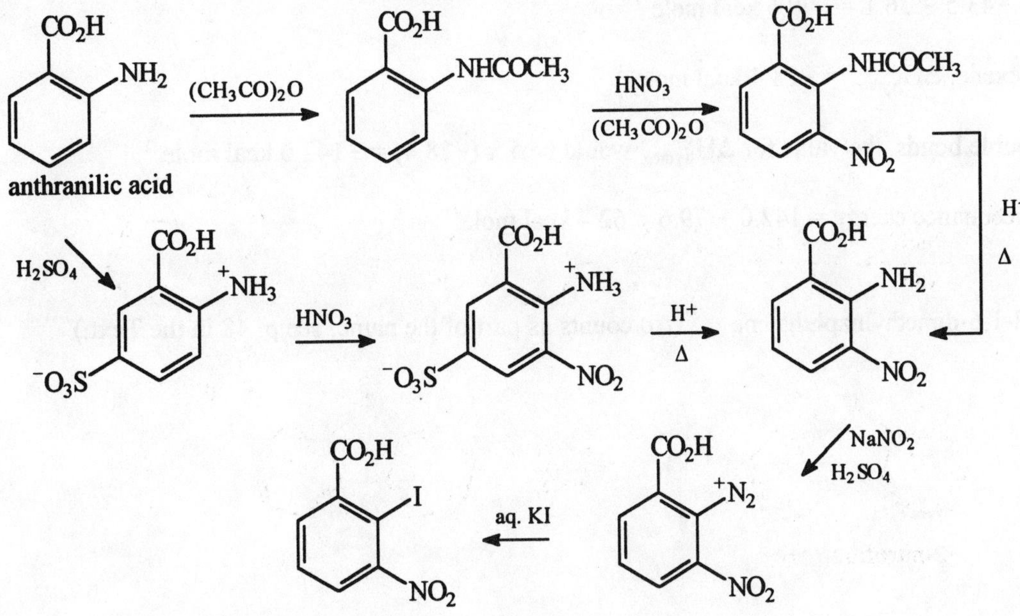

H^+ Δ

H_2SO_4 CO_2H $\overset{+}{N}H_3$ ^-O_3S $\xrightarrow{HNO_3}$ CO_2H $\overset{+}{N}H_3$ ^-O_3S NO_2 $\xrightarrow[\Delta]{H^+}$ CO_2H NH_2 NO_2

$NaNO_2$ H_2SO_4

CO_2H I NO_2 $\xleftarrow{aq.\ KI}$ CO_2H $\overset{+}{N}_2$ NO_2

3. The 6-position conjugates with the 2-methyl. The 7-position does not.

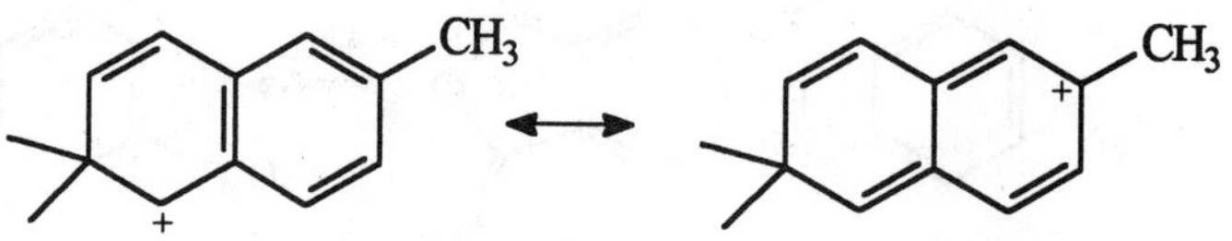

CH_3 ⟷ CH_3

No such tertiary carbocation structure is possible for attack at the 7-position.

4. OH ⇌ O $\underset{}{\overset{HSO_3^-}{\rightleftharpoons}}$ O, SO_3H ⇌ O, SO_3^-

NH_3

$^+NH_2$, SO_3^- $\underset{}{\overset{-OH^-}{\rightleftharpoons}}$ OH, NH_2, SO_3^- ⇌ O^-, $^+NH_3$, SO_3^-

$-H^+$

NH, SO_3H $\underset{}{\overset{-HSO_3^-}{\rightleftharpoons}}$ NH ⇌ NH_2

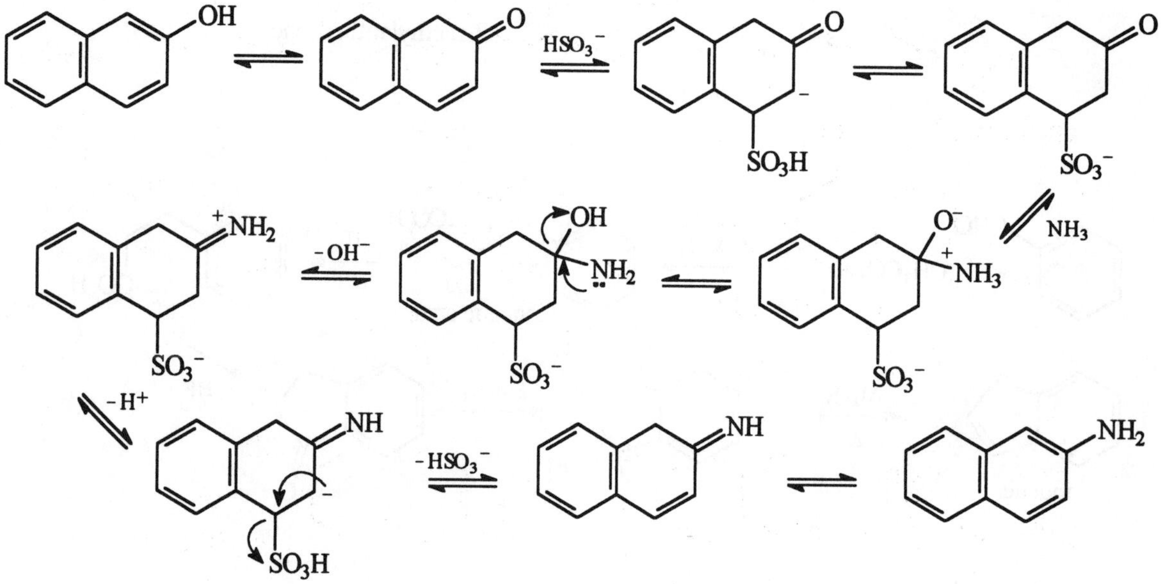

5. a. $\Delta H^\circ_{hydrog.} = -43.5 - 36.1 = -79.6$ kcal mole^{-1}

b. For cyclohexene, $\Delta H^\circ_{hydrog.} = -28.4$ kcal mole^{-1}

For five double bonds, the value for $\Delta H^\circ_{hydrog.}$ would be $5 \times (-28.4) = -142.0$ kcal mole^{-1}

c. Empirical resonance energy = $142.0 - 79.6 = 62.4$ kcal mole^{-1}

6. a. 4-isopropyl-1,6-dimethylnaphthalene (*iso* counts as part of the name; see p. 48 in the Text.)

b.

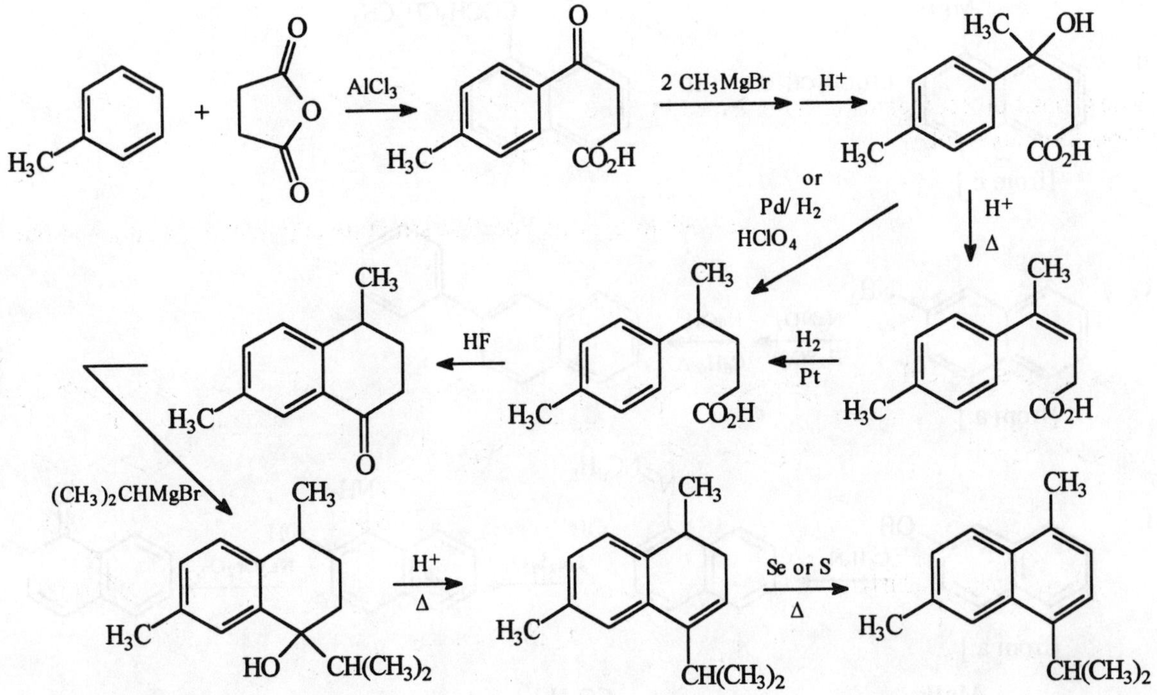

+ $AlCl_3$ → 2 CH_3MgBr, H^+ → H^+, Δ → H_2/Pt ← ; or Pd/ H_2, $HClO_4$; HF ← ; $(CH_3)_2CHMgBr$ → H^+, Δ → Se or S, Δ →

H_3C ; O ; CO_2H ; H_3C, OH ; CH_3 ; HO ; $CH(CH_3)_2$

7. a.

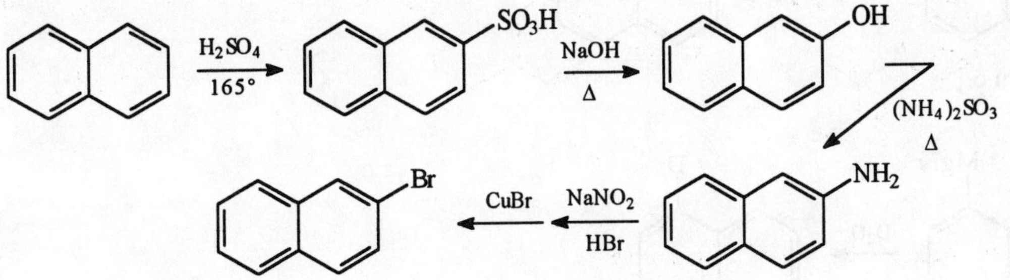

H_2SO_4, 165° → SO_3H ; NaOH, Δ → OH ; $(NH_4)_2SO_3$, Δ → NH_2 ; $NaNO_2$, HBr → CuBr → Br

b.

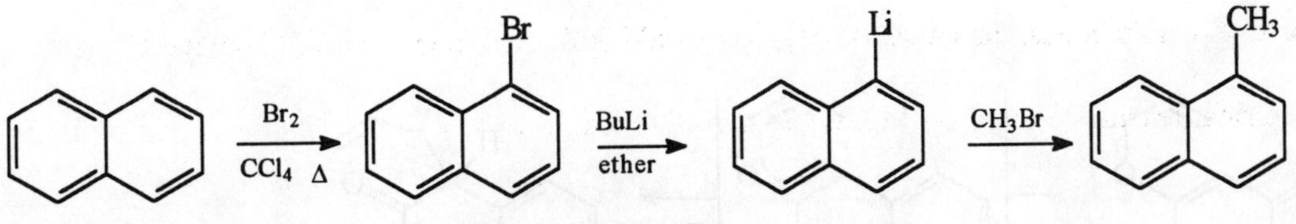

Br_2, CCl_4 Δ → Br ; BuLi, ether → Li ; CH_3Br → CH_3

(see Sect. 30.3.C)

c.

Br ; Mg, ether → MgBr ; 1. $(CH_3)_2CO$ 2. H^+ → OH, CH_3CCH_3 ; Pd/ H_2, $HClO_4$ → CH_3CHCH_3

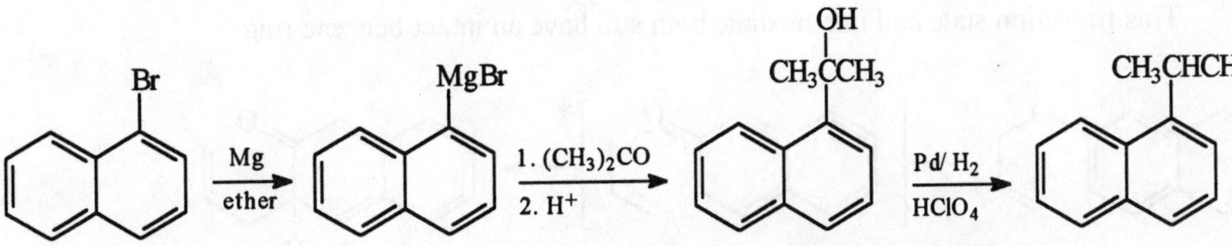

[from b.]

NOTE: benzylic-type alcohol hydrogenolysis

d.

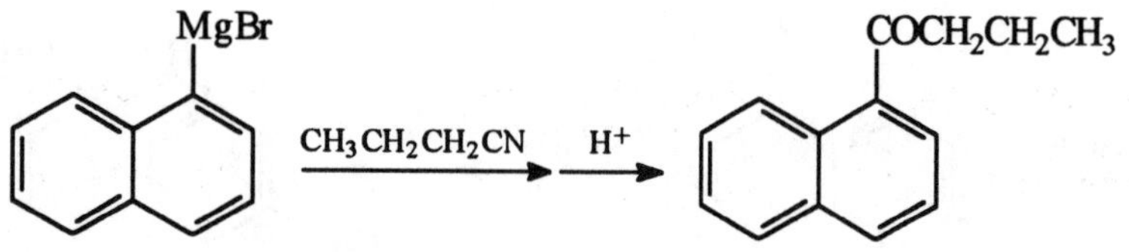

MgBr

[from c.]

$CH_3CH_2CH_2CN$ → H^+ →

$COCH_2CH_2CH_3$

e.

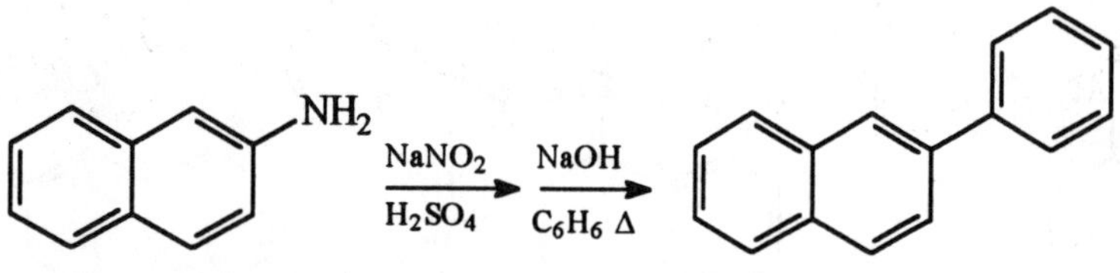

NH_2

[from a.]

$NaNO_2$ / H_2SO_4 → NaOH / C_6H_6 Δ →

f.

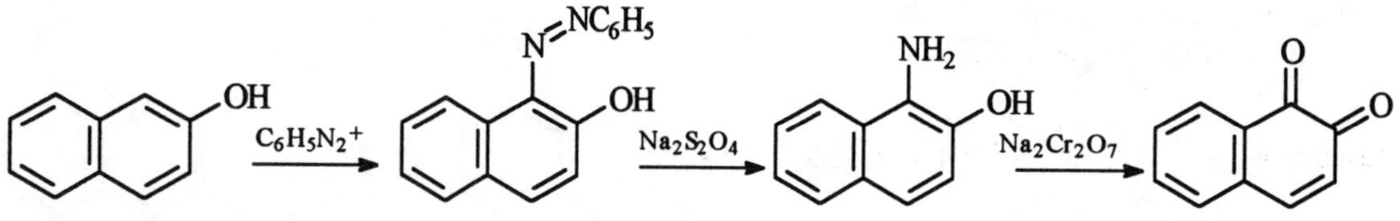

OH

[from a.]

$C_6H_5N_2^+$ →

$N{=}NC_6H_5$, OH

$Na_2S_2O_4$ →

NH_2, OH

$Na_2Cr_2O_7$ →

O, O

g.

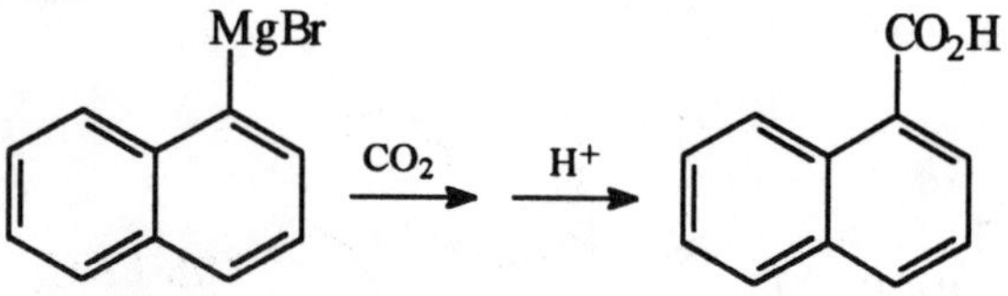

MgBr

[from c.]

CO_2 → H^+ →

CO_2H

h.

MgBr

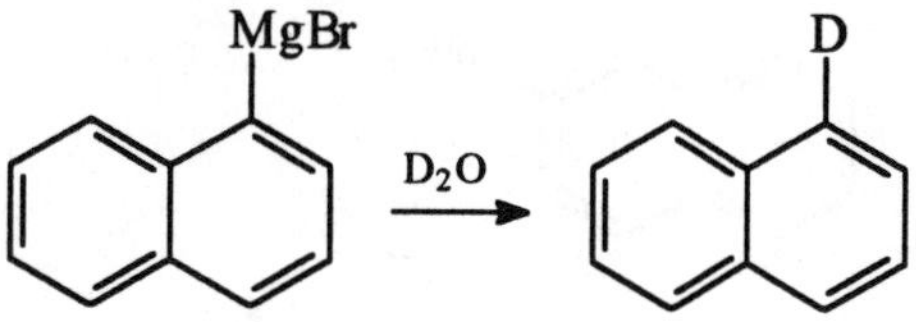

[from c.]

D_2O →

D

8.

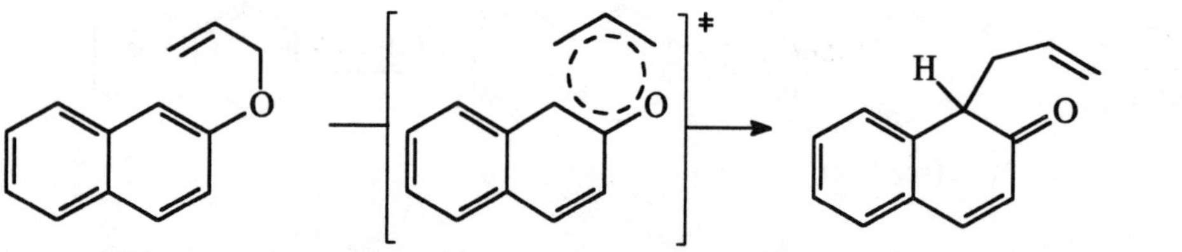

O → [O]‡ → H, O

This transition state and intermediate both still have an intact benzene ring.

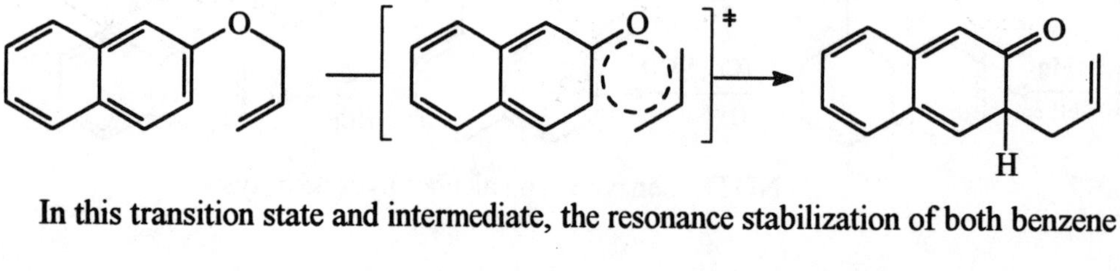

O → [O]‡ → O, H

In this transition state and intermediate, the resonance stabilization of both benzene rings has been lost.

9. a.

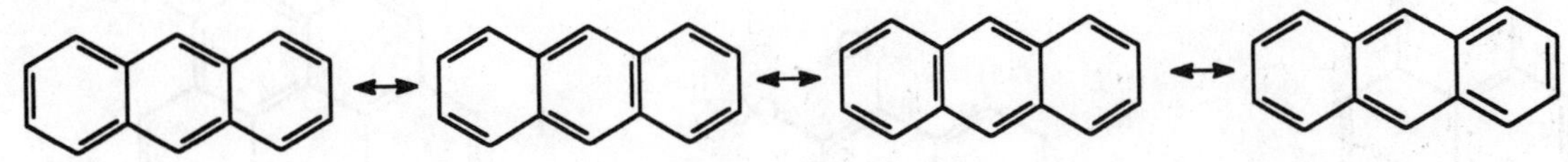

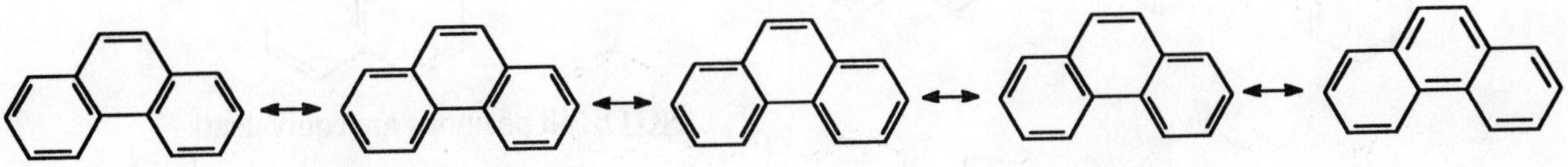

b.

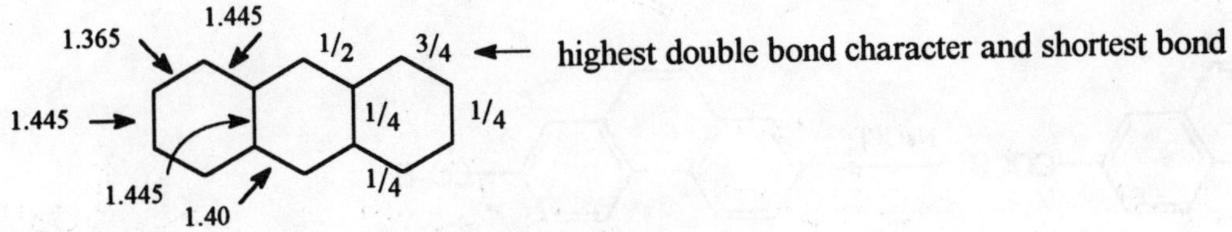

1.365 1.445 1/2 3/4 ← highest double bond character and shortest bond

1.445 → 1/4 1/4

1.445 1/4

1.40

bond lengths (in angstroms) are predicted from curve

10.

a. CH_3 NO_2

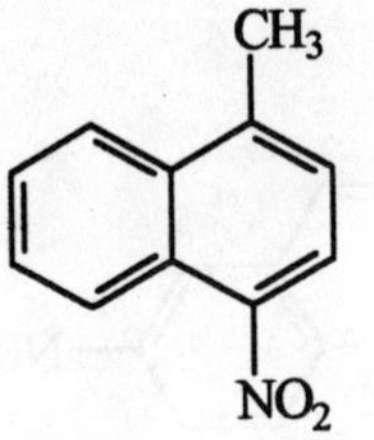

b. NO_2 CN

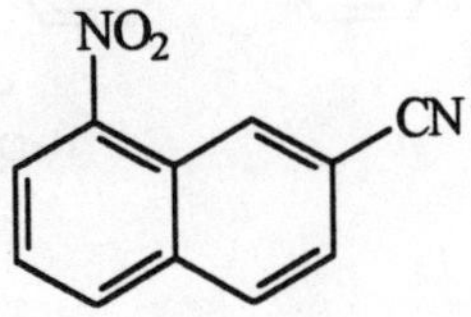

c. H_3C $NHCOCH_3$ NO_2

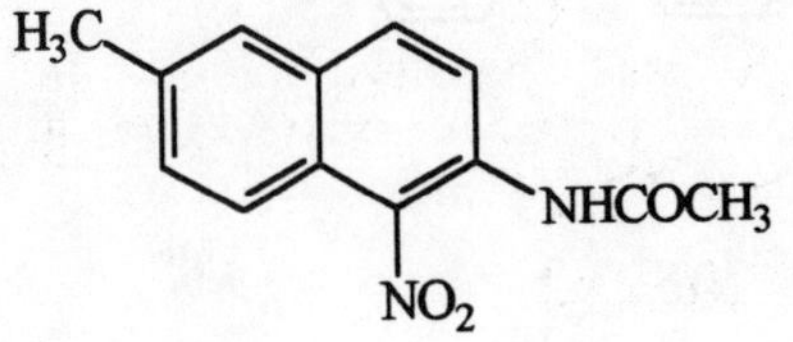

d. $NHCOCH_3$ NO_2 NO_2

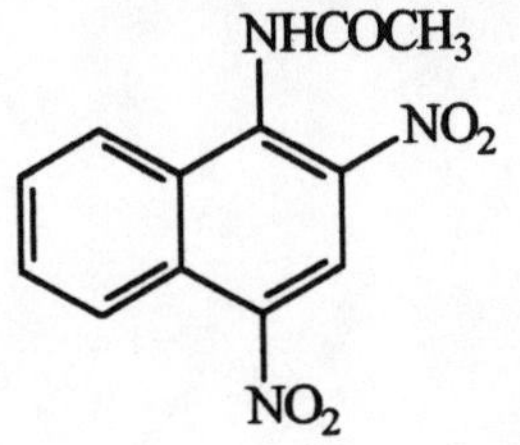

e. NO_2

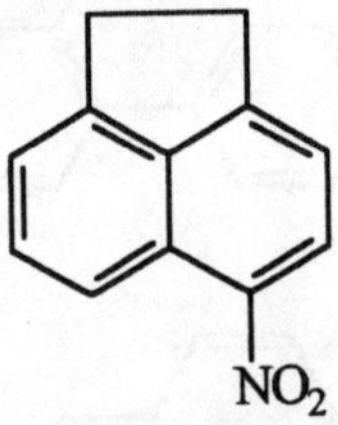

(α-alkylnaphthyl type)

f. NO_2

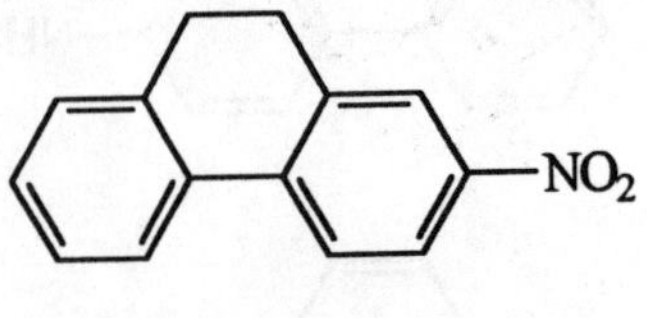

(biphenyl type)

g. HO_3S NO_2

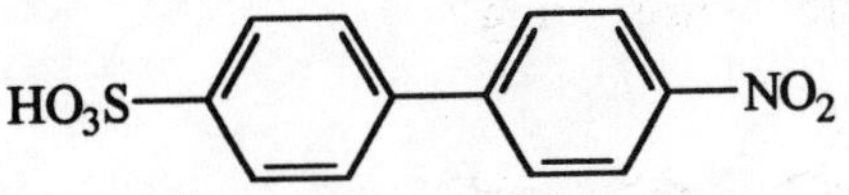

h. NO_2

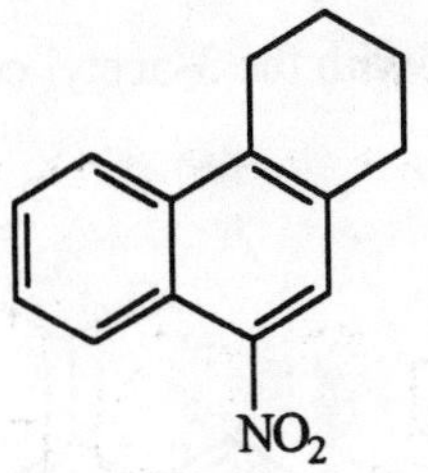

i. O_2N NO_2

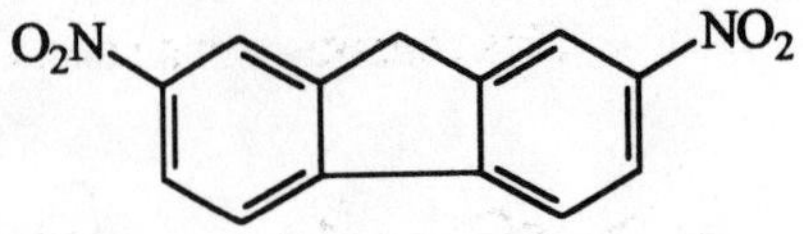

j. 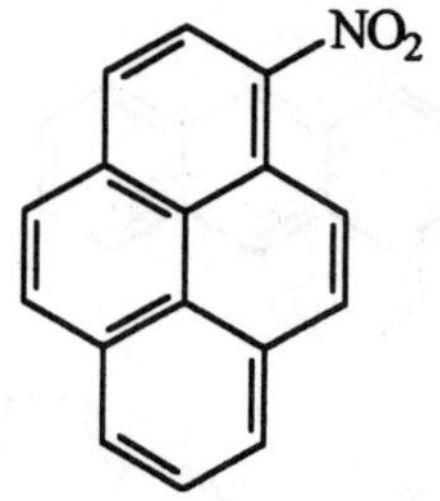 NO_2

k. 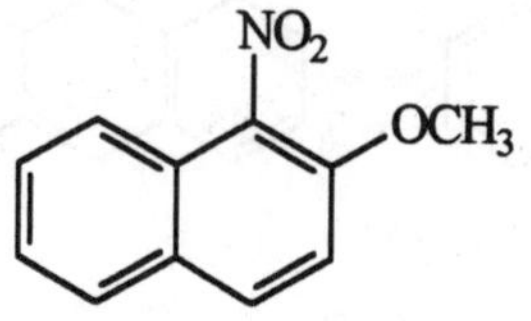 NO_2 OCH_3

l. 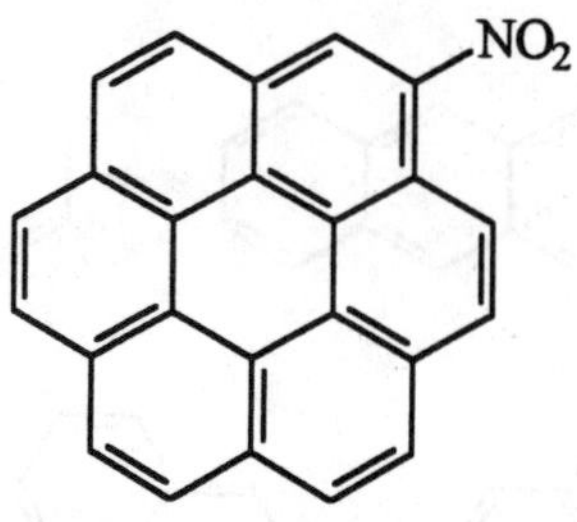 NO_2

NOTE: all positions are equivalent!

11. a. 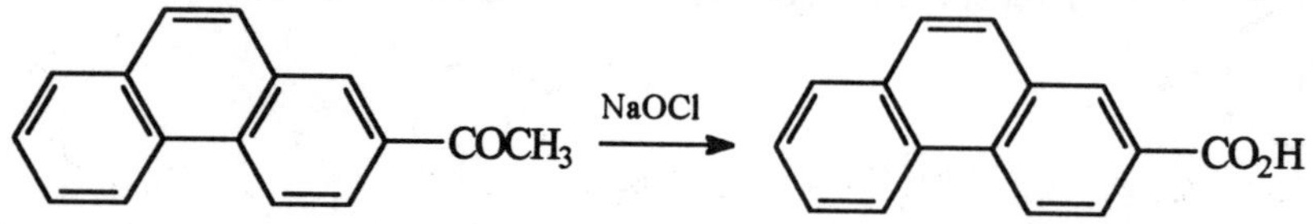 $COCH_3$ $\xrightarrow{NaOCl}$ CO_2H

b. 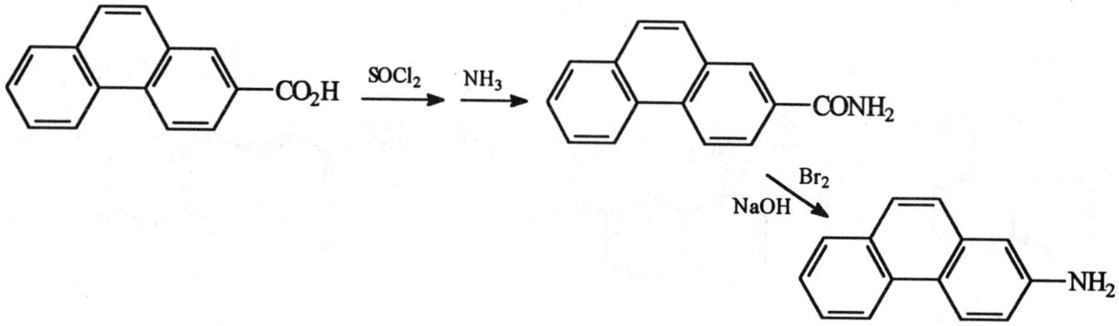 CO_2H $\xrightarrow{SOCl_2}$ $\xrightarrow{NH_3}$ $CONH_2$ $\xrightarrow[NaOH]{Br_2}$ NH_2

c. 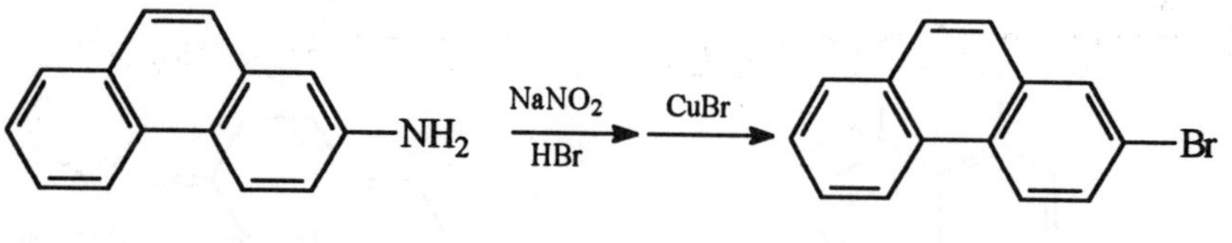 NH_2 $\xrightarrow[HBr]{NaNO_2}$ $\xrightarrow{CuBr}$ Br

d. Br $\xrightarrow[ether]{Mg}$ $\xrightarrow{D_2O}$ D

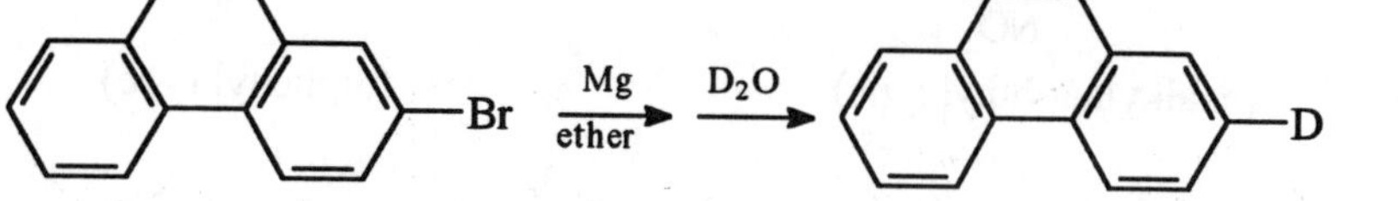

The same sequences apply, starting with the 3-acetyl compound

12. a. 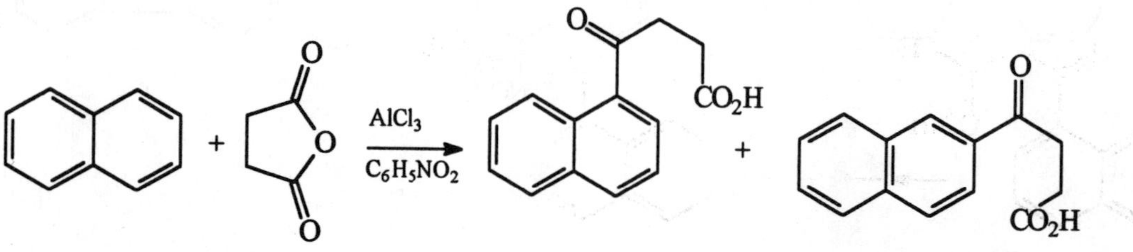 + O O O $\xrightarrow[C_6H_5NO_2]{AlCl_3}$ O CO_2H + O CO_2H

CO_2H H_2NNH_2 KOH glycerol Δ CO_2H HF O

CH_3 Se Δ H^+ Δ CH_3MgBr

O H_2NNH_2 KOH glycerol Δ CO_2H CO_2H HF O etc. H_3C

CH_3 + O O O $AlCl_3$ $C_6H_5NO_2$ O CO_2H CH_3 CH_3

CH_3 + O O O $AlCl_3$ CS_2 CH_3 CO_2H O CH_3

O 1. KH, THF 2. CH_3I CH_3 O H_2NNH_2 KOH glycerol Δ CH_3 Se or S Δ CH_3

(see Section 15.1.B)

b. O CH_3 CO_2H CH_3

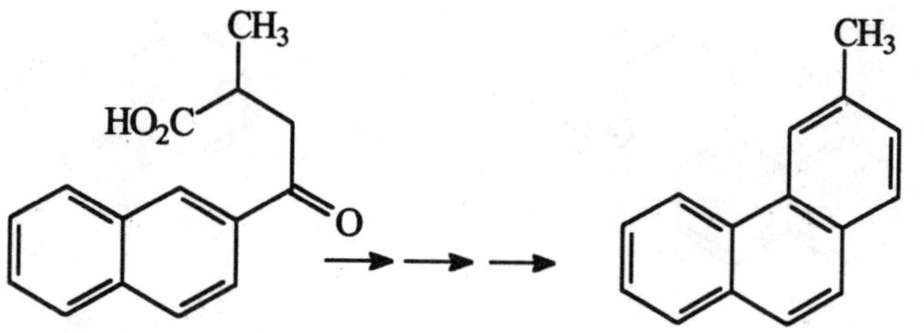

CH_3 HO_2C O → → → CH_3

13.

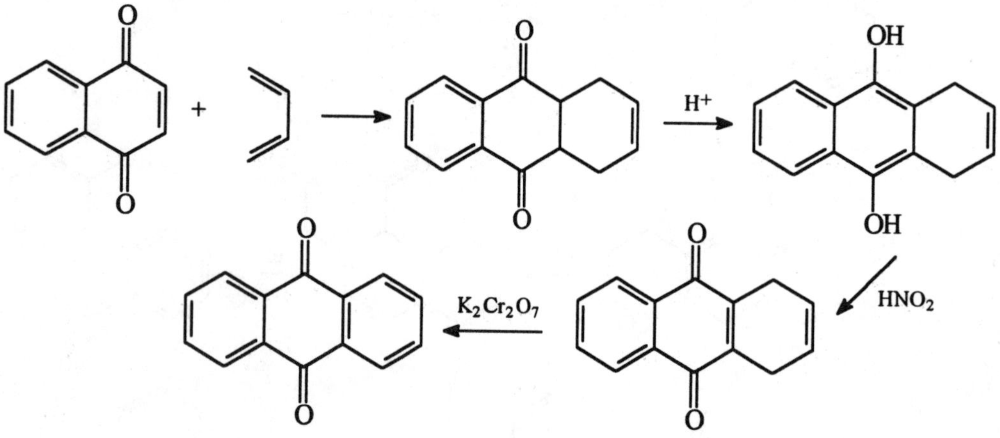

O O + → O O H^+ → OH OH

HNO_2 ↙ O O ← $K_2Cr_2O_7$ O O

14. 1-methylpyrene; 1,2,3,4-tetramethylphenanthrene; 5,6-dimethylchrysene

15. a.

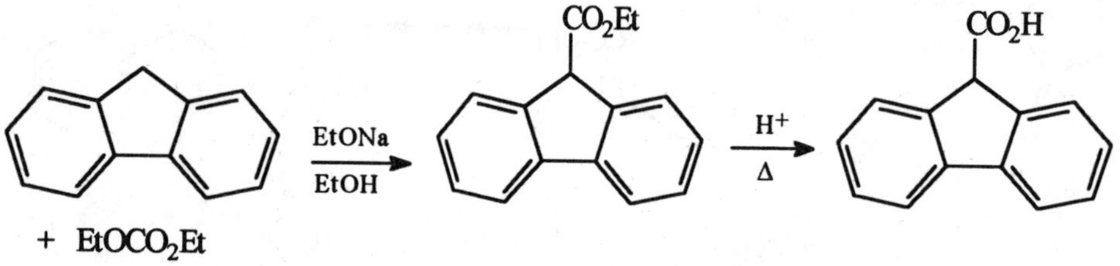

+ $EtOCO_2Et$ $\xrightarrow[EtOH]{EtONa}$ CO_2Et $\xrightarrow[\Delta]{H^+}$ CO_2H

b.

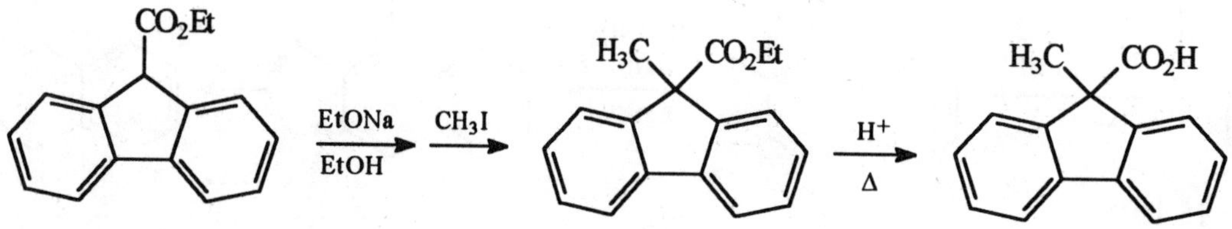

CO_2Et $\xrightarrow[EtOH]{EtONa}$ $\xrightarrow{CH_3I}$ H_3C CO_2Et $\xrightarrow[\Delta]{H^+}$ H_3C CO_2H

c.

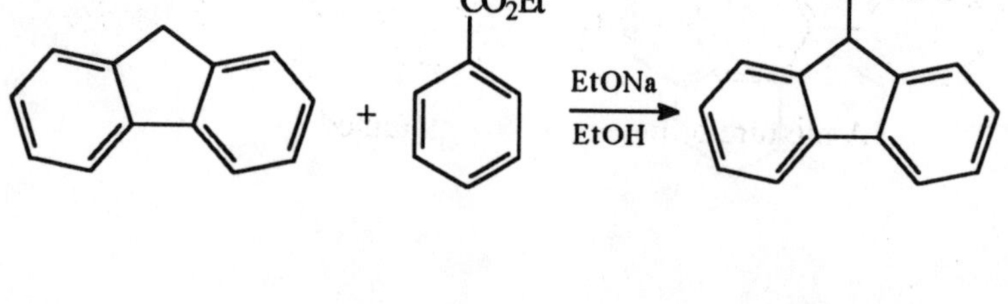

+ CO_2Et $\xrightarrow[EtOH]{EtONa}$ COC_6H_5

d. + HCO_2Et $\xrightarrow[\text{EtOH}]{\text{EtONa}}$ CHO

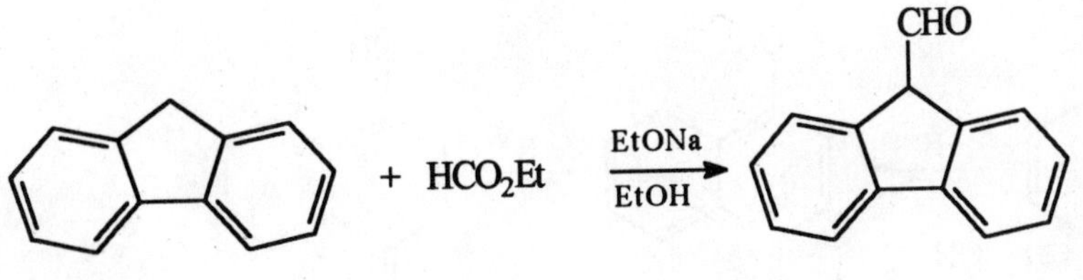

16. + $CH_3CH_2CH_2CH_2Br$ $\xrightarrow[\text{50\% KOH}]{C_6H_5CH_2\overset{+}{N}Et_3Br^-}$ $CH_2CH_2CH_2CH_3$

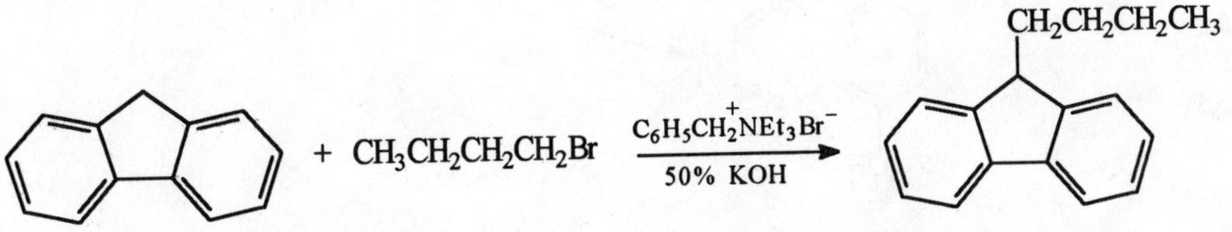

17. a. Cl + NH_2^- ⇌ NH_3 + Cl $\xrightarrow{-Cl^-}$

a benzyne derivative

NH_2 NH_2^-

NH_2 + NH_2^- $\xleftarrow{NH_3}$ NH^- ⇌ NH_2

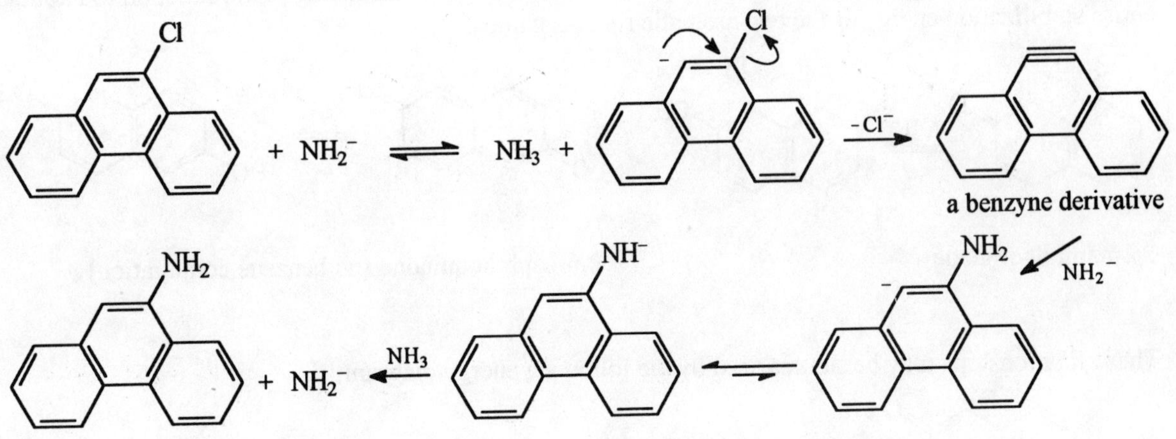

b. $-CH_3$ $\xrightarrow{NH_2^-}$ NH_2 $-CH_3$ + H_2N $-CH_3$

^-NH $-CH_3$ HN^- $-CH_3$

NH_2 $-CH_3$ H_2N $-CH_3$

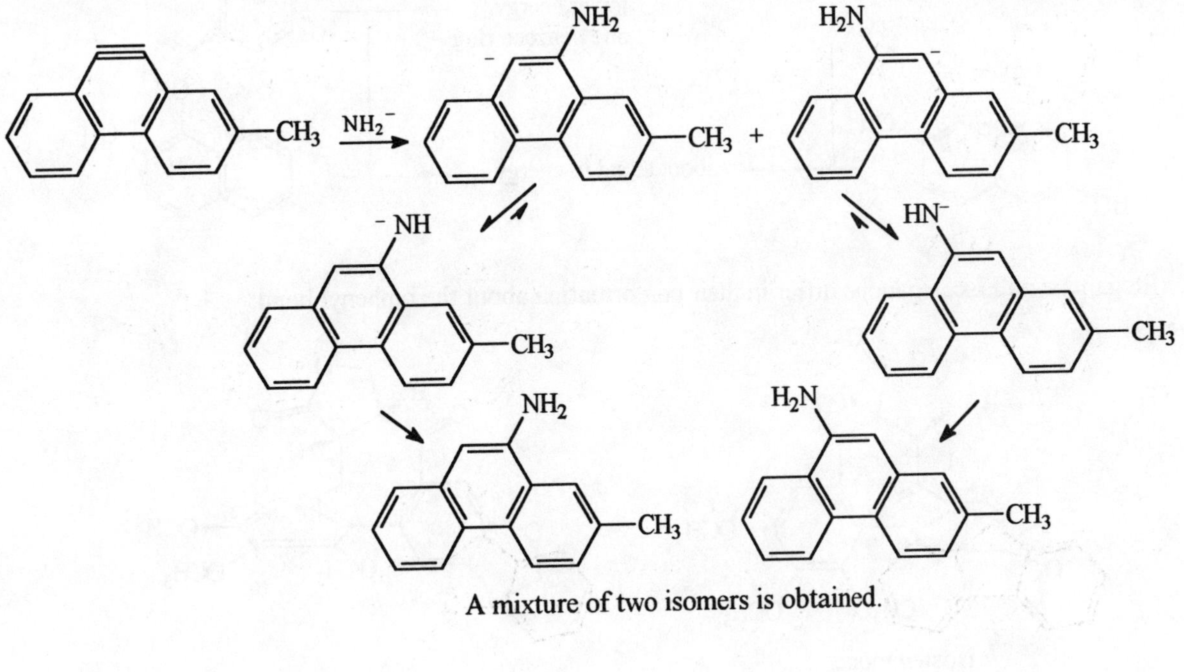

A mixture of two isomers is obtained.

18.

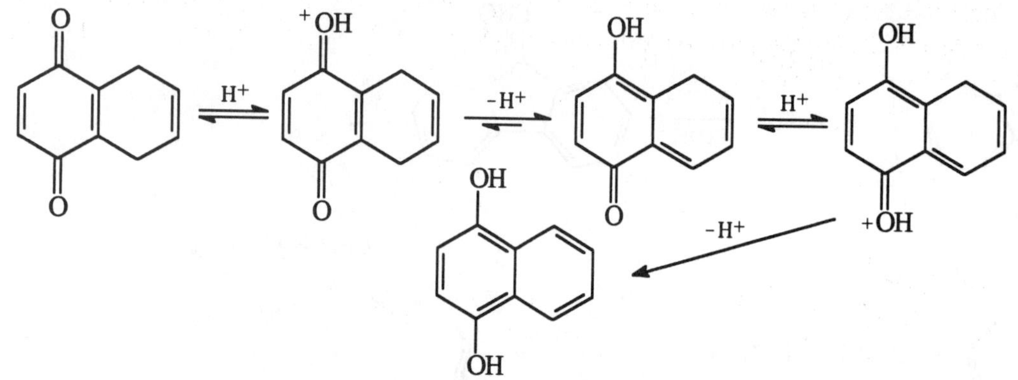

19. 1,2-Naphthoquinone has one benzene ring. On reduction, a naphthalene ring is generated, with a consequent increase in resonance stabilization. 2,6-Naphthoquinone has no benzene ring. On reduction to naphthalene, the entire stabilization energy of the two aromatic rings is gained.

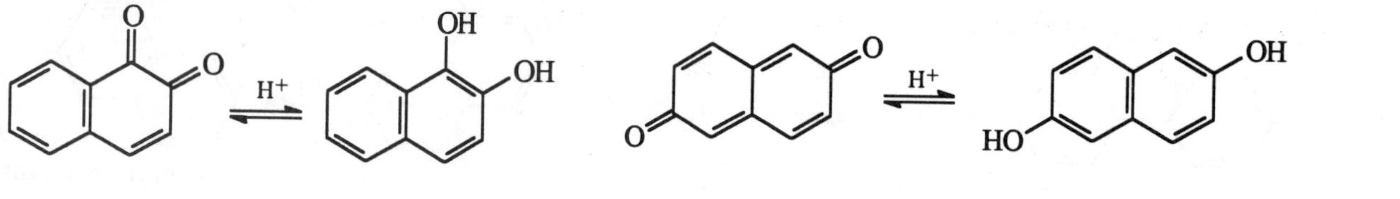

1,2-naphthoquinone

2,6-naphthoquinone (no benzene conjugation)

These relationships may be summarized by the following energy diagram:

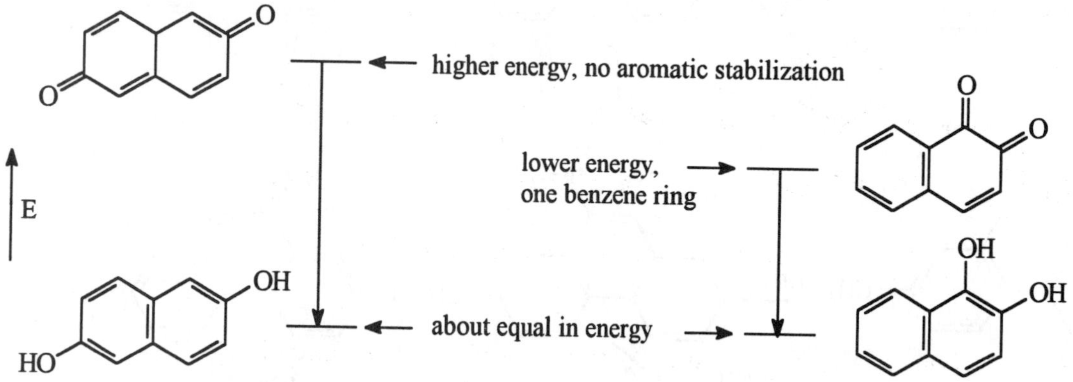

20. Steganone and isosteganone differ in their conformation about the biphenyl bond:

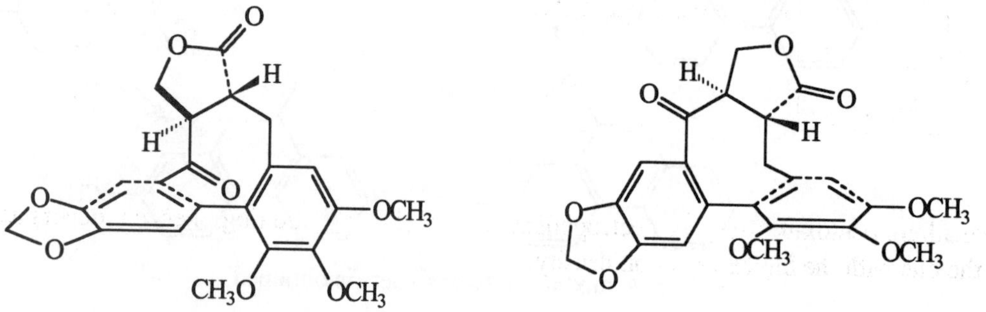

isosteganone

steganone

21. Removal of the bridgehead proton from triptycene places the negative charge in an orbital that cannot overlap with any of the p-orbitals of the aromatic rings. It lies in the nodal plane of all of the π-systems, and so receives no stabilization by resonance:

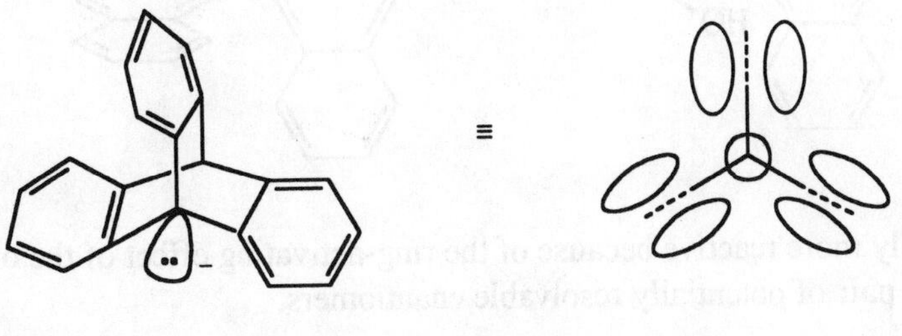

≡

22.

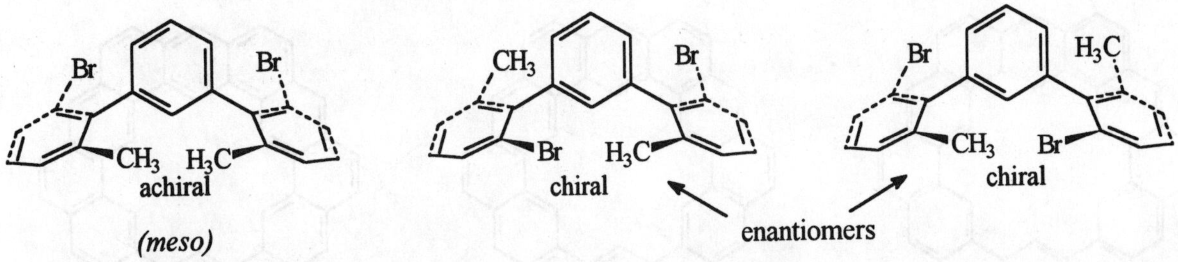

Br Br CH3 H3C achiral (meso)

CH3 Br Br H3C chiral

Br H3C CH3 Br chiral

enantiomers

23.

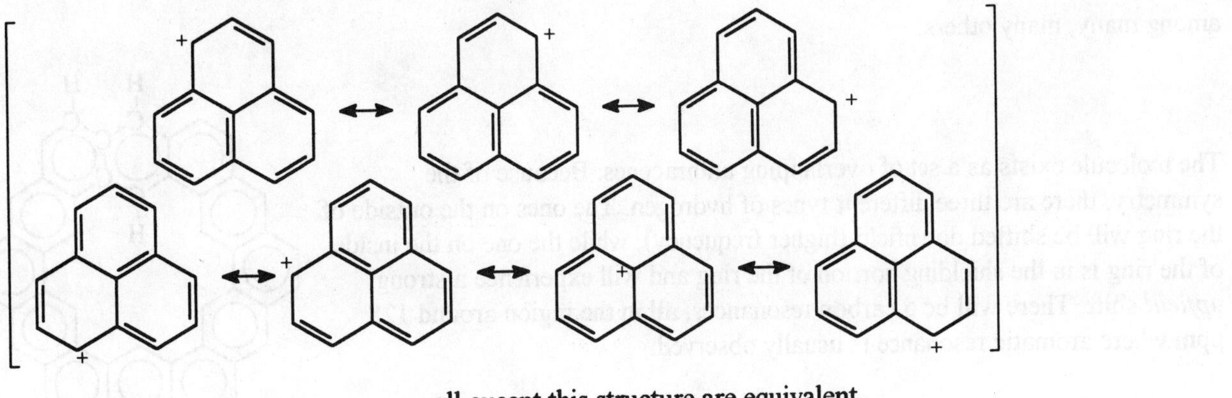

all except this structure are equivalent

Substitution at the 1-positon of pyrene results in the formation of a perinaphthenyl-like cation:

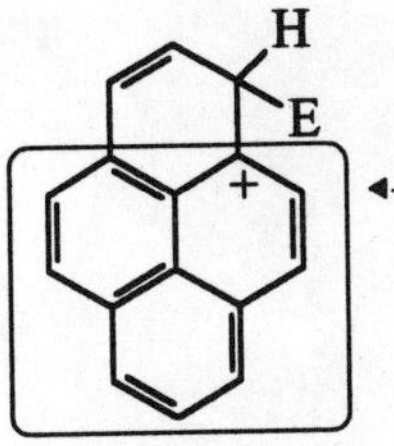

H E + ← perinaphthenyl-like system

24. Oxidation by vanadium pentoxide involves electrophilic attack on the aromatic ring, hence it occurs at the activated ring, the one with the highest electron density.

25. 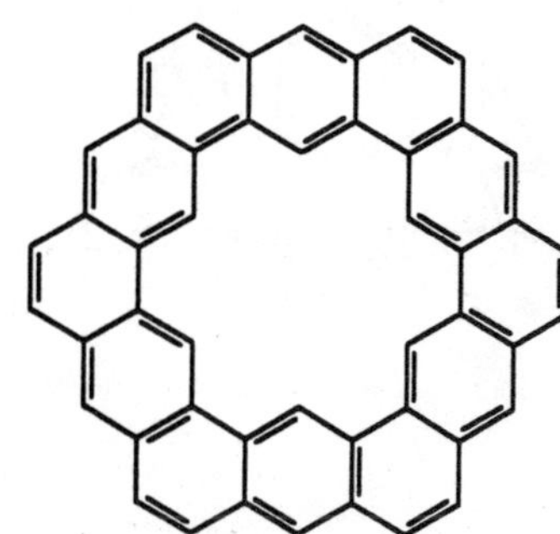

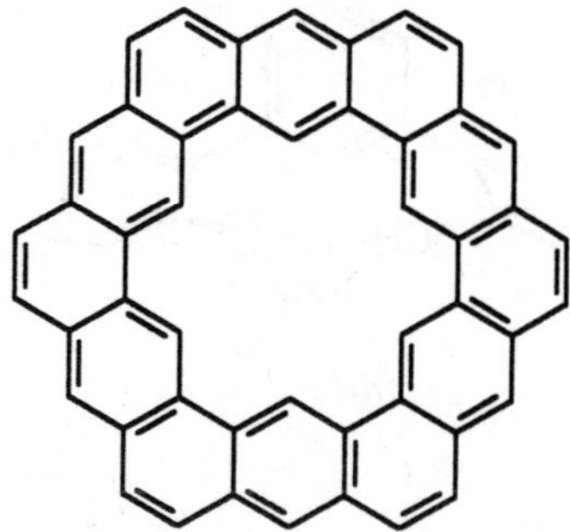

These compounds are generally more reactive because of the ring-activating effect of the of the attached group. The binaphthyl can exist as a pair of potentially resolvable enantiomers.

26.

among many, many others.

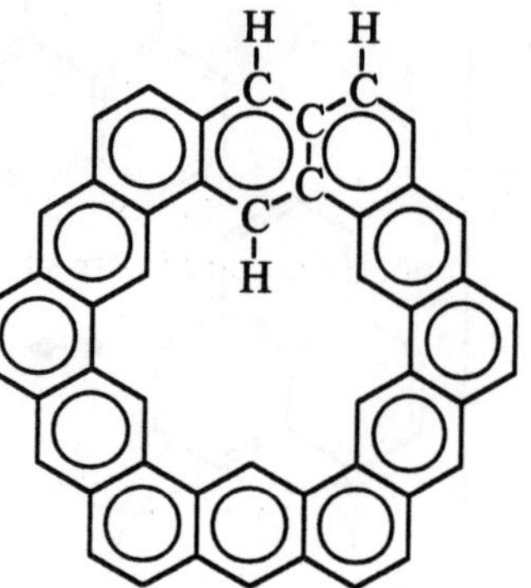

The molecule exists as a set of overlapping anthracenes. Because of the symmetry, there are three different types of hydrogen. The ones on the outside of the ring will be shifted downfield (higher frequency), while the one on the inside of the ring is in the shielding portion of the ring and will experience a strong *upfield* shift. There will be 5 carbon resonances, all in the region around 128 ppm where aromatic resonance is usually observed.

27. Since the wave function is greatest at the C-9 position, that should be the most reactive postion.

31.F. Supplementary Problems

S1. Write the structure of each of the following compounds.

a. 5-dimethylamino-1-naphthalenesulfonic acid
b. 3-bromo-4,4′-dimethylbiphenyl
c. 2,7-dinitrofluorene
d. 1,4-phenanthraquinone
e. benzo[a]chrysene
f. dibenzo[b,e]fluoranthene

S2. Write the major product from each of the following reaction sequences:

a. (succinic anhydride, $AlCl_3$) → H_2NNH_2 KOH glycerol Δ → HF → DDQ Δ

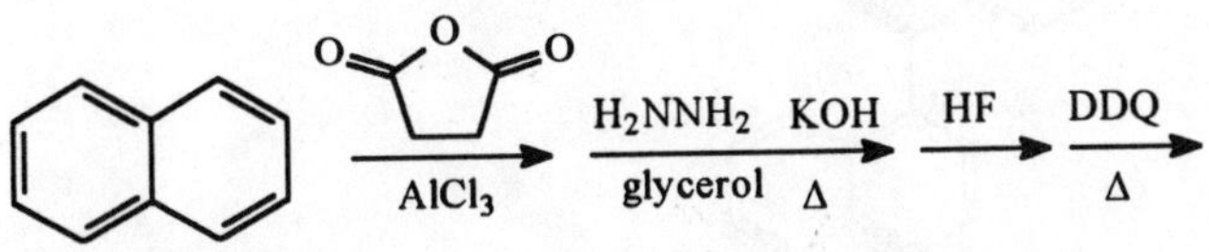

b. (CH_3) → Br_2 / $FeBr_3$ → KNH_2

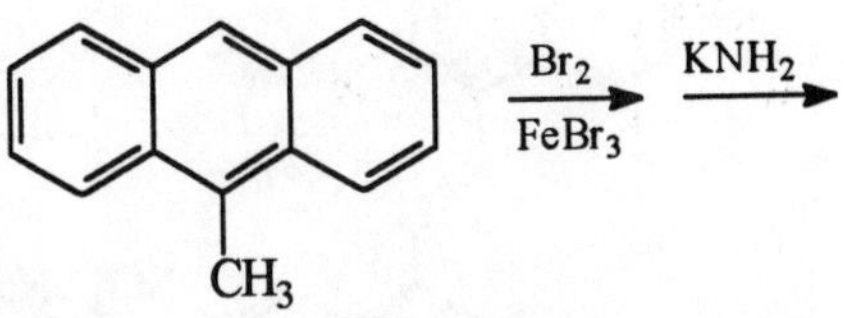

c. (OH) → $(NH_4)_2SO_3$ Δ → 2 CH_3COCl

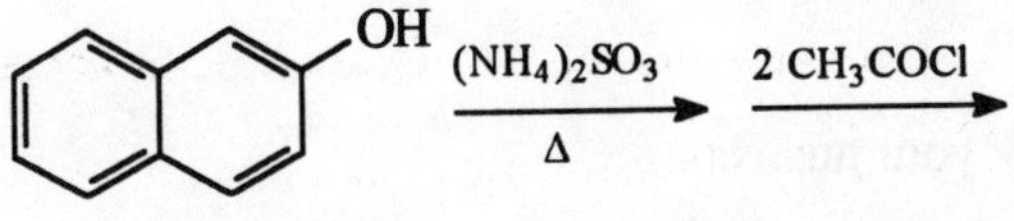

d. (CO_2H) → O_2 / V_2O_5 → Δ → CH_2N_2

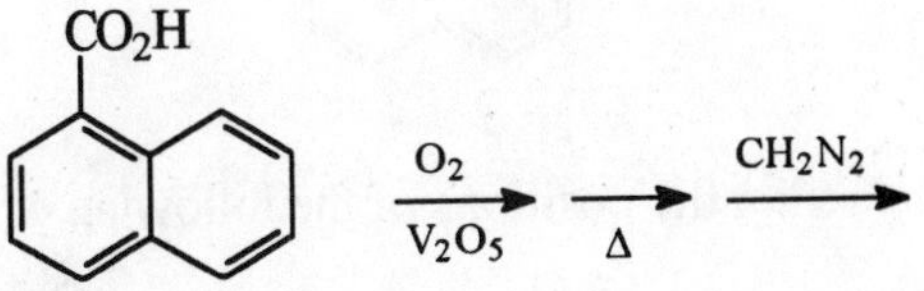

e. (OH, OH) → $K_2Cr_2O_7$ / H_2SO_4 → $(CH_3)_2NH$

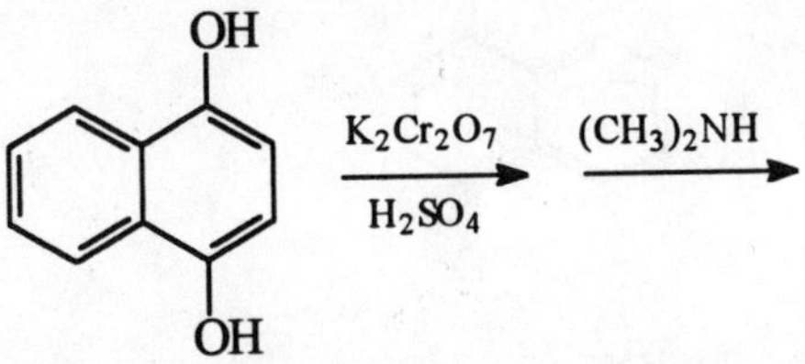

f. (CH_3) → 95% H_2SO_4 165° → KOH Δ

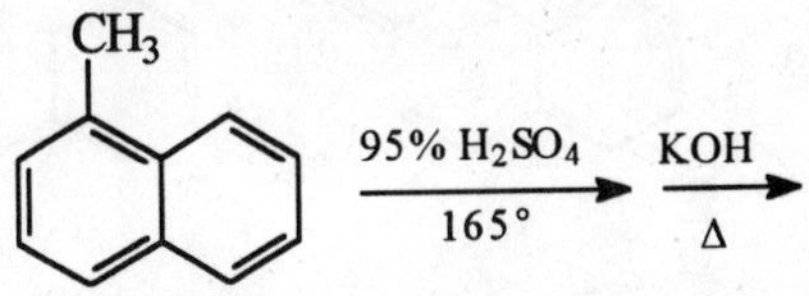

g. → CH_3COCl / $AlCl_3$

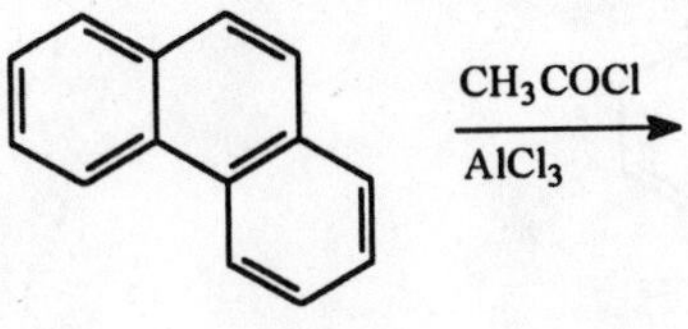

h. → excess Na / NH_3

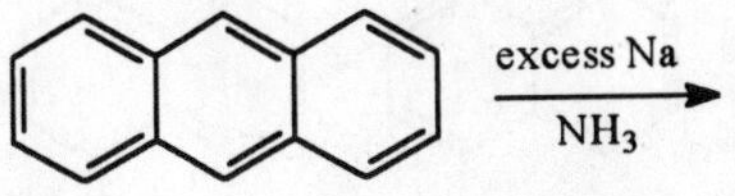

S3. Outline a synthesis for each of the compounds illustrated below, starting from benzene, naphthalene, or any monosubstituted benzene, and any other non-aromatic compounds.

a.

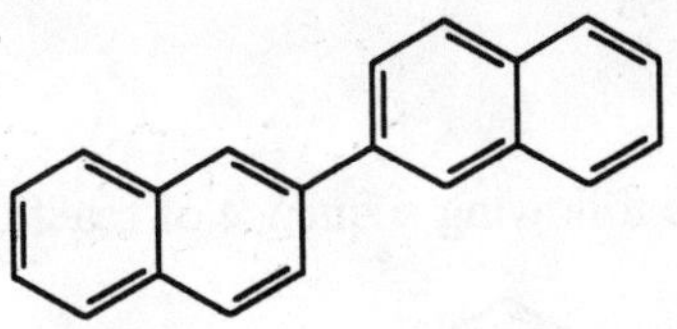

b. D

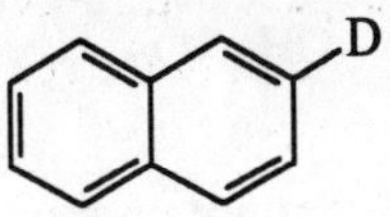

c.

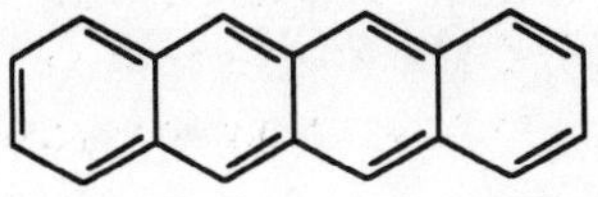

d. CH_2OH HOH_2C

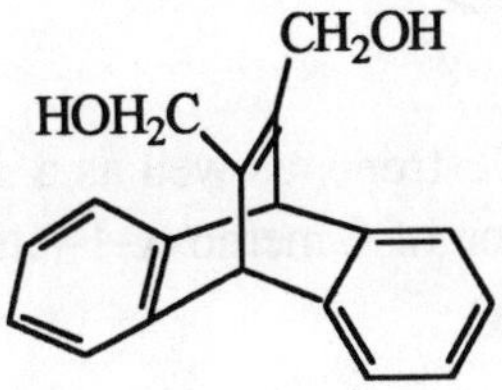

e. O CH_3 O

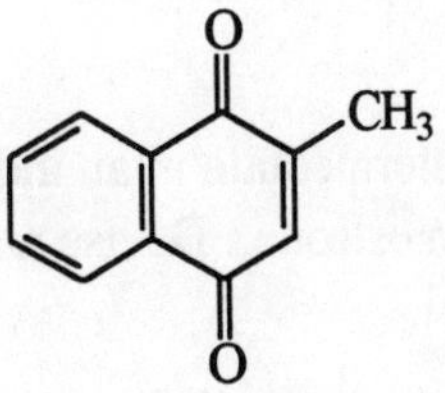

Menadione
vitamin K substitute

f. O O

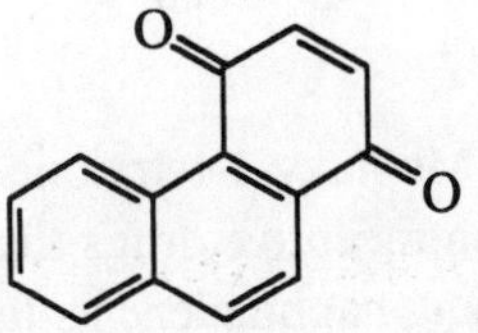

g. OH, N=N, SO_3^- Na^+

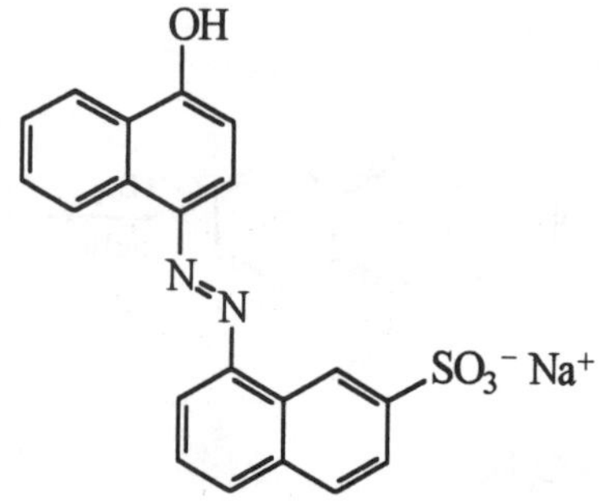

h. OCH_3, OCH_3

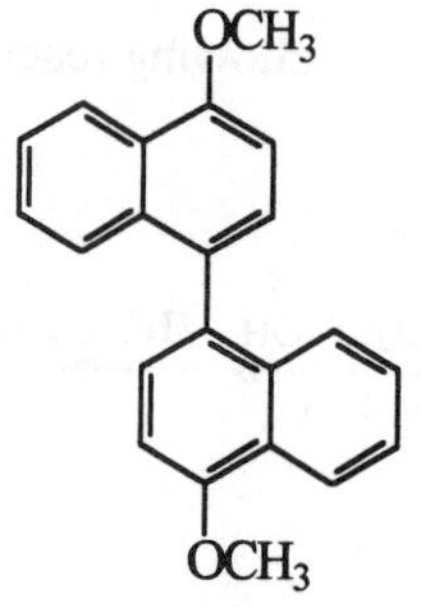

S4. Predict the positions of the following equilibria and justify your answers.

a. O, O + OH, OH ⇌ OH, OH + O, O

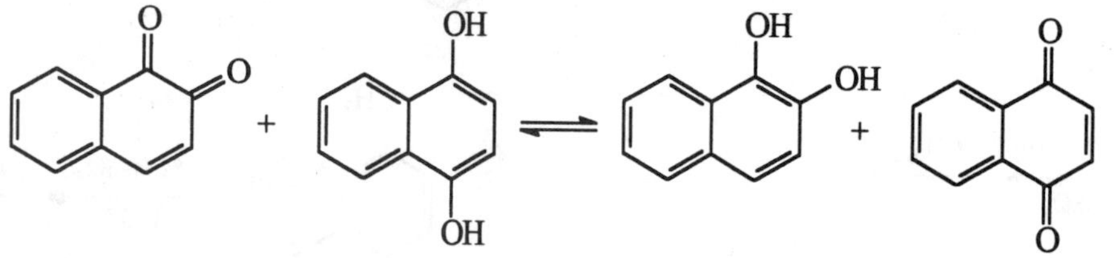

b. OH, OH + O, O ⇌ O, O + OH, OH

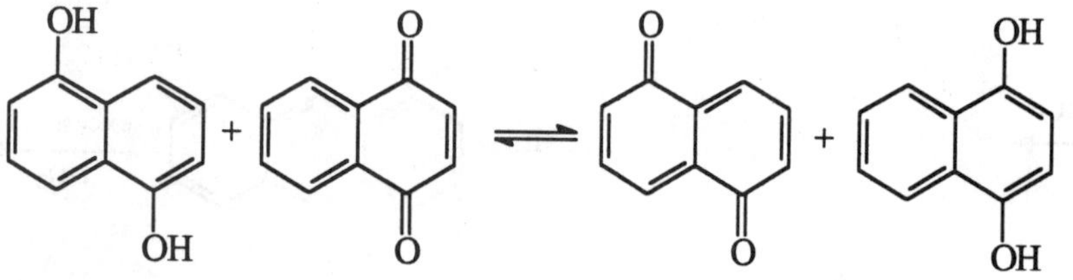

c. O, OH, O, OH ⇌ OH, O, OH, O

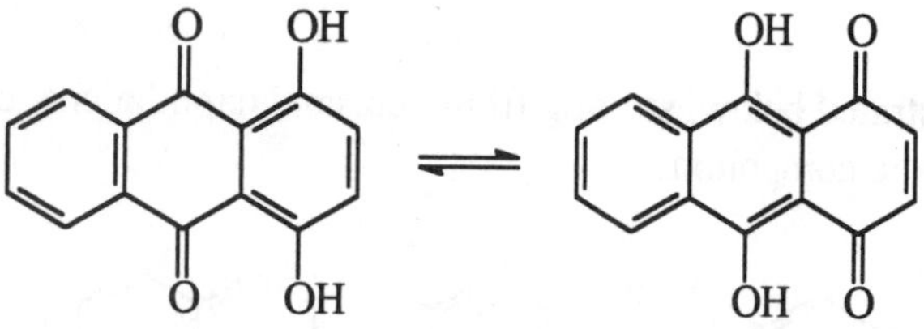

S5. Write the structures of the six intermediates that are produced during the following sequence of reactions.

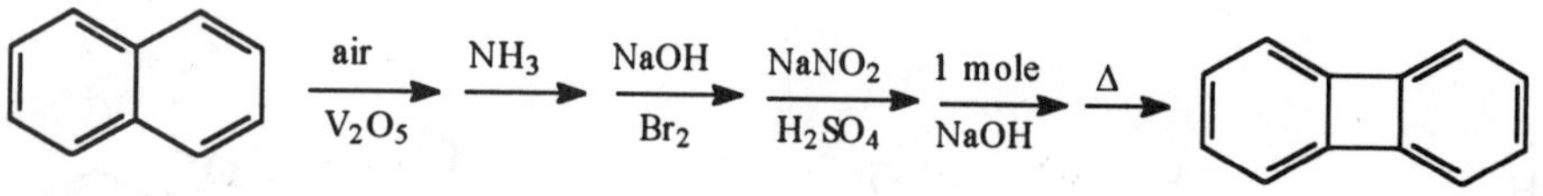

air / V_2O_5 → NH_3 → NaOH / Br_2 → $NaNO_2$ / H_2SO_4 → 1 mole NaOH → Δ →

S6. 6-Methoxy-1-tetralone is an important intermediate in an industrial synthesis of estrone, as well as a number of contraceptive drugs that are derived from estrone. Devise an efficient preparation of 6-methoxy-1-tetralone, using naphthalene as the starting material.

6-methoxy-1-tetralone: O, CH_3O estrone: O, HO

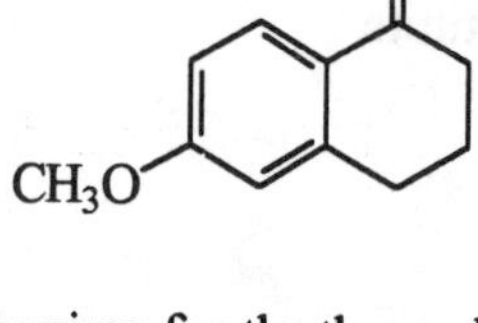

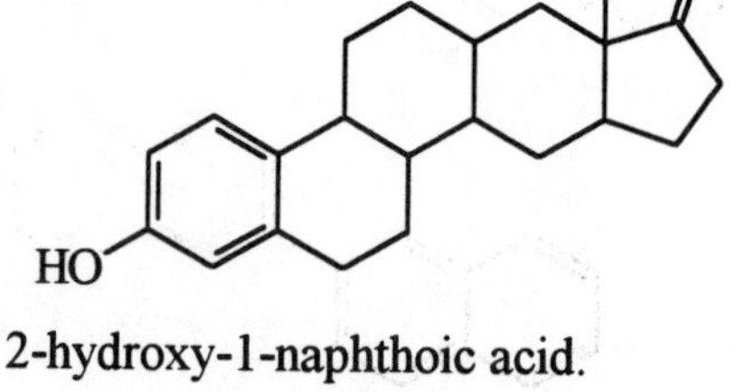

S7. Write a reasonable mechanism for the thermal decarboxylation of 2-hydroxy-1-naphthoic acid.

31.G. Answers to Supplementary Problems

S1. a. 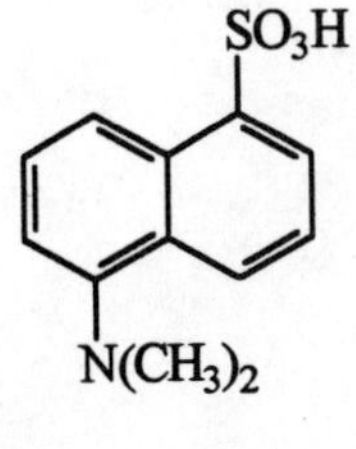 SO_3H $N(CH_3)_2$ b. 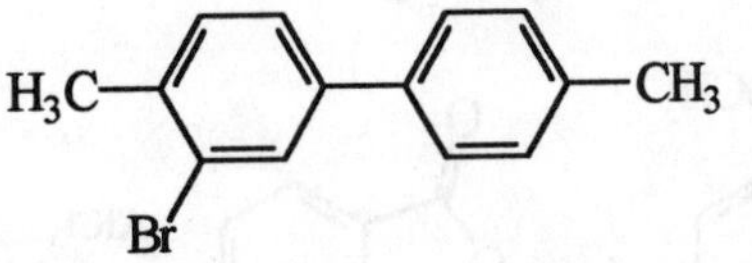 H_3C CH_3 Br c. 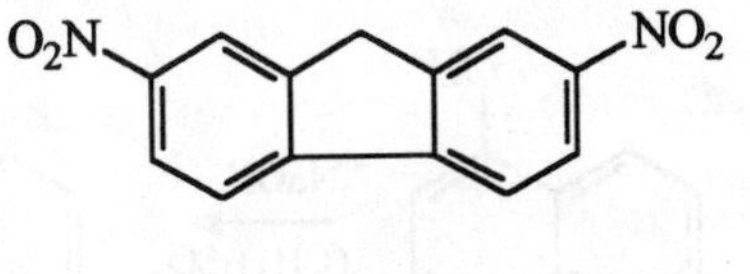 O_2N NO_2

d. 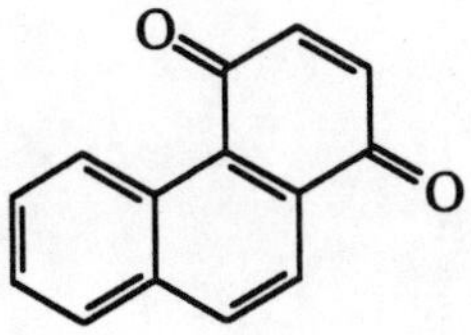 O O e. 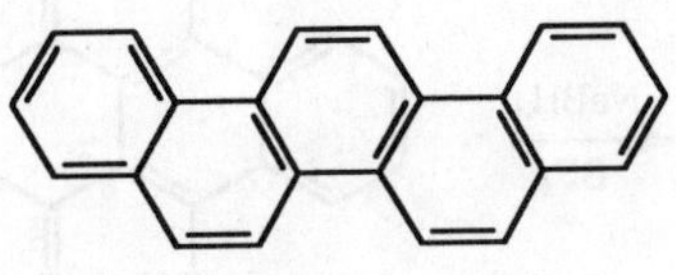 f.

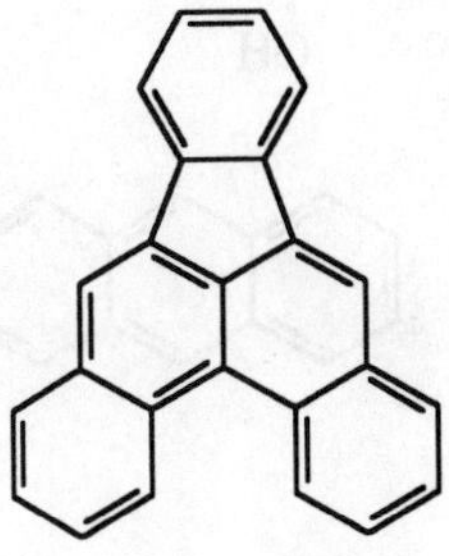

S2. a. 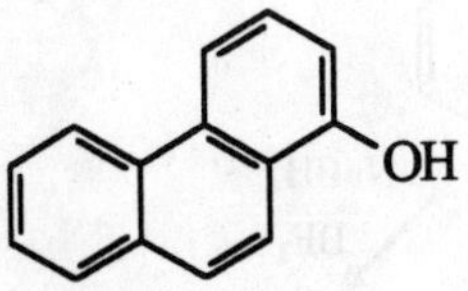 OH b. 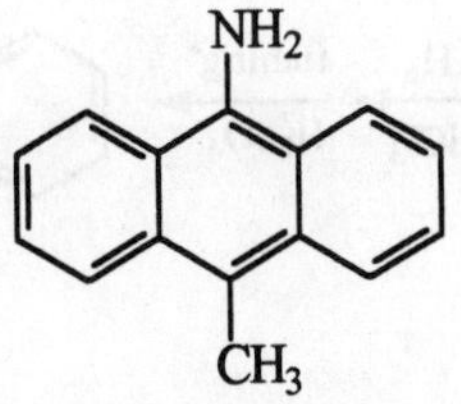 NH_2 CH_3 c. 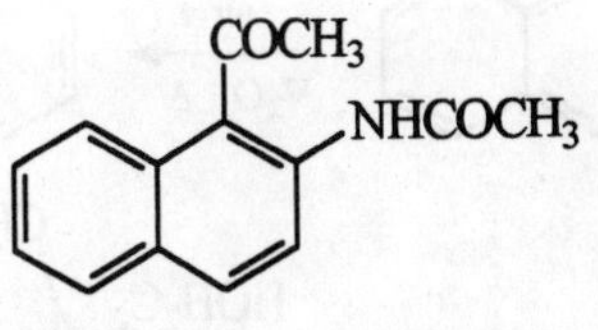 $COCH_3$ $NHCOCH_3$

d. 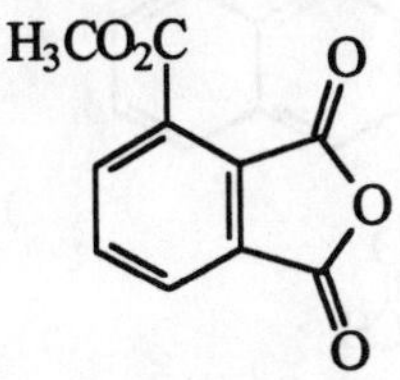 H_3CO_2C O O O e. 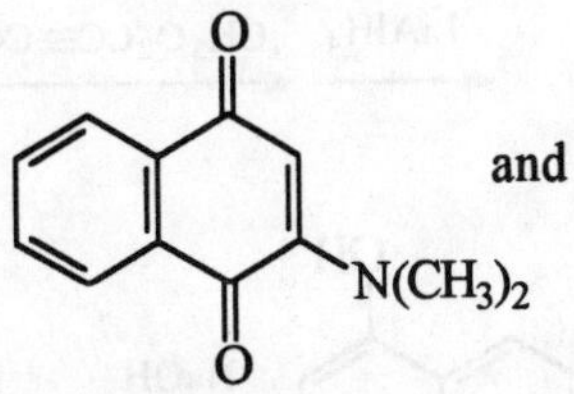 O O $N(CH_3)_2$ and OH OH

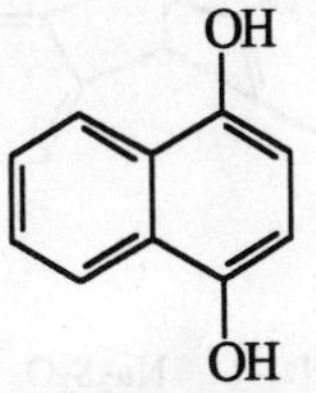

½ mole ½ mole

(see Problem #S9 in Chapter 30 of this Study Guide)

f. 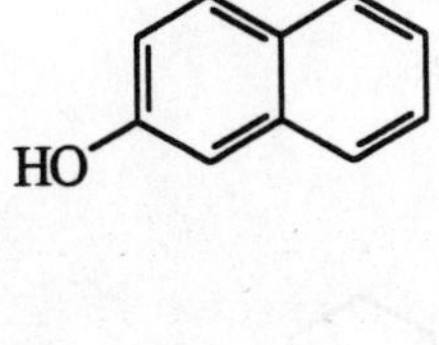 CH_3 HO g. 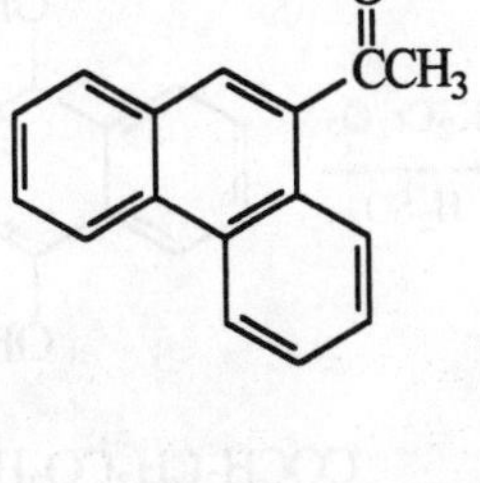 O $\overset{\|}{C}CH_3$ h.

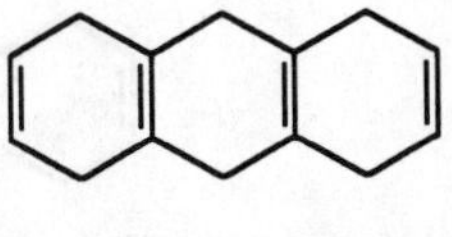

S3. a. 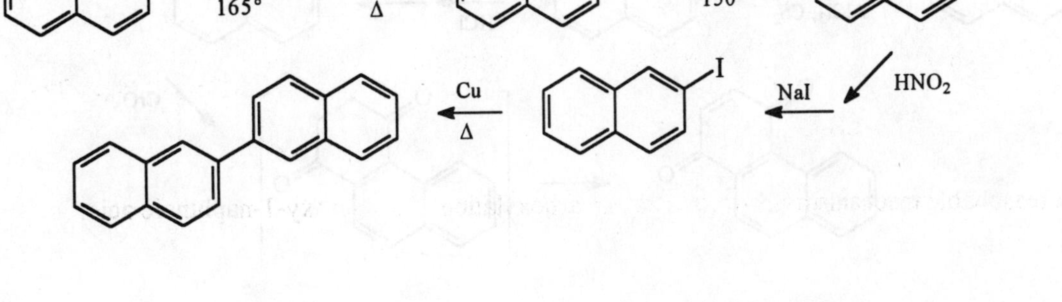 95% H_2SO_4 165° NaOH Δ OH $(NH_4)_2SO_3$ 150° NH_2 HNO_2 NaI I Cu Δ

b. from a. 2-iodonaphthalene $\xrightarrow[\text{ether}]{\text{Mg}}$ $\xrightarrow{D_2O}$ 2-deuterionaphthalene

c. (see e.) 1,4-naphthalenediol $\xrightarrow[(CH_3)_2SO_4]{\text{NaOH}}$ 1,4-dimethoxynaphthalene + phthalic anhydride $\xrightarrow[\Delta]{AlCl_3}$ $\xrightarrow{\text{HBr}}$ 1,4-dihydroxyanthraquinone-type product (OH, OH, O, O) $\xrightarrow[BF_3]{NaBH_4}$ tetracene

d. naphthalene $\xrightarrow[V_2O_5,\ \Delta]{\text{air}}$ phthalic anhydride $\xrightarrow[AlCl_3]{C_6H_6}$ $\xrightarrow[H_2SO_4]{\text{fuming}}$ anthraquinone $\xrightarrow[BF_3]{NaBH_4}$ anthracene $\xrightarrow[\Delta]{CH_3O_2CC{\equiv}CCO_2CH_3}$ $\xrightarrow{LiAlH_4}$ bicyclic adduct with HOH_2C and CH_2OH

e. (see g.) 1-naphthol $\xrightarrow[H_2SO_4]{K_2Cr_2O_7}$ $\xrightarrow{Na_2S_2O_4}$ 1,4-naphthalenediol $\xrightarrow[CH_3Cl]{\text{NaOH}}$ 1,4-dihydroxy-2-naphthaldehyde (OH, $\overset{O}{\|}$CH, OH) $\xrightarrow[\text{HCl}]{\text{Zn}}$ 2-methyl-1,4-naphthalenediol (OH, CH_3, OH) $\xrightarrow[H_2SO_4]{K_2Cr_2O_7}$ 2-methyl-1,4-naphthoquinone (O, CH_3, O)

f. naphthalene $\xrightarrow[AlCl_3,\ CS_2]{\text{succinic anhydride}}$ 1-naphthyl-$COCH_2CH_2CO_2H$ $\xrightarrow[\text{HCl}]{\text{Zn}}$ $\xrightarrow{\text{HF}}$ tetrahydrophenanthrenone (=O) $\xrightarrow{CrO_3}$ [diketone (O, O)] $\longrightarrow$ phenanthrene-1,4-dione-type product (O, O)

g.

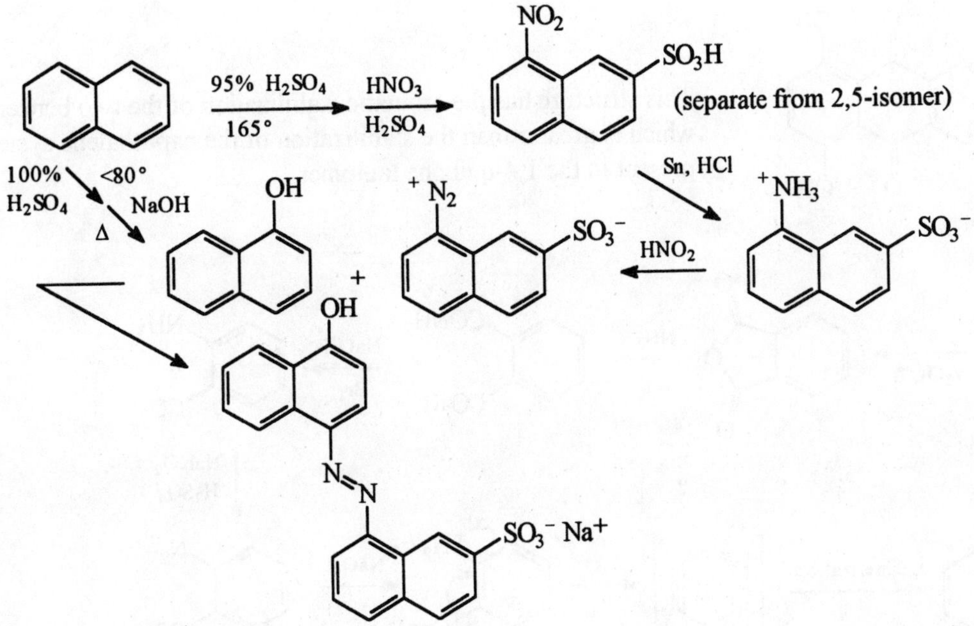

95% H_2SO_4, 165° → ; HNO_3, H_2SO_4 → (1-nitro, 7-SO_3H naphthalene) (separate from 2,5-isomer)

Sn, HCl → $^+NH_3$ … SO_3^-

HNO_2 ← $^+N_2$ … SO_3^-

100% H_2SO_4 <80°; NaOH Δ → OH (1-naphthol) +

OH … N=N … SO_3^- Na^+

h.

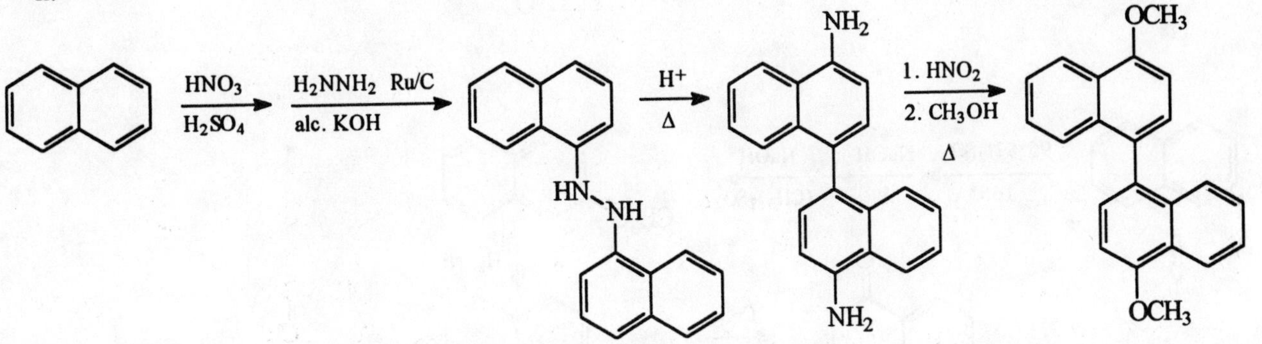

HNO_3, H_2SO_4 → ; H_2NNH_2 Ru/C, alc. KOH → HN–NH ; H^+, Δ → NH_2 … NH_2 ; 1. HNO_2 2. CH_3OH Δ → OCH_3 … OCH_3

S4. a. favors OH OH + O O

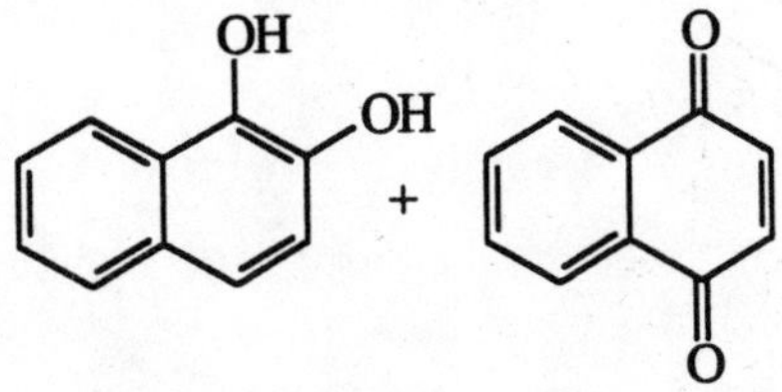

1,2-Naphthoquinone and 1,2-benzoquinone both have higher reduction potentials than their 1,4-isomers (see Table 30.2) because of unfavorable interaction between their aligned dipoles:

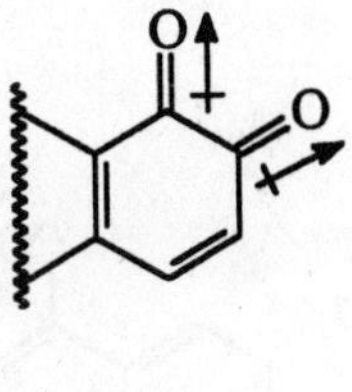

O O

b. favors OH OH + O O

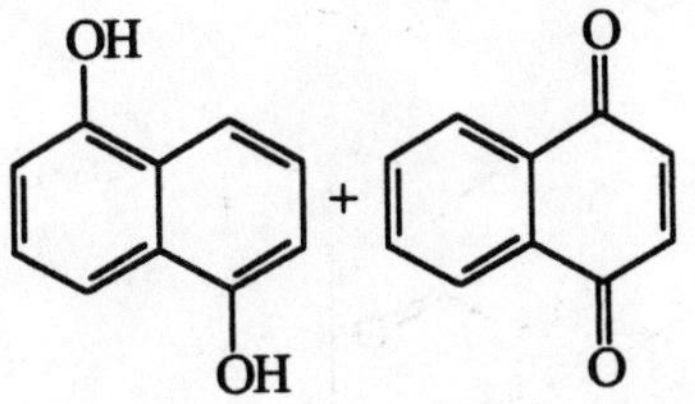

See answer to problem #19 of this chapter.

c. favors 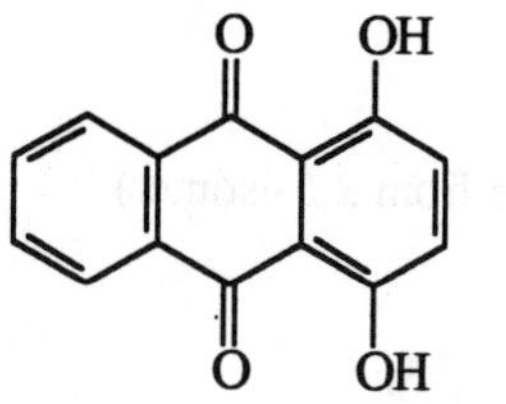

This structure has the aromatic stabilization of the two benzene rings, which is greater than the stabilization of the naphthalene system present in the 1,4-quinone tautomer.

S5.

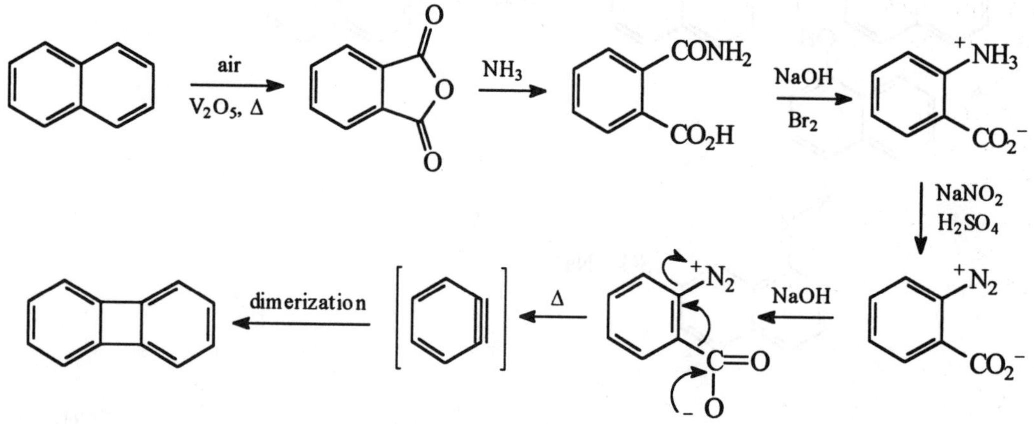

S6.

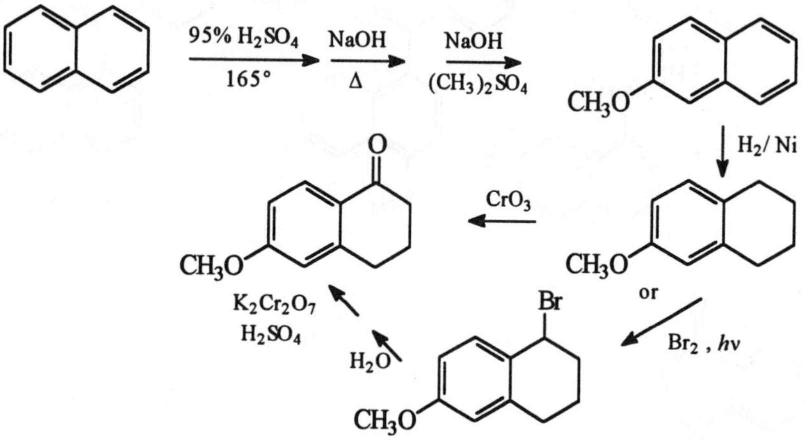

S7.

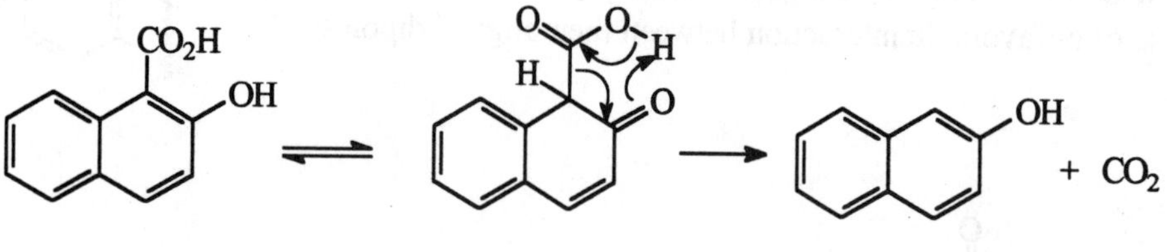

32. HETEROCYCLIC COMPOUNDS

32.A. and 32.B. Chapter Outline and Reactions Discussed

<u>NOTE:</u> *since this Chapter is primarily an outline of reactions, there does not seem to be any point in separating Parts A. and B. in this chapter of the Study Guide.*

32.1 Introduction

definition of heterocycles (-irane (3), -etane (4), -olane 5), -ane (6))
aza- (N), oxa- (O), thia- (S)

32.2. Non-aromatic Heterocycles

A. Nomenclature
B. Three-Membered Rings

epoxides = oxiranes (see Sections 10.11.A and 11.6.E)
aziridines via intramolecular alkylation:

$$HO{-}C{-}C{-}NH_2 \xrightarrow{H_2SO_4} HO_3SO{-}C{-}C{-}\overset{+}{N}H_3 \xrightarrow{OH^-} \text{aziridine (ring N–H)}$$

aziridines via iodo isocyanates:

$$C{=}C + INCO \longrightarrow I{-}C{-}C{-}NCO \xrightarrow{CH_3OH} I{-}C{-}C{-}NHCO_2CH_3 \xrightarrow{OH^-} \text{aziridine (ring N–H)}$$

thiiranes from oxiranes:

$$\text{oxirane (ring O)} \xrightarrow{SCN^-} \text{thiirane (ring S)}$$

reactions: ring opening

$$\text{three-membered ring (ring Y)} \xrightarrow{HX} HY{-}C{-}C{-}X$$

X = halogen, OR; Y = O, S, NR′

C. Four-Membered Rings

oxetane, azetidine, thietane
via ring closure:

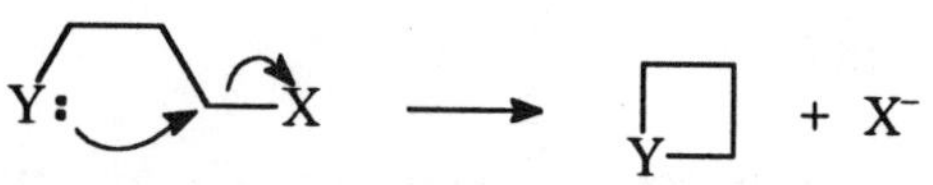

Y: ⌒ X → (four-membered ring with Y) + X^- (poor yields)

X = good leaving group; Y: = O^-, S^-, or NHR

β-lactones and β-lactams via cycloaddition:

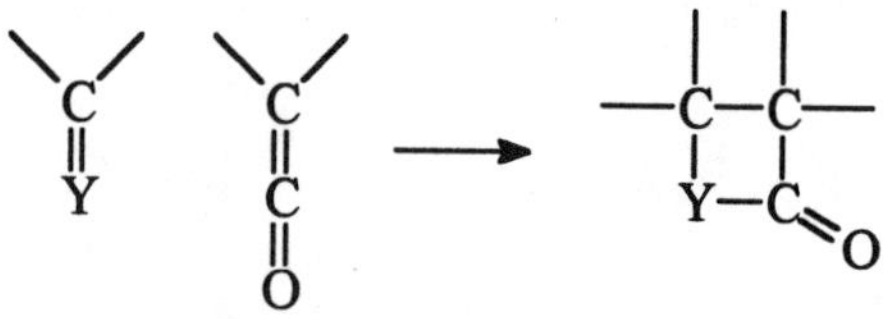

$>C{=}Y \;+\; >C{=}C{=}O \longrightarrow$ four-membered ring –C–C– / Y–C=O

Y = O or NR

ring-opening reactions; similar to three-membered rings

D. Five- and Six-Membered Rings

by hydrogenation of the aromatic heterocycles (furan, pyridine):

e.g., furan $\xrightarrow[\text{cat.}]{H_2}$ tetrahydrofuran

pyridine $\xrightarrow[\text{cat.}]{H_2}$ piperidine

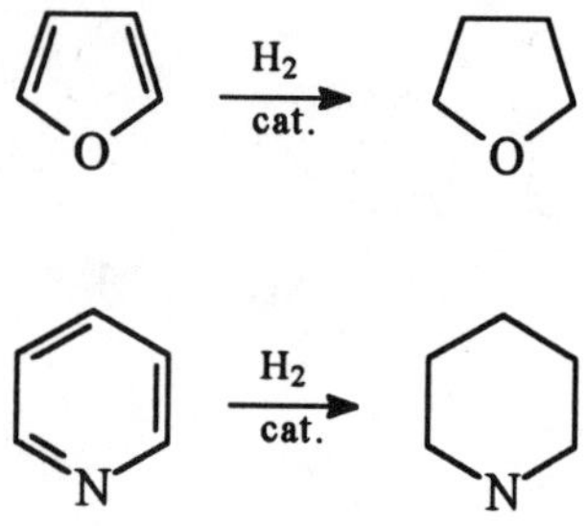

nucleophilic ring closure:

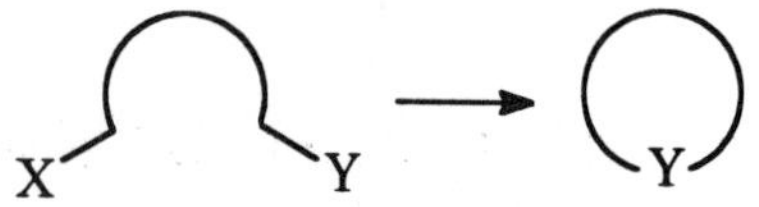

X⌒Y → ring containing Y

32.3 **Furan, Pyrrole, and Thiophene**

A. Structure and Properties
 use of lone pair electrons of heteroatom to attain aromatic six-π-electron system
 effect of aromaticity on pK_a
 position of protonation of pyrrole

B. Synthesis
 furan derivatives from dehydration of pentoses (industrial synthesis)
 pyrrole from distillation of coal-tar
 thiophene from pyrolysis of butenes and sulfur (industrial synthesis)

Paal-Knorr synthesis:

P_2O_5, 100°; NH_3, Δ; P_2S_5, Δ

Knorr pyrrole synthesis:

EtO_2C, NH_2, O, CO_2Et, AcOH, CO_2Et, EtO_2C, N H

C. Reactions

reactions toward electrophiles: pyrrole > furan > thiophene >> benzene

electrophilic aromatic substitution oriented toward 2-positions

hydrolysis (primarily of furans): H_3O^+

polymerization of pyrrole by dilute acid: H^+ → etc.

32.4 Condensed Furans, Pyrroles, and Thiophenes

A. Structure and Nomenclature

indole

B. Synthesis

Fischer indole synthesis:

RH_2C, R', H^+, Δ, R, R'

R′≠H

(use R=H, R′=COOH, then decarboxylate to obtain indole itself)

C. Reactions

electrophilic aromatic substitution

32.5 Azoles

A. Structure and Nomenclature

oxazole, imidazole, thiazole
isoxazole, pyrazole, isothiazole
basicity
occurrence in nature

B. Synthesis

isoxazoles and pyrazoles from 1,3-dicarbonyl compounds:

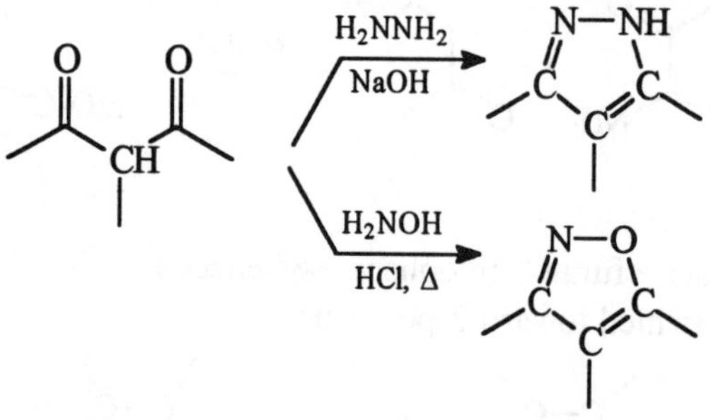

H_2NNH_2, NaOH

H_2NOH, HCl, Δ

isoxazoles via nitrile oxide acetylene 1,3-dipolar cycloaddition:

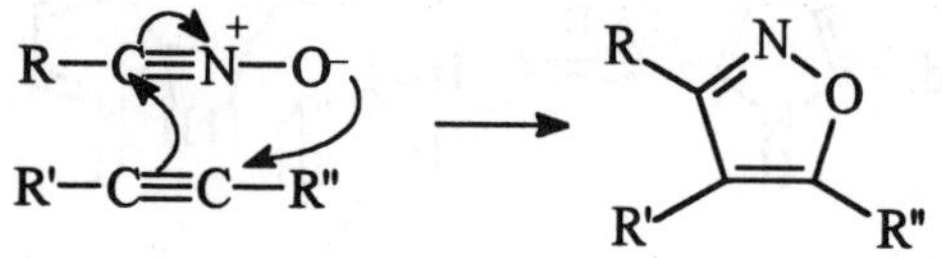

pyrazoles via diazomethane acetylene 1,3-dipolar cycloaddition:

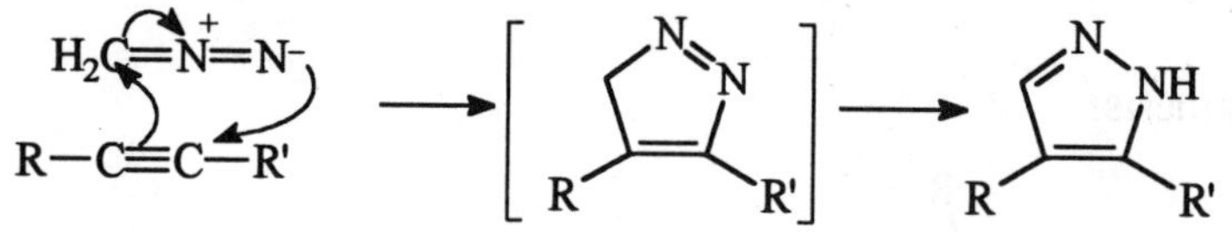

Paal-Knorr-like cyclization:

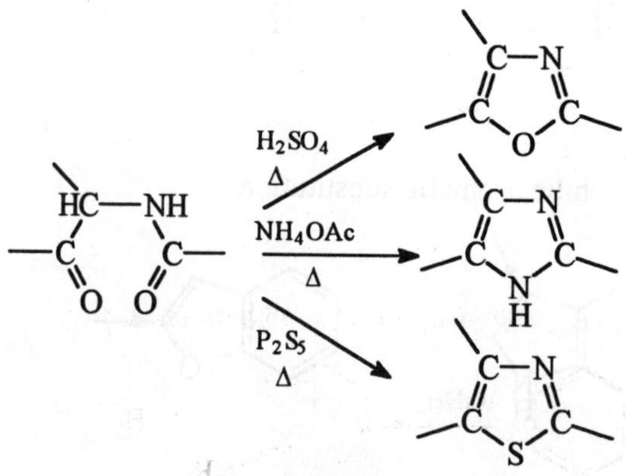

H_2SO_4, Δ

NH_4OAc, Δ

P_2S_5, Δ

C. Reactions

less reactive toward electrophilic aromatic substitution than the monohetero analogs:

pyrazole > isothiazole > isoxazole

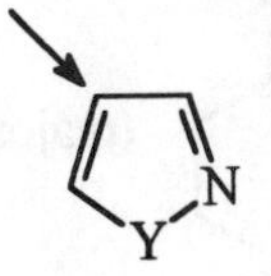

Y = O, S, NH

substitution occurs at C-4

imidazole > thiazole > oxazole

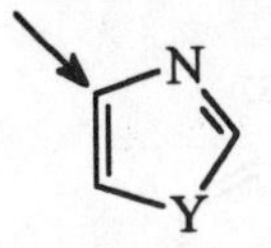

32.6 Pyridine

A. Structure and Physical Properties

basicity (pK_a of pyridinium ion is 5.2)

B. Synthesis

Hantzsch pyridine synthesis:

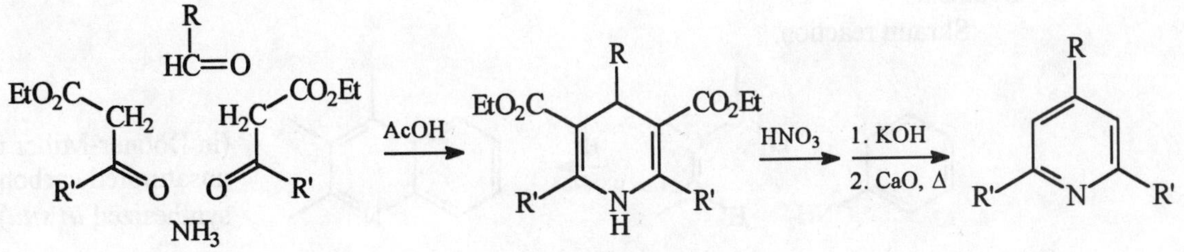

$$RCHO + EtO_2CCH_2COR' + R'COCH_2CO_2Et + NH_3 \xrightarrow{AcOH} \text{1,4-dihydropyridine} \xrightarrow{HNO_3} \xrightarrow{\text{1. KOH; 2. CaO, } \Delta} \text{pyridine}$$

C. Reactions

as a base and nucleophile, and as solvent

resistant to oxidation and electrophilic aromatic substitution:

e.g., pyridine $\xrightarrow[\text{H}_2\text{SO}_4\text{, Fe, }300°]{\text{KNO}_3\text{, HNO}_3}$ 3-nitropyridine (22%)

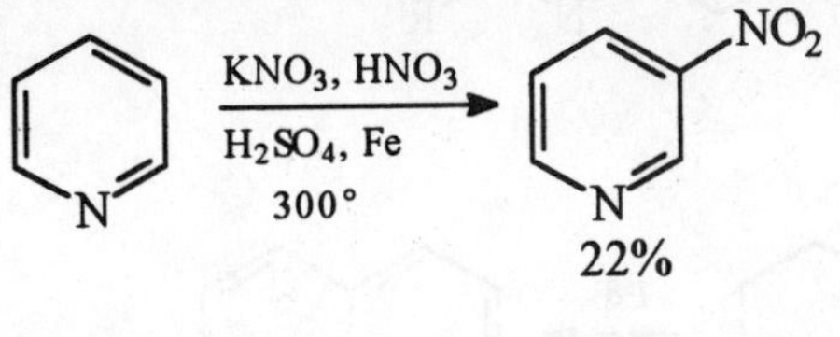

substitution facilitated by activating groups or by N-oxide:

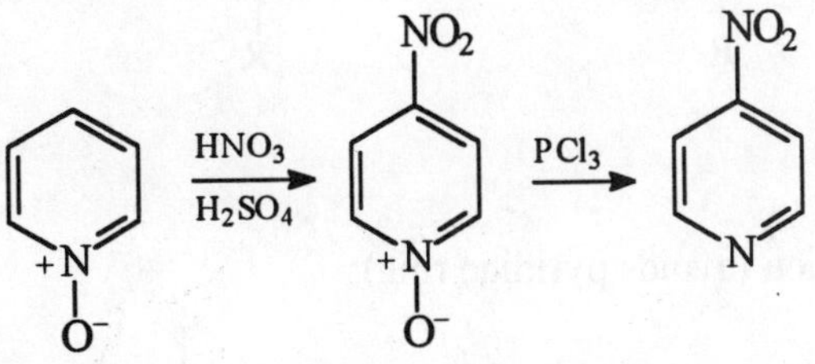

pyridine N-oxide $\xrightarrow[\text{H}_2\text{SO}_4]{\text{HNO}_3}$ 4-nitropyridine N-oxide $\xrightarrow{\text{PCl}_3}$ 4-nitropyridine

nucleophilic substitution: Chichibabin reaction:

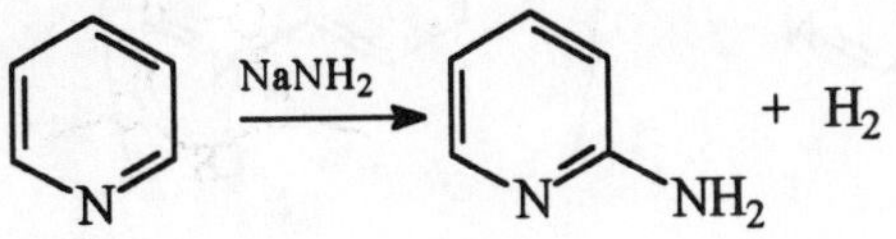

pyridine $\xrightarrow{\text{NaNH}_2}$ 2-aminopyridine + H_2

diazotization of aminopyridines to make pyridones:

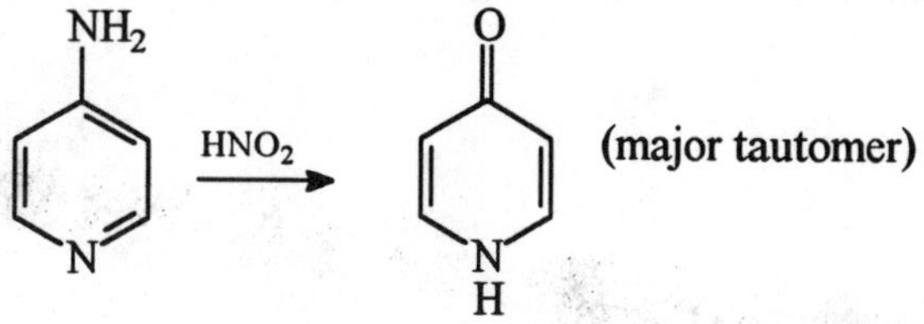

NH_2 HNO$_2$ → O N H (major tautomer)

acidity of α- and γ-alkyl groups:

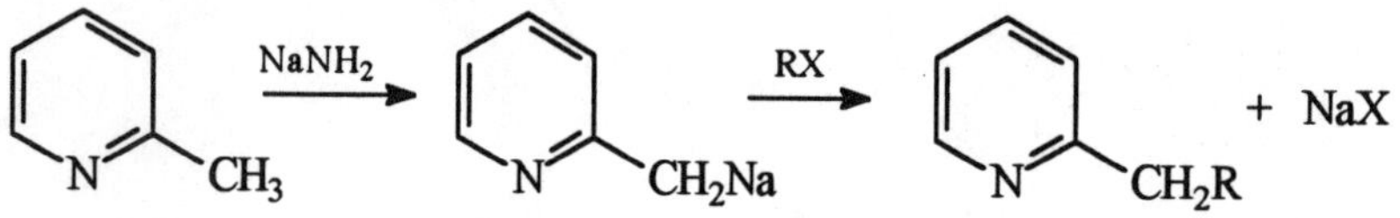

N CH_3 $NaNH_2$ → N CH_2Na RX → N CH_2R + NaX

32.7 Quinoline and Isoquinoline

A. Synthesis and Nomenclature

B. Synthesis

Skraup reaction:

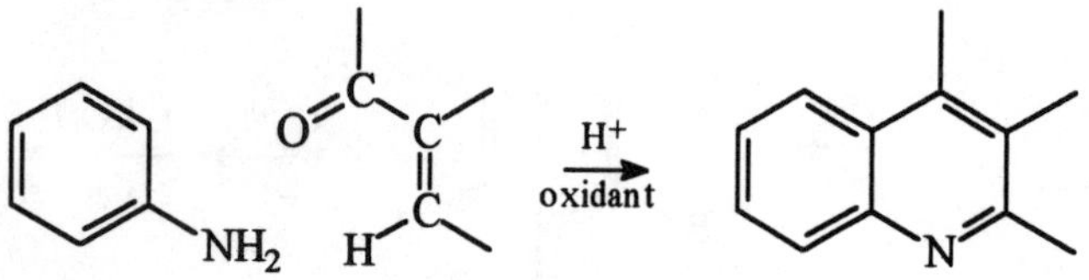

NH_2 O=C C C H H+ oxidant → N

(in Döbner-Miller reaction, the unsaturated carbonyl component is synthesized *in situ*)

Friedländer synthesis:

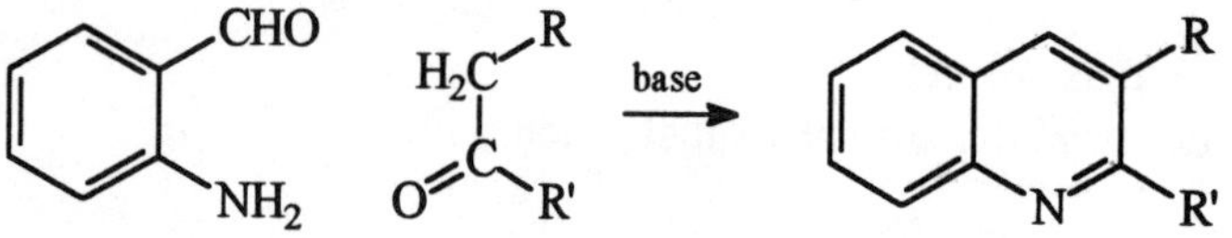

CHO NH_2 H_2C R C O R' base → N R R'

Bischler-Napieralski synthesis:

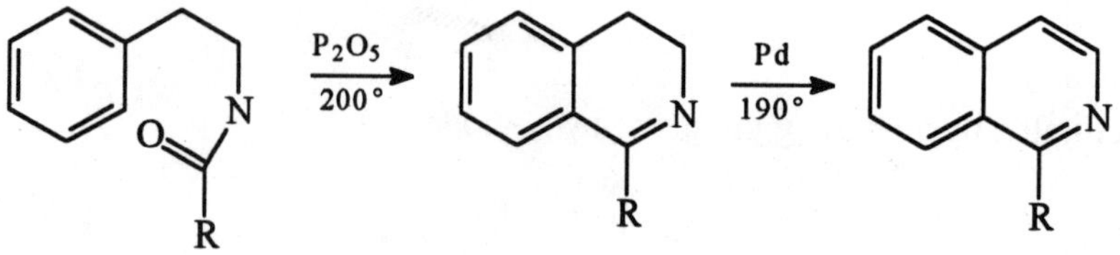

N O R P_2O_5 200° → N R Pd 190° → N R

C. Reactions

electrophilic aromatic substitution (avoids pyridine ring):

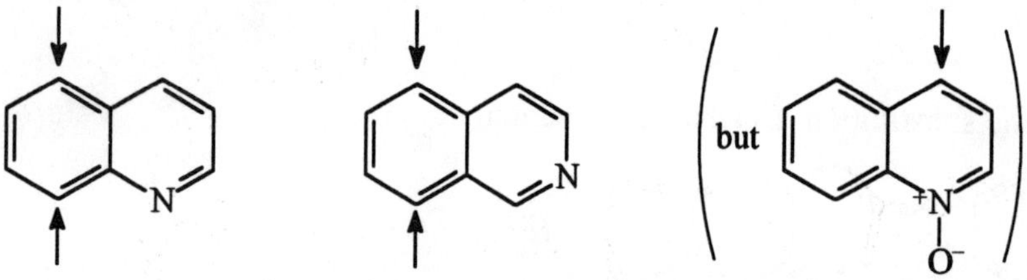

N N (but $^{+}$N O^-)

nucleophilic aromatic substitution:

acidity of alkyl derivatives:

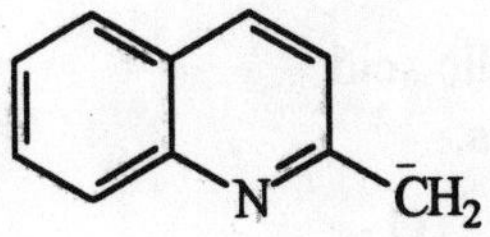

32.8 Diazines

A. Structure and Occurrence
 pyridazine, pyrimidine, pyrazine
 purine
 nucleic acid components

B. Synthesis

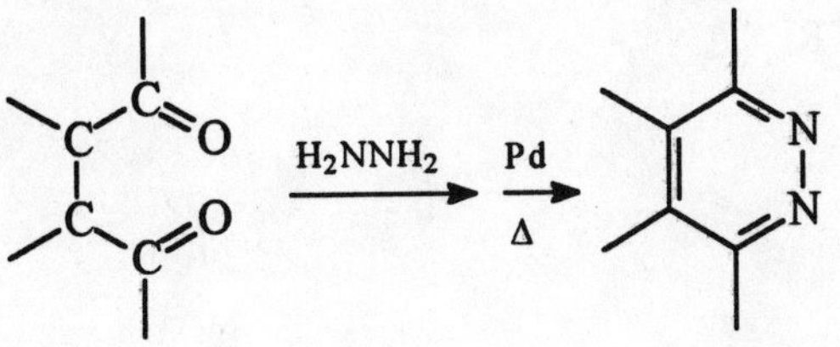

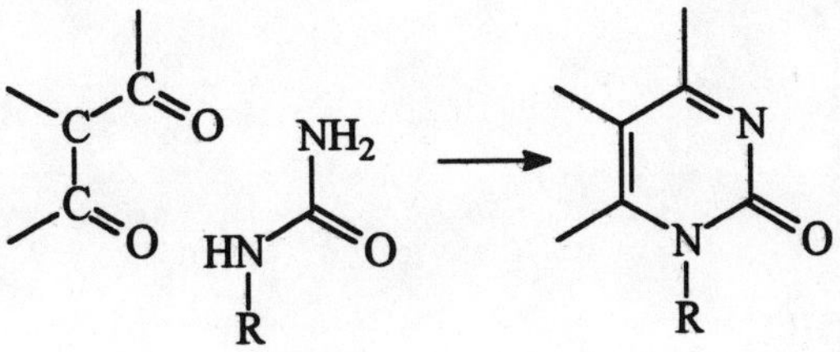

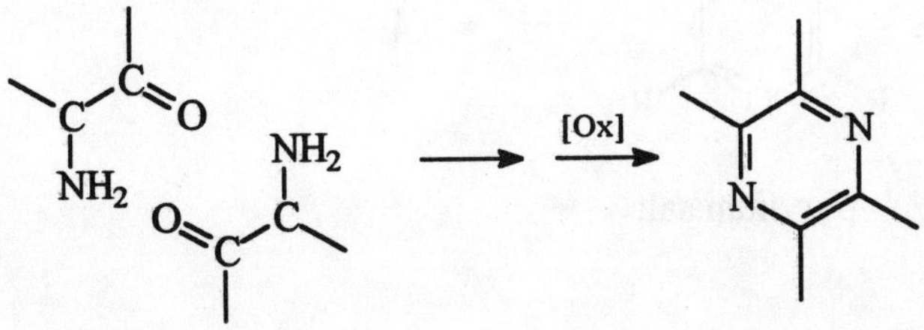

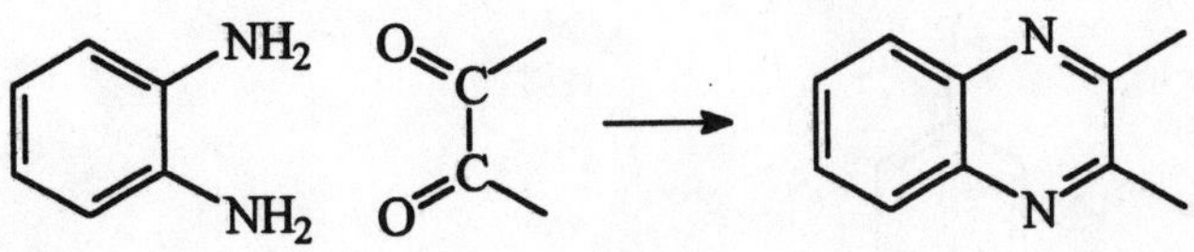

H_2NNH_2 Pd Δ

[Ox]

C. Reactions

electrophilic aromatic substitution requires activating groups
nucleophilic aromatic substitution is reasonably easy

32.9 Pyrones and Pyrylium Salts

A. Pyrones

α-pyrone from pyrolysis of malic acid
γ-pyrones from 1,3,5-triketones:

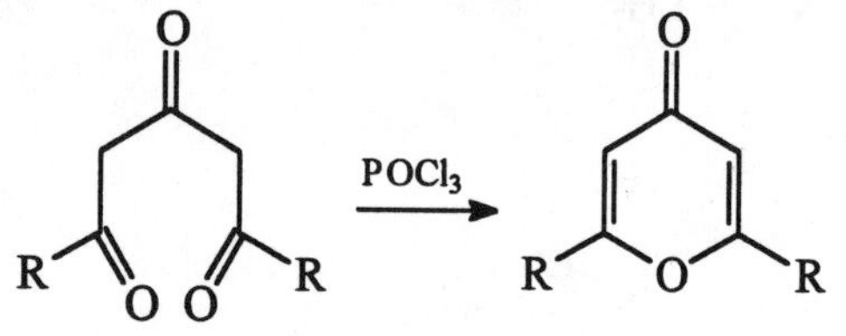

$POCl_3$

α-pyrones as Diels-Alder dienes:

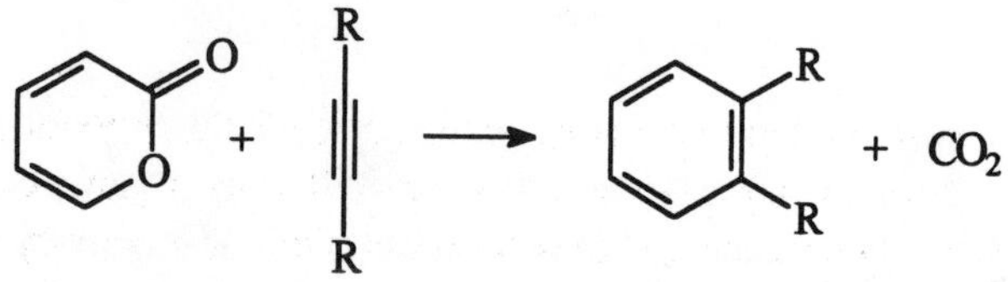

+ CO_2

pyridones from γ-pyrones:

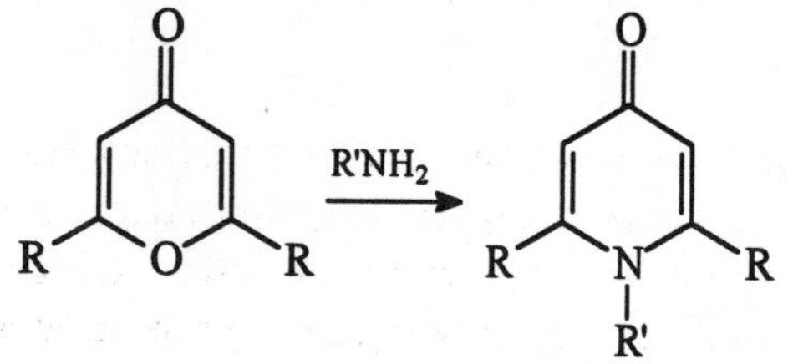

$R'NH_2$

basicity of pyrones:

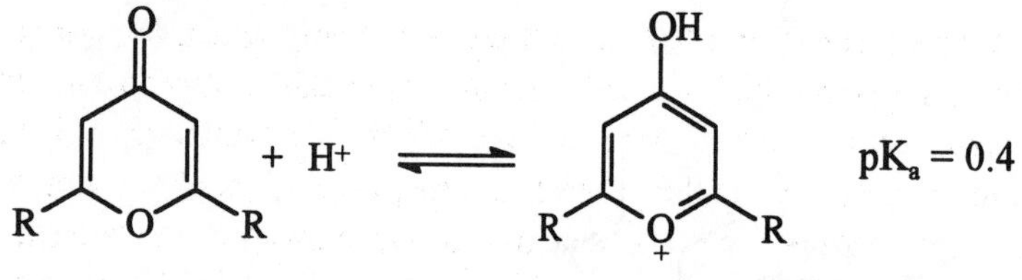

+ H^+

$pK_a = 0.4$

pyrylium salt

B. Pyrylium Salts

from pyrones and a Grignard reagent:

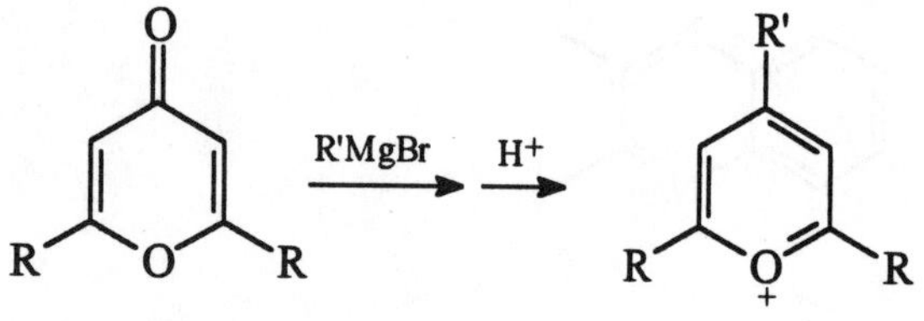

R'MgBr H^+

from enone condensation:

e.g., $(CH_3)_2C{=}CH\overset{O}{\overset{\|}{C}}CH_3$ + $2\ Ac_2O$ $\xrightarrow{H^+}$

reactions with nucleophiles:

+ $^{-}CH_2NO_2$ ⟶

NO_2

32.C. Important Concepts and Hints

There is an astounding amount of material presented in this chapter on heterocyclic compounds. That the chapter is organized as it is reflects both your chemical sophistication as the end of your organic course approaches ("you know more so you can learn more") and the importance and breadth of the field of hetereocyclic chemistry itself. Many biologically significant compounds, both naturally-occurring and man-made, are heterocycles, and most organic chemists encounter heterocyclic compounds either directly or indirectly during the course of their research.

The subject is divided between saturated and unsaturated (usually aromatic) heterocycles. The syntheses and reactions of the saturated heterocycles are almost the same as those of acyclic compounds that have the same functional groups. However, special syntheses and greater reactivity are seen for the three- and four-membered ring compounds. The saturated rings themselves are prepared either by hydrogenation of the aromatic analogues or by intramolecular alkylation reactions. (Peracid epoxidation is an exception.)

The classification of aromatic heterocycles encompasses a vast range of compounds, including five- and six-membered and polycyclic systems, and many combinations and orientations of one or more nitrogens, oxygens, and sulfurs (and others!). There is a correspondingly large number of methods for the synthesis of these compounds. A few can be made by cycloaddition reactions, but by far the greatest number arise from condensation reactions. The single most important characteristic of the condensation reactions that produce unsaturated heterocycles is the following: the ring carbons that are directly attached to the heteroatom were originally either carbonyl carbons or were adjacent to carbonyl carbons. To convince yourself of this fact, go through the chapter outline preceding this section of the Study Guide and look for **exceptions** to this generalization. In a sense, the synthesis of aromatic heterocycles is simply another aspect of the chemistry of carbonyl compounds.

32.D. Answers to Exercises

32.1 a.

b. + INCO → (racemic mixture) MeOH KOH

32.2 NaSCN

(2*S*,3*S*)-2,3-dimethylthiirane

32.3 CH_3OCH_3 → (oxirane, $H_2C{-}CH_2$) + H_2

ΔH_f° = -44.0 -12.6 0 ΔH° = +31.4 kcal mole^{-1}

CH_3SCH_3 → (thiirane, $H_2C{-}CH_2$) + H_2

ΔH_f° = -8.9 +19.7 0 ΔH° = +28.6 kcal mole^{-1}

The thiirane ring is less strained because less distortion of the C-S-C bond angle is required to close the ring. In dimethyl sulfide the C-S-C bond angle is 98.9° (see Section 26.1) vs. the C-O-C bond angle of 111.7° of dimethyl ether.

32.4

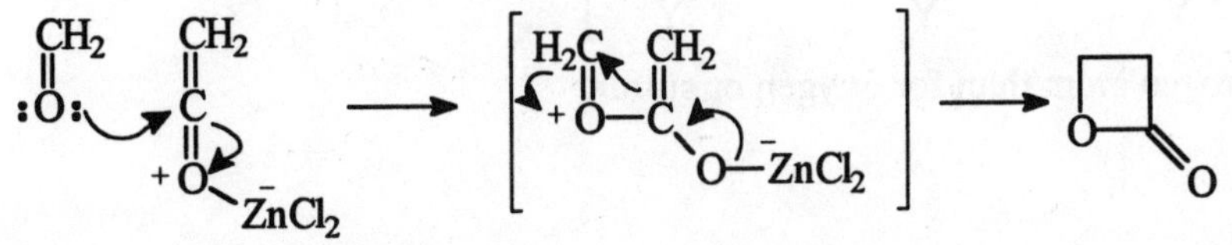

32.5 C-protonation gives a highly delocalized cation: 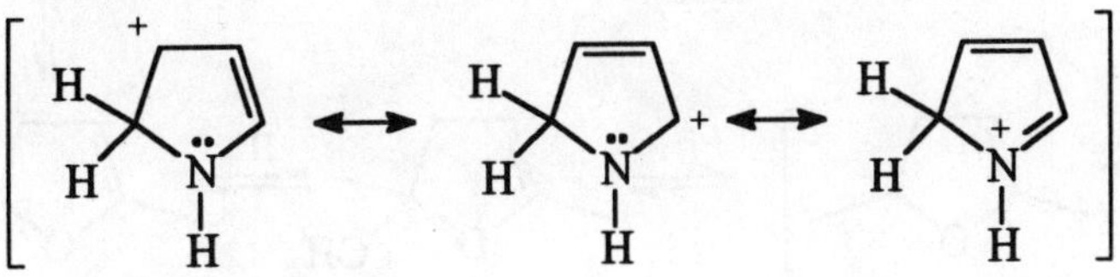 N-protonation gives a localized cation:

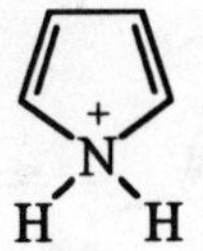

32.6

				$H°$ (kcal mole^{-1})	Aromatic Stabilization (kcal mole^{-1})
cyclopentadiene + 2 H_2 → cyclopentane				-50.3	(0)
ΔH_f° =	31.9	0	-18.4		
furan + 2 H_2 → tetrahydrofuran				-35.7	14.6
ΔH_f° =	-8.3	0	-44.0		
pyrrole + 2 H_2 → pyrrolidine				-26.7	23.6
ΔH_f° =	25.9	0	-0.8		
thiophene + 2 H_2 → tetrahydrothiophene				-35.7	14.6
ΔH_f° =	27.6	0	-8.1		

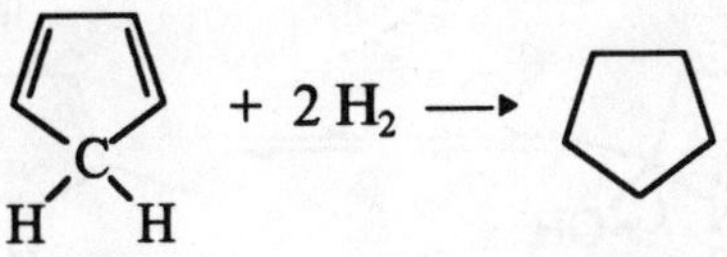

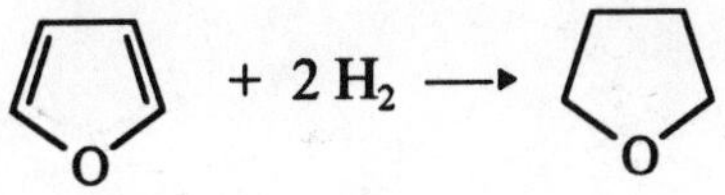

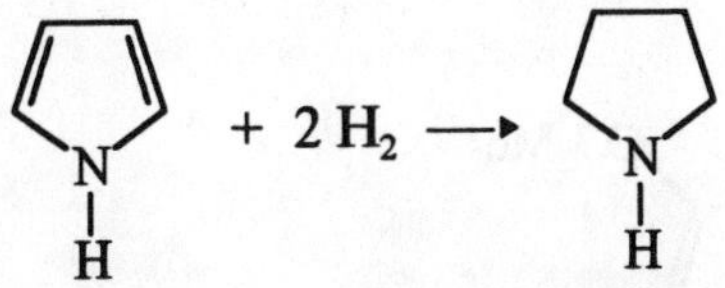

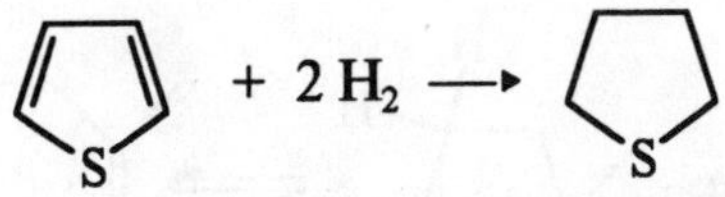

The resonance structures that contribute to aromatic stabilization are those that involve a positive charge on the heteroatom:

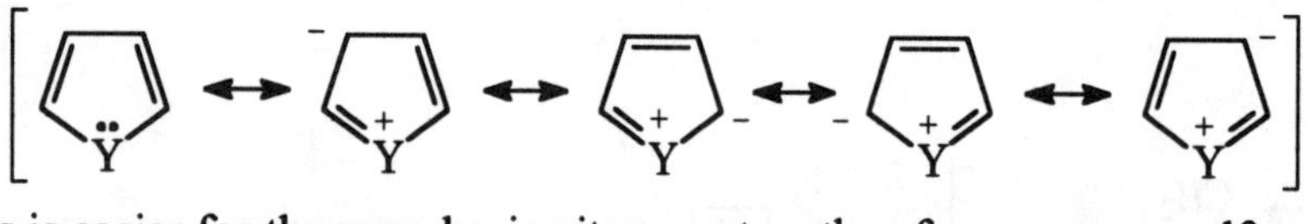

This is easier for the more basic nitrogen atom than for oxygen or sulfur.

32.7

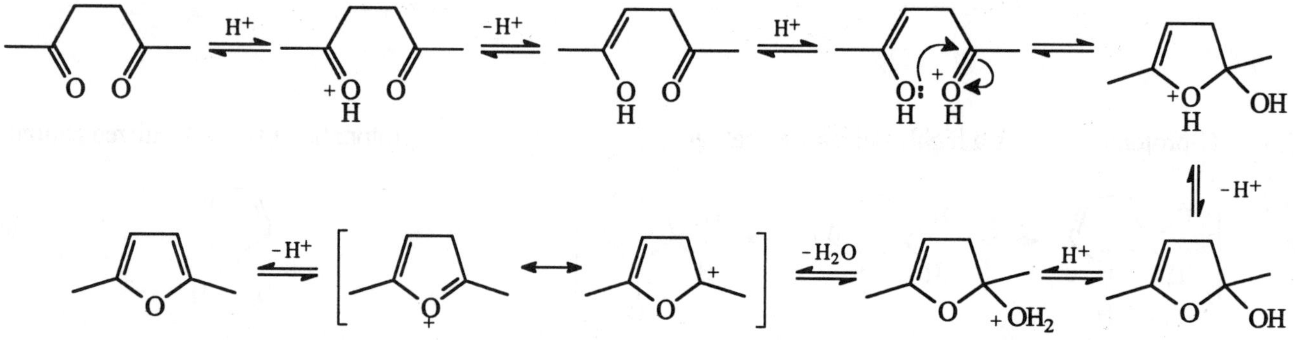

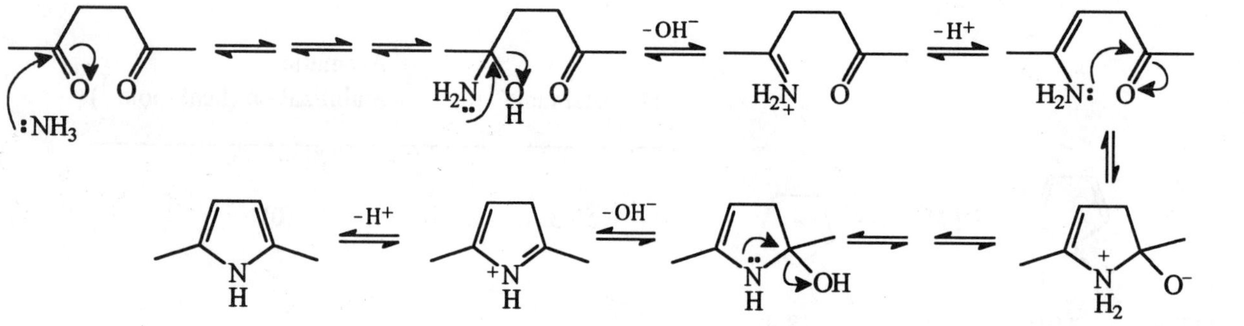

32.8 a. P_2S_5, Δ

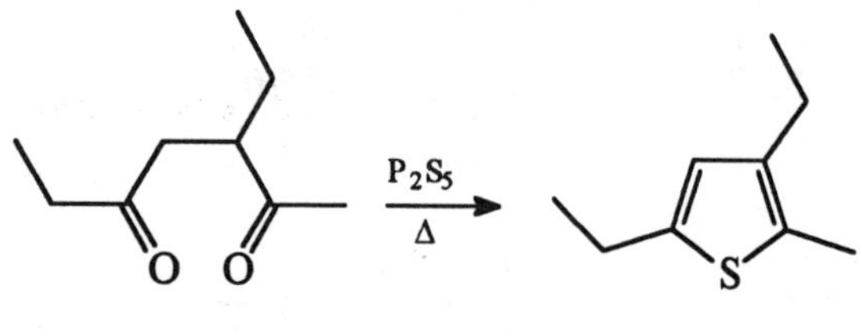

b. AcOH

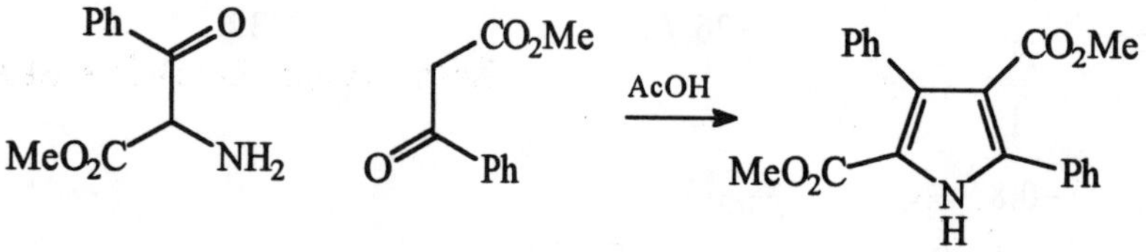

32.9

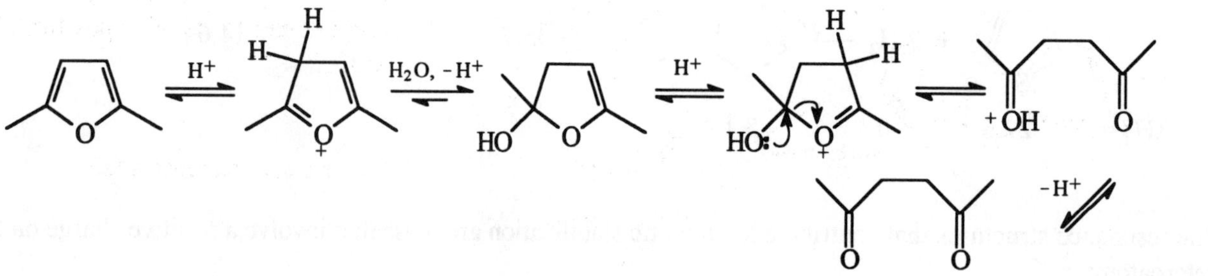

31.10 Pyrrole + $(CH_3CH_2\overset{O}{\overset{\|}{C}})_2O$ $\xrightarrow{BF_3}$ 2-propanoylpyrrole (CCH_2CH_3, C=O)

Furan + Br_2 $\xrightarrow{Fe}$ 2-bromofuran $\xrightarrow{Mg}$ 2-furyl-MgBr $\xrightarrow{D_2O}$ 2-deuteriofuran (D)

Thiophene + $CH_2{=}O$ + HCl $\longrightarrow$ 2-(chloromethyl)thiophene (CH_2Cl)

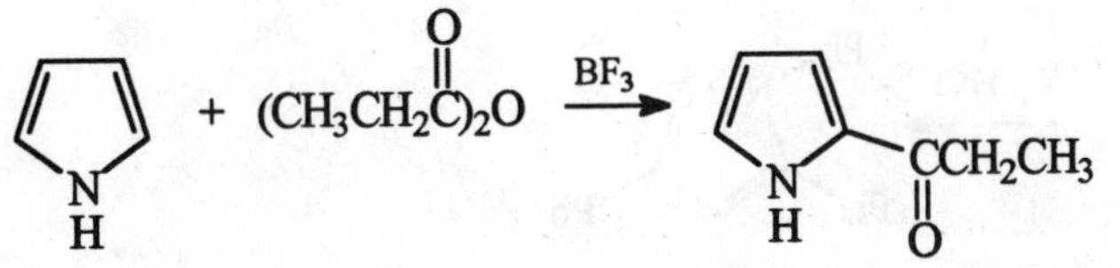

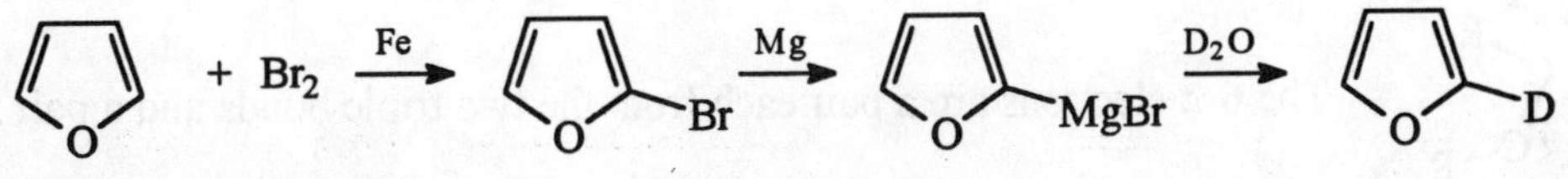

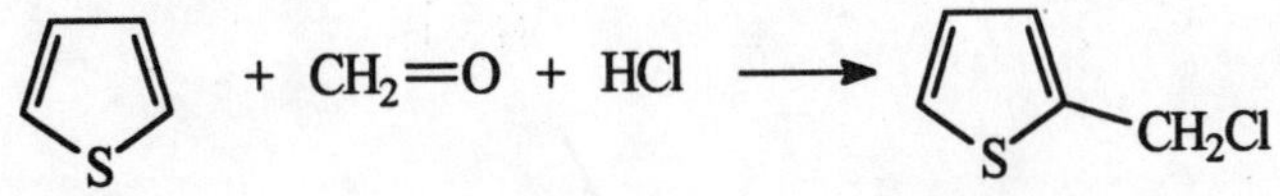

32.11

a. 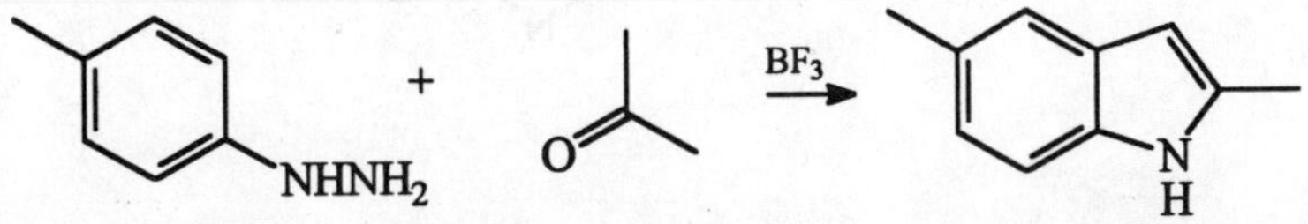 p-tolyl-$NHNH_2$ + acetone $\xrightarrow{BF_3}$ 2,5-dimethylindole (N–H)

b. 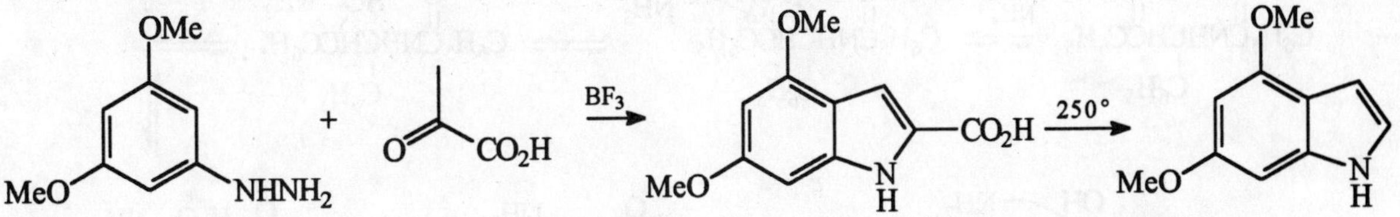 3,5-dimethoxyphenyl-$NHNH_2$ (OMe, MeO) + pyruvic acid (CO_2H) $\xrightarrow{BF_3}$ 4,6-dimethoxyindole-2-carboxylic acid (CO_2H) $\xrightarrow{250°}$ 4,6-dimethoxyindole

c. 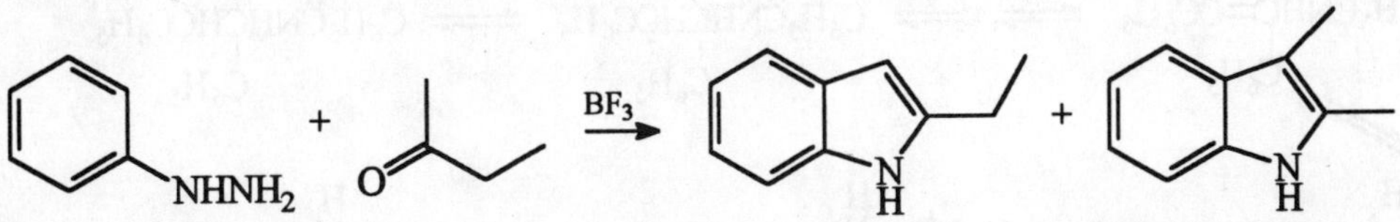 phenyl-$NHNH_2$ + 2-butanone $\xrightarrow{BF_3}$ 2-ethylindole + 2,3-dimethylindole

2,3-dimethylindole is formed as well

d. 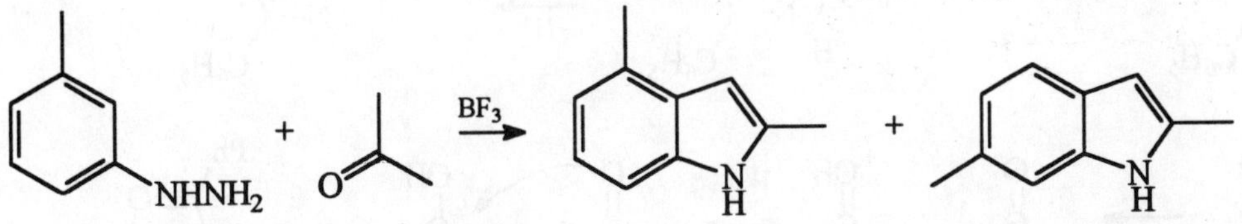 m-tolyl-$NHNH_2$ + acetone $\xrightarrow{BF_3}$ 2,4-dimethylindole + 2,6-dimethylindole

2,6-dimethylindole is formed as well

32.12

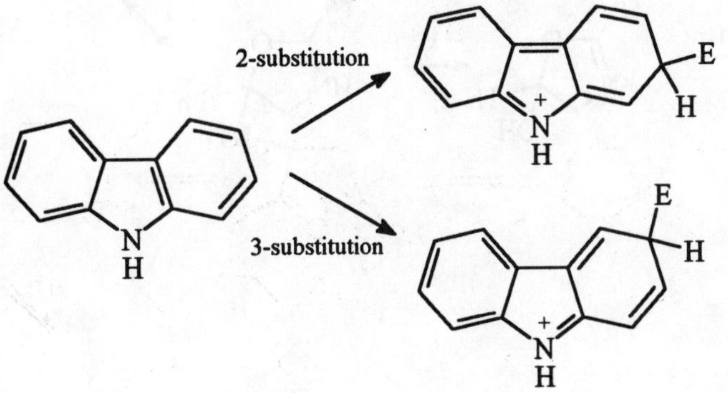

Carbazole (N–H)

2-substitution → cation with C(E)(H) at position 2, $\overset{+}{N}$–H: requires loss of aromaticity of benzene ring in order to stabilize positive charge with nitrogen

3-substitution → cation with C(E)(H) at position 3, $\overset{+}{N}$–H: benzene ring remains intact

32.13 $PhCOCH_2COPh$ + $PhNHNH_2$ $\xrightarrow{\text{aq. HCl}}$ 1,3,5-triphenylpyrazole (Ph, N—N, Ph, Ph)

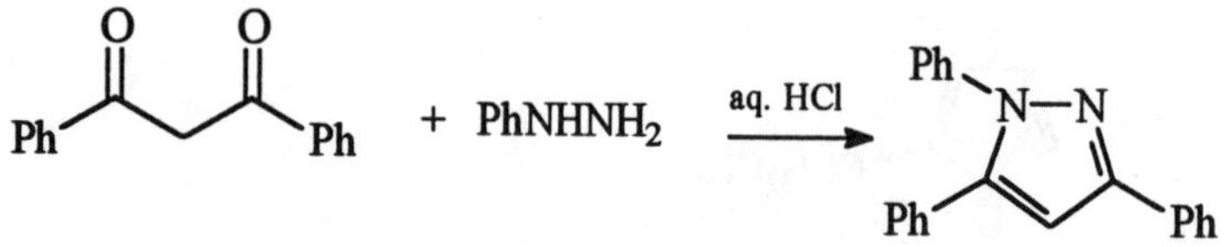

32.14 a. (isoxazole ring: R–C, C–R, N, C–R, O:)

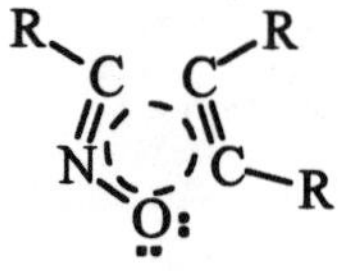

The 6 π electrons are a pair each from the two triple bonds and a pair from O.

b. $CH_3CH_2CH_2NO_2$ $\xrightarrow{\text{PhNCO}}$ $[CH_3CH_2C{\equiv}\overset{+}{N}{-}O^-]$ → (3-ethyl-5-isopropylisoxazole)

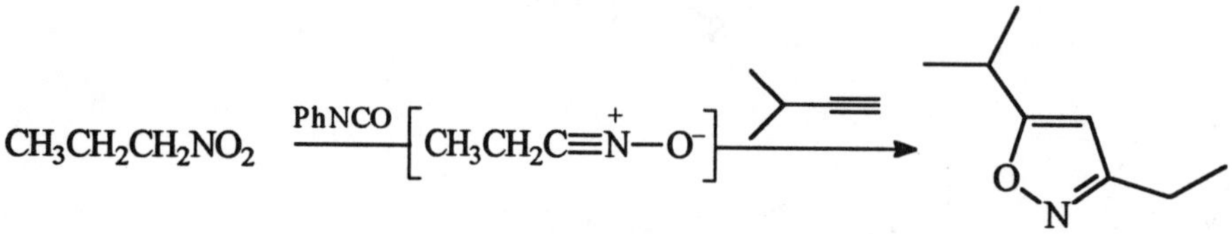

32.15 $C_6H_5\overset{O}{\overset{\|}{C}}NHCH(C_6H_5)\overset{O}{\overset{\|}{C}}C_6H_5$ $\xrightleftharpoons{NH_4^+}$ $C_6H_5\overset{O}{\overset{\|}{C}}NHCH(C_6H_5)\overset{^+OH}{\overset{\|}{C}}C_6H_5$:NH_3 ⇌ $C_6H_5\overset{O}{\overset{\|}{C}}NHCH(C_6H_5)C(HO)(^+NH_3)C_6H_5$ ⇌

⇌ $C_6H_5\overset{O}{\overset{\|}{C}}NHCH(C_6H_5)C(H_2\overset{+}{O})(NH_2)C_6H_5$ $\xrightleftharpoons{-H_2O}$ $C_6H_5\overset{O}{\overset{\|}{C}}NHCH(C_6H_5)\overset{+}{C}(NH_2)C_6H_5$ ⇌⇌ $C_6H_5\overset{+}{C}(OH)NHC(C_6H_5){=}C(NH_2)C_6H_5$

⇌ (HO, C_6H_5, $H\overset{+}{N}H$, C_6H_5, N–H, C_6H_5 ring) ⇌ ⇌ ($H_2\overset{+}{O}$, C_6H_5, N–H, C_6H_5, N–H, C_6H_5 ring) $\xrightarrow[-H^+]{-H_2O}$ 2,4,5-triphenylimidazole (C_6H_5, N–H, C_6H_5, N, C_6H_5)

Ph–C(=O)–NHCH$_2$–C(=O)–Ph (H$^+$) → Ph–C(=O)–NHCH$_2$–C(=$^+$OH)–Ph ↔ Ph–C(=O)–NHCH$_2$–$\overset{+}{C}$(OH)–Ph ⇌ (Ph, =O$^+$, HN, Ph, OH ring)

↔ (Ph, $\overset{+}{}$, O, HN, Ph, OH ring) $\xrightarrow{-H^+}$ (Ph, O, N, Ph, OH ring) $\xrightarrow{H^+}$ (Ph, O, N, Ph, $^+OH_2$ ring) $\xrightarrow[-H^+]{-H_2O}$ 2,5-diphenyloxazole (Ph, O, N, Ph)

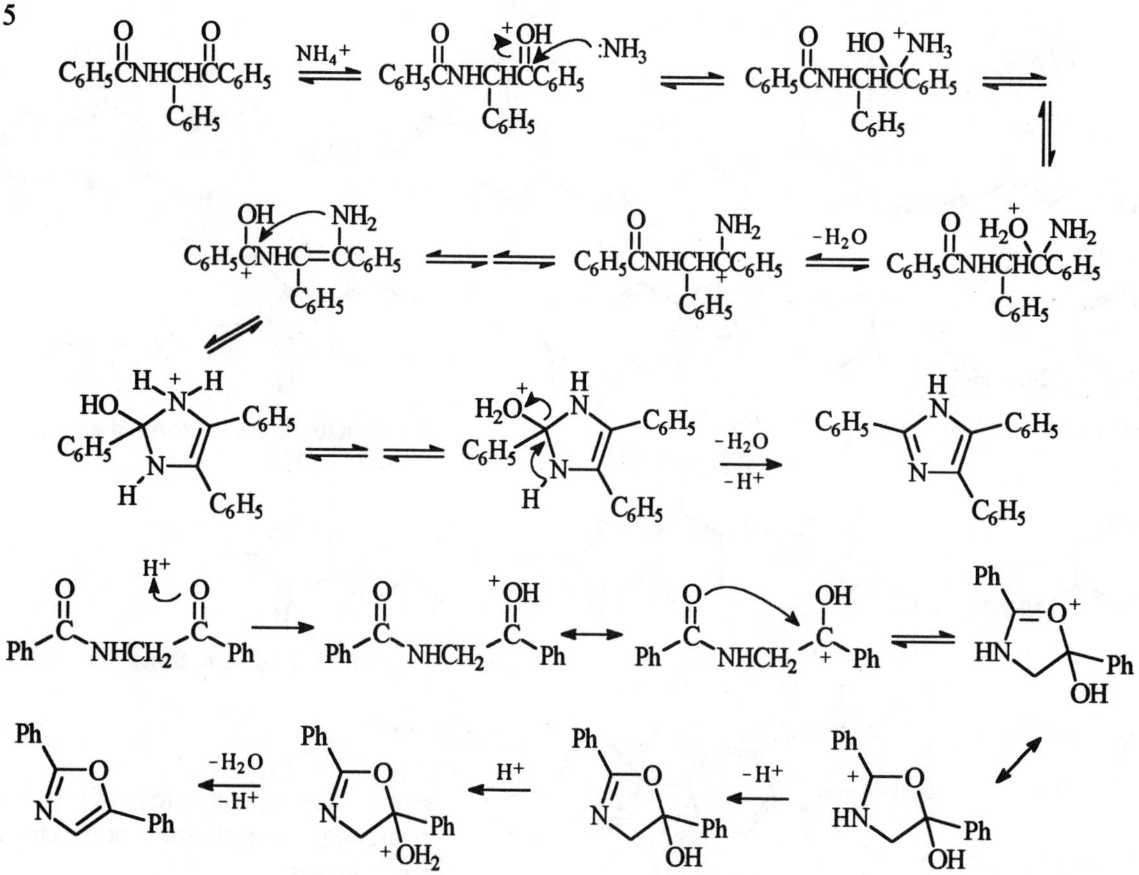

32.16 4-substitution:

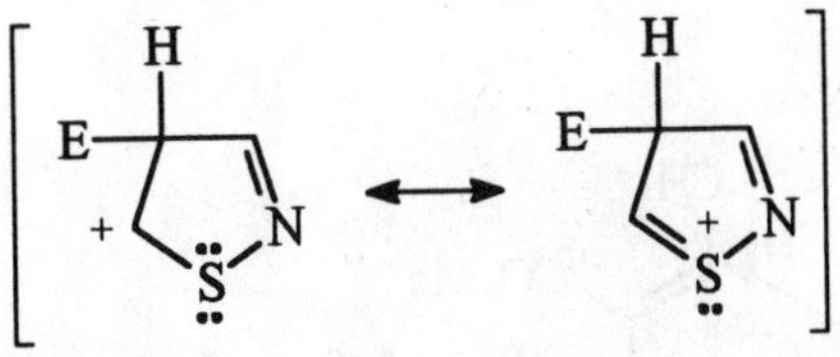

5-substitution:

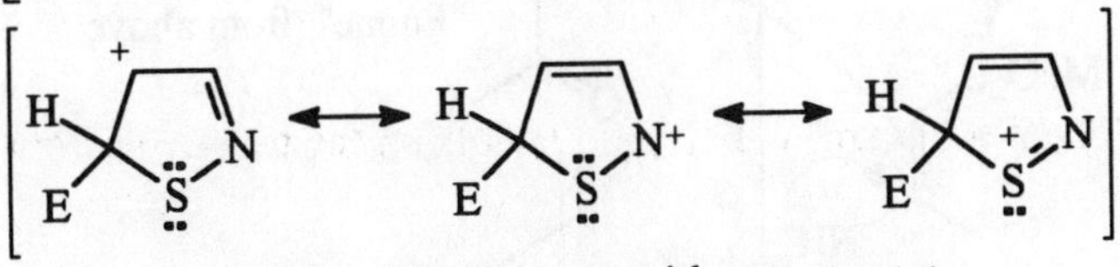

5-Substitution has nitrogen without an octet.

32.17 a. $\xrightarrow[\text{CHCl}_3]{\text{Br}_2}$

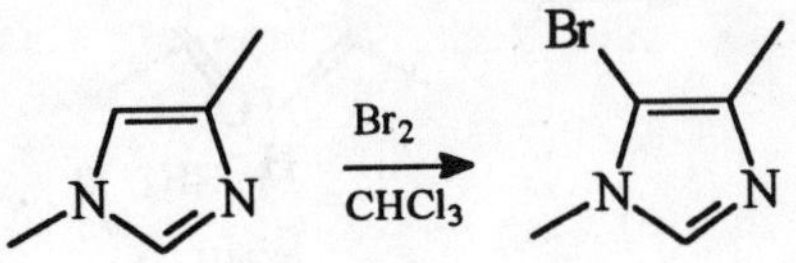

b. $\xrightarrow[\text{H}_2\text{SO}_4]{\text{HNO}_3}$ $\xrightarrow[\text{CHCl}_3]{\text{Br}_2}$

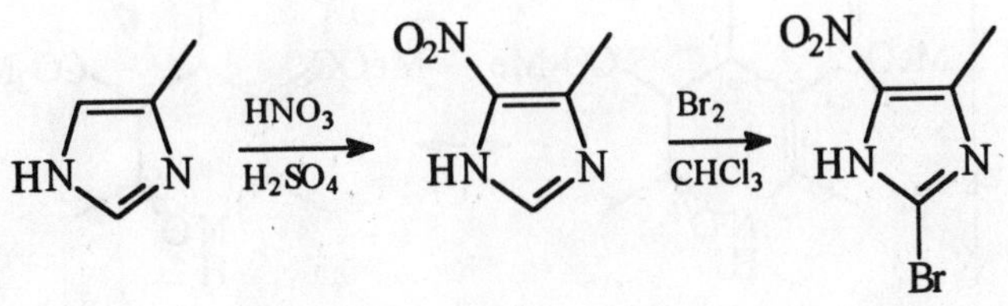

32.18

$MeO_2CCH_2CCH_3$ ⇌ $MeO_2CCH_2CCH_3$ (O⁻, $^+NH_3$) ⇌ ⇌ $MeO_2CCH_2CCH_3$ (OH, NH_2)

The ultimate H^+ source is α-H, transferred as NH_4^+ by the base NH_3;

$MeO_2CCH{=}CCH_3$ (NH_2) "Enamine" ⇌ $MeO_2CC{-}CCH_3$ ($=\overset{+}{N}H_2$) + OH^- → $H_2O + NH_3$ (NH_4^+)

$MeO_2CCH_2CCH_3$ (:NH_3) ⇌ [$MeO_2CCH{=}CCH_3$ (O⁻) ↔ $MeO_2CCHCCH_3$ (O)] + NH_4^+ + $HCCH_3$ (O) ⇌ $MeO_2CC{-}CCH_3$ (H, O; $HCCH_3$, O⁻; H_3N:, $\overset{+}{H}NH_3$) ⇌

$H_2O + NH_3$ ← $^+NH_4$ — OH^- + $MeO_2CC{-}CCH_3$ (O; $HCCH_3$) "Enone" ⇌ [$MeO_2CC{=}CCH_3$ (O⁻; $HCCH_3$, OH) ↔ $MeO_2CC{-}CCH_3$ (O; $HCCH_3$, OH)]

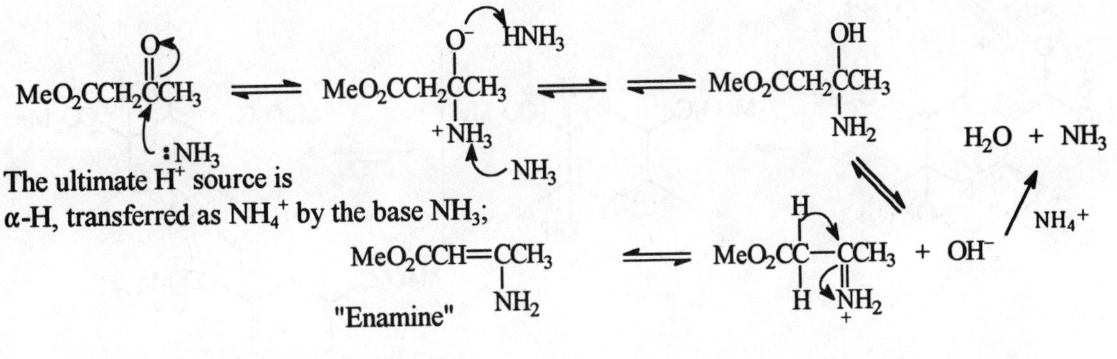

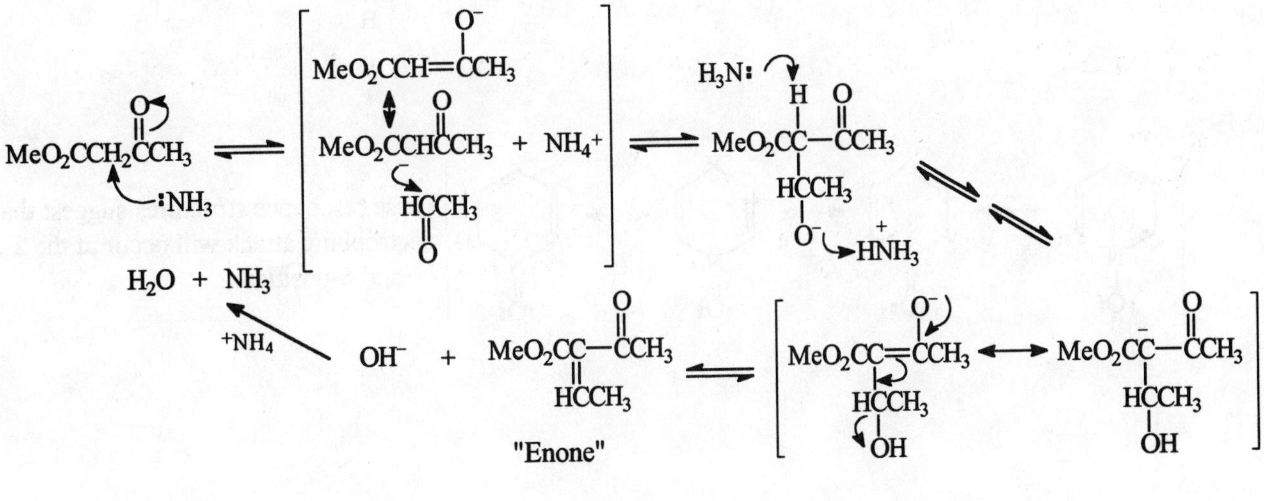

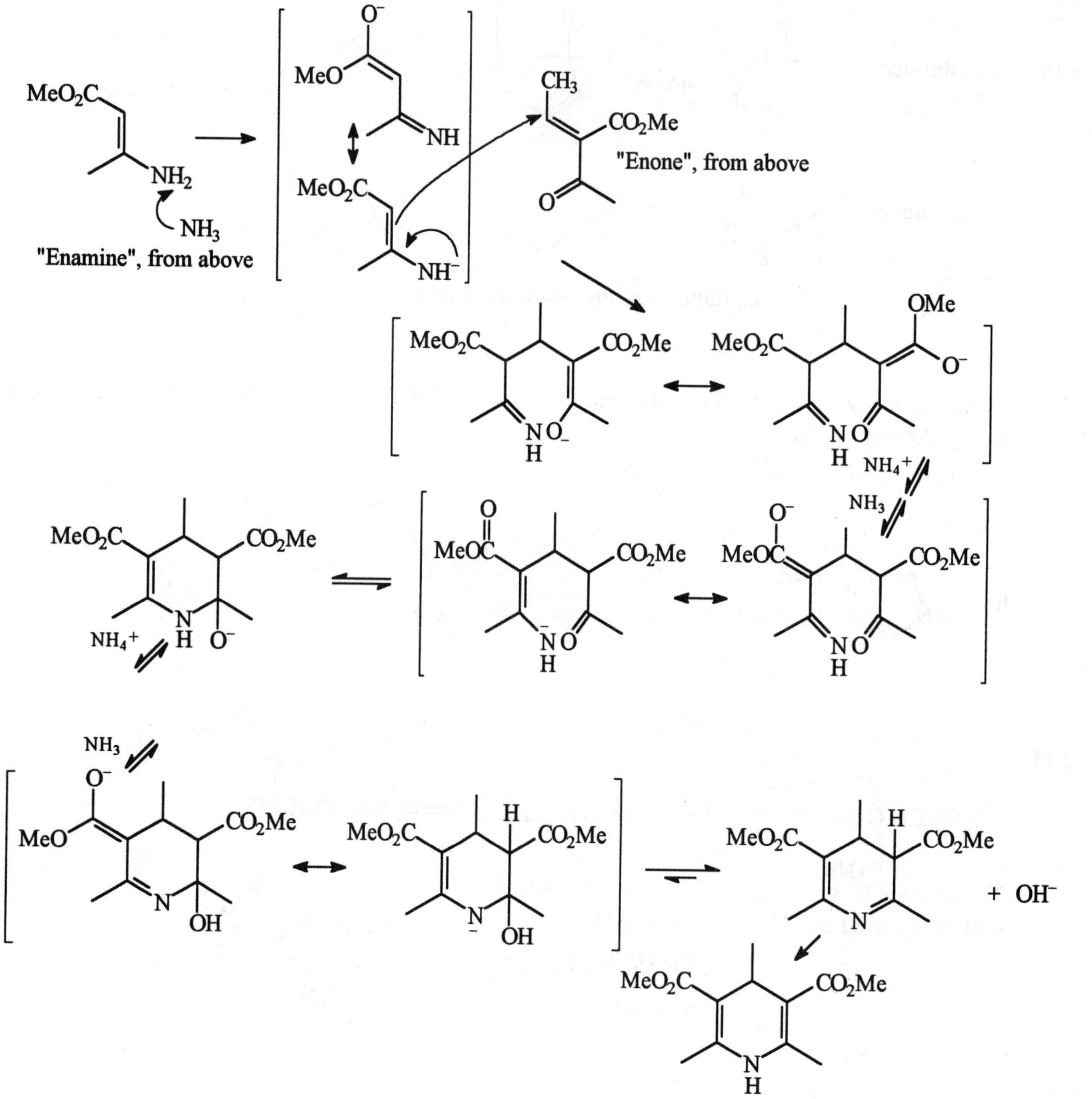

MeO$_2$C NH$_2$ NH$_3$ "Enamine", from above

CH$_3$ CO$_2$Me "Enone", from above

MeO O$^-$ NH MeO$_2$C NH$^-$

MeO$_2$C CO$_2$Me N O$^-$ H

MeO$_2$C OMe O$^-$ N O H

NH$_4^+$ NH$_3$

MeO$_2$C CO$_2$Me N H O$^-$ NH$_4^+$

O MeOC CO$_2$Me N O H

O$^-$ MeOC CO$_2$Me N O H

NH$_3$

O$^-$ MeO CO$_2$Me N OH

MeO$_2$C H CO$_2$Me N OH

MeO$_2$C H CO$_2$Me N + OH$^-$

MeO$_2$C CO$_2$Me N H

32.19

N+ :O: N+ :O: N+ :O: N+ :O:

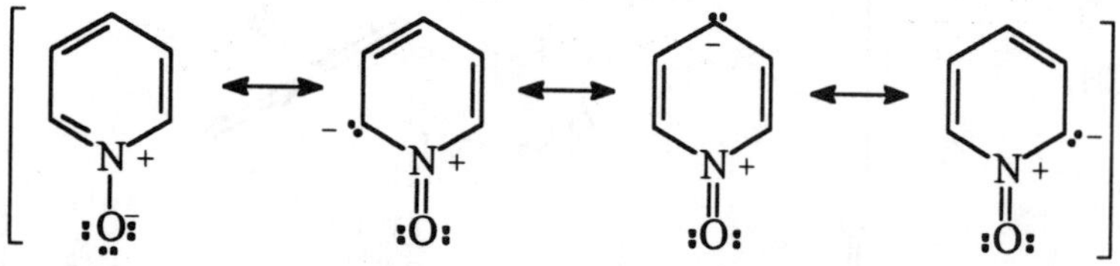

These resonance structures suggest that electrophilic attack will occur at the 2-, 4-, and 6-positions.

32.20

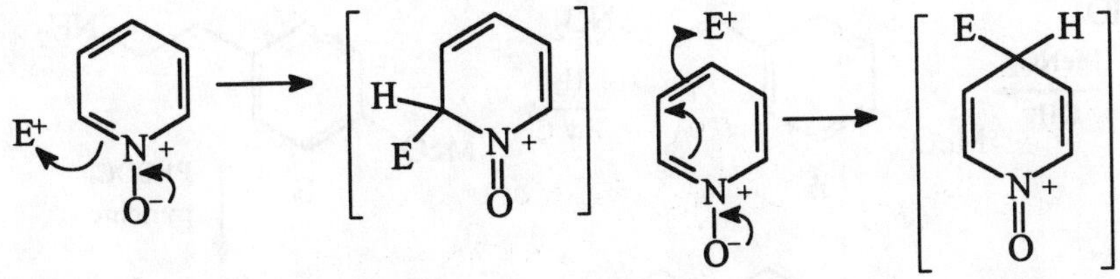

Intermediates with resonance structures having full octets are possible from attack at C-2 and C-4.

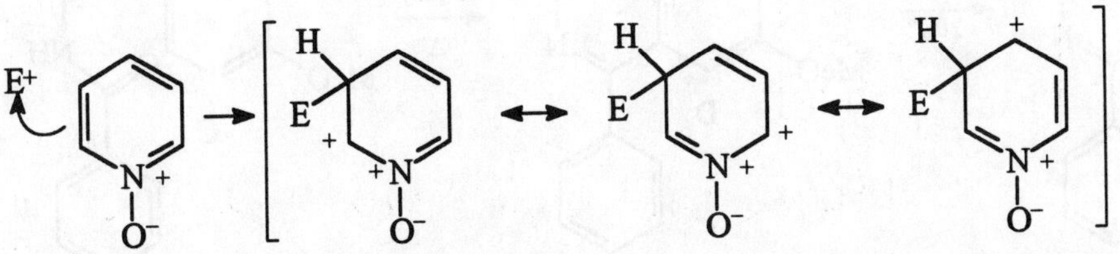

No resonance structure of the intermediate from attack at C-3 has full octets for all atoms. Also, positive charge develops on C's next to N^+.

32.21 PhLi, $C_6H_5CH_3$, 110°; $NaNH_2$, Δ; HNO_2, H_3O^+

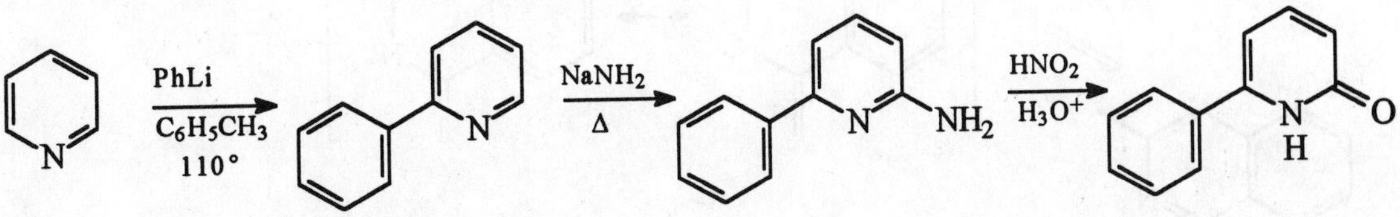

32.22 $^-OCH_3$, CH_3OD; D_2O

32.23 + $HOCH_2CH(OH)CH_2OH$ $\xrightarrow{As_2O_5,\ H_2SO_4\ \Delta}$

+ $HOCH_2CH(OH)CH_2OH$ $\xrightarrow{As_2O_5,\ H_2SO_4\ \Delta}$ +

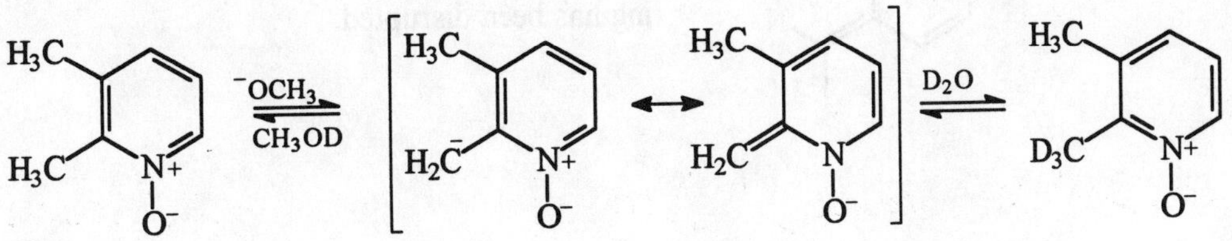

32.24 + $\xrightarrow{NaOH}$

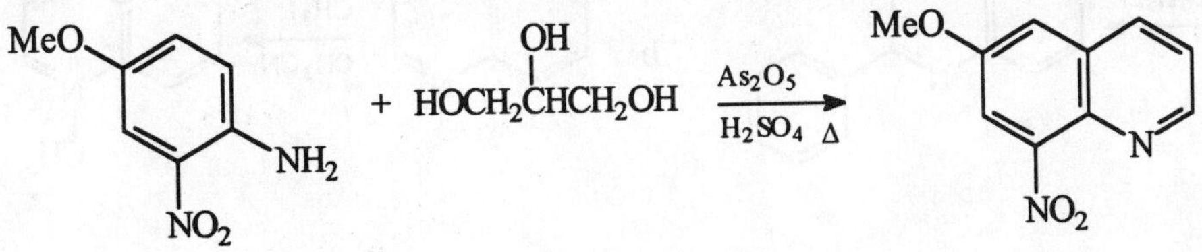

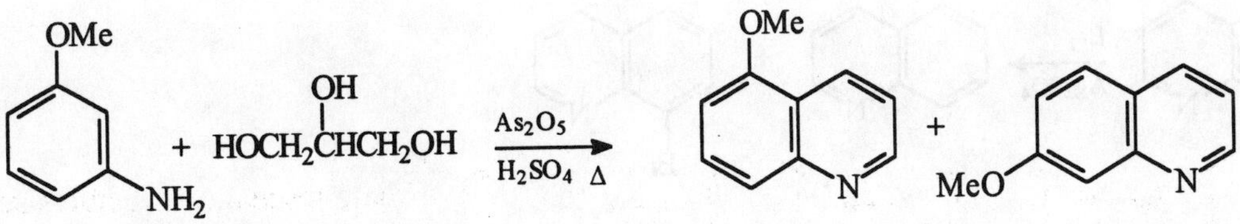

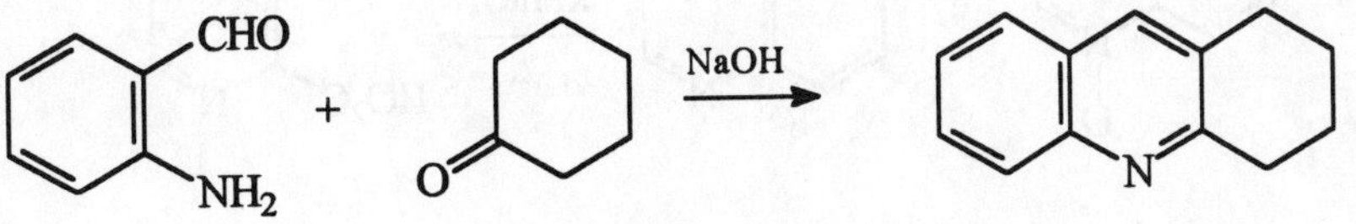

32.25

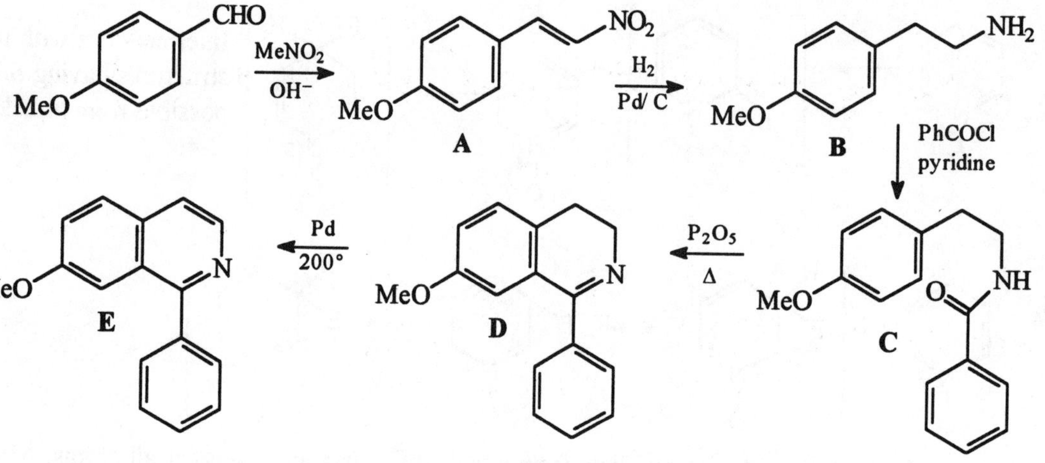

MeO–C₆H₄–CHO $\xrightarrow[OH^-]{MeNO_2}$ **A** $\xrightarrow[Pd/C]{H_2}$ **B** $\xrightarrow[pyridine]{PhCOCl}$ **C** $\xrightarrow[\Delta]{P_2O_5}$ **D** $\xrightarrow[200°]{Pd}$ **E**

MeO, NO_2, NH_2, NH, O, N

32.26

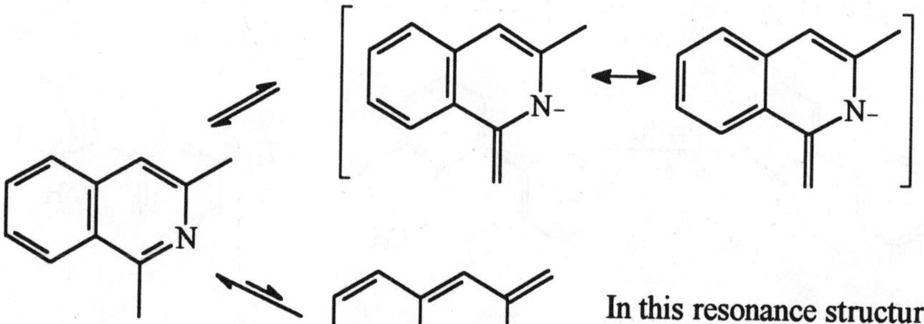

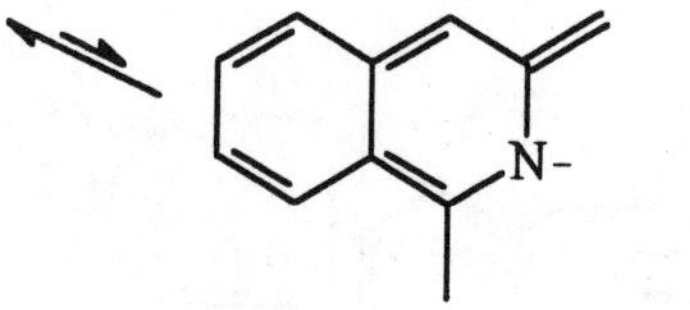

In this resonance structure, the aromaticity of the other ring has been disrupted.

32.27

a. 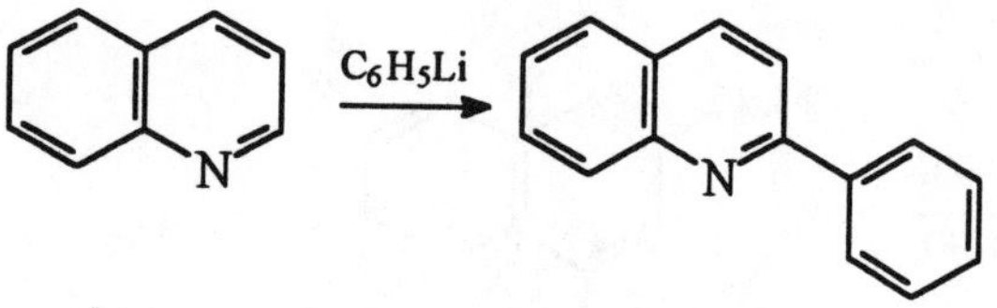 $\xrightarrow{C_6H_5Li}$

b. $\xrightarrow[CH_3CN]{CH_3I}$ CH_3 I^-

c. $\xrightarrow[Ag_2SO_4]{Br_2}$ Br + Br

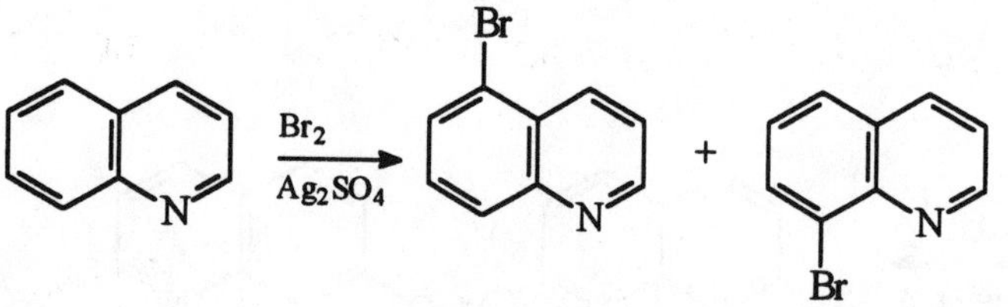

d. $\xrightarrow{30\%\ H_2O_2}$ O^-

e. 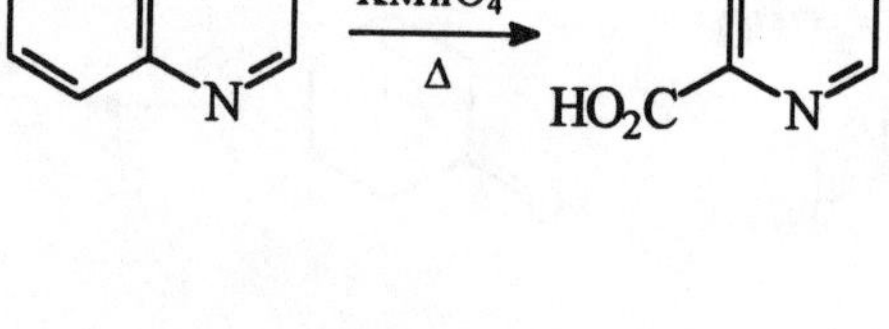 $\xrightarrow[\Delta]{KMnO_4}$ HO_2C HO_2C

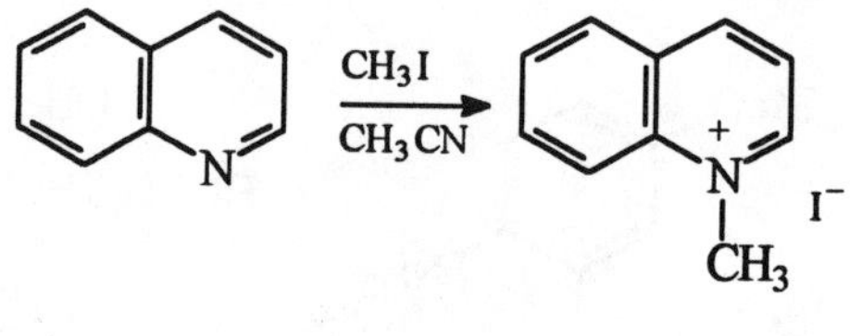

32.28 a. Ph–CO–CH$_2$CH$_2$–CO–Ph + H_2NNH_2 $\xrightarrow{\text{AcOH}}$ Ph–(N–N dihydropyridazine)–Ph $\xrightarrow[\Delta]{\text{Pd}}$ Ph–(N–N pyridazine)–Ph

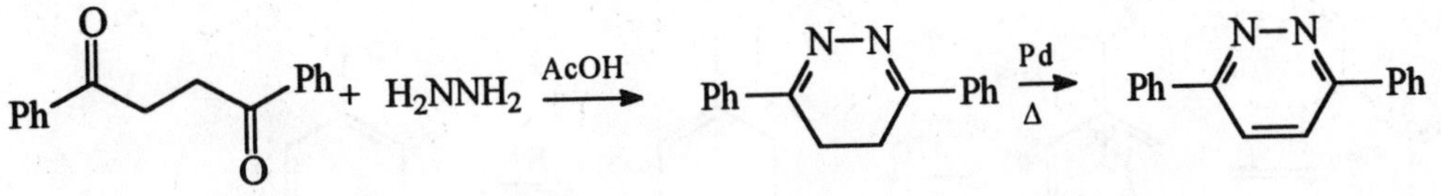

b. Ph–CO–CH$_2$–CO–Ph + $H_2N\overset{O}{\overset{\|}{C}}NH_2$ $\xrightarrow{H^+}$ 4,6-diphenyl-2(1H)-pyrimidinone (N, NH, O, Ph, Ph)

c. 2 Ph–CO–CH$_2$NH$_2$ ⟶ Ph–(dihydropyrazine, N, N)–Ph $\xrightarrow{\text{HgCl}_2}$ Ph–(pyrazine, N, N)–Ph

32.29 2,5-dimethylpyrazine + KOH ⇌ [pyrazine–CH_2^- ⟷ N^- pyrazine=CH_2 ⟷ etc.]

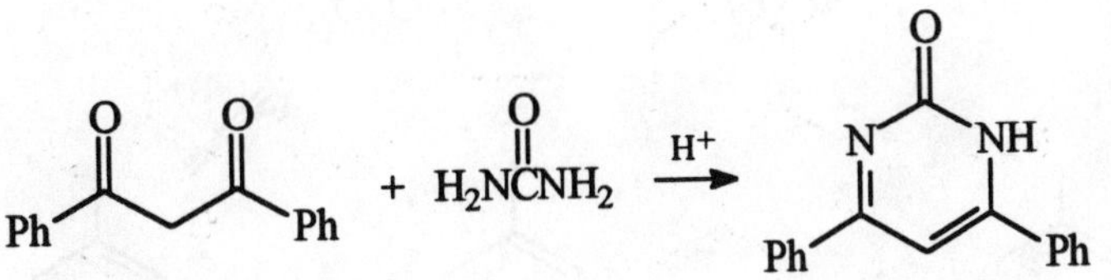

$\searrow$ PhCH=O This is similar to the aldol condensation.

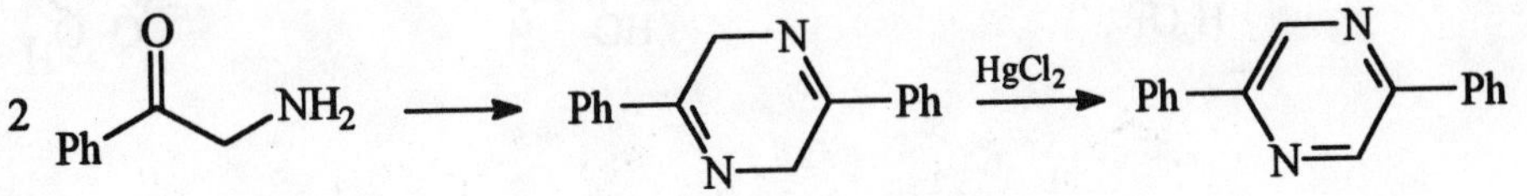

5-methyl-2-styrylpyrazine $\xleftarrow[-H_2O]{\text{KOH}}$ pyrazine–CH$_2$–CH(OH)–Ph

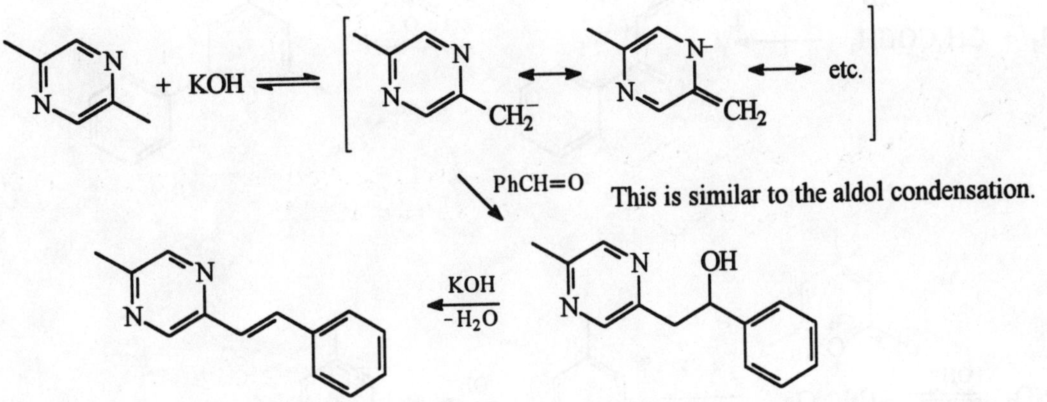

32.30

$HO_2CCH_2\overset{O}{\overset{\|}{C}}H$ ⇌ $HO_2CCH{=}\overset{:OH}{CH}$ + $H\overset{+OH}{C}CH_2CO_2H$ ⇌ HO_2C–CH(CHO)–CH(OH)–CH$_2$CO$_2$H ⇌ cyclic hemiacetal lactone (OH, HO_2C, O, HO, O) $\underset{-H_2O}{\overset{H^+}{\rightleftharpoons}}$ HO_2C-dihydropyranone (HO, O)

$\Downarrow$ H^+

HO_2C–2-pyranone $\underset{}{\overset{-H^+}{\rightleftharpoons}}$ HO_2C–pyranone (+OH) $\underset{-H_2O}{\overset{H^+}{\rightleftharpoons}}$ HO_2C–pyran (HO, OH) $\overset{-H^+}{\rightleftharpoons}$ HO_2C–dihydropyran (HO, +OH)

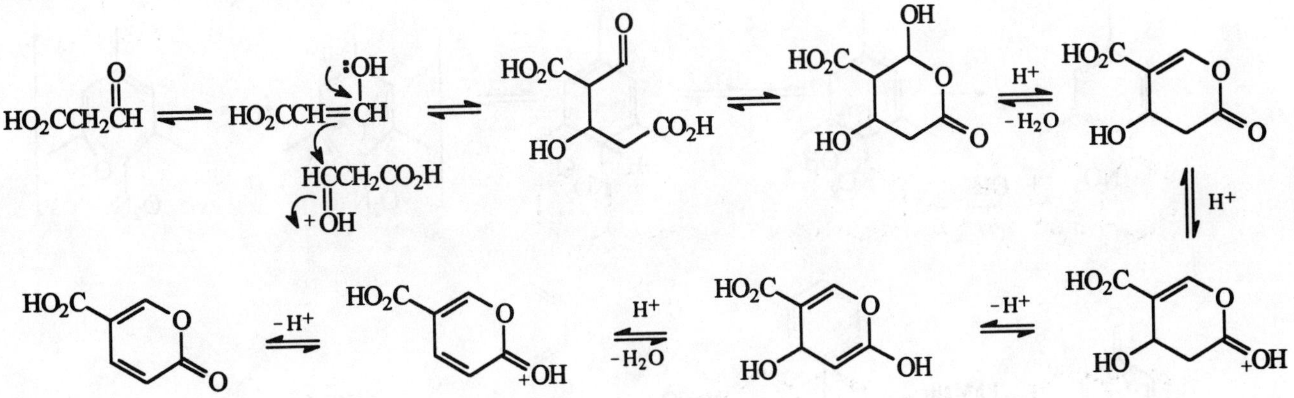

32.31

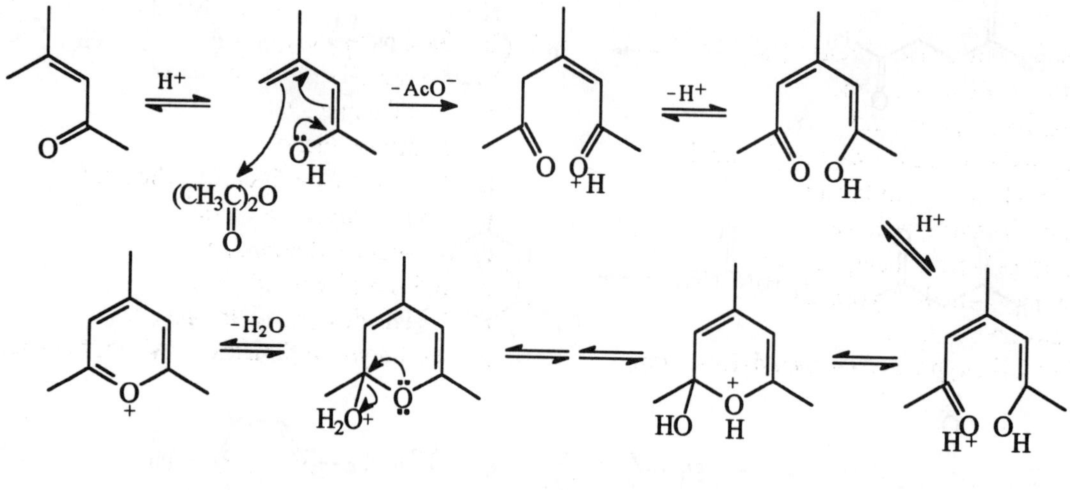

H^+ ; $-AcO^-$; $-H^+$; $(CH_3C(=O))_2O$; H^+ ; $-H_2O$

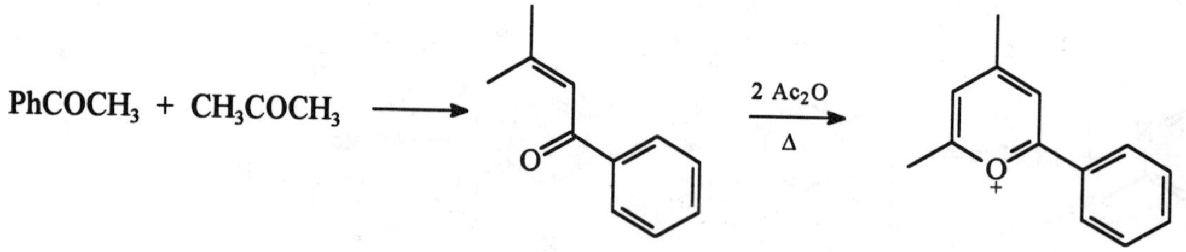

$$PhCOCH_3 + CH_3COCH_3 \longrightarrow \xrightarrow[\Delta]{2\ Ac_2O}$$

32.32

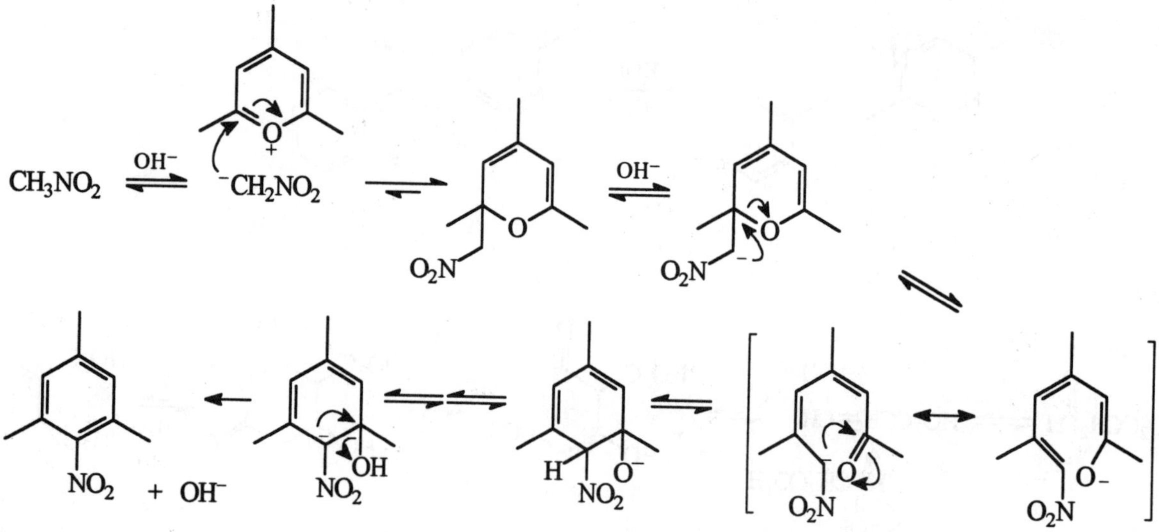

$$CH_3NO_2 \underset{}{\overset{OH^-}{\rightleftharpoons}} {}^-CH_2NO_2$$

OH^- ; O_2N ; $NO_2 + OH^-$

$\xrightarrow[-H_2O]{PhMgBr}$ $\xrightarrow[OH^-]{MeNO_2}$ NO_2

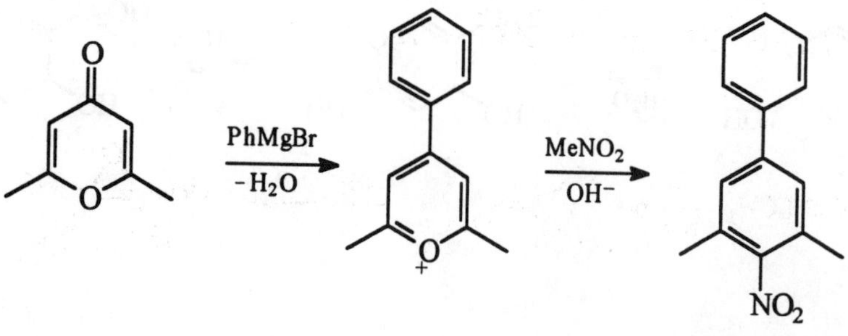

32.E. Answers and Explanations for Problems

1. a. 3-methyltetrahydropyran
   b. 3-azetidinone
   c. 2-ethyl-2-methyloxirane
   d. 3-bromo-2-nitrofuran
   e. 5-chloro-2-furoic acid
   f. 2-aminothiazole
   g. 4-methyl-3-isoxazolecarboxylic acid
   h. 5-nitroisothiazole
   i. 4-nitro-1-phenylimidazole
   j. 6-bromoindole-3-carboxylic acid
   k. 3-pyridinecarboxylic acid
   l. 4-methylpyridine oxide
   m. 7-chloro-1-methylisoquinoline
   n. 2,3-dimethylquinoline
   o. 2-amino-4-methylpyrimidine
   p. 3,6-dimethylpyridazine
   q. 3-chlorobenzofuran
   r. 2-(2-hydroxyethyl)thiophene
   s. 2-(4-methoxyphenyl)-6-phenyl-1,4-pyrone
   t. 4-*t*-butyl-2,6-dimethylpyrylium tetrafluoroborate

2. a. Ph, N, Ph
   b. O
   c. N, O, CH_3
   d. N, Br
   e. NH_2, CH_3, N, N, N, N
   f. N, NH, H_2N, N, N
   g. CH_2CO_2H, N, H
   h. NO_2, N+, O−
   i. Cl, S, CO_2H
   j. Ph, N, N, CH_3
   k. NO_2, N, CO_2H
   l. N, S, NO_2
   m. CN, O, N
   n. CH_3, H_3C, O, O

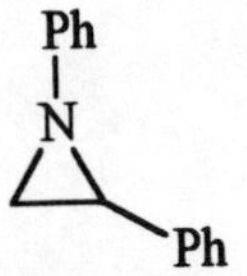

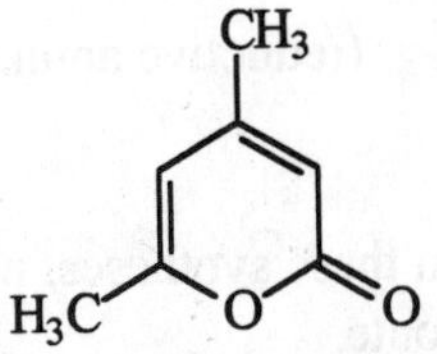

3. a. $PhCH{=}CH_2 \xrightarrow[\text{1. HOCl 2. NaOH}]{C_6H_5CO_3H \text{ or}} PhCH{-}CH_2 \text{ (epoxide, O)} \xrightarrow{^{-}SCN} PhCH{-}CH_2 \text{ (episulfide, S)}$

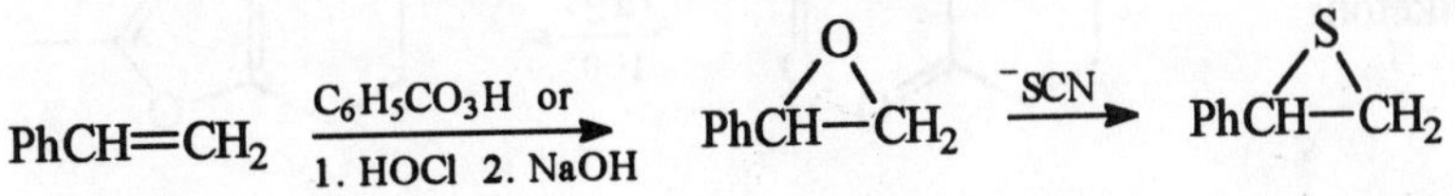

   b. $PhC{\equiv}CH \xrightarrow[\text{liq. } NH_3]{NaNH_2} \xrightarrow{CH_3I} PhC{\equiv}CCH_3 \xrightarrow[NH_3]{Na}$ (Ph)(H)C=C(H)(CH_3) $\xrightarrow{C_6H_5CO_3H}$ Ph, H, O, H, CH_3 (epoxide)

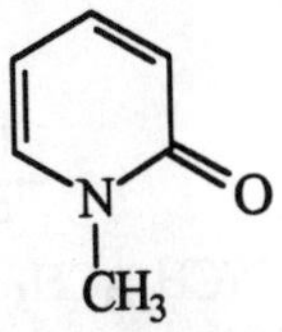

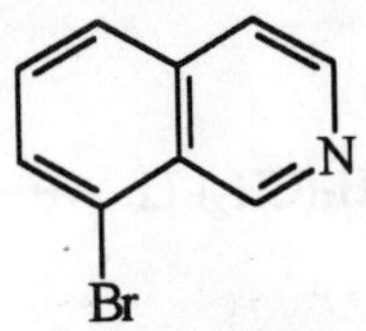

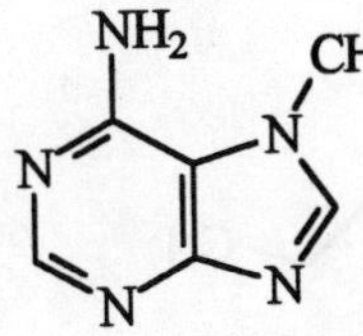

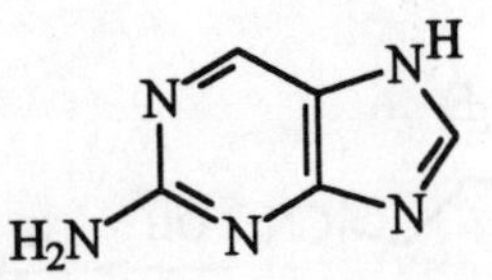

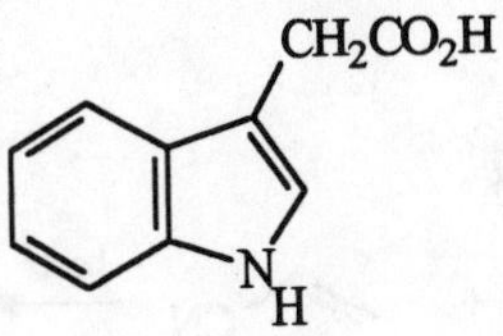

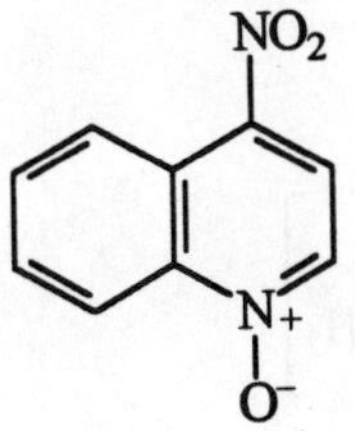

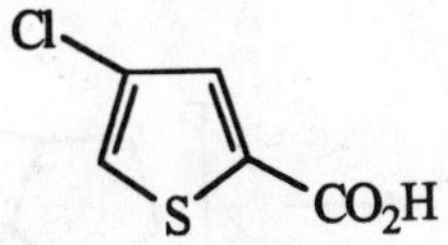

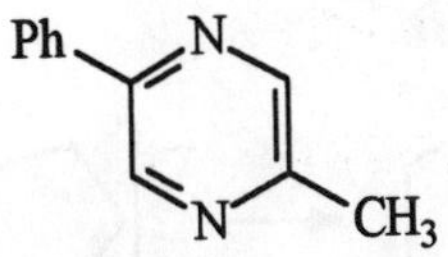

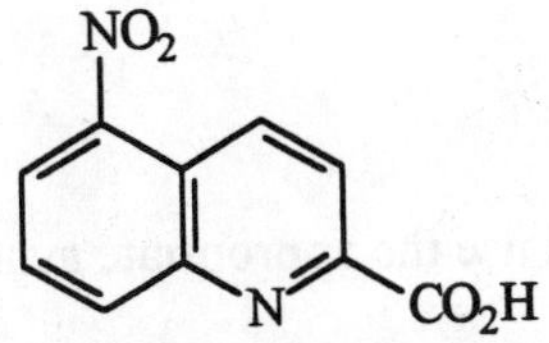

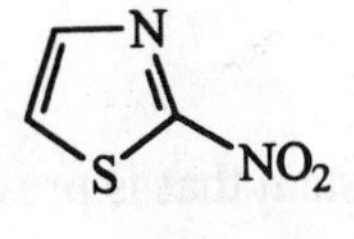

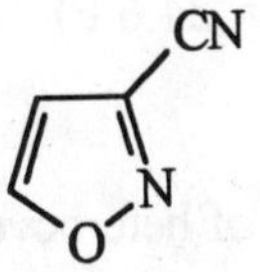

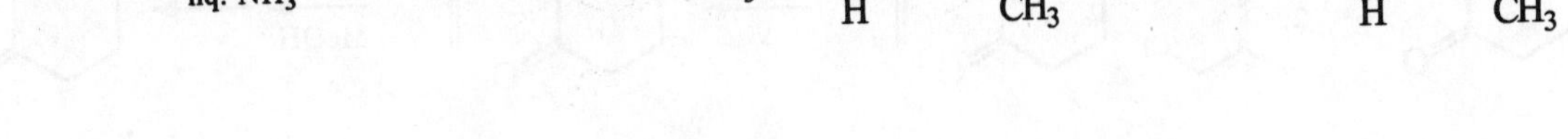

c. (methylenecyclohexane) $=CH_2$ + INCO ⟶ (1-isocyanato-1-(iodomethyl)cyclohexane) NCO, I $\xrightarrow{MeOH}$ NCO_2Me, I $\xrightarrow[MeOH]{KOH}$ (spiro aziridine) NH

methylenecyclohexane via Wittig reaction 14.7.D

d. H_3C, CH_3, C=C=O + CHPh=N–Ph ⟶ (β-lactam) CH_3, H_3C, Ph, N, Ph, O

NOTE: $(CH_3)_2CBrCOBr \xrightarrow{Zn} (CH_3)_2C{=}C{=}O$ *and* $C_6H_5NH_2 + O{=}CHC_6H_5 \longrightarrow C_6H_5N{=}CHC_6H_5$
(see Section 20.3.C) (see Section 14.6.C)

e. (pyridine) N + $CH_3(CH_2)_3Li$ ⟶ (2-butylpyridine) N, $(CH_2)_3CH_3$ $\xrightarrow[cat.]{H_2}$ (2-butylpiperidine) N–H, $(CH_2)_3CH_3$

f. (ethyl 2-oxocyclopentanecarboxylate) CO_2Et, O + (acrylonitrile) CN $\xrightarrow{base}$ CO_2Et, CN, O $\xrightarrow{OH^-}$ $\xrightarrow[\Delta]{H^+}$ CN, O

$\downarrow$ H_2 cat.

[O, NH_2] ⟶ [N (cyclic imine)] ⟶ (bicyclic amine) N–H

(reductive amination, Section 24.6.F)

4. In these syntheses, note the type of heterocyclic ring system that is present and use the appropriate synthetic route.

a. This problem requires a 1,4-diketone: (2-(2-oxopropyl)cyclohexanone) O, O $\xrightarrow[100°]{P_2O_5}$ (2-methyl-4,5,6,7-tetrahydrobenzofuran) O

The 1,4-diketone may be prepared in several ways. One possibility is:

(ethyl 2-oxocyclohexanecarboxylate) CO_2Et, O $\xrightarrow[2.\ CH_2{=}C(CH_3)CH_2Cl]{1.\ NaOEt}$ CO_2Et, O $\xrightarrow[\Delta]{H_3O^+}$ O $\xrightarrow[MeOH]{O_3}$ $\xrightarrow{Me_2S}$ O, O

b. Knorr pyrrole synthesis (Section 32.3.B)

MeO_2C ... Ph ... O + O ... Ph ... H_2N ... CO_2Me $\xrightarrow{\text{AcOH}}$ MeO_2C ... Ph ... Ph ... N H ... CO_2Me

$PhCO_2Me + CH_3CO_2Me \xrightarrow[\text{MeOH}]{\text{NaOMe}} PhCCH_2CO_2Me$ (C=O) $\xrightarrow{\text{HONO}}$ $PhCCCO_2Me$ (C=O, C=NOH)

(Section 29.5.C)

$\xrightarrow[(CH_3CO)_2O]{H_2/\,Pt}$ $PhCCHCO_2Me$ (C=O, $NHCOCH_3$) $\xrightarrow[H^+\ \Delta]{\text{MeOH}}$ $PhCCHCO_2Me$ (C=O, NH_2)

c. Fischer indole synthesis (Section 32.4.B)

Cl ... $NHNH_2$ + O ... $\xrightarrow[100°]{\text{PPA}}$ Cl ... N H

Cl ... NH_2 $\xrightarrow[\text{HCl}]{NaNO_2}$ Cl ... $^+N_2$ $\xrightarrow{Na_2SO_3}$ Cl ... $NHNH_2$

d. OH ... NO_2 $\xrightarrow{(CH_3)_2SO_4}$ OCH_3 ... NO_2 $\xrightarrow[\text{cat.}]{H_2}$ OCH_3 ... NH_2 $\xrightarrow[\text{HCl}]{NaNO_2}$ OCH_3 ... $^+N_2$

$\xrightarrow{Na_2SO_3}$ OCH_3 ... $NHNH_2$ $\xrightarrow{(CH_3)_2CO}$ OCH_3 ... NHN=C(CH_3)CH_3 $\xrightarrow[100°]{\text{PPA}}$ N H ... OCH_3

e. $NHNH_2$ + CH_2CH_3 ... O=C ... CH_2CH_3 $\xrightarrow[100°]{\text{PPA}}$ CH_3 ... CH_2CH_3 ... N H

f. hydroxylamine + 1,3-dicarbonyl compound (Section 32.5.B);
β-Diketones are prepared from esters + ketones:

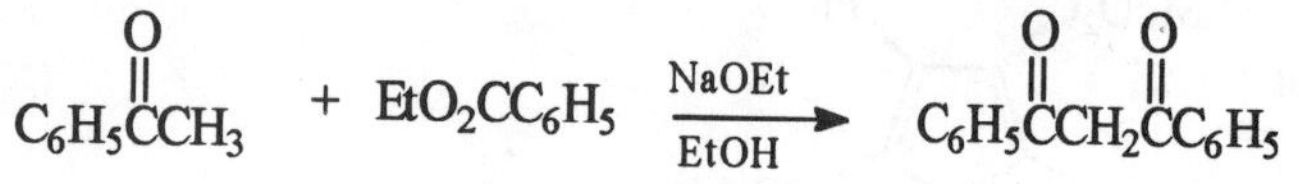

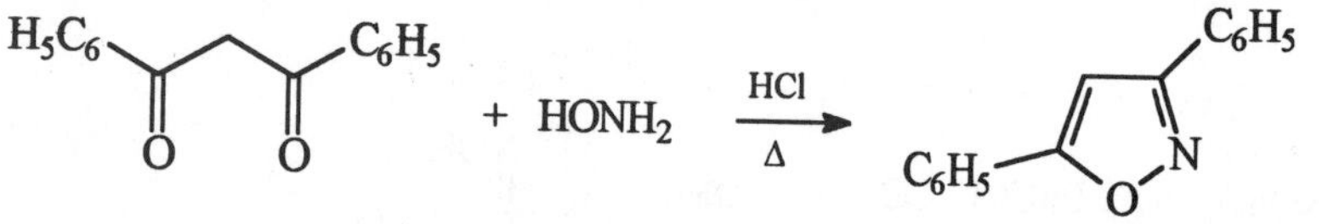

$$C_6H_5\overset{O}{\overset{\|}{C}}CH_3 + EtO_2CC_6H_5 \xrightarrow[EtOH]{NaOEt} C_6H_5\overset{O}{\overset{\|}{C}}CH_2\overset{O}{\overset{\|}{C}}C_6H_5$$

$$H_5C_6COCH_2COC_6H_5 + HONH_2 \xrightarrow[\Delta]{HCl} \text{3,5-diphenylisoxazole } (C_6H_5\text{ at C-3 and C-5})$$

g. β-diketone + hydrazine:

$$H_5C_6COCH_2COCH_3 + H_2NNH_2 \xrightarrow[\Delta]{HCl} \text{3-methyl-5-phenylpyrazole } (C_6H_5\text{, }CH_3\text{, N-H})$$

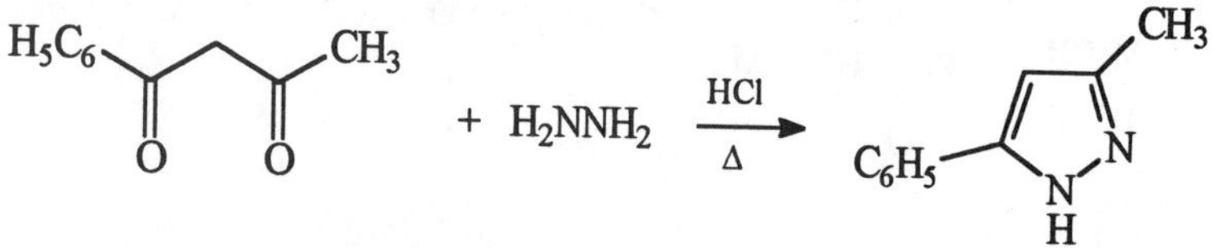

Note that unsymmetrical pyrazoles can be prepared because hydrazine is symmetrical. With hydroxylamine, this β-diketone would give a mixture of isoxazoles.

The β-diketone can be made by: $C_6H_5COCH_3 + CH_3CO_2Et \xrightarrow{EtO^-} C_6H_5COCH_2COCH_3$

or: $C_6H_5CO_2Et + CH_3COCH_3 \xrightarrow{EtO^-} C_6H_5COCH_2COCH_3$

h. This isoxazole is unsymmetrical, and the required β-diketone is hard to make. An alternative preparation is a cycloaddition with nitrile oxides (see Section 32.5.B):

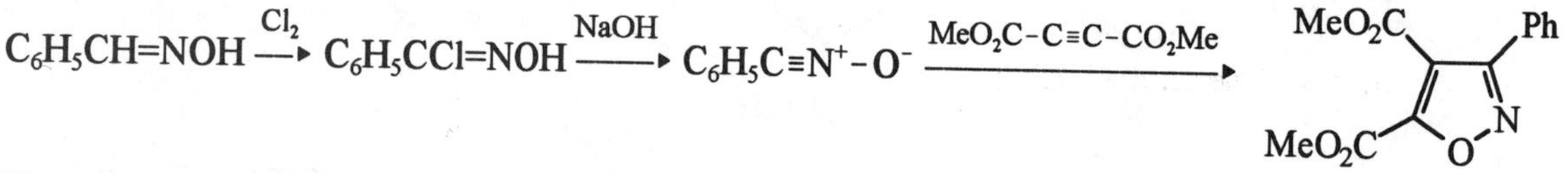

$$C_6H_5CH{=}NOH \xrightarrow{Cl_2} C_6H_5CCl{=}NOH \xrightarrow{NaOH} C_6H_5C{\equiv}N^+{-}O^- \xrightarrow{MeO_2C-C\equiv C-CO_2Me} \text{isoxazole with } MeO_2C\text{, }MeO_2C\text{, Ph}$$

i. Here also, the β-dicarbonyl approach does not look promising. An alternative preparation uses diazomethane (Section 32.5.B):

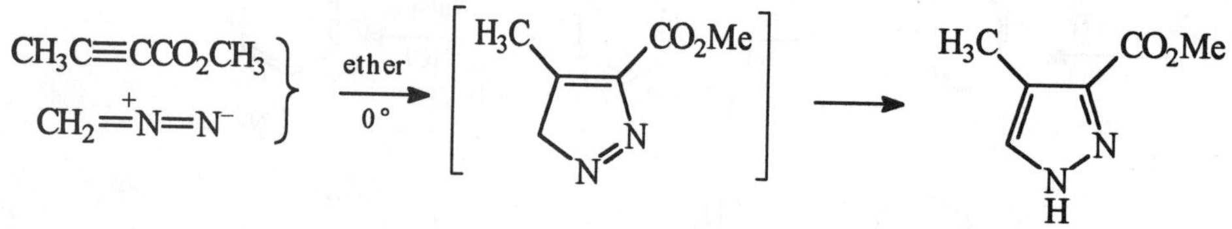

$$\left.\begin{matrix} CH_3C{\equiv}CCO_2CH_3 \\ CH_2{=}\overset{+}{N}{=}N^- \end{matrix}\right\} \xrightarrow[0°]{ether} \left[\text{pyrazoline with } H_3C\text{, }CO_2Me\text{, N=N}\right] \longrightarrow \text{pyrazole with } H_3C\text{, }CO_2Me\text{, N-H}$$

The acetylene compound may be prepared by:

$$CH_3C{\equiv}CH \xrightarrow{RMgX} CH_3C{\equiv}CMgX \xrightarrow{CO_2} \xrightarrow{H^+} CH_3C{\equiv}CCO_2H \xrightarrow[H^+]{MeOH}$$

j. Don't be fooled by the way this compound is written. Imidazoles are in rapid tautomeric equilibrium (remember their basicity):

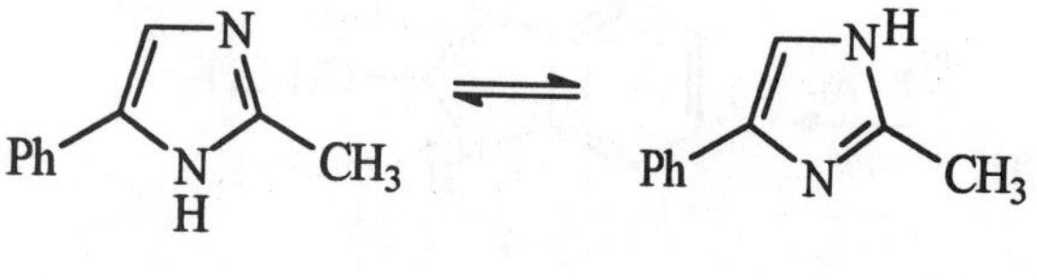

$$\text{Ph, N-H, }CH_3\text{ imidazole} \rightleftharpoons \text{Ph, NH, }CH_3\text{ imidazole}$$

The Paal-Knorr cyclization can be designed in two ways (Section 32.5.B):

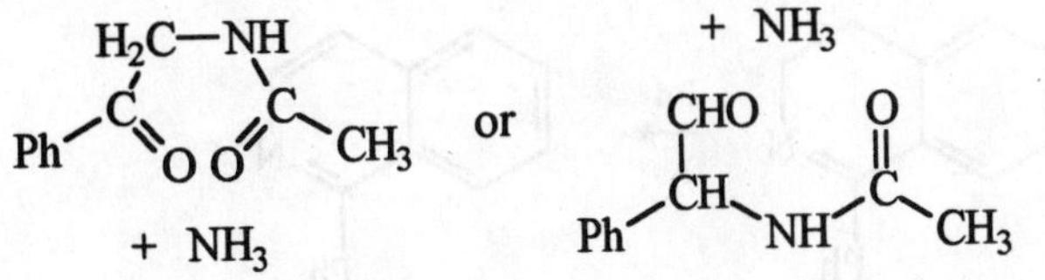

$H_2C{-}NH$ … Ph–C(=O) … O=C–CH_3 + NH_3 or CHO, Ph–CH–NH–C(=O)–CH_3 + NH_3

The first is a better approach since it does not involve a sensitive aldehyde. Ketones are better than aldehydes in all of these cyclizations.

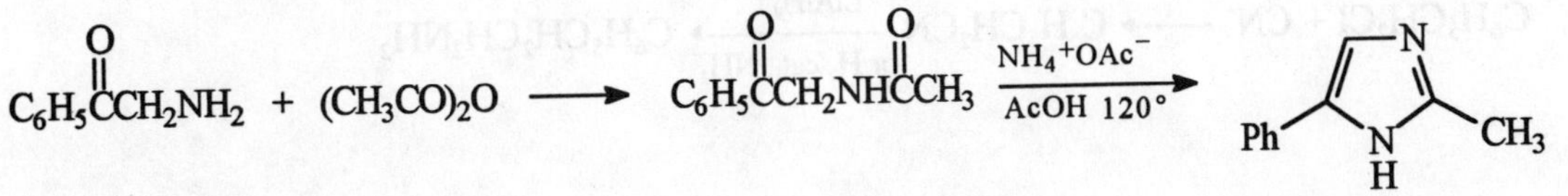

$$C_6H_5\overset{O}{\overset{\|}{C}}CH_2NH_2 + (CH_3CO)_2O \longrightarrow C_6H_5\overset{O}{\overset{\|}{C}}CH_2NH\overset{O}{\overset{\|}{C}}CH_3 \xrightarrow[\text{AcOH } 120°]{NH_4{}^+OAc^-}$$ (Ph, CH_3, N, N–H imidazole)

k. Hantzsch pyridine synthesis (Section 32.6.B):

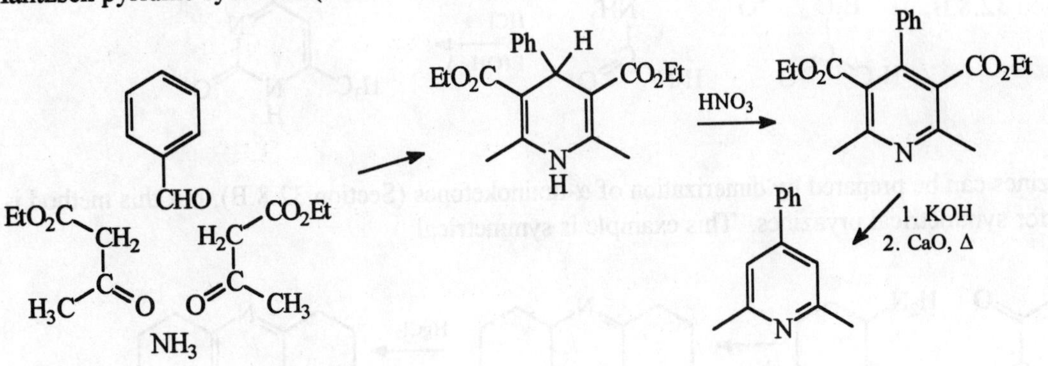

(Benzaldehyde CHO; EtO_2C–CH_2–C(=O)–H_3C; H_2C(CO_2Et)–C(=O)–CH_3; NH_3) → (Ph, H, EtO_2C, CO_2Et, N–H dihydropyridine) $\xrightarrow{HNO_3}$ (Ph, EtO_2C, CO_2Et, N pyridine) $\xrightarrow{\text{1. KOH; 2. CaO, } \Delta}$ (Ph, N pyridine)

l. Skraup reaction (Section 32.7.B):

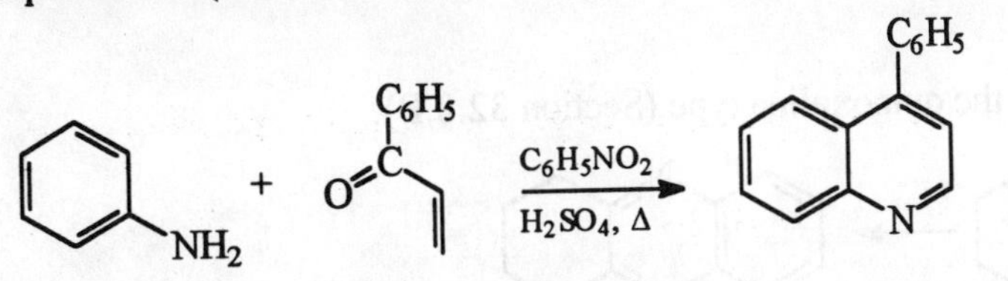

(aniline, NH_2) + C_6H_5–C(=O)–CH=CH_2 $\xrightarrow[H_2SO_4,\ \Delta]{C_6H_5NO_2}$ (C_6H_5, N quinoline)

m. Skraup reaction with:

(H_3C, NH_2 aniline) + $\overset{OH}{CH_2}CHOH\overset{OH}{CH_2}$ $\xrightarrow[As_2O_5]{H_2SO_4}$ (H_3C, N quinoline)

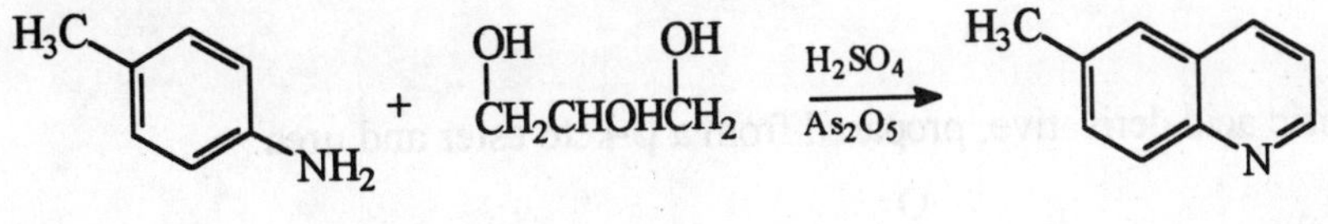

n. This kind of quinoline is best prepared by the Friedländer method (Section 32.7.B):

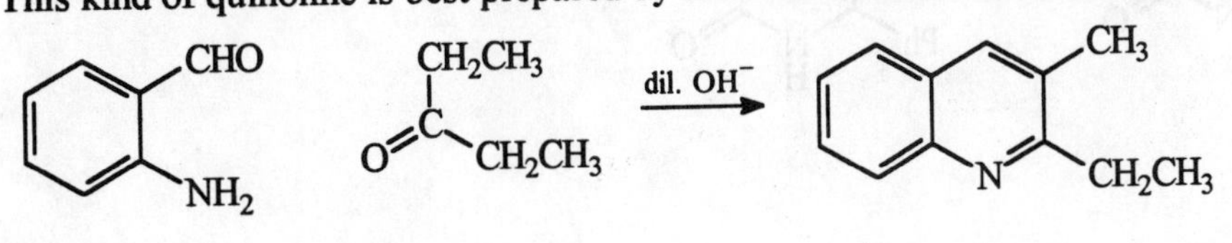

(CHO, NH_2 benzene) CH_2CH_3–C(=O)–CH_2CH_3 $\xrightarrow{\text{dil. } OH^-}$ (CH_3, N, CH_2CH_3 quinoline)

o. Isoquinolines are prepared by Bischler-Napieralski synthesis (Section 32.7.B):

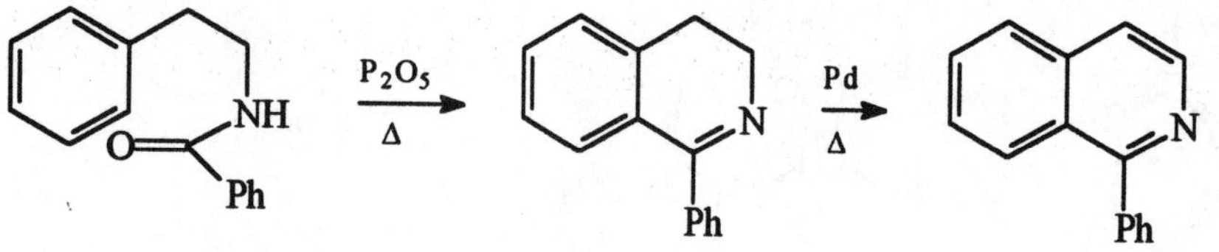

$\xrightarrow[\Delta]{P_2O_5}$ $\xrightarrow[\Delta]{Pd}$

The amide is prepared from $C_6H_5CH_2CH_2NH_2$ + C_6H_5COCl (benzoyl chloride):

$$C_6H_5CH_2Cl + CN^- \longrightarrow C_6H_5CH_2CN \xrightarrow[\text{or } H_2/\text{cat., } NH_3]{LiAlH_4} C_6H_5CH_2CH_2NH_2$$

p. Section 32.8.B:

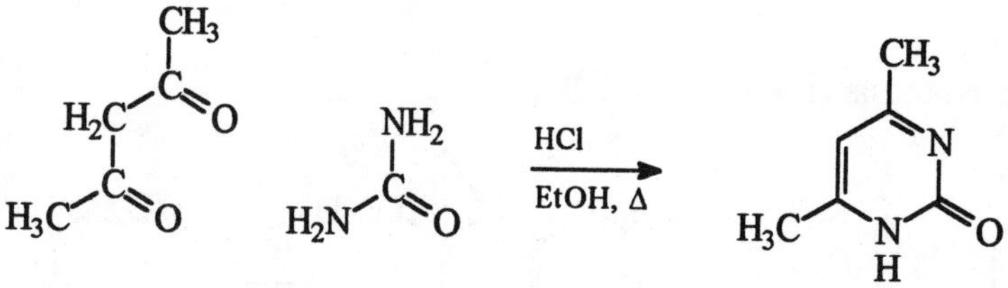

$\xrightarrow[\text{EtOH, }\Delta]{HCl}$

q. Pyrazines can be prepared by dimerization of α-aminoketones (Section 32.8.B), but this method is useful only for symmetrical pryazines. This example is symmetrical:

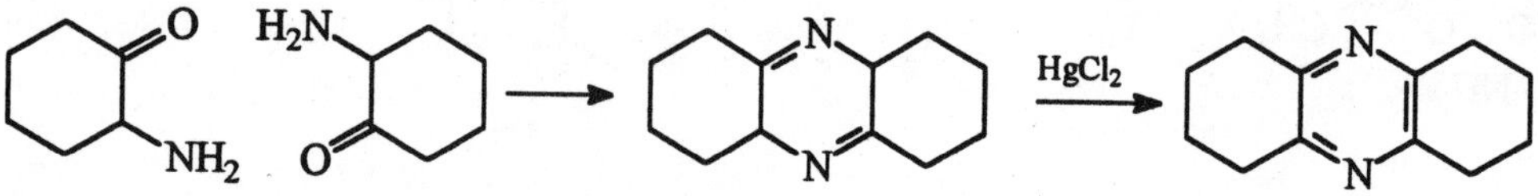

$\xrightarrow{HgCl_2}$

r. This unsymmetrical pyrazine is of the quinoxaline type (Section 32.8.B):

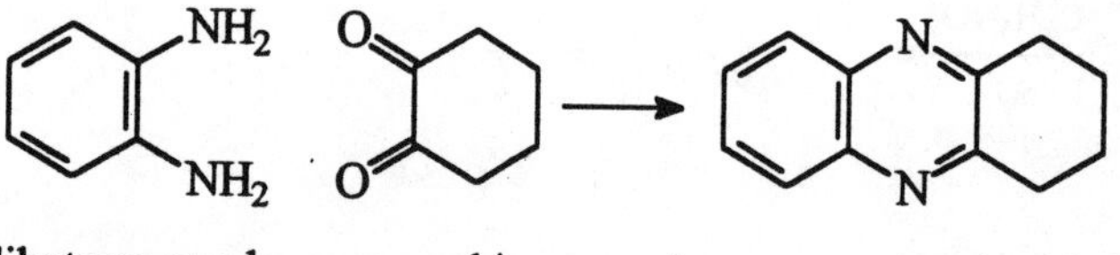

The diketone can be prepared in several ways; one method is given in Section 27.7.A.

s. This compound is a barbituric acid derivative, prepared from a β-keto ester and urea.

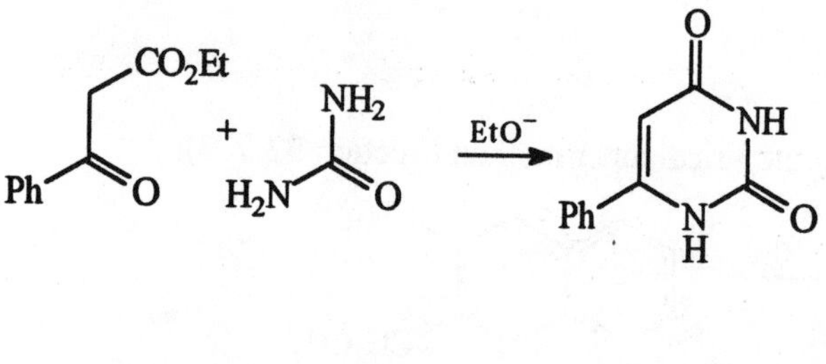

$\xrightarrow{EtO^-}$

t.

$$CH_3\overset{O}{\overset{\|}{C}}CH_3 + 2\,CH_3CO_2Et \xrightarrow{NaOEt} CH_3\overset{O}{\overset{\|}{C}}CH_2\overset{O}{\overset{\|}{C}}CH_2\overset{O}{\overset{\|}{C}}CH_3 \xrightarrow{POCl_3}$$ 2,6-dimethyl-4-pyrone (H_3C, O, CH_3) $\xrightarrow{1.\ EtMgBr;\ 2.\ H^+}$ 4-ethyl-2,6-dimethylpyrylium ion (CH_2CH_3, H_3C, O^+, CH_3)

u.

$$CH_3\overset{O}{\overset{\|}{C}}CH_3 + 2\,C_6H_5CO_2Et \xrightarrow{NaOEt} C_6H_5\overset{O}{\overset{\|}{C}}CH_2\overset{O}{\overset{\|}{C}}CH_2\overset{O}{\overset{\|}{C}}C_6H_5 \xrightarrow{POCl_3}$$ 2,6-diphenyl-4-pyrone (H_5C_6, O, C_6H_5)

5. a. N-methylpyrrole (CH_3) + $H\overset{O}{\overset{\|}{C}}N(CH_3)_2 \xrightarrow{POCl_3}$ [N-methylpyrrol-2-yl-$CH{=}\overset{+}{N}(CH_3)_2$] $\xrightarrow{H_2O}$ 1-methylpyrrole-2-CHO (CH_3)

Vilsmeier reaction

b. thiophene + $I_2 \xrightarrow[benzene]{HgO}$ 2-iodothiophene $\xrightarrow[ether]{Mg}\ \xrightarrow{CH_3COCH_3}$ 2-thienyl–$C(CH_3)_2$–OH

(Section 32.3.C)

c. isoquinoline $\xrightarrow{C_6H_5Li}$ 1-phenylisoquinoline (C_6H_5)

d. furan + $I_2 \xrightarrow[benzene]{HgO}$ 2-iodofuran $\xrightarrow{Mg}$ 2-furyl–MgI $\xrightarrow[H^+]{\text{cyclohexanone}}$ 1-(2-furyl)cyclohexanol (HO)

e. Friedel-Crafts acylations occur readily on furan; only mild Lewis acids are required, if any:

furan + $ClCOC_6H_5 \xrightarrow{BF_3}$ 2-furyl phenyl ketone (C=O)

f. [benzofuran] $\xrightarrow{Br_2}$ [2-bromobenzofuran] $\xrightarrow{Mg}$ $\xrightarrow{D_2O}$ [2-deuteriobenzofuran]

g. [indole] + Br_2 $\xrightarrow{AcOH}$ [3-bromoindole]

h. [pyridine] $\xrightarrow[H_2SO_4,\ Fe,\ 300°]{KNO_3,\ HNO_3}$ [3-nitropyridine]

vigorous nitrating conditions

i. α-Picoline must first be converted into the N-oxide so that nitration will occur at the γ-position:

[2-methylpyridine] $\xrightarrow{H_2O_2}$ [2-methylpyridine N-oxide] $\xrightarrow[H_2SO_4]{\text{fuming } HNO_3}$ [2-methyl-4-nitropyridine N-oxide] $\xrightarrow[\Delta]{PCl_3}$ [2-methyl-4-nitropyridine]

j. Chichibabin reaction:

[pyridine] + CH_3NHNa $\xrightarrow{\Delta}$ [2-($NHCH_3$)pyridine] (Section 32.6.C)

k. [pyridine] + CH_3CH_2Br $\xrightarrow{\Delta}$ [N-ethylpyridinium] $\xrightarrow{OH^-}$ [N-ethyl-2-hydroxy-1,2-dihydropyridine] $\xrightarrow{Fe(CN)_6^{3-}}$ [N-ethyl-2-pyridone]

(See Problem 11.)

l. [2-methylquinoline] $\xrightarrow{H_2O_2}$ [2-methylquinoline N-oxide] $\xrightarrow[H_2SO_4,\ \Delta]{HNO_3}$ [2-methyl-4-nitroquinoline N-oxide] $\xrightarrow{PCl_3}$ [2-methyl-4-nitroquinoline]

m. (2-methylpyridine) $\xrightarrow{EtO^-}$ [2-pyridyl–CH_2^-] + furan-2-CHO ⟶ [2-pyridyl–CH_2–C(–O^-)(H)–2-furyl] ⟶ 2-pyridyl–CH=CH–2-furyl

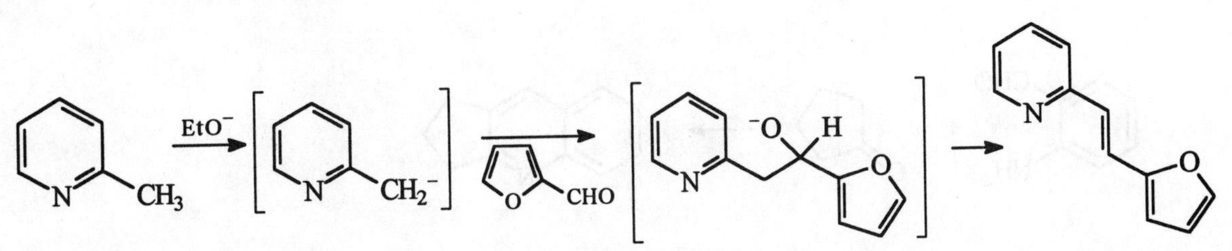

n. (phthalazine) $\xrightarrow[H_2SO_4\ \Delta]{HNO_3}$ 5-nitrophthalazine (NO_2)

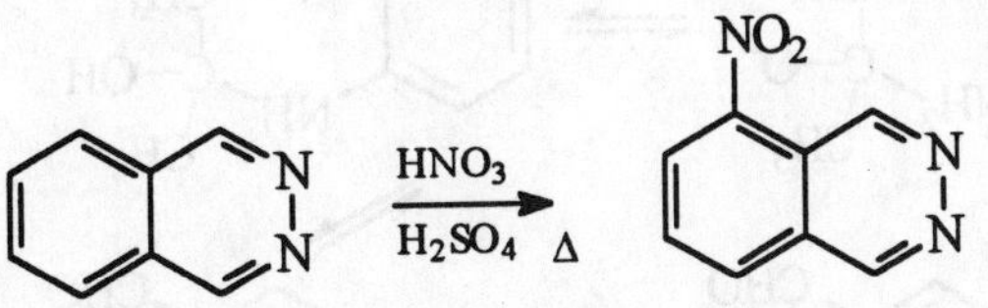

6. **Michael addition reaction:**

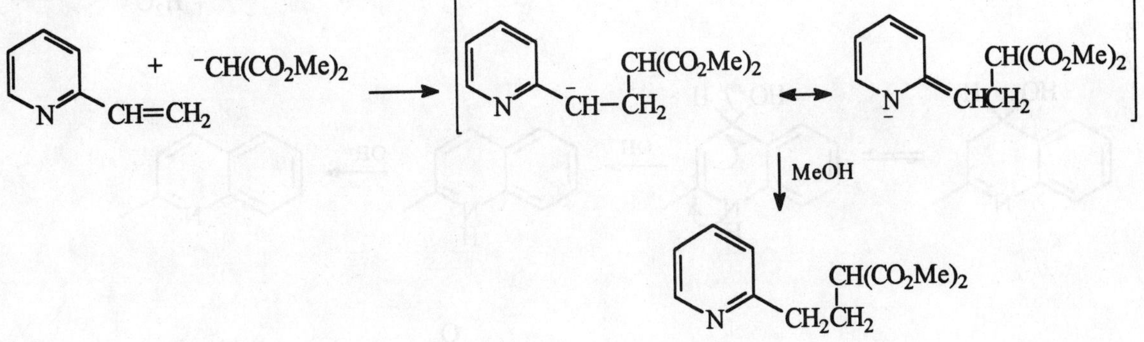

2-pyridyl–CH=CH_2 + $^-CH(CO_2Me)_2$ ⟶ [2-pyridyl–$\bar{C}H$–CH_2–$CH(CO_2Me)_2$ ⟷ (N$^-$)pyridylidene=CH–CH_2–$CH(CO_2Me)_2$] $\xrightarrow{MeOH}$ 2-pyridyl–CH_2CH_2–$CH(CO_2Me)_2$

The corresponding reaction does not occur with 3-vinylpyridine because the negative charge cannot be delocalized onto the nitrogen via resonance.

7. a. (CH_3, NO_2-benzene) $\xrightarrow[H_2SO_4,\ 10°]{Ac_2O,\ CrO_3}$ ($CH(OAc)_2$, NO_2) $\xrightarrow[\Delta]{HCl}$ (CHO, NO_2) $\xrightarrow[NH_3]{FeSO_4}$ (CHO, NH_2)

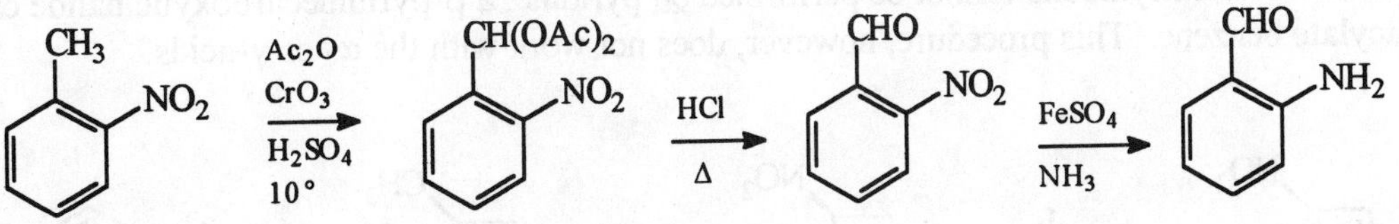

b. (CHO, NH_2-benzene) + CH_3–C(=O)–Ph ⟶ 2-Ph-quinoline (N, Ph)

(CHO, NH_2-benzene) + CH_3–C(=O)–CH_2–CO_2Et ⟶ quinoline (N, CH_3, CO_2Et)

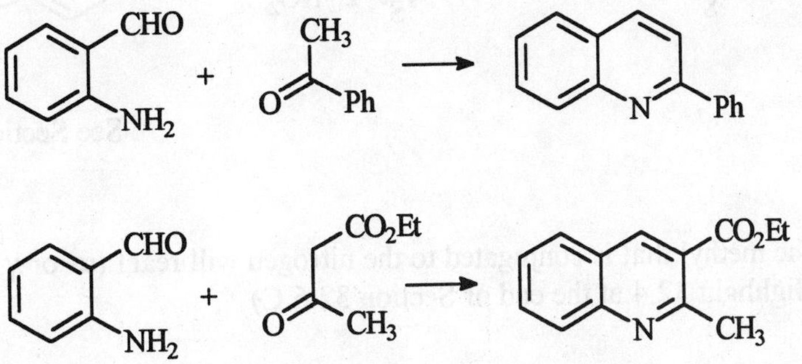

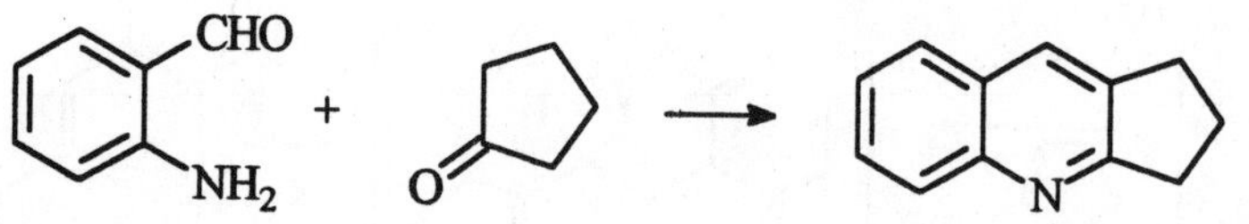

c.

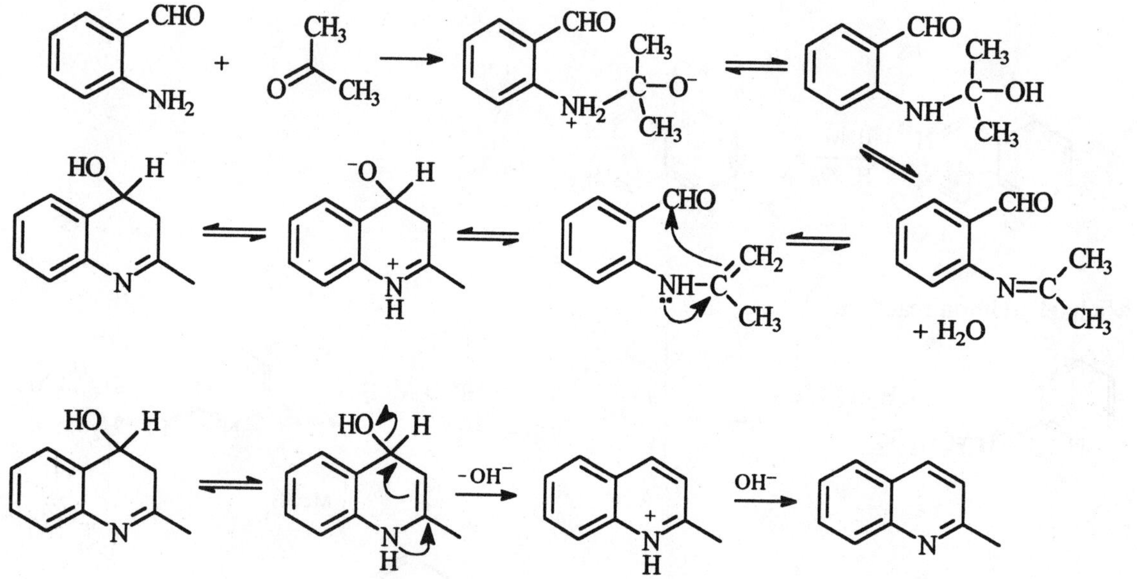

8.

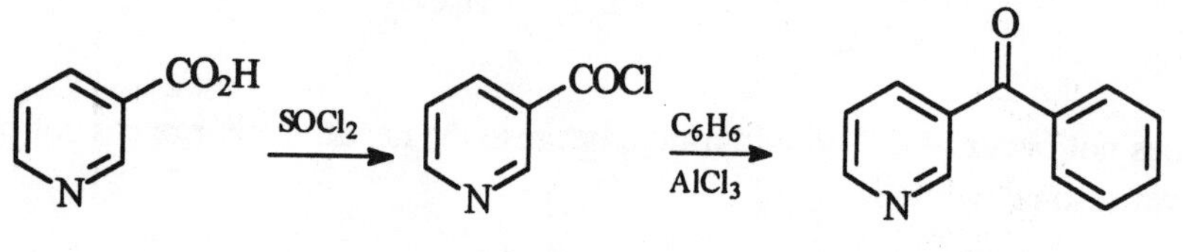

Although Friedel-Crafts acylations cannot be performed on pyridine, a β-pyridinecarboxylic halide can be utilized to acylate benzene. This procedure, however, does not work with the α- or γ-acids.

9. a. b. 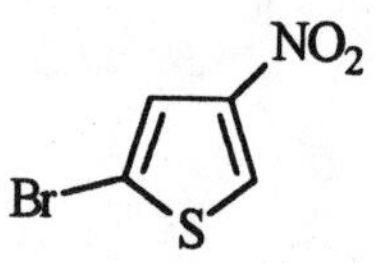 c. 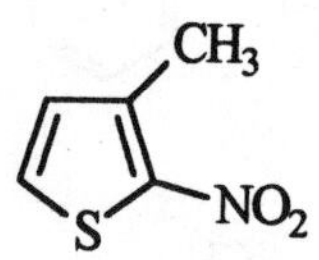 d.

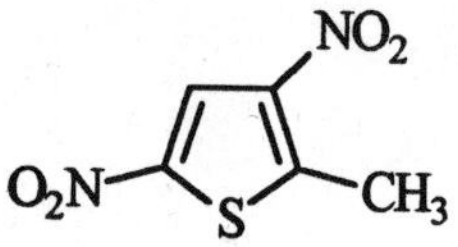

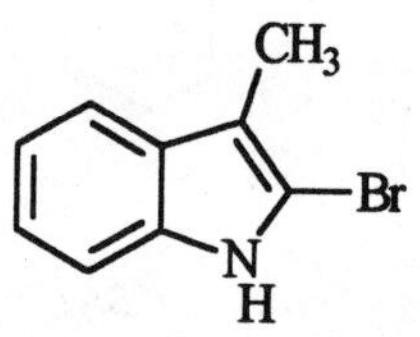

For (a)-(c), see Section 32.3.C

See Section 32.4.C

e. only the methyl that is conjugated to the nitrogen will react (α- or γ-methyl) (see Highlight 32.4 at the end of Section 32.6.C)

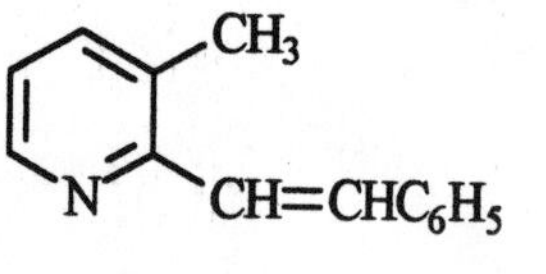

f. 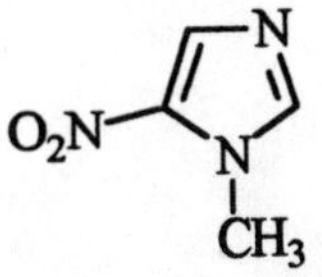 O_2N-substituted 1-methylimidazole (CH_3 on N)

g. a Fischer indole synthesis;

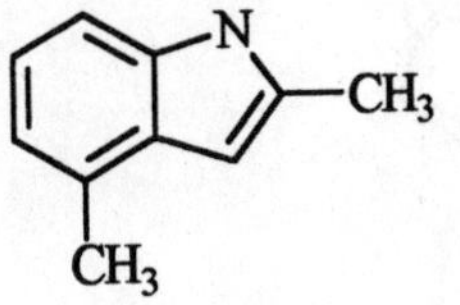

4-methyl-2-methylindole (CH_3, CH_3) and 2,6-dimethylindole (H_3C, CH_3)

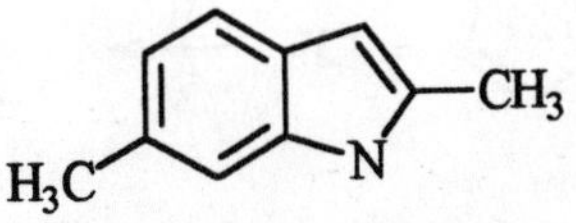

We expect a greater amount of this product, since there is less steric hindrance for the cyclization reaction.

10. 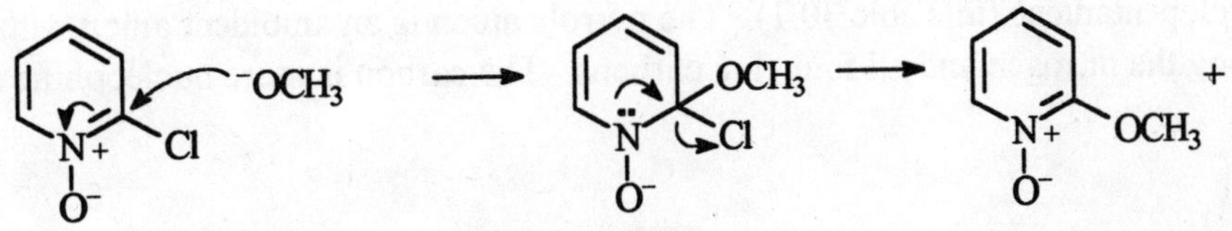 $^{-}OCH_3$ + 2-chloropyridine N-oxide (N+, O^-, Cl) ⟶ addition intermediate (OCH_3, Cl, N, O^-) ⟶ 2-methoxypyridine N-oxide (N+, OCH_3, O^-) + Cl^-

11. 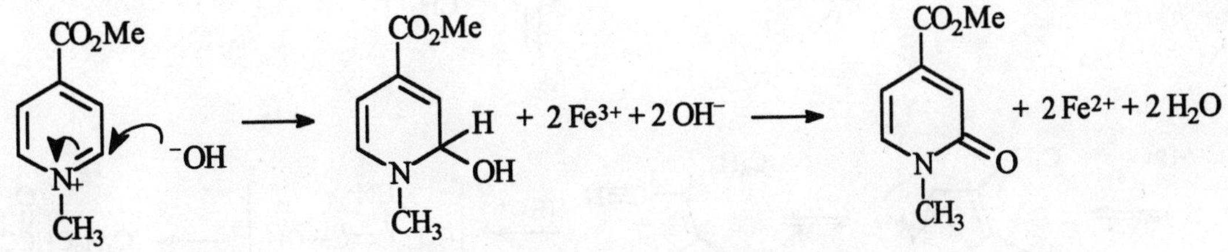 CO_2Me-substituted N-methylpyridinium (N+, CH_3) + ^{-}OH ⟶ (CO_2Me, H, OH, N, CH_3) + $2\,Fe^{3+} + 2\,OH^-$ ⟶ (CO_2Me, N, O, CH_3) + $2\,Fe^{2+} + 2\,H_2O$

12. 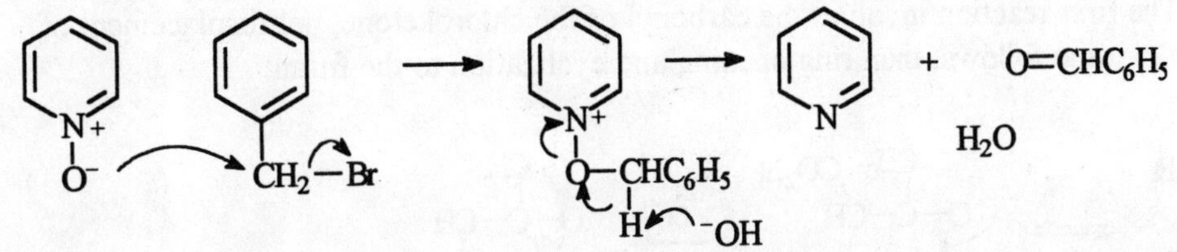 pyridine N-oxide (N+, O^-) + $C_6H_5CH_2$—Br ⟶ (N+, O—CHC_6H_5, H, ^{-}OH) ⟶ pyridine (N) + $O{=}CHC_6H_5$ + H_2O

Pyridine serves as a leaving group in a reaction that is essentially an E2 reaction, and which constitutes a new aldehyde synthesis.

13.

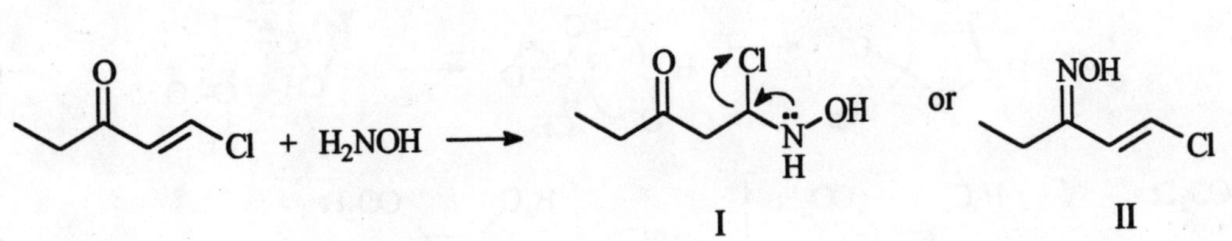

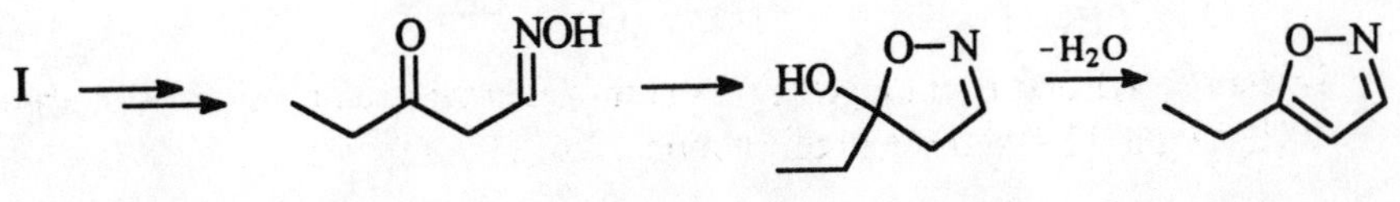

(O, Cl) + H_2NOH ⟶ (O, Cl, N—OH, H) I or (NOH, Cl) II

I ⟶⟶ (O, NOH) ⟶ HO (O—N) $\xrightarrow{-H_2O}$ (O—N)

II → [protonated isoxazolium, N–O⁺H] + Cl^- $\xrightarrow{-H^+}$ [isoxazole]

14. [pyrrole] + EtMgBr $\xrightarrow{-EtH}$ [pyrrole anion resonance structures ↔ etc.] ^+MgBr

Pyrrole is rather acidic; recall cyclopentadiene (in Table 30.1). The pyrrole anion is an ambident anion with negative charges distributed among the nitrogen and all four ring carbons. The carbon is more nucleophilic and displaces on CH_3I.

[pyrrole anion] + CH_3–I → [2-methyl-2H-pyrrole] ⇌ [2-methylpyrrole]

15.

C_4H_9, CH_3 isoxazole + $^-$O-t-Bu ⇌ [anion] ⇌ C_4H_9(C≡N)C=C(O$^-$)CH_3 $\xrightarrow{H_3O^+}$ [$C_4H_9C(CN)=C(OH)-CH_3$] → $C_4H_9CH(CN)C(=O)CH_3$

16. This one is rather subtle. The first reaction involves the carbonyl of the chloroketone, not displacement of chloride. Cyclization to an oxirane follows, then ring opening and cyclization to the furan:

$CH_3C(=O)CH_2-Cl$ + $^-CH(CO_2Et)C(=O)CH_3$ ⇌ $^-O-C(CH_3)(CH_2Cl)-CH(CO_2Et)C(=O)CH_3$ $\xrightarrow{-Cl^-}$ [oxirane] $\xrightarrow{\text{pyridine}}$ [oxirane carbanion] → $H_2C(O^-)-C(CH_3)=C-C(=O)CH_3$ → [cyclic alkoxide] ⇌ [resonance forms] $\xrightarrow{-OH^-}$ [ethyl 2,4-dimethylfuran-3-carboxylate]

17. a. H_2 + [cyclohexene] → [cyclohexane] $\Delta H^\circ = -28.4\ \text{kcal mole}^{-1}$

-1.1 -29.5

H_2 + [tetrahydropyridine] → [piperidine, N–H] $\Delta H° = -21$ kcal mole^{-1}

Therefore, for pyridine (two C=C bonds and one C=N bond), we would expect:

[pyridine] $\xrightarrow{3\ H_2}$ [piperidine, N–H]

predicted $\Delta H° = 2 \times (-28.4) - 21 = -77.8$ kcal mole^{-1}

actual $\Delta H° = -11.8 - (34.6) = -46.4$ kcal mole^{-1}

The resonance energy is $-46.4 - (-77.8) = 31.4$ kcal mole^{-1}

This empirical resonance energy is similar to that of benzene.

b. [pyridine] → 5 C + 5 H + N $\Delta H°_{atomiz} = 1193.4$ kcal mole^{-1}

+ 34.6 5 x 170.9 5 x 52.1 113.0

Bond energies: 5 C–H + 2 C=C + 2 C–C + N–C + N=C

5 x 99 + 2 x 146 + 2 x 83 + 73 + 147 = 1173 kcal mole^{-1}

Therefore, the empirical resonance energy is 1193 – 1173 = 20 kcal mole^{-1}

Note that using bond energies,

H_2 + [tetrahydropyridine] → [piperidine, N–H] (C–H) + (N–H) + (C–N) – (C=N) – (H–H)

= 99 + 93 + 73 – 147 – 104

$\Delta H° = -14$ kcal mole^{-1} The experimental value used in part (a) is -21 kcal mole^{-1}.

[piperidine, N–H] → 5 C + 11 H + N experimental $\Delta H° = 1552.4$ kcal mole^{-1}

calculated from Bond Energy Table: 1561 kcal mole^{-1}

The Table of Bond Energies gives $\Delta H°_{atomiz}$ that are accurate to ±1% or less, but this still amounts to several kcal mole^{-1}; i.e., in practice, the $\Delta H°$ we calculate only amount to a few percent of the total atomization energies. In general, energy differences derived from average bond energies can have substantial errors.

18. Compare the bond dipole of C–Br in (3-bromofuran) with that for (2-bromofuran)

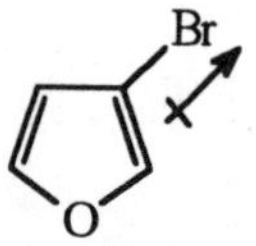

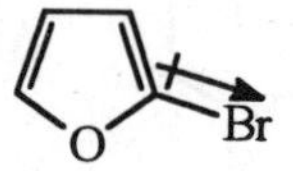

0.91 D < 1.46 D; therefore (furan dipole)

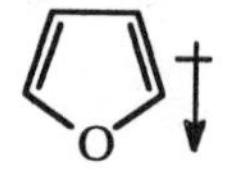

The similarity between (2,5-dibromothiophene) - (thiophene) = 0.6 D

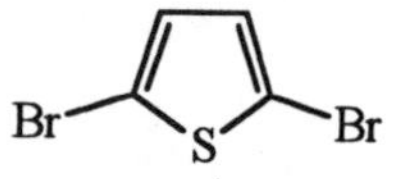

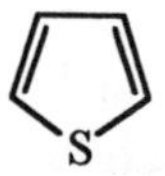

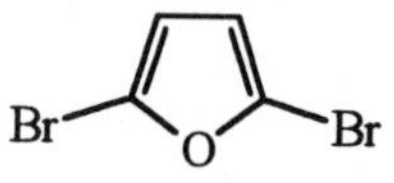

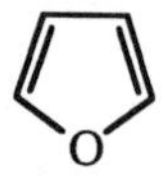

1.1D 0.51 D

and (2,5-dibromofuran) - (furan) = 0.9 D

1.63 D 0.70 D

suggests (thiophene dipole) for thiophene. If the thiophene dipole were in the opposite direction, the effect of the bromines could not be rationalized.

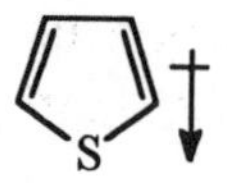

In the case of pyrrole, (oxazole) net 2.8 D, but (imidazole) 6.2D strongly suggests (pyrrole) 1.81 D

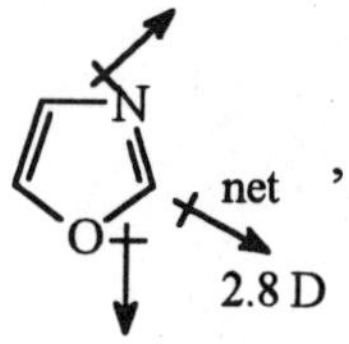

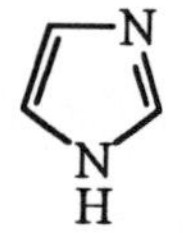

19. 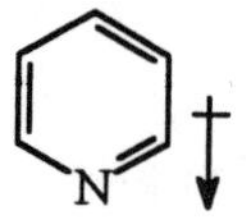 (pyridine); (piperidine) is also (dipole), and pyridine is expected to be enhanced in this direction by polarization, as suggested by resonance structures such as: (pyridine resonance structure with + and −)

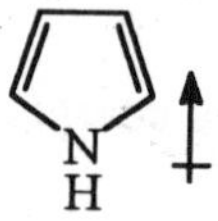

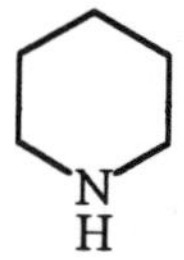

20.

HO–C₆H₄–OH $\xrightarrow[POCl_3]{HOCON(CH_3)_2}$ (2,4-dihydroxybenzaldehyde, CHO) $\xrightarrow[NaO_2CCH_3]{(CH_3CO)_2O}$ AcO-coumarin $\xrightarrow[\Delta]{H^+}$ HO-coumarin

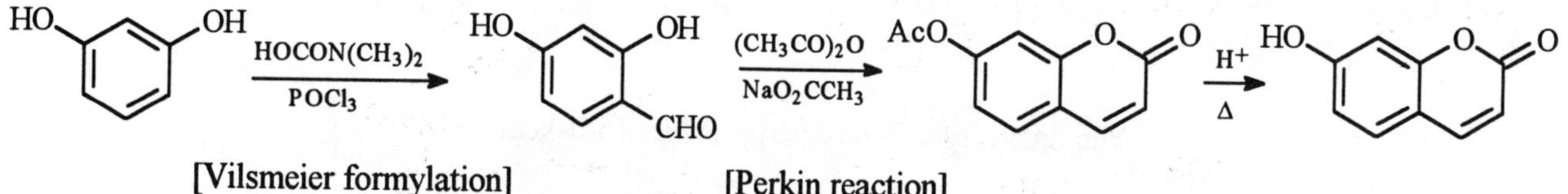

[Vilsmeier formylation]
[or Reimer-Tiemann: $CHCl_3$, OH^-]

[Perkin reaction]

NOTE: the lactone ring hydrolyzes less readily than a normal phenol ester, especially in acid.

21. For $C_{20}H_{21}NO_4$, Zeisel determination gives the partial formula $C_{16}H_9N(OCH_3)_4$.
The oxidation to a ketone indicates: $C_{15}H_7N(CH_2)(OCH_3)_4 \longrightarrow C_{15}H_7N(CO)(OCH_3)_4$
The oxidation products from the ketone can be explained on the basis that the initial products undergo further reaction:

$$C_{15}H_7N(CO)(OCH_3)_4 \xrightarrow{[O]} C_9H_4N(CO_2H)(OCH_3)_2 + C_6H_3(CO_2H)(OCH_3)_2$$

$$C_9H_4N(CO_2H)(OCH_3)_2 \longrightarrow \text{pyridine-2,3,4-tricarboxylic acid (HOOC, HOOC, } CO_2H) + \text{4,5-dimethoxyphthalic acid } (CH_3O, CH_3O, CO_2H, CO_2H)$$

Note that the two primary products are $C_{15} \longrightarrow C_9 + C_6$; this is the only combination that can give the molecular formula of the ketone by working backwards. Furthermore, the CO_2H of both primary product carboxylic acids must come from the same ketone carbonyl. Thus, the structures of the ketone and of papaverine must be:

ketone — 6,7-dimethoxyisoquinolin-1-yl–C=O–(3,4-dimethoxyphenyl); *papaverine* — 6,7-dimethoxyisoquinolin-1-yl–CH_2–(3,4-dimethoxyphenyl)

22.

$$C_6H_5C(=NOH)H \underset{}{\overset{OH^-}{\rightleftharpoons}} \left[C_6H_5C(=N\text{–}O^-)H \longleftrightarrow C_6H_5\bar{C}(N{=}O)H \right] \xrightarrow{Cl\text{–}Cl} C_6H_5CH(N{=}O)\text{–}Cl + Cl^-$$

$$C_6H_5CH(N{=}O)Cl \overset{OH^-}{\rightleftharpoons} \left[C_6H_5C(=N\text{–}O^-)Cl \longleftrightarrow C_6H_5\bar{C}(N{=}O)Cl \right] \overset{H_2O}{\rightleftharpoons} C_6H_5C(=NOH)Cl + {}^-OH$$

This mechanism is actually similar to the chlorination of phenol in basic solution (Section 30.6.A).

23. acting as a base

$C_5H_5^-$ H O^-

-OH^-

OH

24. CO_2Me / N^+ / CH_3 —Na→ MeO_2C (radical) / N / CH_3 + CO_2Me / N (radical) / CH_3 —$ClCH_2Br$→ MeO_2C, Cl / N / CH_3 + CO_2Me / N / Cl, H / CH_3

25.

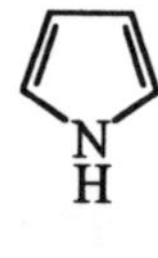

	Pyrrole	Pyridine	Imidazole
pK_a:	-4.4	5.2	7.0

The lack of basicity in the nitrogen in pyrrole reflects the aromatic character of the molecule. The electron pair on the nitrogen is part of the 6π electron system of the molecule. If the electron pair were to be used to accept H^+, the aromatic system would be destroyed. Pyridine is a moderately good base; since the nitrogen is sp^2 hybridized, it is less basic than an aliphatic amine. In imidizole, the pyrrole nitrogen is still not basic, but the sp^2 nitrogen is more basic than the one in pyridine probably reflecting resonance stabilization of the symmetric cation:

26.

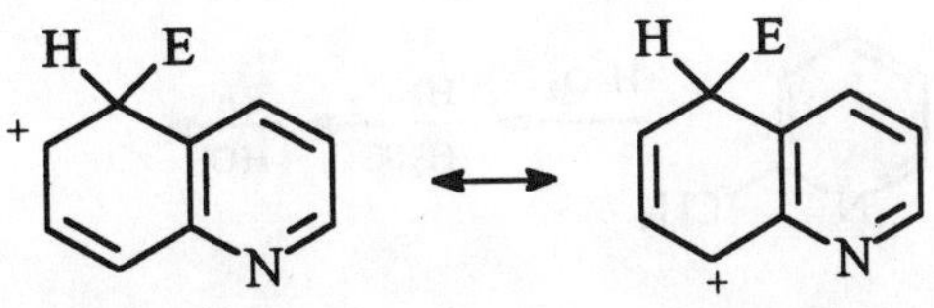

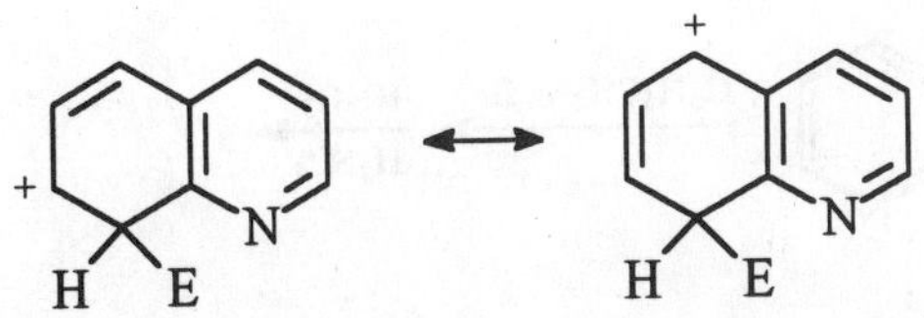

Resonance structures can be written for 5- and 8- substitution that do not disturb the aromatic character of the pyridine ring. This is not true for substitution at the 6- and 7- positions:

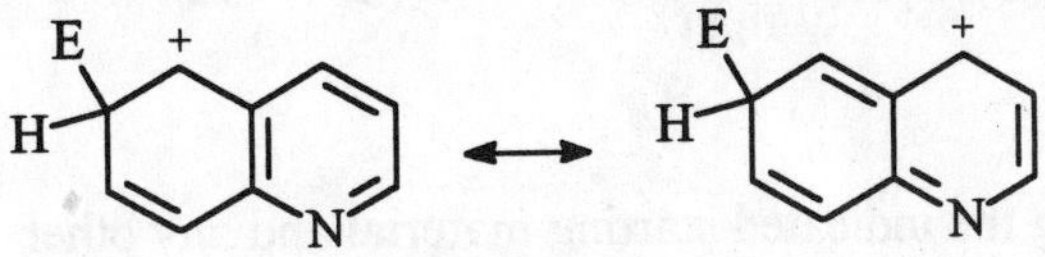

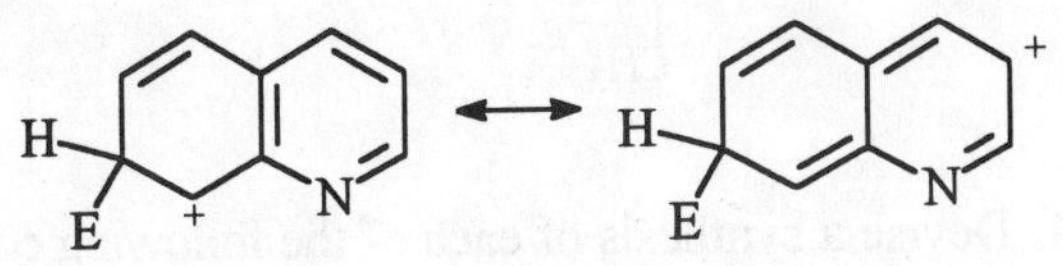

32.F. Supplementary Problems

S1. Name each of the following compounds

a. 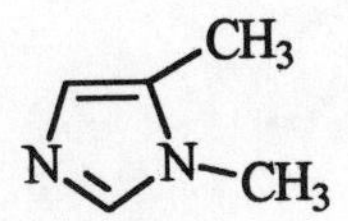 b. 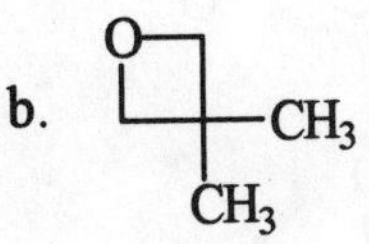 c. 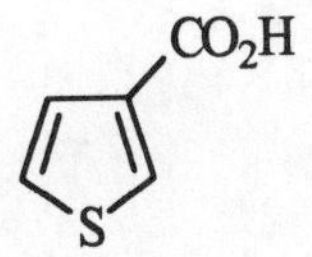 d.

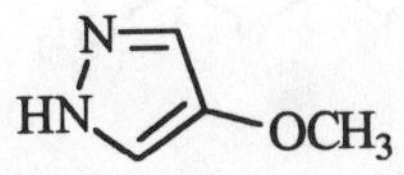

e. 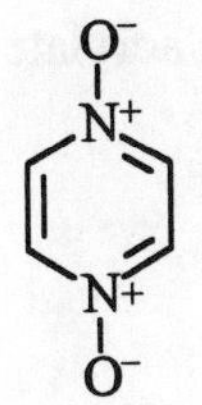 f. 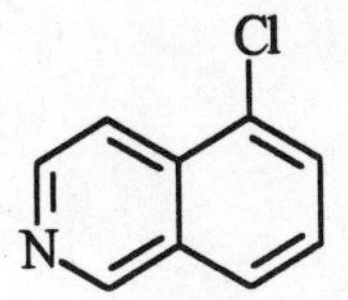 g. 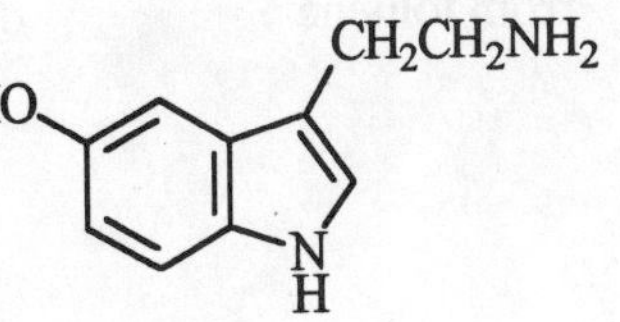 h.

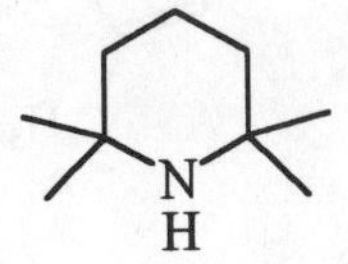

S2. Write the structure of the following compounds:

a. 4-bromofuran-3-carboxaldehyde
b. 3-nitrobenzofuran
c. *trans*-2,3-diphenyloxirane
d. N-acetylpyrrole
e. 2-chloroquinoline
f. 5-nitroisoxazole
g. 6-methylthiopurine
h. 2-methylthietane

S3. What is the major product to result from each of the following reaction sequences?

a. 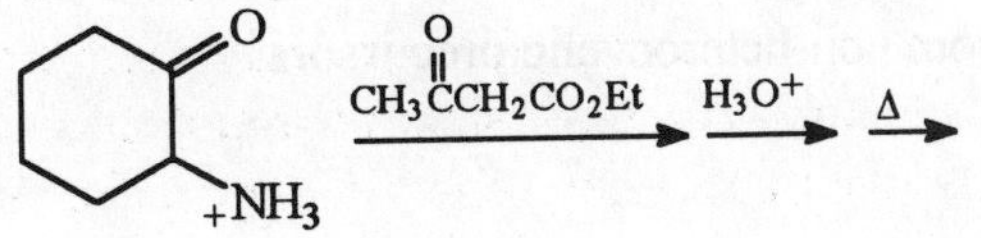 $\xrightarrow{CH_3\overset{O}{\overset{\|}{C}}CH_2CO_2Et} \xrightarrow{H_3O^+} \xrightarrow{\Delta}$

b. $\xrightarrow{Cl_2} \xrightarrow{AcNO_2}$

c. $\xrightarrow[HCl]{CH_2O} \xrightarrow{NaCN} \xrightarrow{LiAlH_4}$

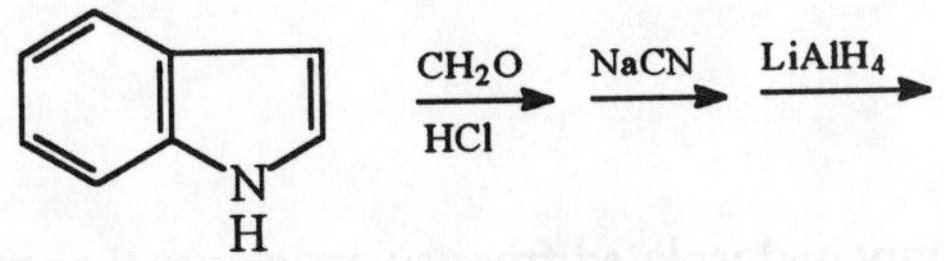

d. 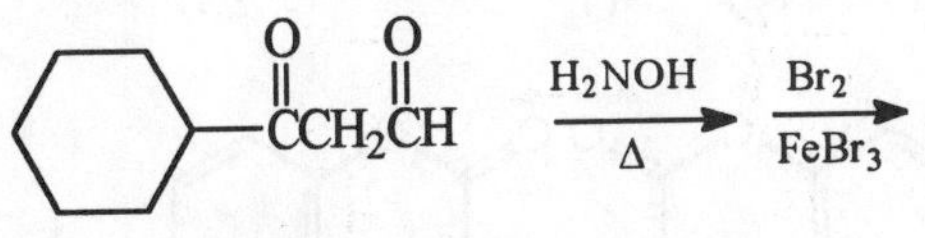 $\xrightarrow[\Delta]{H_2NOH} \xrightarrow[FeBr_3]{Br_2}$

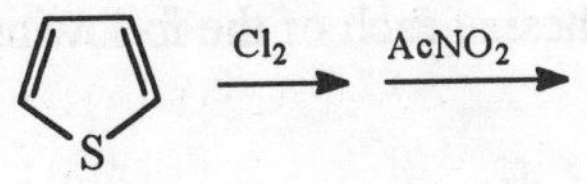

e. 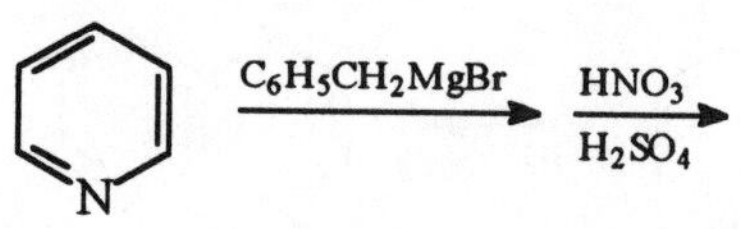 pyridine $\xrightarrow{C_6H_5CH_2MgBr}$ $\xrightarrow[H_2SO_4]{HNO_3}$

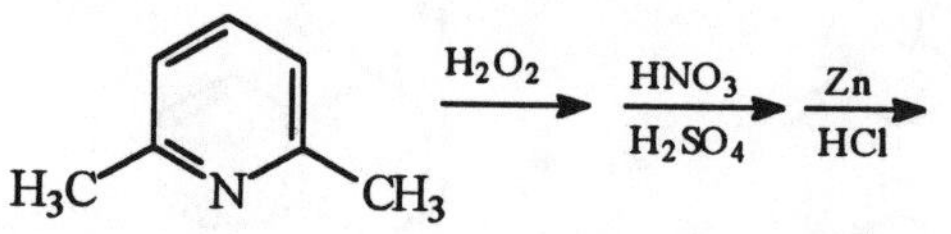

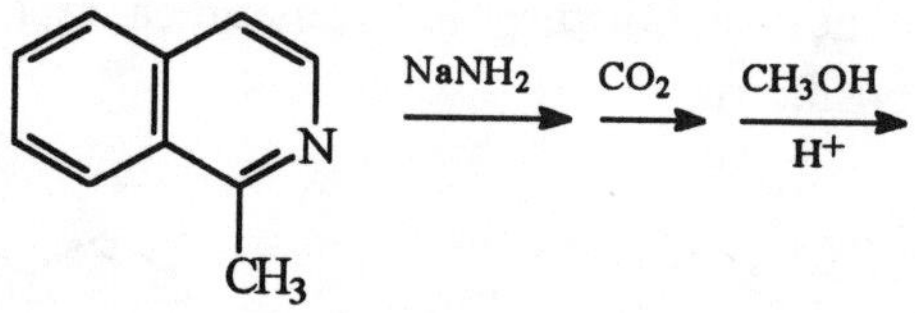

f. 2,6-dimethylpyridine (H_3C, N, CH_3) $\xrightarrow{H_2O_2}$ $\xrightarrow[H_2SO_4]{HNO_3}$ $\xrightarrow[HCl]{Zn}$

g. 1-methylisoquinoline (N, CH_3) $\xrightarrow{NaNH_2}$ $\xrightarrow{CO_2}$ $\xrightarrow[H^+]{CH_3OH}$

h. $C_6H_5CH_2\overset{O}{\overset{\|}{C}}H$ $\xrightarrow[H_2/Pt]{NH_3}$ $\xrightarrow{\text{3,4-methylenedioxybenzoyl chloride (COCl)}}$ $\xrightarrow[\Delta]{P_2O_5}$

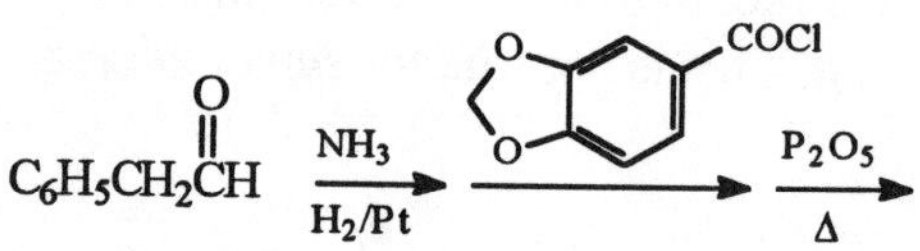

S4. Devise a synthesis of each of the following compounds, using the indicated starting material and any other reagents.

a. 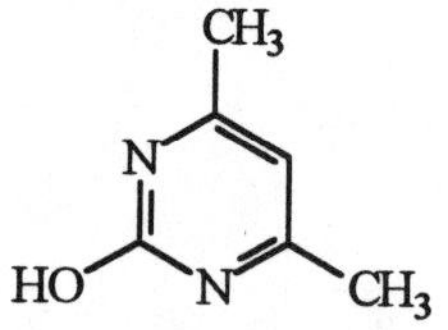 pyrimidine (CH_3, N, HO, N, CH_3) from urea

b. 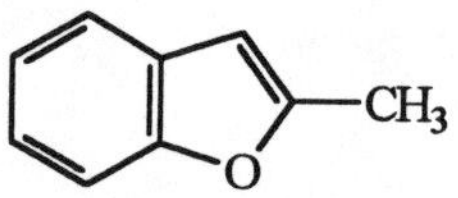 2-methylbenzofuran (O, CH_3) from phenol

c. 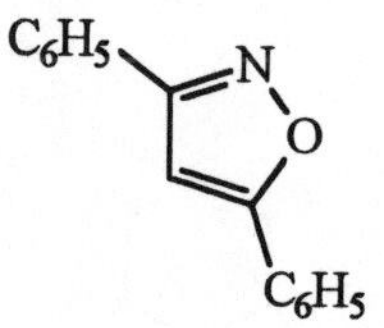 isoxazole (C_6H_5, N, O, C_6H_5) from toluene

d. 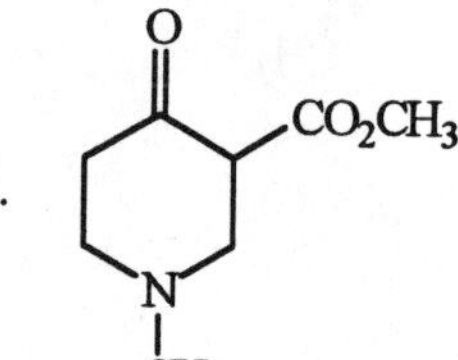

from methyl acrylate

e. 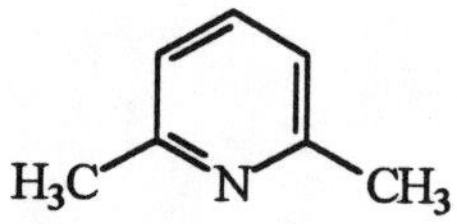 2,6-dimethylpyridine (H_3C, N, CH_3) from acetic acid

f. 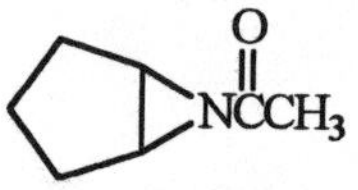 bicyclic aziridine ($N\overset{O}{\overset{\|}{C}}CH_3$) from cyclopentane

g. 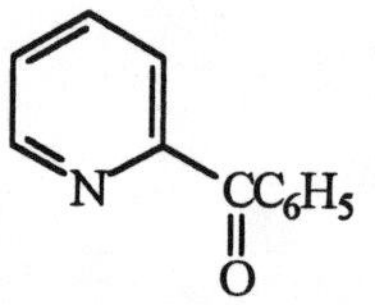 pyridine ring (N) with $\overset{}{C}C_6H_5$ (C=O) from pyridine

h. 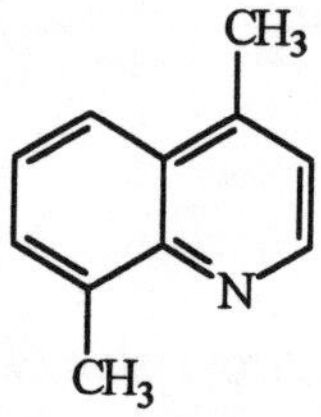 quinoline (CH_3, N, CH_3) from toluene

S5. Show how to synthesize each of the following compounds from non-heterocyclic precursors.

a. 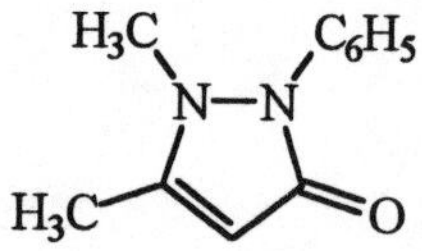 (H_3C, C_6H_5, N—N, H_3C, O) antipyrine, an ingredient in many commercial headache remedies

b. 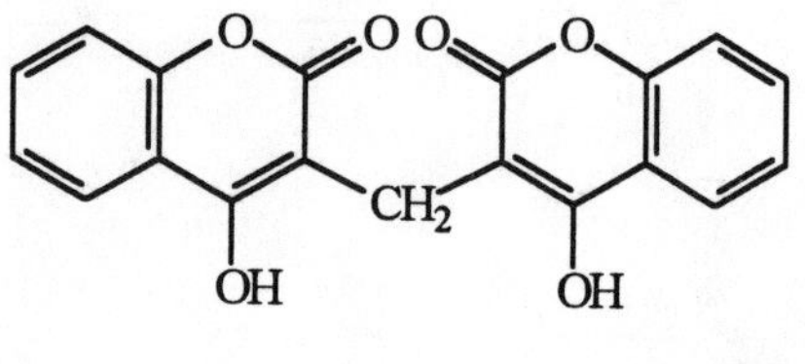 (O, O, O, O, CH_2, OH, OH) dicumarol, a compound isolated from sweet clover that causes a severe bleeding tendency in cattle

c. 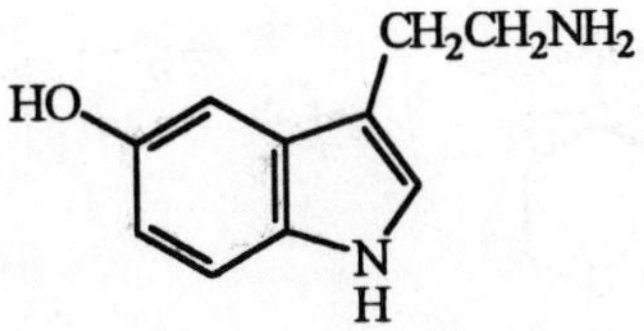 [structure: 5-HO-indole with 3-$CH_2CH_2NH_2$] serotonin, one of the molecules involved in the transmission of nerve impulses in the brain

d. 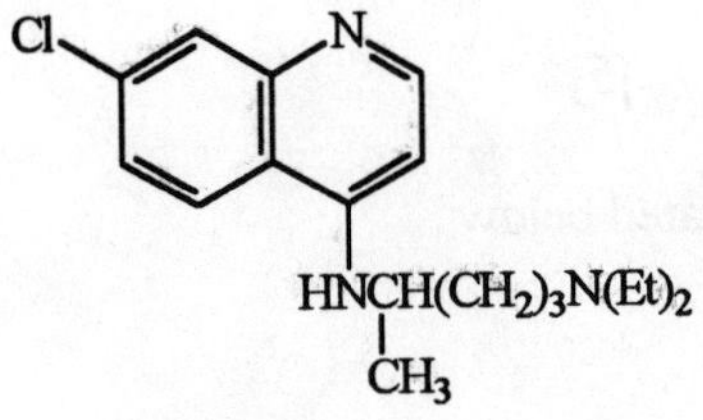 [structure: 7-Cl-quinoline with 4-$HNCH(CH_3)(CH_2)_3N(Et)_2$] chloroquinone, an important antimalarial drug

e. 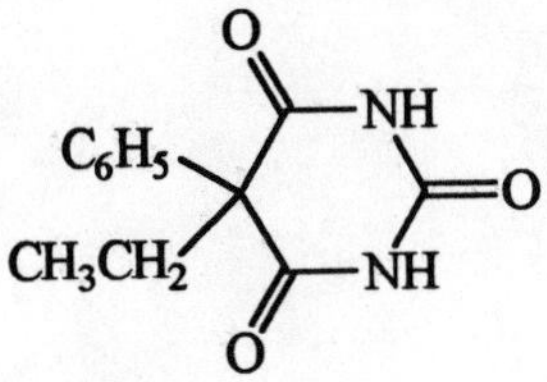 [structure: 5-C_6H_5-5-CH_3CH_2-barbituric acid] phenobarbital, one of the barbiturates (sedatives, depressants)

S6. Write a reasonable mechanism for each of the following transformations:

a. 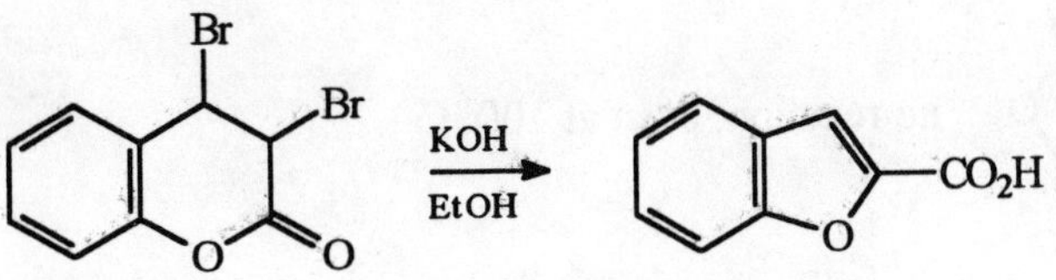 3,4-dibromo-dihydrocoumarin $\xrightarrow[EtOH]{KOH}$ benzofuran-2-CO_2H

b. 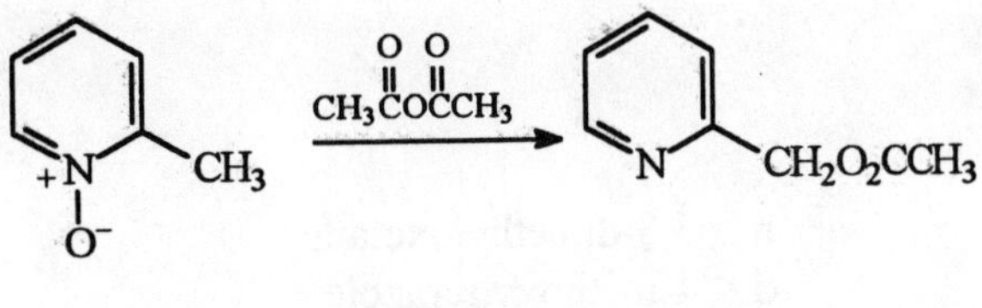 2-methylpyridine N-oxide ($+N$–O^-, CH_3) $\xrightarrow{CH_3\overset{O}{\overset{\|}{C}}O\overset{O}{\overset{\|}{C}}CH_3}$ 2-pyridyl–$CH_2O_2CCH_3$

c. 2-fluoro-3-methylbenzoxazolium ($+N$, CH_3, BF_4^-) + $NaOCH_2CH_2CH_3$ ⟶ 3-methylbenzoxazol-2-one (=O, N–CH_3) + $FCH_2CH_2CH_3$ + $NaBF_4$

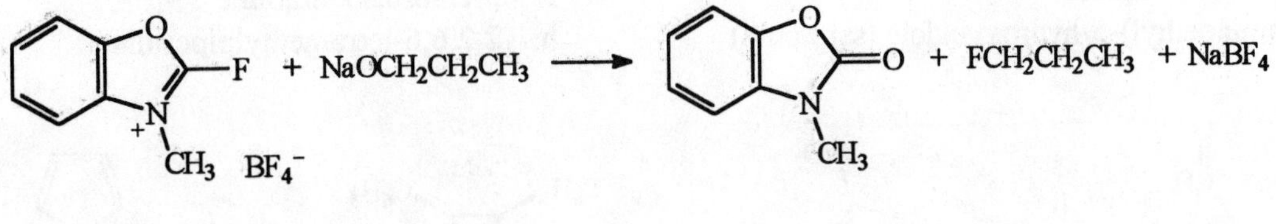

d. 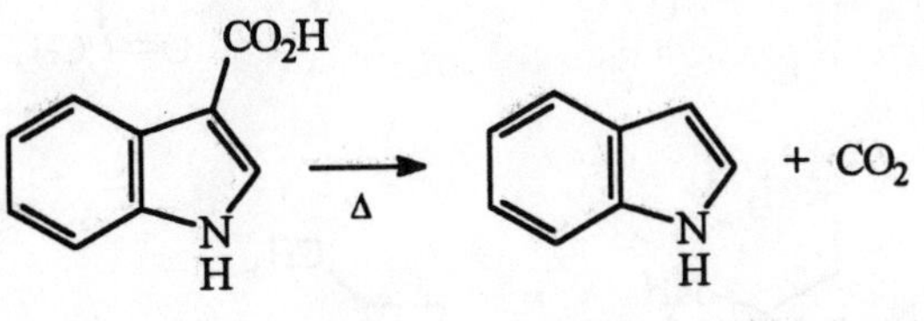 indole-3-CO_2H (N–H) $\xrightarrow{\Delta}$ indole (N–H) + CO_2

S7. Rank the following compounds in order of **basicity**.

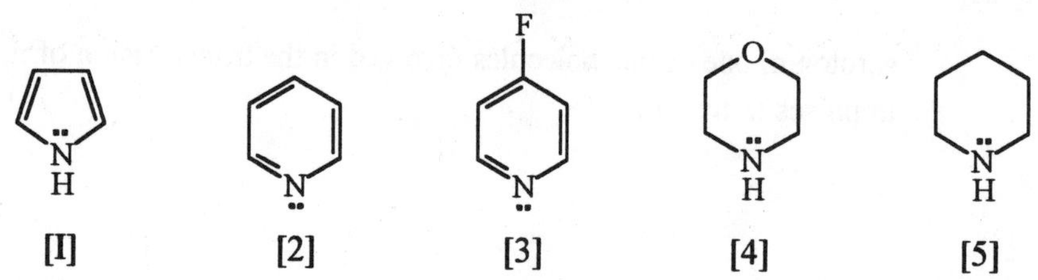

S8. Explain the differences in reactivity of the three furan derivatives illustrated below:

very fast reaction, even at −50°C

reacts at room temperature

no reaction, even at 100°C

32.G. Answers to Supplementary Problems

S1. a. 1,5-dimethylimidazole
b. 3,3-dimethyloxetane
c. 3-thiophenecarboxylic acid
d. 4-methoxypyrazole
e. pyrazine di-N-oxide
f. 5-chloroisoquinoline
g. 3-(2-aminoethyl)-5-hydroxyindole (serotonin)
h. 2,2,6,6-tetramethylpiperidine

S2. a. b. c. d.

e. f. g. h.

S3. a. (structure: 2-methyl-4,5,6,7-tetrahydroindole; $-CH_3$, N–H)

b. (structure: O_2N–thiophene(S)–Cl)

c. (structure: indole with $CH_2CH_2NH_2$ at C-3; N–H)

d. (structure: 5-cyclohexyl-4-bromoisoxazole; O–N, Br)

e. (structure: 2-(4-nitrobenzyl)pyridine; NO_2, N)

f. (structure: 4-amino-2,6-dimethylpyridine; NH_2, H_3C, N, CH_3)

g. (structure: isoquinoline with $CH_2CO_2CH_3$ at C-1; N)

h. (structure: 1-(3,4-methylenedioxyphenyl)isoquinoline; N, O, O)

S4. a. H_2NCNH_2 (C=O) + $CH_3CCH_2CCH_3$ (two C=O) $\xrightarrow[EtOH]{HCl}$ (pyrimidine with OH at C-2, N, N, H_3C, CH_3)

b. C_6H_5OH $\xrightarrow[CH_2=C(CH_3)CH_2Br]{NaOH}$ $\xrightarrow{\Delta}$ (2-(2-methylallyl)phenol; OH, CH_2) $\xrightarrow[2.\ Zn,\ HOAc]{1.\ O_3}$ $\xrightarrow{P_2O_5}$ (2-methylbenzofuran; O, CH_3)

c. $C_6H_5CH_3$ $\xrightarrow[h\nu]{Br_2}$ $\xrightarrow{NaNO_2}$ $C_6H_5CH_2NO_2$ $\xrightarrow{C_6H_5NCO}$ $C_6H_5C\equiv\overset{+}{N}-O^-$

$C_6H_5CH_3$ $\xrightarrow[h\nu]{2\ Br_2}$ $\xrightarrow{NaOH}$ $C_6H_5CH=O$ $\xrightarrow{CH_3MgBr}$ $\xrightarrow{H_2SO_4}$ $C_6H_5CH=CH_2$ $\xrightarrow{Br_2}$ $\xrightarrow{NaNH_2}$ $C_6H_5C\equiv CH$

(both combine to give 3,5-diphenylisoxazole; C_6H_5, N, O, C_6H_5)

or

$C_6H_5CH_3$ $\xrightarrow{KMnO_4}$ $C_6H_5CO_2H$ $\xrightarrow[H^+]{CH_3OH}$ $C_6H_5CO_2CH_3$

$C_6H_5CO_2H$ $\xrightarrow{CH_3Li}$ $C_6H_5CCH_3$ (C=O)

$C_6H_5CCH_3$ + $C_6H_5CO_2CH_3$ $\xrightarrow{NaOCH_3}$ $C_6H_5CCH_2CC_6H_5$ (two C=O) $\xrightarrow[H^+\ \Delta]{H_2NOH}$ (3,5-diphenylisoxazole; N–O, C_6H_5, C_6H_5)

d. CH_3NH_2 + 2 $CH_2=CHCO_2CH_3$ $\longrightarrow$ $CH_3N(CH_2CH_2CO_2CH_3)_2$ (O, OCH_3, CO_2CH_3, N, CH_3) $\xrightarrow[\Delta]{NaOCH_3}$ (1-methyl-3-carbomethoxy-4-piperidone; O, CO_2CH_3, N, CH_3)

e.

$CH_3CO_2H \xrightarrow[H^+, \Delta]{CH_3OH} \xrightarrow[\Delta]{NaOCH_3} CH_3\overset{O}{\overset{\|}{C}}CH_2CO_2CH_3 \xrightarrow[CH_2O]{NH_3}$ H_3CO_2C, CO_2CH_3, H_3C, N, H, CH_3

$\xrightarrow{HNO_3}$ $\xrightarrow[2.\ CaO,\ \Delta]{1.\ KOH}$ H_3C, N, CH_3

f.

$\xrightarrow[h\nu]{Br_2} \xrightarrow{t\text{-}BuO^-} \xrightarrow{INCO} \xrightarrow{NaOH}$ NH $\xrightarrow{Ac_2O}$ $N\overset{O}{\overset{\|}{C}}CH_3$

g.

N + $C_6H_5CH_2Li \longrightarrow$ N, $CH_2C_6H_5$ $\xrightarrow{CrO_3}$ N, O^-, $\underset{\underset{O}{\|}}{C}C_6H_5$ $\xrightarrow{PCl_3}$ N, $\underset{\underset{O}{\|}}{C}C_6H_5$

h.

$C_6H_5CH_3 \xrightarrow[H_2SO_4]{HNO_3} \xrightarrow[HCl]{Zn}$ NH_2, CH_3 $\xrightarrow[FeCl_3,\ ZnCl_2]{CH_2=CH\overset{O}{\overset{\|}{C}}CH_3}$ CH_3, N, CH_3

S5. a.

$CH_3\overset{O}{\overset{\|}{C}}CH_2\overset{O}{\overset{\|}{C}}OCH_3 + CH_3NHNHC_6H_5 \xrightarrow{-H_2O}$ [H_3C, N—NHC_6H_5, CO_2CH_3, H_3C] $\xrightarrow{-CH_3OH}$ H_3C, N—N, C_6H_5, C=O, H_3C

b.

OH, CO_2CH_3 + $CH_3CO_2CH_3 \xrightarrow{NaOCH_3}$ [OH, $COCH_2CO_2CH_3$] $\longrightarrow$ O, O, O

$\rightleftharpoons$ $NaOCH_3$

O, O, O^- $\xrightarrow{CH_2O}$ [O, O, CH_2OH, O $\xrightarrow{-H_2O}$ O, O, CH_2, O ; O, O, O^-] $\longrightarrow$ O, O, O, O, OH, OH

b. $CH_3CO_2CH_3$, NaOCH3, $COCH_2CO_2CH_3$, NaOCH3, CH_2O, $-H_2O$, CH_2OH, CH_2

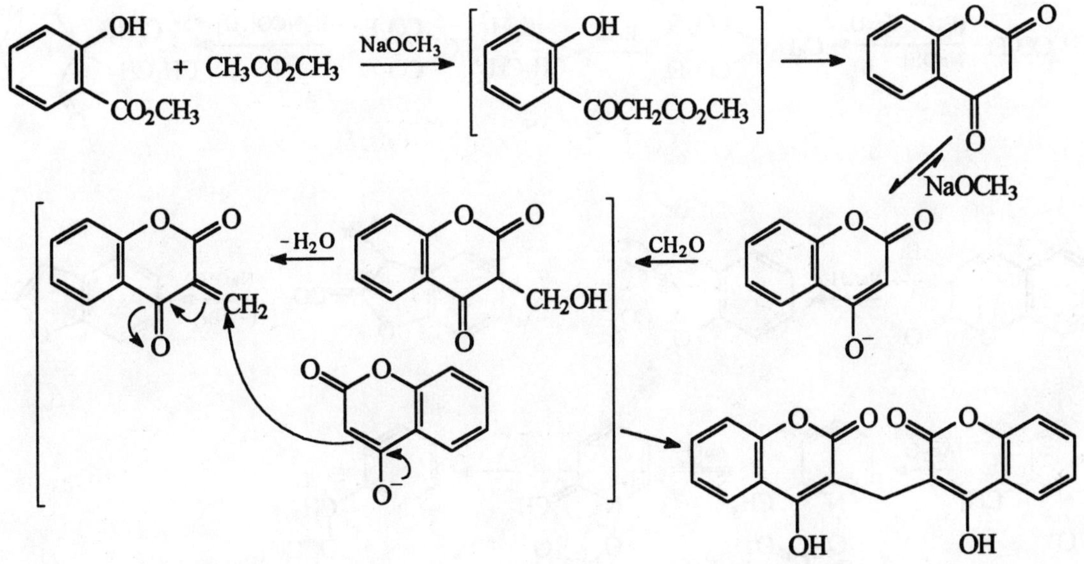

c. CH_3O, NH_2, CH_3CCO_2H, H^+, Δ, CO_2H, $-CO_2$, Δ, 1. DMF 2. $POCl_3$, H_2O, CH, CH_3NO_2 base, $CH{=}CHNO_2$, H_2 Pt, HBr ($-CH_3$), HO, $CH_2CH_2NH_2$

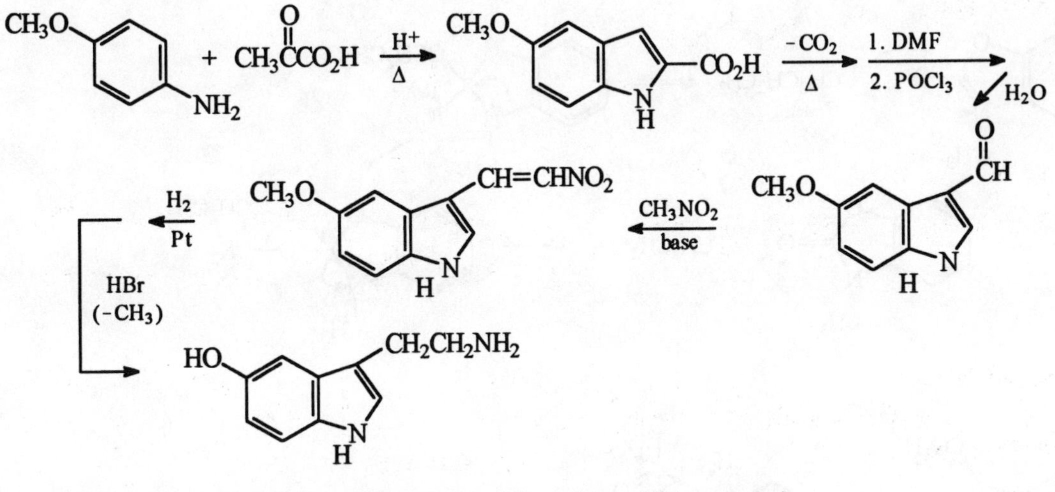

(alternatively, refer to the sequence given in problem #S3 c. above)

d. Cl, NO_2, Zn HCl, $CH_2{=}CHCH$ $FeCl_3$, H_2O_2, HNO_3, O^-, Zn/ HCl, NH_2, $CH_3\overset{O}{C}(CH_2)_3N(Et)_2$ H_2/ Pt, $HNCH(CH_2)_3N(Et)_2$, CH_3

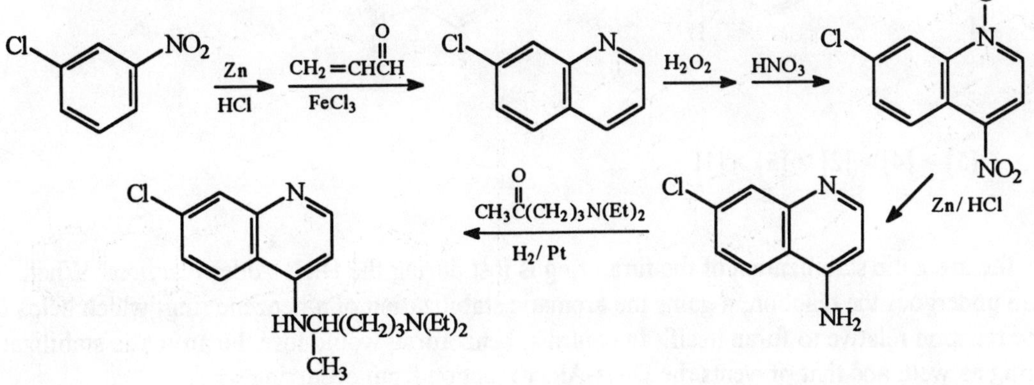

e.

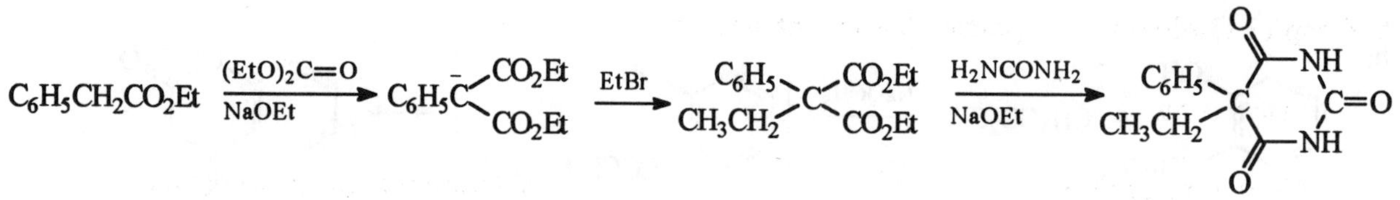

$C_6H_5CH_2CO_2Et \xrightarrow[NaOEt]{(EtO)_2C{=}O} C_6H_5\overset{-}{C}(CO_2Et)_2 \xrightarrow{EtBr} (C_6H_5)(CH_3CH_2)C(CO_2Et)_2 \xrightarrow[NaOEt]{H_2NCONH_2}$ (product: 5-ethyl-5-phenyl barbiturate ring, C_6H_5, CH_3CH_2, C=O, NH, C=O, NH)

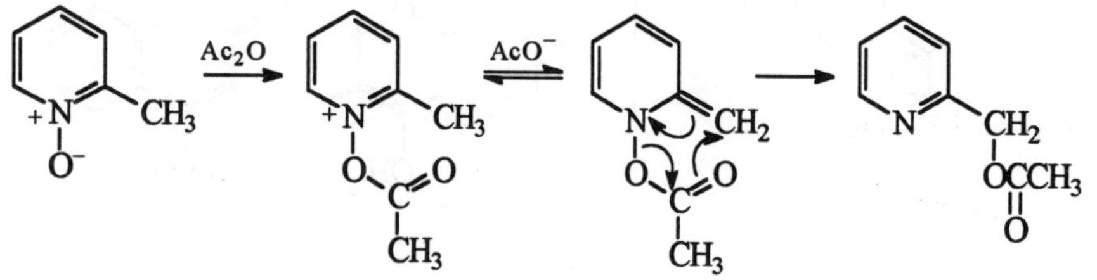

S6. a. Br, Br, O, O $\xrightarrow{NaOH}$ Br, Br, O^-, CO_2^- $\longrightarrow$ Br, O, CO_2^- $\xrightarrow[(-HBr)]{NaOH}$ O, CO_2^-

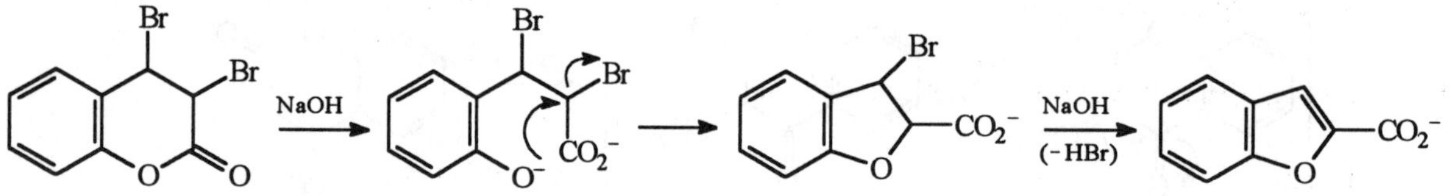

b. $+N$, O^-, CH_3 $\xrightarrow{Ac_2O}$ $+N$, CH_3, O–C=O, CH_3 $\overset{AcO^-}{\rightleftharpoons}$ N, CH_2, O–C=O, CH_3 $\longrightarrow$ N, CH_2–$OCCH_3$ (C=O)

c. O, F, $+N$–CH_3 + $^-OCH_2CH_2CH_3$ $\rightleftharpoons$ O, $OCH_2CH_2CH_3$, F, N–CH_3 $\longrightarrow$ O, $+N$–CH_3, O–CH_2–CH_2CH_3, F^- $\longrightarrow$ O, C=O, N–CH_3 + $FCH_2CH_2CH_3$

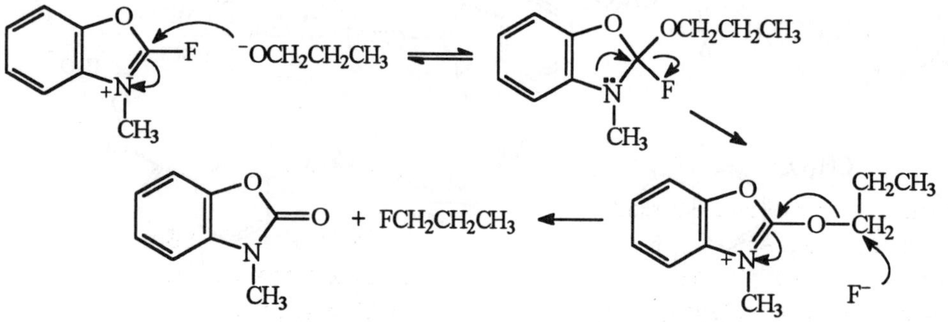

d. CO_2H, N–H $\rightleftharpoons$ H, C(=O)O^-, $+N$–H $\xrightarrow{-CO_2}$ N–H

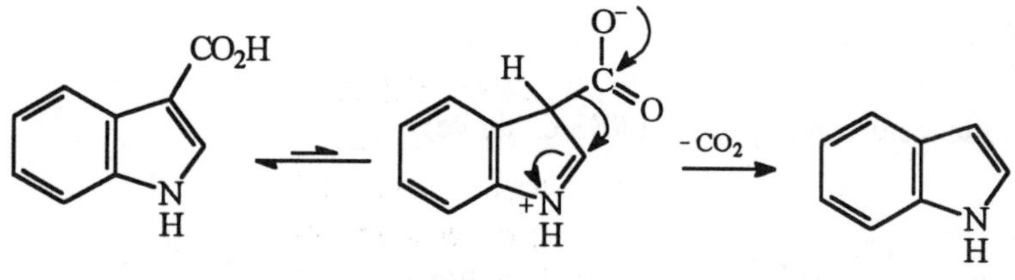

S7. [5] > [4] > [2] > [3] > [1]

S8. In each case, the aromatic stabilization of the furan ring is lost during the Diels-Alder reaction. When isobenzofuran undergoes the reaction, it gains the aromatic stabilization of a benzene ring, which helps to accelerate the reaction relative to furan itself. In contrast, benzofuran would lose the aromatic stabilization of its benzene ring as well, and that prevents the Diels-Alder reaction from occurring.

33. MOLECULAR RECOGNITION; NUCLEIC ACIDS AND SOME BIOLOGICAL CATALYSTS

33.A. Chapter Outline and Important Terms Introduced

33.1 Introduction

molecular recognition
host-guest interaction
specificity of interaction
standard free energy of binding
enthalpy of binding
entropy of binding

33.2 Recognition of Guests by Synthetic Hosts

Kemp's triacid
receptor
preorganization
cryptand
spherand
polar cavity
hydrophobic surface

33.3 Molecular Recognition by an Enzyme

33.4 Catalytically Active Antibodies

antigenic group, epitope
antigen
clonal selection, polyclonal
monoclonal antibodies
hapten
transition state analog
reversible inhibitor

33.5 Nucleic Acids

A. Molecular Components of Nucleic Acids

ribonucleic acid (RNA)
deoxyribonucleic acid (DNA)
nucleotides, ribonucleotides
nucleosides, ribonucleosides
2′-deoxyribonucleotides
2′-deoxyribonucleosides
DNA synthesizer
genetic code

B. Molecular Recognition in Nucleic Acids

base pairing G-C, A-T
Watson-Crick hydrogen bonding
wide(major) grooves
narrow (minor) grooves
base stacking
mRNA, tRNA
exons, introns
triplet code, triad
gene, genetic code
codon, anticodon

C. Determination of the Primary Structure of DNA: Sequencing

restriction nucleases
base pairs
complementary DNA (cDNA)
Maxam-Gilbert method; chemical degradation approach
reverse transcriptase
polymerase chain reaction (PCR)
fluorescence

D. High Specificity Recognition and Cleavage of DNA

base triplets
triple helix
Hoogsteen hydrogen bonding

33.B. Important Reactions Introduced

NOTE: The new reactions introduced in this chapter are enzymatic ones, rather than what we typically think of when we refer to organic reactions. Instead of new reactions, reactions previously encountered are applied to biochemical systems.

33.C. Important Concepts and Hints

This chapter is a distillation of much of what you have learned in organic chemistry. It demonstrates the importance of such diverse topics as bond angles and length, molecular shape and reactivity and how they affect biochemical reactions. Biological chemistry is incredibly complex; this chapter will give you some idea of the means currently employed to help us understand it.

33.D. Answers to Exercises

33.1 $\Delta G° = -RT\ln K_a$ $R = 1.98$ cal K^{-1} mole^{-1} std T = 25°C or 298 K

a. $K_a = 10$ mole^{-1}: $\Delta G° = -1.4$ kcal mole^{-1}

b. $K_a = 5 \times 10^3$ mole^{-1}: $\Delta G° = -5.0$ kcal mole^{-1}

c. $K_a = 8 \times 10^9$ mole^{-1}: $\Delta G° = -13.5$ kcal mole^{-1}

33.2 $\Delta G° = \Delta H° - T\Delta S° = -5$ kcal mole^{-1} $-$ [298K x (-16 cal K^{-1} mole^{-1})]

$\Delta G° = -0.23$ kcal mole^{-1}

At 37°C: $\Delta G° = \Delta H° - T\Delta S° = -5$ kcal mole^{-1} $-$ [310K x (-16 cal K^{-1} mole^{-1})]

$\Delta G° = -0.040$ kcal mole^{-1}

33.3 $K_{Li} = \dfrac{[complex]}{[Li^+]\ [spherand]}$ $K_{Na} = \dfrac{[complex]}{[Na^+]\ [spherand]}$

A measure of the $[Li^+]/[Na^+]$ should be given by K_{Na}/K_{Li}. RT = .59 kcal mole^{-1} at 25°C.

Since $K = \exp(-\Delta G/RT)$: $K_{Li} = \exp(-23$ kcal mole^{-1}/0.59 kcal mole$^{-1}) = 1.2 \times 10^{-17}$

$K_{Na} = \exp(-19.2$ kcal mole^{-1}/0.59 kcal mole$^{-1}) = 7.4 \times 10^{-15}$

$[Li^+]/[Na^+] \sim K_{Na}/K_{Li} = 630:1$; much more Na^+ than Li^+ will be bound as indicated by the large differences in free energy.

33.4 $RCO_2R' + OH^- \leftrightarrows RCO_2H + R'O^- \rightleftharpoons RCO_2^- + R'OH$

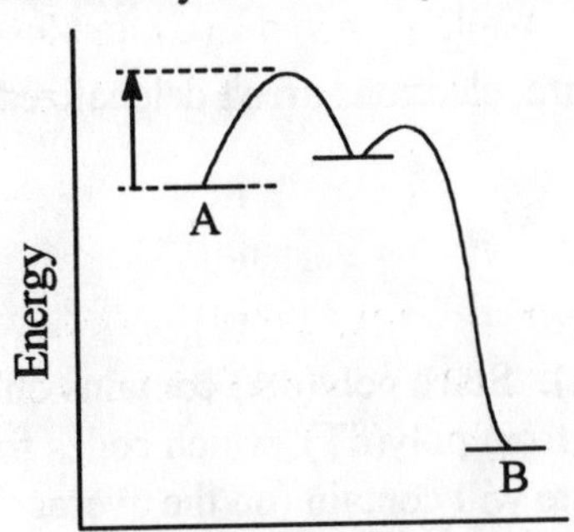

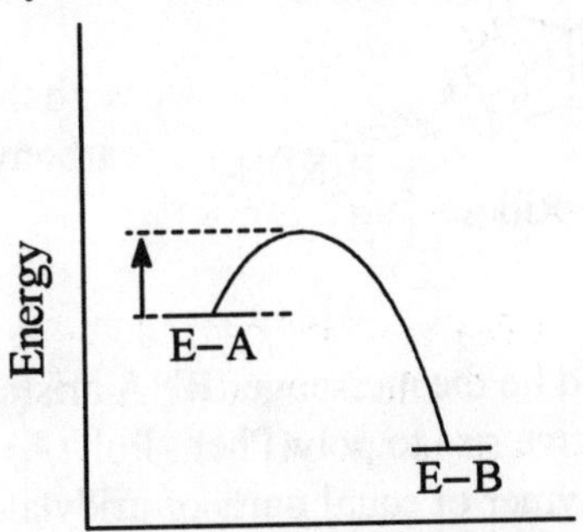

33.5 There are several equilibria involved here: 2 U $\rightleftharpoons$ UU, 2 A $\rightleftharpoons$ AA and U + A $\rightleftharpoons$ UA.

The equilibrium concentration of U = [U] = 0.1 - [UA] - [UU].

The equilibrium concentration of A = [A] = 0.1 - [UA] - [AA].

Since the equilibrium constant for UA formation is so much larger than those for UU or AA, we will make the simplifying assumption that [U] = 0.1 - [UA] = [A].

$K_{UA} = \dfrac{[UA]}{[A]\ [U]}$ = 103. Substitution for [U] and [A] and solution gives:

$[U] = [A] = 2.67 \times 10^{-2}\, M$
$[UA] = .1 - 2.67 \times 10^{-2} = 7.33 \times 10^{-2}\, M$

$[UU] = K_{UU} \times [U]^2 = 6 \times (2.67 \times 10^{-2})^2 = 4.3 \times 10^{-3}\, M$
$[AA] = K_{AA} \times [A]^2 = 3 \times (2.67 \times 10^{-2})^2 = 2.1 \times 10^{-3}\, M$

33.6 Since there are 3 hydrogen bonds in the G-C complex, its density will be greater than that for A-T.

33.7 GTA CCC GAT TGA CAG CAG AAT AAT AAT TAC GAA GAA ATT
His--Gly---Leu---Thr---Val--Val---Leu---Leu---Leu--Met---Leu--Leu--stop

33.8

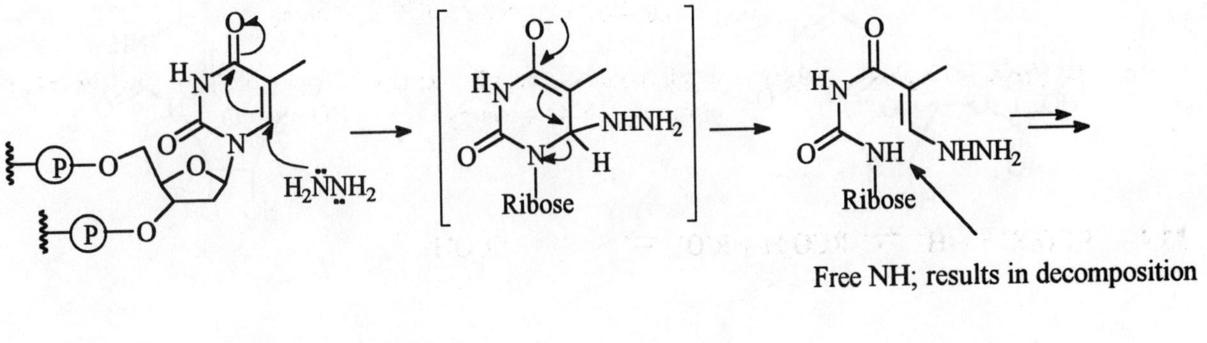

$H_2\ddot{N}NH_2$ → [$NHNH_2$, H, Ribose] → NH, $NHNH_2$, Ribose ⇉

Free NH; results in decomposition

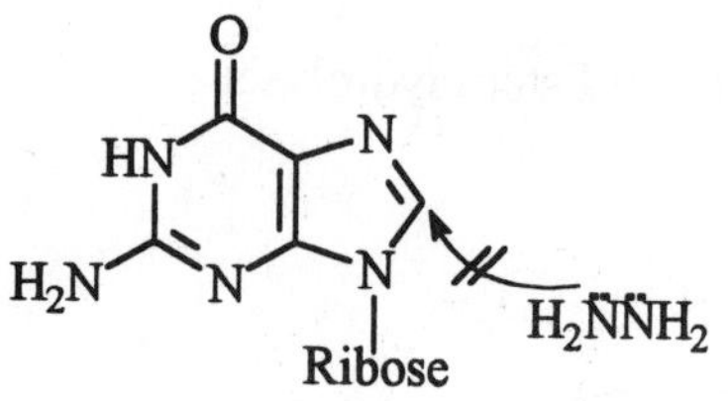

With the purine structure, electrons aren't delocalized onto the carbonyl oxygen.

33.9 Poly(U) would be the messenger RNA arising from poly(dA). Since poly(dA) contains only AAA codons, poly(U) will give rise to poly(Phe). Poly(A) likewise arises from poly(dT), which codes for poly(Lys). A random copolymer of equal parts of uridylate and guanidylate will contain (on the average) equal amounts of the eight triads shown below:

RNA Triad	*Precursor Sequence in Parent DNA (codon)*	*Amino Acid Encoded*
UUU	AAA	Phe
GUU	CAA	Val
UGU	ACA	Cys
UUG	AAC	Leu
GGU	CCA	Gly
GUG	CAC	Val
UGG	ACC	Trp
GGG	CCC	Gly

The polypeptide produced is therefore expected to have the approximate composition: $PheVal_2CysLeuGly_2Trp$

33.E. Answers to Problems

1. a.

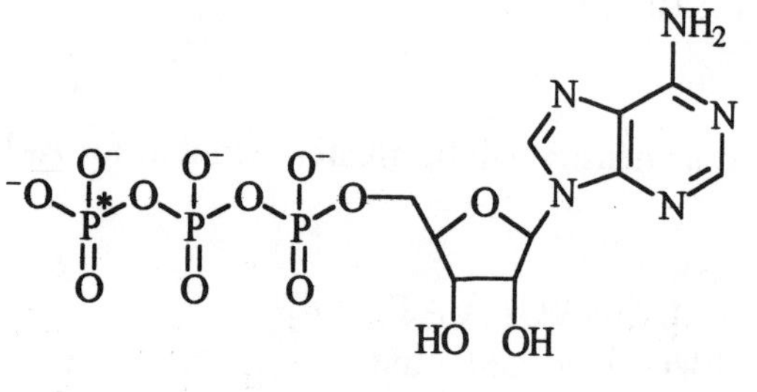

b.

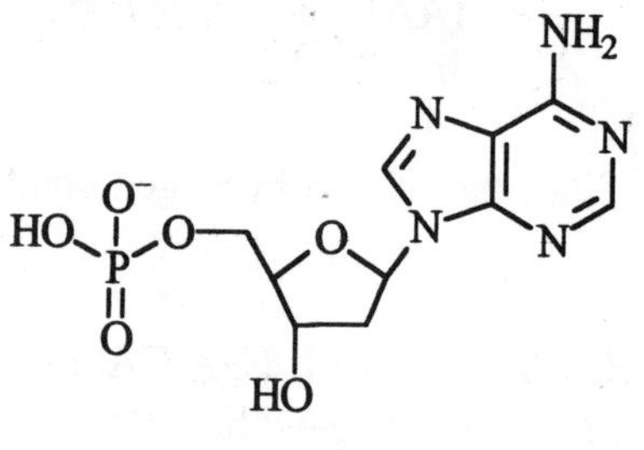

c.

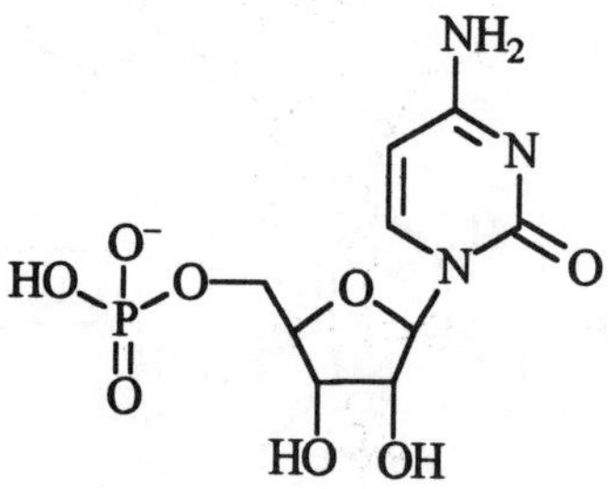

d.

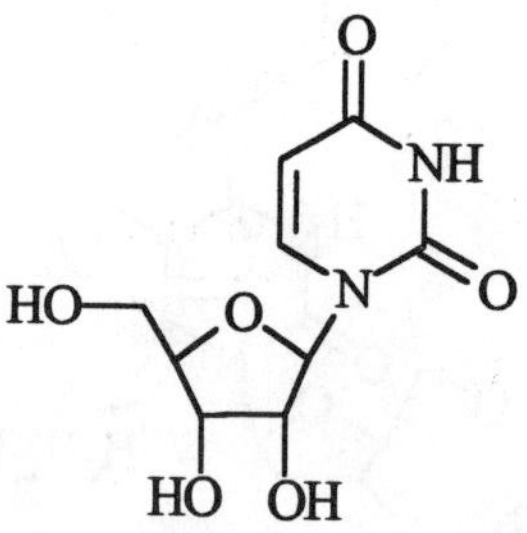

e.

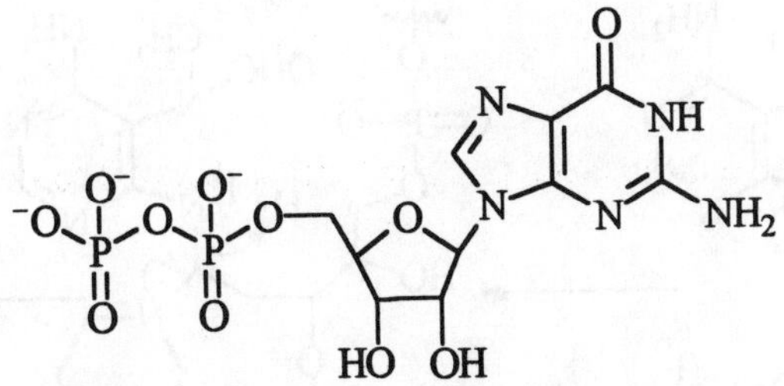

f.

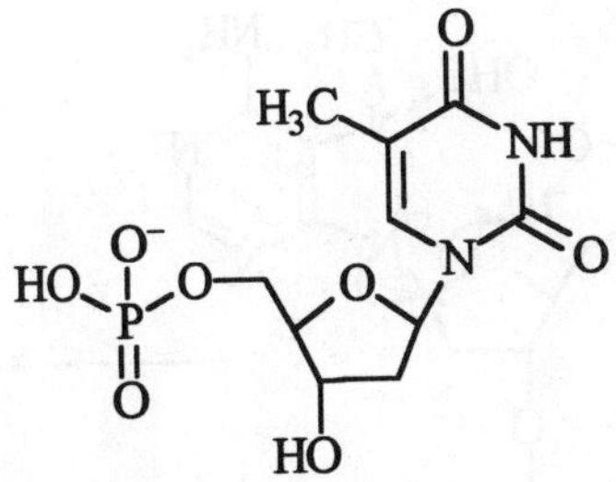

2. a. 5′-γ-^{32}P-ATP b. 5′-dAMP or pdA c. 5′-CMP or pC
d. U e. 5′-GDP f. 5′-dTMP or pdT

3. a. 2′-deoxythymidylic acid, 5′-dTMP, 2′-deoxythymidine 5′-phosphate or pdT
b. uridine 5′triphosphate, 5′-UTP
c. 2′-deoxyguanosine 5′-triphosphate, 5′-dGTP

4. For Ed, Glu-Asp: Glu is coded for by CTC; Asp by CTT, CTG or CTA

Possible combinations: DNA: CTC CTT CTC CTG CTC CTA
mRNA: GAG GAA GAG GAC GAG GAU

5. Of the three -OH's on the phosphate group, one is not bonded to ribose and will be ionized at physiologic pH. This will give a net negative charge to the DNA. Glutathione, γ-Glu-Cys-Gly, also has a free acid group, on the glutamic acid residue, and will also have a net negative charge at physiologic pH. A polypeptide containing cysteine for the radical-scavenging properties of the thiol and a negatively charged side chain, i.e., histidine, lysine or arginine, might be more effective.

6. These steps are outlined in detail on p. 1168.

7.

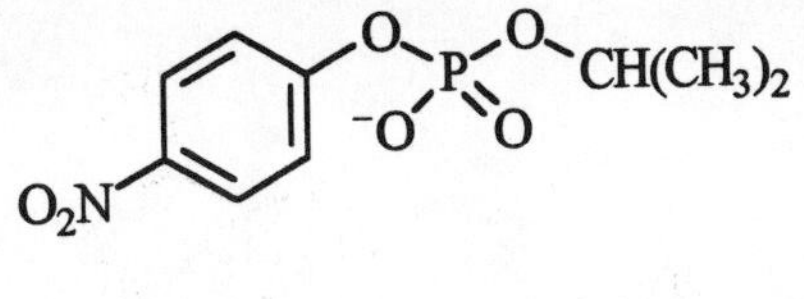

9.

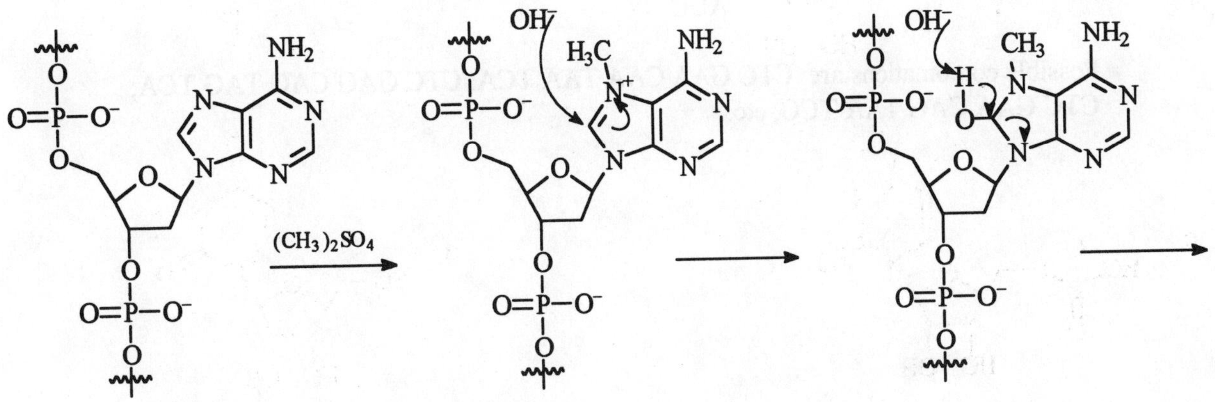

$(CH_3)_2SO_4$

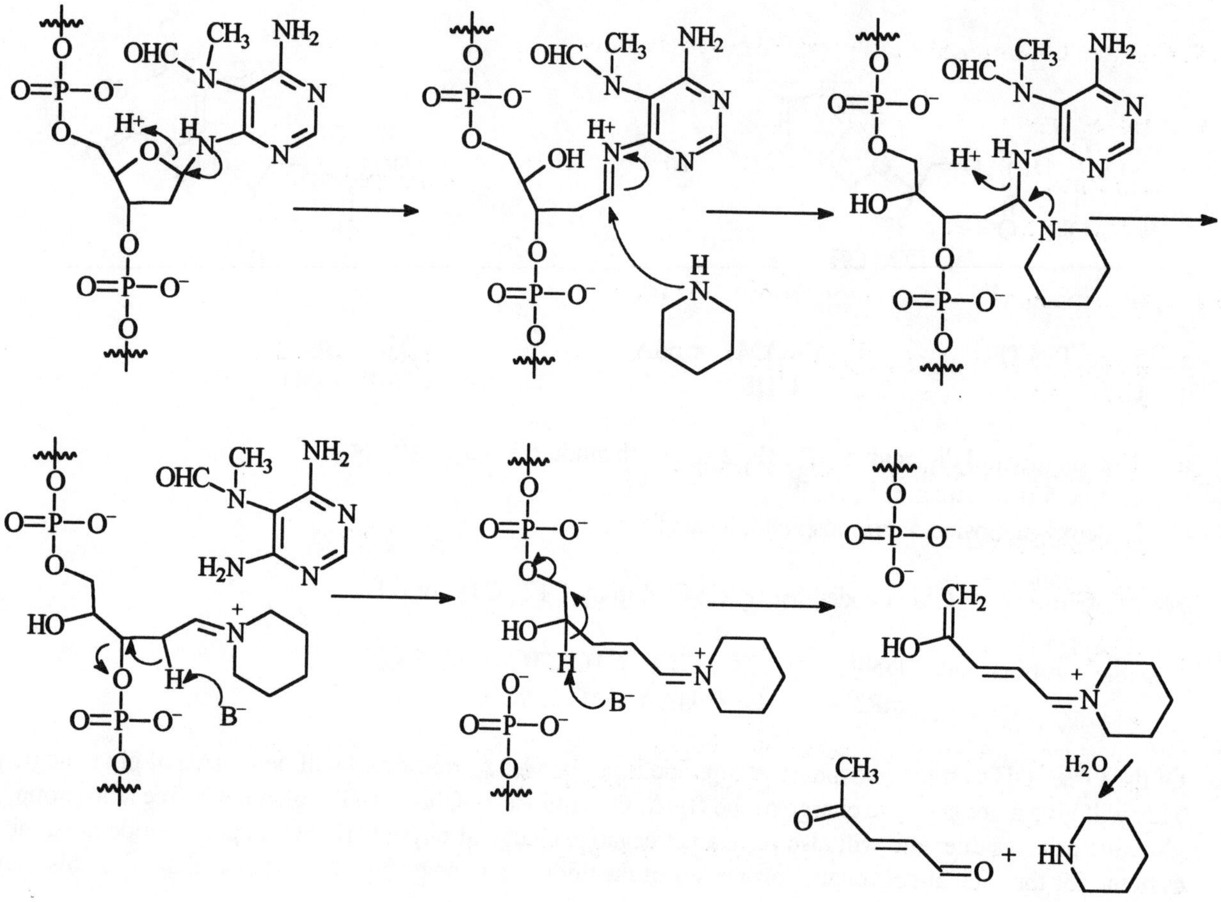

10.

E	L	V	I	S
Glu	Leu	Val	Ile	Ser
CTC	GAA	CAA	TAA	TCA
	GAG	CAG	TAG	TCG
	GAT	CAT	TAT	AGA
	GAC	CAC		AGG
				AGT
				AGC

Possible combinations are: CTC GAA CAA TAA TCA; CTC GAG CAG TAG TCA; CTC GAA CAG TAA TCG, etc.

34. MASS SPECTROMETRY

34.A. Chapter Outline and Important Terms Introduced

34.1 Introduction

radical cation
mass spectrometer
mass spectrum

34.2 Instrumentation (the physics behind the technique)

magnetic sector mass spectrometer
$m/z = B^2r^2e/2V$
magnetic scanning

34.3 The Molecular Ion: Molecular Formula

nominal mass
high resolution
$M + 1$ peaks

34.4. Fragmentation

A. Simple Bond Cleavage
   (to give most stable cationic fragment)
B. Two-bond Cleavage, Elimination of a Neutral Molecule
   alcohols: loss of water
   carbonyl compounds: McLafferty rearrangement

34.B. Important Reactions Introduced (none in this Chapter)

34.C. New Concepts and Hints

HOW TO INTERPRET A MASS SPECTRUM ≡ HOW TO SOLVE MASS SPECTRAL PROBLEMS

All the rules listed below apply to molecules containing only C, H, and O. For halogen- or nitrogen-containing compounds, see item #6, entitled "Complications".

1. Decide whether the particle of highest mass/charge ratio (*m/z*) is the molecular ion (M^+). **(NOTE: The height of the peak (intensity) is something completely different; the tallest peak is usually not M^+.)**
   a. If the formula of the molecule is given, you can determine M^+ right away by calculating the molecular weight.
   b. If *m/z* for the largest particle is **odd**, then it's **not** M^+.
   c. If the next smaller fragment corresponds to loss of an impossible piece (loss of 7, say, or 22 mass units), then the larger fragment is not M^+. The spectrum of 2-methyl-2-butanol (Figure 34.15) provides an example of both of these generalizations.
   d. If the particle of highest *m/z* is **not** M^+ and no hints or formula were provided, try adding water (to the even-numbered fragments) or alkyl groups (to the odd-numbered fragments) to come up with something reasonable. For instance, in the spectrum of 2-methyl-2-butanol, addition of 18 (water) to the next largest fragment and 15 (methyl) to the largest fragment suggest the same molecular ion (*m/z* 88).

2. Look at the major fragment ions, and decide what pieces have been lost (subtract each *m/z* from M^+ and/or from a higher *m/z*).
   a. Odd-mass pieces are radicals: methyl = 15, ethyl = 29, propyl = 43, butyl = 57, etc.
   b. Even-mass pieces are neutral molecules: water = 18, ROH = 17 + alkyl; ethylene (from McLafferty rearrangement) = 28; higher alkenes = 28 + 14 for each additional CH_2 group.

3. Before you attempt to interpret cleavage patterns, see what you can deduce about the molecule from other sources. Is there oxygen in the molecule? Is it a ketone or an alcohol? Because you can *predict* how different classes of compounds will fragment, interpreting an actual cleavage pattern becomes much easier if you know what to expect.

4. Look at even-mass fragments first, the ones that correspond to loss of a neutral molecule.
   a. Loss of 18 (water) is strong evidence for an alcohol; loss of 17 + alkyl (ROH) is evidence for an ether.
   b. Loss of 28 (CH_2=CH_2) or a higher alkene from a carbonyl-containing molecule is the result of McLafferty rearrangement. This means that a carbon γ to the carbonyl group has a hydrogen attached. It also gives you an indication of the type of alkyl group (see problem #11, for example).

5. Finally, look at the odd-mass fragments, which correspond to loss of a radical from the molecule. Such cleavages occur to give the most stable cations and (less importantly) most stable radicals.
   a. Hydrocarbons cleave at branch points so that secondary or tertiary cations can be formed.
   b. Alcohols, ethers, amines, and carbonyl compounds undergo cleavage adjacent to the functional group (so-called α-cleavage) so that oxonium or immonium ions can be formed:

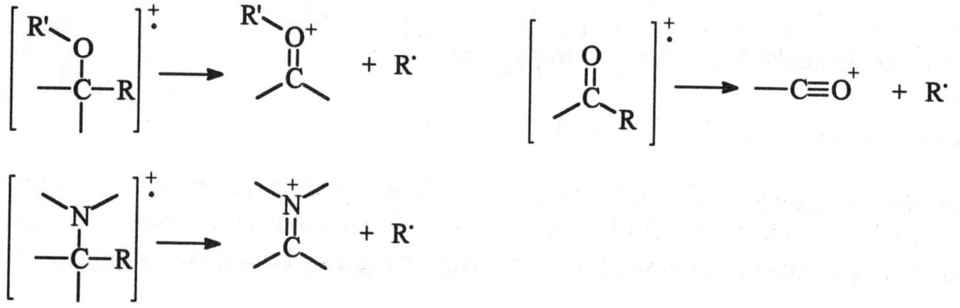

$$[R'-O-C(-R)]^{+\cdot} \longrightarrow R'-O^+=C + R^\cdot \qquad [C(=O)-R]^{+\cdot} \longrightarrow -C\equiv O^+ + R^\cdot$$

$$[N-C(-R)]^{+\cdot} \longrightarrow N^+=C + R^\cdot$$

6. <u>**COMPLICATIONS**</u>:
   a. Chlorine- and bromine-containing ions show doubled peaks because of the presence of two isotopes for each.
   b. The presence of an odd number of nitrogen atoms changes statement 1 (b) above: molecular ions containing an odd number of nitrogens have <u>odd</u> mass.
   c. Molecules that contain several functional groups will obviously produce more complex mass spectra. Cyclic molecules, especially bicyclic molecules, often give fragmentation patterns that are difficult to interpret because more than one bond must be broken to remove a piece.

34.D. Answers to Exercises

34.1 The electron is removed from a π-orbital in forming the radical cation $[CH_2{=}CH_2]^{+\cdot}$. This orbital is higher in energy (= less energy required to **remove** an electron) than the σ-orbitals in methane.

34.2 $CH_3{-}^{+}C{=}O{:} \longleftrightarrow CH_3{-}C{\equiv}O^{+}$

34.3 $$\frac{M+1}{M} = 0.01119\,c + 0.00015\,h + 0.00367\,n + 0.00037\,o + 0.0080\,s$$

a. $C_{10}H_{22}$: $(M+1)/M = 0.1152 = 11.52\%$;

b. $C_{10}H_{22}O$: 11.56;

c. $C_{10}H_{23}N$: 11.90%

34.4

Formula	(Mass)		%Abundance
$C_2H_4{}^{79}Br_2$	186	$(50.69)^2$	= 25.7%
$C_2H_4{}^{79}Br^{81}Br$	188	$(50.69 \times 49.31) \times 2$	= 50.0%
$C_2H_4{}^{81}Br_2$	190	$(49.31)^2$	= 24.3%

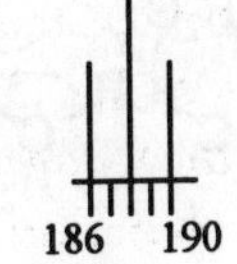

186 190

34.5 $[CH_3CH_2C(CH_3)_2CH_2CH_2CH_2CH_3]^{+\cdot}$ → $CH_3CH_2^{\cdot}$ + $^{+}C(CH_3)_2CH_2CH_2CH_2CH_3$ *m/z* 99

→ $CH_3CH_2C^{+}(CH_3)_2$ *m/z* 71 + $^{\cdot}CH_2CH_2CH_2CH_3$

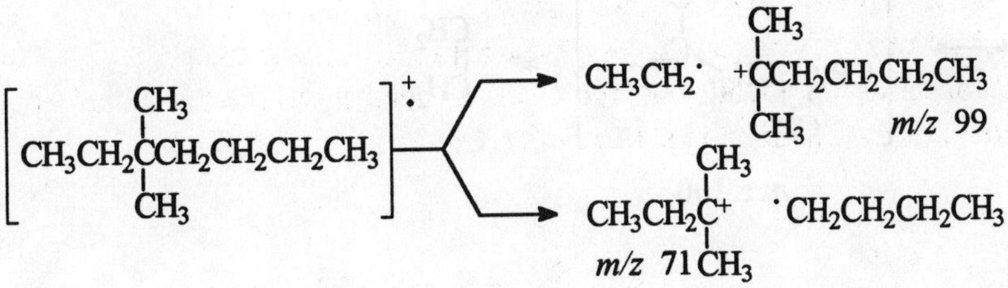

34.6 For molecules containing only C, H, and O, the molecular ions will have even mass and the fragments resulting from single bond cleavage will all have odd mass. The situation is reversed for monoamines: the molecular ions have odd mass and the single-bond cleavage fragments have even mass.

34.7 a.

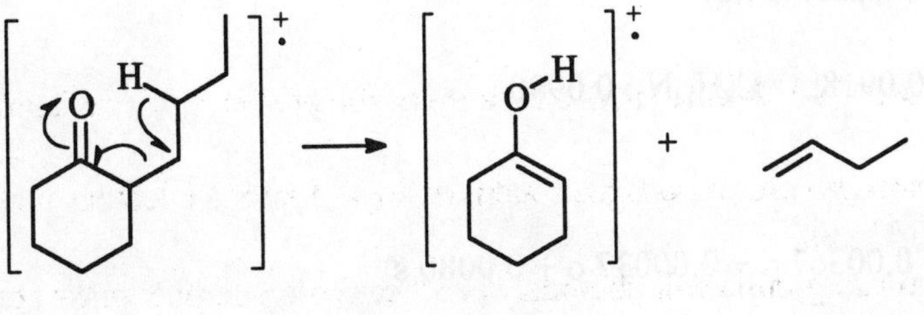

$M^{+} = 154$ → *m/z* 98 +

b.

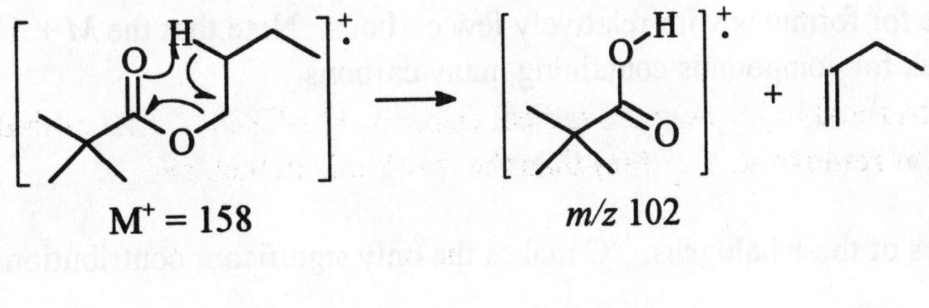

$M^{+} = 158$ → *m/z* 102 +

34.8 $[CH_3CH(CH_3)CH_2 \wr C(=O)-CH_2CH_2CH_3]^{+\cdot} \longrightarrow CH_3CH(CH_3)CH_2^{\cdot} \quad {}^{+}C(=O)CH_2CH_2CH_3$

m/z 71

$[CH_3CH(CH_3)CH_2-C(=O) \wr CH_2CH_2CH_3]^{+\cdot} \longrightarrow CH_3CH(CH_3)CH_2C(=O)^{+} \quad {}^{\cdot}CH_2CH_2CH_3$

m/z 85

$[H_2C(H)-CH(H_3C)-CH_2-C(=O)-CH_2CH_2CH_3]^{+\cdot} \longrightarrow H_3C-CH=CH_2 + [H_2C=C(O-H)-CH_2CH_2CH_3]^{+\cdot}$

m/z 86

$[CH_3CH(CH_3)CH_2-C(=O)-CH_2-CH_2-CH_2(H)]^{+\cdot} \longrightarrow [CH_3CH(CH_3)CH_2-C(O-H)=CH_2]^{+\cdot} + CH_2=CH_2$

m/z 100

34.E. Answers and Explanations to Problems

1. a. Relative probabilities are:

$C_3H_6{}^{35}Cl_2$,	$(0.7577)^2$	= 57%
$C_3H_6{}^{37}Cl^{35}Cl$,	$(0.7577)(0.2423) \times 2$	= 37%
$C_3H_6{}^{37}Cl_2$,	$(0.2423)^2$	= 6%

b. Predicted $(M+1)/M$, according to equation 34-5:

$C_{10}H_{18}$: 0.1146; $C_8H_{10}O_2$: 0.0918; $C_8H_{14}N_2$: 0.0990

2. $$\frac{M+1}{M} = 0.01119\ c + 0.00015\ h + 0.00367\ n + 0.00037\ o + 0.0080\ s$$

a. $C_8H_{14}O_4$: 9.31% b. $C_{10}H_{10}N_2$: 12.1% c. $C_{13}H_{20}$: 14.8% d. $C_{60}H_{122}$: (69.0%)

Equation 34-5 above is only accurate for formulas with relatively few carbons. Note that the $M+1$ peak approaches the intensity of the M peak for compounds containing many carbons.

e. CH_3I: 1.2% f. C_2F_6: 2.2%

Since ^{127}I and ^{19}F are the only isotopes of these halogens, ^{13}C makes the only significant contribution to $M+1$.

3. The positive charge in the $M - 29$ fragment from N-propylaniline is localized on the nitrogen atom as an immonium ion. The presence of a *p*-nitro substituent destabilizes this charge and disfavors the formation of this fragment the same way it decreases the basicity of the aniline itself.

$$\left[\text{HN(CH}_2\text{CH}_2\text{CH}_3\text{)–C}_6\text{H}_4\text{–Y}\right]^{+\cdot} \xrightarrow{-\,C_2H_5^{\cdot}} \text{CH}_2\text{=}\overset{+}{\text{N}}\text{H–C}_6\text{H}_4\text{–Y}$$

4. 1. $M^+ = 128$: $C_{10}H_8$ or C_9H_{20}
   2. weak M^+: likely to be a saturated hydrocarbon, therefore C_9H_{20}
   3. fragment at m/z 113 = $M - 15$: $[R{-}CH_3]^+ \longrightarrow R^+ + {\cdot}CH_3$
   4. no significant fragment at m/z 99 (= $M - C_2H_5$): **not** $R'{-}CH_2CH_3$
   5. *m/z* 43 = $C_3H_7^+$: $(CH_3)_2CH^+$
   6. Structure is 2,6-dimethylheptane: $(CH_3)_2CH \mid CH_2 \mid CH_2 \mid CH_2 \mid CH(CH_3)_2$ (cleavages giving 85, 71, 57, 43)

Notice how the fragmentations correspond with important peaks.

5. Saturated hydrocarbons frequently present problems in mass spectral analyses because of the prevalence of peaks derived from carbocation rearrangements. This example illustrates some of these difficulties. Assignments must be based not just on the presence or absence of given *m/z* peaks, but on their relative intensities.

$(CH_3)_3C \mid CH_2CH_2CH_3$ (57) — This isomer is the only one with a *t*-butyl group and is expected to have the largest *m/z* 57 peak; therefore, b.

$(CH_3)_2CHCH(CH_3) \mid CH_2CH_3$ (71) — This isomer is expected to have the largest *m/z* 71 peak; hence, a.

$(CH_3)_2CH \mid CH_2 \mid CH(CH_3)_2$ (43, 43) — This isomer has two isopropyl ends and is expected to have the largest *m/z* 43 peak; therefore, c.

6. 1. peaks at *m/z* 90 and 92: compound contains Cl (75.8% ^{35}Cl, 24.2% ^{37}Cl)
   2. $M^+ = 90$ (92): formula is C_4H_7Cl
   3. *m/z* 55 = $M - Cl$: $R = C_4H_7$
   4. strength of *m/z* 55 peak suggests that R^+ is quite stable: $CH_3CH{=}CHCH_2^+$ or $CH_2{=}C(CH_3){-}CH_2^+$
   5. IR bands at 1650, 890 cm^{-1} confirm double bond, and suggest that it is $R_2C{=}CH_2$ (Table 17.3).
   6. Structure is 3-chloro-2-methylpropene ("methallyl chloride"): $CH_2{=}C(CH_3){-}CH_2Cl$

7. 1. The IR suggests unsaturation, possibly α,β-unsaturated or C=O next to an aromatic ring since the frequency is less than 1700 cm^{-1}.
   2. The proton NMR shows two isolated methyl groups with chemical shifts indicative of neighboring C=C or C=O.
   3. The M^+ peak indicates $C_7H_2O_3$, precluded by presence of 2 CH_3's; $C_8H_6O_2$ this is at least possible, but would leave 6 carbons with no hydrogen; $C_9H_{10}O$ seems like the most probable formula.
   4. The CMR, coupled with the molecular ion fragment of the mass spectrum, suggests a benzene ring since there are too few carbons for the mass indicated unless there 2 or more equivalent carbons. A *para*-disubstituted benzene will show four signals for six benzene carbons if the substituents are different:

$H_3C-C_6H_4-\overset{O}{\overset{\|}{C}}CH_3$ 4′-methylacetophenone is consistent with the data.

$(M+1)/M = .01119(9) + .00015(10) + .00037(1) = .10$; $M+1$ will be 10.26% the value of M.

8. a. The M^+ peak is 154, the base peak is 136, indicating a loss of mass 18, probably H_2O, indicating an alcohol.

   b. If the peak at 297 is M^+, then the molecule contains nitrogen or an odd-massed heteroatom. Not halogen or the $M+1$ peak would be larger. Consider N. Peak at 282 shows loss of mass 15, CH_3-; base peak at 254 shows loss of 43, C_3H_7.

   c. Mass 142 could be the M^+, but the next large peak, 134, would mean loss of mass 8, an impossibility, therefore 142 is not the molecular ion.

   d. The high mass, coupled with the M, $M+2$, $M+4$ peaks in the ratio of 1:2:1, indicates a dibromide. See Exercise 34.4

9. 1. $M^+ = 254$: $C_{20}H_{14}$, $C_{19}H_{26}$, or $C_{18}H_{38}$
   2. weak M^+ suggests saturated hydrocarbon: $C_{18}H_{38}$
   3. Compound A: major fragments at m/z 239 ($M - 15 = M - CH_3$), 155 ($M - 99 = M - C_7H_{15}$), and 127 ($M - 127 = M - C_9H_{19}$) account for all but one carbon. Since fragmentation of saturated hydrocarbons favors cleavage at branch points, these prominent peaks suggest the structure of 8-methyl-heptadecane:

$-C_9H_{19}^{\cdot}$ → m/z 127

m/z 254

$-C_7H_{15}^{\cdot}$ → m/z 155

   4. Compound B: similarly, the fragments at m/z 239 ($M - CH_3$), 169 ($M - C_6H_{13}$), and 113 ($M - C_{10}H_{21}$) are consistent with 7-methylheptadecane:

$$[\text{C}_{17}\text{H}_{36}]^{+\bullet}\ (m/z\ 254) \xrightarrow{-\text{C}_{10}\text{H}_{21}^{\bullet}} m/z\ 113 \qquad \xrightarrow{-\text{C}_6\text{H}_{13}^{\bullet}} m/z\ 169$$

10. 1. The largest ion (101) has odd mass, and most of the large fragments (m/z = 86, 70, 58) are of even mass, suggesting that the compound contains an odd number of nitrogens.
2. For $M^+ = 101$, a formula of $C_6H_{15}N$ can be proposed.
3. Major fragment is $M - 15$ peak, resulting from loss of a methyl group.
4. If compound is a simple amine, $M - 15$ would be expected to result from α-cleavage:

$$\left[-\ddot{N}-C-CH_3\right]^{+\bullet} \longrightarrow \overset{+}{N}{=}C\langle \; + \;\cdot CH_3$$

5. Absence of other major fragments suggests that only **one** mode of α-cleavage is possible. Four structures are possible with this property:

6. If there are only two resonances in the CMR spectrum, *t*-butylethylamine and *t*-butyldimethylamine are excluded.

11. 1. IR band at 1710 cm^{-1}: ketone
2. $M^+ = 100$: $C_6H_{12}O$
3. m/z 85: $[RCOCH_3]^+ \longrightarrow RC{\equiv}O^+ + \cdot CH_3$
4. m/z 43: $[RCOCH_3]^+ \longrightarrow R\cdot + {}^+O{\equiv}CCH_3$
5. m/z 58: even mass fragment must arise from two bond cleavage process; for a ketone this is most likely the McLafferty rearrangement; the fragment lost (100 – 58 = 42) corresponds to $CH_3CH{=}CH_2$, and could arise from either 2-hexanone and 4-methyl-2-pentanone:

$$[M^+ = 100] \longrightarrow [m/z\ 58]^{+\bullet} + CH_3CH{=}CH_2 \longleftarrow [M^+ = 100]$$

6. To distinguish between these two possibilities is not easy. The fact that the *m/z* 43 fragment is so intense suggests that the structure is 4-methyl-2-pentanone, since this molecule can give rise to fragments of *m/z* 43 in two ways:

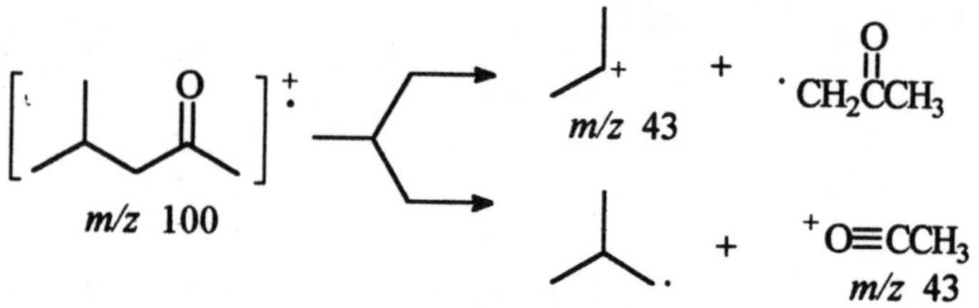

m/z 100 → *m/z* 43 + $\cdot CH_2CCH_3$ (C=O); → isopropyl radical + $^+O{\equiv}CCH_3$ *m/z* 43

12. 1. IR band at 1710 cm^{-1}: ketone
 2. $M^+ = 114$: $C_7H_{14}O$
 3. odd mass fragments of *m/z* 71 and 43 cannot be the two α-cleavage fragments $R{-}C{\equiv}O^+$ and $^+O{\equiv}C{-}R'$, because they don't account for enough carbons (the mass of the two α-cleavage fragments must add up to $M + 28$)
 4. *m/z* 71 must be **both** α-cleavage fragments: ketone is $C_3H_7\overset{O}{\overset{\|}{C}}C_3H_7$
 5. There are **no** even mass peaks, therefore no fragments arising from McLafferty rearrangement
 6. Structure must therefore be diisopropyl ketone:

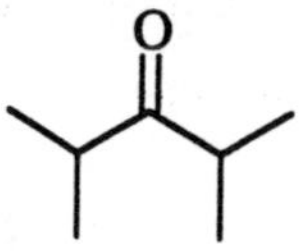

13. 1. IR band at 3400 cm^{-1}: alcohol
 2. $M^+ = 126$ suggests formula of $C_8H_{13}OH$, with two degrees of unsaturation
 3. *m/z* 111 ($M - CH_3$) and 69 ($M - C_4H_9$) suggest 3-methyl-1-heptyn-3-ol as structure:

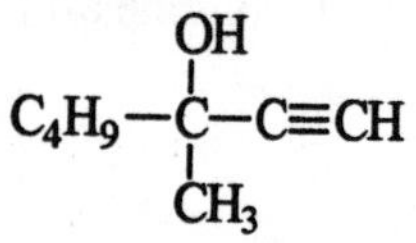

$$C_4H_9{-}\underset{CH_3}{\overset{OH}{C}}{-}C{\equiv}CH$$

 4. But where does *m/z* 87 come from? And why isn't there a strong fragment at *m/z* 108 for loss of H_2O?
 5. Assume that *m/z* 126 is **not** M^+, but represents $M - H_2O$ instead; $M = 144$ would then be a saturated alcohol: $C_9H_{19}OH$
 6. *m/z* 87 ($M - C_4H_9$) is α-cleavage fragment: part structure = R_2C^+OH
 7. *m/z* 69 cannot be an α-cleavage product, because no formula with $C_nH_{2n+1}O = 69$)
 8. with a single α-cleavage product observed, the structure must be $C_4H_9CHOHC_4H_9$
 9. *m/z* 111 $= M - H_2O - CH_3$ suggests that there are methyl branches
 10. absence of *m/z* 97 for $M - H_2O - nC_2H_5$ suggests that there are not any ethyl branches
 11. most reasonable structure is 2,6-dimethyl-4-heptanol: OH
 12. The *m/z* 69 peak is left unexplained, but must result from a more deep-seated rearrangement. This is also a loose end, but the second hypothesis explains much more of the data than the first. Notice how the **absence** of expected peaks can be more significant than the presence of additional ones.

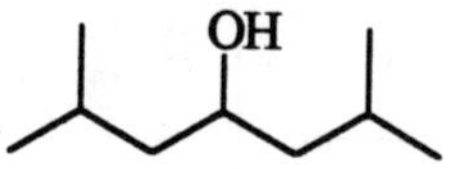

14. 2-Octanone is $C_8H_{16}O$ and has a molecular weight of 128: 113 = 128 − 15 (CH_3), 43 = 128 − 85 (C_6H_{13}):

$CH_3(CH_2)_5$–C(=O)–CH_3 (cleavage labels: 43, 85, 113)

McLafferty rearrangement gives: $[C_3H_7CH(H)–CH_2–CH_2–C(=O)–CH_3]^{+\cdot}$ → $[H_2C{=}C(OH)CH_3]^{+\cdot}$ *m/z* 58

15. 1. Odd M^+, even fragments: an amine
2. $M^+ = 73$: $C_4H_{11}N$
3. *m/z* 58 = *M* − 15: loss of CH_3 by α-cleavage $CH_3–\overset{+}{C}–NR_2 \longrightarrow CH_3^{\cdot} + {>}C{=}\overset{+}{N}R_2$
4. *m/z* 44 = *M* − 29: loss of CH_3CH_2 by α-cleavage $CH_3CH_2–\overset{+}{C}–NR_2 \longrightarrow CH_3H_2^{\cdot} + {>}C{=}\overset{+}{N}R_2$
5. structure must be 1-methylpropanamine: $CH_3CH_2CH(NH_2)CH_3$

16.

		ΔH° (kcal mole^{-1})
$[(CH_3)_2CHCH_2CH_3]^{+\cdot}$ →	$CH_3^+ + (CH_3)_2CHCH_2\cdot$	77
	$CH_3\cdot + (CH_3)_2CHCH_2^+$	38
	$(CH_3)_2CH^+ + CH_3CH_2\cdot$	14
	$(CH_3)_2CH\cdot + CH_3CH_2^+$	38
	$CH_3\cdot + CH_3{}^+CHCH_2CH_3$	25
	$CH_3^+ + CH_3\dot{C}HCH_2CH_3$	76

$(CH_3)_2CH^+ > CH_3{}^+CHCH_2CH_3 > (CH_3)_2CHCH_2^+ \sim CH_3CH_2^+ >> CH_3^+$

34.F. Supplementary Problems

S1. Predict the major peaks in the mass spectra of the following compounds:

a. $CH_3CH_2CH(CH_3)CH_2CH_2CH(CH_3)_2$

b. $(CH_3)_2CHCH(OH)CH_2CH_3$

c. $(CH_3)_3CCH_2CH_2CH_2CH_2OH$

d. $CH_3C(=O)CH_2CH_2CH(CH_2CH_3)_2$

e. $(CH_3)_2CHCH_2OC(CH_3)_3$

f. cyclopentyl$-CH_2C(=O)CH_3$

g. $CH_3CH_2CH_2CH(CH_3)N(CH_3)_2$

h. phenyl$-CH_2CH_2CH_2CH_3$

S2. From the following accurate mass measurements, determine the most likely formula for the molecule or fragment.

a. $m/z = 70.0419$ b. $m/z = 56.0373$ c. $m/z = 81.0861$

S3. Deduce the structure of each of the compounds below, and write a mechanism showing the principal fragments in each mass spectrum.

a. $A \xrightarrow[2.\ H_2O_2,\ OH^-]{1.\ B_2H_6} B + C$ Both B and C show a strong band at 1710 cm^{-1} in the IR.

mass spectrum of B: m/z 114, 86, 85, 57
mass spectrum of C: m/z 114, 99, 58, 43

b. $D \xrightarrow{(C_6H_5)_3P{=}CH_2} E$; $E \xrightarrow[2.\ NaBH_4]{1.\ Hg(OAc)_2,\ H_2O} F$; $E \xrightarrow[\text{dilute } H_2SO_4]{} G$

mass spectrum of D: m/z 86, 58, 29
mass spectrum of F: m/z 102, 87, 84, 45
mass spectrum of G: m/z 87, 84, 73

c. $H \xrightarrow{C_3H_7MgCl} \xrightarrow[\Delta]{H_2SO_4} I \xrightarrow[2.\ H_2O_2,\ OH^-]{1.\ B_2H_6} J \xrightarrow[H_2SO_4]{K_2Cr_2O_7} K$

NMR spectrum of H: δ 0.9 (3H,t), 1.2-1.4 (4H,m), 2.1 (2H,dt), 9.5 (1H, t).

mass spectrum of K: m/z 128, 86, 85, 71

d. $L \xrightarrow[H_2SO_4]{HgSO_4} M \xrightarrow[H_2/Ni]{NH_3} N \xrightarrow[HCO_2H,\ HCl]{CH_2{=}O} O$

mass spectrum of L: m/z 82
IR spectrum of M: 1710 cm^{-1}
mass spectrum of M: m/z 100, 72, 71, 57
mass spectrum of N: m/z 101, 72, 58
mass spectrum of O: m/z 129, 100, 86

34.G. Answers to Supplementary Problems

S1. a. CH_3CH_2–CH(CH_3)–CH_2CH_2–CH(CH_3)–CH_3 (fragments: 99, 57, 43, 113)
$M^+ = 128$

b. $(H_3C)_2$CH–CH(OH)–CH_2CH_3 (fragments: 73, 59)
$M^+ = 102$; $M - H_2O = 84$

c. CH_3–C(CH_3)$_2$–$CH_2CH_2CH_2$–CH_2OH (fragments: 57, 115, 31)
$M^+ = 130$; $M - H_2O = 112$

d. H_3C–CO–CH_2CH_2–CH(CH_2CH_3)$_2$ (fragments: 71, 127, 113, 43) —McLafferty→ $[CH_3C(OH)=CH_2]^{+\cdot}$ (m/z 58) + CH_2=C(CH_2CH_3)$_2$
$M^+ = 142$

e. $(H_3C)_2$CH–CH_2O–C(CH_3)$_3$ (fragments: 87, 11)
($M^+ = 130$); M - (ROH) = 56

f. cyclopentyl–CH_2–CO–CH_3 (fragments: 111, 43) —McLafferty→ cyclopentene + $[H_2C=C(OH)CH_3]^{+\cdot}$ (m/z 58)
$M^+ = 126$

g. $CH_3CH_2CH_2$–CH(CH_3)–N(CH_3)$_2$ (fragments: 100, 72)
$M^+ = 115$

h. C_6H_5–CH_2–CH_2–CH_2–CH_3 (fragment: 91)
$M^+ = 134$

S2. a.

Formula	Calculated Exact Mass
C_5H_{10}	70.07825
C_4H_6O	70.04186 ←
$C_3H_6N_2$	70.05305
$[C_4H_8N]\cdot$	70.06565

b.

Formula	Calculated Exact Mass
C_4H_8	56.0626
C_3H_4O	56.0262
$C_2H_4N_2$	56.0374 ←
$[C_3H_6N]\cdot$	56.0500

c.

Formula	Calculated Exact Mass
$[C_5H_{11}]\cdot$	81.08608 ←
$[C_4H_7O]\cdot$	81.04969
$[C_3H_7N_2]\cdot$	81.06088
C_4H_9N	81.07348

S3. a. $CH_3C\equiv CCH(CH_3)CH_2CH_3$ (A) $\xrightarrow[\text{2. } H_2O_2,\ OH^-]{\text{1. } B_2H_6}$ $CH_3CH_2-C(=O)-CH(CH_3)CH_2CH_3$ (B), $M^+ = 114$

C: $CH_3-C(=O)-CH(CH_3)CH_2CH_3$ (cleavage fragments 43, 99), $M^+ = 114$

McLafferty: $[CH_3CH_2C(OH)=CHCH_3]^{+\cdot}$ *m/z* 86; $[CH_3C(OH)=CH_2]^{+\cdot}$ *m/z* 58

b. $CH_3CH_2CH(CH_3)-CH=O$ (fragment 29) $\xrightarrow{Ph_3P=CH_2}$ $CH_3CH_2CH(CH_3)CH=CH_2$ $\xrightarrow[\text{2. } NaBH_4]{\text{1. } Hg^{2+},\ H_2O}$ $CH_3CH_2CH(CH_3)-CH(OH)-CH_3$ (fragments 87, 43)

M+ = 102; $M - H_2O = 84$

$CH_3CH_2CH(CH_3)CH=CH_2$ $\xrightarrow{\text{dilute } H_2SO_4}$ $[CH_3CH_2CH(CH_3)\overset{+}{C}HCH_3 \longrightarrow CH_3CH_2\overset{+}{C}(CH_3)CH_2CH_3]$ $\xrightarrow{H_2O}$ $CH_3CH_2-C(CH_3)(OH)-CH_2CH_3$ (fragments 73, 87)

McLafferty: $[CH_3CH=CHOH]^{+\cdot}$ *m/z* 58

$M - H_2O = 84$
($M^+ = 102$ not seen)

c. $CH_3CH_2CH_2CH_2CHO \xrightarrow{C_3H_7MgCl} \xrightarrow[\Delta]{H_2SO_4} CH_3CH_2CH_2CH_2CH{=}C(CH_3)_2 \xrightarrow[2.\ H_2O_2,\ OH^-]{1.\ B_2H_6}$

H   I

$CH_3CH_2CH_2CH_2CH_2CH(OH)CH(CH_3)_2 \xrightarrow[H_2SO_4]{K_2Cr_2O_7} CH_3CH_2CH_2CH_2 \mid C(=O) \mid CH(CH_3)_2$

J   K (fragments 85, 71) $M^+ = 128$

McLafferty: $[CH_2{=}C(OH)CH(CH_3)_2]^{+\cdot}$ *m/z* 86

d. $CH_3CH_2C{\equiv}CCH_2CH_3 \xrightarrow[H_2SO_4]{HgSO_4} CH_3CH_2 \mid C(=O) \mid CH_2CH_2CH_3 \xrightarrow[H_2/Ni]{NH_3}$

L   M (fragments 71, 57) $M^+ = 100$

McLafferty: *m/z* 72

$CH_3CH_2 \mid CH(NH_2) \mid CH_2CH_2CH_3 \xrightarrow{CH_2{=}O\ ,\ HCO_2H} CH_3CH_2 \mid CH(N(CH_3)_2) \mid CH_2CH_2CH_3$

N (fragments 72, 58) $M^+ = 101$ O (fragments 100, 86) $M^+ = 129$

36. SPECIAL TOPICS

36.A. Chapter Outline and Important Terms Introduced

36.1 Transition Metal Organometallic Compounds

A. Structure
   - ligand
   1. Lewis Acid Association-Dissociation
   2. Lewis Base Association-Dissociation
   3. Oxidative Addition-Reductive Elimination
   4. Insertion-Deinsertion

coordinatively saturated
coordinatively unsaturated
decarbonylation

36.2 Organic Coloring Matters

A. Color

B. Natural Coloring Matters
   - anthocyanins
   - naphthoquinones, anthraquinones
   - ommochromes
   - porphyrins
   - carotenoids
   - melanins
   - pterins
   - indigoids

C. Dyes and Dyeing
   - vat dyes
   - direct dyes
   - chromophore
   - triphenylmethane dyes
   - indigoid dyes
   - phthalocyanines
   - mordant dyes
   - disperse dyes
   - azo dyes
   - anthraquinone dyes
   - azine dyes

36.3 Photochemistry

A. Electronically Excited States
   - single vs. triplet state
   - vertical transition = Franck Condon transition
   - internal conversion
   - fluorescence
   - phosphorescence
   - Jablonski crossing
   - intersystem crossing
   - triplet energy transfer

B. Photochemical Reactions
   - Norrish type I and II processes
   - triplet sensitizer
   - quantum yield

36.4 Polymer Chemistry

monomers

addition vs. condensation polymers

A. Carbocation Polymerization of Alkenes

B. Free Radical Vinyl Polymerization
   - branched chain vs. linear polyethylene
   - plasticizer
   - cross-linking

C. Anionic and Organometallic Polymerization
   - Ziegler-Natta catalyst
   - isotactic vs. atactic vs. syndiotactic

D. Diene Polymers
   - copolymer vs. homopolymer

E. Condensation Polymers
  nylon
  polyurea

polyurethane

36.5 Natural Products: Terpenes, Steroids, and Alkaloids

metabolism

primary and secondary metabolites

A. Terpenes
  essential oils

mono-, sesqui-, di- and triterpenes

B. Steroids
  estrogens and androgens

cardiac glycosides

C. Alkaloids

36.D Answers to (Selected) Exercises

36.1 a. To satisfy the 18-electron rule, Cr (6 electrons) needs six CO ligands: $Cr(CO)_6$. Fe (8 electrons) needs five CO ligands: $Fe(CO)_5$.

b. $^{-}Mn(CO)_5$ and $^{-}Co(CO)_4$ each have 18 electrons in the valence shell of the metal.

36. 2 The steps involved in this reaction are:

$$Ni(CO)_4 \rightleftharpoons Ni(CO)_3 + CO$$

$$Ni(CO)_3 + (C_6H_5)_3P \rightleftharpoons (C_6H_5)_3PNi(CO)_3$$

The rate expression for formation of product is: $dP/dt = k[Ni(CO)_3][R_3P]$. $Ni(CO)_3$ is in equilibrium with $Ni(CO)_4$, and its concentration is related to that of $Ni(CO)_4$ and CO as shown by the equilibrium constant

$$K = \frac{[Ni(CO)_3][CO]}{[Ni(CO)_4]}.$$

Combining these two expressions gives the following equation: $\frac{dP}{dt} = kK\frac{[Ni(CO)_4][R_3P]}{[CO]}$

and the inverse dependence of rate on CO concentration can be seen. (The concentration of CO in solution is proportional to the pressure of CO in contact with the solution).

36.3

$HCo(CO)_4$ (18 e$^-$) $\xrightarrow[\text{Lewis Base dissociation}]{-CO}$ $HCo(CO)_3$ (16 e$^-$) $\xrightarrow[\text{Lewis Base association}]{CH_2{=}CH_2}$ $CH_2{=}CH_2 \rightarrow HCo(CO)_3$ (18 e$^-$) $\xrightarrow{\text{Insertion}}$ $CH_3CH_2{-}Co(CO)_3$ (16 e$^-$)

$CH_3CH_2{-}Co(CO)_3$ (16 e$^-$) $\xrightarrow[\text{Lewis Base association}]{CO}$ $CH_3CH_2{-}Co(CO)_4$ (18 e$^-$) $\xrightarrow{\text{Insertion}}$ $CH_3CH_2C(=O){-}Co(CO)_3$ (16 e$^-$) $\xrightarrow[\text{Oxidative addition}]{H_2}$ $CH_3CH_2C(=O){-}CoH_2(CO)_3$ (18 e$^-$)

$\xrightarrow{\text{Reductive elimination}}$ CH_3CH_2CHO + $HCo(CO)_3$ (16 e$^-$) $\xrightarrow[\text{Lewis Base association}]{CO}$ $HCo(CO)_4$ (18 e$^-$)

36.4 $[(C_6H_5)_3P]_3RhCl = L_3RhCl$ = Wilkinson's catalyst $CH_3CH_2CH_2CH_2COCl$ = RCOCl

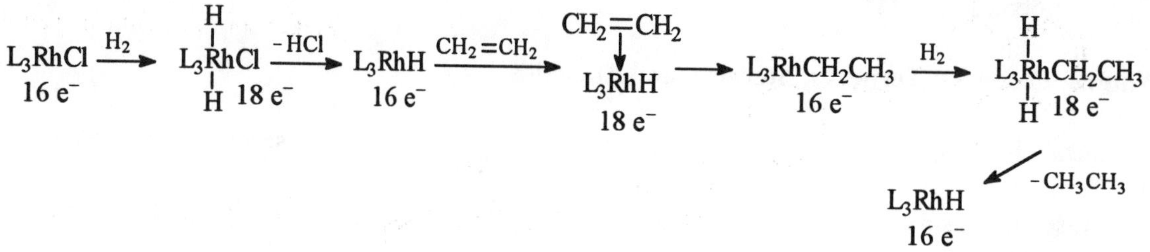

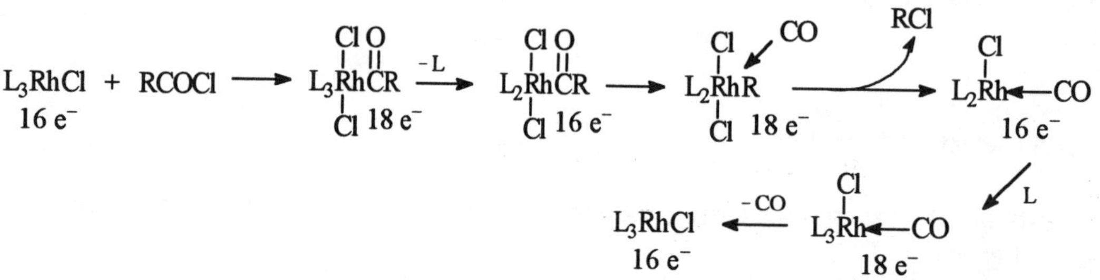

36.5

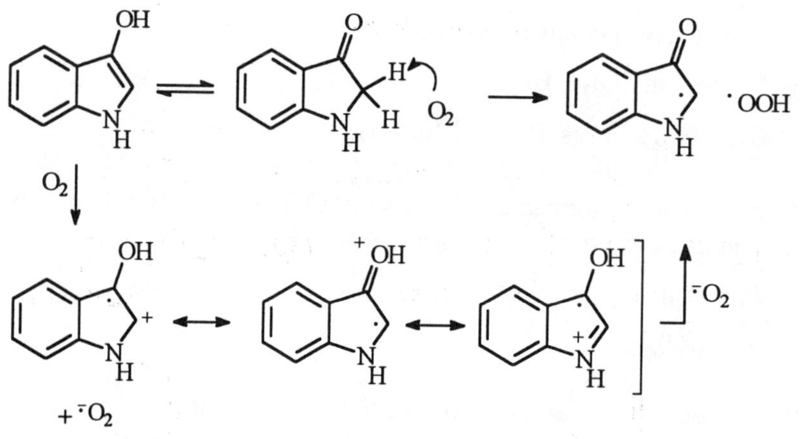

3-hydroxyindole, indoxyl
Very good electron donor

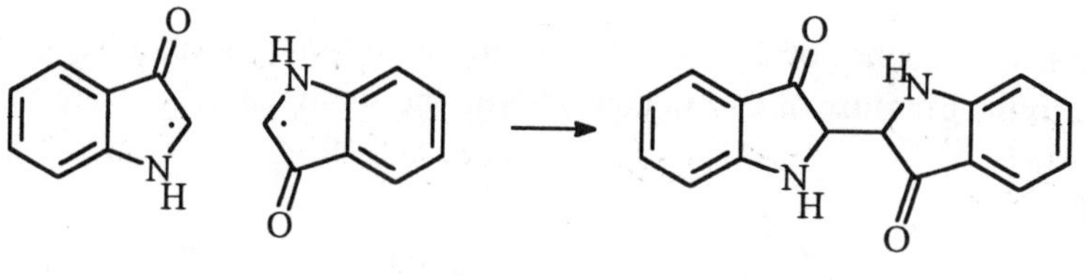

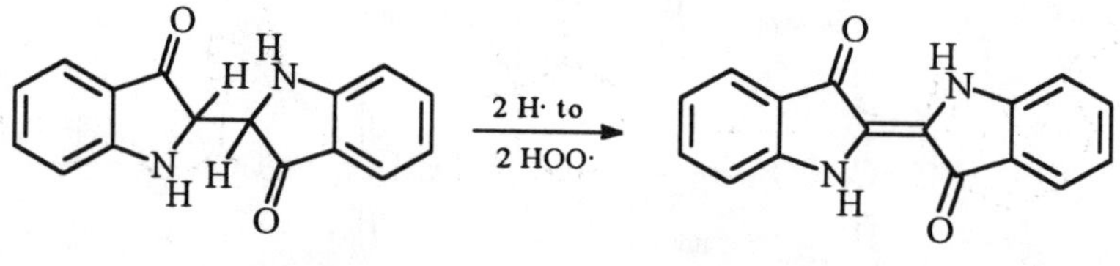

leuco (reduced) form of indigotin — indigotin, **indigo**

36.6

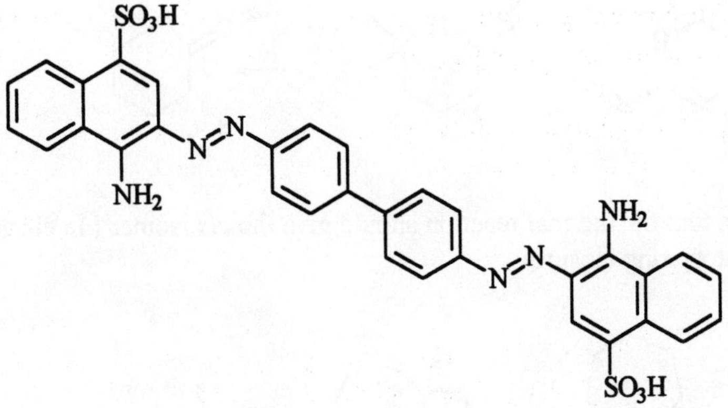

SO_3H, NH_2, N=N, N=N, NH_2, SO_3H

neutral solution: *red*
acidic solution: *blue-green*

36.7

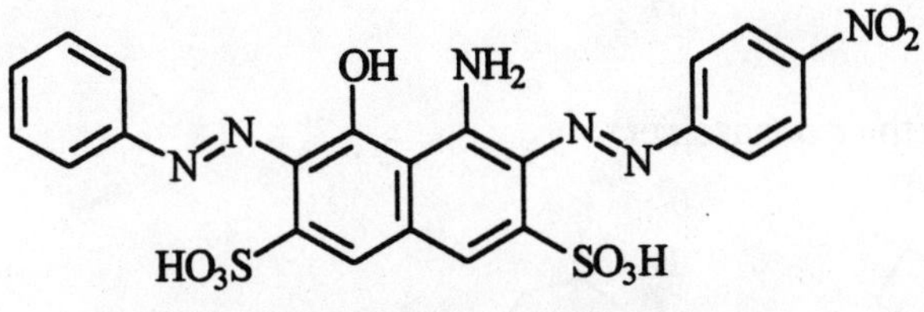

OH, NH_2, NO_2, N=N, N=N, HO_3S, SO_3H

Naphthol Blue Black H

36.8

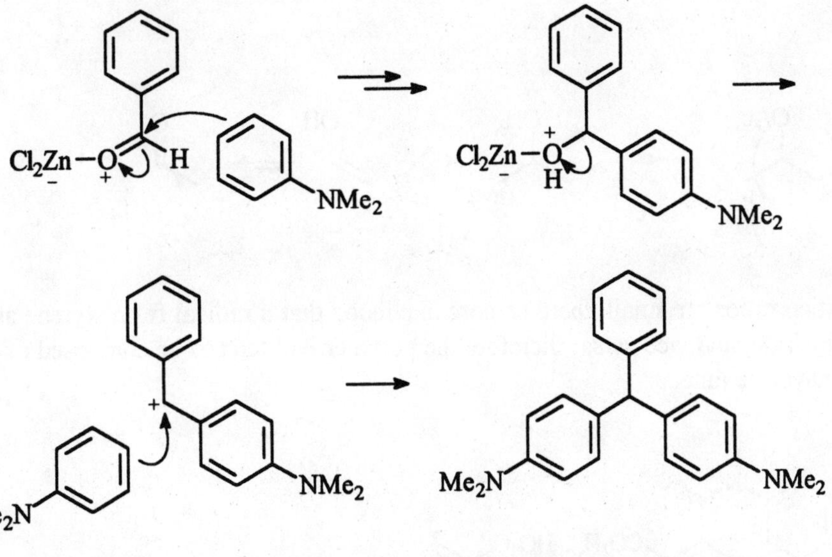

Cl_2Zn-O^+, H, NMe_2, Cl_2Zn-O^+, H, NMe_2

NMe_2, Me_2N, Me_2N, NMe_2

36.9

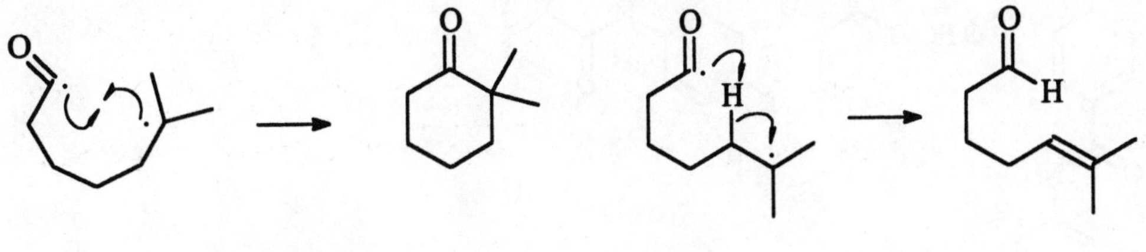

O, O, O, H, O, H

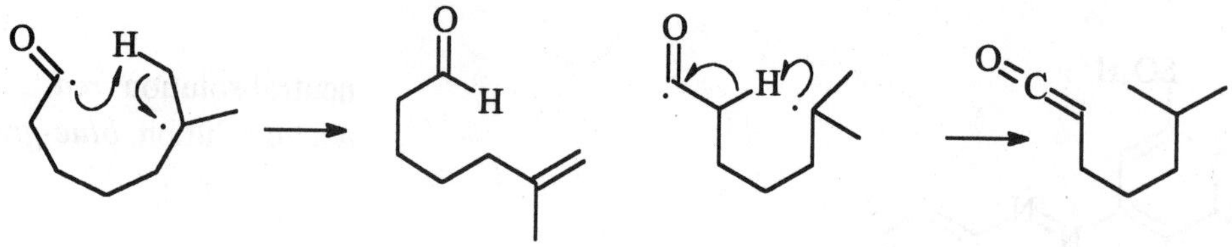

36.10 Molecular orbital theory predicts that the thermal reaction should give the *cis* isomer (4n electrons, Möbius transition state required, conrotatory ring closure):

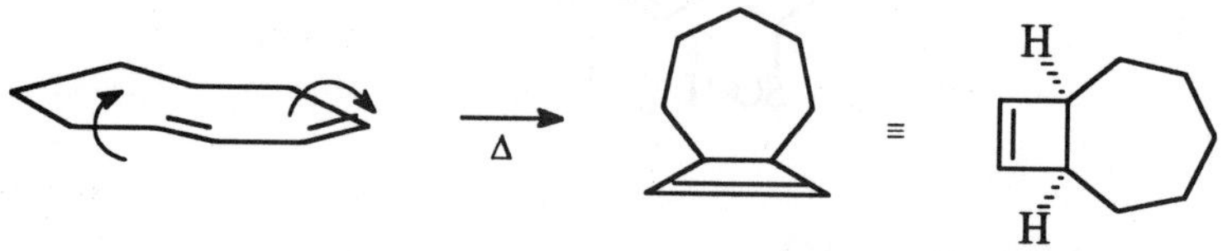

The excited state photochemical process gives the *trans* isomer:

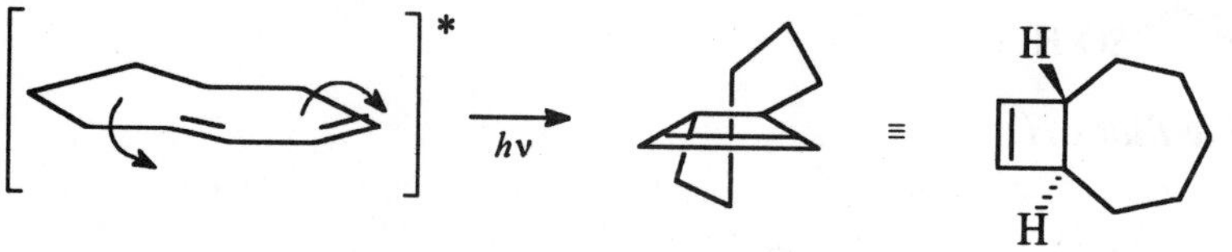

36.11

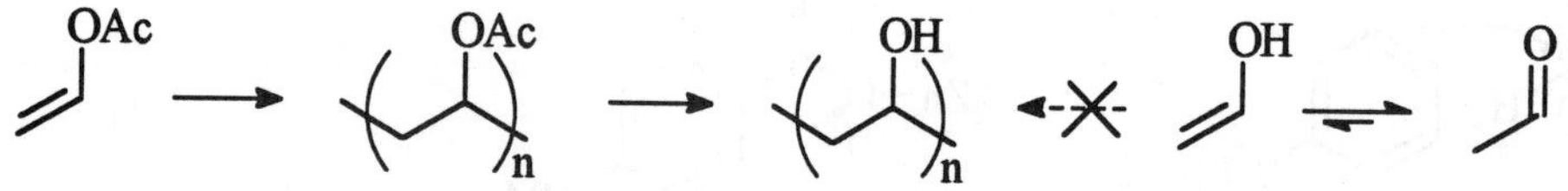

36.12 If both copolymerization ratios are small, there is more likelihood that a radical from styrene addition will react with maleic anhydride, and vice versa; therefore the polymer will tend to be composed of alternating styrene and maleic anhydride units.

36.13

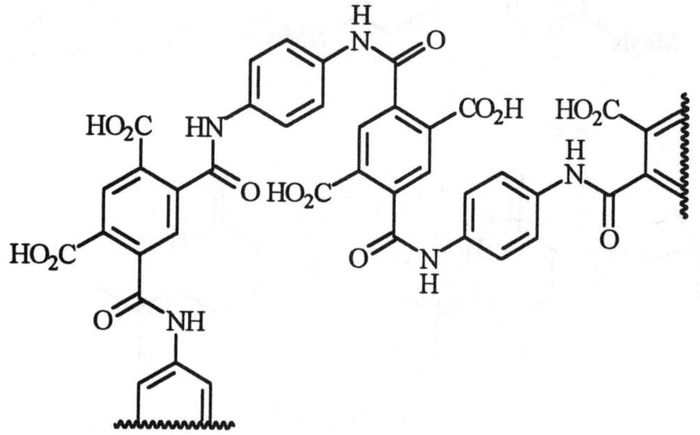

36.14

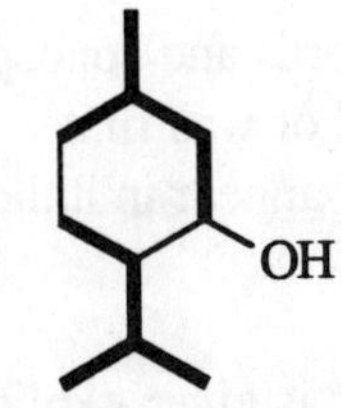

menthol 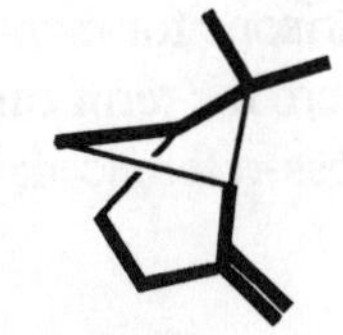 β-pinene 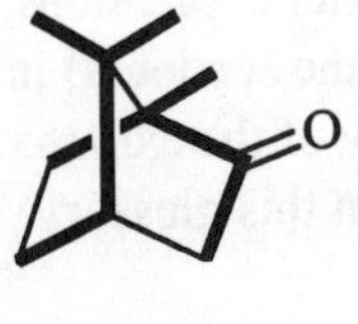 camphor 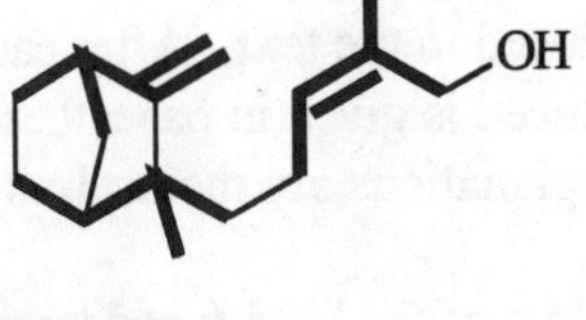 β-santalol

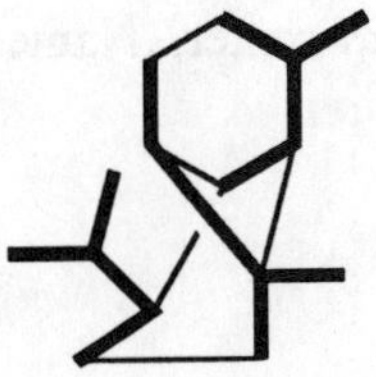

copaene 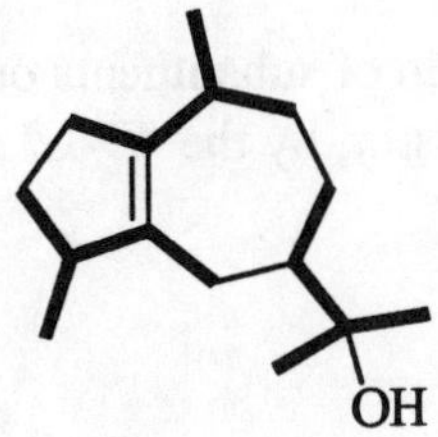 guaiol 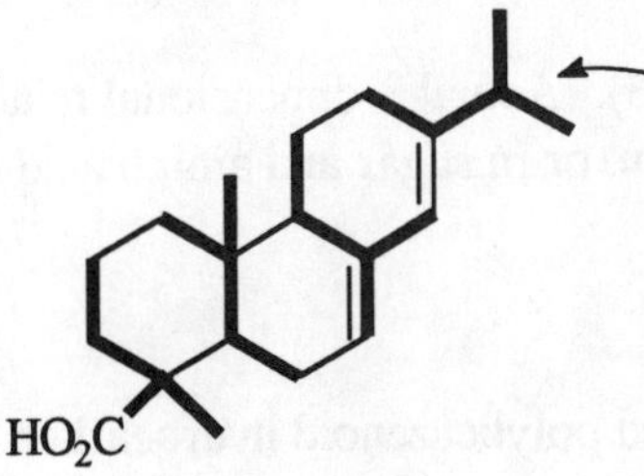 abietic acid

this unit actually arises from rearrangement of:

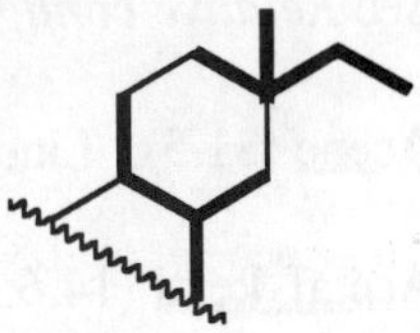

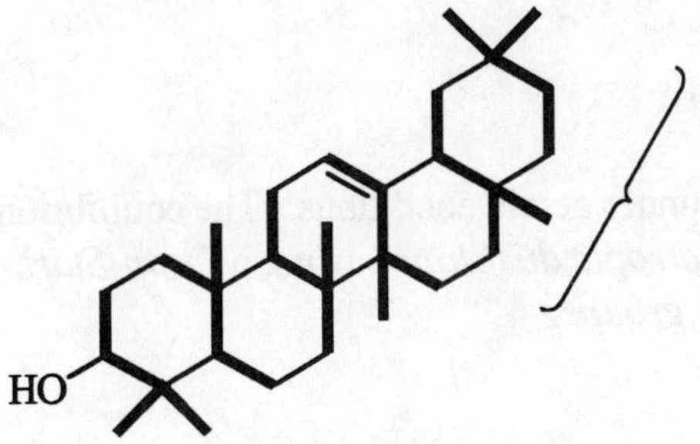

a number of rearrangements have occurred in this region during biosynthesis

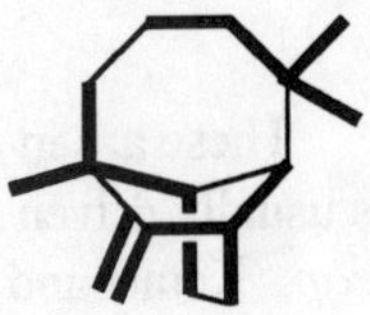

β-amyrin

Longifolene

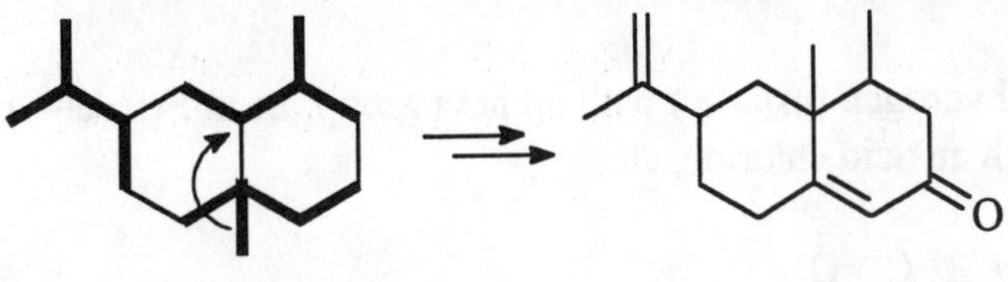

nootkatone

36.15

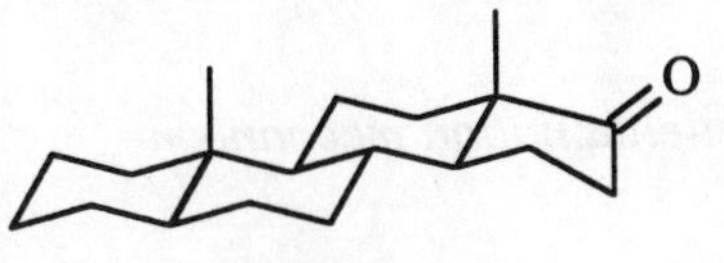

Andosterone

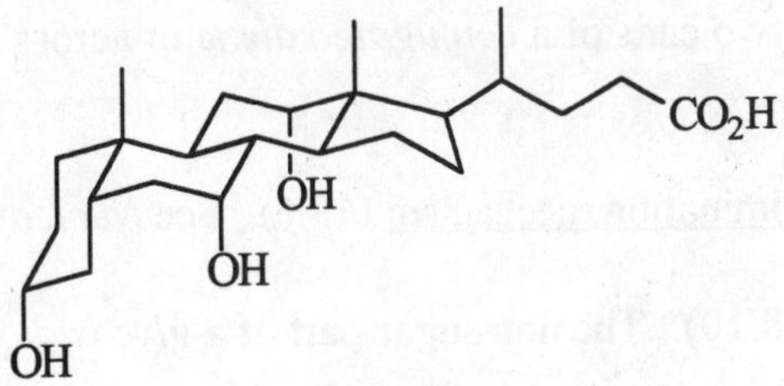

Cholic acid

APPENDIX I: GLOSSARY

This glossary provides definitions, brief explantions, and comparisons for many of the words and concepts introduced in the text. After each entry, the section(s) in the text where the term can be found or was first introduced is given in parentheses. Many of the explanations are cross-referenced; terms that appear in italics in the explanations are themselves defined in this glossary.

Most of the bold-faced terms in the text have been included, except those whose definition is given explicitly in the text and that you can find directly through the index. This glossary also does not include name reactions, which again you can find through the index in the text.

Absolute configuration (7.3). Actual 3-dimensional relationship of substituents on a *chiral* center. Named according to the *R-S convention*, or in sugar and amino acid chemistry, by the *"D-L" convention.*

See *Relative configuration.*

Acene (31.5). Linear, fused polybenzenoid hydrocarbon.

Acetal, ketal (14.6.B).

$$\mathrm{R{-}\overset{\displaystyle OR''}{\overset{|}{\underset{\underset{\displaystyle R'}{|}}{C}}}{-}OR''}$$

acetal: R,R″ = alkyl, R′ = H
ketal: R,R′,R″ = alkyl

These are in equilibrium with the aldehyde or ketone plus alcohol only under acidic conditions. The equilibrium is usually driven to the side of acetal or ketal by removing the water by *azeotropic distillation* using a *Dean-Stark trap*. Ketals and acetals are stable to base and are often used as *protecting groups*.

Achiral (7.1). See *Chiral.*

Activating (23.5). An activating substituent on an aromatic ring increases the electron density of the ring so that it undergoes electrophilic aromatic substitution more readily.

Acylation (23.4.A). The replacement of a hydrogen (usually) with an acyl group, as in Friedel-Crafts acylation of aromatic rings or acylation of an alcohol with an acid chloride, etc.

Acylium ion (19.6, 23.4.A). $[R{-}C{\equiv}O^+ \longleftrightarrow R{-}C^+{=}O]$

1,2-Additions vs. 1,4-additions (conjugate additions) (20.3.A). Additions across one π-bond are called *1,2-additions*, regardless of the actual numbering scheme of the molecule. Addition across a *conjugated* system, for example to the two ends of a *conjugated diene* or across a *conjugated enone*, are called *conjugate* or *1,4-additions*.

Addition-elimination mechanism (19.6). See *Nucleophilic addition-elimination mechanism.*

Aglycon (28.10). The non-sugar part of a *glycoside.*

Alditol (28.5.D). The polyalcohol resulting from reduction of the carbonyl group of a sugar.

<u>Aldonic acid</u> (28.5.E). The monocarboxylic acid resulting from oxidation of the aldehyde carbon of an *aldose*.

<u>Alkylation</u> (23.4.B). The replacement of a hydrogen (usually) with an alkyl group, as in Friedel-Crafts alkylation of an aromatic ring, or alkylation of a ketone via its enolate, etc.

<u>Allyl, allylic</u> (20.1). An *allylic system* is one involving 3 adjacent, overlapping p-orbitals, making 3 molecular orbitals, and filled with 2 *(allyl cation)*, 3 *(allyl radical)*, or 4 *(allyl anion)* electrons. It is an example of *conjugation*.

An *allylic position* is one next to a double bond.

Reactions of *allyl systems* are said to proceed with *allylic rearrangement* if the double bond moves in the course of the reaction.

<u>Amphoteric</u> (29.1, 29.3). Having both acidic and basic character in the same molecule.

<u>Angle of rotation</u>, α (7.2). The amount by which plane polarized light is rotated by an *optically active* sample.

<u>Anisotropic</u> (7.2, 13.10, 21.3.A). Not having the same effect in all directions. *Diamagnetic anisotropy* refers to the *deshielding* effect of electron density that is adjacent to but not surrounding the nucleus being observed in the NMR. (See *chemical shift* and *ring current*)

<u>Anomers, anomeric</u> (28.3). *Anomers* are stereoisomeric sugars that differ in configuration only at the *hemiacetal* or *hemiketal* carbon, also known as the *anomeric carbon*.

<u>Antarafacial</u> (22.5.B). On opposite faces of a π system. See *suprafacial*.

<u>Anti, Gauche, Syn, Eclipsed, Staggered</u> (5.2). These are all terms used to describe three-dimensional relationships of substituents at opposite ends of a single bond.

When the substituents at one end are "lined up" with the substituents at the other end, the molecule is said to have an eclipsed conformation of the single bond. The molecules drawn below are shown in eclipsed conformations.

Eclipsed

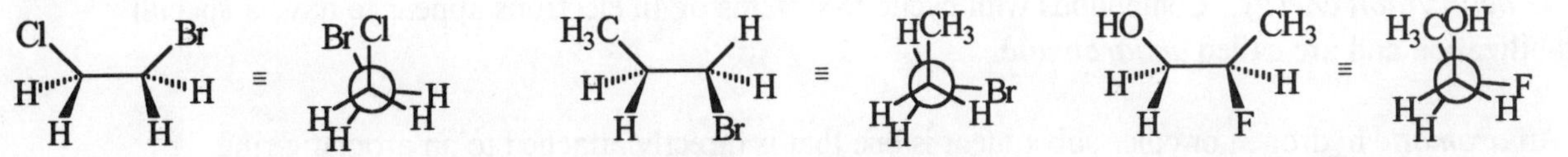

<u>A</u> <u>B</u> <u>C</u>

In the eclipsed conformation <u>A</u>, the chlorine and bromine atoms are *syn* to each other (coplanar and "pointing in the same direction"). Similarly, the methyl and the hydroxy group in the eclipsed conformation <u>C</u> are *syn*.

When the substituents at one end are "in between" those at the other end, the molecule is in a *staggered* conformation, as shown for the molecules below. The staggered conformations are more stable (lower potential energy) than the eclipsed.

Staggered

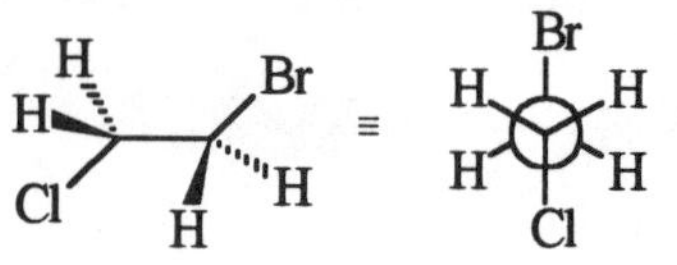

D

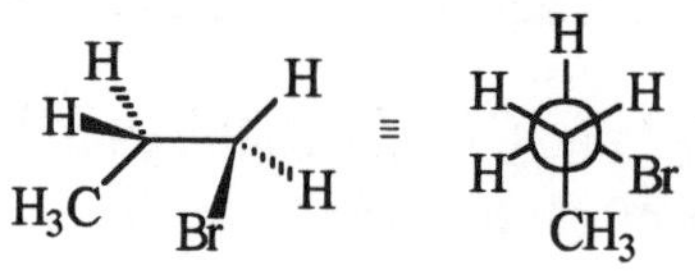

E

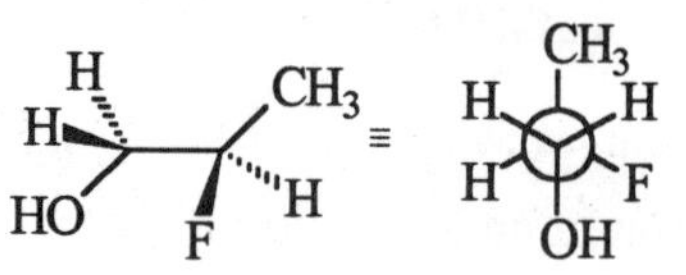

F

In the staggered conformations D and F, the bromine and the chlorine, and the methyl and hydroxy groups, respectively, are *anti* to each other (coplanar, but "pointing in opposite directions"). In the staggered conformation E, the bromine and the methyl group are said to be *gauche* to each other (dihedral angle = 60°); the staggered conformation F has a *gauche* relationship between the fluorine and hydroxy substituents.

Antiaromatic (22.3.A). See *Aromatic.*

Antibonding orbital (2.7). See *Orbital.*

Anti-Markovnikov (11.6.C). An orientation opposite to that predicted by *Markovnikov's rule* for electrophilic addition to an alkene. It is the overall result of the free-radical addition of HBr to an alkene or alkyne, or of their hydration via hydroboration/oxidation (11.6.D).

Annulene (22.3.E). A monocyclic $(CH)_n$ hydrocarbon, for example cyclodecapentaene is [10]annulene.

Apical, Equatorial (26.11.A). In the trigonal bipyramidal conformation of pentavalent phosphorus and silicon compounds, the two substituents that are colinear with the central atom are *apical*, and the three others that are coplanar with the central atom are *equatorial*. These positions are interconverted by *pseudorotation*. The favored conformations are those in which the more electronegative substituents occupy the *apical* positions and the less electronegative groups are *equatorial*.

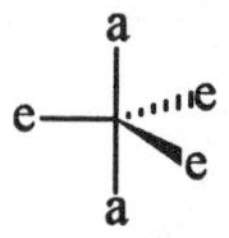

a = apical
e = equatorial

Aromatic (21.1.A, 21.1.C, 22.3). *Aromatic* compounds are those that have a special stabilization of their electronic systems due to a cyclic arrangement of 4n + 2 π-electrons. See *Hückel 4n + 2 rule, resonance energy,* and *delocalization energy.* Compounds with cyclic π-systems of 4n electrons appear to have a special destabilization and are called *antiaromatic.*

An *aromatic* hydrogen or other substituent is one that is directly attached to an aromatic ring.

Aromatization (31.3.B). Formation of an *aromatic* ring from a less unsaturated precursor, usually by a *dehydrogenation* process.

Asymmetric center or carbon (7.1). See *Stereocenter.*

Asymmetric induction (7.8). *Asymmetric induction* is said to occur when a *chiral center* is formed with a preference for one *absolute configuration* over the other.

Atomic orbital (2.5). See *Orbital.*

Autocatalytic (15.1.D). If one of the products of a reaction is a catalyst for the reaction, the reaction is said to be *autocatalytic*. Often such reactions exhibit an *induction period.*

Autoxidation (10.10, 14.8.A). Reaction with atmospheric oxygen, usually to generate an *alkylperoxide*, ROOH, or a peroxycarboxylic acid, RCO_3H.

Average bond energy (6.5). The average energy required to dissociate each of the bonds of a certain type in a molecule. It differs from the *bond dissociation energy*, which refers to a specific, instead of an average, bond of a given type.

Axial and Equatorial (5.6). In the most stable conformation for the cyclohexane ring, called the "chair" form, there are two orientations that substituents can adopt. If they point up and down, perpendicular to the average plane of the six-membered ring, they are *axial*; if they point out somewhat horizontally relative to the ring, they are called *equatorial*, as drawn below:

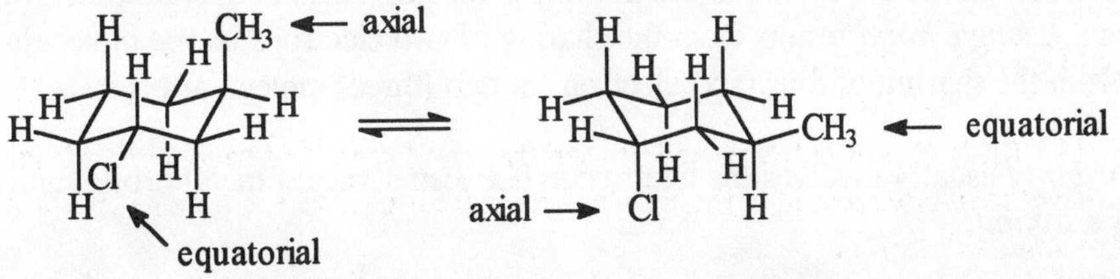

CH_3 ← axial ⇌ CH_3 ← equatorial
equatorial (Cl) · axial → Cl

Note that when the cyclohexane ring flips between the two chair conformations, the axial substituents become equatorial and vice versa. You can see this easily if you have a set of models.

The equatorial positions are less *sterically hindered* and therefore more stable for substituents to occupy than the axial ones.

Azeotropic distillation (14.6.B). Two immiscible solvents will distill as a mixture. This forms the basis for removing water from an equilibrium process by distilling with benzene or toluene. The distillate separates into two phases, with the water being drawn off in a *Dean-Stark trap* and the benzene or toluene returned to the reaction flask.

Beer's law (22.6.C). In ultraviolet *spectroscopy*: $\log \frac{I_0}{I} = \epsilon cd$, where I_0, I = light intensity before and after passing through cell, ϵ = *extinction coefficient*, c = concentration in moles liter^{-1}, and d = path length in cm.

Bent bonds (5.6). When geometric constraints force the bond angles around an atom to be smaller than the preferred angles between the orbitals, the single bonds "bend", which is to say the atomic orbitals do not overlap in their normal end-to-end fashion. Instead, the center of electron density in the bonding orbital lies off to the side of the internuclear axis, and the bond is weaker (see Figure 2.9). This behavior is most important for cyclopropane.

Benzyl, benzylic (21.5). A *benzylic* radical is one involving a carbon radical directly attached to an aromatic ring. Benzylic cations and anions are analogous. A benzylic position is one immediately adjacent to an aromatic ring.

Benzyne (30.3.B). A highly reactive intermediate in the *elimination-addition mechanism* for nucleophilic aromatic substitution: 

Betaine (14.7.D). A *zwitterionic* intermediate. In the Wittig reaction, the betaine may be in equilibrium with an *oxaphosphetane*.

Bimolecular (9.1, 9.2). See *Molecularity*.

Biosynthesis (26.7, 33.3). The synthesis of compounds by living organisms.

Bond (2.7). A net attractive interaction between two atoms. The most important bonds in organic chemistry are *covalent bonds*, which result when two atoms share electrons. The behavior of the shared electrons is described by a *molecular orbital*, and the means by which the sharing is accomplished is overlap of the appropriate *atomic orbitals* on each atom. A *single bond* results from the sharing of two electrons in one molecular orbital; a *double* (triple) bond results from the sharing of four (six) electrons in two (three) molecular orbitals.

A *dative* or *donor bond* usually involves the interaction between a vacant metal orbital and a filled orbital (often a *lone pair*) of a *ligand*.

Bond dissociation energy (D or DH°) (6.1). The energy required to break a bond *homolytically* and separate the two radicals.

α- and β-Branching (9.3). Alkyl groups attached to the carbon undergoing a substitution reaction are α-branches; those on an adjacent carbon are β-branches. Both contribute to *steric hindrance* in *S_N2-reactions*.

Bridged, fused, and spiro ring systems. (20.5; 31.1; 27.3.B)

A *bridged* compound is a bicyclic compound in which the rings share more than two atoms.

A *fused* compound is one in which the two rings share two adjacent atoms.

A *spiro* compound is one in which the rings share only one atom.

Carbenoid (11.6.F). An organometallic complex that behaves like a carbene ($R-\ddot{C}-R'$) source in its reactions.

Carbinolamine (14.6.C). A *hemiaminal*: $R_2C(OH)(NR'_2)$

Carbocation (2.4). A molecule in which a carbon atom is only trivalent and has only six *valence* electrons. The simplest carbocation is the methyl cation. The positively charged carbon is sp^2 *hybridized*, bonding to the hydrogens via the three sp^2 orbitals and leaving the remaining p orbital vacant: CH_3^+

Carbocation rearrangement. (10.6.C). Migration of a substituent, with its bonding electron pair, from an adjacent carbon to a cationic carbon so as to generate a more stable *carbocation*.

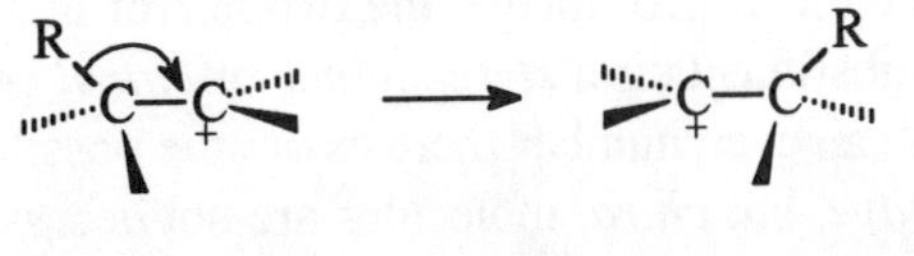

e.g., 2° carbocation ⟶ 3° carbocation, or ring-expansion of a four- to a five-membered ring.

Center of symmetry (7.5). An object has a center of symmetry when the exact same environment is encountered at the same distance in both directions along any line through a particular point, which is called the center of symmetry. Such an object is *achiral*.

Chain reaction (6.3). A chain reaction is one in which (relatively few) *initiation* steps take place, followed by (many) *propagation* steps that convert the starting materials to products, and finally, by (again relatively few) *termination* steps. These reactions usually involve radical intermediates, and transformations involving radical intermediates are usually chain reactions. A chain reaction process is shown below:

Initiation

$$I\text{-}I \xrightarrow{h\nu \text{ or } \Delta} 2\, I\cdot$$

$$I\cdot + A \longrightarrow I' + A\cdot$$

Propagation

$$A\cdot + B \longrightarrow A' + B\cdot$$

$$B\cdot + A \longrightarrow A\cdot + B' \qquad \text{overall process is: } A + B \longrightarrow A' + B'$$

Termination

$$A\cdot + B\cdot,\ A\cdot + A\cdot, \text{ or } B\cdot + B\cdot \longrightarrow AB,\ AA, \text{ or } BB$$

Chair conformation (7.7). The most stable conformation of a cyclohexane ring. In the chair conformation (depicted under *axial*, above), all of the C–C bonds have the staggered conformation and each carbon is free to adopt its preferred tetrahedral geometry.

Characterization (3.4). Determination of the physical and chemical properties of a compound. For example, you could characterize 1-tetradecene ($CH_3(CH_2)_{11}CH{=}CH_2$) by determining its melting point (−12°C), boiling point (232°C), reaction with $KMnO_4$ (to give a carboxylic acid, $CH_3(CH_2)_{11}COOH$), combustion analysis (85.7% C, 14.3% H), and by recording its nmr, ir, and mass *spectra*, etc.

Charge-transfer complex (22.6.D). A complex formed by the face-to-face interaction of two π-systems, one electron-rich and the other electron-poor. There is a certain amount of electron density transferred from the electron-rich (donor) to the electron poor (acceptor) system.

Chemical shift (13.4). The difference in resonance frequency in the nmr between a reference compound (usually tetramethylsilane, *TMS*) and the nucleus of interest, usually measured in parts per million (*ppm*) *downfield* from TMS; this is the so-called "δ scale".
Downfield = *deshielded* = higher frequency = lower field = "to the left" in most spectra.
Upfield = the opposite of downfield.

The presence of electron density *around* a nucleus is *shielding*; the withdrawal of electron density or the presence of electron density *next to* a nucleus (*diamagnetic anisotropy*) is *deshielding*.

Chiral (7.1). A molecule is *chiral* if it is not superimposable on its mirror image; the two mirror image molecules are *enantiomers*. A carbon atom with four different substituents is a *stereocenter* or *chiral center*. A molecule with an odd number of *chiral centers* is *chiral*, but with an even number there exists the possibility of *achiral, meso* compounds. *Achiral* molecules are *optically inactive*, but *chiral* molecules are not necessarily *optically active*, they could be present as a *racemic* mixture (equal amounts of both *enantiomers*).

Chloromethylation (23.4.B). Replacement of an aromatic hydrogen (usually) with a chloromethyl group, usually using CH_2O, HCl, and $ZnCl_2$.

Combination bands (17.2). See *Overtone*.

Condensation (14.6.C). Combination of two molecules with elimination of water or other small molecules, as in the formation of an *imine* or in the aldol condensation (15.2).

Condensed formulas (3.1). See *Structural formulas*.

Configuration:

Electronic: The specific distribution of electrons in *atomic* or *molecular orbitals*.

Stereochemical: The arrangement in 3 dimensions of the substituents on a *chiral center*. See *Absolute configuration*.

Configurational isomers (11.1.A). See *conformation*.

Conformation (5.2), Conformational isomers (11.1.B) = conformers (5.3). There is a fine distinction between *conformers* and other types of *isomers* which is confusing at times. Anytime there are molecules that have the same formula but are put together differently in three dimensions, we can call them isomers. A distinction is drawn between those isomers that interconvert rapidly at ordinary temperatures (without breaking any bonds usually), and those that interconvert very slowly or not at all (and that usually require bonds to be broken and remade during the process). The former are *conformers*, and the different arrangements in three dimensions that a molecule can adopt easily (without breaking bonds) are called different *conformations*. Those that interconvert slowly are called *configurational isomers*.

Conrotatory and disrotatory (22.5.A). In an *electrocyclic* reaction, if the two ends of the *conjugated* system rotate in the same direction (e.g., clockwise) during ring closure the reaction is said to be *conrotatory*; if they rotate in opposite directions (one clockwise, one counterclockwise) it is a *disrotatory* ring closure. *Electrocyclic* ring opening reactions are evaluated the same way.

Coordinatively saturated (36.1.A). A transition metal that has achieved the *18-electron configuration* is *coordinatively saturated*.

Conjugate acid, (conjugate base) (4.5). The species that results from loss of a proton from a molecule is its *conjugate base*; the species that results from protonation of a molecule is its *conjugate acid*;

$$HA \rightleftharpoons A^- + H^+$$

$$B: + H^+ \rightleftharpoons B^+H$$

A^- and B: are conjugate bases of HA and B^+H, respectively: HA and B^+H are conjugate acids of A^- and B:, respectively.

Conjugation (19.5, 20). π-Orbital overlap from more than two atoms. Two double bonds are conjugated if they are adjacent to each other (C=C–C=C, *conjugated diene*) and are *unconjugated* if there is one or more sp^3 hybridized carbons in between (C=C–C–C=C, *unconjugated diene, isolated double bonds*).

A *conjugated enone* is one in which the carbonyl group and double bond are adjacent (C=C–C=O).

Coupling constant, J (13.7). See *Spin-spin splitting.*

Covalent bond (2.7). See *Bond.*

Cracking (6.2). *Pyrolysis* of alkanes to give shorter-chain alkanes and alkenes.

Cross-conjugation (20.3.B). A *conjugated* system that is branched rather than linear is *cross-conjugated.*

Crown ethers (10.11.B). Cyclic polymers of ethylene oxide: 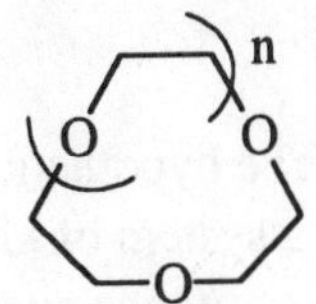

These are important as *phase-transfer catalysts.*

Cumulated double bonds (20.2.C). Double bonds that share the same carbon atom (C=C=C).

Cycloaddition reaction (20.5, 22.5.B, 32.5.B). A reaction in which electron movement occurs in a cyclic manner during the course of addition of a conjugated molecule to a π-bond (for example, the Diels-Alder reaction or *1,3-dipolar cycloaddition*).

D-L convention (28.2). Systems of nomenclature for indicating the *absolute configuration* of *amino acids* and *sugars*: the molecule is drawn in a *Fischer projection* with the carbon chain written vertically and the most oxidized end at the top. If the *heteroatom* substituent (hydroxy or amino group) projects to the right, that *chiral center* is D, if it's on the left it's L. Whether a sugar belongs to the D- or the L-series is determined by the last *chiral center* in the chain. The configuration of an amino acid is determined by the α-carbon.

Deactivating (23.5). The opposite of *Activating*.

Dean-Stark trap (14.6.B). See *Azeotropic distillation.*

Decarbonylation (36.1). Loss of carbon monoxide, usually referring to the reaction catalyzed by transition metal complexes like Wilkinson's catalyst ($[(C_6H_5)_3P]_3RhCl$).

Decarboxylation (27.6.C, 27.7.C). Loss of carbon dioxide.

Decoupling (13.5). Strong electromagnetic irradiation of a nucleus at its resonance frequency in an NMR spectrometer causes it to change its allowed *quantized* orientation in the magnetic field rapidly on the *NMR time scale*, averaging out its effect on nuclei to which it is *coupled.* This reduces the observed *coupling constant* J to 0. It is useful in proton NMR for interpreting splitting patterns, and is routinely used in CMR to simplify the spectra and improve the signal-to-noise ratio.

Off-resonance decoupling in CMR is a partial decoupling of the protons from the carbons, leaving a small, residual coupling so that the number of hydrogens attached to each carbon can be determined.

Degenerate (17.2, 21.1.C). Energy levels that are different but equal in energy are *degenerate*.

Degree of association (15.1.B). How tightly bound an *ion-pair* is. In non-polar solvents, ions are poorly *solvated* and the *ion pairs* are tightly associated. In polar, particularly hydroxylic solvents, *solvation energies* are high, and ions are only loosely associated. The association of ions usually reduces their reactivity.

Dehydration (11.5.B). Loss of water, usually with reference to an alcohol losing a molecule of water to give an alkene.

Dehydrogenation (21.1.D). The opposite of *hydrogenation*. *Dehydrogenation* involves the removal of hydrogen from a molecule and the introduction of *unsaturation*.

Dehydrohalogenation (11.5.A). Removal of HX, usually referring to E2 reaction of an alkyl halide.

Deinsertion (36.1). See *Insertion*.

Delocalization energy (21.1.B). The hypothetical difference in energy between the actual distribution of electrons in a *conjugated* system and a system of identical geometry with electronic isolation of the π-bonds from each other. Contrast with *empirical resonance energy*, (See *Resonance energy*).

Deshielded, shielded (13.4). See *Chemical shift*.

Deuteration (8.9.A). Introduction of deuterium into a molecule. It can be accomplished by *deuterium exchange* or by reaction of an organometallic reagent with D_2O, among other methods.

Deuterium exchange (15.1.A). Equilibration of ionizable hydrogens with deuterium atoms from the solvent, usually used to determine how many α-hydrogens (enolizable hydrogens) are present in a ketone or other carbonyl compound.

Dextrorotatory, levorotatory (7.2). A sample that rotates plane polarized light clockwise (+ direction) is *dextrorotatory* (counterclockwise is *levorotatory* (−)).

Diamagnetic shielding and deshielding (13.4). See *Chemical Shift, Anisotropy,* and *π-Electron circulation*.

Diastereomer (7.5). *Stereoisomers* that are not *enantiomers*. If more than one *chiral center* is present in a molecule, changing the *configuration* of all of them gives the *enantiomer*. Changing the configuration of *less* than all of them gives a *diastereomer*.

Diastereomeric salts (29.5.F). See *Resolution*.

Diaxial (11.6.B). This is equivalent to *anti* in a cyclohexane system.

Diazotization (24.7.B, 25.5). Conversion of an amino into a diazonium group:

$$RNH_2 + HONO + H^+ \longrightarrow RN_2^+ + 2\ H_2O$$

Dienophile (20.5). The "monoene" component of the Diels-Alder reaction. Electron-withdrawing groups on a dienophile usually increase its reactivity.

Dilution principle (20.1.C). A *bimolecular* reaction is slowed down more on dilution than a *unimolecular* reaction.

Dimer, trimer (36.4). See *Polymer*.

1,3-Dipolar cycloaddition (32.5.B). A *cycloaddition reaction* in which the *conjugated* component is 3 atoms long and is *zwitterionic* or with *zwitterionic* character.

Dipole moment, μ (8.1). A dipole moment arises whenever positive and negative charges are separated. The magnitude depends on separation, d, and charge, q: $\mu = q \cdot d$. The general quality of "polarity" depends in part on the presence of a dipole moment in a molecule.

Dispersion force = London force = van der Waals attraction (5.1). These three terms all describe a very weak interaction that results in an attraction between two molecules. It is different from the more easily understood ionic or dipole interactions, and depends instead on differences in the "instantaneous" distribution of electrons. The most important features to remember for this force are: (1) it is the primary force of attraction between hydrocarbon molecules; (2) it decreases very rapidly with distance; and (3) its magnitude is proportional to the "surface area" of a molecule.

Disproportionation (6.2). *Disproportionation* involves two of the same or similar molecules reacting with each other to produce different products: 2 A ⟶ B + C. The *cracking* of alkanes provides an example of this.

Disrotatory (22.5.A). See *Conrotatory*.

Donor-acceptor complex (22.6.D). See *Charge-transfer complex*.

Double bond (2.3). See *Bond*.

Double bond character (19.1). Degree of π-bonding character, often evaluated as amount of contribution from a resonance structure containing a double bond, as in the amide linkage.

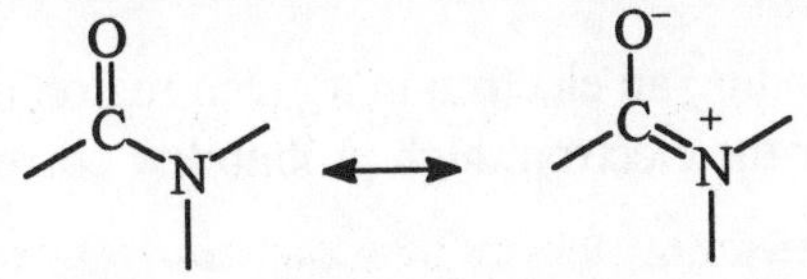

Doublet (13.7). See *Spin-spin splitting*.

Downfield, upfield (13.4). See *Chemical Shift*.

E (entgegen) and Z (zusammen) (11.2). Nomenclature for describing the configuration of a double bond.

E2 (Elimination-bimolecular) (9.6, 11.5.A). The most common mechanism for an elimination reaction. It involves simultaneous removal of the proton by a base, formation of a the π-bond, and departure of the leaving group. All the orbitals involved must be coplanar, and this is usually accomplished in an *anti* relationship:

B^- H c d a b X ⟶ BH + a c b d + X^-

Eclipsed (5.2). See *Anti*.

Edman degradation (29.7.C). A method for the stepwise removal and identification of the N-terminal amino acids from a *peptide* chain.

α, β, and γ-effects (13.5). Characteristic effects on *chemical shift* in cmr spectroscopy that depend on specific substituents and their position relative to the nucleus observed.

Eighteen electron rule (36.1). Just as a 2nd or 3rd period element tries to achieve an octet of *valence electrons* in its bonding arrangement, so does a transition metal try to achieve a filled shell of 2s + 10d + 6p = 18 electrons through its bonds to *ligands* and other groups. A metal that has achieved an *18-electron configuration* is *coordinatively saturated.*

Electrocyclic reaction (22.5). A reaction in which a conjugated molecule undergoes a cyclic rearrangement of electrons to form a molecule with one less π-bond and one more ring, or the reverse of this process.

Electron affinity (2.2). (See definition on page 6 in text).

Electron-attracting and -donating (10.4). See *Inductive effect*.

π-Electron circulation (13.10.A). The motion of all electrons is altered by a magnetic field. The electron motion induced in the π-electrons of alkenes and aromatic rings causes a *downfield (deshielding)* effect on the *chemical shifts* of nuclei attached to them. This effect is known as *diamagnetic deshielding*.

Electron count (2.2, 36.1). The number of electrons in the valence shell of an atom or metal is determined for the purpose of evaluating *formal charges, filled octets, 18-electron* configurations, etc.

Electron density (2.5). Refers to the *probability* of finding an electron in a given region of space. This probability is given by the square of the *wave-function* for the electron; high probability corresponds to high electron density.

Electronegative (2.2). Exerting a strong attraction for electrons. The electronegativity of the elements increases as one goes up and to the right in the periodic table. The more electronegative a group is, the better it is able to stabilize a negative charge, within the constraints of the *octet rule*. Electronegative elements usually need only one or two additional electrons to complete their filled octet.

Electronic transition (22.6). Change of the electronic state of a molecule or atom, usually from the *ground electronic state* to an *excited electronic state* or vice versa. Such transitions are important in ultraviolet *spectroscopy* and photochemistry (22.6.H, 36.3).

Electrophile, electrophilic reagent. (11.6.C, 23 The opposite of *nucleophile* and *nucleophilic reagent*. An electrophile contributes the vacant *orbital* when a bond is formed by the reverse of a *heterolytic* process:

$E^+ + :Nu^- \longrightarrow E\text{-}Nu$ $\quad E^+$ = electrophile $\quad :Nu^-$ = nucleophile

A proton is the simplest electrophile.

Electropositive (2.2). The opposite of *electronegative*. Electropositive elements are those at the left of the periodic table, which achieve a complete *valence* shell (filled *octet*) by losing rather than gaining one or two electrons.

Elimination-addition mechanism (30.3.B). For nucleophilic aromatic substitution via a *benzyne* intermediate.

Empirical formula (3.4). The empirical formula of a compound expresses the ratio of elements present. Compare with *molecular formula*.

Enantiomers (7.1). *Stereoisomers* that differ only by being mirror images of each other. See *Chiral* and *Diastereomer*.

Endo and exo (20.5). In a bicyclic compound, the configuration of a non-bridgehead substituent can be specified by its relationship to the other bridges. If it points in the same direction as the longer bridge, its is *endo*; if it points toward the shorter bridge, it is *exo*; e.g.

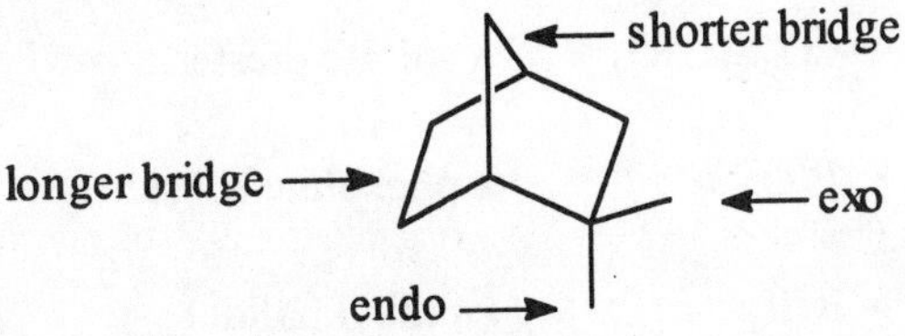

shorter bridge

longer bridge

exo

endo

In the transition state of the Diels-Alder reaction, if a substituent on the *dienophile* points toward the diene, it is *endo*; if it points away it is *exo*.

Envelope (5.6). The most favorable conformation for a cyclopentane ring, in which one carbon is pushed out of the plane of the other four in order to minimize the eclipsing interactions of all of the hydrogens. This conformation is not fixed, and all five carbons of a cyclopentane ring can take the out-of-plane position interchangeably. In this instance, the interconversion is referred to as *pseudorotation*.

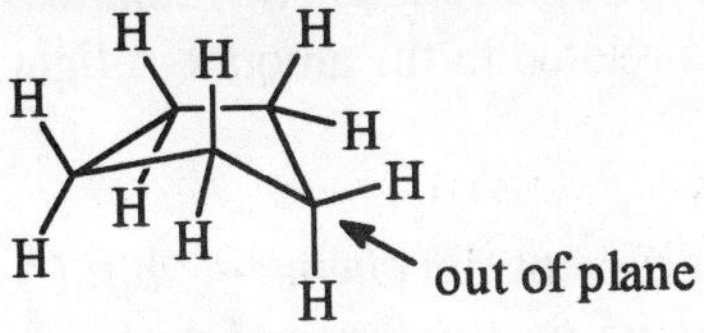

out of plane

Enzyme (28.3, 29.8.E). A *protein* that catalyzes a chemical reaction.

Equatorial (5.6, 26.11.A). See definitions under Axial and Apical.

Equilibrium (4.1). Although the term equilibrium is defined in the text, it is important that you understand the distinction between *equilibrium* and rate. The rate constant for a reaction is a measure of how *fast* it goes; the equilibrium constant is a measure of how *far* it goes. There is not necessarily any connection between these two constants: there are many highly *exothermic* reactions that go very slowly (for instance, the decomposition of TNT in the absence of a detonation), and many reactions that are only slightly exothermic but that go very rapidly (for instance, the neutralization of a weak acid with a weak base).

Erythro and threo (28.2). A controversial nomenclature system for specifying the relative *configuration* of two *chiral centers*. It arose from carbohydrate chemistry, and is used with Fischer projections in the following way: with the carbon chain written vertically, if the "similar" substituents on two chiral centers are on the same side, the relationship between those carbons is *erythro*. If they are on opposite sides, the relationship is *threo*. Complications arise in systems other than carbohydrates and when it is difficult to decide what the "similar" substituents are.

```
      CHO                    CHO
  H———+———OH            HO———+———H
  H———+———OH             H———+———OH
     CH2OH                  CH2OH
   Erythrose               Threose
```

Ester, esterification (10.6.B, 18.7.C, 26.4, 26.7). The compound formed from the loss of water (formally) between an alcohol and an acid is an *ester*.

$ROH + HO_2CR' \longrightarrow RO_2CR'$,	a carboxylic ester
$2\ ROH + O_2S(OH)_2 \longrightarrow O_2S(OR)_2$	a sulfate diester
$3\ ROH + {:}P(OH)_3 \longrightarrow {:}P(OR)_3$	a phosphite triester
(but: $ROH + HBR \longrightarrow RBr$,	an alkyl halide)

Exact mass (34.3). The actual mass, to at least 0.0001 atomic mass unit accuracy, of a molecule or molecular fragment, which allows you to distinguish between *molecular formulas* of the same *nominal mass*.

Excited electronic state (22.6.A). See *Ground electronic state*.

Exo (20.5). See *Endo*.

Extinction coefficient, ϵ (22.6.C). A measure of the probability that a quantum of electromagnetic radiation with the correct energy will be absorbed and result in an *electronic transition*. It is related to the amount of light transmitted by a sample by *Beer's law*.

First-order vs. non-first order nmr spectra (13.7). When the magnitude of the difference in *chemical shift* ($\Delta\nu$) of two nuclei is much greater than their *coupling constant*, J, the spectrum is said to be *first-order*, and the *spin-spin splitting* patterns follow the usual rules. When $\Delta\nu = J$, the spectrum is *non-first-order*, and the splitting patterns are very complex.

First order reaction (4.3). A true first order reaction is one in which the *rate-determining step* involves only *one* molecule. Therefore the equation for the rate of reaction includes only the concentration of that molecule. If the reaction actually involves two molecules, but one is in large excess (for instance, solvent), the reaction is called a *pseudo first order reaction* because the rate equation still has only one concentration as a variable.

Fischer projection (28.2). A system for indicating 3-dimensional structures in two dimensions. In a Fischer projection the horizontal bonds are understood to project forward and the vertical bonds back:

```
     b              b
     |              :
 a———+———c  ≡   a ◀━:━▶ c
     |              :
     d              d
```

In a Fischer projection, exchange of the positions of any pair of substituents or a 90° rotation of the whole picture leads to a representation of the other *configuration* at that chiral center. Any even combination of these changes (for example 2 pair-wise exchanges, 3 pair-wise exchanges and a 90° rotation, or a 180° rotation) lead to a picture of the same molecule again.

Fluorescence (36.3). Loss of a photon form the first *excited singlet state* and transition of the molecule to the *ground state*.

Formal charges (2.2). The difference between the number of *valence* electrons controlled by an atom in the elemental state and in its bonding arrangement in a molecule. The number of valence electrons in the elemental state corresponds to its column in the periodic table; in a molecule, an atom is considered to control all the valence electrons it does not share, and half of thse it does share.

Franck-Condon transition (36.3). An *electronic transition* that occurs more rapidly than atomic (vibrational) motions in a molecule; also called a *vertical transition*.

Free radical (6.1). Any molecule having unpaired electrons. Usually refers to carbon atoms with only three substituents and seven *valence electrons*, such as $\cdot CH_3$

Front side attack (9.1). A possible mode of substitution stereochemistry. It occurs only in rare instances in nucleophilic displacement reactions (which ordinarily involve *inversion* (S_N2) or *racemization* (S_N1)), but is seen in some electrophilic displacement reactions.

Functional groups (3.3). The reactive parts of molecules. These are small, frequently-occurring groups of atoms, such as the hydroxy group or carboxy group, which exhibit a typical reactivity in a wide variety of molecules. For example,

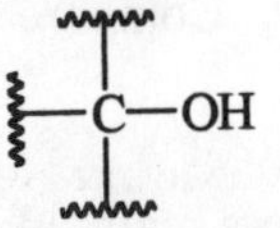

C—OH

hydroxy group

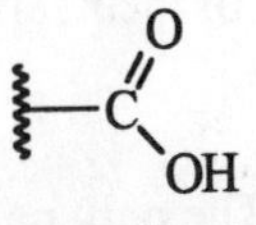

O
C
OH

carboxy group

all molecules having a carboxy group are acidic. A list of the most important functional groups is found in Table 3.1 of the text.

Fundamental vibrational modes (17.2). Modes of vibration that involve more than one bond; all vibrational modes of molecules larger than two atoms are fundamental modes.

Fused ring system (31.1). See *Bridged*.

Gauche (5.2). See definition under *Anti*.

Geminal (12.5.C). Attached to the same carbon. See *Vicinal*.

Gibbs Standard Free Energy Change, ΔG° (4.2). This represents the amount of energy available for work that would be released during a reaction if all of the starting material in its standard state were converted to all of the product in its standard state. (In solution, the standard state is about 1 *M*). Important equations to remember are $\Delta G° = \Delta H° - T\Delta S°$, which indicates how the *enthalpy* (ΔH°) and the *entropy* (ΔS°) of the reaction contribute to the free energy (ΔG°); and $\Delta G° = -RT\ln K$, which relates the free energy change to the equilibrium constant K.

Glycoside (28.3). Cyclic *acetals* or *ketals* of a sugar with another alcohol, called the *aglycon*.

Glyme(s) (10.11.A). Dimethyl ethers of short ethylene oxide polymers: $CH_3(OCH_2CH_2)_nOCH_3$. See *Crown ethers*.

Ground electronic state (22.6.A). The lowest energy distribution of electrons in the molecular orbitals of a molecule or the atomic *orbitals* of an atom. Any higher energy distribution, for example, one resulting from promotion of one of the electrons from a *bonding* to an *antibonding orbital*, is an *excited electronic state*.

Halonium ion (11.6.B). A cyclic halogen cation, usually resulting from the addition of electrophilic halogen to a π-bond.

Harmonic oscillator approximation (17.2). The approximation of a bond vibration as a system that obeys *Hooke's Law*.

Haworth projection (28.3). A semi-perspective drawing of the cyclic hemiacetal or hemiketal form of a sugar to indicate its stereochemistry. Useful for making the transition from *Fischer projections* to perspective drawings of chair conformations.

Heat of combustion (6.4). Enthalpy released on complete oxidation of a compound: $C_nH_m \longrightarrow n\ CO_2 + m/2\ H_2O$. This number provides the same information on the thermodynamic stability of a molecule as the *heat of formation*, but references it to a different standard state. *Heats of combustion* are in fact the values that are obtained experimentally; *heats of formation* are then calculated using the known heats of combustion of the elements.

Heat of formation (5.5). The *enthalpy* released on forming a compound from its elements. Comparison of the heats of formation of isomers is an indication of their relative thermodynamic stability. Compare with *Heat of combustion* (6.4).

Heisenberg Uncertainty Principle (2.5). The only part of chemistry the philosophers really like. It sets a lower limit on the accuracy with which we can know both the position and momentum of a particle. For instance, the more precisely we define how an electron is moving, the less accurately we can know where it is, and vice versa.

Hetero (3.2). Not carbon, hydrogen (or a metal, usually). For example, O, N, F, S etc. are all *heteroatoms*, and cyclic compounds in which there are ring atoms other than carbon are called *heterocycles*.

Heterolysis, heterolytic cleavage (6.3). Cleavage of a bond in which both electrons in the bonding orbital depart with one of the pieces: $A{:}B \longrightarrow A^+ + {:}B^-$. There will always be a change in formal charge on the two atoms involved in such a process. Contrast with *Homolysis*.

Hofmann rule (24.7.E). "In the decomposition of quaternary ammonium hydroxides, the hydrogen is lost most easily from CH_3, next from RCH_2, and least easily from R_2CH."

HOMO = Highest Occupied Molecular Orbital (22.7). In a molecule, the highest energy molecular *orbital* that is occupied by a pair of electrons. Contrast with the *LUMO*, which is the Lowest Unoccupied Molecular Orbital. The interaction between the *HOMO* of one molecule and the *LUMO* of another can be used to predict the manner in which the two molecules might react [*Perturbational molecular orbital theory* (22.7)].

Homolysis, homolytic cleavage (6.3). Cleavage of a bond in which one of the two bonding electrons departs with each of the pieces: A:B ⟶ A· + ·B. There is no change in formal charge when this happens. Contrast with *Heterolysis*.

Hooke's law (17.2). States that the force needed to stretch or compress a spring (or bond) is directly proportional to the distance it is stretched or compressed.

Hückel 4n+2 rule (22.3.A). "Monocyclic π-systems with 4n+2 electrons show relative stability compared to acyclic analogs."

Hückel molecular orbital (22.4). See *Orbital*.

Hybridization (2.8). A recombination of the *orbitals* of a free atom to enable it to make stronger bonds when it is in a molecule. Whereas the atomic state of carbon has one 2s and three 2p orbitals, carbon is hybridized to provide four $2sp^3$ orbitals (sp^3-hybridized), one 2p and three $2sp^2$ orbitals (sp^2-hybridized), or two 2p and two 2sp orbitals (sp-hybridized) in its molecules. The hybridized orbitals have a characteristic spatial relationship that is reflected in the geometry of the molecule. The exponents correspond to the fractional character of a given atomic orbital in the hybrid; for example, in sp^3, the fractional p-character is 3/(3+1), or 0.75. Note that the exponent of s is always unity. (N.B. Don't confuse *hybrid orbitals* with *resonance hybrids*.)

Hydration. The opposite of *dehydration*. Usually the addition of water cross the π-bond of an alkene to give an alcohol (11.6.C), of an alkyne to give a ketone (12.6.B), or of a carbonyl compound to give a gem-diol (carbonyl hydrate) (14.6.A).

Hydrogen bond (4.5, 10.3). A dipole-dipole interaction between a hydrogen atom bonded to an electronegative element and an electron *lone pair* on another electronegative atom. It is a weak bond (~5 kcal mole^{-1}) but important in the chemistry of alcohols, carboxylic acids, amines and similar compounds.

Hydrogenolysis (21.6.B). Cleavage of a single bond by hydrogenation:

$$A-B + H_2 \xrightarrow{\text{catalyst}} AH + HB$$

Usually encountered in Raney-nickel desulfurization or in hydrogenation of benzyl alcohols, etc.

Hydrolysis (10.5, 19.6). See *Solvolysis reaction*.

Hydrophilic (29.8.B). The opposite of *hydrophobic*: polar, often ionic, well-*solvated* by water, water soluble.

Hydrophobic (29.8.B). Repelling water. Nonpolar, hydrocarbon chains are poorly *solvated* by water molecules and are excluded from aqueous solution. They form *lipid* bilayers, *micelles*, or separate phases in which they provide their own solvation.

Hydroxylation (27.3.A). Usually, addition of two hydroxyl groups to a double bond to give a 1,2-diol.

Hyperconjugation (9.7, 22.6.D). The mechanism by which a *carbocation* is stabilized by sharing the electron density of bonds to the adjacent carbon. This results in the stability sequence: $3° > 2° > 1° > CH_3^+$ for carbocations.

It is important in characterizing the interaction of the σ bonds of an alkyl group with an adjacent π-system, as in ultraviolet spectroscopy.

Induction period (15.1.D). If a reaction is *autocatalytic* it will accelerate as it proceeds. The time *before* the reaction gets itself going is called the *induction period.*

Inductive effect (10.4, 18.4.B). The *electron attracting* or *donating* effect of a nearby dipole. For instance, *electronegative* elements like halogens are *electron attracting*; alkyl groups are *electron donating*.

Infrared active, inactive (17.2). See *Selection rule.*

Initiation steps, Initiator (6.3). See *Chain reaction* (6.3).

Inner salt (26.5.B, 29.1). See *Zwitterion.*

Insertion (36.1). A type of reaction in transition metal chemistry in which a donor *ligand* undergoes insertion into a σ-bond between the metal and another group. The reverse of this process is called *deinsertion*. The *electron count* of the metal decreases by two on insertion.

Integrated intensity (13.6). The area under the curve of a given peak in the nmr. For proton nmr, this usually corresponds to the relative number of protons that resonate at that frequency. In *cmr, saturation* and *relaxation* effects affect the integrated intensities of different signals differently, and the area ratios do not correspond to the relative number of carbons.

Interference (2.6). An addition of *wave functions (orbitals)* that results in their cancellation in a certain region of space. This occurs during the formation of an *antibonding* orbital, for instance, in which an atomic orbital wave-function of one sign overlaps with an orbital of opposite sign on the other atom. The opposite of interference is *reinforcement.*

Intermolecular, intramolecular (9.9). *Intermolecular* = between two or more molecules; *intramolecular* = within the same molecule.

Internal conversion (36.3). Relaxation of an *excited* vibrational *state* to a *ground* vibrational *state.*

Intersystem crossing (36.3). Conversion of the first *excited singlet state* to the first *excited triplet state.*

Inversion (9.2). The normal stereochemistry seen in displacement reactions occurring by the S_N2 *mechanism.* It arises from simultaneous bonding of the incoming *nucleophile* and the departing *leaving group* to opposite lobes of the *orbital* on carbon:

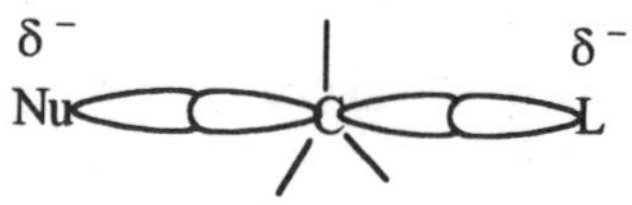

Ionic character (8.6). A highly polarized bond, one between an *electronegative* and an *electropositive* element, has a lot of ionic character. A symmetrical, *covalent* bond between identical groups has no ionic character (e.g, the C-C bond of ethane).

Ionization potential (2.2, 34.4.A). The energy required to remove an electron from an atom or molecule.

Isoelectric point (29.3). The pH at which the average charge on a molecule (usually a *zwitterionic* amino acid or *peptide*) is zero. For the simple amino acids (with only one acidic and one basic group), this is the average of pK_1 and pK_2.

Isolated double bonds (20.2.A). See *Conjugation.*

Isomers (3.6). Compounds that have the same molecular formula but different structures. See *Structural Formulas* and *Stereoisomers*, and the table at the beginning of Chapter 3 in this Study Guide.

J (13.7). See *Coupling Constant.*

Karplus curve (13.9). Figure 13.33. Relates *coupling constant* J to the dihedral angle between the two adjacent C-H bonds.

Kekulé structure (2.2). A structure in which a line between two atoms indicates a *covalent single bond* involving two shared electrons. A double bond is represented by a double line, etc. Contrast with *Lewis* structures.

Ketal (14.6.B). See *Acetal.*

Ketyl (27.3.A). A *radical anion* produced by addition of an electron to a ketone; an intermediate in the pinacol reaction.

Kinetic control (20.2.B). A reaction in which the products are not in equilibrium with the starting materials is under *kinetic control*; that is, the product obtained is the one that is formed the fastest. This is not always the most stable product: see *Thermodynamic control.*

Kinetic order (9.2). Equal to the sum of the exponents of all the concentrations expressed in the rate equation. If the rate depends only on the concentration of one component, the reaction is *first order*; if it depends on the concentration of two components, or on the square of the concentration of one component, the reaction is *second order*.

Leaving Group (9.6). The group that departs in a displacement or an elimination reaction. The nucleophile [or base] attacks and the leaving group leaves, either before (S_N1 [or E1], for example) during (S_N2 [or E2]), or afterwards (reactions of carboxylic acid derivatives, for example). The less basic the departing species is, the better a leaving group it is.

Levorotatory (7.2). See *Dextrorotatory.*

Lewis acid. A Lewis acid is a molecule or ion that furnishes a vacant orbital for bonding. The term is most commonly applied to the strong Lewis acids like $ZnCl_2$, $AlCl_3$, $FeCl_3$, BF_3, etc. However, all *electrophiles* are formally Lewis acids, and the term is used in this sense in the organic chemistry of transition metals (see next entry).

Lewis acid association (36.1). Bond formation between a transition metal with a Lewis acid (*electrophile*). The reverse process is called *Lewis acid dissociation*; the *electron count* of the metal does not change.

Lewis base. A Lewis base is a molecule or ion that has two electrons available for bonding. All nucleophiles are Lewis bases.

Lewis base association (36.1). The formation of a bond to a transition metal in which the two electrons to be shared originate with the *ligand* or other entering group. The reverse process is called *Lewis base dissociation.* The *electron count* of the metal increases by +2 on *Lewis base association.*

Lewis structure (2.2). A structure in which each *valence* electron is indicated by a dot ("electron dot structures"), and a *covalent single bond* is represented by two dots between the atoms. These are helpful in keeping track of electrons, *formal charges*, and valences. Note that "core" electrons are not shown. That is, for carbon only the four valence electrons are depicted; the 1s electrons are understood to be present as well.

Ligand (36.1). Usually refers to a molecule or ion that is stable by itself but that is also capable of forming a *dative* or *donor bond* to a transition metal.

London force (5.1). See *Dispersion force*.

Lone pair electrons (2.2). See *Nonbonded electrons*.

Long range coupling (13.9). Coupling observed over longer distances than the usual H–C–C–H system. It occurs with specific conformations of H–C–C–C–H, with alkynes, H–C–C≡C–H, and frequently with nuclei other than protons (for example, ^{13}C, ^{19}F, ^{31}P).

LUMO = Lowest vacant (Unoccupied) Molecular Orbital (22.7). See HOMO.

Macroscopic isotope (23.2). A non-radioactive isotope used in labeling studies. In contrast to a *tracer isotope*, most of the labeled positions will contain a macroscopic isotope. The presence of a macroscopic isotope is determined by NMR spectroscopy or by mass spectrometry.

Magnetic moment (13.3). Some atomic nuclei behave as if they were spinning (hence the term nuclear "spin") and act like small magnets. Nuclei with a nuclear spin *quantum number* of ½ (^{1}H, ^{13}C, ^{19}F, ^{31}P, among others) are allowed only two orientations in the presence of an external magnetic field: aligned with the field or against it (the α- or β-states). The difference in energy of these two orientations gives rise to the phenomenon of nuclear magnetic resonance spectroscopy (nmr).

Magnetically equivalent (13.4). Having the same *chemical shift*. Equivalent nuclei (for instance the two hydrogens in acetylene) have to be magnetically equivalent; many others can be equivalent on the *nmr time scale* because of rapid conformational changes (for example the hydrogens of a methyl group); still others may be quite different structurally but coincidentally have the same *chemical shift*.

Markovnikov's rule (11.6.C). Original formulation: In the acid-catalyzed addition of water to an alkene (*hydration*), the proton goes on the carbon that already has the greater number of protons (less substituted carbon).

Modern formulation: In electrophilic addition to a π-bond, the *electrophile* adds so as to generate the most stable carbocationic intermediate.

Mass spectrum, mass spectrometer (34.1). A *mass spectrum* is a record of the mass-to-charge ratio of the ions produced when a molecule is subjected to electron impact in a *mass spectrometer*.

McLafferty rearrangement (34.4.B). A two-bond fragmentation mechanism for *radical cations* of carbonyl compounds in the mass spectrometer. It requires the presence of a hydrogen γ to the carbonyl group.

Mechanism (4.1). The details of how the atoms and electrons are reorganized during the course of a reaction.

Meso (7.5). An *achiral* molecule with *chiral* carbon atoms is a *meso* compound. There must be an even number of chiral centers and they must have opposite configurations so that there is a *center* or *plane of symmetry* in the molecule.

Meta-directors (23.5). Substituents on an aromatic ring that destabilize adjacent positive charge and thereby favor electrophilic aromatic substitution at positions *meta* to themselves. See *ortho, para-directors*.

All *m-directors* are *deactivating*.

Metallation (30.3.C). Replacement of a hydrogen or a halogen with a metal cation. Deprotonation is usually accomplished with a strong base like lithium diisopropylamide, alkyllithiums, or sodium amide; halogens are replaced with metals using alkyllithiums (*transmetallation*) or the metals themselves.

Micelle (18.4.D). Molecules with both *hydrophilic* and *hydrophobic* regions often cluster together in aqueous solution in *micelles,* in which the non-polar chains are in the interior and the polar regions are on the exterior.

Migratory aptitude (14.8.A). Relative ease with which a particular substituent moves over to an adjacent atom in rearrangements like carbocation rearrangements, the Baeyer-Villiger reaction, or acylnitrene or carbene rearrangements. Usually H > 3° > 2° > phenyl > 1° > CH_3.

Mixed Claisen condensation (27.7.A). Claisen condensation between a ketone and an ester to give a β-diketone.

Möbius molecular orbital (22.5). See *Orbital.*

Molecular formula (3.4). The molecular formula of a compound expresses the total number of atoms of each element present. It is always the same as, or a multiple of, the *empirical formula*. For instance, the *empirical formula* for glucose is CH_2O, but the *molecular formula* is $C_6H_{12}O_6$.

Molecular ion, M^+ (34.3). The ion produced in a mass spectrometer on ejection of an electron from a molecule.

Molecular orbital (2.7). See *Orbital.*

Molecularity (9.2). The molecularity of a reaction is the number of molecules involved in the rate-determining transition state. For instance, an S_N2 displacement reaction is a typical *bimolecular* reaction; an S_N1 displacement is a typical *unimolecular* reaction. *Termolecular* reactions, and higher, are very rare.

Monochromator (17.1). An instrument that disperses a beam of light into its component frequencies, so that a narrow range of frequencies can be focused on the sample. It is an important component of infrared and ultraviolet/visible spectrophotometers.

Monomer (36.4). See *Polymer.*

Multiplet (13.7). See *Spin-spin splitting.*

Mutarotation (28.3). A change in *optical rotation* that results from the equilibration of *anomers.*

Nitrogen inversion (24.1).

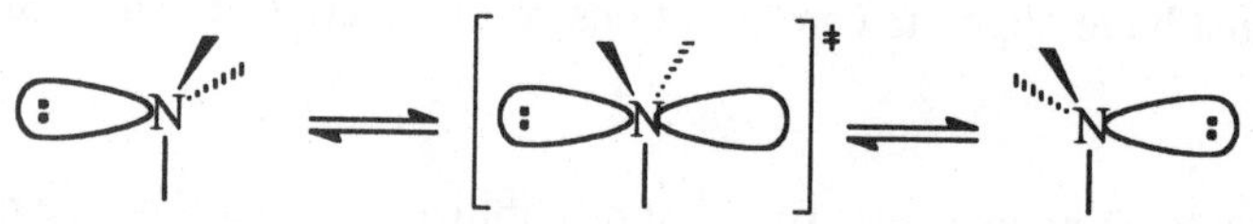

$$\text{:N (pyramidal)} \rightleftharpoons [\text{planar N}]^{\ddagger} \rightleftharpoons \text{N: (pyramidal)}$$

NMR Time Scale (13.11). About 10^{-3} sec, the length of time a molecule must exist in a discrete state for it to be observable by nmr.

Node (2.5). A surface that separates regions of an *orbital* that have opposite signs. For instance, a 2p orbital aligned with the z axis has as its node the xy plane, and the antibonding orbital of a hydrogen molecule has as its node a plane perpendicular to the internuclear axis.

Nominal mass (34.3). The integral mass of a molecule or molecular fragment. See *Exact mass*.

Non-bonded electrons (2.2, 24.1). *Valence electrons* not involved in a *covalent bond*. Oxygen normally has two pairs of non-bonded electrons and nitrogen one, for example: $H\ddot{\underset{..}{O}}H$, $:NH_3$. These *non-bonded electrons* are also called *lone pair* electrons, and they are the electrons involved when these molecules act as *Lewis bases* or *electrophiles*.

Nuclear spin (13.3). See *Magnetic moment*.

Nucleophile, nucleophilic reagent (9.1). A *Lewis base*. In its most general sense, any species that can furnish a pair of electrons to make a bond can be considered a nucleophile: any time a bond is formed by the reverse of a heterolytic process, the species that contributes the electrons is a nucleophile. In discussing displacement reactions, those that involve attack by a nucleophile and loss of a *leaving group* are called nucleophilic displacement reactions.

Nucleophilic addition-elimination mechanism (19.6). The most important mechanism whereby substitution reactions occur in carboxylic acid derivatives. In its general form:

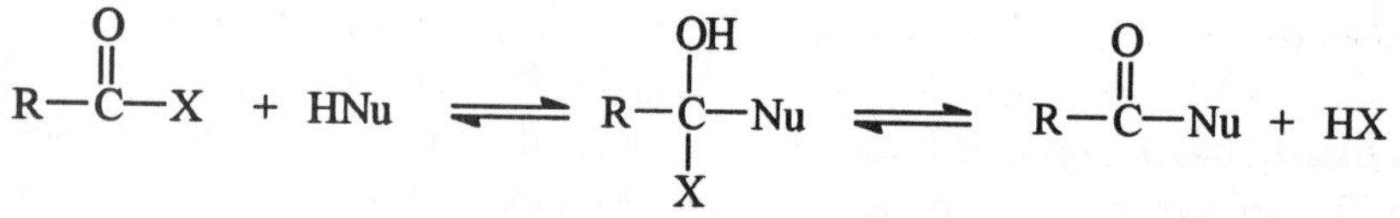

$$R-\overset{O}{\overset{\|}{C}}-X + HNu \rightleftharpoons R-\underset{X}{\overset{OH}{C}}-Nu \rightleftharpoons R-\overset{O}{\overset{\|}{C}}-Nu + HX$$

Nucleophilic aromatic substitution can also occur via an addition-elimination mechanism (30.3.A).

Octet (2.2). A filled *valence* shell for elements in the second and third periods of the periodic table. This arrangement for electrons is very stable, and achieving a filled octet is the most important factor in determining how many and what kinds of bonds an element will form in its molecules.

Off-resonance decoupling (13.5). See *Decoupling*.

Optically active (7.2). Capable of rotating the *plane of polarization* of *plane polarized light*. An *optically active* sample must be composed of *chiral* molecules, but the converse is not always true: a *racemic* mixture of chiral molecules is *optically inactive*.

Orbital (2.5). An equation that describes the behavior of an electron. The term is synonymous with *wave function*, although conceptually chemists envision a volume having specific shape and orientation when they think of orbitals and a mathematical equation when they think of wave functions. An *atomic orbital* describes the behavior of an electron in the vicinity of a single nucleus. *Molecular orbitals* result from the combination (overlap) of two or more atomic orbitals and describe an electron that is shared by several nuclei. Molecular orbitals can be further classified into *bonding, nonbonding,* and *antibonding* orbitals, depending on whether the electron distribution in the molecular orbital is more favorable (lower energy), unchanged (same energy), or less favorable (higher energy) than in the component atomic orbitals. Bonding orbitals result from the *reinforcing* overlap of atomic orbitals, antibonding orbitals from their *interfering* overlap.

Cyclic molecular orbitals are classified as Hückel if they have an even number (or zero) of positive/negative overlaps around the ring; they are classified as Möbius is there is an odd number of such interactions.

Ortho, meta, and para (21.2.A). Positions on a benzene ring relative to a substituent.

R
ortho ortho
meta meta
para

Ortho, para-directors (23.5). Substituents on an aromatic ring that stabilize adjacent positive charge and thereby favor electrophilic aromatic substitution at positions *ortho* and *para* to themselves. See *meta-directors*.

o,p-Directors can be either *activating* or *deactivating*.

Overlap (in NMR spectroscopy) (13.8). When two nuclei resonate at similar *chemical shifts*, their peaks *overlap* (fall on top of one another).

Overtones, combination bands (17.2). Bands that result in an infrared spectrum from simultaneous change of more than one energy level.

Oxaphosphetane (14.7.D). A cyclic intermediate in the Wittig reaction (see *betaine*):

O—PPh_3

Oxidative addition (36.1). A type of reaction in transition metal chemistry in which the two groups at the ends of a σ-bond both become bonded to the metal. The reverse process is called *reductive elimination*. The *electron count* of the metal decreases by two on oxidative addition.

Oxidative coupling (30.3.C). Joining two molecules with overall loss of two protons and two electrons, as in the oxidative coupling of arylcopper compounds to give biphenyls.

Oxonium ion structure (2.4). A structure that involves three bonds to an oxygen atom, and therefore a positive *formal charge* on the oxygen. Although localization of a positive charge on the *electronegative* element oxygen would appear to be difficult, the atom still retains a filled *octet valence* shell (compare with a *carbocation*), and this consideration is the most important. The simplest oxonium ion is the hydronium ion: H_3O^+

Partial rate factors (23.7). In electrophilic aromatic substitution, the partial rate factor is the reactivity of a given position on a substituted aromatic ring relative to benzene, corrected for the number of equivalent positions.

Peptide (29.1). A *peptide bond* is an amide linkage between two amino acids. *Peptides* are dimers, trimers, oligomers, etc. (*condensation polymers*) of amino acids.

"Peroxides" (10.10.G). Usually *alkylperoxides* (ROOH) present in impure materials as a result of *autoxidation*. They are often initiators of free radical chain reactions, and are distinctly different from the reagents hydrogen peroxide (HOOH) or peracids (RCO_3H).

Phase-transfer catalysis (24.5.B). Transfer of ionic reagents from an aqueous or solid phase into an organic solvent, where they show enhanced reactivity. Phase-transfer catalysts are either *quaternary ammonium salts* or *crown ethers*.

Phosphorane (14.7.D). See *Ylide*.

pK = -log K (4.5). pK_a = measure of acidity

$$K_a = \frac{[H^+][A^-]}{[HA]} \qquad K_b = \frac{[HA][OH^-]}{[A^-]}$$

$$pK_a + pK_b = 14$$

The lower the pK_a, the more acidic HA is and the less basic the *conjugate base* A^- is. When pH = pK_a, $[A^-]$ = [HA].

Plane of symmetry (7.5). An object has a plane of symmetry if it can be divided into two halves that are mirror images of each other. Such objects are *achiral*.

Plane polarized light (7.2). Light in which the electric field vectors of all the light waves lies in the same plane, called the *plane* or *polarization*. An *optically active* sample rotates the plane of polarization of a beam of plane polarized light that passes through it.

Poisoned catalyst (12.6.A). A catalyst whose efficiency and activity have been reduced with a "poison". Lindlar's catalyst (Pd/$BaSO_4$, poisoned with quinoline) is an example.

Polarimeter (7.2). An instrument for measuring *optical activity*.

Polarizability (8.2). Often thought of in an intuitive sense as "softness" or "mushiness" of an atom, polarizability refers to the ease of deforming the electron density around an atom. It increases as one descends in the periodic table because the valence orbitals lie further from the nucleus and the electrons are held less tightly. In general, in protic solvents, the more polarizable a *nucleophile* is the more potent it is as a nucleophile.

Polarized light (7.2). See *Plane polarized light*.

Polymer, polymerization (36.4). The sequential linking of *monomers* to produce *dimers, trimers*, molecules of intermediate size (*telomers* or *oligomers*), and then of very large size (polymers) is called *polymerization*.

In the formation of an *addition polymer*, the bonds between the monomers are made by addition to a double bond, by a cationic, radical, or anionic process.

In the formation of *condensation polymers*, the bonds are made with the elimination of a small by-product molecule.

Copolymers incorporate more than one *monomer* in the chain, in contrast to *homopolymers* in which all the units are the same.

Ppm = parts per million (13.4). See *Chemical shift.*

Primary, Secondary, Tertiary, Quaternary (6.1, 24.1, 24.5.A). R is understood to be a carbon substituent, such as an alkyl group.

	Carbon	Hydrogen	Radical	Carbocation	Carbanion	Halide	Alcohol	Amine
Primary:	R–CH_3	R–C(H)(H)–H	$H_2\dot{C}$–R	H_2C^+–R	H_2C^-–R	RCH_2X	RCH_2OH	R–$\ddot{N}H_2$
Secondary:	R–C(H)(H)–R'	R–C(H)(H)–R'	H(R')$\dot{C}$–R	H(R')C^+–R	H(R')C^-–R	R–C(X)(H)–R'	R–C(OH)(H)–R'	R–$\ddot{N}$H–R'
Tertiary:	R–C(H)(R")–R'	R–C(H)(R")–R'	R"(R')$\dot{C}$–R	R"(R')C^+–R	R"(R')C^-–R	R–C(X)(R")–R'	R–C(OH)(R")–R'	R–$\ddot{N}$(R')–R"
Quaternary:	R–C(R''')(R")–R'							R–N^+(R''')(R")–R'

Principle of microscopic reversibility (6.3). "If the easiest way to get from Yosemite Valley to Mono Lake is through Tuolomne Meadows and over Tioga Pass, then the easiest way from Mono Lake to Yosemite Valley is over Tioga Pass and through Tuolomne Meadows", which is to say that the transition state for the forward reaction will be identical to the transition state for the back reaction.

Probability function (2.5). Squaring the value of a *wave function* at each point in space gives a probability function, which describes the likelihood that the electron will be found at that point. Whereas the wave function itself will have regions in which its value is negative and regions in which it is positive, the probability function is always positive (or zero).

Propagation step (6.3). See *Chain reaction.*

Protecting group (16.4). A functional group introduced to mask the reactivity of another functional group so that an otherwise interfering reaction can be carried out elsewhere on the molecule. It is later removed and the original functional group is regenerated.

functional group	*protecting group*
alcohols	*t*-butyl ethers (10.10)
	silyl ethers (26.11.C)
ketones and aldehydes	*acetals* and *ketals* (14.6.B)
anomeric carbons	*glycoside* formation (28.5.B)
amino acids	carbobenzoxy (Cbz) and *t*-butoxycarbonyl (Boc) groups (29.7.B)

Proteins (29.8). High molecular weight poly*peptides*.

Proton decoupled (13.5). See *Decoupling.*

Pseudo first-order reaction (4.3). See *First-order reaction.*

Pseudorotation (26.11.A). Interconversion of the trigonal bipyramidal intermediates that are involved in substitution reactions of phosphorus and silicon derivatives. In one step of *pseudorotation*, the *apical* substituents become *equatorial*, and two (of the three) equatorial substituents become apical; the other equatorial substituent is the pivot. *Pseudorotation* is involved in substitution reactions that occur with *retention* of configuration in these systems.

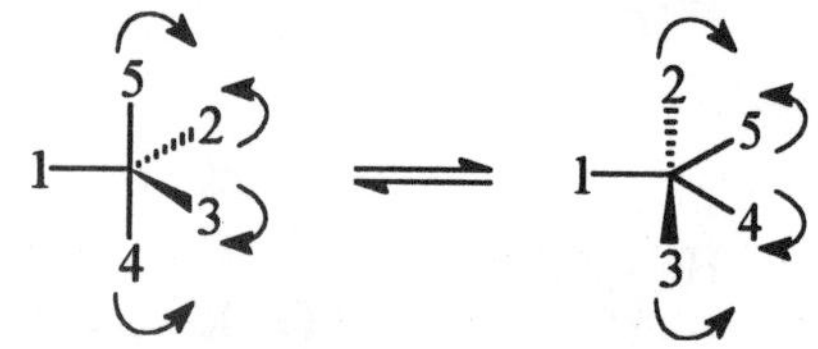

Pseudorotation is also used to describe the interconversions of the cyclopentane *envelope* conformations.

Pyrolysis (6.2). Cleaving a bond or a molecule simply by heating it to a sufficiently high temperature, as in the *cracking* of alkanes or the pyrolytic elimination of sulfoxides (25.3).

Quantum numbers (2.5), Quantized (13.2). At the atomic level, only certain *electronic configurations*, speeds of rotation, degrees of vibration, orientations in a magnetic field, etc., are possible. These are said to be *quantized*, and their specific value is indicated by the *quantum numbers*.

Quantum yield (36.3). The fraction of molecules that proceed to products after absorbing a photon in a photochemical reaction.

Quartet (13.7). See *Spin-spin splitting*.

Quaternary (10.6). Also see *Primary.*

R-S Convention (7.3). System of nomenclature for indicating the *absolute configuration* of a *chiral center.* Involves use of the "*sequence rule*".

Racemic mixture, racemate, racemic compound (7.4). An equimolar mixture of the two *enantiomers* of a compound. It is *optically inactive*. A *racemic compound* is a *racemic mixture* in which both *enantiomers* are present in the same crystal structure.

Racemization (7.4, 15.1.C). The equilibration of one *enantiomer* with the other, to produce a *racemic, optically inactive* mixture consisting of equal amounts of both *enantiomers*.

Radical In chemical reactions, *radicals* are usually present only as short-lived, unstable intermediates, because they have unpaired electrons and lack valence octets. Free radical halogenation (section 6.3) is a good example of a type of reaction that involves radical intermediates.

Radical anion (12.6.A). An anion with an odd number of electrons; usually obtained by addition of an electron to a neutral, even-electron molecule.

Radical cation (34.1). A cation with an odd number of electrons; usually formed from a neutral, even-electron molecule by ejection of an electron, as in the ionization chamber of a *mass spectrometer* on electron impact.

Reducing vs. non-reducing sugar (28.5.E). A sugar in equilibrium with its open chain form in alkaline solution (that is, a *hemiacetal* or *hemiketal*) is a *reducing sugar* because it can be oxidized by Fehling's or Tollen's reagents. If the sugar exists as a *glycoside*, it is a *non-reducing sugar*.

Reduction potential: see *Standard reduction potential*.

Reductive amination (24.6.F). Conversion of a carbon-oxygen double bond to a carbon-nitrogen single bond by reduction of an imine or immonium ion intermediate.

Reductive elimination (36.1). The reverse of *oxidative addition*.

Reinforcement (2.7). The overlap of *orbitals* having the same sign, so that their *wave functions* add together; the opposite of *interference*. The overlap of two atomic orbitals to make a bonding molecular orbital is an example of reinforcement.

Relative configuration (28.6). This terms refers to a *stereochemical* relationship between two or more *chiral centers*. If the *absolute configuration* of two chiral centers is known, so is their *relative configuration*, but it is possible to know their *relative configuration* without knowing their *absolute configuration*.

Terms such as *meso, erythro* and *threo*, and *cis* and *trans* (in cyclic systems) all describe *relative configurations*.

Relaxation (13.6, 14.4). In NMR spectroscopy, *relaxation* is the restoration of the spin distribution of the sample to its equilibrium value. For a ^{13}C nucleus, relaxation is accelerated by hydrogens directly attached to it. If relaxation is slow (for example, for *quaternary* or carbonyl carbons), *saturation* of the NMR signal occurs easily and only a weak resonance peak is seen. See *Integrated intensity*.

Resolution (29.5.F). Separation of the *enantiomers* of a *racemic mixture*, usually by the formation and separation of *diastereomeric salts* or other derivatives, and regeneration of the *enantiomers*.

Resonance (in *Spectroscopy*) (13.4). When the energy of electromagnetic radiation matches the energy difference between two quantum states, as given by the relationship $E = h\nu$, absorption or emission of radiation by the sample is possible, and the system is said to be in *resonance*.

Resonance energy (21.1.B). The special stabilization that an *aromatic* π-system has over a hypothetical system of similar electronic *configuration*, but that lacks the *delocalization* of the aromatic system.

The *empirical resonance energy* is determined by experimental comparison of actual molecules. It differs from *delocalization energy*, which results from comparison with theoretical models.

Resonance hybrid (2.4). A molecule that cannot be adequately represented by a single written structure, but can be understood as a hybrid of two or more *resonance structures*. (N.B. Don't confuse *resonance hybrids* with *hybrid orbitals.*)

Resonance structure (2.4). Structures depicting a molecule that differ only in the distribution of electrons (as opposed to nuclei). They are most important and useful when the alternative bonding arrangements they represent are of similar energy, in which case the molecule they describe is said to be a *resonance hybrid* of the two structures.

Retention (26.11.A). The opposite of *inversion*.

Ring current (21.3.A). The *π-electron circulation* induced in an *aromatic* system by an external magnetic field. See *Anisotropic* and *Chemical shift*..

S_N1 (Substitution$_{Nucleophilic}$Unimolecular (9.8). A description of the mechanism of a displacement reaction that involves only one molecule in the rate determining step and that takes place first with ionization of the carbon-leaving group bond (slow) and then attack by the nucleophile (fast). Usually goes with loss of configuration at the carbon atom, i.e., racemization if that is the only chiral center present. The rate depends on carbocation stability (3° > 2° > 1°).

S_N2 (Substitution$_{Nucleophilic}$Bimolecular) (9.2). A description of a mechanism of a displacement reaction, as the expanded name above implies. These reactions usually occur with inversion of configuration at the carbon undergoing attack. The rate depends on lack of steric hindrance, and is slowed either by α-branching (1° > 2° > 3° in rate) or β-branching (e.g., "neopentyl" systems *very* slow).

Sanger method (29.7.C). A method for determining the N-terminal amino acid of a *peptide* by labeling it with a dinitrophenyl group followed by hydrolysis of the peptide and identification of the derivatized amino acid.

Saponification (19.11.B). Alkaline hydrolysis of esters.

Saturated (3.2). From the point of view of *molecular formula*, *saturated* means that there are 2n + 2 hydrogens for every carbon present in the molecule, after a hydrogen has been added for every halogen atom present and subtracted for every nitrogen atom present. For instance, ethane (C_2H_6), 1,2-dichloroethane ($C_2H_4Cl_2$), and ethylamine (C_2H_7N) are all counted as $C_2H_6 = C_nH_{2n+2}$, where n = 2.

From the point of view of structure, *saturated* means containing only single bonds (σ bonds). It differs from the definition above only for cyclic compounds. For instance, cyclohexane (C_6H_{12}) has only single bonds, but the formula is C_nH_{2n}.

See *Unsaturated*.

Saturation (in the NMR (13.6, 14.4). Equalization of the populations of the α- and β-spin states, which leads to disappearance of the nmr signal. (The strength of the signal depends on the difference in population of these spin states.) Saturation is opposed by *relaxation*. (See *Integrated intensity*).

Schiff base (14.6.C). A substituted *imine*: $\gt C{=}\ddot{N}{-}R$

Second-Order kinetics (9.2). See *Kinetic order*.

Second-Order reaction (4.3). Usually, one in which two molecules are involved in the rate-determining step, and in which, therefore, the rate equation has the concentration of both molecules as variables.

Secondary (6.1). See *Primary*.

Selection rule (17.2). A quantum mechanical requirement for a transition between two energy levels to be allowed. In infrared spectroscopy, the transition must result in a change in the dipole moment of the molecule. Such transitions are *infrared active*.

In ultraviolet-visible spectroscopy, the two electronic states are subject to other quantum mechanical constraints for the *electronic transition* to be allowed.

<u>Sequence rule</u> (7.3). See *R-S convention.*

<u>Sigmatropic rearrangement</u> (21.7). See definition in text.

<u>Single bond</u> (2.3). See *Bond.*

<u>Singlet state</u> (36.3). An electronic state in which all the electron spins are paired, that is, in which the molecule has a net electronic spin of zero.

<u>Solid phase technique</u> (29.7.B). A method for the synthesis of poly*peptides* on a polymer support.

<u>Solvation</u> (4.2), <u>Solvation energy</u> (9.2, 9.5). *Solvation* refers to the interactions between solvent molecules and the molecules dissolved in them (the solute). These interactions often influence the mechanism and rate of a reaction in solution, and therefore frequently make it difficult to compare liquid and gas phase data.

The *solvation energy* is an indication of the strength of these interactions, and is the energy released on transferring a solute from the gas phase into a solvent. It reflects how well the solvent molecules stabilize ions, neutral molecules, or transition states, etc. Difficulties in estimating solvation energies often hinders detailed understanding of reaction kinetics.

<u>Solvolysis reaction</u> (9.8). A substitution reaction in which the nucleophile is the solvent. If the nucleophile is water it is called *hydrolysis*:

$$\text{A-B} + \text{RO-H} \longrightarrow \text{A-OR} + \text{H-B}$$

<u>Specific deuteration</u> (8.9.A). See *Deuteration.*

<u>Specific rotation,</u> [α] (7.2). $[\alpha]_D = \dfrac{\alpha}{l \cdot c}$

α = angle of rotation l = length in decimeters c = concentration in g/mL

<u>Spectra, spectroscopy</u> (3.4). *Spectroscopy* is the experimental evaluation of the way in which a substance interacts with electromagnetic radiation, usually for the purpose of structure determination. More detailed descriptions can be found in the Chapters on Nuclear Magnetic Resonance Spectroscopy (Chapter 13), Infrared Spectroscopy (Chapter 17), and Ultraviolet Spectroscopy (Chapter 22). Mass spectrometry (Chapter 34) is another experimental technique for structure determination, although it is not truly a "spectroscopy".

<u>Spin-spin splitting</u> (13.7). Results from the *coupling* of the *magnetic moments* of two nearby, *magnetically nonequivalent* nuclei, so that the orientation of one of them (with or against an external magnetic field) affects the *chemical shift* of the other. The amount that nucleus A affects the chemical shift of nucleus B is called the *coupling constant*, J, and is equal to the amount that nucleus B affects nucleus A, too.

In simple cases, n adjacent protons cause a splitting into n+1 peaks (*doublets, triplets, quartets,* etc.), but the situation can easily become complicated by nonequivalent coupling constants (two *different* adjacent nuclei) and *non-first-order* effects (*multiplets*). *Magnetically equivalent* nuclei do not split each other.

<u>Spiro ring system</u> (27.3.B). See *Bridged.*

Staggered (5.2). See definition under *Anti*.

Standard reduction potential (30.8.C). An indication of the *electron affinity* of an atom, ion, or molecule: the more *electropositive* a metal is, for example, the more easily it gives up its electron to form the cation, and the more negative the *standard reduction potential* of the cation is. The more positive the *reduction potential* of a quinone is, the stronger an oxidizing agent it is.

Stationary state (36.3). A reaction at equilibrium, usually in reference to a photochemical equilibration.

Stereoaxis (7.1, 20.2.C). *Chiral* molecules such as some allenes and biphenyls, in which formal rotation about a bond interconverts stereoisomers, are said to have a *stereoaxis*.

Stereocenter (7.1). If interchanging two substituents on an atom produces a *stereoisomer*, that atom is said to be a *stereocenter*. Examples of *stereocenters* are carbon atoms with four different substituents, as well as the sp^2-hybridized carbons of an alkene that is capable of *E* and *Z* isomerism. The term *stereocenter* replaces the older term *asymmetric center*.

Stereoisomers (7.1). Molecules that differ only in the three-dimensional relationship between their atoms.

Stereospecific (11.5.A, 16.2, 27.3.A). Requiring or generating a particular three-dimensional relationship. For example, the S_N2 reaction proceeds *stereospecifically* with *inversion*, and the *E2* reaction is usually stereospecific for the *anti* relationship between the *leaving group* and the proton that is being removed.

Stereospecific is also used in the sense of giving only one *stereoisomer*, and is a desirable goal in planning the synthesis of a molecule with chiral centers or *cis* or *trans* double bonds.

Steric hindrance (9.3). An effect that arises when two molecules or parts of molecules try to occupy the same space at the same time. It is responsible for the γ-effect in *CMR* (13.5), preference for *equatorial* versus *axial*-substitution on cyclohexane rings (5.6), the slowing of S_N2 displacement reactions by *α- or β-branching* (9.3), the greater stability of *trans* over *cis* alkenes (11.4), etc.

Structural formulas (3.1). Drawings of molecules that show how the atoms are connected. They contrast with *empirical formulas* and *molecular formulas* because a given structural formula represents only one of all possible structural *isomers*. Even more specific are stereo formulas (Chapter 7). Structural formulas may be very detailed

```
        H   H   H
        |   |   |
(e.g., H—C—C—C—O—H   for 1-propanol) or condensed (CH3CH2CH2OH)
        |   |   |
        H   H   H
```

(e.g., H–C(H)(H)–C(H)(H)–C(H)(H)–O–H for 1-propanol) or condensed ($CH_3CH_2CH_2OH$)

As you become more familiar with organic structures, you will use line formulas to save time. In these simple structural formulas, the C's and H's on carbon are omitted, and only the bonds between carbon and non-hydrogen atoms are drawn (for example, 1-propanol is drawn: ⁄\⁄OH).

Suprafacial (22.5.B). On the same face of a π-system. See *Antarafacial*.

Tautomers, tautomerism (15.1.A). *Isomers* that differ only in the placement of the protons; refers only to systems involving heteroatoms and that interconvert fairly rapidly.

Telomer (36.4). See *Polymer*.

Termination (6.3). See *Chain reaction*.

Tertiary (6.1). See *Primary*.

Thermodynamic control (20.2.B). A reaction in which the products are in equilibrium with the starting materials is said to be under thermodynamic control. That is, the product obtained is the most stable one. See *Kinetic control*.

Three center two-electron bond (8.6). A bond between three atoms, incorporating three *orbitals* in the *bonding molecular orbital*, which is occupied by two electrons.

Tracer isotope (23.2). A radioactive isotope (e.g., tritium (^3H) or ^{14}C) used in labeling studies. Only a small fraction of the atoms at a position labeled with a *tracer isotope* are actually the radioactive isotope itself. The presence of a radioactive isotope is determined with an instrument known as a liquid scintillation counter.

Transesterification (19.7.A). Exchange of alkyl groups in an ester by acid- or base-catalysis, usually via the *nucleophilic addition-elimination mechanism*.

Transmetallation (30.3.B). See *Metallation*.

Triplet (in NMR spectroscopy) (13.7). See *Spin-spin splitting*.

Triplet sensitizer (36.3). A molecule that is readily converted to its excited *triplet state* photochemically, and that will react with a different *ground state* molecule to convert it to its *triplet state*.

Triplet state (in photochemistry) (36.3). A molecule or atom in which two electron spins are unpaired, resulting in a net electronic spin of 1.

Unsaturated (3.5). An *unsaturated* molecule contains double or triple bonds (π-bonds). This reflected in the formula of a hydrocarbon: for every π bond, there are two fewer hydrogens. Compare ethane (C_2H_6), ethylene (C_2H_4), and acetylene (C_2H_2). From the point of view of *molecular* formula, however, a ring results in the same differences: 1-butene ($CH_3CH_2CH{=}CH_2 = C_4H_8$) or cyclobutane $\begin{matrix} CH_2-CH_2 \\ | \quad\quad | \\ CH_2-CH_2 \end{matrix} = C_4H_8$.

Each π-bond *and* ring in a compound, therefore, is considered to be a *degree of unsaturation* from the point of view of molecular formula.

See *Saturated*.

Valence electrons (2.2). Those in the outermost shell of an atom. The attempt by an atom to achieve a filled *octet* of valence electrons is chiefly responsible for the bonding arrangements it undergoes.

van der Waals forces (5.1). See *Dispersion force*.

Vertical transition (36.3). See *Franck-Condon transition*.

Vicinal (11.6.E, 12.5.C). Attached to adjacent carbons. See *Geminal*.

Vinylogy (27.7.B). The similarity of a compound in which two functional groups are separated by a conjugated double bond, with that in which the two groups are directly connected. For example, 4-methoxy-3-penten-2-one is a *vinylogous* ester.

Wave function (2.5). The equation that describes the behavior of an electron. The wavefunction can have regions in which its value is positive and regions in which it is negative, just as an ocean wave has peaks and troughs. These signs have no connection with the electron charge, which is always negative. The square of the wavefunction is a *probability function*.

Ylide (14.7.D, 26.8). A neutral molecule that has a formal negative charge on carbon as in *phosphoranes* ($R_3P^+ - {}^-CH_2$) and sulfur ylides ($R_2S^+ - {}^-CH_2$).

Z (11.2). See *E*.

Zero point energy (6.1). The difference between the lowest point on a potential energy diagram for a bond and the lowest energy attainable (at 0°K) within the constraints of the *Heisenberg Uncertainty Principle*.

Zwitterion (26.5.B, 29.1). A molecule that is an *inner salt*, that is, in which both the cation and anion are part of the same molecule.

APPENDIX II: SUMMARY OF FUNCTIONAL GROUP PREPARATIONS

Functional group interconversions can be discussed as reactions of one function or preparations of another. The following summary lists the preparations of various functional groups discussed in this textbook with reference to each place the reaction is used. Products shown are those for normal work-up of the reaction. Although examples are included with more than one functional group, specific reactions of polyfunctional compounds are not included. Abbreviations used are:

R = alkyl and cycloalkyl; for some cases may also apply to R = H or R = Ar
Ar = aryl
X = halide
Y = leaving group; may be X, sulfonate, and so on
[H] = several reducing agents
[O] = several oxidizing agents

The importance of a given type of reaction of functional group transformation can be gauged roughly by the number of times it is cited.

Acetals and Ketals

Aldehydes or ketones

$$\mathrm{RC(=O)R' + 2\ R''OH \xrightarrow{H^+} RCR'(OR'')_2}$$

394-397, 452, 863, 870, 911-913, 917-919

Acid Anhydrides

Acyl Halides

$$\mathrm{RC(=O)Cl + R'CO_2^- \longrightarrow RC(=O)OC(=O)R'}$$

531

Carboxylic acids

$$\mathrm{(CH_2)_x(CO_2H)_2 \longrightarrow (CH_2)_x(C{=}O)_2O}$$

x = 2 or 3

530, 880, 882

Acyl Halides

Carboxylic acids

$\mathrm{RCOOH + SOCl_2 \longrightarrow RCOCl}$ — 502, 528

$\mathrm{RCOOH + PCl_5 \longrightarrow RCOCl}$ — 502

$\mathrm{RCOOH + PBr_3 \longrightarrow RCOBr}$ — 498, 502, 591

Alcohols

Aldehydes and ketones: Carbanion additions

$\mathrm{RCOR' + R''MgX\ or\ R''Li \longrightarrow \xrightarrow{H^+} RR'R''COH}$ — 401-404, 406-407, 440-441, 446, 451, 531-532, 584-585, 828, 1011, 1059

$RCOR' + R''C{\equiv}C^- \longrightarrow \xrightarrow{H^+} RR'C(OH)C{\equiv}CR''$ — 404, 576

Aldehydes and ketones: Reductions

$RCOR' + LiAlH_4$ or $NaBH_4 \longrightarrow RCHOHR'$ — 412-414, 452, 586, 854, 863, 871, 907, 919, 921, 1067

$RCOR' \xrightarrow{H_2/catalyst} RCHOHR'$ — 414, 632, 921

$RC(=O)R' \xrightarrow{Mg} \xrightarrow{H^+} RC(R')(OH){-}C(R')(OH)R$ — 862

Alkenes

$>C{=}C< + H_2O \xrightarrow{H^+} {-}C(H){-}C(OH){-}$ — 278, 819

$>C{=}C< + HOX \longrightarrow {-}C(X){-}C(OH){-}$ — 236, 277, 286, 443, 1085

$>C{=}C< \xrightarrow[H_2O]{Hg^{2+}} {-}C(Hg^+){-}C(OH){-} \xrightarrow{NaBH_4} {-}C(H){-}C(OH){-}$ — 279

$>C{=}C< \xrightarrow{B_2H_6} \xrightarrow{H_2O_2} {-}C(H){-}C(OH){-}$ — 280-282, 443

$RCH{=}CHR \xrightarrow{O_3} \xrightarrow{NaBH_4} RCH_2OH$ — 285

$>C{=}C< \xrightarrow{KMnO_4} {-}C(OH){-}C(OH){-}$ — 283, 859-860

$>C{=}C< \xrightarrow[H_2O_2]{OsO_4} {-}C(OH){-}C(OH){-}$ — 283-284, 442, 858, 861

Amines

$ArNH_2 \xrightarrow{HNO_2} ArOH$ — 791

Carboxylic acids

$RCOOH \xrightarrow{LiAlH_4} RCH_2OH$ — 413, 504

Esters

$R'COOR \xrightarrow[H_2O]{H^+ \text{ or } OH^-} R'COOH + ROH$ 63-64, 215-216, 243, 546, 628, 875

$R'COOR \xrightarrow[\text{or } NaBH_4]{LiAlH_4} R'CH_2OH + ROH$ 413, 535

$RCOOEt + R'MgX \text{ or } R'Li \longrightarrow \xrightarrow{H^+} RR'_2COH$ 532-533

$RMgX + (EtO)_2C{=}O \longrightarrow R_3COH$ 533

$R'COOR \xrightarrow[\text{or } H_2/\text{cat.}]{Na/EtOH} R'CH_2OH + ROH$ 535

Ethers

$ROR \xrightarrow{HX} ROH + RX$ 239, 1030-1031

Halides and sulfonates

$RY \xrightarrow{H_2O \text{ or } OH^-} ROH$ 56, 171, 172, 197, 215, 243, 559-560, 874

Oxiranes

oxirane $\xrightarrow[\text{or } H^+]{OH^-}$ 1,2-diol (HO, OH) 236-238, 443, 859, 1087

$RM + H_2C{-}CH_2$ (oxirane, O bridging) $\longrightarrow RCH_2CH_2OH$ 239, 440, 447

Silyl ethers

$RO{-}SiR'_3 \xrightarrow{F^- \text{ or } H_3O^+} ROH$ 846-847

Aldehydes

Acetals

$RCH(OR')_2 \xrightarrow{H^+} RCHO + R'OH$ 397, 452, 904, 911-913

Acid derivatives

$RCOCl \xrightarrow[Li(t\text{-}BuO)_3AlH]{H_2/\text{cat. or}} RCHO$ 534

$RCONR'_2 \xrightarrow{LiAlH(OEt)_3} RCHO$ 536

$$RCH(OH)CO_2H \xrightarrow{CaO} \xrightarrow[H_2O_2]{Fe^{3+}} RCHO$$ 928

Alcohols

$$RCH_2OH \xrightarrow{[O]} RCHO$$ 228-230, 387, 581

$$RCH(OH)CH(OH)R' \xrightarrow[Pb(OAc)_2]{HIO_4 \text{ or}} RCHO + R'CHO$$ 866, 871, 923-925

Alkenes

$$RCH{=}CHR \xrightarrow{O_3} \xrightarrow[(CH_3)_2S]{Zn/AcOH \text{ or}} RCHO$$ 284, 388

Alkynes

$$RC{\equiv}CH \xrightarrow{B_2H_6} \xrightarrow[OH^-]{H_2O_2} RCH_2CHO$$ 316, 389

Arenes

$$ArH + HCON(CH_3)_2 \xrightarrow{POCl_3} ArCHO$$ 759-760, 1100

(Vilsmeier reaction)

Enol ethers

$$R_2C{=}CHOR' \xrightarrow[H_2O]{H^+} R_2CHCHO$$ 397

Phenols

$$ArO^- + CHCl_3 \longrightarrow Ar(OH)(CHO)$$ 1023

(Reimer-Tiemann reaction)

Alkanes and Arenes

Alcohols

$$ArC(R)(R')\text{—OH (or COR)} \xrightarrow[HClO_4]{H_2/Pd} ArC(R)(R')\text{—H}$$ 633

Aldehyde

$$RCHO \xrightarrow{L_3RhCl} RH$$ 1223

Summary of Functional Group Preparations

Alkenes

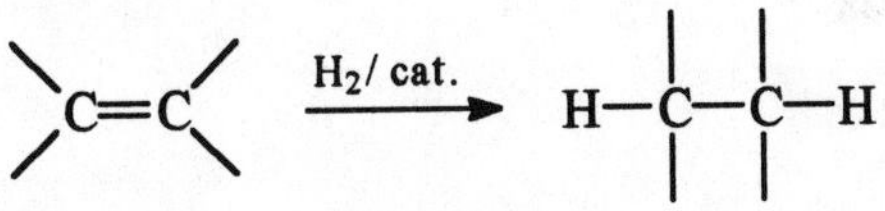

$>C{=}C<$ $\xrightarrow{H_2/\text{ cat.}}$ $H{-}\overset{|}{\underset{|}{C}}{-}\overset{|}{\underset{|}{C}}{-}H$ 271-272, 441, 450, 451, 587, 632, 854

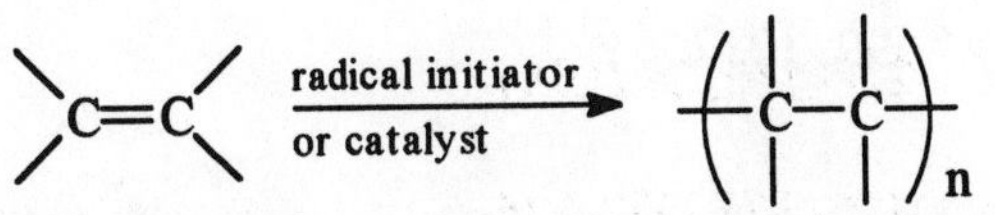

$>C{=}C<$ $\xrightarrow[\text{or catalyst}]{\text{radical initiator}}$ $\left(-\overset{|}{\underset{|}{C}}{-}\overset{|}{\underset{|}{C}}-\right)_n$ 1240-1247

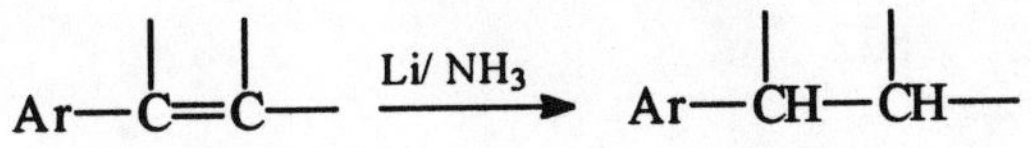

$Ar{-}\overset{|}{C}{=}\overset{|}{C}{-}$ $\xrightarrow{Li/\ NH_3}$ $Ar{-}\overset{|}{C}H{-}\overset{|}{C}H{-}$ 635

$-\overset{O}{\overset{\|}{C}}{-}\overset{|}{C}{=}\overset{|}{C}{-}$ $\xrightarrow{Li/\ NH_3}$ $-\overset{O}{\overset{\|}{C}}{-}\overset{|}{C}H{-}\overset{|}{C}H{-}$ 587-588

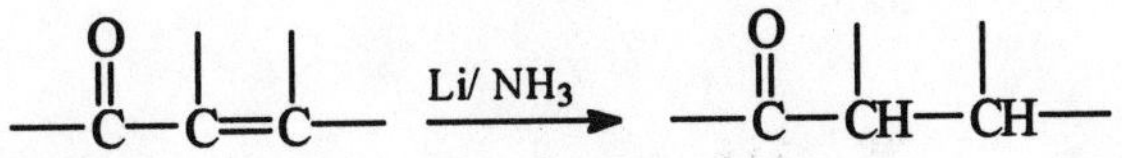

Alkynes

$RC{\equiv}CR' \xrightarrow{H_2/\text{cat.}} RCH_2CH_2R'$ 309

Amines

$ArNH_2 \xrightarrow{HONO} ArN_2^+ \xrightarrow{H_3PO_2} ArH$ 795, 801, 1051

Arenes

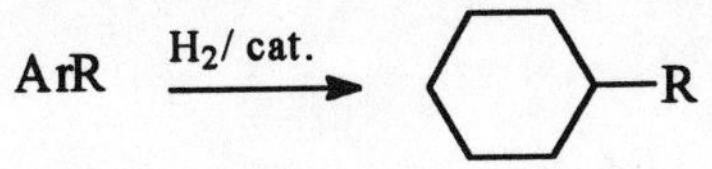

$ArR \xrightarrow{H_2/\text{ cat.}}$ (cyclohexyl)$-R$ 632, 879, 1063, 1069

Cyclohexanes

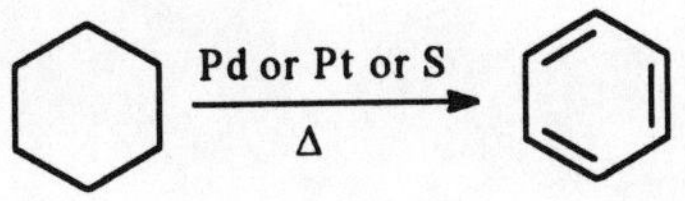

(cyclohexane) $\xrightarrow[\Delta]{\text{Pd or Pt or S}}$ (benzene) 613, 1059, 1068, 1121

Halides and sulfonates

$RX \xrightarrow{LiAlH_4} RH$ 413, 633

$RNa + RX \longrightarrow RR'$ 192

$ArLi$ or $ArMgX + RX \longrightarrow ArR$ 1012

$ArX + RX \xrightarrow{Na} ArR$
(Wurtz-Fitting reaction) 1012

$ArI + Cu \longrightarrow Ar\text{-}Ar$
(Ullmann reaction) 1012

$ArH + RX \xrightarrow{AlCl_3} ArR + ArR'$ 698-700, 718, 824
(Friedel-Crafts alkylation)

$ArH + \ce{>C=C<} \xrightarrow{acid} Ar-\overset{|}{\underset{|}{C}}-\overset{|}{\underset{|}{C}}-H$ 699, 1017

Ar_2CH_2 (or Ar_3CH) $\xrightarrow{NaNH_2} \xrightarrow{RX} Ar_2CHR$ (or Ar_3CR) 632

$ArSO_3H \xrightarrow[100°]{H_3O^+} ArH$ 823

Ketones

$RCOR' \xrightarrow[KOH/diethylene\ glycol,\ \Delta]{H_2NNH_2} RCH_2R'$ 414-415, 441, 451, 715

(Wolff-Kishner reduction)

$RCOR' \xrightarrow[HCl,\ \Delta]{Zn(Hg)} RCH_2R'$ 415, 451, 715

(Clemmensen reduction)

$RCOR' \xrightarrow[BF_3]{HS(CH_2)_3SH}$ (1,3-dithiane, S–C(R)(R)–S) $\xrightarrow{Raney\ Ni} RCH_2R'$ 816

$ArCOR' \xrightarrow{H_2/Pd} ArCH_2R$ 633

Organometallics

$RMgX$, RLi, or $R_3Al \xrightarrow{R'OH} RH$ 168, 451, 453, 633

$R_4Si \xrightarrow{HCl} RH$ 168

Alkenes

Alcohols

$-\overset{H}{\underset{|}{C}}-\overset{OH}{\underset{|}{C}}- \xrightarrow[cold,\ \Delta]{H^+\ or\ Lewis} \ce{>C=C<}$ 227, 233, 267-269, 410, 441, 448, 451, 575, 576, 581, 629, 868-869, 878, 1059 1068

Summary of Functional Group Preparations

Aldehydes and ketones

$RR'C{=}O + Ph_3P{=}CHR'' \longrightarrow RR'C{=}CHR''$ (Wittig reaction)	408-410, 440, 448, 576

Alkynes

$RC{\equiv}CR' \xrightarrow{H_2/cat.}$ *cis*-$RHC{=}CHR'$	309, 311, 442, 449, 576
$RC{\equiv}CR' \xrightarrow[liq.\ NH_3]{Na}$ *trans*-$RHC{=}CHR'$	309-311, 442

Amine oxides

$H{-}\overset{\vert}{\underset{\vert}{C}}{-}\overset{\vert}{\underset{\vert}{C}}{-}{}^{+}N(CH_3)_2O^{-} \xrightarrow{\Delta} \rangle C{=}C\langle$	764-765

Ammonium hydroxides

$H{-}\overset{\vert}{\underset{\vert}{C}}{-}\overset{\vert}{\underset{\vert}{C}}{-}{}^{+}N(CH_3)_3\ \ OH^{-} \xrightarrow{\Delta} \rangle C{=}C\langle$	761-764

Arenes

$ArR \xrightarrow[NH_3]{Na}$ 1-R-cyclohexa-1,4-diene	634-636, 1063, 1069
or 3-R-cyclohexa-1,4-diene (for $R = CO_2H$)	636

Esters

$H{-}\overset{\vert}{\underset{\vert}{C}}{-}\overset{\vert}{\underset{\vert}{C}}{-}OCR({=}O) \xrightarrow{\Delta} \rangle C{=}C\langle + RCO_2H$	575

Ethers

$H{-}\overset{\vert}{\underset{\vert}{C}}{-}\overset{\vert}{\underset{\vert}{C}}{-}OR \xrightarrow[\Delta]{H^{+}} \rangle C{=}C\langle$	234, 453

Halides and sulfonates (X = halogen; Y = X or sulfonate)

$H{-}\overset{\vert}{\underset{\vert}{C}}{-}\overset{\vert}{\underset{\vert}{C}}{-}Y \xrightarrow{base} \rangle C{=}C\langle$	191, 197, 215, 217, 232, 243, 260-267 307, 443, 448, 591, 593

Reaction	Pages
$-\overset{X}{\underset{\vert}{C}}-\overset{Y}{\underset{\vert}{C}}- \xrightarrow[Mg]{Zn\ or} \rangle C=C\langle$	593, 656
$RMgX + CH_2{=}CHCH_2Y \longrightarrow RCH_2CH{=}CH_2$	563, 576

Sulfoxides

Reaction	Pages
$-\overset{H}{\underset{\vert}{C}}-\overset{S(O)R}{\underset{\vert}{C}}- \xrightarrow[\Delta]{} \rangle C=C\langle$	814

Alkynes

Reaction	Pages
$RCX_2CH_2R' \xrightarrow{base} RC{\equiv}CR'$	307-308
$RCHXCHXR' \xrightarrow{base} RC{\equiv}CR'$	307-308
$RC{\equiv}C^- + R'Y \longrightarrow RC{\equiv}CR'$	305-306, 396, 440, 449
$RCH_2C{\equiv}CH \overset{strong\ base}{\rightleftharpoons} RC{\equiv}CCH_3$	307-308

Amides

Acid anhydrides

Reaction	Pages
$(RCO)_2O + R'NH_2 \longrightarrow RCONHR'$	530, 752, 753, 797, 801, 964

Acyl Halides

Reaction	Pages
$RCOCl + R'NH_2 \longrightarrow RCONHR'$	530, 752, 961, 964, 971

Carboxylic acids

Reaction	Pages
$RCOOH + R'NH_2 \xrightarrow{\Delta} RCONHR'$	502-503, 752, 761, 969
$RCOOH + R'NH_2 \xrightarrow{DCC} RCONHR'$	972-973

Esters

Reaction	Pages
$RCOOR' + R''NH_2 \longrightarrow RCONHR''$	530

Ketones

Reaction	Pages
$R_2C{=}O + HN_3 \xrightarrow{H^+} RCONHR$ (Schmidt reaction)	960

Nitriles

Reaction	Pages
$RCN \xrightarrow[or\ H_2O_2/OH^-]{H^+} RCONH_2$	314, 315, 494, 883

Summary of Functional Group Preparations

Amines

Acyl halides

$RCOCl + N_3^- \longrightarrow RCON_3 \xrightarrow[H_2O]{\Delta} RNH_2$ — 750

(Curtius reaction)

Aldehydes and ketones

$RC(=O)R'' + R'NH_2 \xrightarrow{H_2/\text{ cat.}} RCH(NHR')R''$ — 746-747

if NH_3, then $RCH(NH_2)R''$

$RC(=O)H + HCN + NH_3 \longrightarrow RCH(NH_2)CN$ — 959

Amides

$RCONR'R'' \xrightarrow[OH^-]{H^+ \text{ or}} RCOOH + HNR'R''$ — 526, 761, 801, 960, 970, 1086

$RCONR'R'' \xrightarrow{LiAlH_4} RCH_2NR'R''$ — 413, 535-536, 748

$RCONH_2 + Br_2 \xrightarrow{NaOH} RNH_2$ — 750-751

(Hofmann rearrangement)

Amines

$RNH_2 \text{ (or } RR'NH) \xrightarrow[\Delta]{CH_2O,\ HCOOH} RN(CH_3)_2 \text{ (or } RR'NCH_3)$ — 747

(Eschweiler-Clarke reaction)

Carboxylic acids

$RCOOH + NaN_3 \xrightarrow[\Delta]{H^+} RNH_2$ — 751, 960

(Schmidt reaction)

Halides and sulfonates

$RY + R'NH_2 \text{ (or } NH_3) \longrightarrow RR'NH \text{ (or } RNH_2)$ — 172, 197-198, 742-743, 957, 959, 1086, 1087, 1089

RY + (phthalimide anion, N^-) $\longrightarrow$ [N-alkylphthalimide, NR] $\xrightarrow[H_2NNH_2]{OH^- \text{ or}} RNH_2$ — 743

$RY + N_3^- \longrightarrow RN_3 \xrightarrow[\text{or } LiAlH_4]{H_2/\text{cat.}} RNH_2$ — 786

$ArX + NH_2^- \xrightarrow{\text{liq. } NH_3} ArNH_2 + Ar'NH_2$ 1009, 1115

Nitriles

$RCN \xrightarrow[H_2/\text{cat.}]{LiAlH_4 \text{ or}} RCH_2NH_2$ 312, 413, 745

Nitro compounds

$RNO_2 \xrightarrow{LiAlH_4} RNH_2$ 413

$RNO_2 \xrightarrow[H_2/\text{cat.}]{Fe, H^+ \text{ or}} RNH_2$ 744, 780, 800, 801, 1006

Oximes

$RR'C{=}NOH \xrightarrow[\text{or } LiAlH_4]{H_2/\text{cat.}} RR'CHNH_2$ 746, 958

Azides

$RY + N_3^- \longrightarrow RN_3$ 172, 749, 786, 855

$RCH\text{—}CH_2$ (epoxide, O bridging) $+ N_3^- \longrightarrow RCH(OH)CH_2N_3$ 786

Carboxylic Acids

Acid anhydrides

$(RCO)_2O + H_2O \longrightarrow RCOOH$ 522, 526

Acyl halides

$RCOCl + H_2O \longrightarrow RCOOH$ 522, 526

Alcohols

$RCH_2OH \xrightarrow{[O]} RCOOH$ 495-496, 930, 932

Aldehydes

$RCHO \xrightarrow{[O]} RCOOH$ 410-412, 496, 582, 921-923, 930, 932, 933

Alkenes

$RCH{=}CHR' \xrightarrow{KMnO_4} RCOOH + R'COOH$ 741, 880

Summary of Functional Group Preparations

Amides

$$RCONH_2 \xrightarrow[\text{or HONO}]{H^+ \text{ or } OH^-} RCOOH$$ 494, 522, 526

Arenes

$$ArR \xrightarrow{[O]} ArCOOH$$ 630-631, 714, 780, 880, 1063, 1112

Esters

$$RCOOR' \xrightarrow{H^+ \text{ or } OH^-} RCOOH + R'OH$$ 63-64, 215-216, 522-525, 546, 591, 875, 890-891, 894, 971

$$RCOOCH_2Ar \xrightarrow[Pt]{H_2} RCOOH$$ 963, 971

$$RCO_2{-}\overset{|}{\underset{|}{C}}{-}\overset{|}{\underset{|}{C}}{-}H \xrightarrow[\Delta]{} RCO_2H + \;\rangle C{=}C\langle$$ 575

Halides

$$RX \xrightarrow[\text{ether}]{Mg} [RgMX] \xrightarrow{CO_2} RCOOH$$ 495, 1011

Hydroxy ketones

$$R\overset{O}{\overset{\|}{C}}\underset{OH}{\underset{|}{C}}HR' \xrightarrow{HIO_4} RCO_2H + R'\overset{O}{\overset{\|}{C}}H$$ 871

Ketones

$$RCOCH_2R' \xrightarrow{[O]} RCOOH + R'COOH$$ 412

Nitriles

$$RCN \xrightarrow{H^+ \text{ or } OH^-} RCOOH$$ 494, 591, 874, 878, 927, 959

Enamines

$$\rangle CH\overset{O}{\overset{\|}{C}}{-} + R_2NH \longrightarrow \;\rangle C{=}C\langle^{NR_2}$$ 766-767, 1106, 1120

$${-}\overset{|}{C}H{-}\overset{|}{C}H{-}NR_2 \xrightarrow{Hg(OAc)_2} \xrightarrow{OH^-} \;\rangle C{=}C\langle^{NR_2}$$ 767

Epoxides (see Oxiranes)

Esters

Acid anhydrides

$(RCO)_2 + R'OH \longrightarrow RCOOR'$ 528, 919

Acyl halides

$RCOCl + R'OH \longrightarrow RCOOR'$ 189, 528, 591, 871, 1025

Alcohols and phenols

$ROH + R'COOH \xrightarrow{H^+} R'COOR$ 499-501, 875-878, 963

$ROH + R'COCl \longrightarrow R'COOR$ 189, 528, 871, 880, 1025

$ROH + (R'CO)_2O \longrightarrow R'COOR$ 528, 1025

Amides

$RCONH_2 + R'OH \xrightarrow{H^+} RCOOR'$ 592

Carboxylic acids

$RCOOH + CH_2N_2 \longrightarrow RCOOCH_3$ 497, 787

$RCOOH + R'OH \xrightarrow{H^+} RCOOR'$ 499-501, 875-878, 963

$RCO_2^- + R'Y \longrightarrow RCOOR'$ 173-175, 215, 243, 496, 821, 877, 974

Esters

$RCOOR' + R''OH \xrightarrow{H^+ \text{ or } R''O^-} RCOOR'' + R'OH$ 529, 1245

Halides and sulfonates

$RY + R'CO_2^- \longrightarrow R'COOR$ 173-175, 215, 243, 496, 821, 877, 974

Ketones

$R_2C{=}O + R'CO_3H \longrightarrow RCO_2R$ 412, 875

Ethers

Alcohols

$ROH \xrightarrow[\Delta]{H^+} ROR$ 227, 233, 267, 865

Alkenes

$>C{=}C< + ROH \xrightarrow{H^+} -\underset{|}{\overset{H}{\overset{|}{C}}}-\underset{|}{\overset{OR}{\overset{|}{C}}}-$ 452

$>C{=}C< + ROH \xrightarrow{Hg(OAc)_2} \xrightarrow{NaBH_4} -\underset{|}{\overset{H}{\overset{|}{C}}}-\underset{|}{\overset{OR}{\overset{|}{C}}}-$ 279

Summary of Functional Group Preparations

$$>C=C< + ROH + X_2 \longrightarrow -\overset{X}{\underset{|}{\overset{|}{C}}}-\overset{OR}{\underset{|}{\overset{|}{C}}}-$$ 275

Alkynes

$$RC{\equiv}CH \xrightarrow[R'OH]{R'O^-} R\overset{OR'}{\overset{|}{C}}{=}CH_2$$ 315, 397

Halides and sulfonates

$RY + R'O^-$ (or $R'OH$) $\longrightarrow ROR'$ 172, 189, 197, 217, 232, 236-2398 819, 855, 918

$RY + ArO^- \longrightarrow ROAr$ 1020

$ArX + RO^- \longrightarrow ArOR$ 1020

Ketones

$$R\overset{O}{\overset{||}{C}}CH_2R \xrightarrow{LDA} \xrightarrow{Me_2SiCl} R\overset{OSiMe_3}{\overset{|}{C}}{=}CHR$$ 428

Halogen Compounds

Alcohols

$ROH + HX \longrightarrow RX$ 218-219, 224-226, 559

$ROH + SOCl_2 \longrightarrow RCl$ 220, 225-226, 447, 562, 629, 633

$ROH + PBr_3 \longrightarrow RBr$ 220-221, 225-226

$ROH + PI_3 \longrightarrow RI$ 221, 225

Alkanes

$RH + X_2 \xrightarrow{h\nu} RX$ 105-116, 145-147, 242-243, 441, 497-498, 568-569

Alkenes

$$>C=C< + HX \longrightarrow -\overset{H}{\underset{|}{\overset{|}{C}}}-\overset{X}{\underset{|}{\overset{|}{C}}}-$$ 277-278, 288-290, 571

$$>C=C< + X_2 \longrightarrow -\overset{X}{\underset{|}{\overset{|}{C}}}-\overset{X}{\underset{|}{\overset{|}{C}}}-$$ 272-275, 443, 571-573, 583, 689

$$>C=C< + HOX \longrightarrow -\overset{X}{\underset{|}{\overset{|}{C}}}-\overset{OH}{\underset{|}{\overset{|}{C}}}-$$ 236, 275-276, 286, 443, 1085

Reaction	Pages
$>C=C< + CX_4 \longrightarrow -\overset{X}{\underset{\vert}{C}}-\overset{CX_3}{\underset{\vert}{C}}-$	289
$RCH=CH-SiMe_3 + X_2 \longrightarrow RCH=CHX$	844
$CH_2=CH-\overset{H}{\underset{\vert}{C}}- \xrightarrow{NBS} CH_2=CH-\overset{Br}{\underset{\vert}{C}}-$	568-569
Alkynes	
$RC\equiv CR' + HX \longrightarrow RCH=CXR'$	312-313, 316, 317
Amines	
$ArNH_2 \xrightarrow{HONO} ArN_2^+ \xrightarrow{I^-} ArI$	792, 800
$ArNH_2 \xrightarrow{HONO} ArN_2^+ \xrightarrow[\Delta]{CuX} ArX$ (X=Cl,Br)	793, 1006
$ArNH_2 \xrightarrow[BF_3 \text{ (or } HPF_6)]{HONO} ArN_2^+BF_4^-$ (or PF_6^-) $\xrightarrow{\Delta} ArF$	794, 800
Arenes	
$ArH + X_2 \longrightarrow ArX$	689-693, 702-705, 717, 758, 759, 761, 795, 823, 1005-1006, 1021, 1028, 1061, 1070, 1096, 1107, 1114
$ArH + CH_2O + HCl \xrightarrow{ZnCl_2} ArCH_2Cl$	700, 1100
$ArCH_3 + Cl_2 \xrightarrow{h\nu} ArCH_2Cl$	624-626
$ArCHR_2 + Br_2 \xrightarrow{h\nu} ArCBrR_2$	627
$ArH + CCl_4 \xrightarrow{AlCl_3} Ar_3CCl$	700
$Ar\text{-}SiMe_3 + X_2 \longrightarrow Ar\text{-}X + X\text{-}SiMe_3$	845-846
Carboxylic acids	
$RCH_2COOH + Br_2 \xrightarrow{PBr_3} RCHBrCOOH$ (Hell-Volhard-Zelinsky reaction)	498-499, 591, 874
$RCOOH \xrightarrow{Ag^+} \xrightarrow[CCl_4]{Br_2} RBr$	505
$RCOOH \xrightarrow[Br_2]{HgO} RBr$	505

Summary of Functional Group Properties

$RCOOH \xrightarrow[LiCl]{Pb(OAc)_4} RCl$ — 505

Ethers

$ROR' + HX \longrightarrow ROH + R'X \text{ or } RX + R'X$ — 233, 239

Halides and sulfonates

$RY + X^- \longrightarrow RX$ — 172, 191, 222, 562

Ketones

$-\overset{|}{C}H-\overset{O}{\overset{\|}{C}}- \xrightarrow[H^+ \text{ or } OH^-]{X_2} -\overset{X}{\overset{|}{\underset{|}{C}}}-\overset{O}{\overset{\|}{C}}-$ — 430-431

Hemiacetals, Hemiketals: see **Acetals and Ketals**

Imines
Aldehydes and ketones

$RR'C{=}O + R''NH_2 \longrightarrow RR'C{=}NR''$
$R' = \text{alkyl or H}$ — 398-401, 746-747, 925-926, 929, (1104-1105), 1124

Amines

$R_2CHNR_2' \xrightarrow{Hg(OAc)_2} R_2C{=}^+NR_2'$ — 767

Ketones
Acyl Halides

$RCOCl + R'MgX \text{ (or } R'_2CuLi \text{ or } R'_2Cd) \longrightarrow RR'C{=}O$ — 531

$RCOCl + CH_2N_2 \longrightarrow RCOCHN_2 \text{ or } RCOCH_2Cl$ — 787

Alcohols

$RR'CHOH \xrightarrow{[O]} RR'C{=}O$ — 228-230, 242-243, 387-388, 630, 884, 919

$-\underset{|}{\overset{OH}{\overset{|}{C}}}-\underset{|}{\overset{OH}{\overset{|}{C}}}- \xrightarrow[Pb(OAc)_2]{HIO_4 \text{ or}} -\underset{|}{\overset{O}{\overset{\|}{C}}} + \underset{|}{\overset{O}{\overset{\|}{C}}}-$ — 866, 923-925

$R_2\overset{OH}{\overset{|}{C}}-\overset{OH}{\overset{|}{C}}R_2 \xrightarrow{H^+} R_3\overset{O}{\overset{\|}{C}}CR$ — 864

Alkenes

$$\rangle C{=}C\langle \xrightarrow{O_3} \rangle C{=}O + O{=}C\langle$$ 284-285, 388

Alkynes

$$RC{\equiv}CR \xrightarrow[Hg^{2+}]{H^+} RCOCH_2R$$ 313-315, 388-389

$$RC{\equiv}CR \xrightarrow{B_2H_6} \xrightarrow[OH^-]{H_2O_2}$$ 316

Arenes

$$ArH + RCOCl \xrightarrow{\text{Lewis acid}} ArCOR$$ 696-697, 718, 759, 844, 1062

$$ArH + (RCO)_2O \xrightarrow{AlCl_3} ArCOR$$ 1028-1030, 1058, 1065, 1067,

$$ArH + RCOOH \xrightarrow[\text{or Lewis acid}]{H^+} ArCOR$$ 1028, 1068

$$ArCH_2R \xrightarrow{[O]} ArCOR$$ 1055

Dithioacetals

$$\text{(1,3-dithiane, } S{-}C(R)(R'){-}S) \xrightarrow[H_2O]{HgCl_2} RC(=O)R'$$ 836-837

Enol ethers

$$RC(OR'){=}CR_2 \xrightarrow[H_2O]{H^+} RC(=O)CHR_2$$ 397, 635

Esters

$$R'CO_2Et + RCH_2CO_2Et \xrightarrow{\text{base}} R'C(=O)CHRCO_2Et$$ 539-541, 883-885

Ketones

$$RCH_2C(=O)R' \xrightarrow{SeO_2} RC(=O){-}C(=O)R'$$ 884

Nitriles

Aldehydes and ketones

Summary of Functional Group Properties

$$RC(=O)R' + HCN \longrightarrow RR'C(OH)CN$$
R′ = alkyl or H

404-406, 440, 592, 745, 874, 927

$$RC(=O)R' + HCN + NH_3 \longrightarrow RR'C(NH_2)CN$$
R′ = alkyl or H

959

$$-C{=}C-C(=O)- + HCN \longrightarrow -C(CN)-CH-C(=O)-$$

583-584, 878

Amides

$$RCONH_2 \xrightarrow[SOCl_2]{P_2O_5 \text{ or}} RCN$$

543

Amines

$$ArNH_2 \xrightarrow{HONO} ArN_2^+ \xrightarrow{CuCN} ArCN$$
(Sandmeyer reaction)

793, 800

Halides and sulfonates

$$RY + CN^- \text{ (or CuCN)} \longrightarrow RCN$$

173, 276, 304-305, 440, 494, 592, 628, 878

Oximes

$$RCH{=}NOH \xrightarrow{Ac_2O} RCN$$

929

Sulfonic acids

$$ArSO_3H \xrightarrow[\Delta]{NaCN} ArCN$$

825

Nitro Compounds
Amines

$$ArNH_2 \xrightarrow{HONO} ArN_2^+ \xrightarrow[Cu]{NO_2^-} ArNO_2$$

794

Arenes

$$ArH + HNO_3 \longrightarrow ArNO_2$$

695-696, 702-704, 716-717, 758, 760, 800, 801, 1006, 1026-1027, 1054, 1060, 1064, 1095-1096, 1107, 1113-1115, 1121-1122, 1125

Halides and sulfonates

$RY + NO_2^- \longrightarrow RNO_2$ 172, 779

Ketones

$RCOR' + CH_3NO_2 \xrightarrow{\text{base}} RC(OH)(R')CH_2NO_2$ 780

Organometallics

Boron

$RCH{=}CH_2 + B_2H_6 \longrightarrow (RCH_2CH_2)_3B$ 280-282

$RC{\equiv}CR + B_2H_6 \longrightarrow (RCH{=}CR)_3B$ 316

Cadmium

$RMgX + CdCl_2 \longrightarrow R_2Cd$ (R = prim.) 167

Copper

$RLi + CuI \longrightarrow R_2CuLi$ 532, 585, 1012

Lithium

$RX + Li \xrightarrow{\text{ether}} RLi$ 166, 1012

$ArBr + RLi \longrightarrow ArLi$ 1012

Magnesium (Grignard reagents)

$RX + Mg \xrightarrow{\text{ether}} RMgX$ 167, 451, 453, 495, 563-564, 1011, 1070

Mercury

RLi or $RMgX + HgCl_2 \longrightarrow R_2Hg$ 167

$RCH{=}CH_2 + Hg(OAc)_2 \xrightarrow{R'OH} RCH(OR')CH_2HgOAc$ 279

Silicon

$RMgX + SiCl_4 \longrightarrow R_4Si$ 167

Summary of Functional Group Properties

Silver

$$RC{\equiv}CH + AgNO_3 \longrightarrow RC{\equiv}CAg$$ 303

Sodium

$$2\,RX + Na \longrightarrow RR$$ 192

Oxiranes

Aldehydes and ketones

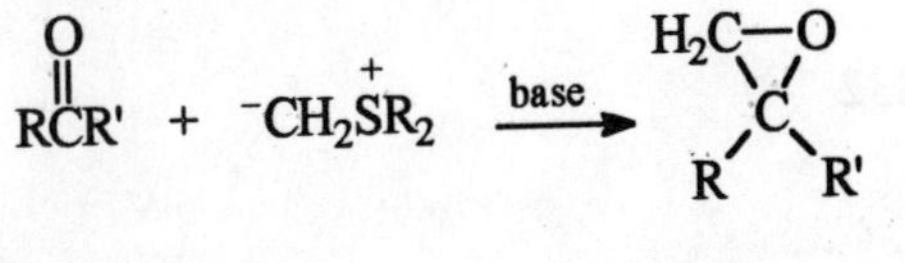

$$RC(=O)R' + {}^{-}CH_2\overset{+}{S}R_2 \xrightarrow{\text{base}} \text{epoxide } H_2C{-}O{-}C(R)(R')$$ 837

Alkenes

$$>C{=}C< \xrightarrow{RCO_3H} \text{epoxide}$$ 285-6, 443, 859, 1085

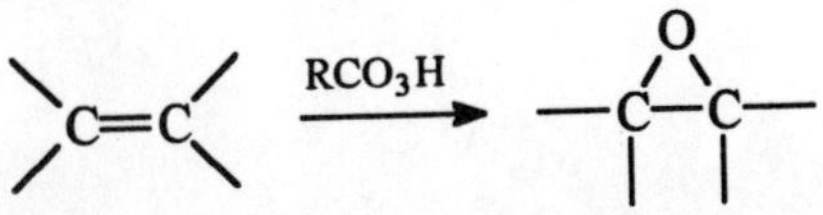

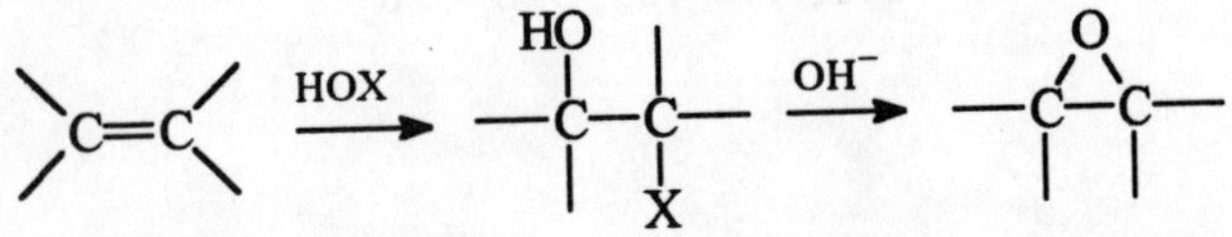

$$>C{=}C< \xrightarrow{HOX} {-}C(OH){-}C(X){-} \xrightarrow{OH^-} \text{epoxide}$$ 236, 286, 443, 1085

Phenols

Amines

$$ArNH_2 \xrightarrow{HONO} ArN_2^+ \xrightarrow[\Delta]{H_2O} ArOH$$ 791, 1017, 1116

Ethers

$$ArOR + HX \longrightarrow ArOH + RX$$ 1030-1032

Halides

$$ArX \xrightarrow{OH^-} ArOH$$ 1007, 1008, 1017

Hydroperoxides

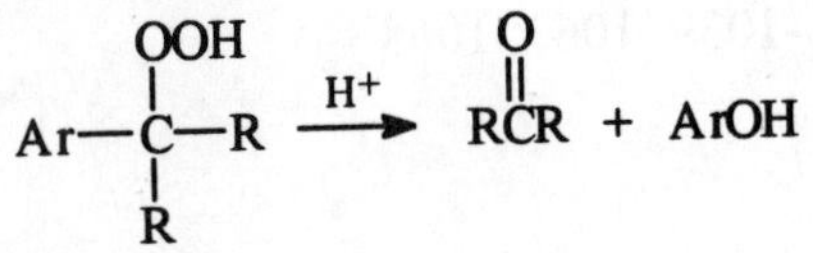

$$Ar{-}C(OOH)(R){-}R \xrightarrow{H^+} RC(=O)R + ArOH$$ 1017-18

Sulfonic acids

$$ArSO_3H \xrightarrow[\Delta]{NaOH} ArOH$$ 7825, 1017, 1063

Phosphorus Compounds

Phosphate esters

$$n\,ROH + Cl_nP(OR')_{3-n} \longrightarrow (RO)_nP(=O)(OR')_{3-n}$$ 830

Phosphines

$$RMgX + PCl_3 \longrightarrow R_3P$$ 828

Phosphite esters

$$ROH + PCl_3 \xrightarrow{\text{pyridine}} (RO)_3P:$$ 832

Phosphonate esters

$$(RO)_3P: + R'X \longrightarrow (RO)_2P(=O)R'$$ 833

Phosphonium salts

$$R_3P: + R'X \longrightarrow R_3P^+R'X^-$$ 172, 408-409, 576, 828

Ylides

$$R_3{}^+PCHR_2' \xrightarrow{nBuLi} R_3{}^+P^-CR_2'$$ 409, 448, 576, 834

Polymers

By addition

$$>C=C< \longrightarrow -(C-C)_n-$$ 1240-1245

By condensation

$$HY'-R-YH \longrightarrow (Y'R) + HY$$ 1245-1246

<u>Quinones</u>

Phenols

$$\text{PhOH or } HO-C_6H_4-YH \xrightarrow{[O]} O=C_6H_4=O$$ 1033-1039, 1063, 1068

Silicon compounds

Silanes

$$RMgX + SiCl_4 \longrightarrow R_4Si$$ 167, 841

$R_3SiCl \xrightarrow{LiAlH_4} R_3SiH$ — 843-844

Silyl ethers

$RO^- + R'_3SiCl \longrightarrow RO{-}SiR'_3$ — 428, 453, 846-847

Sulfur Compounds

Disulfides

$2\ RSH \xrightarrow{[O]} RSSR$ — 810-812

Sulfides

$RS^- + R'Y \longrightarrow RSR'$ — 809, 1087-1089

$>C{=}C< \ + RSH \longrightarrow -\overset{H}{C}-\overset{SR}{C}-$ — 289

$R\overset{O}{\overset{\|}{C}}\bar{C}HR' \xrightarrow{R''SSR''} R\overset{O}{\overset{\|}{C}}\underset{R'}{CH}SR''$ — 812

Sulfinic acids

$ArSO_2Cl \xrightarrow[H_2O]{Zn} ArSO_2H$ — 826

Sulfonamides

$RSO_2Cl + R'NH_2 \longrightarrow RSO_2NHR'$ — 826

Sulfonate esters

$RSO_2Cl + R'OH \xrightarrow{pyridine} RSO_2OR'$ — 222, 562, 821, 1111

Sulfones

$R_2SO \xrightarrow{R'CO_3H} R_2SO_2$ — 812

$R_2S \xrightarrow[or\ R'CO_3H]{KMnO_4} R_2SO_2$ — 812

$ArSO_2Cl + Ar'H \xrightarrow{AlCl_3} ArSO_2Ar'$ — 826

$RSO_2^- + R'X \longrightarrow RSO_2R'$ 827

Sulfonic acids

$RY + SO_3^{2-} \longrightarrow RSO_3^-$ 182, 820

$RCH(=O) + HSO_3^- \longrightarrow RCH(OH)SO_3^-$ 820

$ArSO_2Cl \xrightarrow{H_2O} ArSO_3H$ 812

$ArH + H_2SO_4 \longrightarrow ArSO_3H$ 821-823, 1027-1028, 1062, 1065, 1113

Sulfonium salts

$R_2S + R'X \longrightarrow R_2{}^+SR'X^-$ 172, 815

Sulfonyl halides

$ArSSAr \xrightarrow[HNO_3]{Cl_2} ArSO_2Cl$ 812

$RSO_3Na \xrightarrow{PCl_5} RSO_2Cl$ 820, 826

$ArH \xrightarrow{ClSO_3H} ArSO_2Cl$ 824, 826

Sulfoxides

$R_2S \xrightarrow{[O]} R_2SO$ 812, 813

Thiols

$RY + HS^- \longrightarrow RSH$ 172, 809

$RY + H_2NCSNH_2 \longrightarrow \xrightarrow{OH^-} RSH$ 809

$RX \longrightarrow RMgX \xrightarrow{S} RSH$ 809

$RSSR \xrightarrow[NH_3]{Li} RSH$ 811

Summary of Functional Group Properties

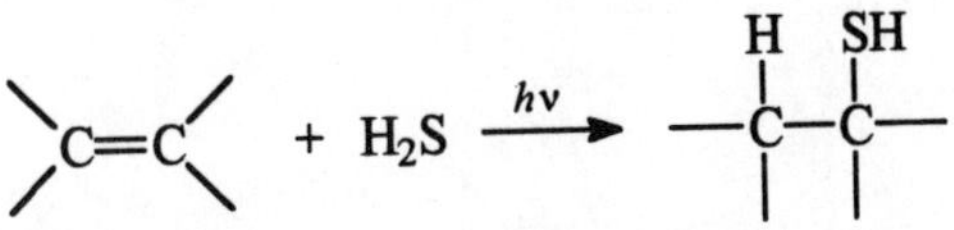

$$>C=C< + H_2S \xrightarrow{h\nu} -\overset{H}{\underset{|}{\overset{|}{C}}}-\overset{SH}{\underset{|}{\overset{|}{C}}}-$$

289

$$ArNH_2 \xrightarrow{HONO} ArN_2^+ \xrightarrow[EtOCS_2^-; OH^-]{HS^- \text{ or}} ArSH$$

791, 800

Thiocyanates

$$RX + {}^-SCN \rightarrow RSCN$$

172, 741

FORMATION OF C-C BONDS

Acyloin and Pinacol Condensations

$$RCOOEt \xrightarrow[ether]{Na} RCOCHOHR$$

866-868

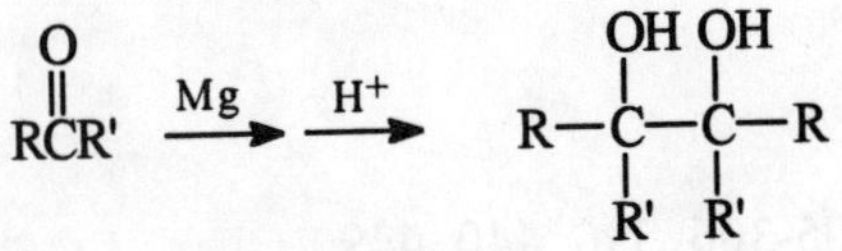

$$R\overset{O}{\overset{\|}{C}}R' \xrightarrow{Mg} \xrightarrow{H^+} R-\underset{R'}{\overset{OH}{C}}-\underset{R'}{\overset{OH}{C}}-R$$

862

Alkene Addition Reactions

Carbenoid additions

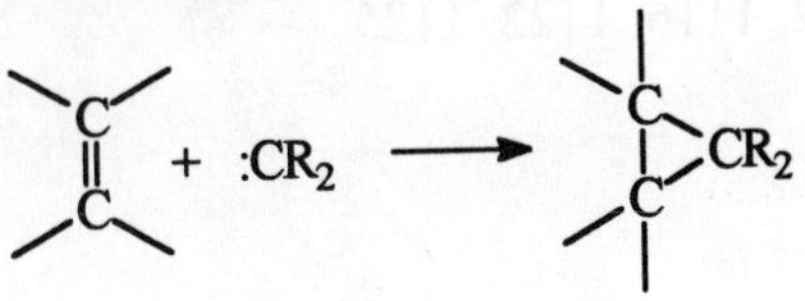

$$>C=C< + :CR_2 \rightarrow \text{cyclopropane } (>C-C<)CR_2$$

287-288, 443, 740, 788

Carbon tetrahalide addition

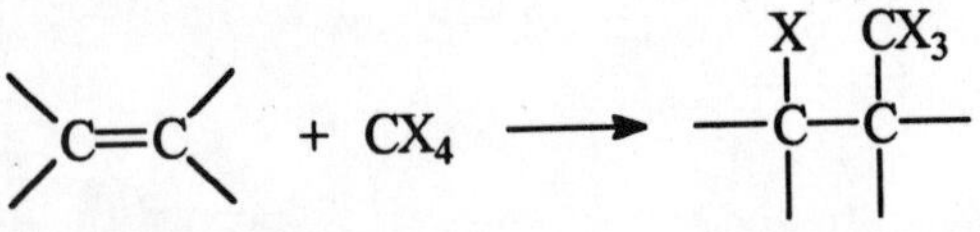

$$>C=C< + CX_4 \rightarrow -\overset{X}{\underset{|}{\overset{|}{C}}}-\overset{CX_3}{\underset{|}{\overset{|}{C}}}-$$

289

Claisen rearrangement

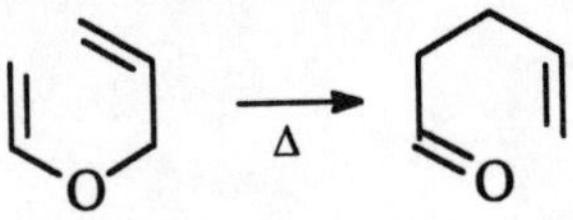

$$\text{allyl vinyl ether} \xrightarrow[\Delta]{} \text{4-pentenal (}\gamma,\delta\text{-unsaturated carbonyl)}$$

638, 1031-1032

Cope rearrangement

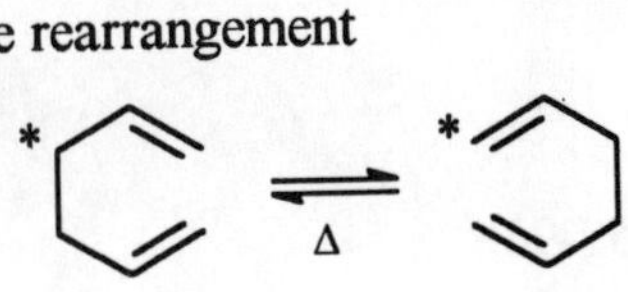

$$\text{*1,5-hexadiene} \underset{\Delta}{\rightleftharpoons} \text{*1,5-hexadiene}$$

637

Cycloaddition reactions
(see also Diels-Alder cycloaddition)

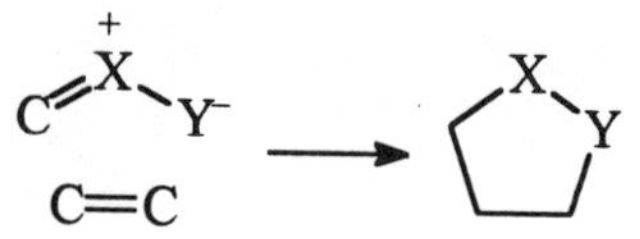

$C{=}\overset{+}{X}{-}Y^-$ + $C{=}C$ ⟶ (five-membered ring containing X–Y) 1105, 1106

Diels-Alder cycloaddition

(diene) + $\|$ $\xrightarrow{\Delta}$ (cyclohexene) 595-599, 662, 668-669, 879, 1042, 1070, 1127

Michael additions

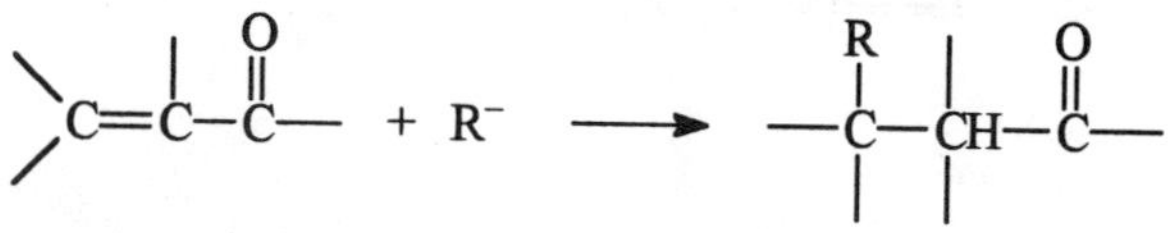

$>C{=}C{-}\overset{O}{\overset{\|}{C}}{-}$ + R^- ⟶ $-\overset{R}{\overset{|}{C}}{-}CH{-}\overset{O}{\overset{\|}{C}}{-}$ 893-895, 1117

(from ketone, β-dicarbonyl or enamine)

Alkylation of Carbon Acids
Acetylides

$RC{\equiv}C^- + R'Y \longrightarrow RC{\equiv}CR'$ 305-306, 396, 440, 449

Benzylic carbanions

Ar_2CH_2 (or Ar_3CH) $\xrightarrow{NaNH_2}\xrightarrow{RX}$ Ar_2CHR(or Ar_3CR) 632, 1116, 1122, 1125

Dithianes

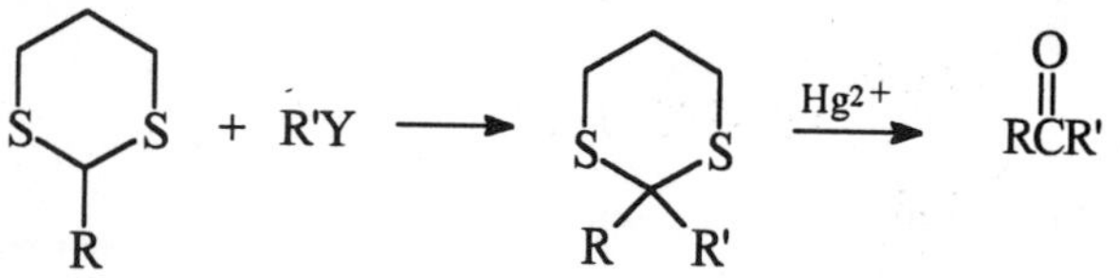

(2-R-1,3-dithiane) + R'Y ⟶ (2-R-2-R'-1,3-dithiane) $\xrightarrow{Hg^{2+}}$ $R\overset{O}{\overset{\|}{C}}R'$ 837

Enamines

$>C{=}CNR_2$ + R'Y ⟶ $R'{-}\overset{|}{C}{-}C{=}\overset{+}{N}R_2$ $\xrightarrow{H_2O}$ $R'{-}\overset{|}{C}{-}\overset{O}{\overset{\|}{C}}{-}$ 767

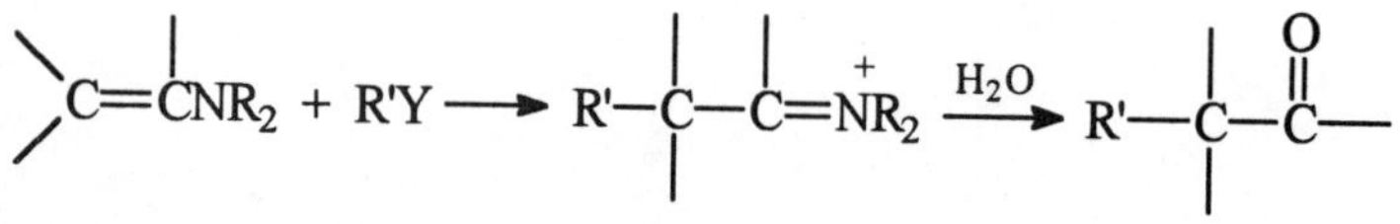

Enolates

$>C{=}\overset{O^-}{\overset{|}{C}}R$ + $R'CH_2Y$ ⟶ $R'CH_2{-}\overset{|}{C}{-}\overset{O}{\overset{\|}{C}}R$ 426-428, 441, 538

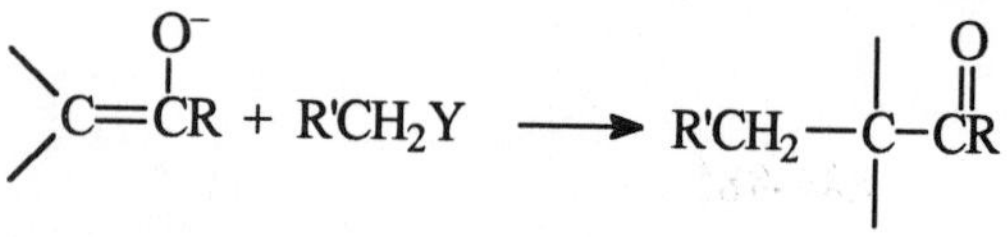

$(EtOOC)_2CH^- + RY \longrightarrow (EtOOC)_2CHR$ 889-891, 957-959

Summary of Functional Group Properties

$RCO^{-}CHCOOEt + R'Y \longrightarrow$ 889-891

$R^{-}CHCOOEt + R'Y \longrightarrow RR'CHCOOEt$ 538

Phosphorus- and sulfur-stabilized carbanions

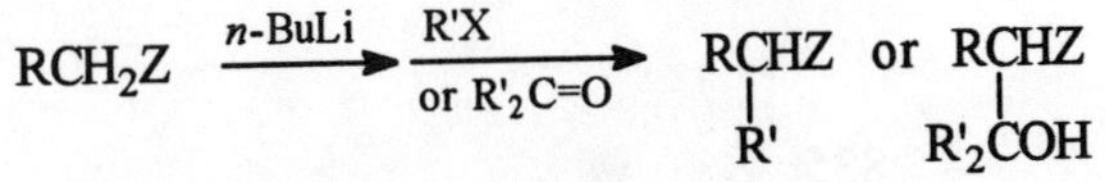

$$RCH_2Z \xrightarrow{n\text{-BuLi}} \xrightarrow[\text{or } R'_2C{=}O]{R'X} \underset{\displaystyle R'}{RCHZ} \text{ or } \underset{\displaystyle R'_2COH}{RCHZ}$$ 834-837

$(Z = PO_3R''_2, SR'', {}^{+}SR''_2, SO_2R'')$

Carbonyl Condensation Reactions

Aldol type

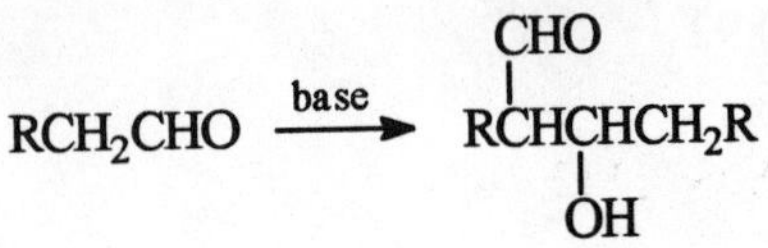

$$RCH_2CHO \xrightarrow{\text{base}} RCH(OH)CH(CHO)CH_2R$$ 432-435, 440, 867

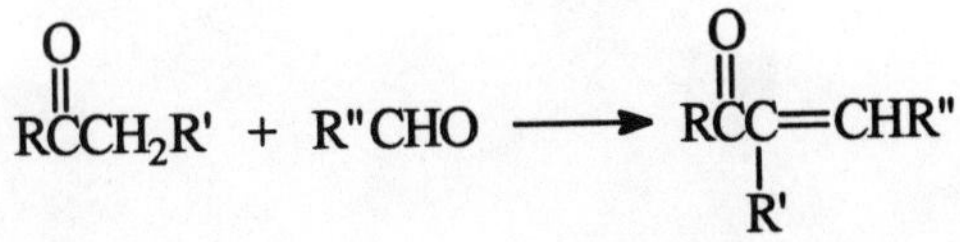

$$RC(=O)CH_2R' + R''CHO \longrightarrow RC(=O)C(R')=CHR''$$ 432-435, 440, 450, 580, 867, 894

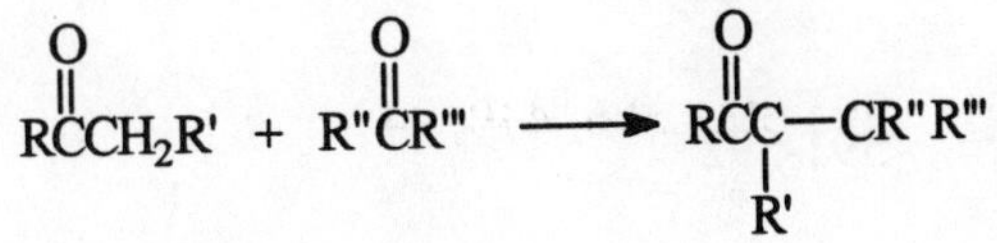

$$RC(=O)CH_2R' + R''C(=O)R''' \longrightarrow RC(=O)C(R')=CR''R'''$$ 580-581, 894

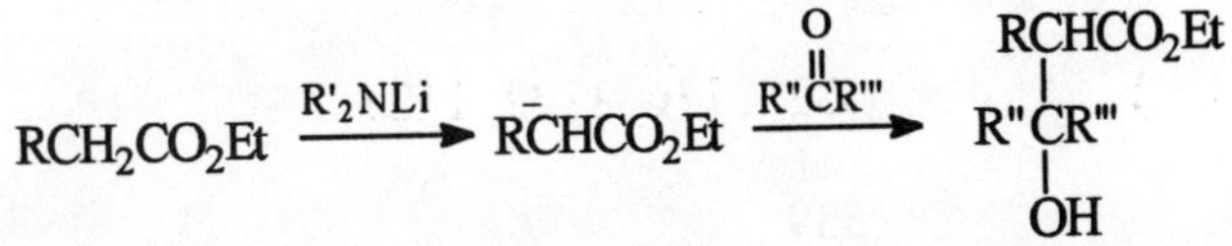

$$RCH_2CO_2Et \xrightarrow{R'_2NLi} R\bar{C}HCO_2Et \xrightarrow{R''C(=O)R'''} R''C(OH)(R''')CH(R)CO_2Et$$ 538, 874

$$RC(=O)R' + R''C(=O)CH_2CO_2Et \longrightarrow RC(OH)(R')CH(CO_2Et)COR'' \text{ or } RR'C{=}C(CO_2Et)(COR'')$$ 891-892, 1110

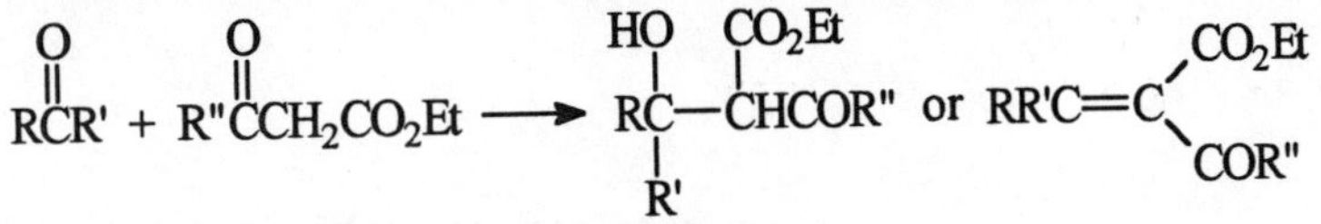

Benzoin

$$ArCHO \xrightarrow{CN^{-}} ArCOCHOHAr$$ 867

Claisen and Dieckmann condensations

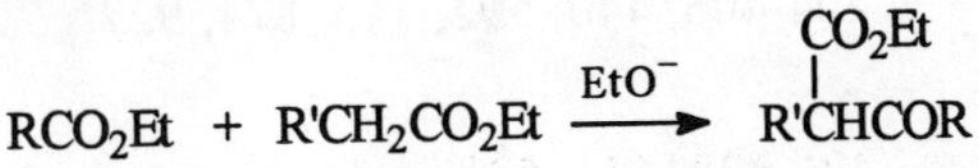

$$RCO_2Et + R'CH_2CO_2Et \xrightarrow{EtO^{-}} R'CH(CO_2Et)COR$$ 539-541, 883-885, 1126

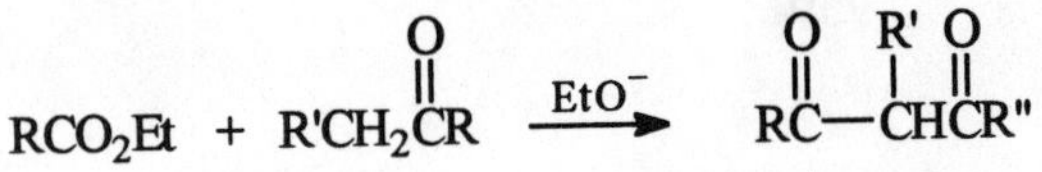

$$RCO_2Et + R'CH_2C(=O)R'' \xrightarrow{EtO^{-}} RC(=O)CH(R')C(=O)R''$$ 884-885

Henry

$$RCH_2NO_2 + R'\overset{O}{\overset{\|}{C}}R' \xrightarrow{\text{base}} R\overset{NO_2}{\overset{|}{C}}\underset{\underset{OH}{|}}{H}CR'_2 \longrightarrow R\overset{NO_2}{\overset{|}{C}}{=}CR'_2$$ 780

Horner-Emmons

$$R\overset{O}{\overset{\|}{C}}R + R'\overset{O}{\overset{\|}{C}}\underset{-}{C}H\overset{O}{\overset{\|}{P}}(OEt)_2 \longrightarrow R_2C{=}CH\overset{O}{\overset{\|}{C}}R'$$ 835

Knoevenagel

$$RCHO + CH_2(COOH)_2 \xrightarrow{\text{amine}} RCH{=}C(COOH)_2$$ 891-892

Mannich

$$R\overset{O}{\overset{\|}{C}}CH_2R' + CH_2O + HNR''_2 \longrightarrow R\overset{O}{\overset{\|}{C}}\underset{\underset{R'}{|}}{C}HCH_2NR''_2$$ 767

Perkin

$$ArCHO + (RCH_2CO)_2O \xrightarrow{RCH_2CO_2^-} ArCH{=}C(R)CO_2H$$ 591-592

Reformatsky

$$R\overset{O}{\overset{\|}{C}}R' + BrCHR''CO_2Et \xrightarrow{Zn} R\underset{\underset{R'}{|}}{\overset{OH}{\overset{|}{C}}}CHR''CO_2Et$$ 539

Wittig

$$Ph_3P{=}CHR + R'_2C{=}O \longrightarrow R'_2C{=}CHR$$ 408-410, 440, 448, 576

Cyanide Reactions

$$RY + CN^- \longrightarrow RCN$$ 172, 183, 276, 304, 440, 494, 591, 628, 879

$$R\overset{O}{\overset{\|}{C}}R' + HCN \longrightarrow RR'\overset{OH}{\overset{|}{C}}CN$$ 404-406, 440, 592, 745, 874, 927

$$R\overset{O}{\overset{\|}{C}}R' + HCN + NH_3 \longrightarrow RR'\overset{NH_2}{\overset{|}{C}}CN$$ 959

Summary of Functional Group Properties

$$-\overset{|}{C}=\overset{|}{C}-\overset{O}{\overset{\|}{C}}- + HCN \longrightarrow -\underset{|}{\overset{CN}{\overset{|}{C}}}-\overset{|}{C}H-\overset{O}{\overset{\|}{C}}-$$ 583, 879

$ArX + CuCN \longrightarrow ArCN$ 793

Grignard and Related Organometallic Reactions

Acyl halides

$RCOCl + R'MgX \longrightarrow RR'C{=}O$ 531-532

$RCOCl + R'_2CuLi \longrightarrow RR'C{=}O$ 531-532

$RCOCl + CH_2N_2 \longrightarrow RCOCHN_2$ or $RCOCH_2Cl$ 788

Aldehydes and ketones

$$RMgX + R'\overset{O}{\overset{\|}{C}}R'' \longrightarrow R\underset{R'}{\underset{|}{\overset{OH}{\overset{|}{C}}}}R''$$ 401-404, 440, 446, 447, 584-585, 1011, 1059, 1127

RLi can substitute for RMgX, R′ = alkyl, H

$$R\overset{O}{\overset{\|}{C}}R' + RC{\equiv}C^- Z^+ \longrightarrow R\underset{R'}{\underset{|}{\overset{OH}{\overset{|}{C}}}}C{\equiv}CR''$$ 404, 440, 576

$Z^+ = {}^+MgX, Li^+, Na^+$; R′ = alkyl, H

Carbon dioxide

$RMgX$ (or RLi) $\xrightarrow{CO_2}$ RCOOH 495-496

Esters

$RCOOEt + R'MgX$ (or R′Li) $\longrightarrow RR'_2COH$ 532

$RMgX + (EtO)_2C{=}O \longrightarrow R_3COH$ 533

Halides

$RMgX + CH_2{=}CHCH_2X \longrightarrow RCH_2CH{=}CH_2$ 563, 576

Oxirane

$RMgX$ + $H_2C\overset{O}{-}CH_2$ (oxirane) $\longrightarrow RCH_2CH_2OH$ 239, 440, 447

Unsaturated ketones

$$\text{RMgX (or } R_2CuLi) + \;>C{=}\overset{|}{C}{-}\overset{O}{\overset{\|}{C}}{-} \longrightarrow -\overset{R}{\overset{|}{\underset{|}{C}}}{-}\overset{|}{C}H{-}\overset{O}{\overset{\|}{C}}{-}$$ 584-585, 854

Formation of C-C Bonds to Aromatic Rings

Amine derivatives

$ArNH_2 \xrightarrow{HONO} ArN_2^+ \xrightarrow{CuCN} ArCN$ 793, 800

$ArNH_2 \xrightarrow{HONO} ArN_2^+ \xrightarrow[OH^-]{Ar'H} Ar\text{-}Ar'$ 796-797, 800

$ArNHNHAr \xrightarrow{H^+} H_2NAr'Ar'NH_2$ 783, 1051
(benzidine rearrangement)

Arenes

$ArH + RCOCl \text{ (or } (RCO)_2O) \xrightarrow{\text{Lewis acid}} ArCOR$ 696-697, 718, 759, 844, 1028-1030, 1062, 1065, 1067, 1096, 1099

$ArH + RCOOH \xrightarrow[\text{Lewis acid}]{H^+ \text{ or}} ArCOR$ 1059, 1067-1068

$ArH + HCON(CH_3)_2 \xrightarrow{POCl_3} ArCHO$ 759, 1100

$ArH + CH_2O + HCl \xrightarrow{ZnCl_2} ArCH_2Cl$ 700, 1100

$ArH + RX \xrightarrow{AlCl_3} ArR \text{ (+ } Ar'R + ArR')$ 698-699, 700

$ArH + \;>C{=}C< \;\xrightarrow{H^+} Ar{-}\overset{|}{\underset{|}{C}}{-}\overset{|}{\underset{|}{C}}{-}H$ 699, 824, 1017

Aryl halides and organometallics

$ArX + RX \xrightarrow{Na} ArR$ 1012

$ArX \xrightarrow{Cu} ArAr$ 1012

$ArX \xrightarrow{Mg} \xrightarrow{CO_2} ArCOOH$ 1011

$ArLi \xrightarrow{CuBr} \xrightarrow{O_2} Ar\text{-}Ar$ 1013

Phenols

$ArO^- + RCHO \longrightarrow HOAr\overset{OH}{\overset{|}{C}}HR$ 1021

$ArO^- + CO_2 \longrightarrow Ar(OH)(CO_2H)$ 1022

$ArO^- + CHCl_3 \longrightarrow Ar(OH)(CHO)$ 1023

$ArOCOR \xrightarrow{AlCl_3} Ar(OH)(COR)$ 1030

$ArOCH_2CH{=}CH_2 \xrightarrow[\Delta]{} Ar(OH)(CH_2CH{=}CH_2)$ 1031-1032

(Claisen rearrangement)

Notes

Notes

Notes

Notes